Annual Review of Biochemistry

Volume 81, 2012

Roger D. Kornberg, *Editor*
Stanford University School of Medicine

James E. Rothman, *Associate Editor*
Yale University School of Medicine

JoAnne Stubbe, *Associate Editor*
Massachusetts Institute of Technology

Jeremy W. Thorner, *Associate Editor*
University of California, Berkeley

www.annualreviews.org • science@annualreviews.org • 650-493-4400

Annual Reviews
4139 El Camino Way • P.O. Box 10139 • Palo Alto, California 94303-0139

Annual Reviews
Palo Alto, California, USA

COPYRIGHT © 2012 BY ANNUAL REVIEWS, PALO ALTO, CALIFORNIA, USA. ALL RIGHTS RESERVED. The appearance of the code at the bottom of the first page of an article in this serial indicates the copyright owner's consent that copies of the article may be made for personal or internal use, or for the personal or internal use of specific clients. This consent is given on the condition that the copier pay the stated per-copy fee of $20.00 per article through the Copyright Clearance Center, Inc. (222 Rosewood Drive, Danvers, MA 01923) for copying beyond that permitted by Section 107 or 108 of the U.S. Copyright Law. The per-copy fee of $20.00 per article also applies to the copying, under the stated conditions, of articles published in any *Annual Review* serial before January 1, 1978. Individual readers, and nonprofit libraries acting for them, are permitted to make a single copy of an article without charge for use in research or teaching. This consent does not extend to other kinds of copying, such as copying for general distribution, for advertising or promotional purposes, for creating new collective works, or for resale. For such uses, written permission is required. Write to Permissions Dept., Annual Reviews, 4139 El Camino Way, P.O. Box 10139, Palo Alto, CA 94303-0139 USA.

International Standard Serial Number: 0066-4154
International Standard Book Number: 978-0-8243-0881-0
Library of Congress Catalog Card Number: 32-25093

All Annual Reviews and publication titles are registered trademarks of Annual Reviews.

∞ The paper used in this publication meets the minimum requirements of American National Standards for Information Sciences—Permanence of Paper for Printed Library Materials, ANSI Z39.48-1992.

Annual Reviews and the Editors of its publications assume no responsibility for the statements expressed by the contributors to this *Annual Review*.

TYPESET BY APTARA
PRINTED AND BOUND BY FRIESENS CORPORATION, ALTONA, MANITOBA, CANADA

Editorial Committee (2012)

Joan W. Conaway, Stowers Institute for Medical Research and Kansas University Medical Center
Christopher Dobson, University of Cambridge
F. Ulrich Hartl, Max Planck Institut für Biochemie
Laura L. Kiessling, University of Wisconsin, Madison
Roger D. Kornberg, Stanford University School of Medicine
Paul Modrich, Duke University Medical Center
James E. Rothman, Yale University School of Medicine
JoAnne Stubbe, Massachusetts Institute of Technology
Jeremy W. Thorner, University of California, Berkeley
Gunnar von Heijne, Stockholm University

Responsible for the Organization of Volume 81 (Editorial Committee, 2010)

Joan W. Conaway
Christopher Dobson
F. Ulrich Hartl
Laura L. Kiessling
Roger D. Kornberg
Paul Modrich
Christian R.H. Raetz
James E. Rothman
JoAnne Stubbe
Jeremy W. Thorner
Gunnar von Heijne
Donald M. Engelman (Guest)
Wayne A. Hendrickson (Guest)
Michele Pagano (Guest)

Production Editor: Jesslyn S. Holombo
Managing Editor: Linley E. Hall
Bibliographic Quality Control: Mary A. Glass
Electronic Content Coordinators: Suzanne K. Moses, Erin H. Lee
Illustration Editor: Glenda Lee Mahoney

Contents

Annual Review of
Biochemistry
Volume 81, 2012

Preface

Preface and Dedication to Christian R.H. Raetz
JoAnne Stubbe xi

Prefatories

A Mitochondrial Odyssey
Walter Neupert 1

The Fires of Life
Gottfried Schatz 34

Chromatin, Epigenetics, and Transcription Theme

Introduction to Theme "Chromatin, Epigenetics, and Transcription"
Joan W. Conaway 61

The COMPASS Family of Histone H3K4 Methylases:
Mechanisms of Regulation in Development
and Disease Pathogenesis
Ali Shilatifard 65

Programming of DNA Methylation Patterns
Howard Cedar and Yehudit Bergman 97

RNA Polymerase II Elongation Control
Qiang Zhou, Tiandao Li, and David H. Price 119

Genome Regulation by Long Noncoding RNAs
John L. Rinn and Howard Y. Chang 145

Protein Tagging Theme

The Ubiquitin System, an Immense Realm
Alexander Varshavsky 167

Ubiquitin and Proteasomes in Transcription
Fuqiang Geng, Sabine Wenzel, and William P. Tansey 177

The Ubiquitin Code
  David Komander and Michael Rape 203

Ubiquitin and Membrane Protein Turnover: From Cradle to Grave
  Jason A. MacGurn, Pi-Chiang Hsu, and Scott D. Emr 231

The N-End Rule Pathway
  Takafumi Tasaki, Shashikanth M. Sriram, Kyong Soo Park, and Yong Tae Kwon ... 261

Ubiquitin-Binding Proteins: Decoders of Ubiquitin-Mediated
Cellular Functions
  Koraljka Husnjak and Ivan Dikic 291

Ubiquitin-Like Proteins
  Annemarthe G. van der Veen and Hidde L. Ploegh 323

Recent Advances in Biochemistry

Toward the Single-Hour High-Quality Genome
  Patrik L. Ståhl and Joakim Lundeberg 359

Mass Spectrometry–Based Proteomics and Network Biology
  Ariel Bensimon, Albert J.R. Heck, and Ruedi Aebersold 379

Membrane Fission: The Biogenesis of Transport Carriers
  Felix Campelo and Vivek Malhotra 407

Emerging Paradigms for Complex Iron-Sulfur Cofactor Assembly
and Insertion
  John W. Peters and Joan B. Broderick 429

Structural Perspective of Peptidoglycan Biosynthesis and Assembly
  Andrew L. Lovering, Susan S. Safadi, and Natalie C.J. Strynadka 451

Discovery, Biosynthesis, and Engineering of Lantipeptides
  Patrick J. Knerr and Wilfred A. van der Donk 479

Regulation of Glucose Transporter Translocation
in Health and Diabetes
  Jonathan S. Bogan 507

Structure and Regulation of Soluble Guanylate Cyclase
  Emily R. Derbyshire and Michael A. Marletta 533

The MPS1 Family of Protein Kinases
  Xuedong Liu and Mark Winey 561

The Structural Basis for Control of Eukaryotic Protein Kinases
  Jane A. Endicott, Martin E.M. Noble, and Louise N. Johnson 587

Measurements and Implications of the Membrane Dipole Potential
Liguo Wang 615

GTPase Networks in Membrane Traffic
Emi Mizuno-Yamasaki, Felix Rivera-Molina, and Peter Novick 637

Roles for Actin Assembly in Endocytosis
Olivia L. Mooren, Brian J. Galletta, and John A. Cooper 661

Lipid Droplets and Cellular Lipid Metabolism
Tobias C. Walther and Robert V. Farese Jr. 687

Adipogenesis: From Stem Cell to Adipocyte
Qi Qun Tang and M. Daniel Lane 715

Pluripotency and Nuclear Reprogramming
Marion Dejosez and Thomas P. Zwaka 737

Endoplasmic Reticulum Stress and Type 2 Diabetes
Sung Hoon Back and Randal J. Kaufman 767

Structure Unifies the Viral Universe
*Nicola G.A. Abrescia, Dennis H. Bamford, Jonathan M. Grimes,
and David I. Stuart* 795

Indexes

Cumulative Index of Contributing Authors, Volumes 77–81 823

Cumulative Index of Chapter Titles, Volumes 77–81 827

Errata

An online log of corrections to *Annual Review of Biochemistry* articles may be found at http://biochem.annualreviews.org/errata.shtml

Related Articles

From the Annual Review of Biophysics, Volume 41 (2012)

    Mechanisms of Sec61/SecY-Mediated Protein Translocation Across Membranes
        Eunyong Park and Tom A. Rapoport

    Racemic Protein Crystallography
        Todd O. Yeates and Stephen B.H. Kent

From the Annual Review of Cell and Developmental Biology, Volume 27 (2011)

    The Coupling of X-Chromosome Inactivation to Pluripotency
        Jane Lynda Deuve and Philip Avner

    The Role of MeCP2 in the Brain
        Jacky Guy, Hélène Cheval, Jim Selfridge, and Adrian Bird

From the Annual Review of Chemical and Biomolecular Engineering, Volume 3 (2012)

    Density of States-Based Molecular Simulations
        Sadanand Singh, Manan Chopra, and Juan J. de Pablo

From the Annual Review of Genetics, Volume 45 (2011)

    V(D)J Recombination: Mechanisms of Initiation
        David G. Schatz and Patrick C. Swanson

    Double-Strand Break End Resection and Repair Pathway Choice
        Lorraine S. Symington and Jean Gautier

From the Annual Review of Genomics and Human Genetics, Volume 12 (2011)

    RNA-Mediated Epigenetic Programming of Genome Rearrangements
        Mariusz Nowacki, Keerthi Shetty, and Laura F. Landweber

From the Annual Review of Immunology, Volume 30 (2012)

    microRNA Regulation of Inflammatory Responses
        Ryan M. O'Connell, Dinesh S. Rao, and David Baltimore

Chromatin Topology and the Regulation of Antigen Receptor Assembly
Claudia Bossen, Robert Mansson, and Cornelis Murre

From the Annual Review of Medicine, Volume 63 (2012)

Nanoparticle Delivery of Cancer Drugs
Andrew Z. Wang, Robert Langer, and Omid C. Farokhzad

From the Annual Review of Microbiology, Volume 65 (2011)

Regulation of Alternative Sigma Factor Use
Sofia Österberg, Teresa del Peso-Santos, and Victoria Shingler

From the Annual Review of Neuroscience, Volume 35 (2012)

The Complement System: An Unexpected Role in Synaptic Pruning During Development and Disease
Alexander H. Stephan, Ben A. Barres, and Beth Stevens

From the Annual Review of Pathology: Mechanisms of Disease, Volume 7 (2012)

RNA Dysregulation in Diseases of Motor Neurons
Fadia Ibrahim, Tadashi Nakaya, and Zissimos Mourelatos

From the Annual Review of Pharmacology and Toxicology, Volume 52 (2012)

Molecular Mechanism of β-Arrestin-Biased Agonism at Seven-Transmembrane Receptors
Eric Reiter, Seungkirl Ahn, Arun K. Shukla, and Robert J. Lefkowitz

From the Annual Review of Plant Biology, Volume 63 (2012)

Epigenetic Mechanisms Underlying Genomic Imprinting in Plants
Claudia Köhler, Philip Wolff, and Charles Spillane

Annual Reviews is a nonprofit scientific publisher established to promote the advancement of the sciences. Beginning in 1932 with the *Annual Review of Biochemistry*, the Company has pursued as its principal function the publication of high-quality, reasonably priced *Annual Review* volumes. The volumes are organized by Editors and Editorial Committees who invite qualified authors to contribute critical articles reviewing significant developments within each major discipline. The Editor-in-Chief invites those interested in serving as future Editorial Committee members to communicate directly with him. Annual Reviews is administered by a Board of Directors, whose members serve without compensation.

2012 Board of Directors, Annual Reviews

Richard N. Zare, *Chairperson of Annual Reviews, Marguerite Blake Wilbur Professor of Natural Science, Department of Chemistry, Stanford University*
Karen S. Cook, *Vice-Chairperson of Annual Reviews, Director of the Institute for Research in the Social Sciences, Stanford University*
Sandra M. Faber, *Vice-Chairperson of Annual Reviews, University Professor of Astronomy and Astrophysics, Astronomer, University of California Observatories/Lick Observatory, University of California, Santa Cruz*
John I. Brauman, *J.G. Jackson-C.J. Wood Professor of Chemistry, Stanford University*
Peter F. Carpenter, *Founder, Mission and Values Institute, Atherton, California*
Susan T. Fiske, *Eugene Higgins Professor of Psychology, Princeton University*
Eugene Garfield, *Emeritus Publisher, The Scientist*
Samuel Gubins, *President and Editor-in-Chief, Annual Reviews*
Steven E. Hyman, *Director, Stanley Center for Psychiatric Research, Broad Institute of MIT and Harvard*
Roger D. Kornberg, *Professor of Structural Biology, Stanford University School of Medicine*
Sharon R. Long, *Wm. Steere-Pfizer Professor of Biological Sciences, Stanford University*
J. Boyce Nute, *Palo Alto, California*
Michael E. Peskin, *Professor of Particle Physics and Astrophysics, SLAC, Stanford University*
Claude M. Steele, *Dean of the School of Education, Stanford University*
Harriet A. Zuckerman, *Senior Fellow, The Andrew W. Mellon Foundation*

Management of Annual Reviews

Samuel Gubins, President and Editor-in-Chief
Paul J. Calvi Jr., Director of Technology
Steven J. Castro, Chief Financial Officer and Director of Marketing & Sales
Jennifer L. Jongsma, Director of Production
Laurie A. Mandel, Corporate Secretary

Annual Reviews of

Analytical Chemistry
Anthropology
Astronomy and Astrophysics
Biochemistry
Biomedical Engineering
Biophysics
Cell and Developmental Biology
Chemical and Biomolecular Engineering
Clinical Psychology
Condensed Matter Physics
Earth and Planetary Sciences
Ecology, Evolution, and Systematics
Economics

Entomology
Environment and Resources
Financial Economics
Fluid Mechanics
Food Science and Technology
Genetics
Genomics and Human Genetics
Immunology
Law and Social Science
Marine Science
Materials Research
Medicine
Microbiology
Neuroscience
Nuclear and Particle Science
Nutrition

Pathology: Mechanisms of Disease
Pharmacology and Toxicology
Physical Chemistry
Physiology
Phytopathology
Plant Biology
Political Science
Psychology
Public Health
Resource Economics
Sociology

SPECIAL PUBLICATIONS
Excitement and Fascination of Science, Vols. 1, 2, 3, and 4

Preface and Dedication to Christian R.H. Raetz

Christian Rudolph Hubert Raetz (Chris to his friends) was taken from us in an untimely fashion on August 16, 2011. He was a faculty member in the Biochemistry Department at Duke University Medical School and past chair of the department. He joined the Editorial Committee of the *Annual Review of Biochemistry* on January 1, 1992, and became an associate editor in 1996, a position held until his death. We all are grateful for his important, insightful contributions to making the *Annual Review of Biochemistry* so successful and for his enthusiasm and passion for science that inspired us all.

Chris was a scientist's scientist. He was dedicated to understanding the details of a problem that captured his attention for his entire independent career: the biosynthesis of lipid A, a glucosamine-based phospholipid that makes up the outer layer of the outer membrane of *Escherichia coli* and most gram-negative bacteria. Lipid A is required for bacterial growth and also plays an important role in the pathogenesis of infections. His elucidation of ten steps in the pathway required an interdisciplinary approach including development of creative screens to identify the new pathway genes, use of bioinformatics to help identify protein function, and a profound knowledge of lipids and their chemical properties to create assays for catalytic activity so that the proteins could be isolated and characterized. He also was an outstanding enzymologist, interested in understanding the detailed chemical mechanisms that in combination with structural biology led to a "rational" design of inhibitors of this pathway. His approach to tackling problems was all inclusive, translating basic science to clinical applications. His fearlessness and depth of knowledge from chemistry to medicine earned him the respect and admiration of his peers throughout his entire career.

Beyond his important scientific contributions, Chris was a very kind and thoughtful individual. He was an outstanding mentor to his graduate students and postdoctoral fellows as well as to his colleagues in academics and industry. He had a playfulness about him, accepting with a smile the several pie-in-the-face bets he lost and delivering with enthusiasm the one pie-in-the-face bet he won. Although he was a self-assured person, he could always laugh at himself: whether commenting on his lack of hair, his inability to drive a stick shift (going 20 mph in fifth gear and 60 mph in first gear), or his distinctive name—Christian Rudolph Hubert. He was excited when good things happened to you and supportive when things were not going well. He was enthusiastic about everyone's science. In fact, he just loved science! It was very challenging for me to write even this short dedication to Chris. I repeatedly found myself teary-eyed thinking about all the good times that I shared with him and his family and wishing I would see him again. I often think I will receive a phone call from him, but alas, I know I will not. However, I have no doubt that he is somewhere at peace, thinking about lipids and reading the next exciting issue of the *Annual Review of Biochemistry*.

JoAnne Stubbe, Associate Editor
April 2, 2012

Christian R.H. Raetz
Photo courtesy of Strawbridge Studios, Durham, North Carolina.

Walter Neupert

A Mitochondrial Odyssey

Walter Neupert

Ludwig-Maximilians-Universität München and Max Planck Institute of Biochemistry, Martinsried D-82152, Germany; email: Neupert@biochem.mpg.de

Keywords

mitochondria, protein transport, protein folding, molecular chaperones, AAA-proteases, membrane biogenesis

Abstract

Good fortune let me be an innocent child during World War II, a hopeful adolescent with encouraging parents during the years of German recovery, and a self-determined adult in a period of peace, freedom, and wealth. My luck continued as a scientist who could entirely follow his fancy. My mind was always set on understanding how things are made. At a certain point, I found myself confronted with the question of how mitochondria and organelles, which cannot be formed de novo, are put together. Intracellular transport of proteins, their translocation across the mitochondrial membranes, and their folding and assembly were the processes that fascinated me. Now, after some 30 years, we have wonderful insights, unimagined views of a complex and at the same time simple machinery and its workings. We have glimpses of how orderly processes are established in the cell to assemble from single molecules our beautiful mitochondria that every day make some 50 kg of ATP for each of us. At the same time, we have learned amazing lessons from the tinkering of evolution that developed mitochondria from bacteria.

Contents

PROLOGUE	2
GROWING UP IN POSTWAR GERMANY	3
GETTING HOOKED ON BIOGENESIS OF MITOCHONDRIA	5
EARLY WORK ON IMPORT OF PROTEINS INTO MITOCHONDRIA	8
DIVERSITY OF IMPORT SIGNALS AND IMPORT PATHWAYS	11
FOLDING AND UNFOLDING OF PROTEINS BEFORE AND DURING IMPORT	13
MITOCHONDRIAL PROCESSING PEPTIDASE	13
MITOCHONRIAL RECEPTORS FOR PRECURSORS AND THE TOM COMPLEX	14
HEAT SHOCK PROTEIN HSP60 AND THE DE NOVO FOLDING OF PROTEINS	16
THE ROLE OF MITOCHONDRIAL HSP70 IN PROTEIN IMPORT	17
THE TIM23 COMPLEX OF THE INNER MEMBRANE	18
SURPRISING ASPECTS OF THE TIM23 TRANSLOCASE	20
THE TIM22 TRANSLOCASE	22
THE DISULFIDE RELAY SYSTEM	23
THE OXA1 COMPLEX	24
THE TOB COMPLEX	25
MITOCHONDRIAL ATP-DEPENDENT PROTEASES	25
THE FUTURE: THE MOLECULAR ARCHITECTURE OF MITOCHONDRIA	27
EPILOGUE	28

ΔAIMΩN, *Dämon*
Wie an dem Tag, der dich der Welt verliehen,
Die Sonne stand zum Grüße der Planeten,
Bist alsobald und fort und fort gediehen
Nach dem Gesetz, wonach du angetreten.
So musst du sein, dir kannst du nicht entfliehen,
So sagten schon Sibyllen, so Propheten;
Und keine Zeit und keine Macht zerstückelt
Geprägte Form, die lebend sich entwickelt.

ΔAIMON, Daemon
As stood the sun to the salute of planets
Upon the day that gave you to the earth,
You grew forthwith, and prospered, in your growing
Heeded the law presiding at your birth.
Sibyls and prophets told it: You must be
None but yourself, from self you cannot flee.
No time there is, no power, can decompose
The minted form that lives and living grows.

(Goethe, Primal Words, Orphic, translated by Christopher Middleton)

PROLOGUE

This poem, written in 1817, has been interpreted many times, mostly by men of letters. For me, it has a firm scientific motivation. Goethe anticipated the ideas and discoveries of Darwin and Mendel made only some 30 to 50 years later. He had observed that nature works according to very strict rules once the individual constellation has been set. The metaphor of the position of the planets at the time when one enters the world means, in terms or our present knowledge: As your set of genes was arranged when egg and sperm fused, you have to be and stay that way for the rest of your life. You are subject to the dualism of chance and law.

I believe my entire life was governed by this principle. I feel in retrospect I was destined for a life of finding solid truth behind things. I realized there was another most enjoyable world full of excitement in humanities, philosophy, and literature, always a source of intellectual curiosity and delight, but I was apparently not made for this world. Now, after so many years thinking along the lines of Mendel and Darwin and the many giants of scientists who built up our present stronghold of wisdom in biology,

I am convinced that we are minted forms not only in our bodies but also in our minds.

GROWING UP IN POSTWAR GERMANY

I was born on October 24, 1939. My first largely unremembered years were the years of terror, ubiquitous misery, fear, and despair. My father was drafted in 1938. Aware of the disaster to come, he volunteered to become a paramedic. He survived the war, but it took him 10 years to return. My first memory of him must have been in 1943 or 1944, when he left after a vacation. I still see him jumping onto the open platform of the last car of the leaving train, which took him back to the Ukraine. He left my mother at home with three little boys. On one of his leaves from service at the eastern front, my father built a shelter for protection against the bombs, which had started to rain on Munich. I remember the family sitting at the radio, a Volksempfänger, listening for the announcement of the approaching bombers mixed with the "Preludes" of Liszt played to report about the great successes of the army when everybody knew that Germany was doomed. I still hear the sirens ringing. Sometimes this was during the day, more often during the night. And then we were running to the shelter, which was a hole in the ground covered with a concrete lid, was open at both sides to avoid the dangerous air pressure, had benches at each side, and had a few steps up to the garden on both sides. I remember in particular the cold during the winters. Later, when the bombing was more and more intense, our mother had just time to drag us out of our beds to the basement of the house.

As far as I can remember, I did not feel the danger and misery around me, like the children now who are playing in the ruins of the wars that still torment the planet. But I remember well the excitement when the war ended, the turmoil just before the Americans arrived, the weapons left behind, and the dark things happening during the last nights. I remember standing at the gate of our house when the American troops moved into Munich, the 1st of May 1945, with an endless chain of jeeps, tanks, and trucks. Later, our curiosity to see these strange people lured us into their camps. And indeed, we saw unbelievable wonders: chocolate, orange juice in gasoline canisters, and, for the first time, African American people. The GIs were very friendly to us kids, and they sometimes shared their Hershey bars.

But life grew darker and darker, and more and more misery was to come. Marauding criminals were a constant threat; there were burglary, looting, and killing. We did not know what had happened to our father. The adults had little hope that there would be better times ever. And starving started. Droves of people from Munich were cycling through the villages in upper Bavaria to exchange valuables for potatoes. Our mother was busy enough looking after her boys, which was necessary. With no fathers around, the boys looked for adventures and formed little gangs. The war had left plenty of exciting things in the woods and in deserted antiaircraft stands. There was a variety of discarded weaponry to play with. One day in the summer of 1945, three boys of families of our neighbors, 10 or so years old, were killed when playing with a hand grenade they picked up somewhere. We grew vegetables in our garden and had a goat and a few hens to provide us with milk and eggs.

In the autumn of 1945, I started elementary school in the village nearby. There was no paper, and there were no pencils. Some had slates and chalks. We had to bring logs to heat the classroom. In the second year, the class consisted of more than 60 kids, since the stream of refugees from the former East Germany had arrived in Bavaria. The school moved to the dancing floor of the village inn. A 20-year-old survivor of the war was our teacher. School was exciting but much more so was our daily march of some two or more miles, depending on our investigative mood. It took us along the road where the GIs were driving with their new miraculously flashy Mercurys and Studebakers between their barracks in Munich and a military airport in the vicinity. Occasionally, the dream happened, and they gave the little hitchhikers a ride to school, unforgettable adventures. 1946 and 1947 were the worst years regarding food. In 1948, after three years of struggling to survive behind the Ural mountains, my father came home. He was lucky not to belong to the 1 million out of 3 million prisoners of war who died somewhere in boundless Siberia. And he was certainly very fortunate not to belong to those survivors who were released only in 1953. He was sent home because he suffered from severe dystrophy. We pushed our fingers into his swollen legs, and it took time until the dents disappeared. He spoke very little about the life in the prison camp, but many years later when he read *One Day in the Life of Ivan Denisovich* by Alexander Solzhenitsyn, he told me that the layout and the regime in the prison camp were exactly as in the gulag. My father immediately got a job, and everything became better. In June 1948, there was the Währungsreform (currency reform), and with this, we had food and a warm kitchen in winter.

My parents and grandparents were not academics. I guess they believed I should learn something useful in times in which survival was the essence. Fortunately, my teacher and the parish priest told my father he should send me to *Gymnasium*. I was the only one of the whole class who was thought to meet the requirements, yet at the same time, my parents and I were told: Many try, but few succeed. Certainly, this was not encouraging to me. I passed the admission examination to be exposed to teachers who were quite demanding, and some were grim or draconic. Most of them had participated in, or at least suffered from, the war and had been part of the prewar white-bread society. They felt school was a Darwinian institution that involved selection of the best. It was not for the fainthearted nor the sensitive. Only about one-third of those admitted to *Gymnasium*, the equivalent of high school and college in Germany, succeeded in getting through. Then, about 3% of an age group qualified for admission to university, compared to a proportion of 40% at present. After the first six years of *Gymnasium* and finding I had survived this insecurity, pressure, and threat, I enjoyed the remaining three years enormously. I began to take to physics, biology, math, and German classes. The American Forces Network in Germany had a daily outpost concert, with the overture to Donna Diana by Emil Nikolaus von Reznicek as the theme music. This started right when we came home from school, and for years, this was a regular pleasure after school stress. The years of the mid-1950s were the time when it became clear that there was a future for Germany. Munich, almost completely destroyed, was largely rebuilt. The impact of American culture, movies, music (including Elvis Presley), parties, and porch swings, which accompanied the reeducation of Germany, was enormous. America was the dream of a new, glorious life. But I also became fascinated by literature; I devoured Molière, Shakespeare, Shaw, Anouilh, even Corneille, and, of course, all the German classical writers, and later realized how little I had understood. When I finished *Gymnasium*, we were confident; life would be good for our generation, there would be jobs, and we would have to work hard, but there was a brave new world waiting for us. We enjoyed skiing in winter and mountain climbing in summer in the Bavarian and Austrian mountains, although everything, compared to present-day standards, happened under very frugal conditions.

When it came to decide what to study, I confess I was confused. I liked literature, but was this something really solid on which to build a profession? After half a year in law school, another one studying botany and zoology, and then one studying geology, I finally decided on chemistry. Rather irritated, not to say guilty, about my wanderings for two years, I rushed through my chemistry classes. I enjoyed the intellectual challenge of the analytical work, of synthesizing components, and the universe of possible novel compounds to be designed and made for the first time by clever experimentation. I considered then to do a PhD in organic chemistry, but when it came to decide on a supervisor, a deep fascination by the question of what really is at the bottom of living organisms made me choose biochemistry.

I should, however, admit what made my decision for biochemistry definite was an accident in my final year in an advanced inorganic chemistry course. This was in a lab that practiced the synthesis of boron tris azide, BN_9, a compound that has the tendency to explode by merely looking at it. Another student in the lab was to prepare lithium azide from sodium azide; both were not so dangerous substances as long as they were not mistreated. This step involved the use of ether, which then was to be removed by mild heating. Unfortunately, the student, after putting half a kilo of lithium azide in an oil bath on a tripod under a Bunsen flame, went for lunch together with his supervisor. The windows, with their frames, were blown into the street; I was grazed by a bullet, the tripod, and thrown out of the lab through the door and was deaf for a whole day. This finally cured my excitement for chemistry. The group leader devoted to BN_9 lost all but three of his fingers in the hood upon minor explosions at the time when he became professor.

GETTING HOOKED ON BIOGENESIS OF MITOCHONDRIA

I went through the biochemistry labs in Munich with complete ignorance of what biochemistry meant. There was one Nobel Prize winner, Adolf Butenandt, with his institute, but they were inaccessible to a boyish chemistry student. Feodor Lynen, not yet a Nobel Prize awardee, offered me a project to isolate the kinase that converted NAD to NADP, and then there was Theodor Bücher, the new holder of the chair of the Institute of Physiological Chemistry in the medical faculty. He was the last of Otto Warburg's PhD students. During the war, he continued to work with Warburg until the end when the Russians occupied Berlin and confiscated Warburg's lab and horses. Bücher's interest was in enzymes of glycolysis, metabolic regulation of the flow of hydrogen, and development of tissues and organs by the study of enzyme patterns. He had just completed a combined electron microscopic and enzymological analysis of the development of the flight muscle of African locusts [see Brosemer et al. (1)].

Bücher showed me the electron micrographs of the developing flight muscle (**Figure 1**). The development of a locust occurs through five instars, and the last molting changes the grasshoppers into little aeroplanes. This process includes the formation of mighty muscles in the thorax with powerful mitochondria densely filled with cristae. The muscle bundles are covered with oily fat, the fuel that enables the locusts to cover large distances. In the 1840s, a swarm of African locusts landed in Brandenburg near Berlin, some 2000 miles from the place where they had started. This high-performance motor, one may call it an insect Ferrari, develops over a period of a few days by an increase of mitochondrial volume by a factor of 20–50. It appeared to be an ideal model to study the biogenesis of mitochondria. I was deeply impressed by the question as to how these beautiful organelles are made and immediately was hooked.

The origin of mitochondria from bacteria had been a matter of speculation and debate for already many decades. Mitochondria were thought to be self-replicating entities in the cell. My supervisor told me to isolate mitochondria and to find conditions to make them grow in the test tube (they were bacteria!) with the unspoken proviso that I should come back when I had succeeded. This way of educating young scientists was the traditional German norm. Nowadays, this attitude has completely disappeared, and the principles of affirmation, close guidance, and supervision, traditionally practiced in the Anglo-American world, have become the gold standard. Looking back, I can see the problematic sides of the old education system, but also the benefits. It was not simply neglect of the young students, but rather respect for their own ways of growing up. By contrast, the old-world principle was clearly not the most efficient and productive one. Many young students got lost on their way sometimes because of minor deficiencies in their development, which could have been easily corrected by some advice and guidance at the right time.

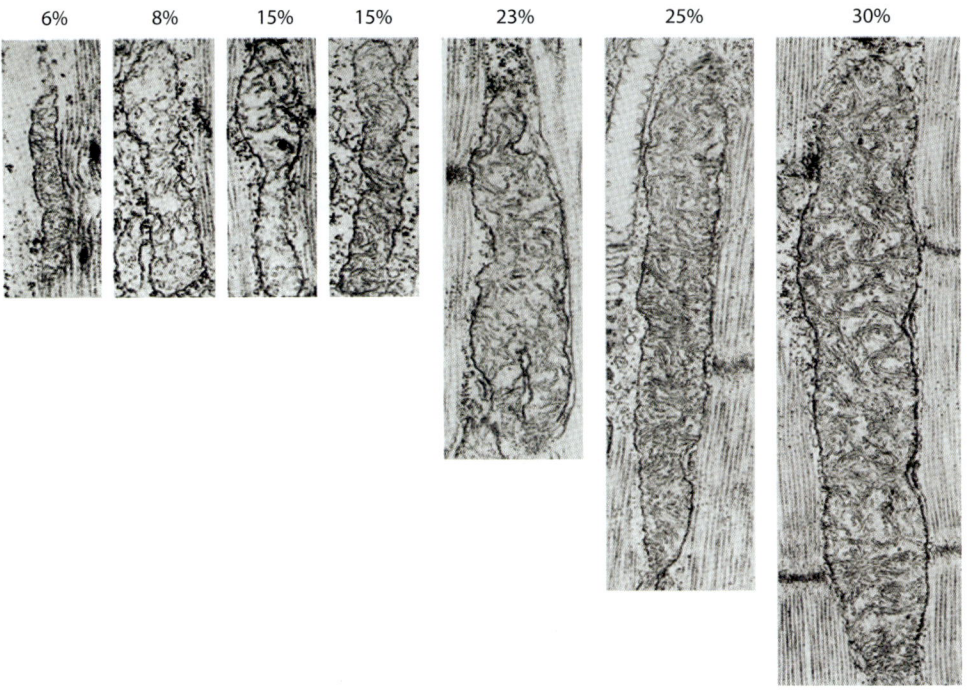

Figure 1

Mitochondria in the developing flight muscle of *Locusta migratoria* before and after the molt that leads it from the grasshopper stage to a flying insect. The sequence of electron micrographs shows mitochondria nine days, six days, three days, and one day before, as well as two hours, three days, and eight days after the imaginal molt. The percent values indicate the relative volume of the muscle occupied by mitochondria. Adapted from Reference 1.

I went to the Anti-Locust Research Center in London to obtain a stock of egg-laying locusts, which were the founders of hundreds of thousands of progeny. I scraped the flight muscles out of the thorax, isolated the mitochondria, and incubated them with radioactive leucine. One year after this start, I convinced myself that mitochondria can make proteins, and it was membrane proteins that were synthesized. This was supported by a few other reports studying the new field of mitochondrial protein synthesis in various organisms. My supervisor passed on to me his invitation to the first "Bari meeting." It was where, for the first time, I met the grand world of mitochondria research and its heroes. This was an exciting event for bioenergetics and biogenesis of mitochondria. The conference, in beautiful and then unspoiled Puglia, left a deep impression on me and so did the lectures. The fights and feuds among the various groups baffled me. There was a group challenging the existence of mitochondrial protein synthesis, arguing that it was an artifact owing to contaminating bacteria. The fiercest storm was on Peter Mitchell who presented his chemiosmotic hypothesis. Argument after argument destroying his hypothesis was thrown at him by the proponents of the "hypothetical intermediate." He listened quietly and after some time took out his hearing aids and put them on the table.

The work with locust mitochondria eventually led to the isolation and characterization of mitochondrial ribosomes. Surprisingly, they had a very small S-value, 55S to 60S, and they contained RNA of a very small size, almost half of that of the cytosolic ribosomes, which for quite some time made us believe they were degradation products (2). But then in several reports, mitochondrial ribosomal RNAs from mammalian cells were reported to be of a similar small size. These ribosomes, during high-resolution electron microscopy, revealed somewhat smaller dimensions than cytosolic ribosomes and a very similar morphology. As it turned out, the extremely small size of the RNA and the lack of 5S RNA are compensated for by an increase of the number of ribosomal proteins.

Somehow, during all this work, I found the time to enlist in medical school and to go through the whole curriculum of preclinical and clinical studies. Studying medicine, in comparison to chemistry, was highly entertaining. The lectures and practical courses were eventful, patients were presented and examined, clinical pictures were complex and their analysis was intellectually demanding. And one got a glimpse of the satisfaction that one can derive from providing help to patients. This was a dangerous temptation, but after I had completed my state exam and received my MD degree and driven by the exhilaration of discovering new things, I dismissed medicine. Yet, I felt enriched by having experienced a completely different exciting world.

A burning question, arising from the finding that mitochondria have the capacity to make proteins, was to find out which proteins are made. What now seems a trivial task was extremely challenging in the late 1960s because there were practically no tools to separate and analyze membrane proteins. So we started to find out which of the mitochondrial membranes contained mitochondrial translation products. These experiments made it necessary to switch from the locust flight muscle to another source that would allow isolation of larger amounts of mitochondria. We chose regenerating rat liver, because mitochondria from this tissue were highly active in synthesizing proteins and new elegant procedures were available to separate outer from inner membranes. The result was that radioactive proteins were found only in the inner but not the outer membrane.

At this stage, it became obvious that further progress would require a different biological system that was more accessible not only to biochemical experimentation but also to genetic analysis. The experiments of Beadle and Tatum with the filamentous fungus *Neurospora* had brought a revolution to eukaryotic genetics. However, not only was this organism accessible to genetic manipulation, but also it was easy to grow on simple inorganic media, and kilogram amounts of cells could be obtained in a very short time. The fungal hyphae could easily be broken, and the content be released to yield morphologically and biochemically intact mitochondria. Moreover, in beautiful experiments, David Luck (3) had described the making of new mitochondria from preexisting ones. In addition, mitochondrial ribosomes were isolated and characterized. We found their proteins to be synthesized by cytosolic ribosomes, as it later turned out, with the exception of one protein of the small subunit (4).

The mid-1970s were a period of great excitement with mitochondrial biogenesis in the Bücher lab in Munich. Walter Sebald and Hanns Weiss had joined the mitochondria group. They were eager to find out which proteins were made by the mitochondrial ribosomes. They worked out methods to isolate cytochrome oxidase and then also cytochrome *b*. Walter Sebald developed the method of differential labeling of proteins in intact *Neurospora* cells by using cycloheximide and chloramphenicol, which specifically inhibited cytosolic and mitochondrial translation, respectively. The differential labeling involved the use of two different isotopes to discriminate between total protein and protein synthesized in the presence of one of the inhibitors. In this way, Sebald discovered the three largest subunits of cytochrome oxidase to be made in the mitochondria, in contrast to the smaller ones [see Sebald et al. (5)]. Hanns Weiss identified cytochrome *b* as a mitochondrial translation product [see Weiss & Ziganke (6)]. These results were confirmed for mitochondria of yeast by the groups of Alex Tzagoloff in New York, Gottfried Schatz in Basel, and Piotr Slonimski in Paris. They had performed pioneering work on mitochondrial genetics in yeast as well as biochemical work on the enzyme complexes of mitochondrial oxidative phosphorylation. The results from all these labs came together in the early 1970s to provide an emerging picture of the genome of mitochondria of yeast and *Neurospora*. These studies were then completed by the discovery of three and two subunits of the ATP synthase in yeast and *Neurospora*, respectively, and one subunit of the mitochondrial ribosomal proteins.

EARLY WORK ON IMPORT OF PROTEINS INTO MITOCHONDRIA

In about 1975, I became interested in how the many mitochondrial proteins encoded in the nucleus and synthesized on cytosolic ribosomes would reach their sites of function in the mitochondria. Astonishingly, very few mitochondrial researchers found this question attractive. There seem to be several reasons for this. There was no immediate genetic approach, and this was the time when molecular biology made a steep rise. Maybe some people were simply put off by the obvious complexity of the process. Molecular geneticists and biologists may not have appreciated the importance of this problem, although it touched on many general aspects of cell biology, such as the mechanisms of movement of proteins across one or two membranes; the folding of proteins before, during, and after membrane passage; the addition of cofactors and assembly into multisubunit complexes; as well as the nature and specificity of signals for intracellular transport.

The processes involved in the intracellular and intraorganellar sorting of nuclear-encoded newly made mitochondrial proteins surfaced during a time when transport of proteins across and insertion into membranes came into the focus of cell biology. One of the major areas that developed at this stage was the secretion of proteins by eukaryotic and prokaryotic cells. Günter Blobel, David Sabatini, and others, inspired by the work of George Palade, studied the processes in the endoplasmic reticulum (ER) that, in 1976, led to the signal hypothesis by Blobel and Dobberstein. Bill Wickner, Jon Beckwith, and others were analyzing protein export and membrane insertion in bacteria with biochemical and genetic methods. Randy Schekman started to identify the proteins of the yeast secretory system by a combined genetic and biochemical approach. Cotranslational translocation and membrane insertion by membrane bound ribosomes were the predominant concepts.

The first proposal for mitochondrial protein import was made by Ron Butow [see Kellems et al. (7)]. He suggested that ribosomes would make mitochondrial proteins that sat at the mitochondrial surface, and therefore translocation would be cotranslational. However, under normal growth conditions, ribosomes were not seen in electron micrographs in association with the mitochondrial surface, in contrast to the ER. This suggested that import of proteins could occur after their synthesis. By contrast, in yeast cells treated with cycloheximide, which arrests translation, the mitochondrial surface was covered by ribosomes. This seemed to argue for a direct release of proteins by active ribosomes into the mitochondria.

Together with Matt Harmey, a sabbatical guest from University College Dublin, and Gerhard Hallermayer, a graduate student, we reasoned that the question of whether import was posttranslational or cotranslational could be solved by performing pulse-chase experiments. The basic assumption was as follows: If proteins are delivered to the mitochondria by ribosomes attached to the mitochondria, the appearance of proteins in mitochondria should occur with the same kinetics as the release of cytosolic proteins from cytosolic ribosomes. In contrast, if proteins were initially released from cytosolic ribosomes into the cytosol and then imported, there should be (*a*) a lag and slower appearance of these proteins in the mitochondria and (*b*) observable extramitochondrial pools of mitochondrial proteins with precursor product kinetics. We designed the experiments in the following way: *Neurospora* cells were grown in the presence of [^{35}S] sulfate to obtain a homogeneous labeling of cellular proteins; were given a pulse of [^{3}H]-labeled leucine for short periods, followed by addition of a large excess of unlabeled leucine to instantly terminate further labeling; were incubated for additional different periods of time (chase); were shock cooled to 0°C; were harvested with breaking of the hyphae; and were separated into cellular fractions; finally, we determined the ^{3}H/^{35}S ratios in proteins. It was necessary to develop methods to rapidly harvest and fractionate the cells, a relatively easy procedure with *Neurospora* hyphae, but this needed at least 10 hands. Total mitochondrial proteins and individual proteins were then analyzed by immune precipitation from cytosolic and mitochondrial fractions and sodium dodecyl sulfate gel electrophoresis. The ^{3}H/^{35}S ratios were determined in gel slices. This was before autoradiography of dried gels was introduced and had the advantage of allowing reliable quantitation. The kinetics varied from 90 to 1,500 s, and newly made proteins showed a distinct time lag for their appearance in mitochondria and continued to accumulate in the mitochondria after blockage by cycloheximide. Among different mitochondrial proteins, the lag phase of labeling varied. Together, this showed the existence of extramitochondrial pools of mitochondrial proteins of different sizes for the various proteins analyzed (**Figure 2**). In addition to these in vivo experiments, protein synthesis in a cell-free homogenate mimicked the delayed kinetics of the labeling of proteins in mitochondria as compared to proteins of the cytosol fraction (**Figure 3**). Altogether, these results demonstrated that transport of proteins into the mitochondria is essentially a posttranslational process. However, it did not exclude the possibility that some mitochondrial proteins could be translocated in a cotranslational manner. The accumulation of cytosolic ribosomes on mitochondria observed by Butow and colleagues (7) indicated the possible location of signals for transport at the N terminus of the nascent chains because translocation could apparently start if elongation was stalled by cycloheximide.

Two of the first proteins whose biogenesis we investigated in more detail were the ADP/ATP carrier and cytochrome *c*. We found that the latter protein was initially synthesized as apocytochrome *c* (8). Luckily, we obtained antibodies that were able to discriminate between both forms of this protein. The apo form was imported into the mitochondria where it was converted to holocytochrome *c* by addition of heme via thioether bridges.

These findings were presented at a conference on the biogenesis of mitochondria and chloroplasts, which we organized in Munich in 1976, and were published in 1977 (9, 10). They raised a large number of doubts and questions. What was the nature of the signals that directed the newly synthesized proteins to the mitochondria? How could completed proteins cross the mitochondrial membranes? What conformation would these proteins have? How could membrane proteins exist as precursors outside the mitochondria? What kind of driving forces would support translocation? How would the different proteins know to which subcompartment of the mitochondria they should go?

In 1979, the labs of Günter Blobel and Gottfried Schatz published a study demonstrating the synthesis of nuclear-encoded subunits of the ATP synthase with N-terminal extensions that were removed upon import (11). These experiments fully confirmed the results of our in vivo and in vitro studies, arriving at the same conclusion that import can be posttranslational. Later in the same year, we reported a study on the synthesis of the matrix protein citrate synthase for which we had observed a large extramitochondrial pool. Upon pulse labeling of intact cells, this protein was present initially as a larger precursor species, which was cleaved when it reached the mitochondria (12). In our initial experiments in 1976 and then in further studies, a difference in the molecular mass of the precursor and mature proteins with cytochrome *c* and the ADP/ATP carrier was not observed (9, 13). This led us to conclude that there is one class of precursor proteins with signals at the N terminus, which are generally cleaved after import, and another class with internal targeting signals. In retrospect, owing to bad luck in

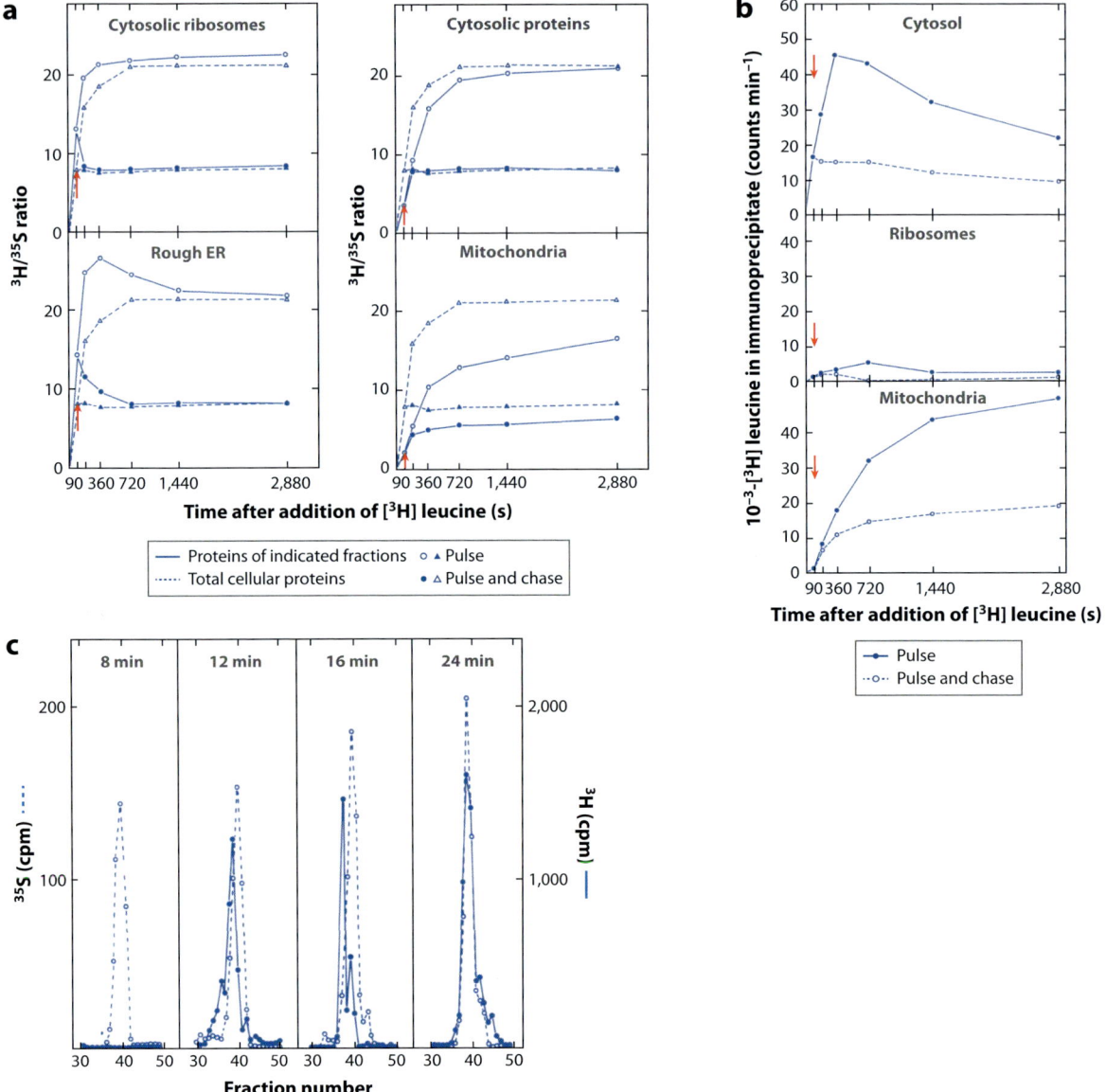

Figure 2

Kinetics of posttranslational protein translocation into mitochondria in intact *Neurospora* cells. (*a*) Kinetics of the appearance of newly synthesized proteins in mitochondria. Release of proteins from ribosomes is much faster than their transfer into the mitochondria, which shows a distinct lag phase. Cells were grown on [^{35}S] sulfate and then subjected to pulse labeling and pulse-chase labeling with [^{3}H] leucine at 8°C. Red arrows indicate the time of addition of a chase with unlabeled leucine; for comparison, the ^{3}H/^{35}S ratios of the total homogenate are included in all panels. (*b*) Translocation of matrix proteins into mitochondria is slow as compared to translocation of total mitochondrial proteins and indicates a large precursor pool in the cytosol. Cells were pulse and pulse-chase labeled with [^{3}H] leucine, and mitochondrial matrix proteins were immunoprecipitated from the indicated fractions. Arrows indicate addition of a chase of unlabeled leucine. (*c*) A larger precursor of citrate synthase appears upon pulse labeling, which is slowly converted to the mature citrate synthase. Cells were prelabeled with [^{3}H] leucine and pulse labeled with [^{35}S] sulfate. Citrate synthase was immunoprecipitated from cell homogenates after various times of labeling with [^{35}S] sulfate and analyzed by sodium dodecyl sulfate gel electrophoresis. Adapted from Reference 9. Abbreviations: cpm, counts per min; ER, endoplasmic reticulum; (s), seconds.

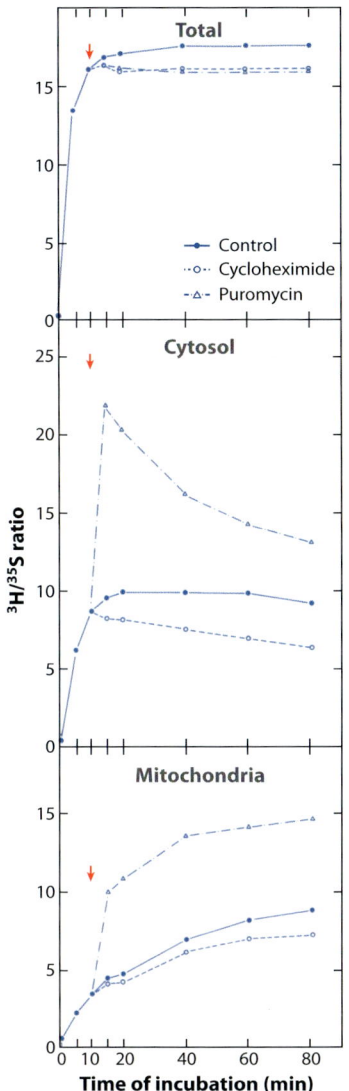

Figure 3

Rapid synthesis of proteins in an in vitro coupled translation-translocation system of *Neurospora* cells and slow uptake into mitochondria. Incorporation into various fractions of a cell-free homogenate were prepared from cells grown on [^{35}S] sulfate. Labeling was performed with [^{3}H] leucine. One aliquot was left untreated (control), or either cycloheximide or puromycin was added after 10 min (*red arrows*). After the indicated time periods, the homogenates were fractionated to separate the cytosol and mitochondria. Adapted from Reference 10.

our earlier experiments, we did not choose for analysis a precursor that belonged to the type with cleavable extensions. But importantly, our initial experiments revealed that it was possible to separate synthesis of mitochondrial proteins from import and to perform these steps in the test tube. Now, an experimental system was available to probe the many burning issues.

DIVERSITY OF IMPORT SIGNALS AND IMPORT PATHWAYS

In 1978, I accepted a professorship at the University of Göttingen. The town was small with some 100,000 inhabitants, including the 30,000 students, who were in the middle of a revolution dominated by Marxist students. The university was in a state of turmoil and paralysis, yet it still carried the historical reputation of a great place for science. At a time when the northern German Mescaleros, which included the progeny of the local professors, burned down buildings around my new lab, I was able to build up a larger group. This began with a few people, in particular Richard Zimmermann, who had started the work on the ADP/ATP carrier and came with me from Munich, and Matt Harmey, a regular guest from Ireland. Also, a number of new students and postdocs contributed to rapid growth of the lab. There was the most enjoyable atmosphere of a new beginning.

We looked at the mechanisms by which various selected proteins of the different mitochondrial subcompartments are imported into the mitochondria. The import signals and pathways of the various precursor proteins turned out to be of unexpected diversity. We studied the import of the ADP/ATP carrier in more detail and found that it required a membrane potential for insertion into the inner membrane (14). Receptor-like components on the outer membrane were required for efficient import. As an example of an outer membrane protein, we chose Porin, the most abundant protein of this membrane. We isolated Porin for the first time, which, as we showed together with Roland Benz at the University of Würzburg, was responsible for the VDAC (voltage-dependent anion channel) activity (15). Porin belongs to the family of β-barrel membrane proteins as was recently confirmed by X-ray crystallography. It required surface components, but not a membrane potential, for insertion into the outer membrane. Like the ADP/ATP carrier, it did not contain a cleavable signal. Therefore, we thought the difference between the precursor and mature forms might be just a matter of the folding state. Helmut Freitag, a postdoc with a very original way of thinking, converted the detergent-purified Porin into a water-soluble form. Magically, this then indeed behaved like a precursor and could compete with Porin synthesized in a cell-free system for import into the outer membrane (16). Thus, we could use it to quantify receptor sites and determine affinities. The nuclear-encoded subunit 9 of the ATP synthase from *Neurospora* served as a model protein for the inner membrane with a mitochondrial targeting sequence, which is exceptionally long and, as it turned out, is cleaved in two steps. This powerful matrix-targeting signal is still used when a passenger protein is to be sent to mitochondria for a variety of experimental purposes.

Along with the availability of efficient methods to study protein import in vivo and in vitro came the ability to analyze energy requirements. Roles for ATP and the electrical membrane potential were identified. However, these requirements were not uniform. Proteins of the mitochondrial matrix required both $\Delta\Psi$ (the electrical membrane potential) and matrix ATP. Some inner membrane proteins required $\Delta\Psi$ only, and outer membrane proteins neither. Some proteins of the intermembrane space needed neither matrix ATP nor a $\Delta\Psi$, others required both. Matrix proteins turned out to require $\Delta\Psi$ only for translocation of the N-terminal targeting signals across the inner membrane and matrix ATP for the translocation of the rest of the polypeptides. This apparently confusing picture was clarified only when we learned more about pathways and the import machineries. The precise role of $\Delta\Psi$ is still not clear; it might be involved in a kind of electrophoretic effect on the positively charged targeting signals or may have a role in gating the opening of the protein-conducting pore.

In 1983, I took over the chair for Physiologische Chemie of the medical faculty in Munich. I was excited by the offer of this prestigious chair. Although the academic atmosphere in Göttingen was very inspiring and the empty country around the town very charming, I knew I would be happier in Munich, where I was born and grew up. Indeed, both my hopes and fears came true. Not only did a number of students and postdocs come along with me to Munich, but I also immediately recruited excellent new people. Soon, my lab included Nikolaus Pfanner, who initially came to perform work for his MD degree and left seven years later to hold the chair of biochemistry at Freiburg University; Ulrich Hartl, who already had obtained his MD; Rupi Pfaller and Thomas Söllner, graduate students; Max Tropschug, a postdoc who introduced the art of cloning; Don Nicholson, a postdoc from London, Ontario; Matt Harmey, who was back from Ireland; and Rosemary Stuart, a graduate student from Dublin. Swiftly, the new lab was set up, the institute was reorganized, and the experimental work became productive. And I enjoyed immensely being back in Munich, where my family was and where my Bavarian heart felt at home.

As expected, my position in Munich brought many commitments on top of organizing the institute, including duties in the medical faculty, which I could avoid for a long time but not entirely. For a number of years, I was vice dean and dean of research. Serious work was also required from me because of my membership in the senate and the governing body of the German Research Council and several subcommittees of this main German funding agency. I served as president of the German Biochemical Society, as chair of the EMBO Council and on the boards of other international scientific organizations. I was sitting on advisory committees and organized conferences. I did my full share in working as a member of editorial boards and as a reviewer. When I cleared my office after retirement, there were some 100 thick folders with reviews, references, and statements to be destroyed. This did not happen without pondering about how much of my life had been consumed. I never really liked all these responsibilities and did not get satisfaction from a single one of them. Still, I felt that as a member of a community that only works if everybody cooperates, I had to make a contribution.

The move to Munich also meant a large teaching load. In the beginning, there were some 500 medical students per year to be taught in lectures, seminars, and practical courses; when I left, there were some 1,000 students. It is a common experience that not all medical students are fond of biochemistry, but there were always enough extremely gifted and interested students who made teaching a pleasure.

FOLDING AND UNFOLDING OF PROTEINS BEFORE AND DURING IMPORT

How are proteins crossing the mitochondrial membranes? Proteins released from cytosolic ribosomes generally undergo rapid folding to their functional conformations. An obvious question was whether this was also true for mitochondrial precursor proteins, and another was whether these proteins would cross mitochondrial membranes in a folded state. As the experiments told us, this was not the case; rather most precursors are not tightly folded. ATP in the cytosol was found as a requirement for many precursors to remain import competent after release from the ribosomes. Further experiments identified a role of cytosolic Hsp70 proteins in interacting with precursors and in preventing them from misfolding and aggregation. Cytosolic Hsp70 and Hsp90 were later found to have an additional function in targeting precursor proteins to the receptors on the outer membrane. These observations hinted at the possibility of precursors crossing the mitochondrial membranes in an unfolded state. Could intermediates arrested in the translocation process be observed? I remember precisely the moment when Manfred Schleyer, a graduate student, showed me a sodium dodecyl sulfate gel in which bands could be seen representing precursor proteins, which were processed by the peptidase in the matrix but were still accessible to added proteases in intact mitochondria (17). Intermediates of this type, which are spanning the outer and inner membranes at the same time, then became invaluable tools for studying the import mechanisms of precursor proteins. Furthermore, binding of antibodies to precursors could lead to a permanent arrest of spanning intermediates, and such complexes were present at sites where outer and inner membranes come close together at contact sites (18). These observations suggested that precursors were crossing the membranes in an unfolded state. Conclusive experiments were then performed by Martin Eilers in the Schatz lab. Cytosolic mouse dihydrofolate reductase (DHFR) was fused to a mitochondrial targeting sequence. This chimeric protein contained a folded DHFR domain but still could be imported into mitochondria (19). This meant it became unfolded, a conclusion that was confirmed by adding the DHFR antagonist methotrexate, which stabilized the domain and completely prevented translocation. These results had important impacts on a number of issues. One immediate problem was that after reaching the matrix, proteins must be folded again. But this was a story that went far beyond the world of mitochondrial biogenesis, and that will be told below.

During the 1980s, the Schatz lab and my lab had very good exchanges of information, and we had regular meetings every year, alternating between Munich and Basel. These meetings were extremely productive and provided new views as well as inspiration for new avenues of research. Our groups were the largest working in this field. In later years, more and more overlap of interests developed, competition was difficult to avoid, and this began to limit free communication. Still, I am convinced these contacts promoted the field enormously, and cooperation is generally much more efficient than competition.

MITOCHONDRIAL PROCESSING PEPTIDASE

The year 1988 was when the mitochondrial processing peptidase was identified. Ulrich

Hartl and Gerd Hawlitschek took a biochemical approach and purified two proteins from *Neurospora*, which together efficiently cleaved mitochondrial targeting signals, initially named MPP and PEP, then α- and β-MPP [see Hawlitschek et al. (20)]. cDNA cloning was used to identify their sequences. In the same year, the Schatz group identified the *MAS* genes, out of a class of yeast mutants (*mas* mutants) that were deficient in mitochondrial assembly. The genes *MAS1* and *MAS2* encoded the yeast homologs of α-MPP and β-MPP, respectively (21, 22). In collaboration with Arthur Horwich at Yale University, who had isolated yeast mutants deficient in mitochondrial biogenesis, we also identified the *mif2* gene, which encodes β-MPP. That the competing groups in the field independently came up (more or less at the same time) with very similar results was a theme that was to prevail in the following years. Sometimes one is inclined to believe in the existence of something akin to parapsychology. But more likely, even if we do not speak openly about our new results, somehow these findings diffuse without our saying anything.

Yet there was one completely unexpected finding. We observed, together with Hanns Weiss in Düsseldorf, who was studying complex III of the *Neurospora* respiratory chain, that β-MPP was also present as core protein I of this complex (23). Core I and II and the MPP proteins belong to one family, and these proteins are apparently derived from bacterial proteases during evolution (24). So there was a protein involved in the processing of precursor proteins, and at the same time, this protein has an essential function in the respiratory chain. Absence of the core proteins leads to a nonfunctional complex III. In yeast, the core proteins and the MPP subunits are related but not identical. MPP proteases are metalloproteases that belong to the Pitrilysin family. X-ray structures showed the relationship to the core proteins in atomic dimensions and yielded profound insights into the enzymatic mechanism of this unusual protease.

MITOCHONRIAL RECEPTORS FOR PRECURSORS AND THE TOM COMPLEX

The approach used to study specific binding on the mitochondrial surface and to determine the numbers and affinity of sites enabled us to find proteins with a receptor-like function. The first one was MOM19, after unification of the nomenclature called Tom20. This protein is anchored to the outer membrane and has a hydrophilic domain facing the cytosol. Specific antibodies to Tom20, when added to isolated mitochondria, were able to inhibit import of a subset of precursors, including those of the outer membrane and the matrix, but very inefficiently inhibit import of the ADP/ATP carrier (25). Soon afterward, we identified another outer membrane protein, MOM72, then Tom70, which like Tom20 was anchored to the outer membrane by an N-terminal transmembrane segment, exposing its large hydrophilic domain into the cytosol. Antibodies against Tom70 strongly inhibited binding to the mitochondrial surface and import of the ADP/ATP carrier, but these antibodies had little effect on the import of precursors, which interacted with Tom20 (26). This work was performed with *Neurospora* mitochondria; Gottfried Schatz's group produced very similar results with yeast, a mutually reassuring progress.

Our previous work had led us to postulate the existence of a "general insertion pore" functioning as a protein-conducting channel in the outer membrane. Having specific antibodies available for the receptors, we isolated a complex, which in addition to Tom20 and Tom70, contained two more proteins, Tom40 and Tom22 (27). Tom40 was identified, and the cDNA sequenced by Michael Kiebler in my group. These results were published back to back with the identification of Tom40 in yeast (28). This exciting work on the translocase of the outer membrane was initiated by Nikolaus Pfanner, Thomas Söllner, and Rupert Pfaller and was one of several lines of research that afterward occupied our group and several others for many years. We found three more subunits in the TOM complex that are important for the stability of this complex (29).

Our work on the TOM complex concentrated on its functional and structural characterization. We stayed mainly with the complex from *Neurospora* because of its accessibility in large amounts and efficient isolation. Frank Nargang of the University of Edmonton, an excellent *Neurospora* geneticist and collaborator, generated mutants and put tags on the individual subunits of the TOM complex and also on those of a number of other complexes of the mitochondrial import machinery. Tom22 was found to cooperate with Tom20 as well as with Tom70 in the passage of precursors to the general insertion pore. These three proteins are essential components required for the formation of intact mitochondria with cristae (29). Tom22 has abundant negative charges on its N-terminal domain, which faces the mitochondrial surface. A mutant analysis suggested the negative charges play an important role in the recognition of the positively charged side of the amphipathic targeting signals. Indeed, as shown recently by the fine structural and biochemical work of Toshi Endo and colleagues (30), this is what takes place when Tom20 and Tom22 together bind matrix-targeting signals. Roland Lill, a postdoc and then group leader together with Andreas Mayer, took over the part of characterizing the TOM complex functionally. They studied the TOM complex in isolated outer membranes and were able to discriminate two different binding sites: the *cis* site on the surface of the outer membrane containing the receptors, and the *trans* site on the inner face of the outer membrane (31, 32). Depending on the experimental conditions, the N-terminal targeting signals of precursors, but not the rest of the polypeptide chains, could transit from *cis* to *trans*. Precursors were observed to move to the *trans* site spontaneously, pointing to a higher affinity of the *trans* site compared with the *cis* site. Interestingly, when a DHFR domain was placed right behind the targeting signal, the DHFR became unfolded upon interaction with mitochondria. Binding of such constructs did not occur, if folding was forced by addition of methotrexate; when they were first bound to the TOM complex and then methotrexate was added, folding occurred as well as release from the mitochondria. Thus, binding and unfolding were equilibrium reactions, an important piece of knowledge for later studies on the mechanism of the import motor.

After a few years of work, Klaus-Peter Künkele and Marcus Dembowski, graduate students, and Roland Lill succeeded in isolating the TOM complex from *Neurospora* in a pure state (33). It came in two forms, (*a*) the core complex containing Tom40, Tom22, and Tom20 and the small subunits Tom7, Tom6, and Tom5, the latter three all C-terminally anchored proteins; and (*b*) the holocomplex containing in addition the receptors Tom20 and Tom70. Single-molecule electron tomography upon negative staining revealed a large structure with three holes in the case of the holocomplex and two holes of ∼2 nm in diameter for the core complex (this work was performed by Stefan Nussberger with the Baumeister group in Martinsried). Still, an X-ray structure of the complex was missing; to our great disappointment, crystals, although easily obtained, did not sufficiently diffract. Together with the group of Michel Thieffry in Paris, we found the reconstituted complex had characteristic cation-selective and voltage-gated pores, which were regulated by mitochondrial targeting peptides (34). The TOM core complex displayed three different levels of conductance, suggesting the existence of two pores. The isolated TOM complex, as then shown by Doron Rapaport, a postdoc from Israel, retained a number of its in vivo properties, such as specific and high-affinity binding of precursor proteins (35). The TOM core complex could be dissociated to yield a subcomplex consisting of a single ring of Tom40 molecules with only two conductance states, demonstrating the central role of Tom40 as the pore-forming entity. Circular dichroism spectral analysis of

Tom40 suggested that β-strands are its main structural elements, which was in agreement with prediction programs (36). Still it is unclear as to whether the pore seen upon electron microscopy and made up by several Tom40 subunits is the protein-conducting channel or whether single Tom40 subunits constitute this channel. Recently, certain proteins that move through the TOM channel were observed to have the ability to exit laterally into the outer membrane. Thus, the TOM complex still poses many intriguing questions—even 20 years after its discovery. Roland Lill moved to the University of Marburg in 1995 to become professor of cell biology and discovered the complex machinery of mitochondria that provides the various cellular compartments with the iron sulfur clusters required for the formation of iron-sulfur proteins, an eminent contribution to understanding mitochondrial biogenesis.

HEAT SHOCK PROTEIN HSP60 AND THE DE NOVO FOLDING OF PROTEINS

Hsp60 and its cofactor Hsp10 appeared on the mitochondrial landscape when we entered into a collaboration with Arthur Horwich, at Yale University, to functionally characterize yeast mutants with defects in mitochondrial import. Ulrich Hartl and Joachim Ostermann analyzed the *mif4* mutant of the Horwich collection of temperature-sensitive mutants [see Cheng et al. (37)]. The phenotype of the *mif4* mitochondria was stunning. They could import precursors into the matrix, but their assembly in the matrix was defective. We proposed that protein-catalyzed assembly takes place in mitochondria and that the *MIF4* gene product had a role as a "workbench" in assisted acquisition of the native conformation of oligomeric proteins. We were particularly excited by this observation when Hsp60 turned out to be the protein encoded by the *MIF4* gene. Hsp60 is a homolog of the bacterial GroEL and the Rubisco subunit-binding protein, RBP60, of chloroplasts. GroEL was known for its role in mediating the assembly of λ-phage. RBP60 was found, by the group of John Ellis in Warwick, to be involved in the assembly of the nuclear-encoded small subunit and the chloroplast-encoded large subunit of Rubisco. They proposed the term molecular chaperones for components facilitating assembly of large protein complexes (38).

Ulrich Hartl then followed the scent of an exciting problem. We knew that proteins are unfolded when they enter the mitochondrial matrix, which is similar to the situation prevailing when polypeptides are emerging from the translocation channel of ribosomes during protein synthesis. Was it primarily the folding of the proteins in the mitochondria that required the action of Hsp60? This question was revolutionary because folding of proteins was generally considered a spontaneous process. As demonstrated by Anfinsen (39), purified ribonuclease A can refold spontaneously after denaturation. Convincing as the Anfinsen experiments were, it was, however, largely ignored that most proteins when unfolded would not refold in the test tube. This was attributed to the experimental conditions, which would be different in the cell. We showed that Hsp60 interacted with unfolded precursor proteins imported into the matrix and that they were released folded in an ATP-dependent manner (**Figure 4**) (40). The conclusion was that folding of proteins in mitochondria is catalyzed by a transient interaction with Hsp60, a process driven by ATP hydrolysis. As it turned out, not all proteins require Hsp60 for folding, but the general principle of chaperone-mediated folding in the cell was established. Within a few years, not without fights and feuds, the existence of catalyzed, chaperone-mediated protein folding was generally accepted. Ulrich Hartl became completely absorbed by protein folding in vivo and in vitro and quickly drifted away from mitochondrial biogenesis into the rapidly emerging new field of protein folding in the cell. He left the Munich lab to join the department of Jim Rothman as an associate member at Sloan-Kettering Institute in New York and returned five years later to become director at the Max Planck Institute of Biochemistry in Martinsried.

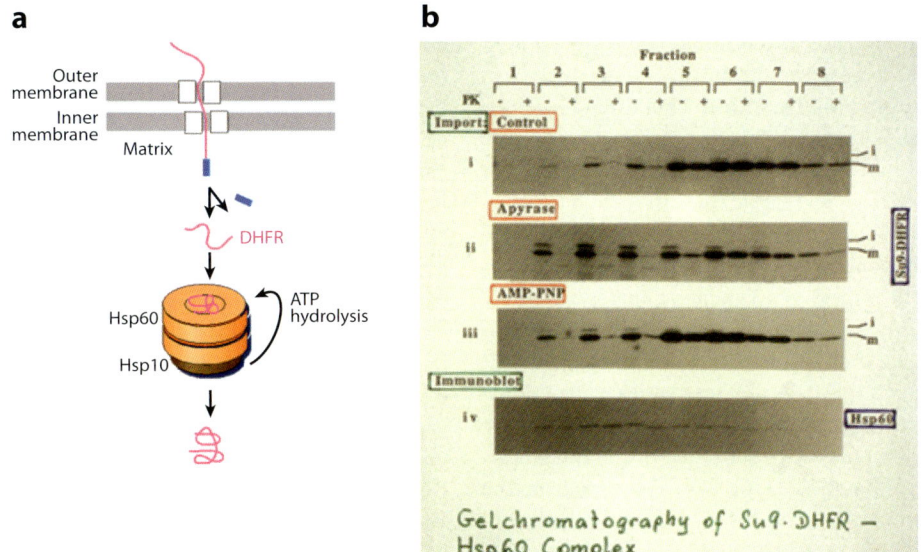

Figure 4

Hsp60-mediated folding in the mitochondria. The unfolded protein dihydrofolate reductase (DHFR) after import into mitochondria is bound to Hsp60 chaperonin in the matrix and released in the presence of ATP in a folded state. The white boxes in the outer and inner membranes symbolize the protein translocases TOM and TIM23. The blue bar represents the matrix-targeting signal. (*b*) The unfolded imported fusion protein Su9-DHFR (Su9, mitochondrial targeting-sequence of subunit 9 of the F_1F_O-ATP synthase of *Neurospora crassa*; DHFR, mouse dihydrofolate reductase) accumulates with Hsp60 in the absence of ATP, but not in its presence, where it is recovered as folded protein. Adapted from Reference 40. Abbreviations: PK, proteinase K; AMP-PNP, adenylyl-imidodiphosphate.

THE ROLE OF MITOCHONDRIAL HSP70 IN PROTEIN IMPORT

Support of protein folding in the mitochondrial matrix is not the only role of ATP in the import and assembly process. As was found early on, precursor proteins would not enter mitochondria at low levels of matrix ATP. So the question was: Which components would transduce the free energy of ATP hydrolysis into the energy for movement of polypeptide chains across both mitochondrial membranes? Experimental access to this question became possible once Betty Craig had created temperature-sensitive mutants of Ssc1, the mitochondrial member of the Hsp70 family in yeast. The mtHsp70 (Ssc1) is an essential protein and the closest relative of bacterial DnaK among all the various Hsp70s (41).

Nikolaus Pfanner and Joachim Ostermann studied the effects of functional depletion of mtHsp70 on mitochondrial protein import [see Kang et al. (42)]. Import was interrupted at a stage in which the N-terminal targeting signals had crossed the inner membrane and were processed by MPP, but the bulk of the precursors were still outside the outer membrane. MtHsp70, therefore, is required for translocation of the mature parts of the precursor proteins, but not for translocation of the targeting signals. Yet, unfolded DHFR as passenger domain was still imported to a low degree in the mutant, apparently owing to residual activity of mtHsp70, whereas folded DHFR was not. We concluded that mtHsp70 is not only required for translocation, but also for the unfolding of folded domains in order to pass through the mitochondrial membrane barrier. Moreover, imported DHFR did not become folded after being completely transferred into the matrix, in contrast to wild-type mitochondria. This indicated that mtHsp70 is also required for folding in the matrix. In further experiments, a direct interaction of mtHsp70 with precursor proteins was demonstrated. In summary, mtHsp70 turned out to be part of a matrix-based motor for the import of polypeptide chains.

Over the years following this initial discovery, the central role of mtHsp70 in protein import was substantiated by a series of functional studies. Precursors arrested by a folded DHFR domain on the cytosolic side of the mitochondria, as found by Doug Cyr, a postdoc from the University of North Carolina at Chapel Hill, and Christian Ungermann, a graduate student, were held in place by mtHsp70 (43, 44). At low levels of ATP in the matrix, they underwent reverse movement as long as the segments having reached the matrix were short, but not when they were long enough to form folded structures. We determined the length of segments of precursors spanning both TOM and TIM complexes, about 45–50 residues, translating into about 15 nm of length of an extended polypeptide chain. The TOM complex measures ~7 nm across the membrane, and the TIM23 complex should not be smaller, meaning the precursor chain must be essentially unfolded. This substantiated our previous conclusion that precursors crossed the both membranes in an extended fashion. Furthermore, we could dissect the roles of the driving forces, ATP hydrolysis and the membrane potential. In the early stages, the electrical membrane potential drives translocation of the matrix-targeting signal, and its binding to mtHsp70 is necessary to make this step irreversible. The precursor chain can oscillate in the import channel, and the cycling of mtHsp70 on and off the incoming polypeptide chain, powered by ATP hydrolysis, then leads to vectorial movement.

The process of reverse movement (or backward slippage) is a salient feature of a specific dual targeting pathway. A number of matrix enzymes are present both in the mitochondria and in the cytosol. In case of fumarase, a single gene encodes a precursor with a matrix-targeting signal. As demonstrated by Ohad Yogev & Ophry Pines (45), a fraction of the precursors end up in the matrix as processed mature enzyme. Another fraction is localized to the cytosol. It is also processed by MPP but then undergoes reverse movement. Folding of part of the precursor in the cytosol and interaction with the Hsp70s in matrix and cytosol appear to determine the degree of the eclipsed distribution.

Like all members of the Hsp70 family, mtHsp70 interacts with cochaperones. Together with Neil Rowley, a graduate student from Cambridge, Benedikt Westermann and Elisabeth Schwarz in my lab identified Mdj1 as the equivalent of bacterial DnaJ [see Rowley et al. (46)]. A surprise was the discovery of a protein we named Hep1, for Hsp70 escort protein. This protein exists in all eukaryotes, and its deletion leads to aggregation of mtHsp70. It appears that Hep1 has a role in stabilizing mtHsp70 in certain stages of its ATP-dependent conformational cycle.

THE TIM23 COMPLEX OF THE INNER MEMBRANE

With the growing insights into the complexity of the import pathways, it was realized that the molecular machinery mediating translocation into and across the inner membrane would have a multiplicity of tasks: Recognition of targeting signals in the intermembrane space after their initial recognition by the TOM complex, opening and closing of a precursor-conducting pore in the inner membrane while protecting the membrane potential, differentiation between precursors destined to the matrix and precursors to be laterally released into the inner membrane, and coordination of the precursor-conducting channel with the TOM complex and with the import motor to prevent idling.

The first three constituents were identified by genetic screening. Michel Meijer and coworkers (47) in Amsterdam discovered a gene encoding Tim44, a hydrophilic protein; this was also found in a biochemical assay by the

Schatz group. Rob Jensen's group (48, 49) identified Tim23 and Tim17 as constituents that were in contact with the translocation intermediates.

At this juncture, Michael Brunner came to our lab after a postdoc period with Jim Rothman in New York. He set out to study how the translocase of the inner membrane works. In a first set of experiments, together with several graduate students, he looked at the interaction of mtHsp70 with the membrane components identified. They found that mtHsp70 interacts with Tim44 in an ATP-dependent manner (50, 51). In the presence of ATP, the Tim44-mtHsp70 complex locked precursor polypeptides in a position in which they cannot undergo reverse movement. When a folded DHFR domain was placed at the C terminus of a precursor, this was held in tight apposition to the TOM complex. These results and others described above led us to propose that the import motor works as a molecular ratchet (52). This view was not shared by others who instead suggested a power stroke model in which mtHsp70 actively pulls on the incoming polypeptide chain. The molecular ratchet relies on the ability of the polypeptide chain to slide in the protein-conducting channel after translocation has been initiated by the effect of the membrane potential on the targeting signal. It further relies on the equilibrium of the (partial) folding and unfolding of precursors at the entry of the TOM-TIM complex. Since the time when these opposing models were put forward, a large amount of experimental data have been accumulated in favor of the molecular ratchet model (52a, 53).

In a next step, we found out that Tim44 forms a complex with the two other components, Tim23 and Tim17. We first termed this complex MIM; later, it became the TIM23 complex. In the isolated complex, we found additional components, in particular a 14-kDa protein, but this had to wait eight more years to be identified. The complex was unfortunately rather unstable in the presence of detergents; it was only with digitonin that it could be isolated together with accumulated precursor polypeptides and the TOM complex, indicating a close cooperation of TOM and TIM23 in the import process.

Tim23 and Tim17 are proteins that are predicted to span the inner membrane four times. These segments show sequence similarity, which points to a common origin of both proteins. In addition, yeast Tim23 has a hydrophilic part at the N-terminal side, comprising ~100 amino acid residues. We performed an analysis of the latter part of Tim23 and found the segment comprising residues 50–100 to be essential for the function of the TIM translocase. We suggested that this part of Tim23 acts as a receptor for mitochondrial targeting signals in the intermembrane space and responds to the membrane potential by changing the conformation and oligomeric state of Tim23 (54). A specific transmembrane helix of Tim23 appears to play an important role in the opening of the translocation channel (55). Altogether, Tim23 seems to be a voltage sensor in mediation of the membrane potential–dependent step of translocation of matrix-targeting signals across the inner membrane. The mechanistic details of the opening and closing of the protein translocation channel, however, are still unclear. Tim17 and Tim23 interact with Tim44 by forming dimers or higher oligomers. On the basis of these findings, we suggested a basic mechanistic property of the import machinery, namely that it acts in a hand-over-hand mechanism in which two neighboring Tim44 and mtHsp70s work together at the outlet of the translocation channel to hold the translocating chain, a plausible element of a ratchet-like import machinery.

A further component of the TIM23 translocase, Tim50, was discovered independently by the groups of Pfanner (56) and Endo (57) in yeast, and in my lab by Kai Hell and Dejana Mokranjac in *Neurospora* [see Mokranjac et al. (58)]. Tim50 is anchored to the inner membrane; its intermembrane space domain interacts with that of Tim23, a process responsible for the recognition of matrix-targeting signals and their passage to the translocation channel. Tim50 is apparently the first component of the TIM machinery to recognize the targeting signals of precursors coming in through the TOM complex. Tim50 can be cross-linked to precursors even in the absence of a membrane potential and stays in their vicinity until the complete polypeptide chain has reached the matrix (58). It took some three years more until the most recently found components of the membrane sector of the TIM23 translocase were discovered, Pam17 and Tim21 (59–61). These are components that are not essential for the viability of yeast. Their association with the other constituents is not a firm one, and they change their interaction with the other subunits depending on the type of precursor translocated. Their precise functions are not known, yet Tim21 has the ability to interact with the TOM complex and thus may have a role in tethering both translocases.

The number of constituents of the import motor of the TIM23 translocase has also expanded since the discovery of Tim44 and mtHsp70. The nucleotide exchange factor Mge1 was found to be absolutely necessary for driving import because it catalyzes an essential step in the binding and release cycle of mtHsp70 to Tim44 and the precursor polypeptide to be imported (62). Then, we identified Tim14, the 14-kDa component observed previously in a complex with Tim44, Tim23, and Tim17 (63). This was again a case where three labs at the same time, in this case at the same meeting, presented a new component of mitochondrial import machinery (63–65). Tim14 (synonym Pam18) is a J domain protein and is membrane anchored. It has the ability to stimulate the ATPase activity of mtHsp70 and can be considered as the "accelerator" of the import motor. Immediately thereafter, we discovered Tim16, which forms a complex with Tim14 (66). The complex, in contrast to the isolated Tim14, did not stimulate mtHsp70 ATPase activity. Thus, it is acting as a "brake" of the motor. In further studies, using *in organello* cross-linking experiments, Tim14 and Tim16 were found in close contact with Tim44 and mtHsp70. These interactions were dependent on the nucleotide present, suggesting that extensive conformational changes take place during the cycles of the motor, involving binding and release of mtHsp70 to and from the precursor polypeptides (67). A more-detailed model of the motor was possible when Dejana Mokranjac and Michael Groll obtained a crystal structure of the Tim14-Tim16 pair. This pair forms a heterotetramer of two subunits each. The structure explains how Tim16 inhibits Tim14 and suggests the way they control the activity of the import motor (**Figure 5**) (68).

With 10 components in total found so far in the TIM23 translocase, models proposed for the mechanisms of action are not are not likely to be convergent. Thus, different views exist about how the divergent processes of complete translocation into the matrix versus lateral release into the inner membrane are triggered and regulated. Likewise, it is a matter of debate whether the membrane part and the import motor act as a single entity or by assembly of these modules upon demand (60, 69–71). The latter model would imply an additional signal in precursor proteins to trigger the assembly reaction for which there is no evidence so far.

And there are more unsolved questions. It is not understood how the electrical membrane potential is involved in the opening of the protein-conducting channel. The roles of Tim17 and Tim23 in forming the protein-conducting channel are still a mystery. The coordination of the various components of the membrane module and the motor as well as the interaction with the TOM complex are also poorly understood. X-ray structural analysis, it is hoped, will shed light on these questions.

SURPRISING ASPECTS OF THE TIM23 TRANSLOCASE

The TIM23 complex surprised us with unexpected variability in its functions. I want to illustrate this by three examples. A first one is the import of Bcs1 type substrates. The AAA-ATPase Bcs1 is involved in the biogenesis of the Rieske FeS protein (72). It is anchored to the inner membrane by a transmembrane domain close to the N terminus, and its

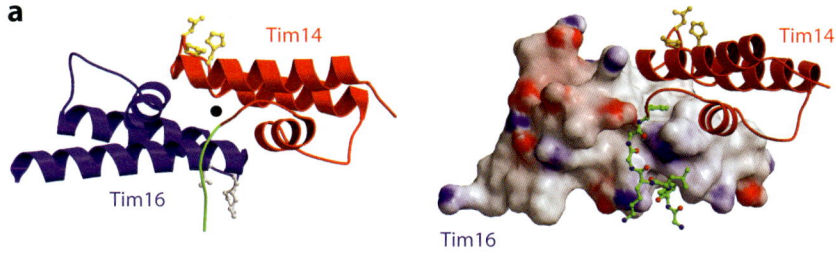

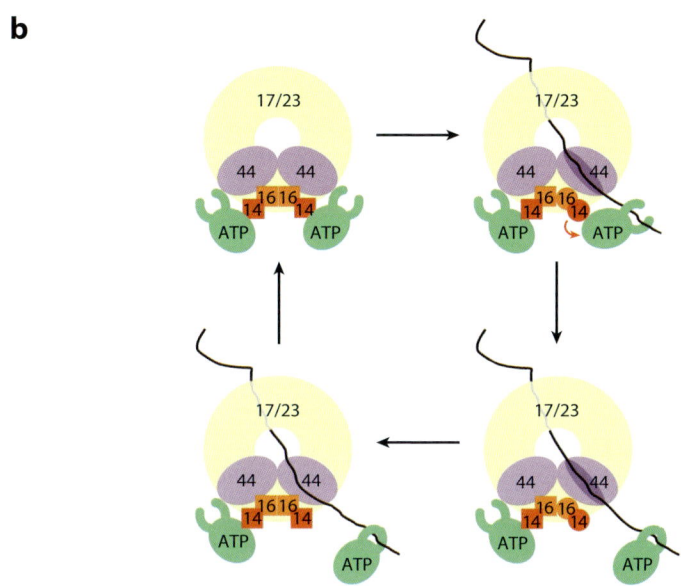

Figure 5

The import motor of the TIM23 translocase. (*a*) X-ray structure of the Tim14-Tim16 subcomplex and (*b*) molecular ratchet model of the motor. Adapted from References 68 and 116. The numbers of the protein components represent affixes to their names as Tim components and reflect their respective molecular mass in kDa.

ATPase domain is exposed to the matrix (73). This protein is synthesized without an N-terminal targeting signal, but such a sequence is present internally right after the transmembrane domain. Taken out of context and put in front of a passenger protein, this sequence works perfectly well as a targeting signal. In Bcs1, however, it works only in conjunction with the transmembrane domain. Rosemary Stuart and Heike Fölsch showed that the Bcs1 precursor threaded into the TIM23 channel as a loop structure (73). This leads to insertion of the membrane segment with an N-out orientation and to the sliding of the rest of the polypeptide through the channel into the matrix.

A second example is the biogenesis of the two forms of the mitochondrial dynamin-like GTPase, Mgm1 (74, 75). This protein plays an important role in the fusion of mitochondria. Mgm1 occurs as a long form, l-Mgm1, anchored by a transmembrane segment some 100 residues from the N terminus, exposing its large GTPase domain to the intermembrane space. About half of the protein exists in this form. The short form, s-Mgm1, lacks the

N-terminal anchor of l-Mgm1 to the inner membrane. It is generated by cleavage at a site that is located at the C-terminal side of the transmembrane anchor present in the l-Mgm1, a reaction catalyzed by the rhomboid protease Pcp1. Pcp1 is an integral membrane protease that cleaves a peptide bond present in or near the lipid phase of the inner membrane. Thus, about half of the Mgm1 precursors are not arrested in the inner membrane when the membrane anchor sequence crosses TIM23 but slips through to allow the more C-terminally located cleavage site to be seen by Pcp1. Because the presence of both forms of Mgm1 is required for mitochondrial fusion, the ratio must be carefully regulated. The factors responsible for this are not completely clear. One factor is the activity of the import motor, which again depends on the ATP levels in the matrix. The homolog of Mgm1 in higher eukaryotes is Opa1, a protein that is involved in crista structure formation and that plays a role in apoptosis. Interestingly, Opa1 is not cleaved by the mitochondrial rhomboid protease, but rather by the m-AAA protease paraplegin, implying a surprising change in the regulation of its synthesis during evolution (76, 77).

A third example is the ability of yeast Tim23 to interact with the outer membrane so that its N terminus becomes exposed on the surface of the outer membrane (78). This reaction was most surprising because in this way a protein is generated that is anchored to the inner membrane by four transmembrane segments, has a domain of ∼50 residues in the intermembrane space, and crosses the outer membrane with a sequence that is not of the usual transmembrane type. Insertion of the N terminus is triggered by the presence of a translocating chain in the TIM23 complex and is reversible (69). The function of this reaction is not clear, but it may help TOM and TIM23 to find each other for import of precursors. Fusion of stably folded protein domains to the N terminus of Tim23 leads to proteins that are permanently spanning outer and inner boundary membranes without impairing the functionality of Tim23. These fusion proteins proved to be extremely useful as they can serve as markers for the inner boundary membranes and contact sites between outer and inner membranes, as described below.

A protein whose sorting pathway was controversial is yeast cytochrome b_2. This protein is present only in yeasts. It is located in the intermembrane space as a soluble protein. Its precursor contains a matrix-targeting signal and a sorting signal, which are successively cleaved off when the precursor, mediated by TIM23, reaches the matrix. Its sorting has been intensely investigated and different pathways were suggested. We had initially proposed a conservative sorting pathway. This turned out not to be the case; rather the protein appears to belong to the class that is laterally released. It is, however, still possible that the precursor makes a partial entry into the matrix, in a way similar to that of Mgm1. In the case of cytochrome b_2, this might be necessary for translocation of the heme-binding domain across the outer membrane, which has a tendency to fold tightly in the presence of heme. Rather frequent observations have convinced me that Goethe was right when he said: *"Die Gelehrten sind meist gehässig, wenn sie widerlegen; einen Irrenden sehen sie gleich als ihren Todfeind an."* (Scholars are usually venomous when they refute somebody; they view an erring peer immediately as a deadly enemy.)

THE TIM22 TRANSLOCASE

Shortly after the initial experiments to identify the TIM23 translocase, Michael Brunner and Christian Sirrenberg detected an open reading frame in yeast showing sequence similarity to Tim23 and Tim17 and the same overall predicted structure of four membrane-spanning segments. Likewise, the encoding gene was essential for viability of cells. Downregulation of the protein led to reduction in the expression of solute carrier proteins, in particular the ADP/ADP carrier, whereas the levels of proteins transported by TIM23 were unaffected. This was the result of a defect in import as observed by in vivo experiments. We named this protein Tim22 (79). Tim22 was present in a large complex not associated with the import motor, which was different from the TIM23 complex. We concluded that mitochondria had two import machineries in the inner membrane with different client precursors. This fitted perfectly with our previous results on the import of carrier proteins. In further studies on the complex, the groups of Jensen (80, 81) and Schatz (82) found additional components of the complex, Tim54 and Tim18, that are not essential but important for the stability of the TIM22 complex.

Following the identification of Tim22, we discovered this membrane protein needs cooperation of other components, Tim10 and Tim12 (83). These proteins were identified through their association with Tim22. In independent studies, they were found also by Carla Koehler and coworkers in the Schatz group (84). Without our groups apparently knowing of each other, the manuscripts were submitted on October 9, 1997, and accepted on November 25, 1997, ours in *Nature*, theirs in *Science*. Both manuscripts reported the interaction of these proteins with the ATP/ADP precursor, a major transport substrate of the pathway, suggesting that these proteins assist translocation of the hydrophobic carriers across the intermembrane space. Then Tim9, present in complex with Tim10, was discovered as another protein required for translocation (85, 86). In this case, we submitted six weeks later to the same journal. Another complex of "small Tim proteins," Tim8 and Tim13, was then described by Koehler and Schatz, similar in its structure to the Tim9-Tim10 complex. It helps, but is not essential for, import of the Tim23 precursor by the TIM22 translocase into the inner membrane. These results revealed an interesting and unexpected interdependence of the TIM23 and TIM22. Both are indispensible for mitochondrial biogenesis and have apparently a related central component responsible for transport across and insertion of precursors into the inner membrane in a membrane potential–dependent manner. Yet, they have a different molecular clientel, and TIM22 is necessary for the biogenesis of TIM23.

THE DISULFIDE RELAY SYSTEM

The discovery of the small Tim proteins in the late 1990s triggered an unforeseen burgeoning area of mitochondrial research. Initially, we were interested in their import pathways and observed that they could enter the intermembrane space via the TOM channel and somehow were retained then in the intermembrane space. We thought some factor would be bound to make translocation vectorial (87). The presence of two CysXXXCys motifs in the small Tim proteins was apparent from the point of their identification. We did not consider that these cysteines could form disulfide bonds. Proteins with S-S bridges so far were not found outside the endomembrane system. However, when other groups provided support for this assumption, we followed this line carefully. An important step was the discovery of Mia40 by the groups of Pfanner (88), Endo (89), and ourselves (90). Clients of this pathway enter the intermembrane space in the reduced state and interact with oxidized Mia40 to form mixed disulfide bonds. Mia40 acts as the initial retention component in the intermembrane space. Then the disulfide bonds are formed, and the proteins are released from Mia40 to undergo oxidative folding, leaving Mia40 in the reduced state. A next important step in our lab was the discovery of the role of Erv1 as disulfide oxidase in the intermembrane space by Nikola Mesecke, Johannes Herrmann, and Kai Hell [see Mesecke et al. (91)]. Erv1 catalyzes the removal of electrons from the reduced Mia40, which thereby returns to the oxidized state for further import cycles. Erv1 delivers the electrons to the respiratory chain to complete the reduction cycle. We termed this pathway the disulfide relay system. This pathway is probably more complex than known so far, as several other components assist in these reactions. Furthermore, this import system works also for other mitochondrial disulfide-containing proteins, such as the CuZn superoxide dismutase together with components mediating the addition of copper ions. This enzyme is of considerable interest to those studying the biology of oxygen radicals and various diseases, such as amyotrophic lateral sclerosis (92–94). In this way, mitochondria became a new important player on the stage of disulfide biochemistry, with many groups exploring these new avenues further.

THE OXA1 COMPLEX

Import of precursor proteins, directed by an N-terminal signal in the majority of cases, implies that the N terminus resides in the matrix. However, early on, we realized that this was not always the case and that the assembled proteins with such a signal expose their N terminus into the intermembrane space. A prominent example is subunit 9 of the F_1F_O-ATP synthase (Su9), which forms the core of this enzyme complex's rotor. Su9 consists of two membrane-spanning helices with a loop on the matrix side. It is encoded in most eukaryotes in the nuclear genome, synthesized with an N-terminal targeting signal, and imported into the mitochondria by TIM23. Su9 of yeast is an exception; it is encoded by mitochondrial DNA, made on mitochondrial ribosomes, and inserted into the inner membrane from the matrix side.

How are these proteins put in place? In 1994, the identification of a nuclear gene termed *OXA1* was reported to be involved in the assembly of cytochrome oxidase (95, 96). Rosemary Stuart and Kai Hell found out that the Oxa1 protein is required for the insertion of cytochrome oxidase subunit II and of nuclear-encoded Su9 into the inner membrane. These observations are a very nice illustration of the events underlying the evolutionary processes involved in the transition from endosymbiosis to present mitochondria. Oxa1 is used both by mitochondrially encoded proteins and by nuclear-encoded proteins. The latter ones, which in the endosymbiont have entered the inner membrane from the inner face, are returned from the cytosol first into the mitochondrial interior, where they then use the "ancient" pathway they used in the bacteria. We have named the Oxa1-mediated pathway "conservative sorting" (97). Tom Fox and his coworkers arrived independently at the same conclusions (98). Ideas and concepts obviously mature at the same time at different places. This finding led to an intense investigation not only of the functional mechanism of Oxa1 but also of the bacterial homolog, YidC. Usually, the mitochondrial researchers have kept and continue to keep a close eye on their bacterial colleagues; this was a case where the latter could learn from the former ones.

We also found that Oxa1 has five transmembrane segments and faces the intermembrane space with its N terminus (99). It is synthesized in the cytosol with a matrix-targeting signal and, intriguingly, undergoes conservative sorting. The Oxa1 precursor can use the assembled preexisting form of Oxa1 for its assembly. Because it can be restored by artificial expression after deletion of its gene, there must be an additional topogenic component involved in its assembly. Perhaps Cox18, a homolog of Oxa1, plays a role, yet its function is not entirely clear. Further studies then showed a number of exciting properties of Oxa1. It exposes a C-terminal helical domain into the matrix, which serves as an anchor for mitochondrial ribosomes to the matrix face of the inner membrane, suggesting a cotranslational insertion pathway of mitochondrially encoded proteins (100). Oxa1 can replace yidC in bacteria, exhibiting an amazing degree of functional conservation during evolution (101).

These findings raised a large number of questions. One of them was: How are the precursors of the hydrophobic membrane proteins that are subject to this conservative sorting imported by the TIM23 translocase without being arrested and laterally released? Some answers are available regarding the distinguishing characteristics of the precursors. As Johannes Herrmann and Stefan Meier observed, the hydrophobicity of the transmembrane segments does not strongly differ between arrested and passing proteins; in both cases, they are considerably lower than those of bacterial or eukaryotic membrane proteins sorted via the endomembrane system. The flanking charges do matter, and charges on the C-terminal side favor arrest. Furthermore, some of the conservatively sorted proteins have an increased number of proline residues in their transmembrane segments, and these residues might prevent formation of a helical arrangement required for lateral release (102).

THE TOB COMPLEX

Mitochondria together with chloroplasts are unique among cellular organelles in that they possess β-barrel proteins in their outer membranes. Mitochondrial Porins, Tom40, and Mdm10 are examples. In 2003, we identified another putative β-barrel protein in *Neurospora*, which we termed Tob55 (103); the yeast counterpart was termed Sam50 (104). They all are apparently inherited from their gram-negative prokaryotic ancestors. How are these proteins put in place? A simple answer would have been by direct insertion from the cytosol. However, Porin and Tom40 use the pore of the TOM complex for entry into the outer membrane, and small Tim proteins assist in their assembly. Tob55 showed sequence similarity to bacterial Omp85 and its homologs. These latter proteins have an essential function in the insertion of β-barrel proteins in prokaryotes. There, these proteins are made in the cytosol, exported by the SecYEG machinery, and inserted after passage through the periplasm into the outer membrane. Important elements of Omp85 are the β-barrel forming 16 β-strands and the Potra domains facing the periplasm. These elements are also present in the mitochondrial counterpart. Tob55 is essential for cell viability in yeast, which is obviously because of its role in the import of the essential protein Tom40. Tob55 forms the central part of a complex that contains two additional proteins on the cytosolic side of the complex, Tob38/Sam35 and Tob37/Sam37 (105). We termed this pathway also as a conservative pathway because the precursor has to return to the side of the membrane from which it was inserted in the prokaryotic ancestor to use the conserved insertion machinery. Many questions regarding this pathway are still open.

The TOB/SAM complex and the disulfide relay system were the last of the mitochondrial import and sorting machineries discovered. The number of protein components known to be involved has increased between 1984 and the present from zero to some 50 (**Figure 6**). Yet, more are to be discovered (105a).

MITOCHONDRIAL ATP-DEPENDENT PROTEASES

For a long time, regulated degradation of proteins has been suspected as playing an important role in mitochondrial biogenesis. When Thomas Langer joined the lab, we became interested in ATP-dependent proteases, which cooperate with molecular chaperones. Mitochondria contain a homolog of the bacterial Lon protease, Pim1. As we observed, the mtHsp70 system, including Mdj1 and Mge1, cooperates with Pim1 in the degradation of misfolded proteins (106). Apparently, also in terms of the quality control of proteins, the mitochondria have retained prokaryotic pathways.

A further example of the same principle came up with the discovery of AAA-ATPase proteins, Yta10/Yta12, proteins of unknown function, and Yme1, described to have a role in DNA escape from mitochondria. We observed that Yta10 is involved in the degradation of incomplete polypeptide chains made by mitochondrial ribosomes. Yta10 and Yta12 were then recognized as subunits of a heterohexameric complex that is able to degrade mitochondrial proteins, in particular mitochondrial membrane proteins (107). The subunits of the complex are anchored to the inner membrane and are composed of two domains, an AAA-ATPase domain proximal to the membrane and a distal metalloprotease domain. The AAA-ATPase has chaperone function, apparently pulling proteins out from the membrane to degrade them with their protease domain. We termed this complex m-AAA protease because it faces the matrix space (108). Yme1 then proved to form a similar, however, homohexameric complex termed i-AAA protease because it faces the intermembrane space. Both complexes are related to the FtsH protein of

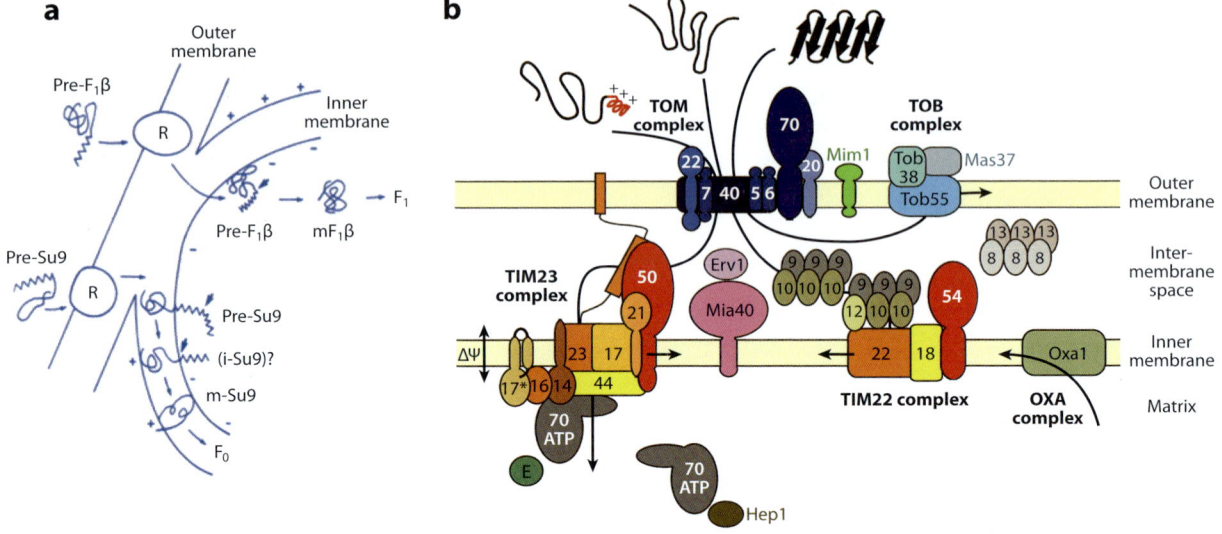

Figure 6

The machinery of mitochondrial protein import. (*a*) Models from 1984 and (*b*) from 2011. Panel (*a*) is adapted from Harmey & Neupert (119). The subunits of the TOM complex, the TOB complex, the TIM23 complex, and the Tim22 complex are indicated by numbers, which represent their apparent molecular mass. Abbreviations: R, receptors for mitochondrial precursor proteins; F_1, F_O, soluble and membrane sectors of the F_1F_O-ATP synthase, respectively; pre-Su9, i-Su9, m-Su9, precursor, intermediate, and mature forms of subunit 9 of the F_1F_O-ATP synthase, respectively; pre-F1β, mF$_1$β, precursor and mature forms of subunit β of the F_1F_O-ATP synthase; ΔΨ, electrical membrane potential; Mia40, a protein that interacts with cysteine-containing client proteins upon import into the intermembrane space; Erv1, mitochondrial disulfide oxidase; Oxa1, a protein that mediates insertion of a group of mitochondrial membrane proteins into the inner membrane from the matrix side; 70 ATP, mitochondrial ATP hydrolyzing heat shock protein Hsp70; Hep1, Hsp70 escort protein; E, nucleotide exchange factor Mge1; Mim1, protein of the outer membrane involved in the assembly of the TOM complex; 17*, Pam17.

bacteria, which was studied at the same time by Koreaki Ito in Kyoto [see Ito & Akiyama (109)].

These findings were the beginning of a long story about the functions of AAA-proteases in mitochondria. Soon after their identification in yeast, mutations in the genes for m-AAA protease in human were described by Andrea Ballabio's group in Milan (110); this work suggested a mechanism involved in the neurodegenerative disorder hereditary spastic tetraplegia. The situation in human is much more complex than in yeast, and deep insights into the genetics of a whole group of related hereditary diseases were obtained in the following years. Thomas Langer, now professor at the University of Cologne, followed up the functions of the AAA-proteases in yeast and mice, which turned out to be quite diverse and surprising [see Tatsuta & Langer (111)]. The m-AAA ATPase was found to play a role in the topogenesis of yeast cytochrome *c* peroxidase (112). It pulls on an intermediate to adjust the sorting signal into a position in which it can be cleaved by the mitochondrial rhomboid protease. For this pulling reaction, the enzyme uses only its chaperone function.

These insights raise many exciting questions as to how evolution works as it, on the one hand, conserves established components and pathways and, on the other hand, finds surprising new solutions. They also taught me how prudent one has to be when drawing general conclusions on the basis of solid and virtually sound concepts arising from evolutionary considerations.

As Goethe said: "*Das Gewisse erfährt man aus der Regel, aber die Regel führt zum Übersehen der Ausnahme und die Natur ist voller Regeln und Ausnahmen, die nach ihrer Aufklärung eine begründbare Variation der Regel sind."* (Certitude is experienced through the rule, but the rule leads to us ignore the exception, and Nature is full of rules and exceptions, which, after their elucidation, are a reasonable variation of the rule.)

THE FUTURE: THE MOLECULAR ARCHITECTURE OF MITOCHONDRIA

I believe my long, enduring fascination by mitochondria has several origins. For me, they presented a continuous journey of discovery, which is as yet not complete. My obsession has also an emotional root, and this is the beauty of mitochondrial architecture, which was triggered when I first saw electron micrographs of mitochondria. I immediately wanted to know how this complex architecture is built and maintained; how cristae are formed; how cristae junctions, the narrow short tube-like structures connecting the cristae and the inner boundary membrane, are shaped; and how the various morphologies of mitochondria in different cells and tissues are generated.

During the time when I was a student, Charles Hackenbrock demonstrated by electron microscopy that outer and inner membranes are held together by contact sites, which became visible when mitochondria underwent condensation of the matrix (113). What are the components responsible for these structures? And what is their biological meaning? Now, 35 years later, these questions are still not answered. But now we have the tools to address these questions, and I am returning to the times of my early scientific steps. As Goethe wrote: *"Der ist der glücklichste Mensch, der das Ende seines Lebens mit dem Anfang in Verbindung setzen kann."* (Happiest is the man who can link the end of his life with its beginning.)

Many years ago, I made attempts to fractionate isolated mitochondria to separate different parts of the mitochondria and to study their composition and biogenesis, but this failed mainly because there were no markers for mitochondrial substructures. In the past years, initially together Andreas Reichert in the group and with Frank Vogel of the Max Delbrück Center Berlin-Buch, we studied the submitochondrial distribution of a series of proteins by immune electron microscopy. A number of components of the respiratory chain were found preferentially, but not exclusively, in the crista membrane as compared to the inner boundary membrane. Conversely, proteins of the mitochondrial import machinery, such as the subunits Tim17 and Tim23 of TIM23, were present preferentially in the inner boundary membrane. The latter distribution depended on active protein import; upon clearance of import sites, the distribution between the cristae and inner boundary membrane became equal. This suggested that the TIM23 complex can be recruited to the inner boundary membrane upon demand. Thus, a specific and dynamic distribution of membrane proteins among different parts of the inner membrane prevails.

Continuing these experiments, we discovered Fcj1, a protein concentrated at crista junctions of yeast. Its deletion resulted in an altered mitochondrial morphology with accumulation of large membrane stacks in the matrix and a virtually complete absence of crista junctions. Fcj1 has low sequence similarity to mitofilin, a mammalian protein; a similar absence of crista junctions was observed upon its downregulation. We then observed that Fcj1 antagonizes the formation of oligomeric F_1F_O-ATP synthase, whereas the subunits e and g of this complex promoted oligomerization. Fcj1 apparently has a key role in the formation of cristae and crista junctions, and oligomers of F_1F_O-ATP synthase play an important role in cristae morphology (113a). Quite recently, we identified a complex of at least six proteins, which is required for the formation of contact sites. One key to these experiments was the establishment of a marker, the above-mentioned GFP-Tim23 (113b). I am confident that these findings are the beginning of what will lead us to a deeper understanding of mitochondrial architecture, particularly to the relationship between the architecture and function of mitochondria.

EPILOGUE

The story told here represents a very personal view and does not describe the work of the many researchers who have contributed to the growth of this field. I apologize to them for missing or only referencing fragments of their contributions. A complete account of all the work would require a whole book and many thousands of quotations. Therefore, I refer the reader to three of my own reviews, two of them in this series (114–116), and a number of others (30, 111, 117, 118). In particular, I have not included references to the work of all of my colleagues, whose names appear in the text, because PubMed will easily lead to their work.

Looking back at almost 50 years of research, I have a sincere feeling of gratitude to the many colleagues who went through my lab as students, postdocs, and group leaders. Without them, nothing would have been achieved. Innumerable concepts, ideas, and practical solutions arose from the continuous open discussion in the lab. The Thursday evening lab seminars were almost always intellectually stimulating, a bouncing back and forth of thoughts and inspirations. My final Goethe quotation: *"Viele Gedanken heben sich erst aus der allgemeinen Kultur hervor, wie die Blüten aus den grünen Zweigen. Zur Rosenzeit sieht man Rosen überall blühen."* (Many ideas emerge just from a common culture, like the blossoms from the green branches. At the time of the roses, one can see roses in flower everywhere.)

It also makes me happy that almost all of the alumni found their place in life and had either a successful or outstandingly successful career. Sad to say, a few got lost. Sometimes, there is no way one can help. I greatly appreciate the collaboration with quite a number of colleagues all over the world. Joining forces has not only been a great support for scientific advancement, but also led to many lasting friendships. And, mercifully, I was saved from deceiving experiments, fabricated data produced in the lab, and their devastating effects. I am also extremely grateful to the Max Planck Society, which after my retirement from university offered me the opportunity to continue research with a smaller group, in a way a "life after death." Finally, I am deeply grateful to my wife Monika, who tolerated my addiction to research and my lack of devotion to life's sensual outlets.

DISCLOSURE STATEMENT

The author is not aware of any affiliations, memberships, funding, or financial holdings that might be perceived as affecting the objectivity of this review.

LITERATURE CITED

1. Brosemer R, Vogell W, Bücher T. 1963. Morphologische und enzymatische Muster bei der Entwicklung indirekter Flugmuskeln von *Locusta migratoria*. *Biochem. Z.* 338:854–910
2. Kleinow W, Neupert W, Miller F. 1974. Electron microscope study of mitochondrial 60S and cytoplasmic 80S ribosomes from *Locusta migratoria*. *J. Cell Biol.* 62:860–75
3. Luck DJL. 1963. Genesis of mitochondria in *Neurospora crassa*. *Proc. Natl. Acad. Sci. USA* 49:233–40
4. Aizawa Y, Xiang Q, Lambowitz AM, Pyle AM. 2003. The pathway for DNA recognition and RNA integration by a group II intron retrotransposon. *Mol. Cell* 11:795–805
5. Sebald W, Machleidt W, Otto J. 1973. Products of mitochondrial protein synthesis in *Neurospora crassa*. Determination of equimolar amounts of three products in cytochrome oxidase on the basis of amino-acid analysis. *Eur. J. Biochem.* 38:311–24
6. Weiss H, Ziganke B. 1974. Cytochrome *b* in *Neurospora crassa* mitochondria. Site of translation of the heme protein. *Eur. J. Biochem.* 41:63–71
7. Kellems RE, Allison VF, Butow RA. 1975. Cytoplasmic type 80S ribosomes associated with yeast mitochondria. *J. Cell Biol.* 65:1–14

8. Korb H, Neupert W. 1978. Biogenesis of cytochrome *c* in *Neurospora crassa*. Synthesis of apocytochrome *c*, transfer to mitochondria and conversion to holocytochrome *c*. *Eur. J. Biochem.* 91:609–20
9. Hallermayer G, Zimmermann R, Neupert W. 1977. Kinetic studies on the transport of cytoplasmically synthesized proteins into the mitochondria in intact cells of *Neurospora crassa*. *Eur. J. Biochem.* 81:523–32
10. Harmey MA, Hallermayer G, Korb H, Neupert W. 1977. Transport of cytoplasmically synthesized proteins into the mitochondria in a cell free system from *Neurospora crassa*. *Eur. J. Biochem.* 81:533–44
11. Maccecchini ML, Rudin Y, Blobel G, Schatz G. 1979. Import of proteins into mitochondria: precursor forms of the extramitochondrially made F_1-ATPase subunits in yeast. *Proc. Natl. Acad. Sci. USA* 76:343–47
12. Harmey MA, Neupert W. 1979. Biosynthesis of mitochondrial citrate synthase in *Neurospora crassa*. *FEBS Lett.* 108:385–89
13. Zimmerman R, Paluch U, Sprinzl M, Neupert W. 1979. Cell-free synthesis of the mitochondrial ADP/ATP carrier protein of *Neurospora crassa*. *Eur. J. Biochem.* 99:247–52
14. Zimmermann R, Hennig B, Neupert W. 1981. Different transport pathways of individual precursor proteins in mitochondria. *Eur. J. Biochem.* 116:455–60
15. Freitag H, Neupert W, Benz R. 1982. Purification and characterisation of a pore protein of the outer mitochondrial membrane from *Neurospora crassa*. *Eur. J. Biochem.* 123:629–36
16. Pfaller R, Freitag H, Harmey MA, Benz R, Neupert W. 1985. A water-soluble form of Porin from the mitochondrial outer membrane of *Neurospora crassa*. Properties and relationship to the biosynthetic precursor form. *J. Biol. Chem.* 260:8188–93
17. Schleyer M, Neupert W. 1985. Transport of proteins into mitochondria: translocational intermediates spanning contact sites between outer and inner membranes. *Cell* 43:339–50
18. Schwaiger M, Herzog V, Neupert W. 1987. Characterization of translocation contact sites involved in the import of mitochondrial proteins. *J. Cell Biol.* 105:235–46
19. Eilers M, Hwang S, Schatz G. 1988. Unfolding and refolding of a purified precursor protein during import into isolated mitochondria. *EMBO J.* 7:1139–45
20. Hawlitschek G, Schneider H, Schmidt B, Tropschug M, Hartl FU, Neupert W. 1988. Mitochondrial protein import: identification of processing peptidase and of PEP, a processing enhancing protein. *Cell* 53:795–806
21. Witte C, Jensen RE, Yaffe MP, Schatz G. 1988. *MAS1*, a gene essential for yeast mitochondrial assembly, encodes a subunit of the mitochondrial processing protease. *EMBO J.* 7:1439–47
22. Yang M, Jensen RE, Yaffe MP, Oppliger W, Schatz G. 1988. Import of proteins into yeast mitochondria: The purified matrix processing protease contains two subunits which are encoded by the nuclear *MAS1* and *MAS2* genes. *EMBO J.* 7:3857–62
23. Schulte U, Arretz M, Schneider H, Tropschug M, Wachter E, et al. 1989. A family of mitochondrial proteins involved in bioenergetics and biogenesis. *Nature* 339:147–49
24. Bolhuis A, Koetje E, Dubois JY, Vehmaanpera J, Venema G, et al. 2000. Did the mitochondrial processing peptidase evolve from a eubacterial regulator of gene expression? *Mol. Biol. Evol.* 17:198–201
25. Sollner T, Griffiths G, Pfaller R, Pfanner N, Neupert W. 1989. MOM19, an import receptor for mitochondrial precursor proteins. *Cell* 59:1061–70
26. Sollner T, Pfaller R, Griffiths G, Pfanner N, Neupert W. 1990. A mitochondrial import receptor for the ADP/ATP carrier. *Cell* 62:107–15
27. Kiebler M, Pfaller R, Sollner T, Griffiths G, Horstmann H, et al. 1990. Identification of a mitochondrial receptor complex required for recognition and membrane insertion of precursor proteins. *Nature* 348:610–16
28. Baker KP, Schaniel A, Vestweber D, Schatz G. 1990. A yeast mitochondrial outer membrane protein essential for protein import and cell viability. *Nature* 348:605–9
29. Kiebler M, Keil P, Schneider H, van der Klei IJ, Pfanner N, Neupert W. 1993. The mitochondrial receptor complex: a central role of MOM22 in mediating preprotein transfer from receptors to the general insertion pore. *Cell* 74:483–92
30. Endo T, Yamano K, Kawano S. 2011. Structural insight into the mitochondrial protein import system. *Biochim. Biophys. Acta* 1808:955–70

31. Mayer A, Lill R, Neupert W. 1993. Translocation and insertion of precursor proteins into isolated outer membranes of mitochondria. *J. Cell Biol.* 121:1233–43
32. Mayer A, Neupert W, Lill R. 1995. Mitochondrial protein import: Reversible binding of the presequence at the *trans* side of the outer membrane drives partial translocation and unfolding. *Cell* 80:127–37
33. Künkele KP, Heins S, Dembowski M, Nargang FE, Benz R, et al. 1998. The preprotein translocation channel of the outer membrane of mitochondria. *Cell* 93:1009–19
34. Künkele KP, Juin P, Pompa C, Nargang FE, Henry JP, et al. 1998. The isolated complex of the translocase of the outer membrane of mitochondria. Characterization of the cation-selective and voltage-gated preprotein-conducting pore. *J. Biol. Chem.* 273:31032–39
35. Stan T, Ahting U, Dembowski M, Künkele KP, Nussberger S, et al. 2000. Recognition of preproteins by the isolated TOM complex of mitochondria. *EMBO J.* 19:4895–902
36. Ahting U, Thieffry M, Engelhardt H, Hegerl R, Neupert W, Nussberger S. 2001. Tom40, the pore-forming component of the protein-conducting TOM channel in the outer membrane of mitochondria. *J. Cell Biol.* 153:1151–60
37. Cheng MY, Hartl FU, Martin J, Pollock RA, Kalousek F, et al. 1989. Mitochondrial heat-shock protein hsp60 is essential for assembly of proteins imported into yeast mitochondria. *Nature* 337:620–25
38. Ellis J. 1987. Proteins as molecular chaperones. *Nature* 328:378–79
39. Anfinsen CB. 1973. Principles that govern the folding of protein chains. *Science* 181:223–30
40. Ostermann J, Horwich AL, Neupert W, Hartl FU. 1989. Protein folding in mitochondria requires complex formation with hsp60 and ATP hydrolysis. *Nature* 341:125–30
41. Craig EA, Kramer J, Shilling J, Werner-Washburne M, Holmes S, et al. 1989. *SSC1*, an essential member of the yeast HSP70 multigene family, encodes a mitochondrial protein. *Mol. Cell. Biol.* 9:3000–8
42. Kang PJ, Ostermann J, Shilling J, Neupert W, Craig EA, Pfanner N. 1990. Requirement for hsp70 in the mitochondrial matrix for translocation and folding of precursor proteins. *Nature* 348:137–43
43. Ungermann C, Neupert W, Cyr DM. 1994. The role of Hsp70 in conferring unidirectionality on protein translocation into mitochondria. *Science* 266:1250–53
44. Ungermann C, Guiard B, Neupert W, Cyr DM. 1996. The delta psi- and Hsp70/MIM44-dependent reaction cycle driving early steps of protein import into mitochondria. *EMBO J.* 15:735–44
45. Yogev O, Pines O. 2011. Dual targeting of mitochondrial proteins: mechanism, regulation and function. *Biochim. Biophys. Acta* 1808:1012–20
46. Rowley N, Prip-Buus C, Westermann B, Brown C, Schwarz E, et al. 1994. Mdj1p, a novel chaperone of the DnaJ family, is involved in mitochondrial biogenesis and protein folding. *Cell* 77:249–59
47. Maarse AC, Blom J, Grivell LA, Meijer M. 1992. *MPI1*, an essential gene encoding a mitochondrial membrane protein, is possibly involved in protein import into yeast mitochondria. *EMBO J.* 11:3619–28
48. Emtage JLT, Jensen RE. 1993. *MAS6* encodes an essential inner membrane component of the yeast mitochondrial protein import pathway. *J. Cell Biol.* 122:1003–12
49. Ryan KR, Menold MM, Garrett S, Jensen RE. 1994. SMS1, a high-copy suppressor of the yeast *mas6* mutant, encodes an essential inner membrane protein required for mitochondrial protein import. *Mol. Biol. Cell* 5:529–38
50. Schneider HC, Berthold J, Bauer MF, Dietmeier K, Guiard B, et al. 1994. Mitochondrial Hsp70/MIM44 complex facilitates protein import. *Nature* 371:768–74
51. Berthold J, Bauer MF, Schneider HC, Klaus C, Dietmeier K, et al. 1995. The MIM complex mediates preprotein translocation across the mitochondrial inner membrane and couples it to the mt-Hsp70/ATP driving system. *Cell* 81:1085–93
52. Okamoto K, Brinker A, Paschen SA, Moarefi I, Hayer-Hartl M, et al. 2002. The protein import motor of mitochondria: a targeted molecular ratchet driving unfolding and translocation. *EMBO J.* 21:3659–71
52a. Neupert W, Brunner M. 2002. The protein import motor of mitochondria. *Nat. Rev. Mol. Cell Biol.* 3:555–65
53. Yamano K, Kuroyanagi-Hasegawa M, Esaki M, Yokota M, Endo T. 2008. Step-size analyses of the mitochondrial Hsp70 import motor reveal the Brownian ratchet in operation. *J. Biol. Chem.* 283:27325–32

54. Bauer MF, Sirrenberg C, Neupert W, Brunner M. 1996. Role of Tim23 as voltage sensor and presequence receptor in protein import into mitochondria. *Cell* 87:33–41
55. Alder NN, Jensen RE, Johnson AE. 2008. Fluorescence mapping of mitochondrial TIM23 complex reveals a water-facing, substrate-interacting helix surface. *Cell* 134:439–50
56. Geissler A, Chacinska A, Truscott KN, Wiedemann N, Brandner K, et al. 2002. The mitochondrial presequence translocase: an essential role of Tim50 in directing preproteins to the import channel. *Cell* 111:507–18
57. Yamamoto H, Esaki M, Kanamori T, Tamura Y, Nishikawa S, Endo T. 2002. Tim50 is a subunit of the TIM23 complex that links protein translocation across the outer and inner mitochondrial membranes. *Cell* 111:519–28
58. Mokranjac D, Paschen SA, Kozany C, Prokisch H, Hoppins SC, et al. 2003. Tim50, a novel component of the TIM23 preprotein translocase of mitochondria. *EMBO J.* 22:816–25
59. van der Laan M, Chacinska A, Lind M, Perschil I, Sickmann A, et al. 2005. Pam17 is required for architecture and translocation activity of the mitochondrial protein import motor. *Mol. Cell. Biol.* 25:7449–58
60. Chacinska A, Lind M, Frazier AE, Dudek J, Meisinger C, et al. 2005. Mitochondrial presequence translocase: Switching between TOM tethering and motor recruitment involves Tim21 and Tim17. *Cell* 120:817–29
61. Mokranjac D, Popov-Celeketic D, Hell K, Neupert W. 2005. Role of Tim21 in mitochondrial translocation contact sites. *J. Biol. Chem.* 280:23437–40
62. Westermann B, Prip-Buus C, Neupert W, Schwarz E. 1995. The role of the GrpE homologue, Mge1p, in mediating protein import and protein folding in mitochondria. *EMBO J.* 14:3452–60
63. Mokranjac D, Sichting M, Neupert W, Hell K. 2003. Tim14, a novel key component of the import motor of the TIM23 protein translocase of mitochondria. *EMBO J.* 22:4945–56
64. Truscott KN, Voos W, Frazier AE, Lind M, Li Y, et al. 2003. A J-protein is an essential subunit of the presequence translocase-associated protein import motor of mitochondria. *J. Cell Biol.* 163:707–13
65. D'Silva PD, Schilke B, Walter W, Andrew A, Craig EA. 2003. J protein cochaperone of the mitochondrial inner membrane required for protein import into the mitochondrial matrix. *Proc. Natl. Acad. Sci. USA* 100:13839–44
66. Kozany C, Mokranjac D, Sichting M, Neupert W, Hell K. 2004. The J domain–related cochaperone Tim16 is a constituent of the mitochondrial TIM23 preprotein translocase. *Nat. Struct. Mol. Biol.* 11:234–41
67. Mokranjac D, Berg A, Adam A, Neupert W, Hell K. 2007. Association of the Tim14•Tim16 subcomplex with the TIM23 translocase is crucial for function of the mitochondrial protein import motor. *J. Biol. Chem.* 282:18037–45
68. Mokranjac D, Bourenkov G, Hell K, Neupert W, Groll M. 2006. Structure and function of Tim14 and Tim16, the J and J-like components of the mitochondrial protein import motor. *EMBO J.* 25:4675–85
69. Popov-Celeketic D, Mapa K, Neupert W, Mokranjac D. 2008. Active remodelling of the TIM23 complex during translocation of preproteins into mitochondria. *EMBO J.* 27:1469–80
70. Wiedemann N, van der Laan M, Hutu DP, Rehling P, Pfanner N. 2007. Sorting switch of mitochondrial presequence translocase involves coupling of motor module to respiratory chain. *J. Cell Biol.* 179:1115–22
71. Popov-Celeketic D, Waegemann K, Mapa K, Neupert W, Mokranjac D. 2011. Role of the import motor in insertion of transmembrane segments by the mitochondrial TIM23 complex. *EMBO Rep.* 12:542–48
72. Nobrega FG, Nobrega MP, Tzagoloff A. 1992. BCS1, a novel gene required for the expression of functional Rieske iron-sulfur protein in *Saccharomyces cerevisiae*. *EMBO J.* 11:3821–29
73. Folsch H, Guiard B, Neupert W, Stuart RA. 1996. Internal targeting signal of the BCS1 protein: a novel mechanism of import into mitochondria. *EMBO J.* 15:479–87
74. Herlan M, Vogel F, Bornhovd C, Neupert W, Reichert AS. 2003. Processing of Mgm1 by the rhomboid-type protease Pcp1 is required for maintenance of mitochondrial morphology and of mitochondrial DNA. *J. Biol. Chem.* 278:27781–88
75. Herlan M, Bornhovd C, Hell K, Neupert W, Reichert AS. 2004. Alternative topogenesis of Mgm1 and mitochondrial morphology depend on ATP and a functional import motor. *J. Cell Biol.* 165:167–73

76. Ishihara N, Fujita Y, Oka T, Mihara K. 2006. Regulation of mitochondrial morphology through proteolytic cleavage of OPA1. *EMBO J.* 25:2966–77
77. Duvezin-Caubet S, Koppen M, Wagener J, Zick M, Israel L, et al. 2007. OPA1 processing reconstituted in yeast depends on the subunit composition of the m-AAA protease in mitochondria. *Mol. Biol. Cell* 18:3582–90
78. Donzeau M, Kaldi K, Adam A, Paschen S, Wanner G, et al. 2000. Tim23 links the inner and outer mitochondrial membranes. *Cell* 101:401–12
79. Sirrenberg C, Bauer MF, Guiard B, Neupert W, Brunner M. 1996. Import of carrier proteins into the mitochondrial inner membrane mediated by Tim22. *Nature* 384:582–85
80. Kerscher O, Holder J, Srinivasan M, Leung RS, Jensen RE. 1997. The Tim54p-Tim22p complex mediates insertion of proteins into the mitochondrial inner membrane. *J. Cell Biol.* 139:1663–75
81. Kerscher O, Sepuri NB, Jensen RE. 2000. Tim18p is a new component of the Tim54p-Tim22p translocon in the mitochondrial inner membrane. *Mol. Biol. Cell* 11:103–16
82. Koehler CM, Murphy MP, Bally NA, Leuenberger D, Oppliger W, et al. 2000. Tim18p, a new subunit of the TIM22 complex that mediates insertion of imported proteins into the yeast mitochondrial inner membrane. *Mol. Cell. Biol.* 20:1187–93
83. Sirrenberg C, Endres M, Folsch H, Stuart RA, Neupert W, Brunner M. 1998. Carrier protein import into mitochondria mediated by the intermembrane proteins Tim10/Mrs11 and Tim12/Mrs5. *Nature* 391:912–15
84. Koehler CM, Jarosch E, Tokatlidis K, Schmid K, Schweyen RJ, Schatz G. 1998. Import of mitochondrial carriers mediated by essential proteins of the intermembrane space. *Science* 279:369–73
85. Koehler CM, Merchant S, Oppliger W, Schmid K, Jarosch E, et al. 1998. Tim9p, an essential partner subunit of Tim10p for the import of mitochondrial carrier proteins. *EMBO J.* 17:6477–86
86. Adam A, Endres M, Sirrenberg C, Lottspeich F, Neupert W, Brunner M. 1999. Tim9, a new component of the TIM22•54 translocase in mitochondria. *EMBO J.* 18:313–19
87. Lutz T, Neupert W, Herrmann JM. 2003. Import of small Tim proteins into the mitochondrial intermembrane space. *EMBO J.* 22:4400–8
88. Chacinska A, Pfannschmidt S, Wiedemann N, Kozjak V, Sanjuan Szklarz LK, et al. 2004. Essential role of Mia40 in import and assembly of mitochondrial intermembrane space proteins. *EMBO J.* 23:3735–46
89. Naoe M, Ohwa Y, Ishikawa D, Ohshima C, Nishikawa S, et al. 2004. Identification of Tim40 that mediates protein sorting to the mitochondrial intermembrane space. *J. Biol. Chem.* 279:47815–21
90. Terziyska N, Lutz T, Kozany C, Mokranjac D, Mesecke N, et al. 2005. Mia40, a novel factor for protein import into the intermembrane space of mitochondria is able to bind metal ions. *FEBS Lett.* 579:179–84
91. Mesecke N, Terziyska N, Kozany C, Baumann F, Neupert W, et al. 2005. A disulfide relay system in the intermembrane space of mitochondria that mediates protein import. *Cell* 121:1059–69
92. Sideris DP, Tokatlidis K. 2007. Oxidative folding of small Tims is mediated by site-specific docking onto Mia40 in the mitochondrial intermembrane space. *Mol. Microbiol.* 65:1360–73
93. Koehler CM, Tienson HL. 2009. Redox regulation of protein folding in the mitochondrial intermembrane space. *Biochim. Biophys. Acta* 1793:139–45
94. Hell K. 2008. The Erv1-Mia40 disulfide relay system in the intermembrane space of mitochondria. *Biochim. Biophys. Acta* 1783:601–9
95. Bonnefoy N, Chalvet F, Hamel P, Slonimski PP, Dujardin G. 1994. *OXA1*, a *Saccharomyces cerevisiae* nuclear gene whose sequence is conserved from prokaryotes to eukaryotes controls cytochrome oxidase biogenesis. *J. Mol. Biol.* 239:201–12
96. Bauer M, Behrens M, Esser K, Michaelis G, Pratje E. 1994. *PET1402*, a nuclear gene required for proteolytic processing of cytochrome oxidase subunit 2 in yeast. *Mol. Gen. Genet.* 245:272–78
97. Hell K, Herrmann JM, Pratje E, Neupert W, Stuart RA. 1998. Oxa1p, an essential component of the N-tail protein export machinery in mitochondria. *Proc. Natl. Acad. Sci. USA* 95:2250–55
98. Bonnefoy N, Fiumera HL, Dujardin G, Fox TD. 2009. Roles of Oxa1-related inner-membrane translocases in assembly of respiratory chain complexes. *Biochim. Biophys. Acta* 1793:60–70
99. Herrmann JM, Neupert W, Stuart RA. 1997. Insertion into the mitochondrial inner membrane of a polytopic protein, the nuclear-encoded Oxa1p. *EMBO J.* 16:2217–26

100. Szyrach G, Ott M, Bonnefoy N, Neupert W, Herrmann JM. 2003. Ribosome binding to the Oxa1 complex facilitates co-translational protein insertion in mitochondria. *EMBO J.* 22:6448–57
101. Preuss M, Ott M, Funes S, Luirink J, Herrmann JM. 2005. Evolution of mitochondrial Oxa proteins from bacterial YidC. Inherited and acquired functions of a conserved protein insertion machinery. *J. Biol. Chem.* 280:13004–11
102. Meier S, Neupert W, Herrmann JM. 2005. Proline residues of transmembrane domains determine the sorting of inner membrane proteins in mitochondria. *J. Cell Biol.* 170:881–88
103. Paschen SA, Waizenegger T, Stan T, Preuss M, Cyrklaff M, et al. 2003. Evolutionary conservation of biogenesis of beta-barrel membrane proteins. *Nature* 426:862–66
104. Kozjak V, Wiedemann N, Milenkovic D, Lohaus C, Meyer HE, et al. 2003. An essential role of Sam50 in the protein sorting and assembly machinery of the mitochondrial outer membrane. *J. Biol. Chem.* 278:48520–23
105. Endo T, Yamano K. 2010. Transport of proteins across or into the mitochondrial outer membrane. *Biochim. Biophys. Acta* 1803:706–14
105a. Wagener N, Ackermann M, Funes S, Neupert W. 2011. A pathway of protein translocation in mitochondria mediated by the AAA-ATPase Bcs1. *Mol. Cell* 44:191–202
106. Wagner I, Arlt H, van Dyck L, Langer T, Neupert W. 1994. Molecular chaperones cooperate with PIM1 protease in the degradation of misfolded proteins in mitochondria. *EMBO J.* 13:5135–45
107. Arlt H, Tauer R, Feldmann H, Neupert W, Langer T. 1996. The YTA10-12 complex, an AAA protease with chaperone-like activity in the inner membrane of mitochondria. *Cell* 85:875–85
108. Leonhard K, Herrmann JM, Stuart RA, Mannhaupt G, Neupert W, Langer T. 1996. AAA proteases with catalytic sites on opposite membrane surfaces comprise a proteolytic system for the ATP-dependent degradation of inner membrane proteins in mitochondria. *EMBO J.* 15:4218–29
109. Ito K, Akiyama Y. 2005. Cellular functions, mechanism of action, and regulation of FtsH protease. *Annu. Rev. Microbiol.* 59:211–31
110. Casari G, De Fusco M, Ciarmatori S, Zeviani M, Mora M, et al. 1998. Spastic paraplegia and OXPHOS impairment caused by mutations in paraplegin, a nuclear-encoded mitochondrial metalloprotease. *Cell* 93:973–83
111. Tatsuta T, Langer T. 2009. AAA proteases in mitochondria: diverse functions of membrane-bound proteolytic machines. *Res. Microbiol.* 160:711–17
112. Tatsuta T, Augustin S, Nolden M, Friedrichs B, Langer T. 2007. m-AAA protease-driven membrane dislocation allows intramembrane cleavage by rhomboid in mitochondria. *EMBO J.* 26:325–35
113. Hackenbrock CR. 1966. Ultrastructural bases for metabolically linked mechanical activity in mitochondria. I. Reversible ultrastructural changes with change in metabolic steady state in isolated liver mitochondria. *J. Cell Biol.* 30:269–97
113a. Rabl R, Soubannier V, Scholz R, Vogel F, Mendl N, et al. 2009. Formation of cristae and crista junctions in mitochondria depends on antagonism between Fcj1 and Su e/g. *J. Cell. Biol.* 185:1047–63
113b. Harner M, Körner C, Walther D, Mokranjac, Kaesmacher J, et al. 2011. The mitochondrial contact site complex, a determinant of mitochondrial architecture. *EMBO J.* 30:4356–70
114. Neupert W, Herrmann JM. 2007. Translocation of proteins into mitochondria. *Annu. Rev. Biochem.* 76:723–49
115. Neupert W. 1997. Protein import into mitochondria. *Annu. Rev. Biochem.* 66:863–917
116. Mokranjac D, Neupert W. 2010. The many faces of the mitochondrial TIM23 complex. *Biochim. Biophys. Acta* 1797:1045–54
117. Lithgow T, Schneider A. 2010. Evolution of macromolecular import pathways in mitochondria, hydrogenosomes and mitosomes. *Philos. Trans. R. Soc. Lond. Ser. B* 365:799–817
118. Chacinska A, Koehler CM, Milenkovic D, Lithgow T, Pfanner N. 2009. Importing mitochondrial proteins: machineries and mechanisms. *Cell* 138:628–44
119. Harmey MA, Neupert W. 1985. Synthesis and intracellular transport of mitochondrial proteins. In *The Enzymes of Biological Membranes*, ed. A Martonosi, 4:431–64. New York: Plenum

The Fires of Life

Gottfried Schatz

Biozentrum, Universität Basel, Basel, CH-4056 Switzerland;
email: gottfried.schatz@unibas.ch

Keywords

European science, mitochondrial formation, oxidative phosphorylation

Abstract

This retrospective recounts the hunt for the mechanism of mitochondrial ATP synthesis, the early days of research on mitochondrial formation, and some of the colorful personalities dominating these often dramatic and emotional efforts. The narrative is set against the backdrop of postwar Austria and Germany and the stream of young scientists who had to leave their countries to receive postdoctoral training abroad. Many of them—including the author—chose the laboratory of a scientist their country had expelled a few decades before. The article concludes with some thoughts on the uniqueness of U.S. research universities and a brief account of the struggles to revive science in Europe.

Contents

CHILDREN OF MARS	36
Beginnings	37
Desert	39
Oasis Years	41
MITOCHONDRIAL DNA	41
Postdoc in New York	45
Promitochondria	48
Cornell Bliss	49
SPELLBOUND BY CYTOCHROME OXIDASE	50
BASEL FOR BIG BIOLOGY	52
TACKLING PROTEIN IMPORT INTO MITOCHONDRIA	53
SCIENCE: A CONTRACT BETWEEN GENERATIONS	56

CHILDREN OF MARS

I am a child of Mars, the god of war. His Second World War overshadowed my early youth and still sways my thoughts and actions. I only discovered this when I no longer lived in Europe and saw my former self in the befogged mirror of early memories: the childhood among barbarians, bombast, and bombs; the teenage years in an impenitent Austria; the tortuous path to science; and the restless life that marks my scientific generation. No wonder I am the eternal misfit. I live in Switzerland, carry an Austrian passport, and was born at the Hungarian border where the sounds of Hungarian, Croatian, and Romani blended into the singing German dialect of the local peasants. Yet my German lacks local color because my mother, a schoolteacher, grew up speaking Hungarian and taught me the textbook German expected of her profession. My wife is Danish but delivered each of our three children in a different country. They converse with her in Danish, with me in English, and with their friends in Swiss German, German, English, or French. As they have chosen Swiss, Russian, and Romanian partners, our panoply of passports would almost do for a domestic poker game. And two universities anointed me Professor of Biochemistry, even though I never attended a course—or passed an exam—on this subject.

That is fine with me. In lopping my roots, father Mars gave me critical distance and freedom. I abhor processions, parades, sermons, uniforms, incense, national celebrations, drum rolls, gun salutes, and official headdresses of any sort. This may not be a shortcut to popularity, but at least it gave me a headstart in spotting the rigid traditions that split the Austria of my youth into a clerical republic and a Marxist counterstate. Rigid traditions stifle innovation, pave the way to prejudice, and are the archenemies of science. Science is an expedition into the unknown where flawed maps and excess baggage can mean danger.

For us children, the war was not all that bad. We often went hungry and spent long nights in cold bomb shelters, but this was a small price to pay for the abandoned steel helmets, live ammunition, and brand-new bayonets we could pick up almost anywhere; they made our war games much more realistic than home-made wooden swords, bows, or arrows. But the ultimate kick was a nocturnal dogfight with its screeching plane engines, barking antiaircraft guns, probing fingers of search lights and—as the eagerly awaited climax—the fiery crash of an interceptor or bomber. But shortly before my ninth birthday, everything was suddenly over. We did not see it as a liberation, but as a defeat. In our war games, being "the Germans" was out and being "the "Americans" was in. Food and coal became even scarcer, and we spent much of our free time begging nearby farmers for bread and milk or ransacking bombed-out houses for firewood. During these years, Mars branded me forever with his coordinates. On leaving a room, I unconsciously turn off all lights, a gourmet shop makes me dizzy, and a restaurant guest griping about his salad dressing makes me angry. To a Mars child, the emotional coordinates of children of peace are out of kilter.

The myths portray Mars as a poor lover, yet he fathered many Austrian and German children. Most of them lead perfectly normal lives; I can only spot them by their age and their home country: We children of Mars lack telltale birthmarks. We neither caused nor fought our father's war and survived it without grasping its horrors. Most new generations dream of a new world; we just tried to patch up the old one. We would have loved to turn back time and watch the bombs and bullets hurtle back into the planes and guns they had come from, and see the rubble reassemble into the houses, schools, and shops it had once been. Our goal was not to invent but to repair, to reconstruct a past that none of us had ever known. During the turbulent first half of the twentieth century, each decade spawned its own generation of Austrians and Germans. The three generations before me created a brilliant culture, proclaimed revolutions, waged civil wars, committed abominable crimes, suffered unspeakable horrors, or laid the foundations for a united Europe once the guns fell silent. The generation after me espoused the visions of my British and French contemporaries who tried to transmute their countries into social utopias. My own generation can offer nothing of this kind. We are the generation without qualities. History seemed to have condemned us to rebuild our ravaged countries, help the next generation back into the saddle, and then drift into oblivion without leaving our trace in the sands of time.

Reconstruction of our cities and roads was astonishingly fast and made everyone forget that the war had also devastated our cultural heritage. Universities were prime examples of this invisible havoc. Having expelled their Jewish or otherwise "politically unacceptable" faculty during the Nazi insanity, they were now mere shadows of their former selves. Their scientific fires were dead. To rekindle them, the children of Mars had to leave their Austrian or German homes to find sparks elsewhere—often with the help of scientists their country had expelled three decades earlier. Many of these fire seekers never returned, and for some of them, their voyage turned into an odyssey whose plights no Muse has ever sung. Those who returned often faced the envy and hostility of those who had sat pretty at home and slipped comfortably into a professorship. Yet, some of those who returned did eventually rise to key positions and revived the scientific fires that now safeguard the future of Austria and Germany. Perhaps, after all, my generation without qualities did leave its mark in the sands of time.

Beginnings

I was born on August 18, 1936, in the tiny village of Strem near the Austrian-Hungarian border. A kind fate had placed the village to the west of this border, which made all the difference when the border turned into a forbidding barrier guarded by dogs, mines, and barbed wire fences. My father was the son of a local peasant whose wife had borne him thirteen children. Six of them died early, five emigrated to the United States, and the oldest surviving son inherited the farm. My father Andreas, "the smart one," studied agricultural engineering because one of his older brothers kept sending him money from New York. My mother descended from a long line of teachers, who, for generations, tried to bring the light of education to border villages such as Strem. Before her marriage in 1935, she was Panika Lantos, talked to her parents in Hungarian, and kissed their hands when she greeted them.

When I was two years old and Austria became part of Hitler's "Thousand Year Reich," the government transferred my father to the Styrian city of Graz, which by then had a well-deserved reputation as a Nazi haven. My grade school teacher, a kind and intelligent woman, genuinely liked us, but also kept telling us with glowing eyes of the glorious advances of "our" armies into Poland and the Soviet Union. After the war, my high school gave me an excellent training in the classics, but essentially none in the sciences. History courses were eclectic: We heard a lot about Greece, Rome, and the crusades, much less about contemporary history up to 1914, and not a peep about what happened later. Books on the Nazis and their war were difficult to come by, but second-hand popular science books were plentiful and cheap—so were chemicals. Today this may be hard to believe, but back then, some drug stores stocked white phosphorus, mercury, and sometimes even potassium metal and sold them to a teenager like me. Soon I spent most of my allowance on chemicals and glassware for the "laboratory" I had set up in our kitchen. I can only guess what today's safety inspectors would have to say about it, but I still recall vividly what my mortified parents said after I had successfully detonated my first device of red phosphorus and potassium chlorate. These quibbles aside, my parents were wonderful. Having left their world at the Hungarian border, they struggled to become city people, but the rigid class structure of the time placed this goal beyond their reach. They stood unflinchingly behind me, showed me their love, and gave me confidence. They also helped me assemble a respectable collection of college textbooks on inorganic chemistry, which I knew forward and backward by the time I finished high school. In the end, they even tolerated my kitchen experiments, which acquainted me with the brilliant yellow of zinc sulfide, the infernal smell of hydrogen selenide, the chameleon-like color changes of freshly precipitated manganese (II) hydroxide, and the lurid glow of a sulfur flame. Chemistry was my enchanted garden.

So was music. When I was about five, my maternal grandfather—a teacher, like all of his relatives—briefly placed his shiny new violin under my chin and got me hooked for life. My first violin teacher was a cobbler, but I soon attended the Graz conservatory as an external student. It was there that it dawned on me that Austria must once have been a much more exciting place. Where had all the great violinists gone? Most of them had funny names such as Menuhin, Heifetz, Szeryng, Goldberg, Stern, Milstein, and Rostal, and they now lived in England or the United States. Why didn't they perform in Austria? There were just too many things I did not understand, and the grown-ups around me were reluctant to explain them. By the time I was sixteen, I had become deeply frustrated and wanted to get away. Thanks to the American Field Service, a private foundation fostering student exchange, I spent the school year of 1952–1953 as an exchange student in Rochester, New York.

That year changed my life. In the early 1950s, the Soviet Union was not yet a credible foil to the United States, which savored its role as the undisputed ruler of the globe. Every day was a new adventure. I valued the freedom I had at school, even though the low academic standards astonished me. Music was a different matter, though. Rochester was home to the first-rate Eastman School of Music, which generously offered me a scholarship for its violin master of arts program. I also tried to conquer the basics of jitterbug and, at the insistence of my American foster family, spent evenings with church groups discussing the deep questions of life such as "Should one kiss a 'date' on the first night out?" The answer was a firm "no," of course, but as I was much too shy for dating, the kissing issue was moot. I had never heard of most of the other things that we were told not to do, but they sounded enticing, and I tried to remember them for later. I was still the diffident Mars child with dislodged coordinates, but now I had another vantage point from which to triangulate the world. I worked as a newspaper boy for the local Rochester newspaper, the *Democrat and Chronicle*, as an usher at a local movie theater, and as a helper at Sears, Roebuck and Co. during the Christmas shopping season. After Christmas, all helpers were laid off; I had no idea what that meant and went back to work the next day. My boss explained to me that I was fired and that I should quit coming back. I did not understand that either, but eventually got the message. This year was not always easy, but it made the United States my second home and was to help me immeasurably during my scientific career.

Returning to Graz felt like being plunked back into a dark hole. My new stereoscopic vision showed me how backward and prejudiced the city was, and I finished my last year in high school as a difficult and rebellious student. But the Eastman School of Music had made me aware of what first-rate violin playing was like, and I began to channel much of my pent-up energy into becoming a better player. Today,

Austrian string players again rank among the best in the world, but at that time, their general level was quite low. I learned only later how effectively the Nazi purges had crippled Austrian and German violin playing for which Jewish artists had set the standard for the past century. As gramophone records were essentially nonexistent, Graz with its imposing opera house, concert halls, and university buildings was not even aware of its intellectual and artistic poverty. The ravages of war and the stifling grip of the Iron Curtain had made it sink into provinciality.

This provinciality could not, however, diminish the music of the great masters. The hours I spent playing as a substitute with the Graz Philharmonic or in the orchestra pit of an opera house were among the happiest of my life. There is no way to describe how a professional orchestra sounds from within or how a successful performance can send a tingling down the spine. Playing string quartets or sonatas with friends transported me into yet another enchanted garden. Yet even this garden was not without blemishes: Most musicians at the time abhorred modern music, which to them was music written after Johannes Brahms. Quite a few of them were also overt anti-Semites, even though the objects of their aversion were gone. Luckily, my violin teacher, a former concertmaster of the Vienna Symphony Orchestra, was a true cosmopolitan untainted by prejudice. He became my first personal role model, and his photograph still comforts me in my office at home.

Desert

"Have fun with them super-duper specials of yours, *Herr Doktor*" growled Aloysius Zacherl across the loan desk of our Graz university library and reluctantly pushed two battered "*Physiologische Chemie*" textbooks a few millimeters in my general direction. According to my well-informed aunt, Herr Zacherl was the illegitimate son of a blue-blooded "*Hofrat*" of local prominence and a decidedly red-blooded peasant's daughter from the nearby hamlet of Sinabelkirchen. This mésalliance of two discordant worlds had spawned an Austrian centaur whose nobly sculpted paternal mouth was condemned to issue the atrocious sounds of the maternal dialect. After his ill-fated attempt at studying law, Zacherl had become a librarian who liked books, but not students, and who did his malicious best to keep the two apart. By facetiously addressing me as "*Herr Doktor*," he was trying to remind me that a lowly undergraduate such as I had no business requesting books that were not required reading.

As a seasoned Austrian, I took his spiteful officialdom in my stride, but the biochemistry textbooks he had dug up for me were pre-World War II vintage. Now, in 1958, they were just useless clunkers. The chemical processes in living cells had fascinated me already as a teenager, but my high school teachers knew nothing about them. I had hoped to study biochemistry at the university, but at that time, the University of Graz had not a single biochemist on its faculty and offered no biochemistry courses of any kind. As my parents could not afford to send me abroad and as international fellowships were essentially nonexistent, I had decided to enroll as a chemistry student and to master biochemistry on my own. Thanks to Herr Zacherl I had now learned that textbooks from our university library were not an option. The bookstores in Graz carried only a single modern, but also outrageously expensive, textbook by the Swiss biochemist Franz Leuthardt and were unable—or unwilling—to find for me what British or American publishers might have to offer.

Such was the intellectual splendor of postwar Graz, which had once been home to such scientific giants as Ludwig Boltzmann, Karl Cori, Karl von Frisch, Viktor Hess, Otto Loewi, Ernst Mach, Erwin Schrödinger, Nikola Tesla, Alfred Wegener, and Richard Zsigmondy. The university faculty still boasted a few first-rate scientists, such as the physicist Adolf Smekal and the physical chemist Otto Kratky, but they were exceptions and could not change the facts that the general level of science in Graz was at best mediocre and that my training as a chemist was rudimentary. The situation with music was not much better. Austria has always prided itself on its musical legacy, yet it held its breath for almost two decades before daring to allow Gustav Mahler's bandmaster music and the degenerate sounds of the New Viennese School back into its concert halls.

I have often wondered why the collapse of the Third Reich left Austria stymied for decades, whereas that of the Austro-Hungarian Empire triggered revolution, civil war, and a dazzling, if tragically brief, cultural flowering. The general hardship after 1945 cannot explain this difference: After 1918, the hardship was even more severe, and the loss of national identity and social fabric much more dramatic than after 1945. But the Third Reich and its aftermath had depleted Austria's intellectual resources twice over: first through the Nazi purges, and then through the efforts of postwar Austria to cleanse its cultural institutions of former Nazis, not all of whom were incompetent. The unhappy outcome was a Golden Age of Mediocrity: Those ready to sail with the prevailing winds landed key positions despite their lack of talent, paralyzing Austria's intellectual life for at least a generation. This period of intellectual, political, and moral anomie robbed my generation of its ideals and laid the foundation for Austria's present political immaturity. Democracy, fairness, minority rights, and tolerance meant little or nothing to us. We wanted to help our relatives and friends, travel to faraway lands, and build ourselves a future.

My future was biochemistry, and I had no choice but to master it on my own. After many false starts, I concocted a six-step course: First, I worked my way through the biochemistry section of *Chemical Abstracts*, a now defunct abstracting periodical that our library did carry. Second, I jotted down the names and addresses of the authors whose articles interested me. Third, I bought several dozen picture postcards of Graz and sent them to these authors with the lapidary handwritten request: "Please send me all your reprints." For the fourth, fifth, and sixth steps, I waited, waited, and waited because I could not afford the luxury of airmail and sent all postcards by surface mail. Looking back, I am amazed that anybody answered me at all. Yet some did, sending me a few of their most recent reprints. Not so David E. Green, a leading researcher in the field of mitochondrial biochemistry, who ran a huge and successful laboratory with several dozen collaborators at the Enzyme Institute of the University of Wisconsin-Madison. Green liked to do things the big way and sent me a hefty package with more than 200 reprints on the function and structure of mammalian mitochondria. Some of these papers are now classics, and all of them bore the mark of Green's polished scientific prose. I devoured these articles, mostly on a solitary bench in our local park, and soon lost myself in an enchanted world of electron-conducting membranes and colorful cytochromes. What could be more exciting and important than the pathways giving energy to life? My private biochemistry course had swung into high gear. Its balance of subjects may have been open to question, but it kindled my lifelong fascination with cellular respiration and mitochondrial biogenesis.

I met Green for the first time on a visit to his Madison laboratory in the early spring of 1965. His generosity as a host, his charm, and his scientific flair immediately impressed me, and it was obvious that his collaborators liked and respected him. After obtaining a biology degree from New York University, he spent eight years at the University of Cambridge, where he acquired not only a doctorate in biochemistry, but also an elegant British accent, which, however, still betrayed his Brooklyn roots. His intellect and experimental skill had quickly won him attention, and when, at the tender age of 27, he published his outstanding monograph *Reconstruction of Chemical Processes in Living Cells*, he established himself as a biochemical *wunderkind*. After his return to the United States in 1940, he briefly joined Harvard University and published his second book entitled *Mechanism of Biological Oxidations*, which brilliantly summarized the knowledge of the time and influenced research on this topic for years to come. He soon left Harvard for Columbia University where he made seminal discoveries on a variety of processes, including the oxidation and transamination of amino acids. His seven years at Columbia, the most productive of his life, propelled him into the top league of biochemists in the United States and the world. In 1948, the University of Wisconsin chose him as one of the three directors of the new Enzyme Institute, where he remained until his premature death in 1983. The Madison campus offered Green fabulous resources, which he exploited to the full. Soon he directed a large team of outstanding collaborators, including Helmut Beinert, Fred L. Crane, Youssef Hatefi, Frank Huennekens, David H. MacLennan, Henry Mahler, Jesse Rabinowitz, Alan Senior, Salih Wakil, Alexander Tzagoloff, and many others. But when I met him in 1965, his had already passed his apogee. Managing his big laboratory occupied so much of his time that he had lost touch with research, acquiring the reputation of someone whose ideas were often no longer rooted in reality. Yet he was one of the great biochemists of his time (1). Future generations may underestimate his scientific contributions because many were not published under his name: Unlike most of his colleagues, he only put his name on a paper if he had participated in the research directly. This exemplary habit was very much appreciated by his collaborators and may partly explain why so many of them rose to international prominence. I remember Green as a generous and engaging man whose magnanimous response paved my way to biochemistry.

Oasis Years

Well before finishing my PhD studies, I decided to leave Graz as soon as possible. But where was I to go? Luckily someone recommended me to Hans Tuppy, who had just been appointed professor of biochemistry at the University of Vienna. Tuppy was a fabulously gifted, imaginative, and dynamic young biochemist who had been a key player in Fred Sanger's Nobel Prize–winning work on the amino acid sequence of insulin. In setting up his department, Tuppy had attracted some of the best young Austrian biochemists. We admired and tried to emulate him, and I am still amazed at how much his small and underfunded laboratory accomplished in the early 1960s.

When I joined Tuppy in the spring of 1961, Vienna was still scarred by the war and the Soviet occupation, and the once famous Vienna Medical School had become a scientific backwater. Tuppy showed us that one man can make a difference. His contacts with Cambridge University kept us in touch with many exciting discoveries in molecular biology, and his impeccable credentials as an anti-Nazi opened the doors for us into the international biochemical community that included many Viennese refugees. We worked day and night for seven days a week, asked cocky questions at seminars, and were both envied and heartily loathed by outsiders as "Tuppy's arrogant Mafia." However, this successful period did not last long. Tuppy had his eyes on a political career and became in rapid succession Rector of the University, President of the Austrian Academy of Sciences, and, finally, Federal Minister for Science and Research. Even for a workaholic like him, a day had only 24 hours, leaving him less and less time to discuss our research with us. Moreover, most of us left the laboratory, one after the other, for postdoctoral training abroad. Rarely has a small laboratory sent out such a large and motivated group of pilgrims, and rarely has such a group been as successful. All of us became professors at universities or research directors at international pharmaceutical companies. Only three of us did not come back: Peter Palese is now a renowned virologist at Mount Sinai Medical School in New York, Manfred Karobath directed drug development at Sandoz in Switzerland and at Rhône Poulenc in France, and I ended up as a member of the Biozentrum in Basel.

MITOCHONDRIAL DNA

By the late 1950s, biochemists already knew that cellular respiration is mediated by a chain of colored cytochromes, flavoproteins, and non-heme iron proteins that pass on electrons from nutrients to dioxygen gas, reducing this gas to water. The best known of these chromoproteins was the intensely red cytochrome *c*. Tuppy and his assistant Günter Kreil, collaborating with Emanuel Margoliash and his group at Northwestern University, had just elucidated the complete amino acid sequence of this cytochrome, awakening Tuppy's interest in mitochondria. He readily agreed with my proposal to work on these organelles. But when he asked me for specifics, he drew a blank: I knew far too little biochemistry to come up with a credible research plan, but I saved my skin by promising him a detailed proposal within a week.

I spent the following days racking my brains trying to come up with a specific and credible research project. Luckily, I did not yet know how crucial the choice of the first project can be for a research career. A clearly defined, safe project promises quick results, but it usually lacks originality and the chance of a scientific coup. An innovative project, by contrast, may be the "launchpad" for a successful career, but it is usually long and risky. The choice is a question of courage. No wonder that the famed biochemist Harold C. Urey once declared courage as the key ingredient of scientific talent. And yet courage alone is not enough; scientific success also demands passion and patience. It takes passion to tackle a problem everybody else considers intractable, and it takes patience to allow courage and passion to wield their force. Innovative research is an expedition into unknown waters. Those who shun them will rarely discover anything new. There is no better advice for a young researcher than the words of John A. Shedd: "A ship in harbor is safe. But that is not what ships are made for."

After several days of soul-searching, I decided to work on how yeast cells form their mitochondria. By felicitous serendipity, David Green had included in his reprint package to Graz a brief paper on this topic by his former postdoctoral fellow Anthony Linnane, and I found the paper fascinating. Few, if any other, biochemists seemed to work on this problem, and my struggles to produce something resembling a doctoral thesis in Graz had at least taught me how to grow and handle yeast cells.

In the end, I let my heart decide. Why do we so often ignore its advice? In science as well as in art, the insouciance of youth is usually keener than the wisdom of old age.

In 1961, the heroic age of mitochondrial research had just drawn to a close (2). During the previous two decades, Albert Claude, George H. Hogeboom, Walter C. Schneider, and George E. Palade had worked out methods for isolating liver mitochondria by differential centrifugation in isotonic sucrose solutions; Eugene P. Kennedy and Albert L. Lehninger had discovered that mitochondria contained the complete enzymic machinery for oxidative phosphorylation, the tricarboxylic acid cycle, and fatty acid oxidation; George E. Palade, Fritjof S. Sjøstrand, and Keith R. Porter had shown the world the first high-resolution electron micrographs of mitochondria; Britton Chance, applying his rapid and ultrasensitive spectroscopic methods, had dissected energy coupling in the mitochondrial respiratory chain into three discrete steps; David E. Green and his colleagues had established the role of mitochondria in fatty acid synthesis and had isolated and characterized four distinct subcomplexes of the respiratory chain; and Harvey S. Penefsky, Maynard E. Pullman, and Efraim Racker had discovered and purified F_1-ATPase, the first defined part of mitochondrial ATP synthase. Mitochondrial researchers ("mitochondriacs") saw themselves, and were also seen by others, as the biochemical elite. As the respiratory chain was no longer a mystery, everybody was convinced that, a few years down the road, the same would be true for oxidative phosphorylation. But the question "How are mitochondria made?" aroused little general interest and, at any rate, seemed beyond the reach of biochemistry. In 1958, Simpson and his colleagues (3) had observed that isolated rat liver mitochondria incorporated labeled amino acids into protein, but for reasons mentioned later, it seemed a hopeless task to identify these proteins.

Enter yeast genetics. In France, the Russian Boris Ephrussi and the Polish man Pjiotr P. Slonimski had spent the preceding

15 years studying strange mutations that abolished the respiration of yeast cells. The mutations never reverted and were not inherited according to Mendel's laws. Because these respiration-deficient yeast mutants utilized glucose less efficiently than respiring cells, they formed smaller colonies on plates containing low glucose levels. Ephrussi baptized them *petite* mutants (the French word for yeast, *levure*, is a feminine noun). Ephrussi and Slonimski were convinced that these mutations reflected the inactivation or the loss of some extrachromosomal factor that controlled the formation of the respiratory system. They took it for granted that the respiratory system of yeast was housed in typical mitochondria and suspected that this might also be true of the mysterious genetic factor. For those of us who studied mitochondral formation in yeast at that time, the slim monographs by Ephrussi (4) and Slonimski (5) on this topic were "holy scripture." But few biochemists knew about them, and their general impact was low. Interest in the genetic control of mitochondrial formation was still an exotic hobby. The reasons for this parochial attitude show how much biochemistry has changed during my lifetime.

In the early 1960s, biology was still fragmented into scientific fiefdoms, and yeast genetics was defended as an arcane calling for the chosen few. Yeast geneticists loved to intimidate outsiders with their obfuscating and often unnecessary patois. On the other side of the border, mitochondriacs were so busy chasing after nonexistent chemical intermediates of oxidative phosphorylation that they would never have deigned to read the *Journal of Molecular Biology* or, God forbid, the journal *Genetics*.

Also, many biochemists did not accept yeast as a model for mammalian cells and refused to believe that the "respiring yeast granules," studied by Slonimski and a few others, had anything to do with mitochondria. In their eyes, yeast was just another microbe, not much different from *Escherichia coli*. There was no general awareness of the profound divide between prokaryotes and eukaryotes. This awareness spread slowly in the late 1960s as more cells were examined with the electron microscope or with the powerful new tools of molecular biology. Today, we know that respiring membrane vesicles isolated from bacteria are vesicular fragments of the bacterial plasma membrane, but at that time, they were often considered to be preexisting intracellular organelles. Finally, although mitochondria had been discovered and first studied in Europe, research on them in the early 1960s was very much an American enterprise. Europeans trying to learn the trade had to embark on a *hadj* to the mitochondrial meccas in New York, Madison, Philadelphia, or Baltimore. The mitochondrial research centers in Amsterdam, Stockholm, and Munich had already begun to flex their muscles, but they were not yet a match for their powerful U.S. rivals.

In sum, when I decided to work on the formation of yeast mitochondria, I had no idea that I would be separated from the mainstream of mitochondrial research by the Atlantic Ocean, the continental divide between yeast and mammals, and the lack of travel funds. Toward the mid-1960s, however, the general ignorance about yeast mitochondria slowly lifted, and the skeptical remarks about "yeast respiratory granules" gradually subsided.

One of the turning points was the discovery of mitochondrial DNA. In about 1962, it became clear that chloroplasts, the green sisters of mitochondria, contained their own DNA. Hans Tuppy and I reasoned that, if this also held for mitochondria, this DNA might well be the mysterious extrachromosomal factor controlling mitochondrial formation in yeast. Together with Ellen Haslbrunner, my first graduate student, we decided to look for DNA in yeast mitochondria. Just at that time, Sidney Brenner and his colleagues at the Medical Research Council in Cambridge devised a simple gadget for generating linear sucrose gradients. By the time-honored Viennese method of offering a bottle of wine, I persuaded our local mechanic to copy this device from a hand-drawn sketch. I then purified yeast mitochondria by isopycnic centrifugation in a sucrose gradient, collected 15 fractions by puncturing first my finger and then the centrifuge tube with a syringe needle, and analyzed each fraction for DNA by the color reaction devised by Karl Diesche. Did I find DNA! Every fraction had lots of it, without any visible peak where the brownish mitochondria had equilibrated as a turbid band. Clearly, DNA had leaked out of the nucleus, and the sucrose gradient had failed to separate it from the mitochondria. But when we replaced sucrose by the X-ray contrasting agent Urografin, the bulk of the DNA sedimented to the bottom of the tube, and a tiny part of it cofractionated precisely with the mitochondria. The DNA in the bottom fraction was easily digested by DNAase and probably represented nuclear DNA. The DNA in the mitochondrial fraction was not readily digested by DNAase unless the fraction was first mistreated with trichloroacetic acid; presumably, it represented DNA enclosed by mitochondrial membranes. Its amount per milligram mitochondrial protein was remarkably constant between different experiments. We submitted our findings under the cautious title "DNA Associated with Yeast Mitochondria" to *Biochemical and Biophysical Research Communications* (6), the newest and "hottest" biochemical journal of the day, and soon, an avalanche of reprint requests and telephone calls descended on our laboratory.

For the first time, I felt the thrill of discovery. But was this discovery important? When a journalist asked me how it would benefit the general public, I answered him that I did not know. What a short-sighted answer! Today, I would not know where to start. I could tell the journalist that the DNA in our own mitochondria carries the blueprints for 13 essential proteins of our body's power plants; that mutations in it can cause terrifying diseases; that it offers unique advantages for tracking the origin of modern humans and their spread across the globe; and that analysis of this DNA from human bones has posthumously exposed the woman by the name of Anna Anderson as a fraud: She was not the "daughter Anastasia of the last Romanov," as she had claimed, but a false Anastasia of Polish peasant stock.

Shortly before our report on mitochondrial DNA appeared, we learned that Margit M.K. Nass and Sylvan Nass, two electron microscopists at the University of Pennsylvania in Philadelphia, had described thread-like inclusions within the matrix of chick embryo mitochondria. As these threads were sensitive to DNAase, but not to RNAase or protease, the authors had correctly identified them as DNA (7). In those days, U.S. journals took up to half a year to reach our Vienna university library, and the electronic grapevine across the Atlantic was still in its infancy. Some reviews had already cited us as the sole discoverers of mitochondrial DNA, and so the two excellent papers by the Nass team left us crestfallen. In retrospect, however, these concordant reports accelerated the general acceptance of mitochondrial DNA and even got some mitochondriacs interested in mitochondrial biogenesis.

What was this mitochondrial DNA doing? My colleague Erhard Wintersberger found that yeast mitochondria also harbored a DNA-dependent RNA polymerase as well as ribosomal RNA hybridizing to mitochondrial DNA. He suspected that this RNA was part of mitochondrial ribosomes, which were discovered by others a few years later. But what about protein-coding genes? The amount of DNA we had measured in yeast mitochondria was insufficient to encode all of the hundreds of mitochondrial proteins. Most of these proteins had to be encoded by nuclear genes, synthesized on cytosolic ribosomes, and then imported into mitochondria—a bizarre scenario!

Yet we knew from the work of Simpson that mitochondria made some proteins themselves, and we suspected that they were encoded by mitochondrial DNA. Simpson's team had gone to great lengths to exclude contaminating microsomes as the source of the observed activity, and a few years later, others found that protein synthesis by isolated mitochondria was insensitive to inhibitors of microsomal protein synthesis, such as cycloheximide, but sensitive to inhibitors of bacterial protein synthesis, such as chloramphenicol and erythromycin. However, efforts to identify these mitochondrially synthesized proteins ran into an insurmountable roadblock. Amino acid incorporation by mitochondria was very low, the labeled proteins were too insoluble to be analyzed by the available methods, and a fraudulent claim that one of them was cytochrome c severely blemished the field's reputation. Perhaps the most frustrating obstacle was the obstinate refusal of the incorporated radioactivity to cofractionate with any of the known mitochondrial enzymes. Because the non-Mendelian *petite* mutants lacked cytochrome oxidase and succinate-cytochrome c reductase, we suspected that these enzymes were synthesized by mitochondria. But when a few heroic researchers labeled isolated mitochondria with radioactive amino acids and then purified cytochrome oxidase or cytochrome b from them, neither enzyme was labeled. In 1964, it was even claimed that mitochondria could not synthesize any proteins at all and that the observed activity merely reflected contaminating bacteria. I should have stayed in the game to exploit our discovery of mitochondrial DNA, but the time had come to leave for my long overdue postdoctoral training in New York City.

Postdoc in New York

My decision to go to New York was born in the summer of 1961 during a visit by the famed biochemist Efraim Racker. It was Racker's first postwar visit to the city he had grown up in and fled after obtaining his MD degree 23 years earlier. After brief sojourns in Denmark and Great Britain, he had finally made it to the United States, where, after brief interludes at the University of Minnesota, the Harlem Hospital in New York City, and Yale University, he was now working at the Public Health Research Institute of the City of New York on biological energy conversion. His discovery of an "energy-rich" thioester intermediate in glycolytic ATP formation by glyceraldehyde-3-phosphate dehydrogenase had propelled him to fame and prompted his conviction that ATP formation by oxidative phosphorylation in mitochondria occurs by a similar mechanism. I had heard that he conducted his research not as a highly organized campaign but as an intuitive and almost artistic endeavor. Indeed, after graduating from high school, he had been accepted into the prestigious Vienna Art Academy, but he had soon left it to study medicine. Later, he used to quip that this academy, by twice rejecting Adolf Hitler, had caused World War II. He remained a passionate painter throughout his life and gave each of his students and postdoctoral fellows at least one of his paintings as a farewell gift.

When I first met Racker, he was only 48 years old. His thinning white hair and wrinkled face made him look older, but his lively demeanor and quick wit soon revealed his true age. As he then still refused to speak his native German, we conversed in English. "How come you speak English so well?" he wanted to know. When I told him that I had spent a year in a U.S. high school, he quickly retorted, alluding to my German accent, "How come you speak English so badly?" When both of us erupted in laughter, we spontaneously shook hands and laid the foundation for our lifelong friendship. We spent the next evening strolling through the streets of Vienna, which evoked in him many long-buried memories. His reminiscences showed me a city I had only known from books: its rich musical life, the legendary public readings by Karl Kraus, and the rising Nazi tide at the university that led to violent attacks on his Jewish student friends. When I told him that I had learned about these events only recently through books by the German publishing house S. Fischer and that most of my colleagues abhorred these crimes as much as I did, he at first would not believe me, but many years later, he confessed to me that our talks that evening had persuaded him to make peace with Austria's young generation. He even started to speak to me in halting German. Science had begun to bridge the gulf between our two generations, but it could not completely heal the wounds from his youth. Shortly before his death, we both attended a scientific meeting in Vienna and, rushing to the conference center, crossed the busy Ringstrasse on a red light. The angry honking immediately attracted a police officer who reprimanded us sternly, but politely. I was relieved to be let off the hook that easily, but Racker became unreasonably aggressive and might have caused both of us considerable grief: Insulting an Austrian official is almost as grave an offense as not paying a Swiss hotel bill.

I had initially planned to train with David Green in Madison, but after that evening with Racker, I spontaneously asked him whether I could work with him. He immediately agreed, but warned me that his laboratory was already full and that I could only join him in 1964. I was more than willing to wait, and so my wife, our one-year-old daughter, and I boarded the little Dutch steamer *Westerdam* on June 27, 1964, for a leisurely 10-day voyage to New York. It was to be the beginning of a restless life that made us change residence 10 times—a tribute Mars exacted from his children who tried to escape his shadow. We arrived in New York with great expectations, but our first impressions were disappointing. My fellowship had seemed lavish if expressed in Austrian Schillings, but now barely met our basic needs. We were as poor in New York as we had been in Vienna. I shall never forget the humiliation of having to postpone the purchase of a larger baby carriage after a New York bank had turned down my application for a $50 loan. Racker's laboratories were another shock. They were housed in the "Public Health Research Institute of the City of New York, Inc.," which, despite its name, was a serious public health threat. It was a ramshackle, dirty building at the foot of East Sixteenth Street, a dead-end slum street abutting the East River. At first I could not even find the place, and when I asked a taxi driver to get me there, he asked suspiciously, "Now why would you want to go *there*?" A coal-fired power plant of "the Confederated Edison Co." supplied our Institute with barely enough electricity; it was the decade of the brownouts but with a surplus of black soot, which settled as a grimy black film on window sills and lab benches. Cockroaches of ghastly dimensions lurked everywhere—even in the circuits of our centrifuges and blenders—where their electrocution often led to mysterious short circuits. But there was nothing ramshackle about Racker's research group in which I spent two and a half exciting years. Racker was then in his early fifties. He looked even older than during our first encounter in Vienna, but he was in superb physical shape: He excelled in competitive sports, such as tennis, ping-pong, and volleyball, and loved to argue about just everything. His presence radiated a competitive and creative restlessness that I found invigorating, even though I soon learned that it was not to everyone's liking. He came as close to reading other people's mind as anyone I have known, and he delighted in getting his coworkers and colleagues off balance. A lack of self-confidence was a dangerous platform from which to approach him. Yet, I have met few human beings who could be as sensitive, helpful, and empathic as he, and none who were as broad-minded, or who had a keener sense of humor. When he felt at ease, he regaled his audience with his quick wit, child-like playfulness, and genuine charm that were irresistible. His humor reflected his youth in Mazzenstadt, the Viennese ghetto district, playing on human folly and life's dark and surrealistic sides. It would do him an injustice to say that he harbored two souls because he had so many. He could be incredibly rude and disarmingly gentle, overbearing and unassuming, stunningly petty and royally generous. His artistic temperament made him approach a scientific problem intuitively rather than methodically, making it often frustrating to discuss science with him. He was the epitome of a scientist-artist and a genuine humanist. In his private universe, the human spirit was both a center of gravity and a point of reference. He became my mentor, then a father figure, and finally one of my closest friends. When he died of a stroke in 1991, so did part of me. I still talk with him, as I do with the ever-growing family of others who are no longer with me. I have learned that the dead can be closer friends than the living. They are always there to offer advice or consolation; they belong to a world that is uniquely mine.

In Racker's lab, I learned how to prepare submitochondrial particles, extract specific proteins from them, and measure oxidative phosphorylation as well as how to keep my calm when challenged in a scientific discussion. This could be quite a trial because the field of oxidative phosphorylation had a well-deserved reputation as a no-holds-barred war zone dominated by weak data and strong personalities. The verbal duels between David E. Green and Efraim Racker were legendary; both lost little time in going straight at each other's jugular, entertaining and shocking everyone by their personal attacks and biting repartees.

When I joined Racker in 1964, it was still a mystery how the energy released during respiration drives ATP synthesis from ADP and inorganic phosphate. Racker's colleagues Harvey S. Penefsky and Maynard E. Pullman had just isolated a key component of this process: By treating submitochondrial particles with ultrasound, they had solubilized a protein that cleaved ATP to ADP and inorganic phosphate and, when added back to the mistreated particles, restored their capacity for oxidative phosphorylation (8, 9). Penefsky and Pullman suspected correctly that this ATPase was just one of many components of the mitochondrial oxidative phosphorylation system and baptized it "factor 1," or F_1. "Ef" Racker was apparently quite satisfied with this designation, as its sound reminded everyone of his nickname and his self-perceived rank in the scientific hierarchy. But this seminal discovery had also led to tensions because Penefsky and Pullman had discovered F_1 on their own and resented Racker's insistence that he be the senior author on their papers. These turbulences aside, all of us expected that the remaining parts of the ATP-forming machine would soon be known and that we would then accompany our master on his well-deserved winter trip to Stockholm.

At first everything progressed well. Racker's postdoctoral fellow Yasuo Kagawa identified a complex of insoluble membrane proteins that tightly binds soluble F_1 and restores its native properties. In mitochondria, F_1 is sensitive to the antibiotic oligomycin and, like most proteins, stable on ice. Once solubilized, however, it is oligomycin insensitive and cold labile. As Kagawa could not separate the insoluble F_1-binding proteins from each other, he called the complex "oligomycin-sensitivity factor" or, in short, F_o. Had we only known that the complex between F_o and F_1 represented the entire ATP-forming machine! But we did not, and so we lost precious years hunting for additional factors that would stimulate oxidative phosphorylation by mistreated submitochondrial particles. This approach led nowhere. Like most of our competitors, we were looking for a protein that, by reversibly adopting an energy-rich state, could transfer energy from the respiratory chain to F_1, allowing it to make ATP. We were all mesmerized by Slater's "chemical hypothesis," which predicted such an intermediate. Racker was particularly partial to this hypothesis, as it corresponded to his earlier discovery of an energy-rich thioester intermediate in the first ATP-forming step of glycolysis. This *fata morgana* led hundreds of researchers—us included—down the garden path, seduced one of Green's collaborators to falsify data, and tarnished the reputation of our field. The British microbiologist Peter Mitchell had warned as early as 1961 that the link between respiration and ATP synthesis was not an energy-rich substance, but a proton electrochemical potential gradient across the mitochondrial inner membrane (10). Most biochemists, however, ignored this heretical view. Maynard E. Pullman and I published the first comprehensive review of Mitchell's "chemiosmotic hypothesis" in 1967 (11), yet it took 10 more years for the leading researchers in the field to publish their famous "peace treaty" in which they declared their agreement with Mitchell's concept (12).

In New York, I lost almost an entire year trying to confirm reports from David Green's laboratory on protein factors stimulating oxidative phosphorylation at specific regions of the respiratory chain. Using an existing assay for ATP synthesis in the cytochrome oxidase region and developing a novel one for the NADH dehydrogenase region, I proved these claims to be incorrect. In the early spring of 1965, my postdoctoral colleague June M.

Fessenden and I took a plane to snowed-in Madison in a last attempt to confirm the published results in the lion's den—again without success. Some of them were the result of wishful thinking, but one turned out to be a deliberate fraud. Sadly, fraud is no longer a rarity in today's hypercompetitive atmosphere, but in those more tranquil days, it was a shocking exception that seriously shook our field. Luckily, my next project was successful: Together with Harvey Penefsky and Ef Racker, I purified F_1 from yeast and used it to show that F_1 was not only a catalyst for ATP synthesis, but also a structural "plug" for the proton pore of F_o, thereby preventing a short-circuit of the respiration-driven proton flow (13).

With his 17-year-long lonely battle against virtually the entire mitochondrial community, Peter Mitchell wrote one of the most bizarre chapters in the history of science. He confirmed that fundamentally new ideas emanate from creative individuals rather than from groups or institutions. But Mitchell also acknowledged his intellectual debt to Robert Robertson, Robert Davies, Alexander George Ogston, and others. Scientific discoveries, like works of art, are children of solitude, yet are not born in isolation.

Promitochondria

When I returned to Vienna in the fall of 1966, I could not wait to do what my move to New York had put on hold: to identify the proteins encoded by mitochondrial DNA. On the hunch that some of them might be part of Kagawa's $F_o.F_1$ complex—the "ATP synthase"—I checked whether this complex was altered in the extrachromosomal *petite* mutants of yeast. The result was striking: The mutants still had F_1, but it was oligomycin insensitive and cold labile even in the intact mitochondria. F_1 itself seemed to be normal because it became oligomycin sensitive and cold stable when I mixed it with F_o from beef heart mitochondria (14). As it was already known that the *petite* mutation caused massive alterations, or even the complete loss of mitochondrial DNA, I reasoned that mitochondrial DNA encoded at least one of the F_o proteins. I was getting close! My next steps should have been to isolate F_o from wild-type yeast, to characterize its protein subunits, and to check whether they were made by mitochondria. But two things got in the way: my desire to settle the controversial fate of mitochondria during anaerobic growth of yeast and my decision to leave Europe for good and settle with my family in the United States.

The fate of mitochondria during anaerobic growth of yeast had been a long-standing and emotional issue. The Australian biochemist Anthony Linnane had reported that yeast cells growing by fermentation in the absence of oxygen lose mitochondria, but regain them upon aeration (15, 16). I believed that anaerobically grown cells could not respire and that this defect was reversed by aeration, but I thought it very unlikely that this physiological change reflected the disappearance and reappearance of mitochondrial organelles. Shortly before leaving for my postdoctoral U.S. stay in 1964, I had shown that the anaerobically grown cells still had mitochondria-like structures, which contained succinate dehydrogenase and an oligomycin-sensitive ATPase. I had termed these structures "promitochondria" but had failed to convince others in the field. Now, however, I had better insight and better tools. I showed that the ATPase activity of promitochondria was inhibited by an antiserum against purified F_1 and that it was oligomycin sensitive in promitochondria from wild-type cells, but not in those from *petite* mutants. Together with my old school pal and colleague Fritz Paltauf from the University of Graz, I then went on to show with Richard S. Criddle (17) that promitochondria contain the mitochondria-specific lipid cardiolipin, and Helmut Plattner from the University of Innsbruck helped me (18) to identify promitochondria as double-walled structures with typical cristae in electron micrographs of freeze-etched anaerobically grown cells. Independently and at about the same time, Hewson Swift and colleagues (19) at the University of Chicago published similar electron micrographs of anaerobically grown yeast cells. Later on, my coworkers and I (20) used pulse-chase experiments and quantitative electron microscope autoradiography to prove that promitochondria from wild-type cells were structural precursors of the respiring mitochondria that arose upon respiratory adaptation. This put an end to the claim that mitochondria can arise de novo, even though many details of promitochondrial maturation remained to be worked out. It seems that a key event is the oxygen-requiring formation of heme that stimulates not only the transcription of many nuclear genes for mitochondrial proteins, but also the assembly of oligomeric cytochrome complexes of the respiratory chain. Mitochondria contain so many vital enzymes that a cell cannot lose these organelles—even when respiration is not necessary. Many years later, Michael P. Yaffe and I (21) exploited this fact in a screen for yeast mutants defective in mitochondrial protein import.

Cornell Bliss

In 1966, Racker had accepted a prestigious Albert Einstein professorship at Cornell University in Ithaca, New York, and had persuaded many prominent membrane biochemists to join him there. This cast included internationally known senior figures such as Quentin H. Gibson, Leon A. Heppel, and André T. Jagendorf as well as younger scientists such as Richard E. McCarty, Peter C. Hinkle, June Fessenden-Raden, and David C. Wharton. He had also invited me, but I had already planned to return to my old Public Health Research Institute in New York City. I soon changed my mind, however, and, in the fall of 1968, moved with my family to Ithaca. Like most of the other young staff members there, I was eager to make my mark and cared little about Cornell traditions. Not only did I not lunch at the Faculty Club, I did not even know that such a Club existed. The address of my small basement laboratory, Wing Wing G-1, would today suggest a *Drosophila* mutant or an oncogene. After I received my own grant, I hired my first technician, Jo Saltzgaber, and, in a spell of hubris, even bought an automatic pencil sharpener and an electric typewriter.

The excellent facilities at Cornell renewed my determination to track down the proteins made by mitochondria. My previous attempts at incorporating ^{14}C-labeled leucine into isolated yeast mitochondria had taught me two things. First, I would never get the protein products hot enough without five additional National Institutes of Health grants. Second, the methods for preparing yeast mitochondria were so lengthy, and our common centrifuges at Wing Wing so dirty, that I would always have to worry about protein synthesis by contaminating bacteria. Why not label the mitochondrially synthesized proteins within their proper environment—the living yeast cell? The protocol was simple enough: Inhibit the cytosolic ribosomes of yeast cells with cycloheximide, incubate the cells with glucose as an energy source and labeled leucine, and then isolate and analyze the mitochondria. The results made us dance; virtually the entire incorporated label was in the mitochondrial fraction, and all labeling was abolished when we inhibited the cells with both chloramphenicol and cycloheximide (22). This in vivo labeling protocol was fast, minimized bacterial contamination, and labeled the mitochondria at least one hundred times more efficiently than protocols using isolated mitochondria.

With this new labeling tool, we showed that mitochondria of extrachromosomal *petite* mutants of yeast could no longer synthesize proteins. Stefan Kužela et al. (23) at the University of Bratislava, then in Czechoslovakia, had reached the same conclusion by examining the incorporation of labeled amino acids into the isolated mitochondria. This could only mean that the *petite* mutation prevented expression of all the genes that were still left on the defective mitochondrial DNA molecules. Now, we finally understood why all *petite* mutants lacked the same set of enzymes even though they had widely different deletions in their mitochondrial DNA, or no mitochondrial DNA at all.

SPELLBOUND BY CYTOCHROME OXIDASE

When we analyzed the labeled mitochondria by electrophoresis in sodium dodecyl sulfate-polyacrylamide gels, which the virologist Jacob V. Maizel pioneered in 1966, we found six broad radioactivity peaks, but none of them coincided with a major stained protein band. The mitochondrial protein products were clearly not major mitochondrial proteins. But what were their properties and their functions? It was at this critical stage that Thomas L. Mason joined my laboratory as a postdoctoral fellow. Tom knew about my earlier work on the biosynthesis of the $F_1.F_o$ complex and was disappointed to hear from me that this project was already well advanced in Alexander Tzagoloff's laboratory in New York City. He therefore wanted to check whether cytochrome oxidase—another enzyme missing from *petite* mutants—was made by mitochondria. Cytochrome oxidase (also termed cytochrome aa_3) is the last enzyme of the mitochondrial respiratory chain. It had already been purified by detergent extraction of mammalian mitochondria, but its subunit composition was unknown. Tom and I decided to ignore earlier reports that cytochrome oxidase was not labeled by isolated mitochondria. We expected to see things others had missed because we now had better eyes. Our Cornell colleague David C. Wharton, an expert on cytochrome oxidase, helped us purify the enzyme from yeast mitochondria and resolve it into three large and four small subunits. The three large ones were labeled in cycloheximide-poisoned yeast cells and, therefore, made by mitochondria; the four small ones were labeled in chloramphenicol-poisoned yeast cells and, therefore, made in the cytosol. Presumably, these small subunits were imported into the mitochondria and only then assembled with the three large subunits (24, 25). Once again, we were not alone. Hanns Weiss and his colleagues (26) at the University of Munich concluded that only a single large subunit of cytochrome oxidase from the mold *Neurospora crassa* was made by mitochondria. This result turned out to be an artifact of electrophoresis. After Meryl S. Rubin & Alexander Tzagoloff (27) had confirmed our results, many subsequent studies by different groups (28) established that mitochondria of most eukaryotes synthesize three large subunits of cytochrome oxidase. As discussed below, these subunits are the catalytic "heart" of the enzyme.

Tzagoloff (29) also unraveled the biogenesis of the yeast $F_1.F_o$ complex, now commonly referred to as ATP synthase. He established its subunit composition and showed that the smallest subunit was made by mitochondria. He published this finding in 1971 (29), the same year we first reported our results on cytochrome oxidase at a Gordon Conference. His landmark contribution did not immediately receive the attention it deserved, perhaps because ATP synthase was not yet considered as well-defined an entity as cytochrome oxidase.

For a while, many colleagues believed that these mitochondrially made proteins were not functional parts of either cytochrome oxidase or ATP synthase, but hydrophobic contaminants. Hydrophobicity was the angel with the flaming sword that protected the mitochondrial magic garden from unworthy intruders. We tried to purify these hydrophobic subunits; they precipitated like balls of steel. We tried to sequence them; they merely scoffed. Strange legends started to grow around them. Some claimed that the large cytochrome oxidase subunits were in reality tight aggregates of smaller proteins. Others considered them as structural scaffolds without specific function.

Gradually the fog lifted (29). When my postdoctoral fellows Eberhard Ebner and Tom Mason isolated nuclear yeast mutants that specifically lacked cytochrome oxidase activity, they noted with delight that these mutants also lacked one or two mitochondrially made subunits of the enzyme (30). Then Robert O. Poyton showed that an antiserum specific for the second largest cytochrome oxidase subunit inhibited the activity of the purified enzyme [see Poyton & Schatz (31)]. Both findings implied that these large subunits were essential for the function of the enzyme. A few years later,

Thomas D. Fox used the nuclear yeast mutants to uncover a translational control mechanism by which nuclear genes govern the expression of mitochondrial genes (32). And once several groups worked out the primary sequence of the large cytochrome oxidase subunits, these turned out to be normal, albeit extremely hydrophobic proteins. Perhaps the most decisive breakthrough was the discovery of specific mitochondrial DNA mutations that altered only a single mitochondrially made protein (33). Some of these mutations made ATP synthase or cytochrome *b* resistant to antibiotics; others selectively inactivated one of the enzyme complexes of oxidative phosphorylation; and still others had no physiological effect, but altered the electrophoretic mobility of a mitochondrially made protein. Analysis of these specific mutations led to the first genetic map of yeast mitochondrial DNA (33). This map, however, did not tell whether a gene affecting a mitochondrially made protein was regulatory or structural, but Slonimski's laboratory and ours showed that one of the genes affecting cytochrome oxidase function was, in fact, the structural gene of the second largest cytochrome oxidase subunit (34). All of this became history when Fred Sanger's group (35) at Cambridge deciphered the complete nucleotide sequence of human mitochondrial DNA, Giuseppe Attardi's group (36) provided the matching transcript map, and several groups sequenced most of the remaining mitochondrially made proteins. When the three-dimensional atomic structure of bovine cytochrome oxidase arrived on the scene (37), it showed that subunit I carries the two heme *a* groups as well as the two copper atoms, whereas subunit II binds cytochrome *c*. The function of the other subunits, particularly that of subunit III, is still unclear. Today, we know that human mitochondria make 13 proteins, all of them encoded by mitochondrial DNA: 3 subunits of cytochrome oxidase, 1 subunit of the cytochrome bc_1 complex, 7 subunits of the NADH dehydrogenase complex, and 2 subunits of the F_o part of ATP synthase. Mitochondria of the yeast *Saccharomyces cerevisiae* make only 8 stable proteins, those of plants at least twice as many and mitochondria of the protozoon *Reclinomonas americana* make no fewer than 62. Most or all of these mitochondrial genes bear a striking resemblance to their counterparts in bacteria—dramatic evidence for the long-held view that mitochondria evolved from bacterial endosymbionts.

My Cornell years showed me what a first-rate university can offer. By comparison, most European universities were grossly underfunded, granted their young researchers little independence, and clung to antiquated hierarchies and rigid faculty divisions. My colleagues in Europe had much more power than I at Cornell, but wielding this power left them little time for research. Running a modern university with public funds has become so complex that Wilhelm von Humboldt's ideal of a "scholars' republic" should be abandoned in favor of an enlightened presidential system, which frees researchers from the burden of administrative and political duties. The prime goals of a university president should be to select the best faculty and then to guard its precious time, but neither goal seemed to be a priority at the European university I visited at that time. I am afraid that this situation has not changed much.

My years at Cornell also dropped me smack in the midst of the hippie revolution. This proved to be a challenge—particularly for someone coming straight from conservative Austria. Although most of my students badly needed a haircut, a shave, or sartorial advice, they were outgoing and idealistic. With my friend Stuart J. Edelstein, I taught the introductory biochemistry course to a class of several hundred unruly students, which called for the equanimity of the Dalai Lama and the guts of a lion tamer. We stopped at nothing to keep our students attentive. To explain the role of ATP, I held up a huge wooden model of the molecule, which could fire a spring-loaded red Styrofoam™ ball representing the γ-phosphoryl group into the audience. To make the Krebs cycle less boring, I went through its steps by rotating the hand of a monstrous clock-like device. Whenever the hand reached a decarboxylation reaction, teaching assistants hidden behind this "Krebs cycle clock" released a green helium-filled balloon and fired a starter's pistol. I had heard that, in the United States, shooting always helps. The students seemed to appreciate our efforts and even took my German accent in good humor. Their comment "Hey Prof, you sound just like Henry Kissinger" was a compliment because the German-born U.S. secretary of state was notorious for his political craftiness and alleged allure for the opposite sex. Everything was questioned; everything tried. It seemed that a new society was just around the corner—until the flowers of this exuberant revolution wilted in the chill of the rising political violence, the oil shock, and the bitterness of the Vietnam defeat.

BASEL FOR BIG BIOLOGY

My wife and I had intended to move up the evolutionary tree from resident aliens to full-fledged U.S. citizens, but evolution has its own ways. In 1972, it made me accept a visiting professorship in Zurich, which, to most Austrians, conjures up hard-working burghers, discreet bankers wearing gold watches, and a nightlife reminiscent of Vienna's municipal cemetery. But our very first evening stroll along the elegant Bahnhofstrasse and a brief glance at the calendar of events in the venerable *Neue Zürcher Zeitung* quickly showed us the city's wealth, its cosmopolitan flair, and its rich cultural life. Later, I was equally impressed by the level of science at the university and at the Swiss Federal Institute of Technology. This visit made me realize that my image of Europe still reflected my years in postwar Austria and that the tranquil life in a small U.S. college town was worlds apart from the splendor and the intellectual excitement of Zurich. When the University of Zurich offered me a professorship, I was about to accept but evolution intervened again in the guise of the molecular biologist Thomas Hohn, who invited me to visit the newly created Biozentrum at the University of Basel. I had heard of this new institute through the grapevine and a British science magazine, which had run a highly laudatory article entitled "Basel for Big Biology." Indeed, this visit convinced me that the Biozentrum was the place for me. Even though it had just started operation, it radiated the informal and international atmosphere I so loved at Cornell. When my future Swiss colleague Max M. Burger phoned me at my Cornell office shortly upon my return and asked me to join the Biozentrum as a professor, I forewent the time-honored mating dance of academic courtship and simply said "yes"—provided my wife would say the same. She said nothing of this sort, however, and could not understand why we should now move our family across the Atlantic for the fifth time. Her loaded question "Is this move really necessary?" dominated our evening discussions for several weeks until I finally convinced her that this move would benefit us all.

In setting up the Biozentrum, the penny-wise citizens of Basel and their venerable university had outdone themselves—but not without insistent prodding by Basel's pharmaceutical giants F. Hoffmann-La Roche, Sandoz, Ciba, and Geigy. When these companies got wind of the university's plan to appoint a new professor of biochemistry, they pressed for an entirely new multidisciplinary institution in which biochemists, cell biologists, physical chemists, and pharmacologists would attack biological problems in a concerted manner. At first Basel's city fathers thought this idea to be plain wacky, but once they had decided to go ahead, they did so with their customary thoroughness. In no time, they erected an attractive building in which wide-open staircases invited casual interactions between scientists from different floors. By offering very generous financial support, they also had little difficulty hiring first-rate scientists. Most of them were either foreigners or Swiss nationals who had spent many years abroad. The cell biologists, Max M. Burger and Walter J. Gehring, came from Princeton and Yale, respectively; the crystallographer, Johan N. Jansonius, left the University of Groningen; the biophysicists, Jürgen Engel, Kasper Kirschner, Joachim Seelig, and Gerhard Schwarz, had all been trained at the Max-Planck-Institute in Göttingen; and the microbiologists, Eduard Kellenberger and Werner Arber, came from the University of Geneva and, together with the Roche Research Director, Albert Pletscher, gave the Biozentrum the legal statutes and the career structures of a modern and open research institution.

It is a rare privilege to belong to the founding generation of a scientific institute. All of us were eager to put our international experience into practice and make the Biozentrum one of the best research centers in the world. Although we did have our occasional squabbles about teaching and budgets, we never knew the internecine feuds common at European universities. Yet our presence did not please everyone: Colleagues from the university's other departments reproached us for wasting money, flouting university traditions, skipping faculty meetings, and—the ultimate sin— embracing American manners. Even our habit of working late into the night raised some eyebrows: An anonymous resident from across the street complained to the government that the nocturnal lights from our laboratory windows "interfered with his marital life." This being Switzerland, the government made us install outrageously expensive time-controlled Venetian blinds to protect the marital bliss of its population. But when Werner Arber received the Nobel Prize in 1978, Basel staged a city-wide celebration replete with a fearsome battalion of *Fasnacht* drums. After such an event, our critics could no longer denounce us as a bunch of money-squandering show-offs. As the Biozentrum's international reputation grew and we collected more and more scientific honors, Basel's citizens started to take pride in their Biozentrum. Today, it has become a firm part of the city's scientific and cultural scene. Still, if Basel University were to conduct a popularity poll among its faculty, most of us at the Biozentrum would rank near the bottom. Old resentments die hard.

TACKLING PROTEIN IMPORT INTO MITOCHONDRIA

During our first years in Basel, I continued to study the mitochondrial genetic system and its protein products. The Biozentrum proved to be a magnet for young scientists from around the world, and once again, I was lucky to have many outstanding students and postdoctoral fellows join my laboratory. The first one was Thomas D. Fox with whom I never coauthored a paper because he preferred to follow his own nose. This nose led him to the startling discovery that the genetic code of yeast mitochondrial DNA diverges from the general code (38) and that the nuclear genes affected in Ebner's cytochrome oxidase-less yeast mutants (30) encoded translational activators of specific mitochondrial mRNAs (39).

As the size of my research group grew, I felt the itch to tackle a new problem, but which one? In the fall of 1977, my good fairy sent me Maria-Luisa Maccecchini who became my first Swiss PhD student. At that time, my laboratory was already full, but she would not take "no" for an answer. "Maria-Luisa," I said finally, "if you will teach me how to spell your last name, I will teach you how to study protein import into mitochondria." It was a deal, and the start of a long trek into new territory.

Several laboratories had already described systems for studying uptake of proteins by isolated mitochondria. In particular, Walter Neupert and his colleagues (40, 41) at the University of Munich had translated total mRNA from *Neurospora crassa* in a cell-free extract, then added mitochondria, and found that some mitochondrial proteins (notably cytochrome c and the ADP/ATP carrier) were taken up by the organelles. However, uptake was difficult to prove convincingly as the measured radioactivities were low, and neither cytochrome c nor the ADP/ATP carrier was proteolytically altered upon import. Aided by a visit to Günter Blobel's laboratory at Rockefeller University, Maccecchini approached the problem with a highly efficient reticulocyte lysate. She translated total mRNA from yeast in the presence of radiolabeled methionine, then added yeast mitochondria, and checked whether these would import the α- and β-subunits of yeast F_1. The choice of these two proteins was a stroke of luck: Both were made as larger precursors and then cleaved in a single step to their mature size upon import. This modification was readily seen as a mobility shift in sodium dodecyl sulfate-polyacrylamide gels and offered a fast and reliable way to monitor import (42). Susan Gasser, my next graduate student, then established many key features of this import process (43). When she graduated with honors in 1982, Bruce Alberts and his colleagues handed her the best graduation gift a student could desire: almost two pages of their magisterial textbook *Molecular Biology of the Cell* describing the results of her thesis. Susan is now the successful director of the Friedrich Miescher Institute for Biomedical Research in Basel and a widely admired role model for young women scientists.

The discovery of mitochondrial precursor proteins marked the beginning of our 20-year-long effort to unravel the complex mitochondrial protein import system. In 1980, my Swiss student, Peter C. Böhni, identified a soluble metalloprotease in the matrix that removed the transient N-terminal sequences of precursors destined to cross the mitochondrial inner membrane. He partially purified the enzyme, but we obtained the pure enzyme only eight years later after a circuitous detour involving yeast genetics. In 1981, Michael Yaffe and I decided to isolate yeast mutants defective in mitochondrial protein import, but how were we to find such mutants? We reasoned that mitochondrial protein import should be necessary for life because it is a prerequisite for mitochondrial function, and our earlier work had shown this function to be vital. We therefore screened for conditional mutants. During the Christmas recess of 1982, I produced a collection of about 2,000 temperature-sensitive yeast mutants, which Yaffe then tested for the accumulation of an uncleaved precursor to the F_1 β-subunit at the nonpermissive temperature. This brute force dragnet netted us two mutants (21), which later proved to be defective in the genes for the two different subunits of the matrix protease. In 1988, Gerd Hawlitschek in Walter Neupert's group at Munich finally obtained the homogeneous enzyme from *Neurospora crassa* [see Hawlitschek et al. (44)], and my student Meija Yang isolated it from yeast [see Yang et al. (45)].

This protease was not the only one participating in the import process. Akira Ohashi et al. (46) and Susan Gasser and coworkers (47) found that the precursors to the imported mitochondrial proteins cytochrome c_1 and cytochrome b_2 are cleaved twice: first by the metalloprotease just discussed, and then by another enzyme that was tightly bound to the outer surface of the inner membrane. Why would mitochondria resort to such proteolytic extravagance? As cytochrome c_1 and cytochrome b_2 are both located in the space between the two mitochondrial membranes, we suspected that their import requires two different signals: (*a*) a matrix-targeting signal at the extreme N terminus that, by itself, would direct the protein into the matrix, where it is cleaved off by the matrix protease; and (*b*) a downstream "stop-transfer" signal that makes the once-cleaved precursor get stuck across the inner membrane. Cleavage of the stuck intermediate by a protease on the outer face of the inner membrane then generates the mature protein. The doubly cleaved cytochrome b_2 is released into the intermembrane space, whereas cytochrome c_1 remains tethered to the inner membrane through its hydrophobic C terminus. This stop-transfer model sparked a drawn-out and emotional controversy, but it is now generally accepted (48, 49). A similar combination of signals also directs many proteins to the mitochondrial outer membrane. By attaching the corresponding targeting sequences, Eduard C. Hurt, Dolf van Loon, and others in my group (50, 51) could direct non-mitochondrial "passenger proteins," such as dihydrofolate reductase, to any of the four major mitochondrial compartments or redirect mitochondrial proteins between compartments. Following up on this work, Martin Eilers made the puzzling observations that import of fusion proteins containing dihydrofolate reductase as the passenger protein was blocked by the dihydrofolate reductase ligand methotrexate and that it was dramatically accelerated by denaturing the protein. As methotrexate stabilizes the native conformation of dihydrofolate reductase, denaturation destroys it. Thus, Eilers concluded that proteins must enter mitochondria in the unfolded state [see Eilers & Schatz (52)]. But how do they find the mitochondria? Howard Riezman discovered that they bind to specific protein receptors on the mitochondrial surface [see Riezman et al. (53)], but it took several years before Walter Neupert's group in Munich and we could identify these receptors.

I will always remember the moment when my German postdoctoral fellow Eduard C. Hurt burst into my office and told me that even a targeting sequence as short as a dozen amino acids could direct a protein into the mitochondrial matrix [see Hurt et al. (54)]. How was this information encrypted? David A. Roise found the equally startling answer: All it takes is a positively charged amphiphilic helix [see Roise et al. (55)]. The exact amino acid sequence did not matter—even randomly generated sequences would do the job, as long as they could fold into such a helix. What then prevented cytosolic proteins from entering mitochondria? Our computers told us that evolution had apparently selected against such helices at the N termini of cytosolic proteins (56).

By the mid-1980s, our laboratory basked in the glow of a golden period, and it would take too long to recount all of what we found. But I cannot resist telling how Dietmar Vestweber tracked down the first subunit of a mitochondrial protein import channel [see Vestweber et al. (57)]. Vestweber's capacity for work has become a Biozentrum legend, and rumors that he once left the laboratory to get some sleep were never verified. To identify proteins of the import channel, he constructed an artificial precursor protein that gets stuck in this channel; he then photocross-linked the stuck precursor to any mitochondrial protein nearby; and he then identified the cross-linked protein. This may sound simple, but it required the construction of an artificial precursor stitched together by chemistry and genetic engineering from three different proteins as well as the synthesis of a trifunctional and photoactivatable cross-linker. The fish Dietmar finally caught was an integral outer membrane protein, the deletion of which proved to be lethal (58). We called it ISP42, but it was later renamed Tom40 and identified as the key subunit of the protein import channel across the mitochondrial outer membrane.

Vestweber's departure from our laboratory in early 1989 marked a new phase in our research. Having spent a decade sketching the broad outlines of the mitochondrial protein import system, the time had now come to flesh out the sketch. This phase saw again gifted young people and new hunts, as well as success and disappointment. But finishing a painting rarely matches the excitement of drawing the sketch. The sketch is an exclamation mark calling for attention, whereas the finishing phase is a more private battle, often too subtle for a riveting story. Yet it was this final phase that helped me become a mature scientist. Most of the insecurities of my early years had faded; I finally felt at ease leading a research group; and the honors that started to come my way in rapid succession sweetened the departure *d'un certain âge*. Just before I closed my laboratory, Carla Koehler and Sabeeha Merchant opened a new door by showing that a devastating human disease inflicting deafness, muscle weakness, dementia, or blindness is caused by a defective mitochondrial protein import system [see Koehler et al. (59)]. Our sketch of the protein import system showed us many nuts and bolts, but did not tell us how they functioned. To me as a chemist, this question was at the heart of the problem, but not everybody saw it this way. In the hothouse of today's science, discovering a new protein is great; cloning its gene a must; documenting its intracellular distribution by "red-green-merge" fluorescence imaging is politically correct; and coaxing the editor of a "high-impact" journal to pick the image for the journal cover is the ultimate victory. But deciphering the mechanism by which the protein works is a long uphill struggle with an uncertain outcome. Yet, when everything is said and done, it is the mechanism that counts. Unless we know it, we know nothing. And so we tried to identify the forces that move proteins into mitochondria as well as the mechanisms by which these proteins acquire their mature structure within the proper compartment (60). These questions are still not fully resolved, but the fleshed-out picture revealed a fiendishly complex network of interacting receptors and channels, aided by ATP-powered chaperones and enzymes. On their long voyage from endosymbionts to well-behaved cellular citizens, mitochondria have left no stone unturned. They are not quite there yet because the oxidizing sparks from their furnaces damage their host, and when this damage exceeds a certain threshold, mitochondria order the host to kill itself. Mitochondria are not only obedient servants, but also angels of death.

SCIENCE: A CONTRACT BETWEEN GENERATIONS

In 1998, Swiss activists mounted a massive attack on genetic engineering. This "Gene Protection Initiative" was mainly orchestrated by educated, professionally successful women from the German-speaking part of Switzerland (61), and even though it was roundly defeated, it made me aware of the fragility of science in a democratic society and the rising tide of political and bureaucratic controls that was threatening the country's scientific innovation. Everyone bemoaned this danger—but why not fight it? In the fall of 1999, I decided to stop research, close down my laboratory and office, and follow the call of the federal government in Bern to head the Swiss Science and Technology Council. As the country's top science advisor, I hoped to persuade its political leaders to adopt a more enlightened science policy. I was then only 63 years old, and my decision surprised and even shocked many of my friends and colleagues. But my four years in science politics brought me face to face with the plight of European science. Switzerland is one of the world's top scientific nations, yet many of its universities are poorly governed and unaware of the damage their antiquated hierarchies inflict on young researchers. Universities should be ticking intellectual time bombs, yet many are among our most conservative institutions. Elsewhere in Europe, matters are even worse. Descending into the netherworld of science politics was a daunting experience, but at least our Science Council could stem the planned increase of politically inspired project grants, reaffirm the importance of long-term basic research, and persuade some universities to adopt a "tenure track" system for their young faculty. And these years of reflecting, lecturing, and writing on these issues gradually transformed me into the author of essays and books that I am today (62, 65).

Looking at science from the outside made me aware of how much it has given me. Mars had done its best to stunt me, but science had rescued me through the scientists who helped me on my way—foremost among them Hans Tuppy, Efraim Racker, and David E. Green. By introducing me to the fires of life, they lit the fires of science in me and let me witness at close range one of the greatest biological discoveries of the twentieth century. Science is a contract between generations. Rarely was a generation as dependent on this contract as mine. And rarely was a generation as entitled to renege on it as the Jewish refugees. In honoring this contract so generously, they changed my life.

DISCLOSURE STATEMENT

The author is not aware of any affiliations, memberships, funding, or financial holdings that might be perceived as affecting the objectivity of this review.

ACKNOWLEDGMENTS

This Recollection draws on the previous autobiographical articles "Interplanetary Travels" (66), "From 'Granules' to Organelles: How Yeast Mitochondria Became Respectable" (67), "The Hunt for Mitochondrially Synthesized Proteins" (28), "Coming in from the Cold: How Answering a Postcard Can Launch a Scientific Career" (68), as well as on my book *Feuersucher* (65). All required permissions were given. I wish to thank Michael P. Murphy and Heimo Brunetti for helpful comments and Jesslyn Holombo for her exceptional care in copyediting the manuscript.

LITERATURE CITED

1. Beinert H, Stumpf PK, Wakil SJ. 2003. David Ezra Green. In *Biographical Memoirs*, 84:1–34. Washington, DC: Natl. Acad. Sci.
2. Ernster L, Schatz G. 1981. Mitochondria: a historical review. *J. Cell Biol.* 91:s227–55
3. McLean JR, Cohn GL, Brandt IK, Simpson MV. 1958. Incorporation of amino acids into the protein of isolated mitochondria. *J. Biol. Chem.* 233:657–63
4. Ephrussi B. 1953. *Nucleo-Cytoplasmic Relationships in Microorganisms*. Oxford: Clarendon
5. Slonimski PP. 1953. *La Formation des Enzymes Respiratoires chez la Levure*. Paris: Masson
6. Schatz G, Haslbrunner E, Tuppy H. 1964. Deoxyribonucleic acid associated with yeast mitochondria. *Biochem. Biophys. Res. Commun.* 15:127–32
7. Nass S, Nass MMK. 1963. Intramitochondrial fibers with DNA characteristics II. Enzymatic and other hydrolytic treatments. *J. Cell Biol.* 19:613–29
8. Pullman ME, Penefsky HS, Datta A, Racker E. 1960. Partial resolution of the enzymes catalyzing oxidative phosphorylation. I. Purification and properties of soluble dinitrophenol-stimulated adenosine triphosphatase. *J. Biol. Chem.* 235:3322–29
9. Penefsky HS, Pullman ME, Datta A, Racker E. 1960. Partial resolution of the enzymes catalyzing oxidative phosphorylation. II. Participation of a soluble adenosine triphosphatase in oxidative phosphorylation. *J. Biol. Chem.* 235:3330–36
10. Mitchell P. 1961. Coupling of phosphorylation to electron and hydrogen transfer by a chemi-osmotic type of mechanism. *Nature* 191:144–48
11. Pullman ME, Schatz G. 1967. Mitochondrial oxidation and energy coupling. *Annu. Rev. Biochem.* 36:530–610
12. Boyer PD, Chance B, Ernster L, Mitchell P, Racker E, Slater EC. 1977. Oxidative phosphorylation and photophosphorylation. *Annu. Rev. Biochem.* 46:955–1026
13. Schatz G, Penefsky HS, Racker E. 1967. Partial resolution of the enzymes catalyzing oxidative phosphorylation. XIV. Interaction of purified mitochondrial adenosine triphosphatase from baker's yeast with submitochondrial particles from beef heart. *J. Biol. Chem.* 242:2552–60
14. Schatz G. 1968. Impaired binding of mitochondrial adenosine triphosphatase in the cytoplasmic "petite" mutant of *Saccharomyces cerevisiae*. *J. Biol. Chem.* 243:2192–99
15. Linnane AW, Vitols E, Nowland PG. 1962. Studies on the origin of yeast mitochondria. *J. Cell Biol.* 13:345–50
16. Wallace PG, Linnane AW. 1964. Oxygen-induced synthesis of yeast mitochondria. *Nature* 201:1191–94
17. Criddle RS, Schatz G. 1969. Promitochondria of anaerobically grown yeast. I. Isolation and biochemical properties. *Biochemistry* 8:322–34
18. Plattner H, Schatz G. 1969. Promitochondria of anaerobically grown yeast. III. Morphology. *Biochemistry* 8:339–43
19. Swift H, Rabinowitz M, Getz G. 1968. Cytochemical studies on mitochondrial nucleic acids. In *Biochemical Aspects of the Biogenesis of Mitochondria*, ed. EC Slater, JM Tager, S Papa, E Quagliarello, pp. 3–19. Bari, Italy: Adriatica Editrice
20. Plattner H, Salpeter MM, Saltzgaber J, Schatz G. 1970. Promitochondria of anaerobically grown yeast. IV. Conversion into respiring mitochondria. *Proc. Natl. Acad. Sci. USA* 66:1252–59
21. Yaffe M, Schatz G. 1984. Two nuclear mutations which block mitochondrial protein import in yeast. *Proc. Natl. Acad. Sci. USA* 81:4819–23
22. Schatz G, Saltzgaber J. 1969. Protein synthesis by yeast promitochondria in vivo. *Biochem. Biophys. Res. Commun.* 37:996–1001

23. Kužela S, Smigan P, Kováč L. 1969. Biochemical characteristics of respiration-deficient yeast mutants differing in buoyant densities of mitochondrial DNA. *Experientia* 25:1042–43
24. Mason TL, Ebner E, Poyton RO, Wharton DC, Mennucci L, Schatz G. 1972. Participation of mitochondrial and cytoplasmic protein synthesis in mitochondrial formation. In *Mitochondria: Biogenesis and Bioenergetics*, ed. P Borst, EC Slater, GS van den Bergh, pp. 53–69. Amsterdam: North Holland
25. Mason TL, Schatz G. 1973. Cytochrome *c* oxidase of baker's yeast. II. Site of translation of the protein components. *J. Biol. Chem.* 248:1355–60
26. Weiss H, Sebald W, Bücher T. 1971. Cycloheximide-resistant incorporation of amino acids into a polypeptide of the cytochrome oxidase of *Neurospora crassa*. *Eur. J. Biochem.* 22:19–26
27. Rubin MS, Tzagoloff A. 1973. Assembly of the mitochondrial membrane system. X. Mitochondrial synthesis of three of the subunit proteins of yeast cytochrome oxidase. *J. Biol. Chem.* 248:4275–79
28. Schatz G. 1997. The hunt for mitochondrially synthesized proteins. *Protein Sci.* 6:728–34
29. Tzagoloff A. 1971. Assembly of the mitochondrial membrane system. IV. Role of mitochondrial and cytoplasmic protein synthesis in the biosynthesis of the rutamycin-sensitive adenosine triphosphatase. *J. Biol. Chem.* 246:3050–56
30. Ebner E, Mason TL, Schatz G. 1973. Mitochondrial assembly in respiration-deficient mutants of *Saccharomyces cerevisiae*. II. Effect of nuclear and extrachromosomal mutations on the formation of cytochrome oxidase. *J. Biol. Chem.* 248:5369–78
31. Poyton RO, Schatz G. 1975. Cytochrome *c* oxidase from baker's yeast: immunological evidence for the participation of a mitochondrially synthesized subunit in enzymatic activity. *J. Biol. Chem.* 250:762–66
32. Costanzo MC, Poutre CG, Strick CA, Fox TD. 1985. Yeast nuclear gene products required for translation of specific mitochondrial messenger RNAs. In *Achievements and Perspectives in Mitochondrial Research*, ed. F Palmieri, C Saccone, AM Kroon, pp. 355–60. Amsterdam: Elsevier
33. Tzagoloff A, Akai A, Needleman RB, Zulch G. 1975. Assembly of the mitochondrial membrane system. Cytoplasmic mutants of *Saccharomyces cerevisiae* with lesions in enzymes of the respiratory chain and in the mitochondrial ATPase. *J. Biol. Chem.* 250:8236–42
34. Cabral F, Solioz M, Rudin Y, Schatz G, Clavilier L, Slonimski PP. 1978. Identification of the structural gene for yeast cytochrome *c* oxidase subunit II on mitochondrial DNA. *J. Biol. Chem.* 253:297–304
35. Anderson S, Bankier AT, Barrell BG, de Bruijn MHL, Coulson AR, et al. 1981. Sequence and organization of the human mitochondrial genome. *Nature* 290:457–65
36. Attardi G, Chomyn A, Doolittle RF, Mariottini P, Ragan CI. 1986. Seven unidentified reading frames of human mitochondrial DNA encode subunits of the respiratory chain NADH dehydrogenase. *Cold Spring Harb. Symp. Quant. Biol.* 51:103–14
37. Tsukihara T, Aoyama H, Yamashita E, Tomizaki T, Yamaguchi H, et al. 1996. The whole structure of the 13-subunit oxidized cytochrome *c* oxidase at 2.8 Å. *Science* 272:1136–44
38. Fox TD. 1979. Five TGA "stop" codons occur within the translated sequence of the yeast mitochondrial gene for cytochrome *c* oxidase subunit II. *Proc. Natl. Acad. Sci. USA* 76:6534–38
39. Fox TD. 1986. Nuclear gene products required for translation of specific mitochondrially coded mRNAs in yeast. *Trends Genet.* 2:97–99
40. Harmey MA, Hallermayer G, Korb H, Neupert W. 1977. Transport of cytoplasmically synthesized proteins into the mitochondria in a cell free system from *Neurospora crassa*. *Eur. J. Biochem.* 81:533–44
41. Hallermayer G, Zimmermann R, Neupert W. 1977. Kinetic studies on the transport of cytoplasmically synthesized proteins into the mitochondria in intact cells of *Neurospora crassa*. *Eur. J. Biochem.* 81:323–33
42. Macchecchini M-L, Rudin Y, Blobel G, Schatz G. 1979. Import of proteins into mitochondria: precursor forms of the extra mitochondrially made F_1-ATPase subunits in yeast. *Proc. Natl. Acad. Sci. USA* 76:343–47
43. Gasser SM, Daum G, Schatz G. 1982. Import of proteins into mitochondria. Energy-dependent uptake of precursors by isolated mitochondria. *J. Biol. Chem.* 257:13034–41
44. Hawlitschek G, Schneider H, Schmidt B, Tropschug M, Hartl FU, Neupert W. 1988. Mitochondrial protein import: identification of processing peptidase and of PEP, a processing enhancing protein. *Cell* 53:795–806
45. Yang M, Jensen RE, Yaffe MP, Schatz G. 1988. Import of proteins into yeast mitochondria: The purified matrix processing protease contains two subunits which are encoded by the nuclear *MAS1* and *MAS2* genes. *EMBO J.* 7:3857–62

46. Ohashi A, Gibson J, Gregor I, Schatz G. 1982. Import of proteins into mitochondria. The precursor of cytochrome c_1 is processed in two steps, one of them heme-dependent. *J. Biol. Chem.* 257:13042–47
47. Gasser SM, Ohashi A, Daum G, Böhni P, Gibson J, et al. 1982. The imported mitochondrial proteins cytochrome b_2 and cytochrome c_1 are processed in two steps. *Proc. Natl. Acad. Sci. USA* 79:267–71
48. Glick BS, Brand A, Cunningham K, Müller S, Hallberg RL, Schatz G. 1992. Cytochromes c_1 and b_2 are sorted to the intermembrane space of yeast mitochondria by a stop-transfer mechanism. *Cell* 69:809–22
49. Bömer U, Meijer M, Guiard B, Dietmeier K, Pfanner N, Rassow J. 1997. The sorting route of cytochrome b_2 branches from the general mitochondrial import pathway at the preprotein translocase of the inner membrane. *J. Biol. Chem.* 272:30439–46
50. Hurt EC, Pesold-Hurt B, Schatz G. 1984. The cleavable prepiece of an imported mitochondrial protein is sufficient to direct cytosolic dihydrofolate reductase into the mitochondrial matrix. *FEBS Lett.* 178:306–10
51. van Loon APGM, Brändli A, Schatz G. 1986. The cleavable presequences of two imported mitochondrial precursor proteins contain information for intracellular targeting and for intramitochondrial sorting. *Cell* 44:801–12
52. Eilers M, Schatz G. 1986. Binding of a specific ligand inhibits import of a purified precursor protein into mitochondria. *Nature* 322:228–32
53. Riezman H, Hay R, Witte C, Nelson N, Schatz G. 1983. Yeast mitochondrial outer membrane specifically binds cytoplasmically-synthesized precursors of mitochondrial proteins. *EMBO J.* 2:1113–18
54. Hurt EC, Pesold-Hurt B, Suda K, Oppliger W, Schatz G. 1985. The first twelve amino acids (less than half of the presequence) of an imported mitochondrial protein can direct mouse cytosolic dihydrofolate reductase into the yeast mitochondrial matrix. *EMBO J.* 4:2061–68
55. Roise D, Horvath SJ, Tomich JM, Richards JH, Schatz G. 1986. A chemically synthesized mitochondrial signal peptide can form an amphipathic helix and perturb natural and artificial phospholipid bilayers. *EMBO J.* 5:1327–34
56. Lemire BD, Fankhauser C, Baker A, Schatz G. 1989. The mitochondrial targeting function of randomly generated peptide sequences correlates with predicted helical amphiphilicity. *J. Biol. Chem.* 264:20206–15
57. Vestweber D, Brunner J, Baker A, Schatz G. 1989. A 42K outer-membrane protein is a component of the yeast mitochondrial protein import site. *Nature* 341:205–9
58. Baker KP, Schaniel A, Vestweber D, Schatz G. 1990. ISP42, a protein of the yeast mitochondrial outer membrane, is essential for protein import and cell viability. *Nature* 348:605–9
59. Koehler CM, Leuenberger D, Merchant S, Renold A, Junne T, Schatz G. 1999. Human deafness dystonia syndrome is a mitochondrial disease. *Proc. Nat. Acad. Sci. USA* 96:2141–46
60. Schatz G. 1998. The Swiss vote on gene technology. *Science* 281:1810–11
61. Schatz G. 2005. *Jeff's View on Science and Scientists*. Amsterdam: Elsevier
62. Schatz G. 2008. *Jenseits der Gene*. Zürich: NZZ Libro
63. Schatz G. 2008. *Jenseits der Gene*. Zürich: Kein & Aber. Audio version.
64. Schatz G. 2011. *A Matter of Wonder*. Basel: Karger
65. Schatz G. 2011. *Feuersucher. Die Jagd nach dem Rätsel der Lebensenergie*. Zürich: Wiley VCH/Weinheim/NZZ Libro
66. Schatz G. 2000. Interplanetary travels. In *Comprehensive Biochemistry*, ed. G Semenza, R Jaenicke, 41:449–530. Amsterdam: Elsevier Sci.
67. Schatz G. 1993. From 'granules' to organelles: how yeast mitochondria became respectable. In *The Early Days of Yeast Genetics*, ed. MN Hall, pp. 241–46. Cold Spring Harbor, NY: Cold Spring Harb. Lab.
68. Schatz G. 2009. Coming in from the cold: how answering a postcard can launch a scientific career. *Nat. Cell Biol.* 11:364

Introduction to Theme "Chromatin, Epigenetics, and Transcription"

Joan W. Conaway[1,2]

[1]Stowers Institute for Medical Research, Kansas City, Missouri 64110;
email: JLC@stowers.org

[2]Department of Biochemistry and Molecular Biology, Kansas University Medical Center, Kansas City, Kansas 66160

Keywords

DNA methylation, noncoding RNA, histone modification, RNA polymerase II, elongation

Abstract

Transcriptional regulation in eukaryotes depends on a complex network of interactions between RNA polymerases and a host of transcription factors and coregulators that control their activity during transcription initiation and elongation. Among these are an enormous variety of enzymes and proteins that modulate chromatin structure via changes in DNA methylation, histone modifications, and nucleosome location. This volume of the *Annual Review of Biochemistry* contains a set of four reviews addressing the interplay between mechanisms that regulate DNA methylation, chromatin structure, and transcription.

The development of a multicellular organism from a fertilized egg requires proper execution of a transcription program that results in synthesis of many thousands of messenger RNAs (mRNAs), at exactly the right time, in the right set of cells, and in the right amount. Misregulation of the transcriptional program is associated not only with developmental defects but also with many disease states. Accordingly, major goals of research over the past several decades have been to define the components of the transcriptional regulatory apparatus and to work out the details of how they function together to attain the exquisitely precise control of the transcription output required for normal development and function. This work has led over time to the revelation that regulation of eukaryotic mRNA synthesis is a multi-tiered process governed by an intricate interplay between chromatin structure and the RNA polymerase II (Pol II) transcription machinery. Chromatin structure is controlled locally, at the gene level, and globally through the action of a withering collection of enzymes and proteins that regulate both the methylation status of chromosomal DNA and the distribution and modification of nucleosomes along it. Like the cell's machinery for regulating chromatin structure, the RNA Pol II transcription machinery is elaborate and composed of a large collection of transcription factors that control not only transcript initiation, but also transcript elongation through chromatin. This volume of the *Annual Review of Biochemistry* includes four articles addressing the interplay between chromatin structure and RNA Pol II transcription.

At its most fundamental level, the transcription of a gene by any RNA polymerase can be described in terms of a transcription cycle that can be divided into multiple, mechanistically distinct stages. In the first stage, the enzyme must be able to identify, bind to, and initiate transcription correctly at promoters in the background of the large amount of nonpromoter DNA in the cell. Having initiated transcription, an RNA polymerase must elongate the nascent transcript until it reaches the gene's 3′ end, a process that in higher eukaryotes can require it to traverse hundreds of kilobases before it terminates transcription and is released from the gene.

For many years, it was assumed that the majority of Pol II transcriptional regulation was accomplished at the earliest stages of the transcription cycle by processes that affect the efficiency with which the enzyme is recruited to promoters and initiates transcription; however, hints that transcript elongation could also be a key site for regulation came from evidence that most Pol II elongation in vitro or in cells is sensitive to inhibition by the protein kinase inhibitor 5,6-dichloro-1-β-D-ribofuranosylbenzimidazole or DRB (1, 2). Subsequently, seminal studies of heat shock gene expression in *Drosophila* and of *c-myc*, *c-fos*, and other oncogenes in human cells revealed that transcription of some genes can be regulated by promoter-proximal pausing in which Pol II pauses shortly after initiating transcription and is released into productive elongation only upon gene activation (3–5). Over the next decade or so, biochemical studies identified a collection of transcription elongation factors, including negatively acting factors such as DRB sensitivity-inducing factor (DSIF) and the negative elongation factor (NELF), which induce transcriptional pausing, and positively acting factors, such as the protein kinase P-TEFb and eleven-nineteen lysine rich in leukemia (ELL), which promote efficient elongation. With the advent of genome-wide methods for measuring nascent transcript synthesis as well as the location of Pol II and elongation factors on chromatin, it has become clear that control at the level of transcript elongation is a general feature of Pol II transcription and plays a critical role in development and disease. In their review, entitled "RNA Polymerase II Elongation Control," Zhou and colleagues (6) discuss our current understanding of the mechanisms and transcription factors that control transcript elongation and associated processes, such as RNA capping, splicing, and polyadenylation.

In eukaryotic cells, chromosomal DNA is packaged into arrays of nucleosomes, each of which contains ∼146 bp of DNA wrapped around a histone octamer containing two copies each of histones H2A, H2B, H3, and H4 (7). These arrays are, in turn, folded into chromatin fibers made up of higher-order nucleosomal structures, allowing the several meters of DNA that make up the human genome to fit into nuclei with diameters of just a few microns. Although this compaction presents a serious impediment to the RNA Pol II transcription machinery, as well as to enzymes involved in other nuclear processes such as DNA replication and repair, it also provides remarkable opportunities for gene regulation. Indeed, it is now clear that establishment and maintenance of distinct patterns of gene expression in different cell lineages are due in significant part to epigenetic mechanisms that involve alterations in chromatin structure driven in part by posttranslational modifications of histones and by DNA methylation.

Regions of transcriptionally silenced chromosomal DNA are enriched in CpG dinucleotides containing cytosine residues methylated at position 5 (8). In addition, the chromatin of silenced regions contains high levels of histone H3 dimethylated on lysine 9 (H3K9me2) and histone H3 trimethylated on lysine 27 (H3K27me3) (9). In contrast, transcriptionally active regions are associated with hypomethylated DNA and with chromatin with a different constellation of histone marks, including histone H4 trimethylated on lysine 4 (H3K4me3) (9). Two reviews in this series, "Programming of DNA Methylation Patterns," by Cedar & Bergman (10), and "The COMPASS Family of Histone H3K4 Methylases: Mechanisms of Regulation in Development and Disease Pathogenesis," by Shilatifard (11), review our current understanding of how these epigenetic marks are established, maintained, and function to control chromatin structure and gene regulation.

In the final review in the series, "Genome Regulation by Long Noncoding RNAs," Rinn & Chang (12) describe the identification and functional analysis of a large set of long noncoding RNAs (lncRNAs), at least some of which are proving to have roles in establishing gene expression patterns important for stem cell maintenance, cell fate determination during development, and diseases such as cancer. Until recently, it was thought that the vast majority of higher eukaryotic genomes are made up of nontranscribed, so-called junk DNA that lacks known protein coding sequences and has no clear function. With the advent of high-throughput methods for analyzing the entire population of RNAs present in a given cell type, however, it has become clear that a much larger fraction of the genome is transcribed than previously appreciated. Indeed, mammalian cells express thousands of lncRNAs that are encoded by previously unannotated regions of the genome. That these lncRNAs often exhibit sequence conservation across species suggested that they have some function linked to their structures (13, 14), and the challenges are now to determine which of them contribute to regulatory processes and how. In their review, Rinn & Chang discuss emerging evidence that lncRNAs can assemble into ribonucleoprotein complexes and contribute to gene regulation by mechanisms almost as diverse as those employed by more conventional protein regulators. Some lncRNAs serve as decoys that mimic the DNA-binding sites of transcriptional regulators and prevent them from binding to their normal targets in promoter DNA. Among these are the lncRNA Gas5, which binds the the glucocorticoid receptor, and an another called PANDA, which binds the transcription factor NF-Y. Others can serve as scaffolds or adapters to promote interactions between multiple proteins or protein complexes that regulate chromatin structure. For example, an lncRNA called HOTAIR can bind to both the histone H3 K27 methyltransferase polycomb repressive complex 2 (PRC2) and the histone H3 K4 demethylase LSD1-CoREST to coordinate a repressive chromatin methylation state. In addition, lncRNAs can serve as guides that target their protein partners to specific chromosomal locations, much like the DNA-binding domain of a sequence-specific DNA-binding protein. Thus, HOTAIR as well as a host of other lncRNAs have been shown to bind PRC2, and in several cases, these have been shown to contribute to PRC2 recruitment and histone H3K27 methylation at specific loci.

A few decades ago, we knew virtually nothing about the the enzymes and proteins that coordinate epigenetic modifications of chromatin and DNA with transcription or about how their functions, and malfunctions, contribute to normal development and disease. Taken together, these four reviews highlight just how much has been learned. But as they also make clear, each new answer raises new and unexpected questions, and research in this area should be fruitful for many years to come.

DISCLOSURE STATEMENT

The author is not aware of any affiliations, memberships, funding, or financial holdings that might be perceived as affecting the objectivity of this review.

ACKNOWLEDGMENT

The author thanks Ronald C. Conaway for helpful suggestions during the preparation of this introduction.

LITERATURE CITED

1. Tamm I, Kikuchi T. 1979. Early termination of heterogeneous nuclear RNA transcripts in mammalian cells: accentuation by 5,6-dichloro 1-beta-D-ribofuranosylbenzimidazole. *Proc. Natl. Acad. Sci. USA* 76:5750–54
2. Zandomeni R, Bunick D, Ackerman S, Mittleman B, Weinmann R. 1983. Mechanism of action of DRB. III. Effect on specific in vitro initiation of transcription. *J. Mol. Biol.* 167:561–74
3. Bentley DL, Groudine M. 1986. A block to elongation is largely responsible for decreased transcription of c-*myc* in differentiated HL60 cells. *Nature* 321:702–6
4. Fort P, Rech J, Vie A, Piechaczyk M, Bonnieu A, et al. 1987. Regulation of c-*fos* gene expression in hamster fibroblasts: initiation and elongation of transcription and mRNA degradation. *Nucleic Acids Res.* 15:5657–67
5. Rougvie AE, Lis JT. 1988. The RNA polymerase II molecule at the 5′ end of the uninduced *hsp70* gene of *D. melanogaster* is transcriptionally engaged. *Cell* 54:795–804
6. Zhou Q, Li T, Price DH. 2012. RNA polymerase II elongation control. *Annu. Rev. Biochem.* 81:119–43
7. Kornberg RD. 1977. Structure of chromatin. *Annu. Rev. Biochem.* 46:931–54
8. Yisraeli J, Szyf M. 1984. Gene methylation patterns and expression. In *DNA Methylation: Biochemistry and Biological Significance*, ed. A Razin, H Cedar, AD Riggs, pp. 352–70. New York: Springer-Verlag
9. Li B, Carey M, Workman JL. 2007. The role of chromatin during transcription. *Cell* 128:707–19
10. Cedar H, Bergman Y. 2012. Programming of DNA methylation patterns. *Annu. Rev. Biochem.* 81:97–117
11. Shilatifard A. 2012. The COMPASS family of histone H3K4 methylases: mechanisms of regulation in development and disease pathogenesis. *Annu. Rev. Biochem.* 81:65–95
12. Rinn JL, Chang HY. 2012. Genome regulation by long noncoding RNAs. *Annu. Rev. Biochem.* 81:145–66
13. Ponjavic J, Ponting CP, Lunter G. 2007. Functionality or transcriptional noise? Evidence for selection within long noncoding RNAs. *Genome Res.* 17:556–65
14. Guttman M, Amit I, Garber M, French C, Lin MF, et al. 2009. Chromatin signature reveals over a thousand highly conserved large non-coding RNAs in mammals. *Nature* 458:223–27

The COMPASS Family of Histone H3K4 Methylases: Mechanisms of Regulation in Development and Disease Pathogenesis

Ali Shilatifard

Stowers Institute for Medical Research, Kansas City, Missouri 64110; email: ASH@Stowers.org

Keywords

histone methylation, MLL, Set1, Rad6/Bre1, histone monoubiquitination, chromosomal translocations in leukemia

Abstract

The *Saccharomyces cerevisiae* Set1/COMPASS was the first histone H3 lysine 4 (H3K4) methylase identified over 10 years ago. Since then, it has been demonstrated that Set1/COMPASS and its enzymatic product, H3K4 methylation, is highly conserved across the evolutionary tree. Although there is only one COMPASS in yeast, *Drosophila* possesses three and humans bear six COMPASS family members, each capable of methylating H3K4 with nonredundant functions. In yeast, the histone H2B monoubiquitinase Rad6/Bre1 is required for proper H3K4 and H3K79 trimethylations. The machineries involved in this process are also highly conserved from yeast to human. In this review, the process of histone H2B monoubiquitination-dependent and -independent histone H3K4 methylation as a mark of active transcription, enhancer signatures, and developmentally poised genes is discussed. The misregulation of histone H2B monoubiquitination and H3K4 methylation result in the pathogenesis of human diseases, including cancer. Recent findings in this regard are also examined.

Contents

1. INTRODUCTION 67
2. MOLECULAR AND BIOCHEMICAL PROPERTIES OF SET1/COMPASS 69
   - 2.1. Biochemical Purification of Set1/COMPASS as the First Histone H3K4 Methylase 69
   - 2.2. Kinetic Properties and Contribution of Set1/COMPASS Subunits in H3K4 Methylation 70
   - 2.3. Transcription Elongation and the Pattern of H3K4 Methylation by Set1/COMPASS 72
3. THE METAZOANS' COMPASS FAMILY 73
   - 3.1. Biochemical Purification of the COMPASS family from *Drosophila* to Human 73
   - 3.2. Composition of the Metazoan COMPASS Family of H3K4 Methylases 73
   - 3.3. Structural Studies of the COMPASS Family from Yeast to Human 74
   - 3.4. Functional Diversity of the Metazoans' COMPASS-Like Complexes 74
   - 3.5. Trx in *Drosophila* and Its Mammalian Homolog MLL1/MLL2 Are Positive Regulators of Homeotic Gene Expression 75
   - 3.6. Mammalian MLL3/MLL4: The Homologs of *Drosophila* Trr 76
   - 3.7. A Histone H3K27 Demethylase as a Component of the Trr/MLL3/MLL4 COMPASS-Like Complexes ... 77
   - 3.8. The Metazoan COMPASS Subunit DPY-30 in Dosage Compensation 77
4. A NOVEL YEAST BIOCHEMICAL SCREEN DEFINING THE H3K4 METHYLATION PATHWAY .. 78
   - 4.1. The Histone H2B Monoubiquitinase Rad6 Is Required for H3K4 Trimethylation by Set1/COMPASS 78
   - 4.2. Yeast Bre1, an E3 Ligase Required for Rad6 Recruitment to Chromatin 78
   - 4.3. The Conserved Role for Bre1 in Set1/COMPASS Function from Yeast to Human 79
   - 4.4. Rad6/Bre1, H2B Monoubiquitination, and Transcription Elongation Control 79
   - 4.5. A Role for Rad6/Bre1 in DNA Double-Strand Break 80
   - 4.6. The PAF1 Complex as a Platform for Histone-Modifying Enzymes on Transcribing Polymerase 81
5. HISTONE H3K4 METHYLASES IN DEVELOPMENT AND DISEASE 81
   - 5.1. Licensed to Elongate: A Model for Leukemic Pathogenesis Through MLL1 81
   - 5.2. The COMPASS Family in the Regulation of Life Span and Aging 82
   - 5.3. The Trr/COMPASS Family in Nuclear Receptor Transactivation and Cancer 83
6. CONCLUDING REMARKS 84

1. INTRODUCTION

The very long linear sequences of nucleotides within DNA possess the genetic blueprint required for the creation of all organisms, with the exception of RNA viruses. The several meters of DNA constituting our genome must remain functional and accessible to the transcriptional machinery and yet be protected and packaged within a very minute space within the nucleus of our cells. To achieve such dynamic packaging of this genomic information, DNA is found in complex with an equal mass of histone proteins to form nucleosomes, the basic unit of chromatin (**Figure 1**).

The eukaryotic genome is compacted at several levels. The first level requires the wrapping of DNA around the outside of an octamer of histone proteins, i.e., H2A, H2B, H3, and H4, or some variant of these canonical histones to form nucleosomes (1–3). Using electron microscopy, the array of nucleosomes on DNA was observed as a series of "beads on a string" or the 11-nm model. This beads-on-a-string structure of nucleosomes can further interact through

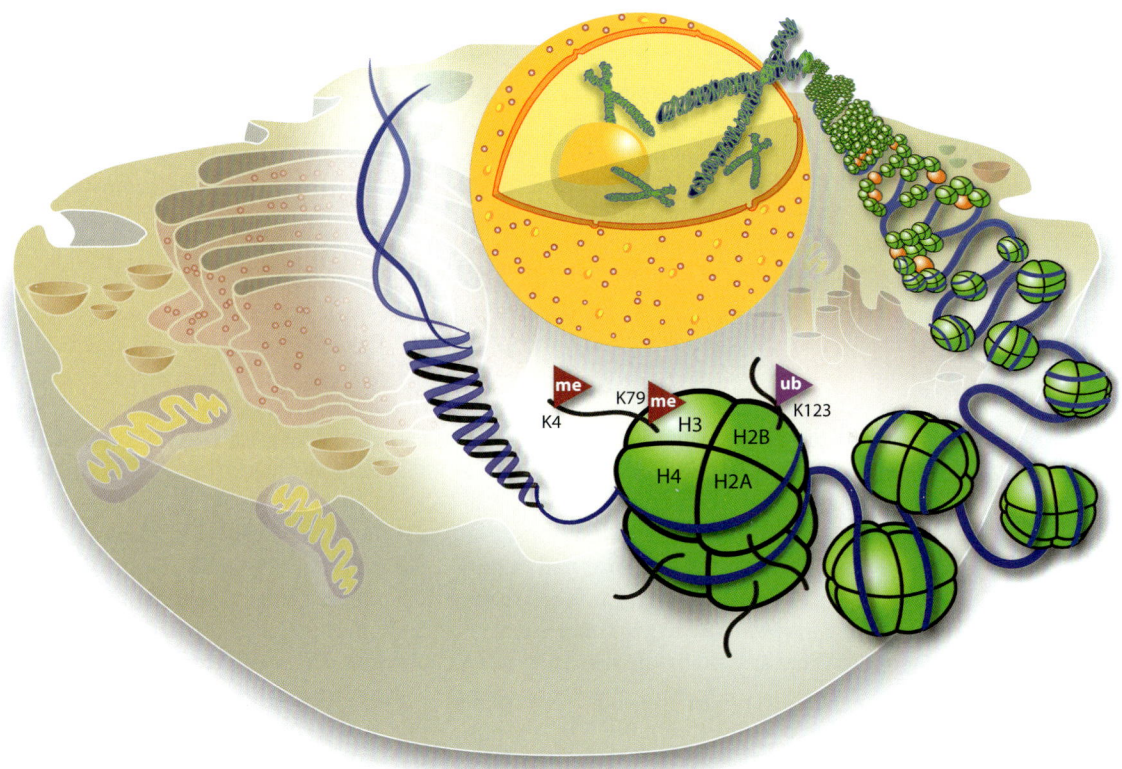

Figure 1

Chromatin, histone modifications, and gene expression. The very long eukaryotic DNA is compacted within the cell nucleus through its interactions with histone proteins, forming the nucleosomes. Structural studies of nucleosomes demonstrated that the histone N-terminal tails protrude outward beyond the gyres of DNA. Many of the amino acid residues within the histone tails can be posttranslationally modified, providing a landing pad for a diverse array of transcription factors, chromatin remodelers, and DNA-interacting proteins to regulate gene expression and other DNA-dependent processes. In this figure, sites of posttranslational modifications on histone tails for histone H3 lysine 4 (H3K4) methylation and histone H2B monoubiquitination by the COMPASS family of H3K4 methylases and the Rad6/Bre1 complex, respectively, are shown. Abbreviations: me, methylation; ub, ubiquitin.

short-range internucleosomal exchanges to compact the nucleosomal array into a structure known as the 30-nm fiber. To form the fully packaged metaphase chromosomes, linker histone proteins, such as histone H1, help to stabilize and further compact the 30-nm fiber into stable chromatin fibers, which can then be further packaged into metaphase chromosomes (**Figure 1**).

Upon packaging, the DNA within the nucleus is found in both lightly and tightly packed regions of chromatin. On the basis of a series of cytogenetic observations by light microscopy, Heitz (4) reported the presence of darkly stained chromatin in the nucleus, which he referred to as "heterochromatin." Heitz (4) found that the heterochromatin remained condensed throughout the cell cycle. This observation was in contrast to the behavior of the euchromatic regions (the lightly packed regions), which are subjected to cycles of condensation and decondensation at different stages of the cell cycle. Initial studies by Heitz suggested that heterochromatin is devoid of genes; however, functional studies in *Drosophila* established that some genes, such as *rolled* and *light*, reside within the heterochromatin (5, 6).

In the early 1930s, Muller & Altenberg (7) described the phenomenon of position-effect variegation (PEV) of gene expression in which genes expressed near heterochromatin are silenced in *Drosophila*. PEV, originally observed in insects, has since been demonstrated to exist in other organisms, including plants and mammals. The variegation in gene expression is thought to be regulated by the spreading effect of factors from the adjacent heterochromatin. This explanation for the molecular mechanism of PEV presumes that the juxtaposed gene locus is condensed to form a transcriptionally inactive state through the movement of factors or posttranslational modifications of factors from heterochromatic regions to regions within its vicinity. In the late 1970s, studies by Grigliatti and colleagues (8, 9) demonstrated that the variegation in gene expression in *Drosophila* is a function of histone gene multiplicity or expression levels, suggesting for the first time that chromatin and its basic components, the histones, may have a fundamental role in the process of regulating gene expression.

Nucleosomes and histones also represent an analogous inhibitory effect to transcription in other cell types in addition to *Drosophila*. For example, attenuation of histone levels, and thereby nucleosome levels, by genetic means in yeast cells resulted in the activation of genes that were otherwise inactive (10, 11). Not only were nucleosomes found to be inhibitory to transcription, but also their presence on DNA can control the interaction of DNA-binding factors. Although DNA sequences recognized by DNA-binding proteins are found throughout the genomes of organisms, the actual in vivo interaction of DNA-binding proteins with chromatin is highly specific and is only found on precise regions. For example, the repressor-activator protein 1 (Rap1) is associated with certain limited sites on chromatin that have regions with the potential to act as promoters, although Rap1's primary DNA sequence consensus binding site is found throughout the genome (12). Similar to these in vivo studies, fundamental in vitro biochemical studies demonstrated that nucleosomes are inhibitory to the initiation of transcription by RNA polymerase II (Pol II) (13–15). Biochemical studies also demonstrated that even if Pol II is initiated, the presence of nucleosomes, or just histone tetramers, provide a blockage to productive transcription-elongating Pol II (16, 17). The above findings in agreement with other findings in the field suggested that not only does chromatin regulate the initiation and elongation steps of gene expression in vivo but also that chromatin is pivotal in modulating the interaction of DNA-binding proteins with their consensus binding sites on DNA. Indeed, using DNaseI hypersensitivity mapping, Weintraub & Groudine (18) provided the seminal concept that nucleosomes can block the access of transcription factors to DNA and the regulatory regions, thereby controlling gene expression, initiation of DNA replication and recombination, as well as other processes requiring DNA access.

To define the mechanism of heterochromatin formation and the molecular machinery involved in the regulation of gene expression from heterochromatin and other developmental loci, such as the homeotic gene-containing regions, fundamental genetic studies were performed in *Drosophila* and other organisms. From these studies, the *trithorax* (*trx*) group and the *Polycomb* (*Pc*) group of genes were identified as playing opposing roles in homeotic gene expression (19–21). The suppressor and enhancer screens of PEV in *Drosophila* identified suppressors and enhancers of PEV [Su(var)] and [E(var)], respectively. Many of the genes found from the above screens encode for proteins bearing a 130- to 140-amino acid motif called the SET domain (22, 23). This domain takes its name from the *Drosophila* genes *Su(var)3-9*, *Enhancer of zeste* [*E(z)*], and *trx* (22, 23). Many of the SET domain-containing proteins have now been demonstrated to possess histone (H) or lysine (K) methyltransferase (HMTase or KMTase) activity.

Detailed structural studies of nucleosomes demonstrated that the N-terminal tails of each of the histones protrude outward beyond the gyres of DNA (**Figure 1**) (24). Many amino acid residues within the histone tails can be post-translationally modified; and the modifications of histone tails could provide a landing pad for a diverse array of transcription factors, chromatin remodelers, and DNA-interacting proteins to regulate gene expression. The covalent modifications of histones to date include acetylation, phosphorylation, ADP-ribosylation, biotinylation, ubiquitination, and methylation (3, 25–29). Although, the list of SET domain-containing proteins with identified KMTase activity toward histones is growing, so far, this list only includes eight classes of enzymes, KMT1-KMT8 (30).

The KMT2 class and its first member, Set1, in yeast *Saccharomyces cerevisiae* were isolated within the Set1/COMPASS functioning as histone H3K4 methylases. Given the limitation in space, this review only focuses on the description of the KMT2 class and its histone substrate. I discuss our current knowledge on (*a*) how histone H3K4 methylation by the KMT2 class is regulated by the histone H2B monoubiquitinase Rad6/Bre1 and Paf1 complexes, (*b*) how these marks function as a sign of active transcription and enhancer signatures, and (c) how these modifications function on the developmentally poised genes in mammalian cells.

2. MOLECULAR AND BIOCHEMICAL PROPERTIES OF SET1/COMPASS

2.1. Biochemical Purification of Set1/COMPASS as the First Histone H3K4 Methylase

Early genetic studies in *Drosophila* identified two families of proteins, the trithorax group (trxG) and the Polycomb group (PcG), as playing a central role in the regulation of gene expression throughout development (31). The clustered homeotic (*Hox*) genes in the *Bithorax* and *Antennapedia* gene complexes in *Drosophila* are some of the known targets of the trxG and PcG proteins. The *Drosophila trx* gene was discovered by the identification of several mutations in this gene that resulted in the partial transformation of the halteres into wings (32). The halteres are small knobbed-like structures, functioning as a gyroscope and informing the insect about position of the body throughout flight, found behind the wings in *Drosophila*. Mutations in *trx* and *Pc* genes resembled mutations in the *Hox* genes, leading to the suggestion that TrxG proteins function as positive effectors, whereas the PcG proteins function in the repression of homeotic gene expression (33, 34).

Most of the trxG and PcG proteins are conserved in mammals and function within similar pathways to those of their *Drosophila* counterparts (35, 36). The first mammalian homolog of *Drosophila trx*, the mixed lineage leukemia (*MLL*) gene, was cloned by the identification of its random translocations found in patients suffering from hematological malignancies, including acute myeloid and lymphoid leukemia (37–40). Despite the critical role of the MLL gene product in the pathogenesis of hematological malignancies, very little was known about the molecular and biochemical properties of MLL until the genetic and biochemical studies of its yeast homolog, Set1.

To begin to define the biochemical function(s) of MLL in the hope of a better understanding of its role in leukemic pathogenesis, initial studies from our laboratory in the late 1990s set out to purify MLL-containing complexes. Our approach was to fractionate nuclear extracts and follow MLL by immunoblotting. Cloning and sequencing of MLL predicted a 400-kDa polypeptide on Western blotting, but it was much later that it was determined that MLL is proteolytically cleaved in vivo, resulting in N-terminal (~150-kDa) and C-terminal (~200-kDa) fragments that can associate noncovalently in cells (41). Although we were unable to detect full-length MLL at its expected electrophoretic mobility (~400 kDa), and therefore could not identify MLL complexes from mammalian cells, we had identified Set1 of yeast *S. cerevisiae* as a MLL homolog (42). We used conventional chromatographic methods to isolate a yeast Set1-containing complex from nuclear extracts obtained from ~300 liters of yeast culture (42). This complex was named COMPASS, for complex of proteins associated with Set1, and demonstrated to be a histone H3 lysine 4 (H3K4) methylase. Set1/COMPASS was the first H3K4 methylase identified and is capable of catalyzing the mono-, di-, and trimethylation of histone H3K4 (42–45).

2.2. Kinetic Properties and Contribution of Set1/COMPASS Subunits in H3K4 Methylation

Set1 alone is not active as a KMTase, as Set1 within COMPASS is the active form of the enzyme. Indeed, any alteration to the C-terminal domain of Set1 containing its SET domain, such as insertion of tags (including TAP, HA, or Flag), results in the full loss of the enzyme's activity (43, 44). Set1/COMPASS consists of eight subunits, including Set1, Cps60, Cps50, Cps40, Cps35, Cps30, Cps25, and Cps15 (**Figure 2; Table 1**), with each of the subunits having a specific function in the assembly and regulation of the pattern of H3K4 mono-, di-, and trimethylation by the enzyme. Several subunits of Set1/COMPASS (including Set1, Cps50, and Cps30) are essential for complex formation/stability and for full KMTase activity of the complex, as yeast cells lacking any of these subunits are defective in H3K4 mono-, di-, and trimethylation. The Cps35 subunit of COMPASS (also known as SWD2) is the only essential subunit of the complex in yeast, and it is shared with other complexes, including the cleavage and polyadenylation factor, functioning in transcription termination on snoRNA genes (46–48). The essentiality

Figure 2

The subunit composition of the COMPASS family from yeast to human. The yeast Set1/COMPASS is the founding member of the family of COMPASS histone H3 lysine 4 (H3K4) methyltransferases. There is only one Set1/COMPASS in yeast capable of methylating histone H3 on its fourth lysine. From yeast to *Drosophila*, COMPASS is divided into three branches. The Set1/COMPASS of *Drosophila* is the direct descendent of the yeast Set1 complex (shown by *solid arrows*). There are two COMPASS-like complexes in *Drosophila* (*dotted arrows*); one trithorax (Trx)-containing complex and the other trithorax-related containing complex (Trr). The SET domain-containing enzymes from yeast to *Drosophila* are shown (*red*), and the conserved subunits between yeast and *Drosophila* are shown (*green*). The complex-specific subunits are shown (*blue* and *purple*). From *Drosophila* to mammalian cells, each branch of the COMPASS family is further divided into two members for the total of six COMPASS members in mammals. The mammalian Set1A and -B are direct homologs of yeast and *Drosophila* Set1 and are found in Set1A-B/COMPASS (*solid arrows*). The subunit composition and function of Trx of *Drosophila* is divided between MLL1 and MLL2 in the mammalian cells. Both MLL1 and MLL2 are found in COMPASS-like complexes. The subunit composition and function of Trr of *Drosophila* is divided between MLL3 and MLL4 in the mammalian cells, both of which are also found in COMPASS-like complexes. All six mammalian COMPASS family members are capable of methylating H3K4. The known common subunits shared between yeast, *Drosophila*, and the human complexes are shown (*green*). Cps35 in yeast and Wdr82 in *Drosophila* and mammals are restricted to the Set1/COMPASS branch. Menin (*blue*) is found in complex with Trx and the MLL1/MLL2/COMPASS-like complexes. The shared subunits among the trr and the MLL3/MLL4 complexes (UTX, PTIP, PA1, and NCOA6) are shown (*purple*).

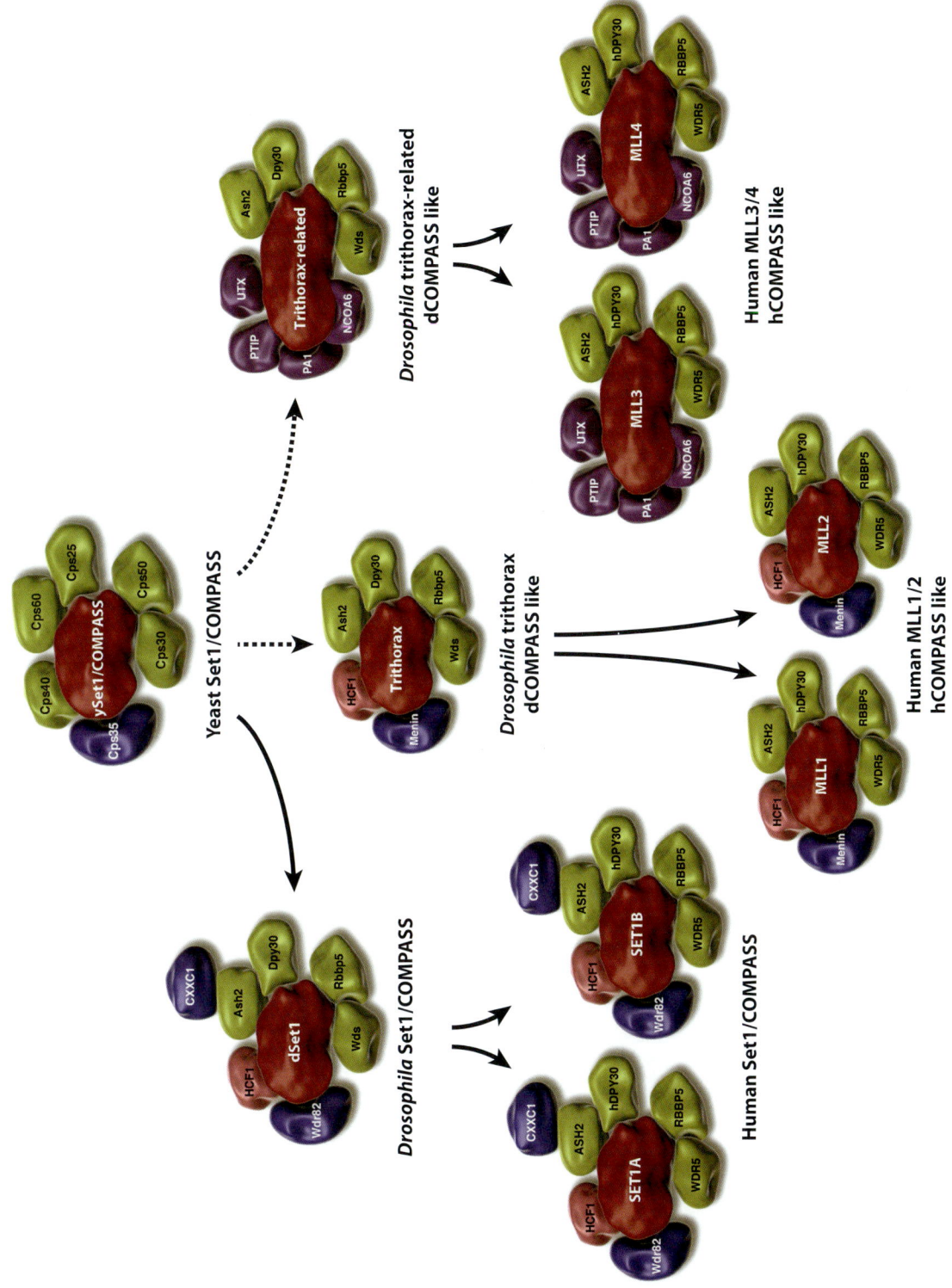

Table 1 Subunit composition and biological activities of COMPASS family from yeast to human

Yeast Set1/COMPASS	Drosophila COMPASS/COMPASS-like complexes	Mammalian COMPASS/COMPASS-like complexes	Biological/biochemical properties
Set1	Set1, Trx, Trr	Set1A/B, MLL1–4	The catalytic subunits
Cps60 (Bre2)	Ash2	Ash2L	Required for H3K4me3
Cps50 (Swd1)	RbBP5	RbBP5	Required for assembly
Cps40 (Spp1)	CXXC1 (dCfp1)	CXXC1 (Cfp1)	Components of Set1 complexes
Cps35 (Swd2)	Wdr82	Wdr82	Required for proper H3K4 di- and trimethylation
Cps30 (Swd3)	Wds	Wdr5	Required for assembly
Cps25 (Sdc1)	Dpy30	Dpy30	Required for H3K4me2/3
	HCF1	HCF1	Components of Set1 complexes
	Menin	Menin	Trx and MLL1/MLL2 specific
	PTIP, PA1, NCOA6, UTX	PTIP, PA1, NCOA6	Trr and MLL3/MLL4 specific

of Cps35 for growth viability is because of its requirement within the cleavage and polyadenylation factor complex and not in COMPASS. The lethality in yeast cells carrying deletion of Cps35 is suppressed by the overexpression of the C-terminal fragment of Sen1, a superfamily I helicase functioning in snoRNA termination (47). The Cps35 null strains in a Sen1-overexpression background demonstrate a significant reduction in H3K4 di- and trimethylation levels, with no detectable change in H3K4 monomethylation. Therefore, the Cps35 subunit of Set1/COMPASS is required for proper H3K4 di- and trimethylation, but not monomethylation (**Table 1**). The Cps40 subunit of the complex is required for proper H3K4 trimethylation, as in its absence, yeast cells lack over 80% of H3K4 trimethylation with very little effect on H3K4 mono- and dimethylation levels (**Table 1**) (45). The Cps25 and Cps60 subunits of the complex appear to be required for proper H3K4 di- and trimethylation but not for monomethylation (45, 49, 50).

2.3. Transcription Elongation and the Pattern of H3K4 Methylation by Set1/COMPASS

Biochemical screens in yeast identified a role for the RNA Pol II C-terminal domain kinases, Ctk1, Ctk2, and Ctk3 (the Ctk complex), in the regulation of H3K4 monomethylation by Set1/COMPASS (51, 52). These studies demonstrated that the loss of the Ctk complex kinase activity in yeast cells results in reduced histone H3K4 monomethylation levels, followed by a global increase in the histone H3K4 trimethylation levels (51). Given that Set1/COMPASS associates with transcribing Pol II through its interactions with the Paf1 complex (53, 54) and that there is an established role for the Ctk complex in transcriptional elongation control through RNA polymerase II's C-terminal domain phosphorylation, a possible role for the rate of transcription elongation by Pol II in dictating the pattern of H3K4 mono-, di-, and trimethylation by Set1/COMPASS was tested (51). These in vivo studies demonstrated that, as Set1/COMPASS travels with transcribing Pol II during promoter escape, where Pol II transcribes at a slower rate, the chromatin within such regions are mostly trimethylated as Set1/COMPASS has a longer window of opportunity to methylate its substrate, the nucleosomes. However, in regions where Pol II transcribes at a faster rate, during productive elongation, Set1/COMPASS spends less time with its substrate, and therefore, the chromatin within such areas is monomethylated (51). In support of such observations, in vitro enzymological studies demonstrated that the onset of monomethylation on an unmethylated histone H3 by Set1/COMPASS is virtually immediate, whereas the onset of trimethylation occurs upon an extended time of association between the histone H3 tail and Set1/COMPASS, further supporting the notion that the pattern of H3K4 trimethylation by the enzyme requires an extended interaction with its substrate (51).

3. THE METAZOANS' COMPASS FAMILY

3.1. Biochemical Purification of the COMPASS family from *Drosophila* to Human

The COMPASS family of H3K4 methylases is highly conserved from yeast to plant and to human (42, 43, 55–58). Three independent genes in *Drosophila*, *dSet1*, *trithorax (trx)* and *trithorax-related (trr)*, are homologs to yeast Set1 (**Figure 2**; **Table 1**) (59). The *dSet1* is the direct descendent of yeast Set1, whereas *trx* and *trr* genes are more distantly related to yeast Set1 (**Figure 2**). Although it was initially reported that Trx is found in a heterotrimeric complex with CBP and SBF1 (60), our studies have demonstrated that in *Drosophila* dSet1, Trx, and Trr are all three found in COMPASS-like complexes capable of methylating histone H3K4 (59). Each enzyme localizes to hundreds of sites on polytene chromosomes, but the deletion or RNAi-mediated knockdown of either one of these genes results in lethality in *Drosophila*, indicating that the H3K4 methylase function of each complex is nonredundant. Loss of dSet1 results in global losses of H3K4 di- and trimethylation (59, 61), indicating that Trx and Trr may have more specialized functions. dSet1, Trx, and Trr each interact with a set of core COMPASS subunits, but there are also subunits unique to each complex (59). Importantly, each of these unique subunits has an ortholog in the mammalian version of its COMPASS-like complex; mammals, however, have two homologous complexes for each of the *Drosophila* complexes.

In human cells, Set1A, Set1B, MLL1, MLL2, MLL3, and MLL4 are homologs of yeast Set1 (**Figure 2**; **Table 1**) (62, 63). Following the identification of Set1/COMPASS in yeast as the first H3K4 methylase (42–45), the affinity purification of the menin tumor suppressor protein, which is the product of the *MEN1* gene, resulted in the isolation of MLL2 and many of the human homologs of the yeast COMPASS subunits within the same complex (56). This biochemical study demonstrated that not only is MLL2 associated within a COMPASS-like complex, but it can also function as a histone H3K4 methylase similar to its founding member yeast Set1/COMPASS (56). Subsequent studies demonstrated that the MLL1–MLL4 are all found in COMPASS-like complexes capable of mono-, di-, and trimethylating H3K4 (**Figure 2**; **Table 1**) (64–71). The Set1A and Set1B proteins in human cells were also identified in COMPASS-like compositions functioning as H3K4 methylases (58, 72, 73). These studies established that Set1/COMPASS, identified orginally in yeast (42), is highly conserved from yeast to human, both in composition and function. Detailed biochemical studies have demonstrated that each COMPASS-like complex in metazoans carries shared subunits that are similar to their yeast counterpart. Additionally, the metazoan complexes also possess complex-specific subunits bringing about target specificity and functioning as interaction modules for each complex (**Figure 2**; **Table 1**). Details of the composition of the human COMPASS-like complexes are described below.

3.2. Composition of the Metazoan COMPASS Family of H3K4 Methylases

The shared subunits between yeast COMPASS and the metazoan Set1/MLL/COMPASS-like complexes include Ash2 (related to Cps60), RbBp5 (related to Cps50), Wdr5 (related to Cps30), and Dpy30 (related to Cps25)

(**Figure 2**; **Table 1**) (62, 74). Metazoan Set1/COMPASS possesses CXXC1 and Wdr82, which are related to yeast Cps40 and Cps35, respectively; however, these proteins are not detected stoichiometrically within the MLL1–MLL4 complexes (62, 66, 73, 75, 76). The menin tumor suppressor protein, whose gene, *MEN1*, is mutated in familial multiple endocrine neoplasia type 1, together with the lens epithelium-derived growth factor are associated with both the MLL1 and MLL2 complexes. However, these proteins are absent in the MLL3 and MLL4 complexes (56, 66). The histone H3K27 demethylase UTX, the Pax transactivation domain-interacting protein (PTIP), nuclear receptor coactivator (NCOA6), and PTIP-associated 1 (PA1) are specifically found within the MLL3–4/COMPASS-like complexes similar to the Trr complex in *Drosophila* (59, 64–66, 69, 71, 77–79).

3.3. Structural Studies of the COMPASS Family from Yeast to Human

Recently, we have reconstituted in vitro fully functional yeast Set1/COMPASS and a human MLL/COMPASS-like complex (80). This study resulted in the identification of a minimum set of COMPASS family subunits required for proper histone H3K4 methylation. These conserved subunits include the methyltransferase C-terminal SET domain of Set1/MLL, Cps60/Ash2L, Cps50/RbBp5, Cps30/Wdr5, and Cps25/Dpy30. Reconstitution of the active Set1/COMPASS and MLL/COMPASS-like complexes and analyses of their structures by cryoelectron microscopy reconstructions combined with immunolabeling and two-dimensional electron microscopy analysis of the individual subcomplexes revealed that the COMPASS family members have a Y-shaped architecture with Cps30 and Cps50 localizing on the top two adjacent lobes and Cps60-Cps25 forming the base. Our studies in yeast demonstrated that the SET domain of Set1 is localized at the juncture of Cps30 and Cps50 and the Cps60–Cps25 module, lining the walls of a central channel. This channel may act as the platform for catalysis and regulative processing of various degrees of H3K4 methylation.

These structural studies suggested that the centrally located active site within the COMPASS channel may only be reached by flexible peptide termini such as the fourth lysine of the H3 tail. Based on this observation, we hypothesized that the COMPASS family may function primarily as exo- and not endomethylases. This hypothesis was tested by the addition of a few amino acids to the N terminus of histone H3 in yeast. Although wild-type H3 can be methylated on K4 by Set1/COMPASS in vivo, the addition of a few amino acids extending the N terminus of H3 and moving the K4 site to a more internal one results in H3 no longer being methylated by the COMPASS family (80). This observation was also tested in vitro. By providing an N-terminally extended histone H3 to either a Set1/COMPASS or a MLL/COMPASS-like complex, H3 can no longer be methylated by the COMPASS family (80). Therefore, the COMPASS family appears to function as exomethylases under these experimental settings. Given this observation, the identification of new COMPASS family substrates (based on antibody studies), which bear an internal methylation site(s), need to be further confirmed enzymologically and using mass spectrometry [for analysis of the chemical composition of the substrate(s)] with appropriate controls, such as a purified complex with a catalytically dead SET domain.

3.4. Functional Diversity of the Metazoans' COMPASS-Like Complexes

Following the identification of MLL1 within a COMPASS-like complex functioning as an H3K4 trimethylase (56), the vast majority of H3K4 methylation studies were focused on MLL1's activity. However, biochemical and cell biological studies in mammalian cells demonstrated that the attenuation of Wdr82 (a Set1A/B-specific subunit) levels in cells results in the reduction of Set1A/B protein levels, but not the core components or MLL1 (66, 75). This loss in the Wdr82 levels is followed by a global loss in H3K4 trimethylation levels with very little alteration in the H3K4 di- and trimethylation levels (66). This study demonstrated that (*a*) mammalian Wdr82, the Set1A/B-specific factor, is required for the majority of proper H3K4 trimethylation by Set1, and not for the MLL COMPASS-like complexes; (*b*) Set1A/B complexes in mammalian cells are the major H3K4 methylases; and (*c*) MLL1–MLL4 do not function globally as the major regulators of H3K4 trimethylation but must have gene-specific functions. Recently, the generality of the role of Set1/COMPASS as a major histone H3K4 methylase in other organisms was confirmed (59, 61).

Studies in *Caenorhabditis elegans* confirmed that many of the components of the COMPASS family are also highly conserved in *C. elegans* (81). Similar to studies in mammalian cells (58), the loss of the Set1 homolog in *C. elegans* resulted in a major decrease in bulk H3K4 trimethylation; however, removal of the only MLL1–MLL4 homolog in *C. elegans*, Set-16, did not have a global effect on the pattern of H3K4 methylation (81). Recent studies have demonstrated a specific role for the components of Set1/COMPASS in *C. elegans* in the regulation of the expression of the gustatory neurons ASEL and ASER, a bilaterally symmetric neuron pair (known as ASE left and right) that is functionally lateralized and occurs stereotypically (82). This study demonstrated that Set1/COMPASS has a role in the specification of a left-right asymmetry in this sensory neuron pair. Several shared components of the COMPASS family in *C. elegans*, including Ash2, Wdr5, and Rbbp5, are required for ASE laterality defects (82). Furthermore, Wdr82, which is specific to Set1/COMPASS and not the MLL1–MLL4/COMPASS-like complexes, is also required for ASE laterality (82), suggesting a specific role for Set1/COMPASS in this process. *C. elegans* lacks the Trx/MLL1/MLL2 complexes (please see below) (83), therefore this function of the COMPASS family in *C. elegans* could be shared by either Trx or Trr complexes in *Drosophila*.

3.5. Trx in *Drosophila* and Its Mammalian Homolog MLL1/MLL2 Are Positive Regulators of Homeotic Gene Expression

Early studies by Korsmeyer and colleagues (84) demonstrated that MLL1 is required for proper segment identity in mammals and that it positively regulates *Hox* gene expression. The targeted deletion of both copies of *MLL1* in mouse embryonic stem (ES) cells by homologous recombination resulted in embryonic lethality (84, 85). *MLL1* heterozygous ($+/-$) mice demonstrated retarded growth, hematopoietic abnormalities, and, similar to the loss of the *trx* gene in *Drosophila*, bidirectional homeotic transformations (84). Reduction of MLL1 levels in *MLL1* heterozygous ($+/-$) animals also demonstrated abnormalities in the anterior boundaries of *Hoxa7* and *Hoxc9* expression, whereas their posterior expression was lost in $MLL^{-/-}$ embryos. Subsequent cellular studies confirmed a role for MLL1 as a regulator of *Hox* gene expression and linked MLL's specific role in this process to its role in leukemic pathogenesis (86–89).

Given the importance of the HMTase activity of MLL1 in leukemogenesis, and its existence in just one of the six COMPASS-like complexes, the global pattern of H3K4 methylation was tested in null- and wild-type mouse embryonic fibroblasts for MLL1 (77). Surprisingly, these studies demonstrated that MLL1 is required for the H3K4 trimethylation of only less than 5% of the promoters carrying this modification (77). The gene ontology analysis of the MLL1 H3K4 methylation targets demonstrated that many of these genes include developmental regulators, such as *Hox* genes; however, not all *Hox* genes require MLL1 for their proper H3K4 methylation (77). Also, the loss of MLL1 resulted in decreased levels of Pol II recruitment, decreased expression, and a concomitant loss of H3K4 methylation at these genes. Analysis of the entire *Hox* cluster established that MLL1 is only required for the methylation of a subset of *Hox* genes. However, the deletion of menin, which is a component of both the MLL1 and MLL2 complexes (**Figure 2**), resulted in the abolishment of the majority of H3K4 methylation at the entire *Hox* loci. This finding established that MLL1 and MLL2 are the functional homologs of the *Drosophila trx* gene, as the loss of the MLL3/MLL4 and/or the Set1 complexes demonstrated little to no effect on the H3K4 methylation or expression of this homeotic cluster (77). Similar findings in *Drosophila* also suggest that Set1/COMPASS is the major H3K4 methylase and that Trx/COMPASS-like and Trr/COMPASS-like complexes have limited and specific genomic targets (59).

3.6. Mammalian MLL3/MLL4: The Homologs of *Drosophila* Trr

The *Drosophila trithorax-related (trr)* gene was identified on the basis of the homology of its SET domain with *trx* and was cloned using a degenerate polymerase chain reaction strategy (90, 91). Similar to *trx*, mutations or deletions of the *trr* gene result in embryonic lethality, indicating that the functions of these two related genes are not redundant. Indeed, the loss of Trr function does not result in homeotic transformations as seen with the loss of function of Trx, and no dominant interaction has been reported between trr^1 and the hypomorphic alleles of either *Pc* or *trx*, revealing that Trr is not a major regulator of the expression of homeotic genes (74, 90).

The major steroid hormone pathway receptor in *Drosophila*, the ecdysone receptor (EcR), colocalizes extensively with Trr on polytene chromosomes (92). Furthermore, Trr and EcR can be coimmunoprecipitated in an ecdysone-dependent manner (92). Following the demonstration that the SET domain of Trr can also function as a histone H3K4 methylase (92), it was demonstrated that association of Trr and EcR on chromatin in an ecdysone-dependent manner is followed by the trimethylation of histone H3K4 of EcR target promoters. More importantly, in *trr* alleles lacking the SET domain of Trr, the H3K4 trimethylation levels on EcR target genes are significantly abridged, suggesting that the catalytic activity of Trr is required for ecdysone-dependent gene activation through H3K4 methylation in vivo (92).

Cloning of one of the human homologs of *Drosophila trr*, the *MLL4/ALR* gene (93), was followed by the purification of its complex from human cells in a COMPASS-like composition capable of interacting with the estrogen receptor (ER) (94). The ER α directly binds to MLL4's two LXXLL motifs near its C terminus in a ligand-dependent manner (94). Analogously to EcR and Trr in *Drosophila*, the MLL4 complex is recruited to the promoters of the ER α target genes following estrogen stimulation, and the expression of the ER α target genes is abridged upon inhibition of MLL4 expression. Subsequently, other complexes containing MLL3/MLL4 were also identified, and it was demonstrated that MLL3/MLL4 exists in COMPASS-like complexes (**Figure 2**) (64–66, 69, 71, 78). Overall, studies on the MLL3/MLL4 complexes have demonstrated that these COMPASS family members play an important role as coactivators of nuclear receptor signaling, adipogenesis, and immunoglobulin class switching (71, 94–98). The prominent role of the MLL3/MLL4 and Trr complexes in nuclear receptor coactivation indicates specialization among different classes of the MLL1–MLL4 proteins. Indeed, *C. elegans*, which has a reduced homeotic gene cluster (99), lacks a MLL1/MLL2/Trx homolog but does have Set1 and MLL3/MLL4/Trr proteins. Interestingly, *C. elegans* has an expanded number of nuclear receptors encoded in its genome, perhaps explaining why this organism has a *MLL3/4*-related gene *Set-16*, but not *MLL1/2* genes (83, 100).

3.7. A Histone H3K27 Demethylase as a Component of the Trr/MLL3/MLL4 COMPASS-Like Complexes

As the histone H3K4 methylase function of Trx in *Drosophila* and MLL1/MLL2 in mammalian cells is considered a positive regulator of homeotic gene expression, histone H3K27 methylation by the PcG proteins is considered a negative regulator of this process. The Jumonji C domain-containing histone demethylase, UTX, functions as a histone H3K27 demethylase (78, 79, 101–104). Surprisingly, UTX was identified copurifying with the MLL3/MLL4 complexes (64, 65, 69, 78) and functioning during retinoic acid signaling events (78). Similar complexes containing Trr and UTX have also been observed in *Drosophila* (59) and function in cell growth control, where the Trr/COMPASS-like complex interacts with the Notch and Rbf pathways to suppress tumorigenesis (104). A similar phenomenon was observed in *C. elegans* and in human cells (105). It is not clear at this time why the Trr/MLL3/MLL4-containing complexes carry H3K27 demethylase activity within their complexes and why the Set1A/B and Trx/MLL1/MLL2 complexes lack this activity/subunit. The H3K27 methylation machinery and a MLL3-like protein can be found even in unicellular organisms, such as ciliated protozoa, indicating an ancient role for antagonism between H3K4 and H3K27 methylation that goes beyond regulating *Hox* gene patterning (83). Recent genome-wide studies of PcG proteins indicate that these proteins play a much larger role than repression of *Hox* genes (106–108). Perhaps the MLL3/MLL4 H3K4 methyltransferases together with H3K27 demethylase activity represent a distinct form of antagonizing PcG function that differs from the Trx-Hox paradigm. *MLL3* and *MLL4*, as well as *UTX*, are frequently mutated genes in human cancers (109–112). Therefore, determining the way that these two classes of enzymes interact with each other to regulate gene expression is an important area of future research.

3.8. The Metazoan COMPASS Subunit DPY-30 in Dosage Compensation

Dosage compensation is a process by which the expression of X-linked genes is equalized between males and females. In mammals, this is achieved by silencing one of the two X chromosomes in females. In *Drosophila*, genes on the single X chromosome in males are transcribed at twice the rate to achieve the same transcriptional output from two X chromosomes in females. In *C. elegans*, the two X chromosomes in hermaphrodites are transcribed at half the level of the X chromosome in XO males (113). In each lineage, along with the evolution of an autosome into a sex chromosome, there was also the co-option of global chromatin-modifying machinery for the purpose of dosage compensation. In mammals, the machinery includes proteins involved with heterochromatin formation that are co-opted for silencing one of the X chromosomes (forming the cytologically visible Barr body) (114). In *Drosophila*, the MSL complex implements H4K16 acetylation on the male X chromosome to achieve dosage compensation. In both *Drosophila* and mammals, a related H4K16 acetylation complex modifies histones in both males and females on all chromosomes for proper gene expression genome wide (115). In *C. elegans*, a condensin complex (called the DCC for dosage compensation complex) shares subunits with the chromosome compaction machinery necessary for the faithful segregation of the mitotic and meiotic chromosomes (113, 116–119). Although mutations in many of the DCC genes cause hermaphrodite-specific lethality, rare survivors have a shortened or dumpy morphology. Another gene whose mutation gives a dumpy phenotype is *dpy-30* (120, 121). However, DPY-30 is also essential for the proper development of XO males, indicating an additional, more general function. *C. elegans* DPY-30 is homologous to Cps25 found in yeast COMPASS and to DPY-30 found in *Drosophila* and the mammalian COMPASS family. Coimmunoprecipitation experiments and genome-wide profiling studies demonstrate that *C. elegans* DPY-30 associates not only with COMPASS and COMPASS-like complexes, where it is essential for the H3K4me3 mark, but also that it functions in the DCC independently of other COMPASS subunits (122). As a member of the DCC, DPY-30 plays an essential role in the loading of DCC components onto X chromosomes of hermaphrodites. Thus, the same subunit that is required for H3K4me3, a mark of active transcription from yeast to mammals, is also a component of a complex that reduces transcription through chromatin condensation. The existence of DPY-30 in the context of the DCC has the potential to provide clues for the function of DPY-30 and its homologs of the COMPASS family, owing to unique genetic tools developed over the past 25 years in studies of *C. elegans* dosage compensation.

4. A NOVEL YEAST BIOCHEMICAL SCREEN DEFINING THE H3K4 METHYLATION PATHWAY

4.1. The Histone H2B Monoubiquitinase Rad6 Is Required for H3K4 Trimethylation by Set1/COMPASS

A proteomic screen called global proteomic analysis in *S. cerevisiae* (GPS) was devised to identify factors and molecular pathway(s) required for proper histone H3K4 methylation by Set1/COMPASS (123). This method takes advantage of the availability of the collection of the yeast nonessential deletion mutants (124, 125) and antibodies, generated toward H3K4 mono-, di-, and trimethylated H3K4, to identify factors required for implementation of this histone modification in cells (123). In GPS, extracts from strains, deleted for each of the nonessential yeast genes from the collection, are generated and analyzed by sodium dodecyl sulfate-polyacrylamide gel electrophoresis followed by Western blot analysis. This method allows the identification of the protein-coding genes in the yeast genome required for Set1/COMPASS function.

The first hit from this biochemical screen was the demonstration that the histone H2B monoubiquitinase Rad6 is required for proper H3K4 methylation by Set1/COMPASS (126). At the same time, another independent study also reported the role of H2B monoubiquitination by Rad6 in H3K4 methylation (127). These studies suggested that H2B monoubiquitination by Rad6 is required for H3K4 di- and trimethylation by Set1/COMPASS and that methylation of H3K4 is downstream of Rad6 (45, 126, 127). The histone H2B monoubiquitination cross talk is also required for proper histone H3K79 trimethylation by the non-Set domain-containing enzyme, Dot1 (128–130). Indeed, global genomic studies have demonstrated that promoters within the yeast genome bearing H2B monoubiquitination possess H3K79 trimethylation and that promoters lacking H2B monoubiquitination only bear H3K79 mono- and dimethylation, further suggesting a link for this *trans*-histone tail cross talk throughout the yeast genome (131).

4.2. Yeast Bre1, an E3 Ligase Required for Rad6 Recruitment to Chromatin

Following the identification of the role of histone H2B monoubiquitination by Rad6 in regulating H3K4 methylation, investigators were searching to identify the E3-conjugating enzyme required for Rad6 function in chromatin modification and transcription. Our GPS screen identified the C3HC4 RING finger–containing protein, Bre1, as a factor required for proper H3K4 di- and trimethylation by COMPASS and H3K79 methylation by Dot1 (130). Deletion or generation of a single point mutation in the RING finger of Bre1 rendered this E3 ligase inactive in H2B monoubiquitination and resulted in the loss of H3K4 and H3K79 methylations in strains bearing such mutations (130). In support of this observation, biochemical purification of Rad6 demonstrated that this enzyme exists in complexes with several different E3 ligases, including Bre1 (130). This study demonstrated the existence of physical interactions between Rad6 and Bre1 (130).

In strains lacking Bre1, Rad6 is not able to interact with chromatin. Therefore, the levels of H2B monoubiquitination and H3K4 and H3K79 trimethylation are lost, suggesting that Bre1 is required for the recruitment of Rad6 to chromatin on genes (53, 130). Genetic studies also demonstrated that Bre1 and its interacting factor, Lge1, are required for proper H2B monoubiquitination and H3K4 methylation (132). This study also suggested a function for Bre1/Lge1-dependent H2B monoubiquitination in the control of cell size in yeast (132).

Studies in yeast demonstrated that a subunit of Set1/COMPASS, Cps35 (also known as Swd2), associates with the enzyme and chromatin in an H2B monoubiquitination-dependent manner and that the interaction between Cps35 and monoubiquitinated nucleosomes appears to be indirect as Cps35 is not capable of interacting with free ubiquitin (76, 133). Given that Cps35 is required for proper H3K4 di- and trimethylation (47, 48, 76), such studies suggest that Cps35 can function in regulating histone H2B monoubiquitination/H3K4 methylation cross talk. Similarly, other studies have also found a role for Cps35/Swd2 as a factor within Set1/COMPASS regulating monoubiquitination/H3K4 methylation cross talk (134). In support of these observations, the *Drosophila* and human homolog of Cps35, a protein known as Wdr82, is also required for Set1/COMPASS function in H3K4 trimethylation (58, 59).

4.3. The Conserved Role for Bre1 in Set1/COMPASS Function from Yeast to Human

Both Rad6 and its E3-conjugating enzyme, Bre1, are highly conserved in structure and function from yeast to human cells, demonstrating the power of genetic and biochemical screens in yeast in identifying the molecular machinery required for proper H3K4 methylation (3, 83, 135–139). Following the identification of yeast Bre1 as the E3-conjugating enzyme for Rad6 in H2B monoubiquitination, *Drosophila* Bre1 was identified through genetic screens as a factor that is required for the proper expression of Notch target genes during development (139). Similar to yeast cells (130), depletion of Bre1 levels in *Drosophila* results in a reduction of H2B monoubiquitination levels, followed by a reduction in both histone H3K4 and H3K79 methylation levels (59, 130, 138). In addition to a role in the regulation of the Notch signaling pathway, Bre1 and H2B monoubiquitination also function in Wnt signaling and the maintenance of multiple types of adult stem cells (138, 140, 141).

Although initial studies in mammalian cells suggested that the RING domain-containing E3 ubiquitin ligase, Mdm2, which ubiquitinates the p53 tumor suppressor protein, can also monoubiquitinate histone H2B both in vivo and in vitro (142), it has been firmly established now that, similar to the original studies in yeast (130), human homolog of yeast Bre1 is the E3 ligase functioning with human Rad6 to regulate the global pattern of H2B monoubiquitination (135–137). These studies demonstrated that the machinery required for the proper regulation of H2B monoubiquitination is highly conserved from yeast to human (3). Indeed, when using recombinant nucleosomes containing a wild-type or monoubiquitinated H2B version, the human Set1/COMPASS containing Wdr82 (mammalian homolog of yeast Cps35) used the H2B monoubiquitinated substrate with higher kinetics and efficacy than the nonmonoubiquitinated version, indicating the conservation of function of H2B monoubiquitination in H3K4 trimethylation by Set1/COMPASS from yeast to human (137).

4.4. Rad6/Bre1, H2B Monoubiquitination, and Transcription Elongation Control

Although the Rad6/Bre1 complex and histone H2B monoubiquitination are required for proper histone H3K4 and H3K79 trimethylation from yeast to human, this modification and its machinery have recently been linked to other pathways. The global analysis of the pattern of H2B monoubiquitination in yeast demonstrated that this mark is found on other regions of chromatin lacking some or all of the histone marks mentioned above (131). For example, histone H3K4 trimethylation is only associated with the promoters of active genes; however, histone H2B monoubiquitination is found covering the entire body of actively transcribed genes (131). Histone H3K79 trimethylation appears to be associated with genes that are on average larger in length versus H3K79 dimethylation, which is found on the promoters and associates with genes that are on average shorter in length (131). At the same time, H3K79 dimethylation was demonstrated to be associated with cell cycle regulation and the G1/S transition, and H3K79 trimethylation does not appear to play a role in these processes. Subsequent studies demonstrated that histone H3K79 trimethylation is required for Wnt signaling in metazoans and gene expression in yeast (138, 143).

In vitro studies using reconstituted transcription systems demonstrated that histone H2B monoubiquitination works cooperatively with the transcription elongation factor FACT to regulate the elongation properties of RNA Pol II (144). In support of this observation, studies in the fission yeast, *Schizosaccharomyces pombe*, established that the loss of H2B monoubiquitination is associated with defects in nuclear structure, cell growth, and septation. Chromatin immunoprecipitation experiments analyzing the localization properties of RNA Pol II in *S. pombe* cells lacking H2B monoubiquitination confirmed that the loss of H2B monoubiquitination results in an altered pattern of Pol II distribution and histone occupancy in the wake of Pol II transcription in gene-coding regions (145). These in vivo findings further suggest that H2B monoubiquitination could regulate the rate and the process of transcriptional elongation control (145). Whether H2B monoubiquitination and/or Rad6/Bre1 can directly regulate the elongation properties of Pol II, or whether H2B monoubiquitination of nucleosomes alters transcription elongation resulting from changes in the positioning and/or occupancy of nucleosomes, in front of and/or in the wake of transcribing polymerase, remains to be resolved.

4.5. A Role for Rad6/Bre1 in DNA Double-Strand Break

Histone H2B monoubiquitination by Rad6/Bre1 has also been proposed to be required for DNA double-strand break repair by homologous recombination (146, 147). The protein kinase ATM is central to the DNA damage response as it phosphorylates many of the factors found within the DNA damage response network. In human cells, double-stranded DNA breaks induce histone H2B monoubiquitination catalyzed by Rad6/Bre1, and this process is regulated by the ATM-dependent phosphorylation of the Bre1A (Rnf20) and Bre1B (Rnf40) subunits of the human Bre1 complex to form a heterodimer (146). Histone H2B monoubiquitination was demonstrated to be involved in the two major double-stranded repair pathways: nonhomologous end-joining and homologous recombination repair (146).

In another independent study, the Bre1A subunit of human Bre1 was shown to be recruited to double-stranded DNA break sites independently of H2AX. This study showed that the histone H2B monoubiquitination-dependent methylation of H3K4 at double-stranded break sites is involved in the recruitment of SNF2h, a chromatin remodeling factor, which functions in homologous recombination repair, and in enhanced sensitivity to radiation damage (147). However, these studies did not identify which COMPASS family member is involved in this process. Given the requirement of H2B monoubiquitination for Set1A-B/COMPASS function in human cells, Set1A or B/COMPASS could be the appropriate methyltransferase over the MLL1–MLL4/COMPASS family.

Although the exact molecular mechanism for Rad6/Bre1 and histone H2B monoubiquitination in DNA damage repair is still in its infancy, these studies point to the diverse implementation, regulation, and biological outcomes for this highly conserved process. Given that genetic and biochemical studies in yeast have been very fruitful in this regard, studies concentrating on defining the role of the Rad6/Bre1 complex in DNA double-stranded break repair and in homologous recombination using yeast as a model system would be highly informative.

4.6. The PAF1 Complex as a Platform for Histone-Modifying Enzymes on Transcribing Polymerase

Biochemical affinity purifications, with yeast RNA Pol II as the bait, identified the polymerase-associated factor 1 (PAF1) as a Pol II-interacting protein (148). Deletion of *PAF1* in yeast resulted in several general transcriptional phenotypes, including a defect in the induction of the galactose-regulated genes, suggesting a general role for Paf1 in transcriptional control (148). Another factor, Rtf1, was genetically identified to function as a RNA Pol II elongation factor in yeast (149). Further biochemical studies demonstrated that both Paf1 and Rtf1 and a few other factors, including Cdc73, Ctr9, and Leo1, interact within a same complex, named the Paf1 complex, controlling transcriptional initiation and elongation in yeast (150–153).

Very little was known about the exact molecular function of the Paf1 complex in transcription until the subunits of this complex appeared as hits in the GPS biochemical screen when exploring for the factors required for proper H3K4 and H3K79 methylation (54). Several components of the Paf1 complex, including Paf1, Rtf1, Cdc73, and Ctr9, were identified as factors whose deletion resulted in a reduction in H3K4 and H3K79 trimethylation levels (54). The Paf1 complex regulates H3K4 methylation at several levels. The components of the Paf1 complex are required for proper H2B monoubiquitination by Rad6/Bre1 because in the absence of the Paf1 complex both Rad6 and Bre1 are recruited to chromatin. However, Rad6/Bre1 are not functional in H2B monoubiquitination until Paf1 is recruited to the same site (53). Not only does the Paf1 complex function by regulating H2B monoubiquitination through Rad6/Bre1, but it also plays a role as a landing pad for Set1/COMPASS on RNA Pol II promoting H3K4 methylation on chromatin within the transcribed regions of the genes (53, 54). These early studies proposed that the Paf1 complex plays a role as a "platform" for the association of the histone-modifying machinery with the transcribing RNA Pol II (53, 54, 154). Indeed, subsequent studies identified other factors, such as Set2 in addition to COMPASS, that required the Paf1 complex for their association and methylase function (155).

Similar to the subunits of Set1/COMPASS and Rad6/Bre1, which were initially identified in yeast to play a role in histone H3 and H2B modifications and shown to be functionally conserved in mammalian cells (42, 43, 53, 126, 127, 130, 135, 137, 144), the Paf1 complex is also highly conserved in structure and function from yeast to mammalian cells (144, 156–159). Even the platform function of the Paf1 complex, as a site of recruitment for Set1/COMPASS (53, 154), is conserved in mammalian cells. The Paf1 complex in human cells can associate and recruit the MLL1/COMPASS-like complex to transcribing RNA Pol II (160, 161). However, given that the Paf1 complex associates with all transcriptionally active sites in mammalian cells, this finding does not describe how MLL, which is only required for H3K4 methylation on a small subset of genes (77), is specifically recruited by the Paf1 complex to such loci.

5. HISTONE H3K4 METHYLASES IN DEVELOPMENT AND DISEASE

5.1. Licensed to Elongate: A Model for Leukemic Pathogenesis Through MLL1

Among the four genes encoding for MLLs, the *MLL1* gene is the subject of frequent chromosomal translocations associated with acute myeloid and lymphoid leukemia (162–164). Chromosomal translocations involving the

MLL1 gene in childhood malignancies account for ~10% of all detectable translocation events, and many of the patients bearing such chromosomal translocations are at high risk of relapse and require aggressive treatment (162, 164). Upon translocation, the C-terminal half of MLL is replaced by the translocating chromosomes, and the entire SET domain of MLL is lost as a result of such translocations. Although the chromosomal copy of MLL within the translocations loses its SET domain, the SET domain of MLL from the other chromosomal copy of MLL appears to be required for leukemic pathogenesis (165). A role for MLL1 translocations as a cause of leukemic pathogenesis is well established, as knockin or immortalization models of cancer have demonstrated that MLL1 translocations result in the pathogenesis of hematological malignancies with varying latencies of disease depending on the translocation partners of MLL (162, 166–169). Although MLL's contribution to leukemic pathogenesis was known for a long time, the exact molecular mechanism and contribution of so many of the seemingly unrelated MLL partners in leukemic pathogenesis were not clear until recently.

There are a large number of MLL translocation partners with very few sequence similarities (170). As diverse as the number of MLL partners in leukemia, so are the numbers of distinct models proposed that describe how MLL1 translocations into so many unrelated genes result in the pathogenesis of leukemia (171–174). The ELL protein, which is one of the major translocation partners of MLL, was the first MLL-translocation partner in leukemia for which a molecular and biochemical function was determined (175, 176). Over 15 years ago, ELL was biochemically purified and demonstrated to function as a RNA Pol II elongation factor (175). ELL increases the catalytic rate of transcription elongation by RNA Pol II by suppressing transient pausing by the enzyme (175). Given that biochemically ELL functions as a Pol II elongation factor and that ELL is one of the many MLL partners found in leukemia, it was proposed over fifteen years ago that the misregulation of the elongation stage of transcription could be central to leukemic pathogenesis through MLL translocations (175–177). Biochemical and genetic studies with human ELL1–3 and the sole copy of ELL in *Drosophila* established that ELL can function as a RNA Pol II elongation factor both in vitro and in vivo (175, 178–183). However, it was not clear how MLL translocations, to so many unrelated genes, resulted in leukemia.

Following the biochemical purifications of many of the MLL chimeras, including MLL-ELL, MLL-ENL, MLL-AF9, and MLL-AFF1, it was demonstrated that AFF4, itself a partner of MLL in leukemia, is a common factor among the complexes purified from these translocations (184). Further biochemical purification of AFF4 resulted in the demonstration that many of the MLL translocation partners are found associated with the elongation factors ELL and the positive transcription elongation factor (P-TEFb) within a complex called the super elongation complex (SEC) (**Figure 3***a*) (184). It is the translocation of MLL into SEC that is involved in the misrecruitment of SEC to MLL target genes, perturbing transcription elongation checkpoints at these loci and resulting in leukemic pathogenesis (**Figure 3***b,c*) (83, 162, 184, 185). Not only is SEC involved in the regulation of transcription of the MLL target genes in leukemia, but this elongation complex also plays a central role in regulating transcription elongation control and gene expression during development, in response to environmental stimuli and in human immunodeficiency virus Tat-dependent transactivation (186–188).

5.2. The COMPASS Family in the Regulation of Life Span and Aging

In addition to marking actively transcribed regions and the regulation of developmentally controlled genes, histone H3K4 methylation machineries have recently been linked with life span in a germ line–dependent manner in *C. elegans* (189). Using a directed RNAi screen in fertile worms, the Ash2L subunit of the

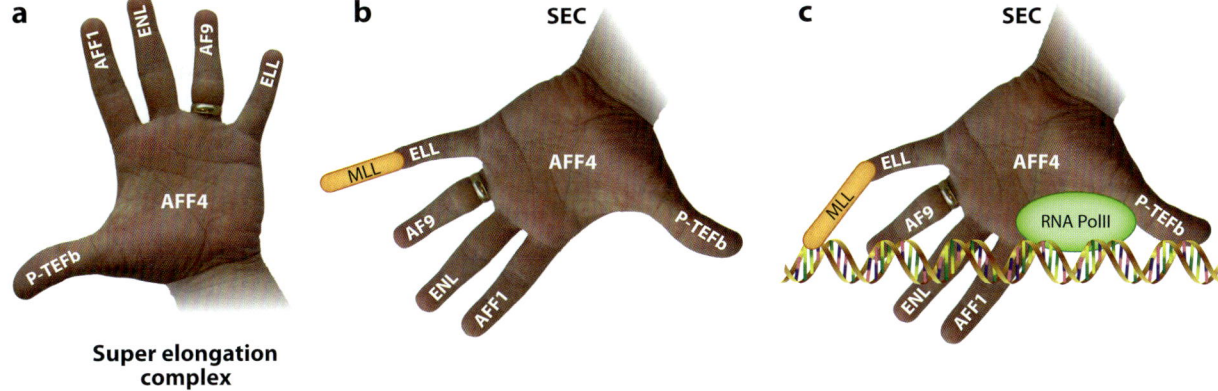

Figure 3

MLL translocation into super elongation complex (SEC) in leukemic pathogenesis. The *MLL* gene is found in a variety of chromosomal translocations associated with childhood leukemia. The ELL protein was the first translocation partner of MLL for which a biological function was demonstrated. Biochemically, ELL was identified as a RNA polymerase II (Pol II) elongation factor over fifteen years ago, and it was proposed then that the elongation stage of transcription could be central in the pathogenesis of leukemia through MLL translocations. We now know that many of the MLL translocation partners, such as ENL, AF9, AFF1, and AFF4, are found in a SEC with ELL and the positive transcription elongation factors (P-TEFb). It has been proposed that the translocation of MLL into SEC results in the mistargeting of SEC to MLL-regulated genes and the disruption elongation stage of transcription without appropriate transcriptional elongation checkpoints. The hand image represents the fact that all fingers (MLL translocation partners) are linked through the palm (AFF4) and that without the palm, the fingers (translocation partners of MLL) are not held together. The translocations of many of the factors (on the fingers) into MLL (*yellow*) results in the association of SEC with MLL target genes misregulating the elongation checkpoint on these genes and therefore resulting in the pathogenesis of childhood leukemia.

COMPASS complexes of *C. elegans* was demonstrated to play a role in the regulation of life span. On the basis of searching for homology, it has been proposed that *C. elegans* possesses two of the three branches of the COMPASS family of H3K4 methylases (83). One is Set-2 of *C. elegans*, which is very similar to the *Drosophila* Set1 and human Set1A/B proteins. The other is *C. elegans* Set-16, which bears resemblance to *Drosophila* Trr and human MLL3/MLL4. The loss of Trx/MLL1/MLL2 in *C. elegans* can be explained by the fact that this organism does not possess a large number of homeotic genes for development.

Along this line, reductions in the levels of other members of Set1/COMPASS in *C. elegans*, including Wdr5 and Set-2 (the Set1 homolog), resulted in extension of life span, suggesting that an excess of H3K4 methylation is detrimental for longevity (189). In support of this observation, loss of the histone H3K4 demethylase RBR-2 in *C. elegans* was demonstrated to be also required for normal life span. Given that histone H3K4 trimethylation by Set1/COMPASS requires the H2B monoubiquitinase Rad6/Bre1 from yeast to human (3, 76, 190), it will be of great interest to determine whether this pathway is also required for longevity in *C. elegans*, and whether di- and/or trimethylation of histone H3K4 is required for this process. This pathway can be further dissected by identification of the conserved role of Wdr82, which is specifically required for H3K4 trimethylation by Set1/COMPASS. The determination of the role of these factors in the specific regulation of di- and trimethylation in the regulation of longevity will also be of great interest.

5.3. The Trr/COMPASS Family in Nuclear Receptor Transactivation and Cancer

The Trr family of H3K4 methylases in *Drosophila* and their human homologs, the

MLL3/MLL4 complexes, have been shown to function in diverse processes. Initially for *Drosophila* Trr and later for mammalian MLL3/MLL4, it was demonstrated that they function in the nuclear hormone transactivation pathway. The EcR, which is the major steroid hormone pathway receptor in *Drosophila*, was demonstrated to colocalize extensively with Trr in an ecdysone-dependent manner (92). Furthermore, Trr and the EcR can be isolated within a common complex in an ecdysone-dependent manner (92). Following these initial observations in *Drosophila*, it was demonstrated that the human homolog of Trr, the MLL3/MLL4 family of the H3K4 methylases, also functions within the nuclear hormone pathway (94). This class of H3K4 methylases was demonstrated to work in the cross talk between the ATPase-dependent chromatin remodeling complexes and efficient nuclear receptor transactivation.

In other studies, the MLL3/MLL4 family members have been linked to cancer through interactions with p53 and the DNA damage pathway (191). Recently, frequent mutations in MLL4 were associated with non-Hodgkin's lymphoma (110). Follicular lymphoma (FL) and diffuse large B-cell lymphoma (DLBCL) are the most common non-Hodgkin's lymphomas. DNA sequence analysis from samples from patients suffering from FL and DLBCL demonstrated that about 32% of DLBCL and 89% of FL cases had somatic mutations in the MLL4 gene (110). Given the widespread role of MLL1 in leukemic pathogenesis and recent findings about MLL3/MLL4 in cancer, further investigation of the roles of the COMPASS family of H3K4 methylases in human will further our understanding of the diverse mechanistic role of H3K4 methyltransferases in development and during disease pathogenesis.

6. CONCLUDING REMARKS

It has been almost 10 years since publication of the original report on the biochemical purification of Set1/COMPASS as the yeast homolog of *Drosophila* Trx and mammalian MLL (42) and since its biochemical characterization as the first histone H3K4 methylase (42–44, 192). As described in this review, there has been an exponential increase in our knowledge regarding the COMPASS family of H3K4 methylases and the biological role of this histone modification from yeast to humans. We now know that Set1/COMPASS is highly conserved from yeast to human both enzymatically and structurally. Histone H3K4 trimethylation in yeast was shown to mark the promoters of the actively transcribed regions (53, 54, 193), and this mark is highly conserved from yeast to human (194). Although there is only one Set1/COMPASS in yeast, there are three COMPASS-like complexes in *Drosophila*, which are the Set1, Trx, and Trr/COMPASS-like complexes (**Figure 2**; **Table 1**) (59). The activities of these three family members of the H3K4 methylases of *Drosophila* are expanded in mammalian cells. The activity of Set1/COMPASS is shared by the Set1A-B/COMPASS complexes. The Trx and Trr activities are shared by the MLL1/MLL2 and MLL3/MLL4 COMPASS-like complexes, respectively (**Figure 2**). The H3K4 methylase activities of Set1, Trx, and Trr of *Drosophila* and Set1A-B, MLL1/MLL2, and MLL3/MLL4 of mammals are highly specific and nonredundant as the deletion of any of these methylases results in lethality, indicating that these enzymes cannot compensate for one another (84, 195–197).

Although we have learned an extensive amount of information about the process of implementation and the removal of the different patterns of histone H3K4 methylation during the past 10 years, there is still much to be learned about their biology and the reasons behind the diversity of the COMPASS family in metazoan cells. For example, Set1/COMPASS of yeast can implement the mono-, di-, and trimethylation of H3K4, and several subunits of COMPASS are required to regulate this process (**Table 1**). However, in mammalian cells the different patterns of H3K4 methylations have been attributed to different biological functions. For example, as with yeast, H3K4 trimethylation has become a universal mark of actively transcribed regions in metazoans. And yet, other patterns of H3K4 methylation by the metazoan COMPASS family have been associated with other possible biological processes. Histone H3K4 monomethylation is becoming a landmark for the identification of enhancers (198, 199). It is not clear which COMPASS family member is involved in the implementation of H3K4 monomethylation on the enhancers and what the role of this mark is in enhancer organization and function within the genomes.

Another example of diversity and variation in the H3K4 methylation patterns is its coexistence with histone H3K27 trimethylation, which is referred to as bivalency (200). Promoters for developmentally regulated genes in mammalian stem cells are marked by both H3K4 trimethylation and H3K27 methylation. Such bivalently marked regions are thought to be required for proper regulation and for the stem cells' commitment to one fate or another. However, it is not clear which COMPASS family member is involved in the creation of bivalently marked promoters on developmentally controlled genes or how such regions function during development (201). Furthermore, other organisms, such as *Xenopus tropicalis*, which share similar Trithorax and Polycomb group proteins, do not appear to use bivalently marked promoters throughout development (201, 202). Are bivalently marked promoters a mammalian-specific phenomenon? Future research on COMPASS family members should include the molecular characterization of the mechanism of the recruitment of the COMPASS family to different loci within the genome. Because of the lack of reliable antibodies toward Set1A-B, MLL1/MLL2, and MLL3/MLL4, it is not clear at this time how different COMPASS family members are recruited to different loci. Early studies demonstrated that, in the absence of MLL1/MLL2, the Hox cluster in mouse embryonic fibroblasts lose their H3K4 methylation and concomitantly cease to express these loci (77). This finding suggested that unlike Set1/COMPASS, which implements H3K4 methylation subsequent to transcriptional activation, H3K4 methylation, implemented by the MLL1/MLL2/COMPASS family, is instructive to transcription. How is it that the MLL1/MLL2 complexes, which are found within almost identical complexes, are recruited to different loci within the Hox cluster? How is it that Set1A-B and MLL3/MLL4 activities are excluded from these loci? Future studies addressing the above questions should employ molecular genetics as well as biochemical and cell biological approaches on the COMPASS family in diverse model systems. The results from these studies will greatly add to our knowledge of the molecular complexity and functional diversity of this highly conserved ancient machinery.

SUMMARY POINTS

1. Yeast Set1/COMPASS was the first histone H3K4 methyltransferase identified, containing Set1 and six other subunits. Unlike most other SET domain-containing proteins, Set1 alone is not active and requires association with components of COMPASS to be active.

2. *Drosophila* has three COMPASS-like complexes (Set1, Trx, and Trr), and mammals have six (Set1A/B, MLL1/MLL2, and MLL3/MLL4), which together form the COMPASS family of H3K4 methylases.

3. Set1/COMPASS and H3K4 methylation are involved in a number of processes including regulation of gene expression, homologous recombination, and life span.

4. *Drosophila* Trithorax (Trx) and the human MLL1 and MLL2 complexes are a part of the same COMPASS branch, positively regulating homeotic gene expression.

5. *Drosophila* Trithorax-related (Trr) and the human MLL3 and MLL4 complexes are part of another COMPASS branch that is involved in regulating hormone-responsive genes. UTX, an H3K27-demethylating enzyme, is restricted to this COMPASS branch.

6. The regulation of different methylation states by Set1/COMPASS is a highly controlled process involving interactions with the PAF complex and requiring the H2B monoubiquitination machinery.

7. Structural studies have demonstrated that the COMPASS family has a conserved Y-shaped architecture with Cps50/RbBp5 and Cps30/Wdr5 localizing on the top two adjacent lobes and Cps60/Ash2-Cps25/Dpy30 forming the base. The catalytic SET domain is located at the juncture, where a central channel may act as the platform for regulative processing of various degrees of H3K4 methylation.

8. The most common translocation partners of MLL1 in leukemia are part of a super elongation complex (SEC) required for *Hox* and *Myc* gene overexpression found in MLL translocation-based leukemia.

FUTURE ISSUES

1. The biochemical function(s) of both the common subunits and the subunits restricted to each branch of the COMPASS family need to be better defined.

2. Although it is known that different members of the COMPASS family have different gene specificities, the mechanisms for gene target selectivity for COMPASS family members is poorly understood.

3. Because COMPASS can implement H3K4 mono-, di-, and trimethylation, structural and genetic studies will be needed to understand the contribution of COMPASS subunits in the regulation of H3K4 mono-, di-, and trimethylation states and the functional consequence of each mark on chromatin.

4. The requirement for the H2B monoubiquitination machinery in H3K4 trimethylation by COMPASS is conserved from yeast to humans, but the mechanistic connection has remained elusive.

5. Do the different members of the COMPASS family work together or separately? Could these complexes work at distinct locations within a gene, or regulate distinct steps of gene regulation? What is the role of H3K4 trimethylation at promoters and H3K4 monomethylation at enhancers?

6. Genes that are bivalently marked with H3K4 and H3K27 trimethylation are prevalent in mammalian stem cells, but the mechanism for how this state is set, and which COMPASS family member(s) works with the PcG family, is unknown.

7. It is still poorly understood how translocations of MLL into subunits of SEC result in the pathogenesis of childhood leukemia. The identification of functional targets of MLL-SEC translocations will allow development of targeted therapeutic approaches.

8. The development of COMPASS-specific small molecular inhibitors could be used to help define the H3K4 methylation machinery in vitro and in vivo and could be used as possible drugs in situations where H3K4 methylation is misregulated.

DISCLOSURE STATEMENT

The author is not aware of any affiliations, memberships, funding, or financial holdings that might be perceived as affecting the objectivity of this review.

ACKNOWLEDGMENTS

I am indebted to many of my colleagues for their conversations and suggestions while writing this article. I am grateful to my colleague, Edwin Smith, for his critical reading of this review and valuable comments and input. Also, I thank Laura Shilatifard for editorial assistance and Julia Schulze for help with the illustrations. I apologize to my many colleagues whose work I was not able to cite in this review owing to space limitations. Studies in my laboratory on the COMPASS family of histone H3K4 methylases and childhood leukemia are supported by the National Institutes of Health grants R01CA150265, R01GM069905, and R01CA89455; the Alex's Lemonade Stand Foundation for Childhood Cancer; and the Stowers Institute for Medical Research.

LITERATURE CITED

1. Kornberg RD. 1974. Chromatin structure: a repeating unit of histones and DNA. *Science* 184:868–71
2. Kornberg RD, Lorch Y. 1999. Twenty-five years of the nucleosome, fundamental particle of the eukaryote chromosome. *Cell* 98:285–94
3. Smith E, Shilatifard A. 2010. The chromatin signaling pathway: diverse mechanisms of recruitment of histone-modifying enzymes and varied biological outcomes. *Mol. Cell* 40:689–701
4. Heitz E. 1928. Das Heterochromatin der Moose. *Jahrb. Wiss. Bot.* 69:762–818
5. Hilliker AJ, Holm DG. 1975. Genetic analysis of the proximal region of chromosome 2 of *Drosophila melanogaster*. I. Detachment products of compound autosomes. *Genetics* 81:705–21
6. Yasuhara JC, DeCrease CH, Wakimoto BT. 2005. Evolution of heterochromatic genes of *Drosophila*. *Proc. Natl. Acad. Sci. USA* 102:10958–63
7. Muller HJ, Altenburg E. 1930. The frequency of translocations produced by X-rays in *Drosophila*. *Genetics* 15:283–311
8. Moore GD, Procunier JD, Cross DP, Grigliatti TA. 1979. Histone gene deficiencies and position-effect variegation in *Drosophila*. *Nature* 282:312–14
9. Moore GD, Sinclair DA, Grigliatti TA. 1983. Histone gene multiplicity and position effect variegation in *Drosophila melanogaster*. *Genetics* 105:327–44
10. Han M, Grunstein M. 1988. Nucleosome loss activates yeast downstream promoters in vivo. *Cell* 55:1137–45
11. Han M, Kim UJ, Kayne P, Grunstein M. 1988. Depletion of histone H4 and nucleosomes activates the PHO5 gene in *Saccharomyces cerevisiae*. *EMBO J.* 7:2221–28
12. Lieb JD, Liu X, Botstein D, Brown PO. 2001. Promoter-specific binding of Rap1 revealed by genome-wide maps of protein-DNA association. *Nat. Genet.* 28:327–34
13. Lorch Y, LaPointe JW, Kornberg RD. 1987. Nucleosomes inhibit the initiation of transcription but allow chain elongation with the displacement of histones. *Cell* 49:203–10
14. Knezetic JA, Luse DS. 1986. The presence of nucleosomes on a DNA template prevents initiation by RNA polymerase II in vitro. *Cell* 45:95–104

15. Workman JL, Roeder RG. 1987. Binding of transcription factor TFIID to the major late promoter during in vitro nucleosome assembly potentiates subsequent initiation by RNA polymerase II. *Cell* 51:613–22
16. Chang CH, Luse DS. 1997. The H3/H4 tetramer blocks transcript elongation by RNA polymerase II in vitro. *J. Biol. Chem.* 272:23427–34
17. Ujvari A, Hsieh FK, Luse SW, Studitsky VM, Luse DS. 2008. Histone N-terminal tails interfere with nucleosome traversal by RNA polymerase II. *J. Biol. Chem.* 283:32236–43
18. Weintraub H, Groudine M. 1976. Chromosomal subunits in active genes have an altered conformation. *Science* 193:848–56
19. Pirrotta V. 1998. Polycombing the genome: PcG, trxG, and chromatin silencing. *Cell* 93:333–36
20. Mahmoudi T, Verrijzer CP. 2001. Chromatin silencing and activation by Polycomb and trithorax group proteins. *Oncogene* 20:3055–66
21. Orlando V. 2003. Polycomb, epigenomes, and control of cell identity. *Cell* 112:599–606
22. Stassen MJ, Bailey D, Nelson S, Chinwalla V, Harte PJ. 1995. The *Drosophila* trithorax proteins contain a novel variant of the nuclear receptor type DNA binding domain and an ancient conserved motif found in other chromosomal proteins. *Mech. Dev.* 52:209–23
23. Tschiersch B, Hofmann A, Krauss V, Dorn R, Korge G, Reuter G. 1994. The protein encoded by the *Drosophila* position-effect variegation suppressor gene Su(var)3–9 combines domains of antagonistic regulators of homeotic gene complexes. *EMBO J.* 13:3822–31
24. Luger K, Mader AW, Richmond RK, Sargent DF, Richmond TJ. 1997. Crystal structure of the nucleosome core particle at 2.8 Å resolution. *Nature* 389:251–60
25. Bhaumik SR, Smith E, Shilatifard A. 2007. Covalent modifications of histones during development and disease pathogenesis. *Nat. Struct. Mol. Biol.* 14:1008–16
26. Workman JL, Kingston RE. 1998. Alteration of nucleosome structure as a mechanism of transcriptional regulation. *Annu. Rev. Biochem.* 67:545–79
27. Kouzarides T. 2007. Chromatin modifications and their function. *Cell* 128:693–705
28. Berger SL. 2007. The complex language of chromatin regulation during transcription. *Nature* 447:407–12
29. Shilatifard A. 2006. Chromatin modifications by methylation and ubiquitination: Implications in the regulation of gene expression. *Annu. Rev. Biochem.* 75:243–69
30. Allis CD, Berger SL, Cote J, Dent S, Jenuwien T, et al. 2007. New nomenclature for chromatin-modifying enzymes. *Cell* 131:633–36
31. Ringrose L, Paro R. 2004. Epigenetic regulation of cellular memory by the Polycomb and Trithorax group proteins. *Annu. Rev. Genet.* 38:413–43
32. Ingham PW, Whittle R. 1980. Trithorax: a new homeotic mutation of *Drosophila melanogaster* causing transformations of abdominal and thoracic imaginal segments. *Mol. Gen. Genet.* 179:607–14
33. Ingham PW. 1985. Genetic control of the spatial pattern of selector gene expression in *Drosophila*. *Cold Spring Harb. Symp. Quant. Biol.* 50:201–8
34. Lewis EB. 1978. A gene complex controlling segmentation in *Drosophila*. *Nature* 276:565–70
35. Schuettengruber B, Chourrout D, Vervoort M, Leblanc B, Cavalli G. 2007. Genome regulation by polycomb and trithorax proteins. *Cell* 128:735–45
36. Margueron R, Reinberg D. 2011. The Polycomb complex PRC2 and its mark in life. *Nature* 469:343–49
37. Gu Y, Nakamura T, Alder H, Prasad R, Canaani O, et al. 1992. The t(4;11) chromosome translocation of human acute leukemias fuses the *ALL-1* gene, related to *Drosophila trithorax*, to the *AF-4* gene. *Cell* 71:701–8
38. Tkachuk DC, Kohler S, Cleary ML. 1992. Involvement of a homolog of *Drosophila* trithorax by 11q23 chromosomal translocations in acute leukemias. *Cell* 71:691–700
39. Ziemin-van der Poel S, McCabe NR, Gill HJ, Espinosa R III, Patel Y, et al. 1991. Identification of a gene, *MLL*, that spans the breakpoint in 11q23 translocations associated with human leukemias. *Proc. Natl. Acad. Sci. USA* 88:10735–39
40. Mohan M, Lin C, Guest E, Shilatifard A. 2010. Licensed to elongate: a molecular mechanism for MLL-based leukaemogenesis. *Nat. Rev. Cancer* 10:721–28

41. Hsieh JJ, Ernst P, Erdjument-Bromage H, Tempst P, Korsmeyer SJ. 2003. Proteolytic cleavage of MLL generates a complex of N- and C-terminal fragments that confers protein stability and subnuclear localization. *Mol. Cell. Biol.* 23:186–94
42. Miller T, Krogan NJ, Dover J, Erdjument-Bromage H, Tempst P, et al. 2001. COMPASS: a complex of proteins associated with a trithorax-related SET domain protein. *Proc. Natl. Acad. Sci. USA* 98:12902–7
43. Krogan NJ, Dover J, Khorrami S, Greenblatt JF, Schneider J, et al. 2002. COMPASS, a histone H3 (lysine 4) methyltransferase required for telomeric silencing of gene expression. *J. Biol. Chem.* 277:10753–55
44. Roguev A, Schaft D, Shevchenko A, Pijnappel WW, Wilm M, et al. 2001. The *Saccharomyces cerevisiae* Set1 complex includes an Ash2 homologue and methylates histone 3 lysine 4. *EMBO J.* 20:7137–48
45. Schneider J, Wood A, Lee JS, Schuster R, Dueker J, et al. 2005. Molecular regulation of histone H3 trimethylation by COMPASS and the regulation of gene expression. *Mol. Cell* 19:849–56
46. Nedea E, He X, Kim M, Pootoolal J, Zhong G, et al. 2003. Organization and function of APT, a subcomplex of the yeast cleavage and polyadenylation factor involved in the formation of mRNA and small nucleolar RNA 3′-ends. *J. Biol. Chem.* 278:33000–10
47. Nedea E, Nalbant D, Xia D, Theoharis NT, Suter B, et al. 2008. The Glc7 phosphatase subunit of the cleavage and polyadenylation factor is essential for transcription termination on snoRNA genes. *Mol. Cell* 29:577–87
48. Cheng H, He X, Moore C. 2004. The essential WD repeat protein Swd2 has dual functions in RNA polymerase II transcription termination and lysine 4 methylation of histone H3. *Mol. Cell. Biol.* 24:2932–43
49. Schlichter A, Cairns BR. 2005. Histone trimethylation by Set1 is coordinated by the RRM, autoinhibitory, and catalytic domains. *EMBO J.* 24:1222–31
50. Dehe PM, Dichtl B, Schaft D, Roguev A, Pamblanco M, et al. 2006. Protein interactions within the Set1 complex and their roles in the regulation of histone 3 lysine 4 methylation. *J. Biol. Chem.* 281:35404–12
51. Wood A, Shukla A, Schneider J, Lee JS, Stanton JD, et al. 2007. Ctk complex-mediated regulation of histone methylation by COMPASS. *Mol. Cell. Biol.* 27:709–20
52. Xiao T, Shibata Y, Rao B, Laribee RN, O'Rourke R, et al. 2007. The RNA polymerase II kinase Ctk1 regulates positioning of a 5′ histone methylation boundary along genes. *Mol. Cell. Biol.* 27:721–31
53. Wood A, Schneider J, Dover J, Johnston M, Shilatifard A. 2003. The Paf1 complex is essential for histone monoubiquitination by the Rad6-Bre1 complex, which signals for histone methylation by COMPASS and Dot1p. *J. Biol. Chem.* 278:34739–42
54. Krogan NJ, Dover J, Wood A, Schneider J, Heidt J, et al. 2003. The Paf1 complex is required for histone H3 methylation by COMPASS and Dot1p: linking transcriptional elongation to histone methylation. *Mol. Cell* 11:721–29
55. Jiang D, Kong NC, Gu X, Li Z, He Y. 2011. *Arabidopsis* COMPASS-like complexes mediate histone H3 lysine-4 trimethylation to control floral transition and plant development. *PLoS Genet.* 7:e1001330
56. Hughes CM, Rozenblatt-Rosen O, Milne TA, Copeland TD, Levine SS, et al. 2004. Menin associates with a trithorax family histone methyltransferase complex and with the *Hoxc8* locus. *Mol. Cell* 13:587–97
57. Lee JH, Tate CM, You JS, Skalnik DG. 2007. Identification and characterization of the human Set1B histone H3-Lys4 methyltransferase complex. *J. Biol. Chem.* 282:13419–28
58. Wu M, Wang PF, Lee JS, Martin-Brown S, Florens L, et al. 2008. Molecular regulation of H3K4 trimethylation by Wdr82, a component of human Set1/COMPASS. *Mol. Cell. Biol.* 28:7337–44
59. Mohan M, Herz HM, Smith ER, Zhang Y, Jackson J, et al. 2011. The COMPASS family of H3K4 methylases in *Drosophila*. *Mol. Cell. Biol.* 31:4310–18
60. Petruk S, Sedkov Y, Smith S, Tillib S, Kraevski V, et al. 2001. Trithorax and dCBP acting in a complex to maintain expression of a homeotic gene. *Science* 294:1331–34
61. Ardehali MB, Mei A, Zobeck KL, Caron M, Lis JT, Kusch T. 2011. *Drosophila* Set1 is the major histone H3 lysine 4 trimethyltransferase with role in transcription. *EMBO J.* 30:2817–28
62. Shilatifard A. 2008. Molecular implementation and physiological roles for histone H3 lysine 4 (H3K4) methylation. *Curr. Opin. Cell Biol.* 20:341–48
63. Eissenberg JC, Shilatifard A. 2010. Histone H3 lysine 4 (H3K4) methylation in development and differentiation. *Dev. Biol.* 339:240–49

64. Cho YW, Hong T, Hong S, Guo H, Yu H, et al. 2007. PTIP associates with MLL3- and MLL4-containing histone H3 lysine 4 methyltransferase complex. *J. Biol. Chem.* 282:20395–406
65. Issaeva I, Zonis Y, Rozovskaia T, Orlovsky K, Croce CM, et al. 2007. Knockdown of ALR (MLL2) reveals ALR target genes and leads to alterations in cell adhesion and growth. *Mol. Cell. Biol.* 27:1889–903
66. Wu M, Wang PF, Lee JS, Martin-Brown S, Florens L, et al. 2008. Molecular regulation of H3K4 trimethylation by Wdr82, a component of human Set1/COMPASS. *Mol. Cell. Biol.* 28:7337–44
67. Wysocka J, Myers MP, Laherty CD, Eisenman RN, Herr W. 2003. Human Sin3 deacetylase and trithorax-related Set1/Ash2 histone H3-K4 methyltransferase are tethered together selectively by the cell-proliferation factor HCF-1. *Genes Dev.* 17:896–911
68. Yokoyama A, Wang Z, Wysocka J, Sanyal M, Aufiero DJ, et al. 2004. Leukemia proto-oncoprotein MLL forms a SET1-like histone methyltransferase complex with menin to regulate *Hox* gene expression. *Mol. Cell. Biol.* 24:5639–49
69. Patel SR, Kim D, Levitan I, Dressler GR. 2007. The BRCT-domain containing protein PTIP links PAX2 to a histone H3, lysine 4 methyltransferase complex. *Dev. Cell* 13:580–92
70. Dou Y, Milne TA, Tackett AJ, Smith ER, Fukuda A, et al. 2005. Physical association and coordinate function of the H3 K4 methyltransferase MLL1 and the H4 K16 acetyltransferase MOF. *Cell* 121:873–85
71. Goo YH, Sohn YC, Kim DH, Kim SW, Kang MJ, et al. 2003. Activating signal cointegrator 2 belongs to a novel steady-state complex that contains a subset of trithorax group proteins. *Mol. Cell. Biol.* 23:140–49
72. Lee JH, Tate CM, You JS, Skalnik DG. 2007. Identification and characterization of the human Set1B histone H3-Lys4 methyltransferase complex. *J. Biol. Chem.* 282:13419–28
73. Lee JH, Skalnik DG. 2005. CpG-binding protein (CXXC finger protein 1) is a component of the mammalian Set1 histone H3-Lys4 methyltransferase complex, the analogue of the yeast Set1/COMPASS complex. *J. Biol. Chem.* 280:41725–31
74. Eissenberg JC, Shilatifard A. 2010. Histone H3 lysine 4 (H3K4) methylation in development and differentiation. *Dev. Biol.* 339:240–49
75. Lee JH, Skalnik DG. 2008. Wdr82 is a C-terminal domain-binding protein that recruits the Setd1A histone H3-Lys4 methyltransferase complex to transcription start sites of transcribed human genes. *Mol. Cell. Biol.* 28:609–18
76. Lee JS, Shukla A, Schneider J, Swanson SK, Washburn MP, et al. 2007. Histone crosstalk between H2B monoubiquitination and H3 methylation mediated by COMPASS. *Cell* 131:1084–96
77. Wang P, Lin C, Smith ER, Guo H, Sanderson BW, et al. 2009. Global analysis of H3K4 methylation defines MLL family member targets and points to a role for MLL1-mediated H3K4 methylation in the regulation of transcriptional initiation by RNA polymerase II. *Mol. Cell. Biol.* 29:6074–85
78. Lee MG, Villa R, Trojer P, Norman J, Yan KP, et al. 2007. Demethylation of H3K27 regulates polycomb recruitment and H2A ubiquitination. *Science* 318:447–50
79. Agger K, Cloos PA, Christensen J, Pasini D, Rose S, et al. 2007. UTX and JMJD3 are histone H3K27 demethylases involved in *HOX* gene regulation and development. *Nature* 449:731–34
80. Takahashi YH, Westfield GH, Oleskie AN, Trievel RC, Shilatifard A, Skiniotis G. 2011. *Proc. Natl. Acad. Sci. USA* 108:20526–31
81. Li T, Kelly WG. 2011. A role for Set1/MLL-related components in epigenetic regulation of the *Caenorhabditis elegans* germ line. *PLoS Genet.* 7:e1001349
82. Poole RJ, Bashllari E, Cochella L, Flowers EB, Hobert O. 2011. A genome-wide RNAi screen for factors involved in neuronal specification in *Caenorhabditis elegans*. *PLoS Genet.* 7:e1002109
83. Smith E, Lin C, Shilatifard A. 2011. The super elongation complex (SEC) and MLL in development and disease. *Genes Dev.* 25:661–72
84. Yu BD, Hess JL, Horning SE, Brown GA, Korsmeyer SJ. 1995. Altered Hox expression and segmental identity in *Mll*-mutant mice. *Nature* 378:505–8
85. Yu BD, Hanson RD, Hess JL, Horning SE, Korsmeyer SJ. 1998. MLL, a mammalian *trithorax*-group gene, functions as a transcriptional maintenance factor in morphogenesis. *Proc. Natl. Acad. Sci. USA* 95:10632–36
86. Ayton PM, Cleary ML. 2003. Transformation of myeloid progenitors by MLL oncoproteins is dependent on Hoxa7 and Hoxa9. *Genes Dev.* 17:2298–307

87. Zeisig BB, Milne T, Garcia-Cuellar MP, Schreiner S, Martin ME, et al. 2004. Hoxa9 and Meis1 are key targets for MLL-ENL-mediated cellular immortalization. *Mol. Cell. Biol.* 24:617–28
88. Milne TA, Briggs SD, Brock HW, Martin ME, Gibbs D, et al. 2002. MLL targets SET domain methyltransferase activity to *Hox* gene promoters. *Mol. Cell* 10:1107–17
89. Terranova R, Agherbi H, Boned A, Meresse S, Djabali M. 2006. Histone and DNA methylation defects at Hox genes in mice expressing a SET domain-truncated form of Mll. *Proc. Natl. Acad. Sci. USA* 103:6629–34
90. Sedkov Y, Benes JJ, Berger JR, Riker KM, Tillib S, et al. 1999. Molecular genetic analysis of the *Drosophila trithorax*-related gene which encodes a novel SET domain protein. *Mech. Dev.* 82:171–79
91. Tillib S, Petruk S, Sedkov Y, Kuzin A, Fujioka M, et al. 1999. Trithorax- and Polycomb-group response elements within an *Ultrabithorax* transcription maintenance unit consist of closely situated but separable sequences. *Mol. Cell. Biol.* 19:5189–202
92. Sedkov Y, Cho E, Petruk S, Cherbas L, Smith ST, et al. 2003. Methylation at lysine 4 of histone H3 in ecdysone-dependent development of *Drosophila*. *Nature* 426:78–83
93. Prasad R, Zhadanov AB, Sedkov Y, Bullrich F, Druck T, et al. 1997. Structure and expression pattern of human *ALR*, a novel gene with strong homology to ALL-1 involved in acute leukemia and to *Drosophila trithorax*. *Oncogene* 15:549–60
94. Mo R, Rao SM, Zhu YJ. 2006. Identification of the MLL2 complex as a coactivator for estrogen receptor alpha. *J. Biol. Chem.* 281:15714–20
95. Lee J, Saha PK, Yang QH, Lee S, Park JY, et al. 2008. Targeted inactivation of MLL3 histone H3-Lys-4 methyltransferase activity in the mouse reveals vital roles for MLL3 in adipogenesis. *Proc. Natl. Acad. Sci. USA* 105:19229–34
96. Daniel JA, Santos MA, Wang Z, Zang C, Schwab KR, et al. 2010. PTIP promotes chromatin changes critical for immunoglobulin class switch recombination. *Science* 329:917–23
97. Kim DH, Lee J, Lee B, Lee JW. 2009. ASCOM controls farnesoid X receptor transactivation through its associated histone H3 lysine 4 methyltransferase activity. *Mol. Endocrinol.* 23:1556–62
98. Lee S, Kim DH, Goo YH, Lee YC, Lee SK, Lee JW. 2009. Crucial roles for interactions between MLL3/4 and INI1 in nuclear receptor transactivation. *Mol. Endocrinol.* 23:610–19
99. Aboobaker AA, Blaxter ML. 2003. Hox gene loss during dynamic evolution of the nematode cluster. *Curr. Biol.* 13:37–40
100. Robinson-Rechavi M, Maina CV, Gissendanner CR, Laudet V, Sluder A. 2005. Explosive lineage-specific expansion of the orphan nuclear receptor HNF4 in nematodes. *J. Mol. Evol.* 60:577–86
101. Lan F, Bayliss PE, Rinn JL, Whetstine JR, Wang JK, et al. 2007. A histone H3 lysine 27 demethylase regulates animal posterior development. *Nature* 449:689–94
102. Seenundun S, Rampalli S, Liu QC, Aziz A, Palii C, et al. 2010. UTX mediates demethylation of H3K27me3 at muscle-specific genes during myogenesis. *EMBO J.* 29:1401–11
103. Smith ER, Lee MG, Winter B, Droz NM, Eissenberg JC, et al. 2008. *Drosophila* UTX is a histone H3 Lys27 demethylase that colocalizes with the elongating form of RNA polymerase II. *Mol. Cell. Biol.* 28:1041–46
104. Herz HM, Madden LD, Chen Z, Bolduc C, Buff E, et al. 2010. The H3K27me3 demethylase dUTX is a suppressor of Notch- and Rb-dependent tumors in *Drosophila*. *Mol. Cell. Biol.* 30:2485–97
105. Wang JK, Tsai MC, Poulin G, Adler AS, Chen S, et al. 2010. The histone demethylase UTX enables RB-dependent cell fate control. *Genes Dev.* 24:327–32
106. Schwartz YB, Kahn TG, Nix DA, Li XY, Bourgon R, et al. 2006. Genome-wide analysis of Polycomb targets in *Drosophila melanogaster*. *Nat. Genet.* 38:700–5
107. Schwartz YB, Kahn TG, Stenberg P, Ohno K, Bourgon R, Pirrotta V. 2010. Alternative epigenetic chromatin states of polycomb target genes. *PLoS Genet.* 6:e1000805
108. Enderle D, Beisel C, Stadler MB, Gerstung M, Athri P, Paro R. 2011. Polycomb preferentially targets stalled promoters of coding and noncoding transcripts. *Genome Res.* 21:216–26
109. Ashktorab H, Schaffer AA, Daremipouran M, Smoot DT, Lee E, Brim H. 2010. Distinct genetic alterations in colorectal cancer. *PLoS ONE* 5: e8879
110. Morin RD, Mendez-Lago M, Mungall AJ, Goya R, Mungall KL, et al. 2011. Frequent mutation of histone-modifying genes in non-Hodgkin lymphoma. *Nature* 476:298–303

111. Parsons DW, Li M, Zhang X, Jones S, Leary RJ, et al. 2011. The genetic landscape of the childhood cancer medulloblastoma. *Science* 331:435–39
112. van Haaften G, Dalgliesh GL, Davies H, Chen L, Bignell G, et al. 2009. Somatic mutations of the histone H3K27 demethylase gene UTX in human cancer. *Nat. Genet.* 41:521–23
113. Lieb JD, Capowski EE, Meneely P, Meyer BJ. 1996. DPY-26, a link between dosage compensation and meiotic chromosome segregation in the nematode. *Science* 274:1732–36
114. Escamilla-Del-Arenal M, da Rocha ST, Heard E. 2011. Evolutionary diversity and developmental regulation of X-chromosome inactivation. *Hum. Genet.* 130:307–27
115. Laverty C, Lucci J, Akhtar A. 2010. The MSL complex: X chromosome and beyond. *Curr. Opin. Genet. Dev.* 20:171–78
116. Chuang PT, Albertson DG, Meyer BJ. 1994. DPY-27: a chromosome condensation protein homolog that regulates *C. elegans* dosage compensation through association with the X chromosome. *Cell* 79:459–74
117. Chuang PT, Lieb JD, Meyer BJ. 1996. Sex-specific assembly of a dosage compensation complex on the nematode X chromosome. *Science* 274:1736–39
118. Lieb JD, Albrecht MR, Chuang PT, Meyer BJ. 1998. MIX-1: an essential component of the *C. elegans* mitotic machinery executes X chromosome dosage compensation. *Cell* 92:265–77
119. Mets DG, Meyer BJ. 2009. Condensins regulate meiotic DNA break distribution, thus crossover frequency, by controlling chromosome structure. *Cell* 139:73–86
120. Hsu DR, Chuang PT, Meyer BJ. 1995. DPY-30, a nuclear protein essential early in embryogenesis for *Caenorhabditis elegans* dosage compensation. *Development* 121:3323–34
121. Hsu DR, Meyer BJ. 1994. The *dpy-30* gene encodes an essential component of the *Caenorhabditis elegans* dosage compensation machinery. *Genetics* 137:999–1018
122. Pferdehirt RR, Kruesi WS, Meyer BJ. 2011. An MLL/COMPASS subunit functions in the *C. elegans* dosage compensation complex to target X chromosomes for transcriptional regulation of gene expression. *Genes Dev.* 25:499–515
123. Schneider J, Dover J, Johnston M, Shilatifard A. 2004. Global proteomic analysis of *S. cerevisiae* (GPS) to identify proteins required for histone modifications. *Methods Enzymol.* 377:227–34
124. Giaever G, Chu AM, Ni L, Connelly C, Riles L, et al. 2002. Functional profiling of the *Saccharomyces cerevisiae* genome. *Nature* 418:387–91
125. Goffeau A, Barrell BG, Bussey H, Davis RW, Dujon B, et al. 1996. Life with 6000 genes. *Science* 274:546, 563–67
126. Dover J, Schneider J, Tawiah-Boateng MA, Wood A, Dean K, et al. 2002. Methylation of histone H3 by COMPASS requires ubiquitination of histone H2B by Rad6. *J. Biol. Chem.* 277:28368–71
127. Sun ZW, Allis CD. 2002. Ubiquitination of histone H2B regulates H3 methylation and gene silencing in yeast. *Nature* 418:104–8
128. Ng HH, Xu RM, Zhang Y, Struhl K. 2002. Ubiquitination of histone H2B by Rad6 is required for efficient Dot1-mediated methylation of histone H3 lysine 79. *J. Biol. Chem.* 277:34655–57
129. Briggs SD, Xiao T, Sun ZW, Caldwell JA, Shabanowitz J, et al. 2002. Gene silencing: *trans*-histone regulatory pathway in chromatin. *Nature* 418:498
130. Wood A, Krogan NJ, Dover J, Schneider J, Heidt J, et al. 2003. Bre1, an E3 ubiquitin ligase required for recruitment and substrate selection of Rad6 at a promoter. *Mol. Cell* 11:267–74
131. Schulze JM, Jackson J, Nakanishi S, Gardner JM, Hentrich T, et al. 2009. Linking cell cycle to histone modifications: SBF and H2B monoubiquitination machinery and cell-cycle regulation of H3K79 dimethylation. *Mol. Cell* 35:626–41
132. Hwang WW, Venkatasubrahmanyam S, Ianculescu AG, Tong A, Boone C, Madhani HD. 2003. A conserved RING finger protein required for histone H2B monoubiquitination and cell size control. *Mol. Cell* 11:261–66
133. Zheng S, Wyrick JJ, Reese JC. 2010. Novel *trans*-tail regulation of H2B ubiquitylation and H3K4 methylation by the N terminus of histone H2A. *Mol. Cell. Biol.* 30:3635–45
134. Vitaliano-Prunier A, Menant A, Hobeika M, Géli V, Gwizdek C, Dargemont C. 2008. Ubiquitylation of the COMPASS component Swd2 links H2B ubiquitylation to H3K4 trimethylation. *Nat. Cell Biol.* 10:1365–71

135. Zhu B, Zheng Y, Pham AD, Mandal SS, Erdjument-Bromage H, et al. 2005. Monoubiquitination of human histone H2B: the factors involved and their roles in *HOX* gene regulation. *Mol. Cell* 20:601–11
136. Kim J, Hake SB, Roeder RG. 2005. The human homolog of yeast BRE1 functions as a transcriptional coactivator through direct activator interactions. *Mol. Cell* 20:759–70
137. Kim J, Guermah M, McGinty RK, Lee JS, Tang Z, et al. 2009. RAD6-mediated transcription-coupled H2B ubiquitylation directly stimulates H3K4 methylation in human cells. *Cell* 137:459–71
138. Mohan M, Herz HM, Takahashi YH, Lin C, Lai KC, et al. 2010. Linking H3K79 trimethylation to Wnt signaling through a novel Dot1-containing complex (DotCom). *Genes Dev.* 24:574–89
139. Bray S, Musisi H, Bienz M. 2005. Bre1 is required for Notch signaling and histone modification. *Dev. Cell* 8:279–86
140. Smith E, Shilatifard A. 2009. Developmental biology. Histone cross-talk in stem cells. *Science* 323:221–22
141. Buszczak M, Paterno S, Spradling AC. 2009. *Drosophila* stem cells share a common requirement for the histone H2B ubiquitin protease scrawny. *Science* 323:248–51
142. Minsky N, Oren M. 2004. The RING domain of Mdm2 mediates histone ubiquitylation and transcriptional repression. *Mol. Cell* 16:631–39
143. Takahashi YH, Schulze JM, Jackson J, Hentrich T, Seidel C, et al. 2011. Dot1 and histone H3K79 methylation in natural telomeric and HM silencing. *Mol. Cell* 42:118–26
144. Pavri R, Zhu B, Li G, Trojer P, Mandal S, et al. 2006. Histone H2B monoubiquitination functions cooperatively with FACT to regulate elongation by RNA polymerase II. *Cell* 125:703–17
145. Tanny JC, Erdjument-Bromage H, Tempst P, Allis CD. 2007. Ubiquitylation of histone H2B controls RNA polymerase II transcription elongation independently of histone H3 methylation. *Genes Dev.* 21:835–47
146. Moyal L, Lerenthal Y, Gana-Weisz M, Mass G, So S, et al. 2011. Requirement of ATM-dependent monoubiquitylation of histone H2B for timely repair of DNA double-strand breaks. *Mol. Cell* 41:529–42
147. Nakamura K, Kato A, Kobayashi J, Yanagihara H, Sakamoto S, et al. 2011. Regulation of homologous recombination by RNF20-dependent H2B ubiquitination. *Mol. Cell* 41:515–28
148. Shi X, Finkelstein A, Wolf AJ, Wade PA, Burton ZF, Jaehning JA. 1996. Paf1p, an RNA polymerase II-associated factor in *Saccharomyces cerevisiae*, may have both positive and negative roles in transcription. *Mol. Cell. Biol.* 16:669–76
149. Costa PJ, Arndt KM. 2000. Synthetic lethal interactions suggest a role for the *Saccharomyces cerevisiae* Rtf1 protein in transcription elongation. *Genetics* 156:535–47
150. Mueller CL, Jaehning JA. 2002. Ctr9, Rtf1, and Leo1 are components of the Paf1/RNA polymerase II complex. *Mol. Cell. Biol.* 22:1971–80
151. Pokholok DK, Hannett NM, Young RA. 2002. Exchange of RNA polymerase II initiation and elongation factors during gene expression in vivo. *Mol. Cell* 9:799–809
152. Chang M, French-Cornay D, Fan HY, Klein H, Denis CL, Jaehning JA. 1999. A complex containing RNA polymerase II, Paf1p, Cdc73p, Hpr1p, and Ccr4p plays a role in protein kinase C signaling. *Mol. Cell. Biol.* 19:1056–67
153. Squazzo SL, Costa PJ, Lindstrom DL, Kumer KE, Simic R, et al. 2002. The Paf1 complex physically and functionally associates with transcription elongation factors in vivo. *EMBO J.* 21:1764–74
154. Gerber M, Shilatifard A. 2003. Transcriptional elongation by RNA polymerase II and histone methylation. *J. Biol. Chem.* 278:26303–6
155. Krogan NJ, Kim M, Tong A, Golshani A, Cagney G, et al. 2003. Methylation of histone H3 by Set2 in *Saccharomyces cerevisiae* is linked to transcriptional elongation by RNA polymerase II. *Mol. Cell. Biol.* 23:4207–18
156. Zhu B, Mandal SS, Pham AD, Zheng Y, Erdjument-Bromage H, et al. 2005. The human PAF complex coordinates transcription with events downstream of RNA synthesis. *Genes Dev.* 19:1668–73
157. Wang P, Bowl MR, Bender S, Peng J, Farber L, et al. 2008. Parafibromin, a component of the human PAF complex, regulates growth factors and is required for embryonic development and survival in adult mice. *Mol. Cell. Biol.* 28:2930–40
158. Kim J, Guermah M, Roeder RG. 2010. The human PAF1 complex acts in chromatin transcription elongation both independently and cooperatively with SII/TFIIS. *Cell* 140:491–503

159. Tenney K, Gerber M, Ilvarsonn A, Schneider J, Gause M, et al. 2006. *Drosophila* Rtf1 functions in histone methylation, gene expression, and Notch signaling. *Proc. Natl. Acad. Sci. USA* 103:11970–74
160. Muntean AG, Tan J, Sitwala K, Huang Y, Bronstein J, et al. 2010. The PAF complex synergizes with MLL fusion proteins at *HOX* loci to promote leukemogenesis. *Cancer Cell* 17:609–21
161. Milne TA, Kim J, Wang GG, Stadler SC, Basrur V, et al. 2010. Multiple interactions recruit MLL1 and MLL1 fusion proteins to the *HOXA9* locus in leukemogenesis. *Mol. Cell* 38:853–63
162. Mohan M, Lin C, Guest E, Shilatifard A. 2010. Licensed to elongate: a molecular mechanism for MLL-based leukaemogenesis. *Nat. Rev. Cancer* 10:721–28
163. Rowley JD. 1998. The critical role of chromosome translocations in human leukemias. *Annu. Rev. Genet.* 32:495–519
164. Meyer C, Kowarz E, Hofmann J, Renneville A, Zuna J, et al. 2009. New insights to the MLL recombinome of acute leukemias. *Leukemia* 23:1490–99
165. Thiel AT, Blessington P, Zou T, Feather D, Wu X, et al. 2010. MLL-AF9-induced leukemogenesis requires coexpression of the wild-type *Mll* allele. *Cancer Cell* 17:148–59
166. Corral J, Lavenir I, Impey H, Warren AJ, Forster A, et al. 1996. An *Mll-AF9* fusion gene made by homologous recombination causes acute leukemia in chimeric mice: a method to create fusion oncogenes. *Cell* 85:853–61
167. Daser A, Rabbitts TH. 2004. Extending the repertoire of the mixed-lineage leukemia gene *MLL* in leukemogenesis. *Genes Dev.* 18:965–74
168. DiMartino JF, Miller T, Ayton PM, Landewe T, Hess JL, et al. 2000. A carboxy-terminal domain of ELL is required and sufficient for immortalization of myeloid progenitors by MLL-ELL. *Blood* 96:3887–93
169. Lavau C, Szilvassy SJ, Slany R, Cleary ML. 1997. Immortalization and leukemic transformation of a myelomonocytic precursor by retrovirally transduced HRX-ENL. *EMBO J.* 16:4226–37
170. Tenney K, Shilatifard A. 2005. A COMPASS in the voyage of defining the role of trithorax/MLL-containing complexes: linking leukemogensis to covalent modifications of chromatin. *J. Cell Biochem.* 95:429–36
171. Mueller D, Bach C, Zeisig D, Garcia-Cuellar MP, Monroe S, et al. 2007. A role for the MLL fusion partner ENL in transcriptional elongation and chromatin modification. *Blood* 110:4445–54
172. Martin ME, Milne TA, Bloyer S, Galoian K, Shen W, et al. 2003. Dimerization of MLL fusion proteins immortalizes hematopoietic cells. *Cancer Cell* 4:197–207
173. So CW, Lin M, Ayton PM, Chen EH, Cleary ML. 2003. Dimerization contributes to oncogenic activation of MLL chimeras in acute leukemias. *Cancer Cell* 4:99–110
174. Bitoun E, Oliver PL, Davies KE. 2007. The mixed-lineage leukemia fusion partner AF4 stimulates RNA polymerase II transcriptional elongation and mediates coordinated chromatin remodeling. *Hum. Mol. Genet.* 16:92–106
175. Shilatifard A, Lane WS, Jackson KW, Conaway RC, Conaway JW. 1996. An RNA polymerase II elongation factor encoded by the human *ELL* gene. *Science* 271:1873–76
176. Shilatifard A. 1998. Factors regulating the transcriptional elongation activity of RNA polymerase II. *FASEB J.* 12:1437–46
177. Shilatifard A, Conaway RC, Conaway JW. 2003. The RNA polymerase II elongation complex. *Annu. Rev. Biochem.* 72:693–715
178. Shilatifard A, Duan DR, Haque D, Florence C, Schubach WH, et al. 1997. ELL2, a new member of an ELL family of RNA polymerase II elongation factors. *Proc. Natl. Acad. Sci. USA* 94:3639–43
179. Miller T, Williams K, Johnstone RW, Shilatifard A. 2000. Identification, cloning, expression, and biochemical characterization of the testis-specific RNA polymerase II elongation factor ELL3. *J. Biol. Chem.* 275:32052–56
180. Gerber M, Ma J, Dean K, Eissenberg JC, Shilatifard A. 2001. *Drosophila* ELL is associated with actively elongating RNA polymerase II on transcriptionally active sites in vivo. *EMBO J.* 20:6104–14
181. Gerber MA, Shilatifard A, Eissenberg JC. 2005. Mutational analysis of an RNA polymerase II elongation factor in *Drosophila melanogaster*. *Mol. Cell. Biol.* 25:7803–11
182. Eissenberg JC, Ma J, Gerber MA, Christensen A, Kennison JA, Shilatifard A. 2002. dELL is an essential RNA polymerase II elongation factor with a general role in development. *Proc. Natl. Acad. Sci. USA* 99:9894–99

183. Smith ER, Winter B, Eissenberg JC, Shilatifard A. 2008. Regulation of the transcriptional activity of poised RNA polymerase II by the elongation factor ELL. *Proc. Natl. Acad. Sci. USA* 105:8575–79
184. Lin C, Smith ER, Takahashi H, Lai KC, Martin-Brown S, et al. 2010. AFF4, a component of the ELL/P-TEFb elongation complex and a shared subunit of MLL chimeras, can link transcription elongation to leukemia. *Mol. Cell* 37:429–37
185. Yokoyama A, Lin M, Naresh A, Kitabayashi I, Cleary ML. 2010. A higher-order complex containing AF4 and ENL family proteins with P-TEFb facilitates oncogenic and physiologic MLL-dependent transcription. *Cancer Cell* 17:198–212
186. He N, Liu M, Hsu J, Xue Y, Chou S, et al. 2010. HIV-1 Tat and host AFF4 recruit two transcription elongation factors into a bifunctional complex for coordinated activation of HIV-1 transcription. *Mol. Cell* 38:428–38
187. Sobhian B, Laguette N, Yatim A, Nakamura M, Levy Y, et al. 2010. HIV-1 Tat assembles a multifunctional transcription elongation complex and stably associates with the 7SK snRNP. *Mol. Cell* 38:439–51
188. Lin C, Garrett AS, De Kumar B, Smith ER, Gogol M, et al. 2011. Dynamic transcriptional events in embryonic stem cells mediated by the super elongation complex (SEC). *Genes Dev.* 25:1486–98
189. Greer EL, Maures TJ, Hauswirth AG, Green EM, Leeman DS, et al. 2010. Members of the H3K4 trimethylation complex regulate lifespan in a germline-dependent manner in *C. elegans*. *Nature* 466:383–87
190. Lee JS, Smith E, Shilatifard A. 2010. The language of histone crosstalk. *Cell* 142:682–85
191. Lee J, Kim DH, Lee S, Yang QH, Lee DK, et al. 2009. A tumor suppressive coactivator complex of p53 containing ASC-2 and histone H3-lysine-4 methyltransferase MLL3 or its paralogue MLL4. *Proc. Natl. Acad. Sci. USA* 106:8513–18
192. Shilatifard A. 2006. Chromatin modifications by methylation and ubiquitination: implications in the regulation of gene expression. *Annu. Rev. Biochem.* 75:243–69
193. Ng HH, Robert F, Young RA, Struhl K. 2003. Targeted recruitment of Set1 histone methylase by elongating Pol II provides a localized mark and memory of recent transcriptional activity. *Mol. Cell* 11:709–19
194. Bernstein BE, Kamal M, Lindblad-Toh K, Bekiranov S, Bailey DK, et al. 2005. Genomic maps and comparative analysis of histone modifications in human and mouse. *Cell* 120:169–81
195. Andreu-Vieyra CV, Chen R, Agno JE, Glaser S, Anastassiadis K, et al. 2010. MLL2 is required in oocytes for bulk histone 3 lysine 4 trimethylation and transcriptional silencing. *PLoS Biol.* 8:pii:e1000453
196. Glaser S, Lubitz S, Loveland KL, Ohbo K, Robb L, et al. 2009. The histone 3 lysine 4 methyltransferase, Mll2, is only required briefly in development and spermatogenesis. *Epigenet. Chromatin* 2:5
197. Daniel JA, Santos MA, Wang Z, Zang C, Schwab KR, et al. 2010. PTIP promotes chromatin changes critical for immunoglobulin class switch recombination. *Science* 329:917–23
198. Heintzman ND, Stuart RK, Hon G, Fu Y, Ching CW, et al. 2007. Distinct and predictive chromatin signatures of transcriptional promoters and enhancers in the human genome. *Nat. Genet.* 39:311–18
199. Heintzman ND, Ren B. 2009. Finding distal regulatory elements in the human genome. *Curr. Opin. Genet. Dev.* 19:541–49
200. Bernstein BE, Mikkelsen TS, Xie X, Kamal M, Huebert DJ, et al. 2006. A bivalent chromatin structure marks key developmental genes in embryonic stem cells. *Cell* 125:315–26
201. Herz HM, Nakanishi S, Shilatifard A. 2009. The curious case of bivalent marks. *Dev. Cell* 17:301–3
202. Akkers RC, van Heeringen SJ, Jacobi UG, Janssen-Megens EM, Francoijs KJ, et al. 2009. A hierarchy of H3K4me3 and H3K27me3 acquisition in spatial gene regulation in *Xenopus* embryos. *Dev. Cell* 17:425–34

Programming of DNA Methylation Patterns

Howard Cedar and Yehudit Bergman

Department of Developmental Biology and Cancer Research, Hebrew University Medical School, Ein Kerem, Jerusalem, Israel 91120; email: cedar@cc.huji.ac.il, yehuditb@ekmd.huji.ac.il

Keywords

maintenance, repression, chromatin, development, reprogramming, imprinting

Abstract

DNA methylation represents a form of genome annotation that mediates gene repression by serving as a maintainable mark that can be used to reconstruct silent chromatin following each round of replication. During development, germline DNA methylation is erased in the blastocyst, and a bimodal pattern is established anew at the time of implantation when the entire genome gets methylated while CpG islands are protected. This brings about global repression and allows housekeeping genes to be expressed in all cells of the body. Postimplantation development is characterized by stage- and tissue-specific changes in methylation that ultimately mold the epigenetic patterns that define each individual cell type. This is directed by sequence information in DNA and represents a secondary event that provides long-term expression stability. Abnormal methylation changes play a role in diseases, such as cancer or fragile X syndrome, and may also occur as a function of aging or as a result of environmental influences.

Contents

INTRODUCTION	98
PRINCIPLES OF METHYLATION	98
Methylation Metabolism	98
Maintenance of DNA Methylation Patterns	99
Methylation and Gene Repression	100
DNA Methylation and Chromatin Structure	100
DNA METHYLATION PATTERNS DURING DEVELOPMENT	101
Erasure	101
Generation of a Bimodal Methylation Pattern	102
Mechanism of CpG Island Protection	102
Sequence Information Generates Methylation Patterns	103
Global Repression	103
Postimplantation Methylation Changes	104
Tissue-Specific Methylation Patterns	105
ABNORMAL DNA METHYLATION	105
De Novo Methylation in Cancer	105
Role of Methylation in Cancer	106
Fragile X Syndrome	107
DEMETHYLATION	107
Active Demethylation	107
Demethylation by Repair	108
Reprogramming	109
METHYLATION DURING GAMETOGENESIS	109
Erasure of Methylation Patterns	109
Imprinting	110
NONPROGRAMMED INFLUENCES ON DNA METHYLATION	110

5mC:
5-methylcytidine

INTRODUCTION

Although the existence of methylated nucleotides has been known for a long time, studies of its function in vivo mostly concentrated on its role as a restriction modification mechanism in bacteria, where 6-methyladenine or 5-methylcytidine (5mC) appear at fixed nucleotide sequences throughout the genome. Nucleotide analysis of animal and plant DNA indicated that these organisms also have a fair amount of 5mC. Strikingly, when this DNA was subjected to nearest-neighbor analysis, a method that detects and quantifies the bases located 5' adjacent to any labeled nucleotide, it was determined that almost all of this methylation is concentrated in the dinucleotide sequence CpG (1). Furthermore, in any particular cell type, only a portion of the CpGs is actually methylated (2). Taken together, these observations on the placement and distribution of methylation strongly suggested that this modification must play a different biological role in animals and plants as compared to bacteria.

PRINCIPLES OF METHYLATION

Methylation Metabolism

In order to gain some insight into the significance of this modification system, it was important to ask whether DNA methyl groups are freely metabolized throughout the genome or, alternatively, are located at fixed positions in each cell type. This was accomplished by using bacterial restriction enzymes to assay methylation at specific sites in the DNA (3). Enzymes such as HpaII (CCGG) or HhaI (GCGC), which have a CpG dinucleotide in their recognition sequences, are inhibited by methylation on the internal C residue. This unmethylated site can be cut, whereas the same site in a methylated form remains undigested. Using this assay, it was shown that different CpG residues in the genome are either highly methylated or present in an unmethylated form, strongly suggesting that methyl groups have fixed positions (4).

One of the major steps in understanding how methyl groups are organized and managed in the genome came about through the use of DNA-mediated gene transfer to stably insert foreign sequences into the endogenous genome of fibroblast cells in culture. Plasmid DNA derived from *Escherichia coli* is unmethylated at all its HpaII (CCGG) recognition sequences, but these same sites could be artificially methylated in vitro using the specific methylases found in the same *Haemophilus* parainfluenza bacteria (5). Strikingly, unmethylated DNA remained unmodified in these cells, but methylated DNA retained its original methyl groups, even after many generations of growth in culture. These studies clearly showed that methylation is not the result of ongoing transient metabolism, but rather methylation marks must be located at specific sites in the genome. Furthermore, these patterns appeared to be stably maintained through cell division (6–8).

Maintenance of DNA Methylation Patterns

In addition to demonstrating that methylation patterns can be faithfully maintained, these transfection experiments also revealed important principles about the mechanism for this process. Because methylation was added artificially in vitro prior to the introduction of DNA into cells, maintenance of the methylation state clearly had nothing to do with the DNA sequence. This clearly suggested that modification patterns in somatic cells do not come about by the recruitment of factors through specific local sequence motifs. Rather, it appeared that maintenance must be accomplished by some sort of autonomous mechanism that can actually read and copy the modification pattern per se at the time of replication.

The key to understanding this process evolved from studies of the enzymatic DNA methylation activity found in crude extracts from somatic cells. These experiments showed that, although completely unmodified DNA is a rather poor substrate for methylation, hemimethylated DNA (i.e., methylated on one strand) works extremely efficiently with a 100-fold better K_m for this reaction (9). As previously noted, methyl moieties are located within CpG residues, and because these sites have strand symmetry, these results implied that DNA must be normally methylated on both strands. During replication, synthesis of the new strand generates a hemimethylated site that is then recognized by the maintenance methylase and therefore is methylated on the opposite strand, thereby regenerating the original bimethylated state originally present in the mother cell. In contrast, unmethylated CpG residues still appear completely unmethylated during replication and therefore do not constitute a substrate for the enzyme. In this way, the methylation pattern on the native strands serves as a template for regenerating methylation patterns during replication (10).

It is clear that the basis for this semiconservative mechanism derives from the fact that methyl groups in CpG residues are symmetrically disposed on both strands of the DNA. Strong support for this idea came from studies of plant DNA. Nearest-neighbor analysis of this DNA originally indicated that all C-containing dinucleotides are partially methylated. Upon closer examination, however, it could be seen that all of these methyl groups are actually located in CpNpG trinucleotide symmetrical sequences where N can be any of the four bases. Thus, every instance of modified C on one strand of the DNA is opposed by an equivalent methyl group two nucleotides over on the other strand (11). At the time of replication, methylation on the native DNA strand can thus serve as a template for complementing the methylation pattern on the newly synthesized strand, thus reproducing the methyl profile present in the mother cell.

It is now known that in animal cells the maintenance reaction is carried out by the enzyme Dnmt1 (12), which is perpetually localized to replication foci (13) and therefore is constantly available to provide CpG maintenance function. Exactly how hemimethylated DNA is recognized is not well understood (14, 15), but recent studies on Dnmt1 indicate that it does not operate alone. Rather, it probably works as part of a larger complex that includes other essential factors, such as Np95 (UHRF1) (16–18). Ultimately, these complexes constitute the biochemical basis for epigenetic

CpG island: a region that contains a high density of CpGs and is located at the promoters of many genes

memory by providing an enzymatic platform for copying methylation patterns in a semiconservative manner in the same way that DNA sequence itself is reproduced from generation to generation.

Methylation and Gene Repression

Once it became possible to use restriction enzymes for analyzing the DNA methylation patterns of specific endogenous genes, it became immediately obvious that this modification is correlated with gene repression. Tissue-specific genes were found to be highly methylated in most tissue samples but undermethylated in their tissue of expression (19). At the same time, housekeeping genes were shown to have a unique CpG island promoter structure, which is constitutively unmethylated in every cell (20, 21). Furthermore, early genomic studies based on analysis of total mRNA even demonstrated that active genes, in general, are undermethylated as compared to inactive DNA (22). This correlation has also been confirmed over and over again using more comprehensive assays, such as bisulfite analysis to measure endogenous gene DNA methylation (23).

Although these studies demonstrate a strong association with repression, it was still necessary to actually test whether DNA methylation has a causal effect on gene expression, and this was ultimately accomplished by employing DNA-mediated gene transfer. These experiments showed that unmethylated genes are actively transcribed when inserted into the genome, whereas the exact same sequences are repressed if the inserted DNA had been premethylated in vitro (6, 7, 24). Because the only difference between these templates is the presence of methyl groups, these studies provided convincing proof that DNA methylation itself is responsible for gene inhibition. Even though these experiments were based exclusively on the analysis of exogenous gene sequences, the implication of these results is that DNA methylation can explain why the homologous endogenous genes are completely silenced in these same exact cells, and this concept has been reinforced by subsequent transfection studies in vitro (25) and transgenic studies in vivo (26, 27), demonstrating that methylated DNA molecules are repressed.

When DNA methylation was first being explored as a mechanism of gene repression, it was commonly thought that methyl groups work by preventing the binding of key transcription factors much in the same way that bacterial restriction enzymes cannot cut when their recognition site is methylated. Although this may be true for a small number of specific regulatory factors, this is certainly not a general phenomenon. Many transcription factors do not have CpGs in their binding sites, and even when present, as is the case for Sp1 (28), DNA methylation does not necessarily inhibit their binding.

DNA Methylation and Chromatin Structure

A great deal of evidence points to the idea that the major effect of methyl groups is to model chromatin structure, and this may be carried out at many different levels. Microinjection experiments, for example, have demonstrated that a gene template inserted directly into the nucleus of cells in culture is initially unaffected by DNA methylation. Only after these substrates have had a chance to get packaged into a chromosomal structure does one begin to see the effects of this modification on transcription (29). The most convincing evidence for this idea comes from DNA-mediated cell transfection studies showing that unmethylated substrates are packaged into an open chromatin structure following their integration into the genome, whereas the exact same DNA remains completely resistant to DNase I if it is methylated (30). Because these experiments were carried out using bacterial DNA sequences that do not harbor any eukaryotic regulatory information, one can conclude that methyl moieties themselves must be responsible for generating a closed chromatin structure regardless of sequence context.

Although the precise mechanisms by which DNA methylation affects chromatin packaging have not been completely elucidated, many studies have concentrated on nucleosome structure. Early experiments indicated that most methylated CpGs are concentrated within the central core of nucleosomes, as opposed to internucleosomal regions, suggesting that the positioning of this modification may have an intrinsic effect on where nucleosomes reside on the DNA (31, 32), and these results have recently been confirmed by sophisticated genome-wide analysis (33). In addition to this intrinsic effect, methylation may also play a role in regulating other factors that ultimately impinge on nucleosome displacement.

Another way that methylation mediates gene repression is through methyl-binding proteins, such as MeCP2, MBD2, and MBD3 (34). These factors can specifically recognize methyl moieties and, once bound, may model local chromatin structure, perhaps by recruiting modifying enzymes that bring about overall histone deacetylation (35, 36) or the methylation of specific lysine residues (37). Alternatively, methylation may actually prevent the binding of chromatin proteins, such as boundary-forming CTCF (38) or Cfp1 (39), which is known to be exclusively located on unmethylated CpG islands.

These observations on chromatin structure provide important insight on the significance of DNA methylation as a mechanism for stable gene silencing. Although gene expression profiles are mediated by the initial interactions with transcriptional regulatory factors, it is most likely the state of gene accessibility at the level of chromatin structure that actually determines gene activity or repression. In principle then, the chromatin conformation could provide a mold for preserving long-term gene expression profiles. The problem with this idea is that these basic structural features get disrupted in the wake of DNA polymerase advancement during every round of replication, and the newly made DNA must then be repackaged into chromatin (40). There is, as yet, no well-established mechanism for reproducing chromatin profiles at the replication fork (see Reference 41). Because the underlying DNA methylation pattern is autonomously copied following replication, it is possible that this modification pattern may play a role both in the generation of active chromatin over undermethylated regions as well as in the formation of closed chromatin over methylated DNA (42). It is probably in this manner that DNA methylation mediates stable gene repression.

DNA METHYLATION PATTERNS DURING DEVELOPMENT

Erasure

Studies in tissue culture have been very helpful in characterizing basic concepts of DNA methylation metabolism and function, but these experiments do not reveal very much about the precise role of this modification during normal development in vivo. To understand this process, it was necessary to first map the dynamic pattern of DNA methylation as a function of embryogenesis and organogenesis (**Figure 1**). Early studies using restriction enzymes indicated that methyl groups inherited from parental gametes are largely erased in the preimplantation morula and blastula (43–45). This process of erasure appears to take place in two distinct stages, with much of the paternal genome undergoing active demethylation, which begins in the zygote (46), and further demethylation takes place during the first few early embryonic replication cycles, perhaps passively as a result of Dnmt1 relocation from the nucleus to the cytoplasm (47). Although the level of DNA methylation in the blastula is very low, the exact pattern of this modification in these preimplantation cells has not yet been accurately determined. Differentially methylated regions located at imprinting centers must have a mechanism to preserve them through this erasure stage (48, 49), and other studies indicate that additional specific regions and a number of different repeated sequences may also be partially protected from this process (50, 51).

Imprinting centers: regulatory regions that are marked in the gametes and that control domain-wide allele-specific expression

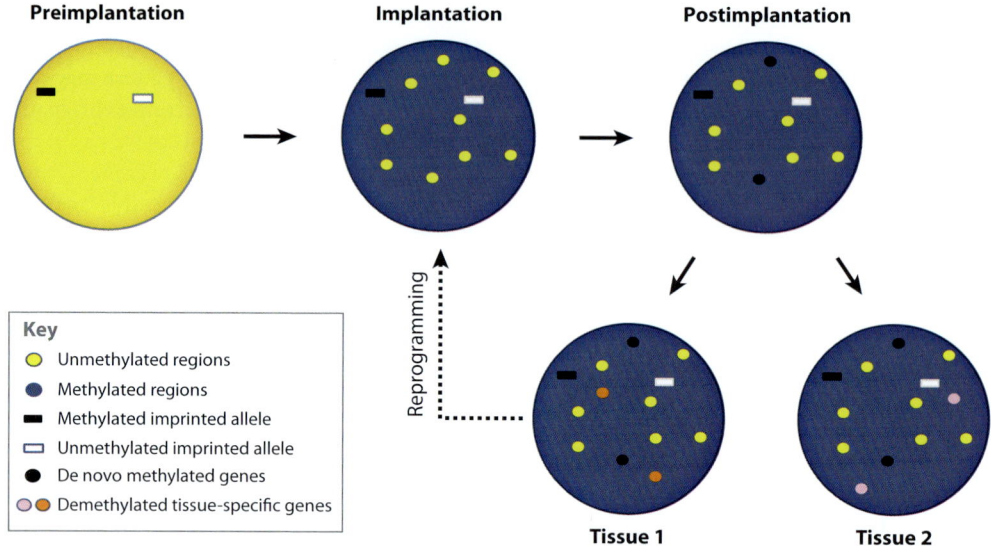

Figure 1

The generation of DNA methylation patterns during development. Almost all methylation in the gametes is erased (*yellow*) in the preimplantation embryo, but imprinting centers retain methylation on one allele (*black*). At the time of implantation, the entire genome gets methylated (*blue*), but the CpG islands are protected (*yellow circles*). Postimplantation, pluripotency genes are de novo methylated (*black*). Tissue-specific genes undergo demethylation (*orange* in Tissue 1, *pink* in Tissue 2) in their cell type of expression. Imprinting centers remain differentially methylated throughout development. Somatic cell reprogramming by induced pluripotent stem cells or fusion resets the methylation pattern of somatic cells to the stage of implantation.

Generation of a Bimodal Methylation Pattern

In the next stage of embryogenesis at about the time of implantation, the entire genome gets remodified through a dramatic wave of de novo methylation (**Figure 1**), and genetic analyses indicate that this is mediated by Dnmt3a and -3b, which are present at high concentrations at this stage of development (52). Strikingly, this process appears to generate a bimodal pattern of methylation, with most sequences becoming methylated to high levels (>80%), while CpG island-like windows are protected and therefore remain unmethylated (23). This concept of what happens in the early embryo derives from the observation that CpG island-like windows are constitutively unmodified in a large number of adult cell types, whereas other regions are constitutively methylated (53). This includes repeated sequences in satellite DNA that make up over 50% of all methyl moieties in the nucleus (32). Because all of these tissues are ultimately derived from the early inner cell mass at the time of implantation, it is reasonable to assume that the bimodal pattern is generated at this early stage and then maintained largely intact over succeeding cell divisions. This overall picture is also supported by the observation that mouse and human embryonic stem (ES) cells derived directly from this stage of development already show this same bimodal profile of DNA methylation (53, 54).

Mechanism of CpG Island Protection

Although the precise mechanism for CpG island protection is not known, it is quite clear that this is mediated by common sequence motifs present within the islands. This was initially deduced from transfection studies in ES cells. Unlike somatic cells, these cells can actually de novo methylate exogenously introduced DNA sequences and, at the same time, also have the ability to recognize and protect CpG islands (55). Experiments using these cells in vitro indicated that Sp1 motifs play a role in this protection process, and this was further confirmed in vivo using transgenic mice (56, 57). Furthermore, these elements were even able to protect a non-CpG island sequence from de novo methylation at the time of implantation after being introduced as a transgene in one-cell embryos (26).

Although these studies indicate, in principle, that *cis* acting sequences are involved in protecting specific CpG islands from de novo methylation at the time of implantation, the general sequence rules for this mechanism have not been deciphered. Genome-wide studies indicate that almost all unmethylated CpG islands contain transcription start sites, are marked with H3K4me3 in ES cells (58, 59), and harbor many transcription factor–binding motifs (53). This suggests a model whereby sites of RNA polymerase binding in the blastocyst may serve as a mark for preventing de novo methylation during implantation. According to this idea, generation of a bimodal DNA methylation pattern in early development essentially serves to perpetuate the factor-mediated basal transcription profile of the preimplantation embryo.

Sequence Information Generates Methylation Patterns

By comparing constitutively unmethylated and methylated DNA sequences, it has been possible to derive a well-defined accurate algorithm that can distinguish between these two types of sequences, and by means of this mathematical formula, it is possible to accurately predict the basal methylation state of any sequence window in the genome (53). These studies serve to emphasize the idea that, although DNA methylation patterns represent an epigenetic mark that is not in itself inherited from the parents, the ultimate modification profile is fully determined by underlying sequence information within the DNA itself.

Once this epigenetic code is deciphered, it should be possible to predict the full dynamics of DNA methylation during development.

Global Repression

De novo establishment of a bimodal methylation pattern in the early embryo has far-reaching implications for the management of gene regulation in the organism. Unlike simple biological systems where almost the entire genome is transcribed, even though only a relatively small number of genes are specifically recognized and repressed, animals have a large genome in which over 50% of the genes may be silenced in any given cell type. This type of repression pattern clearly requires a different mechanism that can carry out gene repression in a global manner without the need to recognize specific motifs on every target gene. According to this scheme, at the time of implantation, the entire genome undergoes global modification while CpG islands are protected. This insures that housekeeping genes that have CpG island promoters will be kept unmethylated and may also provide a mechanism for setting up regulatory modules, such as enhancers, in a permanent active configuration so they can bind key transcription factors at later stages of development (60).

At the same time, this wave of methylation guarantees that genes with nonisland promoters will automatically get modified and therefore be repressed in most tissues of the body (26, 27). While this relatively sparse background methylation may only have a small influence on the transcription levels of tissue-specific genes in their nonexpressing cell types, it also brings about a dramatic repression of endogenous viral sequences and foreign elements throughout the genome (61). It may also cause blanket inactivation of many cryptic promoters. Because, at this early stage, the entire genome undergoes DNA modification in a nonspecific manner, this event probably represents the only time in development when methylation itself actually serves as the primary cause of silencing. Even in this case, however, DNA methylation may

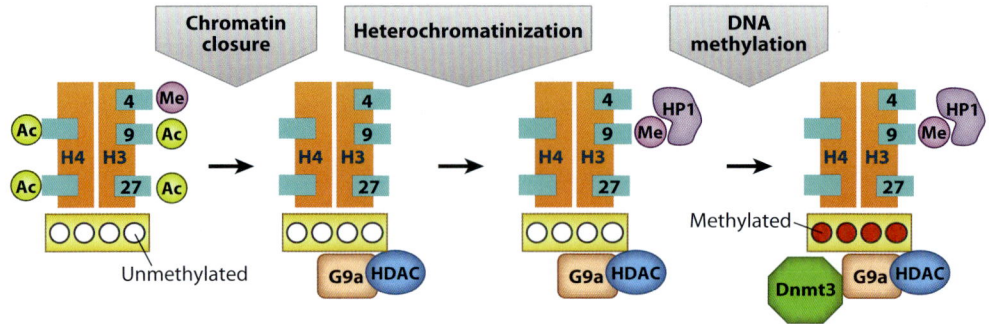

Figure 2

Inactivation of pluripotency genes. The *Oct-3/4* gene is unmethylated (*white circles*) at the implantation stage (ES cells) and active. With the onset of differentiation, transcription is first turned off by repressor proteins. The histone methylase G9a, together with a histone deacetylase (HDAC) and other chromatin modifying enzymes, is then recruited and brings about chromatin inactivation. In the next step, G9a methylates K9, and this serves as a binding site for heterochromatin protein 1 (HP1), thereby generating local heterochromatin. In the last step, G9a recruits de novo methylases (Dnmt3a and -3b) to cause promoter methylation (*red circles*).

represent only one of multiple factors that can mediate long-term repression through its effects on chromatin structure (62).

Postimplantation Methylation Changes

Following implantation, the animal genome can undergo additional changes in methylation, but these events are all of a tissue-specific or gene-specific nature. Perhaps one of the most significant of these modifications is that involved in the silencing of pluripotency genes (**Figure 2**). Genes, such as *Oct-3/4* and *Nanog*, for example, are active in the early embryo and still maintain an unmethylated transcribed promoter at the time of implantation. Following this stage, however, these genes undergo inactivation, thereby setting the stage for embryonic differentiation (63). Using ES cells as a model system, it has been possible to learn about the mechanism of this repression process, which occurs in three steps.

With the onset of differentiation, *Oct-3/4* transcription is initially turned off through a simple repression-factor mechanism. In the second step, the histone methyltransferase G9a is specifically recruited to the promoter of these genes, thus facilitating histone deacetylation and subsequent methylation of H3K9, which then binds HP1, forming heterochromatin (64). Finally, this same G9a complex can also recruit the de novo methylases Dnmt3a and -3b, thus bringing about methylation of the pluripotency-gene promoters themselves (**Figure 2**) (65–67). Although this de novo methylation represents a secondary event, it still plays an important role in stabilizing the silent state. Indeed, experiments in vitro, using mutant ES cells, clearly show that both the transcriptional repression and heterochromatinization steps are easily reversed in culture, but once DNA methylation has occurred, it is no longer possible to return to the pluripotent state (64, 65). In a similar manner, it has been demonstrated that viruses, initially inactivated by transient mechanisms in ES cells, may then become permanently silenced following differentiation-mediated de novo methylation (68).

Another major event that occurs soon after implantation throughout all cells of the embryo is the inactivation of one X chromosome in female animals. Here too, repression takes place as a multistep process, beginning with rapid chromosome-wide changes in replication timing, gene expression, and chromatin structure (69), followed by de novo methylation of CpG island promoters (70). In this case, as well, DNA

Pluripotency genes: genes, such as *Oct-3/4* and *Nanog*, that are required for increased potency in the early embryo

modification is probably mediated by histone methylases capable of generating heterochromatin and then recruiting Dnmts that carry out targeted local methylation many days after the initial inactivation event (71). These findings once again serve to emphasize the important concept that DNA methylation often plays a role as a secondary mechanism programmed to insure long-term silencing.

Tissue-Specific Methylation Patterns

Other alterations of the basic bimodal methylation pattern occur in a cell-type-specific manner. Many genes that are silenced throughout the organism and expressed specifically in a single tissue have non-CpG island promoters that automatically undergo de novo methylation at the time of implantation. During tissue development, these genes have to be specifically recognized by cell-type-specific factors that apparently recruit the molecules needed for demethylating their promoters (72), decondensing the overlying chromatin structures and making them accessible to the transcription machinery (**Figure 1**). A number of experiments in different cell types have demonstrated that this demethylation occurs in an active manner that does not require DNA replication (73, 74) and is mediated by specific *cis* acting sequences (75) and *trans* acting factors (76). Because this type of demethylation is specific and requires prior recognition, it cannot be considered the primary underlying cause of gene activation. Once demethylation has occurred, however, it is possible that this serves to stably maintain chromatin accessibility.

In a manner similar to demethylation, genes may also undergo tissue-specific de novo methylation at CpG island sequences that were originally protected at the time of implantation (77). Interestingly, these targets are not necessarily associated with promoters, and many are actually located within coding regions, where their methylation is associated with gene activation (53). Although the mechanism for this reverse effect is not known, one possibility is that these internal CpG islands carry promoters for antisense transcripts whose methylation would inhibit the production of these repressive RNA molecules. Alternatively, these regions may simply contain binding sites for transcriptional repressors. A large number of these targets have been shown to undergo de novo methylation in brain tissue, and genetic studies demonstrate that this may have profound effects on gene expression and function (78).

Although the precise mechanism for targeting de novo methylation in vivo is not yet known, it appears that this may be largely mediated by the polycomb complex (**Figure 3**). Almost all of the sites that undergo this type of modification are known polycomb targets (53), and it has been shown that the polycomb complex has an inherent ability to recruit Dnmt3a and -3b (79). It is still not known, however, which *cis* acting elements and *trans* acting factors may be involved in determining cell-type and gene-specific modification.

ABNORMAL DNA METHYLATION

De Novo Methylation in Cancer

In the same way that normal development is dependent on the proper programming of methylation, abnormal cell behavior is also associated with alterations in DNA modification patterns. A prime example of this phenomenon is cancer. All tumors that have ever been examined show changes in DNA methylation, suggesting that this may represent a basic element of cancer biology that has a significant effect on tumor pathology. Early studies indicated, for example, that cancer is characterized by widespread demethylation (80), and using more advanced technologies, it has been possible to map this effect to specific blocks covering a large portion of the genome (81). In addition, almost all tumors undergo de novo methylation at specific sites (82). It has been commonly assumed that this modification process occurs in a completely random manner, with the final pattern seen in the tumor being determined mainly by

> **Polycomb complex:** a complex of proteins bound at specific genes where they cause repression, probably through histone modification

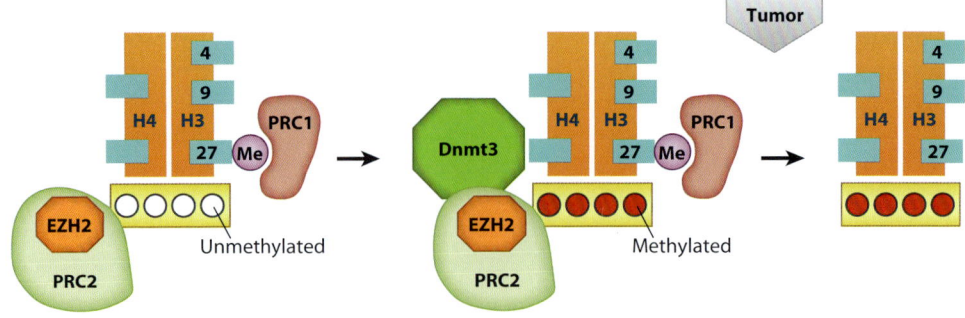

Figure 3
Targeted de novo methylation. Genes targeted by the polycomb complex, PRC2, normally have unmethylated CpG island promoters (*white circles*) but are repressed by virtue of the histone methylase EZH2-mediated methylation of H3K27, which then recruits the chromodomain-containing complex, PRC1, generating a form of heterochromatin. Targeted de novo methylation (*red circles*) can occur at specific sites during normal development or abnormally in cancer, and this probably occurs because EZH2 itself recruits Dnmt3a and Dnmt3b. In some tumors, it has been found that PRC2 may no longer remain bound to the methylated islands, leading to a situation where flexible polycomb repression is replaced with a more stable form of methylation-mediated repression (139).

selection, thus explaining why many tumor suppressor genes are found methylated in various cancer types. Although this may represent a small part of the methylation picture, more recent results have shown that DNA methylation in cancer is actually quite extensive, with hundreds of CpG islands in the genome getting methylated in a largely biallelic manner (83).

According to this selection idea, DNA methylation in the tumor serves to inactivate genes that would normally be expressed in the normal tissue, but transcription analysis showed that most methylation targets are actually inactive at the beginning (84). It appears that many of these sites harbor polycomb, and it may be this complex that actually recruits the methylases that then bring about this abnormal modification (**Figure 3**) (85–87). From this point of view, de novo methylation in cancer behaves in a programmed manner, in the sense that target sites are already predetermined independent of whether these genes are active or inactive or whether they actually play some role in tumorigenesis (88). Studies on human colon cancer suggest that de novo methylation occurs very early in the process of tumor evolution and that some of the polycomb target sites are associated with genes essential for epithelial cell differentiation (85). It is thus possible that DNA methylation plays a role in tumorigenesis by replacing the normally flexible polycomb repression mechanism with a more permanent silencing mode, thus inhibiting crypt cells from undergoing final differentiation and promoting the survival of proliferating precursors.

Role of Methylation in Cancer

The overall part played by DNA methylation in cancer may best be studied by using a genetics approach in mouse model systems. Min⁻ mice carry a deletion in one allele of *Apc*, a key tumor suppresser gene that is deleted in almost every case of human colon cancer. These animals develop thousands of intestinal microadenomas owing to spontaneous deletion of the second *Apc* allele, but only about 5% of these actually develop into full-blown adenomas (89). When these mice are genetically manipulated to express lower levels of Dnmts, or when they are treated from birth with low doses of 5 azacytidine, the number of adenomas that is generated is reduced by almost 100-fold (90), whereas the number of microadenomas is unaffected (91). These observations suggest that by preventing de novo methylation after birth, it is possible to reduce the incidence of tumor formation. Similar results have been obtained with other mouse tumor models (92).

When taken together, these observations support a model for intestinal cancer that involves two separate and independent molecular events; both of these are necessary for tumor formation. Spontaneous deletion of *Apc* is what brings about the generation of microadenomas, but abnormal de novo methylation is required to convert this into an authentic adenoma. In keeping with this concept, min$^-$ mice that overexpress the de novo methylase, Dnmt3b, have increased de novo methylation at tumor target genes in several of their tissues and develop twofold more adenomas than are normally seen in this model system (89).

It has been known for a long time that cells growing in culture have abnormal methylation patterns, largely characterized by excess de novo methylation at CpG islands (93, 94). This is much different than the profile observed in ES cells growing in culture, probably because these stem cells have the molecular machinery required for setting up and preserving the correct bimodal pattern characteristic of the early embryo. Indeed, once these stem cells undergo differentiation, thereby losing this ability to protect CpG islands, they also become subject to abnormal sequence-directed de novo methylation (59, 95). In these cases, as well, polycomb-binding sites represent the major targets for new methylation (59, 96).

Fragile X Syndrome

Abnormal de novo methylation also plays a role in the pathogenesis of fragile X syndrome, causing the *FMR1* gene, carrying an expanded triplet repeat, to become repressed at a very early stage in development. Although the precise mechanism for this process has not been elucidated, inactivation is known to be accompanied by H3K9me3 heterochromatinization (97, 98). Taken together, these findings suggest that programming of abnormal modification may involve a common mechanism (71) whereby various histone methylases mediate unscheduled DNA methylation and repression.

DEMETHYLATION

Demethylation is a common event that occurs at a number of different stages during normal development. Early in preimplantation development, for example, the entire genome is subject to a wave of demethylation that apparently erases almost all of the methyl marks inherited from the parents, and a very similar process also takes place during early stages of gametogenesis (43, 44). In addition to this global form, there are also many instances of site-specific demethylation. This usually involves tissue-specific genes that become automatically methylated at the time of implantation and then undergo demethylation in their cell type of expression (19).

Active Demethylation

Extensive efforts over several decades have been invested in deciphering the biochemical mechanisms that may be involved in this demethylation process. Considering the way DNA methylation patterns are maintained throughout cell division, it was originally suggested that demethylation may occur through a passive mechanism whereby selected modified sites may not be efficiently recapitulated during replication. Although this appears to occur in an artificial manner when cells are treated with 5 azacytidine, early studies already demonstrated that demethylation in vivo largely occurs in an active manner, even in the absence of replication and cell division (73, 74).

The most obvious biochemical mechanism for demethylation would be the reverse of DNA methylation, involving direct removal of methyl groups from the 5′ position of cytosine. This type of demethylation has indeed been observed in extracts from cancer cells, where it was proposed that MBD2 may have an enzymatic activity that can combine methyl groups with H_2O to generate methanol (99). Even though this type of reaction may be chemically and

> **Fragile X syndrome:** a developmental disease caused by a triplet repeat (CGG) expansion in the *FMR1* gene that brings bout mental retardation

energetically feasible (100), other laboratories have not yet been able to reproduce these studies (101).

Demethylation by Repair

An alternative mechanism for demethylation was suggested by experiments showing that 5mC may be a weak substrate for glycosylases. On this basis, it was proposed that the apyrimidinic base product of this reaction may then undergo DNA patch repair, which would result in the replacement of this damaged nucleotide with unmethylated cytosine, thereby completing the demethylation process (102). This idea was originally confirmed in studies showing nucleotide substitution both in vivo (103) and in vitro (104) and has now received wide support from a number of different cellular and biological systems showing that glycosylases (105–107) and excision-repair enzymes may indeed be essential for demethylation (108, 109).

One of the major problems with this proposed demethylation pathway is that 5mC is actually only a very weak substrate for the known glycosylases. It now appears, however, that demethylation may take place through a more complex process that involves first modifying 5mC to prepare it for glycosylation. 5-hydroxymethylcytidine (5hmC), for example, has been detected at a number of sites known to be undergoing demethylation in vivo and in vitro (110–112), suggesting that this may represent a necessary intermediate in the demethylation pathway. It has also been shown that the cytosine deaminase, Aid, may be required for demethylation during spermatogenesis (113).

Taken together, these studies suggest that demethylation may occur through a series of biochemical steps, ultimately leading to cytosine substitution by repair (**Figure 4**). In one possible scenario, 5mC is first hydroxylated, thereby generating 5hmC (114). This modified nucleoside may then be recognized by a deaminase, which coverts it to 5-hydroxymethyluridine (5hmU), a nucleotide variant that is normally removed by specific glycosylases to generate apyrimidinic acid sites, which are subsequently repaired. In keeping with this idea, many of the enzymatic components involved in this pathway have been found together in distinct complexes, apparently making up a type of demethylation machine (107, 115). The recent discovery of other ten-eleven-translocation (Tet)-catalyzed chemical changes at the 5′ position raises the possibility that demethylation may even involve additional intermediate reaction steps (116).

Evidence for this type of multistep repair process has been derived from a number of different demethylation examples in vivo, including the global removal of methyl groups seen during gametogenesis (108), the general paternal-genome demethylation that takes

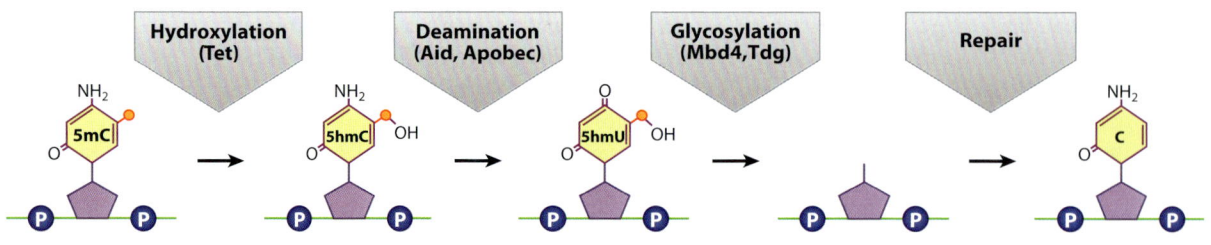

Figure 4

The demethylation pathway. Active demethylation may take place through a series of biochemical steps that modify 5-methylcytidine (5mC) to make it a recognized substrate for removal and replacement by repair with unmethylated cytosine. In the first step, a Tet enzyme brings about hydroxylation of the methyl group to form 5-hydroxymethylcytidine (5hmC). Deamination then occurs through the involvement of activation-induced deaminase (Aid) or Apobec family proteins to generate 5-hydroxymethyluridine (5hmU), which then becomes a substrate for a glycosylase (Mbd4 or Tdg). The resulting apyrimidinic acid residue is then replaced with C by means of path repair base or nucleotide excision repair (BER or NER).

place in the zygote (111), as well as tissue-specific demethylation (105, 114). This process evidently works in a modular manner. Thus, even though all demethylation events may occur through the same sequence of biochemical reactions, the enzymes that mediate these steps may vary in different cell types and at different stages of development. Indeed, hydroxymethylation (Tet1, Tet2, Tet3), deamination (Aid, Apobec), and glycosylation (Tdg, Mdb4) can all be carried out by different family members.

Reprogramming

As noted above, genes required for pluripotency, such as *Oct-3/4* or *Nanog*, undergo targeted methylation postimplantation and, in this way, prevent differentiated cells from undergoing dedifferentiation back to their pluripotent state. There is no question that the methylated state of these genes is what provides stability to the differentiated phenotype during normal development (65). Despite this layer of protection, it is still possible to reprogram somatic cells artificially either in vitro (117) or by nuclear transfer in vivo (118). How does this occur? A wide variety of studies in a number of different organisms have shown that somatic cell nuclei transplanted into primed denucleated oocytes undergo reprogramming to totipotency and then serve as the genetic source for generating an entire organism, including extraembryonic tissues (118). It appears that this is made possible by the global demethylation that takes place in the preimplantation embryo, which first enables trophectoderm differentiation by turning on the initially silenced *Elf5* gene (119), and then activates key early embryonic genes, such as *Oct-3/4* and *Nanog* (120, 121).

Reprogramming of somatic nuclei in vitro can be carried out in two different ways. The most efficient method is to fuse somatic cells with ES cells. When this is done, *Oct-3/4*, *Nanog*, and other pluripotency CpG island-like genes undergo rapid demethylation, thus setting in motion the regulatory network that defines the early ES cell phenotype (122).

Previous experiments have shown that mouse ES cells have an activity capable of demethylating a wide variety of CpG islands in a reaction that is targeted by specific *cis* acting sequences, such as Sp1 (56, 57), and this probably includes the regulatory regions of *Oct-3/4* and other pluripotency genes (123). A similar process may occur in induced pluripotent stem (iPS) cells, where the addition of key stem-cell transcription factors apparently serve to turn on a set of endogenous ES master genes, including the factors needed to carry out CpG island-specific demethylation (**Figure 4**). It is interesting that, although demethylation of CpG islands generates a pluripotency phenotype, it is not sufficient for totipotency. This may be because genes like *Elf5* that are required for trophoblast formation (119) in the preimplantation embryo have a non-CpG island promoter, which remains methylated in ES cells.

METHYLATION DURING GAMETOGENESIS

Erasure of Methylation Patterns

In addition to the DNA methylation changes that occur during early embryonic development, the process of gametogenesis is also characterized by an elaborate program of modification dynamics. Germ line cells probably emerge from the postimplantation embryo carrying the full bimodal methylation pattern and migrate along the genital ridge to ultimately take part in forming the gamete (124). These primitive germ cells undergo global demethylation at about 10–12 days post coitum (44). One of the important consequences of this event is that it facilitates the removal of all methyl sites associated with imprinted genes, and it is this erasure that allows reestablishment of new parent-specific methyl imprints at late stages of gametogenesis, either in the oocytes or sperm (125).

Following this global demethylation, the entire genome gets remethylated, while CpG islands are protected in a manner reminiscent of the process that occurs during implantation

Induced pluripotent stem (iPS) cells: these cells are generated by adding pluripotency factors to somatic cells, causing them to be reprogrammed

in the early embryo (44). It thus appears that the cycle of erasure followed by reestablishment of a new methylation pattern is a basic aspect of epigenetic regulation. Both in man and in mouse, there are a large number of genes that emerge from spermatogenesis in an unmethylated form, even though they are fully methylated in all somatic cells, and many of these indeed have a testes-specific pattern of expression (53, 58). This profile of undermethylation may be generated by one of two possible mechanisms. Some genes clearly get methylated as part of the remethylation process and then undergo tissue-specific demethylation later in spermatogenesis (126). It is also possible, however, that some genes are actually protected from de novo methylation during early gametogenesis and then remain this way during the formation of a haploid genome in sperm. Recent evidence indicates that CpG islands also undergo specific methylation changes during oogenesis (51).

Imprinting

One of the important events that occurs during gametogenesis is the establishment of imprinting. Over 100 imprinted genes have been identified in mammals, and these are all clustered at well-defined loci in the genome, where *cis*-acting imprinting centers regulate their allele-specific expression. DNA methylation plays a major role in orchestrating this process. At most imprinted loci, the center undergoes de novo methylation during late oogenesis but remains unmodified during spermatogenesis. However, there are also a few loci where the imprinting center actually becomes methylated postmitotically in the testis, while remaining unmethylated in the oocytes. These parental-specific methylation profiles are generated by a combination of paternal or maternal gametic factors that either promote or prevent de novo methylation (127).

Once formed, differential methyl imprints are preserved in an allele-specific form throughout early embryogenesis and then serve as an epigenetic mark that either turns on or turns off the imprinting center's regulatory activity controlling all of the genes in the entire locus. It is not yet clear why these methylation patterns do not get erased in the preimplantation embryo (49) and do not lose their allele specificity during the wave of de novo methylation at the time of implantation, but it is possible that other maintenance mechanisms, such as replication timing (128) or histone modification (129), may play a supporting role. Following implantation, however, there is no question that allele-specific methylation is carried on through every replication cycle by the maintenance methylase, Dnmt1 (130).

NONPROGRAMMED INFLUENCES ON DNA METHYLATION

It has already been demonstrated that many CpG islands are subject to creeping de novo methylation as a function of aging in a variety of different cell types (131), suggesting that, in addition to built-in developmental events, the basic pattern of modification may also be influenced by nonprogrammed changes during the lifetime of the organism. The major question, however, is whether methylation patterns can also be influenced by environmental cues, either during embryogenesis or even afterward in the newborn or adult organism. A number of pioneering studies on this topic appear to indicate that this may indeed be possible.

Using a mouse model that contains a retrotransposon integrated upstream of the agouti gene, it was shown that offspring from each litter have variable coat color, ranging from dark brown to yellow, even though all the animals are isogenic, and molecular analysis indicated that this probably occurs because the agouti gene promoter is methylated to different degrees in each individual (132). This methylation variability appears to occur stochastically and probably takes place at about the time of implantation. The degree of methylation appears to be under the control of a number of different modifier genes located throughout the genome (133, 134), but the coat color in the offspring can also be modulated by exposing pregnant mothers to ethanol (135) or to a methyl-rich diet (136). In general, there appear to be many natural sequences in the genome that are also subject to variable stochastic methylation during development, and it is clear that this effect can generate trait variability in isogenic mice (137). Similar stochastic effects may also be responsible for causing tumor variability (81).

These results on a mouse model confirm the idea that DNA methylation patterns can be influenced by environment, either during gestation or perhaps even after birth. Nonetheless, these changes do not seem to be fully inherited through the germ line into the next generation (136). In the case of the agouti trait, for example, all parent mice, regardless of coat color, still generate approximately the same range of variable offspring. This is consistent with the idea that methylation patterns are largely erased between generations first during gametogenesis, and then again in early embryogenesis (44). This picture is very different from the *trans*-generational epigenetic effects observed in plants, and this is undoubtedly caused by the lack of embryonic erasure in this organism (138).

SUMMARY POINTS

1. DNA methylation patterns are erased during preimplantation and then re-established throughout development via sequence information in the DNA.
2. Once established, DNA methylation patterns can be maintained autonomously through many cell divisions.
3. DNA methylation inhibits gene expression by affecting chromatin structure.
4. Changes in methylation during postimplantation development are usually secondary to factor-mediated gene activation or repression, but this subsequent methylation pattern provides long-term stability.
5. Active demethylation takes place through a multistep biochemical pathway that involves hydroxylation, deamination, glycosylation, and subsequent repair.
6. Somatic cell reprogramming involves resetting the methylation pattern to that of an early implantation embryo.
7. Abnormal methylation in cell lines and cancer takes place through a programmed process that involves the recruitment of de novo methylases by the polycomb complex.

FUTURE ISSUES

1. What is the mechanism involved in setting up methylation patterns? How is local sequence information translated in epigenetic information?
2. How does methylation play a role in lineage determination during development?
3. What are the roles of environment and aging on DNA methylation? What are methylation's effects on long-term physiology and disease susceptibility?

DISCLOSURE STATEMENT

The authors are not aware of any affiliations, memberships, funding, or financial holdings that might be perceived as affecting the objectivity of this review.

ACKNOWLEDGMENTS

This work was supported by research grants from the Israel Academy of Sciences (H.C., Y.B.), the National Institutes of Health (Y.B.), the Israel Cancer Research Foundation (H.C., Y.B.), the European Community 5th Framework Quality of Life Program (Y.B.), and the European Research Council (H.C.).

LITERATURE CITED

1. Sinsheimer RL. 1955. The action of pancreatic deoxyribonuclease. II. Isomeric dinucleotides. *J. Biol. Chem.* 215:579–83
2. Cedar H, Solage A, Glaser G, Razin A. 1979. Direct detection of methylated cytosine in DNA by use of the restriction enzyme MspI. *Nucleic Acids Res.* 6:2125–32
3. Bird AP. 1978. Use of restriction enzyme to study eukaryotic DNA methylation. II. The symmetry of methylation sites supports semiconservative copying of the methylation pattern. *J. Mol. Biol.* 118:49–60
4. van der Ploeg LHT, Flavell RA. 1980. DNA methylation in the human γ-globin locus in erythroid and nonerythroid tissues. *Cell* 19:947–58
5. Quint A, Cedar H. 1981. In vitro methylation of DNA with Hpa II methylase. *Nucleic Acids Res.* 9:633–46
6. Pollack Y, Stein R, Razin A, Cedar H. 1980. Methylation of foreign DNA sequences in eukaryotic cells. *Proc. Natl. Acad. Sci. USA* 77:6463–67
7. Wigler M, Levy D, Perucho M. 1981. The somatic replication of DNA methylation. *Cell* 24:33–40
8. Stein R, Gruenbaum Y, Pollack Y, Razin A, Cedar H. 1982. Clonal inheritance of the pattern DNA methylation in mouse cells. *Proc. Natl. Acad. Sci. USA* 79:61–65
9. Gruenbaum Y, Cedar H, Razin A. 1982. Substrate and sequence specificity of a eukaryotic DNA methylase. *Nature* 295:620–22
10. Razin A, Riggs AD. 1980. DNA methylation and gene function. *Science* 210:604–10
11. Gruenbaum Y, Naveh-Many T, Cedar H, Razin A. 1981. Sequence specificity of methylation in higher plant DNA. *Nature* 292:860–62
12. Li E, Bestor TH, Jaenisch R. 1992. Targeted mutation of the DNA methyltransferase gene results in embryonic lethality. *Cell* 69:915–26
13. Leonhardt H, Page AW, Weier HU, Bestor TH. 1992. A targeting sequence directs DNA methyltransferase to sites of DNA replication in mammalian nuclei. *Cell* 71:865–73
14. Song J, Rechkoblit O, Bestor TH, Patel DJ. 2011. Structure of DNMT1-DNA complex reveals a role for autoinhibition in maintenance DNA methylation. *Science* 331:1036–40
15. Rajakumara E, Wang Z, Ma H, Hu L, Chen H, et al. 2011. PHD finger recognition of unmodified histone H3R2 links UHRF1 to regulation of euchromatic gene expression. *Mol. Cell* 43:275–84
16. Bostick M, Kim JK, Esteve PO, Clark A, Pradhan S, Jacobsen SE. 2007. UHRF1 plays a role in maintaining DNA methylation in mammalian cells. *Science* 317:1760–64
17. Sharif J, Koseki H. 2011. Recruitment of Dnmt1: roles of the SRA protein Np95 (Uhrf1) and other factors. In *Progress in Molelcular Bioliolgy and Translational Science*, Vol. 101: *Modifications of Nuclear DNA and its Regulatory Proteins*, ed. X Cheng, RM Blumenthal, pp. 289–310. London: Acad. Press
18. Achour M, Jacq X, Ronde P, Alhosin M, Charlot C, et al. 2008. The interaction of the SRA domain of ICBP90 with a novel domain of DNMT1 is involved in the regulation of *VEGF* gene expression. *Oncogene* 27:2187–97
19. Yisraeli J, Szyf M. 1984. Gene methylation patterns and expression. In *DNA Methylation: Biochemistry and Biological Significance*, ed. A Razin, H Cedar, AD Riggs, pp. 352–70. New York: Springer-Verlag
20. Stein R, Sciaky-Gallili N, Razin A, Cedar H. 1983. Pattern of methylation of two genes coding for housekeeping functions. *Proc. Natl. Acad. Sci. USA* 80:2422–26
21. Bird AP, Taggart M, Frommer M, Miller OJ, Macleod D. 1985. A fraction of the mouse genome that is derived from islands of nonmethylated CpG-rich DNA. *Cell* 40:91–99
22. Naveh-Many T, Cedar H. 1981. Active gene sequences are undermethylated. *Proc. Natl. Acad. Sci. USA* 78:4246–50

23. Laurent L, Wong E, Li G, Huynh T, Tsirigos A, et al. 2010. Dynamic changes in the human methylome during differentiation. *Genome Res.* 20:320–31
24. Stein R, Razin A, Cedar H. 1982. In vitro methylation of the hamster adenine phosphoribosyltransferase gene inhibits its expression in mouse L cells. *Proc. Natl. Acad. Sci. USA* 79:3418–22
25. Schubeler D, Lorincz MC, Cimbora DM, Telling A, Feng YQ, et al. 2000. Genomic targeting of methylated DNA: influence of methylation on transcription, replication, chromatin structure, and histone acetylation. *Mol. Cell. Biol.* 20:9103–12
26. Siegfried Z, Eden S, Mendelsohn M, Feng X, Tzubari B, Cedar H. 1999. DNA methylation represses transcription in vivo. *Nat. Genet.* 22:203–6
27. Goren A, Simchen G, Fibach E, Szabo PE, Tanimoto K, et al. 2006. Fine tuning of globin gene expression by DNA methylation. *PLoS ONE* 1:e46
28. Höller M, Westin G, Jiricny J, Schaffner W. 1988. Sp1 transcription factor binds DNA and activates transcription even when the binding site is CpG methylated. *Genes Dev.* 2:1127–35
29. Buschhausen G, Wittig B, Graessmann M, Graessmann A. 1987. Chromatin structure is required to block transcription of the methylated herpes simplex virus thymidine kinase gene. *Proc. Natl. Acad. Sci. USA* 84:1177–81
30. Keshet I, Lieman-Hurwitz J, Cedar H. 1986. DNA methylation affects the formation of active chromatin. *Cell* 44:535–43
31. Razin A, Cedar H. 1977. Distribution of 5-methylcytosine in chromatin. *Proc. Natl. Acad. Sci. USA* 74:2725–28
32. Solage A, Cedar H. 1978. Organization of 5-methylcytosine in chromosomal DNA. *Biochemistry* 17:2934–38
33. Chodavarapu RK, Feng S, Bernatavichute YV, Chen PY, Stroud H, et al. 2010. Relationship between nucleosome positioning and DNA methylation. *Nature* 466:388–92
34. Klose RJ, Bird AP. 2006. Genomic DNA methylation: the mark and its mediators. *Trends Biochem. Sci.* 31:89–97
35. Nan X, Ng H-H, Johnson CA, Laherty CD, Turner BM, et al. 1998. Transcriptional repression by the methyl-CpG-binding protein MeCP2 involves a histone deacetylase complex. *Nature* 393:386–89
36. Jones PL, Veenstra GJC, Wade PA, Vermaak D, Kass SU, et al. 1998. Methylated DNA and MeCP2 recruit histone deacetylase to repress transcription. *Nat. Genet.* 19:187–91
37. Fuks F, Hurd PJ, Wolf D, Nan X, Bird AP, Kouzarides T. 2003. The methyl-CpG-binding protein MeCP2 links DNA methylation to histone methylation. *J. Biol. Chem.* 278:4035–40
38. Bell AC, Felsenfeld G. 2000. Methylation of a CTCF-dependent boundary controls imprinted expression of the *Igf2* gene. *Nature* 405:482–85
39. Thomson JP, Skene PJ, Selfridge J, Clouaire T, Guy J, et al. 2010. CpG islands influence chromatin structure via the CpG-binding protein Cfp1. *Nature* 464:1082–86
40. Lucchini R, Sogo JM. 1995. Replication of transcriptionally active chromatin. *Nature* 374:276–80
41. Hansen KH, Bracken AP, Pasini D, Dietrich N, Gehani SS, et al. 2008. A model for transmission of the H3K27me3 epigenetic mark. *Nat. Cell. Biol.* 10:1291–300
42. Hashimshony T, Zhang J, Keshet I, Bustin M, Cedar H. 2003. The role of DNA methylation in setting up chromatin structure during development. *Nat. Genet.* 34:187–92
43. Monk M, Boubelik M, Lehnert S. 1987. Temporal and regional changes in DNA methylation in the embryonic, extraembryonic and germ cell lineages during mouse embryo development. *Development* 99:371–82
44. Kafri T, Ariel M, Brandeis M, Shemer R, Urven L, et al. 1992. Developmental pattern of gene-specific DNA methylation in the mouse embryo and germline. *Genes Dev.* 6:705–14
45. Chaillet JR, Vogt TF, Beier DR, Leder P. 1991. Parental-specific methylation of an imprinted transgene is established during gametogenesis and progressively changes during embryogenesis. *Cell* 66:77–83
46. Mayer W, Niveleau A, Walter J, Fundele R, Haaf T. 2000. Demethylation of the zygotic paternal genome. *Nature* 403:501–2
47. Carlson LL, Page AW, Bestor TH. 1992. Properties and localization of DNA methyltransferase in preimplantation mouse embryos: implications for genomic imprinting. *Genes Dev.* 6:2536–41

48. Birger Y, Shemer R, Perk J, Razin A. 1999. The imprinting box of the mouse *Igf2r* gene. *Nature* 397:84–88
49. Bourc'his D, Xu GL, Lin CS, Bollman B, Bestor TH. 2001. Dnmt3L and the establishment of maternal genomic imprints. *Science* 294:2536–39
50. Sanford JP, Clark HJ, Chapman VM, Rossant J. 1987. Differences in DNA methylation during oogenesis and spermatogenesis and their persistence during early embryogenesis in the mouse. *Genes Dev.* 1:1039–46
51. Smallwood SA, Tomizawa SI, Krueger F, Ruf N, Carli N, et al. 2011. Dynamic CpG island methylation landscape in oocytes and preimplantation embryos. *Nat. Genet.* 43:811–14
52. Okano M, Bell DW, Haber DA, Li E. 1999. DNA methyltransferases Dnmt3a and Dnmt3b are essential for de novo methylation and mammalian development. *Cell* 99:247–57
53. Straussman R, Nejman D, Roberts D, Steinfeld I, Blum B, et al. 2009. Developmental programming of CpG island methylation profiles in the human genome. *Nat. Struct. Mol. Biol.* 16:564–71
54. Mohn F, Weber M, Schübeler D, Roloff TC. 2009. Methylated DNA immunoprecipitation (MeDIP). *Methods Mol. Biol.* 507:55–64
55. Frank D, Keshet I, Shani M, Levine A, Razin A, Cedar H. 1991. Demethylation of CpG islands in embryonic cells. *Nature* 351:239–41
56. Brandeis M, Frank D, Keshet I, Siegfried Z, Mendelsohn M, et al. 1994. Sp1 elements protect a CpG island from de novo methylation. *Nature* 371:435–38
57. Macleod D, Charlton J, Mullins J, Bird AP. 1994. Sp1 sites in the mouse *aprt* gene promoter are required to prevent methylation of the CpG island. *Genes Dev.* 8:2282–92
58. Weber M, Hellmann I, Stadler MB, Ramos L, Pääbo S, et al. 2007. Distribution, silencing potential and evolutionary impact of promoter DNA methylation in the human genome. *Nat. Genet.* 39:457–66
59. Meissner A, Mikkelsen TS, Gu H, Wernig M, Hanna J, et al. 2008. Genome-scale DNA methylation maps of pluripotent and differentiated cells. *Nature* 454:766–70
60. Xu J, Watts JA, Pope SD, Gadue P, Kamps M, et al. 2009. Transcriptional competence and the active marking of tissue-specific enhancers by defined transcription factors in embryonic and induced pluripotent stem cells. *Genes Dev.* 23:2824–38
61. Walsh CP, Chaillet JR, Bestor TH. 1998. Transcription of IAP endogenous retrovirus is constrained by cytosine methylation. *Nat. Genet.* 20:116–17
62. Lande-Diner L, Zhang J, Ben-Porath I, Amariglio N, Keshet I, et al. 2007. Role of DNA methylation in stable gene repression. *J. Biol. Chem.* 282:12194–200
63. Scholer HR. 1991. Octamania: the POU factors in murine development. *Trends Genet.* 7:323–29
64. Feldman N, Gerson A, Fang J, Li E, Zhang Y, et al. 2006. G9a-mediated irreversible epigenetic inactivation of *Oct-3/4* during early embryogenesis. *Nat. Cell Biol.* 8:188–94
65. Epsztejn-Litman S, Feldman N, Abu-Remaileh M, Shufaro Y, Gerson A, et al. 2008. De novo DNA methylation promoted by G9a prevents reprogramming of embryonically silenced genes. *Nat. Struct. Mol. Biol.* 15:1176–83
66. Dong KB, Maksakova IA, Mohn F, Leung D, Appanah R, et al. 2008. DNA methylation in ES cells requires the lysine methyltransferase G9a but not its catalytic activity. *EMBO J.* 27:2691–701
67. Tachibana M, Matsumura Y, Fukuda M, Kimura H, Shinkai Y. 2008. G9a/GLP complexes independently mediate H3K9 and DNA methylation to silence transcription. *EMBO J.* 27:2681–90
68. Gautsch JW, Wilson MC. 1983. Delayed de novo methylation in teratocarcinoma suggests additional tissue-specific mechanisms for controlling gene expression. *Nature* 301:32–37
69. Keohane AM, Lavender JS, O'Neill LP, Turner BM. 1998. Histone acetylation and X inactivation. *Dev. Genet.* 22:65–73
70. Lock LF, Takagi N, Martin GR. 1987. Methylation of the *Hprt* gene on the inactive X occurs after chromosome inactivation. *Cell* 48:39–46
71. Cedar H, Bergman Y. 2009. Linking DNA methylation and histone modification: patterns and paradigms. *Nat. Rev. Genet.* 10:295–304
72. Yisraeli J, Adelstein RS, Melloul D, Nudel U, Yaffe D, Cedar H. 1986. Muscle-specific activation of a methylated chimeric actin gene. *Cell* 46:409–16
73. Sullivan CH, Grainger RM. 1986. δ-Crystallin genes become hypomethylated in postmitotic lens cells during chicken development. *Proc. Natl. Acad. Sci. USA* 83:329–33

74. Paroush Z, Keshet I, Yisraeli J, Cedar H. 1990. Dynamics of demethylation and activation of the α-actin gene in myoblasts. *Cell* 63:1229–37
75. Lichtenstein M, Keini G, Cedar H, Bergman Y. 1994. B cell-specific demethylation: a novel role for the intronic κ-chain enhancer sequence. *Cell* 76:913–23
76. Kirillov A, Kistler B, Mostoslavsky R, Cedar H, Wirth T, Bergman Y. 1996. A role for nuclear NF-κB in B-cell-specific demethylation of the Igκ locus. *Nat. Genet.* 13:435–41
77. Illingworth R, Kerr A, Desousa D, Jørgensen H, Ellis P, et al. 2008. A novel CpG island set identifies tissue-specific methylation at developmental gene loci. *PLoS Biol.* 6:e22
78. Wu H, Coskun V, Tao J, Xie W, Ge W, et al. 2010. Dnmt3a-dependent nonpromoter DNA methylation facilitates transcription of neurogenic genes. *Science* 329:444–48
79. Viré E, Brenner C, Deplus R, Blanchon L, Fraga M, et al. 2006. The Polycomb group protein EZH2 directly controls DNA methylation. *Nature* 439:871–74
80. Feinberg AP, Vogelstein B. 1983. Hypomethylation distinguishes genes of some human cancers from their normal counterparts. *Nature* 301:89–92
81. Hansen KD, Timp W, Bravo HC, Sabunciyan S, Langmead B, et al. 2011. Increased methylation variation in epigenetic domains across cancer types. *Nat. Genet.* 43:768–75
82. Baylin SB, Höppener JW, de Bustros A, Steenbergh PH, Lips CJ, Nelkin BD. 1986. DNA methylation patterns of the calcitonin gene in human lung cancers and lymphomas. *Cancer Res.* 46:2917–22
83. Zardo G, Tiirikainen MI, Hong C, Misra A, Feuerstein BG, et al. 2002. Integrated genomic and epigenomic analyses pinpoint biallelic gene inactivation in tumors. *Nat. Genet.* 32:453–58
84. Keshet I, Schlesinger Y, Farkash S, Rand E, Hecht M, et al. 2006. Evidence for an instructive mechanism of de novo methylation in cancer cells. *Nat. Genet.* 38:149–53
85. Schlesinger Y, Straussman R, Keshet I, Farkash S, Hecht M, et al. 2007. Polycomb mediated histone H3(K27) methylation pre-marks genes for de novo methylation in cancer. *Nat. Genet.* 39:232–36
86. Widschwendter M, Fiegl H, Egle D, Mueller-Holzner E, Spizzo G, et al. 2007. Epigenetic stem cell signature in cancer. *Nat. Genet.* 39:157–58
87. Ohm JE, McGarvey KM, Yu X, Cheng L, Schuebel KE, et al. 2007. A stem cell-like chromatin pattern may predispose tumor suppressor genes to DNA hypermethylation and heritable silencing. *Nat. Genet.* 39:237–42
88. Luckow B, Schütz G. 1987. CAT constructions with multiple unique restriction sites for the functional analysis of eukaryotic promoter and regulatory elements. *Nucleic Acids Res.* 15:5490
89. Linhart HG, Lin H, Yamada Y, Moran E, Steine EJ, et al. 2007. Dnmt3b promotes tumorigenesis in vivo by gene-specific de novo methylation and transcriptional silencing. *Genes Dev.* 21:3110–22
90. Laird PW, Jackson-Grusby L, Fazeli A, Dickinson SL, Jung WE, et al. 1995. Suppression of intestinal neoplasia by DNA hypomethylation. *Cell* 81:197–205
91. Lin H, Yamada Y, Nguyen S, Linhart H, Jackson-Grusby L, et al. 2006. Suppression of intestinal neoplasia by deletion of Dnmt3b. *Mol. Cell. Biol.* 26:2976–83
92. McCabe MT, Low JA, Daignault S, Imperiale MJ, Wojno KJ, Day ML. 2006. Inhibition of DNA methyltransferase activity prevents tumorigenesis in a mouse model of prostate cancer. *Cancer Res.* 66:385–92
93. Jones PA, Wolkowicz MJ, Rideout WM 3rd, Gonzales FA, Marziasz CM, et al. 1990. De novo methylation of the MyoD1 CpG island during the establishment of immortal cell lines. *Proc. Natl. Acad. Sci. USA* 87:6117–21
94. Antequera F, Boyes J, Bird A. 1990. High levels of de novo methylations and altered chromatin structure at CpG islands in cell lines. *Cell* 62:503–14
95. Shen Y, Chow J, Wang Z, Fan G. 2006. Abnormal CpG island methylation occurs during in vitro differentiation of human embryonic stem cells. *Hum. Mol. Genet.* 15:2623–35
96. Mohn F, Weber M, Rebhan M, Roloff TC, Richter J, et al. 2008. Lineage-specific polycomb targets and de novo DNA methylation define restriction and potential of neuronal progenitors. *Mol. Cell* 30:755–66
97. Pietrobono R, Tabolacci E, Zalfa F, Zito I, Terracciano A, et al. 2005. Molecular dissection of the events leading to inactivation of the *FMR1* gene. *Hum. Mol. Genet.* 14:267–77
98. Coffee B, Zhang F, Ceman S, Warren ST, Reines D. 2002. Histone modifications depict an aberrantly heterochromatinized *FMR1* gene in fragile X syndrome. *Am. J. Hum. Genet.* 71:923–32

99. Bhattacharya SK, Ramchandani S, Cervoni N, Szyf M. 1999. A mammalian protein with specific demethylase activity for mCpG DNA. *Nature* 397:579–83
100. Cedar H, Verdine GL. 1999. Gene expression. The amazing demethylase. *Nature* 397:568–69
101. Ng HH, Zhang Y, Hendrich B, Johnson CA, Turner BM, et al. 1999. MBD2 is a transcriptional repressor belonging to the MeCP1 histone deacetylase complex. *Nat. Genet.* 23:58–61
102. Jost JP. 1993. Nuclear extracts of chicken embryos promote an active demethylation of DNA by excision repair of 5-methyldeoxycytidine. *Proc. Natl. Acad. Sci. USA* 90:4684–88
103. Razin A, Szyf M, Kafri T, Roll M, Giloh H, et al. 1986. Replacement of 5-methylcytosine by cytosine: a possible mechanism for transient DNA demethylation during differentiation. *Proc. Natl. Acad. Sci. USA* 83:2827–31
104. Weiss A, Keshet I, Razin A, Cedar H. 1996. DNA demethylation in vitro: involvement of RNA. *Cell* 86:709–18
105. Kim MS, Kondo T, Takada I, Youn MY, Yamamoto Y, et al. 2009. DNA demethylation in hormone-induced transcriptional derepression. *Nature* 461:1007–12
106. Cortazar D, Kunz C, Selfridge J, Lettieri T, Saito Y, et al. 2011. Embryonic lethal phenotype reveals a function of TDG in maintaining epigenetic stability. *Nature* 470:419–23
107. Cortellino S, Xu J, Sannai M, Moore R, Caretti E, et al. 2011. Thymine DNA glycosylase is essential for active DNA demethylation by linked deamination-base excision repair. *Cell* 146:67–79
108. Hajkova P, Jeffries SJ, Lee C, Miller N, Jackson SP, Surani MA. 2010. Genome-wide reprogramming in the mouse germ line entails the base excision repair pathway. *Science* 329:78–82
109. Barreto G, Schafer A, Marhold J, Stach D, Swaminathan SK, et al. 2007. Gadd45a promotes epigenetic gene activation by repair-mediated DNA demethylation. *Nature* 445:671–75
110. Wu H, D'Alessio AC, Ito S, Xia K, Wang Z, et al. 2011. Dual functions of Tet1 in transcriptional regulation in mouse embryonic stem cells. *Nature* 473:389–93
111. Wossidlo M, Nakamura T, Lepikhov K, Marques CJ, Zakhartchenko V, et al. 2011. 5-Hydroxymethylcytosine in the mammalian zygote is linked with epigenetic reprogramming. *Nat. Commun.* 2:241
112. Iqbal K, Jin SG, Pfeifer GP, Szabo PE. 2011. Reprogramming of the paternal genome upon fertilization involves genome-wide oxidation of 5-methylcytosine. *Proc. Natl. Acad. Sci. USA* 108:3642–47
113. Popp C, Dean W, Feng S, Cokus SJ, Andrews S, et al. 2010. Genome-wide erasure of DNA methylation in mouse primordial germ cells is affected by AID deficiency. *Nature* 463:1101–5
114. Guo JU, Su Y, Zhong C, Ming GL, Song H. 2011. Hydroxylation of 5-methylcytosine by TET1 promotes active DNA demethylation in the adult brain. *Cell* 145:423–34
115. Rai K, Huggins IJ, James SR, Karpf AR, Jones DA, Cairns BR. 2008. DNA demethylation in zebrafish involves the coupling of a deaminase, a glycosylase, and gadd45. *Cell* 135:1201–12
116. Ito S, Shen L, Dai Q, Wu SC, Collins LB, et al. 2011. Tet proteins can convert 5-methylcytosine to 5-formylcytosine and 5-carboxylcytosine. *Science* 333:1300–3
117. Yamanaka S. 2009. A fresh look at iPS cells. *Cell* 137:13–17
118. Jaenisch R, Hochedlinger K, Blelloch R, Yamada Y, Baldwin K, Eggan K. 2004. Nuclear cloning, epigenetic reprogramming, and cellular differentiation. *Cold Spring Harb. Symp. Quant. Biol.* 69:19–27
119. Ng RK, Dean W, Dawson C, Lucifero D, Madeja Z, et al. 2008. Epigenetic restriction of embryonic cell lineage fate by methylation of Elf5. *Nat. Cell. Biol.* 10:1280–90
120. Boiani M, Eckardt S, Scholer HR, McLaughlin KJ. 2002. Oct4 distribution and level in mouse clones: consequences for pluripotency. *Genes Dev.* 16:1209–19
121. Bortvin A, Eggan K, Skaletsky H, Akutsu H, Berry DL, et al. 2003. Incomplete reactivation of Oct4-related genes in mouse embryos cloned from somatic nuclei. *Development* 130:1673–80
122. Cowan CA, Atienza J, Melton DA, Eggan K. 2005. Nuclear reprogramming of somatic cells after fusion with human embryonic stem cells. *Science* 309:1369–73
123. Gidekel S, Bergman Y. 2002. A unique developmental pattern of *Oct-3/4* DNA methylation is controlled by a *cis*-demodification element. *J. Biol. Chem.* 277:34521–30
124. McCarrey JR. 1993. Development of the germ cell. In *Cell and Molecular Biology of the Testis*, ed. C Desjardins, LL Ewing, pp. 58–89. New York: Oxford Univ. Press

125. Shemer R, Razin A. 1996. Establishment of imprinted methylation patterns during development. In *Epigenetic Mechanisms of Gene Regulation*, ed. VEA Russo, RA Martienssen, AD Riggs, pp. 215–29. Cold Spring Harbor, NY: Cold Spring Harb. Lab. Press
126. Ariel M, Cedar H, McCarrey JR. 1994. Developmental changes in methylation of spermatogenesis-specific genes include reprogramming in the epididymis. *Nat. Genet.* 7:59–63
127. Hudson QJ, Kulinski TM, Huetter SP, Barlow DP. 2010. Genomic imprinting mechanisms in embryonic and extraembryonic mouse tissues. *Heredity* 105:45–56
128. Simon I, Tenzen T, Reubinoff BE, Hillman D, McCarrey JR, Cedar H. 1999. Asynchronous replication of imprinted genes is established in the gametes and maintained during development. *Nature* 401:929–32
129. Ng RK, Gurdon JB. 2008. Epigenetic inheritance of cell differentiation status. *Cell Cycle* 7:1173–77
130. Li E, Beard C, Jaenisch R. 1993. Role for DNA methylation in genomic imprinting. *Nature* 366:362–65
131. Maegawa S, Hinkal G, Kim HS, Shen L, Zhang L, et al. 2010. Widespread and tissue specific age-related DNA methylation changes in mice. *Genome Res.* 20:332–40
132. Morgan HD, Sutherland HG, Martin DI, Whitelaw E. 1999. Epigenetic inheritance at the agouti locus in the mouse. *Nat. Genet.* 23:314–18
133. Chong S, Vickaryous N, Ashe A, Zamudio N, Youngson N, et al. 2007. Modifiers of epigenetic reprogramming show paternal effects in the mouse. *Nat. Genet.* 39:614–22
134. Blewitt ME, Vickaryous NK, Hemley SJ, Ashe A, Bruxner TJ, et al. 2005. An *N*-ethyl-*N*-nitrosourea screen for genes involved in variegation in the mouse. *Proc. Natl. Acad. Sci. USA* 102:7629–34
135. Kaminen-Ahola N, Ahola A, Maga M, Mallitt KA, Fahey P, et al. 2010. Maternal ethanol consumption alters the epigenotype and the phenotype of offspring in a mouse model. *PLoS Genet.* 6:e1000811
136. Daxinger L, Whitelaw E. 2010. Transgenerational epigenetic inheritance: more questions than answers. *Genome Res.* 20:1623–28
137. Whitelaw NC, Chong S, Whitelaw E. 2010. Tuning in to noise: epigenetics and intangible variation. *Dev. Cell* 19:649–50
138. Feng S, Jacobsen SE, Reik W. 2010. Epigenetic reprogramming in plant and animal development. *Science* 330:622–27
139. Gal-Yam EN, Egger G, Iniguez L, Holster H, Einarsson S, et al. 2008. Frequent switching of Polycomb repressive marks and DNA hypermethylation in the PC3 prostate cancer cell line. *Proc. Natl. Acad. Sci. USA* 105:12979–84

RNA Polymerase II Elongation Control

Qiang Zhou,[1] Tiandao Li,[2] and David H. Price[2]

[1]Department of Molecular and Cell Biology, University of California, Berkeley, California 94720; email: qzhou@berkeley.edu

[2]Department of Biochemistry, University of Iowa, Iowa City, Iowa 52242; email: tiandao-li@uiowa.edu, david-price@uiowa.edu

Keywords

P-TEFb, NELF, DSIF, SEC, 7SK snRNP

Abstract

Regulation of the elongation phase of transcription by RNA polymerase II (Pol II) is utilized extensively to generate the pattern of mRNAs needed to specify cell types and to respond to environmental changes. After Pol II initiates, negative elongation factors cause it to pause in a promoter proximal position. These polymerases are poised to respond to the positive transcription elongation factor P-TEFb, and then enter productive elongation only under the appropriate set of signals to generate full-length properly processed mRNAs. Recent global analyses of Pol II and elongation factors, mechanisms that regulate P-TEFb involving the 7SK small nuclear ribonucleoprotein (snRNP), factors that control both the negative and positive elongation properties of Pol II, and the mRNA processing events that are coupled with elongation are discussed.

Contents

1. INTRODUCTION 120
2. PROMOTER PROXIMALLY PAUSED POLYMERASES 122
3. THE POSITIVE TRANSCRIPTION ELONGATION FACTOR, P-TEFb 123
   - 3.1. Structure and Function of P-TEFb 124
   - 3.2. Regulation of P-TEFb 125
4. THE PRODUCTIVE ELONGATION PHASE OF RNA POLYMERASE II TRANSCRIPTION 127
   - 4.1. TFIIS and TFIIF 127
   - 4.2. ELL1, -2, and -3 128
   - 4.3. PAFc 129
   - 4.4. Super Elongation Complexes ... 129
5. EVENTS COUPLED TO TRANSCRIPTION ELONGATION 131
   - 5.1. Capping 132
   - 5.2. Splicing 132
   - 5.3. Cleavage and Polyadenylation .. 133

erly processed mRNA. All of these steps are connected to each other to varying extents, and our goal is to review recent findings concerning postinitiation transcriptional events. Results from mammalian systems are emphasized, and findings in yeast and *Drosophila* are denoted as such because there is some species specificity.

It is now clear that many transcriptional choices are available to fine-tune the repertoire of genes expressed. Most are not a matter of yes or no decisions, but rather to what frequency the regulated events are allowed. On some genes, in any particular cell, or cell type, there is no initiation by Pol II, and the consequences on gene expression are obvious. Recently, a choice was uncovered by looking at Pol II occupation by chromatin immunoprecipitation followed by sequencing (ChIP-Seq) or by sequencing nascent transcripts. From many promoters in yeast (1, 2) and mammals (3, 4), transcription can proceed in either direction (**Figure 1**), but in general, only one direction is allowed to proceed into productive elongation (5). The same

1. INTRODUCTION

Normal development and cellular responses to environmental challenges are accomplished by dynamic regulation of the expression of tens of thousands of genes in mammals. Transcription is the obvious first step in gene expression and, overall, provides many mechanisms that are essential to achieving the appropriate mix of proteins and RNAs in each cell. Access to promoter regions encoded in the DNA template as well as the body of genes is complicated because much of each organism's genome is buried in highly repressive nucleosomes. A host of factors are required to deliver RNA polymerase II (Pol II) to the appropriate promoters, allow initiation, cause the transition into productive elongation, and couple the splicing and polyadenylation machinery that is needed to generate a prop-

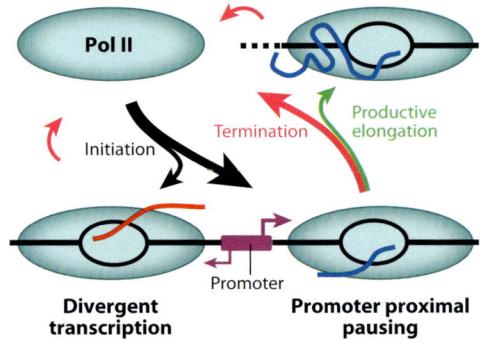

Figure 1

Divergent transcription, promoter proximally paused polymerases and productive elongation. Many promoters allow Pol II to initiate and elongate in both directions using two transcription start sites. Transcription upstream of the gene gives rise to short unstable transcripts, and transcription into the gene produces paused polymerases that are poised for entry into productive elongation or termination. The thickness of the arrows represents the relative flow of polymerases: initiation (*black*), termination (*pink*), transition into productive elongation (*green*). RNA is blue in the sense direction, and the divergent transcript is red.

techniques have also demonstrated that many metazoan genes experience initiation to a fairly high extent with little correlation to the level of expression achieved. The resulting promoter proximally paused polymerases are a prevalent feature on *Drosophila* (6, 7) and mammalian (8, 9) chromosomes, and accordingly, these poised polymerases permit a choice of either terminating or entering productive elongation through the action of the positive transcription elongation factor P-TEFb (**Figure 1**) (10). Finally, there is evidence that all long Pol II transcripts are not necessarily processed into functional mRNAs (11), arguing that the coupled RNA processing events are not guaranteed by default.

ChIP-Seq has provided an unprecedented view of the transcription of metazoan genomes. A number of high-quality ChIP-Seq data sets examining the chromosomal distribution of Pol II and elongation factors in mouse embryonic stem (MES) cells were recently published (9). As an example of what was found, a very active gene (*RPL11*) and a neighboring gene that is much less active (*TCEB3*) are presented in **Figure 2**. Information from the MES cell data sets was used to generate a plot showing the relative positions for Pol II before and after a one-hour flavopiridol treatment of cells, which inhibits P-TEFb function (12). Also shown are the distributions of two phosphorylated forms of Pol II that are found in elongation complexes [Ser5 and Ser2 of the C-terminal domain (CTD) of the large subunit] (13). The total signals for each of the four tracks for the two genes were normalized so that the relative distributions across the regions can be viewed. The height of the signals between data sets cannot be compared because each is the end result of a complicated protocol that starts with a potentially variable immunoprecipitation step. However, the relative position of signals within a data set is informative. To interpret the graphs presented, it is important to realize that for a protein located precisely at a specific point on the DNA, the signal detecting that protein will be about 400-bp wide with the center of the peak exactly over the site of binding.

The patterns of Pol II found over the two example genes are different and lead to significant insight into the transcription elongation process. For an active gene like the ribosomal protein L11, there is a peak of Pol II about 50 bp downstream of the transcription start site (TSS) owing to promoter proximally paused polymerases (**Figure 2a**). Downstream of the poised polymerase a low but significant signal is found over the body of the gene, as well as a large peak downstream of the poly(A) addition site. Treatment of cells with flavopiridol for one hour eliminated almost all of the downstream signals, indicating that productive elongation and the production of mRNAs were blocked by inhibiting P-TEFb. Ser5 phosphorylation tracks with Pol II over the 5′ gene regions but is greatly reduced in the region downstream of the poly(A) site. Ser2 phosphorylation is found predominately over the polymerases downstream of the poly(A) site. This pattern is typical for all highly transcribed genes (9). A different pattern is found over the more inactive genes (**Figure 2b**). In this case, only poised polymerases are found before and after flavopiridol treatment of the cells. Ser5, but not Ser2, phosphorylation is found over the poised polymerases. A third type of gene exists that has no polymerase at all (not shown), but most mammalian and many *Drosophila* genes are similar to the lowly expressed gene, *TCEB3* (6, 9).

The distribution of Pol II across genes in budding yeast has some similarities as well as some significant differences. Importantly, there are no prominent poised polymerases on the closely packed yeast genes, and this is likely because of the lack of the negative elongation factor (NELF). Overall, the pattern of CTD phosphorylation is similar, with Pol II near the 5′ end of genes having more Ser5 phosphorylation and Pol II gaining more Ser2 phosphorylation at the 3′ end of genes (14). The lack of poised polymerases does not necessarily mean that elongation is not controlled in a related way in yeast compared to metazoans. Indeed a number of factors are exchanged on and off the polymerase during elongation (15).

P-TEFb: positive transcription elongation factor

NELF: negative elongation factor

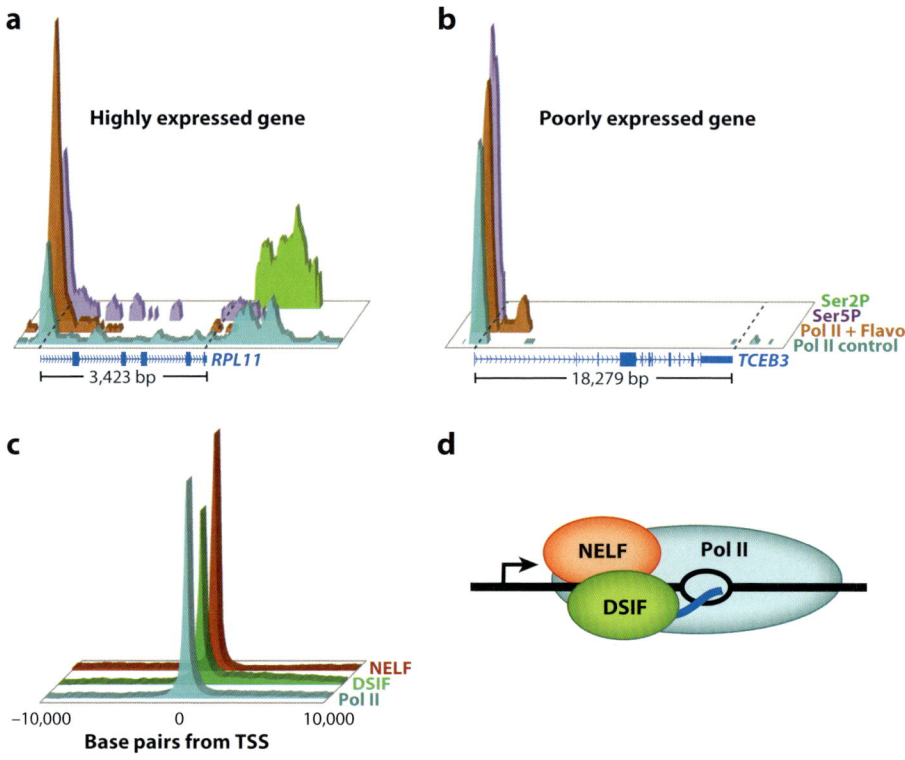

Figure 2

Pol II and elongation factor occupancy from chromatin immunoprecipitation sequencing (ChIP-Seq). (*a,b*) Mouse embryonic stem (MES) cell ChIP-Seq data sets from Rahl et al. (9) were used to generate the views of Pol II from control or flavopiridol-treated cells as well as Ser5 and Ser2 phosphorylation of the C-terminal domain of the large subunit of Pol II across two neighboring genes on mouse chromosome 4. The data for each track for each gene were normalized so that the area under all curves was equal (except for the Ser2P for TCEB3 for which there was no data). (*b*) Metagene analysis of MES cell ChIP-Seq data (9) for Pol II, Spt5 subunit of the DSIF and NELFe subunit of NELF. The relative distributions of the polymerase and factors were compiled for the region from −10,000 to +10,000 bp around the transcription start sites (TSSs) for ∼20,000 RefSeq genes. In (*c*), background signals from the lowest 10% of the 20,000 data points in each distribution were averaged and subtracted from all data points, and the resulting curves were normalized so that the total area under each was equal. (*d*) Diagram of the engaged polymerase with nascent transcript (*blue line*) under the influence of the DSIF and NELF.

2. PROMOTER PROXIMALLY PAUSED POLYMERASES

To control any pathway, it is first necessary to place a restriction in the flow through that pathway, which can then be modulated. As demonstrated by the examples in **Figure 2a,b**, in the production of mRNA, a major restriction point occurs shortly after Pol II initiates. The polymerase becomes stably engaged by elongating the nascent transcript past about 12 nt and then comes under the control of factors that significantly slow or halt elongation (16). These promoter proximately paused polymerases are the major form of polymerase found bound to metazoan chromosomes and are poised for entry into productive elongation (9, 17). Two factors, the 5,6-dichloro-1-β-D-ribofuranosylbenzimidazole (DRB) sensitivity-inducing factor (DSIF) (18) and NELF (19), have been clearly implicated in generating poised polymerases (**Figure 2**) (20, 21). Published ChIP-Seq data for Pol II, DSIF, and NELF in MES cells (9) were reanalyzed to generate a genome-wide average (**Figure 2c**). A set of 20,000 genes containing all genes except for those with TSSs within 1,000 bp of another TSS were aligned by their TSSs, and for each protein, the ChIP-Seq signal at each base pair position from −10,000 to +10,000 bp was summed. To correct for the non-specific background of DNA bound during the immunoprecipitation, the average of the lowest 10% of the signals for each base pair was subtracted from all other positions, and then the total area under each curve was normalized as in **Figure 2a,b**. The predominant peak of Pol II is found at about +50 bp (**Figure 2c**). Pol II is virtually absent in the −10,000 to −2,000 region and is found at low but significant levels downstream of the poised polymerase peak (**Figure 2c**). The distribution of DSIF is very similar to that of the polymerase except that the regions downstream of the poised polymerase are slightly more populated, but NELF is found only over the poised polymerase peak (**Figure 2c**). The data support the idea that poised polymerases are under the control of DSIF and NELF (**Figure 2d**).

The positional information, described above, supports what was learned about DSIF and NELF functions from earlier biochemical and molecular analyses. In vitro transcription experiments initially provided the tools to identify both the positive and negative factors involved in controlling the elongation phase of transcription (16). A key feature of transcription of virtually any promoter in vitro using a crude nuclear extract is the inhibition of the generation of long transcripts by the ATP analog DRB (16, 22). P-TEFb (23), NELF (19), and DSIF (18) were purified on the basis of their activities during reconstitution of DRB-sensitive transcription using fractionated nuclear extracts. Individually, the three factors have almost no effect on elongation of pure Pol II, but together, NELF and DSIF reduce the elongation rate of Pol II significantly, and P-TEFb can reverse the negative effect of the two factors (20, 21). Pol II in a crude nuclear extract in which P-TEFb has been inhibited elongates significantly slower than Pol II in the presence of DSIF and NELF (21), and the search for the other negative factors is ongoing.

Although poised polymerases in the bodies of genes have garnered the most attention, the polymerases synthesizing divergent transcripts also seem to be paused (5). Divergent transcripts arise by initiation in an antisense direction about 200 bp upstream from the TSS driving mRNA production. The location of the bulk of polymerases, determined using ChIP-Seq as well as short transcripts from RNA-Seq, ranges from about +50 to −250 bp from the orienting TSS (5). This suggests that polymerases encounter a block to elongation in both directions. Supporting this idea, NELF and DSIF are found over both regions, and knockdown of either factor increased transcription in the divergent direction (5). Even though most divergent transcripts are degraded rapidly by the RNA exosome, some can be extended for more than a thousand nucleotides (24). The longer divergent transcripts are sensitive to flavopiridol, indicating that they are dependent on P-TEFb just like mRNAs derived from the sense direction (24). ChIP-Seq data demonstrate that, even for promoters of highly active genes, polymerases synthesizing the associated divergent transcripts do not experience much productive elongation (9). It is important to determine what directs P-TEFb to function primarily on poised polymerases transcribing in the sense direction. The purpose of this divergent transcription is not clear, but it could be involved in altering the promoter architecture by helping to maintain a nucleosome-free region between the two TSSs (sense and antisense) or by generating strong torsional strain on the promoter region caused by the helical tension (negative supercoiling) imparted by polymerases traveling away from each other (5).

3. THE POSITIVE TRANSCRIPTION ELONGATION FACTOR, P-TEFb

By default, negative elongation factors associate with Pol II during initiation leading to the

DSIF: 5,6-dichloro-1-β-D-ribofuranosyl-benzimidazole sensitivity-inducing factor

generation of poised polymerases. P-TEFb is required to reverse their effects and allow the polymerases to enter into productive elongation (10, 25). P-TEFb is an essential cyclin-dependent kinase, which has been identified in eukaryotes from yeast to humans, and is regulated in many standard ways and in metazoans by an unusual mechanism involving reversible association with a small nuclear ribonucleoprotein (snRNP) complex (10, 26). Because poised polymerases are prevalent across the genomes of metazoans, the function of P-TEFb must be accurately directed at the correct time to the appropriate genes. There are a number of comprehensive reviews (10, 25, 27–35) on the history of discovery and details of P-TEFb and its function, and accordingly, we mainly focus on general mechanisms, recent results, and new directions.

3.1. Structure and Function of P-TEFb

P-TEFb is a nuclear-localized, cyclin-dependent kinase (CDK) that has some properties similar to all CDKs, others that are similar only to CDKs involved in transcription, and a few that are totally unique to P-TEFb. A common property to all is the two-subunit arrangement. A CDK subunit contains the active site, and a cyclin subunit is needed to invoke a conformational change that, in conjunction with phosphorylation of the T-loop in the CDK subunit, leads to activation of the kinase. Phosphorylation of Thr186 is critical for activation of the catalytic subunit of P-TEFb, CDK9 (36, 37), although other sites can be phosphorylated with other functional consequences (38, 39). The cyclin subunit in mammals is usually either cyclin T1 or cyclin T2 (40), although cyclin K can also be found associated with CDK9 (41–43). Similar to other CDKs involved in transcription (CDK7, -8, -12, -13), cyclin T1, cyclin T2, and CDK9 do not change their concentration significantly during the cell cycle (44). X-ray structures are available for P-TEFb containing most of CDK9 and the N-terminal third of cyclin T1

(45) and for the human immunodeficiency virus (HIV) Tat•P-TEFb complex containing a similarly truncated cyclin T1 and 86 amino acid Tat (46). It should be noted that the P-TEFb structure alone contains three accidental mutations in CDK9, one of which likely causes a conformational change in the active site compared to the Tat•P-TEFb structure (Tahir Tahirov, personal communication). The Tat•P-TEFb structure should prove useful in the design of anti-HIV drugs that target the transcription of the virus.

P-TEFb has been demonstrated to phosphorylate a number of proteins. The CTD of the largest subunit of mammalian Pol II comprises 52 tandemly repeated copies of a heptapeptide with the consensus Tyr-Ser-Pro-Thr-Ser-Pro-Ser ($Y_1S_2P_3T_4S_5P_6S_7$). Like CDK7•cyclin H, P-TEFb phosphorylates the CTD (47), but, although CDK7 phosphorylates Ser5 of the CTD repeats, P-TEFb can phosphorylate primarily Ser2 (48). As described below, CTD phosphorylation is likely to be more involved in the coupling of RNA processing with transcription than in the release of promoter proximally paused polymerases. The role of P-TEFb in CTD phosphorylation has been challenged by the discovery that CDK12 and CDK13, two human kinases evolutionarily related to the Ser2 kinase Ctk1 in yeast, can also phosphorylate Ser2 (49, 50). Treatment of cells with the potent P-TEFb inhibitor, flavopiridol, does lead to loss of Ser2 phosphorylation, but this could be the result of a loss of Pol II in the 3′ regions of genes where Ser2 phosphorylation occurs (**Figure 2a**), and this could explain why Ser2 phosphorylation was lost in *Caenorhabditis elegans* when CDK9 or cyclin T was knocked down (51). For P-TEFb's main function in releasing poised polymerases, two negative factors, DSIF and NELF, are targeted (10). Phosphorylation of the C-terminal region of the larger of the two DSIF subunits, hSpt5, by P-TEFb is required for the transition into productive elongation (52, 53), and the factor then tracks with Pol II throughout the gene and in the region downstream of the poly(A) site (9). Phosphorylation of one of the four NELF subunits (NELFe) has also been linked with its removal from Pol II during the transition into productive elongation (54). Even though yeast do not have NELF or poised polymerases, the Bur1/2 kinase has been demonstrated to phosphorylate DSIF and seems to be performing a similar function to transition the polymerase into productive elongation (55, 56).

3.2. Regulation of P-TEFb

Because poised polymerases that are waiting for P-TEFb to begin mRNA production are prevalent across genomes, regulating the activity of the kinase is essential for proper gene expression. Most mechanisms used to control protein levels have been found to be involved in controlling P-TEFb subunits; these mechanisms include expression of alternative forms (40, 57), as well as transcriptional (58), translational (58), and posttranslational (27) control, some of which involves miRNAs (59), and regulated turnover of the proteins (60). However, one unique mechanism, reversible association with the 7SK snRNP, plays the major role (**Figure 3**). In rapidly growing cells, such as HeLa cells, up to 90% of P-TEFb is sequestered in an inactive form by an RNA-binding protein, HEXIM1 or HEXIM2 (HEXIM, hexamethylene bisacetamide-inducible proteins), which associates with the 7SK snRNP (36, 61–67). The bound CDK9 has been activated by T-loop phosphorylation (36, 37) but is held in an inactive state by an inhibitory domain of HEXIM that is exposed when it is bound to 7SK RNA (36, 68). The La-related protein (LARP7) (69–71) and 7SK methyl phosphate capping enzyme (MePCE) (72, 73) are constitutive components of the 7SK snRNP. Together, they stabilize the RNA and may be involved in regulation of the release of P-TEFb. Interestingly, the C-terminal domain of LARP7 can bind to the active site of MePCE and inactivate its methyl transferase activity, which prevents the reverse reaction that removes the 7SK cap structure (72).

The biochemical properties of the 7SK snRNP with HEXIM and the sequestered

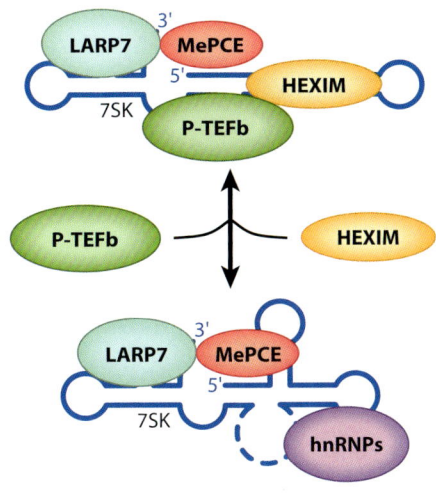

Figure 3

7SK snRNA is depicted in a cartoon view in which the secondary structures known to be involved in its function are shown. The La-related protein, LARP7, is constitutively associated with 7SK. The methyl phosphate capping enzyme, MePCE, which methylates the triphosphate at the 5′ end of 7SK is also bound. After binding to the major 5′ stem and loop, HEXIM1 or HEXIM2 (HEXIM) undergoes a conformational change and binds to and inhibits P-TEFb. When P-TEFb is released from the 7SK snRNP, HEXIM is also released, and there is a structural change in 7SK. hnRNPs then replace P-TEFb and HEXIM.

P-TEFb are ideal for a role in delivering P-TEFb to genes to be expressed while protecting poised polymerases on many other genes from inappropriate activation. In nuclei treated with mild detergents, the 7SK snRNP is the only snRNP that is readily extracted under low-salt conditions (74). This indicates that it is not tightly anchored anywhere in the nucleus, although transient interactions of the 7SK snRNP with chromatin have been found over the HIV promoter found in the long terminal repeat (LTR) (75). Diffusion of the 7SK snRNP would provide a means to supply P-TEFb to any location. Higher salt is needed to extract the 7SK-free P-TEFb, which is likely associated with the genes undergoing productive elongation at the time of extraction (74). Thus, an important mechanism for achieving selective P-TEFb function is the extraction

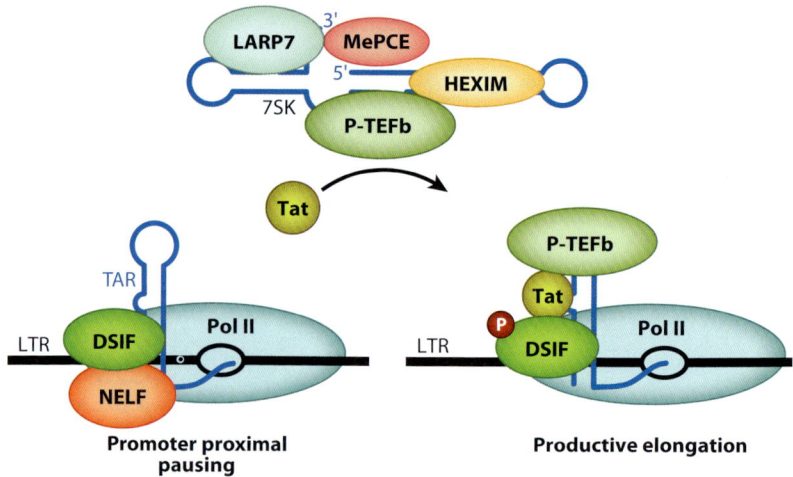

Figure 4

P-TEFb-mediated transition into productive elongation. Regulation of HIV transcription is used as an example of how P-TEFb is released from the 7SK snRNP and specifically recruited to its site of action. HIV Tat interacts with P-TEFb in the 7SK snRNP, leading to extraction of P-TEFb and the formation of a Tat•P-TEFb complex. P-TEFb is recruited to the poised polymerase on the HIV LTR by an interaction with the HIV nascent transcript TAR, where it phosphorylates DSIF (P) as well as NELF and the polymerase (not shown). Also not shown is the SEC, which can accompany and potentially help recruit P-TEFb.

SEC: super elongation complex

of P-TEFb from the 7SK snRNP, and recent evidence has demonstrated that this is possible. The HIV Tat protein and the P-TEFb-binding domain of the cellular bromodomain protein BRD4 can both individually directly release the P-TEFb from the 7SK snRNP without any energy or covalent modifications of the complex (**Figure 4**) (76–83). On the HIV LTR, the released P-TEFb is recruited to the nascent transcript held by the poised polymerase (**Figure 4**). Extraction of the P-TEFb by Tat does not require the HIV LTR (77, 79) and may be a special case because HIV has evolved to take over the P-TEFb control process. In contrast, BRD4 is generally found associated with acetylated histone tails on active chromatin, and its release of P-TEFb from 7SK snRNP could represent a more normal example of selective extraction at the site of activation (32, 84). Many cellular genes that employ BRD4 to recruit P-TEFb to their promoters are primary response genes, whose expression proceeds through signal-induced transcriptional elongation (85). A large number of transcription factors can bind to P-TEFb, but it is not yet clear if any or all of these can extract P-TEFb (10, 86). Because Tat has been found bound to P-TEFb in the super elongation complex (SEC) that also contains several other transcription factors/cofactors (81, 87), it is likely that the overall recruitment of P-TEFb, especially at the HIV LTR, is a multistep process. The interactions of SEC with the polymerase-associated factor complex (PAFc) through the PAF1 subunit and the human Mediator complex through the MED26 subunit of Mediator may provide additional means for delivering P-TEFb to cellular genes (88–92). Many details of the recruitment of P-TEFb to specific sites of action are currently unknown and this is an important area of future research.

What happens to P-TEFb after it has completed its function on a specific gene is just as important as its initial recruitment. Because free P-TEFb might inappropriately trigger the release of poised polymerases into productive elongation on undesired genes, it is critical to resequester it in the 7SK snRNP. During the extraction of P-TEFb, 7SK undergoes a conformational change that causes the release of

HEXIM (**Figure 3**) (76). This conformation is likely stabilized by the binding of several heterogeneous nuclear ribonucleoproteins (hnRNPs) (93, 94). For P-TEFb to be picked back up by the 7SK snRNP, HEXIM must first rebind. Because the concentration of HEXIM is about five times the concentration of P-TEFb in HeLa cells and half of the 7SK snRNPs do not contain HEXIM or P-TEFb (36, 61, 66), it is likely that entry of HEXIM into the 7SK snRNP is a regulated step. How this is accomplished is not clear. Support for the general model of P-TEFb's extraction and critical resequestration was recently obtained by observing the movement of 7SK in living cells containing an inducible gene array (95). During activation of the array, no specific recruitment of 7SK was observed, but when the genes were shut down, the 7SK concentration increased in that region. One speculative model would be that the 7SK snRNP containing hnRNPs might be temporarily recruited to the genes as they are shutting down to transfer the hnRNPs to the mRNA, and this might activate 7SK by changing the conformation of the RNA, allowing rebinding of HEXIM and then P-TEFb. Evidence for some of this model was provided by studies indicating the importance of the hnRNPs in 7SK dynamics (94).

4. THE PRODUCTIVE ELONGATION PHASE OF RNA POLYMERASE II TRANSCRIPTION

The properties of Pol II after P-TEFb triggers its release from the poised position are dictated by the factors that are now allowed to function. Prior to the transition into productive elongation, NELF and DSIF are primary elongation factors (9). After phosphorylation of DSIF by P-TEFb, NELF is ejected from the elongation complex, whereas DSIF is retained (9). Other elongation factors come into the complex and provide it with the remarkable ability to elongate at a steady ~3.8 kb/min rate through all obstacles for up to 2 million base pairs (96). The continued kinase activity of P-TEFb is not needed to maintain this rate (97). The identity and exact function of all of these factors are not known, but a number of candidate factors have been described. Some have been identified by their ability to affect Pol II elongation through activity-based biochemical approaches or by assays that detect their presence in elongation complexes in vivo. For example, transcription factor IIS (TFIIS) aids in restarting arrested Pol II after it falls into a backtracked state (98), and transcription factor IIF (TFIIF) has been shown to stimulate the elongation rate of Pol II in the absence of other factors (99). The PAFc, the SEC, and the SEC subunits ELL1 and -2 (eleven-nineteen lysine-rich leukemia 1 and 2 proteins), which, like TFIIF, stimulate elongation of pure Pol II, have recently gained much attention for their intricate roles in transcription, as well as transcription-coupled pre-mRNA processing. An excellent review of the Pol II elongation factors was published in 2004 (100), so we focus more heavily on recent work on the factors implicating them in productive elongation.

4.1. TFIIS and TFIIF

During transcription in vitro, Pol II has a strong propensity to pause or enter an arrested state at frequent sites along the template, and the first two factors discovered that affect elongation, TFIIS and TFIIF, affect pausing and arrest. The difference between pausing and arrest is that Pol II pausing is transient, and if given time, the polymerase will return to the elongation mode, but arrest cannot be overcome with time. If pausing persists, such as when nucleoside triphosphates are removed or a physical roadblock is imposed, the pause gradually decays into arrest (101), which is characterized by the "backtracking" of Pol II such that the active site within the polymerase is no longer aligned with the 3' end of the nascent RNA. Irreversibly backtracked Pol II is frequently a target for ubiquitylation/degradation (102). In eukaryotic cells, the transcript cleavage stimulatory factor TFIIS (originally S-II) is necessary to rescue Pol II from this arrested state by allowing Pol II to endonucleolytically cleave the RNA to generate a new 3′ end, which is properly aligned with the catalytic site (98). This type of activity is present throughout evolution as evidenced by the presence of two TFIIS-like factors, GreA and GreB in *E. coli* (103).

In cells, Pol II likely oscillates between the paused and arrested states, especially at promoter-proximal regions where negative factors restrict the forward momentum of the polymerase (104–106). Thus, TFIIS is expected to play a critical role in promoting the activity of early elongation complexes. Indeed, global analysis of paused genes in *Drosophila* reveals that Pol II backtracking and TFIIS-induced transcript cleavage are widespread among early elongation complexes (105). Depletion of TFIIS via RNAi from *Drosophila* cells delays the induction of heat-shock genes (107); however, global gene expression profiles are not significantly affected in TFIIS-depleted *Drosophila* cells (105). The yeast *ppr2* gene encoding the TFIIS ortholog is not essential, although *ppr2* mutants display sensitivity to 6-azauracil (108), a phenotype that is frequently associated with elongation defects (109).

The second factor discovered that affects elongation by Pol II is TFIIF. Identified by its strong affinity for Pol II (110) and by its requirement for initiation of transcription, it is also found to greatly increase the rate of elongation of Pol II in vitro (99, 111). Although it associates with free Pol II and pauses elongation complexes, it does not maintain this tight association during elongation (21, 99). Its positive action during elongation can be ascribed to its ability to bind to paused elongation complexes and to subsequently convert them into an elongation-competent form. Because elongation rates in vitro are dictated by the dwell time at many pause sites and TFIIF reduces this time, it has a concentration-dependent effect on elongation with the maximum rate stimulation of about 20-fold. Although ChIP experiments indicate that TFIIF is mostly found in promoter proximal regions in yeast (112), it has been found downstream, including at the 3′ end of genes in humans (113).

4.2. ELL1, -2, and -3

ELL1, the founding member of the ELL family that also contains ELL2 and ELL3, was originally purified from rat liver extracts as a factor that promoted Pol II elongation in vitro (114). Employing a mechanism different from that of TFIIS, ELL1 directly increases the catalytic rate of Pol II by suppressing transient pausing of Pol II at multiple sites along the DNA template (114). Prior to this discovery, ELL1 had been identified as a fusion partner of the mixed lineage leukemia (MLL) protein in acute myeloid leukemia (115).

MLL plays an important role in maintaining the appropriate expression of the *Hox* gene loci in hematopoietic cells and is a key regulator of hematopoietic stem cell development and maintenance (88). However, the *MLL* gene is frequently involved in aberrant chromosomal translocations with a large number of other genes to create fusion proteins, leading to the development of acute and aggressive myeloid and lymphoblastic leukemia. Wild-type MLL has histone methyltransferase activity capable of methylating histone 3 lysine 4 (H3K4), and this function is lost after the translocation. However, the myriad of DNA- and protein-binding domains present in the MLL N-terminal portion, which is retained in the chimeras, contribute to leukemogenesis. At the time when little was known about the functions of most of the MLL fusion partners, the discovery of the elongation stimulatory activity of ELL1 provided one of the earliest clues to how a fusion protein may induce aberrant expression of MLL target genes to cause leukemia (see below for further discussions).

On the basis of their homology to ELL1, ELL2 and ELL3 were identified and shown to have similar elongation stimulatory activity in vitro (116, 117). Although ELL2 is expressed in many of the same tissues as ELL1, ELL3 is a testis-specific factor. In *Drosophila*, there is only one ELL ortholog, dELL, that is detected at numerous sites of active transcription on polytene chromosomes, and dELL colocalizes with Pol II at heat-shock gene loci upon heat shock

(118). Interestingly, mutations in dELL preferentially affect the expression of large genes (119), implicating a critical role for this factor in facilitating promoter-distal elongation by Pol II, a process that is tightly coupled to mRNA 3′ end formation. Indeed, human ELL2 functions in this coupling (see below).

4.3. PAFc

Composed of five subunits Paf1, Ctr9, Cdc73, Rtf1, and Leo1, the PAFc is a multifunctional factor that contributes to transcriptional elongation, the generation of transcription-associated histone modifications, and the coupling of elongation with downstream events of mRNA biogenesis. The dependence on several, but not all, of the PAFc subunits for histone H2B monoubiquitylation and H3K4 and H3K79 methylation on active genes has been demonstrated (100). The PAFc-mediated recruitment of the ubiquitin-conjugating enzyme Rad6 and the SET domain histone methyltransferases to coding regions is responsible for these modifications. Although the modified nucleosomes are expected to influence the Pol II elongation on chromatin templates, PAFc has been shown to employ other modification-independent mechanisms to further control the elongation process. Indeed, PAFc has been demonstrated to promote Pol II elongation on both naked DNA and chromatin templates in the presence of other factors (120–122). Notably, the effect on chromatin is mechanistically different from the established role of PAFc in histone monoubiquitylation and methylation (122).

Like most other elongation factors, PAFc does not possess any enzymatic activity. Furthermore, the isolated complex has very little ability to stimulate the Pol II elongation rate in the absence of other factors. Instead, it functions as a mediator or adaptor to facilitate other elongation factors to bind and affect Pol II. Recent studies have revealed both genetic and physical interactions between the PAFc and a number of elongation factors, including DSIF, Tat-SF1, TFIIS, and SEC, as well as their cooperative interactions with Pol II (81, 120, 122–124). Through these interactions, PAFc and its partners coordinate the various aspects of transcriptional elongation and modulate the elongation complex at multiple stages. In agreement with its broad involvement in elongation control, PAF1c has been detected along the entire length of several actively transcribed genes in yeast (112, 125). A recent ChIP-seq analysis in MES cells further corroborates this view and shows that the PAFc subunit Ctr9 cross-links throughout coding regions of active genes and extends to the 3′ end together with the Ser2-phosphorylated Pol II (9). The peak Ctr9 occupancy is at the 3′ end of genes, reflecting the fact that PAFc also plays a key role in transcription-coupled mRNA 3′ end formation (see below).

4.4. Super Elongation Complexes

4.4.1. Compositions of human and *Drosophila* super elongation complexes.
In addition to being sequestered in the 7SK snRNP and the BRD4-containing complex, P-TEFb is also found in several closely related complexes that are collectively referred to as the SECs (81, 87, 126–128). Although only a small fraction of P-TEFb is present in the SECs (88), it is highly active as a CTD kinase (126). SECs also contain AFF1 (AF4/FMR2 family, member 1), AFF4 (AF4/FMR2 family, member 4), ELL1, ELL2, ENL, AF9 (ALL1-fused gene from chromosome 9), and probably additional subunits. Not all of them exist in a single complex. For example, the highly homologous ENL and AF9 were recently shown to exist in separate SECs that displayed similar but non-identical functions (124). It is possible that the homologous ELL1 and ELL2 as well as the similar AFF1 and AFF4 proteins are also present in different but closely related complexes (88, 128a). Thus, the number of possible combinations among these homologous subunits could be fairly large, which can greatly increase the regulatory diversity and gene-control options by a family of SECs.

Unlike mammals, *Drosophila* has only one dELL, one AFF-like protein encoded by the *lilliputian* (*lilli*) gene, and a single ENL/AF9-like protein called Ear. Together with *Drosophila* CDK9 and cyclin T, they exist in a fly version of the SEC (88). Both Lilli and dELL are well-known transcriptional regulators that exhibit strong genetic interactions with genes that contain paused Pol II and are essential for normal embryonic and eye development (reviewed in References 88 and 129). As expected from the functional coupling between dELL and P-TEFb within the *Drosophila* SEC, CDK9 depletion in larvae reduces the amount of dELL detected on the chromosomes (130), and dELL knockdown decreases Ser2 phosphorylation on the Pol II CTD (131).

4.4.2. Super elongation complexes and human diseases.
Human SECs were initially identified because of their close connection to two important diseases: HIV/AIDS (acquired immunodeficiency syndrome) and leukemia. As discussed above, HIV Tat can efficiently capture P-TEFb from the 7SK snRNP, raising the question of whether Tat delivers P-TEFb alone or together with other factors to the HIV LTR to activate viral transcription. To address this question, a tandem affinity-purification strategy was employed to isolate the complex that contains both CDK9 and Tat. This has led to the identification of the SEC components ELL2, AFF4, ENL, and AF9 as additional subunits of the Tat-P-TEFb complex (87). Through directly isolating proteins associated with Tat, the same set of factors plus a few others (e.g., AFF1, ELL1, and the PAFc subunits) were also found to interact with Tat and P-TEFb (81). The bromodomain protein, BRD4, is noticeably absent from the list, which is consistent with earlier demonstrations that BRD4 and Tat compete for binding to P-TEFb and that the BRD4-P-TEFb complex is incompatible with Tat transactivation (82, 83).

The interaction between Tat and SECs serves two purposes: (*a*) It recruits SECs to the HIV LTR through cooperative bindings to the viral TAR RNA, thereby allowing P-TEFb and ELL, elongation factors of two distinct classes, to act on the same polymerase enzyme to synergistically activate transcription (81, 87); and (*b*) it enhances the SEC formation through increasing the half-life of ELL2, which otherwise would be a short-lived protein targeted by the 26S proteasome for degradation (87). In HeLa and HEK293 cells, ~40% of total ELL2 proteins are in SECs, whereas a much smaller percentage of ELL1, which is stable in the absence of Tat, is sequestered in these complexes (87).

Not only are SECs targeted and regulated by the HIV-1 Tat protein, but they also play an important role in causing "runaway" transcription of genes that are normally subject to stringent control by wild-type MLL in developing lymphocytes. AFF1, AFF4, AF9, ENL, and ELL1, the five integral SEC components, are all well-known translocation partners of MLL (132). Their fusions to the N-terminal chromatin-targeting portion of MLL result in the recruitment of SECs and their associated elongation stimulatory activity to the MLL target genes to induce leukemic transformation (126–128).

4.4.3. Targeting SECs to RNA polymerase II for general elongation.
SECs are not made just for HIV-1 Tat or the MLL fusions. Rather, they are required for general transcriptional elongation of many non-HIV and non-MLL-target genes. In both mouse and human cells, they are recruited to many rapidly induced genes in response to differentiation signals to release paused Pol II for dynamic induction of transcription (89). Interestingly, most of these genes contain paused Pol II in their undifferentiated, unstimulated state. Even for a gene, *Cyp26a1*, which displays no detectable paused Pol II at the promoter, SECs are still required for rapid induction, although the precise role of SECs and their targets are unknown in this case (89). Nevertheless, these data underscore the general importance of SECs during developmental control of gene transcription.

In the absence of gene-specific recruitment factors, such as Tat and MLL, how are SECs recruited to the Pol II elongation complex? One potential mechanism relies on the interaction of the conserved YEATS domain in the N-terminal regions of ENL/AF9 with PAFc and, through PAFc, the elongating Pol II on chromatin templates (**Figure 5**) (124). This model is consistent with the previous demonstrations of an interaction between the PAFc and SECs (81) and also the requirement of PAFc for efficient Ser2 phosphorylation of the Pol II CTD (133). Moreover, it also explains why the YEATS domain is dispensable for leukemogenesis when ENL/AF9 is translocated to MLL (134). Apparently, the interactions with PAFc and DNA by the N-terminal portion of MLL in the fusion proteins (134, 135) can effectively substitute for the PAFc/chromatin-targeting function of the ENL/AF9 YEATS domain, which is frequently missing in the MLL chimeras. Notably, PAFc may not be the only means to recruit SECs to the elongation complex. A recent proteomic study has identified the human Mediator subunit MED26 as a docking site for SECs and proposed MED26 as a molecular switch for interacting first with TFIID during transcription initiation and then with SECs during the transition into early elongation (90). This observation agrees well with the previous data that indicate a role for the Mediator in controlling transcriptional elongation of some genes (136).

5. EVENTS COUPLED TO TRANSCRIPTION ELONGATION

The Pol II elongation complex does not merely extend a nascent mRNA chain to produce the full-length mRNA precursor. It also serves as a central coordinator to facilitate efficient integration of multiple nuclear processes that collectively control the proper expression of a gene. The coupling of transcription with key nuclear events, such as mRNA surveillance and export, has been reviewed elsewhere (100, 137). This review focuses on the roles of Pol II and several key elongation factors in modulating mRNA processing and maturation, which consist of the addition of the 7-methylguanylate cap structure to the 5′ end of nascent mRNA, the removal of intronic sequences, and the

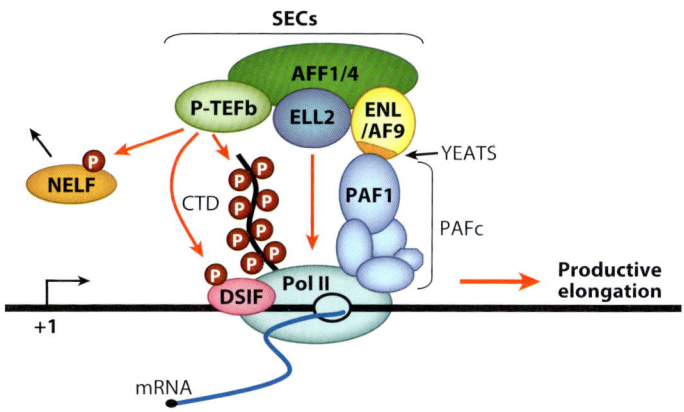

Figure 5

The super elongation complex, SEC. In the absence of sequence-specific recruitment factors such as HIV-1 Tat and the MLL fusions, the SEC complex, which is assembled around the scaffolding protein AFF4, is recruited to the elongating Pol II through the interaction of the YEATS domain of either ENL or AF9 with the PAF1 subunit of PAFc. This allows the SEC to use its P-TEFb and ELL2 functional modules to exert a multitude of effects (e.g., direct stimulation of the Pol II catalytic rate by ELL2, phosphorylation of Ser2 on the Pol II CTD, and DSIF by P-TEFb, and phosphorylation and release of NELF by P-TEFb) that result in the synergistic activation of Pol II elongation.

cleavage of RNA at the 3′ end followed by the addition of a poly(A) tail. Importantly, being cotranscriptional not only allows these events to occur more efficiently than if they do so completely independently, but it is also a necessity for fast and controlled expression of many, especially large, genes. Recently, using an assay in which elongation of long genes was synchronized by manipulating P-TEFb activity in vivo, it was found that splicing of exons, even when the intervening intron was several hundred thousand base pairs in length, occurred within 10 min of the time the polymerase passed the downstream exon (96). In many of the genes studied, the elongation complex was still in productive elongation moving at about 3.8 kb/min as the splicing machinery efficiently removed the intron (96).

An essential player in the coordination of these processing events is the CTD of Pol II, which is conveniently located next to the pre-mRNA exit channel (138) and functions as a platform for the assembly and actions of the various enzymes and proteins responsible for mRNA maturation (100). Additional posttranslational modifications of the CTD consensus sequences (e.g., serine and threonine glycosylation and proline isomerization) have also been reported (139, 140). These modifications, in conjunction with the many degenerate heptads in the CTD, offer tremendous combinatorial possibilities to modulate the interactions as well as the activities of the myriad of RNA processing factors, which in turn influence the many steps during the conversion of a gene into its corresponding mature, translatable mRNA.

5.1. Capping

Capping of the 5′ ends of nascent pre-mRNAs, which occurs when the transcripts reach a length of 25 to 30 nt, is influenced by the phosphorylation status of the CTD and accomplished by the sequential actions of the capping enzyme, which contains both RNA triphosphatase and guanylyltransferase activities, and then the guanine-7-methyltransferase (141). mRNAs produced by yeast Pol II mutants lacking an intact CTD are not efficiently capped (142, 143). Yeast guanylyltransferase interacts tightly and specifically with the CTD phosphorylated on Ser5 (141, 144). Phosphorylation of the CTD on Ser5, but not Ser2, stimulates the guanylyltransferase activity (145, 146). Kin28, the TFIIH-associated kinase responsible for Ser5 phosphorylation in yeast, is required for targeting guanylyltransferase to the Pol II transcription complex (147). The Spt5 subunit of DSIF, which is part of the early elongation complex, also stimulates capping together with the Ser5-phosphorylated CTD (142, 148). The human capping enzyme is stimulated 100,000-fold in vitro when the transcript is present in an isolated elongation complex compared to free RNA; however, unlike the situation in yeast, there is only a threefold dependence on Ser5 phosphorylation of the CTD (146).

These findings suggest the existence of a "checkpoint" mechanism to control the tight coupling between capping and early elongation. As Pol II containing the phospho-Ser5 CTD and bound by DSIF is specifically recognized by NELF, which serves to pause the elongation complex (19, 149), a window of opportunity is created for the CTD and DSIF to recruit and stimulate the capping machinery to cap a nascent transcript. Because the capping enzyme is also capable of relieving NELF-mediated repression of elongation (150), it may signal and cooperate with other enzymes, such as P-TEFb, to convert a paused elongation complex into a productive one once capping is completed. The second step of capping carried out by the guanine-7-methyltransferase has also been shown to be a control point during Myc activation of genes (11, 151, 152). Control of the fraction of caps that are methylated regulates the subsequent translation of mRNAs of Myc-dependent genes (151).

5.2. Splicing

Like capping, both constitutive and alternative splicing also maintain close cross talk with mammalian transcription. Furthermore, the Pol II CTD, especially the hyperphosphorylated form, is intimately involved in this cross talk (100). Although the exact mechanism is yet to be defined, it is suspected that the hyperphosphorylated CTD of elongating Pol II acts by recruiting key splicing factors to the elongation complex. For instance, the phospho-Ser2 CTD has been shown to interact with the Ser/Arg-rich (SR) proteins (e.g., SF2/ASF and SC35), Spt6, and snRNP particles (153–157). Providing visual support for this recruitment model, Pol II with the complete, but not a truncated, CTD was found to recruit splicing factors to sites of active transcription in living cells (158). In addition to the recruitment of splicing factors, the CTD may also assist the formation of early splicing intermediates, such as the U1 snRNP-containing complex assembled at the 5′ splice site and the U2 snRNP complex at the branch point adenosine (159).

Given the importance of the Ser2-phosphorylated CTD in the coupling between splicing and transcript elongation, it does not come as a surprise that recent studies have revealed a key role for P-TEFb in controlling this process. The release of P-TEFb from the 7SK snRNP as a result of LARP7 or MePCE depletion is shown to promote the inclusion of an alternative exon containing an exonic splicing enhancer element and a suboptimal 3′ splice site (154). This is likely achieved through the increased occupancy of P-TEFb at the promoter and exonic regions of the gene, which leads to a chain of events: enhanced elongation and CTD phosphorylation on Ser2, elevated associations of the SR proteins SF2/ASF and probably others with the CTD, the SR protein-mediated assembly of the so-called cross exon recognition complex (160) at the alternative exon, and ultimately the inclusion of this exon in the mature mRNA. In addition to the P-TEFb-dependent CTD phosphorylation on Ser2, reducing the rate of Pol II elongation by mutation or drugs also results in different splicing patterns in several other genes, suggesting a general role for the Pol II elongation complex in controlling splice-site selection during alternative splicing (161).

The P-TEFb-induced cotranscriptional mRNA splicing is also important for controlling the inducible inflammatory gene expression program in response to Toll-like receptor signaling in macrophages (85). In the absence of stimulation, Pol II at many of the GC-rich primary response genes generates low levels of full-length but unspliced and untranslatable transcripts. Gene induction is accomplished through signal-dependent recruitment of P-TEFb by BRD4, which recognizes the inducibly acquired histone H4 acetylation on lysines 5, 8, and 12. This results in robust CTD phosphorylation on Ser2 and production of high levels of fully spliced mature mRNA transcripts.

Not only does the splicing machinery take advantage of the Pol II elongation complex to efficiently remove introns from pre-mRNAs, but splicing factors can also promote elongation, providing yet another "checkpoint control" to ensure that a pre-mRNA is not synthesized unless the machinery for its processing is properly assembled and positioned. Demonstrating reciprocal cross talk between SR proteins and P-TEFb, RNAi-mediated depletion of SF2/ASF or SC35 decreases the levels of P-TEFb and phospho-Ser2 CTD at the promoter and exonic regions of some genes (154, 162). Furthermore, SC35 depletion causes Pol II accumulation within the gene body and attenuated transcriptional elongation in a gene-specific manner (162). Another example illustrating the stimulatory effect of the splicing machinery on Pol II elongation concerns the spliceosomal U snRNPs, which stimulate elongation when directed to an intron-free DNA template by the elongation factor Tat-SF1 (163). When bound to the U snRNPs, the Tat-SF1 can also interact with P-TEFb (163) and probably also with DSIF and PAFc (120); all are components of the Pol II elongation complex.

5.3. Cleavage and Polyadenylation

The proper processing of the human mRNA 3′ end is carried out by coordinated actions of a set of proteins and proceeds in two steps: (*a*) cleavage of the RNA precursor at a position 10–35 nt downstream of the consensus AAUAAA signal and (*b*) addition of 200–300 adenosine nucleotides to the newly generated 3′-OH terminus of the transcript by poly(A) polymerase. Multiple lines of evidence indicate that the Pol II CTD, especially the Ser2-phosphorylated form, promotes transcription-coupled 3′ end formation through recruiting key components of the processing machinery to their target sites. For instance, the Pcf11 subunit of the yeast cleavage/polyadenylation factor IA complex specifically recognizes the phospho-Ser2 CTD (164, 165). Similarly, the binding of human cleavage stimulation factor 64 (CstF-64) to Pol II depends on Ser2 phosphorylation in the CTD (166).

Although components of the cleavage/polyadenylation machinery can bind to the phospho-Ser2 CTD in vitro (e.g., see References 164 and 165), the bindings are likely enhanced in vivo by transcription elongation factors that track along with Pol II, resulting in more efficient recognition

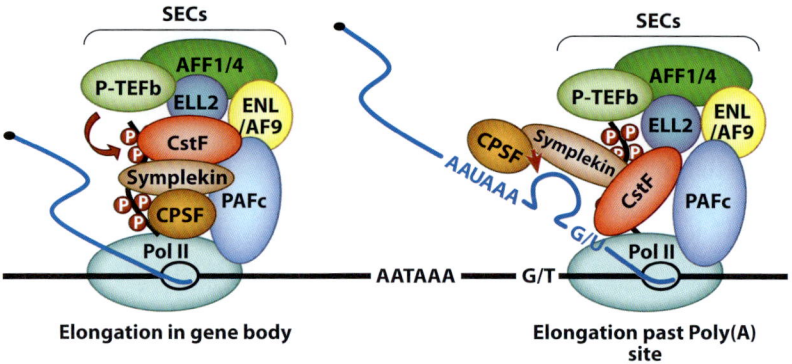

Figure 6

Coupling of 3′ end processing with transcription. Polyadenylation factors such as CstF, CPSF and symplekin are recruited to the Pol II elongation complex through the concerted actions of the phospho-Ser2 CTD and transcription elongation factors ELL2 and PAFc, which track along with Pol II during productive elongation. Once the cleavage/polyadenylation signals (AAUAAA followed by a G/U-rich sequence) emerge in the nascent mRNA, they are recognized by the processing machinery to result in efficient polyadenylation.

and cleavage at poly(A) sites (**Figure 6**). For example, the multifunctional PAFc was found to be required for transcription-coupled polyadenylation stimulated by the prototypical transcriptional activator GAL4-VP16 (167). Moreover, the PAFc subunit Cdc73 is essential for proper histone mRNA 3′ processing as well as export to the cytoplasm (168). PAFc likely accomplishes these tasks through recruiting key polyadenylation factors, such as the cleavage/polyadenylation specificity factor and the CstF, as well as symplekin, a scaffolding protein for 3′ processing, to transcribing Pol II (169, 170). Consistent with a critical role for PAFc in this process, the occupancy of the PAFc subunit Ctr9 peaks at the 3′ end of actively transcribed genes (9).

Another elongation factor that also facilitates the association of mRNA 3′ processing machinery with elongating Pol II is the SEC component ELL2 (171). It enhances the association of CstF-64 with Pol II across the gene encoding the immunoglobulin heavy-chain complex in plasma cells. As a result, ELL2 accelerates the use of a weak promoter-proximal secretory-specific poly(A) site and enhances exon skipping of a non-consensus splice signal that is in direct competition with the poly(A) site (171). Utilization of such sites for alternative polyadenylation is emerging as a common mechanism for controlling gene expression in development and disease (167). Like ELL2, PAFc has also been implicated in the control of alternative polyadenylation site usage. The usage of the promoter-proximal poly(A) site of *Ints6*, a target gene of the PAFc component Cdc73, is stimulated by Cdc73 in human cells (169).

The ELL2-promoted 3′ processing of mRNA correlates with a significant increase in Ser2 phosphorylation in the Pol II CTD (166). Notably, PAFc is also required for this modification (133). It had been difficult to explain these results prior to the recent discovery that ELL2 and P-TEFb are components of SECs, which depend on PAFc to interact with elongating Pol II (87, 91). With this new information, many seemingly unrelated events that have previously been attributed separately to ELL2 and PAFc, such as the promotion of Pol II elongation, phosphorylation of the CTD on Ser2, enhanced recruitment of polyadenylation factors to Pol II containing the phospho-Ser2 CTD, and finally accelerated mRNA 3′ formation, can now be placed into a complete and coherent picture (**Figure 6**).

SUMMARY POINTS

1. Pol II elongation control is essential for the regulation of gene expression.
2. Promoter proximally paused polymerases under the control of NELF and DSIF are prevalent across metazoan genomes.
3. P-TEFb controls the transition into productive elongation, generating mature mRNAs in a process at the 5′ end of genes.
4. The 7SK snRNP allows regulated release of active P-TEFb by specific transcription factors.
5. Productive elongation complexes are influenced by DSIF, PAFc, SEC, and other factors.
6. RNA processing is functionally coupled with transcription elongation.

FUTURE ISSUES

1. Are there other negative factors in addition to DSIF and NELF that are involved in promoter proximal pausing?
2. Does divergent transcription have a functional consequence?
3. What regulates directional preference in productive elongation at divergent promoters?
4. What are the mechanisms for delivery of active P-TEFb to specific genes?
5. How is P-TEFb re-sequestered in the 7SK snRNP?
6. What are all the factors needed for productive elongation? How are they sequentially recruited and then removed at the 3′ end of genes?

DISCLOSURE STATEMENT

The authors are not aware of any affiliations, memberships, funding, or financial holdings that might be perceived as affecting the objectivity of this review.

ACKNOWLEDGMENTS

This work was supported by National Institutes of Health grants GM35500 and AI074392 to D.H.P. and AI41757 and AI095057 to Q.Z. We thank Jiannan Guo for helpful comments on the manuscript.

LITERATURE CITED

1. Neil H, Malabat C, d'Aubenton-Carafa Y, Xu Z, Steinmetz LM, Jacquier A. 2009. Widespread bidirectional promoters are the major source of cryptic transcripts in yeast. *Nature* 457:1038–42
2. Xu Z, Wei W, Gagneur J, Perocchi F, Clauder-Munster S, et al. 2009. Bidirectional promoters generate pervasive transcription in yeast. *Nature* 457:1033–37
3. Seila AC, Calabrese JM, Levine SS, Yeo GW, Rahl PB, et al. 2008. Divergent transcription from active promoters. *Science* 322:1849–51
4. Core LJ, Waterfall JJ, Lis JT. 2008. Nascent RNA sequencing reveals widespread pausing and divergent initiation at human promoters. *Science* 322:1845–48

5. Seila AC, Core LJ, Lis JT, Sharp PA. 2009. Divergent transcription: a new feature of active promoters. *Cell Cycle* 8:2557–64
6. Muse GW, Gilchrist DA, Nechaev S, Shah R, Parker JS, et al. 2007. RNA polymerase is poised for activation across the genome. *Nat. Genet.* 39:1507–11
7. Zeitlinger J, Stark A, Kellis M, Hong JW, Nechaev S, et al. 2007. RNA polymerase stalling at developmental control genes in the *Drosophila melanogaster* embryo. *Nat. Genet.* 39:1512–16
8. Guenther MG, Levine SS, Boyer LA, Jaenisch R, Young RA. 2007. A chromatin landmark and transcription initiation at most promoters in human cells. *Cell* 130:77–88
9. Rahl PB, Lin CY, Seila AC, Flynn RA, McCuine S, et al. 2010. c-Myc regulates transcriptional pause release. *Cell* 141:432–45
10. Peterlin BM, Price DH. 2006. Controlling the elongation phase of transcription with P-TEFb. *Mol. Cell* 23:297–305
11. Cowling VH, Cole MD. 2007. The Myc transactivation domain promotes global phosphorylation of the RNA polymerase II carboxy-terminal domain independently of direct DNA binding. *Mol. Cell. Biol.* 27:2059–73
12. Chao SH, Price DH. 2001. Flavopiridol inactivates P-TEFb and blocks most RNA polymerase II transcription in vivo. *J. Biol. Chem.* 276:31793–99
13. Egloff S, Murphy S. 2008. Cracking the RNA polymerase II CTD code. *Trends Genet.* 24:280–88
14. Komarnitsky P, Cho EJ, Buratowski S. 2000. Different phosphorylated forms of RNA polymerase II and associated mRNA processing factors during transcription. *Genes Dev.* 14:2452–60
15. Mayer A, Lidschreiber M, Siebert M, Leike K, Söding J, Cramer P. 2010. Uniform transitions of the general RNA polymerase II transcription complex. *Nat. Struct. Mol. Biol.* 17:1272–78
16. Marshall NF, Price DH. 1992. Control of formation of two distinct classes of RNA polymerase II elongation complexes. *Mol. Cell. Biol.* 12:2078–90
17. Nechaev S, Adelman K. 2008. Promoter-proximal Pol II: when stalling speeds things up. *Cell Cycle* 7:1539–44
18. Wada T, Takagi T, Yamaguchi Y, Ferdous A, Imai T, et al. 1998. DSIF, a novel transcription elongation factor that regulates RNA polymerase II processivity, is composed of human Spt4 and Spt5 homologs. *Genes Dev.* 12:343–56
19. Yamaguchi Y, Takagi T, Wada T, Yano K, Furuya A, et al. 1999. NELF, a multisubunit complex containing RD, cooperates with DSIF to repress RNA polymerase II elongation. *Cell* 97:41–51
20. Renner DB, Yamaguchi Y, Wada T, Handa H, Price DH. 2001. A highly purified RNA polymerase II elongation control system. *J. Biol. Chem.* 276:42601–9
21. Cheng B, Price DH. 2007. Properties of RNA polymerase II elongation complexes before and after the P-TEFb-mediated transition into productive elongation. *J. Biol. Chem.* 282:21901–12
22. Chodosh LA, Fire A, Samuels M, Sharp PA. 1989. 5,6-Dichloro-1-beta-D-ribofuranosylbenzimidazole inhibits transcription elongation by RNA polymerase II in vitro. *J. Biol. Chem.* 264:2250–57
23. Marshall NF, Price DH. 1995. Purification of P-TEFb, a transcription factor required for the transition into productive elongation. *J. Biol. Chem.* 270:12335–38
24. Flynn RA, Almada AE, Zamudio JR, Sharp PA. 2011. Antisense RNA polymerase II divergent transcripts are P-TEFb dependent and substrates for the RNA exosome. *Proc. Natl. Acad. Sci. USA* 108:10460–65
25. Price DH. 2000. P-TEFb, a cyclin-dependent kinase controlling elongation by RNA polymerase II. *Mol. Cell. Biol.* 20:2629–34
26. Diribarne G, Bensaude O. 2009. 7SK RNA, a non-coding RNA regulating P-TEFb, a general transcription factor. *RNA Biol.* 6:122–28
27. Choo S, Schroeder S, Ott M. 2010. CYCLINg through transcription: Post-translational modifications of P-TEFb regulate transcription elongation. *Cell Cycle* 9:1697–705
28. Lenasi T, Barboric M. 2010. P-TEFb stimulates transcription elongation and pre-mRNA splicing through multilateral mechanisms. *RNA Biol.* 7:145–50
29. Wang Y, Liu XY, De Clercq E. 2009. Role of the HIV-1 positive elongation factor P-TEFb and inhibitors thereof. *Mini Rev. Med. Chem.* 9:379–85
30. Canduri F, Perez PC, Caceres RA, de Azevedo WF Jr. 2008. CDK9 a potential target for drug development. *Med. Chem.* 4:210–18

31. Romano G, Giordano A. 2008. Role of the cyclin-dependent kinase 9-related pathway in mammalian gene expression and human diseases. *Cell Cycle* 7:3664–68
32. Zhou Q, Yik JH. 2006. The Yin and Yang of P-TEFb regulation: implications for human immunodeficiency virus gene expression and global control of cell growth and differentiation. *Microbiol. Mol. Biol. Rev.* 70:646–59
33. Garriga J, Grana X. 2004. Cellular control of gene expression by T-type cyclin/CDK9 complexes. *Gene* 337:15–23
34. Rice AP, Herrmann CH. 2003. Regulation of TAK/P-TEFb in CD4+ T lymphocytes and macrophages. *Curr. HIV Res.* 1:395–404
35. De Falco G, Giordano A. 2002. CDK9: from basal transcription to cancer and AIDS. *Cancer Biol. Ther.* 1:342–47
36. Li Q, Price JP, Byers SA, Cheng D, Peng J, Price DH. 2005. Analysis of the large inactive P-TEFb complex indicates that it contains one 7SK molecule, a dimer of HEXIM1 or HEXIM2, and two P-TEFb molecules containing CDK9 phosphorylated at threonine 186. *J. Biol. Chem.* 280:28819–26
37. Chen R, Yang Z, Zhou Q. 2004. Phosphorylated positive transcription elongation factor b (P-TEFb) is tagged for inhibition through association with 7SK snRNA. *J. Biol. Chem.* 279:4153–60
38. Fong YW, Zhou Q. 2000. Relief of two built-in autoinhibitory mechanisms in P-TEFb is required for assembly of a multicomponent transcription elongation complex at the human immunodeficiency virus type 1 promoter. *Mol. Cell. Biol.* 20:5897–907
39. Garber ME, Mayall TP, Suess EM, Meisenhelder J, Thompson NE, Jones KA. 2000. CDK9 autophosphorylation regulates high-affinity binding of the human immunodeficiency virus type 1 Tat-P-TEFb complex to TAR RNA. *Mol. Cell. Biol.* 20:6958–69
40. Peng J, Zhu Y, Milton JT, Price DH. 1998. Identification of multiple cyclin subunits of human P-TEFb. *Genes Dev.* 12:755–62
41. Fu TJ, Peng J, Lee G, Price DH, Flores O. 1999. Cyclin K functions as a CDK9 regulatory subunit and participates in RNA polymerase II transcription. *J. Biol. Chem.* 274:34527–30
42. Chang PC, Li M. 2008. Kaposi's sarcoma–associated herpesvirus K-cyclin interacts with CDK9 and stimulates CDK9-mediated phosphorylation of p53 tumor suppressor. *J. Virol.* 82:278–90
43. Yu DS, Zhao R, Hsu EL, Cayer J, Ye F, et al. 2010. Cyclin-dependent kinase 9-cyclin K functions in the replication stress response. *EMBO Rep.* 11:876–82
44. Garriga J, Peng J, Parreno M, Price DH, Henderson EE, Grana X. 1998. Upregulation of cyclin T1/CDK9 complexes during T cell activation. *Oncogene* 17:3093–102
45. Baumli S, Lolli G, Lowe ED, Troiani S, Rusconi L, et al. 2008. The structure of P-TEFb (CDK9/cyclin T1), its complex with flavopiridol and regulation by phosphorylation. *EMBO J.* 27:1907–18
46. Tahirov TH, Babayeva ND, Varzavand K, Cooper JJ, Sedore SC, Price DH. 2010. Crystal structure of HIV-1 Tat complexed with human P-TEFb. *Nature* 465:747–51
47. Marshall NF, Peng J, Xie Z, Price DH. 1996. Control of RNA polymerase II elongation potential by a novel carboxyl-terminal domain kinase. *J. Biol. Chem.* 271:27176–83
48. Ramanathan Y, Rajpara SM, Reza SM, Lees E, Shuman S, et al. 2001. Three RNA polymerase II carboxyl-terminal domain kinases display distinct substrate preferences. *J. Biol. Chem.* 276:10913–20
49. Bartkowiak B, Greenleaf AL. 2011. Phosphorylation of RNAPII: To P-TEFb or not to P-TEFb? *Transcription* 2:115–19
50. Bartkowiak B, Liu P, Phatnani HP, Fuda NJ, Cooper JJ, et al. 2010. CDK12 is a transcription elongation-associated CTD kinase, the metazoan ortholog of yeast Ctk1. *Genes Dev.* 24:2303–16
51. Shim EY, Walker AK, Shi Y, Blackwell TK. 2002. CDK-9/cyclin T (P-TEFb) is required in two postinitiation pathways for transcription in the *C. elegans* embryo. *Genes Dev.* 16:2135–46
52. Yamada T, Yamaguchi Y, Inukai N, Okamoto S, Mura T, Handa H. 2006. P-TEFb-mediated phosphorylation of hSpt5 C-terminal repeats is critical for processive transcription elongation. *Mol. Cell* 21:227–37
53. Chen H, Contreras X, Yamaguchi Y, Handa H, Peterlin BM, Guo S. 2009. Repression of RNA polymerase II elongation in vivo is critically dependent on the C-terminus of Spt5. *PLoS ONE* 4:e6918

54. Fujinaga K, Irwin D, Huang Y, Taube R, Kurosu T, Peterlin BM. 2004. Dynamics of human immunodeficiency virus transcription: P-TEFb phosphorylates RD and dissociates negative effectors from the transactivation response element. *Mol. Cell. Biol.* 24:787–95
55. Zhou K, Kuo WH, Fillingham J, Greenblatt JF. 2009. Control of transcriptional elongation and co-transcriptional histone modification by the yeast BUR kinase substrate Spt5. *Proc. Natl. Acad. Sci. USA* 106:6956–61
56. Pei Y, Shuman S. 2003. Characterization of the *Schizosaccharomyces pombe* CDK9/Pch1 protein kinase: Spt5 phosphorylation, autophosphorylation, and mutational analysis. *J. Biol. Chem.* 278:43346–56
57. Shore SM, Byers SA, Dent P, Price DH. 2005. Characterization of CDK9(55) and differential regulation of two CDK9 isoforms. *Gene* 350:51–58
58. Herrmann CH, Carroll RG, Wei P, Jones KA, Rice AP. 1998. Tat-associated kinase, TAK, activity is regulated by distinct mechanisms in peripheral blood lymphocytes and promonocytic cell lines. *J. Virol.* 72:9881–88
59. Sung TL, Rice AP. 2009. miR-198 inhibits HIV-1 gene expression and replication in monocytes and its mechanism of action appears to involve repression of cyclin T1. *PLoS Pathog.* 5:e1000263
60. Kiernan RE, Emiliani S, Nakayama K, Castro A, Labbe JC, et al. 2001. Interaction between cyclin T1 and SCF(SKP2) targets CDK9 for ubiquitination and degradation by the proteasome. *Mol. Cell. Biol.* 21:7956–70
61. Michels AA, Fraldi A, Li Q, Adamson TE, Bonnet F, et al. 2004. Binding of the 7SK snRNA turns the HEXIM1 protein into a P-TEFb (CDK9/cyclin T) inhibitor. *EMBO J.* 23:2608–19
62. Yik JH, Chen R, Pezda AC, Samford CS, Zhou Q. 2004. A human immunodeficiency virus type 1 Tat-like arginine-rich RNA-binding domain is essential for HEXIM1 to inhibit RNA polymerase II transcription through 7SK snRNA-mediated inactivation of P-TEFb. *Mol. Cell. Biol.* 24:5094–105
63. Michels AA, Nguyen VT, Fraldi A, Labas V, Edwards M, et al. 2003. MAQ1 and 7SK RNA interact with CDK9/cyclin T complexes in a transcription-dependent manner. *Mol. Cell. Biol.* 23:4859–69
64. Nguyen VT, Kiss T, Michels AA, Bensaude O. 2001. 7SK small nuclear RNA binds to and inhibits the activity of CDK9/cyclin T complexes. *Nature* 414:322–25
65. Yang Z, Zhu Q, Luo K, Zhou Q. 2001. The 7SK small nuclear RNA inhibits the CDK9/cyclin T1 kinase to control transcription. *Nature* 414:317–22
66. Byers SA, Price JP, Cooper JJ, Li Q, Price DH. 2005. HEXIM2, a HEXIM1-related protein, regulates positive transcription elongation factor b through association with 7SK. *J. Biol. Chem.* 280:16360–67
67. Yik JH, Chen R, Pezda AC, Zhou Q. 2005. Compensatory contributions of HEXIM1 and HEXIM2 in maintaining the balance of active and inactive positive transcription elongation factor b complexes for control of transcription. *J. Biol. Chem.* 280:16368–76
68. Barboric M, Kohoutek J, Price JP, Blazek D, Price DH, Peterlin BM. 2005. Interplay between 7SK snRNA and oppositely charged regions in HEXIM1 direct the inhibition of P-TEFb. *EMBO J.* 24:4291–303
69. He N, Jahchan NS, Hong E, Li Q, Bayfield MA, et al. 2008. A La-related protein modulates 7SK snRNP integrity to suppress P-TEFb-dependent transcriptional elongation and tumorigenesis. *Mol. Cell* 29:588–99
70. Krueger BJ, Jeronimo C, Roy BB, Bouchard A, Barrandon C, et al. 2008. LARP7 is a stable component of the 7SK snRNP while P-TEFb, HEXIM1 and hnRNP A1 are reversibly associated. *Nucleic Acids Res.* 36:2219–29
71. Markert A, Grimm M, Martinez J, Wiesner J, Meyerhans A, et al. 2008. The La-related protein LARP7 is a component of the 7SK ribonucleoprotein and affects transcription of cellular and viral polymerase II genes. *EMBO Rep.* 9:569–75
72. Xue Y, Yang Z, Chen R, Zhou Q. 2010. A capping-independent function of MePCE in stabilizing 7SK snRNA and facilitating the assembly of 7SK snRNP. *Nucleic Acids Res.* 38:360–69
73. Jeronimo C, Forget D, Bouchard A, Li Q, Chua G, et al. 2007. Systematic analysis of the protein interaction network for the human transcription machinery reveals the identity of the 7SK capping enzyme. *Mol. Cell* 27:262–74

74. Biglione S, Byers SA, Price JP, Nguyen VT, Bensaude O, et al. 2007. Inhibition of HIV-1 replication by P-TEFb inhibitors DRB, seliciclib and flavopiridol correlates with release of free P-TEFb from the large, inactive form of the complex. *Retrovirology* 4:47
75. D'Orso I, Frankel AD. 2010. RNA-mediated displacement of an inhibitory snRNP complex activates transcription elongation. *Nat. Struct. Mol. Biol.* 17:815–21
76. Krueger BJ, Varzavand K, Cooper JJ, Price DH. 2010. The mechanism of release of P-TEFb and HEXIM1 from the 7SK snRNP by viral and cellular activators includes a conformational change in 7SK. *PLoS One* 5:e12335
77. Barboric M, Yik JH, Czudnochowski N, Yang Z, Chen R, et al. 2007. Tat competes with HEXIM1 to increase the active pool of P-TEFb for HIV-1 transcription. *Nucleic Acids Res.* 35:2003–12
78. Schulte A, Czudnochowski N, Barboric M, Schonichen A, Blazek D, et al. 2005. Identification of a cyclin T-binding domain in Hexim1 and biochemical analysis of its binding competition with HIV-1 Tat. *J. Biol. Chem.* 280:24968–77
79. Sedore SC, Byers SA, Biglione S, Price JP, Maury WJ, Price DH. 2007. Manipulation of P-TEFb control machinery by HIV: recruitment of P-TEFb from the large form by Tat and binding of HEXIM1 to TAR. *Nucleic Acids Res.* 35:4347–58
80. Muniz L, Egloff S, Ughy B, Jády BE, Kiss T. 2010. Controlling cellular P-TEFb activity by the HIV-1 transcriptional transactivator Tat. *PLoS Pathog.* 6:e1001152
81. Sobhian B, Laguette N, Yatim A, Nakamura M, Levy Y, et al. 2010. HIV-1 Tat assembles a multifunctional transcription elongation complex and stably associates with the 7SK snRNP. *Mol. Cell* 38:439–51
82. Yang Z, Yik JH, Chen R, He N, Jang MK, et al. 2005. Recruitment of P-TEFb for stimulation of transcriptional elongation by the bromodomain protein BRD4. *Mol. Cell* 19:535–45
83. Bisgrove DA, Mahmoudi T, Henklein P, Verdin E. 2007. Conserved P-TEFb-interacting domain of BRD4 inhibits HIV transcription. *Proc. Natl. Acad. Sci. USA* 104:13690–95
84. Wu SY, Chiang CM. 2007. The double bromodomain-containing chromatin adaptor BRD4 and transcriptional regulation. *J. Biol. Chem.* 282:13141–45
85. Hargreaves DC, Horng T, Medzhitov R. 2009. Control of inducible gene expression by signal-dependent transcriptional elongation. *Cell* 138:129–45
86. Marshall RM, Grana X. 2006. Mechanisms controlling CDK9 activity. *Front. Biosci.* 11:2598–613
87. He N, Liu M, Hsu J, Xue Y, Chou S, et al. 2010. HIV-1 Tat and host AFF4 recruit two transcription elongation factors into a bifunctional complex for coordinated activation of HIV-1 transcription. *Mol. Cell* 38:428–38
88. Smith E, Lin C, Shilatifard A. 2011. The super elongation complex (SEC) and MLL in development and disease. *Genes Dev.* 25:661–72
89. Lin C, Garrett AS, De Kumar B, Smith ER, Gogol M, et al. 2011. Dynamic transcriptional events in embryonic stem cells mediated by the super elongation complex (SEC). *Genes Dev.* 25:1486–98
90. Takahashi H, Parmely TJ, Sato S, Tomomori-Sato C, Banks CA, et al. 2011. Human mediator subunit MED26 functions as a docking site for transcription elongation factors. *Cell* 146:92–104
91. He N, Zhou Q. 2011. New insights into the control of HIV-1 transcription: when Tat meets the 7SK snRNP and super elongation complex (SEC). *J. Neuroimmune Pharmacol.* 6:260–68
92. Mohan M, Lin C, Guest E, Shilatifard A. 2010. Licensed to elongate: a molecular mechanism for MLL-based leukaemogenesis. *Nat. Rev. Cancer* 10:721–28
93. Barrandon C, Bonnet F, Nguyen VT, Labas V, Bensaude O. 2007. The transcription-dependent dissociation of P-TEFb-HEXIM1-7SK RNA relies upon formation of hnRNP-7SK RNA complexes. *Mol. Cell. Biol.* 27:6996–7006
94. Van Herreweghe E, Egloff S, Goiffon I, Jady BE, Froment C, et al. 2007. Dynamic remodelling of human 7SK snRNP controls the nuclear level of active P-TEFb. *EMBO J.* 26:3570–80
95. Prasanth KV, Camiolo M, Chan G, Tripathi V, Denis L, et al. 2010. Nuclear organization and dynamics of 7SK RNA in regulating gene expression. *Mol. Biol. Cell* 21:4184–96
96. Singh J, Padgett RA. 2009. Rates of in situ transcription and splicing in large human genes. *Nat. Struct. Mol. Biol.* 16:1128–33

97. Egyhazi E, Ossoinak A, Pigon A, Holmgren C, Lee JM, Greenleaf AL. 1996. Phosphorylation dependence of the initiation of productive transcription of Balbiani ring 2 genes in living cells. *Chromosoma* 104:422–33

98. Fish RN, Kane CM. 2002. Promoting elongation with transcript cleavage stimulatory factors. *Biochim. Biophys. Acta* 1577:287–307

99. Price DH, Sluder AE, Greenleaf AL. 1989. Dynamic interaction between a *Drosophila* transcription factor and RNA polymerase II. *Mol. Cell. Biol.* 9:1465–75

100. Sims RJ 3rd, Belotserkovskaya R, Reinberg D. 2004. Elongation by RNA polymerase II: the short and long of it. *Genes Dev.* 18:2437–68

101. Gu W, Reines D. 1995. Identification of a decay in transcription potential that results in elongation factor dependence of RNA polymerase II. *J. Biol. Chem.* 270:11238–44

102. Sigurdsson S, Dirac-Svejstrup AB, Svejstrup JQ. 2010. Evidence that transcript cleavage is essential for RNA polymerase II transcription and cell viability. *Mol. Cell* 38:202–10

103. Laptenko O, Lee J, Lomakin I, Borukhov S. 2003. Transcript cleavage factors GreA and GreB act as transient catalytic components of RNA polymerase. *EMBO J.* 22:6322–34

104. Rasmussen EB, Lis JT. 1995. Short transcripts of the ternary complex provide insight into RNA polymerase II elongational pausing. *J. Mol. Biol.* 252:522–35

105. Nechaev S, Fargo DC, dos Santos G, Liu L, Gao Y, Adelman K. 2010. Global analysis of short RNAs reveals widespread promoter-proximal stalling and arrest of Pol II in *Drosophila*. *Science* 327:335–38

106. Ujvari A, Pal M, Luse DS. 2002. RNA polymerase II transcription complexes may become arrested if the nascent RNA is shortened to less than 50 nucleotides. *J. Biol. Chem.* 277:32527–37

107. Adelman K, Marr MT, Werner J, Saunders A, Ni Z, et al. 2005. Efficient release from promoter-proximal stall sites requires transcript cleavage factor TFIIS. *Mol. Cell* 17:103–12

108. Nakanishi T, Shimoaraiso M, Kubo T, Natori S. 1995. Structure-function relationship of yeast S-II in terms of stimulation of RNA polymerase II, arrest relief, and suppression of 6-azauracil sensitivity. *J. Biol. Chem.* 270:8991–95

109. Mason PB, Struhl K. 2005. Distinction and relationship between elongation rate and processivity of RNA polymerase II in vivo. *Mol. Cell* 17:831–40

110. Sopta M, Carthew RW, Greenblatt J. 1985. Isolation of three proteins that bind to mammalian RNA polymerase II. *J. Biol. Chem.* 260:10353–60

111. Flores O, Maldonado E, Reinberg D. 1989. Factors involved in specific transcription by mammalian RNA polymerase II: Factors IIE and IIF independently interact with RNA polymerase II. *J. Biol. Chem.* 264:8913–21

112. Krogan NJ, Kim M, Ahn SH, Zhong G, Kobor MS, et al. 2002. RNA polymerase II elongation factors of *Saccharomyces cerevisiae*: a targeted proteomics approach. *Mol. Cell. Biol.* 22:6979–92

113. Cojocaru M, Jeronimo C, Forget D, Bouchard A, Bergeron D, et al. 2008. Genomic location of the human RNA polymerase II general machinery: evidence for a role of TFIIF and Rpb7 at both early and late stages of transcription. *Biochem. J.* 409:139–47

114. Shilatifard A, Lane WS, Jackson KW, Conaway RC, Conaway JW. 1996. An RNA polymerase II elongation factor encoded by the human *ELL* gene. *Science* 271:1873–76

115. Thirman MJ, Levitan DA, Kobayashi H, Simon MC, Rowley JD. 1994. Cloning of *ELL*, a gene that fuses to *MLL* in a t(11;19)(q23;p13.1) in acute myeloid leukemia. *Proc. Natl. Acad. Sci. USA* 91:12110–14

116. Shilatifard A, Duan DR, Haque D, Florence C, Schubach WH, et al. 1997. ELL2, a new member of an ELL family of RNA polymerase II elongation factors. *Proc. Natl. Acad. Sci. USA* 94:3639–43

117. Miller T, Williams K, Johnstone RW, Shilatifard A. 2000. Identification, cloning, expression, and biochemical characterization of the testis-specific RNA polymerase II elongation factor ELL3. *J. Biol. Chem.* 275:32052–56

118. Gerber M, Ma J, Dean K, Eissenberg JC, Shilatifard A. 2001. *Drosophila* ELL is associated with actively elongating RNA polymerase II on transcriptionally active sites in vivo. *EMBO J.* 20:6104–14

119. Eissenberg JC, Ma J, Gerber MA, Christensen A, Kennison JA, Shilatifard A. 2002. dELL is an essential RNA polymerase II elongation factor with a general role in development. *Proc. Natl. Acad. Sci. USA* 99:9894–99

120. Chen Y, Yamaguchi Y, Tsugeno Y, Yamamoto J, Yamada T, et al. 2009. DSIF, the Paf1 complex, and Tat-SF1 have nonredundant, cooperative roles in RNA polymerase II elongation. *Genes Dev.* 23:2765–77

121. Rondón AG, Gallardo M, Garcia-Rubio M, Aguilera A. 2004. Molecular evidence indicating that the yeast PAF complex is required for transcription elongation. *EMBO Rep.* 5:47–53

122. Kim J, Guermah M, Roeder RG. 2010. The human PAF1 complex acts in chromatin transcription elongation both independently and cooperatively with SII/TFIIS. *Cell* 140:491–503

123. Bai X, Kim J, Yang Z, Jurynec MJ, Akie TE, et al. 2010. TIF1gamma controls erythroid cell fate by regulating transcription elongation. *Cell* 142:133–43

124. He N, Chan CK, Sobhian B, Chou S, Xue Y, et al. 2011. Human polymerase-associated factor complex (PAFc) connects the super elongation complex (SEC) to RNA polymerase II on chromatin. *Proc. Natl. Acad. Sci. USA* 108:E636–45

125. Pokholok DK, Hannett NM, Young RA. 2002. Exchange of RNA polymerase II initiation and elongation factors during gene expression in vivo. *Mol. Cell* 9:799–809

126. Lin C, Smith ER, Takahashi H, Lai KC, Martin-Brown S, et al. 2010. AFF4, a component of the ELL/P-TEFb elongation complex and a shared subunit of MLL chimeras, can link transcription elongation to leukemia. *Mol. Cell* 37:429–37

127. Yokoyama A, Lin M, Naresh A, Kitabayashi I, Cleary ML. 2010. A higher-order complex containing AF4 and ENL family proteins with P-TEFb facilitates oncogenic and physiologic MLL-dependent transcription. *Cancer Cell* 17:198–212

128. Mueller D, Garcia-Cuéllar MP, Bach C, Buhl S, Maethner E, Slany RK. 2009. Misguided transcriptional elongation causes mixed lineage leukemia. *PLoS Biol.* 7:e1000249

128a. Biswas D, Milne TA, Basrur V, Kim J, Elenitoba-Johnson KS, et al. 2011. Function of leukemogenic mixed lineage leukemia 1 (MLL) fusion proteins through distinct partner protein complexes. *Proc. Natl. Acad. Sci. USA* 108:15751–56

129. Levine M. 2011. Paused RNA polymerase II as a developmental checkpoint. *Cell* 145:502–11

130. Eissenberg JC, Shilatifard A, Dorokhov N, Michener DE. 2007. CDK9 is an essential kinase in *Drosophila* that is required for heat shock gene expression, histone methylation and elongation factor recruitment. *Mol. Genet. Genomics* 277:101–14

131. Smith ER, Winter B, Eissenberg JC, Shilatifard A. 2008. Regulation of the transcriptional activity of poised RNA polymerase II by the elongation factor ELL. *Proc. Natl. Acad. Sci. USA* 105:8575–79

132. Harper DP, Aplan PD. 2008. Chromosomal rearrangements leading to MLL gene fusions: clinical and biological aspects. *Cancer Res.* 68:10024–27

133. Jaehning JA. 2010. The Paf1 complex: platform or player in RNA polymerase II transcription? *Biochim. Biophys. Acta* 1799:379–88

134. Slany RK, Lavau C, Cleary ML. 1998. The oncogenic capacity of HRX-ENL requires the transcriptional transactivation activity of ENL and the DNA binding motifs of HRX. *Mol. Cell. Biol.* 18:122–29

135. Muntean AG, Tan J, Sitwala K, Huang Y, Bronstein J, et al. 2010. The PAF complex synergizes with MLL fusion proteins at *HOX* loci to promote leukemogenesis. *Cancer Cell* 17:609–21

136. Balamotis MA, Pennella MA, Stevens JL, Wasylyk B, Belmont AS, Berk AJ. 2009. Complexity in transcription control at the activation domain-mediator interface. *Sci. Signal.* 2:ra20

137. Kohler A, Hurt E. 2007. Exporting RNA from the nucleus to the cytoplasm. *Nat. Rev. Mol. Cell Biol.* 8:761–73

138. Cramer P, Bushnell DA, Kornberg RD. 2001. Structural basis of transcription: RNA polymerase II at 2.8 angstrom resolution. *Science* 292:1863–76

139. Xu YX, Manley JL. 2007. The prolyl isomerase Pin1 functions in mitotic chromosome condensation. *Mol. Cell* 26:287–300

140. Kelly WG, Dahmus ME, Hart GW. 1993. RNA polymerase II is a glycoprotein. Modification of the COOH-terminal domain by O-GlcNAc. *J. Biol. Chem.* 268:10416–24

141. Shuman S. 2001. Structure, mechanism, and evolution of the mRNA capping apparatus. *Prog. Nucleic Acid Res. Mol. Biol.* 66:1–40

142. Cho EJ, Takagi T, Moore CR, Buratowski S. 1997. mRNA capping enzyme is recruited to the transcription complex by phosphorylation of the RNA polymerase II carboxy-terminal domain. *Genes Dev.* 11:3319–26
143. McCracken S, Fong N, Yankulov K, Ballantyne S, Pan G, et al. 1997. The C-terminal domain of RNA polymerase II couples mRNA processing to transcription. *Nature* 385:357–61
144. Bentley DL. 2005. Rules of engagement: co-transcriptional recruitment of pre-mRNA processing factors. *Curr. Opin. Cell Biol.* 17:251–56
145. Ho CK, Shuman S. 1999. Distinct roles for CTD Ser-2 and Ser-5 phosphorylation in the recruitment and allosteric activation of mammalian mRNA capping enzyme. *Mol. Cell* 3:405–11
146. Moteki S, Price D. 2002. Functional coupling of capping and transcription of mRNA. *Mol. Cell* 10:599–609
147. Rodriguez CR, Cho EJ, Keogh MC, Moore CL, Greenleaf AL, Buratowski S. 2000. Kin28, the TFIIH-associated carboxy-terminal domain kinase, facilitates the recruitment of mRNA processing machinery to RNA polymerase II. *Mol. Cell. Biol.* 20:104–12
148. Wen Y, Shatkin AJ. 1999. Transcription elongation factor hSPT5 stimulates mRNA capping. *Genes Dev.* 13:1774–79
149. Yamaguchi Y, Inukai N, Narita T, Wada T, Handa H. 2002. Evidence that negative elongation factor represses transcription elongation through binding to a DRB sensitivity-inducing factor/RNA polymerase II complex and RNA. *Mol. Cell. Biol.* 22:2918–27
150. Mandal SS, Chu C, Wada T, Handa H, Shatkin AJ, Reinberg D. 2004. Functional interactions of RNA-capping enzyme with factors that positively and negatively regulate promoter escape by RNA polymerase II. *Proc. Natl. Acad. Sci. USA* 101:7572–77
151. Cowling VH, Cole MD. 2010. Myc regulation of mRNA cap methylation. *Genes Cancer* 1:576–79
152. Cole MD, Cowling VH. 2009. Specific regulation of mRNA cap methylation by the c-Myc and E2F1 transcription factors. *Oncogene* 28:1169–75
153. Das R, Yu J, Zhang Z, Gygi MP, Krainer AR, et al. 2007. SR proteins function in coupling RNAP II transcription to pre-mRNA splicing. *Mol. Cell* 26:867–81
154. Barboric M, Lenasi T, Chen H, Johansen EB, Guo S, Peterlin BM. 2009. 7SK snRNP/P-TEFb couples transcription elongation with alternative splicing and is essential for vertebrate development. *Proc. Natl. Acad. Sci. USA* 106:7798–803
155. Yoh SM, Cho H, Pickle L, Evans RM, Jones KA. 2007. The Spt6 SH2 domain binds Ser2-P RNAPII to direct Iws1-dependent mRNA splicing and export. *Genes Dev.* 21:160–74
156. Mortillaro MJ, Blencowe BJ, Wei X, Nakayasu H, Du L, et al. 1996. A hyperphosphorylated form of the large subunit of RNA polymerase II is associated with splicing complexes and the nuclear matrix. *Proc. Natl. Acad. Sci. USA* 93:8253–57
157. Yuryev A, Patturajan M, Litingtung Y, Joshi RV, Gentile C, et al. 1996. The C-terminal domain of the largest subunit of RNA polymerase II interacts with a novel set of serine/arginine-rich proteins. *Proc. Natl. Acad. Sci. USA* 93:6975–80
158. Misteli T, Spector DL. 1999. RNA polymerase II targets pre-mRNA splicing factors to transcription sites in vivo. *Mol. Cell* 3:697–705
159. Hirose Y, Manley JL. 2000. RNA polymerase II and the integration of nuclear events. *Genes Dev.* 14:1415–29
160. Maniatis T, Tasic B. 2002. Alternative pre-mRNA splicing and proteome expansion in metazoans. *Nature* 418:236–43
161. Muñoz MJ, de la Mata M, Kornblihtt AR. 2010. The carboxy terminal domain of RNA polymerase II and alternative splicing. *Trends Biochem. Sci.* 35:497–504
162. Lin S, Coutinho-Mansfield G, Wang D, Pandit S, Fu XD. 2008. The splicing factor SC35 has an active role in transcriptional elongation. *Nat. Struct. Mol. Biol.* 15:819–26
163. Fong YW, Zhou Q. 2001. Stimulatory effect of splicing factors on transcriptional elongation. *Nature* 414:929–33
164. Barilla D, Lee BA, Proudfoot NJ. 2001. Cleavage/polyadenylation factor IA associates with the carboxyl-terminal domain of RNA polymerase II in *Saccharomyces cerevisiae*. *Proc. Natl. Acad. Sci. USA* 98:445–50

165. Licatalosi DD, Geiger G, Minet M, Schroeder S, Cilli K, et al. 2002. Functional interaction of yeast pre-mRNA 3′ end processing factors with RNA polymerase II. *Mol. Cell* 9:1101–11
166. Shell SA, Martincic K, Tran J, Milcarek C. 2007. Increased phosphorylation of the carboxyl-terminal domain of RNA polymerase II and loading of polyadenylation and cotranscriptional factors contribute to regulation of the Ig heavy chain mRNA in plasma cells. *J. Immunol.* 179:7663–73
167. Nagaike T, Logan C, Hotta I, Rozenblatt-Rosen O, Meyerson M, Manley JL. 2011. Transcriptional activators enhance polyadenylation of mRNA precursors. *Mol. Cell* 41:409–18
168. Farber LJ, Kort EJ, Wang P, Chen J, Teh BT. 2010. The tumor suppressor parafibromin is required for posttranscriptional processing of histone mRNA. *Mol. Carcinog.* 49:215–23
169. Rozenblatt-Rosen O, Nagaike T, Francis JM, Kaneko S, Glatt KA, et al. 2009. The tumor suppressor Cdc73 functionally associates with CPSF and CstF 3′ mRNA processing factors. *Proc. Natl. Acad. Sci. USA* 106:755–60
170. Takagaki Y, Manley JL. 2000. Complex protein interactions within the human polyadenylation machinery identify a novel component. *Mol. Cell. Biol.* 20:1515–25
171. Martincic K, Alkan SA, Cheatle A, Borghesi L, Milcarek C. 2009. Transcription elongation factor ELL2 directs immunoglobulin secretion in plasma cells by stimulating altered RNA processing. *Nat. Immunol.* 10:1102–9

Genome Regulation by Long Noncoding RNAs

John L. Rinn[1] and Howard Y. Chang[2]

[1]Department of Stem Cell and Regenerative Biology, Harvard University, Cambridge, Massachusetts 02138; email: john_rinn@harvard.edu

[2]Howard Hughes Medical Institute and Program in Epithelial Biology, Stanford University School of Medicine, Stanford, California 94305; email: howchang@stanford.edu

Keywords

functional genomics, chromatin, histone modifications, epigenetics

Abstract

The central dogma of gene expression is that DNA is transcribed into messenger RNAs, which in turn serve as the template for protein synthesis. The discovery of extensive transcription of large RNA transcripts that do not code for proteins, termed long noncoding RNAs (lncRNAs), provides an important new perspective on the centrality of RNA in gene regulation. Here, we discuss genome-scale strategies to discover and characterize lncRNAs. An emerging theme from multiple model systems is that lncRNAs form extensive networks of ribonucleoprotein (RNP) complexes with numerous chromatin regulators and then target these enzymatic activities to appropriate locations in the genome. Consistent with this notion, lncRNAs can function as modular scaffolds to specify higher-order organization in RNP complexes and in chromatin states. The importance of these modes of regulation is underscored by the newly recognized roles of long RNAs for proper gene control across all kingdoms of life.

Contents

- INTRODUCTION 146
- GENOMIC DISCOVERY OF LONG
  NONCODING RNAs 147
  - Tiling Microarrays 148
  - Chromatin Marks 149
  - RNA Sequencing 150
- GENOMIC CHARACTERIZATION
  OF LONG NONCODING
  RNAS 150
  - Excluding Protein-Coding
    Potential 151
  - Inference of Long Noncoding RNA
    Functions by Coexpression: Guilt
    by Association 151
  - High-Throughput Loss of Function
    by RNA Inference 152
- LONG NONCODING RNAS IN
  GENE REGULATION 152
  - Long Noncoding RNAs Bind to and
    Target Chromatin Regulators ... 152
  - Enhancer RNAs 155
  - The RNA-Chromatin Interface 155
  - Emerging Mechanistic Themes:
    Decoys, Scaffolds, and Guides ... 156
- THE GLOBAL RNA-CHROMATIN
  NETWORK 158
  - Sequencing of lncRNAs Bound to
    Proteins: RNA
    Immunoprecipitation and
    Cross-Linked
    Immunoprecipitation 158
  - Long Noncoding RNA Structure
    and Function 158
  - Long Noncoding RNAs and
    Disease 159

Long noncoding RNA (lncRNA): an RNA that functions as a large RNA gene

INTRODUCTION

The centrality of RNA in the flow of genetic information came to light in Jacob & Monod's 1961 paper "Genetic Regulatory Mechanisms in the Synthesis of Proteins," establishing the concept of messenger RNAs (1). In the 50 years since this landmark paper, numerous regulatory RNAs of all shapes and sizes have been discovered (2, 3). Long noncoding RNAs (lncRNAs) biochemically resemble mRNAs posited by Jacob & Monod, yet do not template protein synthesis. Rather, lncRNAs function as RNA genes to orchestrate genetic regulatory outputs. Today, lncRNA transcripts have emerged as a cryptic, but critical, layer in the genetic regulatory code (**Figures 1** and **2**).

Studies over the past several decades have pointed to the presence of large amounts of RNA that was transcribed but did not encode proteins (4–8). Some of this RNA was later explained by mRNA splicing and RNA genes comprising translation machinery and its regulation (i.e., ribosomal RNA, tRNA, RNase P, SRP-7S), yet the vast majority was still unexplained. Biochemical experiments were able to characterize many abundant structural and regulatory RNAs by cellular localization and sequence similarity (5, 8–12), and genetic studies identified a few lncRNA genes involved in imprinting and other cellular processes (*XIST*, *H19*, *AIR*) (13, 14). Additional genetic studies also pointed to an emerging class of small regulatory RNAs, such as microRNAs (miRNAs) (12, 15–18), that regulate the translation of mRNAs to fine-tune key genetic pathways. Collectively, these classical studies identified a diverse repertoire of RNAs but may have only scratched the surface of RNAs' functions in the cell.

The advent of full genome sequences enabled an unprecedented survey of the genomic landscape for new genes. Surprisingly, this prospecting for genes led to the discovery of numerous lncRNA genes, but not many more protein genes. At the same time, DNA microarray technology revealed that the genome encodes at least as many lncRNAs as the known protein-coding genes (19–22). In fact, further advancements in RNA sequencing and microarray technology allowed a consortium-wide effort to define all the transcribed bases in the genome. At present, lncRNAs are operationally defined as RNA genes larger than 200 bp that do not appear to have coding potential. Although this working definition is somewhat arbitrary, the size cutoff clearly distinguishes lncRNAs

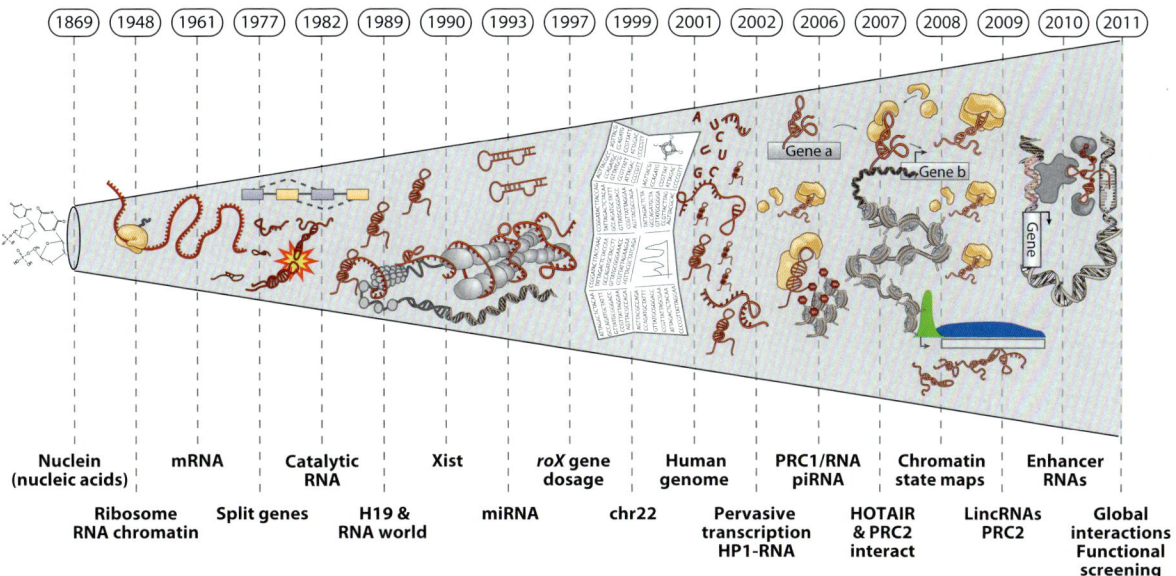

Figure 1

Timeline of discoveries of RNAs in biological regulation.

from small regulatory RNAs, such as miRNAs or piRNAs. Some classically defined small nuclear RNAs are in fact greater than 200 nucleotides, but the lncRNA designation is only prospectively applied to newly recognized transcripts. The conclusion was that a vast majority of the genome was transcribed (23).

In contrast to the substantial progress made in mapping lncRNAs, the functional roles for lncRNAs remained mostly elusive. In fact, the notion of such a widespread abundance of transcription was becoming controversial (23–25). More recently, dozens of functional examples have emerged implicating lncRNAs in numerous cellular processes ranging from embryonic stem cell (ESC) pluripotency, cell-cycle regulation, and diseases, such as cancer. Although lncRNAs exert a diverse spectrum of regulatory mechanisms across a variety of cellular pathways, a common theme is emerging: lncRNAs drive the formation of ribonucleic-protein complexes, which in turn influence the regulation of gene expression.

Here, we discuss multiple lines of evidence that point to lncRNAs as key regulatory layers in global gene regulation. We review the the technological approaches for genome-wide discovery and characterization of lncRNAs, as well as highlight emerging mechanistic themes based on well-studied examples from diverse model systems. We further evaluate several emerging studies indicating important roles for lncRNAs in the etiology of a wide spectrum of diseases.

GENOMIC DISCOVERY OF LONG NONCODING RNAS

At the turn of the twenty-first century, the scientific community was abuzz with great anticipation of the human genome project (26, 27). Perhaps at center stage was the burning question: How many genes are there in the human genome? Can the complexity of different organisms be explained by the sheer number of classic protein-coding genes, their splicing diversity, or perhaps new types of regulation? This simple yet profound question drove the progress of many technologies, such as microarray and DNA sequencing,

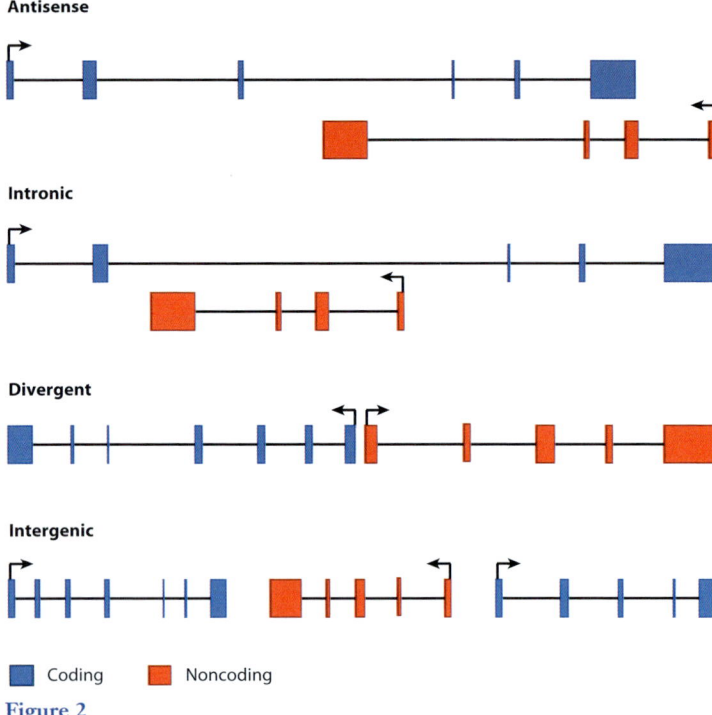

Figure 2

Anatomy of long noncoding RNA (lncRNA) loci. lncRNAs are often defined by their location relative to nearby protein-coding genes. Antisense lncRNAs are lncRNAs that initiate inside or 3′ of a protein-coding gene, are transcribed in the opposite direction of protein-coding genes, and overlap at least one coding exon. Intronic lncRNAs are lncRNAs that initiate inside of an intron of a protein-coding gene in either direction and terminate without overlapping exons. Bidirectional lncRNAs are transcripts that initiate in a divergent fashion from the promoter of a protein-coding gene; the precise distance cutoff that constitutes bidirectionality is not defined but is generally within a few hundred base pairs. Finally, intergenic lncRNAs (also termed large intervening noncoding RNAs or lincRNAs) are lncRNAs with separate transcriptional units from protein-coding genes. One definition required lincRNAs to be 5 kb away from protein-coding genes (44).

Pervasive transcription: the phenomenon, recognized in the early twenty-first century, showing that a vast majority of the genome is transcriptionally active

at an unprecedented rate. One of the first applications of automated Sanger sequencing in the mid-1990s was the mapping of expressed sequence tags (28, 29) that identify fragments of genomic regions that were being actively transcribed. This first glimpse into the transcriptome in 1996 revealed an intriguing new notion that many genes would be lurking in yet undefined regions of the human genome (29). Yet limited by short sequence reads, lower coverage, and an incomplete reference human genome to align expressed sequence tags, what these new genes may encode remained elusive.

Tiling Microarrays

In addition to sequencing advances, new technologies were emerging to understand the regulation of gene expression and de novo identification of new genes. In particular, the advent of DNA microarray technology provided the ability to survey on the order of 20,000 gene or genomic loci. In parallel, the first complete human chromosome 22 sequence was released in 1999 (30). The combined power of microarrays and draft genome sequences provided the first glimpse into pervasive transcription of noncoding RNAs. Specifically, two independent studies reported initial estimates that there may be as many lncRNA genes as protein-coding genes (21, 22). These studies used DNA microarrays with tiled or nested target sequences comprising the entirety of a chromosomal DNA sequence and allowed an unbiased survey of transcribed regions. Some limitations of tiling array studies included the potential for cross hybridization, the lack of strand-specific information if cDNAs were hybridized to the array, and the connectivity between transcribed regions not being known. Nonetheless, both studies were able to confirm expression from numerous noncoding loci by reverse transcription polymerase chain reaction, RNA-blot analysis, and evolutionary conservation studies, yet these findings were met with healthy skepticism that they may simply represent transcriptional noise. In a potentially interesting historical parallel, the discovery of "DNA-like RNA" upon phage infection in 1956 was a critical clue leading to the "mRNA hypothesis" (31). Yet the significance of this finding was not recognized for at least five years because the DNA-like RNA was less than 1% of total RNA (mostly rRNA) and thus assumed to be irrelevant noise.

Thus, one of the fruits of the Human Genome Project was the discovery of numerous new RNA genes, but not new protein-coding genes. For example, the number of human miRNAs quickly rose from a handful to nearly 1,000 (32–34). In fact, the further advancements in RNA sequencing, cDNA cloning, and

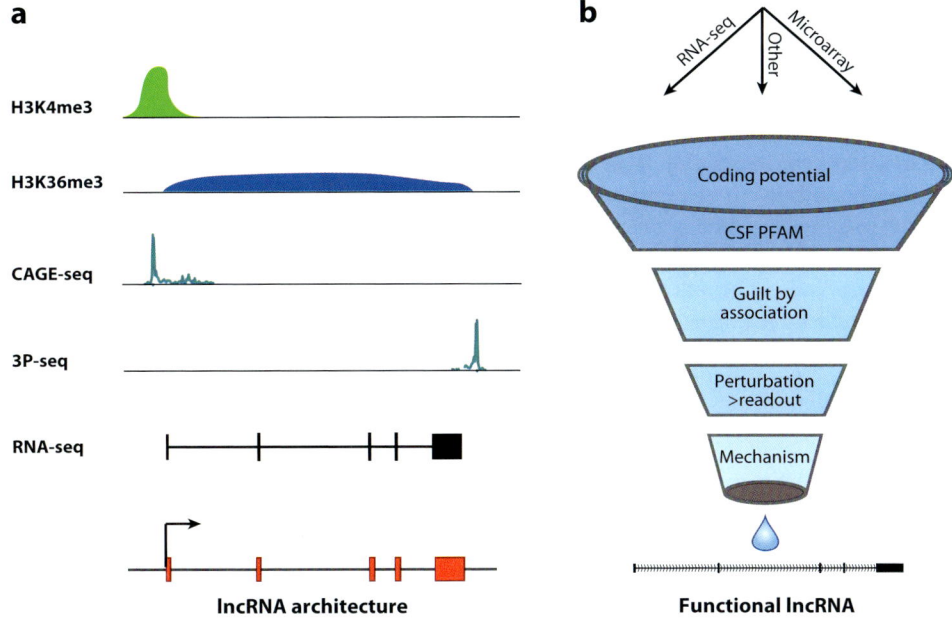

Figure 3

Functional discovery pipeline of long noncoding RNAs (lncRNAs). (*a*) Genome-wide discovery of lncRNAs. Chromatin marks of transcription initiation (histone H3 lysine 4 trimethylation, H3K4me3) and elongation (H3 lysine 36 trimethylation, H3K36me3) define transcribed regions of the genome, and sequencing of capped RNA fragments (CAGE-seq) or polyadenylation ends (3P-seq) define the precise beginnings and ends of transcripts. RNA sequencing (RNA-seq) can directly define the primary structure of lncRNAs. (*b*) Multiple bioinformatic tools and functional studies refine the set of lncRNAs associated with specific biological functions. Abbreviations: CSF, codon substitution frequency analysis; PFAM, Protein Family Database.

microarray technology of the next decade allowed a consortium-wide effort to define all the transcribed bases in the human genome. The conclusion was that a vast majority of the genome was transcribed (23). Despite the observed pervasive transcription throughout the genome, pinpointing functional RNA molecules was tantamount to finding needles in a haystack. The notion of such a widespread abundance of transcription was becoming increasingly controversial (24, 25, 35–38).

Chromatin Marks

A critical clue for hunting RNA genes came from chromatin, the DNA-protein complex where all eukaryotic genes reside (**Figure 3***a*). With the full genome sequence in hand, chromatin immunoprecipitation followed by massively parallel sequencing generated genomic maps of the chromatin architecture that have been termed the epigenome (39–41). Massively parallel sequencing of DNA sites occupied by histones and their modifications revealed numerous interesting domains of genomic architecture (39, 42). This included a clear signature of polymerase II–transcribed genes occupied by histone H3 lysine 4 trimethylation (H3K4me3) at the promoters of genes, followed by histone H3 lysine 36 trimethylation along the transcribed unit (K4-K36 domains) (42–44).

Surprisingly, surveying the entire mouse and human genomes by chromatin marks in several cell types revealed that approximately 5,000 K4-K36 domains represented lncRNAs (44). These lncRNAs had discrete gene loci that reside in previously unannotated intergenic

Large intergenic noncoding RNA (lincRNA): a lncRNA that does not overlap protein-coding genes

regions between protein-coding genes; hence, these RNAs were named large intergenic noncoding RNAs (lincRNAs). Further analysis of these loci revealed highly conserved promoter regions that recruit the binding and direct regulation of key transcription factors (24, 44, 45). lincRNAs show more sequence conservation throughout evolution over introns or untranscribed intergenic sequences, suggestive of their functionality (24, 44, 45), and syntenic lincRNAs in different species can have conserved functions (45a). Moreover, expression patterns of lncRNAs are associated with numerous key cellular processes, such as pluripotency, immune response, and regulation of the cell cycle (44, 46–48). More recently, approximately a third of lincRNAs were found to associate with chromatin-modifying complexes (49) and to modulate key cellular pathways (48, 50–52).

RNA Sequencing

The advent of deep sequencing technology led to the ability to sequence cDNA at an unprecedented scale and throughput, termed RNA-seq (53–57). These approaches have been coupled to computational methods, allowing the reconstruction of transcripts and their isoforms at single-nucleotide resolution (55–57). These studies have provided an unbiased identification of noncoding transcripts across many cell types and tissues (56, 58).

In addition to full-length reconstruction algorithms, several applications have emerged to perform RNA-seq. For example, a method termed 3-seq targets the polyadenylated tail of cDNA to quantitatively measure the abundance of transcripts using more affordable short sequence reads (59). Moreover, a variant of this method can be employed to precisely map 3′ ends of transcripts (60). Recently, metabolically labeled mRNAs have been utilized to measure nascent transcription, thereby providing insights into the pausing of polymerase and transcriptional dynamics (61). These and many other emerging technologies are providing ever deeper insights into the dynamic transcriptome (**Figure 3a**).

Recent studies have utilized RNA sequencing and transcript abundance estimations to identify specific properties of distinct classes of large RNA genes. For example, one study (58) identified 8,000 lincRNAs in the human genome by integrating numerous annotation sources in combination with RNA sequencing. This study revealed several global properties of lncRNAs, including a tendency for location next to developmental regulators, enrichment of tissue-specific expression patterns, identification of thousands of orthologous lincRNAs between human and mouse, and localization of hundreds of lincRNAs in gene deserts associated with genetic traits (58). Leveraging the ever increasing depth of sequencing and read lengths has allowed some of the first steps toward characterizing lncRNAs on a global scale.

GENOMIC CHARACTERIZATION OF LONG NONCODING RNAS

We next discuss experimental and computational approaches to identify, map, and derive hypotheses for lncRNA function (**Figure 3**). By combining the above technical approaches, it is now possible to identify all transcribed loci (K4-K36 chromatin domains) as well as to precisely map the primary structure of the RNA products (RNA-seq) (44, 45, 56). These combined layers of information are synergistic where chromatin modifications identify stably transcribed gene loci, and RNA sequencing allows detection of even very low-abundance transcripts that alone could be argued as transcriptional noise. The additional chromatin information also indicates the promoter region of a given locus (H3K4me3) and the transcribed unit (H3K36me3), thereby assisting in the mapping of RNA transcript 5′ and 3′ ends. Through the successive addition of additional layers of information (such as conservation, coding potential patterns, and anatomical properties) progress is being made toward identifying lncRNA gene families (43–45).

On the basis of the anatomical properties of their gene loci, lncRNAs have been further classified: for example, antisense lncRNAs that overlap known protein-coding genes, intronic lncRNAs that are encoded within introns of protein-coding genes, lncRNAs that overlap protein-coding genes termed overlapping transcripts, and lincRNAs that are encoded completely within the intergenic genomic space between protein-coding loci (**Figure 2**). Although this anatomic characterization has been used initially, it is likely that research will show many of these lncRNAs share similar mechanistic and functional roles.

Excluding Protein-Coding Potential

Whether an RNA transcript functions by coding for protein is fundamental to the definition of lncRNA, but obtaining the answer is a challenging task. Many studies have assessed lncRNA-coding potential by translating each lncRNA in all 3′ frames and performing homology queries (i.e., BLASTX) across large protein family and domain databases (i.e., Swissprot and PFAM). These informatic analyses are good initial indications of protein-coding capacity but may miss newly evolved protein sequences or very small open reading frames (<50 amino acids). To address the former issue, codon substitution frequency analyses have been used to determine if codons for amino acids are preferentially conserved through evolution, indicating preservation of protein-coding potential (62, 63). Yet even these two methods combined could still miss small open reading frames buried in these long transcripts. Ribosomal profiling, an experimental menthod, which identifies putative RNAs that are bound and scanned by the ribosome, has provided additional insights into those RNAs that may encode small peptides (64). Moreover, this method identifies the region of ribosomal occupancy, thereby further homing in on potential translated regions that can be used as refined input into informatics predictions, such as codon substitution frequency and BLASTX. Although some portion of lncRNAs may encode small peptides (65), we note this does not rule out the potentially dual nature of lncRNAs acting through RNA and their protein products. This has been exemplified by numerous mRNAs that contain regulatory noncoding RNA elements (i.e., p53, Sgrs, Oskar, VegT, and others) (66–71).

Inference of Long Noncoding RNA Functions by Coexpression: Guilt by Association

With the mapping of thousands of lncRNA loci, the next challenge is to determine what lncRNAs do. A first step in hypothesis generation is to use the expression patterns of lncRNAs to identify specific cell types or biological processes associated with each candidate lncRNA. Some of the first expression studies of lncRNAs identified lncRNAs that are highly expressed in certain brain regions. In situ hybridization studies further confirmed these expression patterns, revealing exquisite patterns of expression in specific substructures of the mouse brain [see Mercer et al. (72)]. A similar study by this group identified numerous lncRNAs that were tightly correlated with pluripotency transcription factors, suggesting that many lncRNAs may function in stem cell pluripotency transcriptional networks (73).

More recently, an informatic method termed guilt by association allowed a global understanding of lncRNAs and protein-coding genes that are tightly coexpressed and thus presumably coregulated (44). This method, using gene-expression analyses, identifies protein-coding genes and pathways significantly correlated with a given lncRNA. Thus, on the basis of the known functions of the coexpressed protein-coding genes, hypotheses are generated for the functions and potential regulators of the candidate lncRNA. Moreover, this analysis revealed families of lncRNAs on the basis of the pathways with which they do and do not associate. This approach has predicted diverse roles for lncRNAs, ranging from stem cell pluripotency to cancer (44). For example, numerous lncRNAs that were tightly correlated with p53 were induced in a p53-dependent manner, many more than would be expected by chance (44, 47, 48). These lncRNAs also were enriched with the p53-binding motif

Codon substitution frequency: determines the evolutionary pressure to preserve synonymous amino acid content

Guilt by association: hypothesis generation of lncRNA function by the coexpression of lncRNAs with protein-coding mRNAs; can group lncRNAs into the pathways they may regulate

Loss-of-function (LOF) approaches: define lncRNA functional roles

in their promoters. Moreover, one of these lncRNAs, termed lincRNA-p21 and predicted to be associated with p53, was found to be directly regulated by p53 and subsequently formed a lncRNA-ribonucleoprotein (RNP) with a nuclear factor to act as a global transcriptional repressor, facilitating p53-mediated apoptosis (48). Similarly, several lncRNAs predicted to be associated with adipogenesis and pluripotency have recently been identified as requirements for maintaining these cellular states (46; M. Guttman and L. Sun, personal communications).

Other expression correlation analyses have revealed additional functional roles of lncRNAs. For example, a recent study profiling lncRNAs across more than 130 breast cancers contained varying grades of the tumor and clinical information (74). This study identified numerous lncRNAs that are specifically up- or downregulated in tumor subtypes. For example, it was identified that a lncRNA termed HOTAIR, encoded in the HOXC cluster, was a strong predictor of breast cancer metastasis. In fact, enforced expression of HOTAIR was sufficient to drive breast cancer metastasis. More global expression studies of lncRNAs overlapping promoter regions of protein-coding genes identified numerous lncRNAs associated with cell-cycle regulation (47). This led to the functional characterization of a lncRNA named PANDA, which plays a critical role in inhibiting p53-mediated apoptosis. The guilt-by-association methods are universally applicable to any biological system. For example, a family of telomere-encoded lncRNAs in the malaria parasite (*Plasmodium falciparum*) was identified by their stage-specific coexpression with PfsiP2, a key virulence transcription factor (75).

These and other correlative studies have started to identify specific roles of lncRNAs in global transcriptional regulation. Homing in on the pathways with which lncRNAs are associated has allowed for (or provided) hypothesis-driven experiments that identified functional lncRNAs. Yet the full scope of lncRNA transcriptional regulation and function is far from understood. To understand the more global regulatory roles of lncRNAs, comprehensive gain- or loss-of-function (LOF) experiments need to be performed.

High-Throughput Loss of Function by RNA Inference

Indeed, a very recent study performed a LOF study across most (237) of the lincRNAs expressed in mouse ESCs and characterized the resulting effects on global gene expression (52). The authors demonstrated that knockdowns of lincRNAs have major consequences on gene expression patterns, comparable to knockdown of well-known ESC regulators. Intriguingly, this global screen determined that lincRNAs primarily affect gene expression in *trans*. Perhaps more importantly, dozens of lincRNAs were found to be functionally required in the maintenance of the pluripotent state. Further investigation into the molecular circuitry of ESCs showed that lincRNA genes are regulated by key transcription factors and that lincRNA transcripts physically bind to multiple chromatin regulatory proteins to affect shared gene expression programs (52). This study provided the first glimpse of global lincRNA functional properties and mechanisms and highlights their key role in the circuitry controlling the ESC state.

LONG NONCODING RNAS IN GENE REGULATION

Long Noncoding RNAs Bind to and Target Chromatin Regulators

The intimate connection between RNA and chromatin—the DNA-protein complex where all eukaryotic genes reside—was recognized over 40 years ago (76). In 1975, Paul & Duerksen (5) made the surprising finding that biochemically purified chromatin contained twice as much RNA as DNA, raising the idea that RNA may influence chromatin structure and gene regulation. Through the years, it has been demonstrated that RNA is required for proper chromatin structure and recruitment of the chromatin-modifying complexes to DNA (14). Yet the specific RNA species associated remained elusive. Genetic studies in the ensuing decades revealed a few lncRNAs that were associated with heterochromatin formation and imprinting [i.e., Xist (77), Air (78), H19 (79)]. Breakthroughs over the past few years have revealed numerous examples of lncRNAs in controlling the access or dismissal of regulatory proteins from chromatin (**Table 1**). Here, we first focus on the protein-binding partners of lncRNAs, next review the targeting mechanism of lncRNAs, and lastly discuss the emerging mechanistic themes of lncRNAs in gene regulation.

Several studies have shown that lncRNAs can target several chromatin modification complexes involved in gene silencing (**Table 1**). One of the most dramatic examples of lncRNA-mediated chromatin regulation occurs during X chromosome dosage compensation in mammals. Briefly, dosage compensation refers to the process whereby the gene expression level of the two X chromosomes in female cells is made equal to the single X in male cells. The lncRNA Xist is expressed from one of the two X chromosomes in female cells and results in altering the chromatin structure of an entire chromosome—the inactive X—where most genes are transcriptionally silenced (reviewed in Reference 80). Importantly, Xist physically associates with the Polycomb repressive complex 2 (PRC2) through a structured domain termed Repeat A, resulting in the localization of the PRC2 and its cognate histone mark histone H3 lysine 27 trimethylation (H3K27me3) to the inactive X chromosome (81). In an analogous fashion, plants control the seasonal timing of flowering (a process termed vernalization) by a cold-inducible intronic lncRNA termed COLDAIR; COLDAIR recruits the PRC2 in *cis* to silence the flowering regulator gene *FLC* (82). PRC2 recruitment by lncRNA can also regulate distantly located genes throughout the genome. Human lncRNA HOTAIR was the first of such RNAs recognized. HOTAIR physically associates with the PRC2 and modulates the PRC2 and H3K27me3 localization of hundreds of sites throughout the genome (83, 84). Several additional studies have identified HOTAIR and Xist as interfacing with the PRC2 via the catalytic methyltransferase subunit EZH2 (81, 85), although other proteins are likely also involved (86). The precise molecular interactions between lncRNAs and the Polycomb complex have yet to be defined.

In addition, lncRNAs can target diverse chromatin regulators. In imprinting, the paternally and maternally inherited alleles are differentially expressed, and lncRNAs are often involved in distinguishing the two alleles. Both Air and Kcnq1ot1 are lncRNAs that are transcribed from the silenced paternal allele, and they specifically bind to and recruit the histone H3 lysine 9 methylase G9a in *cis* to mediate H3K9me3 and transcriptional silencing of *Kcnq1* or *Igf2r* loci, respectively (87, 88). DNA methylation can be regulated by lncRNA. In plants, the interplay of small interfering RNAs and nascent lncRNAs in targeting DNA methylation is well known (89), but a different mechanism, apparently independent of small regulatory RNAs, also operates in mammalian cells. Transcriptional repression of the repetitive ribosomal RNA gene loci (*rDNA*) depends in part on an ncRNA termed pRNA, which recruits DNMT3b to mediate cystosine methylation (90). Additionally, AN-RIL, a lncRNA that is associated with cardiac disease, associates with CBX7 of the PRC1, facilitating the H3K27me3-based silencing of the *INK4a* locus (91). Two p53-regulated lncRNAs, lincRNA-p21 and PANDA, have been recently identified as interfacing with DNA-binding proteins such as hnRNP-K and NF-YA, and these interactions result in transcriptional repression at specific genomic loci (47, 48). Finally, beyond chromatin modifications, SRA, a lncRNA, can interact with and enhance the function of the insulator protein CTCF (92); CTCF can control higher-order chromosomal looping and "insulate" specific genes from the effects of long-range enhancers and regulatory elements. On the basis of these numerous examples, it is clear that many chromatin regulatory complexes moonlight as

PRC1 and -2: Polycomb repressive complex 1 and 2

Table 1 Protein partners of long noncoding RNAs[a]

Proteins[b]	Long noncoding RNAs	Long noncoding RNA functions	References
CTCF	SRA	Enhances insulator function of CTCF	92
DNMT3b	pRNA	Targets DNMT3b in *cis* to the rRNA locus via an RNA:DNA:DNA triplex for cytosine methylation and gene silencing	90
G9a	Kcnq1ot1, Air	Targets H3K9 methylase G9a in *cis* for imprinting	87, 88
Glucocorticoid receptor	Gas5	Binds to glucocorticoid receptor as a decoy and titrates GR away from target genes	102
hnRNP-K	lincRNA-p21	Targets hnRNP-K in *trans* to mediate p53-dependent gene repression	48
LSD1-CoREST	HOTAIR, many others	Targets the LSD1 complex to demethylate H3K4me2 to enforce gene silencing	49, 84
MLL-WDR5	HOTTIP, some eRNAs?	Binds to and localizes the MLL complex and H3K4me3 via chromosomal looping for gene activation	95, 96
NF-YA	PANDA	p53 inducible and titrates away NF-YA to favor survival over cell death during DNA damage	47
PRC1	ANRIL, Xist	Targets PRC1 in *cis* for gene silencing. ANRIL influences p16INK4a expression and cell senescence	9, 91
PRC2	Xist, HOTAIR, ANRIL, COLDAIR, Gtl2, Kcnq1ot1, many others	Targets PRC2 either in *cis* or *trans* to mediate H3K27 methylation and gene silencing for dosage compensation, imprinting, and developmental gene expression	49, 81, 83, 88, 109
Serine/arginine-rich splicing factors	MALAT1	Sequesters serine/arginine splicing factors to regulate alternative splicing	125
Staufen	1/2 SBS RNAs	Pairs with mRNAs via Alu repeats and targets them into a nonsense-mediated decay pathway	126
Set1 and Hda1/2/3 HDACs	CUTs, XUTs	Antisense RNAs repress sense transcription via control of H3K4me3 and histone deacetylation	127–130
hnRNP-A	TERRA	Controls telomerase access to telomeres in a cell-cycle phase-specific manner	131
TFIIB	DHFR minor	Titrates away TFIIB during cell quiescence to decrease *DHFR* transcription	99
TLS	CCND1 promoter ncRNA	Allosterically binds TLS to inhibit CREB binding protein and p300 activity, leads to repression of the *CCND1* gene	132
YY1	Xist	YY1 binding nucleates Xist on the inactive X chromosome	100

[a]Many of these long noncoding RNA-protein complexes function at chromatin.
[b]Abbreviations: 1/2 SBS, half Staufen binding site; CUTs, cryptic unstable transcripts; MLL, mixed lineage leukemia protein; TLS, translocated in liposarcoma; XUTs, Xrn1-sensitive unstable transcripts.

RNA-binding proteins; the ability to bind lncRNAs endows them with condition- or allele-specific recognition of target gene chromatin.

Enhancer RNAs

For historical reasons, many of the initial studies focused on RNAs associated with repressive chromatin-modifying complexes. Yet several other studies have also demonstrated that active chromatin states are associated with lncRNAs. Genome-scale mapping of histone modifications and enhancer-binding proteins have provided an additional layer of information to identify lncRNAs involved in gene activation. Chromatin immunoprecipitation followed by massively parallel sequencing analysis of H3K4me1, H3K27ac, and p300, several marks associated with gene-activating enhancers, showed these regions also produce lncRNA transcripts. Many such enhancer RNAs were bidirectional, lacked a polyA tail, and had a very low copy number (93, 94). Although many of these transcripts were initially thought to be by-products of polymerase II transcription or enhancer-promoter interaction, more evidence is pointing to functional roles of the lncRNAs. A recent study performed LOF experiments and found 7 of 12 lncRNAs affected expression of their cognate neighboring genes (95). The authors continued to demonstrate it was not the act of transcription rather the RNA itself that was important for gene enhancer activation. Although this trend of lncRNAs affecting transcription of neighboring genes is not a universal phenomenon (47, 52), these studies clearly demonstrate a functional role for the RNA molecule beyond that of a simple by-product of transcription in enhancer regions.

More recently, HOTTIP, an enhancer-like lncRNA, has been discovered to directly interact with the WDR5 protein, a key component of the mixed lineage leukemia/Trx complex, which catalyzes the activating H3K4me3 mark (96). HOTTIP is encoded on the distal 5′ end of the *HOXA* gene cluster. Chromosomal looping of the 5′ end of *HOXA* in an enhancer-like manner brings HOTTIP into spatial proximity with multiple *HOXA* genes, enforcing the maintenance of H3K4me3 and gene activation. Remarkably, in vivo HOTTIP LOF experiments silenced *HOXA* expression and altered limb morphology, consistent with its role in activating *HOXA* genes (96). Collectively, these studies demonstrate the critical importance of lncRNAs interfacing with chromatin-modifying machinery, resulting in enhancer-based gene activation, and these findings raise the possibility that many other enhancer-like RNAs may operate via similar mechanisms. Thus, a critical path forward for understanding the functions of lncRNAs is to understand the repertoire of lncRNA-binding proteins.

The RNA-Chromatin Interface

How does a lncRNA interface with selective regions of the genome? Several hypotheses have been forwarded, including (*a*) formation of an RNA:DNA:DNA triplex; (*b*) RNA binding to a sequence-specific DNA-binding protein; (*c*) an RNA:DNA hybrid that displaces a single strand of DNA (so-called R-loop); and (*d*) an RNA:RNA hybrid of lncRNA with a nascent transcript (97, 98). Mechanisms (*a*) and (*b*) have been experimentally demonstrated in several systems (**Figure 4**).

Two studies have demonstrated that the association of lncRNAs and chromatin complexes can also include recruitment to DNA. The first of such studies demonstrated that a lncRNA encoded upstream of the *DHFR* gene forms a triplex structure with the promoter, which binds to and sequesters the general transcription factor IIB and prevents transcription of *DHFR* (99). More recently, another study identified a 150–250-nucleotide species of ncRNA, termed pRNA, which also forms a triplex at rDNA loci to recruit DNMT3b to this location through the DNA-RNA triplex (90). In each of these cases, purified ncRNA is able to bind to the cognate DNA sequence to form a triplex structure in vitro, but it is difficult to demonstrate that such a triplex forms in living cells. Nonetheless, on the basis of these precedents,

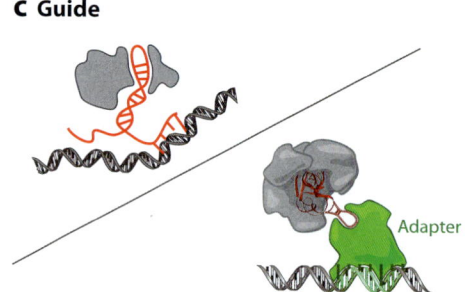
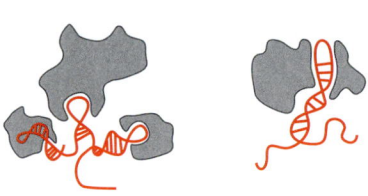
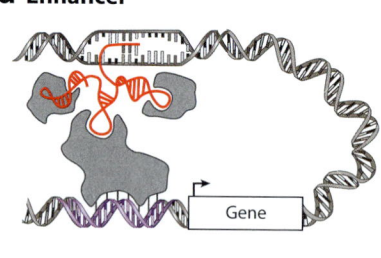

Figure 4

Models of long noncoding RNA (lncRNA) mechanisms of action. (*a*) The lncRNAs can act as decoys that titrate away DNA-binding proteins, such as transcription factors. (*b*) These lncRNAs may act as scaffolds to bring two or more proteins into a complex or spatial proximity and (*c*) may also act as guides to recruit proteins, such as chromatin modification enzymes, to DNA; this may occur through RNA-DNA interactions or through RNA interaction with a DNA-binding protein. (*d*) Such lncRNA guidance can also be exerted through chromosome looping in an enhancer-like model, where looping defines the *cis* nature and spread of the lncRNA effect.

it is likely that many more DNA-RNA interactions will be identified that serve as molecular beacons to recruit specific protein complexes.

By contrast, the ability of Xist to localize to the inactive X chromosome depends on the ability of Xist to bind to the sequence-specific transcription factor, YY1 (100). When nascent Xist is transcribed, the interaction of the Xist repeat C region with YY1 on the inactive X chromosome captures and nucleates Xist. Conversely, ectopic insertion of multiple copies of the YY1 motif can mobilize Xist from the inactive X chromosome to the ectopic sites. Why Xist is not captured by the numerous YY1 sites on autosomes remains a mystery.

In addition to the four hypothesized targeting strategies, the recent example of HOTTIP introduced a new concept for lncRNA targeting via chromosomal looping

(**Figure 4**) (96). Mature HOTTIP RNA appears to have no ability to seek out the *HOXA* locus if HOTTIP is ectopically produced elsewhere in the genome. But endogenous, nascent HOTTIP RNA is brought to its target genes via chromosomal looping. In this way, the lncRNA can serve as a faithful conduit to transform spatial information in chromosome conformation into chemical information in histone modifications.

Emerging Mechanistic Themes: Decoys, Scaffolds, and Guides

The ability of lncRNAs to bind to protein partners endows them with several regulatory capacities. Despite our limited knowledge from just dozens of characterized examples, several mechanistic themes of lncRNAs' functions have emerged (101). Three main themes encompass many of the examples discussed thus far (**Figure 4**).

1. Decoys: First, and at the simplest level, lncRNAs can serve as decoys that preclude the access of regulatory proteins to DNA. For example, the lncRNA Gas5 is induced upon growth factor starvation; Gas5 contains a hairpin sequence motif that resembles the DNA-binding site of the glucocorticoid receptor (102). Thus, upon starvation conditions, Gas5 is induced and serves as a decoy to release the receptor from DNA to prevent transcription of metabolic genes. A more recent lncRNA decoy example was identified, termed PANDA, which associates with the transcription factor NF-YA to prevent p53-mediated apoptosis (47). NF-YA transactivates several key genes for apoptosis, but PANDA binding to NF-YA titrates NF-YA away from target gene chromatin.

2. Scaffold: The lncRNAs can serve as adaptors to bring two or more proteins into discrete complexes (103). The telomerase RNA TERC is a classic example of an RNA scaffold that assembles the telomerase complex (104). Another prime example of lncRNA scaffolds is HOTAIR, which can simultaneously bind both the PRC2 and the LSD1-CoREST complex via specific domains of the RNA structure (84). This combination of interactions coordinates H3K27 methylation and H3K4me2 demethylation, ensuring gene silencing. Additional examples include ANRIL, which combines the PRC2 and the PRC1 (91, 105); and Kcnq1ot1, which interacts with both the PRC2 and G9a to promote H3K27me3 and H3K9me3, two different silencing histone marks (88). Both of these combinations are likely to reinforce the transcriptionally silent state. Importantly, the concept of RNA as molecular scaffold is likely to generalize more globally as hundreds of lncRNAs have been identified to form ribonucleic-protein interactions with multiple protein partners (49, 52).

3. Guides: As described above (**Table 1**), many lncRNAs are individually required for the proper localization of specific protein complexes. The lncRNAs involved in dosage compensation and imprinting (Xist, Kcnq1ot1, Air) serve as guides to target gene silencing activity in an allele-specific fashion. HOTAIR also serves as a guide to localize the PRC2 in developmental and cancer-related gene expression (74, 83). As another example, lincRNA-p21 is directly induced by p53 upon DNA damage and, in turn, physically associates with nuclear factor hnRNP-K to reroute this protein to specific promoters (48). Guide lncRNAs thereby combine two basic molecular functions—binding of a protein partner plus a mechanism to interface with selective regions of the genome.

Decoy: the notion that lncRNAs can associate with DNA-binding proteins to prevent their binding to DNA recognition elements

Scaffold: the formation of lncRNA-RNPs where the RNA joins several proteins together in a complex

Guide: the formation of a lncRNA-RNP that imparts specificity to genomic locations through either DNA-protein or RNA-DNA recognition rules

The concept of guide lncRNAs has previously been parsed by whether the guidance occurs in *cis* (on neighboring genes) or in *trans* (on distantly located genes). The *cis* actors have been assumed to occur in a cotranscriptional manner, leading to the analogy of lncRNAs as tethers (106). But recent experiments, where ectopically supplied lncRNAs are shown to seek out their cognate target sites, show that even *cis*-acting lncRNAs have the capacity to act in *trans* (90, 99, 100). The *cis* actions also do not simply correlate with distance from the site of lncRNA synthesis (87, 96), perhaps reflecting the important role of chromosomal looping in so-called *cis* effects. Interestingly, all of the classic long and small ncRNAs that have been well characterized also work in *trans* through interactions with proteins (i.e., rRNA, tRNA, snoRNA, RNase P, TERC). Two groups have recently developed methods to map the genomic binding sites of specific lncRNAs in a comprehensive manner (106a, 106b). Future studies that allow global mapping of lncRNA binding sites may better define the *cis* versus *trans* nature of lncRNA actions.

These three modes of action encompass many of the recently discovered lncRNA mechanistic themes, yet there are likely many other mechanisms to be uncovered. It is clear from the above examples that, as additional protein partners and targeting mechanisms (not just to DNA but perhaps also to other cellular structures) are discovered, it is possible to build complex regulatory scripts out of lncRNAs (107). Essential to understanding the meaning of such scripts will be a systematic understanding of the individual parts of lncRNA and their relevant interactions—akin to deciphering the codons of messenger RNAs.

THE GLOBAL RNA-CHROMATIN NETWORK

Sequencing of lncRNAs Bound to Proteins: RNA Immunoprecipitation and Cross-Linked Immunoprecipitation

Could the above described examples of Xist, HOTAIR, lincRNA-p21 and HOTTIP simply be a few quirky examples of lncRNAs interacting with chromatin-modifying complexes or are lncRNAs a more global phenomenon? Several recent studies point to the latter. These studies employ protein immunoprecipitation followed by microarray or deep sequencing to enumerate all RNAs associated with a protein complex of interest (RNA immunoprecipitation sequencing, RIP-seq) (49, 108, 109); often, cross-linking is performed to trap the relevant interactions in living cells (CLIP-seq) (110). For example, human PRC2 is associated with approximately 20% of lincRNAs expressed in a given cell type. Moreover, depletion of several PRC2-bound lincRNAs resulted in altered expression of PRC2-regulated gene loci in *trans* (49). Several of these PRC2-bound lincRNAs were independently identified as bound to the chromatin fraction (111), and a subset of these same lncRNAs are also associated with

RNA immunoprecipitation (RIP) and cross-linked immunoprecipitation (CLIP): procedures to identify lncRNAs associated with protein coding genes such as PRC2

LSD1 complexes (49), raising the possibility of numerous lncRNA scaffolds. A similar study in mouse ESCs identified approximately 9,000 lncRNAs associated with PRC2 (109). Thus, numerous lncRNAs are associated with PRC2 and may serve as guides as do HOTAIR, Xist, Kcnq1ot1, and Air.

To address lncRNA-protein interactions at a more global level, a recent study systematically performed RIP in combination with LOF experiments by depleting the lincRNAs bound to a given chromatin-modifying complex, as well as depleting the chromatin-modifying complex itself (52). Interestingly, depletion of the lincRNAs associated with a given complex collectively phenocopied the depletion of the complex itself for PRC2 and several other complexes. These results strongly suggest that lincRNAs serve to modulate the targeting of the chromatin-modifying complex to specific genomic loci, which is an emerging mechanistic theme. Yet future studies will need to investigate the directness of these interactions and determine how lncRNAs may alter the activities of highly dynamic chromatin-modifying complexes.

Long Noncoding RNA Structure and Function

One of the most the intriguing features of RNA is the malleable adoption of secondary and tertiary structures that relate to function. Classic chemical probing and structural studies have resulted in a structural understanding of several lncRNAs, including the atomic structure of the largest known RNP complex–the ribosome (112). Several recent studies have developed genome-scale approaches to measure RNA secondary structures; these approaches also apply to lncRNAs (113). These studies use either chemical probes to acylate flexible RNA bases that are not participating in structural interactions or use specific enzymes that cleave structured and unstructured regions of RNAs. For example, a large-scale application of chemical probing followed by sequencing of reverse-transcription products, termed SHAPE (selective 2′-hydroxyl acylation analyzed by primer extension), revealed the secondary structure of the entire RNA genome of the human immunodeficiency virus (114). More recently, deep sequencing of RNA fragments generated by enzymes that cleave single-stranded and double-stranded regions of RNA mapped the secondary structure of the entire budding yeast transcriptome, revealing several global structural properties (115). Notably, this study observed a triplet-based structural motif across gene bodies that correlates with translational efficiency. By contrast, 5′ and 3′ untranslated regions were observed to be much more lowly structured. In addition to the *Saccharomyces cerevisiae* transcriptome, this study also confirmed the known structural motifs and structural properties of the HOTAIR lncRNA. Two other studies successfully mapped the secondary structures of mouse small nuclear RNAs and compared wild-type and mutant RNase P (116, 117). With these new technologies in hand, it will be possible to gain a much needed understanding of the relationship between lncRNA structure and function, perhaps revealing common motifs of RNA structure that result in specific protein interactions or other functional properties.

Long Noncoding RNAs and Disease

Underscoring the importance of lncRNAs' regulatory roles is their emergence as key players in the etiology of several disease states (118). The strongest connection at present is with cancer (119). Dozens of lncRNAs have been documented to have altered expression in human cancers and are regulated by specific oncogenic and tumor-suppressor pathways, such as p53, MYC, and NF-κB (44, 47, 48). Hung et al. (47) recently described a class of lncRNAs that show periodic expression during the human cell cycle, and many of these are dysregulated in expression in human cancer samples. The lncRNA HOTAIR is highly induced in approximately one-quarter of human breast cancers, and HOTAIR expression is strongly predictive of eventual metastasis and death (74). HOTAIR overexpression in fact drives breast cancer metastasis in vivo, in part by relocalizing Polycomb occupancy patterns genome wide to alter the positional identity of cancer cells (74). Elevated HOTAIR level is also predictive of metastasis or progression in colon and liver cancers, suggesting a general oncogenic trait (120, 121). In effect, cancer cells reprogram themselves to act as if they belong in other anatomic sites (74). The concept of lncRNAs as disease markers is strongly bolstered by the notable discovery that lncRNAs, perhaps owing to their secondary structures, are stable in body fluids and enable noninvasive diagnoses (122). Chinnaiyan and colleagues (122) discovered a large set of lncRNAs in human prostate cancers by RNA-seq and also identified PCAT-1, a lncRNA involved in gene repression that can identify poor-prognosis patients on the basis of its level in urine. The human lncRNA ANRIL is located upstream of the *CDKN2A* tumor-suppressor locus encoding the p16 CDK inhibitor. Mutations in ANRIL are associated with cancer and cardiovascular disease and also lead to aberrant ANRIL transcripts and loss of p16 repression (123). Together, these examples illustrate diverse pathogenic mechanisms—from altering the epigenetic landscape (HOTAIR and ANRIL), modulation of the p53 pathway (lincRNA-p21 and PANDA), and alternative splicing that increases oncogenic protein production (Zeb2 antisense RNA) (124) to controlling the DNA damage response (*CCND1* promoter transcript) (132).

Although cancer has been the most studied, it is likely that lncRNAs are involved in the pathogenesis of many other diseases. Consistent with this notion, hundreds of genomic regions that do not contain protein-coding genes are strongly associated with a wide spectrum of human diseases. Future studies will need to pinpoint potential lncRNA transcripts in these regions and discern whether and how the noncoding genome contributes to human diseases.

SUMMARY POINTS

1. Several examples of functional lncRNAs have been identified that play key roles from pluripotency to cancer.
2. A common emerging theme of lncRNAs is that they form ribonucleic acid-protein interactions to carry out their functions by modulating chromatin-modifying complexes, by interaction with transcription factors, and likely by many additional mechanisms.
3. Enhancers transcribe RNAs. To date, two classes of enhancer RNAs have been identified: those that are by-products of transcription and lncRNAs that play a role in forming enhancer contacts to promote gene expression.

FUTURE ISSUES

1. Very little is understood about how specific lncRNAs seek out selective sites in the genome for interaction, the nature of lncRNA-chromatin interactions, and their possible functional roles in lncRNA biology.
2. Numerous annotation resources are available that will need to be compiled into parsimonious lncRNA transcript databases.
3. A deeper understanding of the sequence and structural elements that relate to lncRNA function will allow classification, or even prediction, of lncRNA families as has occurred for protein families with similar structural domains.
4. Research should identify potential ways of distinguishing *cis*-acting enhancer-like lncRNAs from lncRNAs that function in *trans*.
5. There is a need for large-scale LOF or gain-of-function studies to causally demonstrate lncRNA functions.
6. There is a need for detailed mapping and structural studies to understand the sequence and or structural basis of RNA-protein and RNA-DNA interactions.
7. The clear difference in conservation between protein-coding and lncRNA genes raises the question: How are these lncRNAs rapidly evolving?
8. Genomic regions genetically associated with disease contain only lncRNAs, pointing to their genetic importance to disease, yet the functional roles of lncRNA remain largely unresolved.
9. There is a clear need to develop genetic model systems to understand lncRNAs' function in vivo.

DISCLOSURE STATEMENT

The authors are not aware of any affiliations, memberships, funding, or financial holdings that might be perceived as affecting the objectivity of this review.

ACKNOWLEDGMENTS

We apologize to colleagues whose work could not be discussed or cited owing to space limitations. J.L.R. is supported by National Institutes of Health (NIH) grants 1DP2OD00667-01 and 1P01GM099117-01. J.L.R. is a Damon Runyon-Rachleff, Searle, Smith Family, and Merkin Foundation scholar. H.Y.C. is supported by NIH grants R01-HG004361 and R01-CA118750 and by the California Institute for Regenerative Medicine. H.Y.C. is an Early Career Scientist of the Howard Hughes Medical Institute.

LITERATURE CITED

1. Jacob F, Monod J. 1961. Genetic regulatory mechanisms in the synthesis of proteins. *J. Mol. Biol.* 3:318–56
2. Amaral PP, Dinger ME, Mercer TR, Mattick JS. 2008. The eukaryotic genome as an RNA machine. *Science* 319:1787–89
3. Ponting CP, Oliver PL, Reik W. 2009. Evolution and functions of long noncoding RNAs. *Cell* 136:629–41
4. Weinberg RA, Penman S. 1968. Small molecular weight monodisperse nuclear RNA. *J. Mol. Biol.* 38:289–304
5. Paul J, Duerksen JD. 1975. Chromatin-associated RNA content of heterochromatin and euchromatin. *Mol. Cell. Biochem.* 9:9–16
6. Salditt-Georgieff M, Harpold MM, Wilson MC, Darnell JE Jr. 1981. Large heterogeneous nuclear ribonucleic acid has three times as many 5′ caps as polyadenylic acid segments, and most caps do not enter polyribosomes. *Mol. Cell. Biol.* 1:179–87
7. Salditt-Georgieff M, Darnell JE Jr. 1982. Further evidence that the majority of primary nuclear RNA transcripts in mammalian cells do not contribute to mRNA. *Mol. Cell. Biol.* 2:701–7
8. Nickerson JA, Krochmalnic G, Wan KM, Penman S. 1989. Chromatin architecture and nuclear RNA. *Proc. Natl. Acad. Sci. USA* 86:177–81
9. Bernstein E, Duncan EM, Masui O, Gil J, Heard E, Allis CD. 2006. Mouse polycomb proteins bind differentially to methylated histone H3 and RNA and are enriched in facultative heterochromatin. *Mol. Cell. Biol.* 26:2560–69
10. Maison C, Bailly D, Peters AH, Quivy JP, Roche D, et al. 2002. Higher-order structure in pericentric heterochromatin involves a distinct pattern of histone modification and an RNA component. *Nat. Genet.* 30:329–34
11. Warner JR, Soeiro R, Birnboim HC, Girard M, Darnell JE. 1966. Rapidly labeled HeLa cell nuclear RNA. I. Identification by zone sedimentation of a heterogeneous fraction separate from ribosomal precursor RNA. *J. Mol. Biol.* 19:349–61
12. Bartel DP. 2009. MicroRNAs: target recognition and regulatory functions. *Cell* 136:215–33
13. Nagano T, Fraser P. 2011. No-nonsense functions for long noncoding RNAs. *Cell* 145:178–81
14. Bernstein E, Allis CD. 2005. RNA meets chromatin. *Genes Dev.* 19:1635–55
15. Lagos-Quintana M, Rauhut R, Lendeckel W, Tuschl T. 2001. Identification of novel genes coding for small expressed RNAs. *Science* 294:853–58
16. Lau NC, Lim LP, Weinstein EG, Bartel DP. 2001. An abundant class of tiny RNAs with probable regulatory roles in *Caenorhabditis elegans*. *Science* 294:858–62
17. Lee RC, Ambros V. 2001. An extensive class of small RNAs in *Caenorhabditis elegans*. *Science* 294:862–64
18. Lee RC, Feinbaum RL, Ambros V. 1993. The *C. elegans* heterochronic gene *lin-4* encodes small RNAs with antisense complementarity to *lin-14*. *Cell* 75:843–54
19. Bertone P, Stolc V, Royce TE, Rozowsky JS, Urban AE, et al. 2004. Global identification of human transcribed sequences with genome tiling arrays. *Science* 306:2242–46
20. Carninci P, Kasukawa T, Katayama S, Gough J, Frith MC, et al. 2005. The transcriptional landscape of the mammalian genome. *Science* 309:1559–63
21. Kapranov P, Cawley SE, Drenkow J, Bekiranov S, Strausberg RL, et al. 2002. Large-scale transcriptional activity in chromosomes 21 and 22. *Science* 296:916–19
22. Rinn JL, Euskirchen G, Bertone P, Martone R, Luscombe NM, et al. 2003. The transcriptional activity of human Chromosome 22. *Genes Dev.* 17:529–40

23. Birney E, Stamatoyannopoulos JA, Dutta A, Guigo R, Gingeras TR, et al. 2007. Identification and analysis of functional elements in 1% of the human genome by the ENCODE pilot project. *Nature* 447:799–816
24. Ponjavic J, Ponting CP, Lunter G. 2007. Functionality or transcriptional noise? Evidence for selection within long noncoding RNAs. *Genome Res.* 17:556–65
25. Shoemaker DD, Schadt EE, Armour CD, He YD, Garrett-Engele P, et al. 2001. Experimental annotation of the human genome using microarray technology. *Nature* 409:922–27
26. Lander ES, Linton LM, Birren B, Nusbaum C, Zody MC, et al. 2001. Initial sequencing and analysis of the human genome. *Nature* 409:860–921
27. Venter JC, Adams MD, Myers EW, Li PW, Mural RJ, et al. 2001. The sequence of the human genome. *Science* 291:1304–51
28. Kawai J, Shinagawa A, Shibata K, Yoshino M, Itoh M, et al. 2001. Functional annotation of a full-length mouse cDNA collection. *Nature* 409:685–90
29. Schuler GD, Boguski MS, Stewart EA, Stein LD, Gyapay G, et al. 1996. A gene map of the human genome. *Science* 274:540–46
30. Dunham I, Shimizu N, Roe BA, Chissoe S, Hunt AR, et al. 1999. The DNA sequence of human chromosome 22. *Nature* 402:489–95
31. Volkin E, Astrachan L. 1956. Intracellular distribution of labeled ribonucleic acid after phage infection of *Escherichia coli*. *Virology* 2:433–37
32. Ambros V, Bartel B, Bartel DP, Burge CB, Carrington JC, et al. 2003. A uniform system for microRNA annotation. *RNA* 9:277–79
33. Bartel DP. 2004. MicroRNAs: genomics, biogenesis, mechanism, and function. *Cell* 116:281–97
34. Bentwich I, Avniel A, Karov Y, Aharonov R, Gilad S, et al. 2005. Identification of hundreds of conserved and nonconserved human microRNAs. *Nat. Genet.* 37:766–70
35. Clark MB, Amaral PP, Schlesinger FJ, Dinger ME, Taft RJ, et al. 2011. The reality of pervasive transcription. *PLoS Biol.* 9:e1000625
36. Struhl K. 2007. Transcriptional noise and the fidelity of initiation by RNA polymerase II. *Nat. Struct. Mol. Biol.* 14:103–5
37. van Bakel H, Nislow C, Blencowe BJ, Hughes TR. 2010. Most "dark matter" transcripts are associated with known genes. *PLoS Biol.* 8:e1000371
38. van Bakel H, Nislow C, Blencowe BJ, Hughes TR. 2011. Response to "the reality of pervasive transcription." *PLoS Biol.* 9:e1001102
39. Barski A, Cuddapah S, Cui K, Roh TY, Schones DE, et al. 2007. High-resolution profiling of histone methylations in the human genome. *Cell* 129:823–37
40. Bernstein BE, Mikkelsen TS, Xie X, Kamal M, Huebert DJ, et al. 2006. A bivalent chromatin structure marks key developmental genes in embryonic stem cells. *Cell* 125:315–26
41. Rando OJ, Chang HY. 2009. Genome-wide views of chromatin structure. *Annu. Rev. Biochem.* 78:245–71
42. Mikkelsen TS, Ku M, Jaffe DB, Issac B, Lieberman E, et al. 2007. Genome-wide maps of chromatin state in pluripotent and lineage-committed cells. *Nature* 448:553–60
43. Marson A, Levine SS, Cole MF, Frampton GM, Brambrink T, et al. 2008. Connecting microRNA genes to the core transcriptional regulatory circuitry of embryonic stem cells. *Cell* 134:521–33
44. Guttman M, Amit I, Garber M, French C, Lin MF, et al. 2009. Chromatin signature reveals over a thousand highly conserved large non-coding RNAs in mammals. *Nature* 458:223–27
45. Khalil AM, Guttman M, Huarte M, Garber M, Raj A, et al. 2009. Many human large intergenic noncoding RNAs associate with chromatin-modifying complexes and affect gene expression. *Proc. Natl. Acad. Sci. USA* 106:11667–72
45a. Ulitsky I, Shkumatava A, Jan CH, Sive H, Bartel DP. 2011. Conserved function of lincRNAs in vertebrate embryonic development despite rapid sequence evolution. *Cell* 147:1537–50
46. Loewer S, Cabili MN, Guttman M, Loh YH, Thomas K, et al. 2010. Large intergenic non-coding RNA-RoR modulates reprogramming of human induced pluripotent stem cells. *Nat. Genet.* 42:1113–17

47. Hung T, Wang Y, Lin MF, Koegel AK, Kotake Y, et al. 2011. Extensive and coordinated transcription of noncoding RNAs within cell-cycle promoters. *Nat. Genet.* 43:621–29
48. Huarte M, Guttman M, Feldser D, Garber M, Koziol MJ, et al. 2010. A large intergenic noncoding RNA induced by p53 mediates global gene repression in the p53 response. *Cell* 142:409–19
49. Khalil AM, Guttman M, Huarte M, Garber M, Raj A, et al. 2009. Many human large intergenic noncoding RNAs associate with chromatin-modifying complexes and affect gene expression. *Proc. Natl. Acad. Sci. USA* 106:11667–72
50. Koziol MJ, Rinn JL. 2010. RNA traffic control of chromatin complexes. *Curr. Opin. Genet. Dev.* 20:142–48
51. Rinn JL, Huarte M. 2011. To repress or not to repress: This is the guardian's question. *Trends Cell Biol.* 21:344–53
52. Guttman M, Donaghey J, Carey BW, Garber M, Grenier JK, et al. 2011. lincRNAs act in the circuitry controlling pluripotency and differentiation. *Nature* 477:295–300
53. Mortazavi A, Williams BA, McCue K, Schaeffer L, Wold B. 2008. Mapping and quantifying mammalian transcriptomes by RNA-Seq. *Nat. Methods* 5:621–28
54. Marioni JC, Mason CE, Mane SM, Stephens M, Gilad Y. 2008. RNA-seq: an assessment of technical reproducibility and comparison with gene expression arrays. *Genome Res.* 18:1509–17
55. Trapnell C, Pachter L, Salzberg SL. 2009. TopHat: discovering splice junctions with RNA-Seq. *Bioinformatics* 25:1105–11
56. Guttman M, Garber M, Levin JZ, Donaghey J, Robinson J, et al. 2010. Ab initio reconstruction of cell type–specific transcriptomes in mouse reveals the conserved multi-exonic structure of lincRNAs. *Nat. Biotechnol.* 28:503–10
57. Garber M, Grabherr MG, Guttman M, Trapnell C. 2011. Computational methods for transcriptome annotation and quantification using RNA-seq. *Nat. Methods* 8:469–77
58. Cabili MN, Trapnell C, Goff L, Koziol M, Tazon-Vega B, et al. 2011. Integrative annotation of human large intergenic noncoding RNAs reveals global properties and specific subclasses. *Genes Dev.* 25:1915–27
59. Beck AH, Weng Z, Witten DM, Zhu S, Foley JW, et al. 2010. 3′-end sequencing for expression quantification (3SEQ) from archival tumor samples. *PLoS ONE* 5:e8768
60. Jan CH, Friedman RC, Ruby JG, Bartel DP. 2011. Formation, regulation and evolution of *Caenorhabditis elegans* 3′UTRs. *Nature* 469:97–101
61. Core LJ, Waterfall JJ, Lis JT. 2008. Nascent RNA sequencing reveals widespread pausing and divergent initiation at human promoters. *Science* 322:1845–48
62. Lin MF, Carlson JW, Crosby MA, Matthews BB, Yu C, et al. 2007. Revisiting the protein-coding gene catalog of *Drosophila melanogaster* using 12 fly genomes. *Genome Res.* 17:1823–36
63. Lin MF, Deoras AN, Rasmussen MD, Kellis M. 2008. Performance and scalability of discriminative metrics for comparative gene identification in 12 *Drosophila* genomes. *PLoS Comput. Biol.* 4:e1000067
64. Pueyo JI, Couso JP. 2011. Tarsal-less peptides control Notch signalling through the Shavenbaby transcription factor. *Dev. Biol.* 355:183–93
65. Ingolia NT, Lareau LF, Weissman JS. 2011. Ribosome profiling of mouse embryonic stem cells reveals the complexity and dynamics of mammalian proteomes. *Cell* 147:789–802
66. Wadler CS, Vanderpool CK. 2007. A dual function for a bacterial small RNA: SgrS performs base pairing-dependent regulation and encodes a functional polypeptide. *Proc. Natl. Acad. Sci. USA* 104:20454–59
67. Dinger ME, Pang KC, Mercer TR, Mattick JS. 2008. Differentiating protein-coding and noncoding RNA: challenges and ambiguities. *PLoS Comput. Biol.* 4:e1000176
68. Leygue E. 2007. Steroid receptor RNA activator (SRA1): unusual bifaceted gene products with suspected relevance to breast cancer. *Nucl. Recept. Signal.* 5:e006
69. Jenny A, Hachet O, Zavorszky P, Cyrklaff A, Weston MD, et al. 2006. A translation-independent role of *oskar* RNA in early *Drosophila* oogenesis. *Development* 133:2827–33
70. Kloc M, Wilk K, Vargas D, Shirato Y, Bilinski S, Etkin LD. 2005. Potential structural role of noncoding and coding RNAs in the organization of the cytoskeleton at the vegetal cortex of *Xenopus* oocytes. *Development* 132:3445–57

71. Candeias MM, Malbert-Colas L, Powell DJ, Daskalogianni C, Maslon MM, et al. 2008. *p53* mRNA controls p53 activity by managing Mdm2 functions. *Nat. Cell Biol.* 10:1098–105
72. Mercer TR, Dinger ME, Sunkin SM, Mehler MF, Mattick JS. 2008. Specific expression of long noncoding RNAs in the mouse brain. *Proc. Natl. Acad. Sci. USA* 105:716–21
73. Dinger ME, Amaral PP, Mercer TR, Pang KC, Bruce SJ, et al. 2008. Long noncoding RNAs in mouse embryonic stem cell pluripotency and differentiation. *Genome Res.* 18:1433–45
74. Gupta RA, Shah N, Wang KC, Kim J, Horlings HM, et al. 2010. Long non-coding RNA HOTAIR reprograms chromatin state to promote cancer metastasis. *Nature* 464:1071–76
75. Broadbent KM, Park D, Wolf AR, Van Tyne D, Sims JS, et al. 2011. A global transcriptional analysis of *Plasmodium falciparum* malaria reveals a novel family of telomere-associated lncRNAs. *Genome Biol.* 12:R56
76. Britten RJ, Davidson EH. 1969. Gene regulation for higher cells: a theory. *Science* 165:349–57
77. Brown CJ, Ballabio A, Rupert JL, Lafreniere RG, Grompe M, et al. 1991. A gene from the region of the human X inactivation centre is expressed exclusively from the inactive X chromosome. *Nature* 349:38–44
78. Barlow DP, Stoger R, Herrmann BG, Saito K, Schweifer N. 1991. The mouse insulin-like growth factor type-2 receptor is imprinted and closely linked to the *Tme* locus. *Nature* 349:84–87
79. Bartolomei MS, Zemel S, Tilghman SM. 1991. Parental imprinting of the mouse *H19* gene. *Nature* 351:153–55
80. Wutz A. 2011. Gene silencing in X-chromosome inactivation: advances in understanding facultative heterochromatin formation. *Nat. Rev. Genet.* 12:542–53
81. Zhao J, Sun BK, Erwin JA, Song JJ, Lee JT. 2008. Polycomb proteins targeted by a short repeat RNA to the mouse X chromosome. *Science* 322:750–56
82. Heo JB, Sung S. 2010. Vernalization-mediated epigenetic silencing by a long intronic noncoding RNA. *Science* 331:76–79
83. Rinn JL, Kertesz M, Wang JK, Squazzo SL, Xu X, et al. 2007. Functional demarcation of active and silent chromatin domains in human *HOX* loci by noncoding RNAs. *Cell* 129:1311–23
84. Tsai MC, Manor O, Wan Y, Mosammaparast N, Wang JK, et al. 2010. Long noncoding RNA as modular scaffold of histone modification complexes. *Science* 329:689–93
85. Kaneko S, Li G, Son J, Xu CF, Margueron R, et al. 2010. Phosphorylation of the PRC2 component Ezh2 is cell cycle–regulated and up-regulates its binding to ncRNA. *Genes Dev.* 24:2615–20
86. Maenner S, Blaud M, Fouillen L, Savoye A, Marchand V, et al. 2010. 2-D structure of the A region of Xist RNA and its implication for PRC2 association. *PLoS Biol.* 8:e1000276
87. Nagano T, Mitchell JA, Sanz LA, Pauler FM, Ferguson-Smith AC, et al. 2008. The Air noncoding RNA epigenetically silences transcription by targeting G9a to chromatin. *Science* 322:1717–20
88. Pandey RR, Mondal T, Mohammad F, Enroth S, Redrup L, et al. 2008. Kcnq1ot1 antisense noncoding RNA mediates lineage-specific transcriptional silencing through chromatin-level regulation. *Mol. Cell* 32:232–46
89. Law JA, Jacobsen SE. 2010. Establishing, maintaining and modifying DNA methylation patterns in plants and animals. *Nat. Rev. Genet.* 11:204–20
90. Schmitz KM, Mayer C, Postepska A, Grummt I. 2010. Interaction of noncoding RNA with the rDNA promoter mediates recruitment of DNMT3b and silencing of rRNA genes. *Genes Dev.* 24:2264–69
91. Yap KL, Li S, Muñoz-Cabello AM, Raguz S, Zeng L, et al. 2010. Molecular interplay of the noncoding RNA ANRIL and methylated histone H3 lysine 27 by polycomb CBX7 in transcriptional silencing of INK4a. *Mol. Cell* 38:662–74
92. Yao H, Brick K, Evrard Y, Xiao T, Camerini-Otero RD, Felsenfeld G. 2010. Mediation of CTCF transcriptional insulation by DEAD-box RNA-binding protein p68 and steroid receptor RNA activator SRA. *Genes Dev.* 24:2543–55
93. Kim TK, Hemberg M, Gray JM, Costa AM, Bear DM, et al. 2010. Widespread transcription at neuronal activity-regulated enhancers. *Nature* 465:182–87
94. De Santa F, Barozzi I, Mietton F, Ghisletti S, Polletti S, et al. 2010. A large fraction of extragenic RNA pol II transcription sites overlap enhancers. *PLoS Biol.* 8:e1000384

95. Ørom UA, Derrien T, Beringer M, Gumireddy K, Gardini A, et al. 2010. Long noncoding RNAs with enhancer-like function in human cells. *Cell* 143:46–58
96. Wang KC, Yang YW, Liu B, Sanyal A, Corces-Zimmerman R, et al. 2011. A long noncoding RNA maintains active chromatin to coordinate homeotic gene expression. *Nature* 472:120–24
97. Hung T, Chang HY. 2010. Long noncoding RNA in genome regulation: prospects and mechanisms. *RNA Biol.* 7:582–85
98. Bonasio R, Tu S, Reinberg D. 2010. Molecular signals of epigenetic states. *Science* 330:612–16
99. Martianov I, Ramadass A, Serra Barros A, Chow N, Akoulitchev A. 2007. Repression of the human dihydrofolate reductase gene by a non-coding interfering transcript. *Nature* 445:666–70
100. Jeon Y, Lee JT. 2011. YY1 tethers Xist RNA to the inactive X nucleation center. *Cell* 146:119–33
101. Wang KC, Chang HY. 2011. Molecular mechanisms of long noncoding RNAs. *Mol. Cell* 43:904–14
102. Kino T, Hurt DE, Ichijo T, Nader N, Chrousos GP. 2010. Noncoding RNA Gas5 is a growth arrest- and starvation-associated repressor of the glucocorticoid receptor. *Sci. Signal.* 3:ra8
103. Spitale RC, Tsai MC, Chang HY. 2011. RNA templating the epigenome: long noncoding RNAs as molecular scaffolds. *Epigenetics* 6:539–43
104. Zappulla DC, Cech TR. 2006. RNA as a flexible scaffold for proteins: yeast telomerase and beyond. *Cold Spring Harb. Symp. Quant. Biol.* 71:217–24
105. Kotake Y, Nakagawa T, Kitagawa K, Suzuki S, Liu N, et al. 2010. Long non-coding RNA ANRIL is required for the PRC2 recruitment to and silencing of $p15^{INK4B}$ tumor suppressor gene. *Oncogene* 30:1956–62
106. Lee JT. 2009. Lessons from X-chromosome inactivation: long ncRNA as guides and tethers to the epigenome. *Genes Dev.* 23:1831–42
106a. Chu C, Qu K, Zhong FL, Artandi SE, Chang HY. 2011. Genomic maps of long noncoding RNA occupancy reveal principles of RNA-chromatin interactions. *Mol. Cell* 44:667–78
106b. Simon MD, Wang CI, Kharchenko PV, West JA, Chapman BA, et al. 2011. The genomic binding sites of a noncoding RNA. *Proc. Natl. Acad. Sci. USA* 08:20497–502
107. Delebecque CJ, Lindner AB, Silver PA, Aldaye FA. 2011. Organization of intracellular reactions with rationally designed RNA assemblies. *Science* 333:470–74
108. Gerber AP, Herschlag D, Brown PO. 2004. Extensive association of functionally and cytotopically related mRNAs with Puf family RNA-binding proteins in yeast. *PLoS Biol.* 2:E79
109. Zhao J, Ohsumi TK, Kung JT, Ogawa Y, Grau DJ, et al. 2010. Genome-wide identification of polycomb-associated RNAs by RIP-seq. *Mol. Cell* 40:939–53
110. Licatalosi DD, Mele A, Fak JJ, Ule J, Kayikci M, et al. 2008. HITS-CLIP yields genome-wide insights into brain alternative RNA processing. *Nature* 456:464–69
111. Mondal T, Rasmussen M, Pandey GK, Isaksson A, Kanduri C. 2010. Characterization of the RNA content of chromatin. *Genome Res.* 20:899–907
112. Steitz TA. 2008. A structural understanding of the dynamic ribosome machine. *Nat. Rev. Mol. Cell Biol.* 9:242–53
113. Wan Y, Kertesz M, Spitale RC, Segal E, Chang HY. 2011. Understanding the transcriptome through RNA structure. *Nat. Rev. Genet.* 12:641–55
114. Watts JM, Dang KK, Gorelick RJ, Leonard CW, Bess JW Jr, et al. 2009. Architecture and secondary structure of an entire HIV-1 RNA genome. *Nature* 460:711–16
115. Kertesz M, Wan Y, Mazor E, Rinn JL, Nutter RC, et al. 2010. Genome-wide measurement of RNA secondary structure in yeast. *Nature* 467:103–7
116. Underwood JG, Uzilov AV, Katzman S, Onodera CS, Mainzer JE, et al. 2010. FragSeq: transcriptome-wide RNA structure probing using high-throughput sequencing. *Nat. Methods* 7:995–1001
117. Lucks JB, Mortimer SA, Trapnell C, Luo S, Aviran S, et al. 2011. Multiplexed RNA structure characterization with selective 2′-hydroxyl acylation analyzed by primer extension sequencing (SHAPE-Seq). *Proc. Natl. Acad. Sci. USA* 108:11063–68
118. Wapinski O, Chang HY. 2011. Long noncoding RNAs and human disease. *Trends Cell Biol.* 21:354–61
119. Tsai MC, Spitale RC, Chang HY. 2011. Long intergenic noncoding RNAs: new links in cancer progression. *Cancer Res.* 71:3–7

120. Kogo R, Shimamura T, Mimori K, Kawahara K, Imoto S, et al. 2011. Long non-coding RNA HOTAIR regulates Polycomb-dependent chromatin modification and is associated with poor prognosis in colorectal cancers. *Cancer Res.* 71:6320–26
121. Yang Z, Zhou L, Wu LM, Lai MC, Xie HY, et al. 2011. Overexpression of long non-coding RNA HOTAIR predicts tumor recurrence in hepatocellular carcinoma patients following liver transplantation. *Ann. Surg. Oncol.* 18:1243–50
122. Prensner JR, Iyer MK, Balbin OA, Dhanasekaran SM, Cao Q, et al. 2011. Transcriptome sequencing across a prostate cancer cohort identifies PCAT-1, an unannotated lincRNA implicated in disease progression. *Nat. Biotechnol.* 29:742–49
123. Burd CE, Jeck WR, Liu Y, Sanoff HK, Wang Z, Sharpless NE. 2010. Expression of linear and novel circular forms of an INK4/ARF-associated non-coding RNA correlates with atherosclerosis risk. *PLoS Genet.* 6:e1001233
124. Beltran M, Puig I, Peña C, Garcia JM, Alvarez AB, et al. 2008. A natural antisense transcript regulates Zeb2/Sip1 gene expression during Snail1-induced epithelial-mesenchymal transition. *Genes Dev.* 22:756–69
125. Tripathi V, Ellis JD, Shen Z, Song DY, Pan Q, et al. 2010. The nuclear-retained noncoding RNA MALAT1 regulates alternative splicing by modulating SR splicing factor phosphorylation. *Mol. Cell* 39:925–38
126. Gong C, Maquat LE. 2011. lncRNAs transactivate STAU1-mediated mRNA decay by duplexing with 3′ UTRs via Alu elements. *Nature* 470:284–88
127. van Dijk EL, Chen CL, d'Aubenton-Carafa Y, Gourvennec S, Kwapisz M, et al. 2011. XUTs are a class of Xrn1-sensitive antisense regulatory non-coding RNA in yeast. *Nature* 475:114–17
128. Berretta J, Pinskaya M, Morillon A. 2008. A cryptic unstable transcript mediates transcriptional *trans*-silencing of the Ty1 retrotransposon in *S. cerevisiae*. *Genes Dev.* 22:615–26
129. Camblong J, Beyrouthy N, Guffanti E, Schlaepfer G, Steinmetz LM, Stutz F. 2009. *Trans*-acting antisense RNAs mediate transcriptional gene cosuppression in *S. cerevisiae*. *Genes Dev.* 23:1534–45
130. Camblong J, Iglesias N, Fickentscher C, Dieppois G, Stutz F. 2007. Antisense RNA stabilization induces transcriptional gene silencing via histone deacetylation in *S. cerevisiae*. *Cell* 131:706–17
131. Flynn RL, Centore RC, O'Sullivan RJ, Rai R, Tse A, et al. 2011. TERRA and hnRNPA1 orchestrate an RPA-to-POT1 switch on telomeric single-stranded DNA. *Nature* 471:532–36
132. Wang X, Arai S, Song X, Reichart D, Du K, et al. 2008. Induced ncRNAs allosterically modify RNA-binding proteins in *cis* to inhibit transcription. *Nature* 454:126–30

The Ubiquitin System, an Immense Realm

Alexander Varshavsky

Division of Biology, California Institute of Technology, Pasadena, California 91125; email: avarsh@caltech.edu

Keywords

proteolysis, ubiquitylation, proteasome, N-end rule, arginylation

Among the functions of intracellular proteolysis are the elimination of misfolded or otherwise abnormal proteins; the maintenance of amino acid pools in cells affected by stresses, such as starvation; and the generation of protein fragments that act as hormones, antigens, or other effectors. Many intracellular proteins are either conditionally or constitutively short-lived, with in vivo half-lives that can be as brief as a few minutes. In some cases, a proteolytic pathway targets and destroys a protein cotranslationally, i.e., an emerging polypeptide chain can be degraded while it is still a ribosome-associated peptidyl-tRNA (1, 2). The regulated and processive degradation of intracellular proteins is carried out largely by the ubiquitin (Ub)-proteasome system (Ub system), in conjunction with molecular chaperones, autophagy, and lysosomal proteolysis. Other mediators of intracellular protein degradation include proteases such as caspases, calpains, and separases. These and other nonprocessive proteases can function as "upstream" components of the Ub system, generating protein fragments that are targeted and degraded to short peptides by Ub-mediated pathways. Proteins that are damaged, misfolded, or otherwise abnormal are often recognized as such and selectively destroyed by the Ub system. Physiologically important exceptions include conformationally perturbed proteins and/or their aggregates that are harmful but cannot be efficaciously repaired or removed. The resulting proteotoxicity underlies both aging and specific diseases, including neurodegeneration.

One major role of the Ub system is the regulation of proteins whose concentrations must vary with time and alterations in the state of a cell. Short in vivo half-lives of such proteins provide a way to generate their spatial gradients and to rapidly adjust their concentration or subunit composition through changes in the rate of their degradation. In addition, a short half-life of a protein would lead to a rapid decrease in its concentration upon cessation of its synthesis. This way, transcriptional or translational regulation of specific proteins can acquire switch-like properties, because a short-lived protein that is no longer made would not persist in a cell, in contrast to a metabolically stable protein. Proteolysis can also serve to activate or otherwise modulate protein molecules and specific circuits, e.g., by cleaving off and destroying an autoinhibitory domain of a protein or by selectively eliminating an inhibitory subunit of a protein complex. These and other properties of the Ub system make it, among other things, a major regulator of gene expression. For example, most transcriptional activators and repressors in eukaryotes are conditionally short-lived proteins that are destroyed by the Ub system at spatiotemporally regulated rates that underlie the finely tuned physiological functions of these proteins.

Ub is a 76-residue protein that mediates proteolysis through the enzymatic conjugation of Ub to proteins that contain primary degradation signals, called degrons (3). A primary degron is a feature of a protein (a region of its amino acid sequence and/or a conformational determinant) that makes the protein metabolically unstable. Ub-protein conjugation marks proteins for their recognition and degradation by the 26S proteasome, a processive, ATP-dependent protease. Ub is conjugated to proteins either as a single moiety or as a poly-Ub chain that is linked (in most cases) to the ε-amino group of an internal Lys residue in a substrate protein. Ub is a "secondary" degron, in that Ub is conjugated to proteins because they contain primary degradation signals. Ub has nonproteolytic functions as well. The design of the Ub system is summarized in **Figure 1**.

The field of Ub and regulated protein degradation was created in the 1980s, largely through the complementary discoveries by the laboratory of A. Hershko at the Technion (Haifa, Israel) and by my laboratory, then at the Massachusetts Institute of Technology (Cambridge, Massachusetts). In 1978–1985, the elegant biochemical insights by Hershko and coworkers produced the initial understanding of Ub-mediated protein degradation in cell extracts, including the isolation of enzymes that mediate Ub conjugation. In 1984–1990, these mechanistic (enzymological) advances with cell-free systems were complemented by our genetic and biochemical discoveries with mammalian cells and the yeast *Saccharomyces cerevisiae*. These discoveries revealed the singularly important biology of the Ub system, including the first demonstration that the bulk of protein degradation in a living cell requires Ub conjugation, and the identification of the first Ub-conjugating (E2) enzymes with specific physiological functions, in the cell cycle (Cdc34 E2) and DNA repair (Rad6 E2). These advances initiated the understanding of the massive, multilevel involvement of the Ub system in the regulation of the cell cycle and DNA damage response. We also discovered the critical roles of the Ub system in stress resistance, protein synthesis, and transcriptional regulation. In 1990, we identified and cloned an E3 ligase termed Ubr1, the first molecularly cloned and analyzed E3 Ub ligase. Together with the Rad6 E2 and Cdc34 E2 results, the cloning and characterization of the Ubr1 E3 opened up a particularly large field, because we now know that the mammalian genome encodes at least 1,000 distinct E3s. The targeting of many distinct degrons in cellular proteins by this astounding diversity of specific E3 Ub ligases underlies the unprecedented functional reach of the Ub system. The term Ub ligase denotes either an E3-E2 holoenzyme or its E3 component. In 1986, we discovered the first primary degradation signals in short-lived proteins. (Specific signals that mark proteins for conjugation to Ub were presumed to exist, but their nature was a mystery.) These signals included degrons that give rise to the N-end rule, which relates the in vivo half-life of a protein to the identity of its N-terminal residue (**Figure 2**). The N-end rule pathway (it mediates the N-end rule) was the first specific pathway of the Ub system. Other discoveries by our laboratory in the 1980s included the first poly-Ub chains, their specific topology, and their necessity for proteolysis; the subunit selectivity of protein degradation (this fundamental capability of the Ub system underlies most of its functions, as it makes possible the subunit-specific remodeling of oligomeric proteins); the first physiological

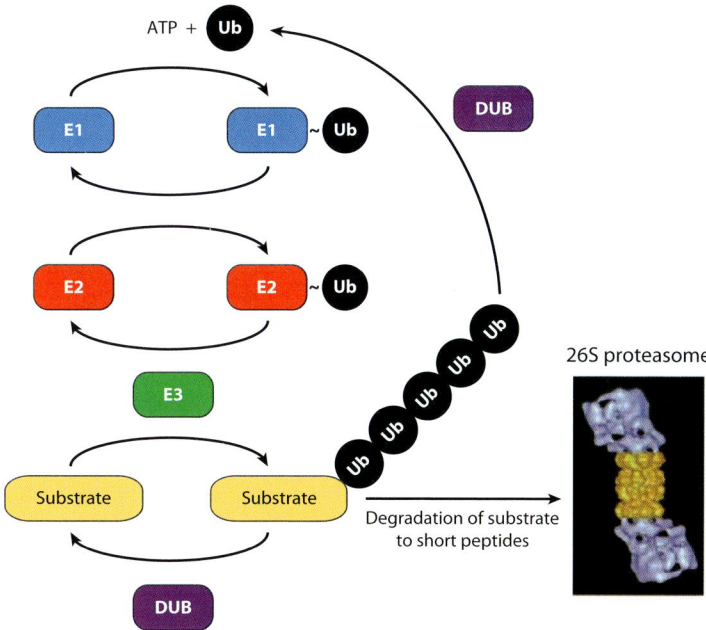

Figure 1

The ubiquitin-proteasome system (Ub system). This diagram illustrates the fundamental design of Ub-mediated processes. The conjugation of Ub to other proteins involves a preliminary ATP-dependent step in which the last residue of Ub (Gly76) is joined, via a thioester bond, to a Cys residue of the E1 (Ub-activating) enzyme. The "activated" Ub moiety is transferred to a Cys residue in one of several Ub-conjugating (E2) enzymes and, from there, through an isopeptide bond to a Lys residue of an ultimate acceptor protein ("substrate"). E2 enzymes function as subunits of E2-E3 Ub ligase holoenzymes that can produce substrate-linked poly-Ub chains. Such chains have distinct topologies, i.e., specific Lys residues of Ub are used to form the isopeptide Gly-Lys bonds between adjacent Ub moieties in a chain. Substrate-linked poly-Ub chains with specific Ub-Ub isopeptide bonds mediate either the processive degradation of a substrate by the 26S proteasome or other metabolic fates of chain-linked substrates. Monoubiquitylation of specific substrates also occurs and has specific functions. The term Ub ligase denotes either an E3-E2 holoenzyme or its E3 component. An individual mammalian genome encodes at least a thousand distinct E3s. One role of E3 is the initial recognition of a substrate's degradation signal (degron). Ub has nonproteolytic functions as well. For accounts of the early history of the Ub field, see References 4–6. Abbreviations: DUB, deubiquitylating enzyme; E1-E3, enzymes of Ub conjugation.

substrate of the Ub system (the MATα2 transcriptional repressor; before this advance, the Ub system was examined using artificial substrates); and the first genes that encode Ub precursors (a linear poly-Ub chain and Ub fusions to specific ribosomal proteins). For accounts of the early history of the Ub field, see References 1, 4–6.

By the end of the 1980s, our studies had revealed the major biological functions of the Ub system as well as the basis for its specificity, i.e., the first degradation signals in short-lived proteins. The resulting discovery of physiological regulation by intracellular protein degradation has transformed the understanding of biological circuits, as it became clear that control through regulated protein degradation rivals, and often surpasses in significance, the classical regulation through transcription and translation (6). Just how strikingly broad and elaborate Ub functions are was understood more systematically and in great detail over the next two decades, through studies by many laboratories that began entering this field in the 1990s, an expansion that continues to the present day. One of the new directions in later studies involved a family of Ub-like proteins. Some of these proteins, referred to as Ub-family modifiers, are Ub-like not only in their structural (spatial) similarity to Ub, but also

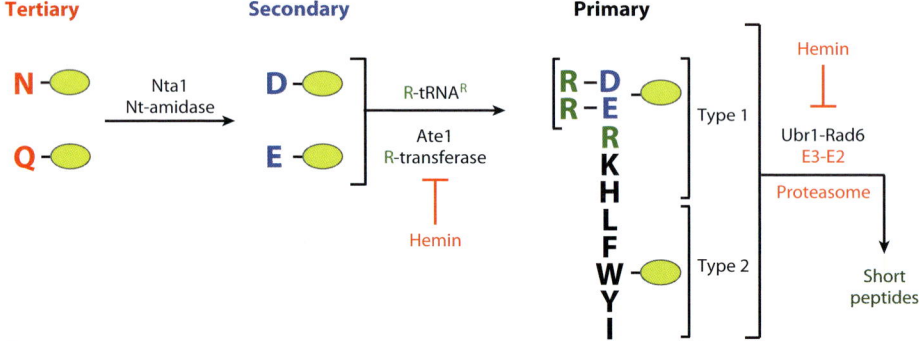

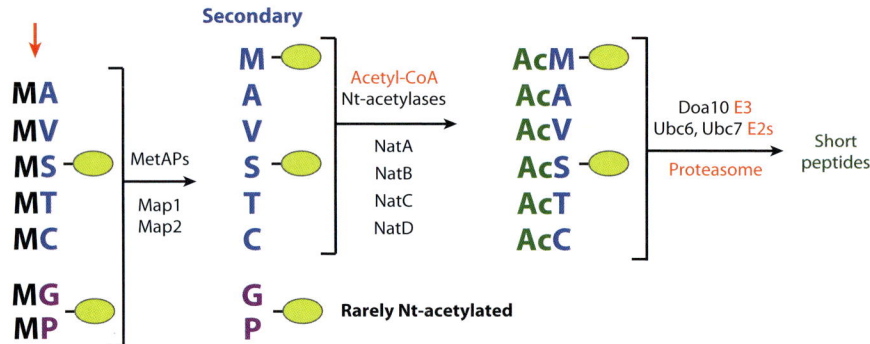

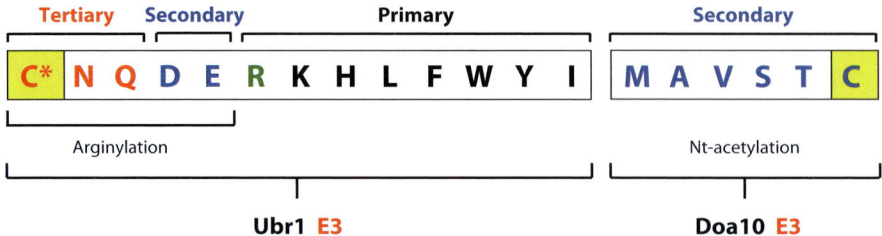

in their ability to be conjugated to other proteins, in reactions catalyzed by enzymes that are similar (though not identical) to the E1-E3 enzymes of the Ub system. The functions of Ub-family modifiers other than Ub itself encompass a broad variety of processes, including the control of autophagy, the nuclear transport of specific proteins, the cohesion/segregation of replicated chromosomes, the repair of DNA, and a multitude of signal transduction pathways. For recent reviews of the Ub and Ub-like systems, see References 1 and 6–48.

How did the Ub system emerge in the course of evolution? Prokaryotes, i.e., bacteria

Figure 2

The Arg/N-end rule and the Ac/N-end rule pathways in the yeast *Saccharomyces cerevisiae*. (*a*) The Arg/N-end rule pathway (59–62). N-terminal residues are indicated by single-letter abbreviations for amino acids. Yellow ovals denote the rest of a protein substrate. In contrast to the directly recognized (by the Ubr1/Rad6 E3-E2 Ub ligase) primary destabilizing N-terminal residues (Arg, Lys, His, Leu, Phe, Tyr, Trp, and Ile), the N-terminal residues Asp, Glu, Asn, and Gln can be targeted by Ubr1/Rad6 only after their N-terminal arginylation (Nt-arginylation) by the Ate1 Arg-tRNA-protein transferase (R-transferase). These destabilizing residues are called secondary or tertiary, depending on the number of steps (arginylation of Asp and Glu; deamidation/arginylation of Asn and Gln) that precede the targeting and polyubiquitylation by the Ubr1/Rad6 Ub ligase of Nt-arginylated N-end rule substrates. One aspect of the Arg/N-end rule pathway that is not illustrated in this diagram is a physical and functional interaction between the Ubr1 E3 of the Arg/N-end rule pathway and the Ufd4 E3 of the Ub-fusion-degradation pathway (59). Specifically, the targeting apparatus of the Arg/N-end rule pathway comprises a physical complex of the RING-type Ubr1 E3 (called N-recognin) and the HECT-type Ufd4 E3, together with their cognate E2 enzymes, Rad6 and Ubc4 (or Ubc5), respectively. The 2010 discovery (59) of the Ubr1-Ufd4 complex unified two proteolytic pathways that have been studied separately over two decades. (*b*) The Ac/N-end rule pathway (63). The red arrow on the left indicates the removal of N-terminal Met by Met aminopeptidases (MetAPs). This Met residue is retained if a residue at position 2 is nonpermissive (too large) for MetAPs (Reference 63 and the references therein). If the retained N-terminal Met or N-terminal Ala, Val, Ser, Thr, and Cys are followed by acetylation-permissive residues, the above N-terminal residues are usually N-terminally acetylated (Nt-acetylated) by the indicated Nt-acetylases (64). The resulting N-degrons are called AcN-degrons (63). The term "secondary" refers to the necessity of modification (Nt-acetylation) of a destabilizing N-terminal residue before a protein can be recognized by a cognate Ub ligase. Although second-position Gly or Pro can be made N-terminal by MetAPs, and although the Doa10 E3 can recognize Nt-acetylated Gly and Pro (63), few proteins with N-terminal Gly or Pro are Nt-acetylated (64). (*c*) The Arg/N-end rule pathway and the Ac/N-end rule pathway. Both of these branches of the N-end rule pathway target, through different mechanisms, the N-terminal Cys residue (*yellow squares*). The oxidized Cys residue (Cys-sulfinate or Cys-sulfonate) is marked by an asterisk. Physiological substrates of the Ac/N-end rule pathway comprise a majority of proteins in a cell. For example, more than 80% of human proteins are cotranslationally Nt-acetylated (References 64–67 and references therein). Thus, most eukaryotic proteins harbor a specific degradation signal from the moment of their birth (63). The discovery of the Ac/N-end rule pathway has also revealed the main physiological roles of two classes of enzymes, Nt-acetylases and MetAPs. Specifically, Nt-acetylases produce AcN-degrons, whereas the upstream MetAps, by cotranslationally cleaving off the N-terminal Met residue, make possible these degradation signals, all of them except for those AcN-degrons that contain the (nonremoved) Nt-acetylated N-terminal Met (*b*). Nt-acetylases and MetAps are universally present and essential enzymes (64, 66, 68–72) whose physiological roles were largely unknown. These enzymes are now components of the Ac/N-end rule pathway (63). Regulated degradation of specific proteins by the Arg/N-end rule pathway mediates a legion of biological functions, including the sensing of heme, nitric oxide, oxygen, and short peptides; selective elimination of misfolded proteins; regulation of DNA repair and cohesion/segregation of chromosomes; signaling by G proteins; regulation of peptide import, apoptosis, meiosis, immunity, fat metabolism, cell migration, actin filaments, cardiovascular development, spermatogenesis, neurogenesis, and memory; the functioning of adult organs, including the brain, muscle, testis, and pancreas; and regulation of leaf and shoot development, leaf senescence, and seed germination in plants (References 6, 60, 61, 63, and 73–86 and the references therein). In addition, the recently discovered Ac/N-end rule pathway is likely to mediate, among other things, protein quality control, the regulation of in vivo stoichiometries of proteins that form multisubunit complexes, and the degradation of long-lived proteins (63). Abbreviations: Map1, Map2, Met-aminopeptidases (MetAps); Nta1, N-terminal amidase.

and archaea, contain Ub-like proteolytic pathways but lack the bona fide Ub system (49, 50). Although possible evolutionary routes that resulted in Ub, ubiquitylation, Ub-mediated proteolysis, and other Ub functions are somewhat constrained by existing evidence, the specific steps in early Ub evolution are unknown. One idea is that primordial Ub, i.e., a protein containing the characteristic β-grasp Ub fold (45, 47), emerged and functioned as a small, rapidly folding N-terminal cotranslational chaperone of early polypeptides, during the (dimly understood) transition from the RNA world toward translation-endowed organisms (51). This function of primordial Ub would be analogous to the role that modern Ub appears to play as a cotranslational N-terminal chaperone of two specific ribosomal proteins (52). The chaperone function of a primordial N-terminal Ub moiety (51) would be expected to result in its initial spread, through DNA recombination and positive selection, among open reading frames of early polypeptides, before the appearance of deubiquitylases and other Ub-specific enzymes. This model addresses the "chicken-and-egg" problem by endowing Ub (more accurately, a Ub-like polypeptide) with a positively selected (fitness-increasing) property before the emergence of present-day Ub functions, i.e., before the appearance of enzymes that mediate the thioester and isopeptide conjugation chemistry of Ub and Ub-like proteins. In this interpretation, the extant genes that encode N-terminal fusions of Ub or Ub-like moieties to unrelated proteins are derived, at least in part, from the early era of Ub evolution, when Ub and Ub-like fusions may have been encoded by many more genes than today. In this scenario (51), both deubiquitylases and Ub ligases (but not necessarily a primordial proteasome) were later arrivals.

Processive proteolytic systems that lack Ub, for example, the bacterial Leu/N-end rule pathway (1), are significantly less elaborate, both compositionally and mechanistically, than the analogous Ub-dependent pathways in eukaryotes. The remarkable complexity (including compositional complexity) of the eukaryotic Ub system raises the question of how prokaryotes, which are obviously proficient in regard to processive proteolysis (53, 54), get by to a large extent without Ub and ubiquitylation. Possible explanations, including a role that genetic drift and stochastic fixation of mildly deleterious alleles might have played in the evolution of eukaryotes (in contrast to adaptive, i.e., selection-based, Darwinian evolution), are discussed in References 1 and 55.

The Ub and Ub-like systems are of major importance in medicine, given their immense functional range and the multitude of ways in which these systems can malfunction in disease, from cancer and neurodegenerative syndromes to perturbations of immunity and many other illnesses, including birth defects. Both academic laboratories and pharmaceutical companies are developing compounds that target specific components of these systems. The fruits of their labors have already become, or will soon become, clinically useful drugs (56–58). Work in this arena may produce not only "conventional" inhibitors or activators of specific enzymes but also drugs that would direct the Ub system to target, destroy, and thereby downregulate any specific protein. After three decades of ever-expanding studies in this vast biomedical realm, new directions of inquiry, new problems, and new applications of fundamental discoveries continue unabated. Advances in the understanding of the Ub and Ub-like systems are being published at a clip that exceeds anyone's ability to follow these studies in their entirety, a state of affairs that is frustrating and exhilarating at the same time. It is a safe bet that new and major breakthroughs in this arena will continue to occur and will be accompanied by momentous advances in the application of accumulated fundamental understanding to problems in clinical medicine.

This informal overview of the subject and its early history introduces the timely and detailed reviews, by leading researchers in the field, of the Ub and Ub-like systems. I greatly benefited from reading their excellent disquisitions. It is now your turn to enjoy these scholarly, pithy reviews.

DISCLOSURE STATEMENT

The author is not aware of any affiliations, memberships, funding, or financial holdings that might be perceived as affecting the objectivity of this review.

ACKNOWLEDGMENTS

I am grateful to Christopher Brower, Roger Kornberg, Anna Shemorry, and Brandon Wadas for their helpful comments on the manuscript. Studies in our laboratory are supported by grants from the National Institutes of Health and the March of Dimes Foundation.

LITERATURE CITED

1. Varshavsky A. 2011. The N-end rule pathway and regulation by proteolysis. *Protein Sci.* 20:1298–345
2. Turner GC, Varshavsky A. 2000. Detecting and measuring cotranslational protein degradation in vivo. *Science* 289:2117–20
3. Varshavsky A. 1991. Naming a targeting signal. *Cell* 64:13–15
4. Hershko A, Ciechanover A, Varshavsky A. 2000. The ubiquitin system. *Nat. Med.* 10:1073–81
5. Varshavsky A. 2006. The early history of the ubiquitin field. *Protein Sci.* 15:647–54
6. Varshavsky A. 2008. Discovery of cellular regulation by protein degradation. *J. Biol. Chem.* 283:34469–89
7. Schulman BA. 2011. Twists and turns in ubiquitin-like conjugation cascades. *Protein Sci.* 20:1941–54
8. Wolf DH. 2011. The ubiquitin clan: a protein family essential for life. *FEBS Lett.* 585:2769–71
9. Thomas LR, Tansey WP. 2011. Proteolytic control of the oncoprotein transcription factor Myc. *Adv. Cancer Res.* 110:77–106
10. Weissman AM, Shabek N, Ciechanover A. 2011. The predator becomes the prey: regulating the ubiquitin system by ubiquitylation and degradation. *Nat. Rev. Mol. Cell Biol.* 12:605–20
11. Dou H, Huang C, Nguyen TV, Lu L-S, Yeh ETH. 2011. SUMOylation and de-SUMOylation in response to DNA damage. *FEBS Lett.* 585:2891–96
12. Watson IR, Irwin MS, Ohh M. 2011. NEDD8 pathways in cancer, sine quibus non. *Cancer Cell* 19:168–76
13. Xie Y. 2010. Structure, assembly and homeostatic regulation of the 26S proteasome. *J. Mol. Cell Biol.* 2:308–17
14. Stadtmueller BM, Hill CP. 2011. Proteasome activators. *Mol. Cell* 41:8–19
15. Wenzel DM, Stoll KE, Klevit RE. 2011. E2s: structurally economical and functionally replete. *Biochem. J.* 433:31–42
16. Malynn BA, Ma A. 2010. Ubiquitin makes its mark on immune regulation. *Immunity* 33:843–52
17. Liu F, Walters KJ. 2010. Multitasking with ubiquitin through multivalent interactions. *Trends Biochem. Sci.* 35:352–60
18. Gallastegui N, Groll M. 2010. The 26S proteasome: assembly and function of a destructive machine. *Trends Biochem. Sci.* 35:634–42
19. Bohn S, Beck F, Sakata E, Walzthoeni T, Beck M, et al. 2010. Structure of the 26S proteasome from *Schizosaccharomyces pombe* at subnanometer resolution. *Proc. Natl. Acad. Sci. USA* 107:20992–97
20. Gareau JR, Lima CD. 2010. The SUMO pathway: emerging mechanisms that shape specificity, conjugation and recognition. *Nat. Rev. Mol. Cell Biol.* 11:861–71
21. Ulrich HD, Walden H. 2010. Ubiquitin signalling in DNA replication and repair. *Nat. Rev. Mol. Cell Biol.* 11:479–89
22. Duda DM, Scott DC, Calabrese MF, Zimmerman ES, Zheng N, et al. 2011. Structural regulation of cullin-RING ubiquitin ligase complexes. *Curr. Opin. Struct. Biol.* 21:257–64
23. Stolz A, Wolf DH. 2010. Endoplasmic reticulum-associated protein degradation: a chaperone-assisted journey to hell. *Biochim. Biophys. Acta* 1803:694–705
24. Scott DC, Monda JK, Grace CRR, Duda DM, Kriwacki RW, et al. 2010. A dual mechanism for Rub1 ligation to Cdc53. *Mol. Cell* 39:784–96

25. Rubenstein EM, Hochstrasser M. 2010. Redundancy and variation in the ubiquitin-mediated proteolytic targeting of a transcription factor. *Cell Cycle* 9:4282–85
26. Geoffroy M-C, Hay RT. 2010. An additional role for SUMO in ubiquitin-mediated proteolysis. *Nat. Rev. Mol. Cell Biol.* 10:564–68
27. Skaug B, Chen ZJ. 2010. Emerging role of ISG15 in antiviral immunity. *Cell* 143:187–90
28. Bawa-Khalfe T, Yeh ET. 2010. SUMO losing balance: SUMO proteases disrupt SUMO homeostasis to facilitate cancer development and progression. *Genes Cancer* 1:748–52
29. Hochstrasser M. 2009. Origin and function of ubiquitin-like proteins. *Nature* 458:422–29
30. Bergink S, Jentsch S. 2009. Principles of ubiquitin and SUMO modifications in DNA repair. *Nature* 458:461–67
31. Lu Z, Hunter T. 2009. Degradation of activated protein kinases by ubiquitination. *Annu. Rev. Biochem.* 78:435–75
32. Hampton RY, Garza RM. 2009. Protein quality control as a strategy for cellular regulation: lessons from ubiquitin-mediated regulation of the sterol pathway. *Chem. Rev.* 109:1561–74
33. Grabbe C, Dikic I. 2009. Functional roles of ubiquitin-like domain (ULD) and ubiquitin-binding domain (UBD) containing proteins. *Chem. Rev.* 109:1481–94
34. Daulni A, Tansey WP. 2009. Damage control: DNA repair, transcription, and the ubiquitin-proteasome system. *DNA Repair* 8:444–48
35. Deshaies RJ, Joazeiro CAP. 2009. RING domain E3 ubiquitin ligases. *Annu. Rev. Biochem.* 78:399–434
36. Finley D. 2009. Recognition and processing of ubiquitin-protein conjugates by the proteasome. *Annu. Rev. Biochem.* 78:477–513
37. Reyes-Turcu FE, Ventii KH, Wilkinson KD. 2009. Regulation and cellular roles of ubiquitin-specific deubiquitinating enzymes. *Annu. Rev. Biochem.* 78:363–97
38. Hirsch C, Gauss R, Horn SC, Neuber O, Sommer T. 2009. The ubiquitylation machinery of the endoplasmic reticulum. *Nature* 458:453–60
39. Marques AJ, Palanimurugan R, Matias AC, Ramos PC, Dohmen RJ. 2009. Catalytic mechanism and assembly of the proteasome. *Chem. Rev.* 109:1509–36
40. Ravid T, Hochstrasser M. 2008. Diversity of degradation signals in the ubiquitin-proteasome system. *Nat. Rev. Mol. Cell Biol.* 9:679–89
41. Vembar SS, Brodsky JL. 2008. One step at a time: endoplasmic reticulum-associated degradation. *Nat. Rev. Mol. Cell Biol.* 9:944–58
42. Dye BT, Schulman BA. 2007. Structural mechanisms underlying posttranslational modification by ubiquitin-like proteins. *Annu. Rev. Biophys. Biomol. Struct.* 36:131–50
43. Scheffner M, Staub O. 2007. HECT E3s and human disease. *BMC Biochem.* 8(Suppl. I):S6
44. Merlet J, Burger J, Gomes JE, Pintard L. 2009. Regulation of cullin-RING E3 ubiquitin-ligases by neddylation and dimerization. *Cell. Mol. Life Sci.* 66:1924–38
45. Burroughs AM, Balaji S, Iyer LM, Aravind L. 2007. Small but versatile: the extraordinary functional and structural diversity of the beta-grasp fold. *Biol. Direct* 2:18
46. Uzunova K, Göttsche K, Miteva M, Weisshaar SR, Glanemann C, et al. 2007. Ubiquitin-dependent proteolytic control of SUMO conjugates. *J. Biol. Chem.* 282:34167–75
47. Iyer LM, Burroughs AM, Aravind L. 2006. The prokaryotic antecedents of the ubiquitin-signaling system and the early evolution of ubiquitin-like beta-grasp domains. *Genome Biol.* 7:R60
48. Johnson ES. 2004. Protein modification by SUMO. *Annu. Rev. Biochem.* 73:355–82
49. Darwin KH. 2009. Prokaryotic ubiquitin-like protein, proteasomes, and pathogenesis. *Nat. Rev. Microbiol.* 7:485–91
50. Humbard MA, Miranda HV, Lim J-M, Krause DJ, Pritz JR, et al. 2010. Ubiquitin-like small archaeal modifier proteins (SAMPS) in *Haloferax volcanii*. *Nature* 463:54–60
51. Graciet E, Hu RG, Piatkov K, Rhee JH, Schwarz EM, et al. 2006. Aminoacyl-transferases and the N-end rule pathway of prokaryotic/eukaryotic specificity in a human pathogen. *Proc. Natl. Acad. Sci. USA* 103:3078–83
52. Finley D, Bartel B, Varshavsky A. 1989. The tails of ubiquitin precursors are ribosomal proteins whose fusion to ubiquitin facilitates ribosome biogenesis. *Nature* 338:394–401

53. Baker TA, Sauer RT. 2006. ATP-dependent proteases of bacteria: recognition logic and operating principles. *Trends Biochem. Sci.* 31:647–53
54. Gottesman S. 2003. Proteolysis in bacterial regulatory circuits. *Annu. Rev. Cell Dev. Biol.* 19:565–87
55. Lynch M. 2007. *The Origins of Genome Architecture*. Sunderland, MA: Sinauer
56. Ablain J, Nasr R, Bazarbachi A, de Thé H. 2011. The drug-induced degradation of oncoproteins: an unexpected Achilles' heel of cancer cells? *Cancer Discov.* 1:117–27
57. Chen SJ, Zhou GB, Zhang XW, Mao JH, de Thé H, et al. 2011. From an old remedy to a magic bullet: molecular mechanisms underlying the therapeutic effects of arsenic in fighting leukemia. *Blood* 117:6425–37
58. Bedford L, Lowe J, Dick LR, Mayer RJ, Brownell JE. 2011. Ubiquitin-like protein conjugation and the ubiquitin-proteasome system as drug targets. *Nat. Rev. Drug Discov.* 10:29–46
59. Hwang C-S, Shemorry A, Varshavsky A. 2010. The N-end rule pathway is mediated by a complex of the RING-type Ubr1 and HECT-type Ufd4 ubiquitin ligases. *Nat. Cell Biol.* 12:1177–85
60. Hwang C-S, Shemorry A, Varshavsky A. 2009. Two proteolytic pathways regulate DNA repair by cotargeting the Mgt1 alkyguanine transferase. *Proc. Natl. Acad. Sci. USA* 106:2142–47
61. Hwang C-S, Varshavsky A. 2008. Regulation of peptide import through phosphorylation of Ubr1, the ubiquitin ligase of the N-end rule pathway. *Proc. Natl. Acad. Sci. USA* 105:19188–93
62. Xia Z, Webster A, Du F, Piatkov K, Ghislain M, Varshavsky A. 2008. Substrate-binding sites of UBR1, the ubiquitin ligase of the N-end rule pathway. *J. Biol. Chem.* 283:24011–28
63. Hwang C-S, Shemorry A, Varshavsky A. 2010. N-terminal acetylation of cellular proteins creates specific degradation signals. *Science* 327:973–77
64. Arnesen T, Van Damme P, Polevoda B, Helsens K, Evjenth R, et al. 2009. Proteomics analyses reveal the evolutionary conservation and divergence of N-terminal acetyltransferases from yeast and humans. *Proc. Natl. Acad. Sci. USA* 106:8157–62
65. Helbig AO, Gauci S, Raijmakers R, van Breukelen B, Slijper M, et al. 2010. Profiling of N-acetylated protein termini provides in-depth insights into the N-terminal nature of the proteome. *Mol. Cell. Proteomics* 9:928–39
66. Polevoda B, Sherman F. 2003. N-terminal acetyltransferases and sequence requirements for N-terminal acetylation of eukaryotic proteins. *J. Mol. Biol.* 325:595–622
67. Goetze S, Qeli E, Mosimann C, Staes A, Gerrits B, et al. 2009. Identification and functional characterization of N-terminally acetylated proteins in *Drosophila melanogaster*. *PLoS Biol.* 7:e1000236
68. Moerschell RP, Hosokawa Y, Tsunasawa S, Sherman F. 1990. The specificities of yeast methionine aminopeptidase and acetylation of amino-terminal methionine in vivo. Processing of altered iso-1-cytochromes *c* created by oligonucleotide transformation. *J. Biol. Chem.* 265:19638–43
69. Frottin F, Martinez A, Peynot P, Mitra S, Holz RC, et al. 2006. The proteomics of N-terminal methionine cleavage. *Mol. Cell. Proteomics* 5:2336–49
70. Mullen JR, Kayne PS, Moerschell RP, Tsunasawa S, Gribskov M, et al. 1989. Identification and characterization of genes and mutants for an N-terminal acetyltransferase from yeast. *EMBO J.* 8:2067–75
71. Park EC, Szostak JW. 1992. ARD1 and NAT1 proteins form a complex that has N-terminal acetyltransferase activity. *EMBO J.* 11:2087–93
72. Gautschi M, Just S, Mun A, Ross S, Rücknagel P, et al. 2003. The yeast N-alpha-acetyltransferase NatA is quantitatively anchored to the ribosome and interacts with nascent polypeptides. *Mol. Cell. Biol.* 23:7403–14
73. Tasaki T, Kwon YT. 2007. The mammalian N-end rule pathway: new insights into its components and physiological roles. *Trends Biochem. Sci.* 32:520–28
74. Mogk A, Schmidt R, Bukau B. 2007. The N-end rule pathway of regulated proteolysis: prokaryotic and eukaryotic strategies. *Trends Cell Biol.* 17:165–72
75. Eisele F, Wolf DH. 2008. Degradation of misfolded proteins in the cytoplasm by the ubiquitin ligase Ubr1. *FEBS Lett.* 582:4143–46
76. Heck JW, Cheung SK, Hampton RY. 2010. Cytoplasmic protein quality control degradation mediated by parallel actions of the E3 ubiquitin ligases Ubr1 and San1. *Proc. Natl. Acad. Sci. USA* 107:1106–11
77. Hu R-G, Wang H, Xia Z, Varshavsky A. 2008. The N-end rule pathway is a sensor of heme. *Proc. Natl. Acad. Sci. USA* 105:76–81

78. Hu R-G, Sheng J, Xin Q, Xu Z, Takahashi TT, et al. 2005. The N-end rule pathway as a nitric oxide sensor controlling the levels of multiple regulators. *Nature* 437:981–6
79. Wang H, Piatkov KI, Brower CS, Varshavsky A. 2009. Glutamine-specific N-terminal amidase, a component of the N-end rule pathway. *Mol. Cell* 34:686–95
80. Graciet E, Wellmer F. 2010. The plant N-end rule pathway: structure and functions. *Trends Plant Sci.* 15:447–53
81. Brower CS, Varshavsky A. 2009. Ablation of arginylation in the mouse N-end rule pathway: loss of fat, higher metabolic rate, damaged spermatogenesis, and neurological perturbations. *PLoS ONE* 4:e7757
82. Zenker M, Mayerle J, Lerch MM, Tagariello A, Zerres K, et al. 2005. Deficiency of UBR1, a ubiquitin ligase of the N-end rule pathway, causes pancreatic dysfunction, malformations and mental retardation (Johanson-Blizzard syndrome). *Nat. Genet.* 37:1345–50
83. Hwang C-S, Sukalo M, Batygin O, Addor MC, Brunner H, et al. 2011. Ubiquitin ligases of the N-end rule pathway: assessment of mutations in *UBR1* that cause the Johanson-Blizzard syndrome. *PLoS ONE* 6:e24925
84. Prasad R, Kawaguchi S, Ng DTW. 2010. A nucleus-based quality control mechanism for cytosolic proteins. *Mol. Biol. Cell* 21:2117–27
85. Kurosaka S, Leu NA, Zhang F, Bunte R, Saha S, et al. 2010. Arginylation-dependent neural crest cell migration is essential for mouse development. *PLoS Genet.* 6:e1000878
86. Zhang F, Saha S, Shabalina SA, Kashina A. 2010. Differential arginylation of actin isoforms is regulated by coding sequence-dependent degradation. *Science* 329:1534–7

Ubiquitin and Proteasomes in Transcription

Fuqiang Geng, Sabine Wenzel, and William P. Tansey

Department of Cell and Developmental Biology, Vanderbilt University School of Medicine, Nashville, Tennessee 37232-8240; email: fuqiang.geng@vanderbilt.edu, sabine.a.wenzel@vanderbilt.edu, william.p.tansey@vanderbilt.edu

Keywords

chaperone, chromatin, coactivators, gene expression, proteolysis

Abstract

Regulation of gene transcription is vitally important for the maintenance of normal cellular homeostasis. Failure to correctly regulate gene expression, or to deal with problems that arise during the transcription process, can lead to cellular catastrophe and disease. One of the ways cells cope with the challenges of transcription is by making extensive use of the proteolytic and nonproteolytic activities of the ubiquitin-proteasome system (UPS). Here, we review recent evidence showing deep mechanistic connections between the transcription and ubiquitin-proteasome systems. Our goal is to leave the reader with a sense that just about every step in transcription—from transcription initiation through to export of mRNA from the nucleus—is influenced by the UPS and that all major arms of the system—from the first step in ubiquitin (Ub) conjugation through to the proteasome—are recruited into transcriptional processes to provide regulation, directionality, and deconstructive power.

Contents

INTRODUCTION	178
CHALLENGES OF EUKARYOTIC GENE TRANSCRIPTION	178
FLEXIBILITY AND UTILITY OF THE UBIQUITIN-PROTEASOME SYSTEM	179
Protein Ubiquitylation	180
The Proteasome	180
REGULATING TRANSCRIPTIONAL ACTIVATORS AND COACTIVATORS	181
Nonproteolytic Functions of Ubiquitin in Controlling Transcriptional Activators	182
Proteolytic Control of Activators and Coactivators	183
REGULATION OF HISTONES BY THE UBIQUITIN-PROTEASOME SYSTEM	186
CONNECTIONS BETWEEN THE TRANSCRIPTION AND UBIQUITIN-PROTEASOME SYSTEMS	187
THE PROTEASOME IN TRANSCRIPTION	190
CONCLUSIONS	192

Ubiquitin-proteasome system (UPS): includes all proteins required for ubiquitylation, deubiquitylation, recognition, and processing (proteoltyic or otherwise) of ubiquitylated proteins

Ub: ubiquitin

RNA polymerase II (Pol II): a 12-subunit enzyme responsible for mRNA synthesis

INTRODUCTION

Eukaryotic gene transcription is a phenomenally complex process. For a gene to be transcribed, a large cohort of proteins need to descend upon DNA in a highly orchestrated manner, dealing not only with the physical constraints of transcribing a template that is encased in chromatin, but also ensuring that genes are turned on only at the right place, at the right time, and in response to the right signals. Cells employ a myriad of processes to ensure that these challenges are met appropriately; one of the most recently discovered of these processes is by engaging the proteolytic and nonproteolytic capabilities of the ubiquitin-proteasome system (UPS).

The aim of this review is discuss how ubiquitin (Ub), Ub-dependent transactions, and the proteasome impact eukaryotic transcription by RNA polymerase II (Pol II). This is an area of intense investigation, and one in which many surprising and counterintuitive discoveries have been made. Our review centers on the UPS—other reviews have dealt with roles of Ub-like modifications like SUMO (1)—and focuses on advances that have been made in this arena in the past 10 years (for reviews of earlier work see References 2–4). We highlight four major arms of intersection between the transcription and ubiquitin-proteasome systems, discuss the mechanisms that are at work, and present working models of how the UPS intervenes at key stages in transcriptional processes.

CHALLENGES OF EUKARYOTIC GENE TRANSCRIPTION

At its simplest level, mRNA transcription occurs when Pol II engages DNA upstream of the open reading frame of a gene and synthesizes a complementary strand of RNA that serves as a template for protein synthesis. But numerous elaborations on this scheme both present obstacles that must be overcome and provide critical regulatory opportunities (**Figure 1**). Unlike bacterial RNA polymerases, Pol II cannot initiate promoter-specific transcription but relies on interactions with a host of general transcription factors (GTFs) that engage polymerase at core promoter elements and facilitate both initiation and elongation of transcription (5). In vivo Pol II and the GTFs are recruited to chromatin by the action of transcriptional activators (TAs), proteins that possess a DNA-binding domain that tethers them to promoter DNA, and a transcriptional activation domain (TAD) that either directly or via coactivators brings GFTs and Pol II to sites of transcription. Transcriptional activators and coactivators (and conversely repressors and corepressors) are a major point of regulation of gene expression, and these proteins are heavily regulated by processes that control their abundance, localization, and activity.

The compaction of DNA into chromatin also has a major impact on transcription (6). Nucleosomes present a barrier to the binding of transcriptional regulators and are incompatible with passage of Pol II (7), meaning that histones must either be locally remodeled for transcription to occur or evicted from DNA ahead of the elongating Pol II complex (8). Importantly, because nucleosomes also act to conceal cryptic promoter elements present throughout the genome (9, 10), histones must be redeposited appropriately in the wake of Pol II to prevent transcription of nonauthentic RNA from occurring. The ability of nucleosomes to control template accessibility affords tremendous regulatory potential, and numerous chromatin modifying and remodeling factors (6, 11) act to not only control transcriptional processes but also to signal to the cell the transcriptional state of a given piece of chromatin.

Finally, it is worth emphasizing that each transcript is synthesized by a single Pol II complex and that events such as premessenger RNA processing and mRNA export are tightly coupled to transcription. The net effect of this arrangement is that Pol II has to continually change its entourage of interacting proteins: one set for initiation; another for elongation and RNA processing; and yet another for termination, 3′ processing, and mRNA export. Many of these events are coordinated by the so-called CTD code (12), which corresponds to distinct patterns of phosphorylation within the C-terminal domain of the largest subunit of Pol II and acts as a molecular beacon to broadcast specific points in the transcription process. The ability of transcriptional proteins to "sense" the activity of Pol II and other factors involved in transcription is a key mechanism through which order is brought to each of the stages in gene expression.

FLEXIBILITY AND UTILITY OF THE UBIQUITIN-PROTEASOME SYSTEM

The most recognizable role of the UPS is in protein turnover. Ub-mediated proteolysis

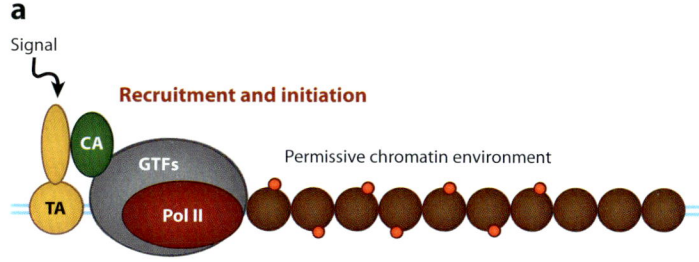
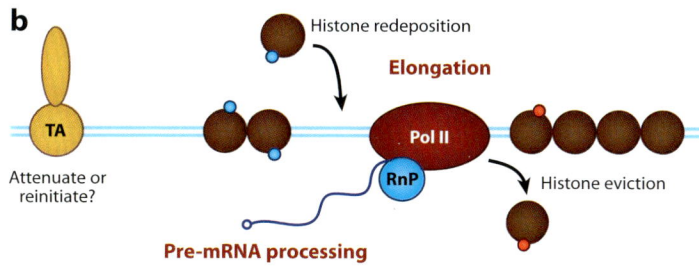
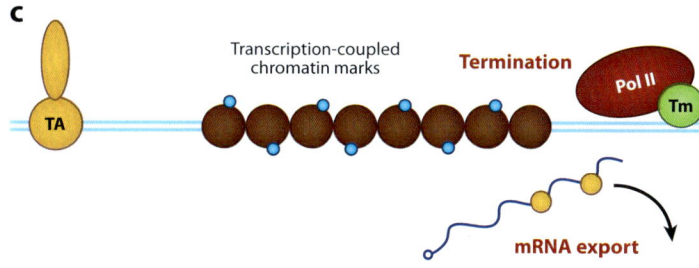

Figure 1

Transcriptional regulation. (*a*) Activation of transcription. Within a permissive chromatin environment, a transcriptional activator (TA) binds to promoter sequences, often in response to environmental signals. The activator typically requires a coactivator (CA), and together they recruit general transcription factors (GTFs) and RNA polymerase II (Pol II), leading to the initiation of transcription. (*b*) As Pol II enters the elongation phase, histones are evicted ahead of the transcribing Pol II and redeposited in its wake. At this stage, components of the RNA processing (RnP) machinery join Pol II, carrying out premessenger RNA maturation. This is also a stage when transcription can reinitiate, or the signal to initiate is terminated. (*c*) Transcription-coupled chromatin marks persist on the recently transcribed gene, and transcription terminates by association of Pol II with the termination machinery (Tm). mRNA is then exported from the nucleus. Blue lines represent the template DNA duplex.

is a highly specific process in which covalent linkage of substrates to the small protein molecule Ub signals their destruction by the 26S proteasome. Regulated protein destruction by the UPS is instrumental in processes ranging

General transcription factors (GTFs): a group of ancillary proteins necessary for promoter-driven transcription by Pol II

TA: transcriptional activator

Transcriptional activation domain (TAD): a transferable element present in transcriptional activators that is responsible for recruiting Pol II and GTFs to chromatin

C-terminal domain (CTD) of RNA polymerase II: a repeating motif in Pol II that is phosphorylated depending on the transcription stage

from cell cycle regulation through to antigen presentation, but it is important to remember that proteolysis is only one of many possible outcomes for a ubiquitylated protein. First and foremost, ubiquitylation is a posttranslational modification that provides a complex and nuanced way to alter the localization, activity, or stability of a linked protein substrate. Similarly, the complexities of the proteasome endow it with capabilities that extend well beyond simple protein breakdown. As we describe below, these noncanonical functions of the UPS feature prominently in the control of gene activity, and transcription makes full use of all of the capabilities of the UPS to insure appropriate transcriptional outcomes.

Protein Ubiquitylation

The conjugation of Ub is a multistep process (**Figure 2**) that begins when Ub forms a high-energy thioester linkage with a Ub-activating enzyme, also known as an E1. The "activated" Ub is then transferred to the active-site cysteine of a Ub-conjugating enzyme (E2), which functions in concert with a Ub-protein ligase (E3) to conjugate Ub to an amino group on the

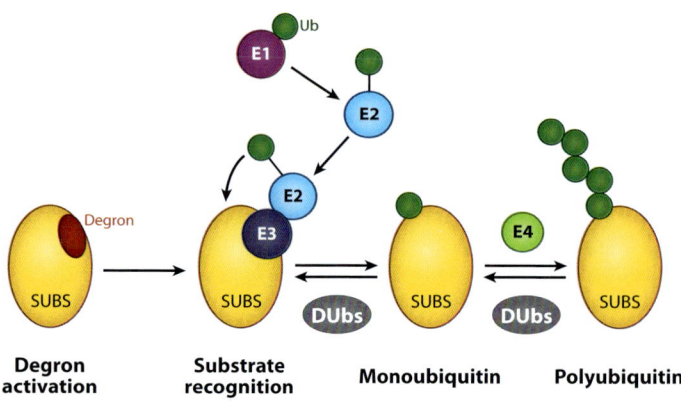

Figure 2

Ubiquitylation. A destruction element in the substrate (a degron) is activated, often by phosphorylation in response to a specific signal. The degron is then bound by a ubiquitin (Ub)-protein ligase (E3), which acts in conjunction with a Ub-activating enzyme (E1) and Ub-conjugating enzyme (E2) to transfer Ub (*green circle*) to the substrate (SUBS), typically at a lysine residue. Repeated rounds of this process, perhaps catalyzed by a Ub chain elongation factor (E4), give rise to a polyubiquitylated substrate. Ubiquitylation can be opposed by the action of deubiquitylating enzymes (DUbs), which can remove all Ub or trim the length of the Ub chain.

substrate, typically at a lysine residue. Specificity in this process is governed by the interaction between an E3 and its target site on the substrate, known as a degron. There are many hundreds of E3s in the mammalian genome, and degrons themselves are often activated by posttranslational modifications, providing a high level of selectivity and control to the process of ubiquitylation. Although numerous proteins are monoubiquitylated, most substrates undergo additional rounds of ubiquitylation on Ub (perhaps catalyzed by an additional class of enzymes known as E4s), leaving them in a heavily modified and polyubiquitylated state.

A few characteristics of protein ubiquitylation are worth highlighting. First, Ub carries eight potential sites of ubiquitylation, meaning that Ub chains acquire unique characteristics not just via their length, but also by their topology (13). It is generally thought that proteins targeted for proteasomal destruction must carry at least a tetra-Ub chain in which Ub groups are linked via lysine 48 of Ub [K48-linkage (14)]; whereas other linkages, such as lysine 63, adopt distinct conformations that are recognized by Ub-binding receptor proteins for nonproteolytic outcomes (15). This diversity in chain recognition generates a range of signals far more diverse than might be expected from simple iteration of the Ub moiety itself. Second, ubiquitylation is a dynamic process that can be reversed by the action of deubiquitylating enzymes (DUbs) that either trim Ub chains or remove them entirely (16). The dynamic nature of ubiquitylation means that Ub conjugation is not a terminal process and that DUbs can act to modify the fate of a ubiquitylated substrate. Finally, ubiquitylation is unusual in that it is a very large posttranslational modification, and it is likely that the bulkiness of Ub can itself serve to alter the inherent characteristics of a ubiquitylated substrate protein.

The Proteasome

The proteasome is a self-compartmentalized protease (17) that carries its proteolytic activities deep within its interior. This arrangement means that for a protein to be destroyed by the proteasome, it not only typically needs to be ubiquitylated, but also presented in an appropriate manner to gain access to the central proteolytic chamber. Delivery of substrates to the proteasome is often facilitated by Ub-dependent chaperones and shuttling factors (18, 19), but once a substrate arrives on its doorstep, much of the actions of the proteasome are devoted to feeding substrates to the inner protease sites.

The 26S proteasome (**Figure 3**) consists of a 20S proteolytic core particle, capped at one or both ends by a 19S regulatory particle, the latter of which can be further subdivided into lid and base subcomplexes (20). The lid of the 19S regulatory particle is composed of ~10 subunits, whereas the base is composed of six AAA-type ATPases and two non-ATPase subunits. Ubiquitylated proteins delivered to the proteasome are recognized by receptors in the 19S complex, deubiquitylated, unfolded by ATPases in the base, and translocated into the 20S complex for destruction. Within the 20S complex, three distinct proteoltyic activities—chymotryptic-, tryptic-, and caspase-like—attack the target, typically cleaving it into peptides of between 4–25 amino acids in length (17), but occasionally, these activities clip the substrate at specific sites, generating functional protein fragments (21). The proteasome thus carries at least six distinct biochemical activities: (*a*) recognition of ubiquitylated proteins, (*b*) deubiquitylation, (*c*) protein unfolding/chaperone, (*d*) protein translocation, (*e*) protein destruction, and (*f*) protein processing, which are typically devoted to proteolysis but could theoretically be diverted to other processes as well. As we discuss below, the potential moonlighting of these functions has received particular attention in the field of transcriptional regulation.

REGULATING TRANSCRIPTIONAL ACTIVATORS AND COACTIVATORS

Given their importance in setting gene expression levels, it is not surprising that TAs and coactivators are a major point of intervention of the UPS in transcriptional control. The most obvious way in which the UPS controls these proteins is by fine-tuning their steady-state levels or by destroying them when their function is no longer appropriate (2), but examples also exist in which the UPS controls the localization of activators (22) or processes them into a functional state via limited proteolysis (23). These actions are in line with traditional views of how the UPS regulates protein activity, acting either upstream or downstream of the point at which its substrates function. In transcription, however, the UPS has inserted itself in a prominent way directly into the heart of gene regulatory mechanisms, controlling activators and coactivators on chromatin and during the process of transcriptional regulation. The ability of the UPS to modulate activators at their workplace endows the system with tremendous potential to micromanage the proteins that control gene activity.

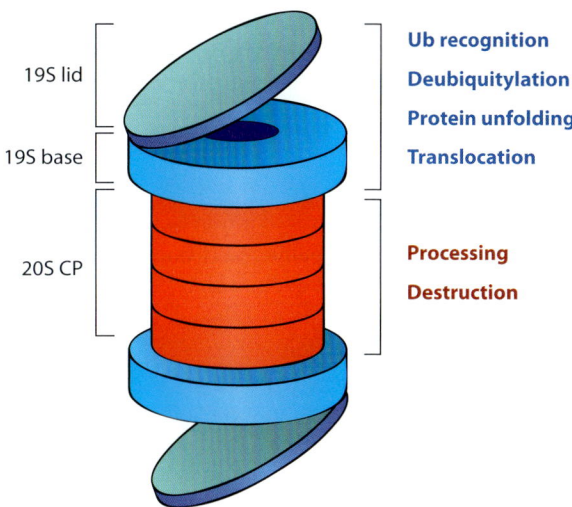

Figure 3

Architecture of the 26S proteasome. The proteasome is a self-compartmentalized protease that consists of three main subcomplexes: a 20S core particle (CP) that houses three distinct proteolytic activities in its inner core, a 19S lid that recognizes and deubiquitylates substrates, and a 19S base structure that uses the energy of ATP hydrolysis to unfold proteins and pass them into the 20S CP for destruction. 26S proteasomes can be capped at one or both ends by a 19S complex.

Ubiquitylation: covalent linkage of Ub onto an accessible amino group of an acceptor protein

Ubiquitin-activating enzyme (E1): the enzyme that carries out the first stage in protein ubiquitylation

Ubiquitin-conjugating enzyme (E2): accepts Ub from E1 and carries out the chemistry of protein ubiquitylation, usually with an E3

Ubiquitin-conjugating enzyme (E3): recognizes substrates and works with an E2 to promote substrate ubiquitylation

Deubiquitylating enzyme (DUb): a class of enzymes that remove Ub modification from substrates

Nonproteolytic Functions of Ubiquitin in Controlling Transcriptional Activators

Many TAs and coactivators can be regulated by nonproteolytic ubiquitylation. A consensus has yet to emerge on the effects of ubiquitylation on transcription factor activity, but on the basis of published examples, we can define two general ways in which Ub acts, independent of proteolysis, to influence activator and coactivator function.

The first group is the least mechanistically developed and encompasses examples where monoubiquitylation stimulates TA activity through unknown means (**Figure 4a**). A collection of natural (24–27) and synthetic (28, 29) transcription factors are regulated by this mechanism, and although exactly how is unclear, the phenomenon does appear to be both biologically relevant and a point of regulation. In the case of FOXO4, for example, a transcription factor important for the cellular response to stress, oxidative stress induces both monoubiquitylation and deubiquitylation of the activator. Monoubiquitylation of FOXO4 causes it to enter the nucleus and stimulates its transcriptional activity, whereas deubiquitylation by the DUb Usp7 reverses

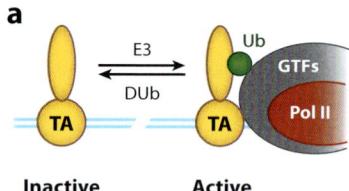

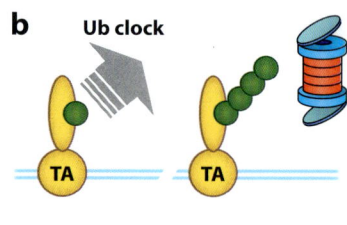

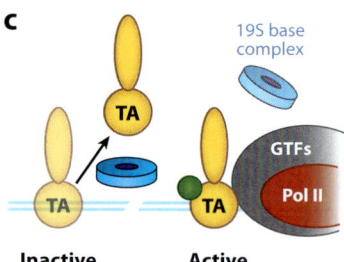

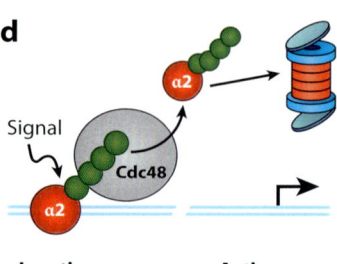

Figure 4

Roles for ubiquitylation in controlling transcriptional regulators on chromatin. (*a*) Activation by monoubiquitylation. In this view, monoubiquitylation of DNA-bound activators stimulates their inherent activation properties, a process that is promoted by Ub ligases (E3s) and antagonized by deubiquitylating enzymes (DUbs). (*b*) The Ub clock model. In this view, the period of activity of a transcriptional regulator is governed by the length of time it takes to transition from the monoubiquitylated and active form to a polyubiquitylated form that is rapidly destroyed by the proteasome. (*c*) Control of promoter stripping by activator ubiquitylation. In this model, activators are aggressively removed from chromatin by resident ATPases in the 19S base complex (*blue ring*). The presence of Ub on the activator blocks this activity, allowing the TA to stably associate with its cognate DNA element. (*d*) Extraction of ubiquitylated transcription factors from chromatin by Ub-selective chaperones. Here, polyubiquitylation of the α2 repressor causes it to be extracted from promoter DNA by the ATPase activities of Cdc48 (p97), a process that allows immediate cessation of function before α2 is subsequently destroyed by the proteasome. Note that these models are not mutually exclusive, and it is possible that multiple mechanisms contribute to transcription factor control at a specific promoter. Abbreviations: GTFs, general transcription factors; Pol II, RNA polymerase II; TA, transcriptional activator. Blue lines represent the template DNA duplex.

this process. By having both ubiquitylation and deubiquitylation under the control of oxidative stress, cells inherently limit FOXO4 activity, ensuring that ongoing rounds of FOXO4-driven transcription are responsive to the continual presence of the activating signal. Another interesting example centers on SRC-3, a coactivator for hormone-liganded transcription factors (**Figure 4b**). SRC-3 is activated by monoubiquitylation (30) but destroyed by the proteasome once the Ub chain on SRC-3 is extended beyond a certain threshold. The time it takes for SRC-3 to transition from a monoubiquitylated to a polyubiquitylated species thus acts as a "molecular clock," defining a specific period of time during which SRC-3 can function before it is destroyed. In this way, ubiquitylation acts as a self-limiting mechanism that prevents renegade transcription factors from uncontrollably activating transcription. In the section titled Proteolytic Control of Activators and Coactivators, we discuss an even more intimate coupling of activator turnover and function that likely achieves the same objective.

The second group includes a growing number of examples in which ubiquitylation affects how a transcription factor binds to its target sites in the genome, both positively and negatively. At one end of the spectrum, monoubiquitylation of the yeast activator Gal4 appears to prevent it from being stripped off chromatin by ATPases resident in the 19S base complex (**Figure 4c**) (31), suggesting that ubiquitylation could act to lock an activator onto chromatin in an active configuration. At the other extreme, ubiquitylation can also play a proactive role in dissociating transcriptional regulators from their target sites (**Figure 4d**) (32, 33). The best-understood example in this group is the yeast transcriptional repressor $\alpha 2$, which controls the transition between different mating types. Although Ub-mediated proteolysis of $\alpha 2$ is important for mating-type switching (34), the critical point of regulation in this instance occurs at the level of $\alpha 2$ binding to chromatin and precedes proteolysis. When the signal to switch mating types is received, $\alpha 2$ is ubiquitylated on chromatin, and this ubiquitylation recruits the Ub-selective chaperone Cdc48 (p97), a AAA-type ATPase that mobilizes ubiquitylated proteins for destruction by the proteasome (19). Once Cdc48 encounters ubiquitylated $\alpha 2$, it uses the energy of ATP hydrolysis to extract $\alpha 2$ from chromatin, immediately terminating its function and subsequently directing it to the proteasome for permanent inhibition of $\alpha 2$ activity. The $\alpha 2$ study is particularly important because it demonstrates that steps in the linear sequence of Ub-mediated proteolysis may be both temporally and spatially separated to effect immediate versus long-term outcomes and that just because $\alpha 2$ is destroyed by the UPS does not mean that proteolysis per se is the rate-limiting step.

Two final points are worth noting here. First, it is entirely possible that some of the first class of activators, where the mechanism of activation by ubiquitylation is unknown, are regulated by Ub at the level of promoter occupancy. Second, the principal that ubiquitylation can control transcription factor activity separate from proteolysis may not be restricted to monoubiquitylation, nor to TAs. Atypical poly-Ub chain topologies, which do not lead to proteasomal proteolysis, have been found to stimulate transcription factor activity under some circumstances (35), and the activity of the repressive histone deacetylase complex mSin3 is stimulated by formation of lysine 63-linked Ub chains on its Sds3 subunit (36). Thus, the rules that are distilled from transcriptional regulators may be broadly applicable to other components of the transcriptional apparatus, and similar mechanisms may be at work.

Proteolytic Control of Activators and Coactivators

The concept that transcriptional regulators can be made to accumulate or disappear by manipulating their Ub-mediated proteolysis is well established (2, 3). There is, however, another more intriguing side to this story: Ub-mediated proteolysis can promote the activity of the transcriptional regulators it destroys. This counterintuitive role of the

UPS in controlling transcription has only been appreciated in recent years, but this role appears to apply to a growing number of transcriptional regulators and provides insight into how other processes may be controlled by an "activation by destruction" mechanism (37).

The first clue to the existence of this phenomenon came from the discovery that TADs often overlap with degrons (38). This overlap is widespread—there are now nearly 30 transcription factors that have overlapping TADs and degrons (**Figure 5**)—and is intimate, as manipulations that affect protein turnover typically have corresponding effects on TAD activity [e.g., (28, 37, 39, 40)]. The coupling of these two functions within TAs [and coactivators (41)] raises an intriguing question: Why are two apparently opposing functions—activation and destruction—linked in these proteins?

Since this overlap was first noted, significant progress has been made in both expanding the range of transcription factors that fall into this category and exposing its significance. Although deep mechanistic insight is still lacking, a few important conclusions can be made. Destruction of activators is likely to be a direct consequence of their ability to activate transcription (39, 42). Activator turnover requires DNA binding and ongoing transcriptional activity (43), and this turnover is often signaled by kinases that are integral parts of the transcriptional apparatus (44–46). Ub ligases are recruited to sites of activator function (2, 44, 46–50). And the ongoing activity of these proteins is dependent on their ubiquitylation and on the proteolytic capacity of the proteasome (2, 3, 37, 39, 41, 44, 51–58). Taken together with the widespread overlap of activation and destruction elements, these observations indicate that Ub-mediated destruction of activators is inherently linked to the way in which they stimulate transcription.

A number of models have been proposed to explain the connection between activator destruction and function (2, 3, 59); the majority of these models invoke a "suicide" theme in which destruction of the activator is an obligate part of the activation mechanism that both drives the process forward and is guaranteed to terminate the activation signal. This notion is closely aligned with the Ub clock hypothesis, and it is possible that a similar mechanism underlies both processes, the only difference being the number of rounds of transcription an activator can stimulate before it is terminated. There are, however, a number of issues that need to be noted in any critical evaluation of a suicide-type

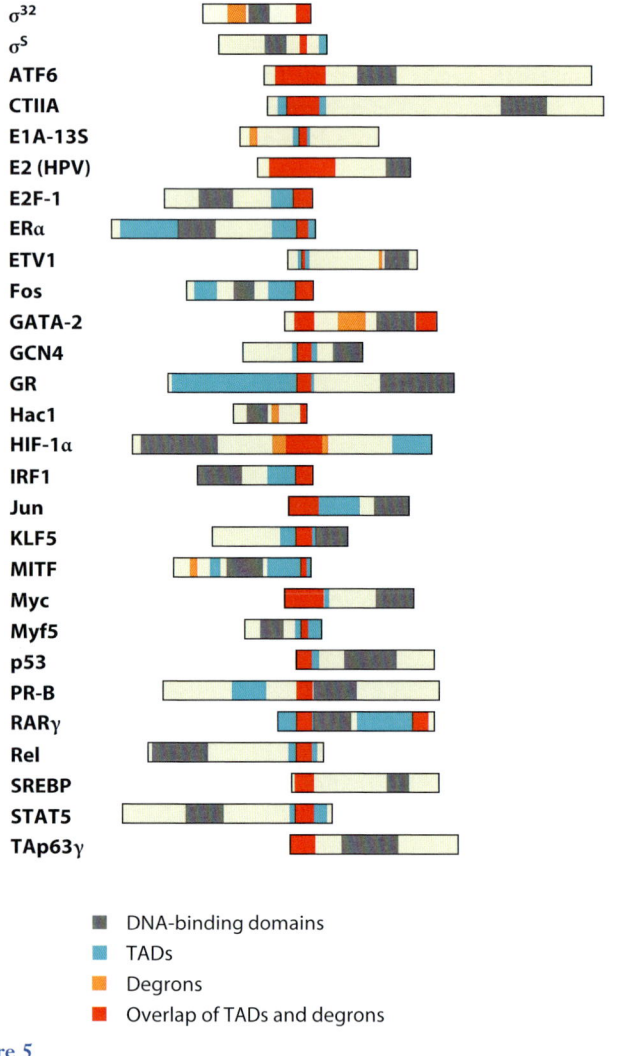

Figure 5
Transcriptional activation domains (TADs) often overlap with degrons. The domain structure of 28 transcription factors is shown. Although the overlap of TADs and degrons is not always perfect, it is worth remembering that different studies have typically defined each type of element and that different sets of mutations (which define the boundaries) have been used in these studies.

model. First, the mechanism must apply only to a certain mode of transcriptional activation, as there are TADs that stimulate robust levels of transcription without triggering proteolysis (38). Second, the mechanism must apply only in certain instances, as there are clear examples of individual transcription factors that can be both activated and inhibited by Ub-mediated proteolysis (44). Third, there must be a way for the UPS to attack activators only after they have functioned; otherwise, proteolysis could precede activation and would obligatorily inhibit TA activity. Finally, transcription-coupled destruction of an activator needs to serve a functional purpose such that, if it is blocked, repeated rounds of transcriptional activation cannot occur.

Taking these issues into account, we suggest a revised model (**Figure 6**) in which TAs are regulated by two distinct modes of Ub-mediated proteolysis. Mode 1 occurs off chromatin, and in this case, the UPS limits the activity of a transcription factor by restricting its abundance. If proteolysis does not occur, the TA accumulates, and its function is enhanced. Mode 2, however, occurs on chromatin and during the process of gene induction. We posit that kinases associated with the transcriptional machinery (44–46) serve to mark an activator as "spent," trapping it in an inactive state that cannot stimulate further rounds of transcription. At the same time, these phosphorylation events bring in components of the UPS that destroy the activator in situ, clearing the deck for promoter association with a fresh activator molecule. In this mode, proteolysis serves a positive role in transcription by allowing pristine activators to access the promoter and stimulate additional rounds of transcription. This model, which is a refinement of one originally proposed by Lipford et al. (3, 37), predicts that the UPS is only required when the activator is locked in its phosphorylated state. Thus, if the mode of activation does not lead to activator phosphorylation, or if the sites of phosphorylation on the TA are blocked (37), the UPS becomes dispensable for activator function.

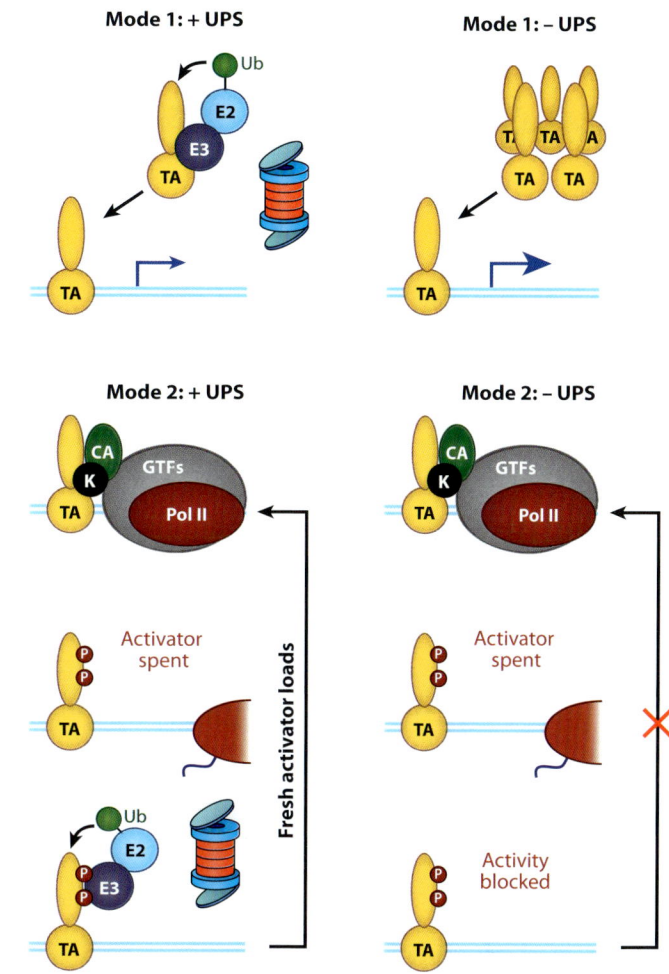

Figure 6

Two modes of activator regulation by ubiquitin-mediated proteolysis. The figure shows how a transcription factor can be regulated in disparate ways by proteolysis and the consequences of disrupting the ubiquitin-proteasome system (UPS) on those modes of regulation. Mode 1 occurs off chromatin, and the UPS is used to limit the concentration of available activators. When the UPS is disrupted in this case, the transcription activator (TA) accumulates, and transcription is induced. Mode 2 occurs on promoter DNA and during the course of transcriptional activation. We posit that kinases (K) associated with the general transcriptional machinery phosphorylate activators at some point during transcriptional activation. This phosphorylation (P, *red circles*) has two functions. Its marks the activator as spent and incapable of stimulating further rounds of transcription. Concurrently, it also recruits a Ub ligase that ubiquitylates the transcription activator (TA), allowing it to be destroyed and a fresh activator to reach the promoter. In this mode, if the UPS is perturbed, the inactive TA remains on chromatin, and subsequent rounds of transcription are blocked. Abbreviations: CA, coactivator; E2, ubiquitin-conjugating enzyme; E3, ubiquitin-protein ligase. Blue lines represent the template DNA duplex.

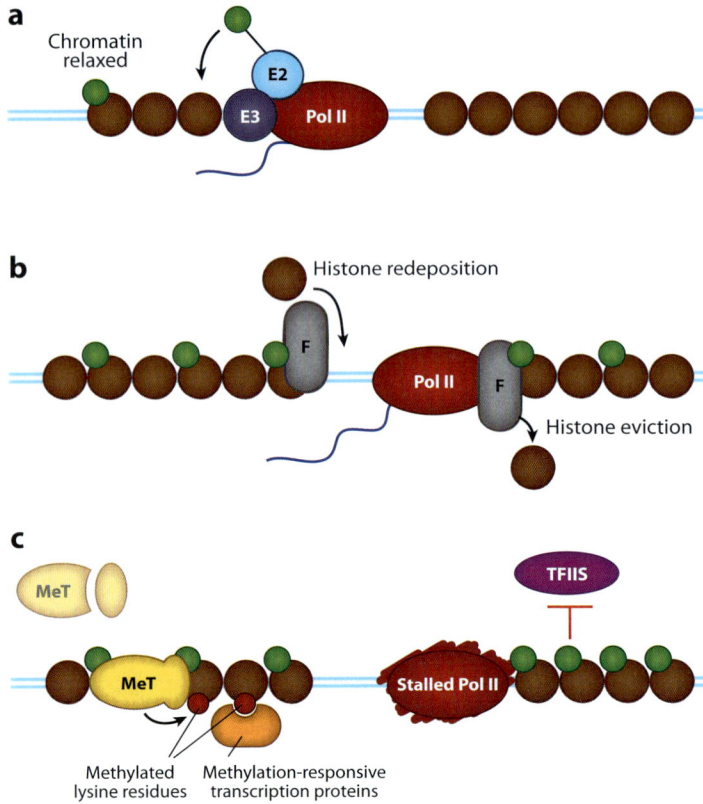

Figure 7

Functions of H2B ubiquitylation. (*a*) During transcriptional elongation, the ubiquitylation machinery for H2B (Rad6/Bre1) is recruited to transcriptional complexes, where it ubiquitylates H2B, most prominently on lysine 120. This modification does not impact nucleosome structure, but it does relax higher-order chromatin configurations. (*b*) As transcription proceeds, ubiquitylation of H2B selectively recruits the FACT elongation complex (F), which not only displaces H2A/H2B dimers ahead of RNA polymerase II (Pol II), but also insures that histones are redeposited afterward. (*c*) H2B ubiquitylation recruits and activates Dot1- and Set1-containing histone methyltransferase complexes (MeT), which promote the di- and trimethylation of histone H3 at lysine residues 4 and 79. These methylation events, in turn, recruit other factors to chromatin and repel binding of transcriptional silencing complexes (not shown). If levels of ubiquitylated H2B accumulate, elongation factors such as TFIIS are excluded from chromatin, preventing the rescue of stalled polymerase molecules. Abbreviations: E2, ubiquitin-conjugating enzyme; E3, ubiquitin-protein ligase. Blue lines represent the DNA duplex. Brown circles represent nucleosomes. Green circles represent ubiquitylated histone H2B.

transcription factor activity. Our model gives mechanistic insight into how this may occur, but it also raises some thought-provoking questions. Does this type of regulation apply to activators only, or are coactivators and repressors also subject to this type of control? What are the steps in transcription that depend on activator phosphorylation and proteolysis, and are other transcription proteins, besides coactivators, also destroyed in the process? And—more broadly—does this type of inherently self-limiting mechanism apply to other cellular events in which the UPS has a regulatory foothold?

REGULATION OF HISTONES BY THE UBIQUITIN-PROTEASOME SYSTEM

One of the most high-profile intersections between the transcription and ubiquitin-proteasome systems centers on ubiquitylation of histones and the impact this has on chromatin structure and function. It is now clear that all histones can be ubiquitylated (60), that this modification is abundant [almost 10% of H2B, for example, is present in the ubiquitylated form (61)], and that histone ubiquitylation acts nonproteolytically to control gene activity. The best-studied examples of histone ubiquitylation are H2A and H2B. H2A ubiquitylation is typically associated with chromatin compaction and transcriptional repression, whereas H2B ubiquitylation is associated with gene activation (60). Numerous excellent reviews have been written on the subject of histone modifications, so rather than cover this subject comprehensively here, we focus on just one example, H2B, that illustrates some of the ways in which the presence of Ub on a core histone can influence gene activities (**Figure 7**).

Although H2B can be ubiquitylated at multiple sites (62), the major modification of H2B is monoubiquitylation at lysine 120, a process carried out by the Rad6 Ub-conjugating enzyme (63) and the Bre1 Ub-protein ligase (64). This modification is tightly coupled to active chromatin (65), is likely signaled by

Linking the destruction of TAs directly to their activity installs a set of mandatory checkpoints to the function of any particular transcription factor molecule and provides a powerful way for the cell to tightly control combinatorial processes that are linked to transcriptional elongation (66–68), and depends on transcription-coupled alterations to chromatin structure (66, 69). Thus, ubiquitylation of H2B marks the template as having recently been transcribed and sets forth a process that impacts transcriptional events. But what does H2B ubiquitylation do?

Inherently, modification of H2B by Ub does not disrupt nucleosome architecture (70), but does impair chromatin fiber compaction, and can work with other histone modifications to relax higher-order chromatin structure (71). This phenomenon is consistent with the bulky nature of the Ub moiety, which presumably prevents tight compaction of chromatinized DNA strands, and has obvious ramifications for influencing the ability of Pol II to access its template. In vivo, however, H2B ubiquitylation is unlikely to act solely through this mechanism. The effects of H2B ubiquitylation cannot be recapitulated by replacing Ub with another bulky Ub-like modification, such as SUMO (72) or HUb1 (71), suggesting that components in the transcriptional machinery recognize the Ub modification on H2B to elicit a biological response.

Ubiquitylation of H2B is closely connected to the function of the FACT nucleosome-remodeling complex, which is important for displacing H2A/H2B dimers ahead of Pol II and for reassembling them after polymerase has passed (73). H2B ubiquitylation stimulates the activity of FACT (74) and appears to negatively act against other nucleosome-restoring complexes, making FACT the preferred nucleosome rebuilder at transcribed genes where H2B ubiquitylation occurs (75). The ability of ubiquitylated H2B to select for FACT implies that a specific molecular recognition event, signaled by Ub on H2B, acts to stimulate FACT activity at specific sites in the genome. In a similar vein, recognition of Ub by histone methyltransferase complexes (76–78) is likely involved in another essential function of H2B ubiquitylation—signaling processive methylation of histone H3 (79). In this instance, the presence of H2B ubiquitylation at active sites of transcription allows localized assembly and activation of methyltransferase complexes (80–83), which then go on to modify their respective sites in H3. The linking in *trans* of these two sets of histone modifications may seem like a lot of effort, but it provides the cell with a way to amplify and extend the actions of H2B ubiquitylation by tying it into a host of processes that respond to H3 methylation (84–87).

Finally, like many ways in which the UPS controls transcription, H2B ubiquitylation is a dynamic process that acts both positively and negatively, and the timing and extent of this modification determines the biological outcome. Removal of Ub from H2B, which is mediated by a DUb that is an integral part of the SAGA coactivator complex, is important for optimal levels of transcription (88). If ubiquitylated H2B accumulates, Pol II cannot recruit kinases important for transcriptional elongation (89), and stalled Pol II complexes cannot be reactivated by factors such as TFIIS (90). It seems likely, therefore, that H2B ubiquitylation is not a static event but that H2B is cycling between its modified and unmodified states, giving the cell a set of constant updates on transcriptional processes and providing critical opportunities for regulatory intervention.

CONNECTIONS BETWEEN THE TRANSCRIPTION AND UBIQUITIN-PROTEASOME SYSTEMS

Given the growth of this field in recent years, we cannot hope to cover all of the various examples that have surfaced of how Ub-dependent processes impact transcription. We can, however, make two important points that illustrate the extent to which the UPS regulates gene activity and the physical connections between the two processes.

The first point is that activators, coactivators, and histones are not the sole venue of intervention of the UPS in transcriptional processes. Termination of transcription depends on proteasome function (91). Core components of the transcriptional machinery

Table 1 Proteins that physically link the transcription and ubiquitin-proteasome systems

Component	Linkage between the transcription and ubiquitin-proteasome systems	Reference
Asr1	Ub ligase that binds directly to the C-terminal domain of RNA polymerase II in response to S5 phosphorylation	98
Atf1	Sequence-specific DNA-binding transcription factor that activates the anaphase-promoting complex (APC) Ub ligase complex	100
BAF250	Component of the SWI/SNF-A chromatin-remodeling complex that associates with elongin C to form an E3 that ubiquitylates histone H2B	103
CBP/p300	Transcriptional coactivator and coactivator of the APC Ub ligase complex	101
CCR4–Not	Transcriptional repressor complex containing an E3	143
CSN	COP9 signalosome; prevalent Ub ligase complex and transcriptional regulator	144
Elongin B,C	Ub ligase complex that associates with RNA polymerase II (Pol II); isolated as a factor that stimulates transcriptional elongation	145
GCN5	Histone acetyltransferase and Ub ligase cofactor	146
Med8	Component of the Pol II-associated mediator complex; forms a Ub ligase complex with elongins B,C	145
Met4	Transcriptional activator that acts as a substrate-specificity factor for the SCF^{Met30} Ub ligase	104
Rpn4	Substrate, ligand, and transcriptional regulator of the proteasome	147
Ssl1	Ub ligase and core component of the TFIIH general transcription factor	96
TAF_I	Part of the basal transcription factor TFIID, carries on both E1 and E2 activities; ubiquitylates histones and activators	95, 148
TBL1/TBLR1	Ubiquitin ligases that are part of the N-CoR corepressor complex	52
Tfb3	RING-finger component of TFIIH, stimulates Ub-like modification on the Cullin-type Ub ligase	97
TIF1γ	Ub ligase that is activated by binding to modified histone tails; destroys Smad4 on chromatin	102
Tom1	Ub ligase associated with the SAGA chromatin-remodeling complex; ubiquitylates the mRNA export factor Yra1	149, 150
Ubp8, SGf11, Sus1, Sgf73	Deubiquitylating enzyme module of the SAGA complex; deubiquitylates histone H2B	88
Uch37	Deubiquitylating enzyme that is an integral part of the Ino80 chromatin remodeler; activated by the proteasome	151

are regulated by Ub-dependent processes (36, 47). And events that occur commensurate with transcription, such as premRNA splicing and mRNA export, are also impacted by Ub and the proteasome (92, 93). Indeed, given that Ub-dependent processes are also important for translation (94), it seems that the UPS has inserted itself into just about every stage in the expression of the genetic information.

The second point is that the extensive functional links between the transcription and ubiquitin-proteasome systems are now supported by a host of physical interactions between components in the two pathways (**Table 1**). Molecules that directly connect the transcription and ubiquitin-proteasome systems influence processes ranging from gene repression through to chromatin modifications and to elongation of transcription. The understanding that has come from studying these molecules gives important insight into how the UPS is such an efficient regulator of transcriptional processes.

Notably, many proteins that connect the transcription and ubiquitin-proteasome systems are integral components of the transcriptional apparatus. TAF_I, for example, is a core component of the TFIID complex and is unusual in that it has combined E1 and E2 activities (95), allowing it to single-handedly activate

Ub and conjugate it to a substrate. The basal factor TFIIH has two potential ways it can influence ubiquitylation, acting as a direct Ub ligase (96) and by modulating the activity of other ligases that are regulated by the Ub-like modification Nedd8 (97). And as mentioned above, the SAGA chromatin-remodeling complex has a built in DUb (88) that deubiquitylates H2B, and potentially other substrates. The setting of Ub-conjugating and deconjugating enzymes within components of the transcriptional apparatus gives the UPS extraordinary opportunities to influence gene expression mechanisms.

One of the more interesting features to emerge from the study of proteins that connect these systems is that many of them have evolved the ability to sense the activity or environment of molecules involved in gene regulation. In this regard, three general mechanisms have emerged to explain how the UPS zeros in on its active transcriptional targets. First, the UPS can read activity-dependent modifications or configurations of transcription proteins (**Figure 8a**). This concept is at the heart of our model of how Ub-mediated proteolysis controls TAs (**Figure 6**), but examples also exist in which an E3 can directly interpret the CTD code on Pol II (98) or in which the CTD is read together with other structural changes in polymerase that only occur when the molecule is active (99). Second, components of

a Context-dependent substrate activation

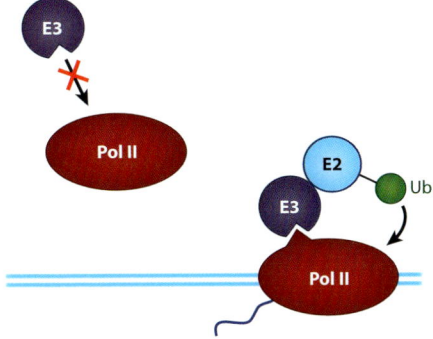

b Context-dependent UPS activation

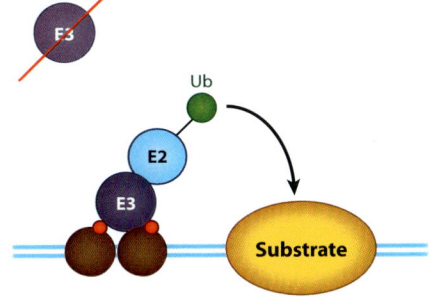

c Reprogramming Ub-ligase selectivity

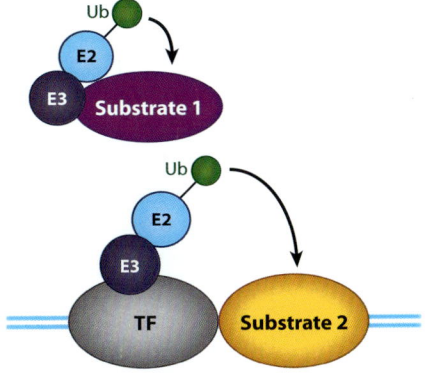

Figure 8

Ways in which the ubiquitin-proteasome system (UPS) can sense the activity of the transcriptional machinery. (*a*) By recognizing the specific context of the substrate. This example shows how the ubiquitylation machinery detects RNA polymerase II (Pol II) only when it is engaged in transcription, via a combination of phosphorylation events and structural changes in polymerase that are induced by transcription. (*b*) By context-specific activation of the ubiquitylation machinery. In this example, the ubiquitin-protein ligase, E3, is not active off chromatin, but it is activated by histone tails, insuring that it ubiquitylates its target only within a transcription setting. (*c*) By reprogramming Ub ligase selectivity. Here, a transcription factor (TF) serves as a substrate adapter for the ubiquitylation machinery, reprogramming E3 specificity from substrate 1 to substrate 2 and allowing selective modification of the second substrate in a transcriptional context. Abbreviation: E2, ubiquitin-conjugating enzyme. Blue lines represent the DNA duplex. Green circles represent ubiquitin. Brown circles represent nucleosomes. Red circles represent covalent histone modifications.

the UPS can be activated specifically within a transcriptional context (**Figure 8b**) (100, 101). One of the best examples of this is the Ub ligase TIF1γ, which is activated by modified histone tails (102), allowing it to select only its chromatin-bound pool of substrate (Smad4) for ubiquitylation. Finally, Ub ligase selectivity can be directly reprogrammed by transcription proteins to target new substrates (**Figure 8c**) (103, 104). In one surprising example, the transcription factor Met4 inserts itself within a Ub ligase complex (104), changing E3 selectivity and allowing Met4 to direct the proteolysis of its transcriptional coregulators. By these three mechanisms, the activity of the UPS can be co-opted and tuned with laser-like precision toward transcriptional processes. The close interplay between molecules important for transcription and Ub-dependent transactions, and the ability of transcription proteins to inherently alter the function of proteins such as Ub ligases, makes the UPS an effective tool for controlling transcriptional processes.

THE PROTEASOME IN TRANSCRIPTION

Implicit in much of our discussion is the concept that the proteasome is available to process ubiquitylated substrates within the immediate vicinity of chromatin, either directly or via chaperones, such as Cdc48. This indeed appears to be the case. Proteasome subunits are clearly recruited to sites of transcription (105–112), and genome-wide studies suggest that one or more components of the proteasome interact with the majority of highly active genes in yeast (109), raising the intriguing possibility that few Pol II–transcribed genes, and perhaps those transcribed by RNA polymerase I (113), are expressed without proteasomes being present. Studies have implicated the proteasome in just about every step in gene transcription (**Table 2**), from control of activators and how they associate with chromatin through to regulated exchange of coactivators (52, 114), elongation (115, 116), termination (91), covalent histone modifications (114, 117–120), and repressing cryptic transcription (111, 121). However, the nature of the proteasome that is involved in gene regulation and the exact contributions of 19S versus 20S activities are unclear.

One school of thought posits that the transcriptionally relevant form of the proteasome is an independent 19S base complex (122), which uses the energy of ATP hydrolysis to regulate dynamic exchanges of key proteins at sites of transcription. Support for this notion comes from genetic evidence linking 19S base

Table 2 Transcriptionally relevant activities of the proteasome

Component of the proteasome	Activity	Reference
19S	Regulates transcriptional activator loading onto chromatin	152
	Strips activators off chromatin	31, 153
	Promotes activator/coactivator association	106, 114
	Promotes transcriptional elongation	115, 116
	Regulates histone modifications	107, 114, 117–120
26S	Promotes dynamic association of activators with chromatin	154–157
	Activity-coupled destruction of transcription factors	37, 44, 47, 54, 55, 57
	Promotes coactivator exchange	52, 158
	Terminates transcription appropriately	91
	Represses cryptic transcription	111, 121
	Resolves permanently stalled RNA polymerase II complexes at sites of DNA damage	99, 129

components—but not those in 20S—to activities of the yeast Gal4 activator (123–126), from the finding that TADs and GTFs can selectively interact with 19S proteins in vitro (117, 127, 128), and from some notable discrepancies in the distribution of 19S and 20S components on chromatin, as measured by chromatin immunoprecipitation (105, 106, 108, 109, 111). Importantly, biochemical studies have clearly indicated that 19S complexes are sufficient for promoting events such as transcriptional elongation (116), and stimulating coactivator binding (114), lending further support to the notion that ATPases in the 19S base are all that is needed for transcriptionally relevant processes.

By contrast, an equally compelling case can be made for the role of 20S proteins in critical transcriptional events. 20S components bind active genes (91, 109, 111, 112); are required for activator and coactivator function (described above), as well as for transcriptional termination (91) and repression of cryptic transcription (111, 121); and are recruited to chromatin in response to DNA damage (129), where they facilitate removal of Pol II complexes that have terminally stalled at DNA lesions. It appears, therefore, that the proteolytic activity of the proteasome is required for normal transcriptional processes and also to cope with problems that arise during expression of the genetic information.

How can we reconcile these apparently disparate observations? It needs to be stated that evidence for the existence of free 19S base complexes is lacking and is confounded by the inherent instability of proteasome complexes during extraction (130). Most in vivo evidence supporting a 20S-free role of proteasome components in transcription is based on chromatin immunoprecipitation studies, which are susceptible to issues of epitope accessibility, and on studies using pharmacological inhibitors of the chymotryptic site of the proteasome, which can leave other proteoltyic activities intact (57). It should also be stressed that there is no conceptual need to physically segregate the proteasome for its nonproteolytic activities to act; protein unfolding occurs on the surface of the 19S complex (131), and 26S proteasomes can disrupt protein complexes independent of proteolysis (132). Depending on their characteristics, some proteins are inexorably shuttled into the proteolytic chamber, whereas others escape degradation (133). Thus, canonical 26S proteasomes certainly have the capability to act both proteolytically and nonproteolytically during the transcription process.

We propose that transcriptionally active genes recruit the entire proteasome and use it as a kind of "Swiss army knife" that carries an integrated set of biological functions (**Figure 9**). During the early stages of transcription, ATPases in 19S complex regulate activator occupancy (134) and coactivator recruitment (52, 114), and the entire proteasome is poised to destroy the activator once its suicide mechanism is activated. After Pol II has initiated transcription, proteasomes enter the transcribed portion of the gene and again put their proteolytic and nonproteolytic activities to work, remodeling the chromatin template to facilitate transcriptional elongation (115, 116) and linked histone modifications (107, 114, 117–120), as well as serving to remove failed transcriptional complexes or those that have initiated at inappropriate sites in the genome (109, 111, 135). And as transcription nears completion, the proteasome uses its activities to drive changes in the composition of Pol II complexes necessary for appropriate termination of transcription (91). Thus, just as one might carry a Swiss army knife—both for routine tasks and to cope with emergencies—transcription complexes carry proteasomes and sample their various activities depending on the stage or challenge to transcription.

Notably, our model does not exclude the possibility that individual functions of the proteasome may be transiently inhibited or that proteasomes may be fragmented to achieve gene-specific transcriptional outcomes. For example, the human immunodeficiency virus 1 (HIV-1) promoter recruits both 19S and 20S components when in the process of synthesizing short, nonproductive transcripts but, in response to the activator Tat, recruits a

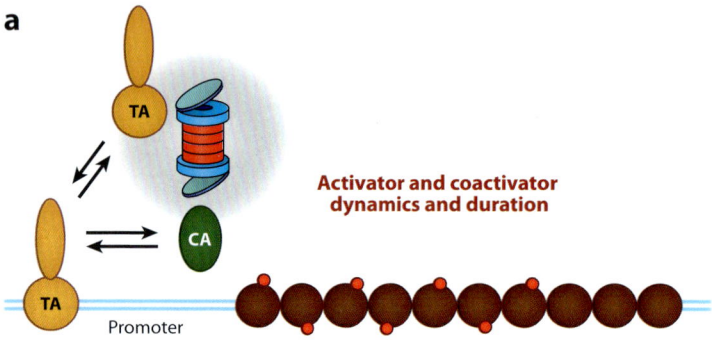

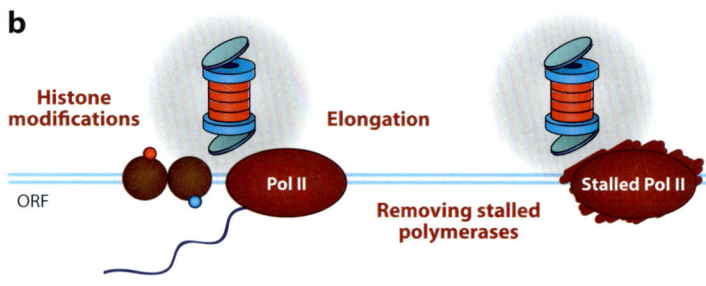

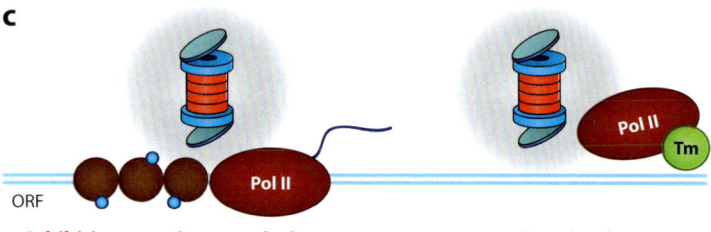

Figure 9

The Swiss army knife model of proteasome function in transcription. This model posits that the entire 26S proteasome is recruited into transcriptional processes, and its various activities are used depending on the molecular requirements. (*a*) The proteasome uses a combination of proteolytic and nonproteolytic functions to control activator binding, residency time, and coactivator (CA) exchange on promoter DNAs. (*b*) We posit that proteasomes enter the open reading frame (ORF) of transcribed genes, where their ATPase activities stimulate elongation, while their proteolytic functions serve to remove terminally stalled RNA polymerase II (Pol II) complexes that may arise. (*c*) Proteolytic activity of the proteasome is also important for attenuating cryptic, nonauthentic, and transcription complexes as well as for processes required for appropriate termination of the transcription event. Abbreviation: Tm, termination machinery. Blue lines represent the DNA duplex. Brown circles represent nucleosomes. Red circles represent covalent histone modifications present before transcription. Cyan circles represent covalent histone modifications present after transcription. The spool is the 26S proteasome (as in **Figure 3**).

proteasome disassembly protein, PAAF1, that displaces 20S proteins, allowing the 19S functions to predominate in synthesis of long, productive mRNAs (136). These results lend support to the notion that proteasomes are recruited *en masse* into transcriptional processes, but also demonstrate that proteasome subcomplexes can be tailored to meet specific transcription requirements on select promoter DNAs.

How proteasomes are recruited into transcriptional complexes is largely unresolved. This could occur by direct interaction with TADs (105, 117, 137–140), by histone modifications specific to active chromatin (107), or by specific adapter proteins (112, 141). Alternatively, if the canonical 26S proteasome is the form that is involved in transcription, there is no reason to believe that proteasomes could not simply come in response to the presence of specific ubiquitylated substrates—a notion supported by the finding that the Ub ligase activity of the CCR4-Not complex is required for specific recruitment of proteasome proteins to the active *PMA1* gene in yeast (142).

The Swiss army knife model makes a number of frank predictions about how proteasomes associate with chromatin, and these predictions need to be challenged experimentally, the most important of which is the tracking of proteasome subunit association across genomes. This research will likely require combined efforts that use multiple approaches to systematically monitor proteasome interaction with chromatin as well as functional approaches that compare the effects of disrupting both 19S and 20S functions on patterns of authentic and nonauthentic transcription. Only once unambiguous assignment of how these subunits interact with active genes is made—or once a transcriptionally relevant activity of the 19S base complex that cannot function within a 26S setting has been found—can these issues be resolved with certainty.

CONCLUSIONS

We hope to make it clear in this review that the UPS is intimately involved, physically and functionally, in gene regulatory mechanisms. Both proteolytic and nonproteolytic activities of the system impact transcriptional processes by targeting multiple steps in gene activity, from controlling activators through to export of a finished transcript from the nucleus, and possibly beyond. We emphasize that, because of the nature of the UPS and the linkage between many of its activities, it is likely that the examples we have cited here represent points on a spectrum of possibilities rather than absolute ways in which the UPS is always involved in gene regulation. It is easy to imagine, for example, how lines between proteolytic versus nonproteolytic regulation of a specific transcription protein could be blurred by events that change the type of Ub linkage or that transiently inactivate one or more functions in downstream UPS components, such as the proteasome. Indeed, it may very well be that transcription is a perfect venue in which to expose the various intricacies of the UPS in protein regulation. On a related note, it will be very interesting to learn whether processes other than transcription can be regulated in so many interesting ways by Ub, Ub-dependent processes, and the proteasome.

SUMMARY POINTS

1. The ubiquitin-proteasome system (UPS) interacts with chromatin and regulates multiple steps in gene transcription, from controlling activators through to mRNA export.

2. Proteolytic and nonproteolytic actions of the UPS are important for transcriptional regulation. This includes proteolytic and nonproteolytic roles for Ub as well as the proteasome.

3. Transcriptional regulators are a major point of intervention of the UPS in transcriptional processes. Nonproteolytic ubiquitylation can modulate activator binding and activity, whereas proteolytic ubiquitylation can both inhibit and promote activator function.

4. The UPS can paradoxically be required for the function of transcription proteins it destroys. This likely reflects the action of the UPS in promoting activator turnover on chromatin and provides the cell with a way to maintain tight control over activators by inexorably linking their destruction to their activity.

5. Histone H2B ubiquitylation is a dynamic process that controls higher-order chromatin compaction; that influences transcription by recruiting effectors, promoting nucleosome disassembly and reassembly; and that signals additional histone modifications.

6. The extraordinary ability of the UPS to impact transcription stems from close physical connections between the two processes and from the ability of the UPS to detect the activity of proteins that are functioning in a transcription context.

7. The proteasome associates with chromatin and uses chaperone functions in the ATPase base, as well as its proteolytic capabilities, to modulate activator binding, coactivator recruitment, transcriptional elongation, and transcriptional termination. The proteasome also plays an important role in clearing RNA polymerase complexes that have stalled or initiated at incorrect sites in the genome.

8. The proteasome is an integrated multifunctional machine that has the potential to bring at least six distinct biochemical activities to bear in transcriptional processes.

FUTURE ISSUES

1. The mechanism through which monoubiquitylation regulates the binding and activity of TAs and coactivators needs to be exposed.
2. Suicide models for activator function need to be challenged biochemically to determine the steps in transcriptional activation that require the UPS, the role of phosphorylation in this process, and whether regulatory proteins are the sole molecules being destroyed during transcriptional induction.
3. The role of ubiquitylation on histones H1, H3, and H4 needs to be understood both it terms of its impact on chromatin organization and its influence on transcriptional processes.
4. Receptors for ubiquitylated proteins in the transcriptional apparatus need to be identified. Given the large number of transcription proteins regulated by Ub, it seems likely that Ub-specific receptors exist, but few examples are known.
5. The influence of Ub chain length and topology on the function of transcription proteins needs to be understood. Understanding how differences in Ub chains are established and recognized could help resolve instances where it is uncertain whether proteolytic or nonproteolytic ubiquitylation is at work.
6. Clear resolution of the relative contribution of 19S versus 20S activities of the proteasome in each step of transcription is needed. The unambiguous assignment of the association of various components of the proteasome on chromatin, and their mode of recruitment to transcription sites, also needs to be resolved.
7. Biochemical analysis of transcriptionally relevant functions of the proteasome is required to reveal the steps in transcription that are directly impacted and to understand the molecular actions of chaperone proteins in the 19S base.
8. Other processes heavily controlled by the UPS need to be carefully examined to see if similar mechanisms, such as activation by destruction, recognition on the basis of activity, and functionally independent 19S chaperone action, are at work.

DISCLOSURE STATEMENT

The authors are not aware of any affiliations, memberships, funding, or financial holdings that might be perceived as affecting the objectivity of this review.

ACKNOWLEDGMENTS

We thank members of the Tansey laboratory, past and present, for useful discussions. Research on connections between the transcription and ubiquitin-proteasome systems in the laboratory is supported by a grant from the National Institutes of Health (GM067728).

LITERATURE CITED

1. Ouyang J, Valin A, Gill G. 2009. Regulation of transcription factor activity by SUMO modification. *Methods Mol. Biol.* 497:141–52
2. Muratani M, Tansey WP. 2003. How the ubiquitin-proteasome system controls transcription. *Nat. Rev. Mol. Cell Biol.* 4:192–201

3. Lipford JR, Deshaies RJ. 2003. Diverse roles for ubiquitin-dependent proteolysis in transcriptional activation. *Nat. Cell Biol.* 5:845–50
4. Dennis AP, O'Malley BW. 2005. Rush hour at the promoter: how the ubiquitin-proteasome pathway polices the traffic flow of nuclear receptor-dependent transcription. *J. Steroid Biochem. Mol. Biol.* 93:139–51
5. Baumann M, Pontiller J, Ernst W. 2010. Structure and basal transcription complex of RNA polymerase II core promoters in the mammalian genome: an overview. *Mol. Biotechnol.* 45:241–47
6. Li B, Carey M, Workman JL. 2007. The role of chromatin during transcription. *Cell* 128:707–19
7. Hartzog GA. 2003. Transcription elongation by RNA polymerase II. *Curr. Opin. Genet. Dev.* 13:119–26
8. Park YJ, Luger K. 2008. Histone chaperones in nucleosome eviction and histone exchange. *Curr. Opin. Struct. Biol.* 18:282–89
9. Carrozza MJ, Li B, Florens L, Suganuma T, Swanson SK, et al. 2005. Histone H3 methylation by Set2 directs deacetylation of coding regions by Rpd3S to suppress spurious intragenic transcription. *Cell* 123:581–92
10. Kaplan CD, Laprade L, Winston F. 2003. Transcription elongation factors repress transcription initiation from cryptic sites. *Science* 301:1096–99
11. Suganuma T, Workman JL. 2011. Signals and combinatorial functions of histone modifications. *Annu. Rev. Biochem.* 80:473–99
12. Egloff S, Murphy S. 2008. Cracking the RNA polymerase II CTD code. *Trends Genet.* 24:280–88
13. Behrends C, Harper JW. 2011. Constructing and decoding unconventional ubiquitin chains. *Nat. Struct. Mol. Biol.* 18:520–28
14. Thrower JS, Hoffman L, Rechsteiner M, Pickart CM. 2000. Recognition of the polyubiquitin proteolytic signal. *EMBO J.* 19:94–102
15. Ikeda F, Dikic I. 2008. Atypical ubiquitin chains: new molecular signals. 'Protein Modifications: Beyond the Usual Suspects' review series. *EMBO Rep.* 9:536–42
16. Komander D, Clague MJ, Urbe S. 2009. Breaking the chains: structure and function of the deubiquitinases. *Nat. Rev. Mol. Cell Biol.* 10:550–63
17. Voges D, Zwickl P, Baumeister W. 1999. The 26S proteasome: a molecular machine designed for controlled proteolysis. *Annu. Rev. Biochem.* 68:1015–68
18. Madura K. 2004. Rad23 and Rpn10: Perennial wallflowers join the melee. *Trends Biochem. Sci.* 29:637–40
19. Stolz A, Hilt W, Buchberger A, Wolf DH. 2011. Cdc48: a power machine in protein degradation. *Trends Biochem. Sci.* 36:515–23
20. Sauer RT, Baker TA. 2011. AAA+ proteases: ATP-fueled machines of protein destruction. *Annu. Rev. Biochem.* 80:587–612
21. Liu CW, Corboy MJ, DeMartino GN, Thomas PJ. 2003. Endoproteolytic activity of the proteasome. *Science* 299:408–11
22. Hoppe T, Matuschewski K, Rape M, Schlenker S, Ulrich HD, Jentsch S. 2000. Activation of a membrane-bound transcription factor by regulated ubiquitin/proteasome-dependent processing. *Cell* 102:577–86
23. Palombella VJ, Rando OJ, Goldberg AL, Maniatis T. 1994. The ubiquitin-proteasome pathway is required for processing the NF-kappa B1 precursor protein and the activation of NF-kappa B. *Cell* 78:773–85
24. Burgdorf S, Leister P, Scheidtmann KH. 2004. TSG101 interacts with apoptosis-antagonizing transcription factor and enhances androgen receptor-mediated transcription by promoting its monoubiquitination. *J. Biol. Chem.* 279:17524–34
25. Brès V, Kiernan RE, Linares LK, Chable-Bessia C, Plechakova O, et al. 2003. A non-proteolytic role for ubiquitin in Tat-mediated transactivation of the HIV-1 promoter. *Nat. Cell Biol.* 5:754–61
26. van der Horst A, de Vries-Smits AM, Brenkman AB, van Triest MH, van den Broek N, et al. 2006. FOXO4 transcriptional activity is regulated by monoubiquitination and USP7/HAUSP. *Nat. Cell Biol.* 8:1064–73
27. Moren A, Hellman U, Inada Y, Imamura T, Heldin CH, Moustakas A. 2003. Differential ubiquitination defines the functional status of the tumor suppressor Smad4. *J. Biol. Chem.* 278:33571–82
28. Salghetti SE, Caudy AA, Chenoweth JG, Tansey WP. 2001. Regulation of transcriptional activation domain function by ubiquitin. *Science* 293:1651–53

29. Kurosu T, Peterlin BM. 2004. VP16 and ubiquitin; binding of P-TEFb via its activation domain and ubiquitin facilitates elongation of transcription of target genes. *Curr. Biol.* 14:1112–16
30. Wu RC, Feng Q, Lonard DM, O'Malley BW. 2007. SRC-3 coactivator functional lifetime is regulated by a phospho-dependent ubiquitin time clock. *Cell* 129:1125–40
31. Archer CT, Delahodde A, Gonzalez F, Johnston SA, Kodadek T. 2008. Activation domain-dependent monoubiquitylation of Gal4 protein is essential for promoter binding in vivo. *J. Biol. Chem.* 283:12614–23
32. Wilcox AJ, Laney JD. 2009. A ubiquitin-selective AAA-ATPase mediates transcriptional switching by remodelling a repressor-promoter DNA complex. *Nat. Cell Biol.* 11:1481–86
33. Wang B, Suzuki H, Kato M. 2008. Roles of monoubiquitinated Smad4 in the formation of Smad transcriptional complexes. *Biochem. Biophys. Res. Commun.* 376:288–92
34. Laney JD, Hochstrasser M. 2003. Ubiquitin-dependent degradation of the yeast Mat(alpha)2 repressor enables a switch in developmental state. *Genes Dev.* 17:2259–70
35. Adhikary S, Marinoni F, Hock A, Hulleman E, Popov N, et al. 2005. The ubiquitin ligase HectH9 regulates transcriptional activation by Myc and is essential for tumor cell proliferation. *Cell* 123:409–21
36. Ramakrishna S, Suresh B, Lee EJ, Lee HJ, Ahn WS, Baek KH. 2011. Lys-63-specific deubiquitination of SDS3 by USP17 regulates HDAC activity. *J. Biol. Chem.* 286:10505–14
37. Lipford JR, Smith GT, Chi Y, Deshaies RJ. 2005. A putative stimulatory role for activator turnover in gene expression. *Nature* 438:113–16
38. Salghetti SE, Muratani M, Wijnen H, Futcher B, Tansey WP. 2000. Functional overlap of sequences that activate transcription and signal ubiquitin-mediated proteolysis. *Proc. Natl. Acad. Sci. USA* 97:3118–23
39. Wang X, Muratani M, Tansey WP, Ptashne M. 2010. Proteolytic instability and the action of nonclassical transcriptional activators. *Curr. Biol.* 20:868–71
40. Bhat KP, Greer SF. 2011. Proteolytic and non-proteolytic roles of ubiquitin and the ubiquitin proteasome system in transcriptional regulation. *Biochim. Biophys. Acta* 1809:150–55
41. Amazit L, Roseau A, Khan JA, Chauchereau A, Tyagi RK, et al. 2011. Ligand-dependent degradation of SRC-1 is pivotal for progesterone receptor transcriptional activity. *Mol. Endocrinol.* 25:394–408
42. Leung A, Geng F, Daulny A, Collins G, Guzzardo P, Tansey WP. 2008. Transcriptional control and the ubiquitin-proteasome system. *Ernst Schering Found. Symp. Proc.* 2008:75–97
43. Sundqvist A, Ericsson J. 2003. Transcription-dependent degradation controls the stability of the SREBP family of transcription factors. *Proc. Natl. Acad. Sci. USA* 100:13833–38
44. Muratani M, Kung C, Shokat KM, Tansey WP. 2005. The F box protein Dsg1/Mdm30 is a transcriptional coactivator that stimulates Gal4 turnover and cotranscriptional mRNA processing. *Cell* 120:887–99
45. Chi Y, Huddleston MJ, Zhang X, Young RA, Annan RS, et al. 2001. Negative regulation of Gcn4 and Msn2 transcription factors by Srb10 cyclin-dependent kinase. *Genes Dev.* 15:1078–92
46. Chymkowitch P, Le May N, Charneau P, Compe E, Egly JM. 2011. The phosphorylation of the androgen receptor by TFIIH directs the ubiquitin/proteasome process. *EMBO J.* 30:468–79
47. Andress EJ, Holic R, Edelmann MJ, Kessler BM, Yu VP. 2011. Dia2 controls transcription by mediating assembly of the RSC complex. *PLoS ONE* 6:e21172
48. Gammoh N, Gardiol D, Massimi P, Banks L. 2009. The Mdm2 ubiquitin ligase enhances transcriptional activity of human papillomavirus E2. *J. Virol.* 83:1538–43
49. Punga T, Bengoechea-Alonso MT, Ericsson J. 2006. Phosphorylation and ubiquitination of the transcription factor sterol regulatory element-binding protein-1 in response to DNA binding. *J. Biol. Chem.* 281:25278–86
50. von der Lehr N, Johansson S, Wu S, Bahram F, Castell A, et al. 2003. The F-box protein Skp2 participates in c-Myc proteosomal degradation and acts as a cofactor for c-Myc-regulated transcription. *Mol. Cell* 11:1189–200
51. Greer SF, Zika E, Conti B, Zhu XS, Ting JP. 2003. Enhancement of CIITA transcriptional function by ubiquitin. *Nat. Immunol.* 4:1074–82
52. Perissi V, Aggarwal A, Glass CK, Rose DW, Rosenfeld MG. 2004. A corepressor/coactivator exchange complex required for transcriptional activation by nuclear receptors and other regulated transcription factors. *Cell* 116:511–26
53. Higazi A, Abed M, Chen J, Li Q. 2011. Promoter context determines the role of proteasome in ligand-dependent occupancy of retinoic acid responsive elements. *Epigenetics* 6:202–11

54. Kaluz S, Kaluzova M, Stanbridge EJ. 2006. Proteasomal inhibition attenuates transcriptional activity of hypoxia-inducible factor 1 (HIF-1) via specific effect on the HIF-1alpha C-terminal activation domain. *Mol. Cell. Biol.* 26:5895–907
55. Spoel SH, Mou Z, Tada Y, Spivey NW, Genschik P, Dong X. 2009. Proteasome-mediated turnover of the transcription coactivator NPR1 plays dual roles in regulating plant immunity. *Cell* 137:860–72
56. Chae E, Tan QK, Hill TA, Irish VF. 2008. An *Arabidopsis* F-box protein acts as a transcriptional co-factor to regulate floral development. *Development* 135:1235–45
57. Collins GA, Gomez TA, Deshaies RJ, Tansey WP. 2010. Combined chemical and genetic approach to inhibit proteolysis by the proteasome. *Yeast* 27:965–74
58. Yin H, Jiang Y, Li H, Li J, Gui Y, Zheng XL. 2011. Proteasomal degradation of myocardin is required for its transcriptional activity in vascular smooth muscle cells. *J. Cell. Physiol.* 226:1897–906
59. Thomas D, Tyers M. 2000. Transcriptional regulation: kamikaze activators. *Curr. Biol.* 10:R341–43
60. Weake VM, Workman JL. 2008. Histone ubiquitination: triggering gene activity. *Mol. Cell* 29:653–63
61. Osley MA. 2006. Regulation of histone H2A and H2B ubiquitylation. *Brief. Funct. Genomics Proteomics* 5:179–89
62. Geng F, Tansey WP. 2008. Polyubiquitylation of histone H2B. *Mol. Biol. Cell* 19:3616–24
63. Robzyk K, Recht J, Osley MA. 2000. Rad6-dependent ubiquitination of histone H2B in yeast. *Science* 287:501–4
64. Hwang WW, Venkatasubrahmanyam S, Ianculescu AG, Tong A, Boone C, Madhani HD. 2003. A conserved RING finger protein required for histone H2B monoubiquitination and cell size control. *Mol. Cell* 11:261–66
65. Minsky N, Shema E, Field Y, Schuster M, Segal E, Oren M. 2008. Monoubiquitinated H2B is associated with the transcribed region of highly expressed genes in human cells. *Nat. Cell Biol.* 10:483–88
66. Kim J, Roeder RG. 2009. Direct Bre1-Paf1 complex interactions and RING finger-independent Bre1-Rad6 interactions mediate histone H2B ubiquitylation in yeast. *J. Biol. Chem.* 284:20582–92
67. Kim J, Guermah M, McGinty RK, Lee JS, Tang Z, et al. 2009. RAD6-mediated transcription-coupled H2B ubiquitylation directly stimulates H3K4 methylation in human cells. *Cell* 137:459–71
68. Zhang F, Yu X. 2011. WAC, a functional partner of RNF20/40, regulates histone H2B ubiquitination and gene transcription. *Mol. Cell* 41:384–97
69. Zheng S, Wyrick JJ, Reese JC. 2010. Novel *trans*-tail regulation of H2B ubiquitylation and H3K4 methylation by the N terminus of histone H2A. *Mol. Cell. Biol.* 30:3635–45
70. Davies N, Lindsey GG. 1994. Histone H2B (and H2A) ubiquitination allows normal histone octamer and core particle reconstitution. *Biochim. Biophys. Acta* 1218:187–93
71. Fierz B, Chatterjee C, McGinty RK, Bar-Dagan M, Raleigh DP, Muir TW. 2011. Histone H2B ubiquitylation disrupts local and higher-order chromatin compaction. *Nat. Chem. Biol.* 7:113–19
72. Chandrasekharan MB, Huang F, Sun ZW. 2009. Ubiquitination of histone H2B regulates chromatin dynamics by enhancing nucleosome stability. *Proc. Natl. Acad. Sci. USA* 106:16686–91
73. Formosa T. 2012. The role of FACT in making and breaking nucleosomes. *Biochim. Biophys. Acta.* 1819:247–55
74. Pavri R, Zhu B, Li G, Trojer P, Mandal S, et al. 2006. Histone H2B monoubiquitination functions cooperatively with FACT to regulate elongation by RNA polymerase II. *Cell* 125:703–17
75. Fleming AB, Kao CF, Hillyer C, Pikaart M, Osley MA. 2008. H2B ubiquitylation plays a role in nucleosome dynamics during transcription elongation. *Mol. Cell* 31:57–66
76. Chatterjee C, McGinty RK, Fierz B, Muir TW. 2010. Disulfide-directed histone ubiquitylation reveals plasticity in hDot1L activation. *Nat. Chem. Biol.* 6:267–69
77. Wu L, Zee BM, Wang Y, Garcia BA, Dou Y. 2011. The RING finger protein MSL2 in the MOF complex is an E3 ubiquitin ligase for H2B K34 and is involved in crosstalk with H3 K4 and K79 methylation. *Mol. Cell* 43:132–44
78. Oh S, Jeong K, Kim H, Kwon CS, Lee D. 2010. A lysine-rich region in Dot1p is crucial for direct interaction with H2B ubiquitylation and high level methylation of H3K79. *Biochem. Biophys. Res. Commun.* 399:512–17
79. Sun ZW, Allis CD. 2002. Ubiquitination of histone H2B regulates H3 methylation and gene silencing in yeast. *Nature* 418:104–8

80. Lee JS, Shukla A, Schneider J, Swanson SK, Washburn MP, et al. 2007. Histone crosstalk between H2B monoubiquitination and H3 methylation mediated by COMPASS. *Cell* 131:1084–96
81. Vitaliano-Prunier A, Menant A, Hobeika M, Geli V, Gwizdek C, Dargemont C. 2008. Ubiquitylation of the COMPASS component Swd2 links H2B ubiquitylation to H3K4 trimethylation. *Nat. Cell Biol.* 10:1365–71
82. McGinty RK, Kohn M, Chatterjee C, Chiang KP, Pratt MR, Muir TW. 2009. Structure-activity analysis of semisynthetic nucleosomes: mechanistic insights into the stimulation of Dot1L by ubiquitylated histone H2B. *ACS Chem. Biol.* 4:958–68
83. McGinty RK, Kim J, Chatterjee C, Roeder RG, Muir TW. 2008. Chemically ubiquitylated histone H2B stimulates hDot1L-mediated intranucleosomal methylation. *Nature* 453:812–16
84. Pinskaya M, Gourvennec S, Morillon A. 2009. H3 lysine 4 di- and tri-methylation deposited by cryptic transcription attenuates promoter activation. *EMBO J.* 28:1697–707
85. Chandrasekharan MB, Huang F, Sun ZW. 2010. Histone H2B ubiquitination and beyond: Regulation of nucleosome stability, chromatin dynamics and the *trans*-histone H3 methylation. *Epigenetics* 5:460–68
86. Leung A, Cajigas I, Jia P, Ezhkova E, Brickner JH, et al. 2011. Histone H2B ubiquitylation and H3 lysine 4 methylation prevent ectopic silencing of euchromatic loci important for the cellular response to heat. *Mol. Biol. Cell* 22:2741–53
87. Terzi N, Churchman LS, Vasiljeva L, Weissman J, Buratowski S. 2011. H3K4 trimethylation by Set1 promotes efficient termination by the Nrd1-Nab3-Sen1 pathway. *Mol. Cell. Biol.* 31:3569–83
88. Ingvarsdottir K, Krogan NJ, Emre NC, Wyce A, Thompson NJ, et al. 2005. H2B ubiquitin protease Ubp8 and Sgf11 constitute a discrete functional module within the *Saccharomyces cerevisiae* SAGA complex. *Mol. Cell. Biol.* 25:1162–72
89. Wyce A, Xiao T, Whelan KA, Kosman C, Walter W, et al. 2007. H2B ubiquitylation acts as a barrier to Ctk1 nucleosomal recruitment prior to removal by Ubp8 within a SAGA-related complex. *Mol. Cell* 27:275–88
90. Shema E, Kim J, Roeder RG, Oren M. 2011. RNF20 inhibits TFIIS-facilitated transcriptional elongation to suppress pro-oncogenic gene expression. *Mol. Cell* 42:477–88
91. Gillette TG, Gonzalez F, Delahodde A, Johnston SA, Kodadek T. 2004. Physical and functional association of RNA polymerase II and the proteasome. *Proc. Natl. Acad. Sci. USA* 101:5904–9
92. Song EJ, Werner SL, Neubauer J, Stegmeier F, Aspden J, et al. 2010. The Prp19 complex and the Usp4Sart3 deubiquitinating enzyme control reversible ubiquitination at the spliceosome. *Genes Dev.* 24:1434–47
93. Gwizdek C, Iglesias N, Rodriguez MS, Ossareh-Nazari B, Hobeika M, et al. 2006. Ubiquitin-associated domain of Mex67 synchronizes recruitment of the mRNA export machinery with transcription. *Proc. Natl. Acad. Sci. USA* 103:16376–81
94. Shcherbik N, Pestov DG. 2010. Ubiquitin and ubiquitin-like proteins in the nucleolus: multitasking tools for a ribosome factory. *Genes Cancer* 1:681–89
95. Pham AD, Sauer F. 2000. Ubiquitin-activating/conjugating activity of TAFII250, a mediator of activation of gene expression in *Drosophila*. *Science* 289:2357–60
96. Takagi Y, Masuda CA, Chang WH, Komori H, Wang D, et al. 2005. Ubiquitin ligase activity of TFIIH and the transcriptional response to DNA damage. *Mol. Cell* 18:237–43
97. Rabut G, Le Dez G, Verma R, Makhnevych T, Knebel A, et al. 2011. The TFIIH subunit Tfb3 regulates cullin neddylation. *Mol. Cell* 43:488–95
98. Daulny A, Geng F, Muratani M, Geisinger JM, Salghetti SE, Tansey WP. 2008. Modulation of RNA polymerase II subunit composition by ubiquitylation. *Proc. Natl. Acad. Sci. USA* 105:19649–54
99. Somesh BP, Sigurdsson S, Saeki H, Erdjument-Bromage H, Tempst P, Svejstrup JQ. 2007. Communication between distant sites in RNA polymerase II through ubiquitylation factors and the polymerase CTD. *Cell* 129:57–68
100. Ors A, Grimaldi M, Kimata Y, Wilkinson CR, Jones N, Yamano H. 2009. The transcription factor Atf1 binds and activates the APC/C ubiquitin ligase in fission yeast. *J. Biol. Chem.* 284:23989–94
101. Turnell AS, Stewart GS, Grand RJ, Rookes SM, Martin A, et al. 2005. The APC/C and CBP/p300 cooperate to regulate transcription and cell-cycle progression. *Nature* 438:690–95

102. Agricola E, Randall RA, Gaarenstroom T, Dupont S, Hill CS. 2011. Recruitment of TIF1gamma to chromatin via its PHD finger-bromodomain activates its ubiquitin ligase and transcriptional repressor activities. *Mol. Cell* 43:85–96
103. Li XS, Trojer P, Matsumura T, Treisman JE, Tanese N. 2010. Mammalian SWI/SNF-A subunit BAF250/ARID1 is an E3 ubiquitin ligase that targets histone H2B. *Mol. Cell. Biol.* 30:1673–88
104. Ouni I, Flick K, Kaiser P. 2010. A transcriptional activator is part of an SCF ubiquitin ligase to control degradation of its cofactors. *Mol. Cell* 40:954–64
105. Gonzalez F, Delahodde A, Kodadek T, Johnston SA. 2002. Recruitment of a 19S proteasome subcomplex to an activated promoter. *Science* 296:548–50
106. Malik S, Shukla A, Sen P, Bhaumik SR. 2009. The 19S proteasome subcomplex establishes a specific protein interaction network at the promoter for stimulated transcriptional initiation in vivo. *J. Biol. Chem.* 284:35714–24
107. Ezhkova E, Tansey WP. 2004. Proteasomal ATPases link ubiquitylation of histone H2B to methylation of histone H3. *Mol. Cell* 13:435–42
108. Sikder D, Johnston SA, Kodadek T. 2006. Widespread, but non-identical, association of proteasomal 19 and 20 S proteins with yeast chromatin. *J. Biol. Chem.* 281:27346–55
109. Auld KL, Brown CR, Casolari JM, Komili S, Silver PA. 2006. Genomic association of the proteasome demonstrates overlapping gene regulatory activity with transcription factor substrates. *Mol. Cell* 21:861–71
110. Tran K, Mahr JA, Spector DH. 2010. Proteasome subunits relocalize during human cytomegalovirus infection, and proteasome activity is necessary for efficient viral gene transcription. *J. Virol.* 84:3079–93
111. Szutorisz H, Georgiou A, Tora L, Dillon N. 2006. The proteasome restricts permissive transcription at tissue-specific gene loci in embryonic stem cells. *Cell* 127:1375–88
112. Morris MC, Kaiser P, Rudyak S, Baskerville C, Watson MH, Reed SI. 2003. Cks1-dependent proteasome recruitment and activation of CDC20 transcription in budding yeast. *Nature* 424:1009–13
113. Fatyol K, Grummt I. 2008. Proteasomal ATPases are associated with rDNA: the ubiquitin proteasome system plays a direct role in RNA polymerase I transcription. *Biochim. Biophys. Acta* 1779:850–59
114. Lee D, Ezhkova E, Li B, Pattenden SG, Tansey WP, Workman JL. 2005. The proteasome regulatory particle alters the SAGA coactivator to enhance its interactions with transcriptional activators. *Cell* 123:423–36
115. Ferdous A, Kodadek T, Johnston SA. 2002. A nonproteolytic function of the 19S regulatory subunit of the 26S proteasome is required for efficient activated transcription by human RNA polymerase II. *Biochemistry* 41:12798–805
116. Ferdous A, Gonzalez F, Sun L, Kodadek T, Johnston SA. 2001. The 19S regulatory particle of the proteasome is required for efficient transcription elongation by RNA polymerase II. *Mol. Cell* 7:981–91
117. Truax AD, Koues OI, Mentel MK, Greer SF. 2010. The 19S ATPase S6a (S6′/TBP1) regulates the transcription initiation of class II transactivator. *J. Mol. Biol.* 395:254–69
118. Koues OI, Mehta NT, Truax AD, Dudley RK, Brooks JK, Greer SF. 2010. Roles for common MLL/COMPASS subunits and the 19S proteasome in regulating CIITA pIV and MHC class II gene expression and promoter methylation. *Epigenetics Chromatin* 3:5
119. Koues OI, Dudley RK, Truax AD, Gerhardt D, Bhat KP, et al. 2008. Regulation of acetylation at the major histocompatibility complex class II proximal promoter by the 19S proteasomal ATPase Sug1. *Mol. Cell. Biol.* 28:5837–50
120. Koues OI, Dudley RK, Mehta NT, Greer SF. 2009. The 19S proteasome positively regulates histone methylation at cytokine inducible genes. *Biochim. Biophys. Acta* 1789:691–701
121. Cheung V, Chua G, Batada NN, Landry CR, Michnick SW, et al. 2008. Chromatin- and transcription-related factors repress transcription from within coding regions throughout the *Saccharomyces cerevisiae* genome. *PLoS Biol.* 6:e277
122. Kodadek T. 2010. No splicing, no dicing: non-proteolytic roles of the ubiquitin-proteasome system in transcription. *J. Biol. Chem.* 285:2221–26
123. Russell SJ, Johnston SA. 2001. Evidence that proteolysis of Gal4 cannot explain the transcriptional effects of proteasome ATPase mutations. *J. Biol. Chem.* 276:9825–31

124. Russell SJ, Sathyanarayana UG, Johnston SA. 1996. Isolation and characterization of SUG2. A novel ATPase family component of the yeast 26 S proteasome. *J. Biol. Chem.* 271:32810–17
125. Xu Q, Singer RA, Johnston GC. 1995. Sug1 modulates yeast transcription activation by Cdc68. *Mol. Cell. Biol.* 15:6025–35
126. Swaffield JC, Melcher K, Johnston SA. 1995. A highly conserved ATPase protein as a mediator between acidic activation domains and the TATA-binding protein. *Nature* 374:88–91
127. Sun L, Johnston SA, Kodadek T. 2002. Physical association of the APIS complex and general transcription factors. *Biochem. Biophys. Res. Commun.* 296:991–99
128. Rasti M, Grand RJ, Yousef AF, Shuen M, Mymryk JS, et al. 2006. Roles for APIS and the 20S proteasome in adenovirus E1A-dependent transcription. *EMBO J.* 25:2710–22
129. Verma R, Oania R, Fang R, Smith GT, Deshaies RJ. 2011. Cdc48/p97 mediates UV-dependent turnover of RNA Pol II. *Mol. Cell* 41:82–92
130. Verma R, Chen S, Feldman R, Schieltz D, Yates J, et al. 2000. Proteasomal proteomics: identification of nucleotide-sensitive proteasome-interacting proteins by mass spectrometric analysis of affinity-purified proteasomes. *Mol. Biol. Cell* 11:3425–39
131. Navon A, Goldberg AL. 2001. Proteins are unfolded on the surface of the ATPase ring before transport into the proteasome. *Mol. Cell* 8:1339–49
132. Nishiyama A, Tachibana K, Igarashi Y, Yasuda H, Tanahashi N, et al. 2000. A nonproteolytic function of the proteasome is required for the dissociation of Cdc2 and cyclin B at the end of M phase. *Genes Dev.* 14:2344–57
133. Fishbain S, Prakash S, Herrig A, Elsasser S, Matouschek A. 2011. Rad23 escapes degradation because it lacks a proteasome initiation region. *Nat. Commun.* 2:192
134. Ferdous A, Sikder D, Gillette T, Nalley K, Kodadek T, Johnston SA. 2007. The role of the proteasomal ATPases and activator monoubiquitylation in regulating Gal4 binding to promoters. *Genes Dev.* 21:112–23
135. Daulny A, Tansey WP. 2009. Damage control: DNA repair, transcription, and the ubiquitin-proteasome system. *DNA Repair* 8:444–48
136. Lassot I, Latreille D, Rousset E, Sourisseau M, Linares LK, et al. 2007. The proteasome regulates HIV-1 transcription by both proteolytic and nonproteolytic mechanisms. *Mol. Cell* 25:369–83
137. Chang C, Gonzalez F, Rothermel B, Sun L, Johnston SA, Kodadek T. 2001. The Gal4 activation domain binds Sug2 protein, a proteasome component, in vivo and in vitro. *J. Biol. Chem.* 276:30956–63
138. Archer CT, Burdine L, Kodadek T. 2005. Identification of Gal4 activation domain-binding proteins in the 26S proteasome by periodate-triggered cross-linking. *Mol. BioSyst.* 1:366–72
139. Satoh T, Ishizuka T, Tomaru T, Yoshino S, Nakajima Y, et al. 2009. Tat-binding protein-1 (TBP-1), an ATPase of 19S regulatory particles of the 26S proteasome, enhances androgen receptor function in cooperation with TBP-1-interacting protein/Hop2. *Endocrinology* 150:3283–90
140. Schwarz T, Sohn C, Kaiser B, Jensen ED, Mansky KC. 2010. The 19S proteasomal lid subunit POH1 enhances the transcriptional activation by Mitf in osteoclasts. *J. Cell. Biochem.* 109:967–74
141. Chaves S, Baskerville C, Yu V, Reed SI. 2010. Cks1, Cdk1, and the 19S proteasome collaborate to regulate gene induction-dependent nucleosome eviction in yeast. *Mol. Cell. Biol.* 30:5284–94
142. Laribee RN, Shibata Y, Mersman DP, Collins SR, Kemmeren P, et al. 2007. CCR4/NOT complex associates with the proteasome and regulates histone methylation. *Proc. Natl. Acad. Sci. USA* 104:5836–41
143. Yao T, Song L, Jin J, Cai Y, Takahashi H, et al. 2008. Distinct modes of regulation of the Uch37 deubiquitinating enzyme in the proteasome and in the Ino80 chromatin-remodeling complex. *Mol. Cell* 31:909–17
144. Saleh A, Collart M, Martens JA, Genereaux J, Allard S, et al. 1998. TOM1p, a yeast hect-domain protein which mediates transcriptional regulation through the ADA/SAGA coactivator complexes. *J. Mol. Biol.* 282:933–46
145. Iglesias N, Tutucci E, Gwizdek C, Vinciguerra P, Von Dach E, et al. 2010. Ubiquitin-mediated mRNP dynamics and surveillance prior to budding yeast mRNA export. *Genes Dev.* 24:1927–38
146. Boutet SC, Biressi S, Iori K, Natu V, Rando TA. 2010. Taf1 regulates Pax3 protein by monoubiquitination in skeletal muscle progenitors. *Mol. Cell* 40:749–61

147. Xie Y, Varshavsky A. 2001. RPN4 is a ligand, substrate, and transcriptional regulator of the 26S proteasome: a negative feedback circuit. *Proc. Natl. Acad. Sci. USA* 98:3056–61
148. Brower CS, Sato S, Tomomori-Sato C, Kamura T, Pause A, et al. 2002. Mammalian mediator subunit mMED8 is an Elongin BC-interacting protein that can assemble with Cul2 and Rbx1 to reconstitute a ubiquitin ligase. *Proc. Natl. Acad. Sci. USA* 99:10353–58
149. Mao X, Gluck N, Li D, Maine GN, Li H, et al. 2009. GCN5 is a required cofactor for a ubiquitin ligase that targets NF-kappaB/RelA. *Genes Dev.* 23:849–61
150. Chamovitz DA. 2009. Revisiting the COP9 signalosome as a transcriptional regulator. *EMBO Rep.* 10:352–58
151. Albert TK, Hanzawa H, Legtenberg YI, de Ruwe MJ, van den Heuvel FA, et al. 2002. Identification of a ubiquitin-protein ligase subunit within the CCR4-NOT transcription repressor complex. *EMBO J.* 21:355–64
152. Ostendorff HP, Peirano RI, Peters MA, Schluter A, Bossenz M, et al. 2002. Ubiquitination-dependent cofactor exchange on LIM homeodomain transcription factors. *Nature* 416:99–103
153. Zhang H, Sun L, Liang J, Yu W, Zhang Y, et al. 2006. The catalytic subunit of the proteasome is engaged in the entire process of estrogen receptor-regulated transcription. *EMBO J.* 25:4223–33
154. Reid G, Hübner MR, Métivier R, Brand H, Denger S, et al. 2003. Cyclic, proteasome-mediated turnover of unliganded and liganded ERalpha on responsive promoters is an integral feature of estrogen signaling. *Mol. Cell* 11:695–707
155. Walsh HE, Shupnik MA. 2009. Proteasome regulation of dynamic transcription factor occupancy on the GnRH-stimulated luteinizing hormone beta-subunit promoter. *Mol. Endocrinol.* 23:237–50
156. Stavreva DA, Müller WG, Hager GL, Smith CL, McNally JG. 2004. Rapid glucocorticoid receptor exchange at a promoter is coupled to transcription and regulated by chaperones and proteasomes. *Mol. Cell. Biol.* 24:2682–97
157. Kim YC, Wu SY, Lim HS, Chiang CM, Kodadek T. 2009. Non-proteolytic regulation of p53-mediated transcription through destabilization of the activator·promoter complex by the proteasomal ATPases. *J. Biol. Chem.* 284:34522–30
158. Satoh T, Ishizuka T, Yoshino S, Tomaru T, Nakajima Y, et al. 2009. Roles of proteasomal 19S regulatory particles in promoter loading of thyroid hormone receptor. *Biochem. Biophys. Res. Commun.* 386:697–702

The Ubiquitin Code

David Komander[1] and Michael Rape[2]

[1]Division of Protein and Nucleic Acid Chemistry, Medical Research Council Laboratory of Molecular Biology, Cambridge, CB2 0QH, United Kingdom; email: dk@mrc-lmb.cam.ac.uk

[2]Department of Molecular and Cell Biology, University of California, Berkeley, California 94720-3202; email: mrape@berkeley.edu

Keywords

ubiquitin chain, linkage specificity

Abstract

The posttranslational modification with ubiquitin, a process referred to as ubiquitylation, controls almost every process in cells. Ubiquitin can be attached to substrate proteins as a single moiety or in the form of polymeric chains in which successive ubiquitin molecules are connected through specific isopeptide bonds. Reminiscent of a code, the various ubiquitin modifications adopt distinct conformations and lead to different outcomes in cells. Here, we discuss the structure, assembly, and function of this ubiquitin code.

Contents

1. INTRODUCTION 204
2. STRUCTURE OF THE
   UBIQUITIN CODE 205
   2.1. Ubiquitin 205
   2.2. Ubiquitin Chain Structure 206
3. WRITING THE UBIQUITIN
   CODE 207
   3.1. Monoubiquitylation 207
   3.2. Ubiquitin Chain Assembly by
        RING Domain and U-Box
        Ligase Enzymes 209
   3.3. Chain Formation by HECT
        E3s 211
   3.4. Chain Formation by
        RING-In-Between-RING E3
        Ubiquitin Ligases 212
4. READING THE CODE:
   CONCEPTS IN UBIQUITIN
   BINDING 212
   4.1. Exploiting the Distance Between
        Ubiquitin Molecules 212
   4.2. Exploiting Chain Flexibility 213
   4.3. Recognizing the Linkage
        Context 214
   4.4. Combining Binding Sites 214
   4.5. Detecting the Free C Terminus
        of Unanchored Chains 214
5. ERASING THE CODE 215
   5.1. Housekeeping and
        Substrate-Specific
        Deubiquitinating Enzymes 215
   5.2. Linkage-Specific
        Deubiquitinating Enzymes 215
   5.3. Ubiquitin Chain Editing 216
6. CELLULAR FUNCTIONS
   OF THE UBIQUITIN
   CODE 216
   6.1. Proteolytic Functions of the
        Ubiquitin Code 216
   6.2. Nonproteolytic Functions of the
        Ubiquitin Code 218
7. CONCLUSIONS 220

1. INTRODUCTION

When in 1532 Spanish conquistadores set foot on the Inca Empire, they found a highly organized society that did not utilize a system of writing. Instead, the Incas recorded tax payments or mythology with quipus, devices in which pieces of thread were connected through specific knots. Although the quipus have not been fully deciphered, it is thought that the knots between threads encode most of the quipus' content. Intriguingly, cells use a regulatory mechanism—ubiquitylation—that is reminiscent of quipus: During this reaction, proteins are modified with polymeric chains in which the linkage between ubiquitin molecules encodes information about the substrate's fate in the cell.

Ubiquitylation is brought about by ubiquitin-activating enzymes (E1s), ubiquitin-conjugating enzymes (E2s), and ubiquitin ligase enzymes (E3s) (1–3). These enzymes first catalyze the formation of an isopeptide bond between the C terminus of ubiquitin and usually a substrate lysine, leading to monoubiquitylation (**Figure 1a**). Monoubiquitylation can occur at a defined residue, such as Lys164 in proliferating cell nuclear antigen (PCNA) (4), or it might be confined to a domain, as in the transcription factor p53 (5). It is possible that multiple lysine residues become modified with one ubiquitin each during multimonoubiquitylation (**Figure 1b**), with the epidermal growth factor receptor (EGFR) as an example (6).

Modification of the N terminus or one of the seven lysine residues of a substrate-attached ubiquitin leads to formation of polymeric chains. These chains can be short and contain only two ubiquitin molecules or long and incorporate more than ten moieties. Ubiquitin chains are homogenous if the same residue is modified during elongation, as in Met1- (or linear), Lys11-, Lys48-, or Lys63-linked chains (**Figure 1c**). Chains have mixed topology if different linkages alternate at succeeding positions of the chain (**Figure 1d**), as seen in NF-κB signaling or protein trafficking (7–10).

If a single ubiquitin is modified with multiple molecules, branched chains of unknown function are generated (**Figure 1e**).

All possible linkages have been detected in cells (11, 12). For chains linked through Lys6, Lys27, Lys29, or Lys33, few substrates are known, and their significance is poorly understood. However, it has been well established that monoubiquitylation and four homogenous chain types trigger distinct outcomes in the cell, suggesting that ubiquitylation can act as a code to store and transmit information. In this review, we discuss the structure, assembly, and function of this ubiquitin code.

2. STRUCTURE OF THE UBIQUITIN CODE

2.1. Ubiquitin

Ubiquitin is a highly stable protein that adopts a compact β-grasp fold with a flexible six-residue C-terminal tail (**Figure 2a**) (13). Most of its core residues are rigid, but the β1/β2 loop containing Leu8 shows flexibility that is important for recognition by ubiquitin-binding proteins (**Figure 2b**) (14). With three conservative changes, ubiquitin is almost invariant from yeast to man. This suggests high evolutionary pressure to conserve the structure of ubiquitin and implies that many of its surfaces are recognized by ubiquitin-binding domains (UBDs).

Ubiquitin is often recognized through a hydrophobic surface that consists of Ile44, Leu8, Val70, and His68 (**Figure 2a–c**) (15). The Ile44 patch is bound by the proteasome and most UBDs, rendering it essential for cell division (15–17). Another hydrophobic surface is centered on Ile36 and involves Leu71 and Leu73 of the ubiquitin tail (**Figure 2c**). The Ile36 patch can mediate interactions between ubiquitin molecules in chains, and it is recognized by HECT E3s (18), DUBs (19), and UBDs (20). A surface comprising Gln2, Phe4, and Thr12 is required for cell division in yeast (**Figure 2c**) (17). This Phe4 patch might function in trafficking (17), and it interacts with the UBAN domain (21) and the ubiquitin-specific protease (USP) domain of DUBs (19). The divergence between Phe4 patches of ubiquitin and its closest homolog Nedd8 enables DUBs to distinguish between these modifiers (22). In higher eukaryotes, the TEK-box of ubiquitin, a three-dimensional motif that includes Thr12, Thr14, Glu34, Lys6, and Lys11, is required for mitotic degradation (**Figure 2c**) (23). As deamidation of Gln40 by the bacterial protein Cif blocks chain assembly (24), additional surfaces might fulfill as yet unidentified functions.

With respect to the ubiquitin code, the most important features of ubiquitin are its N terminus and its seven lysines, which are the attachment sites for chain assembly. These residues cover all surfaces of ubiquitin and point into distinct directions (**Figure 2d**). Lys6 and Lys11 are located in the most dynamic region

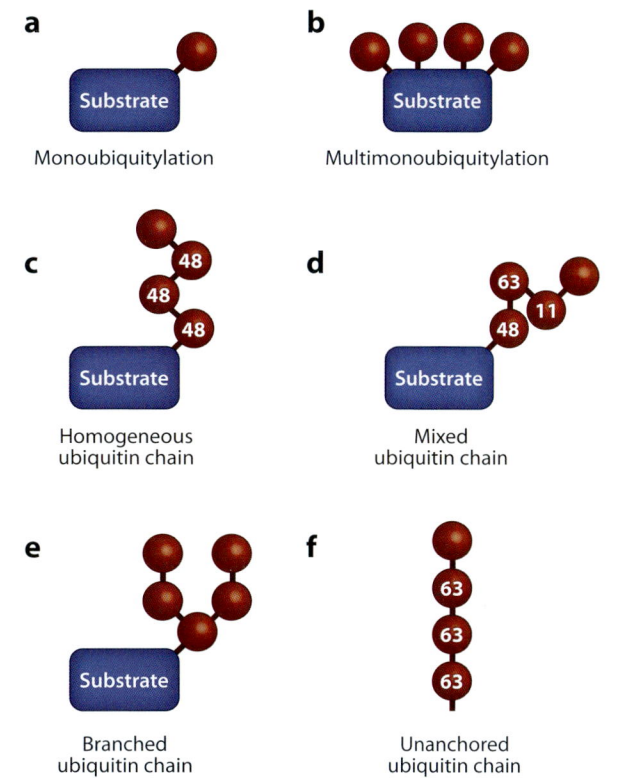

Figure 1

The different topologies of ubiquitylation. (*a*) Monoubiquitylation. (*b*) Multimonoubiquitylation. (*c*) Homogenous ubiquitin chain. (*d*) Mixed ubiquitin chain. (*e*) Branched ubiquitin chain. (*f*) Unanchored ubiquitin chain.

E2: ubiquitin-conjugating enzyme

E3: ubiquitin ligase enzyme

UBD: ubiquitin-binding domain

Homologous to E6AP C terminus (HECT): a class of E3s ubiquitin ligases

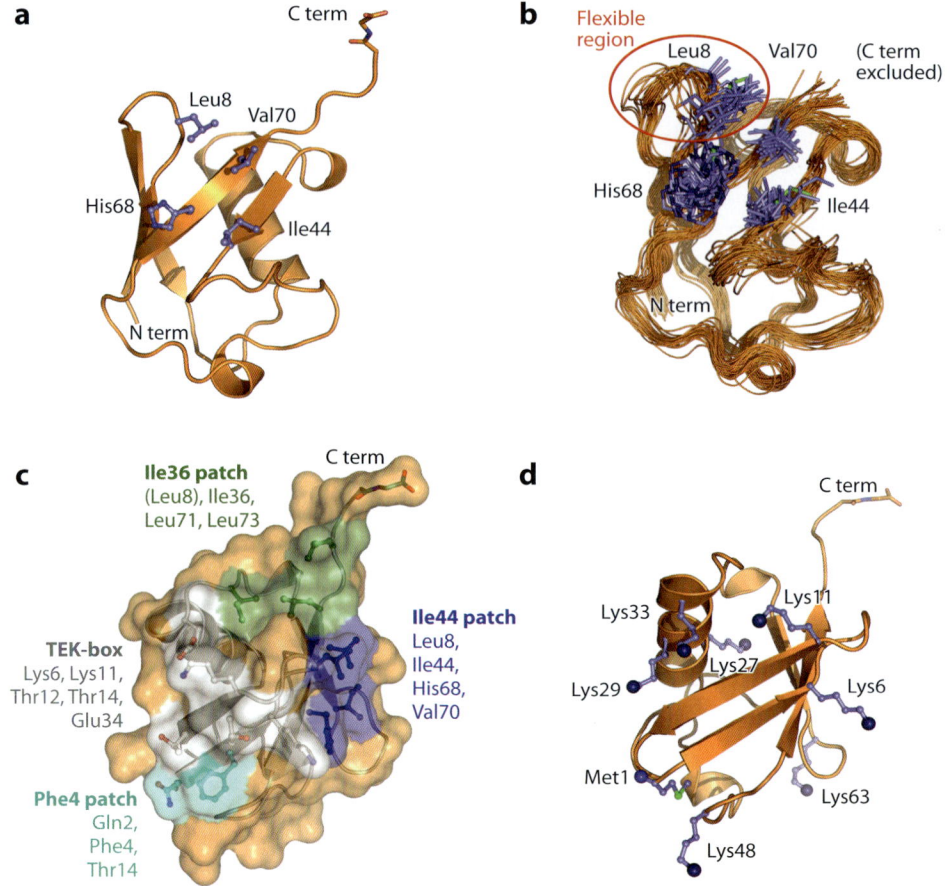

Figure 2

Structural features of ubiquitin. (*a*) Structure of ubiquitin, indicating the C-terminal (C term) tail and residues of the Ile44 patch [Protein Data Bank (pdb) code 1ubq] (13). (*b*) NMR ensemble of ubiquitin on the basis of residual dipolar couplings (14). The first 30 structures of the ensemble (pdb 2k39) are shown. The Ile44 residues are indicated, and the flexible region is highlighted. (*c*) The ubiquitin surface is shown with Ile44 (*blue*), Ile36 (*green*), Phe4 patches (*cyan*), and TEK-box (*white*) highlighted. (*d*) Structure of ubiquitin showing the seven Lys residues and Met1. Blue spheres indicate amino groups used in ubiquitin chain formation. Abbreviation: N term, N terminus.

Deubiquitinating enzyme or deubiquitinase (DUB): an enzyme that cleaves the isopeptide bond between a lysine and the C terminus of ubiquitin

of ubiquitin that may undergo conformational changes in the context of a chain or upon association with UBDs. As Lys27 is buried, linkage assembly through this residue would require localized changes in ubiquitin structure.

2.2. Ubiquitin Chain Structure

Structural characterization of five chain types revealed that different linkages result in distinct chain conformations. Ubiquitin chains adopt either "compact" conformations, where adjacent moieties interact with each other, or "open" conformations, where no interfaces are present except for the linkage site. The canonical Lys48-linked chains adopt compact conformations (**Figure 3***a*) (25–28). In the prevalent model for Lys48-linked diubiquitin, the ubiquitin moieties interact via their Ile44 patches (25–27), and two such diubiquitin modules pack tightly in tetraubiquitin (28). NMR analysis using residual dipolar couplings has identified a minor population of Lys48-linked diubiquitin in which the Ile36 patch of the distal ubiquitin interacts with the Ile44 patch of the proximal unit (27, 29). This structural flexibility might give binding partners of Lys48-linked chains access to the Ile44 patch, a hot spot for ubiquitin recognition.

Similar to Lys48 linkages, Lys6- and Lys11-linked chains adopt compact conformations, with Lys11-linked chains also displaying structural flexibility (**Figure 3*b,c***) (30–32). In one structure of Lys11-linked diubiquitin (30), an asymmetric interface covering the α-helix of ubiquitin is involved, and in another study, the ubiquitin moieties interact symmetrically via Ile36 patches (31). Both conformations are consistent with NMR analysis, suggesting that they coexist in equilibrium (30). Indeed, an analysis of crystal packing revealed a higher-order assembly of Lys11-linked chains that encompasses both conformations (33). In all Lys11-linked chain models, the Ile44 patch is solvent exposed and ready to interact with binding partners.

In contrast to the aforementioned linkages, Met1- and Lys63-linked chains mostly display open conformations (**Figure 3*d,e***), as shown by NMR analysis of Lys63-linked ubiquitin (26, 34) and crystal structures of both chain types (35–37). Reminiscent of beads on a string, the extended open conformation endows Lys63 and Met1 linkages with high conformational freedom. Most binding partners of these chains, therefore, likely exploit the distance and flexibility between chain moieties, rather than recognizing a defined geometric assembly of different ubiquitin surfaces (38).

Together, the various structures revealed a large array of geometries that can be utilized by binding partners to distinguish between modifications. As described below, linkage-specific binding proteins might recognize the distance between chain entities or sense the relative orientation of ubiquitin surfaces at successive chain positions. The conformational flexibility of some chain types raises the possibility that UBDs remodel chains to increase interaction interfaces or improve specificity. The structural diversity of the various modifications, therefore, forms the foundation of the ubiquitin code.

HOW TO DISSECT THE UBIQUITIN CODE

Ubiquitin modifications can be analyzed by a plethora of approaches. Reconstitution of cellular pathways in extracts supplemented with ubiquitin mutants revealed the roles of Lys11-linked chains in mitotic degradation (23) and Lys63-linked chains in kinase activation (60, 97). The biochemical and structural analysis of enzymes led researchers to discover Met1-linked chains as regulators of NF-κB signaling (21, 82). In cells, monoubiquitylation is often studied by analyzing linear fusions between ubiquitin and the candidate substrate (141). To dissect roles of chains in vivo, recombinant ubiquitin mutants can be injected into cells or *Xenopus* embryos (23). Mutant ubiquitin can also be overexpressed in cells, which, owing to the tight regulation of endogenous ubiquitin levels, leads only to a modest excess of mutant ubiquitin and often results in weak phenotypes. In a more careful approach, the genes encoding ubiquitin and ubiquitin-ribosome fusions are deleted or their mRNAs depleted by siRNA, and mutant ubiquitin is expressed as a ribosomal fusion (161, 179). The abundance of ubiquitin chain types can be analyzed with antibodies that specifically detect Met1, Lys11, Lys48, or Lys63 linkages (31, 82, 90) or by quantitative proteomics (11, 12, 55, 128). In all cases, assigning a function to a ubiquitin modification requires a combination of these experimental approaches.

3. WRITING THE UBIQUITIN CODE

Any functional code requires its specific assembly—just as review articles make sense only if letters are arranged in a sequence that gives meaning to the resulting words. In a similar manner, ubiquitylation will trigger specific outcomes only if the responsible enzymes catalyze formation of largely the same product each time they act on their substrate.

3.1. Monoubiquitylation

The enzymes catalyzing monoubiquitylation have to recognize substrate lysine residues, while sparing those of ubiquitin from modification, a specificity that can be

Distal ubiquitin: last ubiquitin moiety in a chain without a modified lysine

a Lys48 diUb
Dist
Prox
Dist
Prox
Prox
Dist
Dist
Prox
Lys48 tetraUb (Xtal, NMR)
b Lys11 diUb
Lys11 octaUb (Xtal)
c Lys6 diUb
Lys6 tetraUb ?
d Lys63 diUb
Lys63 tetraUb (Xtal, NMR)
e Met1 diUb
Met1 tetraUb ?
Ile44 hydrophobic patch
Ile36 hydrophobic patch
Free C terminus (substrate)

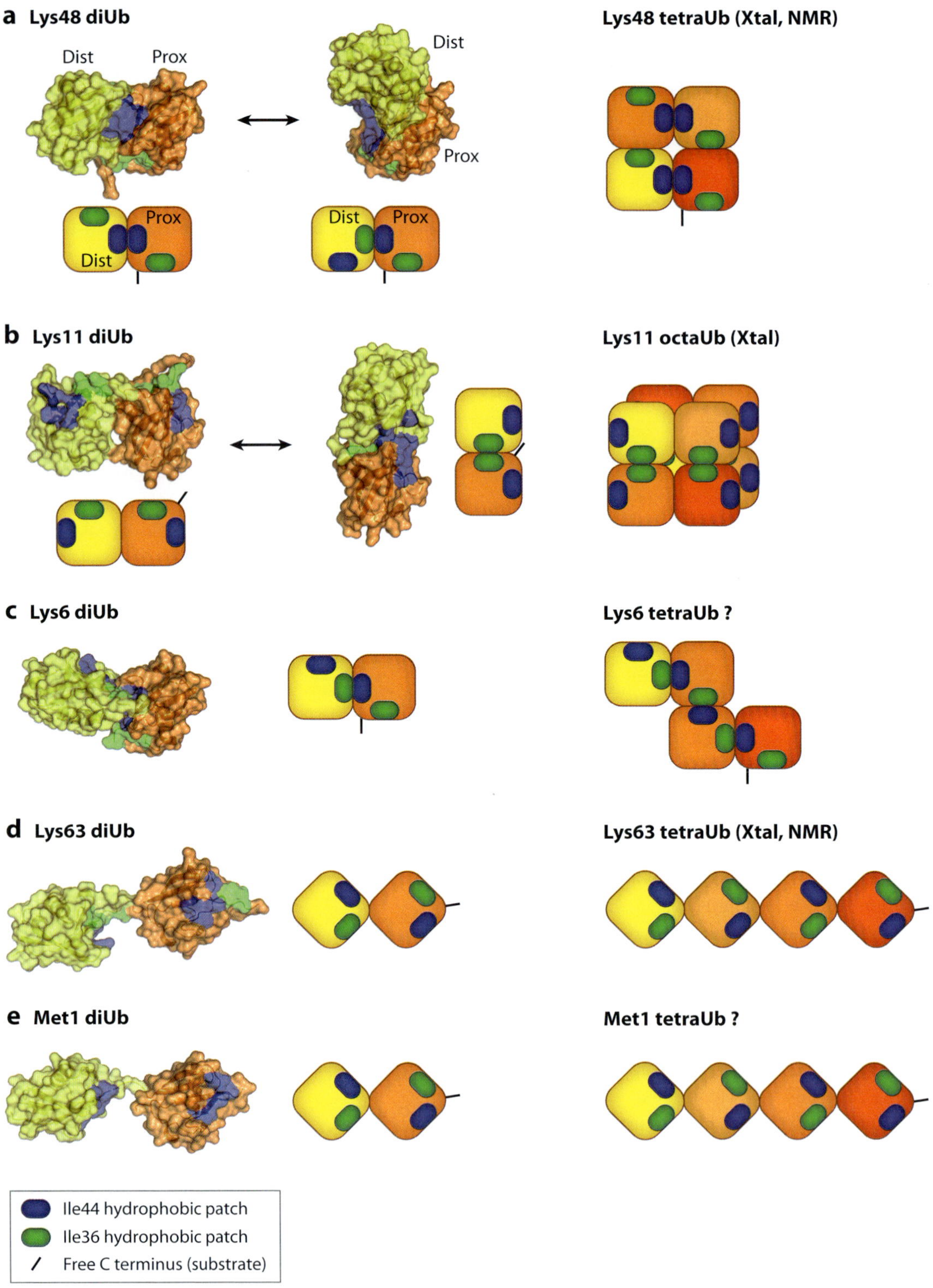

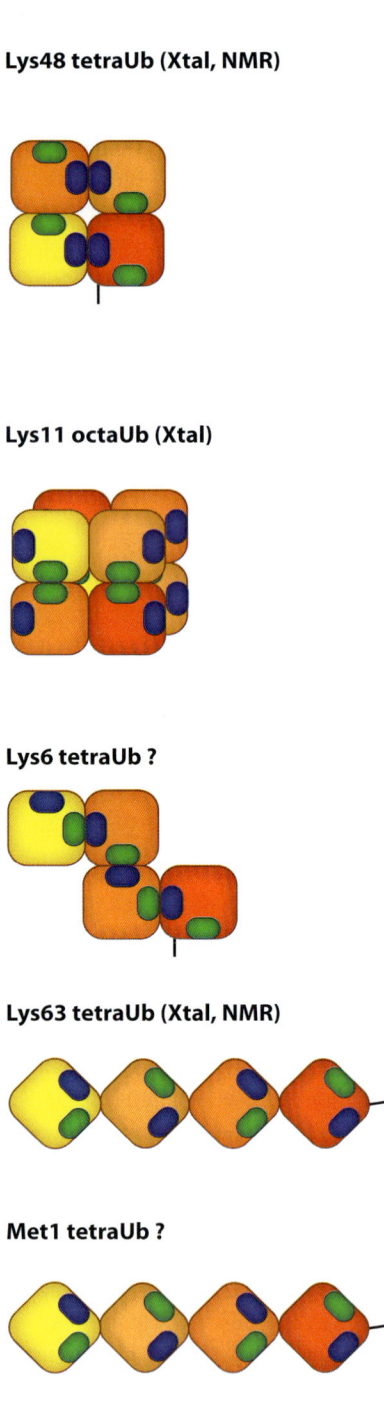

determined by the E2, the E3, or a particular substrate-E3 complex.

In an example of the latter approach, the polycomb E3 ligase complex Bmi1-RING1 monoubiquitylates histone H2A on Lys119 (39), even though it collaborates with Ube2D/UbcH5, a nonspecific E2 that usually modifies multiple substrate and ubiquitin lysine residues (40). Bmi1-RING1 binds to both DNA and nucleosomes, which results in a stiff substrate-E3 complex that exposes the active site of Ube2D toward Lys119 of H2A. Owing to the rigidity of this assembly, ubiquitylation of Lys119 introduces a steric impediment to further modification, thereby restricting the reaction to monoubiquitylation.

Alternatively, E3 enzymes can block the ability of E2s to catalyze chain formation, as seen with Rad18 and its E2 Rad6. Although Rad6 can synthesize mixed or Lys48-linked chains (41, 42), it promotes monoubiquitylation of PCNA when collaborating with Rad18 (4, 43). Similar to other E2s, Rad6 depends on a noncovalent ubiquitin-binding site for chain formation; Rad18 occupies this site, thereby blocking chain formation without interfering with monoubiquitylation (41).

In some cases, the E2 determines monoubiquitylation, yet the molecular basis for this specificity is poorly understood. For example, the E2s Ube2W and Ube2T, together with the E3 FANCL, decorate the DNA repair protein FANCD2 with a single ubiquitin (44, 45). Ube2W also catalyzes monoubiquitylation with other E3s, such as Brca1-Bard1 and CHIP (46, 47). If these E3s utilize the nonspecific Ube2D instead of Ube2W, substrates are modified with ubiquitin chains, showing that in this case it is the E2, Ube2W, that encodes the information for monoubiquitylation.

3.2. Ubiquitin Chain Assembly by RING Domain and U-Box Ligase Enzymes

The enzymes that catalyze chain formation face a different specificity issue: They need to modify specific lysine residues of ubiquitin. For E3s containing a RING or U-box domain, this linkage specificity is likely determined by the E2 (3). This hypothesis is supported by the observation that RING or U-box E3s can synthesize different chain types depending on the E2: Brca1-Bard1 or Murf, for example, assembles Lys63 linkages with the heterodimeric E2 enzyme Ube2N-Uev1A, but Lys48 linkages when bound to Ube2K (46, 48). Similarly, CHIP synthesizes Lys63-linked chains with Ube2N-Uev1A but is unspecific with Ube2D (49). Conversely, RING E3s that interact with a single E2 generally display the specificity of this E2: Multi-subunit E3 ligases of the SCF family decorate substrates with Lys48-linked chains by using the Lys48-specific E2 Ube2R1 (50); gp78, a regulator of endoplasmic reticulum–associated degradation, assembles Lys48-linked chains with the Lys48-specific E2 Ube2G2 (51); and the anaphase promoting complex (APC/C) produces Lys11-linked chains using the Lys11-specific Ube2S (52).

RING E3s and their E2s initiate chain formation on a substrate lysine (**Figure 4***a*), which can occur at random positions or in preferred sequence environments, referred to as chain initiation motifs (53). The initiating E2s often assemble short chains, which can

Really interesting new gene (RING): a class of E3 enzymes

Chain initiation: modification of a substrate lysine residue with the first ubiquitin

---

Figure 3

Ubiquitin chain structure. Diubiquitin (diUb) molecules of different linkages are shown with the distal (dist) molecule in yellow and the proximal (prox) molecule in orange, alongside a schematic representation. Ile44 (*blue*) and Ile36 (*green*) patches are indicated.
(*a*) Lys48-linked diUb, Protein Data Bank (pdb) code 1aar (*left*) (25), pdb 2pe9 (*right*) (185). The tetramer model (tetraUb) is based on data in Reference 28. (*b*) Lys11-linked diubiquitin, pdb 2xew (*left*) (30), pdb 3nob (*right*) (31). The model of Lys11-linked octaubiquitin (octaUb) is based on crystal packing of both structures (33). (*c*) Lys6-linked diubiquitin, pdb 2xk5 (32). (*d*) Lys63-linked polyubiquitin, pdb 2jf5 (35). The model of a longer chain is based on Reference 186. (*e*) Met1-linked polyubiquitin, pdb 2w9n (35).

be connected nonspecifically as for Ube2D (54); contain a favored linkage, as seen for the Lys11-preferring Ube2C (23, 55); or are homogenous, as for Lys48-specific Ube2R1 (50). In most cases, the initiating E2s cooperate with a specific chain-elongating E2 (**Figure 4b**). This allows for assembly of Lys11-linked chains by Ube2S (52, 56), Lys48-linked chains by Ube2K/Ubc1 or Ube2R1 (46, 50, 57), and Lys63-linked chains by Ube2N-Uev1A

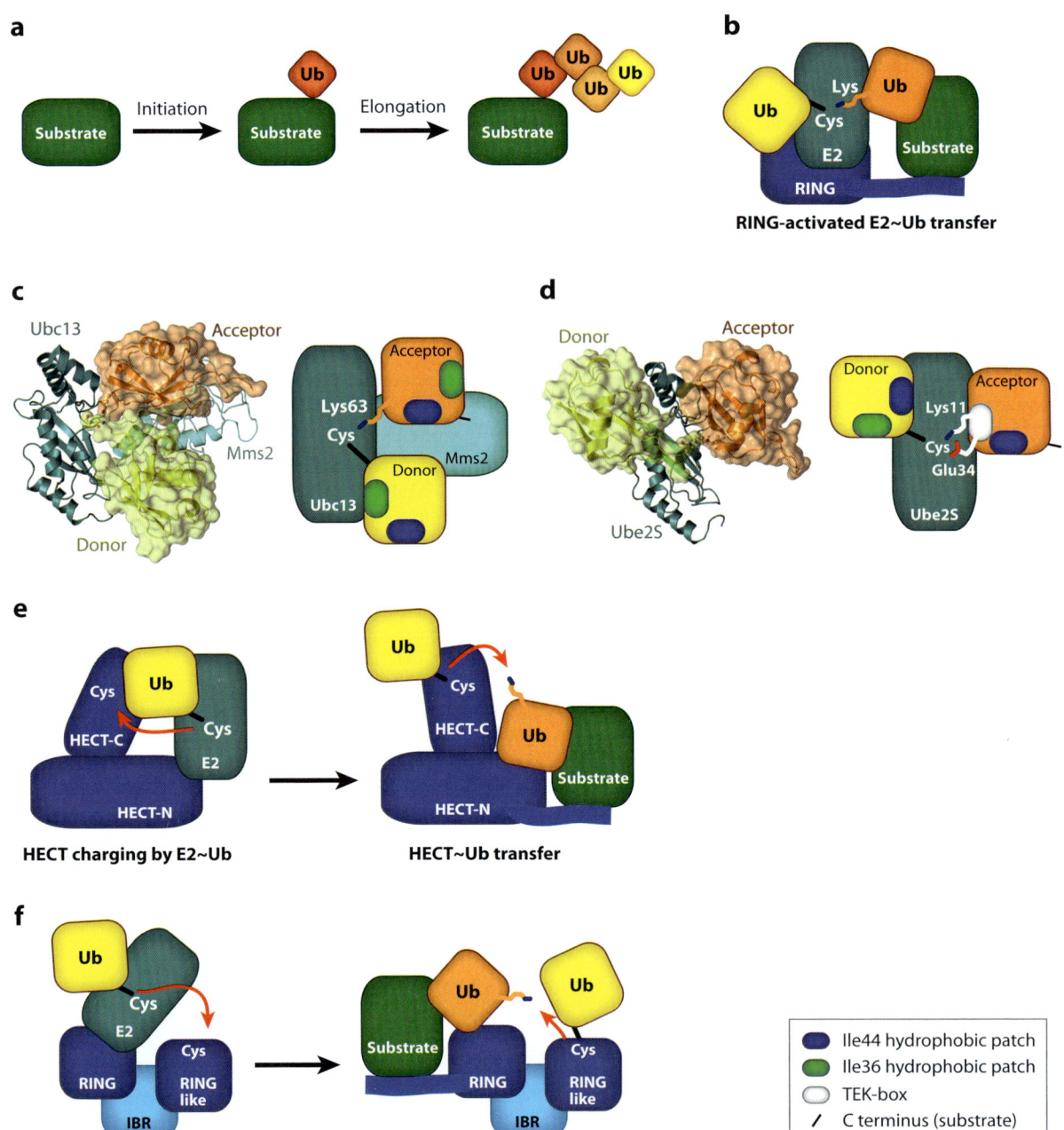

(46, 58). E2s with specificity for Lys6, Lys27, Lys29, or Lys33 have not been reported (59).

Selection of the appropriate lysine used for chain formation requires recognition of a specific acceptor ubiquitin surface by the E2 donor ubiquitin complex. The Lys63-specific Ube2N achieves this feat by teaming up with an auxiliary subunit, Uev1A (**Figure 4c**) (60). Uev1A contains a UBC-E2 variant domain that has lost its catalytic cysteine but has retained its capacity to noncovalently bind ubiquitin (58). This interaction orients the acceptor ubiquitin such that Lys63 faces the active site of charged Ube2N. By contrast, monomeric E2s directly recognize the acceptor ubiquitin. Ube2S, for example, binds the TEK-box of ubiquitin through a surface close to its active site (**Figure 4d**) (61). A similar interaction is required for Lys11 linkage formation by Ube2C (23) and Ube2D (62), suggesting that the TEK-box is of broad importance for Lys11 linkage formation. The Lys48-specific E2s, Cdc34, Ubc1, or Ube2G2, employ an acidic loop or residues close to their active site for acceptor recognition, yet the corresponding ubiquitin surface is not well defined (50, 51, 57, 63).

In most cases, the acceptor ubiquitin is bound with very low affinity: Even Ubc13-Mms2 binds acceptor ubiquitin with a K_m of only 437 μM (64). Such low affinities suggest that additional mechanisms contribute to the linkage specificity of E2s. Previous work with the SUMO-E2 Ube2I/Ubc9 identified an aspartate of the E2 that deprotonates the acceptor lysine, thereby turning it into an efficient nucleophile (65). An acidic residue is also required in ubiquitin E2s that show activity toward lysine, while it is not required for activity toward cysteine, as the thiol group does not need to be deprotonated at physiological pH (66). Importantly, the Lys11-specific Ube2S lacks an acidic residue in its active site and, instead, utilizes the Glu34 of the acceptor ubiquitin for Lys11 activation (**Figure 4d**) (61). Other lysine residues of ubiquitin are not paired up with properly oriented acidic residues, explaining why Ube2S does not modify them with high efficiency. As the Lys48-specific yeast Ubc1 requires Tyr59 of ubiquitin for linkage formation, similar mechanisms of substrate-assisted catalysis might also contribute to formation of other chain types (57).

3.3. Chain Formation by HECT E3s

A different class of enzymes, the HECT E3s, contain a catalytic cysteine (67). E2s charge this cysteine with ubiquitin before this ubiquitin is used for modification. HECT E3s display a wide range of linkage specificities: Yeast Rsp5 and human Nedd4 assemble Lys63-linked chains (68, 69); E6AP (E6-associated protein) synthesizes Lys48 linkages (68, 70); KIAA10/UBE3C promotes formation of Lys29 and Lys48 linkages (70); a bacterial HECT-like E3 triggers assembly of Lys6- and Lys48 linkages (71); and HUWE1 appears to be nonspecific (68).

Figure 4

Mechanism of linkage-specific ubiquitin (Ub) chain assembly by E2s. (*a*) Ubiquitin chain formation proceeds through an initiation step, during which a substrate lysine residue is modified, and elongation, during which ubiquitin molecules are added to the growing chain. (*b*) Proposed mechanism of RING E3-catalyzed ubiquitin transfer from a charged E2 to a substrate or ubiquitin lysine. ∼Ub indicates a covalent ubiquitin thioester intermediate. (*c*) Heterodimeric Ubc13-Mms2 recognizes acceptor ubiquitin through the UBC-variant domain of Mms2. This positions the acceptor Lys63 to the active site of charged Ubc13. (*d*) Monomeric Ube2S catalyzes linkage formation through substrate-assisted catalysis. Ube2S recognizes the TEK-box of acceptor ubiquitin. Activation of the acceptor Lys11 requires a ubiquitin residue, Glu34. Thus, the correct surface of the acceptor ubiquitin, which includes Lys11 and Glu34, must be exposed to the catalytic cysteine of charged Ube2S in order for the active site to be completed and for linkage formation to occur. (*e*) Mechanism of HECT ubiquitin chain formation. The E2 charges a cysteine in the HECT domain C-terminal (HECT-C) lobe forming a HECT-ubiquitin thioester. The HECT domain presumably positions the acceptor for linkage-specific chain assembly. (*f*) Mechanism of RING-in-between-RING (RBR) ubiquitin chain formation. The RING domain binds to and discharges the E2 to a cysteine in the C-terminal RING-like domain. Abbreviations: HECT-N, HECT N-terminal lobe; IBR, in-between-RING.

The catalytic domain of HECT E3s consists of an N-terminal lobe, which binds the E2, and a C-terminal lobe, which contains the active-site cysteine (**Figure 4e**) (72, 73). As the acceptor lysine attacks the thioester between the cysteine of the E3 and ubiquitin, linkage specificity should be determined by the HECT E3 and not the E2. Indeed, Rsp5 and E6AP assemble Lys63- or Lys48-linked chains, respectively, despite using the nonspecific E2 Ube2D; many HECT E3s promote chain formation with Ube2L3/UbcH7, a thiol-reactive E2 that does not modify lysine residues (66), and HECT domain swaps are sufficient to change linkage specificity despite using the same E2s (68).

To determine linkage specificity, HECT domains must orient and activate the acceptor lysine. Indeed, Rsp5, Smurf2, and Nedd4 engage in noncovalent interactions with an acceptor ubiquitin through residues of their N-terminal lobe (69, 74, 75). KIAA10 binds to acceptor ubiquitin through sequences that are N-terminal to the HECT domain (70), and E6AP requires both E2 and ubiquitin surfaces for acceptor recognition (70). However, although this acceptor binding turned out to be required for processive chain formation, it does not determine linkage specificity (69, 74). Thus, how HECT domains synthesize chains of defined topologies remains poorly understood.

3.4. Chain Formation by RING-In-Between-RING E3 Ubiquitin Ligases

A distinct set of E3s contains a sequence of a RING, a RING-in-between-RING (RBR), and a RING-like domain. These RBR E3s are RING/HECT hybrids: They utilize the RING domain to recruit an E2, which transfers ubiquitin to a cysteine in the RING-like domain, to form a thioester intermediate (**Figure 4f**) (66). RBR E3s display linkage specificity: The linear ubiquitin chain assembly complex (LUBAC) assembles Met1-linked chains (8, 76–78), and parkin catalyzes monoubiquitylation as well as Lys63-, Lys48-, and Lys27-linked chain formation (79, 80). As RBR E3s function with Ube2L3, an E2 without reactivity against lysine (66), linkage specificity of chain formation must be determined by the E3, and not the E2. Consistent with this notion, RBR E3s can synthesize chains that differ from the inherent specificity of a cooperating E2: Whereas Ube2K usually assembles Lys48-linked chains (81), LUBAC and Ube2K produce Met1-linked chains (82). How RBR E3s determine their linkage specificity, however, is not known.

4. READING THE CODE: CONCEPTS IN UBIQUITIN BINDING

Once the code has been written, effector proteins with UBDs translate the modifications into specific outcomes (15). Although many UBDs have been described, their potential for linkage-specific ubiquitin recognition has rarely been assessed comprehensively, and limited structural and mechanistic insight is available. For example, a crystal structure of Lys48-linked chains in complex with a protein has not been reported. Fortunately, Lys63-linked diubiquitin has been caught in complex with UBDs, revealing several concepts of ubiquitin recognition.

4.1. Exploiting the Distance Between Ubiquitin Molecules

In the various chain types, the distance between successive ubiquitin moieties can differ considerably. Proteins with multiple UBDs exploit this property by introducing a defined spacer between the UBDs, as seen in proteins that contain tandem repeats of ubiquitin-interacting motifs (UIMs) (38, 83). UIMs consist of an α-helix with a hydrophobic binding site for the Ile44 patch of ubiquitin (84). In Rap80, a subunit of the Brca1-E3, two UIMs are separated by a seven-residue helix that positions the UIMs to recognize extended Lys63-, but not compact Lys48-, linked chains (**Figure 5a**) (38, 83). Conversely, the two UIMs of Ataxin-3 are separated by a short linker of two residues, which

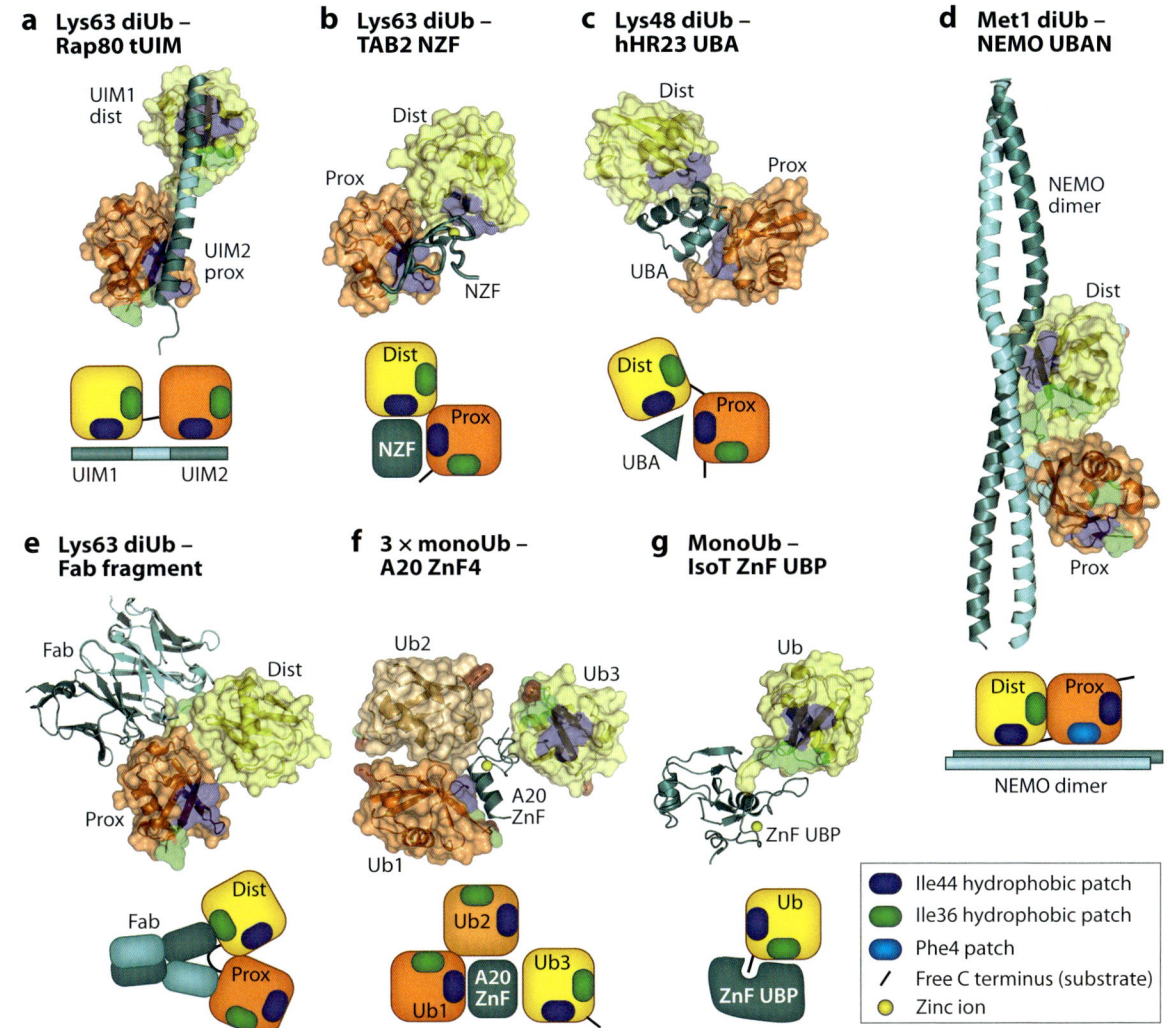

Figure 5

Concepts in ubiquitin binding. Ubiquitin binding domains are shown in cyan/teal, and the ubiquitin chains appear as in **Figure 3**. Zinc ions are shown as yellow spheres. (*a*) Crystal structure of the RAP80 tandem ubiquitin-interacting motif (tUIM) bound to Lys63-linked diubiquitin (diUb), Protein Data Bank (pdb) code 3a1q (86). (*b*) Crystal structure of the TAB2 NZF domain bound to Lys63-linked diubiquitin, pdb code 2wwz (87). (*c*) NMR model of the hHR23 UBA domain bound to Lys48-linked diubiquitin, pdb code 1zo6 (89). (*d*) Crystal structure of the NEMO UBAN domain bound to Met1-linked diubiquitin, pdb code 2zvn (21). (*e*) Crystal structure of the Lys63-specific antibody bound to Lys63-linked diubiquitin, pdb code 3dvg (90). Only part of the Fab fragment is shown. (*f*) Crystal structure of the A20 ZnF domain bound to three ubiquitin molecules, pdb code 3oj3 (94). (*g*) Crystal structure of the Usp5/IsoT ZnF UBP domain bound to ubiquitin, pdb code 2g45 (20). Abbreviations: dist, distal; prox, proximal.

results in preferential recognition of more compact Lys48 linkages. Swapping the linkers made Ataxin-3 Lys63 and Rap80 Lys48 selective, emphasizing how the length of the linker, the "ruler," determines binding specificity (38).

4.2. Exploiting Chain Flexibility

TAB2 and TAB3, adaptors of the TAK1 kinase complex, contain Npl4-like zinc fingers (NZFs) (85). These NZF domains are able to distinguish between structurally similar

Proximal ubiquitin: the ubiquitin moiety attached to a substrate or with a free C terminus in unanchored chains

Unanchored ubiquitin chains: ubiquitin chains that are not attached to substrates

Lys63- and Met1-linked chains, a specificity enabled by the dynamic nature of these chain types (**Figure 5b**). As seen in crystal structures of NZF domains bound to Lys63-linked diubiquitin (86, 87), the Ile44 patches of each ubiquitin interact with a perpendicular surface on the same NZF domain. This "bending" of Lys63-linked dimers displaces Met1 of the proximal ubiquitin from the C terminus of the distal unit, thereby preventing an equivalent binding mode for Met1-linked chains (86, 87).

A different type of chain flexibility can be important for recognition of compact chains, as seen for UBA domains of proteasomal shuttling factors that bind Lys48-linked ubiquitin. These UBA domains slot into a Lys48-linked ubiquitin dimer to interact with Ile44 patches of both ubiquitin molecules (**Figure 5c**) (88, 89), an event that requires dynamic opening of compact Lys48-linked chains (88, 89). Thus, similar to NZF domains, certain UBA domains require ubiquitin chain flexibility to allow for recognition of a particular linkage.

4.3. Recognizing the Linkage Context

The simplest way of recognizing a particular ubiquitin modification is to directly bind the isopeptide bond or the sequence context of the linkage. As an example, the Lys63-linkage-specific antibody interacts with the C terminus of the distal ubiquitin as well as with its Ile36 patch (**Figure 5e**) (90). Although Lys63 on the proximal Ub is not contacted, residues closely preceding Lys63 mediate antibody binding. Interestingly, this interaction requires a compact Lys63-linked diubiquitin in which the linkage is accessible to the antibody (90).

The linkage context is also recognized by the UBAN domain of NEMO, a subunit of IκB kinase (91), which binds linear diubiquitin (21, 35, 92). NEMO forms a symmetric dimer that has two adjacent ubiquitin-binding sites, with one recognizing the distal Ile44 patch and the second binding to the proximal Phe4 patch (**Figure 5d**) (21). Although the peptide bond between ubiquitin molecules is not directly contacted, Gln2 of the proximal ubiquitin makes key interactions (21). Similar interactions are not accessible for the proximal unit in Lys63-linked diubiquitin, which can only bind the distal site with weaker affinity (93). NEMO, therefore, illustrates how UBDs can recognize the linkage context to distinguish between structurally related chain topologies.

4.4. Combining Binding Sites

A single UBD can also achieve specificity by recognizing distinct surfaces in multiple ubiquitin molecules of a chain. This interaction mode is illustrated by a regulator of inflammatory processes, A20, which binds Lys63-linked chains. NMR and crystallographic analysis revealed distinct interactions between one A20 zinc-finger domain and three ubiquitin molecules (**Figure 5f**) (94). A20 binds the Ile44 patch, the TEK-box, and a surface surrounding Asp58 of distinct ubiquitin moieties. Although the ubiquitin molecules were not covalently linked, all Lys63 side chains were in proximity to an adjacent ubiquitin C-terminus, suggesting that this structure represents a model for the interaction of Lys63-linked tri-ubiquitin with one A20 zinc-finger domain.

4.5. Detecting the Free C Terminus of Unanchored Chains

Unanchored ubiquitin chains are generated during ubiquitin biosynthesis as products of the *UBB* and *UBC* genes (95), by DUBs that internally cleave ubiquitin chains or release entire chains from substrates (96), or by ubiquitylation enzymes that synthesize unanchored polymers for signal transduction (97, 98). Unanchored chains contain at their proximal end a free ubiquitin C terminus, which is normally masked by attachment of ubiquitin to substrates. Using its N-terminal ZnF-UBP domain, the DUB USP5/IsoT binds this free C terminus of ubiquitin, including the Gly-Gly sequence, with nanomolar affinity (**Figure 5g**) (20, 99). This interaction mediates substrate binding, converts the active site of USP5 into a catalytically competent state, and allows USP5 to preferentially disassemble unanchored chains from their proximal end (20, 99). Because of the latter property, USP5 has been used to validate the role of unanchored chains in signaling (97, 98); however, it should be noted that USP5 can act on attached chains with an activity that is comparable to other USP enzymes (30).

5. ERASING THE CODE

Any useful code should be carefully employed only at times of need. Indeed, to prevent ubiquitylation from being constitutively on, modifications are reversed by DUBs. Human cells contain ~55 USPs, 14 ovarian tumor DUBs (OTUs), 10 JAMM family DUBs, 4 ubiquitin C-terminal hydrolases (UCHs) and 4 Josephin domain DUBs (96). To specifically control ubiquitin-dependent signaling, these enzymes have to deal with chains of distinct linkage, topology, and length.

5.1. Housekeeping and Substrate-Specific Deubiquitinating Enzymes

Several DUBs, referred to as housekeeping enzymes, play important roles in establishing the ubiquitin code. For example, proteasome-bound DUBs, such as USP14, UCH37/UCH-L5, and RPN11/POH1, protect ubiquitin from degradation (100). This process is vital for keeping sufficient levels of free ubiquitin that can be used for chain assembly. Similar functions might be performed by DUBs that interact with ubiquitin-processing complexes, such as the COP9 signalosome (USP15) (101), or the p97 segregase [YOD1 (102), VCIP135 (103), Ataxin-3 (104)].

Another large group of DUBs disassembles chains independently of the linkage, yet these enzymes gain specificity by being targeted to a select set of substrates. These DUBs include most members of the USP family, which regulate many cellular reactions, including splicing, protein trafficking, or chromatin remodeling. Many USP DUBs are recruited to substrates through interaction domains (96) or adaptor subunits (105). Although a comprehensive analysis has not been reported, most USPs are active against all linkages (22, 32, 35) and also hydrolyze the isopeptide bond between the substrate and the first ubiquitin. An exception from this nonspecificity is CYLD, which prefers Met1- and Lys63-linked chains (35, 98, 106). Hence, most USPs can be considered nonspecific with regard to the ubiquitin code but specific with respect to their substrates.

5.2. Linkage-Specific Deubiquitinating Enzymes

In contrast to the aforementioned examples, several DUBs respond to the ubiquitin code and display specificity toward one or a few linkages. JAMM family DUBs are often Lys63 specific, as seen for AMSH (107), AMSH-LP (108), BRCC36, and POH1 (109). In addition, linkage-specific OTU DUBs have been described; these descriptions showed that OTUB1 is specific for Lys48 linkages (110, 111), Cezanne is specific for Lys11 linkages (30), and Trabid is specific for Lys29 and Lys33 linkages (32, 112). As linkage-specific DUBs may not be able to cleave off the last ubiquitin (30), their activity might generate monoubiquitylated substrates with distinct signaling properties.

The structure of AMSH-LP with Lys63-linked diubiquitin revealed the basis for the linkage specificity displayed by JAMM family DUBs (108). AMSH-LP binds to the open conformation of the Lys63-linked diubiquitin and contacts Gln62 and Glu64 of the proximal ubiquitin. Thus, reminiscent of some UBDs, JAMM DUBs might recognize the sequence context of the Lys63-isopeptide bond. The structure of Trabid revealed a different mechanism to achieve specificity; this enzyme uses an Ankyrin-repeat UBD directly upstream of the catalytic OTU domain to position the proximal ubiquitin (112). However, for the OTU and remaining DUB families, structures are available only in complex with a single ubiquitin (19, 113–115), which revealed high-affinity binding sites for the distal ubiquitin, while leaving the interaction sites for the proximal ubiquitin that provides the modified lysine unclear. These structures did imply, however, that compact chain conformations should not be recognized by DUBs unless they undergo extensive remodeling to expose the isopeptide bond.

5.3. Ubiquitin Chain Editing

Ubiquitin chain editing is perhaps the most sophisticated utilization of the ubiquitin code. During this process, one chain type is replaced by a chain of different topology, which changes the fate of the modified substrate. Editing could be achieved by complexes between sequentially acting DUBs and E3s, as they are often observed in cells (105). Alternatively, single proteins might combine DUB and E3 activity, as seen in a pathway that regulates the transcription factor NF-κB (116). Activation of NF-κB relies on Lys63-specific ubiquitylation of proteins, whereas its inactivation includes a negative feedback loop centered on A20, a protein that combines DUB and E3 domains. It has been proposed that A20 first deubiquitylates Lys63-modified proteins and then modifies them with Lys48-linked chains to trigger their degradation. In this manner, A20 not only stops signaling through Lys63-linked chains, but it also removes the signal transducers that need to be resynthesized before signaling can resume. Although many experiments, including linkage-specific antibodies, support this view of A20 action (90, 94, 116), additional layers of regulation might contribute. Indeed, A20 prefers to cleave Lys48-linked ubiquitin chains in vitro (117, 118), and it was shown to interact with ubiquitin-binding adaptors, such as TAX1BP1 (119) or ABINs (120); E3 ligases, such as ITCH (121) and RNF11 (122); and E2 enzymes (123).

6. CELLULAR FUNCTIONS OF THE UBIQUITIN CODE

The combinatorial action of ubiquitylating, deubiquitylating, and ubiquitin-binding proteins determines the modified protein's fate; it carries the meaning of the code. Historically, the idea of a ubiquitin code emerged from the distinct consequences of proteolytic Lys48-linked and nonproteolytic Lys63-linked chains, a view that might be too simplified: Multiple chain types, including Lys63-linked chains, are now known to drive degradation, whereas Lys48-linked chains can function nonproteolytically, for example, in transcription factor regulation (124). This suggests that the functions of ubiquitylation depend on chain topology, but also on other factors, such as the timing and reversibility of the reaction, enzyme or substrate localization, or interactions between E3s and effectors. Proteomic analyses found thousands of proteins that act in almost all signaling pathways to be modified with ubiquitin (125, 126). To pay tribute to this complexity, we focus our discussion on selected roles of ubiquitylation that are brought about by distinct chain types and influenced by additional cellular inputs.

6.1. Proteolytic Functions of the Ubiquitin Code

6.1.1. Regulation of proteasomal degradation.
It is well established that ubiquitin chains can target proteins to the 26S proteasome, a protease required for cell division in all eukaryotes (100). Consistent with Lys48 being the only essential lysine of ubiquitin in yeast, the role in proteasomal targeting was first assigned to Lys48-linked chains (127). Many E3s, including the SCF, gp78, or E6AP, trigger substrate turnover by synthesizing Lys48-linked chains (50, 51, 68). As a result, Lys48 linkages are the most abundant linkage in all organisms subjected to quantitative proteomic analysis, and their levels increase rapidly when the proteasome is inhibited (11, 12, 125, 128).

However, early experiments had already indicated that other linkages could also be recognized by the proteasome (129, 130). These atypical linkages accumulate upon proteasome inhibition (12), suggesting that they also contribute to protein degradation. Indeed, Lys11-linked chains bind proteasomal receptors and trigger degradation of cell cycle regulators during mitosis (23, 31, 52). In human cells, Lys11 linkages accumulate dramatically upon activation of the responsible E3, the APC/C, and inhibition of Lys11-linked chain formation stabilizes APC/C-substrates and leads to cell cycle arrest (23, 31). Lys11-linked chains are, therefore, proteolytic signals that are particularly important during mitosis.

Other chain types mediate proteasomal degradation less frequently. Lys29-linked chains contribute to substrate turnover in the ubiquitin-fusion-degradation pathway (131, 132), and in a few cases, Lys63-linked or mixed chains were held accountable for triggering degradation (48, 55, 133). Thus, Lys11-, Lys29-, Lys48-, and Lys63-linked chains might all have roles in proteasomal degradation, a diversity in targeting signals that is reflected by the plasticity in substrate recognition by proteasomal subunits: Rpn13 binds monoubiquitin and Lys48-linked diubiquitin with similar affinity (134), S5a/Rpn10 interacts with chains of multiple topologies (135), and various proteasomal shuttling factors show only modest preference for Lys48 compared to Lys63 linkages (136).

Why Lys11- and Lys48-linked chains trigger degradation more frequently than other modifications is not entirely clear. It is possible that enzymes synthesizing Lys11- and Lys48-linked chains are less likely to introduce branches, which can impede degradation (48). In addition, Lys11- and Lys48-specific enzymes, such as the APC/C or SCF, often interact with the proteasome to efficiently couple ubiquitylation and degradation (137, 138). For some E3s, binding to the proteasome is required for sending substrates to degradation; deletion of a proteasome-binding domain in Ufd4 does not affect ubiquitylation, but inhibits substrate turnover (139). It is also possible that atypical linkages are more prone to deubiquitylation, although our limited understanding of DUB biology has not allowed this hypothesis to be rigorously tested.

6.1.2. Regulation of lysosomal degradation.
The degradation of plasma membrane proteins occurs in lysosomes, and substrates are targeted to this proteolytic compartment through monoubiquitylation or Lys63-linked chains (140). Ubiquitylation can be initiated at the membrane and lead to endocytosis, as seen for yeast membrane receptors that are substrates of the HECT E3 Rsp5 (141). In such cases, an in-frame fusion of ubiquitin to the substrate is often sufficient for internalization, even if the ubiquitylation machinery is disrupted (142). Alternatively, ubiquitylation can occur at endosomal membranes to control localization after internalization. This was demonstrated with EGFR for which mutation of all ubiquitylation sites did not block endocytosis but strongly affected its routing to lysosomes (143, 144). As a result, deubiquitylation by Usp8 or AMSH can lead to recycling of EGFR to the plasma membrane (145, 146).

Ubiquitylated membrane proteins are recognized by different ESCRT complexes, which bind ubiquitin with a modest preference for Lys63 linkages (147, 148). In addition to ubiquitin-binding motifs, ESCRT-complexes recognize the coat of endocytic vesicles or lipids of endosomal membranes (147). Thus, the UBDs that read out the ubiquitin modification are enriched in proximity of their substrates, suggesting that colocalization of substrates and effectors helps determine the consequences of a ubiquitylation event.

Although ubiquitylated protein aggregates are also degraded in lysosomes, they pass through autophagosomes on their route to elimination. The aggregates are coupled to autophagosomes by adaptor molecules, which bind to substrates with a moderate preference for Lys63-linked chains (149). Reminiscent of ESCRT-complexes, autophagy receptors are enriched in proximity to their substrates through interactions with autophagosomal membranes. Thus, Lys63-linked chains can trigger proteolysis, yet efficient targeting to lysosomes may also require additional inputs, such as the specific localization of effector proteins.

6.2. Nonproteolytic Functions of the Ubiquitin Code

Ubiquitylation is also able to regulate signaling nonproteolytically as it can be used to recruit proteins to participate in particular signaling pathways, to attract trafficking factors that change a substrate's localization, or to control a substrate's activity. In most cases, the ubiquitin conjugate is recognized with low affinity, but the multivalent recognition of both substrate and ubiquitin allows for tight regulation. These nonproteolytic functions of the ubiquitin code are often the consequence of monoubiquitylation or Met1- and Lys63-linked chain formation.

6.2.1. Regulation of protein interactions.

The attachment of a single ubiquitin often suffices to recruit binding partners as seen for PCNA, a processivity factor for DNA polymerases (4, 150). In response to DNA damage, PCNA is monoubiquitylated (4, 151), which recruits Y family DNA polymerases (152, 153). These polymerases recognize PCNA through a PCNA-interaction motif, the PIP-box, and ubiquitin through UBZ or UBM domains, leading to a high-affinity interaction that replaces replicative polymerases from PCNA. In this manner, monoubiquitylation of PCNA contributes to a ubiquitin-dependent polymerase switch that rescues stalled replication forks from collapsing. Following the successful repair of the damaged DNA, the recruitment signal for Y family polymerases is turned off by Usp1-dependent deubiquitylation, allowing the replication machinery to return to its normal state (154). Similarly, monoubiquitylation of FANCD2 and FANCI, two proteins involved in DNA repair, recruits the FAN1 nuclease, and this signaling event is also turned off by Usp1 (155–158). Thus, monoubiquitylation is a tool for the reversible recruitment of an enzyme to a particular cellular location.

Protein interactions can also be regulated by Lys63-linked chains: The modification of a spliceosomal protein with Lys63-linked chains stabilizes an snRNP complex, which is reorganized upon Usp4-dependent deubiquitylation (159, 160). Similarly, modification of a ribosomal protein with Lys63-linked chains stabilizes polysomes to promote translation (161). The scaffolding role of Lys63-linked chains is most apparent during the response that cells mount to DNA damage, an event that is dependent on a series of E3 enzymes (162–166). The recruitment of several E3s to sites of DNA damage depends on Lys63-linked chains that are probably attached to histone proteins. The E3s Rnf8 or Rnf168 directly recognize Lys63 linkages through MIU or UBZ domains, whereas Brca1 depends on a binding partner, Rap80. Together, these E3s generate ubiquitin-rich foci that act as stable recruitment platforms for DNA repair enzymes and for checkpoint molecules that inhibit cell cycle progression in the face of damage.

An interesting concept has been introduced with unanchored Lys63-linked chains acting as transient mediators of protein interactions (98). Rather than being attached to a substrate, unanchored chains function as recruitment platforms that attract and cluster multiple recognition factors. Because of the high activity of DUBs that recognize unanchored chains, such as USP5, the half-life of these signaling intermediates is likely short, and some of their specificity might be gained by being synthesized in proximity to their binding partners. Unanchored Lys63-linked ubiquitin chains mediate the activation of TAK1 and IKK kinases and the RIG-I antiviral protein (97, 98, 167). RIG-I binds Lys63-linked, but not Lys48- or Met1-linked, chains through tandem CARD domains (97), which leads to RIG-I dimerization and facilitates downstream signaling events (167).

Ubiquitylation can also impair interactions. For example, monoubiquitylation of Smad4 blocks its association with the transcriptional cofactor Smad2 (168). By deubiquitylating Smad4, USP9X relieves the impediment for cofactor binding and triggers transcriptional activation. Similar results are achieved in EGFR signaling through coupled monoubiquitylation (169–171). Monoubiquitylation of substrate adaptors of the endocytic machinery leads to an intramolecular interaction between the conjugate and the adaptor's UBD and blocks the ability of the adaptor to recognize its ubiquitylated cargo (170). The transcription factor Met4 shows that such regulation does not rely on monoubiquitylation. Met4 is modified by SCF^{Met30} with Lys48-linked chains, which bind to an internal UIM in Met4 and block its ability to engage with coactivators (124). These examples underscore that the topology of the conjugate and the context of the modification can determine the outcome of ubiquitylation.

6.2.2. Regulation of protein activity.
Ubiquitylation can affect a protein's activity by different means. In a straightforward mechanism of activation, an inhibitor is sent for degradation. Among many examples, the inhibitor of the NF-κB transcription factor, IκBα, is degraded after modification with Lys48-linked chains by SCF^{βTrCP} (172, 173). The proteasome can also activate a protein by cleaving off inhibitory domains, as seen for proteasomal cleavage of an NF-κB precursor (174). Similar reactions are observed in budding yeast, where proteasomal processing of the transcription factor SPT23 is a prerequisite for its release from the endoplasmic reticulum membrane (175), or in fission yeast, where activation of the membrane-bound transcription factor SREBP requires ubiquitin-dependent cleavage (176). Proteasomal processing might involve atypical linkages, as cleavage of NF-κB might require multimonoubiquitylation (177), whereas SPT23-processing is brought about by Rsp5, an E3 that catalyzes monoubiquitylation or Lys63-linked chain formation (175). The substrates also have to withstand unfolding through proteasomal ATPases (178), again underscoring how the outcome of ubiquitylation can be determined, at least in part, by the sequence context of the modification.

Activation of NF-κB usually occurs in response to external stimuli, such as the tumor necrosis factor α. Binding of the tumor necrosis factor α to its membrane receptor initiates a variety of ubiquitylation events, such as formation of Lys63-linked chains by TRAF6 (60), mixed Lys11/Lys63-linked chains by the RING E3 ligase cIAP1 (7, 179), or Met1-linked chains by LUBAC (82). LUBAC modifies NEMO, a subunit of the IκBα kinase (IKK) complex. Intriguingly, the Met1-linked chains on NEMO are recognized by the UBAN domain of NEMO itself, which may cause a conformational change in the intertwined helices of NEMO dimers (21). As NEMO is a core regulatory subunit of IKK, these conformational changes might lead to allosteric activation of IKK. NEMO also associates with Lys63-linked chains (180) and with mixed Lys11/Lys63-linked chains that are detected on the receptor interacting protein 1, RIP1 (7). Lys63-linked chains activate IKK by promoting its binding to the upstream TAK1 kinase complex (181). These findings suggest that distinct types of ubiquitin topologies, i.e., Met1-, Lys11/Lys63-, Lys63-, or Lys48-linked chains, regulate signaling through inhibitor degradation, proteasomal processing, allosteric activation, or recruitment of upstream activating enzymes.

6.2.3. Regulation of protein localization.
The role of ubiquitylation in regulating localization serves as a final example for the diverse functions of the ubiquitin code. Ubiquitin-dependent changes in localization had originally been observed in yeast, where internalization of plasma membrane proteins can be brought about by monoubiquitylation (141, 142). Interestingly, ubiquitin can determine the intracellular location, even if it is not connected to a substrate through an isopeptide bond; the E2 Ube2E3/UbcM2 is transported into the nucleus only when it is charged with a thioester-linked ubiquitin (182). The ubiquitin-loaded Ube2E3 is bound by a transport factor of the importin family, but whether these importins also interact with other cargoes in a ubiquitin-dependent manner has not been determined.

Ubiquitylation can also indirectly affect protein localization. Following its multimonoubiquitylation, the transcription factor p53 is exported out of the nucleus (183). p53 is modified on several C-terminal lysine residues, and in-frame fusions of ubiquitin to p53 are

Chain elongation: extension of chains by addition of additional ubiquitin molecules

sufficient to drive its nuclear export (5). Because cytoplasmic accumulation of these fusions depends on a nuclear export signal in p53, monoubiquitylation likely changes the accessibility of the nuclear export sequence in p53 to the export machinery, rather than being an export signal itself. As opposed to p53 degradation, monoubiquitylation of p53 can be reversed by USP10, which allows reimport and reactivation (184). Thus, by affecting intra- or intermolecular binding events, ubiquitylation can lead to many different consequences that result from a combinatorial recognition of the ubiquitin conjugate and additional factors, such as membrane lipids, binding partners, or substrate domains.

7. CONCLUSIONS

Groundbreaking work with Lys48- and Lys63-linked chains suggested that different ubiquitin linkages result in unique consequences for the modified proteins. During the past years, much of this ubiquitin code hypothesis has been confirmed: Different chain topologies adopt unique compact or open conformations; they are synthesized by enzymes that assemble the specific modifications; they are recognized by linkage-specific ubiquitin-binding proteins that couple the modification to a particular outcome; they are disassembled by enzymes that act as erasers of the code; and they function in a wide range of different processes.

Several questions, however, remain open. We know little about the physiological relevance of Lys6, Lys27, Lys29, and Lys33 linkages or more complex structures, such as branched chains. Moreover, it is possible that substrates are modified with multiple chain types at the same time, but whether this is associated with a particular function has not been tested. Even for better-understood topologies, fundamental questions need to be addressed: Why, for example, does the APC/C assemble Lys11- instead of Lys48-linked chains to drive degradation? Is there physiological relevance to differences in chain length? Do proteins that restrict chain elongation serve roles other than preventing the waste of ubiquitin? How dynamic are ubiquitin chains? And can the flexibility of chain structures be regulated to modulate recognition by ubiquitin-binding proteins?

In a broader sense, the systems biology of the ubiquitin code has to be analyzed in more detail. Given the importance of E2s in determining specificity, it is surprising how little we know about physiological E2-E3 pairs. The same holds true for DUBs, which are often found in complexes with E3s and, hence, have a lot of potential in modulating the ubiquitin code. Finally, only a few E3s have been studied in sufficient detail to allow a somewhat comprehensive assessment of their substrates. More insight into substrate modifications and their cellular consequences will undoubtedly uncover more functions for the ubiquitin code.

A major breakthrough in understanding the Inca quipus came with the discovery that certain combinations of knots encode numbers, yet most of the information found between those numeric knots remains mysterious. It is the numbers of the ubiquitin code, i.e., the residues used for assembling chains, that became better understood in recent years; what all topologies mean, however, has not been completely unraveled. Thus, as it is the case for those trying to decipher the code of an ancient empire, there is much to be learned before we can crack the ubiquitin code in its entirety.

SUMMARY POINTS

1. Proteins can be modified with a single ubiquitin or with polymeric chains that differ in the connection between ubiquitin molecules.
2. The different ubiquitin modifications adopt distinct structures.

3. Ubiquitin-binding proteins exploit various strategies to specifically interact with particular types of ubiquitin modifications.

4. Ubiquitin chains can be disassembled by nonspecific or linkage-specific DUBs.

5. The various ubiquitin modifications trigger a wide range of biological reactions, including protein degradation, activation, and localization.

6. The consequences of ubiquitylation are determined by the chain topology in combination with additional factors, such as substrate localization or sensitivity to deubiquitylation.

FUTURE ISSUES

1. What are the functions of Lys6, Lys27, Lys29, and Lys33 linkages or branched ubiquitin chains?

2. Are there more complex structures, such as multiple chain types, attached to a single substrate, and what are their functions?

3. How important is chain length for signaling?

4. Can the dynamics of ubiquitin chains be regulated to modulate signaling?

5. Can we dissect the network of enzymes and substrates to discover novel functions of the ubiquitin code?

DISCLOSURE STATEMENT

The authors are not aware of any affiliations, memberships, funding, or financial holdings that might be perceived as affecting the objectivity of this review.

ACKNOWLEDGMENTS

We apologize to all whose work could not be cited owing to page limitations. D.K. would like to thank Yogesh Kulathu for comments on the manuscript. D.K. is funded by the Medical Research Council and the EMBO Young Investigator Program. M.R. is grateful to Julia Schaletzky for many discussions and for critically reading this manuscript, and to Gary Urton of Harvard University for his introduction into Inca archeology. M.R. is funded by grants from the National Institutes of Health, the Pew Foundation, the March of Dimes Foundation, and the California Institute for Regenerative Medicine. Both authors thank their labs for continued discussions and for their enthusiasm toward unraveling the ubiquitin code.

LITERATURE CITED

1. Deshaies RJ, Joazeiro CA. 2009. RING domain E3 ubiquitin ligases. *Annu. Rev. Biochem.* 78:399–434
2. Schulman BA, Harper JW. 2009. Ubiquitin-like protein activation by E1 enzymes: the apex for downstream signalling pathways. *Nat. Rev. Mol. Cell Biol.* 10:319–31
3. Ye Y, Rape M. 2009. Building ubiquitin chains: E2 enzymes at work. *Nat. Rev. Mol. Cell Biol.* 10:755–64
4. Hoege C, Pfander B, Moldovan GL, Pyrowolakis G, Jentsch S. 2002. RAD6-dependent DNA repair is linked to modification of PCNA by ubiquitin and SUMO. *Nature* 419:135–41

5. Carter S, Bischof O, Dejean A, Vousden KH. 2007. C-terminal modifications regulate MDM2 dissociation and nuclear export of p53. *Nat. Cell Biol.* 9:428–35
6. Haglund K, Sigismund S, Polo S, Szymkiewicz I, Di Fiore PP, Dikic I. 2003. Multiple monoubiquitination of RTKs is sufficient for their endocytosis and degradation. *Nat. Cell Biol.* 5:461–66
7. Dynek JN, Goncharov T, Dueber EC, Fedorova AV, Izrael-Tomasevic A, et al. 2010. c-IAP1 and UbcH5 promote K11-linked polyubiquitination of RIP1 in TNF signalling. *EMBO J.* 29:4198–209
8. Gerlach B, Cordier SM, Schmukle AC, Emmerich CH, Rieser E, et al. 2011. Linear ubiquitination prevents inflammation and regulates immune signalling. *Nature* 471:591–96
9. Boname JM, Thomas M, Stagg HR, Xu P, Peng J, Lehner PJ. 2010. Efficient internalization of MHC I requires lysine-11 and lysine-63 mixed linkage polyubiquitin chains. *Traffic* 11:210–20
10. Goto E, Yamanaka Y, Ishikawa A, Aoki-Kawasumi M, Mito-Yoshida M, et al. 2010. Contribution of lysine 11-linked ubiquitination to MIR2-mediated major histocompatibility complex class I internalization. *J. Biol. Chem.* 285:35311–19
11. Peng J, Schwartz D, Elias JE, Thoreen CC, Cheng D, et al. 2003. A proteomics approach to understanding protein ubiquitination. *Nat. Biotechnol.* 21:921–26
12. Xu P, Duong DM, Seyfried NT, Cheng D, Xie Y, et al. 2009. Quantitative proteomics reveals the function of unconventional ubiquitin chains in proteasomal degradation. *Cell* 137:133–45
13. Vijay-Kumar S, Bugg CE, Cook WJ. 1987. Structure of ubiquitin refined at 1.8 A resolution. *J. Mol. Biol.* 194:531–44
14. Lange OF, Lakomek NA, Fares C, Schroder GF, Walter KF, et al. 2008. Recognition dynamics up to microseconds revealed from an RDC-derived ubiquitin ensemble in solution. *Science* 320:1471–75
15. Dikic I, Wakatsuki S, Walters KJ. 2009. Ubiquitin-binding domains—from structures to functions. *Nat. Rev. Mol. Cell Biol.* 10:659–71
16. Shih SC, Sloper-Mould KE, Hicke L. 2000. Monoubiquitin carries a novel internalization signal that is appended to activated receptors. *EMBO J.* 19:187–98
17. Sloper-Mould KE, Jemc JC, Pickart CM, Hicke L. 2001. Distinct functional surface regions on ubiquitin. *J. Biol. Chem.* 276:30483–89
18. Kamadurai HB, Souphron J, Scott DC, Duda DM, Miller DJ, et al. 2009. Insights into ubiquitin transfer cascades from a structure of a UbcH5B~ubiquitin-HECT^{NEDD4L} complex. *Mol. Cell* 36:1095–102
19. Hu M, Li P, Li M, Li W, Yao T, et al. 2002. Crystal structure of a UBP-family deubiquitinating enzyme in isolation and in complex with ubiquitin aldehyde. *Cell* 111:1041–54
20. Reyes-Turcu FE, Horton JR, Mullally JE, Heroux A, Cheng X, Wilkinson KD. 2006. The ubiquitin binding domain ZnF UBP recognizes the C-terminal diglycine motif of unanchored ubiquitin. *Cell* 124:1197–208
21. Rahighi S, Ikeda F, Kawasaki M, Akutsu M, Suzuki N, et al. 2009. Specific recognition of linear ubiquitin chains by NEMO is important for NF-kappaB activation. *Cell* 136:1098–109
22. Ye Y, Akutsu M, Reyes-Turcu F, Enchev RI, Wilkinson KD, Komander D. 2011. Polyubiquitin binding and cross-reactivity in the USP domain deubiquitinase USP21. *EMBO Rep.* 12:350–57
23. Jin L, Williamson A, Banerjee S, Philipp I, Rape M. 2008. Mechanism of ubiquitin-chain formation by the human anaphase-promoting complex. *Cell* 133:653–65
24. Cui J, Yao Q, Li S, Ding X, Lu Q, et al. 2010. Glutamine deamidation and dysfunction of ubiquitin/NEDD8 induced by a bacterial effector family. *Science* 329:1215–18
25. Cook WJ, Jeffrey LC, Carson M, Chen Z, Pickart CM. 1992. Structure of a diubiquitin conjugate and a model for interaction with ubiquitin conjugating enzyme (E2). *J. Biol. Chem.* 267:16467–71
26. Tenno T, Fujiwara K, Tochio H, Iwai K, Morita EH, et al. 2004. Structural basis for distinct roles of Lys63- and Lys48-linked polyubiquitin chains. *Genes Cells* 9:865–75
27. Varadan R, Walker O, Pickart C, Fushman D. 2002. Structural properties of polyubiquitin chains in solution. *J. Mol. Biol.* 324:637–47
28. Eddins MJ, Varadan R, Fushman D, Pickart CM, Wolberger C. 2007. Crystal structure and solution NMR studies of Lys48-linked tetraubiquitin at neutral pH. *J. Mol. Biol.* 367:204–11
29. Ryabov Y, Fushman D. 2006. Interdomain mobility in di-ubiquitin revealed by NMR. *Proteins* 63:787–96
30. Bremm A, Freund SM, Komander D. 2010. Lys11-linked ubiquitin chains adopt compact conformations and are preferentially hydrolyzed by the deubiquitinase Cezanne. *Nat. Struct. Mol. Biol.* 17:939–47

31. Matsumoto ML, Wickliffe KE, Dong KC, Yu C, Bosanac I, et al. 2010. K11-linked polyubiquitination in cell cycle control revealed by a K11 linkage-specific antibody. *Mol. Cell* 39:477–84
32. Virdee S, Ye Y, Nguyen DP, Komander D, Chin JW. 2010. Engineered diubiquitin synthesis reveals Lys29-isopeptide specificity of an OTU deubiquitinase. *Nat. Chem. Biol.* 6:750–57
33. Bremm A, Komander D. 2011. Emerging roles for Lys11-linked polyubiquitin in cellular regulation. *Trends Biochem. Sci.* 36:355–63
34. Varadan R, Assfalg M, Haririnia A, Raasi S, Pickart C, Fushman D. 2004. Solution conformation of Lys63-linked di-ubiquitin chain provides clues to functional diversity of polyubiquitin signaling. *J. Biol. Chem.* 279:7055–63
35. Komander D, Reyes-Turcu F, Licchesi JD, Odenwaelder P, Wilkinson KD, Barford D. 2009. Molecular discrimination of structurally equivalent Lys 63-linked and linear polyubiquitin chains. *EMBO Rep.* 10:466–73
36. Weeks SD, Grasty KC, Hernandez-Cuebas L, Loll PJ. 2009. Crystal structures of Lys-63-linked tri- and di-ubiquitin reveal a highly extended chain architecture. *Proteins* 77:753–59
37. Datta AB, Hura GL, Wolberger C. 2009. The structure and conformation of Lys63-linked tetraubiquitin. *J. Mol. Biol.* 392:1117–24
38. Sims JJ, Cohen RE. 2009. Linkage-specific avidity defines the lysine 63-linked polyubiquitin-binding preference of rap80. *Mol. Cell* 33:775–83
39. Wang H, Wang L, Erdjument-Bromage H, Vidal M, Tempst P, et al. 2004. Role of histone H2A ubiquitination in Polycomb silencing. *Nature* 431:873–78
40. Bentley ML, Corn JE, Dong KC, Phung Q, Cheung TK, Cochran AG. 2011. Recognition of UbcH5c and the nucleosome by the Bmi1/Ring1b ubiquitin ligase complex. *EMBO J.* 30:3285–97
41. Hibbert RG, Huang A, Boelens R, Sixma TK. 2011. E3 ligase Rad18 promotes monoubiquitination rather than ubiquitin chain formation by E2 enzyme Rad6. *Proc. Natl. Acad. Sci. USA* 108:5590–95
42. Hwang CS, Shemorry A, Auerbach D, Varshavsky A. 2010. The N-end rule pathway is mediated by a complex of the RING-type Ubr1 and HECT-type Ufd4 ubiquitin ligases. *Nat. Cell Biol.* 12:1177–85
43. Kim J, Guermah M, McGinty RK, Lee JS, Tang Z, et al. 2009. RAD6-mediated transcription-coupled H2B ubiquitylation directly stimulates H3K4 methylation in human cells. *Cell* 137:459–71
44. Alpi AF, Pace PE, Babu MM, Patel KJ. 2008. Mechanistic insight into site-restricted monoubiquitination of FANCD2 by Ube2t, FANCL, and FANCI. *Mol. Cell* 32:767–77
45. Machida YJ, Machida Y, Chen Y, Gurtan AM, Kupfer GM, et al. 2006. UBE2T is the E2 in the Fanconi anemia pathway and undergoes negative autoregulation. *Mol. Cell* 23:589–96
46. Christensen DE, Brzovic PS, Klevit RE. 2007. E2-BRCA1 RING interactions dictate synthesis of mono- or specific polyubiquitin chain linkages. *Nat. Struct. Mol. Biol.* 14:941–48
47. Scaglione KM, Zavodszky E, Todi SV, Patury S, Xu P, et al. 2011. Ube2w and ataxin-3 coordinately regulate the ubiquitin ligase CHIP. *Mol. Cell* 43:599–612
48. Kim HT, Kim KP, Lledias F, Kisselev AF, Scaglione KM, et al. 2007. Certain pairs of ubiquitin-conjugating enzymes (E2s) and ubiquitin-protein ligases (E3s) synthesize nondegradable forked ubiquitin chains containing all possible isopeptide linkages. *J. Biol. Chem.* 282:17375–86
49. Zhang M, Windheim M, Roe SM, Peggie M, Cohen P, et al. 2005. Chaperoned ubiquitylation-crystal structures of the CHIP U box E3 ubiquitin ligase and a CHIP-Ubc13-Uev1a complex. *Mol. Cell* 20:525–38
50. Petroski MD, Deshaies RJ. 2005. Mechanism of lysine 48-linked ubiquitin-chain synthesis by the cullin-RING ubiquitin-ligase complex SCF-Cdc34. *Cell* 123:1107–20
51. Li W, Tu D, Brunger AT, Ye Y. 2007. A ubiquitin ligase transfers preformed polyubiquitin chains from a conjugating enzyme to a substrate. *Nature* 446:333–37
52. Williamson A, Wickliffe KE, Mellone BG, Song L, Karpen GH, Rape M. 2009. Identification of a physiological E2 module for the human anaphase-promoting complex. *Proc. Natl. Acad. Sci. USA* 106:18213–18
53. Williamson A. 2011. Regulation of ubiquitin chain initiation to determine the timing of substrate degradation. *Mol. Cell* 42:744–57
54. Wu K, Kovacev J, Pan ZQ. 2010. Priming and extending: a UbcH5/Cdc34 E2 handoff mechanism for polyubiquitination on a SCF substrate. *Mol. Cell* 37:784–96

55. Kirkpatrick DS, Hathaway NA, Hanna J, Elsasser S, Rush J, et al. 2006. Quantitative analysis of in vitro ubiquitinated cyclin B1 reveals complex chain topology. *Nat. Cell Biol.* 8:700–10
56. Wu T, Merbl Y, Huo Y, Gallop JL, Tzur A, Kirschner MW. 2010. UBE2S drives elongation of K11-linked ubiquitin chains by the anaphase-promoting complex. *Proc. Natl. Acad. Sci. USA* 107:1355–60
57. Rodrigo-Brenni MC, Foster SA, Morgan DO. 2010. Catalysis of lysine 48-specific ubiquitin chain assembly by residues in E2 and ubiquitin. *Mol. Cell* 39:548–59
58. Eddins MJ, Carlile CM, Gomez KM, Pickart CM, Wolberger C. 2006. Mms2-Ubc13 covalently bound to ubiquitin reveals the structural basis of linkage-specific polyubiquitin chain formation. *Nat. Struct. Mol. Biol.* 13:915–20
59. David Y, Ziv T, Admon A, Navon A. 2010. The E2 ubiquitin-conjugating enzymes direct polyubiquitination to preferred lysines. *J. Biol. Chem.* 285:8595–604
60. Deng L, Wang C, Spencer E, Yang L, Braun A, et al. 2000. Activation of the IkappaB kinase complex by TRAF6 requires a dimeric ubiquitin-conjugating enzyme complex and a unique polyubiquitin chain. *Cell* 103:351–61
61. Wickliffe KE, Lorenz S, Wemmer DE, Kuriyan J, Rape M. 2011. The mechanism of linkage-specific ubiquitin chain elongation by a single-subunit E2. *Cell* 144:769–81
62. Bosanac I, Phu L, Pan B, Zilberleyb I, Maurer B, et al. 2011. Modulation of K11-linkage formation by variable loop residues within UbcH5A. *J. Mol. Biol.* 408:420–31
63. Sadowski M, Suryadinata R, Lai X, Heierhorst J, Sarcevic B. 2010. Molecular basis for lysine specificity in the yeast ubiquitin-conjugating enzyme Cdc34. *Mol. Cell. Biol.* 30:2316–29
64. Hofmann RM, Pickart CM. 2001. In vitro assembly and recognition of Lys-63 polyubiquitin chains. *J. Biol. Chem.* 276:27936–43
65. Yunus AA, Lima CD. 2006. Lysine activation and functional analysis of E2-mediated conjugation in the SUMO pathway. *Nat. Struct. Mol. Biol.* 13:491–99
66. Wenzel DM, Lissounov A, Brzovic PS, Klevit RE. 2011. UBCH7 reactivity profile reveals parkin and HHARI to be RING/HECT hybrids. *Nature* 474:105–8
67. Rotin D, Kumar S. 2009. Physiological functions of the HECT family of ubiquitin ligases. *Nat. Rev. Mol. Cell Biol.* 10:398–409
68. Kim HC, Huibregtse JM. 2009. Polyubiquitination by HECT E3s and the determinants of chain type specificity. *Mol. Cell. Biol.* 29:3307–18
69. Maspero E, Mari S, Valentini E, Musacchio A, Fish A, et al. 2011. Structure of the HECT:ubiquitin complex and its role in ubiquitin chain elongation. *EMBO Rep.* 12:342–49
70. Wang M, Pickart CM. 2005. Different HECT domain ubiquitin ligases employ distinct mechanisms of polyubiquitin chain synthesis. *EMBO J.* 24:4324–33
71. Lin DY, Diao J, Zhou D, Chen J. 2011. Biochemical and structural studies of a HECT-like ubiquitin ligase from *Escherichia coli* O157:H7. *J. Biol. Chem.* 286:441–49
72. Scheffner M, Nuber U, Huibregtse JM. 1995. Protein ubiquitination involving an E1-E2-E3 enzyme ubiquitin thioester cascade. *Nature* 373:81–83
73. Huang L, Kinnucan E, Wang G, Beaudenon S, Howley PM, et al. 1999. Structure of an E6AP-UbcH7 complex: insights into ubiquitination by the E2-E3 enzyme cascade. *Science* 286:1321–26
74. Kim HC, Steffen AM, Oldham ML, Chen J, Huibregtse JM. 2011. Structure and function of a HECT domain ubiquitin-binding site. *EMBO Rep.* 12:334–41
75. Ogunjimi AA, Wiesner S, Briant DJ, Varelas X, Sicheri F, et al. 2010. The ubiquitin binding region of the Smurf HECT domain facilitates polyubiquitylation and binding of ubiquitylated substrates. *J. Biol. Chem.* 285:6308–15
76. Ikeda F, Deribe YL, Skanland SS, Stieglitz B, Grabbe C, et al. 2011. SHARPIN forms a linear ubiquitin ligase complex regulating NF-kappaB activity and apoptosis. *Nature* 471:637–41
77. Tokunaga F, Nakagawa T, Nakahara M, Saeki Y, Taniguchi M, et al. 2011. SHARPIN is a component of the NF-kB activating linear ubiquitin chain assembly complex. *Nature* 471:633–36
78. Tokunaga F, Iwai K. 2009. [Involvement of LUBAC-mediated linear polyubiquitination of NEMO in NF-kappaB activation]. *Tanpakushitsu Kakusan Koso* 54:635–42
79. Doss-Pepe EW, Chen L, Madura K. 2005. Alpha-synuclein and parkin contribute to the assembly of ubiquitin lysine 63-linked multiubiquitin chains. *J. Biol. Chem.* 280:16619–24

80. Geisler S, Holmstrom KM, Skujat D, Fiesel FC, Rothfuss OC, et al. 2010. PINK1/Parkin-mediated mitophagy is dependent on VDAC1 and p62/SQSTM1. *Nat. Cell Biol.* 12:119–31
81. Haldeman MT, Xia G, Kasperek EM, Pickart CM. 1997. Structure and function of ubiquitin conjugating enzyme E2-25K: The tail is a core-dependent activity element. *Biochemistry* 36:10526–37
82. Tokunaga F, Sakata S, Saeki Y, Satomi Y, Kirisako T, et al. 2009. Involvement of linear polyubiquitylation of NEMO in NF-kappaB activation. *Nat. Cell Biol.* 11:123–32
83. Sato Y, Yoshikawa A, Mimura H, Yamashita M, Yamagata A, Fukai S. 2009. Structural basis for specific recognition of Lys 63-linked polyubiquitin chains by tandem UIMs of RAP80. *EMBO J.* 28:2461–68
84. Swanson KA, Kang RS, Stamenova SD, Hicke L, Radhakrishnan I. 2003. Solution structure of Vps27 UIM-ubiquitin complex important for endosomal sorting and receptor downregulation. *EMBO J.* 22:4597–606
85. Kanayama A, Seth RB, Sun L, Ea C-K, Hong M, et al. 2004. TAB2 and TAB3 activate the NF-kappaB pathway through binding to polyubiquitin chains. *Mol. Cell* 15:535–48
86. Sato Y, Yoshikawa A, Yamashita M, Yamagata A, Fukai S. 2009. Structural basis for specific recognition of Lys 63-linked polyubiquitin chains by NZF domains of TAB2 and TAB3. *EMBO J.* 28:3903–9
87. Kulathu Y, Akutsu M, Bremm A, Hofmann K, Komander D. 2009. Two-sided ubiquitin binding explains specificity of the TAB2 NZF domain. *Nat. Struct. Mol. Biol.* 16:1328–30
88. Trempe JF, Brown NR, Lowe ED, Gordon C, Campbell ID, et al. 2005. Mechanism of Lys48-linked polyubiquitin chain recognition by the Mud1 UBA domain. *EMBO J.* 24:3178–89
89. Varadan R, Assfalg M, Raasi S, Pickart C, Fushman D. 2005. Structural determinants for selective recognition of a Lys48-linked polyubiquitin chain by a UBA domain. *Mol. Cell* 18:687–98
90. Newton K, Matsumoto ML, Wertz IE, Kirkpatrick DS, Lill JR, et al. 2008. Ubiquitin chain editing revealed by polyubiquitin linkage-specific antibodies. *Cell* 134:668–78
91. Skaug B, Jiang X, Chen ZJ. 2009. The role of ubiquitin in NF-kappaB regulatory pathways. *Annu. Rev. Biochem.* 78:769–96
92. Ivins FJ, Montgomery MG, Smith SJ, Morris-Davies AC, Taylor IA, Rittinger K. 2009. NEMO oligomerization and its ubiquitin-binding properties. *Biochem. J.* 421:243–51
93. Yoshikawa A, Sato Y, Yamashita M, Mimura H, Yamagata A, Fukai S. 2009. Crystal structure of the NEMO ubiquitin-binding domain in complex with Lys 63-linked di-ubiquitin. *FEBS Lett.* 583:3317–22
94. Bosanac I, Wertz IE, Pan B, Yu C, Kusam S, et al. 2010. Ubiquitin binding to A20 ZnF4 is required for modulation of NF-kappaB signaling. *Mol. Cell* 40:548–57
95. Pickart CM. 2001. Mechanisms underlying ubiquitination. *Annu. Rev. Biochem.* 70:503–33
96. Komander D, Clague MJ, Urbé S. 2009. Breaking the chains: structure and function of the deubiquitinases. *Nat. Rev. Mol. Cell Biol.* 10:550–63
97. Zeng W, Sun L, Jiang X, Chen X, Hou F, et al. 2010. Reconstitution of the RIG-I pathway reveals a signaling role of unanchored polyubiquitin chains in innate immunity. *Cell* 141:315–30
98. Xia ZP, Sun L, Chen X, Pineda G, Jiang X, et al. 2009. Direct activation of protein kinases by unanchored polyubiquitin chains. *Nature* 461:114–19
99. Reyes-Turcu FE, Shanks JR, Komander D, Wilkinson KD. 2008. Recognition of polyubiquitin isoforms by the multiple ubiquitin binding modules of isopeptidase T. *J. Biol. Chem.* 283:19581–92
100. Finley D. 2009. Recognition and processing of ubiquitin-protein conjugates by the proteasome. *Annu. Rev. Biochem.* 78:477–513
101. Hetfeld BK, Helfrich A, Kapelari B, Scheel H, Hofmann K, et al. 2005. The zinc finger of the CSN-associated deubiquitinating enzyme USP15 is essential to rescue the E3 ligase Rbx1. *Curr. Biol.* 15:1217–21
102. Ernst R, Mueller B, Ploegh HL, Schlieker C. 2009. The otubain YOD1 is a deubiquitinating enzyme that associates with p97 to facilitate protein dislocation from the ER. *Mol. Cell* 36:28–38
103. Uchiyama K, Jokitalo E, Kano F, Murata M, Zhang X, et al. 2002. VCIP135, a novel essential factor for p97/p47-mediated membrane fusion, is required for Golgi and ER assembly in vivo. *J. Cell Biol.* 159:855–66
104. Doss-Pepe EW, Stenroos ES, Johnson WG, Madura K. 2003. Ataxin-3 interactions with Rad23 and valosin-containing protein and its associations with ubiquitin chains and the proteasome are consistent with a role in ubiquitin-mediated proteolysis. *Mol. Cell. Biol.* 23:6469–83

105. Sowa ME, Bennett EJ, Gygi SP, Harper JW. 2009. Defining the human deubiquitinating enzyme interaction landscape. *Cell* 138:389–403
106. Komander D, Lord CJ, Scheel H, Swift S, Hofmann K, et al. 2008. The structure of the CYLD USP domain explains its specificity for Lys63-linked polyubiquitin and reveals a B box module. *Mol. Cell* 29:451–64
107. McCullough J, Clague MJ, Urbé S. 2004. AMSH is an endosome-associated ubiquitin isopeptidase. *J. Cell Biol.* 166:487–92
108. Sato Y, Yoshikawa A, Yamagata A, Mimura H, Yamashita M, et al. 2008. Structural basis for specific cleavage of Lys 63-linked polyubiquitin chains. *Nature* 455:358–62
109. Cooper EM, Cutcliffe C, Kristiansen TZ, Pandey A, Pickart CM, Cohen RE. 2009. K63-specific deubiquitination by two JAMM/MPN +complexes: BRISC-associated Brcc36 and proteasomal Poh1. *EMBO J.* 28:621–31
110. Edelmann MJ, Iphofer A, Akutsu M, Altun M, di Gleria K, et al. 2009. Structural basis and specificity of human otubain 1-mediated deubiquitination. *Biochem. J.* 418:379–90
111. Wang T, Yin L, Cooper EM, Lai MY, Dickey S, et al. 2009. Evidence for bidentate substrate binding as the basis for the K48 linkage specificity of otubain 1. *J. Mol. Biol.* 386:1011–23
112. Licchesi JD, Mieszczanek J, Mevissen TET, Rutherford TJ, Akutsu M, et al. 2011. An ankyrin-repeat ubiquitin-binding domain determines TRABID's specificity for atypical ubiquitin chains. *Nat. Struct. Mol. Biol.* 19:62–71
113. Weeks SD, Grasty KC, Hernandez-Cuebas L, Loll PJ. 2011. Crystal structure of a Josephin-ubiquitin complex: evolutionary restraints on ataxin-3 deubiquitinating activity. *J. Biol. Chem.* 286:4555–65
114. Johnston SC, Riddle SM, Cohen RE, Hill CP. 1999. Structural basis for the specificity of ubiquitin C-terminal hydrolases. *EMBO J.* 18:3877–87
115. Messick TE, Russell NS, Iwata AJ, Sarachan KL, Shiekhattar R, et al. 2008. Structural basis for ubiquitin recognition by the Otu1 ovarian tumor domain protein. *J. Biol. Chem.* 283:11038–49
116. Wertz IE, O'Rourke KM, Zhou H, Eby M, Aravind L, et al. 2004. De-ubiquitination and ubiquitin ligase domains of A20 downregulate NF-kappaB signalling. *Nature* 430:694–99
117. Lin SC, Chung JY, Lamothe B, Rajashankar K, Lu M, et al. 2008. Molecular basis for the unique deubiquitinating activity of the NF-kappaB inhibitor A20. *J. Mol. Biol.* 376:526–40
118. Komander D, Barford D. 2008. Structure of the A20 OTU domain and mechanistic insights into deubiquitination. *Biochem. J.* 409:77–85
119. Shembade N, Harhaj NS, Liebl DJ, Harhaj EW. 2007. Essential role for TAX1BP1 in the termination of TNF-alpha-, IL-1- and LPS-mediated NF-kappaB and JNK signaling. *EMBO J.* 26:3910–22
120. Heyninck K, De Valck D, Vanden Berghe W, Van Criekinge W, Contreras R, et al. 1999. The zinc finger protein A20 inhibits TNF-induced NF-kappaB-dependent gene expression by interfering with an RIP- or TRAF2-mediated transactivation signal and directly binds to a novel NF-kappaB-inhibiting protein ABIN. *J. Cell Biol.* 145:1471–82
121. Shembade N, Harhaj NS, Parvatiyar K, Copeland NG, Jenkins NA, et al. 2008. The E3 ligase Itch negatively regulates inflammatory signaling pathways by controlling the function of the ubiquitin-editing enzyme A20. *Nat. Immunol.* 9:254–62
122. Shembade N, Parvatiyar K, Harhaj NS, Harhaj EW. 2009. The ubiquitin-editing enzyme A20 requires RNF11 to downregulate NF-kappaB signalling. *EMBO J.* 28:513–22
123. Shembade N, Ma A, Harhaj EW. 2010. Inhibition of NF-kappaB signaling by A20 through disruption of ubiquitin enzyme complexes. *Science* 327:1135–39
124. Flick K, Raasi S, Zhang H, Yen JL, Kaiser P. 2006. A ubiquitin-interacting motif protects polyubiquitinated Met4 from degradation by the 26S proteasome. *Nat. Cell Biol.* 8:509–15
125. Kim W, Bennett EJ, Huttlin EL, Guo A, Li J, et al. 2011. Systematic and quantitative assessment of the ubiquitin-modified proteome. *Mol. Cell* 44:325–40
126. Emanuele MJ, Elia AE, Xu Q, Thoma CR, Izhar L, et al. 2011. Global identification of modular cullin-RING ligase substrates. *Cell* 147:459–74
127. Chau V, Tobias JW, Bachmair A, Marriott D, Ecker DJ, et al. 1989. A multiubiquitin chain is confined to specific lysine in a targeted short-lived protein. *Science* 243:1576–83

128. Kaiser SE, Riley BE, Shaler TA, Trevino RS, Becker CH, et al. 2011. Protein standard absolute quantification (PSAQ) method for the measurement of cellular ubiquitin pools. *Nat. Methods* 8:691–96
129. Baboshina OV, Haas AL. 1996. Novel multiubiquitin chain linkages catalyzed by the conjugating enzymes E2EPF and RAD6 are recognized by 26 S proteasome subunit 5. *J. Biol. Chem.* 271:2823–31
130. Thrower JS, Hoffman L, Rechsteiner M, Pickart CM. 2000. Recognition of the polyubiquitin proteolytic signal. *EMBO J.* 19:94–102
131. Johnson ES, Ma PC, Ota IM, Varshavsky A. 1995. A proteolytic pathway that recognizes ubiquitin as a degradation signal. *J. Biol. Chem.* 270:17442–56
132. Koegl M, Hoppe T, Schlenker S, Ulrich HD, Mayer TU, Jentsch S. 1999. A novel ubiquitination factor, E4, is involved in multiubiquitin chain assembly. *Cell* 96:635–44
133. Saeki Y, Kudo T, Sone T, Kikuchi Y, Yokosawa H, et al. 2009. Lysine 63-linked polyubiquitin chain may serve as a targeting signal for the 26S proteasome. *EMBO J.* 28:359–71
134. Husnjak K, Elsasser S, Zhang N, Chen X, Randles L, et al. 2008. Proteasome subunit Rpn13 is a novel ubiquitin receptor. *Nature* 453:481–88
135. Zhang D, Chen T, Ziv I, Rosenzweig R, Matiuhin Y, et al. 2009. Together, Rpn10 and Dsk2 can serve as a polyubiquitin chain-length sensor. *Mol. Cell* 36:1018–33
136. Sims JJ, Haririnia A, Dickinson BC, Fushman D, Cohen RE. 2009. Avid interactions underlie the Lys63-linked polyubiquitin binding specificities observed for UBA domains. *Nat. Struct. Mol. Biol.* 16:883–89
137. Verma R, Chen S, Feldman R, Schieltz D, Yates J, et al. 2000. Proteasomal proteomics: identification of nucleotide-sensitive proteasome-interacting proteins by mass spectrometric analysis of affinity-purified proteasomes. *Mol. Biol. Cell* 11:3425–39
138. Seeger M, Hartmann-Petersen R, Wilkinson CR, Wallace M, Samejima I, et al. 2003. Interaction of the anaphase-promoting complex/cyclosome and proteasome protein complexes with multiubiquitin chain-binding proteins. *J. Biol. Chem.* 278:16791–96
139. Xie Y, Varshavsky A. 2002. UFD4 lacking the proteasome-binding region catalyses ubiquitination but is impaired in proteolysis. *Nat. Cell Biol.* 4:1003–7
140. Mukhopadhyay D, Riezman H. 2007. Proteasome-independent functions of ubiquitin in endocytosis and signaling. *Science* 315:201–5
141. Terrell J, Shih S, Dunn R, Hicke L. 1998. A function for monoubiquitination in the internalization of a G protein–coupled receptor. *Mol. Cell* 1:193–202
142. Stringer DK, Piper RC. 2011. A single ubiquitin is sufficient for cargo protein entry into MVBs in the absence of ESCRT ubiquitination. *J. Cell Biol.* 192:229–42
143. Huang F, Kirkpatrick D, Jiang X, Gygi S, Sorkin A. 2006. Differential regulation of EGF receptor internalization and degradation by multiubiquitination within the kinase domain. *Mol. Cell* 21:737–48
144. Huang F, Goh LK, Sorkin A. 2007. EGF receptor ubiquitination is not necessary for its internalization. *Proc. Natl. Acad. Sci. USA* 104:16904–9
145. Mizuno E, Iura T, Mukai A, Yoshimori T, Kitamura N, Komada M. 2005. Regulation of epidermal growth factor receptor down-regulation by UBPY-mediated deubiquitination at endosomes. *Mol. Biol. Cell* 16:5163–74
146. Clague MJ, Urbe S. 2006. Endocytosis: the DUB version. *Trends Cell Biol.* 16:551–59
147. Raiborg C, Stenmark H. 2009. The ESCRT machinery in endosomal sorting of ubiquitylated membrane proteins. *Nature* 458:445–52
148. Ren X, Hurley JH. 2010. VHS domains of ESCRT-0 cooperate in high-avidity binding to polyubiquitinated cargo. *EMBO J.* 29:1045–54
149. Kirkin V, McEwan DG, Novak I, Dikic I. 2009. A role for ubiquitin in selective autophagy. *Mol. Cell* 34:259–69
150. Moldovan GL, Pfander B, Jentsch S. 2007. PCNA, the maestro of the replication fork. *Cell* 129:665–79
151. Freudenthal BD, Gakhar L, Ramaswamy S, Washington MT. 2010. Structure of monoubiquitinated PCNA and implications for translesion synthesis and DNA polymerase exchange. *Nat. Struct. Mol. Biol.* 17:479–84
152. Bienko M, Green CM, Crosetto N, Rudolf F, Zapart G, et al. 2005. Ubiquitin-binding domains in Y-family polymerases regulate translesion synthesis. *Science* 310:1821–24

153. Bienko M, Green CM, Sabbioneda S, Crosetto N, Matic I, et al. 2010. Regulation of translesion synthesis DNA polymerase η by monoubiquitination. *Mol. Cell* 37:396–407
154. Huang TT, Nijman SM, Mirchandani KD, Galardy PJ, Cohn MA, et al. 2006. Regulation of monoubiquitinated PCNA by DUB autocleavage. *Nat. Cell Biol.* 8:339–47
155. Huang M, D'Andrea AD. 2010. A new nuclease member of the FAN club. *Nat. Struct. Mol. Biol.* 17:926–28
156. Joo W, Xu G, Persky NS, Smogorzewska A, Rudge DG, et al. 2011. Structure of the FANCI-FANCD2 complex: insights into the Fanconi anemia DNA repair pathway. *Science* 333:312–16
157. Moldovan GL, D'Andrea AD. 2009. How the Fanconi anemia pathway guards the genome. *Annu. Rev. Genet.* 43:223–49
158. Nijman SM, Huang TT, Dirac AM, Brummelkamp TR, Kerkhoven RM, et al. 2005. The deubiquitinating enzyme USP1 regulates the Fanconi anemia pathway. *Mol. Cell* 17:331–39
159. Bellare P, Small EC, Huang X, Wohlschlegel JA, Staley JP, Sontheimer EJ. 2008. A role for ubiquitin in the spliceosome assembly pathway. *Nat. Struct. Mol. Biol.* 15:444–51
160. Song EJ, Werner SL, Neubauer J, Stegmeier F, Aspden J, et al. 2010. The Prp19 complex and the Usp4Sart3 deubiquitinating enzyme control reversible ubiquitination at the spliceosome. *Genes Dev.* 24:1434–47
161. Spence J, Gali RR, Dittmar G, Sherman F, Karin M, Finley D. 2000. Cell cycle-regulated modification of the ribosome by a variant multiubiquitin chain. *Cell* 102:67–76
162. Al-Hakim A, Escribano-Diaz C, Landry MC, O'Donnell L, Panier S, et al. 2010. The ubiquitous role of ubiquitin in the DNA damage response. *DNA Repair* 9:1229–40
163. Doil C, Mailand N, Bekker-Jensen S, Menard P, Larsen DH, et al. 2009. RNF168 binds and amplifies ubiquitin conjugates on damaged chromosomes to allow accumulation of repair proteins. *Cell* 136:435–46
164. Huang J, Huen MS, Kim H, Leung CC, Glover JN, et al. 2009. RAD18 transmits DNA damage signalling to elicit homologous recombination repair. *Nat. Cell Biol.* 11:592–603
165. Sobhian B, Shao G, Lilli DR, Culhane AC, Moreau LA, et al. 2007. RAP80 targets BRCA1 to specific ubiquitin structures at DNA damage sites. *Science* 316:1198–202
166. Stewart GS, Panier S, Townsend K, Al-Hakim AK, Kolas NK, et al. 2009. The RIDDLE syndrome protein mediates a ubiquitin-dependent signaling cascade at sites of DNA damage. *Cell* 136:420–34
167. Hou F, Sun L, Zheng H, Skaug B, Jiang QX, Chen ZJ. 2011. MAVS forms functional prion-like aggregates to activate and propagate antiviral innate immune response. *Cell* 146:448–61
168. Dupont S, Mamidi A, Cordenonsi M, Montagner M, Zacchigna L, et al. 2009. FAM/USP9x, a deubiquitinating enzyme essential for TGFbeta signaling, controls Smad4 monoubiquitination. *Cell* 136:123–35
169. Polo S, Sigismund S, Faretta M, Guidi M, Capua MR, et al. 2002. A single motif responsible for ubiquitin recognition and monoubiquitination in endocytic proteins. *Nature* 416:451–55
170. Hoeller D, Crosetto N, Blagoev B, Raiborg C, Tikkanen R, et al. 2006. Regulation of ubiquitin-binding proteins by monoubiquitination. *Nat. Cell Biol.* 8:163–69
171. Hoeller D, Hecker CM, Wagner S, Rogov V, Dotsch V, Dikic I. 2007. E3-independent monoubiquitination of ubiquitin-binding proteins. *Mol. Cell* 26:891–98
172. Winston JT, Strack P, Beer-Romero P, Chu CY, Elledge SJ, Harper JW. 1999. The SCFbeta-TRCP-ubiquitin ligase complex associates specifically with phosphorylated destruction motifs in IkappaBalpha and beta-catenin and stimulates IkappaBalpha ubiquitination in vitro. *Genes Dev.* 13:270–83
173. Margottin-Goguet F, Hsu JY, Loktev A, Hsieh HM, Reimann JDR, Jackson PK. 2003. Prophase destruction of Emi1 by the SCF$^{\beta TrCP/Slimb}$ ubiquitin ligase activates the anaphase promoting complex to allow progression beyond prometaphase. *Dev. Cell* 4:813–26
174. Palombella VJ, Rando OJ, Goldberg AL, Maniatis T. 1994. The ubiquitin-proteasome pathway is required for processing the NF-kappa B1 precursor protein and the activation of NF-kappa B. *Cell* 78:773–85
175. Hoppe T, Matuschewski K, Rape M, Schlenker S, Ulrich HD, Jentsch S. 2000. Activation of a membrane-bound transcription factor by regulated ubiquitin/proteasome-dependent processing. *Cell* 102:577–86
176. Stewart EV, Nwosu CC, Tong Z, Roguev A, Cummins TD, et al. 2011. Yeast SREBP cleavage activation requires the Golgi Dsc E3 ligase complex. *Mol. Cell* 42:160–71

177. Kravtsova-Ivantsiv Y, Cohen S, Ciechanover A. 2009. Modification by single ubiquitin moieties rather than polyubiquitination is sufficient for proteasomal processing of the p105 NF-kappaB precursor. *Mol. Cell* 33:496–504
178. Rape M, Hoppe T, Gorr I, Kalocay M, Richly H, Jentsch S. 2001. Mobilization of processed, membrane-tethered SPT23 transcription factor by CDC48$^{UFD1/NPL4}$, a ubiquitin-selective chaperone. *Cell* 107:667–77
179. Xu M, Skaug B, Zeng W, Chen ZJ. 2009. A ubiquitin replacement strategy in human cells reveals distinct mechanisms of IKK activation by TNFalpha and IL-1beta. *Mol. Cell* 36:302–14
180. Laplantine E, Fontan E, Chiaravalli J, Lopez T, Lakisic G, et al. 2009. NEMO specifically recognizes K63-linked poly-ubiquitin chains through a new bipartite ubiquitin-binding domain. *EMBO J.* 28:2885–95
181. Wang C, Deng L, Hong M, Akkaraju GR, Inoue J, Chen ZJ. 2001. TAK1 is a ubiquitin-dependent kinase of MKK and IKK. *Nature* 412:346–51
182. Plafker SM, Plafker KS, Weissman AM, Macara IG. 2004. Ubiquitin charging of human class III ubiquitin-conjugating enzymes triggers their nuclear import. *J. Cell Biol.* 167:649–59
183. Li M, Brooks CL, Wu-Baer F, Chen D, Baer R, Gu W. 2003. Mono- versus polyubiquitination: differential control of p53 fate by Mdm2. *Science* 302:1972–75
184. Yuan J, Luo K, Zhang L, Cheville JC, Lou Z. 2010. USP10 regulates p53 localization and stability by deubiquitinating p53. *Cell* 140:384–96
185. Ryabov YE, Fushman D. 2007. A model of interdomain mobility in a multidomain protein. *J. Am. Chem. Soc.* 129:3315–27
186. Datta AB, Hura GL, Wolberger C. 2009. The structure and conformation of Lys63-linked tetraubiquitin. *J. Mol. Biol.* 392:1117–24

Ubiquitin and Membrane Protein Turnover: From Cradle to Grave

Jason A. MacGurn, Pi-Chiang Hsu, and Scott D. Emr

Weill Institute for Cell and Molecular Biology, Cornell University, Ithaca, New York 14853; email: sde26@cornell.edu

Keywords

endocytosis, quality control, protein trafficking, ubiquitin ligase

Abstract

From the moment of cotranslational insertion into the lipid bilayer of the endoplasmic reticulum (ER), newly synthesized integral membrane proteins are subject to a complex series of sorting, trafficking, quality control, and quality maintenance systems. Many of these processes are intimately controlled by ubiquitination, a posttranslational modification that directs trafficking decisions related to both the biosynthetic delivery of proteins to the plasma membrane (PM) via the secretory pathway and the removal of proteins from the PM via the endocytic pathway. Ubiquitin modification of integral membrane proteins (or "cargoes") generally acts as a sorting signal, which is recognized, captured, and delivered to a specific cellular destination via specialized trafficking events. By affecting the quality, quantity, and localization of integral membrane proteins in the cell, defects in these processes contribute to human diseases, including cystic fibrosis, circulatory diseases, and various neuropathies. This review summarizes our current understanding of how ubiquitin modification influences cargo trafficking, with a special emphasis on mechanisms of quality control and quality maintenance in the secretory and endocytic pathways.

Contents

INTRODUCTION	232
THE CRADLE: UBIQUITIN AND QUALITY CONTROL IN THE SECRETORY PATHWAY	235
Quality Control at the Endoplasmic Reticulum	235
Ubiquitination and Cargo Sorting at the Golgi Complex	236
Quality Control at the Golgi Complex	237
TO THE GRAVE: UBIQUITIN-MEDIATED TURNOVER OF PLASMA MEMBRANE PROTEINS	237
Ubiquitin-Mediated Endocytosis	237
Sorting of Ubiquitinated Cargoes on the Endosome	239
Ubiquitin Editing at the Plasma Membrane and Endosome	240
THE ROLE OF UBIQUITIN IN CELL SURFACE REMODELING	241
Regulation of Signaling Receptors by Ubiquitin	241
Ubiquitin-Mediated Downregulation of Ion Channels	243
Ubiquitin-Mediated Downregulation of Transporters	245
QUALITY MAINTENANCE AT THE PLASMA MEMBRANE	248

INTRODUCTION

The abundance and localization of integral membrane proteins—including nutrient transporters, ion channels, and signaling receptors—play a critical role in the growth, differentiation, and survival of eukaryotic cells. The complex process of remodeling cell surface protein composition involves both biosynthetic addition of new integral membrane proteins (or cargoes) to the plasma membrane (PM) and removal of cargoes from the PM by endocytosis. A key regulator of the sorting, trafficking, and turnover of integral membrane proteins is ubiquitination.

Ubiquitin is a 76-amino acid peptide that can be conjugated to other proteins by the formation of an isopeptide bond between the C-terminal glycine carboxy group of ubiquitin and the ε-amino group of a lysine residue on the recipient protein (1, 2). Ubiquitin conjugation, or ubiquitination, is the end result of three sequential enzymatic reactions catalyzed by a ubiquitin-activating enzyme (E1), a ubiquitin-conjugating enzyme (E2), and a ubiquitin ligase (E3) (3–5). A key feature of ubiquitination is that it is potentially recursive: Ubiquitin can modify itself on any one of its seven lysine residues (K6, K11, K27, K29, K33, K48, K63) or on its N terminus, leading to the formation of polyubiquitin chains that have different structures and properties depending on how the chains are assembled (1, 6).

E3 ubiquitin ligases mediate the addition of ubiquitin to substrates and thus determine the type and extent of ubiquitin modification. There are two major ubiquitin ligase families: (*a*) the RING (really interesting new gene) domain E3 ligases (616 in humans, 40 in yeast), which bind to E2-ubiquitin conjugates and substrates and provide optimal steric orientation for direct ubiquitin transfer to substrate (4, 7); and (*b*) the HECT (homologous to E6-AP C terminus) domain E3 ligases (28 in humans, 5 in yeast), which accept ubiquitin from E2 enzymes on catalytic cysteine residues and catalyze direct transfer of ubiquitin to substrates (2, 8, 9). The activity of ubiquitin ligases can be antagonized by deubiquitinating enzymes, or DUBs. DUBs are proteases that catalyze the cleavage of ubiquitin-lysine isopeptide bonds. There are roughly 100 DUBs encoded in the human genome (less than 20 in yeast), which can be classified on the basis of five known DUB protein families: the Ubiquitin-Specific Protease (USP) family, the Ubiquitin C-Terminal Hydrolase (UCH) family, the Ovarian Tumor domain (OTU) family, the Machado-Josephin domain (MJD) family, and the JAMM (JAB1/MPN/MOV34) domain family (10).

DUB: deubiquitinating enzyme

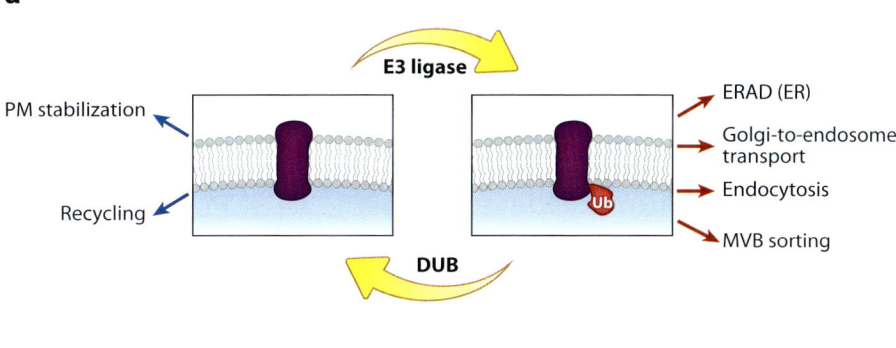
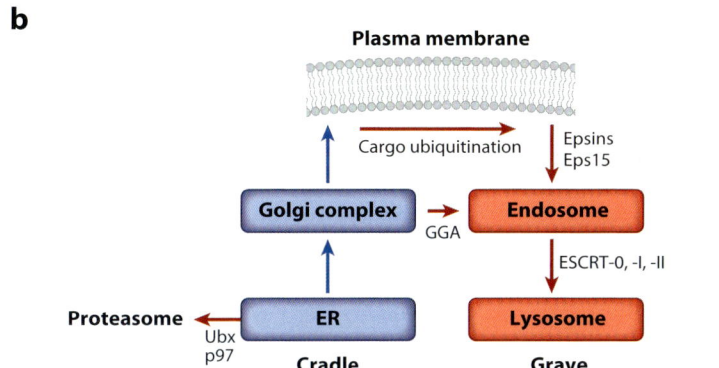

Figure 1

Common themes in ubiquitin-mediated membrane trafficking and turnover events. (*a*) The ubiquitin modification cycle of membrane protein sorting and trafficking. (*b*) Ubiquitin-binding proteins in the membrane protein trafficking circuit. Abbreviations: DUB, deubiquitinating enzyme; Eps15, epidermal growth factor receptor pathway 15; ER, endoplasmic reticulum; ERAD, endoplasmic reticulum–associated degradation; ESCRT, endosomal sorting complex required for transport; GGA, Golgi-localized, γ-ear-containing, ARF-binding protein; MVB, multivesicular body; PM, plasma membrane; Ubx, ubiquitin regulatory X domain-containing proteins; p97, an AAA ATPase that recognizes ubiquitinated substrates at the ER (via Ubx protein adaptors) and is thought to extract substrates from the ER membrane.

DUBs from the USP, UCH, OTU, and MJD families are cysteine proteases, whereas JAMM family DUBs are metalloproteases. The reversible nature of ubiquitination allows it to function as an important regulatory switch in the cell.

It is well established that protein K48-linked polyubiquitination of soluble, cytosolic proteins results in degradation by targeting to the proteasome and thus regulates protein stability (1, 11). However, integral membrane proteins are not directly accessible to the proteasome, and thus, eukaryotic cells have evolved elaborate mechanisms to monitor, sort, and degrade them. These mechanisms rely on membrane trafficking events in the cell that are often guided by ubiquitin modification of substrates (**Figure 1***a*). There is an emerging consensus that, although K48-linked polyubiquitin chains function in proteasomal degradation, ubiquitin-mediated membrane trafficking events in the cell are largely governed by monoubiquitination and K63-linked polyubiquitination (12, 13). Here, we review what is known about the role ubiquitination plays in the sorting, trafficking, and degradation of integral membrane proteins. Although autophagy can target organelles or protein aggregates

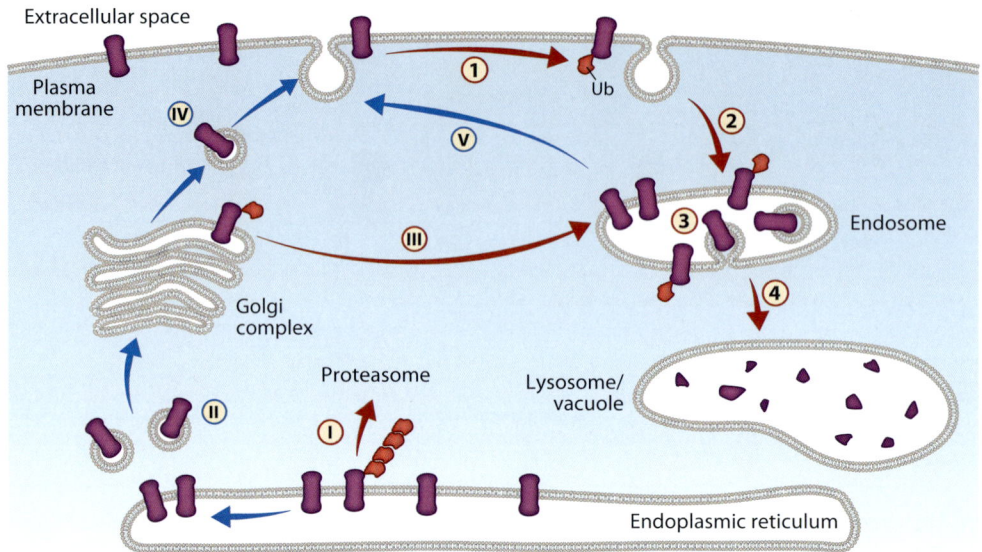

Secretion and quality control
I. ERAD of misfolded proteins
II. ER exit and traffic to Golgi complex
III. Golgi-to-endosome traffic
IV. Exocytosis
V. Endosomal recycling to cell surface

Targeting for lysosomal degradation
1. Cargo recognition and ubiquitination
2. Sorting and ubiquitin-mediated endocytosis
3. ESCRT-mediated sorting and MVB formation
4. Degradation in the lysosome

Figure 2
Ubiquitin-mediated membrane protein trafficking processes in the cell. Abbreviations: ER, endoplasmic reticulum; ERAD, endoplasmic reticulum–associated degradation; ESCRT, endosomal sorting complex required for transport; MVB, multivesicular body; PM, plasma membrane.

for lysosomal degradation in a ubiquitin-dependent manner (14–16), a specific role for autophagy in the turnover of PM proteins is less clear and thus is not discussed in this review. Instead, we place special emphasis on the role of ubiquitin in the quality control circuitry of integral membrane proteins from the cradle to the grave (**Figures 1b** and **2** and **Table 1**).

Table 1 The modular architecture of ubiquitin-dependent trafficking and turnover processes[a]

Processes	Endoplasmic reticulum[b]	Golgi complex	Plasma membrane	Endosome
Detection	BiP Hrd1 complex	Unknown	Hsp70/Hsp90 (m) ARTs (y)	Unknown
Ubiquitination	Hrd1, gp78 (m) Hrd1, Doa10 (y)	Rsp5, Tul1 (y)	Nedd4, Cbl, CHIP (m) Rsp5 (y)	Nedd4, Cbl (m) Rsp5 (y)
Recognition and sorting	Ubx proteins	GGAs	Epsins/Eps15 (m) Ents/Ede1 (y)	ESCRT-0
Degradation	Proteasome	Lysosome	Lysosome	Lysosome

[a]For mammalian (m) and yeast (y) systems, this table highlights the molecular machinery responsible for the detection, ubiquitination, sorting, and degradation of integral membrane proteins at different compartments in the cell.
[b]Abbreviations: ARTs, arrestin-related trafficking adaptors; GGAs, γ-ear-containing ARF-binding proteins; Ubx, ubiquitin regulatory X domain-containing proteins.

THE CRADLE: UBIQUITIN AND QUALITY CONTROL IN THE SECRETORY PATHWAY

Quality Control at the Endoplasmic Reticulum

In general, proteins that do not contain *cis*-acting endoplasmic reticulum (ER) retention sequences are packaged into coat protein complex II (COPII) vesicles and transported to the Golgi complex, provided such proteins are properly folded and are not retained in the ER by *trans*-acting factors. Unlike downstream compartments in the secretory and endocytic pathways, ubiquitin modification does not seem to play a direct role in regulating cargo sorting for ER exit. However, lumenal or integral membrane proteins that do not fold properly are retained in the ER and are subject to ER–associated degradation (ERAD). This elaborate process is initiated by substrate recognition, which can occur through several mechanisms, including prolonged association with ER chaperones or modified glycan processing [the Parodi enzyme (17, 18)]. Selected substrates are then retrotranslocated from the ER to the cytoplasm by a mechanism that is still poorly understood but requires the activity of multiprotein ER membrane complexes that have ubiquitin ligase activity (**Table 1**). During retrotranslocation, ERAD substrates are ubiquitinated and recognized by specialized adaptors called Ubx proteins (19, 20), which contain a ubiquitin-binding domain (UBD) for substrate recognition and a UBX domain to recruit the AAA ATPase p97 (Cdc48 in yeast). Engagement of p97 with the substrate and E3 ligase at the ER membrane is critical for retrotranslocation, and its ATPase activity likely supplies most of the energy for membrane extraction. Once the polyubiquitinated substrate is extracted, it can be degraded by the proteasome. Thus, ubiquitination of ERAD substrates is not a determinant of cargo selection or sorting but rather provides a recognition signal for cytosolic recruitment of the molecular motor that supplies energy for retrotranslocation and subsequently targets the protein for proteasomal degradation. This review does not attempt to cover the molecular mechanisms of ERAD, which have been extensively studied and are the subject of several recent reviews (21–25).

Because ERAD dysfunction results in escape of misfolded proteins into the secretory pathway (26–28), ERAD provides a crucial quality assurance step for proteins exiting the ER. In various human diseases, ERAD prevents surface expression of mutant proteins with decreased stability (see the sidebar titled Human Diseases Associated with Membrane Protein Ubiquitination and Turnover). Indeed, secretion of many proteins, even wild-type proteins, can be enhanced by the addition of chemical chaperones (29), suggesting that ERAD may sometimes degrade nascent integral membrane proteins to limit surface expression. Furthermore, some human pathogens, such as cytomegalovirus, hijack ERAD to prevent

ERAD: endoplasmic reticulum–associated degradation

Parodi enzyme: UDP-glucose: glycoprotein glucosyltransferase in the ER that detects misfolded glycoproteins and tags them by attaching a single glucose moiety

HUMAN DISEASES ASSOCIATED WITH MEMBRANE PROTEIN UBIQUITINATION AND TURNOVER

ERAD contributes to various human diseases including cystic fibrosis, an autosomal recessive genetic disorder caused by mutations in the cystic fibrosis transmembrane conductance regulator (CFTR) protein. CFTR is a chloride channel responsible for maintaining proper salt homeostasis in lung epithelia. The most common mutation linked to cystic fibrosis is ΔF508, a folding mutant that is subject to ERAD in physiological conditions and cannot traffic to the cell surface. Several chemical chaperones or "trafficking correctors" have been described, which allow ΔF508 to evade ERAD, traffic to the cell surface, and function as an active chloride channel (188–191). The trafficking corrector approach has been explored for a number of ERAD-mediated human diseases, especially lysosomal storage disorders, such as Gaucher disease and Fabry disease (192). Defects in ubiquitination of ion channels at the PM contribute to hereditary hypertension and to channelopathies that cause various disorders, including congenital heart failure and epilepsy (155, 193).

surface expression of key immune response regulators, such as major histocompatibility complex class I (30–32). Thus, it is tempting to speculate that, in addition to its role in ER protein quality control, ERAD may also function in protein quantity control to regulate PM protein composition.

Although ERAD has a clearly defined role in ER protein quality control, its role as a regulator of protein traffic is less clear. In some instances, ERAD may promote ER-to-Golgi traffic by degrading ER retention factors. One example involves the Insig proteins (Insig-1 and Insig-2, for insulin-induced genes), which interact with SREBP (sterol response element-binding protein)-SCAP (SREBP cleavage-activating protein) complexes in the presence of cholesterol, resulting in retention of SREBP-SCAP in the ER (33–35). However, in cholesterol-depleted cells, Insigs are subject to ubiquitination (by the ER-localized ubiquitin ligase gp78, see **Table 1**) and degradation, which results in SCAP-SREBP exit from the ER (36). Thus, by destabilizing ER retention factors, ERAD can promote ER-to-Golgi traffic.

Ubiquitination and Cargo Sorting at the Golgi Complex

In yeast, ubiquitin modification is important for Golgi-to-endosome traffic of resident vacuole proteins, such as carboxypeptidase S (37–39), and thus represents an important regulatory step that determines whether individual cargo molecules are delivered to the PM or the endosome. Ubiquitin-mediated Golgi-to-endosome traffic has been described for many different cargoes, including the yeast general amino acid transporter, Gap1 (40–43); the proton pump, Pma1 (44, 45); the uracil transporter, Fur4 (46); the siderophore iron transporter, Sit1 (47); and the ferrichrome transporter, Arn1 (48). In most cases, Rsp5 is the responsible ubiquitin ligase, although the mechanisms that regulate substrate targeting of Rsp5 on the Golgi complex remain unclear.

How ubiquitinated cargoes on the Golgi complex are recognized and sorted into vesicles is still poorly understood. Currently, the best candidate mediators of such Golgi-to-endosome traffic are the Golgi-localized, γ-ear-containing ARF-binding proteins, or GGAs. As their name suggests, the GGA proteins localize to the Golgi complex and contain four critical domains: a VHS domain, a GAT (GGA and TOM) domain that binds to GTP-Arf, a γ-adaptin ear homology domain, and a hinge region that binds to clathrin. GGA proteins are required for Golgi-to-endosome trafficking of many cargoes in both yeast and mammalian cells and sort cargo based on the binding of the VHS domain to acidic dileucine motifs (mannose 6-phosphate receptors) (49) or based on binding of the GAT domain to ubiquitin. Interestingly, the GAT domain consists of a three-helical bundle (50) capable of binding the same surface of ubiquitin at two different interfaces, each sufficient for Golgi-to-endosome transport in yeast (51). The VHS domain of GGA3 was also shown to bind ubiquitin (52, 53), indicating that GGA3 is capable of binding at least three ubiquitin moieties. GGA proteins also interact with phosphatidylinositol 4-phosphate on the Golgi complex (54, 55), an interaction that was shown to increase the affinity of the GAT domain for ubiquitin (55). For many Golgi-to-endosome trafficking cargoes studied in both yeast and mammalian cells, disruption of GGA protein function results in aberrant trafficking phenotypes, such as localization to the PM. Indeed, fusion of cargo to ubiquitin often results in constitutive vacuolar trafficking that is GGA dependent but independent of endocytosis (56), indicating that ubiquitin modification of cargo is sufficient to drive GGA-dependent Golgi-to-endosome traffic. Despite the general consensus that ubiquitinated cargoes on the Golgi complex are recognized by GGA proteins and sorted into clathrin-coated vesicles that traffic directly to endosomes, some evidence is emerging that cargo ubiquitination on the Golgi complex may not be required for

Golgi-to-endosome trafficking (48, 57, 58). Future studies need to clarify the role of cargo ubiquitination and GGA protein function in Golgi-to-endosome traffic.

Quality Control at the Golgi Complex

Compared to the ER, relatively little is known about mechanisms of protein quality control at the Golgi complex. In some cases, misfolded proteins that escape ERAD are targeted for degradation by a Golgi complex quality control (GQC) system. Part of the GQC system in yeast involves the function of Vps10 (the yeast homolog of mammalian sortilins), which can recognize misfolded proteins in the Golgi complex (generated by fusing a thermolabile λ-repressor domain to cargo) and target them for trafficking to endosomes (59, 60), but this pathway is not known to be dependent on ubiquitin. By contrast, some GQC mechanisms are ubiquitin dependent. The Golgi-localized RING E3 ubiquitin ligase, Tul1, was shown to mediate the ubiquitination and subsequent vacuolar trafficking of mutant cargoes in yeast with polar residue substitutions in transmembrane domains (61), suggesting that the GQC system can sense intramembrane protein misfolding. Recently, the GQC system was also described to mediate the vacuolar trafficking of misfolding mutants of the Wsc1 protein. Although the vacuolar trafficking of misfolded Wsc1 was Rsp5 dependent, the results of this study suggest that ubiquitination of the Wsc1 protein is required for endosomal sorting but not for Golgi-to-endosome traffic (62). By contrast, exit of misfolded Wsc1 variants from the Golgi complex was shown to depend on the yeast sortilin Vps10 (63). Ultimately, the relative contribution of the GQC system to the turnover of misfolded proteins compared to quality control mechanisms at the ER and PM needs to be assessed, but it is becoming clear that quality control mechanisms at all three locations in the cell may act in parallel to limit the delivery and accumulation of misfolded proteins at the cell surface.

TO THE GRAVE: UBIQUITIN-MEDIATED TURNOVER OF PLASMA MEMBRANE PROTEINS

Ubiquitin-Mediated Endocytosis

Ubiquitin-mediated endocytosis first came into focus with studies in yeast demonstrating a ubiquitin requirement for endocytosis of various PM cargoes, including Ste6 (64), Ste2 (65, 66), Pdr5 (67), and Fur4 (68). Indeed, for many cargoes studied in yeast, ubiquitin modifications are both necessary and sufficient for endocytosis (65, 69), although ubiquitin-independent endocytosis of cargoes has also been described (70, 71). The general consensus is that ubiquitin-mediated endocytosis is the dominant mechanism for internalization of most cargoes studied in yeast. However, in mammalian cells, the role of ubiquitin in endocytosis is somewhat more complicated. For many endocytic cargoes in mammalian cells, including receptor tyrosine kinases (RTKs) and G protein–coupled receptors (GPCRs), ubiquitin modification appears to be sufficient for endocytic uptake (72, 73). Strikingly, although many of these cargoes exhibit ligand-dependent ubiquitin modification, they also exhibit ubiquitin-independent endocytosis. Thus, in mammalian cells, ubiquitination is often sufficient but not required for internalization by endocytosis. This is indicative of multiple redundant yet distinct mechanisms of endocytosis (74), which may coordinately influence the fate of internalized cargo on the endosome.

A fundamental question of ubiquitin-mediated endocytosis is how ubiquitin-modified cargoes are recognized and sorted into endocytic vesicles. One prevailing hypothesis is that various UBDs located within the endocytic machinery may recruit ubiquitinated cargoes for internalization. For example, epsin family proteins (Ent1 and Ent2 in yeast) interact with clathrin, the AP-2 adaptor, and other endocytic proteins and have long been studied for their role in endocytosis.

Epsins interact with ubiquitin via tandem ubiquitin-interacting motif (UIM) domains, which have been proposed to function in the sorting of ubiquitinated cargoes for endocytosis. Similarly, the Eps15 and Eps15R proteins (similar to yeast Ede1) interact with other endocytic proteins via N-terminal EH (Eps15 homology) domains and also interact with ubiquitin via C-terminal tandem UIM domains (75). In both yeast and mammals, epsins and Eps15 family members exhibit redundant functions for endocytosis (72, 76) and have been shown to bind to ubiquitinated cargo (72, 77).

Many UBD-containing proteins, including Eps15 proteins (75), are subject to coupled monoubiquitination (78), fueling speculation that UBDs in epsins and Eps15 family proteins may function in the assembly of the endocytic protein interaction network. It is conceivable that ubiquitin modification of endocytic proteins can "seed" the assembly of endocytic protein interaction networks. Perhaps the best evidence that ubiquitin modification regulates the function of endocytic proteins comes from the investigation of photoreceptor cell differentiation during compound eye development in *Drosophila*. One *Drosophila* mutant called *fat facets* exhibits excess photoreceptor cell differentiation, indicating that the protein encoded by *fat facets* (Faf) is a differentiation inhibitor. The Faf protein was shown to be a DUB (79) that deubiquitinates the protein encoded by a gene called *liquid facets* (*Lqf*), a *Drosophila* epsin family protein (80, 81). Faf regulates the stability and abundance of Lqf (and perhaps other endocytic proteins) by antagonizing its ubiquitination and subsequent proteasomal degradation (80). Both Faf and Lqf are required for the endocytosis of the Notch ligand Delta (82) (see below), demonstrating how ubiquitination/deubiquitination dynamics of an endocytic protein is critical to cell differentiation and patterning during development.

In addition to epsins and Eps15 proteins, other endocytic proteins have been shown to bind ubiquitin, including Cin85 (yeast Sla1) (83), amphiphysin (83), and the yeast protein Lsb5 (84). The molecular function of these ubiquitin-binding activities is not clear, but the myriad of ubiquitin-binding elements within the endocytic machinery, possibly including UBDs that have not yet been described, suggests either significant redundancy or elaborate avidity affects, making it difficult to dissect precise biochemical events that govern ubiquitin-mediated endocytosis. Because the affinity of interactions between individual UBDs and monoubiquitin is low, it is possible that efficiency of endocytic sorting might be determined by the number of ubiquitin modifications, the extent and linkage types of polyubiquitin chains, lateral interaction with other ubiquitinated cargoes at the PM, as well as orthogonal, ubiquitin-independent sorting signals. Thus, despite a clear regulatory role of ubiquitination in targeting cargo for endocytosis, it has been difficult to establish an exclusive role for ubiquitin in endocytosis for the following reasons:

1. In mammalian cells, there are different kinds of endocytosis, which have different sorting determinants and often act redundantly for internalization of specific cargo, although trafficking outcomes may be determined by the route of entry (i.e., lysosomal sorting versus recycling).
2. The endocytic protein interaction network contains many UBDs, which may act in a redundant manner or serve as avidity sensors to determine the efficiency of sorting for endocytosis.
3. Some endocytic proteins are ubiquitinated, suggesting that some UBDs may mediate interactions within the endocytic protein network and that others may mediate recognition of ubiquitinated cargo.
4. PM cargoes participate in multiple interactions with other PM cargoes, suggesting that cargo homo- or hetero-oligomerization at the PM may allow nonubiquitinated cargo to piggyback during ubiquitin-mediated endocytosis.

Sorting of Ubiquitinated Cargoes on the Endosome

Endosomes are complex, multifunctional organelles that serve as sorting platforms for many different kinds of membrane traffic in the cell. They are the destination of endocytic vesicles derived from the PM, the *trans*-Golgi network, and other endosomes. It is at endosomes where the fate of cargo is decided: Some proteins are recycled back to the *trans*-Golgi network or to the PM, and others are sorted for lysosomal delivery where protein degradation occurs. In order for lysosomal degradation to occur, transmembrane proteins must be sorted into the lumen of the endosome through a complex sorting and vesiculation process known as multivesicular body (MVB) biogenesis. MVBs are formed when the limiting membrane of the endosome buds into the lumen, forming intralumenal vesicles (ILVs) loaded with cargo. ILV formation is the topological antithesis of other vesiculation events in which vesicles bud into the cytoplasm, such as clathrin-mediated endocytosis. Cargo sorting on endosomal membranes and ILV formation require the activity of five evolutionarily conserved protein complexes: ESCRT-0, ESCRT-I, ESCRT-II, ESCRT-III, and the Vps4 AAA-ATPase complex. Following ESCRT-mediated cargo sorting and ILV formation, MVBs can fuse with lysosomes, depositing their contents into the lumen of the lysosome (or yeast vacuole), where cargoes are degraded and their amino acids can be recycled (see the sidebar titled ESCRT Proteins and Human Disease).

The structure and function of the five ESCRT complexes have been the subject of intense investigation and several recent reviews (85–87), and thus, we only briefly discuss their roles in the capture and sorting of ubiquitinated cargoes. The ESCRT-0 complex is responsible for the capture of ubiquitinated cargo on the endosome. ESCRT-0 is a heterodimer consisting of two subunits: Hrs and STAM1/2 (Vps27 and Hse1 in yeast, respectively). Interaction with ESCRT-I is mediated by the PTAP-like motif at the C terminus of Hrs. Five UBDs have been characterized in the ESCRT-0 complex: Hrs can interact with ubiquitin via two UIM domains and a VHS domain, whereas STAM can interact with ubiquitin via a VHS and a UIM domain. The yeast and human ESCRT-0 structures reveal a coiled-coil core with unstructured, flexible linkers connecting the UBDs, FYVE domain, and the PTAP-like motif (88–90). In vivo ESCRT-0 assembles into a tetramer (two heterodimers) (91), which could facilitate binding of up to 10 distinct ubiquitin moieties. In vitro studies using a synthetic cargo, consisting of ubiquitin fused to GFP and tethered to giant unilamellar vesicles, demonstrated that ESCRT-0 was sufficient to cluster ubiquitinated cargo on membranes (92). The parallels between ESCRT-0-mediated sorting of ubiquitinated cargoes on the endosome and epsin/Eps15-mediated sorting of cargoes on the PM are worth noting. Both sorting modules exhibit similar domain architecture, with multiple UBDs for cargo capture, lipid-binding domains for subcellular targeting, and an interface for interaction with vesiculation machinery. On the

ESCRT PROTEINS AND HUMAN DISEASE

Given their role in receptor downregulation, ion homeostasis, and degradation of misfolded membrane proteins, it is not surprising that mutations in various ESCRT proteins are associated with various human diseases. It has long been known that certain retroviruses, including HIV, recruit ESCRT machinery to the PM to mediate retroviral budding (194), a reaction topologically similar to MVB biogenesis. Decreased expression of ESCRT proteins is associated with aberrant growth factor receptor signaling, which leads to hyperproliferation and cancer (195). Indeed, *Tsg101*, the human equivalent of yeast Vps23, was originally identified as a tumor suppressor gene, although it is still not clear if this is because of its role in the downregulation of growth signaling pathways. More recently, mutations in human ESCRT-III subunits have been linked to neurodegenerative disorders, including frontotemporal dementia (196), amyotrophic lateral sclerosis (197–199), and spongiform neurodegeneration (200). Thus, some neurodegenerative disorders are most likely caused, at least in part, by the accumulation of misfolded membrane proteins in cells with decreased ESCRT function.

MVB: multivesicular body

ILV: intralumenal vesicle

ESCRT: endosomal sorting complex required for transport

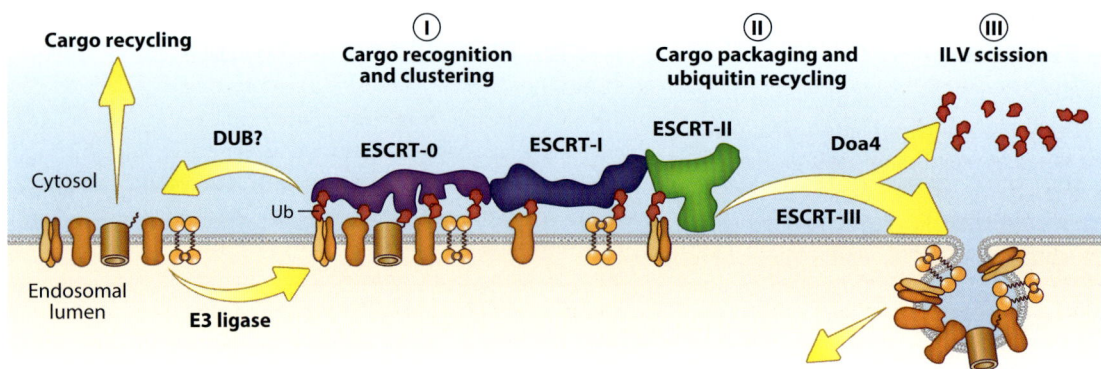

Figure 3

Model for sorting of ubiquitinated cargo on the endosome. Abbreviations: Doa4, deubiquitinating enzyme that recycles ubiquitin from cargo prior to MVB biogenesis; DUB, deubiquitinating enzyme; ESCRT, endosomal sorting complex required for transport; ILV, intralumenal vesicle.

basis of these similarities of structure and function, it is tempting to compare these modules to flexible "nets" that can capture ubiquitinated cargo and interact with downstream components of the vesiculation machinery (**Figure 3**).

In addition to the myriad of UBDs on ESCRT-0, several UBDs on both ESCRT-I and ESCRT-II also have been characterized. In ESCRT-I, an elongated stalk-like protein complex consisting of TSG101 (Vps23 in yeast), Vps28, Vps37, and Mvb12, the UEV (ubiquitin-conjugating enzyme E2 variant) domain of TSG101/Vps23 and a novel unstructured sequence on Mvb12 have been shown to bind ubiquitin (38, 93). In ESCRT-II, the GLUE domain of Vps36 was shown to contain a UBD (94, 95). No UBDs have been described for ESCRT-III, which is thought to function primarily in ILV formation and scission. Interestingly, although individually dispensable for cargo sorting, mutation of all three UBDs in ESCRT-I and ESCRT-II abrogated cargo sorting into ILVs, highlighting UBD redundancy in the system (93). It has been proposed that the multitude of UBDs may favor recognition of polyubiquitinated cargo. Indeed, many endocytic cargoes are subject to multimonoubiquitination or K63-linked polyubiquitination in vivo (57, 96–98). However, a recent study demonstrated that K63 linkages

are not required for MVB sorting of cargo (56). Furthermore, even though the UBDs of human ESCRT-0 exhibit a much higher affinity for polyubiquitin than monoubiquitin, they exhibit only a slight bias for K63 over K48 linkages (52), suggesting that ESCRTs do not specifically recognize K63 linkages. The emerging model of endosomal cargo sorting thus involves a variety of different UBDs in ESCRT-0, ESCRT-I, and ESCRT-II, which function with overlapping specificities in the recognition of various ubiquitinated cargoes.

Ubiquitin Editing at the Plasma Membrane and Endosome

The proteasome associates with both ubiquitin ligases and DUBs, which edit ubiquitin chains on potential substrates, an effect that can either rescue a substrate or hasten its demise. As has been observed with the proteasome, the ESCRT machinery associates with various E3 ubiquitin ligases and DUBs that can influence cargo fate on the endosome. In general, the association of E3 ubiquitin ligases with ESCRTs is thought to promote cargo sorting for lysosomal degradation (99–102), although there are some cases where such association negatively regulates ESCRT stability (103–105). In yeast, Rsp5 clearly localizes to endosomal structures

(39, 106) and has been shown to interact with a C-terminal PY motif in Hse1 (ESCRT-0), an interaction required for efficient sorting of some cargoes (99). Thus, even as cargoes are being sorted by ESCRTs, last-minute ubiquitin conjugation may influence sorting decisions.

Just as E3 ubiquitin ligases on the endosome can nudge cargo toward the grave, so can endosomal DUBs rescue cargo by removing ubiquitin, antagonizing their association with ESCRTs, and promoting their recycling back to the cell surface. For example, UCH-L3 is a DUB that promotes epithelial sodium channel (ENaC) recycling to the cell surface by catalyzing its deubiquitination (107). The DUB Usp10 was shown to mediate the deubiquitination of cystic fibrosis transmembrane conductance regulator (CFTR) on endosomes and thus promote recycling to the surface (108). Two DUBs, Usp33 and Usp20, were recently shown to deubiquitinate the β_2-adrenergic receptor on endosomes and thus promote recycling to the cell surface (109). AMSH (associated molecule with the SH3 domain of STAM), a JAMM family DUB, was shown to interact with STAM (110) and antagonize the lysosomal degradation of the epidermal growth factor receptor (EGFR) (111). In yeast, the UBP family DUB Ubp2 was shown to interact with Rsp5 (indirectly via the adaptor protein Rup1) and to associate with ESCRT-0 components on endosomes (99, 112, 113), although a role for Ubp2 in cargo recycling has not been rigorously demonstrated.

In addition to their role in rescuing ubiquitinated cargo from incorporation into ILVs, DUBs play a major role in recycling of ubiquitin from cargo prior to the ILV scission event. The recycling of ubiquitin from cargo significantly impacts ubiquitin homeostasis in the cell, and mutants defective for ubiquitin recycling exhibit MVB sorting defects (114–116). This function has been well characterized for the yeast DUB Doa4 (117). Doa4 is recruited to Snf7, part of the ESCRT-III machinery, indirectly by its ability to bind the Snf7-interacting protein Bro1 (118–120). In mammalian cells, recycling of ubiquitin from cargo may be mediated by UBPY, a UBP family DUB. Like Doa4, UBPY is required for efficient lysosomal degradation of trafficking cargo, and loss of UBPY function results in the accumulation of ubiquitinated proteins, particularly on endosomes (121, 122). Consistent with a function analogous to Doa4 in yeast, several reports have provided evidence that UBPY function is required for efficient lysosomal degradation of the EGFR (121–124). Thus, AMSH and UBPY may each contribute to both cargo stability and degradation by deubiquitinating cargo at different stages of ESCRT sorting.

ENaC: epithelial sodium channel

CFTR: cystic fibrosis transmembrane conductance regulator

THE ROLE OF UBIQUITIN IN CELL SURFACE REMODELING

Cells remodel protein composition at the PM to attenuate signaling responses, regulate nutrient uptake, and control ion flux across the PM. For single-celled eukaryotic organisms, such as yeast, surface remodeling plays an important role in nutrient homeostasis, stress responses, and pheromone signaling. In metazoans, surface remodeling plays a critical role in cell differentiation and fate decisions. A classic example involves T-cell maturation, which utilizes cell surface remodeling to convert thymocytes from double positive (CD4$^+$8$^+$) to single positive (CD4$^-$8$^+$ or CD4$^+$8$^-$) with only a single major histocompatibility complex coreceptor at the cell surface (125). A key aspect of cell surface remodeling involves the targeting of specific proteins for endocytosis, and one such targeting determinant involves ubiquitin modification of trafficking cargo. Therefore, to understand how cells regulate PM protein composition, it is important to understand how specific PM proteins are targeted for ubiquitination and how these ubiquitinated cargoes are recognized, sorted into buds, and internalized by endocytosis.

Regulation of Signaling Receptors by Ubiquitin

One of the first PM cargoes demonstrated to undergo ubiquitin-mediated endocytosis was the yeast pheromone receptor Ste2, a GPCR (65). Ubiquitin modification is both necessary

ARRDC protein: arrestin domain-containing protein

ART: arrestin-related trafficking adaptor

and sufficient for Ste2 endocytosis (69), leading to the hypothesis that endocytosis of GPCRs in mammalian cells would likewise be ubiquitin dependent. Given their pharmacological significance (126), mammalian GPCRs and the mechanisms that mediate their downregulation have been studied extensively. Although several GPCRs studied in mammalian cells undergo ligand-dependent ubiquitination, these ubiquitination events are not required for internalization but are critical for MVB sorting on the endosome (127–129).

Although GPCR ubiquitination is dispensable for endocytosis in mammalian cells, ubiquitin modification still plays an important role in the regulation of GPCR endocytosis. Following ligand binding, GPCRs activate associated heterotrimeric G proteins by promoting nucleotide exchange, leading to subunit release and dissociation. A ligand-bound GPCR can subsequently be recognized by a GPCR kinase, which phosphorylates active GPCRs, causing them to recruit specialized adaptor proteins called β-arrestins. β-arrestins are multifunctional scaffold proteins that serve a variety of functions during receptor downregulation: They desensitize receptor signaling by preventing reassociation of G proteins with the GPCR; they promote GPCR endocytosis by binding to endocytosis proteins, including clathrin and AP-2; and they function as adaptors for E3 ubiquitin ligases. Interestingly, β-arrestins are ubiquitinated by the E3 ligase Mdm2, and this modification is critical for GPCR endocytosis (128). Furthermore, ubiquitination of β-arrestins enhances receptor interaction and thus promote endocytosis (130). Recently, the deubiquitinating enzyme Usp33 was shown to promote β-arrestin deubiquitination and to antagonize arrestin function (109). Thus, even though GPCR ubiquitination itself is dispensable for endocytosis, β-arrestin ubiquitination is an important regulator of receptor internalization and downregulation.

It is worth noting that the paradigm of arrestin-mediated downregulation has recently expanded beyond GPCRs to many additional types of PM proteins (131, 132). Furthermore, in addition to two β-arrestins and two visual arrestins, the human genome encodes at least five additional arrestin domain-containing (ARRDC) proteins. The human ARRDC family and the yeast family of arrestin-related trafficking adaptors (ARTs) exhibit striking similarity; both families contain N-terminal arrestin domains and multiple C-terminal PY motifs (133). Although the function of the ARRDC family of proteins is still under investigation, ARRDC3 was recently shown to function as an adaptor for the Nedd4 ubiquitin ligase and is required for the ubiquitination and endocytic downregulation of the β_2-adrenergic receptor (134). One recent study demonstrated that *Arrdc3* null mice exhibit resistance to obesity and increased energy expenditure, phenotypes that suggest an inability to downregulate β_2-adrenergic receptor signaling in adipose tissues (135). Another recent study demonstrated that decreased ARRDC3 expression is associated with certain types of breast cancer and that ARRDC3 may suppress cancer progression by promoting the endocytic downregulation of β-4 integrin (136). Thus, the growing family of arrestins and arrestin-related proteins may generally function as a family of ubiquitin ligase adaptors involved in the targeted ubiquitination and endocytic downregulation of GPCRs as well as other classes of PM cargo.

Attenuation of signaling processes from ligand-bound RTKs, like the EGFR, have been studied intensely because these signals facilitate cell growth and proliferation and can contribute to the onset of various cancers. A crucial determinant of the internalization of active RTKs is Cbl, a RING domain-containing E3 ubiquitin ligase. There are three Cbl family RING E3 ligases encoded in the human genome—c-Cbl, Cbl-b, and Cbl-c. In the case of the EGFR, ligand binding promotes receptor dimerization and autophosphorylation, which result in the recruitment of the adaptor protein Grb2 complexed with Cbl. The RING domain of Cbl, in turn, can recruit an E2 enzyme and catalyze ubiquitin transfer to the EGFR. Although ubiquitin modification is sufficient to trigger EGFR endocytosis (73, 137) and the ubiquitinated EGFR interacts with epsin UIM domains (77), EGFR ubiquitination is not required for its internalization (**Figure 4a**) (137). The EGFR is subject to at least two modes of internalization: (*a*) ubiquitin-independent, clathrin-dependent endocytosis at low extracellular concentrations of the EGF and (*b*) ubiquitin-dependent, clathrin-independent internalization at high extracellular concentrations of the EGF (72, 138, 139). Although the mechanism of ubiquitin-dependent internalization has not been elucidated, it requires epsins and Eps15 family proteins (72), consistent with the possibility that UIM domains in these proteins may recognize and sort the ubiquitinated EGFR.

Importantly, interaction of the EGFR with Cbl is required for ubiquitin-independent endocytosis of the EGFR (137), suggesting that Cbl-mediated EGFR internalization may not be related to receptor ubiquitination. Such a function may be related to Cbl-mediated recruitment of additional factors to active RTK signaling complexes, including Cin85 and endophilin, which are part of the endocytic protein interaction network. Furthermore, many such endocytic proteins, including Cin85, epsins, and Eps15 family proteins, are known to undergo coupled monoubiquitination (75, 140) and as such may be modified by Cbl. Thus, it is tempting to speculate that EGFR interaction with Cbl may trigger local assembly of endocytic machinery, although future work must determine how Cbl regulates interactions within the endocytic protein network.

Although ubiquitin-mediated endocytosis is often associated with attenuation of signaling processes, there are also examples where it is required to activate signaling receptors. Perhaps the best example of this involves the *Drosophila* Notch ligand Delta (141). Surface expression of Delta activates Notch signaling in adjacent cells, which triggers specific differentiation programs. The RING-type E3 ligase *neuralized* (*neur*) was found to promote Delta endocytosis in signal-sending cells but also to promote Notch signaling in signal-receiving cells. These observations are paradoxical: How is removal of a Notch ligand from the cell surface required for Notch activation in a neighboring cell? The precise mechanism is still under investigation, but recent studies have indicated that *neur*-mediated ubiquitination may trigger transcytosis of Delta from the basolateral membrane of the signal-sending cell to the apical membrane of the cell (142), thus delivering the ligand to sites that can activate Notch on signal-receiving cells. This mode of Notch signal regulation is a common theme in vertebrate development (143–145), making this system an excellent example of how ubiquitin-mediated endocytosis can remodel the cell surface to drive cell differentiation.

Ubiquitin-Mediated Downregulation of Ion Channels

Regulation of the ENaC, a three-subunit complex at the PM, is critical for cellular ion homeostasis (see the sidebar titled Human Diseases Associated with Membrane Protein Ubiquitination and Turnover). ENaC undergoes ubiquitin-dependent endocytosis, which is mediated by its interaction with the E3 ubiquitin ligase Nedd4-2. Nedd4 family proteins are HECT-type E3 ubiquitin ligases with a distinct domain architecture: They contain an N-terminal C2 domain, a series of two to four WW domains, and a C-terminal HECT ubiquitin ligase domain. WW domains typically mediate protein-protein interactions responsible for substrate targeting and are known to bind PY motifs, although some can also interact with phosphopeptides (146). The interaction of Nedd4-2 WW domains with C-terminal PY motifs in each of the ENaC subunits is required for ENaC ubiquitination and endocytosis. Ubiquitin-modified ENaC has been shown to undergo clathrin-mediated endocytosis, which requires the activity of epsins (**Figure 4b**) (147). Regulatory factors can also influence Nedd4 function: The Sgk1 kinase was shown to phosphorylate Nedd4-2 and abrogate its interaction with ENaC by promoting its interaction with a 14-3-3 protein (148–150), whereas the AMP sensor kinase was shown to destabilize ENaC

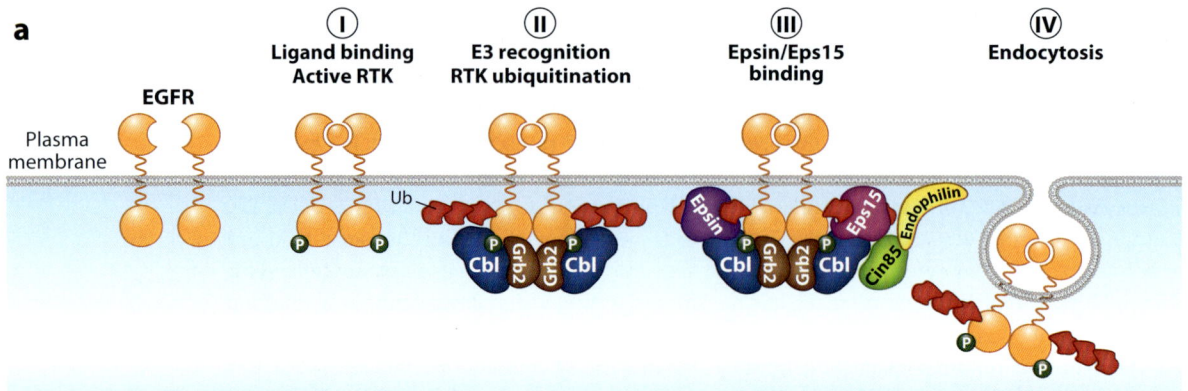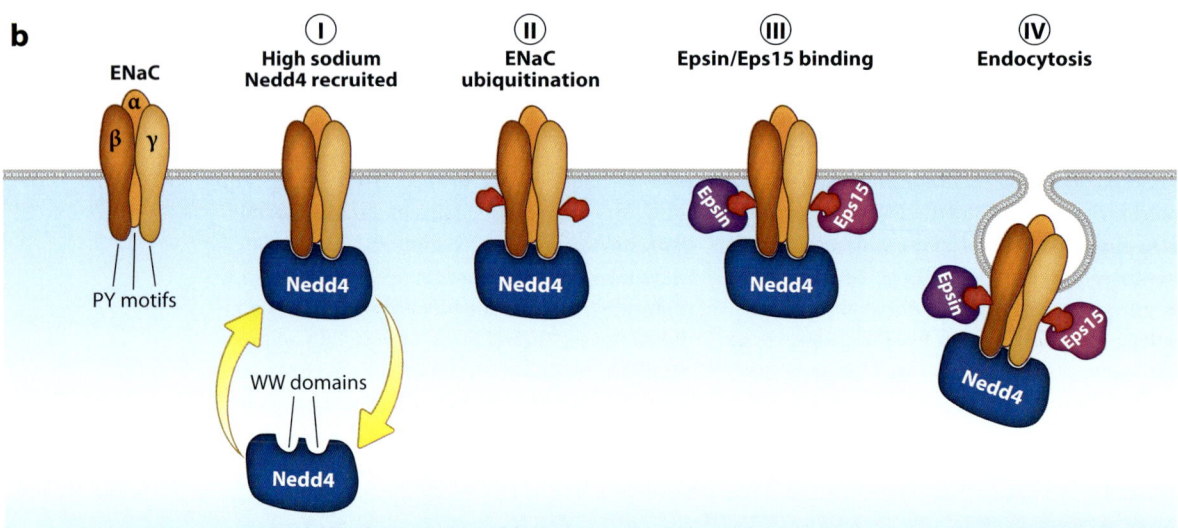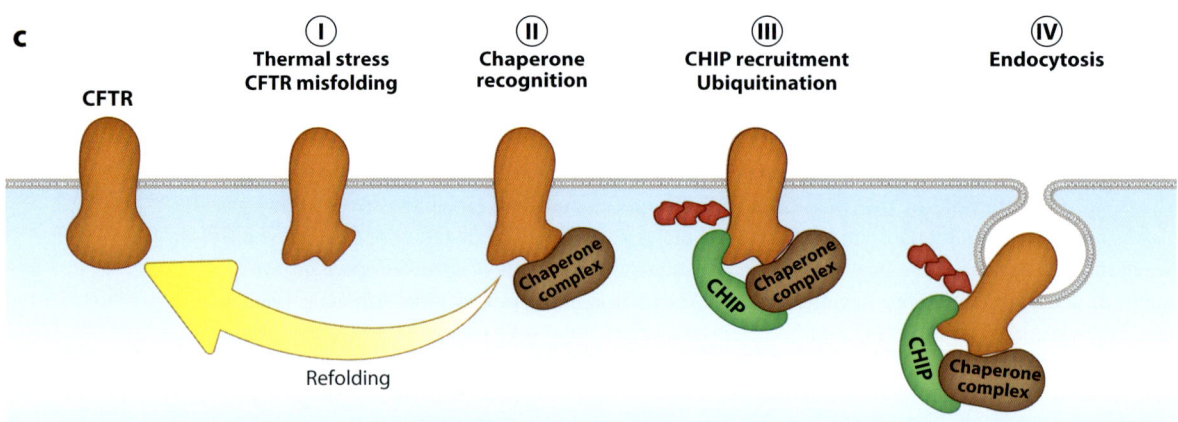

at the cell surface by phosphorylating Nedd4-2 and enhancing its association with ENaC (151).

Aside from ENaC, Nedd4 family members have been implicated in the ubiquitin-mediated endocytosis of many other ion channels (152–154). Importantly, many ion channels linked to channelopathies (see the sidebar titled Human Diseases Associated with Membrane Protein Ubiquitination and Turnover), including the cardiac voltage-gated sodium channel, Na_V 1.5, and the potassium channels, KCNQ1 and hERG1, all encode PY motifs and may be regulated by Nedd4 family E3 ligases (155). Additionally, connexin43, a gap junction protein that facilitates ion exchange between cells, contains a PY motif and has been shown to undergo Nedd4-dependent ubiquitination and endocytosis (156). There are nine Nedd4 family ubiquitin ligases encoded in the human genome, and although not all have been characterized, several Nedd4 family members have been implicated in the ubiquitin-mediated endocytosis of various other PM cargoes, including amino acid transporters (157, 158), RTKs (159–161), and glucose transporters (162). Thus, Nedd4 family E3 ligases may function generally to target a variety of PM cargoes for ubiquitin-mediated endocytosis, and their role is discussed throughout the remainder of this review.

Ubiquitin-Mediated Downregulation of Transporters

A clear example of ubiquitin-dependent endocytosis of a transporter in mammalian cells involves the dopamine transporter (DAT). DAT is a sodium chloride ion symporter responsible for dopamine reuptake at synapses, a process critical to control of dopamine signaling and thus synaptic function. DAT is ubiquitinated at the PM and internalized in a protein kinase C-dependent manner (163). DAT ubiquitination requires the activity of Nedd4-2, and internalization of DAT is clathrin dependent and requires epsins and Eps15 family proteins (164). Importantly, DAT endocytosis involves ubiquitin modification at three N-terminal lysine residues. Although DAT ubiquitination is dependent on the WW4 domain of Nedd4-2 (98), it is unclear exactly how this interaction is mediated because DAT lacks a PY motif. Further studies are needed to determine (*a*) if this interaction is direct or adaptor driven and (*b*) how this interaction is regulated in a protein kinase C-dependent manner.

Unlike DAT, some transporters constitutively cycle between the PM and endosomes via a ubiquitin-independent recycling pathway. This allows nutrient receptors, such as the transferrin receptor (TfR) and the low-density lipoprotein receptor, to scavenge nutrients from the environment, deposit them in the lumen of the endosome, and then return to the surface to collect more nutrients. However, ubiquitination of these nutrient receptors is sufficient to divert such cargo away from the recycling pathway and instead target them for sorting into MVBs and lysosomal degradation. For example, instead of recycling back to the surface, the TfR fused to ubiquitin is targeted for lysosomal degradation (165). Furthermore, excess iron was shown to stimulate TfR ubiquitination and lysosomal degradation (166), although the E3 ligase responsible is not known. Similarly, in cholesterol-replete conditions, the low-density lipoprotein receptor is ubiquitinated and targeted for lysosomal degradation. This is mediated by the RING E3 ligase,

Channelopathies: human diseases caused by mutations that affect ion channel function

Figure 4

Examples of ubiquitin-mediated endocytosis. (*a*) Endocytosis of receptor tyrosine kinases (RTKs) can be ubiquitin mediated, but these are not always ubiquitin dependent. (*b*) Endocytic downregulation of the epithelial sodium channel (ENaC) is ubiquitin dependent and is important for ion homeostasis in epithelial cells. (*c*) Plasma membrane quality control is mediated by chaperone recruitment of the CHIP (C terminus of Hsp70-interacting protein) ubiquitin ligase. Abbreviations: Cbl, a RING E3 ubiquitin ligase; CFTR, cystic fibrosis transmembrane conductance regulator; Cin85, ubiquitin-binding endocytic adaptor protein; Eps15, epidermal growth factor receptor substrate 15; Grb2, growth factor receptor-bound protein 2; Nedd4, HECT-type E3 ubiquitin ligase; P, phosphate.

IDOL (167), which is transcriptionally induced by a sterol-binding transcription factor (168). Thus, cells upregulate ubiquitination of nutrient receptors, such as the TfR and low-density lipoprotein receptor, which are normally recycled, to reduce the cellular level of these receptors and thereby limit nutrient accumulation.

For most cargoes studied in yeast, the primary determinant of endocytosis appears to be ubiquitin modification, although other endocytic targeting signals have been described (70, 71, 169). Thus, cargo ubiquitination at the PM represents a critical point of regulation for internalization. The primary E3 ubiquitin ligase that mediates cargo ubiquitination and targeting for endocytosis is Rsp5, the lone Nedd4 family member encoded in the yeast genome. Because most yeast PM proteins do not contain PY motifs and thus cannot interact directly with Rsp5, this presents a conundrum: How does one ubiquitin ligase selectively regulate the abundance of perhaps hundreds of different proteins at the PM? The answer appears to lie in the existence of an elaborate network of adaptor proteins, each capable of targeting Rsp5 ubiquitin ligase activity to specific substrates at specific locations in the cell in a highly regulated manner. At the core of this adaptor network is a family of yeast ART proteins. There are 10 ARTs encoded in the yeast genome; each ART contains an N-terminal arrestin-like domain and multiple C-terminal PY motifs that interact with the WW domains of Rsp5. Like F-box proteins, which function in the recruitment of specific substrates to SCF ubiquitin ligase complexes, ARTs recruit Rsp5 ubiquitin ligase activity to specific proteins at the PM and thus play a critical role in regulating which PM proteins are selectively targeted for endocytosis. Other proteins, including Bul1, Bul2, Ear1, and Ssh4, also act as specificity adaptors for Rsp5. It has been proposed that Bul1 and Bul2 may in fact be bona fide members of the ART protein family (170), although these proteins do not appear to contain significant primary sequence homology with ARTs. Despite their divergence, it is tempting to speculate that Bul1 and Bul2 could contain arrestin-like domains, but confirmation will require structural analysis of these proteins.

The mechanism of ART-mediated regulation of endocytosis was recently elucidated in two studies directed at understanding how PM proteins in yeast are ubiquitinated. *ART1* was identified in a chemical-genetic screen for mutants with defects in the endocytosis of Can1, the arginine transporter in yeast (171). Importantly, Δ*art1* cells are defective for endocytosis of a few specific cargoes, including Can1 and the methionine transporter Mup1, but are not defective for endocytosis of most other cargoes (171). Art1 interacts with Rsp5 via two C-terminal PY motifs, an interaction required for cargo ubiquitination (171). Additionally, Art1 localizes to the PM and interacts with cargo, confirming its role as a ubiquitin ligase adaptor (**Figure 5***a*). Bioinformatic analysis identified additional ART proteins encoded in the yeast genome (Art1-Art10), each capable of binding to Rsp5. Endocytosis of the lysine transporter Lyp1 is dependent on Art2, and Can1-Lyp1 cargo chimera experiments revealed that ART specificity for these cargoes is mediated through the N-terminal cytosolic tail (171). A parallel study revealed that ART family proteins function redundantly in the ubiquitination and endocytosis of the yeast manganese transporter, Smf1. In this investigation, simultaneous deletion of seven ART family proteins, as well as the Rsp5-interacting protein, Bsd2, was required to block Smf1 ubiquitination and endocytosis (172). This endocytic defect could be complemented by reintroduction of wild-type *ART2* but not by an allele of *art2* with point mutations in its C-terminal PY motifs (172). Both of these studies contributed to a model of ARTs as Rsp5 adaptors that function in targeting PM proteins for endocytosis. Since then, other reports have described similar endocytic functions for other ART family proteins (170, 173). It has also recently been reported that Art3 and Art6 may function in Golgi-to-endosome traffic, but it is unclear if this relates to their role as ubiquitin ligase adaptors (174). This study is noteworthy because it suggests that ART proteins have nonendocytic trafficking functions,

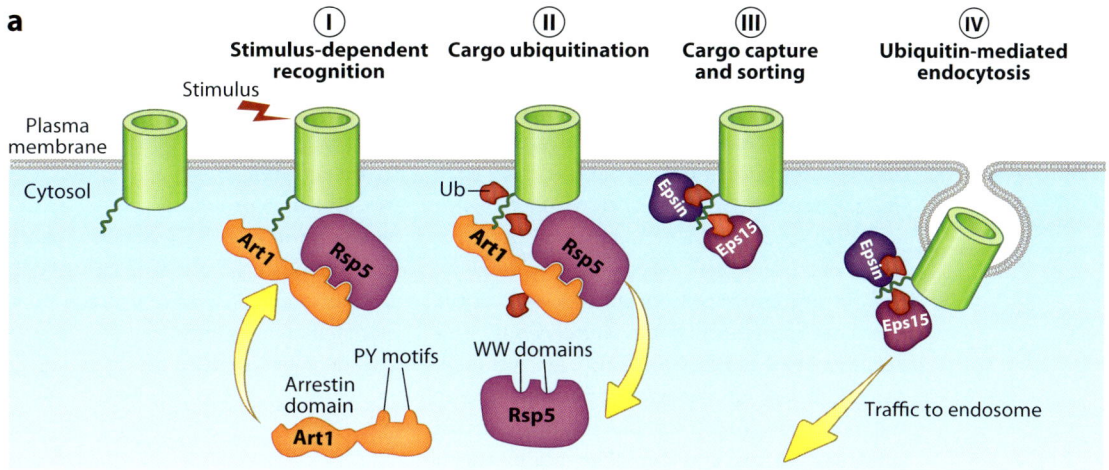

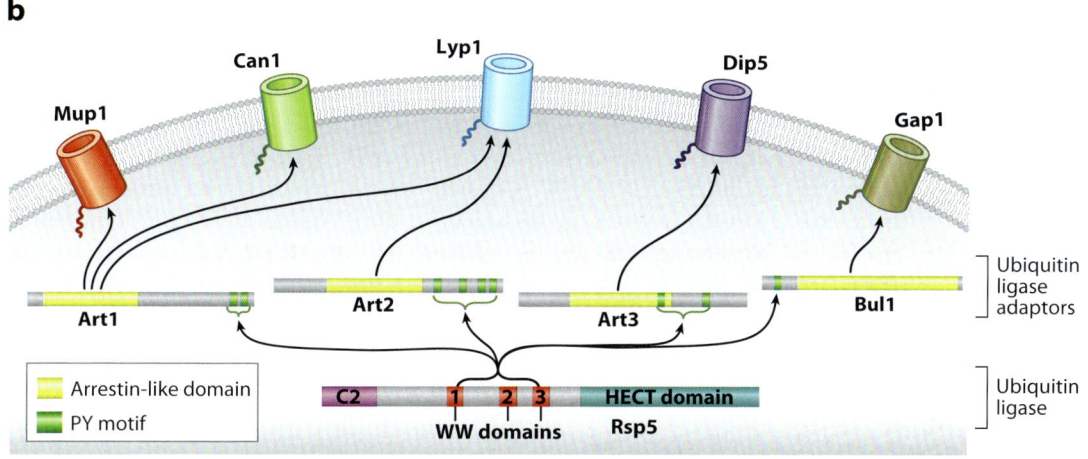

Figure 5

(*a*) Model of arrestin-related trafficking adaptor (ART) family protein function as Rsp5 adaptors in yeast. (*b*) Model for adaptor-driven substrate targeting of the Rsp5 ubiquitin ligase in yeast. Abbreviations: Bul1, another known adaptor for Rsp5; Can1, yeast arginine transporter; C2, a domain that binds to calcium and lipids; Dip5, yeast dicarboxylic amino acid transporter; Gap1, yeast general amino acid transporter; Lyp1, yeast lysine transporter; Mup1, yeast methionine transporter.

and it demonstrated that Art3 and Art6 interact with clathrin and AP-1. The later observation is reminiscent of β-arrestins in mammalian cells, which serve as endocytic adaptors by binding to clathrin and AP-2.

To drive Rsp5 specificity, ART family proteins must be capable of interacting with specific PM proteins in a regulated manner (**Figure 5***b*). ART-cargo interactions have been demonstrated (171–173), and although the precise cargo-binding domain has not yet been mapped, the best candidate is the arrestin domain. More experimentation is required to map and characterize cargo specificity determinants within the ART family of proteins. Importantly, ART function is tightly regulated and often appears to be "activated" by specific stimuli. For example, stimulation

Coupled ubiquitination: a phenomenon whereby ubiquitin-binding proteins and ubiquitin ligase adaptors are often observed to be ubiquitinated

of yeast cells with methionine activates Art1 to target the methionine transporter Mup1, resulting in rapid ubiquitination and endocytic downregulation (171). One major mechanism of regulation may involve stimulus-induced cargo phosphorylation, as is the case with the recognition of GPCRs by β–arrestins in mammalian cells. Indeed, Smf1 phosphorylation correlates with increased Art2 binding, and substrate-induced phosphorylation of the dicarboxylic amino acid transporter, Dip5, triggers increased binding to Art3 (173), although the regulatory kinases and/or phosphatases in these cases are not known. It is also possible that substrate binding and transport might trigger conformational changes in cargo proteins that are recognized by ARTs. Recognition of either conformational changes or cargo phosphorylation represents a local regulatory mechanism that is based on sensing cargo activity.

An alternative regulatory mechanism for ART-cargo recognition may involve posttranslational modifications of the ART proteins themselves. Most ARTs appear to be ubiquitinated, and in the case of Art1, this is Rsp5 dependent (171). Art1 ubiquitination occurs on a single lysine residue (175); mutation of this residue prevents ubiquitin modification and results in a loss-of-function phenotype (171). Such coupled ubiquitination events are not uncommon for ubiquitin ligase adaptors, and although its functional significance is clear, further experimentation is required to determine precisely how ubiquitin modification regulates Art1 function. Recently, a TORC1-Npr1 negative kinase cascade was shown to regulate Art1 function to tune the abundance of Can1 (the arginine transporter) at the cell surface (176). In this mechanism, TORC1 phosphorylates and inhibits the Npr1 kinase, which, in turn, phosphorylates and inhibits Art1. This example demonstrates how stress- and nutrient-sensitive signaling pathways regulate Art1 activity (176). Other ART proteins are also extensively phosphorylated (177) and are likely subject to phosphoregulation.

Although human homologs of ARTs are not obvious on the basis of their primary sequence, domain architecture suggests that yeast ARTs are similar to the family of ARRDC proteins in mammalian cells (134, 171). Like ARTs, ARRDC proteins have an N-terminal arrestin domain and C-terminal PY motifs, which have been shown to mediate interactions with Nedd4 family proteins (134). Thus, elucidating mechanisms of ART function in yeast may have important implications for understanding arrestin and ARRDC protein function in mammalian cells.

QUALITY MAINTENANCE AT THE PLASMA MEMBRANE

Unlike cell surface remodeling, which is driven by the cellular response to environmental dynamics, PM protein quality maintenance is a protective mechanism driven by protein-intrinsic factors and is essential for maintaining cellular homeostasis. By contrast to ERAD, which degrades proteins that do not acquire a native fold in the ER, quality maintenance at the PM limits the accumulation of proteins that have exceeded their functional lifetime after being damaged or misfolded at their site of action. Thus, PM quality maintenance (PMQM) involves both the structural maintenance of integral membrane proteins at the cell surface and the removal of misfolded or damaged PM proteins. PMQM is critical to cell survival because accumulation of such proteins at the PM can generate toxic protein aggregates and lead to loss of PM integrity. Compared to quality control mechanisms at the ER and a recently described quality control pathway in the nucleus (178–180), very little is known about the molecular mechanisms of quality maintenance at the PM. However, any quality maintenance system should incorporate certain key features, which include (*a*) the ability to distinguish between native and misfolded/damaged proteins, (*b*) the ability to refold or restore function of misfolded/damaged proteins or target them for degradation, and (*c*) upregulation in response to misfolding stress. In contrast to established quality control systems like ERAD and the N-end rule pathway, which clearly meet these criteria, the molecular mechanisms that govern PM quality control are only now coming into focus. Here, we describe recent efforts to characterize PM quality control mechanisms and discuss some critical questions that will drive future research.

The existence of machinery dedicated to PM protein quality surveillance and maintenance has long been inferred on the basis of various observations consistent with the targeting of misfolded or damaged PM proteins for endocytosis and lysosomal degradation. The first evidence of PM quality control came from studies in yeast using an allele of the pheromone receptor Ste2, which misfolds at high temperature. The misfolding of this Ste2 allele (as well as a temperature-sensitive allele of the arginine transporter Can1) at a restrictive temperature correlated with rapid endocytic removal from the PM and lysosomal degradation (66, 181). Similarly, a temperature-sensitive allele of the PM H^+ ATPase Pma1 is misfolded and targeted for endocytosis and vacuolar degradation at the restrictive temperature (182). This allele of Pma1 acquires ubiquitin modifications at the restrictive temperature, and internalization requires both the E3 ubiquitin ligase Rsp5 (yeast Nedd4 homolog) and the UIM domain of Ent1 (yeast epsin) (183), indicating that endocytosis of this misfolded protein is ubiquitin mediated. Quality control and proper maintenance of protein conformation at the PM are also tightly linked to the lipid microenvironments in the membrane. For example, proper folding of the yeast general amino acid transporter, Gap1, does not occur in the absence of sphingolipids, resulting in an inactive transporter (184). This inactive Gap1 can still be delivered to the PM but is rapidly ubiquitinated, internalized, and targeted for vacuolar degradation (184), underscoring the important role of lipids in PMQM.

Although the phenomenon of PMQM has been observed in yeast and appears to be ubiquitin mediated, very little is known about the mechanisms for detection of misfolded PM proteins. One study in yeast demonstrated that misfolding of PM protein cytosolic domains was not sufficient to induce endocytosis (185). However, recent studies in mammalian cells have elucidated a mechanism for the detection and removal of misfolded proteins from the cell surface. Using the ΔF508 CFTR mutant protein, which can only traffic to the PM at a low temperature, one study demonstrated that raising the temperature resulted in ΔF508 CFTR misfolding and ubiquitin-mediated endocytosis (**Figure 4c**) (186). The authors identified CHIP (C terminus of Hsp70-interacting protein) as the E3 ubiquitin ligase responsible for ubiquitinating misfolded CFTR at the PM (186). CHIP contains a RING E3 ligase domain as well as an N-terminal tetratricopeptide repeat domain that can interact with several chaperones, including Hsc70 and Hsp90. The authors demonstrated that the misfolded ΔF508 CFTR exhibits increased interaction with CHIP, Hsc70, and Hsp90 and that this interaction is responsible for ubiquitination of misfolded ΔF508 CFTR (186). A similar study fused a thermolabile bacteriophage λ-domain to CD4 and demonstrated that λ-domain misfolding conferred CHIP-mediated ubiquitination and endocytosis (187). These studies have elucidated a mechanism by which misfolded cytosolic domains of PM proteins are recognized and ubiquitinated, effectively targeting them for endocytosis and lysosomal trafficking. Whether this represents the only or primary mechanism of PMQM is not yet known.

Several key questions regarding PMQM remain to be elucidated. For example, although CHIP-mediated ubiquitination can explain detection of misfolded cytosolic domains of PM proteins, detection of misfolded transmembrane domains or damaged extracellular domains would likely require a different mechanism. Furthermore, because CHIP is not conserved in yeast, it remains unclear what system would perform an analogous function in yeast. Thus, it seems likely that other mechanisms might exist for detection of misfolded proteins at the PM. Finally, it will be important to dissect the molecular mechanisms of sorting and endocytosis of misfolded PM proteins in mammalian cells. The ubiquitin-mediated endocytosis mechanism reserved for misfolded proteins may be distinct from clathrin-mediated endocytosis, which might explain why cargo ubiquitination of many PM proteins in mammalian cells is often observed as sufficient, but not required, for endocytosis.

SUMMARY POINTS

1. Ubiquitination of integral membrane proteins at the PM is often sufficient, but not always required, for endocytic targeting of PM proteins. On the endosome, cargo ubiquitination is typically both necessary and sufficient for MVB sorting and lysosomal degradation.

2. Although the human genome encodes >600 E3 ubiquitin ligases, it encodes <95 DUBs. DUBs are versatile regulators of membrane trafficking events in the cell with roles in cargo recycling, ubiquitin recycling, and possibly the regulated assembly of transport machinery.

3. Ubiquitin recognition and sorting machinery at the Golgi complex (GGAs), PM (epsins, Eps15), and endosome (ESCRT-0) share similar characteristics, including multiple UBDs, lipid-binding domains, and flexible structures with the potential to multimerize. It is conceivable that other protein complexes with similar characteristics may serve similar functions at different locations in the cell.

4. Modification of cargo with K63-linked ubiquitin chains enhances the efficiency of MVB sorting on the endosome but is not a strict requirement for ESCRT-mediated turnover of integral membrane proteins.

5. A wide substrate range for E3 ubiquitin ligases can be facilitated through the use of a modular substrate-targeting adaptor network.

6. Ubiquitination systems in the ER, Golgi complex, and PM work sequentially to limit the toxic accumulation of misfolded proteins at the PM. Although ERAD substrates are degraded by the proteasome, post-ER quality maintenance systems operate by targeting misfolded/damaged proteins for lysosomal degradation.

FUTURE ISSUES

1. How are Golgi-localized E3 ubiquitin ligases targeted to specific substrates? How are misfolded proteins at the Golgi complex detected and targeted for ubiquitination?

2. Does ubiquitination of cargo or transport machinery (i.e., endocytic proteins, ESCRT complex proteins) seed the assembly of protein interaction networks that drive localized trafficking events at specific locations in the cell (i.e., PM, endosomes)? What do UBDs in the endocytic and ESCRT machinery recognize?

3. How do DUBs regulate cargo recycling and assembly of transport machinery?

4. What structural features of ubiquitin ligase adaptors (i.e., ARTs) determine substrate selectivity? What posttranslational modifications regulate the function of E3 ubiquitin ligase adaptors?

5. How are various integral membrane protein quality control systems in the cell (ERAD, GQC, PMQM, and autophagy) regulated and coordinated to protect cells from the toxic accumulation of misfolded membrane proteins?

DISCLOSURE STATEMENT

The authors are not aware of any affiliations, memberships, funding, or financial holdings that might be perceived as affecting the objectivity of this review.

ACKNOWLEDGMENTS

We thank N.J. Buchkovich for thoughtful discussions and critical reading of the manuscript. P.H. is supported by a fellowship from the Taiwanese Ministry of Education. J.A.M. is supported by a Fleming Research Fellowship.

LITERATURE CITED

1. Komander D. 2009. The emerging complexity of protein ubiquitination. *Biochem. Soc. Trans.* 37:937–53
2. Ye Y, Rape M. 2009. Building ubiquitin chains: E2 enzymes at work. *Nat. Rev. Mol. Cell Biol.* 10:755–64
3. Wenzel DM, Stoll KE, Klevit RE. 2011. E2s: structurally economical and functionally replete. *Biochem. J.* 433:31–42
4. Deshaies RJ, Joazeiro CA. 2009. RING domain E3 ubiquitin ligases. *Annu. Rev. Biochem.* 78:399–434
5. Nagy V, Dikic I. 2010. Ubiquitin ligase complexes: from substrate selectivity to conjugational specificity. *Biol. Chem.* 391:163–69
6. Behrends C, Harper JW. 2011. Constructing and decoding unconventional ubiquitin chains. *Nat. Struct. Mol. Biol.* 18:520–28
7. Duda DM, Scott DC, Calabrese MF, Zimmerman ES, Zheng N, Schulman BA. 2011. Structural regulation of cullin-RING ubiquitin ligase complexes. *Curr. Opin. Struct. Biol.* 21:257–64
8. Kee Y, Huibregtse JM. 2007. Regulation of catalytic activities of HECT ubiquitin ligases. *Biochem. Biophys. Res. Commun.* 354:329–33
9. Bernassola F, Karin M, Ciechanover A, Melino G. 2008. The HECT family of E3 ubiquitin ligases: multiple players in cancer development. *Cancer Cell* 14:10–21
10. Komander D, Clague MJ, Urbé S. 2009. Breaking the chains: structure and function of the deubiquitinases. *Nat. Rev. Mol. Cell Biol.* 10:550–63
11. Xu P, Duong D, Seyfried N, Cheng D, Xie Y, et al. 2009. Quantitative proteomics reveals the function of unconventional ubiquitin chains in proteasomal degradation. *Cell* 137:133–45
12. Hoeller D, Dikic I. 2010. Regulation of ubiquitin receptors by coupled monoubiquitination. *Subcell. Biochem.* 54:31–40
13. Lauwers E, Erpapazoglou Z, Haguenauer-Tsapis R, André B. 2010. The ubiquitin code of yeast permease trafficking. *Trends Cell Biol.* 20:196–204
14. Chen Y, Klionsky DJ. 2011. The regulation of autophagy—unanswered questions. *J. Cell Sci.* 124:161–70
15. Menzies FM, Moreau K, Rubinsztein DC. 2011. Protein misfolding disorders and macroautophagy. *Curr. Opin. Cell Biol.* 23:190–97
16. Yang Z, Klionsky DJ. 2010. Eaten alive: a history of macroautophagy. *Nat. Cell Biol.* 12:814–22
17. Parodi AJ. 2000. Protein glucosylation and its role in protein folding. *Annu. Rev. Biochem.* 69:69–93
18. Trombetta ES, Parodi AJ. 2003. Quality control and protein folding in the secretory pathway. *Annu. Rev. Cell Dev. Biol.* 19:649–76
19. Hänzelmann P, Buchberger A, Schindelin H. 2011. Hierarchical binding of cofactors to the AAA ATPase p97. *Structure* 19:833–43
20. Schuberth C, Buchberger A. 2008. UBX domain proteins: major regulators of the AAA ATPase Cdc48/p97. *Cell Mol. Life Sci.* 65:2360–71
21. Brodsky JL, Skach WR. 2011. Protein folding and quality control in the endoplasmic reticulum: recent lessons from yeast and mammalian cell systems. *Curr. Opin. Cell Biol.* 23:464–75
22. Bagola K, Mehnert M, Jarosch E, Sommer T. 2011. Protein dislocation from the ER. *Biochim. Biophys. Acta* 1808:925–36
23. Tsai YC, Weissman AM. 2011. Ubiquitylation in ERAD: reversing to go forward? *PLoS Biol.* 9:e1001038

24. Vembar SS, Brodsky JL. 2008. One step at a time: endoplasmic reticulum-associated degradation. *Nat. Rev. Mol. Cell Biol.* 9:944–57
25. Claessen JH, Kundrat L, Ploegh HL. 2011. Protein quality control in the ER: balancing the ubiquitin checkbook. *Trends Cell Biol.* 22:22–32
26. Taxis C, Vogel F, Wolf DH. 2002. ER-Golgi traffic is a prerequisite for efficient ER degradation. *Mol. Biol. Cell* 13:1806–18
27. Kincaid MM, Cooper AA. 2007. Misfolded proteins traffic from the endoplasmic reticulum (ER) due to ER export signals. *Mol. Biol. Cell* 18:455–63
28. Schaheen B, Dang H, Fares H. 2009. Derlin-dependent accumulation of integral membrane proteins at cell surfaces. *J. Cell Sci.* 122:2228–39
29. Bernier V, Lagacé M, Bichet DG, Bouvier M. 2004. Pharmacological chaperones: potential treatment for conformational diseases. *Trends Endocrinol. Metab.* 15:222–28
30. Stagg HR, Thomas M, van den Boomen D, Wiertz EJ, Drabkin HA, et al. 2009. The TRC8 E3 ligase ubiquitinates MHC class I molecules before dislocation from the ER. *J. Cell Biol.* 186:685–92
31. Lilley BN, Ploegh HL. 2004. A membrane protein required for dislocation of misfolded proteins from the ER. *Nature* 429:834–40
32. Lilley BN, Ploegh HL. 2005. Viral modulation of antigen presentation: manipulation of cellular targets in the ER and beyond. *Immunol. Rev.* 207:126–44
33. Flury I, Garza R, Shearer A, Rosen J, Cronin S, Hampton RY. 2005. INSIG: a broadly conserved transmembrane chaperone for sterol-sensing domain proteins. *EMBO J.* 24:3917–26
34. Dong XY, Tang SQ. 2010. Insulin-induced gene: a new regulator in lipid metabolism. *Peptides* 31:2145–50
35. Bengoechea-Alonso MT, Ericsson J. 2007. SREBP in signal transduction: cholesterol metabolism and beyond. *Curr. Opin. Cell Biol.* 19:215–22
36. Lee JN, Song B, DeBose-Boyd RA, Ye J. 2006. Sterol-regulated degradation of Insig-1 mediated by the membrane-bound ubiquitin ligase gp78. *J. Biol. Chem.* 281:39308–15
37. Cowles CR, Snyder WB, Burd CG, Emr SD. 1997. Novel Golgi to vacuole delivery pathway in yeast: identification of a sorting determinant and required transport component. *EMBO J.* 16:2769–82
38. Katzmann DJ, Babst M, Emr SD. 2001. Ubiquitin-dependent sorting into the multivesicular body pathway requires the function of a conserved endosomal protein sorting complex, ESCRT-I. *Cell* 106:145–55
39. Katzmann DJ, Sarkar S, Chu T, Audhya A, Emr SD. 2004. Multivesicular body sorting: Ubiquitin ligase Rsp5 is required for the modification and sorting of carboxypeptidase S. *Mol. Biol. Cell* 15:468–80
40. Risinger AL, Kaiser CA. 2008. Different ubiquitin signals act at the Golgi and plasma membrane to direct GAP1 trafficking. *Mol. Biol. Cell* 19:2962–72
41. Rubio-Texeira M, Kaiser CA. 2006. Amino acids regulate retrieval of the yeast general amino acid permease from the vacuolar targeting pathway. *Mol. Biol. Cell* 17:3031–50
42. Soetens O, De Craene JO, Andre B. 2001. Ubiquitin is required for sorting to the vacuole of the yeast general amino acid permease, Gap1. *J. Biol. Chem.* 276:43949–57
43. Helliwell SB, Losko S, Kaiser CA. 2001. Components of a ubiquitin ligase complex specify polyubiquitination and intracellular trafficking of the general amino acid permease. *J. Cell Biol.* 153:649–62
44. Pizzirusso M, Chang A. 2004. Ubiquitin-mediated targeting of a mutant plasma membrane ATPase, Pma1-7, to the endosomal/vacuolar system in yeast. *Mol. Biol. Cell* 15:2401–9
45. Liu Y, Sitaraman S, Chang A. 2006. Multiple degradation pathways for misfolded mutants of the yeast plasma membrane ATPase, Pma1. *J. Biol. Chem.* 281:31457–66
46. Blondel MO, Morvan J, Dupré S, Urban-Grimal D, Haguenauer-Tsapis R, Volland C. 2004. Direct sorting of the yeast uracil permease to the endosomal system is controlled by uracil binding and Rsp5p-dependent ubiquitylation. *Mol. Biol. Cell* 15:883–95
47. Erpapazoglou Z, Froissard M, Nondier I, Lesuisse E, Haguenauer-Tsapis R, Belgareh-Touzé N. 2008. Substrate- and ubiquitin-dependent trafficking of the yeast siderophore transporter Sit1. *Traffic* 9:1372–91
48. Kim Y, Deng Y, Philpott CC. 2007. GGA2- and ubiquitin-dependent trafficking of Arn1, the ferrichrome transporter of *Saccharomyces cerevisiae*. *Mol. Biol. Cell* 18:1790–802

49. Puertollano R, Aguilar RC, Gorshkova I, Crouch RJ, Bonifacino JS. 2001. Sorting of mannose 6-phosphate receptors mediated by the GGAs. *Science* 292:1712–16
50. Suer S, Misra S, Saidi LF, Hurley JH. 2003. Structure of the GAT domain of human GGA1: a syntaxin amino-terminal domain fold in an endosomal trafficking adaptor. *Proc. Natl. Acad. Sci. USA* 100:4451–56
51. Bilodeau PS, Winistorfer SC, Allaman MM, Surendhran K, Kearney WR, et al. 2004. The GAT domains of clathrin-associated GGA proteins have two ubiquitin binding motifs. *J. Biol. Chem.* 279:54808–16
52. Ren X, Hurley JH. 2010. VHS domains of ESCRT-0 cooperate in high-avidity binding to polyubiquitinated cargo. *EMBO J.* 29:1045–54
53. Puertollano R, Bonifacino JS. 2004. Interactions of GGA3 with the ubiquitin sorting machinery. *Nat. Cell Biol.* 6:244–51
54. Demmel L, Gravert M, Ercan E, Habermann B, Müller-Reichert T, et al. 2008. The clathrin adaptor Gga2p is a phosphatidylinositol 4-phosphate effector at the Golgi exit. *Mol. Biol. Cell* 19:1991–2002
55. Wang J, Sun HQ, Macia E, Kirchhausen T, Watson H, et al. 2007. PI4P promotes the recruitment of the GGA adaptor proteins to the *trans*-Golgi network and regulates their recognition of the ubiquitin sorting signal. *Mol. Biol. Cell* 18:2646–55
56. Stringer DK, Piper RC. 2011. A single ubiquitin is sufficient for cargo protein entry into MVBs in the absence of ESCRT ubiquitination. *J. Cell Biol.* 192:229–42
57. Lauwers E, Jacob C, André B. 2009. K63-linked ubiquitin chains as a specific signal for protein sorting into the multivesicular body pathway. *J. Cell Biol.* 185:493–502
58. Merhi A, Gérard N, Lauwers E, Prévost M, André B. 2011. Systematic mutational analysis of the intracellular regions of yeast Gap1 permease. *PLoS ONE* 6:e18457
59. Hong E, Davidson AR, Kaiser CA. 1996. A pathway for targeting soluble misfolded proteins to the yeast vacuole. *J. Cell Biol.* 135:623–33
60. Jørgensen MU, Emr SD, Winther JR. 1999. Ligand recognition and domain structure of Vps10p, a vacuolar protein sorting receptor in *Saccharomyces cerevisiae*. *Eur. J. Biochem.* 260:461–69
61. Reggiori F, Pelham HR. 2002. A transmembrane ubiquitin ligase required to sort membrane proteins into multivesicular bodies. *Nat. Cell Biol.* 4:117–23
62. Wang S, Thibault G, Ng DT. 2011. Routing misfolded proteins through the MVB pathway protects against proteotoxicity. *J. Biol. Chem.* 286:29376–87
63. Wang S, Ng DT. 2010. Evasion of endoplasmic reticulum surveillance makes Wsc1p an obligate substrate of Golgi quality control. *Mol. Biol. Cell* 21:1153–65
64. Kölling R, Hollenberg CP. 1994. The ABC-transporter Ste6 accumulates in the plasma membrane in a ubiquitinated form in endocytosis mutants. *EMBO J.* 13:3261–71
65. Hicke L, Riezman H. 1996. Ubiquitination of a yeast plasma membrane receptor signals its ligand-stimulated endocytosis. *Cell* 84:277–87
66. Jenness DD, Li Y, Tipper C, Spatrick P. 1997. Elimination of defective alpha-factor pheromone receptors. *Mol. Cell. Biol.* 17:6236–45
67. Egner R, Kuchler K. 1996. The yeast multidrug transporter Pdr5 of the plasma membrane is ubiquitinated prior to endocytosis and degradation in the vacuole. *FEBS Lett.* 378:177–81
68. Galan JM, Moreau V, Andre B, Volland C, Haguenauer-Tsapis R. 1996. Ubiquitination mediated by the Npi1p/Rsp5p ubiquitin-protein ligase is required for endocytosis of the yeast uracil permease. *J. Biol. Chem.* 271:10946–52
69. Shih SC, Sloper-Mould KE, Hicke L. 2000. Monoubiquitin carries a novel internalization signal that is appended to activated receptors. *EMBO J.* 19:187–98
70. Chen L, Davis NG. 2002. Ubiquitin-independent entry into the yeast recycling pathway. *Traffic* 3:110–23
71. Tan PK, Howard JP, Payne GS. 1996. The sequence NPFXD defines a new class of endocytosis signal in *Saccharomyces cerevisiae*. *J. Cell Biol.* 135:1789–800
72. Sigismund S, Woelk T, Puri C, Maspero E, Tacchetti C, et al. 2005. Clathrin-independent endocytosis of ubiquitinated cargos. *Proc. Natl. Acad. Sci. USA* 102:2760–65
73. Haglund K, Sigismund S, Polo S, Szymkiewicz I, Di Fiore PP, Dikic I. 2003. Multiple monoubiquitination of RTKs is sufficient for their endocytosis and degradation. *Nat. Cell Biol.* 5:461–66
74. Goh LK, Huang F, Kim W, Gygi S, Sorkin A. 2010. Multiple mechanisms collectively regulate clathrin-mediated endocytosis of the epidermal growth factor receptor. *J. Cell Biol.* 189:871–83

75. Polo S, Sigismund S, Faretta M, Guidi M, Capua MR, et al. 2002. A single motif responsible for ubiquitin recognition and monoubiquitination in endocytic proteins. *Nature* 416:451–55
76. Shih SC, Katzmann DJ, Schnell JD, Sutanto M, Emr SD, Hicke L. 2002. Epsins and Vps27p/Hrs contain ubiquitin-binding domains that function in receptor endocytosis. *Nat. Cell Biol.* 4:389–93
77. Kazazic M, Bertelsen V, Pedersen KW, Vuong TT, Grandal MV, et al. 2009. Epsin 1 is involved in recruitment of ubiquitinated EGF receptors into clathrin-coated pits. *Traffic* 10:235–45
78. Hoeller D, Dikic I. 2010. Regulation of ubiquitin receptors by coupled monoubiquitination. *Subcell. Biochem.* 54:31–40
79. Huang Y, Baker RT, Fischer-Vize JA. 1995. Control of cell fate by a deubiquitinating enzyme encoded by the *fat facets* gene. *Science* 270:1828–31
80. Chen X, Zhang B, Fischer JA. 2002. A specific protein substrate for a deubiquitinating enzyme: Liquid facets is the substrate of Fat facets. *Genes Dev.* 16:289–94
81. Cadavid ALM, Ginzel A, Fischer JA. 2000. The function of the *Drosophila* Fat facets deubiquitinating enzyme in limiting photoreceptor cell number is intimately associated with endocytosis. *Development* 127:1727–36
82. Overstreet E, Fitch E, Fischer JA. 2004. Fat facets and Liquid facets promote Delta endocytosis and Delta signaling in the signaling cells. *Development* 131:5355–66
83. Stamenova SD, French ME, He Y, Francis SA, Kramer ZB, Hicke L. 2007. Ubiquitin binds to and regulates a subset of SH3 domains. *Mol. Cell* 25:273–84
84. Costa R, Warren DT, Ayscough KR. 2005. Lsb5p interacts with actin regulators Sla1p and Las17p, ubiquitin and Arf3p to couple actin dynamics to membrane trafficking processes. *Biochem. J.* 387:649–58
85. Hurley JH. 2010. The ESCRT complexes. *Crit. Rev. Biochem. Mol. Biol.* 45:463–87
86. Shields SB, Piper RC. 2011. How ubiquitin functions with ESCRTs. *Traffic* 12:1306–17
87. Henne WM, Buchkovich NJ, Emr SD. 2011. The ESCRT pathway. *Dev. Cell* 21:77–91
88. Ren X, Kloer DP, Kim YC, Ghirlando R, Saidi LF, et al. 2009. Hybrid structural model of the complete human ESCRT-0 complex. *Structure* 17:406–16
89. Ren X, Hurley JH. 2011. Structural basis for endosomal recruitment of ESCRT-I by ESCRT-0 in yeast. *EMBO J.* 30:2130–39
90. Prag G, Watson H, Kim YC, Beach BM, Ghirlando R, et al. 2007. The Vps27/Hse1 complex is a GAT domain–based scaffold for ubiquitin-dependent sorting. *Dev. Cell* 12:973–86
91. Mayers JR, Fyfe I, Schuh AL, Chapman ER, Edwardson JM, Audhya A. 2011. ESCRT-0 assembles as a heterotetrameric complex on membranes and binds multiple ubiquitinylated cargoes simultaneously. *J. Biol. Chem.* 286:9636–45
92. Wollert T, Hurley JH. 2010. Molecular mechanism of multivesicular body biogenesis by ESCRT complexes. *Nature* 464:864–69
93. Shields SB, Oestreich AJ, Winistorfer S, Nguyen D, Payne JA, et al. 2009. ESCRT ubiquitin-binding domains function cooperatively during MVB cargo sorting. *J. Cell Biol.* 185:213–24
94. Hirano S, Suzuki N, Slagsvold T, Kawasaki M, Trambaiolo D, et al. 2006. Structural basis of ubiquitin recognition by mammalian Eap45 GLUE domain. *Nat. Struct. Mol. Biol.* 13:1031–32
95. Slagsvold T, Aasland R, Hirano S, Bache KG, Raiborg C, et al. 2005. Eap45 in mammalian ESCRT-II binds ubiquitin via a phosphoinositide-interacting GLUE domain. *J. Biol. Chem.* 280:19600–6
96. Duncan LM, Piper S, Dodd RB, Saville MK, Sanderson CM, et al. 2006. Lysine-63-linked ubiquitination is required for endolysosomal degradation of class I molecules. *EMBO J.* 25:1635–45
97. Paiva S, Vieira N, Nondier I, Haguenauer-Tsapis R, Casal M, Urban-Grimal D. 2009. Glucose-induced ubiquitylation and endocytosis of the yeast Jen1 transporter: role of lysine 63-linked ubiquitin chains. *J. Biol. Chem.* 284:19228–36
98. Vina-Vilaseca A, Sorkin A. 2010. Lysine 63-linked polyubiquitination of the dopamine transporter requires WW3 and WW4 domains of Nedd4-2 and UBE2D ubiquitin-conjugating enzymes. *J. Biol. Chem.* 285:7645–56
99. Ren J, Kee Y, Huibregtse JM, Piper RC. 2007. Hse1, a component of the yeast Hrs-STAM ubiquitin-sorting complex, associates with ubiquitin peptidases and a ligase to control sorting efficiency into multivesicular bodies. *Mol. Biol. Cell* 18:324–35

100. Zhou R, Kabra R, Olson DR, Piper RC, Snyder PM. 2010. Hrs controls sorting of the epithelial Na$^+$ channel between endosomal degradation and recycling pathways. *J. Biol. Chem.* 285:30523–30
101. Bhandari D, Trejo J, Benovic JL, Marchese A. 2007. Arrestin-2 interacts with the ubiquitin-protein isopeptide ligase atrophin-interacting protein 4 and mediates endosomal sorting of the chemokine receptor CXCR4. *J. Biol. Chem.* 282:36971–79
102. Malik R, Marchese A. 2010. Arrestin-2 interacts with the endosomal sorting complex required for transport machinery to modulate endosomal sorting of CXCR4. *Mol. Biol. Cell* 21:2529–41
103. McDonald B, Martin-Serrano J. 2008. Regulation of Tsg101 expression by the steadiness box: a role of Tsg101-associated ligase. *Mol. Biol. Cell* 19:754–63
104. Kim BY, Olzmann JA, Barsh GS, Chin LS, Li L. 2007. Spongiform neurodegeneration-associated E3 ligase Mahogunin ubiquitylates TSG101 and regulates endosomal trafficking. *Mol. Biol. Cell* 18:1129–42
105. Kim GH, Park E, Kong YY, Han JK. 2006. Novel function of POSH, a JNK scaffold, as an E3 ubiquitin ligase for the Hrs stability on early endosomes. *Cell Signal.* 18:553–63
106. Wang G, McCaffery JM, Wendland B, Dupré S, Haguenauer-Tsapis R, Huibregtse JM. 2001. Localization of the Rsp5p ubiquitin-protein ligase at multiple sites within the endocytic pathway. *Mol. Cell. Biol.* 21:3564–75
107. Butterworth MB, Edinger RS, Ovaa H, Burg D, Johnson JP, Frizzell RA. 2007. The deubiquitinating enzyme UCH-L3 regulates the apical membrane recycling of the epithelial sodium channel. *J. Biol. Chem.* 282:37885–93
108. Bomberger JM, Barnaby RL, Stanton BA. 2009. The deubiquitinating enzyme USP10 regulates the post-endocytic sorting of cystic fibrosis transmembrane conductance regulator in airway epithelial cells. *J. Biol. Chem.* 284:18778–89
109. Berthouze M, Venkataramanan V, Li Y, Shenoy SK. 2009. The deubiquitinases USP33 and USP20 coordinate beta2 adrenergic receptor recycling and resensitization. *EMBO J.* 28:1684–96
110. McCullough J, Clague MJ, Urbé S. 2004. AMSH is an endosome-associated ubiquitin isopeptidase. *J. Cell Biol.* 166:487–92
111. McCullough J, Row PE, Lorenzo O, Doherty M, Beynon R, et al. 2006. Activation of the endosome-associated ubiquitin isopeptidase AMSH by STAM, a component of the multivesicular body-sorting machinery. *Curr. Biol.* 16:160–65
112. Kee Y, Lyon N, Huibregtse JM. 2005. The Rsp5 ubiquitin ligase is coupled to and antagonized by the Ubp2 deubiquitinating enzyme. *EMBO J.* 24:2414–24
113. Kee Y, Muñoz W, Lyon N, Huibregtse JM. 2006. The deubiquitinating enzyme Ubp2 modulates Rsp5-dependent Lys63-linked polyubiquitin conjugates in *Saccharomyces cerevisiae*. *J. Biol. Chem.* 281:36724–31
114. Amerik AY, Nowak J, Swaminathan S, Hochstrasser M. 2000. The Doa4 deubiquitinating enzyme is functionally linked to the vacuolar protein-sorting and endocytic pathways. *Mol. Biol. Cell* 11:3365–80
115. Amerik A, Sindhi N, Hochstrasser M. 2006. A conserved late endosome-targeting signal required for Doa4 deubiquitylating enzyme function. *J. Cell Biol.* 175:825–35
116. Swaminathan S, Amerik AY, Hochstrasser M. 1999. The Doa4 deubiquitinating enzyme is required for ubiquitin homeostasis in yeast. *Mol. Biol. Cell* 10:2583–94
117. Dupré S, Haguenauer-Tsapis R. 2001. Deubiquitination step in the endocytic pathway of yeast plasma membrane proteins: crucial role of Doa4p ubiquitin isopeptidase. *Mol. Cell. Biol.* 21:4482–94
118. Luhtala N, Odorizzi G. 2004. Bro1 coordinates deubiquitination in the multivesicular body pathway by recruiting Doa4 to endosomes. *J. Cell Biol.* 166:717–29
119. Nikko E, Marini AM, André B. 2003. Permease recycling and ubiquitination status reveal a particular role for Bro1 in the multivesicular body pathway. *J. Biol. Chem.* 278:50732–43
120. Odorizzi G, Katzmann DJ, Babst M, Audhya A, Emr SD. 2003. Bro1 is an endosome-associated protein that functions in the MVB pathway in *Saccharomyces cerevisiae*. *J. Cell Sci.* 116:1893–903
121. Row PE, Prior IA, McCullough J, Clague MJ, Urbé S. 2006. The ubiquitin isopeptidase UBPY regulates endosomal ubiquitin dynamics and is essential for receptor down-regulation. *J. Biol. Chem.* 281:12618–24
122. Mizuno E, Kobayashi K, Yamamoto A, Kitamura N, Komada M. 2006. A deubiquitinating enzyme UBPY regulates the level of protein ubiquitination on endosomes. *Traffic* 7:1017–31
123. Alwan HA, van Leeuwen JE. 2007. UBPY-mediated epidermal growth factor receptor (EGFR) deubiquitination promotes EGFR degradation. *J. Biol. Chem.* 282:1658–69

124. Row PE, Liu H, Hayes S, Welchman R, Charalabous P, et al. 2007. The MIT domain of UBPY constitutes a CHMP binding and endosomal localization signal required for efficient epidermal growth factor receptor degradation. *J. Biol. Chem.* 282:30929–37
125. Zamoyska R, Lovatt M. 2004. Signalling in T-lymphocyte development: integration of signalling pathways is the key. *Curr. Opin. Immunol.* 16:191–96
126. Overington J, Al-Lazikani B, Hopkins A. 2006. How many drug targets are there? *Nat. Rev. Drug Discov.* 5:993–96
127. Martin NP, Lefkowitz RJ, Shenoy SK. 2003. Regulation of V2 vasopressin receptor degradation by agonist-promoted ubiquitination. *J. Biol. Chem.* 278:45954–59
128. Shenoy SK, McDonald PH, Kohout TA, Lefkowitz RJ. 2001. Regulation of receptor fate by ubiquitination of activated beta 2-adrenergic receptor and beta-arrestin. *Science* 294:1307–13
129. Tanowitz M, von Zastrow M. 2002. Ubiquitination-independent trafficking of G protein–coupled receptors to lysosomes. *J. Biol. Chem* 277:50219–22
130. Shenoy SK, Lefkowitz RJ. 2003. Trafficking patterns of beta-arrestin and G protein–coupled receptors determined by the kinetics of beta-arrestin deubiquitination. *J. Biol. Chem.* 278:14498–506
131. Simonin A, Fuster D. 2010. Nedd4-1 and beta-arrestin-1 are key regulators of Na+/H+exchanger 1 ubiquitylation, endocytosis, and function. *J. Biol. Chem.* 285:38293–303
132. Lefkowitz RJ, Rajagopal K, Whalen EJ. 2006. New roles for beta-arrestins in cell signaling: not just for seven-transmembrane receptors. *Mol. Cell* 24:643–52
133. Lin C, MacGurn J, Chu T, Stefan C, Emr S. 2008. Arrestin-related ubiquitin-ligase adaptors regulate endocytosis and protein turnover at the cell surface. *Cell* 135:714–25
134. Nabhan JF, Pan H, Lu Q. 2010. Arrestin domain-containing protein 3 recruits the NEDD4 E3 ligase to mediate ubiquitination of the β2-adrenergic receptor. *EMBO Rep.* 11:605–11
135. Patwari P, Emilsson V, Schadt EE, Chutkow WA, Lee S, et al. 2011. The arrestin domain-containing 3 protein regulates body mass and energy expenditure. *Cell Metab.* 14:671–83
136. Draheim KM, Chen HB, Tao Q, Moore N, Roche M, Lyle S. 2010. ARRDC3 suppresses breast cancer progression by negatively regulating integrin beta4. *Oncogene* 29:5032–47
137. Huang F, Goh LK, Sorkin A. 2007. EGF receptor ubiquitination is not necessary for its internalization. *Proc. Natl. Acad. Sci. USA* 104:16904–9
138. Sigismund S, Argenzio E, Tosoni D, Cavallaro E, Polo S, Di Fiore PP. 2008. Clathrin-mediated internalization is essential for sustained EGFR signaling but dispensable for degradation. *Dev. Cell* 15:209–19
139. Chen H, De Camilli P. 2005. The association of epsin with ubiquitinated cargo along the endocytic pathway is negatively regulated by its interaction with clathrin. *Proc. Natl. Acad. Sci. USA* 102:2766–71
140. Bezsonova I, Bruce MC, Wiesner S, Lin H, Rotin D, Forman-Kay JD. 2008. Interactions between the three CIN85 SH3 domains and ubiquitin: implications for CIN85 ubiquitination. *Biochemistry* 47:8937–49
141. Weinmaster G, Fischer JA. 2011. Notch ligand ubiquitylation: What is it good for? *Dev. Cell* 21:134–44
142. Benhra N, Vignaux F, Dussert A, Schweisguth F, Le Borgne R. 2010. Neuralized promotes basal to apical transcytosis of Delta in epithelial cells. *Mol. Biol. Cell* 21:2078–86
143. Koo BK, Lim HS, Song R, Yoon MJ, Yoon KJ, et al. 2005. Mind bomb 1 is essential for generating functional Notch ligands to activate Notch. *Development* 132:3459–70
144. Koo BK, Yoon KJ, Yoo KW, Lim HS, Song R, et al. 2005. Mind bomb-2 is an E3 ligase for Notch ligand. *J. Biol. Chem.* 280:22335–42
145. Hansson EM, Lanner F, Das D, Mutvei A, Marklund U, et al. 2010. Control of Notch-ligand endocytosis by ligand-receptor interaction. *J. Cell Sci.* 123:2931–42
146. Verdecia MA, Bowman ME, Lu KP, Hunter T, Noel JP. 2000. Structural basis for phosphoserine-proline recognition by group IV WW domains. *Nat. Struct. Biol.* 7:639–43
147. Wang H, Traub LM, Weixel KM, Hawryluk MJ, Shah N, et al. 2006. Clathrin-mediated endocytosis of the epithelial sodium channel. Role of epsin. *J. Biol. Chem.* 281:14129–35
148. Debonneville C, Flores SY, Kamynina E, Plant PJ, Tauxe C, et al. 2001. Phosphorylation of Nedd4-2 by Sgk1 regulates epithelial Na^+ channel cell surface expression. *EMBO J.* 20:7052–59

149. Ichimura T, Yamamura H, Sasamoto K, Tominaga Y, Taoka M, et al. 2005. 14-3-3 Proteins modulate the expression of epithelial Na$^+$ channels by phosphorylation-dependent interaction with Nedd4-2 ubiquitin ligase. *J. Biol. Chem.* 280:13187–94
150. Nagaki K, Yamamura H, Shimada S, Saito T, Hisanaga S, et al. 2006. 14-3-3 Mediates phosphorylation-dependent inhibition of the interaction between the ubiquitin E3 ligase Nedd4-2 and epithelial Na$^+$ channels. *Biochemistry* 45:6733–40
151. Bhalla V, Oyster NM, Fitch AC, Wijngaarden MA, Neumann D, et al. 2006. AMP-activated kinase inhibits the epithelial Na$^+$ channel through functional regulation of the ubiquitin ligase Nedd4-2. *J. Biol. Chem.* 281:26159–69
152. Lin A, Hou Q, Jarzylo L, Amato S, Gilbert J, et al. 2011. Nedd4-mediated AMPA receptor ubiquitination regulates receptor turnover and trafficking. *J. Neurochem.* 119:27–39
153. He Y, Hryciw DH, Carroll ML, Myers SA, Whitbread AK, et al. 2008. The ubiquitin-protein ligase Nedd4-2 differentially interacts with and regulates members of the Tweety family of chloride ion channels. *J. Biol. Chem.* 283:24000–10
154. Hryciw DH, Ekberg J, Lee A, Lensink IL, Kumar S, et al. 2004. Nedd4-2 functionally interacts with ClC-5: involvement in constitutive albumin endocytosis in proximal tubule cells. *J. Biol. Chem.* 279:54996–5007
155. Rougier JS, Albesa M, Abriel H. 2010. Ubiquitylation and SUMOylation of cardiac ion channels. *J. Cardiovasc. Pharmacol.* 56:22–28
156. Leithe E, Rivedal E. 2007. Ubiquitination of gap junction proteins. *J. Membr. Biol.* 217:43–51
157. Hatanaka T, Hatanaka Y, Setou M. 2006. Regulation of amino acid transporter ATA2 by ubiquitin ligase Nedd4-2. *J. Biol. Chem.* 281:35922–30
158. Vina-Vilaseca A, Bender-Sigel J, Sorkina T, Closs EI, Sorkin A. 2011. Protein kinase C–dependent ubiquitination and clathrin-mediated endocytosis of the cationic amino acid transporter CAT-1. *J. Biol. Chem.* 286:8697–706
159. Li Y, Zhou Z, Alimandi M, Chen C. 2009. WW domain containing E3 ubiquitin protein ligase 1 targets the full-length ErbB4 for ubiquitin-mediated degradation in breast cancer. *Oncogene* 28:2948–58
160. Lin Q, Wang J, Childress C, Sudol M, Carey DJ, Yang W. 2010. HECT E3 ubiquitin ligase Nedd4-1 ubiquitinates ACK and regulates epidermal growth factor (EGF)-induced degradation of EGF receptor and ACK. *Mol. Cell. Biol.* 30:1541–54
161. Arévalo JC, Waite J, Rajagopal R, Beyna M, Chen ZY, et al. 2006. Cell survival through Trk neurotrophin receptors is differentially regulated by ubiquitination. *Neuron* 50:549–59
162. Dieter M, Palmada M, Rajamanickam J, Aydin A, Busjahn A, et al. 2004. Regulation of glucose transporter SGLT1 by ubiquitin ligase Nedd4-2 and kinases SGK1, SGK3, and PKB. *Obes. Res.* 12:862–70
163. Miranda M, Wu CC, Sorkina T, Korstjens DR, Sorkin A. 2005. Enhanced ubiquitylation and accelerated degradation of the dopamine transporter mediated by protein kinase C. *J. Biol. Chem.* 280:35617–24
164. Sorkina T, Miranda M, Dionne KR, Hoover BR, Zahniser NR, Sorkin A. 2006. RNA interference screen reveals an essential role of Nedd4-2 in dopamine transporter ubiquitination and endocytosis. *J. Neurosci.* 26:8195–205
165. Raiborg C, Bache KG, Gillooly DJ, Madshus IH, Stang E, Stenmark H. 2002. Hrs sorts ubiquitinated proteins into clathrin-coated microdomains of early endosomes. *Nat. Cell Biol.* 4:394–98
166. Tachiyama R, Ishikawa D, Matsumoto M, Nakayama KI, Yoshimori T, et al. 2011. Proteome of ubiquitin/MVB pathway: possible involvement of iron-induced ubiquitylation of transferrin receptor in lysosomal degradation. *Genes Cells* 16:448–66
167. Zhang L, Fairall L, Goult BT, Calkin AC, Hong C, et al. 2011. The IDOL-UBE2D complex mediates sterol-dependent degradation of the LDL receptor. *Genes Dev.* 25:1262–74
168. Zelcer N, Hong C, Boyadjian R, Tontonoz P. 2009. LXR regulates cholesterol uptake through Idol-dependent ubiquitination of the LDL receptor. *Science* 325:100–4
169. Howard JP, Hutton JL, Olson JM, Payne GS. 2002. Sla1p serves as the targeting signal recognition factor for NPFX$_{(1,2)}$D-mediated endocytosis. *J. Cell Biol.* 157:315–26
170. Nikko E, Pelham HR. 2009. Arrestin-mediated endocytosis of yeast plasma membrane transporters. *Traffic* 10:1856–67

171. Lin C, MacGurn J, Chu T, Stefan C, Emr S. 2008. Arrestin-related ubiquitin-ligase adaptors regulate endocytosis and protein turnover at the cell surface. *Cell* 135:714–25
172. Nikko E, Sullivan JA, Pelham HRB. 2008. Arrestin-like proteins mediate ubiquitination and endocytosis of the yeast metal transporter Smf1. *EMBO Rep.* 9:1216–21
173. Hatakeyama R, Kamiya M, Takahara T, Maeda T. 2010. Endocytosis of the aspartic acid/glutamic acid transporter Dip5 is triggered by substrate-dependent recruitment of the Rsp5 ubiquitin ligase via the arrestin-like protein Aly2. *Mol. Cell. Biol.* 30:5598–607
174. O'Donnell A, Apffel A, Gardner R, Cyert M. 2010. Alpha-arrestins Aly1 and Aly2 regulate intracellular trafficking in response to nutrient signaling. *Mol. Biol. Cell* 21:3552–66
175. Hitchcock AL, Auld K, Gygi SP, Silver PA. 2003. A subset of membrane-associated proteins is ubiquitinated in response to mutations in the endoplasmic reticulum degradation machinery. *Proc. Natl. Acad. Sci. USA* 100:12735–40
176. MacGurn JA, Hsu PC, Smolka MB, Emr SD. 2011. TORC1 regulates endocytosis via Npr1-mediated phosphoinhibition of a ubiquitin ligase adaptor. *Cell* 147:1104–17
177. Albuquerque C, Smolka M, Payne S, Bafna V, Eng J, Zhou H. 2008. A multidimensional chromatography technology for in-depth phosphoproteome analysis. *Mol. Cell Proteomics* 7:1389–96
178. Fredrickson EK, Rosenbaum JC, Locke MN, Milac TI, Gardner RG. 2011. Exposed hydrophobicity is a key determinant of nuclear quality control degradation. *Mol. Biol. Cell* 22:2384–95
179. Gardner RG, Nelson ZW, Gottschling DE. 2005. Degradation-mediated protein quality control in the nucleus. *Cell* 120:803–15
180. Rosenbaum JC, Fredrickson EK, Oeser ML, Garrett-Engele CM, Locke MN, et al. 2011. Disorder targets misorder in nuclear quality control degradation: a disordered ubiquitin ligase directly recognizes its misfolded substrates. *Mol. Cell* 41:93–106
181. Li Y, Kane T, Tipper C, Spatrick P, Jenness DD. 1999. Yeast mutants affecting possible quality control of plasma membrane proteins. *Mol. Cell. Biol.* 19:3588–99
182. Gong X, Chang A. 2001. A mutant plasma membrane ATPase, Pma1-10, is defective in stability at the yeast cell surface. *Proc. Natl. Acad. Sci. USA* 98:9104–9
183. Liu Y, Chang A. 2006. Quality control of a mutant plasma membrane ATPase: Ubiquitylation prevents cell-surface stability. *J. Cell Sci.* 119:360–69
184. Lauwers E, Grossmann G, André B. 2007. Evidence for coupled biogenesis of yeast Gap1 permease and sphingolipids: essential role in transport activity and normal control by ubiquitination. *Mol. Biol. Cell* 18:3068–80
185. Lewis M, Pelham H. 2009. Inefficient quality control of thermosensitive proteins on the plasma membrane. *PLoS ONE* 4:e5038
186. Okiyoneda T, Barrière H, Bagdány M, Rabeh W, Du K, et al. 2010. Peripheral protein quality control removes unfolded CFTR from the plasma membrane. *Science* 329:805–10
187. Apaja PM, Xu H, Lukacs GL. 2010. Quality control for unfolded proteins at the plasma membrane. *J. Cell Biol.* 191:553–70
188. Sampson HM, Robert R, Liao J, Matthes E, Carlile GW, et al. 2011. Identification of a NBD1-binding pharmacological chaperone that corrects the trafficking defect of F508del-CFTR. *Chem. Biol.* 18:231–42
189. Carlile GW, Robert R, Zhang D, Teske KA, Luo Y, et al. 2007. Correctors of protein trafficking defects identified by a novel high-throughput screening assay. *ChemBioChem* 8:1012–20
190. Pedemonte N, Zegarra-Moran O, Galietta LJ. 2011. High-throughput screening of libraries of compounds to identify CFTR modulators. *Methods Mol. Biol.* 741:13–21
191. Van Goor F, Hadida S, Grootenhuis PD, Burton B, Cao D, et al. 2009. Rescue of CF airway epithelial cell function in vitro by a CFTR potentiator, VX-770. *Proc. Natl. Acad. Sci. USA* 106:18825–30
192. Parenti G. 2009. Treating lysosomal storage diseases with pharmacological chaperones: from concept to clinics. *EMBO Mol. Med.* 1:268–79
193. Kullmann DM. 2010. Neurological channelopathies. *Annu. Rev. Neurosci.* 33:151–72
194. Bieniasz PD. 2009. The cell biology of HIV-1 virion genesis. *Cell Host Microbe* 5:550–58
195. Bache KG, Slagsvold T, Stenmark H. 2004. Defective downregulation of receptor tyrosine kinases in cancer. *EMBO J.* 23:2707–12

196. Urwin H, Ghazi-Noori S, Collinge J, Isaacs A. 2009. The role of CHMP2B in frontotemporal dementia. *Biochem. Soc. Trans.* 37:208–12
197. Cox LE, Ferraiuolo L, Goodall EF, Heath PR, Higginbottom A, et al. 2010. Mutations in CHMP2B in lower motor neuron predominant amyotrophic lateral sclerosis (ALS). *PLoS ONE* 5:e9872
198. Filimonenko M, Stuffers S, Raiborg C, Yamamoto A, Malerød L, et al. 2007. Functional multivesicular bodies are required for autophagic clearance of protein aggregates associated with neurodegenerative disease. *J. Cell Biol.* 179:485–500
199. Parkinson N, Ince PG, Smith MO, Highley R, Skibinski G, et al. 2006. ALS phenotypes with mutations in CHMP2B (charged multivesicular body protein 2B). *Neurology* 67:1074–77
200. Jiao J, Sun K, Walker WP, Bagher P, Cota CD, Gunn TM. 2009. Abnormal regulation of TSG101 in mice with spongiform neurodegeneration. *Biochim. Biophys. Acta* 1792:1027–35

RELATED RESOURCE

2009. *UbiGRID*. **http://biogrid.bio.ed.ac.uk:8080/UbiGRID/**
UbiGRID is a bioinformatic resource for mining reported interactions between proteins involved in ubiquitination.

The N-End Rule Pathway

Takafumi Tasaki,[1] Shashikanth M. Sriram,[1] Kyong Soo Park,[2,3] and Yong Tae Kwon[1,2]

[1]Center for Pharmacogenetics and Department of Pharmaceutical Sciences, School of Pharmacy, University of Pittsburgh, Pittsburgh, Pennsylvania 15261; email: yok5@pitt.edu

[2]World Class University Program, Department of Molecular Medicine and Biopharmaceutical Sciences, Graduate School of Convergence Science and Technology and College of Medicine, and [3]Department of Internal Medicine, College of Medicine, Seoul National University, Seoul 110-799, Korea

Keywords

N-degron, arginylation, ubiquitin, proteolysis

Abstract

The N-end rule pathway is a proteolytic system in which N-terminal residues of short-lived proteins are recognized by recognition components (N-recognins) as essential components of degrons, called N-degrons. Known N-recognins in eukaryotes mediate protein ubiquitylation and selective proteolysis by the 26S proteasome. Substrates of N-recognins can be generated when normally embedded destabilizing residues are exposed at the N terminus by proteolytic cleavage. N-degrons can also be generated through modifications of posttranslationally exposed pro-N-degrons of otherwise stable proteins; such modifications include oxidation, arginylation, leucylation, phenylalanylation, and acetylation. Although there are variations in components, degrons, and hierarchical structures, the proteolytic systems based on generation and recognition of N-degrons have been observed in all eukaryotes and prokaryotes examined thus far. The N-end rule pathway regulates homeostasis of various physiological processes, in part, through interaction with small molecules. Here, we review the biochemical mechanisms, structures, physiological functions, and small-molecule-mediated regulation of the N-end rule pathway.

Contents

INTRODUCTION	262
GENERATION OF N-DEGRONS BY CONJUGATION OF AMINO ACIDS	263
Arginylation in the Eukaryotic N-End Rule Pathway	263
Leucylation and Phenylalanylation in the Prokaryotic N-End Rule Pathway	267
GENERATION OF N-DEGRONS VIA DEAMIDATION	269
GENERATION OF N-DEGRONS BY ACETYLATION	270
RECOGNITION OF N-DEGRONS BY CANONICAL N-RECOGNINS	270
The N-Recognin Family Containing the UBR Box	270
Molecular Principles of N-End Rule Interaction	271
The Classical N-End Rule Pathway in *S. cerevisiae*	271
Redundant Functions of Canonical N-Recognins in Mammalian Development	273
OTHER N-RECOGNIN FAMILY MEMBERS	274
Noncanonical N-Recognins in Mammals	274
The Plant N-End Rule Pathway	276
The Acetylation-Based N-End Rule Pathway	276
The Bacterial N-End Rule Pathway	276
Putative N-Recognins: *Drosophila* Inhibitors of Apoptosis	277
UBIQUITIN ACTIVATION AND CONJUGATION IN THE N-END RULE PATHWAY	277
REGULATION OF THE N-END RULE PATHWAY BY SMALL MOLECULES	278
Regulation of the N-End Rule Pathway by Short Peptides	278
Regulation of the N-End Rule Pathway by Hemin	278
USING ENGINEERED N-DEGRONS AS A MOLECULAR TOOL	279
Engineering N-Degrons to Control Protein Stability	279
Using Engineered N-Degrons as Molecular Probes to Dissect the Function of N-Recognins	280
THE N-DEGRON CODE: EVOLUTION AND BEYOND	280
CONCLUDING REMARKS	281

INTRODUCTION

The selectivity in regulated proteolysis is governed by timely generation and recognition of specific degrons on substrates. Degrons on short-lived proteins in eukaryotes are recognized by ubiquitin (Ub) ligases, which mediate the conjugation of Ub to an internal Lys of the substrate, resulting in ATP-dependent degradation by the 26S proteasome. In bacteria, which lack Ub, specific adaptor proteins recognize and deliver protein substrates to proteolytic machinery. The first degron identified in short-lived proteins is a single N-terminal residue, which is targeted by the N-end rule pathway (1). Substrates of N-recognins are generated through N-terminal Met excision or endoproteolytic cleavages of proteins associated with covalent modifications of posttranslationally exposed N-terminal residues. The N-end rule pathway has long been thought to target a limited number of regulatory proteins of the Ub-proteasome system (UPS). However, recent studies suggest that the majority of cellular proteins may carry N-terminal degradation determinants, at least transiently, through protein-specific and global

N-end rule: a rule that relates the in vivo half-life of a given protein to the destabilizing activity of its N-terminal residue

N-recognin: a recognition component of the N-end rule pathway that recognizes N-degrons

posttranslational modifications under certain physiological states.

The N-end rule pathway was discovered in 1986 when Varshavsky and colleagues (1) found that engineered substrates carrying certain N-terminal residues, which were generated from Ub fusion proteins following cleavage by deubiquitylating enzymes, were rapidly degraded in *Saccharomyces cerevisiae* cells (**Figure 1**). A series of genetic analyses in *S. cerevisiae* identified the N-recognin Ubr1 and proteins involved in the generation of N-degrons. Ubr1, a 200-kDa RING E3 ligase, binds a primary destabilizing residue and mediates protein ubiquitylation and subsequent degradation by the proteasome (2, 3). Substrates of Ubr1 include positively charged (Arg, Lys, and His; type 1) and bulky hydrophobic (Phe, Trp, Tyr, Leu, and Ile; type 2) primary destabilizing residues (**Figure 1**). A destabilizing residue is part of N-degrons, and successful degradation through Ubr1 requires additional sequence features, such as an internal Lys residue (the site of a polyubiquitylation) and an unstructured N-terminal extension (4). In the yeast N-end rule pathway, Arg is the principal degron and can be generated through posttranslational modifications, such as arginylation and deamidation, of pro-N-degrons (Asn, Gln, Asp, and Glu). A recent study identified an alternative N-end rule pathway in *S. cerevisiae* in which acetylated N-terminal residues, which occur in the majority of cellular proteins, act as N-degrons (5, reviewed in References 6–8). Hereafter, we refer to the arginylation-based N-end rule pathway as the classical N-end rule pathway or, simply, the N-end rule pathway.

The N-end rule pathway has been identified in all species examined, ranging from mammals (9) and plants (10, 11) to yeasts (1) and bacteria (12). Several excellent reviews have recently discussed the hierarchical structures and basic mechanisms of the N-end rule pathway in eukaryotes and prokaryotes (6–8, 13, 14). Here, we review the biochemical details, substrates, and physiological functions of posttranslational modifications involved in generation of N-degrons and discuss the functions and mechanisms of various eukaryotic and bacterial N-recognins, with an emphasis on the emerging UBR box N-recognin family in mammalian development. The topics also include how components of the N-end rule pathway sense and react to small molecules, such as oxygen, nitric oxide, heme, and peptides with destabilizing residues, by controlling cellular concentrations of their substrates. Finally, we discuss how N-degrons can be engineered to induce proteolysis of other proteins and to serve as an affinity ligand.

GENERATION OF N-DEGRONS BY CONJUGATION OF AMINO ACIDS

In the classical N-end rule pathway of eukaryotes and bacteria, conjugation of destabilizing amino acids to pro-N-degrons is the major way to generate primary destabilizing residues (**Figure 1**). This modification, mediated by evolutionarily conserved aminoacyl-tRNA transferases (the enzyme EC2.3.2), enables pro-N-degrons to be conditionally recognized by N-recognins (15–19). Most substrates of the classical N-end rule pathway identified thus far carry the N-terminal residues derived from the aminoacyl moiety of aminoacyl-tRNAs or pro-N-degrons, whose activity requires aminoacyl-tRNA transferases. Interestingly, recent studies suggest that mammalian and bacterial N-recognins have been structurally optimized to the degrons derived from aminoacyl-tRNAs (Arg in eukaryotes and Leu and Phe in prokaryotes), highlighting the importance of amino acid conjugation at a licensing step prior to irreversible proteolysis (6, 7, 20, 21). Below, we discuss the catalytic mechanisms, enzymatic specificities, evolution, and functions of aminoacyl-tRNA transferases that generate N-degrons.

Arginylation in the Eukaryotic N-End Rule Pathway

In eukaryotes, the N-terminal Arg is the structurally preferred degron for the UBR box of N-recognins (6, 20, 21). The degron Arg can be generated by *ATE1*-encoded arginyl

N-degron: a class of degrons in which the N-terminal destabilizing residue is the major degradation determinant in substrate recognition

E3 ligase: a protein that recognizes a specific substrate and accelerates the transfer of ubiquitin from an E2 enzyme to the substrate

Pro-N-degron: an N-terminal degradation determinant whose modification can generate an N-degron

UBR box: an ~70-residue zinc-finger motif that acts as a substrate recognition domain for type 1 substrates of the N-end rule pathway

(R)-transferases, which transfer Arg from Arg-tRNA to the N-terminal α-amino group of acceptor substrates (**Figure 2**) (15, 18, 22–25). In *S. cerevisiae*, a single R-transferase, encoded by *Ate1* with no known functions, conjugates Arg to the secondary residues Asp and Glu (**Figure 1*b***) (16). By contrast, the mammalian *ATE1* gene expresses at least six isoforms through alternative splicing of pre-mRNA, including those with either of two homologous exons (18, 26, 27). The physiological importance of protein arginylation has been established by the discovery that ATE1-deficient mouse embryos die owing to defects in cardiac and vascular development (24). Although ATE1 isoforms remain poorly characterized in donor and acceptor specificities, tissue distribution, and physiological functions (18, 26, 27),

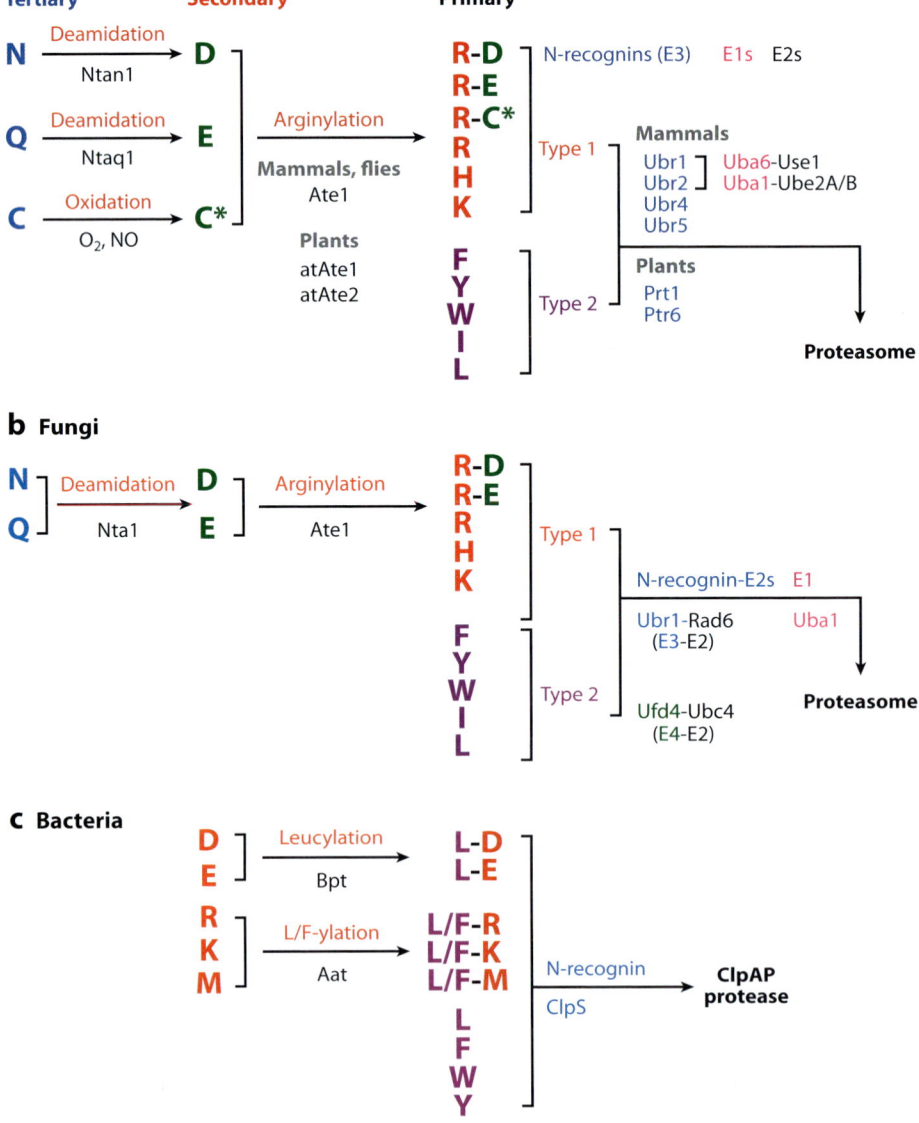

biochemical analyses indicate that Cys as well as Asp and Glu are substrates of arginylation in mammals (reviewed in Reference 4).

Substrates of arginylation include structurally related mammalian RGS (regulator of G-protein signaling) proteins (28–30). These regulators of G protein–signaling function as GTPase-activating proteins for heterotrimeric G protein α-subunits of the i and q classes (**Supplemental Figure 1**; follow the **Supplemental Material link** from the Annual Reviews home page at http://www.annualreviews.org). The degradation of these substrates involves N-terminal Met excision by Met aminopeptidases, which exposes the pro-N-degron Cys2 at the N terminus. The N-terminal Cys2 is subsequently arginylated by ATE1 to generate the degron Arg of N-recognins. Although the exact chemical nature of Cys2 as an arginylation substrate remains murky, various biochemical analyses suggest that the N-terminal Cys2 is not a direct substrate of arginylation but is converted to an arginylation-permissive acceptor following oxidation into Cys sulfinic acid [$CysO_2(H)$] or Cys sulfonic acid [$CysO_3(H)$], a structural mimic of the arginylation substrate Asp (**Supplemental Figure 1**). Interestingly, Cys-dependent degradation of RGS5 is inhibited when molecular oxygen or nitric oxide is deleted (29, 24, 30). On the basis of hypoxia-sensitive proteolysis of RGS proteins and cardiovascular defects of ATE1-deficient mice, it was proposed that oxidation of N-terminal Cys represents an oxygen sensor in cardiovascular development and signaling (4, 24, 30). This conjecture was further supported by recent discoveries that the N-end rule pathway of the plant *Arabidopsis* functions as an oxygen sensor through regulated proteolysis of the hypoxia-sensitive transcription factor family carrying the pro-N-degron Cys2 (31, 32). In normoxia, the ethylene response factor group VII transcription factors, including hypoxia-responsive element 1 and 2 (HRE1 and HRE2) and related to AP2.12

RGS: regulator of G protein–signaling

Figure 1

The classical N-end rule pathway in various eukaryotes and prokaryotes. (*a*) The N-end rule pathway in mammals, flies, and plants. In mammals and other multicellular eukaryotes, the tertiary destabilizing residues Asn and Gln are, respectively, deamidated into the secondary destabilizing residues Asp and Glu by NTAN1 Nt^N-amidase and NTAQ1 Nt^Q-amidase, which in turn are arginylated by *ATE1*-encoded arginyl (R)-transferase isoforms generating the degron Arg. In mammals, N-terminal Cys is converted to a substrate of arginylation through its oxidation into $CysO_2(H)$ or $CysO_3(H)$ prior to arginylation. N-terminal Arg, together with other type 1 and type 2 residues, is recognized and bound by the N-recognin family members, which mediate ubiquitylation and proteasomal degradation, characterized by the UBR box in mammals. Although the components of the plant *Arabidopsis* and fly *Drosophila* N-end rule pathways are not fully characterized, their hierarchical structures appear to be more similar to the mammalian pathway compared to the yeast pathway. In contrast to mammals, the plant *Arabidopsis* genome expresses two distinct R-transferases, AtATE1 and AtATE2, from separate genes. To date, two plant N-recognins, PRT1 and PRT6, have been identified. (*b*) The *S. cerevisiae* N-end rule pathway. A single N-terminal amidohydrolase, Nta1 ($Nt^{N,Q}$-amidase), mediates deamidation of N-terminal Asn and Gln into Asp and Glu, which in turn are arginylated by a single Ate1 R-transferase, generating the degron Arg. N-terminal Arg and other primary degrons are recognized by a single N-recognin Ubr1. (*c*) The bacterial N-end rule pathway. The secondary destabilizing residues Arg, Lys, and Met are conjugated with Leu or Phe by the Aat leucyl/pheylalanyl-tRNA-protein (L/F)-transferase, generating the degrons Leu and Phe. The secondary destabilizing residues Asp and Glu can also be conjugated with Leu by Bpt leucyl-tRNA-protein (L)-transferase, generating the degron Leu. The primary destabilizing residues (Leu, Phe, Trp, and Tyr) are recognized and bound by the N-recognin ClpS, which delivers substrates to the ClpAP protease complex without involving Ub or Ub-like molecules. Abbreviations: Aat, aminoacyl transferase; Bpt, bacterial protein transferase; C*, oxidized Cys; E1, Ub activating enzyme; E2, Ub conjugating enzyme; E3, Ub protein ligase; E4, Ub conjugation factor; L/F-ylation, leucylation/phenylalanylation; N, asparagine; Q, glutamine; Uba1 and Uba6, Ub activating enzymes 1 and 6; Ubc4, Ub conjugating enzyme 4; Ube2A/B, Ub conjugating enzyme E2 A/B; UBR box, Ub ligase N-recognin box; Ubr1, -2, -4, -5, Ub ligase N-recognin1, -2, -4, -5; Ufd4, Ub fusion degradation 4; Use1, Uba6-specific E2 1.

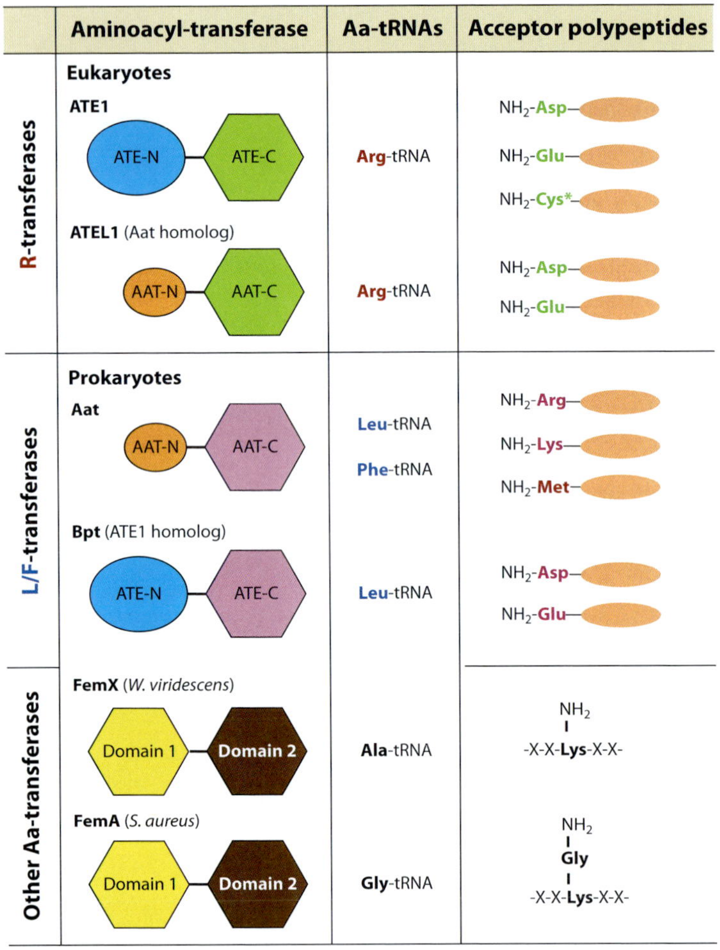

	Aminoacyl-transferase	Aa-tRNAs	Acceptor polypeptides
R-transferases	**Eukaryotes** ATE1 (ATE-N)—(ATE-C)	Arg-tRNA	NH$_2$-Asp— NH$_2$-Glu— NH$_2$-Cys*—
	ATEL1 (Aat homolog) (AAT-N)—(AAT-C)	Arg-tRNA	NH$_2$-Asp— NH$_2$-Glu—
L/F-transferases	**Prokaryotes** Aat (AAT-N)—(AAT-C)	Leu-tRNA Phe-tRNA	NH$_2$-Arg— NH$_2$-Lys— NH$_2$-Met—
	Bpt (ATE1 homolog) (ATE-N)—(ATE-C)	Leu-tRNA	NH$_2$-Asp— NH$_2$-Glu—
Other Aa-transferases	FemX (*W. viridescens*) (Domain 1)—(Domain 2)	Ala-tRNA	NH$_2$ | -X-X-**Lys**-X-X-
	FemA (*S. aureus*) (Domain 1)—(Domain 2)	Gly-tRNA	NH$_2$ | Gly | -X-X-**Lys**-X-X-

Figure 2

Aminoacyl transferases of the N-end rule pathway and structurally related proteins. Eukaryotic and prokaryotic aminoacyl-tRNA (Aa)-transferases can be categorized into arginyl (R)-transferases, leucyl/pheylalanyl-tRNA-protein (L/F)-transferases, and other Aa-transferases on the basis of their enzymatic properties (16, 19, 23). ATE R-transferase and Aat L/F-transferase families can mediate the conjugation of destabilizing amino acids to the N termini of N-end rule substrates, whereas FemX and FemA of the FemABX family mediate the conjugation of amino acids to peptidoglycan pentapeptides, whose residues are shown as the capital letter X. Large (*bright blue*) and small (*orange*) ovals are the conserved ATE-N and AAT-N domains, respectively. Beige ovals represent the protein bodies (substrates) carrying arginylation-permissive N-terminal residues. Hexagons show the evolutionarily conserved GCN5-related N-acetyltransferase (GNAT) fold domains in ATE1 homologs, Aat homologs, and the FemABX family. Structural studies of *Escherichia coli* Aat L/F-transferase and *Weissella viridescens* FemX suggest that this GNAT fold domain is important for recognition of the donor aminoacyl-tRNA and for the enzymatic activity of the transferases (41, 45). Abbreviations: Cys*, the oxidized Cys residue of the acceptor substrate of R-transferase; *S. aureus*, *Staphylococcus aureus*.

(RAP2.12), are downregulated through proteasomal degradation in a manner depending on the pro-N-degron Cys2 (31, 32). In hypoxia, however, these hypoxia-sensitive transcription factors are accumulated, resulting in transcriptional induction of genes that promote anaerobic metabolism and survival of hypoxia. As hypoxia-inducible factor-1 (HIF-1), a known oxygen sensor in animals, is absent in plants, the Cys branch of the N-end rule pathway may represent an oxygen-sensing mechanism in plants. The *Arabidopsis* and human genomes encode at least 206 and 502 proteins, respectively, with the Met-Cys motif (4, 31, 32). Thus, these Met-Cys proteins may represent a unique proteome, whose functions include sensing oxygen and other cellular stresses through oxidation and arginylation of the pro-N-degron Cys.

Polyubiquitination of an ideal N-end rule substrate requires a Lys residue as a site of polyubiquitination and an unstructured N-terminal extension (4). Thus, N-terminal arginylation does not necessarily lead to proteasomal degradation but can serve as a nonproteolytic modification controlling cellular processes. Calreticulin is an endoplasmic reticulum (ER) resident chaperone whose signal peptide is removed upon translocation, exposing the pro-N-degron Asp18 at the N terminus of the mature protein (33). A portion of the mature protein in the ER lumen is arginylated upon retrotranslocation into the cytosol and regulates intracellular relocalization of calreticulin to the stress granule upon cellular stresses, apparently without involving acute proteolysis (33). The protein β-actin, one of most abundant cellular proteins, also undergoes nonproteolytic arginylation at the N-terminal Asp2 or Asp3 following N-terminal Met excision, which plays a role in actin filament properties, actin polymerization, and lamella formation in motile cells (reviewed in Reference 34). In addition to the N terminus, posttranslational arginylation has been observed in the side chain of various residues of many cellular proteins (35). Recent genetic studies revealed various physiological functions of ATE1 in cardiovascular development (24), spermatogenesis (36), metabolism (36), and neural crest migration (37) in animals; seed ripening and germination, shoot and leaf development, and leaf senescence in the plant *Arabidopsis* (10, 38; reviewed in Reference 14); and in apoptosis and viability in the fly *Drosophila* (39). It remains to be determined which of these functions are directly relevant to arginylation-induced proteolysis.

Leucylation and Phenylalanylation in the Prokaryotic N-End Rule Pathway

N-terminal Leu and Phe residues on bacterial proteins are the primary destabilizing residues that can be generated by conjugation of destabilizing amino acids derived from aminoacyl-tRNAs (**Figure 1c**) (16, 19). Two classes of aminoacyl transferases are known to mediate leucylation and phenylalanylation in the N-end rule pathway: leucyl/pheylalanyl-tRNA-protein (L/F)-transferases and leucyl-tRNA-protein (L)-transferases (**Figure 1c**). The *Escherichia coli* L/F-transferase, encoded by *aat*, transfers Leu or Phe to the acceptors Arg and Lys, which are type 1 primary residues in eukaryotes (**Figure 2**) (19). This acceptor specificity was originally determined in the 1971 study by Leibowitz and Soffer (40) and was later confirmed by crystal structures of the L/F-transferase (41). However, recent identification of the first substrate of the bacterial N-end rule pathway, PATase (putrescine aminotransferase), led to an unexpected discovery that the L/F-transferase conjugates Leu or Phe to the N-terminal Met (not Arg or Lys) of PATase (42). Because Met-β-galactosidase is not a substrate of the L/F-transferase and is stable in *E. coli* (12, 42), the substrate specificity of the L/F-transferase needs to be further investigated. The second aminoacyl transferase of the bacterial N-end rule pathway is Bpt L-transferase. In contrast to the Aat L/F-transferase, the Bpt L-transferase transfers Leu to N-terminal Asp and Glu, which are arginylation acceptors in eukaryotes (**Figure 2**) (16). Consistent with R-transferase-like acceptor specificity, the Bpt L-transferase is sequelogous [similar in sequence (43)] to eukaryotic R-transferases but possesses the donor tRNA specificity of prokaryotic Aat L/F-transferase. In the *Vibrio vulnificus* genome, the Aat and Bpt transferases are expressed from a single operon in which *bpt* begins within the stop codon of *aat* (16). The Aat-Bpt tandem gene structure is conserved in a significant portion of β-proteobacteria and γ-proteobacteria but not in *E. coli* and other enterobacteria, a newest group of bacteria where *bpt* has been lost (T. Tasaki and Y.T. Kwon, unpublished data).

The protozoan parasite *Plasmodium falciparum*, which causes malaria in humans, expresses yet another type of transferases, termed ATEL1 (16). ATEL1 is sequelogous to the prokaryotic Aat L/F-transferase but has the enzymatic specificity of eukaryotic ATE1 R-transferases, which conjugates Arg to N-terminal Asp and Glu (**Figure 2**) (16). Thus, eukaryotic R-transferases and prokaryotic Aat L/F-transferase may have a common ancestor but are evolutionarily divergent. Indeed, the C-terminal halves of mammalian R-transferases and bacterial L/F-transferases share a structurally conserved domain, which mediates the synthesis of the interchain peptide of the peptidoglycan and is unique to aminoacyl-tRNA transferases of the FemABX family (**Figure 2**) (44, 45). The conserved GCN5-related N-acetyltransferase folds, which catalyze peptidyltransferase reactions using aminoacyl-tRNA, are confined to the R-transferase, L/F-transferase, and FemABX families (45). The catalytic mechanism of peptide-bond formation by the L/F-transferase was determined using crystal structures of the *E. coli* L/F-transferase in complex with a donor substrate, phenylalanyl adenosine, and an acceptor substrate, the α-casein peptide RYLGYL bearing N-terminal Arg (**Figure 3a,b**) (41). The L/F-transferase mediates phenylalanyl transfer from phenylalanyl adenosine to the peptide RYLGYL, yielding the product peptide FRYLGYL. The N-terminal Phe of the product peptide occupies a hydrophobic pocket with a confined C-shaped edge. The size and shape of the hydrophobic pocket suit hydrophobic residues lacking the branched β-carbon, such as Leu and Phe, but exclude hydrophilic or charged residues (45).

ER: endoplasmic reticulum

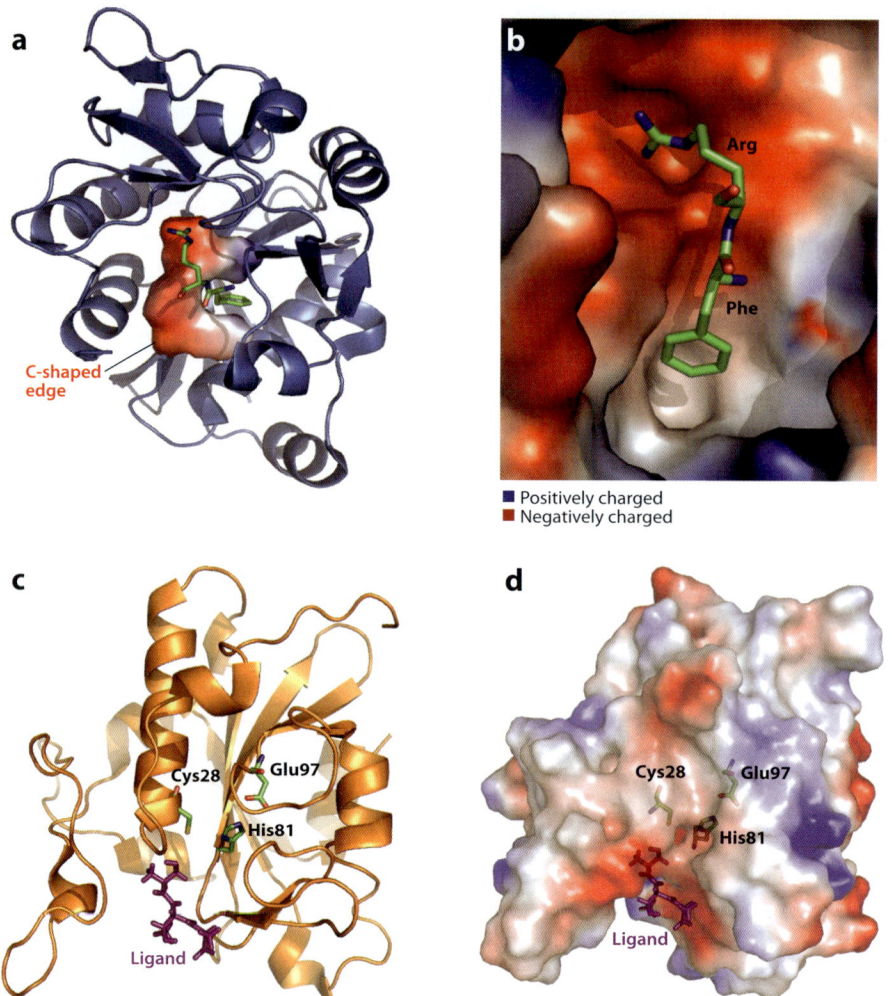

Figure 3

The structures and binding sites of the *E. coli* leucyl/phelyalanyl-tRNA-protein (L/F)-transferase and human NTAQ1. (*a*) Crystal structure of the *E. coli* L/F-transferase complex [Protein Data Bank (PDB) code 2Z3L] with a product peptide. Only the Phe-Arg residues of the bound FRYLG peptide are shown here in a green-colored stick model. The C-shaped structure on the edge of the L/F-transferase hydrophobic pocket is also shown. (*b*) Electrostatic surface potential of L/F-transferase. Blue and red represent positively and negatively charged areas, respectively. The Phe and Arg residues of the FRYLG product peptide are shown in a green-colored stick model. Note that the side chain of Phe resides in the hydrophobic pocket, and the side chain of Arg resides in the negatively charged pocket. (*c*) Crystal structure of human NTAQ1 (PDB code 3C9Q) bound with a peptide-like ligand. The ligand is shown in a purple-colored stick model. NTAQ1 is a monomeric globular protein with a sandwich architecture. The catalytic triad (Cys28, His81, and Glu97) resembles an active site, which is conserved in NTAQ1 proteins. (*d*) Electrostatic surface potential of human NTAQ1. Blue and red colored surfaces depict positively and negatively charged areas, respectively. Note that the catalytic triad residues are in proximity to the binding pocket, bound by the ligand for deamidation.

It has been proposed that the access of the acceptor protein bearing an N-terminal Arg, associated with conformational changes in the L/F-transferase, induces the hydrogen bond breakage between Gln188 and Glu156 and the electron relays between Asp186 and Gln188. This electron relay facilitates the nucleophilic attack of the α-amino group of Arg on the neighboring carbonyl carbon of the esterified aminoacyl-tRNAs, leading to the formation of a peptide bond between Arg and Leu/Phe of the corresponding aminoacyl-tRNA (41).

GENERATION OF N-DEGRONS VIA DEAMIDATION

The tertiary destabilizing residues Asn and Gln on eukaryotic proteins can induce proteasomal degradation of otherwise stable proteins through deamidation into the secondary destabilizing residues Asp and Glu (**Figure 1a,b**) (4, 46, 47). In *S. cerevisiae*, these pro-N-degrons are deamidated by the N-terminal amidohydrolase Nta1 (Nt^{N,Q}-amidase) with no known functions, a member of the Nitrilase superfamily (46, 48). In mammals and other multicellular eukaryotes, N-terminal Asn and Gln are respectively deamidated by *NTAN1*-encoded Nt^N-amidase and *NTAQ1*-encoded Nt^Q-amidase, which are not sequelogous to each other or their yeast counterparts (47–50). Protein sequences and enzymatic properties indicate that these amidases have different evolutionary origins and were independently recruited to the UPS via arginylation (47–50). Despite low sequelogy, they share the principles of substrate recognition and catalysis involving a conserved Cys residue critical for the deamidation activity (46, 49). The physiological function of deamidation was first identified when NTAN1-deficient mice were found to exhibit altered learning, memory, and social behavior (50–52) which were, in part, attributed to misregulation in magnetism-induced proteasomal degradation of the microtubule-associated protein 2 in hippocampal neurons (53, 54).

The cocrystal of human NTAQ1 with a bound peptide (PDB code 3C9Q) illustrated a monomeric globular protein with a three-layer sandwich architecture (**Figure 3c,d**). Its active-site region is structurally similar to that of the human transglutaminase factor XIII, despite lack of sequence similarity (48). Transglutaminase catalyzes formation of an isopeptide bond between the free ε-amino group of a Lys residue and the γ-carboxamide group of a Gln residue (55). The catalytic triad (Cys314, His373, and Asp396) of factor XIII transglutaminase corresponds to the evolutionarily conserved Cys28, His81, and Asp97 of NTAQ1, which are located near the bound peptide (48). It was therefore speculated that NTAQ1 may mediate deamidation of Gln through a transglutamination-like reaction in which a water molecule, instead of the ε-amino group of a Lys, attacks the N-terminal Gln (48).

Substrates of deamidation include the *Drosophila* apoptosis inhibitor, DIAP1. This protein has a short half-life of ∼30 min, in part, through degradation by its RING domain-mediated autoubiquitylation (56, 57). In addition to autoubiquitylation, the N-end rule pathway provides an additional layer of regulation in the turnover of DIAP1 (**Supplemental Figure 2a**) (57). In *Drosophila* cells, DIAP1 can be cleaved after Asp20 by effector caspases (e.g., DrICE and DCP-1), exposing the pre-N-degron Asn21 on Asn21-DIAP1 (57). The N-terminal Asn21 is subsequently deamidated into Asp by Ntan1, which, in turn, is conjugated with Arg by the *Ate1*-encoded R-transferase, resulting in ubiquitylation and proteasomal degradation through N-recognins (57, 58). This pro-N-degron is well conserved in different insect species over an evolutionary distance of ∼300 million years (59). Consistent with N-end rule degradation of DIAP1, inactivation of a single *Ntan1* or *Ate1* allele was found to inhibit head involution defective (HID)-induced apoptosis in the developing eyes of transgenic flies overexpressing the DIAP1 antagonist HID (57). Intriguingly, an analogous analysis resulted in an opposite effect, i.e., accelerated apoptosis when apoptosis was induced by another DIAP1 antagonist, Reaper (57), suggesting that the N-end rule

DIAP1: *Drosophila* inhibitor of apoptosis 1

pathway is part of a complicated network maintaining homeostasis in apoptosis.

The pro-N-degron Asn21 is also required for optimal activity of DIAP1 as an inhibitor of apoptosis (56, 60). For example, whereas overexpression of full-length DIAP1 or Asn21-DIAP1 efficiently inhibits Reaper-induced apoptosis, metabolically more stable Met21-DIAP1 does not show such activity (56, 57). According to the N-end rule, the nonproteolytic activity of Asn21 requires its deamidation into Asp and subsequent arginylation to generate the degron Arg. Ditzel et al. (56) proposed a model in which an N-recognin is recruited by the degron Arg of arginylated DIAP1 to form a heterodimeric E3 complex targeting an effector caspase that also has been recruited to DIAP1 through the BIR domain (**Supplemental Figure 2b**).

GENERATION OF N-DEGRONS BY ACETYLATION

In *S. cerevisiae*, acetylation of N-terminal residues can generate N-degrons, which are targeted by the N-recognin Doa10 in conjunction with the Ubc6 or Ubc7 E2 enzyme (5). N-terminal acetylation is one of the most common protein modifications in eukaryotes, occurring on ~57% of yeast proteins and ~84% human proteins but rarely on prokaryotic proteins (61). This global modification is catalyzed by ribosome-associated N-terminal acetyltransferases (NATs), which transfer acetyl groups from acetyl-coenzyme A to the N-terminal α-amino group. Acetylation starts when the nascent chain with the initiator Met emerges from the ribosome at a length of ~25 residues or somewhat later (~50 residues) if the initiator Met has to be removed by methionine aminopeptidase. Thus, one important factor in N-terminal acetylation is the primary structure of the N-terminal region if NAT and acetyl coenzyme A are not rate limiting (62).

Three major NAT complexes (NatA, NatB, and NatC) conserved from yeast to humans are thought to be responsible for the majority of acetylation, whereas other NATs (NatD, NatE, and NatF) remain poorly characterized (63). NatA acetylates N-terminal Ala, Ser, Thr, Val, Gly, and Cys following N-terminal Met excision by methionine aminopeptidases. NatB or NatC acetylates the N-terminal Met when the second residue is either acidic or hydrophobic, respectively. Eukaryotic proteins with N-terminal Ser (90%) and Ala (50%) are the most frequently acetylated, and these residues along with Met, Gly, and Thr account for over 95% of N-terminal acetylated residues (61, 64). Despite these substrate specificities of NATs, only a subset of proteins with these consensus residues are acetylated (63). Polevoda & Sherman (63, 64) reported that NatA and NatB have a tendency to avoid basic (Lys, Arg, and His) and Pro residues at any positions on the nascent chain during translation, whereas NatC avoids acidic residues. Therefore, the overall efficiency of acetylation may be the net effect of acetylation-permissive and inhibitory residues near the N terminus. It remains unknown how many of N-terminally acetylated proteins undergo proteasomal degradation through N-recognin.

RECOGNITION OF N-DEGRONS BY CANONICAL N-RECOGNINS

Substrate selectivity in the N-end rule pathway is governed by recognition of N-degrons by N-recognins, which induce, in eukaryotes, protein ubiquitylation and proteolysis through the proteasome. The mechanism of substrate selectivity was revealed by the discovery that the UBR box, conserved in many N-recognins, is the substrate recognition domain (65–67). Below, we discuss the enzymatic specificities, structures, functions, and substrates of N-recognins carrying the UBR box, with an emphasis on the mammalian N-recognin family in proteolytic and nonproteolytic physiological processes, including the pathogenesis of human genetic diseases.

The N-Recognin Family Containing the UBR Box

UBR boxes of *S. cerevisiae* Ubr1 and mammalian UBR1 and UBR2 (Ub ligase N-recognin1 and

-2) bind type 1 residues with the dissociation constant of low micromolars (65–67). The mammalian genome encodes at least seven UBR box proteins, UBR1–UBR7 (**Figure 4***a*) (65). With the exception of UBR4, the UBR box family members contain signatures of the substrate recognition components of the UPS (**Figure 4***a*) (65). We here refer to UBR1/E3α, UBR2, and UBR3 as canonical, owing to their sequelogy, size (about 200 kDa), and conserved domains, including the UBR box (type 1 binding site), N domain (type 2 binding site), RING finger (ubiquitylation domain), and autoinhibitory domain (which sterically blocks the UBR box and N domain via intramolecular interaction) (65, 68). UBR4 through UBR7, referred to as noncanonical UBR box proteins, are evolutionarily divergent and nonsequelogous to each other. On the basis of binding and degradation assays, UBR1, UBR2, UBR4, and UBR5 are classified as N-recognins, and UBR3, UBR6, and UBR7 as non-N-recognins (65, 66).

Molecular Principles of N-End Rule Interaction

Recently determined crystal structures of UBR boxes of canonical N-recognins revealed the molecular principles of N-end rule interactions for type 1 degrons (7, 20, 21, 66; reviewed in References 6 and 7). The UBR box consists of two zinc fingers: a typical Cys_2His_2 motif containing a zinc ion and an atypical binuclear Cys_6His_1 motif containing two zinc ions (7, 20, 21). UBR boxes bind to positively charged N-terminal residues through a negatively charged, shallow binding groove (**Figure 5***a,b*). The recognition of N-end rule substrates initiates with hydrogen bonding with the free α-amino group of the N-terminal residue, a unique structure conserved in all proteins. Once engaged with a genuine substrate while scanning the N termini, N-recognin establishes a substrate-selective interaction through hydrogen bonds with the positively charged side chains; this interaction is further supported by additional hydrogen bonds with the side chain of the second residues and the backbone atoms of the first three residues. Overall, N-end rule interactions are largely confined to the first two residues as the side chain at position three stays away from the surface of the UBR box (7, 20, 21), enabling N-recognins to select substrates on the basis of destabilizing N-terminal residues.

Canonical N-recognins have a second, structurally distinct substrate-binding domain, the N domain, which binds type 2 degrons (Phe, Trp, Tyr, Leu, and Ile) (**Figure 4***a*) (66). The N domain appears to be derived from and spalogous (similar in structure) to bacterial N-recognin ClpS, ATP-dependent Clp protease adaptor protein, which recognizes the primary destabilizing residues (Phe, Trp, Tyr, and Leu) (69).

The Classical N-End Rule Pathway in *S. cerevisiae*

In yeasts, a single canonical N-recognin, Ubr1, recognizes type 1 (Arg, Lys, and His) and type 2 (Phe, Trp, Tyr, Leu, and Ile) N termini in concert with the Ub-conjugating enzyme Ubc2/Rad6 (**Figure 1***b*) (2, 3). Substrates of Ubr1 include Scc1, a subunit of the cohesin complex, which holds the sister chromatids connected together during metaphase (70; reviewed in Reference 8). At the metaphase-to-anaphase transition, separase cleavage of Scc1 produces a C-terminal fragment with the degron Arg, which is degraded through Ubr1 for efficient chromatin separation. Notably, analogous cleavage events by separases in many eukaryotes generate destabilizing N-terminal residues on C-terminal Scc1 fragments (Arg in *S. cerevisiae*, Asn in *S. pombe*, Glu in mammals, and Cys in *D. melanogaster*) (8), indicating a strong evolutionary pressure to regulate SCC1 proteins by the N-end rule pathway. Ubr1 can also recognize internal degrons (non-N-degrons). This class of substrates includes the homeodomain protein Cup9 (a transcriptional repressor of the Ptr2 peptide transporter) (71–73), GPA1 (the Gα-subunit that controls signal transduction during mating) (74), and Mgt1 (the O6-alkylguanine-DNA

N domain: the type 2 substrate-recognition domain of N-recognins; this domain has a secondary structure similar to that of the ClpS N-recognin of prokaryotes

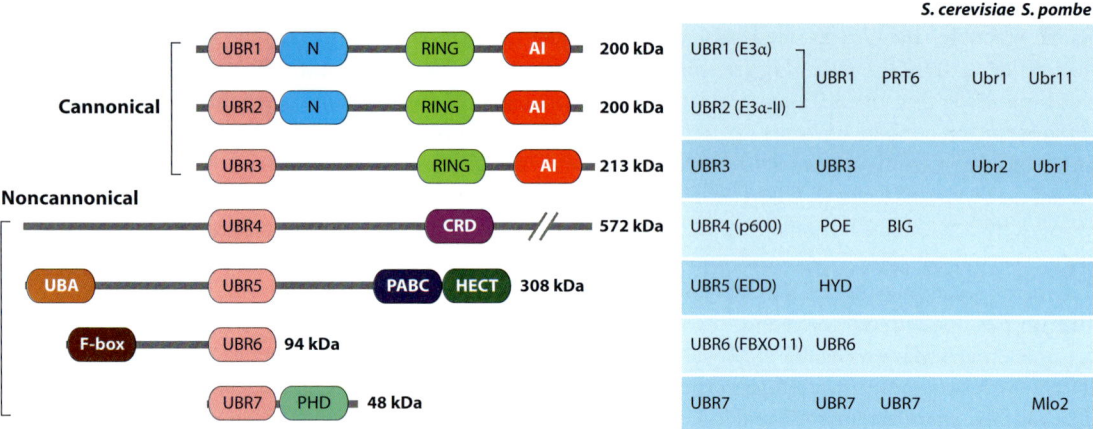

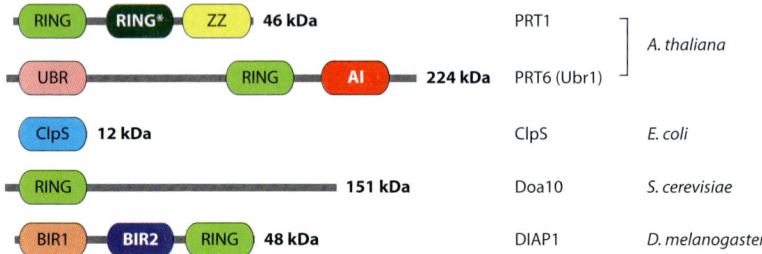

Figure 4

The UBR box protein family and N-recognins. (*a*) The mammalian UBR box protein family. UBR1 and UBR2 (200 kDa) are functionally overlapping canonical UBR box N-recognins. UBR3 (213 kDa) is a canonical UBR box protein but does not show affinity to N-end rule peptides (65, 66, 143). Knockout of *UBR3* in mice resulted in neonatal death associated with female-specific anosmia, a finding consistent with its unique expression in neural tissues of the so-called five senses (olfaction, hearing, vision, touch, and taste). UBR3 mediates degradation of the DNA repair protein APE1 and is required for genomic stability (144). UBR3 is sequelogous and thought to be a functional homolog to *S. cerevisiae* Ubr2 and *Schizosaccharomyces pombe* Ubr1. The homologs in yeasts are involved in transcriptional regulation of the proteasome (through degradation of the transcriptional activator Rpn4), sexual differentiation, nuclear enrichment of the proteasome (through degradation of the nuclear envelope protein Cut8), and the oxidative stress response (145, 146 and references therein; reviewed in References 3, 4, and 8). UBR4 is a sequelog of the *Arabidopsis* BIG, which plays a role in auxin transport, root hair elongation, hormone and light responses, and the regulation of sulfur deficiency-responsive genes (147 and references therein; reviewed in Reference 4). The *Drosophila* homolog of UBR4, POE/PUSH/CALO, has been implicated as an interactor of calmodulin in the retina and the polar granule molecules Vasa and Tudor in germ plasm from early embryos; synaptic transmission; perineurial glial growth; male sterility; and meiotic chromosome pairing, recombination, and segregation in females (reviewed in Reference 4). UBR5 is a sequelog of the *Drosophila* HYD, whose mutations result in imaginal disc hyperplasia associated with uncontrolled cell proliferation through the independent activation of *hedgehog* and *decapentaplegic* (reviewed in Reference 4). UBR6/FBXO11 (94 kDa), a component of a SCF E3 complex, has been implicated in the neddylation of p53 (148) and the human diseases vitiligo (a skin disorder) and otitis media (a common childhood disease characterized by middle ear inflammation following infection) (149). UBR7 (48 kDa) and its sequelogs in multicellular organisms have a PHD domain, which resembles the RING domain. *S. pombe* mlo2, a putative UBR7 homolog, is implicated in chromosome transmission fidelity in mitosis (reviewed in Reference 4). (*b*) Known and putative N-recognins of other species. Abbreviations: AI, autoinhibitory domain; BIR; baculoviral inhibition of apoptosis protein repeat; ClpS, ATP-dependent Clp protease adaptor protein; CRD, cystein-rich domain; Doa10, the ER-localized ubiquitin ligase Doa10; DIAP1, *Drosophila* inhibitor of apoptosis 1; PHD, plant homeodomain finger; HECT, homologous to the E6-AP C terminus; N, N domain; PABC, poly(A)-binding protein C-terminal domain; RING*, composite domain containing RING and CCCH-type Zn fingers; RING, RING finger; UBA domain, Ub association domain; UBR, UBR box; ZZ, a specific zinc finger domain that binds to two zinc ions.

alkyltransferase) (75). In addition to selective proteolysis on the basis of protein-specific degrons of normally folded substrates, Ubr1 mediates protein quality control through degradation of misfolded proteins in conjunction with chaperons, such as Hsp70 and Sse1 (76, 77). Interestingly, the processivity of Ubr1 for some of these substrates was shown to be accelerated by the Ufd4 HECT domain E3 (78, 75, 79). Ufd4 is the recognition component of the Ub fusion degradation pathway that targets the noncleavable N-terminal Ub moiety as a degron (75, 78, 79). Ufd4 in complex with Ubr1 acts as an E4-like processive-enhancement cofactor for Lys48-linked ubiquitylation by Ubr1 once Ubr1 recognizes a substrate (8). This synergistic cotargeting was verified in ubiquitylation on the basis of both N-degrons and internal degrons (75, 78, 79).

Redundant Functions of Canonical N-Recognins in Mammalian Development

UBR1-deficient mice are alive but exhibit pleiotrophic phenotypes, such as defective growth, muscle protein degradation, and fat metabolism, as well as moderate hypoglycemia (80). The mutants also exhibit pancreatic abnormalities with defective secretion of digestive enzymes from exocrine cells, reminiscent of a human genetic disorder, Johanson-Blizzard syndrome (Online Mendelian Inheritance in Man code 243800; **http://www.ncbi.nlm.nih.gov/omim**) (81, 82). In humans, mutations in *UBR1* are the primary cause of this autosomal recessive disorder, characterized by exocrine pancreatic insufficiency and additional clinical features with aplasia or hypoplasia of the alae nasi, oligodontia, sensorineural hearing loss, hypothyroidism, scalp defects, mental retardation, and developmental delay (80, 82). *UBR1* (Online Mendelian Inheritance in Man code 605981) spanning a 161-kb region of chromosome 15q15-q21.1 is highly expressed in acinar cells of the pancreas (82). Several nonsense, splice site, and frameshift mutations as well as missense mutations of *UBR1* were identified (82, 83 and the references therein), among which complete or near-complete null mutations correlate to the Johanson-Blizzard syndrome. Thus, a UBR1-deficiency in humans and mice involves common pathogenic mechanisms with impaired excretion of zymogens and irreversible acinar cell damage.

UBR2-deficient mice develop distinct phenotypes depending on the genetic background and gender (84). In the 129/SvJ genetic background, the mutants of both genders die before adulthood. In contrast, the mutants in a mixed background between the 129/SvJ and C57BL6 strains show infertility in males and partial lethality in females (84). Infertility is caused by arrest of spermatocytes at meiotic prophase I, resulting in apoptosis of spermatocytes and degeneration of testicular tubules (84). Substrates of UBR2 in spermatocytes include histone H2A (85). In vitro UBR2 functions as a scaffolding E3 ligase that mediates monoubiquitylation and polyubiquitylation of H2A by promoting the interaction between the E2 enzyme HR6B/RAD6 and H2A (85). In pachytene spermatocytes, ubiquitylated H2A is implicated in the transcriptional silencing of sex chromosomes. UBR2-deficient spermatocytes fail to maintain a normal level of ubiquitylated H2A and to induce transcriptional silencing of many genes linked to the X and Y chromosomes (85). Thus, in meiotic spermatocytes, a portion of UBR2 molecules (*a*) is associated with meiotic chromatin and (*b*) regulates chromatin dynamics and gene expression through ubiquitylation of H2A.

UBR1 and UBR2 are similar in size (200 kDa), have 46% sequence identity, have conserved domains, and exhibit similar specificity to type 1 and type 2 residues (84). Knockout of *UBR1* in mice resulted in tissue-specific, partial inactivation of the N-end rule pathway (80). Cells lacking both UBR1 and UBR2 still contain significant activities for degradation of N-end rule substrates, perhaps because of functional redundancy between N-recognins (65). In contrast to single-knockout mice, embryos lacking both UBR1 and UBR2 die at about embryonic day 11.0, which is associated with impaired neurogenesis and cardiovascular development (86). During neurogenesis in the neural tube, double-mutant neural progenitors differentiate prematurely, resulting in depletion of self-renewing progenitors, consistent with impaired Notch-1 signaling, which controls the balance between proliferation and differentiation of neural cells (86).

Sequelog: a protein whose nucleotide or amino acid sequence is similar, to a significant extent, to another sequence

OTHER N-RECOGNIN FAMILY MEMBERS

Canonical N-recognins (yeast Ubr1 and its sequelogs) mediate selective proteolysis on the basis of type 1 and type 2 destabilizing N-terminal residues. Recent studies identified new N-recognins that are nonsequelogous to canonical N-recognins but can recognize various N-degrons generated through protein-specific or global modifications. Below, we discuss the biochemical mechanisms, structures, functions, and substrates of recently identified N-recognins from mammals, plants, flies, yeasts, and bacteria.

Noncanonical N-Recognins in Mammals

Mammalian UBR4, an extraordinarily big protein (570 kDa) with no known ubiquitylation domain, has a UBR box and binds to both type 1 and type 2 residues (**Figure 4a**) (65). Mammalian UBR5 [alternatively called EDD (E3 ligase identified by differential display) or hHYD (homolog of the *Drosophila melanogaster* hyperplastic disc)] and its *Drosophila* sequelog selectively bind type 1 residues (65) and have a UBR box, a UBA (Ub association) domain (which can bind to a polyubiquitin chain), a HECT domain, a nuclear localization signal, and PABC domain (65). Knockdown of *UBR4* in cells lacking UBR1 and UBR2 almost completely impaired the type 2 pathway but not the type 1 pathway as determined by degradation of model substrates and the human

Figure 5

Comparison of structures and binding sites of the mammalian UBR box, the bacterial ClpS, and the *Drosophila* BIR2 domain. (*a*) Electrostatic potential of human UBR2 UBR box bound with the peptide Arg-Ile-Phe-Ser [Protein Data Bank (PDB) code 3NY3]. The N-terminal Arg binds to a negatively charged, shallow binding groove, the second residue binds to a hydrophobic pocket, and the other following residues are turned away from the UBR box surface. The white surface color indicates a hydrophobic surface. Note the mononucleate (Cys_2His_2) zinc finger of the UBR box on top of the peptide and the unique binucleate zinc finger directly behind the UBR box contact surface with the peptide. (*b*) Key N-terminal recognition hydrogen bonds (*dotted lines*) of UBR box residues with Arg. Arg atoms are coded by color as follows: green, carbon; red, oxygen; and blue, nitrogen. Note the characteristic interactions, including the electrostatic interaction of the basic Arg side chain with the acidic UBR box surface and the H-bonding interactions from the α-amino group, side chain, and the peptide bond. (*c*) Electrostatic potential of *C.aulobacter crescentus* ClpS bound with the peptide Trp-Leu-Phe (PDB code 3GQ1). The hydrophobic N-terminal residue binds to a deep hydrophobic pocket, and the interaction from the second residue occurs outside of the pocket. (*d*) Key hydrogen bonds (*dotted lines*) of ClpS residues with N-terminal Trp. Trp atoms are coded by color as indicated above. The blue sphere is a water molecule. Note the characteristic interactions, including the hydrophobic interaction of the Trp side chain with the hydrophobic ClpS pocket and the H-bonding interactions from the α-amino group, side chain, and the peptide bond. (*e*) Electrostatic potential of the *Drosophila melanogaster* DIAP1 BIR2 domain with Grim peptide (PDB code 1JD5). The eight-residue peptide (AIAYFIPD) binds the BIR2 surface groove in an extended conformation, and this binding induces a major conformational switch in DIAP1, although the BIR2 binding grove itself is fairly rigid. Note the DIAP1 zinc finger (Cys_3His). (*f*) Key hydrogen bonds (*dotted lines*) of BIR2 residues with the N-terminal Ala of the Grim peptide. Ala atoms are coded by color as indicated above. Note that the N-terminal Ala is positioned in a highly negatively charged BIR2 environment, with the main recognition specificity provided by the H-bonding from the α-amino group and the first peptide bond. This mode of interaction of the N-terminal Ala in Grim peptide to the BIR2 domain of DIAP1 is similar to that of the UBR box interaction with N-terminal Arg, indicating a putative N-recognin behavior of DIAP1.

immunodeficiency virus type 1 integrase bearing the N-degron Phe (65).

Mice lacking UBR4 die during embryogenesis with defects in angiogenesis (T. Tasaki & Y.T. Kwon, unpublished data). Mammalian UBR4 has been implicated in integrin-mediated signaling and cell adhesion (87), E7-mediated cellular immortalization (88, 89), and neuronal migration and development (90). Proteins that interact with mammalian UBR4 include molecules involved in the progression between G1 and S phases, such as the pRb retinoblastoma tumor suppressor (87), the E7 oncoprotein of human and bovine papillomavirus (88, 89), and calmodulin (87). Mice lacking UBR5 also die during

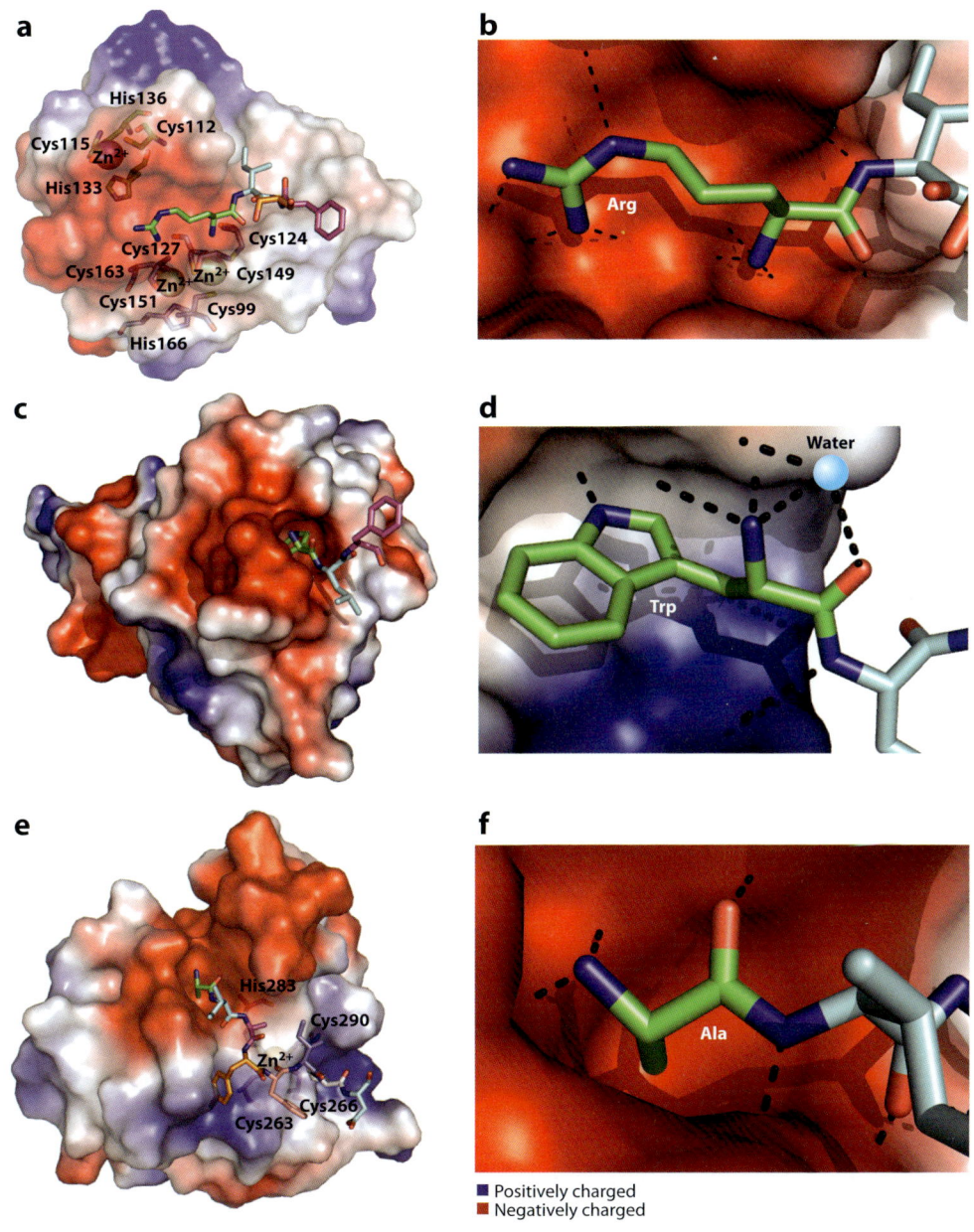

embryogenesis with defects in the development of yolk sac vasculature (91). Mammalian UBR5 plays a role in various processes, such as progesterone-regulated cell proliferation, DNA repair, tumorigenesis, and smooth muscle differentiation (91), in part, through ubiquitylation of its known substrates, TopBP1 (92), the PABP-interacting protein 2 (Paip2) (93), and CDK9 (94), and through interaction with a number of cellular proteins: p53 (95); CIB/KIP (calcium- and integrin-binding protein/DNA-dependent protein kinase-interacting protein) (96); importin $\alpha 5$ (a component of the nuclear import complex) (96); and the progesterone receptor, PR (to potentiate progestin-mediated gene transactivation) (96). Mammalian UBR4 and UBR5 have their homologs in various multicellular eukaryotes but not in fungi (65), and their physiological functions are described in the **Figure 4a** caption.

The Plant N-End Rule Pathway

The hierarchical structure, degrons, and components of the N-end rule pathway in the plant *Arabidopsis* are largely similar to those in mammals except for the genes encoding and specificities of N-recognins (10, 14). Two N-recognins, PRT1 and PRT6, have been identified in plants. PRT1, a 410-residue RING finger protein with a ZZ domain, is nonsequelogous to known N-recognins and does not have the UBR box and the N domain (97). The mutation of *PRT1* stabilizes model substrates bearing aromatic hydrophobic residues (Phe, Trp, and Tyr) at the N termini but not aliphatic hydrophobic residues, such as Leu and Ile (97). PRT6 is a 2006-residue RING finger E3 ligase sequelogous to yeast Ubr1 and has the UBR box but apparently not the N domain (98). *Arabidopsis prt6* mutants are impaired in degradation of type 1, but not type 2, substrates (98). Thus, the mutations in PRT1 and PRT6 selectively stabilize the N-terminal aromatic and basic destabilizing residues, respectively, without affecting substrates with a subset of type 2 degrons, Leu and Ile. It remains unknown whether PRT1 and PRT6 functionally overlap in recognition of the Leu and Ile N-degrons or whether there is yet another N-recognin.

The Acetylation-Based N-End Rule Pathway

A recent study showed that an acetylated N-terminal residue can act as an N-degron in the *S. cerevisiae* N-end rule pathway (5). N-terminal acetylation is irreversible and typically occurs cotranslationally at the retained N-terminal Met residue or at a newly exposed N-terminal amino acid (Ala, Val, Ser, Thr, or Cys) following N-terminal Met excision by Met aminopeptidases (99). The Doa10 E3 ligase in complex with the Ubc6 or Ubc7 E2 enzyme was identified as an N-recognin that targets at least eight substrates carrying acetylated N-degrons (5). Doa10 is a transmembrane protein, which mediates ubiquitination of misfolded proteins on the cytosolic surface of the ER membrane (100). It is unknown whether the acetylated N-degron-based proteolytic system targets the majority of or only a few cellular proteins.

The Bacterial N-End Rule Pathway

In *E. coli*, N-terminal Leu, Phe, Trp, and Tyr are primary destabilizing residues for the bacterial N-recognin ClpS (**Figures 1c** and **4b**) (12, 69, 101–103). ClpS mediates N-end rule degradation without Ub-like molecules (**Figure 4b**) (12, 69, 104); this is in contrast to mammals, where N-recognins use Ub as a secondary degron for proteasomal delivery (**Supplemental Figure 3b**). ClpS recognizes N-end rule substrates through a preformed, deep hydrophobic pocket within which the N-terminal side chain of substrates is completely buried, and the α-amino group and the first peptide bond make additional contacts through hydrogen bonding (**Figure 5c,d**) (102, 103). ClpS-bound substrates are delivered through a multistep process to the ClpAP complex, a ring-shaped proteolytic machinery that functions like the 26S proteasome (**Supplemental Figure 3a**) (104). The ClpP proteolytic complex, a counterpart of the 20S core particle, is a stack of two rings of heptameric complexes in which the active sites are exposed on the internal surface. The size of the chamber is relatively small in diameter (10 Å) so the AAA+ protein ClpA, which forms a hexameric ring, unfolds and feeds ClpS-bound substrates into the ClpP chamber (105). Docking and release of substrate-bound ClpS to ClpAP involve the interaction of ClpS with both the substrate and ClpA on the N- and C-terminal domains of ClpS, respectively (104, 106). While holding the substrate on the N-terminal domain, the C-terminal domain of ClpS docks on a highly mobile N-terminal domain of ClpA. This induces the movement of the ClpA N-terminal domain, placing ClpS-bound substrate in proximity to the ClpAP pore. Substrate-bound ClpS now undergoes a conformational change in the N-terminal region, resulting in its contact with a second unknown site on ClpA. This triggers a conformational change in ClpA and/or ClpS, which in turn results in the delivery of the substrate to the ClpAP complex.

Putative N-Recognins: *Drosophila* Inhibitors of Apoptosis

The apoptosis inhibitors (DIAP1 and DIAP2) and their mammalian homologs (XIAP, cIAPs, and ML-IAP) are RING E3s that suppress undesirable apoptotic activities by inhibiting the functions of initiator and effecter caspases (57, 107, 108). Ditzel & Meier with their coworkers (56, 107, 108) and Yoo et al. (109) proposed that DIAP1 acts as an N-recognin-like component that recognizes N-terminal Ala of peptidase-cleaved caspases as a major determinant in binding and subsequently mediates proteasomal degradation. DIAP1-dependent downregulation of caspases requires the binding of the DIAP1 BIR (baculovirus IAP repeat) domain to the N-terminally exposed IBM (IAP-binding motif) of caspases (110). The IBM is a tetrapeptide consensus sequence [Ala-X-(P/A)-Y, X and Y are hydrophobic] in which N-terminal unmodified Ala is essential for binding with the DIAP1 BIR, contributing to degradation of caspases and IAP antagonists and acting as an N-degron-like element. The IBMs of caspases are typically generated through cleavage by caspases and other proteases during apoptotic induction (111).

The crystal structures of the DIAP1 BIR2 domain in complex with a Hid or Grim peptide show that the BIR2-IBM interaction is similar to that of the UBR box with type 1 residues (**Figure 5***e,f*) (112). The BIR2 binds the IBM on the surface of a hydrophobic groove formed by a ∼70-residue zinc-finger motif (112). The binding involves two hydrogen bonds between the N-terminal α-amino group of the IBM and two negatively charged BIR2 residues (Asp277 and Gln282); this is supported by additional two hydrogen bonds between the first peptide bond of the IBM and surrounding BIR2 residues (112). The N-terminal side chain fits tightly in the binding pocket of the BIR domain. As N-terminal Ala binds to the BIR domain, the next six residues interact with the surface of the shallow groove (112). Thus, N-terminal Ala is an anchoring component in the BIR-IBM interaction, functioning as an N-degron-like determinant. This action mode in ubiquitylation of caspases and IAP antagonists defines DIAP1 and many other IAPs as N-recognin-like components in regulated proteolysis. It remains to be determined whether N-terminal Ala of the IBM is a strong degron.

UBIQUITIN ACTIVATION AND CONJUGATION IN THE N-END RULE PATHWAY

Ub can be activated by two Ub-activating enzymes, UBA1 and UBA6 (113–116) (**Supplemental Figure 4**). UBA1 is the major E1 Ub-activating enzyme responsible for the majority of Ub conjugation to E2 enzymes and substrates (113). UBA6, containing all three conserved domains of UBA1, is an alternative E1 enzyme, representing 5%–10% of UBA1 in abundance. In contrast to UBA1, which works with a broad range of E2s, UBA6 has a designated E2, USE1 (115). Homologs of human UBA6 and USE1 are found in mammals, zebra fish, and sea urchins, but not in worms, flies, and yeast. Mouse embryos lacking UBA6 (also termed E1-L2) die before midgestation (114), indicating the essential role of UBA6 in mammalian development. The physiological E3s and substrates of the UBA6-USE1 system have remained unknown until the recent discovery (116) that UBA6 mediates Ub activation and conjugation for the canonical N-recognins, UBR1 and UBR2, as well as for a non-N-recognin, UBR3. These E3 ligases mediate ubiquitylation with either UBA1-activated UBE2 (alternatively called RAD6) or UBA6-activated USE1. Substrates of UBA6-activated N-recognins include RGS4, whose ubiquitylation can be regulated by both cytosol-specific UBA6 and ubiquitous UBA1 (116).

REGULATION OF THE N-END RULE PATHWAY BY SMALL MOLECULES

The functions of the N-end rule pathway are regulated by various mechanisms, including transcription (80), posttranslational modifications (phosphorylation and ubiquitylation) (117, 118), and interactions with the Ub ligase Ufd4 (78, 75) and small molecules (71, 73, 117, 119). Below, we discuss how components of the N-end rule pathway interact with and react to small peptides and heme by adjusting cellular concentrations of short-lived proteins.

Regulation of the N-End Rule Pathway by Short Peptides

In *S. cerevisiae*, extracellular dipeptides and tripeptides are mainly imported by the transporter Ptr2 whose level is tightly regulated by the N-end rule pathway in response to changing nutritional availability (**Supplemental Figure 5**) (71, 73). When extracellular peptides are limited, the transcriptional repressor Cup9, a substrate of Ubr1, is accumulated and shuts down the transcription of Ptr2 to prevent unnecessary synthesis of the transporter machinery (71–73). However, if cells are exposed to a peptide-rich environment, Ubr1 binds to destabilizing N-terminal residues of constitutively imported small peptides and undergoes an allosteric conformational change, exposing its Cup9-binding site, which is otherwise masked by the C-terminal autoinhibitory domain through intramolecular interaction. Following activation by peptides, Ubr1 mediates the degradation of Cup9 [$t_{1/2}$ (half-life) < 1 min], leading to transcriptional derepression of Ptr2 and accelerated peptide import (72). This allosteric activation of Ubr1 is maximally induced when both type 1 and type 2 peptides bind to Ubr1 (72). Using this positive-feedback circuit, cells sense the levels of extracellular peptides and maintain the homeostasis in peptide uptake.

S. cerevisiae cells sense free amino acids, in particular Trp and Leu, as an indication of a protein-rich environment. Ubr1-dependent degradation of Cup9 links the availability of extracellular free amino acids to transcriptional induction of Ptr2 through the amino acid sensing SPS (Ssy1-Ptr3-Ssy5) pathway (119). Ssy1 is an integral membrane protein that can bind to amino acids. Amino acid–bound Ssy1, together with the peripheral membrane protein Ptr3, activates the chymotrypsin-like endoprotease Ssy5, which requires the casein kinases Yck1 and Yck2 (120). Activated Ssy5 in turn cleaves cytoplasmic retention sequences of the transcription factors Stp1 and Stp2, releasing them to induce the transcription of Ptr2 and other proteins involved in peptide transport. In this process, the kinases Yck1 and Yck2 phosphorylate Ubr1 at Ser300, and this is critical for efficient ubiquitylation of Cup9 and thus the induction of Ptr2 (**Supplemental Figure 5b**) (118).

Regulation of the N-End Rule Pathway by Hemin

Heme is an iron (Fe^{2+})-containing protoporphyrin IX, and hemin is its ferric (Fe^{3+}) counterpart. Hemoproteins, including globins, oxidoreductases (e.g., catalase, peroxidase, P450s, cytochrome oxidases, and nitric oxide synthases), and cytochromes, play a role in electron transfer, redox modification, and sensing of diatomic gases (e.g., oxygen and nitric oxide) (117). A recent study showed that hemin binds to components of the N-end rule pathway and inhibits their functions in the degradation of short-lived proteins (117). Upon binding to hemin, the functions of R-transferases in yeasts and mammals are dually downregulated, in part, by inactivation of enzymatic activities, which involves the formation of a disulfide bond between two adjacent Cys residues (Cys71 and Cys72 in mouse ATE1), and, in part, by metabolic destabilization through the UPS (117). In addition, hemin binds and inhibits in vitro the aminoacylation activity of the human arginyl-tRNA synthetase (the enzyme EC6.1.1.19) that synthesizes Arg-tRNA^{Arg}, a cofactor required for arginylation by R-transferases (121). Hemin also binds and prevents yeast Ubr1 from dipeptide-induced allosteric conformational changes, which would otherwise unmask the Cup9-binding site of Ubr1 from inhibition by the autoinhibitory domain (117). As a consequence, hemin-bound Ubr1 cannot properly ubiquitylate Cup9 even in the presence of dipeptides (117). These results suggest that the N-end rule pathway contributes to homeostasis of heme/hemin and other redox-related molecules by regulating concentrations of proteins involved in redox-dependent processes.

USING ENGINEERED N-DEGRONS AS A MOLECULAR TOOL

Engineered N-degrons have been used to control the metabolic stabilities of various proteins through the N-end rule pathway. Short peptides or small-molecule ligands carrying destabilizing amino acids have been used to inhibit, promote, or probe the interactions of N-recognins with substrates or to identify new N-recognins and their interactors. As the basic principles of N-end rule interactions are conserved across species in eukaryotes and prokaryotes (7, 20, 21, 103), the approaches described below should be generally applicable for a broad range of proteins.

Engineering N-Degrons to Control Protein Stability

The first engineered N-degron, the **X**-e^k extension, is an ∼40-residue *E. coli* Lac repressor fragment bearing an N-terminal destabilizing residue (**X**) (1). The Ub-fusion technique (1, 122) was employed to generate **X** through deubiquitylating enzyme-mediated cotranslational cleavage of Ub-**X**-e^k-protein at the Ub-X junction (**Supplemental Figure 6a**). When fused with β-galactosidase ($t_{1/2} > 20$ h), the Arg-e^k extension induced rapid proteolysis in yeast cells with a half-life of ∼2 min (**Supplemental Figure 6b**) (1). The **X**-e^k extension was used as a portable degron to control the concentration of *S. cerevisiae* ARD1 expressed from the *GAL10* promoter (123). Arg-e^k-ARD, ARD carrying Arg-e^k extension, was normally maintained at a physiologically functional level but was rapidly depleted by turning off the promoter. Ghislain et al. (124) employed a different strategy in which the function of a protein carrying a portable degron was controlled by the expression of Ubr1. Using this strategy, Arg-e^k-Spr54 was accumulated when Ubr1 expression from the GAL1 promoter was repressed and was rapidly depleted when Ubr1 expression was restored. Taxis et al. (125) reported an inducible N-degron using the *Tobacco etch virus* protease, which cleaves the Q**X** junction of the ENLYFQ**X** consensus sequence (X is an N-degron). This cleavage exposed an N-degron on the C-terminal fragment that was conditionally degraded by *Tobacco etch virus* protease expressed from the GAL1 promoter. Dohmen & Varshavsky (126, 127) developed a portable heat-inducible N-degron that can induce temperature-sensitive (ts) degradation through the N-end rule pathway (**Supplemental Figure 6c**). The ts N-degron is based on 21-kDa mouse dihydroforate reductase bearing the Arg degron. The ts Arg dihydroforate reductase is long-lived at 23°C but rapidly degraded at 37°C ($t_{1/2} <$ 10 min) owing to a Pro-to-Leu mutation at position 66, which causes a ts conformational change at the N terminus (126–128). The ts Arg dihydrofolate reductase was shown to induce conditional degradation of other proteins that were fused to their N termini without making individual ts mutants (126–128). Bernal & Venkitaraman (129) applied the heat-inducible degron strategy to avian *Gallus gallus* DT40 cells, which have high homologous recombination efficiency and grow at ambient temperatures. Portable degrons were also used to develop short-lived fluorescent reporters of gene expression and protein localization. Li et al. (130) engineered enhanced green fluorescent protein (EGFP) with the PEST degron of mouse ornithine decarboxylase to produce moderately short-lived EGFP ($t_{1/2} < \sim 2$ h). Dantuma and coworkers (131, 132) tagged an engineered N-degron at the EGFP N terminus to generate fluorescent N-end rule substrates with half-lives of 10 min to 2 h. These substrates were used as reporters of the UPS to study proteasome inhibitors, ER stress, valosin-containing protein (p97), Ran-binding protein-2, and the N-end rule in chloroplasts (131, 133–135). Lee et al. (136) developed a real-time maker of caspase in cultured cells by inserting a caspase cleavage site between the N-degron and EGFP. Finally, Hackett et al. (137) developed cyan fluorescent protein-based fluorescent N-end rule substrates to study various transcriptional activities in *S. cerevisiae*.

Using Engineered N-Degrons as Molecular Probes to Dissect the Function of N-Recognins

The 12-mer X peptides derived from the Sindbis virus polymerase nsP4, an N-end rule substrate (138), were used as an affinity ligand to characterize the interaction of N-recognins with the N termini, leading to elucidation of the peptide-induced allosteric conformational change of yeast Ubr1 in peptide transport (72) and identification of substrate recognition domains of mammalian N-recognins (65, 66). An affinity-based proteomic study using X peptides reported the isolation of endogenous UBR4 and UBR5 from mammalian and *Drosophila* cells (65). The two substrate-binding sites of N-recognins were also exploited as a target of multivalent inhibitors (139–141). In a test-of-concept study (139), a heterotetramer comprising Arg-e^k-βgal and Leu-e^k-βgal, which expose type 1 and type 2 residues from the X-e^k extension, significantly inhibited the degradation of N-end rule substrates in yeast cells. A recent study reported a nonproteinaceous compound, RF-C11 (831 kDa), in which single amino acids (Arg and Phe) were linked by a C11 hydrocarbon chain (140). RF-C11 significantly inhibited the degradation of N-end rule substrates in vivo and in vitro, with higher efficacy compared to monovalent (e.g., dipeptides) or homodivalent controls (e.g., RR-C11 with two Arg ligands), and cardiac hypertrophic responses in cardiomyocytes (140).

THE N-DEGRON CODE: EVOLUTION AND BEYOND

Below, we present a model to reconstitute a series of evolutionary events that led to the current N-degron code comprising 13 amino acids in the genetic code. In this model, the eukaryotic N-end rule pathway was born with the N-recognin Ubr1 that utilizes the UBR box to target type 1 substrates as part of the UPS, whereas ancient bacteria adopted a distinct folding of ClpS to recognize type 2 degrons. Substrate specificities of the N-recognin were expanded when the *ClpS* gene of a prokaryotic endosymbiont was incorporated into *Ubr1*, perhaps about a billion years ago when prokaryotic endosymbionts in eukaryotic cells were converted to organelles, such as the mitochondrion, and possibly the ER as well (142). Because the zinc-finger motif of the UBR box is superior in structure and regulation (e.g., redox modification and allosteric modulation by N-end rule ligands), the *ClpS*-encoded N domain became structurally dependent on the UBR box. In parallel, R-transferase that had evolved from bacterial aminoacyl-tRNA transferases became functionally linked to Ubr1 and thus provided a licensing step prior to irreversible degradation by the proteasome. As mammals evolved, the N-end rule pathway added yet another regulatory mechanism by recruiting

$Nt^{N,Q}$-amidase with a eukaryotic origin, which mediates deamidation of the pro-N-degrons Asn and Gln. An analogous process occurred in mammals with Nt^N-amidase and Nt^Q-amidase through convergence evolution. Although this model explains fairly well the hierarchical structures of the N-end rule pathways, there are several questions remaining to be answered. For example, the majority of known substrates carry the degron Arg or pro-N-degrons (Asn, Gln, Cys, Glu, and Asp) that work through the degron Arg. By sharp contrast, all the primary destabilizing residues (Lys, His, Trp, Phe, Tyr, Leu, and Ile) except for Arg are highly underrepresented in physiological substrates. One intriguing possibility is that these underrepresented primary residues may represent not degrons but ligands (e.g., short peptides) that bind to and modulate the functions of N-recognins. As such, the existence of natural nonpeptide compounds that induce allosteric conformational changes in N-recognins as part of a negative feedback loop is cautiously predicted.

CONCLUDING REMARKS

Since 1986, the N-end rule pathway has long remained as an orphan proteolytic system without clear functions or substrates. Systematic genetic and biochemical dissection over the past decade revealed numerous insights into its components, functions, and mechanisms. The N-end rule pathway is now emerging as a major cellular proteolytic system in which the bulk of cellular proteins potentially carry N-terminal destabilizing residues at least temporarily during their life cycles. What should we focus on in the coming decade? One outstanding issue is the identity of substrates of the classical N-end rule pathway, which lists 13 out of 20 principal amino acids as degradation determinants, and for this reason, the pathway is anticipated to target a large number of substrates. Despite extensive efforts for the past 25 years, only a limited number of substrates have been identified, perhaps because the majority of substrates are conditionally generated under specific physiological processes, for example, apoptotic induction, cellular stresses, or cell cycle transition. Many N-degrons exposed after proteolytic cleavage may exist only briefly and at low concentrations prior to rapid degradation, making it difficult to detect such events under standard assay conditions. A reliable method for genome-wide screening is urgently needed to better understand the N-end rule pathway. One possible approach would be global protein stability profiling with genome-wide cDNA libraries to identify short-lived proteins whose ubiquitylation is sensitive to N-end rule inhibitors (30). It is also important to link the knowledge from known substrates to phenotypes identified from genetic studies in animals. *S. cerevisiae* has only one N-recognin, but mammals have at least four N-recognins whose functions may be overlapping in specific tissues, which makes it challenging to dissect the function of a specific N-recognin. It is still unclear how these mammalian N-recognins cooperate in degradation of N-end rule substrates and whether they recognize N-terminal and second residues of physiological substrates using similar molecular principles. We anticipate that the structures of large fragments containing both the UBR box and the N domain will be elucidated in few years. It is important to understand how small-molecule ligands induce allosteric conformational changes in two binding sites of N-recognins; this information will provide a structural basis for the design of potent inhibitors.

SUMMARY POINTS

1. The N-end rule defines the destabilizing activity of a given N-terminal residue and its posttranslational modification. N-recognins recognize destabilizing N-terminal residues as essential elements of N-degrons and mediate protein degradation.

2. Pro-N-degrons can be converted to N-degrons through conjugation of destabilizing amino acids by aminoacyl-tRNA transferases, which transfer amino acids from aminoacyl-tRNA to the N termini. Eukaryotic R-transferases catalyze arginylation, generating the degron Arg, whereas prokaryotic leucyl/pheylalanyl-tRNA-protein (L/F)-transferases and L-transferases catalyze leucylation and phenylalanylation, generating the degrons Leu and Phe.

3. The tertiary destabilizing residues Asn and Gln are deamidated into the secondary destabilizing residues Asp and Glu, which are subsequently arginylated by R-transferase, generating the primary destabilizing residue Arg.

4. Known mammalian N-recognins bind to type 1 and type 2 substrates through the UBR box and the N domain. Positively charged type 1 residues bind to the negatively charged surface of the UBR box. The N domain is homologous to bacterial ClpS, which binds bulky hydrophobic type 2 residues through a deep hydrophobic pocket.

5. The N-end rule pathway can be regulated by small molecules, such as short peptides and hemin. Short peptides bind the UBR box and the N domain of yeast Ubr1 and synergistically induce the proteolysis of substrates carrying internal degrons, such as Cup9, as a feedback mechanism to maintain the homeostasis of peptide transport.

FUTURE ISSUES

1. Although a number of physiological substrates of the N-end rule pathway have been discovered, many more are likely to remain unknown. A reproducible assay system is urgently needed to screen substrates of the N-end rule pathway.

2. What are the mechanisms that regulate activation of pro-N-degrons, generation of N-degrons, and activity of N-recognins?

3. How are physiological processes in mammals regulated by selective proteolysis of substrates bearing N-degrons? How do mammalian N-recognins cooperate in targeting N-degrons?

4. The structure of the N domain is unknown. It is anticipated that the binding of small-molecule ligands to the UBR box and the N domain of canonical N-recognins synergistically induces a conformational change to expose a site that recognizes internal degrons of substrates.

5. The bulk of eukaryotic proteins (50%–80%) can be acetylated at the N termini. What is the physiological meaning of the acetylation-based N-end rule pathway?

DISCLOSURE STATEMENT

The authors are not aware of any affiliations, memberships, funding, or financial holdings that might be perceived as affecting the objectivity of this review.

ACKNOWLEDGMENTS

We are grateful to Dong Hoon Han and William T. Kwon for editorial assistance. Work in the authors' laboratories was supported by the National Institutes of Health grant (HL083365 to Y.T.K.) and World Class University (R31-2008-000-10103-0 to Y.T.K. and K.S.P.) and World Class Institute (WCI 2009-002 to Bo Yeon Kim) programs through the National Research Foundation funded by the Ministry of Education, Science and Technology, Korea. This article is dedicated to my father who is under post-stroke rehabilitation (Y.T.K.). Takafumi Tasaki is now affiliated with the Medical Research Institute, Kanazawa Medical University, Ishikawa, 920-0293, Japan.

LITERATURE CITED

1. Bachmair A, Finley D, Varshavsky A. 1986. In vivo half-life of a protein is a function of its amino-terminal residue. *Science* 234:179–86
2. Bartel B, Wünning I, Varshavsky A. 1990. The recognition component of the N-end rule pathway. *EMBO J.* 9:3179–89
3. Varshavsky A. 1996. The N-end rule: functions, mysteries, uses. *Proc. Natl. Acad. Sci. USA* 93:12142–49
4. Tasaki T, Kwon YT. 2007. The mammalian N-end rule pathway: new insights into its components and physiological roles. *Trends Biochem. Sci.* 32:520–28
5. Hwang CS, Shemorry A, Varshavsky A. 2010. N-terminal acetylation of cellular proteins creates specific degradation signals. *Science* 327:973–77
6. Sriram SM, Kim BY, Kwon YT. 2011. The N-end rule pathway: emerging functions and molecular principles of substrate recognition. *Nat. Rev. Mol. Cell Biol.* 12:735–47
7. Sriram SM, Kwon YT. 2010. The molecular principles of N-end rule recognition. *Nat. Struct. Mol. Biol.* 17:1164–65
8. Varshavsky A. 2011. The N-end rule pathway and regulation by proteolysis. *Protein Sci.* 20:1298–345
9. Gonda DK, Bachmair A, Wünning I, Tobias JW, Lane WS, Varshavsky A. 1989. Universality and structure of the N-end rule. *J. Biol. Chem.* 264:16700–12
10. Graciet E, Walter F, Maoiléidigh DO, Pollmann S, Meyerowitz EM, et al. 2009. The N-end rule pathway controls multiple functions during *Arabidopsis* shoot and leaf development. *Proc. Natl. Acad. Sci. USA* 106:13618–23
11. Potuschak T, Stary S, Schlögelhofer P, Becker F, Nejinskaia V, Bachmair A. 1998. PRT1 of *Arabidopsis thaliana* encodes a component of the plant N-end rule pathway. *Proc. Natl. Acad. Sci. USA* 95:7904–8
12. Tobias JW, Shrader TE, Rocap G, Varshavsky A. 1991. The N-end rule in bacteria. *Science* 254:1374–77
13. Dougan DA, Truscott KN, Zeth K. 2010. The bacterial N-end rule pathway: expect the unexpected. *Mol. Microbiol.* 76:545–58
14. Graciet E, Wellmer F. 2010. The plant N-end rule pathway: structure and functions. *Trends Plant Sci.* 15:447–53
15. Balzi E, Choder M, Chen WN, Varshavsky A, Goffeau A. 1990. Cloning and functional analysis of the arginyl-tRNA-protein transferase gene *ATE1* of *Saccharomyces cerevisiae*. *J. Biol. Chem.* 265:7464–71
16. Graciet E, Hu RG, Piatkov K, Rhee JH, Schwarz EM, Varshavsky A. 2006. Aminoacyl-transferases and the N-end rule pathway of prokaryotic/eukaryotic specificity in a human pathogen. *Proc. Natl. Acad. Sci. USA* 103:3078–83
17. Kaji H, Novelli GD, Kaji A. 1963. A soluble amino acid–incorporating system from rat liver. *Biochim. Biophys. Acta* 76:474–77
18. Kwon YT, Kashina AS, Varshavsky A. 1999. Alternative splicing results in differential expression, activity, and localization of the two forms of arginyl-tRNA-protein transferase, a component of the N-end rule pathway. *Mol. Cell. Biol.* 19:182–93
19. Shrader TE, Tobias JW, Varshavsky A. 1993. The N-end rule in *Escherichia coli*: cloning and analysis of the leucyl, phenylalanyl-tRNA-protein transferase gene *aat*. *J. Bacteriol.* 175:4364–74
20. Choi WS, Jeong BC, Joo YJ, Lee MR, Kim J, et al. 2010. Structural basis for the recognition of N-end rule substrates by the UBR box of ubiquitin ligases. *Nat. Struct. Mol. Biol.* 17:1175–81

21. Matta-Camacho E, Kozlov G, Li FF, Gehring K. 2010. Structural basis of substrate recognition and specificity in the N-end rule pathway. *Nat. Struct. Mol. Biol.* 17:1182–87
22. Ciechanover A, Ferber S, Ganoth D, Elias S, Hershko A, Arfin S. 1988. Purification and characterization of arginyl-tRNA-protein transferase from rabbit reticulocytes. Its involvement in post-translational modification and degradation of acidic NH2 termini substrates of the ubiquitin pathway. *J. Biol. Chem.* 263:11155–67
23. Kaji H. 1968. Further studies on the soluble amino acid incorporating system from rat liver. *Biochemistry* 7:3844–50
24. Kwon YT, Kashina AS, Davydov IV, Hu RG, An JY, et al. 2002. An essential role of N-terminal arginylation in cardiovascular development. *Science* 297:96–99
25. Soffer RL, Horinishi H. 1969. Enzymic modification of proteins. I. General characteristics of the arginine-transfer reaction in rabbit liver cytoplasm. *J. Mol. Biol.* 43:163–75
26. Hu RG, Brower CS, Wang H, Davydov IV, Sheng J, et al. 2006. Arginyltransferase, its specificity, putative substrates, bidirectional promoter, and splicing-derived isoforms. *J. Biol. Chem.* 281:32559–73
27. Rai R, Kashina A. 2005. Identification of mammalian arginyltransferases that modify a specific subset of protein substrates. *Proc. Natl. Acad. Sci. USA* 102:10123–28
28. Davydov IV, Varshavsky A. 2000. RGS4 is arginylated and degraded by the N-end rule pathway in vitro. *J. Biol. Chem.* 275:22931–41
29. Hu RG, Sheng J, Qi X, Xu Z, Takahashi TT, Varshavsky A. 2005. The N-end rule pathway as a nitric oxide sensor controlling the levels of multiple regulators. *Nature* 437:981–86
30. Lee MJ, Tasaki T, Moroi K, An JY, Kimura S, et al. 2005. RGS4 and RGS5 are in vivo substrates of the N-end rule pathway. *Proc. Natl. Acad. Sci. USA* 102:15030–35
31. Gibbs DJ, Lee SC, Isa NM, Gramuglia S, Fukao T, et al. 2011. Homeostatic response to hypoxia is regulated by the N-end rule pathway in plants. *Nature* 479:415–18
32. Licausi F, Kosmacz M, Weits DA, Giuntoli B, Giorgi FM, et al. 2011. Oxygen sensing in plants is mediated by an N-end rule pathway for protein destabilization. *Nature* 479:419–22
33. Carpio MA, López Sambrooks C, Durand ES, Hallak ME. 2010. The arginylation-dependent association of calreticulin with stress granules is regulated by calcium. *Biochem. J.* 429:63–72
34. Saha S, Kashina A. 2011. Posttranslational arginylation as a global biological regulator. *Dev. Biol.* 358:1–8
35. Wang J, Han X, Saha S, Xu T, Rai R, et al. 2011. Arginyltransferase is an ATP-independent self-regulating enzyme that forms distinct functional complexes in vivo. *Chem. Biol.* 18:121–30
36. Brower CS, Varshavsky A. 2009. Ablation of arginylation in the mouse N-end rule pathway: loss of fat, higher metabolic rate, damaged spermatogenesis, and neurological perturbations. *PLoS ONE* 4:e7757
37. Kurosaka S, Leu NA, Zhang F, Bunte R, Saha S, et al. 2010. Arginylation-dependent neural crest cell migration is essential for mouse development. *PLoS Genet.* 6:e1000878
38. Yoshida S, Ito M, Callis J, Nishida I, Watanabe A. 2002. A delayed leaf senescence mutant is defective in arginyl-tRNA:protein arginyltransferase, a component of the N-end rule pathway in *Arabidopsis*. *Plant J.* 32:129–37
39. Spradling AC, Stern D, Beaton A, Rhem EJ, Laverty T, et al. 1999. The Berkeley *Drosophila* Genome Project gene disruption project: single P-element insertions mutating 25% of vital *Drosophila* genes. *Genetics* 153:135–77
40. Leibowitz MJ, Soffer RL. 1971. Enzymatic modification of proteins. VII. Substrate specificity of leucyl, phenylalanyl-transfer ribonucleic acid-protein transferase. *J. Biol. Chem.* 246:5207–12
41. Watanabe K, Toh Y, Suto K, Shimizu Y, Oka N, et al. 2007. Protein-based peptide-bond formation by aminoacyl-tRNA protein transferase. *Nature* 449:867–71
42. Ninnis RL, Spall SK, Talbo GH, Truscott KN, Dougan DA. 2009. Modification of PATase by L/F-transferase generates a ClpS-dependent N-end rule substrate in *Escherichia coli*. *EMBO J.* 28:1732–44
43. Varshavsky A. 2004. 'Spalog' and 'sequelog': neutral terms for spatial and sequence similarity. *Curr. Biol.* 14:R181–83
44. Rai R, Mushegian A, Makarova K, Kashina A. 2006. Molecular dissection of arginyltransferases guided by similarity to bacterial peptidoglycan synthases. *EMBO Rep.* 7:800–5

45. Suto K, Shimizu Y, Watanabe K, Ueda T, Fukai S, et al. 2006. Crystal structures of leucyl/phenylalanyl-tRNA-protein transferase and its complex with an aminoacyl-tRNA analog. *EMBO J.* 25:5942–50
46. Baker RT, Varshavsky A. 1995. Yeast N-terminal amidase. A new enzyme and component of the N-end rule pathway. *J. Biol. Chem.* 270:12065–74
47. Grigoryev S, Stewart AE, Kwon YT, Arfin SM, Bradshaw RA, et al. 1996. A mouse amidase specific for N-terminal asparagine. The gene, the enzyme, and their function in the N-end rule pathway. *J. Biol. Chem.* 271:28521–32
48. Wang H, Piatkov KI, Brower CS, Varshavsky A. 2009. Glutamine-specific N-terminal amidase, a component of the N-end rule pathway. *Mol. Cell* 34:686–95
49. Cantor JR, Stone EM, Georgiou G. 2011. Expression and biochemical characterization of the human enzyme N-terminal asparagine amidohydrolase. *Biochemistry* 50:3025–33
50. Kwon YT, Balogh SA, Davydov IV, Kashina AS, Yoon JK, et al. 2000. Altered activity, social behavior, and spatial memory in mice lacking the NTAN1p amidase and the asparagine branch of the N-end rule pathway. *Mol. Cell. Biol.* 20:4135–48
51. Balogh SA, Kwon YT, Denenberg VH. 2000. Varying intertrial interval reveals temporally defined memory deficits and enhancements in NTAN1-deficient mice. *Learn. Mem.* 7:279–86
52. Balogh SA, McDowell CS, Kwon YT, Denenberg VH. 2001. Facilitated stimulus-response associative learning and long-term memory in mice lacking the NTAN1 amidase of the N-end rule pathway. *Brain Res.* 892:336–43
53. Goto Y, Taniura H, Yamada K, Hirai T, Sanada N, et al. 2006. The magnetism responsive gene *Ntan1* in mouse brain. *Neurochem. Int.* 49:334–41
54. Hirai T, Taniura H, Goto Y, Ogura M, Sng JC, Yoneda Y. 2006. Stimulation of ubiquitin-proteasome pathway through the expression of amidohydrolase for N-terminal asparagine (Ntan1) in cultured rat hippocampal neurons exposed to static magnetism. *J. Neurochem.* 96:1519–30
55. Pedersen LC, Yee VC, Bishop PD, Le Trong I, Teller DC, Stenkamp RE. 1994. Transglutaminase factor XIII uses proteinase-like catalytic triad to crosslink macromolecules. *Protein Sci.* 3:1131–35
56. Ditzel M, Broemer M, Tenev T, Bolduc C, Lee TV, et al. 2008. Inactivation of effector caspases through nondegradative polyubiquitylation. *Mol. Cell* 32:540–53
57. Ditzel M, Wilson R, Tenev T, Zachariou A, Paul A, et al. 2003. Degradation of DIAP1 by the N-end rule pathway is essential for regulating apoptosis. *Nat. Cell Biol.* 5:467–73
58. Herman-Bachinsky Y, Ryoo HD, Ciechanover A, Gonen H. 2007. Regulation of the *Drosophila* ubiquitin ligase DIAP1 is mediated via several distinct ubiquitin system pathways. *Cell Death Differ.* 14:861–71
59. Tenev T, Ditzel M, Zachariou A, Meier P. 2007. The antiapoptotic activity of insect IAPs requires activation by an evolutionarily conserved mechanism. *Cell Death Differ.* 14:1191–201
60. Yan N, Wu JW, Chai J, Li W, Shi Y. 2004. Molecular mechanisms of DrICE inhibition by DIAP1 and removal of inhibition by Reaper, Hid and Grim. *Nat. Struct. Mol. Biol.* 11:420–28
61. Polevoda B, Arnesen T, Sherman F. 2009. A synopsis of eukaryotic Nalpha-terminal acetyltransferases: nomenclature, subunits and substrates. *BMC Proc.* 3(Suppl. 6):S2
62. Driessen HP, de Jong WW, Tesser GI, Bloemendal H. 1985. The mechanism of N-terminal acetylation of proteins. *CRC Crit. Rev. Biochem.* 18:281–325
63. Polevoda B, Sherman F. 2003. N-terminal acetyltransferases and sequence requirements for N-terminal acetylation of eukaryotic proteins. *J. Mol. Biol.* 325:595–622
64. Polevoda B, Sherman F. 2000. Nalpha-terminal acetylation of eukaryotic proteins. *J. Biol. Chem.* 275:36479–82
65. Tasaki T, Mulder LC, Iwamatsu A, Lee MJ, Davydov IV, et al. 2005. A family of mammalian E3 ubiquitin ligases that contain the UBR box motif and recognize N-degrons. *Mol. Cell. Biol.* 25:7120–36
66. Tasaki T, Zakrzewska A, Dudgeon DD, Jiang Y, Lazo JS, Kwon YT. 2009. The substrate recognition domains of the N-end rule pathway. *J. Biol. Chem.* 284:1884–95
67. Xia Z, Webster A, Du F, Piatkov K, Ghislain M, Varshavsky A. 2008. Substrate-binding sites of UBR1, the ubiquitin ligase of the N-end rule pathway. *J. Biol. Chem.* 283:24011–28
68. Kwon YT, Reiss Y, Fried VA, Hershko A, Yoon JK, et al. 1998. The mouse and human genes encoding the recognition component of the N-end rule pathway. *Proc. Natl. Acad. Sci. USA* 95:7898–903

69. Erbse A, Schmidt R, Bornemann T, Schneider-Mergener J, Mogk A, et al. 2006. ClpS is an essential component of the N-end rule pathway in *Escherichia coli*. *Nature* 439:753–56
70. Rao H, Uhlmann F, Nasmyth K, Varshavsky A. 2001. Degradation of a cohesin subunit by the N-end rule pathway is essential for chromosome stability. *Nature* 410:955–59
71. Byrd C, Turner GC, Varshavsky A. 1998. The N-end rule pathway controls the import of peptides through degradation of a transcriptional repressor. *EMBO J.* 17:269–77
72. Du F, Navarro-Garcia F, Xia Z, Tasaki T, Varshavsky A. 2002. Pairs of dipeptides synergistically activate the binding of substrate by ubiquitin ligase through dissociation of its autoinhibitory domain. *Proc. Natl. Acad. Sci. USA* 99:14110–15
73. Turner GC, Du F, Varshavsky A. 2000. Peptides accelerate their uptake by activating a ubiquitin-dependent proteolytic pathway. *Nature* 405:579–83
74. Madura K, Varshavsky A. 1994. Degradation of Gα by the N-end rule pathway. *Science* 265:1454–58
75. Hwang CS, Shemorry A, Varshavsky A. 2009. Two proteolytic pathways regulate DNA repair by cotargeting the Mgt1 alkylguanine transferase. *Proc. Natl. Acad. Sci. USA* 106:2142–47
76. Eisele F, Wolf DH. 2008. Degradation of misfolded protein in the cytoplasm is mediated by the ubiquitin ligase Ubr1. *FEBS Lett.* 582:4143–46
77. Heck JW, Cheung SK, Hampton RY. 2010. Cytoplasmic protein quality control degradation mediated by parallel actions of the E3 ubiquitin ligases Ubr1 and San1. *Proc. Natl. Acad. Sci. USA* 107:1106–11
78. Hwang CS, Shemorry A, Auerbach D, Varshavsky A. 2010. The N-end rule pathway is mediated by a complex of the RING-type Ubr1 and HECT-type Ufd4 ubiquitin ligases. *Nat. Cell Biol.* 12:1177–85
79. Johnson ES, Ma PC, Ota IM, Varshavsky A. 1995. A proteolytic pathway that recognizes ubiquitin as a degradation signal. *J. Biol. Chem.* 270:17442–56
80. Kwon YT, Xia Z, Davydov IV, Lecker SH, Varshavsky A. 2001. Construction and analysis of mouse strains lacking the ubiquitin ligase UBR1 (E3α) of the N-end rule pathway. *Mol. Cell. Biol.* 21:8007–21
81. Rezaei N, Sabbaghian M, Liu Z, Zenker M. 2011. Eponym: Johanson-Blizzard syndrome. *Eur. J. Pediatr.* 170:179–83
82. Zenker M, Mayerle J, Lerch MM, Tagariello A, Zerres K, et al. 2005. Deficiency of UBR1, a ubiquitin ligase of the N-end rule pathway, causes pancreatic dysfunction, malformations and mental retardation (Johanson-Blizzard syndrome). *Nat. Genet.* 37:1345–50
83. Fallahi GH, Sabbaghian M, Khalili M, Parvaneh N, Zenker M, Rezaei N. 2011. Novel *UBR1* gene mutation in a patient with typical phenotype of Johanson-Blizzard syndrome. *Eur. J. Pediatr.* 170:233–35
84. Kwon YT, Xia Z, An JY, Tasaki T, Davydov IV, et al. 2003. Female lethality and apoptosis of spermatocytes in mice lacking the UBR2 ubiquitin ligase of the N-end rule pathway. *Mol. Cell. Biol.* 23:8255–71
85. An JY, Kim EA, Jiang Y, Zakrzewska A, Kim DE, et al. 2010. UBR2 mediates transcriptional silencing during spermatogenesis via histone ubiquitination. *Proc. Natl. Acad. Sci. USA* 107:1912–17
86. An JY, Seo JW, Tasaki T, Lee MJ, Varshavsky A, Kwon YT. 2006. Impaired neurogenesis and cardiovascular development in mice lacking the E3 ubiquitin ligases UBR1 and UBR2 of the N-end rule pathway. *Proc. Natl. Acad. Sci. USA* 103:6212–17
87. Nakatani Y, Konishi H, Vassilev A, Kurooka H, Ishiguro K, et al. 2005. p600, a unique protein required for membrane morphogenesis and cell survival. *Proc. Natl. Acad. Sci. USA* 102:15093–98
88. DeMasi J, Chao MC, Kumar AS, Howley PM. 2007. Bovine papillomavirus E7 oncoprotein inhibits anoikis. *J. Virol.* 81:9419–25
89. Huh KW, DeMasi J, Ogawa H, Nakatani Y, Howley PM, Münger K. 2005. Association of the human papillomavirus type 16 E7 oncoprotein with the 600-kDa retinoblastoma protein-associated factor, p600. *Proc. Natl. Acad. Sci. USA* 102:11492–97
90. Shim SY, Wang J, Asada N, Neumayer G, Tran HC, et al. 2008. Protein 600 is a microtubule/endoplasmic reticulum-associated protein in CNS neurons. *J. Neurosci.* 28:3604–14
91. Saunders DN, Hird SL, Withington SL, Dunwoodie SL, Henderson MJ, et al. 2004. *Edd*, the murine hyperplastic disc gene, is essential for yolk sac vascularization and chorioallantoic fusion. *Mol. Cell. Biol.* 24:7225–34

92. Honda Y, Tojo M, Matsuzaki K, Anan T, Matsumoto M, et al. 2002. Cooperation of HECT-domain ubiquitin ligase hHYD and DNA topoisomerase II–binding protein for DNA damage response. *J. Biol. Chem.* 277:3599–605
93. Yoshida M, Yoshida K, Kozlov G, Lim NS, De Crescenzo G, et al. 2006. Poly(A) binding protein (PABP) homeostasis is mediated by the stability of its inhibitor, Paip2. *EMBO J.* 25:1934–44
94. Cojocaru M, Bouchard A, Cloutier P, Cooper JJ, Varzavand K, et al. 2011. Transcription factor IIS cooperates with the E3 ligase UBR5 to ubiquitinate the CDK9 subunit of the positive transcription elongation factor B. *J. Biol. Chem.* 286:5012–22
95. Ling S, Lin WC. 2011. EDD inhibits ATM-mediated phosphorylation of p53. *J. Biol. Chem.* 286:14972–82
96. Henderson MJ, Russell AJ, Hird S, Muñoz M, Clancy JL, et al. 2002. EDD, the human hyperplastic discs protein, has a role in progesterone receptor coactivation and potential involvement in DNA damage response. *J. Biol. Chem.* 277:26468–78
97. Stary S, Yin XJ, Potuschak T, Schlögelhofer P, Nizhynska V, Bachmair A. 2003. PRT1 of *Arabidopsis* is a ubiquitin protein ligase of the plant N-end rule pathway with specificity for aromatic amino-terminal residues. *Plant Physiol.* 133:1360–66
98. Garzón M, Eifler K, Faust A, Scheel H, Hofmann K, et al. 2007. *PRT6*/At5g02310 encodes an *Arabidopsis* ubiquitin ligase of the N-end rule pathway with arginine specificity and is not the *CER3* locus. *FEBS Lett.* 581:3189–96
99. Frottin F, Martinez A, Peynot P, Mitra S, Holz RC, et al. 2006. The proteomics of N-terminal methionine cleavage. *Mol. Cell Proteomics* 5:2336–49
100. Swanson R, Locher M, Hochstrasser M. 2001. A conserved ubiquitin ligase of the nuclear envelope/endoplasmic reticulum that functions in both ER-associated and Matalpha2 repressor degradation. *Genes. Dev.* 15:2660–74
101. Schmidt R, Zahn R, Bukau B, Mogk A. 2009. ClpS is the recognition component for *Escherichia coli* substrates of the N-end rule degradation pathway. *Mol. Microbiol.* 72:506–17
102. Schuenemann VJ, Kralik SM, Albrecht R, Spall SK, Truscott KN, et al. 2009. Structural basis of N-end rule substrate recognition in *Escherichia coli* by the ClpAP adaptor protein ClpS. *EMBO Rep.* 10:508–14
103. Wang KH, Roman-Hernandez G, Grant RA, Sauer RT, Baker TA. 2008. The molecular basis of N-end rule recognition. *Mol. Cell* 32:406–14
104. Román-Hernández G, Hou JY, Grant RA, Sauer RT, Baker TA. 2011. The ClpS adaptor mediates staged delivery of N-end rule substrates to the AAA+ ClpAP protease. *Mol. Cell* 43:217–28
105. Effantin G, Ishikawa T, De Donatis GM, Maurizi MR, Steven AC. 2010. Local and global mobility in the ClpA AAA+ chaperone detected by cryo-electron microscopy: functional connotations. *Structure* 18:553–62
106. Zeth K, Ravelli RB, Paal K, Cusack S, Bukau B, Dougan DA. 2002. Structural analysis of the adaptor protein ClpS in complex with the N-terminal domain of ClpA. *Nat. Struct. Biol.* 9:906–11
107. Ditzel M, Meier P. 2005. Ubiquitylation in apoptosis: DIAP1's (N-)en(d)igma. *Cell Death Differ.* 12:1208–12
108. Orme M, Meier P. 2009. Inhibitor of apoptosis proteins in *Drosophila*: gatekeepers of death. *Apoptosis* 14:950–60
109. Yoo SJ, Huh JR, Muro I, Yu H, Wang L, et al. 2002. Hid, Rpr and Grim negatively regulate DIAP1 levels through distinct mechanisms. *Nat. Cell Biol.* 4:416–24
110. Vaux DL, Silke J. 2005. IAPs, RINGs and ubiquitylation. *Nat. Rev. Mol. Cell Biol.* 6:287–97
111. Choi YE, Butterworth M, Malladi S, Duckett CS, Cohen GM, Bratton SB. 2009. The E3 ubiquitin ligase cIAP1 binds and ubiquitinates caspase-3 and -7 via unique mechanisms at distinct steps in their processing. *J. Biol. Chem.* 284:12772–82
112. Wu JW, Cocina AE, Chai J, Hay BA, Shi Y. 2001. Structural analysis of a functional DIAP1 fragment bound to Grim and Hid peptides. *Mol. Cell* 8:95–104
113. McGrath JP, Jentsch S, Varshavsky A. 1991. *UBA 1*: an essential yeast gene encoding ubiquitin-activating enzyme. *EMBO J.* 10:227–36
114. Chiu YH, Sun Q, Chen ZJ. 2007. E1-L2 activates both ubiquitin and FAT10. *Mol. Cell* 27:1014–23

115. Jin J, Li X, Gygi SP, Harper JW. 2007. Dual E1 activation systems for ubiquitin differentially regulate E2 enzyme charging. *Nature* 447:1135–38
116. Lee PC, Sowa ME, Gygi SP, Harper JW. 2011. Alternative ubiquitin activation/conjugation cascades interact with N-end rule ubiquitin ligases to control degradation of RGS proteins. *Mol. Cell* 43:392–405
117. Hu RG, Wang H, Xia Z, Varshavsky A. 2008. The N-end rule pathway is a sensor of heme. *Proc. Natl. Acad. Sci. USA* 105:76–81
118. Hwang CS, Varshavsky A. 2008. Regulation of peptide import through phosphorylation of Ubr1, the ubiquitin ligase of the N-end rule pathway. *Proc. Natl. Acad. Sci. USA* 105:19188–93
119. Xia Z, Turner GC, Hwang CS, Byrd C, Varshavsky A. 2008. Amino acids induce peptide uptake via accelerated degradation of CUP9, the transcriptional repressor of the PTR2 peptide transporter. *J. Biol. Chem.* 283:28958–68
120. Ljungdahl PO. 2009. Amino-acid-induced signalling via the SPS-sensing pathway in yeast. *Biochem. Soc. Trans.* 37:242–47
121. Yang F, Xia X, Lei HY, Wang ED. 2010. Hemin binds to human cytoplasmic arginyl-tRNA synthetase and inhibits its catalytic activity. *J. Biol. Chem.* 285:39437–46
122. Lévy F, Johnsson N, Rümenapf T, Varshavsky A. 1996. Using ubiquitin to follow the metabolic fate of a protein. *Proc. Natl. Acad. Sci. USA* 93:4907–12
123. Park EC, Finley D, Szostak JW. 1992. A strategy for the generation of conditional mutations by protein destabilization. *Proc. Natl. Acad. Sci. USA* 89:1249–52
124. Ghislain M, Dohmen RJ, Lévy F, Varshavsky A. 1996. Cdc48p interacts with Ufd3p, a WD repeat protein required for ubiquitin-mediated proteolysis in *Saccharomyces cerevisiae*. *EMBO J.* 15:4884–99
125. Taxis C, Stier G, Spadaccini R, Knop M. 2009. Efficient protein depletion by genetically controlled deprotection of a dormant N-degron. *Mol. Syst. Biol.* 5:267
126. Dohmen RJ, Varshavsky A. 2005. Heat-inducible degron and the making of conditional mutants. *Methods Enzymol.* 399:799–822
127. Dohmen RJ, Wu P, Varshavsky A. 1994. Heat-inducible degron: a method for constructing temperature-sensitive mutants. *Science* 263:1273–76
128. Lévy F, Johnston JA, Varshavsky A. 1999. Analysis of a conditional degradation signal in yeast and mammalian cells. *Eur. J. Biochem.* 259:244–52
129. Bernal JA, Venkitaraman AR. 2011. A vertebrate N-end rule degron reveals that Orc6 is required in mitosis for daughter cell abscission. *J. Cell Biol.* 192:969–78
130. Li X, Zhao X, Fang Y, Jiang X, Duong T, et al. 1998. Generation of destabilized green fluorescent protein as a transcription reporter. *J. Biol. Chem.* 273:34970–75
131. Dantuma NP, Lindsten K, Glas R, Jellne M, Masucci MG. 2000. Short-lived green fluorescent proteins for quantifying ubiquitin/proteasome-dependent proteolysis in living cells. *Nat. Biotechnol.* 18:538–43
132. Menéndez-Benito V, Heessen S, Dantuma NP. 2005. Monitoring of ubiquitin-dependent proteolysis with green fluorescent protein substrates. *Methods Enzymol.* 399:490–511
133. Yi H, Friedman JL, Ferreira PA. 2007. The cyclophilin-like domain of Ran-binding protein-2 modulates selectively the activity of the ubiquitin-proteasome system and protein biogenesis. *J. Biol. Chem.* 282:34770–78
134. Apel W, Schulze WX, Bock R. 2010. Identification of protein stability determinants in chloroplasts. *Plant J.* 63:636–50
135. Tresse E, Salomons FA, Vesa J, Bott LC, Kimonis V, et al. 2010. VCP/p97 is essential for maturation of ubiquitin-containing autophagosomes and this function is impaired by mutations that cause IBMPFD. *Autophagy* 6:217–27
136. Lee P, Beem E, Segal MS. 2002. Marker for real-time analysis of caspase activity in intact cells. *BioTechniques* 33:1284–87, 1289–91
137. Hackett EA, Esch RK, Maleri S, Errede B. 2006. A family of destabilized cyan fluorescent proteins as transcriptional reporters in *S. cerevisiae*. *Yeast* 23:333–49
138. de Groot RJ, Rümenapf T, Kuhn RJ, Strauss JH. 1991. Sindbis virus RNA polymerase is degraded by the N-end rule pathway. *Proc. Natl. Acad. Sci. USA* 88:8967–71
139. Kwon YT, Lévy F, Varshavsky A. 1999. Bivalent inhibitor of the N-end rule pathway. *J. Biol. Chem.* 274:18135–39

140. Lee MJ, Pal K, Tasaki T, Roy S, Jiang Y, et al. 2008. Synthetic heterovalent inhibitors targeting recognition E3 components of the N-end rule pathway. *Proc. Natl. Acad. Sci. USA* 105:100–5
141. Sriram SM, Banerjee R, Kane RS, Kwon YT. 2009. Multivalency-assisted control of intracellular signaling pathways: application for ubiquitin-dependent N-end rule pathway. *Chem. Biol.* 16:121–31
142. Dyall SD, Brown MT, Johnson PJ. 2004. Ancient invasions: from endosymbionts to organelles. *Science* 304:253–57
143. Tasaki T, Sohr R, Xia Z, Hellweg R, Hörtnagl H, et al. 2007. Biochemical and genetic studies of UBR3, a ubiquitin ligase with a function in olfactory and other sensory systems. *J. Biol. Chem.* 282:18510–20
144. Meisenberg C, Tait PS, Dianova II, Wright K, Edelmann MJ, et al. 2012. Ubiquitin ligase UBR3 regulates cellular levels of the essential DNA repair protein APE1 and is required for genome stability. *Nucleic Acids Res.* 40:701–11
145. Takeda K, Yanagida M. 2005. Regulation of nuclear proteasome by Rhp6/Ubc2 through ubiquitination and destruction of the sensor and anchor Cut8. *Cell* 122:393–405
146. Wang L, Mao X, Ju D, Xie Y. 2004. Rpn4 is a physiological substrate of the Ubr2 ubiquitin ligase. *J. Biol. Chem.* 279:55218–23
147. Yamaguchi N, Suzuki M, Fukaki H, Morita-Terao M, Tasaka M, Komeda Y. 2007. CRM1/BIG-mediated auxin action regulates *Arabidopsis* inflorescence development. *Plant Cell. Physiol.* 48:1275–90
148. Abida WM, Nikolaev A, Zhao W, Zhang W, Gu W. 2007. FBXO11 promotes the neddylation of p53 and inhibits its transcriptional activity. *J. Biol. Chem.* 282:1797–804
149. Rye MS, Bhutta MF, Cheeseman MT, Burgner D, Blackwell JM, et al. 2011. Unraveling the genetics of otitis media: from mouse to human and back again. *Mamm. Genome* 22:66–82

Ubiquitin-Binding Proteins: Decoders of Ubiquitin-Mediated Cellular Functions

Koraljka Husnjak[1] and Ivan Dikic[1,2,3]

[1]Institute of Biochemistry II, School of Medicine, [2]Buchmann Institute for Molecular Life Sciences, Goethe University, 60590 Frankfurt am Main, Germany; email: husnjak@biochem2.de, Ivan.Dikic@biochem2.de

[3]Department of Immunology and Medical Genetics, University of Split, 21000 Split, Croatia

Keywords

ubiquitination, autophagy, NF-κB, 26S proteasome, endocytosis, ubiquitin receptor

Abstract

Ubiquitin acts as a versatile cellular signal that controls a wide range of biological processes including protein degradation, DNA repair, endocytosis, autophagy, transcription, immunity, and inflammation. The specificity of ubiquitin signaling is achieved by alternative conjugation signals (monoubiquitin and ubiquitin chains) and interactions with ubiquitin-binding proteins (known as ubiquitin receptors) that decode ubiquitinated target signals into biochemical cascades in the cell. Herein, we review the current knowledge pertaining to the structural and functional features of ubiquitin-binding proteins and the mechanisms by which they recognize various types of ubiquitin topologies. The combinatorial use of diverse ubiquitin-binding domains (UBDs) in full-length proteins, selective recognition of chains with distinct linkages and length, and posttranslational modifications of ubiquitin receptors or multivalent interactions within protein complexes illustrate a few mechanisms by which a circuitry of signaling networks can be rewired by ubiquitin-binding proteins to control cellular functions in vivo.

Contents

- INTRODUCTION 292
  - Ubiquitin Assembly and Disassembly Machineries 292
  - Ubiquitination Comes in Different Flavors 294
- UBIQUITIN RECEPTORS SPECIFICALLY RECOGNIZE UBIQUITTINATED PROTEINS .. 296
  - Induced Fit or Selection of Ubiquitin-Binding Domain Ensembles 297
- INCREASED BINDING BY MULTIVALENT BINDING SURFACES 300
- COMBINATORIAL USE OF UBIQUITIN-BINDING DOMAINS IN UBIQUITIN-BINDING PROTEINS 303
- LINKAGE-SPECIFIC UBIQUITIN-BINDING PROTEINS 303
  - Preferential Binding to Lys48-Linked Ubiquitin Chain 304
  - Ubiquitin Receptors for Lys63-Linked Ubiquitin Chains 308
  - Decoders of Linear Ubiquitin Chains 309
  - Many Flavors of Zinc-Finger Folds in Ubiquitin-Binding Proteins ... 310
  - E2 and E3 Enzymes Require Binding to Ubiquitin for Their Functions 311
- UBIQUITIN-LIKE MODIFIERS RECOGNIZED BY UBIQUITIN-BINDING PROTEINS 312
- POSTTRANSLATIONAL MODIFICATIONS REGULATE UBIQUITIN RECEPTORS 312
- CONCLUSIONS AND FUTURE PERSPECTIVES 313

INTRODUCTION

Dynamic and reversible posttranslational modification by the small and evolutionarily conserved protein ubiquitin has been recognized in the past 30 years as a critical modulator of numerous cellular processes (reviewed in References 1–9). The exact mechanisms by which these modifications are decoded into the proper cellular response are still a subject of intensive research. Differentially ubiquitinated substrates are recognized by a wide array of ubiquitin-binding proteins, which elicit appropriate responses to these modifications. In this review, we highlight the emerging complexity of the structural and biochemical aspects by which ubiquitin-binding proteins recognize various types of ubiquitin signals and control diverse cellular functions.

Ubiquitin Assembly and Disassembly Machineries

Ubiquitin is a small protein of 76 amino acids that forms a compact globular structure with an exposed C-terminal tail that can be covalently linked to other proteins. The coordinated activity of ubiquitin-activating (E1), ubiquitin-conjugating (E2), and ubiquitin-ligating (E3) enzymes leads to the activation and transfer of ubiquitin to the target proteins: a process known as ubiquitination (**Figure 1**). The last conjugation step involves the formation of an isopeptide bond between the C-terminal Gly carboxyl group of ubiquitin and the primary ε-amino group of Lys residues of the acceptor/substrate protein (**Figure 1**). Initially, ubiquitin needs to be activated by E1 enzymes (two have been described in human proteome), by forming an E1-ubiquitin thioester in an ATP-dependent manner before it can be transferred to E2 enzymes (approximately 40 in human proteome) by forming an E2-ubiquitin thioester. Finally, E3 enzymes (approximately 600 in human proteome) participate in the formation of the isopeptide bond between the Lys of the substrate and the C-terminal tail of ubiquitin. Interestingly, although primarily

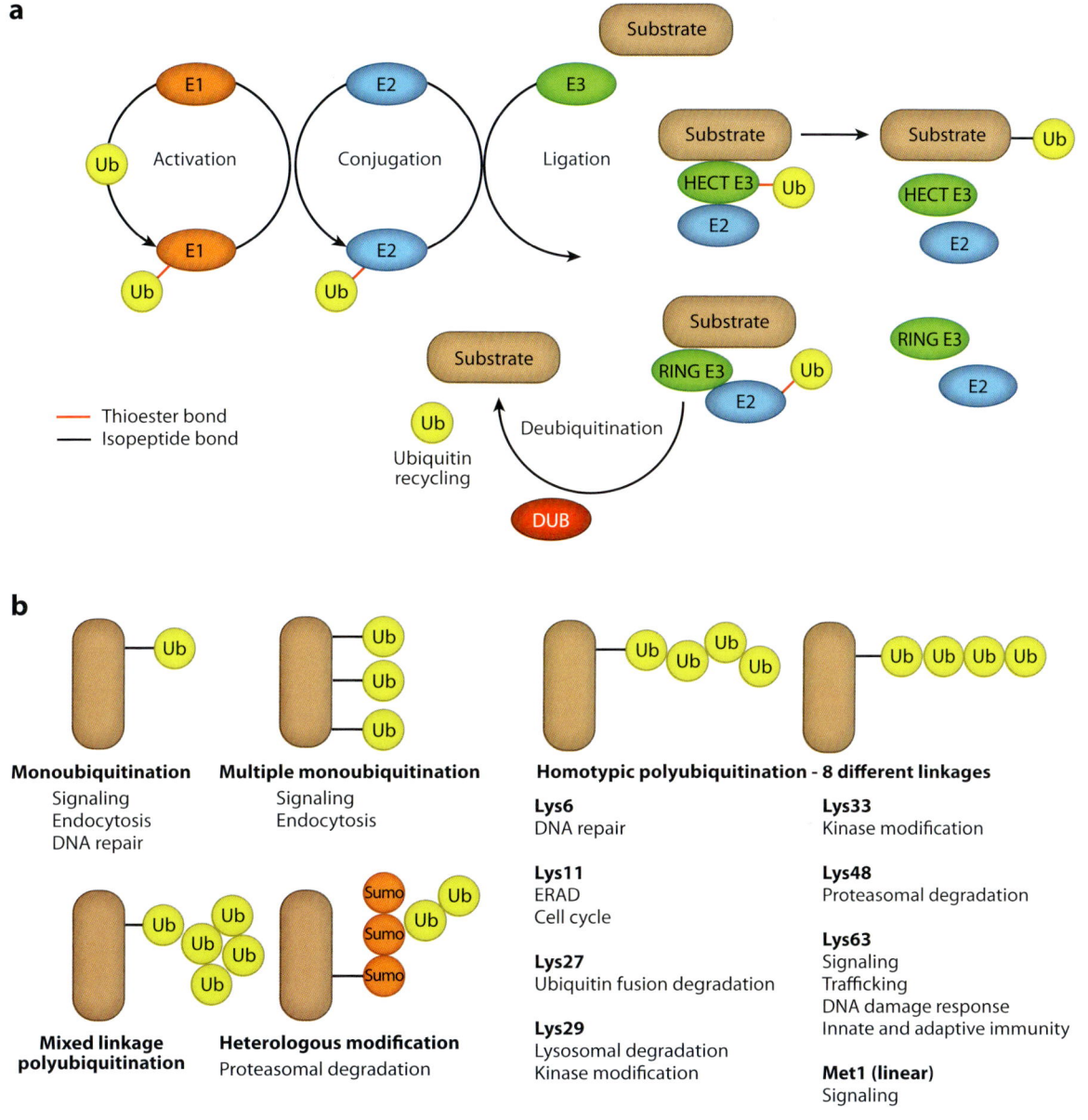

Figure 1

Enzymatic machinery involved in assembly and disassembly of ubiquitin (Ub) chains. (*a*) The coordinated activity of ubiquitin-activating (E1), ubiquitin-conjugating (E2), and ubiquitin-ligating enzyme (E3) is required for ubiquitin attachment to target protein (ubiquitination). Besides modification by single ubiquitin moieties (monoubiquitination), additional ubiquitin molecules can be ligated to a Lys residue of a target protein (multiple monoubiquitination) or a Lys residue of ubiquitin (Lys6, Lys11, Lys27, Lys29, Lys33, Lys48, or Lys63) in a previously attached ubiquitin to form Lys-linked chains. Alternatively, ubiquitin molecules can be linked head to tail forming linear or Met1-linked chains. The modes of action of the two largest groups of E3 ligases (RING and HECT) are shown. Deubiquitinating enzymes (DUBs) remove ubiquitin from modified proteins or disassemble unanchored (*free*) ubiquitin chains. (*b*) Summary of the major cellular processes whereby specific ubiquitin linkages play roles is shown. Abbreviations: ERAD, endoplasmic reticulum-associated degradation; SUMO, small ubiquitin-like modifier.

DUB:
deubiquitinating enzyme

Ubiquitin fusion degradation: a proteolytic system in which nonremovable ubiquitin at the N terminus of a fusion protein is recognized as a degradation signal

Distal ubiquitin: the last ubiquitin unit within a ubiquitin oligomer

modifying Lys residues, ubiquitin can also be attached to the amino group at the N terminus of the substrate (10), as well as to Cys, Ser, and Thr residues of target proteins (11–13). In the budding yeast, *Saccharomyces cerevisiae* ubiquitin modifies more than a thousand different proteins (14), and recent studies have shown that, in the human ubiquitin-modified conjugated proteome, there are more than four to five thousand proteins that are modified by ubiquitin at the steady state (15, 16). The activity of the E1-E2-E3 machinery in the cell is counterbalanced by deubiquitinating enzymes (DUBs), which remove ubiquitin from the targeted proteins (**Figure 1**) (reviewed in Reference 17). More than 100 DUBs have been described that can be further divided into five different subfamilies: ubiquitin-specific proteases (USPs, more than 60 proteins in humans), ubiquitin C-terminal hydrolases (UCHs, four in humans), Machado-Joseph disease protein domain proteases (four in humans), ovarian tumor (Otubain) proteases (14 in humans) and JAB1/MPN/Mov34 metalloenzyme motif proteases (at least four in humans) (18).

Ubiquitination Comes in Different Flavors

In addition to attachment of single (monoubiquitination) or several independent ubiquitin molecules (multiple monoubiquitination) to target proteins, ubiquitin chains (composed of multiple ubiquitin moieties linked via isopeptide bonds) can also be formed on Met1 and all seven internal Lys residues within an ubiquitin molecule (Lys6, Lys11, Lys27, Lys29, Lys33, Lys48, and Lys63) (**Figures 1** and **2**) (19, 20). Although Lys48-linked chains (**Figure 2c**) have been classically studied in the context of protein degradation (reviewed in Reference 21), Lys63-linked chains (**Figure 2e**) have been implicated in a variety of nonproteolytic functions (**Figure 1**) (22). The roles of other types of ubiquitin chains have not been studied in great detail, but their specific roles are just starting to emerge. Mass spectrometry studies revealed that all seven Lys residues and Met1 can participate in the formation of ubiquitin-ubiquitin linkages; however, the percentage of different linkages varied between the individual studies and might reflect technical limitations or differences in cell types or in the model organism employed in the study (i.e., yeast, human) (14, 23–28). In addition, different stages of the cell cycle or intrinsic activity of chain assembly complexes can provide cell-specific ubiquitinome patterns, as recently demonstrated for Lys11 chains (**Figure 2b**). These chains are highly abundant in mitotic cells at a point when the substrates of the E3 ligase anaphase-promoting complex/cyclosome are degraded (29).

Sensitive proteomic approaches provide evidence, at higher resolution, for the role of ubiquitin chains in a variety of cellular functions. For example, all six Lys linkages, except Lys63, are involved in proteasomal degradation of numerous substrates in cells, with Lys48 and Lys11 being the most dominant types of chains (30). By contrast, the same ubiquitin chain is often linked to diverse functions. For example, Lys11 chains promote proteasomal degradation of target proteins during cell-cycle progression (31) and play an important role in endoplasmatic reticulum-associated degradation (30). Furthermore, Lys11 chains have been shown to regulate NF-κB essential modulator (NEMO)-dependent NF-κB activation (32). Both Lys27 and Lys33 linkages can be assembled by U-box E3 ligases during the stress response (33), and Lys29-linked chains play a role in ubiquitin fusion degradation (34). The existence of heterotypic and mixed ubiquitin chains has also been demonstrated in vitro, but their in vivo role is not yet clear (14).

The latest addition to the list of polyubiquitin chains is head-to-tail ubiquitin conjugates, also known as linear or Met1 chains (**Figure 2d**) (35). In linear ubiquitin chains, the terminal Gly76 of the proximal ubiquitin is linked to the Met1 of the distal ubiquitin. To date, a single E3 ligase complex, termed the linear ubiquitin chain assembly complex, has been shown to assemble such chains (35) and plays a crucial role in NF-κB signaling

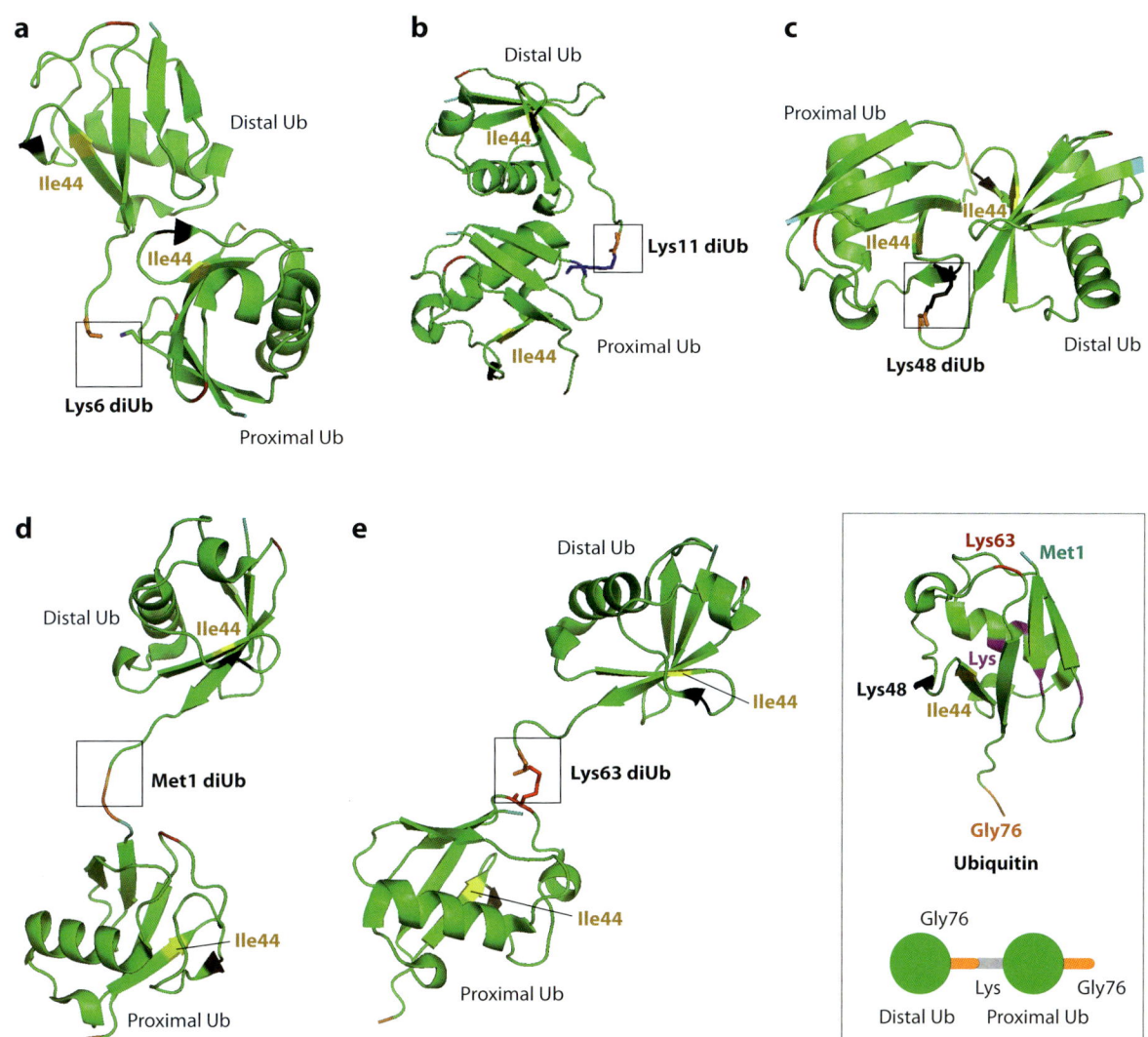

Figure 2

Ubiquitin (Ub) chains have different topologies. Ubiquitin can form eight different homotypic chains; the structure has already been solved for most of these chains. (*a*) Chemically synthesized Lys6-linked diubiquitin (diUb), Protein Data Bank (PDB) code 2XK5, adopts a compact conformation with an asymmetric assembly, whereas (*b*) chemically produced Lys11-linked diUb, PDB code 2XEW, adopts compact conformations with Ile44 patches that are solvent exposed. (*c*) Degradative Lys48-linked diUb adopts a closed conformation, PDB code 1AAR. (*d*) In contrast, Met1-linked diUb chains, PDB code 2W9N, and (*e*) Lys63-linked diUb chains, PDB code 2JF5, form an extended conformation without any contacts between hydrophobic surfaces (the "beads-on-a-string" model). The PDB entry, distal and proximal ubiquitin, and linkage type are shown for each diUb. For clarity, the linkage is shown in a square or rectangle. Crucial residues in ubiquitin structure are colored: Met1 (*cyan*), Ile44 (*yellow*), Lys48 (*black*), Lys63 (*red*), Gly76 (*orange*). In the monoubiquitin structure, all Lys (except for Lys48 and Lys63) are depicted in magenta. This figure was generated from PDB entries by PyMOL software.

UBD: ubiquitin-binding domain

Ubiquitin receptor: an effector protein with UBD(s), which triggers specific cellular responses by recognizing ubiquitinated proteins or free ubiquitin chains

Translesion synthesis polymerase: a DNA polymerase of the Y-family, which performs translesion synthesis across DNA lesions

(25, 26, 35, 36). The limiting factor in our understanding of atypical ubiquitin chains is the inability to synthesize them in vitro, partly because the specific E1-E2-E3 enzymes required to catalyze these chains are not known. Recent developments in the chemical synthesis of diubiquitin chains (37–40) will improve our understanding of these ubiquitin modifications.

UBIQUITIN RECEPTORS SPECIFICALLY RECOGNIZE UBIQUITINATED PROTEINS

Similar to the recognition of phosphorylated proteins via phospho-binding domains recognizing phosphorylated Ser, Thr, or Tyr (41), ubiquitin modification of target proteins is recognized by a variety of ubiquitin-binding domains (UBDs) or ubiquitin receptors that contain at least one UBD within their structure (reviewed in Reference 42). Several distinct modes of UBD interaction with ubiquitin have been described so far (**Figures 3** and **4**), involving approximately 20 different families (**Table 1**). Although structurally quite different, UBDs share a feature of noncovalently binding to ubiquitin signals. UBDs can be divided into several subfamilies, including those containing the following:

1. single or multiple α-helices [ubiquitin-associated (UBA) domain, ubiquitin-interacting motif (UIM), double-sided ubiquitin-interacting motif (DUIM), motif interacting with ubiquitin (MIU), coupling of ubiquitin conjugation to endoplasmic reticulum degradation (CUE), GGA and Tom1 (GAT), Vps27/Hrs/STAM (VHS), ubiquitin binding in ABIN and NEMO (UBAN) protein];
2. zinc fingers [nuclear protein localization 4 zinc finger (NZF), zinc-finger ubiquitin-binding protein (ZnF UBP), zinc finger in A20 protein (ZnF A20), ubiquitin-binding zinc finger (UBZ)];
3. a pleckstrin-homology (PH) fold [gram-like ubiquitin-binding in Eap45 (GLUE), pleckstrin-like receptor for ubiquitin (PRU)];
4. ubiquitin-conjugating-like structures [ubiquitin E2 variant (UEV) and UBC]; or
5. other structures [SH3, ubiquitin-binding motif (UBM), PLAA family ubiquitin-binding domain (PFU), Jab1/MPN] (**Table 1**, **Figures 3** and **4**).

The majority of identified UBDs bind to a hydrophobic patch around Ile44 of ubiquitin (reviewed in Reference 42). The globular ubiquitin surface is mainly polar, with a large hydrophobic area centered on the Leu8, Ile44, and Val70 residues. Adjacent binding surfaces on ubiquitin molecules have also been shown to contribute to the binding specificity of different UBDs. For example, the ubiquitin-binding motif found in the Y-family translesion synthesis polymerases ι and Rev1, binds to the hydrophobic patch centered around Leu8 (**Figure 3***a*) (43, 44), whereas the ubiquitin Asp58 residue binds the ZnF A20 domain of Rabex-5 (**Figure 3***e*) (45, 46). In the case of DUB isopeptidase T (IsoT, USP5), the ZnF UBP domain primarily contacts the C-terminal part of ubiquitin, and the ubiquitin tail penetrates deeply into the hydrophobic pocket of the ZnF UBP domain (**Figure 3***f*) (47). Additionally, ZnF UBP also interacts with a surface centered on Ile36 of ubiquitin. A comprehensive analysis of ubiquitin surface residues revealed two additional areas centered on Ile36/Leu71/Leu73 and Gln2/Phe4/Thr12 residues, which are important for vital functions in yeast (48). Kamadurai et al. (49) showed the importance of the Ile36 site in the E2/E3 complex with ubiquitin (where E3 contacts the Ile36 of the ubiquitin conjugated to the E2); the latter patch was later shown to be involved in ubiquitin binding to UBAN domain of NEMO (36).

Accumulated knowledge about individual UBDs, their major structural features, as well as their binding specificities toward ubiquitin have been summarized in previous reviews (42, 50–52).

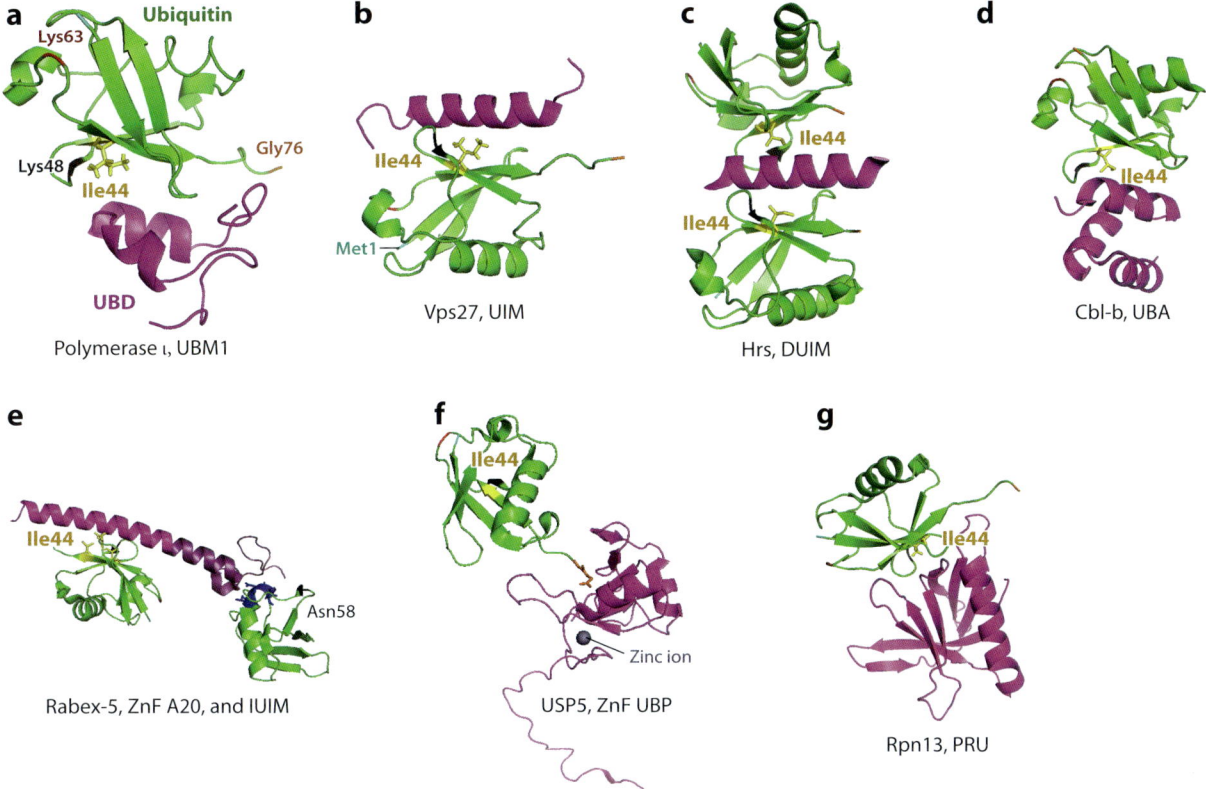

Figure 3

Distinct modes of ubiquitin-binding domains (UBDs) binding to monoubiquitin. Numerous UBDs bind to different surfaces of ubiquitin. UBDs of (*a*) polymerase ι (ubiquitin-binding motif, UBM1), Protein Data Bank (PDB) code 2KWV; (*b*) Vps27 (ubiquitin-interacting motif, UIM) PDB code 1Q0W; (*c*) Hrs, hepatocyte growth factor-regulated tyrosine kinase substrate (double-sided ubiquitin-interacting motif, DUIM), PDB code 2D3G; (*d*) Cbl-b (ubiquitin-associated, UBA, domain), PDB code 2OOB; (*e*) Rabex-5 (ZnF A20 and IUIM), PDB code 2FIF; (*f*) IsoT (ZnF UBP, zinc-finger ubiquitin-binding protein), PDB code 2G45; and (*g*) Rpn13 (pleckstrin-like receptor for ubiquitin, PRU), PDB code 2Z59, are shown. Crucial residues in ubiquitin structure are color coded: Met1 (*cyan*), Ile44 (*yellow*), Lys48 (*black*), Lys63 (*red*), Gly76 (*orange*). Zinc ion in 2G45 structure is shown as a ball. Ubiquitin structures are depicted in green; UBD structures (except for individual residues) are in magenta. This figure was generated from PDB entries by PyMOL software.

Induced Fit or Selection of Ubiquitin-Binding Domain Ensembles

Except for the C-terminal tail, the ubiquitin backbone is relatively rigid. However, recent structural work revealed that UBD:ubiquitin binding changes the conformation of ubiquitin. Lange et al. (53) compared various ubiquitin structures and showed that each UBD:ubiquitin complex corresponds to one of the conformations of the free ubiquitin, thus favoring the conformational selection model. In a follow-up study, Wlodarski & Zagrovic (54) extended the proposed model, suggesting that conformational selection is additionally optimized by residual induced fit close to the binding site. One has to consider that all the modeling was performed on isolated UBDs and not in the context of full-length UBD-containing proteins, which also might bring an additional layer of complexity to the modeling of the interaction mechanisms.

The situation is even more complex when ubiquitin chains, as opposed to monoubiquitin,

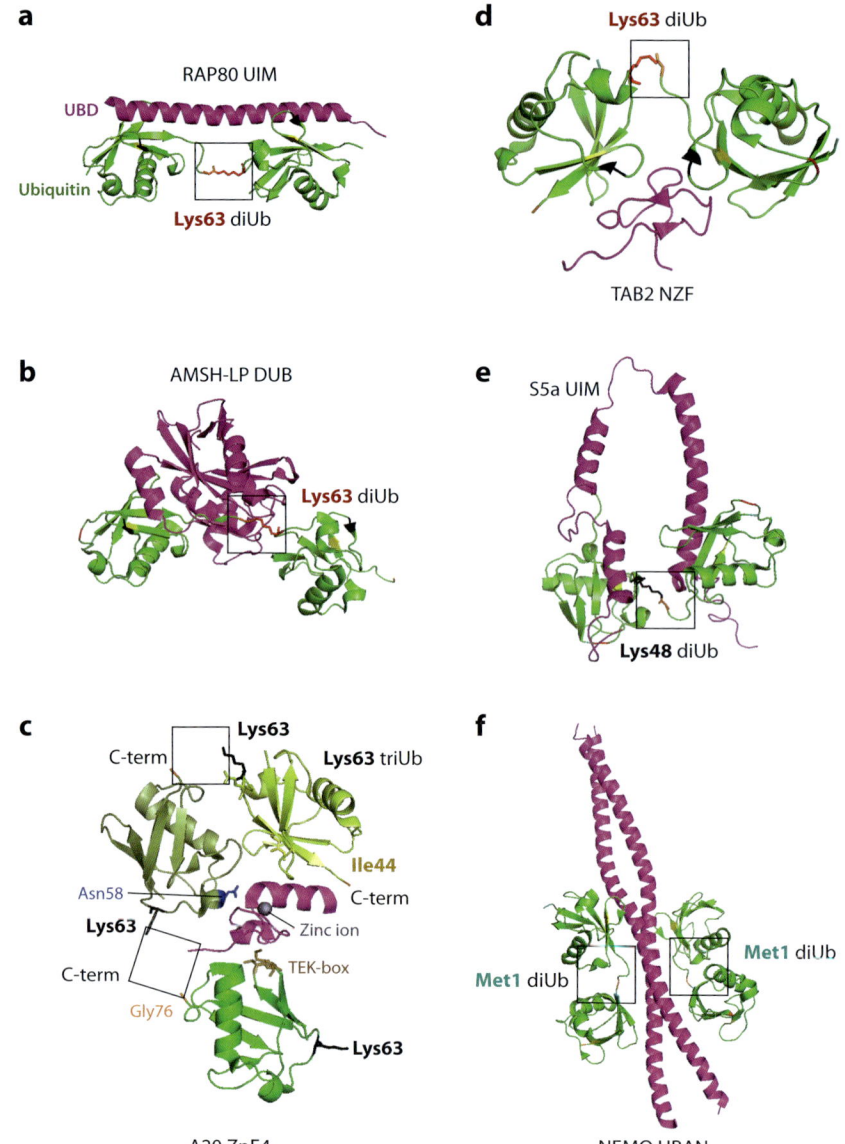

Figure 4

Specificity of ubiquitin receptors is achieved in different ways. Known structures between diubiquitin (diUb) and ubiquitin-binding domains (UBDs) are shown. They include Lys63-specific UBDs: (*a*) RAP80 protein UIMs, Protein Data Bank (PDB) code 3A1Q; (*b*) AMSH-like protease deubiquitinating enzyme (AMSH-LP DUB), PDB code 2ZNV; (*c*) A20 ZnF4, PDB code 3OJ3; (*d*) TAB2 NZF, PDB code 2WWZ; as well as Lys48-specific (*e*) S5a UIMs, PDB code 2KDE; and (*f*) Met1-specific NEMO UBAN, PDB code 2ZVO. Crucial residues in ubiquitin structure are color coded: Met1 (*cyan*), Ile44 (*yellow*), Lys48 (*black*), Lys63 (*red*), Gly76 (*orange*), and the zinc ion in 3OJ3 structure is shown as a ball. Ubiquitin structures are depicted in green; UBD structures (except for individual residues) are in magenta. Residue Asn58 and TEK-box in 3OJ3 structure are depicted in blue (*stick*) and brown (*sticks*), respectively. All Lys63 residues in 3OJ3 (*c*) are depicted as black sticks. The linkage is shown in rectangles, for clarity. This figure was generated from PDB entries by PyMOL software. Abbreviations: C-term, C terminus; TEK-box, Thr-Glu-Lys motif; triUb, chain of 3 ubiquitins.

Table 1 Inventory of ubiquitin-binding domains (UBDs)

UBD abbreviations	Structure	Ubiquitin surface	Type of ubiquitin binding[a]	Proteins with specific UBDs
UBA	Three-helix bundle	Ile44	mUb polyUb (>K48) UBL	PLIC1/2, HHR23A/B, Ddi1, p62/SQSTM1, NBR1, Cbl-b, USP5, UBC1, HERC2
CUE	Three-helix bundle	Ile44	mUb	Cue2 (monomeric), Vps9 (dimeric)
UIM	Single α-helix, often present in tandem	Ile44	mUb pUb (K48, K63) UBL	S5a, Vps27, USP28, ataxin-3, EPS15, STAM, RAP80
MIU/IUIM	Single α-helix[b]	Ile44	mUb	Rabex-5
DUIM	Single α-helix, binds two ubiquitin molecules[c]	Ile44	mUb	Hrs
VHS	Superhelix of eight α-helices	Ile44	mUb	STAM, Hrs
GAT	Three-helix bundle	Ile44	mUb	GGA3, TOM1
NZF	Zinc finger, four β-strands	Ile44	mUb pUb (K63)[d]	Npl4, VPS36, TAB2, TAB3, HOIP, HOIL-1L
ZnF A20	Zinc finger	Asp58[e]	mUb (Rabex-5) Lys63 (A20)	A20, Rabex-5
ZnF UBP	Zinc finger, a globular fold with a deep cleft and pocket to accommodate ubiquitin's tail	Leu8, Ile36, tail	pUb (unanchored)	USP5, HDAC6, BRAP2
UBZ	Zinc finger, ββα-fold	Ile44	mUb pUb	Polymerase η, polymerase κ, FAN1, NDP52, TAX1BP1, WRNIP1
UBC	β-sheet	Ile44	mUb	UbcH5C
UEV	αβ-sequence, lacks E2 catalytic Cys	Ile44	mUb	VPS23, TSG101
UBM	Helix turn helix, helices separated by an invariant Leu-Pro motif	Leu8	mUb	Polymerase ι, polymerase Rev1
GLUE	Split-pleckstrin homology domain	Ile44	mUb	EAP45
PRU	Pleckstrin homology domain, three loops bind ubiquitin	Ile44	mUb (Ile44) pUb (K48 linker) UBL	Rpn13

(Continued)

Table 1 (Continued)

UBD abbreviations	Structure	Ubiquitin surface	Type of ubiquitin binding[a]	Proteins with specific UBDs
Jab1/MPN	Inactive variant of Jab1/MPN domain lacking key residues in the motif	Ile44	mUb	Prp8p
PFU	Four β-strands and two α-helices	Ile44	mUb pUb	Doa1, PLAA
SH3, variant	β-barrel fold, hydrophobic groove binds ubiquitin	Ile44	mUb	Sla1, CIN85, amphiphysin
UBAN	Parallel coiled-coil dimer	Ile44 Phe4 linker	Met1-diUb	NEMO, optineurin, ABIN1-3
WD40 repeat β-propeller	Top surface of WD40 repeat β-propeller	Ile44	mUb	Doa1/UFD3

[a]Abbreviations: (>K48), predominant binding to Lys48-linked ubiquitin chains; Met1-diUb, Met1-linked diubiquitin; mUb, monoubiquitin; pUb, polyubiquitin; UBL, ubiquitin-like domain.
[b]Residues bind to the ubiquitin Ile44 surface in an opposite direction to that of the ubiquitin-interacting motif (UIM).
[c]Key residues of the two binding motifs are repeated and shifted by two residues relative to each other.
[d]Two-sided nuclear protein localization 4 zinc fingers (NZFs) bind Lys63-linked ubiquitin chains (TAB2, TAB3, VPS36).
[e]ZnF A20 of A20 contacts three ubiquitins via Ile44, Asp58 residues and TEK-box (PDB code 3OJ3).

are taken into account (**Figure 2**): Lys48-linked di- and tetraubiquitin chains adopt a closed conformation in solution, resulting in buried hydrophobic patch surfaces around Ile44, which are sequestered at the interdomain interface and tightly packed (**Figure 2c**). Nevertheless, interactions with UBD-containing proteins still occur because this conformation is not rigid under physiological conditions but is dynamic, oscillating between the open and packed structure (55, 56). By contrast, Lys63-linked diubiquitin chains (**Figure 2e**) form an extended conformation without any contacts between hydrophobic surfaces and are fully exposed (56, 57), similar to elongated Met1-linked diubiquitin (**Figure 2d**). By molecular modeling, diubiquitins linked via Lys6, Lys11, Lys27, or Lys48 were predicted to form a closed conformation, whereas chains linked via Lys29, Lys33, Lys63, or Met1 are unable to form such a contact owing to steric occlusion (58). This is consistent with recent structural work showing that Lys6-linked diubiquitin (**Figure 2a**) adopts compact conformation (37).

INCREASED BINDING BY MULTIVALENT BINDING SURFACES

Although ubiquitin networks seem to be involved in virtually all cellular processes, it was initially very surprising that binding affinities of individual UBDs for ubiquitin are generally low, with K_d values in high micromolar ranges (59–61), especially because cellular ubiquitin concentrations can be as high as 85 μM (62). In many instances, protein oligomerization combined with multiple UBDs within single proteins (or within protein subcomplexes) (**Table 2**), as well as the existence of multiple ubiquitin-binding surfaces in single UBDs (as in DUIMs, **Figure 3c**), provide the basis for a multivalent ubiquitin-binding complex (**Figure 5**).

Ubiquitin receptors usually have additional modular structures, which regulate localization/compartmentalization, oligomeric state, enzymatic activity, and additional binding

Table 2 Ubiquitin receptors often contain multiple ubiquitin-binding modules

Protein names	Ubiquitin-binding domains[a]	Function	Function of multiple ubiquitin-binding domains	References
Rad23A/Rad23B	2 UBAs	Shuttle factors for 26S proteasome	Tandem UBDs might prevent deubiquitination and unnecessary chain elongation of substrates during their transfer to the 26S proteasome	89, 174
Ataxin-3	2 N-terminal UBDs 2 C-terminal UIMs	Deubiquitination	Specific binding to Lys48-linked chains	82, 110, 111
RAP80	2 UIMs[b]	DNA repair	Specific binding to Lys63-linked chains	82, 117, 175
USP5/IsoT	ZnF UBP, UBP[c], 2 UBAs	Deubiquitination of unattached ubiquitin chains	Regulation of DUB catalytic activity and processivity	72
USP13/IsoT3	ZnF UBP, 2 UBAs	Deubiquitination of unattached ubiquitin chains	Suggested to stabilize the E3 ligase, SIAH2, by binding to its ubiquitin modifications, independent of DUB activity	176
USP25m	UBA, 2 UIMs	Deubiquitination	Suggested to regulate the ubiquitination state and substrate recognition. Does not seem to be required for catalytic activity or oligomerization	177
Vps27 (yeast)[d]	2 UIMs	Membrane trafficking	Cooperativity to achieve efficient binding to monoubiquitin in sorting cargo into multivesicular bodies	178
Hsj1	2 UIMs[e]	Suggested function as a neuronal shuttling factor for the sorting of chaperone clients to the 26S proteasome	Suggested to prevent cargo aggregation and protects ubiquitinated cargo from being cleaved by DUBs, stimulating its sorting to the 26S proteasomes	179
Epsin1	3 UIMs	Receptor-mediated endocytosis	High-avidity interactions with ubiquitinated proteins and ubiquitin chains	180
Met4 (yeast)	UIM, UIM like	Transcription	Self-protective	76
NEMO	UBAN, ZnF	NF-κB signaling	Suggested to enable binding to long Lys63-linked ubiquitin chains	128

[a]Abbreviations: DUB, deubiquitinating enzyme; UBA domain, ubiquitin-associated domain; UBAN, ubiquitin binding in ABIN and NEMO protein; UBD, ubiquitin-binding domain; UBP, ubiquitin-binding protein; UIM, ubiquitin-interacting motif; ZnF UBP, zinc finger ubiquitin-binding protein.
[b]Two terminal zinc-finger domains in RAP80 have not been shown to bind ubiquitin.
[c]Ubiquitin-specific processing protease.
[d]Human homolog Hrs contains a single double-sided ubiquitin-interacting motif (DUIM).
[e]UIM was bioinformatically predicted to be a DUIM (67).

partners of UBD-containing proteins. For example, many UIM-containing proteins involved in protein trafficking have been shown to associate with each other (63–65), bringing their UBDs in the close proximity and enabling the buildup of large protein complexes (**Figure 5**). Although the binding of individual UIMs to monoubiquitin is weak, ubiquitin chains or oligomerization of ubiquitinated proteins and their receptors leads to high-avidity interactions (**Figure 5**). Many endocytic proteins contain UIMs, including

Avidity: the combined synergistic strength of bond affinities in a protein complex

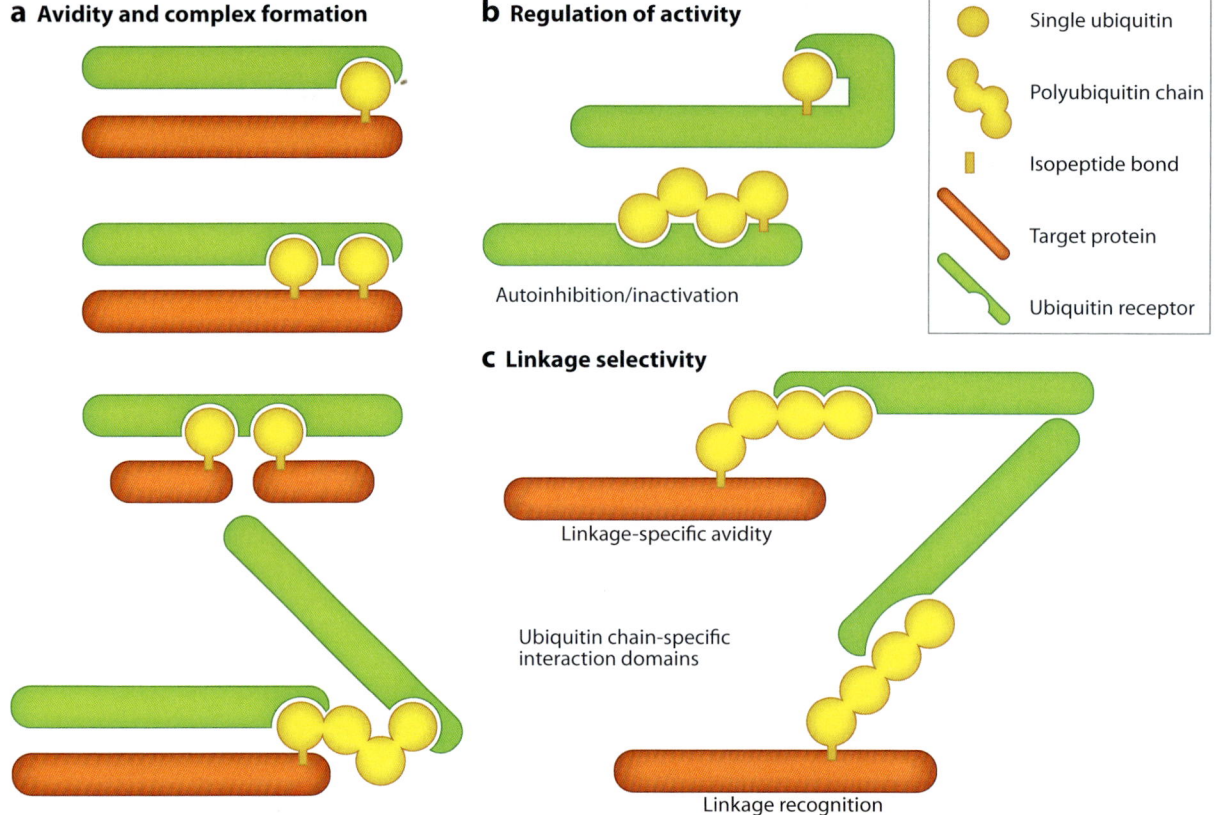

Figure 5

Features of interactions between ubiquitin signals and ubiquitin receptors. (*a*) As exemplified throughout the review, ubiquitin plays important role in protein complex assembly enabling protein oligomerization. (*b*) Combinatorial use of ubiquitin binding and ubiquitination often provides an efficient mechanism for regulation of protein activity, (*c*) whereas diversity of signaling is achieved through linkage-specific ubiquitin receptors whose chain-specific binding is achieved in several distinct ways.

epsins, Eps15, Eps15R, Vps27, and STAM (signal-transducing adaptor molecule). Most UIMs bind monoubiquitin with low affinity (0.2–1 mM) (60); however, the presence of multiple UIMs in UIM-containing proteins (66) implies that cooperative binding might enhance binding affinity, where UIMs could bind to either multiple monoubiquitinated proteins or protein complexes (**Figure 5**). Different structural variations of UIM architecture have been utilized during evolution to expand the binding spectrum to diverse ubiquitin signals. In a DUIM, the UIM forms a single α-helix, which can simultaneously bind to two ubiquitin molecules on the opposite faces of the same helix (67). Both sides of hepatocyte growth factor regulated tyrosine kinase substrate (Hrs) DUIM (**Figure 3c**) are required for Hrs function in the degradative sorting of the internalized epidermal growth factor receptor (67). The predicted DUIM internal repeat consensus sequence was also identified in the endocytic regulator USP25 and MAPK/ERK kinase kinase 1 protein, whereas the UIM-2 of cochaperone heat shock protein J1 (Hsj1), Eps15, and Eps15R have predicted characteristics of the double-sided tandem UIMs (67). Interestingly, the Hrs yeast homolog, Vps27, contains a conventional single-sided UIM (**Figure 3b**), which binds mostly monoubiquitinated cargoes and is unlike mammalian Hrs, which commonly targets multiple monoubiquitinated or oligoubiquitinated receptors (3, 68). Recent work by three independent groups revealed the existence of two novel UBDs in the N-terminal part of Rabex-5 (45, 46, 69), namely the UIM variant MIU (also known as IUIM, inverted UIM) and ZnF A20 (see below and **Figure 3e**) that are essential for coordination of ubiquitin-dependent trafficking of receptor tyrosine kinases. One can speculate that during evolution, multiple variants of UIM conformations (DUIMs, MIU, and their tandem arrangements) have evolved in the context of endocytic receptors to contribute to the increased binding avidity and specificity toward multiple ubiquitinated signals.

COMBINATORIAL USE OF UBIQUITIN-BINDING DOMAINS IN UBIQUITIN-BINDING PROTEINS

In many cases, UBD-containing proteins possess multiple copies of UBDs (**Table 2**, **Figure 5**). In particular, many endocytic adaptors carry more than one UBD, which might explain their ability to build up large protein complexes involved in endocytosis. Rabex-5 is an interesting example because it shows that ubiquitin receptors might simultaneously bind different surfaces of a single ubiquitin molecule (45). STAM1 and STAM2 bind ubiquitin and ubiquitinated proteins by tandemly organized VHS and UIM domains (70), whereas multiple VHS domains within the endosomal sorting complex required for transport (ESCRT)-0 cooperate in high-avidity binding to polyubiquitinated cargo (71). IsoT has four UBDs: a ZnF UBP domain, a ubiquitin-binding protein domain that forms the active site, and two UBA domains involved in linear and Lys48-linked chain binding (72). In addition, tandem use of ubiquitin- and membrane-binding domains is found in endocytic sorting proteins, where UBDs are often accompanied by membrane-binding domains, such as ENTH (in the case of epsins) or a FYVE domain (in the case of Hrs and STAM) (3). Similarly, the ELL-associated protein of 45 kDa (EAP45) involved in endocytic sorting binds ubiquitin via its GLUE domain (73–75), which shares similarities in its primary and predicted secondary structures to phosphoinositide-binding GRAM and PH domains. Because ubiquitin binds far from the proposed phosphoinositide-binding site of EAP45 GLUE, it is assumed that phosphoinositide and ubiquitin binding are independent (75). Such dual binding may affect the function of the full-size EAP45 by controlling its membrane association and, at the same time, positioning the GLUE domain for the most efficient sorting of ubiquitinated cargoes.

One intriguing function of a combinatorial use of UBDs is Met4, a transcriptional activator of the sulfur metabolic network in yeast. Met4 contains tandem UBDs, namely the UIM and UIM-like domains (76), which stabilize it in the Lys48-polyubiquitinated form. Although Met4 is Lys48 polyubiquinated by SCF^{Met30} E3 ligase, it is not degraded by the 26S proteasome, but is restricted from accessing its target promoters (77). Under certain growth conditions, the protective UBD function is lost, which leads to the degradation of Met4 (78, 79). Met4 UBDs restrict the length of its polyubiquitin modification, thus preventing its proteasomal recognition and subsequent degradation. Consequently, UBD inactivation enables the synthesis of longer ubiquitin chains, leading to the quick proteasomal clearance of Met4 (80). What regulates the "off" switch of the protective function of UBDs is still unclear. The choice of Lys48 chains over Lys63 chains may be a consequence of the need for a fast decision-making response in yeast that rapidly turns a protective signal into a death-inducing signal.

LINKAGE-SPECIFIC UBIQUITIN-BINDING PROTEINS

Even though the majority of UBDs can bind to monoubiquitin in vitro (reviewed in Reference 42), there is growing evidence that many ubiquitin-binding proteins interact with selective ubiquitin chains in vivo. Different

Endosomal sorting complex required for transport: a multimeric protein complex required for sorting of ubiquitinated cargoes in the endosome

26S proteasome: a large protease, which degrades proteins modified by polyubiquitin chain, composed of a 19S regulatory particle and a 20S catalytic core

mechanisms have evolved to achieve such in vivo specificity, including increased avidity owing to oligomerization of ubiquitin receptors and multi- or polyubiquitinated substrates, the presence of multiple UBDs in the integral proteins (different UBDs in the same protein or multiplication of the same UBD within a protein), and conformational changes following the UBD:ubiquitin interactions (**Figure 5**) (81). More recently, structural and functional studies have identified a subset of UBDs with high binding selectivity toward ubiquitin chains rather than to monoubiquitin. These UBDs are essential for the regulation of several important physiological processes as summarized in **Figures 6** and **7**.

Linkage-specific UBDs can be divided into two categories (**Figures 4** and **5**): domains recognizing the linker region between two ubiquitin moieties and domains detecting the spatial distribution and positioning of the individual units of the ubiquitin chain without the binding to the actual linkage (42). The latter mechanism was named linkage-specific avidity (82). Yet, the same domains do not have exclusive linkage specificity nor defined recognition of specific chain lengths when present in different ubiquitin receptors.

Preferential Binding to Lys48-Linked Ubiquitin Chain

Hofmann & Bucher (83) identified the UBA domain as a sequence present in several proteins involved in ubiquitin-mediated proteolysis, including shuttle factors, which bind ubiquitinated substrates via their C-terminal UBA domain(s) and deliver them to the 26S proteasome for subsequent degradation. Identified shuttle factors include Rad23 (HHR23A/B in humans), Dsk2 (PLIC1-4 in humans), and Ddi1 and are recruited to the proteasome by their N-terminal ubiquitin-like (UBL) domain (**Figure 6**) (84–87).

Numerous UBA domain–containing proteins bind Lys48-linked chains in strong preference to monoubiquitin (88). In concordance, the UBL-UBA domain–containing protein HHR23A does not bind monoubiquitin, whereas binding to Lys48-linked chains is preferential over Lys63- and Lys29-linked chains (89). Although the extended conformation of Lys63-linked diubiquitin enables each ubiquitin unit to bind to one HHR23A-UBA2 moiety, in a fashion similar to the monoubiquitin:UBA2 interaction (57), binding of Lys48-linked chains is more selective; single UBA molecules bind Lys48-linked chains of up to four ubiquitin moieties (90). The interaction between UBA2 and Lys48-linked diubiquitin requires the classical hydrophobic Ile44 patch on each ubiquitin, as well as at the isopeptide junction while the UBA2 domain is encircled by the hydrophobic surfaces of both ubiquitin molecules (91).

The shuttle factor Rad23 delivers the cargo to the 26S proteasome for degradation (92) without being destroyed by the 26S proteasome. It has been recently shown that successful removal by the 26S proteasome requires that a protein is labeled by a proteasome degron consisting of both a Lys48-linked polyubiquitin chain and an unstructured initiation region that serves as a platform where the 26S proteasome engages the protein to unfold it (93, 94). The C-terminal UBA domain rescues Rad23 from proteasomal degradation (92) by preventing the generation of initiation sites (95, 96). In addition, FAT10 was also shown to act as ubiquitin-independent proteasomal degradation signal in a process accelerated by the UBA-UBL protein NEDD8 ultimate buster 1-long (NUB1L), whose UBL and UBA domains bind the 26S proteasome and FAT10, respectively (97–99).

The 26S proteasome itself has several intrinsic ubiquitin receptors, which reside in the 19S regulatory particle. The very commonly found UIM was actually first identified (and named pUbS) in the 19S proteasomal subunit S5a/Rpn10 as a 20-amino acid motif able to bind ubiquitinated proteins and polyubiquitin chains (100, 101), as well as UBL domain proteins (102). Interestingly, two UIMs of S5a bind polyubiquitin with different affinities (100) and have different specificities for UBL domain proteins. The UIMs of S5a can also bind

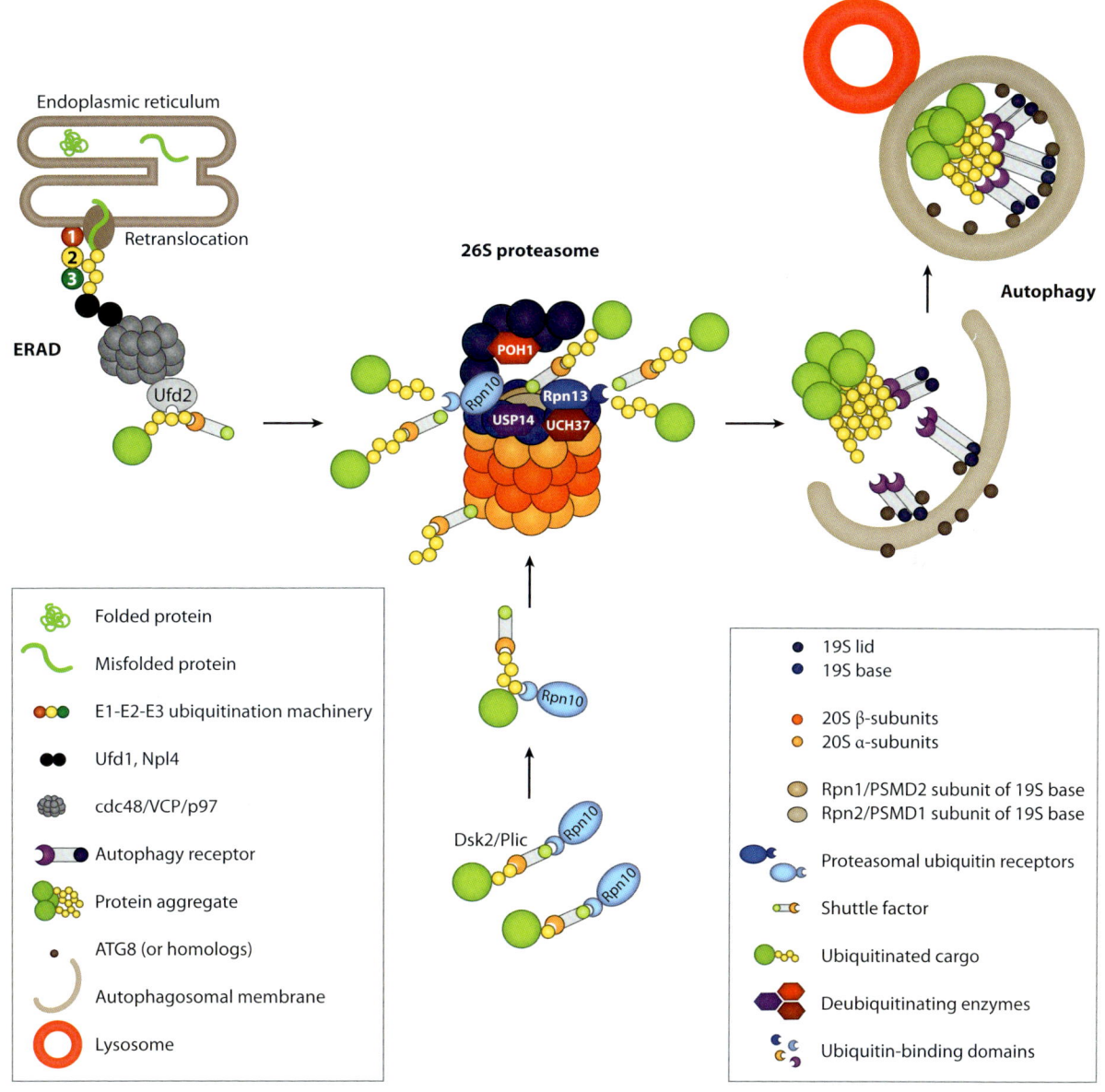

Figure 6

Ubiquitin signaling in quality control pathways. Misfolded proteins are recognized by a cellular quality control system and translocated from the endoplasmic reticulum (ER) to the cytosol, where they are Lys48 ubiquitinated. p97/VCP (Cdc48 in yeast) functions as AAA+ ATPase and is involved in many cellular pathways through interactions with different adaptor proteins. p97 helps in the extraction of misfolded proteins during retranslocation from the ER and its ubiquitin-binding domain (UBD)-containing accessory proteins bind ubiquitinated cargo. The cargo is subsequently transferred to the 26S proteasome, where ubiquitinated cargo is degraded within the 20S catalytic core. The 19S subunit serves as a docking unit for shuttle factors that recruit ubiquitinated cargo to the proteasome and that either directly bind to the 19S base or hand over ubiquitinated proteins to intrinsic proteasomal ubiquitin receptors. Misfolded proteins and damaged organelles are removed from the cell by autophagy, where autophagy receptors, such as p62 and NBR1, bind ubiquitinated cargo via their UBA domains. Autophagosomes are ultimately fused with lysosomes, and their content is degraded. Abbreviations: Dsk2, ubiquitin shuttle factor in yeast (PLIC in humans); ERAD, endoplasmic reticulum–associated degradation; Npl4, nuclear protein localization 4 protein; Ufd1, ubiquitin fusion degradation 1 protein.

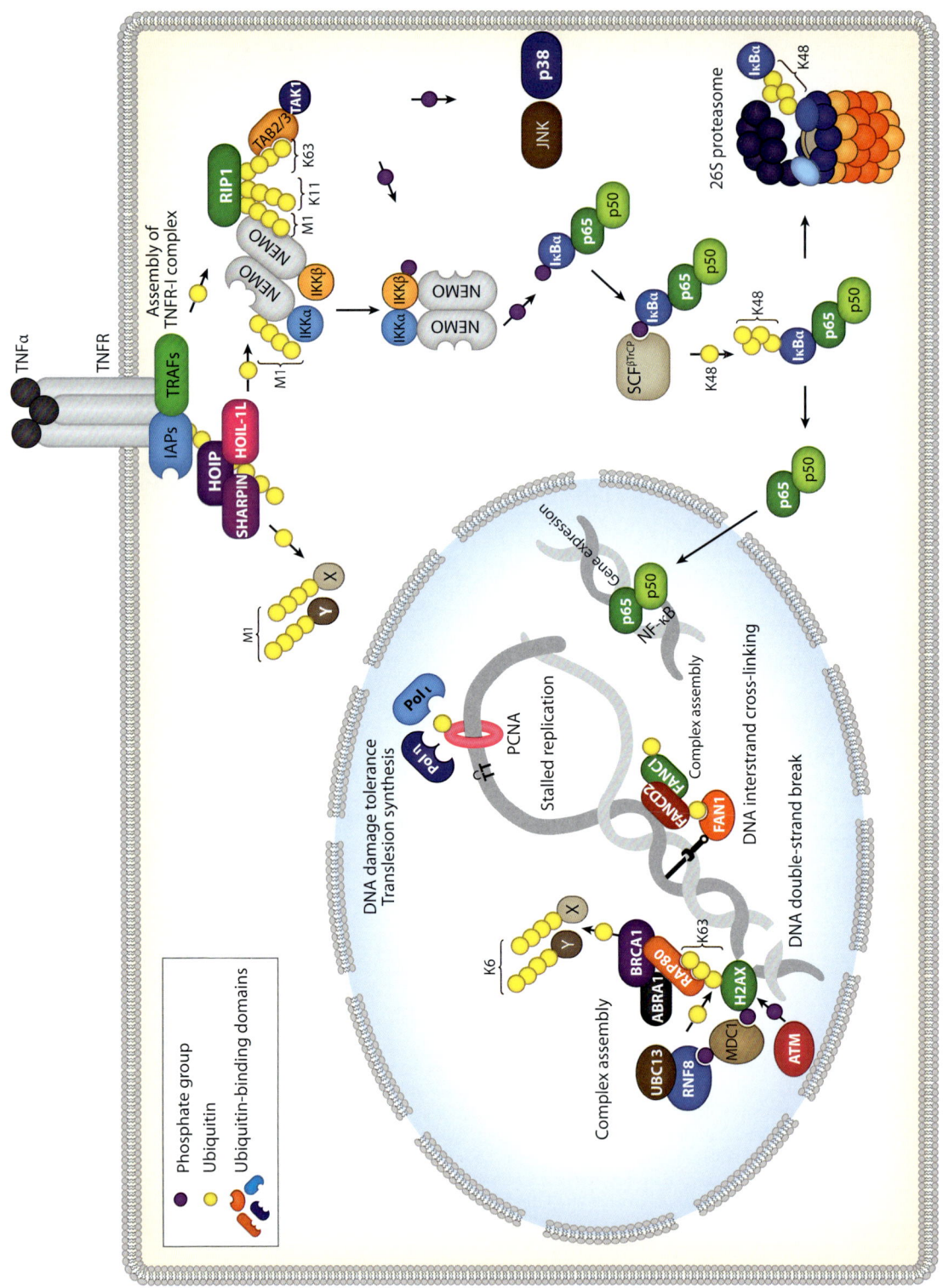

either Lys48-linked (**Figure 4e**) or Lys63-linked polyubiquitin chains (103).

The second intrinsic proteasomal ubiquitin receptor Rpn13 contains the N-terminal UBD with a PH fold, which binds mono- and polyubiquitin, as well as UBL domains of shuttle factors, including HHR23A and Ddi1. Because of its PH domain fold and ubiquitin-binding properties, the domain was named PRU (**Figure 3g**). NMR data show that the PRU domain binds ubiquitin in an unexpected and entirely different way than in the previously identified PH domain GLUE (in EAP45) as loops, rather than secondary structural elements of PRU, are used to capture ubiquitin (**Figure 3g**). Because the loops in the PRU domain form a larger binding surface than the split PH domain of EAP45, the binding to monoubiquitin is also stronger (104, 105). Rpn13 loops bind in the close proximity of Lys48 residue of the proximal ubiquitin, specifically increasing the PRU-binding surface toward Lys48-linked diubiquitin (104, 105). At the same time, preferential binding of PRU to the proximal subunit of Lys48-linked diubiquitin allows DUB UCH37 to mediate a stepwise removal of ubiquitin from the substrate by disassembling the chain from its distal tip (105).

Unlike shuttle factors, which preferentially bind to Lys48-linked ubiquitin chains, purified 26S proteasome components do not exclusively interact with Lys48-linked ubiquitin chains and can bind the entire spectrum of ubiquitin chains (19). However, proteasome inhibitors do not affect the stability of proteins conjugated with Lys63-linked ubiquitin chains (30), arguing in favor of some additional regulatory mechanisms that limit the accessibility of Lys63-linked ubiquitin chains to the 26S proteasome in vivo. It is indeed intriguing to speculate why Lys48 linkages were evolutionarily selected for proteasomal degradation of target

Figure 7

Ubiquitin receptors in DNA repair and NF-κB signaling. In the nucleus, the assembly of DNA damage response complexes at the specific nuclear foci in response to DNA damage, such as DNA double-strand breaks (DSBs), is initiated by regulation of protein-protein interactions through different posttranslational modifications, including phosphorylation, ubiquitination, sumoylation, and acetylation. In a very simplified scenario, phosphorylation of mediator of DNA damage checkpoint 1 (MDC1) by the Ser/Thr kinase ataxia telangiectasia mutated (ATM) leads to the recruitment of the E3 ligase RNF8, which, in collaboration with E2 Ubc13, ubiquitinates histone H2A and its variants. The BRCA1/BARD1 E3 ligase complex is then recruited to Lys63 (K63)-linked chains on histones H2A/H2AX via ubiquitin receptor RAP80, where it ubiquitinates unknown substrates present at the DSB-associated chromatin by Lys6 (K6) chains to initiate DNA repair. Recognition of monoubiquitinated components of Fanconi anemia complex by the UBZ domain of the Fanconi anemia-associated nuclease 1 (FAN1) plays an important role in the repair of DNA cross-links. Stalling of replication forks on DNA leads to monoubiquitination of DNA processivity factor proliferating cell nuclear antigen (PCNA), which, in turn, increases its affinity toward the UBM and UBZ domains of translesion DNA polymerases. Polymerase ι (Pol ι) binds to monoubiquitinated PCNA, enabling DNA synthesis across the lesion. After the lesion is passed, PCNA gets deubiquitinated, and Y-family polymerases disengage from DNA, enabling error-free synthesis to continue. In the membrane, cytosol, and nucleus, the tumor necrosis factor receptor (TNFR) proteins play an important role in immunity and inflammation. Binding of ligands, such as TNFα to TNFR at the plasma membrane, initiates recruitment and ubiquitination of the large protein complex, which includes numerous E3 ligases and scaffolding proteins, to the activated receptor. Several ubiquitin receptors contribute to the activation of the NF-κB pathway. TAK1-binding protein 2 (TAB2) and NEMO bind Lys63 (K63)- and Met1 (M1)-linked chains produced by UBC13/UEV1A/TRAF6 and HOIP/HOIL-1L/SHARPIN (linear ubiquitin chain assembly) complexes, respectively. NEMO can also bind longer ubiquitin chains with variable linkages, including Lys11 (K11) (produced by the UBCH5/c-IAP1 complex) and can also be linearly ubiquitinated. Polyubiquitination of the receptor-interacting protein 1 (RIP1) by E3 ligases TNF receptor-associated factor 2 (TRAF2), TNF receptor-associated factor 5 (TRAF5), and c-IAP1/2 by various ubiquitin chains has also been linked with the activation of the pathway. Activated inhibitor of the NF-κB (IκB) kinase (IKK) complex phosphorylates the inhibitory IκBα, creating a binding site for the Skp-cullin-F-box-βTrCP (SCF$^{β\text{TrCP}}$) ubiquitin E3 ligase complex, which Lys48 ubiquitinates IκBα, targeting it for proteasomal degradation. The NF-κB components p50 and p65 can now freely translocate to the nucleus and activate the transcription of genes promoting cell survival and inflammation. Abbreviations: ABRA1, BRCA1-A complex subunit Abraxas; FANCD2, Fanconi anemia complementation group D2 protein; FANCI, Fanconi anemia complementation group I protein; p38, mitogen-activated protein kinase 14, Jun N-terminal kinase (JNK), transforming growth factor-β-activated kinase 1 (TAK1).

proteins and how Lys63 linkages were saved for degradation-free functions.

Ataxin-3, a deubiquitinating enzyme involved in the development of spinocerebellar ataxia type 3, contains two UIM domains at the C terminus (82), and an additional UIM exists in a splice variant (106). The short two-residue linker between its two UIMs seems to be important for binding to the compactly packed Lys48-linked ubiquitin chains (82, 107) because longer linkers (of the RAP80 type, see below) partially decrease specificity for Lys48-linked chains, shifting the affinity toward Lys63 linkage (82). Although it has also been reported that the isolated ataxin-3 UIMs might bind Lys48- and Lys63-linked diubiquitins (108), binding to longer chains is strongly preferred (107, 109). Importantly, binding kinetics change in the context of the full-length protein. Consistent with this notion, two additional UBDs have been recently identified in the catalytic (Josephin) domain of the protein (110), and a role in regulating specificity for Lys48-linked ubiquitin chains has been suggested (111).

Ubiquitin Receptors for Lys63-Linked Ubiquitin Chains

Lys63-specific ubiquitin-binding proteins have been implicated in the regulation of several cellular functions, including activation of NF-κB, DNA damage repair, and regulation of endosomal sorting pathways. In the NF-κB pathway, activation of the transforming growth factor-β-activated kinase 1 (TAK1) plays an important role in linking receptor proximal events with downstream activation of NF-κB and AP-1 transcription factors. To be activated, TAK1 needs to bind Lys63-linked chains through its subunits Tak1-binding proteins 2 and 3 (TAB2 and TAB3) (112). Although the exact mechanism is still not known, Lys63-linked chain binding is believed to mediate oligomerization and autophosphorylation of TAK1, thus leading to the activation of downstream signaling pathways. Although TAB2 and TAB3 contain two UBDs (CUE domain at the N terminus and NZF at the C terminus), only their C-terminal part specifically interacts with Lys63-linked ubiquitin chains (**Figure 7**). The recently solved cocrystal structures of the NZF domains of TAB2 and TAB3 in a complex with Lys63-linked diubiquitin explain how TAB2 binds neighboring Lys63-linked ubiquitin moieties (**Figure 4d**) (113, 114). The hydrophobic Leu8-Ile44-Val70 patch of each ubiquitin interacts with different surfaces of TAB2, where distal ubiquitin occupies the canonical NZF ubiquitin-binding site, whereas proximal ubiquitin binds to the side of the NZF domain (113, 114). The TAB2 NZF domain does not contact the isopeptide linkage but can actually specifically sense the conformational features of Lys63 linkages for specific recognition. Although TAB2 NZF binding to Lys48-linked diubiquitin was undetectable in pull-down experiments, it was clearly detected in solution (113, 114), which was explained as the consequence of the rotation of a flexible Lys48-isopeptide linker into a conformation similar to the one present in Lys63-linked diubiquitin. Overall affinities for Lys48-linked chains were approximately threefold lower than those for Lys63-linked chains. Because immobilized TAB2 NZF is highly Lys63 chain specific (113, 114), it is assumed that TAB2 binding to protein complexes is a prerequisite for specific linkage recognition and subsequent pathway activation.

The nonproteolytic role of ubiquitination in the coordinated assembly of the DNA repair protein complexes around DNA double-strand breaks has been well recognized (6, 115). The receptor-associated protein 80 (RAP80) recruits the Abraxas-BRCC36 [breast cancer 1/2 (BRCA1/BRCA2)-containing complex subunit 36]-BRCA1-associated RING domain protein 1 (BARD1) complex to DNA damage foci for DNA repair by specifically recognizing the Lys63-linked polyubiquitinated proteins (such as histones H2A and H2AX) (reviewed in Reference 116) by its tandem UIMs (**Figures 4a** and **7**). The length of the linker between two UIMs of RAP80 enables the correct positioning of two ubiquitins within Lys63 dimer, affecting binding avidity (82). Sato et al. (117) solved the cocrystal structure of the RAP80 tandem UIMs in a complex with Lys63-linked diubiquitin, showing that both UIMs and the linker between them form a long α-helix. Each UIM recognizes one Ile44 patch of the neighboring Lys63-linked diubiquitins, without touching the Lys63-linked isopeptide bond (**Figure 4***a*). The length of the inter-UIM linker enables the correct positioning of two ubiquitins within the Lys63 dimer and excludes Lys48 or Met1 chain binding (117). Interestingly, insertion of the ataxin-3 linker (see above) between two UIMs of RAP80 protein shifts RAP80 UIM specificity toward Lys48-linked chains (82). Tandem UIMs of epsin1, which contain the inter-UIM linker of a similar length, also selectively recognize the Lys63-linked diubiquitins and do not bind to Lys48- or Met1-linked chains (117).

Although monoubiquitin was initially described as a sufficient signal for transmembrane receptor internalization and endocytosis (reviewed in Reference 3), subsequent work has shown a critical role for multivalent binding of endocytic adaptor proteins to multiply monoubiquitinated (68) or Lys63-linked polyubiquitinated proteins (118) in regulating the kinetics and fidelity of receptor endocytosis. The pioneering work of Haguenauer-Tsapis's lab (119) has shown that the plasma membrane uracil permease undergoes Lys63 linkage-specific cell-surface ubiquitination, which is dependent on the E3 ligase Npi1/Rsp5, leading to its endocytosis and subsequent vacuolar degradation. Additionally, the AMSH (associated molecule with the SH3 domain of STAM) and AMSH-like protease deubiquitinating enzymes are zinc metalloproteases that regulate receptor trafficking by specifically cleaving Lys63-linked polyubiquitin chains from internalized receptors (120–122). The cocrystal structure of human AMSH-like protease DUB domain, located at the C terminus of the protein, in a complex with Lys63-linked diubiquitin (**Figure 4***b*) (123), reveals how specific recognition of Lys63 linkages is achieved by the DUB domain directly contacting all the residues involved in the Lys63 linkage, including the Lys63 itself.

Decoders of Linear Ubiquitin Chains

The role of linear ubiquitination in the regulation of the innate immunity and inflammation pathways has been recently demonstrated (**Figure 7**) (25–27). However, so far, only a NEMO and RIP1 have been shown to be modified by linear ubiquitin chains (25–27). The decoding of linear ubiquitin signals in cells depends on the presence of the intact UBAN domain, which was initially described in ABIN-1, ABIN-2, ABIN-3, NEMO, and optineurin proteins (124). The high-resolution structure of the NEMO UBAN domain in a complex with linear diubiquitin chains reveals a parallel coiled-coil dimer with a stereo-symmetrical ubiquitin-binding surface that contacts all four ubiquitin moieties of two linear diubiquitins (36). The UBAN domain of NEMO binds the Ile44-surrounding hydrophobic surface on the distal and a novel surface (centered around Phe4) on the proximal ubiquitin. Importantly, the most significant interactions are made through the residues that form the linker region between the linearly linked ubiquitin moieties (**Figure 4***f*). Specific recognition of linear ubiquitin chains by NEMO is important for NF-κB activation in response to the tumor necrosis factor, interleukin 1, and other agonists (36). NEMO residues involved in binding linear ubiquitin chains are found mutated in patients suffering from X-linked ectodermal dysplasia and immunodeficiency (36). Molecular dynamics studies indicated that other types of ubiquitin chains (including Lys63-linked chains) could not adopt the same conformation as linear chains and thus bind NEMO with significantly lower affinity (32, 36, 125). Yet, full-length NEMO is found in complex with Lys63- and Lys11-linked ubiquitin chains in cells (8), which are edited (in length and topology) by E3 ligases and DUBs (126). It was proposed that Lys-linked chains may bind in a circular way, opposite to the perpendicular arrangement previously shown for linear chains, thus surrounding the UBAN dimer and recognizing the patch involved in binding to a distal ubiquitin monomer

Endosome: a small vesicle formed by endocytosis of the plasma membrane; it is responsible for the sorting of internalized proteins

(32, 36, 125, 127). An additional layer of complexity was introduced by showing that the UBAN domain, expressed together with the C-terminal ZnF of NEMO, can bind to longer (four and more) Lys63-linked chains with increased affinity (128). Full-length NEMO thus might act as a multifunctional ubiquitin receptor: a specific linear diubiquitin chain decoder as well as a multivalent, low-affinity receptor for longer Lys63- or Lys11-linked ubiquitin chains. The functional relevance of NEMO binding to ubiquitin chains will likely depend on the local concentration of ubiquitin chains conjugated to substrates in defined subcellular compartments (reviewed in Reference 4).

Many Flavors of Zinc-Finger Folds in Ubiquitin-Binding Proteins

The repeated variation of the structural ZnF fold has also been utilized in numerous proteins to engage in ubiquitin binding. Nuclear protein localization 4 (NPL4) zinc finger (NZF) is a compact zinc-binding module found in many proteins that function in ubiquitin-dependent processes; one of these proteins is the NPL4 subunit of the p97 AAA-ATPase complex (129), where three key NZF residues (Thr13, Phe14, Met25) surrounding the zinc coordination site bind the hydrophobic Ile44 patch of single ubiquitin (61).

Yeast Vps36 protein (also known as EAP45) is involved in sorting of ubiquitinated membrane proteins from endosomes to lysosomes. Mutations in Vps36 NZF domains that inhibit binding to monoubiquitin also inhibit sorting of ubiquitinated proteins into the yeast vacuole (61). Although the NZF domains of yeast Vps36 are regarded as mono-UBDs, it was recently shown that Vps36 contains a two-sided NZF2 domain that can bind Lys63-linked diubiquitin with ~200-fold higher affinity than Met1-linked chains or monoubiquitin (113). NZF domains are also commonly present in proteins involved in the regulation of the NF-κB signaling pathway (i.e., the TAB2, TAB3, HOIP, and HOIL-1L proteins) (**Figure 7**). Interestingly, NZF domains of TAB2/3 proteins specifically bind to Lys63-linked ubiquitin chains (see above). NZF domains of the linear ubiquitin chain assembly complex components HOIP and HOIL-1L bind monoubiquitin (26), although the full-length HOIL-1L preferentially binds to linear ubiquitin chains (130). The HOIL-1L NZF domain consists of a zinc-coordinating NZF core region and an additional α-helical "NZF tail" region, which bind the canonical Ile44 and Phe4 hydrophobic patches on the distal and proximal ubiquitin, respectively, thus explaining how the specific recognition of linear chains is achieved (131).

Bioinformatic analysis has indicated the presence of ubiquitin-binding zinc-finger (UBZ) domains in multiple proteins regulating DNA repair (132), where UBZ domains appear to be critical for the recruitment of UBZ-containing proteins to the sites of DNA damage and local repair, as exemplified by the UBZ-dependent recruitment of polymerase η to monoubiquitinated PCNA upon replication fork stalling and subsequent bypass translesion synthesis (**Figure 7**). The domain was initially identified in the C-terminal part of Y-family DNA polymerases η and κ and was shown to bind to monoubiquitin (43). Another UBZ-containing protein, Fanconi anemia-associated nuclease 1 (FAN1), was shown to bind to monoubiquitinated Fanconi anemia complementation group D2 protein (FANCD2) by its N-terminal UBZ domains, whereas FAN1 C-terminal nuclease activity is required for DNA interstrand cross-link repair (133–136) (**Figure 7**). In addition, the DNA repair protein, Werner helicase-interacting protein 1 (WRNIP1), binds ubiquitin by its N-terminal UBZ domain (137, 138), which can interact with different ubiquitin moieties, including monoubiquitin as well as ubiquitin chains of Lys48 and Lys63 linkages (138). The presence of WRNIP1 in nuclear DNA replication structures requires its intact UBD as well as the oligomerization mediated by its C terminus (138).

The autophagy receptor nuclear dot protein 52 kDa (NDP52, also known as CALCOCO2) (139) and its paralog TAX1BP1 (also known as TXBP151 or T6BP) (140) bind ubiquitin by only their C-terminal UBZ domains. The intact proteins have been shown to bind to multiple polyubiquitinated substrates, indicating that the monoubiquitin-binding ability of UBZ can recognize a wide spectrum of ubiquitin chains. NDP52 is recruited to ubiquitin-coated *Salmonella enterica* in human cells (the identity of linkages on ubiquitinated *Salmonella* is not yet clear) (139), whereas TAX1BP1 binds to polyubiquitinated TRAF6 (most likely conjugated by Lys63- and Lys11-linked chains), resulting in NF-κB inhibition (140).

The ZnF UBP, also known as the PAZ (polyubiquitin-associated zinc binding), BUZ (binder of ubiquitin zinc finger), or DAUP (deacetylase/ubiquitin-specific protease), domain is found in a subfamily of DUBs, as well as in ubiquitin ligase BRAP2 and cytoplasmic deacetylase HDAC6. The ZnF UBP domain of human IsoT is one of the strongest ubiquitin binders with specificity for the free C-terminal tail of ubiquitin (**Figure 3*f***), which could conveniently explain its preference for cleaving unanchored ubiquitin chains (47).

ZnF A20 (45) of Rabex-5 (**Figure 3*e***) binds ubiquitin via its polar interface centered on Asp58, which represents a novel, noncanonical ubiquitin-binding surface (45, 46), whereas the neighboring α-helix (MIU) binds to the canonical Ile44 patch of ubiquitin. Biochemical data suggest that ZnF A20 and MIU can bind ubiquitin independently (45, 46) and that single ubiquitin molecules can bind to two Rabex-5 molecules, which requires two distinct ubiquitin surfaces (Asp58 and Ile44) (45). Rabex-5 acts as a regulator of the small GTPase Rab5, which functions as a master regulator of early endocytic trafficking (141). Despite their different UBD structures, the conserved ubiquitin-binding properties of both Vps9 and Rabex-5 point to the important role of ubiquitin binding for protein function. Rabex-5 binds to the activated and ubiquitinated epidermal growth factor receptor, which requires any of its UBDs (45), thus enabling Rabex-5 recruitment to the membrane where it might, in turn, mediate cargo-dependent activation of Rab5. By simultaneous binding of two ubiquitins, Rabex-5 might be involved in dimerization of ubiquitinated proteins, or alternatively, it might enable transfer of ubiquitinated substrate between two different complexes.

E2 and E3 Enzymes Require Binding to Ubiquitin for Their Functions

In addition to the conserved catalytic αβ-fold core domain (UBC) (142), many E2 enzymes contain a UBD adjacent to the catalytic domain. By contrast, class I UBCs (such as UbcH5C) contain a single UBC catalytic domain with a specific β-sheet that is able to bind to the Ile44 patch of ubiquitin with an affinity of ∼300 μM (143). This UbcH5C region is displaced from the catalytic center, enabling the activated UbcH5C∼ubiquitin (in which ubiquitin Gly76 is covalently linked with the active Cys of UbcH5C) to self-assemble into larger molecular weight complexes (143). This mode of ubiquitin binding is essential for the ability of the E2 enzyme to perform processive ubiquitin transfer reactions in collaboration with BRCA1/BARD1 enzymes, which play an important role in the DNA damage responses (143).

Similarly, several UEV domains have also been shown to bind a single ubiquitin, which can play an important role in determining the specificity of linkage conjugation, as was shown for the UEV domain of Mms2. Mms2 hetero-oligomerizes with Ubc13 and enables selective insertion of Lys63 into the Ubc13 active site to facilitate Lys63-linked chain formation (144). The other example is the UEV domain of the TSG101 endocytic protein, which plays an important role in HIV1 budding and in the vacuolar protein sorting pathway (145).

It was previously shown that yeast Rps5 and human Smurf2 E3 ligases can noncovalently bind ubiquitin by the N-terminal lobe of their HECT domains. Although Rps5-ubiquitin binding was functionally explained as a mechanism for restricting the length of polyubiquitin chains (146), the Smurf2-ubiquitin binding seemed to facilitate chain elongation by stabilizing ubiquitinated substrate binding

Autophagy: an evolutionarily conserved catabolic process by which cells deliver bulk cytosolic components for degradation to the lysosome

Autophagy receptor: a protein that plays a crucial role in selective autophagy by noncovalently binding to autophagy modifiers (e.g., ATG8 family)

to the HECT domain (147). Recent studies favor the latter model in which HECT-ubiquitin binding helps in localizing and positioning the distal ubiquitin to promote further chain elongation, possibly also with a role in the initial recruitment of target proteins (148, 149). Whether a similar mechanism applies to other HECT E3 ligases still needs to be verified.

A subclass of inhibitor of apoptosis proteins (IAPs), characterized by the presence of highly conserved baculoviral IAP repeat domains, functions as RING E3 ligases. By directly binding caspases (key effector proteases in apoptosis), IAPs can inhibit cell death, but they are also linked with inflammatory signaling and immunity, mitogenic kinase signaling, proliferation, and mitosis (150). An evolutionarily conserved UBA domain has recently been identified between the third baculoviral IAP repeat domain and the RING finger in several IAPs, including c-IAPs and XIAP (151, 152). The UBA domain of IAPs can interact with both Lys63- and Met1-linked tetraubiquitin, and RING-mediated IAP dimerization might enable linear or Lys63-linked dimers within tetraubiquitin chains to interact with both UBA domains at the same time (151, 152). The ubiquitin binding of c-IAP1 is not required for its E3 ligase activity but is required for its binding to interacting partners (such as polyubiquitinated NEMO, RIP1, or IAPs, themselves), thus linking IAPs to complexes regulating diverse cellular functions (151, 152).

Single UBA domains can be found at the C terminus of the E3 ligases c-Cbl and Cbl-b, which act as negative regulators of receptor tyrosine kinases by targeting them for ubiquitination and degradation (153, 154). Interestingly, although c-Cbl UBA cannot bind ubiquitin, the UBA domain of Cbl-b strongly binds polyubiquitin chains (155) but weakly binds monoubiquitin (**Figure 3d**) (155). The two UBA domains share 75% sequence similarity as well as the compact three-helix bundle fold (156). However, because the ubiquitin-binding hydrophobic patch on the c-Cbl UBA domain is interrupted by a negatively charged Glu12 residue, c-Cbl UBA cannot bind ubiquitin (156).

UBIQUITIN-LIKE MODIFIERS RECOGNIZED BY UBIQUITIN-BINDING PROTEINS

The additional layer of complexity in the ubiquitin network comes from the fact that a UBL fold can be found in numerous proteins that either exclusively contain a UBL domain (UBL type I proteins) or have an UBL domain as a part of a multidomain structure (UBL type II proteins). UBL type II domains cannot be conjugated to target proteins but serve various functions in the cell. They have been best studied in the ubiquitin-proteasome system, where they recruit polyubiquitinated cargo to the 26S proteasome (**Figure 6**). Additionally, bioinformatic analysis (157) predicted the existence of UBL domains both inside and outside catalytic core domains of USP enzymes. It was initially proposed that internal UBL domain might play a role in the modulation of DUB activity and specificity as well as in recruitment of nonsubstrate proteins that might alter their localization or trafficking, specificity, and enzymatic activity (157). The UBL domain of Ubp6, the yeast homolog of USP14, has already been shown to mediate its recruitment to the 19S proteasome (158). By contrast, USP4 is inhibited by its USP-contained UBL domain, which binds to the catalytic domain and competes with the ubiquitinated substrate (159). E3 ligases may also be regulated by an internal UBL domain (e.g., parkin contains a UBL domain at its N terminus), which inhibits its autoubiquitination (160) owing to interactions with a putative UBD within its C terminus. Binding of UBD-containing proteins, such as Eps15 UIM to parkin's UBL domain cancels autoinhibition and enables its E3 ligase activity (160).

POSTTRANSLATIONAL MODIFICATIONS REGULATE UBIQUITIN RECEPTORS

Many UBD-containing proteins are themselves posttranslationally modified, which can regulate their functions as ubiquitin receptors. The activity of multiple ubiquitin receptors in the endocytic pathway is regulated by their monoubiquitination, an E3-independent process that is based on the interaction between a UBD and an E2-loaded ubiquitin (161–164). This process has been termed coupled monoubiquitination because the self-ubiquitination of the ubiquitin receptor is intimately linked to its ubiquitin-binding ability (161). As a consequence, the *trans* ubiquitin-binding capacity of UBD-containing proteins and their functional role as ubiquitin receptors are impaired (163).

Monoubiquitination also plays positive or negative regulatory roles in the assembly of protein complexes during DNA synthesis (43, 165). Although monoubiquitination of PCNA upon DNA damage enables its interaction with the UBZ domain of polymerase η and subsequently translesion synthesis over DNA lesions (43), monoubiquitination of translesion polymerase η in undamaged cells inhibits its interaction with PCNA, thus keeping the polymerase away from PCNA and enabling error-free DNA synthesis (**Figure 7**). More recent reports indicate the proteasome functions can control protein stability via monoubiquitination of proteasomal receptor Rpn10. Monoubiquitination of Rpn10 controls its capacity to interact with ubiquitinated substrates and thus affects substrate degradation (166). Monoubiquitination is decreased under stress conditions, consequently increasing the receptor availability and capacity of the 19S proteasome to bind ubiquitinated cargo (166).

Other posttranslational modifications, including phosphorylation, play important roles in the regulation of both binding strength and timing of UBD:ubiquitin interactions: Autophagy receptors p62/SQSTM1 and NBR1 mediate the ubiquitinated protein interaction with and sequestration within autophagosomes (reviewed in References 9 and 167). Ubiquitinated proteins are recognized by their C-terminal UBA domain, which was suggested to recognize Lys63-linked chains (168). Interestingly, the isolated UBA domain of NBR1 readily bound to ubiquitin chains, whereas the isolated UBA domain of p62 was unable to bind to Lys48- and Lys63-linked diubiquitins in vitro (169). Recently, Matsumoto and colleagues (170) have shown that casein kinase 2 mediates specific phosphorylation of p62 at Ser403 within its UBA domain. This modification increases the affinity between the UBA domain and Lys63 ubiquitin chains and enhances the ability of p62 to act as an autophagy receptor for ubiquitinated protein aggregates (170). In addition, phosphorylation of a UBAN domain containing protein, optineurin, was shown to promote selective autophagy of ubiquitin-coated cytosolic *S. enterica*. Optineurin is phosphorylated by the TANK-binding kinase 1 on Ser177 within the LC3-interacting region, which increases the binding affinity between optineurin and the ubiquitin-like modifier LC3 and autophagic clearance of cytosolic *Salmonella* (171). In analogy to ubiquitin, phosphorylation by casein kinase 2 controls the binding of SUMO to specific phospho-dependent SUMO interaction motifs (172). Taken together, these examples imply that phosphorylation may represent a general regulatory principle for regulated interactions between ubiquitin or UBL modifiers and their binding domains.

CONCLUSIONS AND FUTURE PERSPECTIVES

The identification of small domains with diverse structural folds, which are able to noncovalently bind to ubiquitin and are an integral part of ubiquitin receptors, provided the core principle by which ubiquitin signals can be recognized and decoded in cells. The number of new UBDs identified in cellular proteins has substantially increased over the past decades with the current list of over 20 families of UBDs present in more than 200 cellular proteins (42). Even though many structural details about UBD:ubiquitin interactions are known, their physiological and functional roles in the context of the intact ubiquitin-binding proteins are less well understood. In many instances, isolated UBDs modulate their binding properties when they are a part of multidomain proteins embedded in their natural protein (sub)complexes.

Autophagosome: a double-membrane vesicle that delivers large cytosolic cargoes such as damaged organelles and protein aggregates to the lysosome

LC3: a mammalian homolog of yeast ATG8 proteins; it can be conjugated to autophagosomal membranes, which play crucial roles in autophagy

We are just beginning to realize a growing spectrum of mechanisms by which ubiquitin-binding proteins recognize and decode ubiquitinated substrates in such complexes in vivo. The major challenge is to obtain structural details of complexes containing ubiquitinated substrates and full-length ubiquitin-binding proteins. In addition, more information is needed to understand how ubiquitin receptors bind and decode ubiquitin chains in cells. Developing ubiquitin sensors, which can be used to monitor the dynamics of ubiquitin chain formation and their length in vivo (recently reviewed in Reference 81), is among the most challenging task in the field. This information will be required for a better understanding of the temporal and spatial control of ubiquitin signaling under physiological conditions, as well as in human diseases that are caused by alterations in the ubiquitin system (173).

SUMMARY POINTS

1. Coordinated activity of E1, E2, and E3 enzymes results in the conjugation of monoubiquitin and ubiquitin chains to target proteins, which are specifically recognized by ubiquitin receptors (i.e., proteins containing UBDs).
2. The majority of structurally diverse UBDs recognize the hydrophobic patch around Ile44 on ubiquitin, although a smaller panel of binding modules can also recognize additional surfaces.
3. The majority of UBDs bind monoubiquitin in the micromolar range, whereas specific chain binding is of higher affinity.
4. UBDs are usually found in multidomain proteins and often exist in multiple copies (tandem UBDs), either of the same or a different structure.
5. Linkage selectivity is achieved by specific types of UBDs, which may specifically recognize either the linkage between two ubiquitins or the positioning of two ubiquitins within a specific linkage.
6. Tandem UBDs perform various functions: enable specific linkage recognition, sense ubiquitin chain length, or protect ubiquitin chains from elongation/degradation.
7. Compartmentalization, ubiquitin receptor dimerization, and tandem UBDs are some of the mechanisms that proteins use to increase binding affinities between a UBD and ubiquitin in vivo.

FUTURE ISSUES

1. The recently reported ability to chemically synthesize atypical ubiquitin linkages should speed up discovery of linkage-specific UBDs and the corresponding E1-E2-E3 machinery.
2. Structural information from full-length ubiquitin receptors in complex with various ubiquitin chains is necessary for a better understanding of linkage and length recognition by UBDs.
3. Ubiquitin-binding properties can change between isolated UBDs and intact proteins, which should be taken into account during data interpretation.

4. Development of various ubiquitin sensors will enable in vivo monitoring of dynamic changes in ubiquitin patterns.

DISCLOSURE STATEMENT

The authors are not aware of any affiliations, memberships, funding, or financial holdings that might be perceived as affecting the objectivity of this review.

ACKNOWLEDGMENTS

We apologize to all scientists whose important contributions were not referenced in this review owing to space limitations. We would like to thank Stefan Mueller, David McEwan, Krishnaraj Rajalingam, and Jaime Lopez-Mosqueda for critical reading and comments on the manuscript. Research in the I.D. laboratory is supported by the Deutsche Forschungsgemeinschaft, the Cluster of Excellence "Macromolecular Complexes" of the Goethe University Frankfurt (EXC115), the LOEWE Centrum for Cell and Gene Therapy Frankfurt, and a European Research Council Advanced Grant. The K.H. laboratory is supported by the LOEWE Centrum for Gene and Cell Therapy Frankfurt.

LITERATURE CITED

1. Hershko A, Ciechanover A. 1998. The ubiquitin system. *Annu. Rev. Biochem.* 67:425–79
2. Varshavsky A. 2005. Regulated protein degradation. *Trends Biochem. Sci.* 30:283–86
3. Mukhopadhyay D, Riezman H. 2007. Proteasome-independent functions of ubiquitin in endocytosis and signaling. *Science* 315:201–5
4. Grabbe C, Husnjak K, Dikic I. 2011. The spatial and temporal organization of ubiquitin networks. *Nat. Rev. Mol. Cell Biol.* 12:295–307
5. Hirsch C, Gauss R, Horn SC, Neuber O, Sommer T. 2009. The ubiquitylation machinery of the endoplasmic reticulum. *Nature* 458:453–60
6. Bergink S, Jentsch S. 2009. Principles of ubiquitin and SUMO modifications in DNA repair. *Nature* 458:461–67
7. Vucic D, Dixit VM, Wertz IE. 2011. Ubiquitylation in apoptosis: a post-translational modification at the edge of life and death. *Nat. Rev. Mol. Cell Biol.* 12:439–52
8. Bianchi K, Meier P. 2009. A tangled web of ubiquitin chains: breaking news in TNF-R1 signaling. *Mol. Cell* 36:736–42
9. Kirkin V, McEwan DG, Novak I, Dikic I. 2009. A role for ubiquitin in selective autophagy. *Mol. Cell* 34:259–69
10. Ciechanover A, Ben-Saadon R. 2004. N-terminal ubiquitination: more protein substrates join in. *Trends Cell Biol.* 14:103–6
11. Cadwell K, Coscoy L. 2005. Ubiquitination on nonlysine residues by a viral E3 ubiquitin ligase. *Science* 309:127–30
12. Ravid T, Hochstrasser M. 2007. Autoregulation of an E2 enzyme by ubiquitin-chain assembly on its catalytic residue. *Nat. Cell Biol.* 9:422–27
13. Wang X, Herr RA, Chua WJ, Lybarger L, Wiertz EJ, Hansen TH. 2007. Ubiquitination of serine, threonine, or lysine residues on the cytoplasmic tail can induce ERAD of MHC-I by viral E3 ligase mK3. *J. Cell Biol.* 177:613–24
14. Peng J, Schwartz D, Elias JE, Thoreen CC, Cheng D, et al. 2003. A proteomics approach to understanding protein ubiquitination. *Nat. Biotechnol.* 21:921–26

15. Wagner SA, Beli P, Weinert BT, Nielsen ML, Cox J, et al. 2011. A proteome-wide, quantitative survey of in vivo ubiquitylation sites reveals widespread regulatory roles. *Mol. Cell Proteomics* 10:M111.013284
16. Kim W, Bennett EJ, Huttlin EL, Guo A, Li J, et al. 2011. Systematic and quantitative assessment of the ubiquitin-modified proteome. *Mol. Cell.* 44:325–40
17. Amerik AY, Hochstrasser M. 2004. Mechanism and function of deubiquitinating enzymes. *Biochim. Biophys. Acta* 1695:189–207
18. Reyes-Turcu FE, Ventii KH, Wilkinson KD. 2009. Regulation and cellular roles of ubiquitin-specific deubiquitinating enzymes. *Annu. Rev. Biochem.* 78:363–97
19. Ikeda F, Dikic I. 2008. Atypical ubiquitin chains: new molecular signals. 'Protein Modifications: Beyond the Usual Suspects' review series. *EMBO Rep.* 9:536–42
20. Iwai K, Tokunaga F. 2009. Linear polyubiquitination: a new regulator of NF-kappaB activation. *EMBO Rep.* 10:706–13
21. Finley D. 2009. Recognition and processing of ubiquitin-protein conjugates by the proteasome. *Annu. Rev. Biochem.* 78:477–513
22. Haglund K, Dikic I. 2005. Ubiquitylation and cell signaling. *EMBO J.* 24:3353–59
23. Ziv I, Matiuhin Y, Kirkpatrick DS, Erpapazoglou Z, Leon S, et al. 2011. A perturbed ubiquitin landscape distinguishes between ubiquitin in trafficking and in proteolysis. *Mol. Cell Proteomics* 10:M111.009753
24. Dammer EB, Na CH, Xu P, Seyfried NT, Duong DM, et al. 2011. Polyubiquitin linkage profiles in three models of proteolytic stress suggest the etiology of Alzheimer disease. *J. Biol. Chem.* 286:10457–65
25. Gerlach B, Cordier SM, Schmukle AC, Emmerich CH, Rieser E, et al. 2011. Linear ubiquitination prevents inflammation and regulates immune signalling. *Nature* 471:591–96
26. Ikeda F, Deribe YL, Skanland SS, Stieglitz B, Grabbe C, et al. 2011. SHARPIN forms a linear ubiquitin ligase complex regulating NF-kappaB activity and apoptosis. *Nature* 471:637–41
27. Tokunaga F, Nakagawa T, Nakahara M, Saeki Y, Taniguchi M, et al. 2011. SHARPIN is a component of the NF-kappaB-activating linear ubiquitin chain assembly complex. *Nature* 471:633–36
28. Tokunaga F, Sakata S, Saeki Y, Satomi Y, Kirisako T, et al. 2009. Involvement of linear polyubiquitylation of NEMO in NF-kappaB activation. *Nat. Cell Biol.* 11:123–32
29. Matsumoto ML, Wickliffe KE, Dong KC, Yu C, Bosanac I, et al. 2010. K11-linked polyubiquitination in cell cycle control revealed by a K11 linkage-specific antibody. *Mol. Cell* 39:477–84
30. Xu P, Duong DM, Seyfried NT, Cheng D, Xie Y, et al. 2009. Quantitative proteomics reveals the function of unconventional ubiquitin chains in proteasomal degradation. *Cell* 137:133–45
31. Jin L, Williamson A, Banerjee S, Philipp I, Rape M. 2008. Mechanism of ubiquitin-chain formation by the human anaphase-promoting complex. *Cell* 133:653–65
32. Dynek JN, Goncharov T, Dueber EC, Fedorova AV, Izrael-Tomasevic A, et al. 2010. c-IAP1 and UbcH5 promote K11-linked polyubiquitination of RIP1 in TNF signalling. *EMBO J.* 29:4198–209
33. Hatakeyama S, Yada M, Matsumoto M, Ishida N, Nakayama KI. 2001. U box proteins as a new family of ubiquitin-protein ligases. *J. Biol. Chem.* 276:33111–20
34. Johnson ES, Ma PC, Ota IM, Varshavsky A. 1995. A proteolytic pathway that recognizes ubiquitin as a degradation signal. *J. Biol. Chem.* 270:17442–56
35. Kirisako T, Kamei K, Murata S, Kato M, Fukumoto H, et al. 2006. **A ubiquitin ligase complex assembles linear polyubiquitin chains.** *EMBO J.* 25:4877–87
36. Rahighi S, Ikeda F, Kawasaki M, Akutsu M, Suzuki N, et al. 2009. **Specific recognition of linear ubiquitin chains by NEMO is important for NF-kappaB activation.** *Cell* 136:1098–109
37. Virdee S, Ye Y, Nguyen DP, Komander D, Chin JW. 2010. Engineered diubiquitin synthesis reveals Lys29-isopeptide specificity of an OTU deubiquitinase. *Nat. Chem. Biol.* 6:750–57
38. Jung JE, Wollscheid HP, Marquardt A, Manea M, Scheffner M, Przybylski M. 2009. Functional ubiquitin conjugates with lysine-epsilon-amino-specific linkage by thioether ligation of cysteinyl-ubiquitin peptide building blocks. *Bioconjug. Chem.* 20:1152–62
39. Yang R, Pasunooti KK, Li F, Liu XW, Liu CF. 2010. Synthesis of K48-linked diubiquitin using dual native chemical ligation at lysine. *Chem. Commun.* 46:7199–201
40. Kumar KS, Spasser L, Erlich LA, Bavikar SN, Brik A. 2010. Total chemical synthesis of di-ubiquitin chains. *Angew. Chem. Int. Ed. Engl.* 49:9126–31

---

35. Presents the first description of linear ubiquitin chains generated by E3 ligases (HOIL-1L and HOIP).

---

36. The first cocrystal structure of diubiquitin in complex with its specific binding domain.

41. Seet BT, Dikic I, Zhou MM, Pawson T. 2006. Reading protein modifications with interaction domains. *Nat. Rev. Mol. Cell Biol.* 7:473–83
42. Dikic I, Wakatsuki S, Walters KJ. 2009. Ubiquitin-binding domains—from structures to functions. *Nat. Rev. Mol. Cell Biol.* 10:659–71
43. **Bienko M, Green CM, Crosetto N, Rudolf F, Zapart G, et al. 2005. Ubiquitin-binding domains in Y-family polymerases regulate translesion synthesis. *Science* 310:1821–24**
44. Bomar MG, D'Souza S, Bienko M, Dikic I, Walker GC, Zhou P. 2010. Unconventional ubiquitin recognition by the ubiquitin-binding motif within the Y family DNA polymerases iota and Rev1. *Mol. Cell* 37:408–17
45. Penengo L, Mapelli M, Murachelli AG, Confalonieri S, Magri L, et al. 2006. Crystal structure of the ubiquitin binding domains of Rabex-5 reveals two modes of interaction with ubiquitin. *Cell* 124:1183–95
46. Lee S, Tsai YC, Mattera R, Smith WJ, Kostelansky MS, et al. 2006. Structural basis for ubiquitin recognition and autoubiquitination by Rabex-5. *Nat. Struct. Mol. Biol.* 13:264–71
47. **Reyes-Turcu FE, Horton JR, Mullally JE, Heroux A, Cheng X, Wilkinson KD. 2006. The ubiquitin binding domain ZnF UBP recognizes the C-terminal diglycine motif of unanchored ubiquitin. *Cell* 124:1197–208**
48. Sloper-Mould KE, Jemc JC, Pickart CM, Hicke L. 2001. Distinct functional surface regions on ubiquitin. *J. Biol. Chem.* 276:30483–89
49. Kamadurai HB, Souphron J, Scott DC, Duda DM, Miller DJ, et al. 2009. Insights into ubiquitin transfer cascades from a structure of a UbcH5B~ubiquitin-HECT^{NEDD4L} complex. *Mol. Cell* 36:1095–102
50. Hicke L, Schubert HL, Hill CP. 2005. Ubiquitin-binding domains. *Nat. Rev. Mol. Cell Biol.* 6:610–21
51. Hurley JH, Lee S, Prag G. 2006. Ubiquitin-binding domains. *Biochem. J.* 399:361–72
52. Winget JM, Mayor T. 2010. The diversity of ubiquitin recognition: hot spots and varied specificity. *Mol. Cell* 38:627–35
53. Lange OF, Lakomek NA, Fares C, Schroder GF, Walter KF, et al. 2008. Recognition dynamics up to microseconds revealed from an RDC-derived ubiquitin ensemble in solution. *Science* 320:1471–75
54. Wlodarski T, Zagrovic B. 2009. Conformational selection and induced fit mechanism underlie specificity in noncovalent interactions with ubiquitin. *Proc. Natl. Acad. Sci. USA* 106:19346–51
55. Varadan R, Walker O, Pickart C, Fushman D. 2002. Structural properties of polyubiquitin chains in solution. *J. Mol. Biol.* 324:637–47
56. Ryabov Y, Fushman D. 2006. Interdomain mobility in di-ubiquitin revealed by NMR. *Proteins* 63:787–96
57. Varadan R, Assfalg M, Haririnia A, Raasi S, Pickart C, Fushman D. 2004. Solution conformation of Lys63-linked di-ubiquitin chain provides clues to functional diversity of polyubiquitin signaling. *J. Biol. Chem.* 279:7055–63
58. Fushman D, Walker O. 2010. Exploring the linkage dependence of polyubiquitin conformations using molecular modeling. *J. Mol. Biol.* 395:803–14
59. Kang RS, Daniels CM, Francis SA, Shih SC, Salerno WJ, et al. 2003. Solution structure of a CUE-ubiquitin complex reveals a conserved mode of ubiquitin binding. *Cell* 113:621–30
60. Fisher RD, Wang B, Alam SL, Higginson DS, Robinson H, et al. 2003. Structure and ubiquitin binding of the ubiquitin-interacting motif. *J. Biol. Chem.* 278:28976–84
61. Alam SL, Sun J, Payne M, Welch BD, Blake BK, et al. 2004. Ubiquitin interactions of NZF zinc fingers. *EMBO J.* 23:1411–21
62. Kaiser SE, Riley BE, Shaler TA, Trevino RS, Becker CH, et al. 2011. Protein standard absolute quantification (PSAQ) method for the measurement of cellular ubiquitin pools. *Nat. Methods* 8:691–96
63. Asao H, Sasaki Y, Arita T, Tanaka N, Endo K, et al. 1997. Hrs is associated with STAM, a signal-transducing adaptor molecule. Its suppressive effect on cytokine-induced cell growth. *J. Biol. Chem.* 272:32785–91
64. Bache KG, Raiborg C, Mehlum A, Stenmark H. 2003. STAM and Hrs are subunits of a multivalent ubiquitin-binding complex on early endosomes. *J. Biol. Chem.* 278:12513–21
65. Chen H, Fre S, Slepnev VI, Capua MR, Takei K, et al. 1998. Epsin is an EH-domain-binding protein implicated in clathrin-mediated endocytosis. *Nature* 394:793–97
66. Hofmann K, Falquet L. 2001. A ubiquitin-interacting motif conserved in components of the proteasomal and lysosomal protein degradation systems. *Trends Biochem. Sci.* 26:347–50

43. Novel UBDs in translesion synthesis polymerases mediate their recruitment to the nuclear foci on monoubiquitinated PCNA.

47. The deep binding pocket of USP5 ZnF UBP accommodates ubiquitin's tail and enables binding to unanchored ubiquitin chains.

67. Hirano S, Kawasaki M, Ura H, Kato R, Raiborg C, et al. 2006. Double-sided ubiquitin binding of Hrs-UIM in endosomal protein sorting. *Nat. Struct. Mol. Biol.* 13:272–77
68. Haglund K, Di Fiore PP, Dikic I. 2003. Distinct monoubiquitin signals in receptor endocytosis. *Trends Biochem. Sci.* 28:598–603
69. Mattera R, Tsai YC, Weissman AM, Bonifacino JS. 2006. The Rab5 guanine nucleotide exchange factor Rabex-5 binds ubiquitin (Ub) and functions as a Ub ligase through an atypical Ub-interacting motif and a zinc finger domain. *J. Biol. Chem.* 281:6874–83
70. Mizuno E, Kawahata K, Kato M, Kitamura N, Komada M. 2003. STAM proteins bind ubiquitinated proteins on the early endosome via the VHS domain and ubiquitin-interacting motif. *Mol. Biol. Cell* 14:3675–89
71. Ren X, Hurley JH. 2010. VHS domains of ESCRT-0 cooperate in high-avidity binding to polyubiquitinated cargo. *EMBO J.* 29:1045–54
72. Reyes-Turcu FE, Shanks JR, Komander D, Wilkinson KD. 2008. Recognition of polyubiquitin isoforms by the multiple ubiquitin binding modules of isopeptidase T. *J. Biol. Chem.* 283:19581–92
73. Slagsvold T, Aasland R, Hirano S, Bache KG, Raiborg C, et al. 2005. Eap45 in mammalian ESCRT-II binds ubiquitin via a phosphoinositide-interacting GLUE domain. *J. Biol. Chem.* 280:19600–6
74. Alam SL, Langelier C, Whitby FG, Koirala S, Robinson H, et al. 2006. Structural basis for ubiquitin recognition by the human ESCRT-II EAP45 GLUE domain. *Nat. Struct. Mol. Biol.* 13:1029–30
75. Hirano S, Suzuki N, Slagsvold T, Kawasaki M, Trambaiolo D, et al. 2006. Structural basis of ubiquitin recognition by mammalian Eap45 GLUE domain. *Nat. Struct. Mol. Biol.* 13:1031–32
76. Tyrrell A, Flick K, Kleiger G, Zhang H, Deshaies RJ, Kaiser P. 2010. Physiologically relevant and portable tandem ubiquitin-binding domain stabilizes polyubiquitylated proteins. *Proc. Natl. Acad. Sci. USA* 107:19796–801
77. Kaiser P, Flick K, Wittenberg C, Reed SI. 2000. Regulation of transcription by ubiquitination without proteolysis: Cdc34/SCF(Met30)-mediated inactivation of the transcription factor Met4. *Cell* 102:303–14
78. Kuras L, Rouillon A, Lee T, Barbey R, Tyers M, Thomas D. 2002. Dual regulation of the Met4 transcription factor by ubiquitin-dependent degradation and inhibition of promoter recruitment. *Mol. Cell* 10:69–80
79. Chandrasekaran S, Deffenbaugh AE, Ford DA, Bailly E, Mathias N, Skowyra D. 2006. Destabilization of binding to cofactors and SCFMet30 is the rate-limiting regulatory step in degradation of polyubiquitinated Met4. *Mol. Cell* 24:689–99
80. Flick K, Ouni I, Wohlschlegel JA, Capati C, McDonald WH, et al. 2004. Proteolysis-independent regulation of the transcription factor Met4 by a single Lys48-linked ubiquitin chain. *Nat. Cell Biol.* 6:634–41
81. Ikeda F, Crosetto N, Dikic I. 2010. What determines the specificity and outcomes of ubiquitin signaling? *Cell* 143:677–81
82. **Sims JJ, Cohen RE. 2009. Linkage-specific avidity defines the lysine 63–linked polyubiquitin-binding preference of Rap80.** ***Mol. Cell*** **33:775–83**
83. Hofmann K, Bucher P. 1996. The UBA domain: a sequence motif present in multiple enzyme classes of the ubiquitination pathway. *Trends Biochem. Sci.* 21:172–73
84. Chen L, Shinde U, Ortolan TG, Madura K. 2001. Ubiquitin-associated (UBA) domains in Rad23 bind ubiquitin and promote inhibition of multi-ubiquitin chain assembly. *EMBO Rep.* 2:933–38
85. Funakoshi M, Sasaki T, Nishimoto T, Kobayashi H. 2002. Budding yeast Dsk2p is a polyubiquitin-binding protein that can interact with the proteasome. *Proc. Natl. Acad. Sci. USA* 99:745–50
86. Elsasser S, Finley D. 2005. Delivery of ubiquitinated substrates to protein-unfolding machines. *Nat. Cell Biol.* 7:742–49
87. Grabbe C, Dikic I. 2009. Functional roles of ubiquitin-like domain (ULD) and ubiquitin-binding domain (UBD) containing proteins. *Chem. Rev.* 109:1481–94
88. Wilkinson CR, Seeger M, Hartmann-Petersen R, Stone M, Wallace M, et al. 2001. Proteins containing the UBA domain are able to bind to multi-ubiquitin chains. *Nat. Cell Biol.* 3:939–43
89. Raasi S, Pickart CM. 2003. Rad23 ubiquitin-associated domains (UBA) inhibit 26 S proteasome-catalyzed proteolysis by sequestering lysine 48–linked polyubiquitin chains. *J. Biol. Chem.* 278:8951–59

82. Describes the linkage-specific binding of RAP80 toward Lys63 ubiquitin chains via two UIM domains.

90. Raasi S, Orlov I, Fleming KG, Pickart CM. 2004. Binding of polyubiquitin chains to ubiquitin-associated (UBA) domains of HHR23A. *J. Mol. Biol.* 341:1367–79
91. Varadan R, Assfalg M, Raasi S, Pickart C, Fushman D. 2005. Structural determinants for selective recognition of a Lys48-linked polyubiquitin chain by a UBA domain. *Mol. Cell* 18:687–98
92. Heessen S, Masucci MG, Dantuma NP. 2005. The UBA2 domain functions as an intrinsic stabilization signal that protects Rad23 from proteasomal degradation. *Mol. Cell* 18:225–35
93. Prakash S, Tian L, Ratliff KS, Lehotzky RE, Matouschek A. 2004. An unstructured initiation site is required for efficient proteasome-mediated degradation. *Nat. Struct. Mol. Biol.* 11:830–37
94. Inobe T, Fishbain S, Prakash S, Matouschek A. 2011. Defining the geometry of the two-component proteasome degron. *Nat. Chem. Biol.* 7:161–67
95. Fishbain S, Prakash S, Herrig A, Elsasser S, Matouschek A. 2011. Rad23 escapes degradation because it lacks a proteasome initiation region. *Nat. Commun.* 2:192
96. Heinen C, Acs K, Hoogstraten D, Dantuma NP. 2011. C-terminal UBA domains protect ubiquitin receptors by preventing initiation of protein degradation. *Nat. Commun.* 2:191
97. Schmidtke G, Kalveram B, Weber E, Bochtler P, Lukasiak S, et al. 2006. The UBA domains of NUB1L are required for binding but not for accelerated degradation of the ubiquitin-like modifier FAT10. *J. Biol. Chem.* 281:20045–54
98. Schmidtke G, Kalveram B, Groettrup M. 2009. Degradation of FAT10 by the 26S proteasome is independent of ubiquitylation but relies on NUB1L. *FEBS Lett.* 583:591–94
99. Hipp MS, Raasi S, Groettrup M, Schmidtke G. 2004. NEDD8 ultimate buster-1L interacts with the ubiquitin-like protein FAT10 and accelerates its degradation. *J. Biol. Chem.* 279:16503–10
100. Young P, Deveraux Q, Beal RE, Pickart CM, Rechsteiner M. 1998. Characterization of two polyubiquitin binding sites in the 26 S protease subunit 5a. *J. Biol. Chem.* 273:5461–67
101. Deveraux Q, Ustrell V, Pickart C, Rechsteiner M. 1994. A 26 S protease subunit that binds ubiquitin conjugates. *J. Biol. Chem.* 269:7059–61
102. Walters KJ, Kleijnen MF, Goh AM, Wagner G, Howley PM. 2002. Structural studies of the interaction between ubiquitin family proteins and proteasome subunit S5a. *Biochemistry* 41:1767–77
103. Wang Q, Young P, Walters KJ. 2005. Structure of S5a bound to monoubiquitin provides a model for polyubiquitin recognition. *J. Mol. Biol.* 348:727–39
104. **Husnjak K, Elsasser S, Zhang N, Chen X, Randles L, et al. 2008. Proteasome subunit Rpn13 is a novel ubiquitin receptor. *Nature* 453:481–88**
105. Schreiner P, Chen X, Husnjak K, Randles L, Zhang N, et al. 2008. Ubiquitin docking at the proteasome through a novel pleckstrin-homology domain interaction. *Nature* 453:548–52
106. Goto J, Watanabe M, Ichikawa Y, Yee SB, Ihara N, et al. 1997. Machado-Joseph disease gene products carrying different carboxyl termini. *Neurosci. Res.* 28:373–77
107. Chai Y, Berke SS, Cohen RE, Paulson HL. 2004. Poly-ubiquitin binding by the polyglutamine disease protein ataxin-3 links its normal function to protein surveillance pathways. *J. Biol. Chem.* 279:3605–11
108. Song AX, Zhou CJ, Peng Y, Gao XC, Zhou ZR, et al. 2010. Structural transformation of the tandem ubiquitin-interacting motifs in ataxin-3 and their cooperative interactions with ubiquitin chains. *PLoS One* 5:e13202
109. Burnett B, Li F, Pittman RN. 2003. The polyglutamine neurodegenerative protein ataxin-3 binds polyubiquitylated proteins and has ubiquitin protease activity. *Hum. Mol. Genet.* 12:3195–205
110. Nicastro G, Masino L, Esposito V, Menon RP, De Simone A, et al. 2009. Josephin domain of ataxin-3 contains two distinct ubiquitin-binding sites. *Biopolymers* 91:1203–14
111. Nicastro G, Todi SV, Karaca E, Bonvin AM, Paulson HL, Pastore A. 2010. Understanding the role of the Josephin domain in the polyUb binding and cleavage properties of ataxin-3. *PLoS One* 5:e12430
112. Chen ZJ, Sun LJ. 2009. Nonproteolytic functions of ubiquitin in cell signaling. *Mol. Cell* 33:275–86
113. Kulathu Y, Akutsu M, Bremm A, Hofmann K, Komander D. 2009. Two-sided ubiquitin binding explains specificity of the TAB2 NZF domain. *Nat. Struct. Mol. Biol.* 16:1328–30
114. **Sato Y, Yoshikawa A, Yamashita M, Yamagata A, Fukai S. 2009. Structural basis for specific recognition of Lys63-linked polyubiquitin chains by NZF domains of TAB2 and TAB3. *EMBO J.* 28:3903–9**

104. Explains the identification and functional characterization of Rpn13 as a high-affinity ubiquitin receptor at the proteasome.

114. Presents the structural basis for specific recognition of Lys63-linked ubiquitin chains by TAB2/TAB3 NZF domains.

115. Ulrich HD, Walden H. 2010. Ubiquitin signalling in DNA replication and repair. *Nat. Rev. Mol. Cell Biol.* 11:479–89
116. Yan J, Jetten AM. 2008. RAP80 and RNF8, key players in the recruitment of repair proteins to DNA damage sites. *Cancer Lett.* 271:179–90
117. Sato Y, Yoshikawa A, Mimura H, Yamashita M, Yamagata A, Fukai S. 2009. Structural basis for specific recognition of Lys63-linked polyubiquitin chains by tandem UIMs of RAP80. *EMBO J.* 28:2461–68
118. Huang F, Kirkpatrick D, Jiang X, Gygi S, Sorkin A. 2006. Differential regulation of EGF receptor internalization and degradation by multiubiquitination within the kinase domain. *Mol. Cell* 21:737–48
119. Galan JM, Haguenauer-Tsapis R. 1997. Ubiquitin Lys63 is involved in ubiquitination of a yeast plasma membrane protein. *EMBO J.* 16:5847–54
120. McCullough J, Clague MJ, Urbe S. 2004. AMSH is an endosome-associated ubiquitin isopeptidase. *J. Cell Biol.* 166:487–92
121. Mizuno E, Iura T, Mukai A, Yoshimori T, Kitamura N, Komada M. 2005. Regulation of epidermal growth factor receptor down-regulation by UBPY-mediated deubiquitination at endosomes. *Mol. Biol. Cell* 16:5163–74
122. Kim MS, Kim JA, Song HK, Jeon H. 2006. STAM-AMSH interaction facilitates the deubiquitination activity in the C-terminal AMSH. *Biochem. Biophys. Res. Commun.* 351:612–18
123. **Sato Y, Yoshikawa A, Yamagata A, Mimura H, Yamashita M, et al. 2008. Structural basis for specific cleavage of Lys63-linked polyubiquitin chains. *Nature* 455:358–62**

> 123. Presents evidence of how deubiquitinating enzyme AMSH-LP specifically recognizes and cleaves Lys63-linked ubiquitin chains.

124. Wagner S, Carpentier I, Rogov V, Kreike M, Ikeda F, et al. 2008. Ubiquitin binding mediates the NF-kappaB inhibitory potential of ABIN proteins. *Oncogene* 27:3739–45
125. Lo YC, Lin SC, Rospigliosi CC, Conze DB, Wu CJ, et al. 2009. Structural basis for recognition of diubiquitins by NEMO. *Mol. Cell* 33:602–15
126. Wertz IE, O'Rourke KM, Zhou H, Eby M, Aravind L, et al. 2004. De-ubiquitination and ubiquitin ligase domains of A20 downregulate NF-kappaB signalling. *Nature* 430:694–99
127. Ivins FJ, Montgomery MG, Smith SJ, Morris-Davies AC, Taylor IA, Rittinger K. 2009. NEMO oligomerization and its ubiquitin-binding properties. *Biochem. J.* 421:243–51
128. Laplantine E, Fontan E, Chiaravalli J, Lopez T, Lakisic G, et al. 2009. NEMO specifically recognizes K63-linked poly-ubiquitin chains through a new bipartite ubiquitin-binding domain. *EMBO J.* 28:2885–95
129. Meyer HH, Wang Y, Warren G. 2002. Direct binding of ubiquitin conjugates by the mammalian p97 adaptor complexes, p47 and Ufd1-Npl4. *EMBO J.* 21:5645–52
130. Haas TL, Emmerich CH, Gerlach B, Schmukle AC, Cordier SM, et al. 2009. Recruitment of the linear ubiquitin chain assembly complex stabilizes the TNF-R1 signaling complex and is required for TNF-mediated gene induction. *Mol. Cell* 36:831–44
131. Sato Y, Fujita H, Yoshikawa A, Yamashita M, Yamagata A, et al. 2011. Specific recognition of linear ubiquitin chains by the Npl4 zinc finger (NZF) domain of the HOIL-1L subunit of the linear ubiquitin chain assembly complex. *Proc. Natl. Acad. Sci. USA* 108:20520–25
132. Hofmann K. 2009. Ubiquitin-binding domains and their role in the DNA damage response. *DNA Repair* 8:544–56
133. Smogorzewska A, Desetty R, Saito TT, Schlabach M, Lach FP, et al. 2010. A genetic screen identifies FAN1, a Fanconi anemia–associated nuclease necessary for DNA interstrand crosslink repair. *Mol. Cell* 39:36–47
134. Kratz K, Schopf B, Kaden S, Sendoel A, Eberhard R, et al. 2010. Deficiency of FANCD2-associated nuclease KIAA1018/FAN1 sensitizes cells to interstrand crosslinking agents. *Cell* 142:77–88
135. MacKay C, Declais AC, Lundin C, Agostinho A, Deans AJ, et al. 2010. Identification of KIAA1018/FAN1, a DNA repair nuclease recruited to DNA damage by monoubiquitinated FANCD2. *Cell* 142:65–76
136. Liu T, Ghosal G, Yuan J, Chen J, Huang J. 2010. FAN1 acts with FANCI-FANCD2 to promote DNA interstrand cross-link repair. *Science* 329:693–96
137. Bish RA, Myers MP. 2007. Werner helicase-interacting protein 1 binds polyubiquitin via its zinc finger domain. *J. Biol. Chem.* 282:23184–93

138. Crosetto N, Bienko M, Hibbert RG, Perica T, Ambrogio C, et al. 2008. Human Wrnip1 is localized in replication factories in a ubiquitin-binding zinc finger-dependent manner. *J. Biol. Chem.* 283:35173–85
139. Thurston TL, Ryzhakov G, Bloor S, von Muhlinen N, Randow F. 2009. The TBK1 adaptor and autophagy receptor NDP52 restricts the proliferation of ubiquitin-coated bacteria. *Nat. Immunol.* 10:1215–21
140. Iha H, Peloponese JM, Verstrepen L, Zapart G, Ikeda F, et al. 2008. Inflammatory cardiac valvulitis in TAX1BP1-deficient mice through selective NF-kappaB activation. *EMBO J.* 27:629–41
141. Zerial M, McBride H. 2001. Rab proteins as membrane organizers. *Nat. Rev. Mol. Cell Biol.* 2:107–17
142. Pickart CM. 2001. Mechanisms underlying ubiquitination. *Annu. Rev. Biochem.* 70:503–33
143. Brzovic PS, Lissounov A, Christensen DE, Hoyt DW, Klevit RE. 2006. A UbcH5/ubiquitin noncovalent complex is required for processive BRCA1-directed ubiquitination. *Mol. Cell* 21:873–80
144. Eddins MJ, Carlile CM, Gomez KM, Pickart CM, Wolberger C. 2006. Mms2-Ubc13 covalently bound to ubiquitin reveals the structural basis of linkage-specific polyubiquitin chain formation. *Nat. Struct. Mol. Biol.* 13:915–20
145. Sundquist WI, Schubert HL, Kelly BN, Hill GC, Holton JM, Hill CP. 2004. Ubiquitin recognition by the human TSG101 protein. *Mol. Cell* 13:783–89
146. French ME, Kretzmann BR, Hicke L. 2009. Regulation of the RSP5 ubiquitin ligase by an intrinsic ubiquitin-binding site. *J. Biol. Chem.* 284:12071–79
147. Ogunjimi AA, Wiesner S, Briant DJ, Varelas X, Sicheri F, et al. 2010. The ubiquitin binding region of the Smurf HECT domain facilitates polyubiquitylation and binding of ubiquitylated substrates. *J. Biol. Chem.* 285:6308–15
148. Kim HC, Steffen AM, Oldham ML, Chen J, Huibregtse JM. 2011. Structure and function of a HECT domain ubiquitin-binding site. *EMBO Rep.* 12:334–41
149. Maspero E, Mari S, Valentini E, Musacchio A, Fish A, et al. 2011. Structure of the HECT:ubiquitin complex and its role in ubiquitin chain elongation. *EMBO Rep.* 12:342–49
150. Gyrd-Hansen M, Meier P. 2010. IAPs: from caspase inhibitors to modulators of NF-kappaB, inflammation and cancer. *Nat. Rev. Cancer* 10:561–74
151. Blankenship JW, Varfolomeev E, Goncharov T, Fedorova AV, Kirkpatrick DS, et al. 2009. Ubiquitin binding modulates IAP antagonist-stimulated proteasomal degradation of c-IAP1 and c-IAP2(1). *Biochem. J.* 417:149–60
152. Gyrd-Hansen M, Darding M, Miasari M, Santoro MM, Zender L, et al. 2008. IAPs contain an evolutionarily conserved ubiquitin-binding domain that regulates NF-kappaB as well as cell survival and oncogenesis. *Nat. Cell Biol.* 10:1309–17
153. Thien CB, Langdon WY. 2005. c-Cbl and Cbl-b ubiquitin ligases: substrate diversity and the negative regulation of signalling responses. *Biochem. J.* 391:153–66
154. Schmidt MH, Dikic I. 2005. The Cbl interactome and its functions. *Nat. Rev. Mol. Cell Biol.* 6:907–18
155. Davies GC, Ettenberg SA, Coats AO, Mussante M, Ravichandran S, et al. 2004. Cbl-b interacts with ubiquitinated proteins; differential functions of the UBA domains of c-Cbl and Cbl-b. *Oncogene* 23:7104–15
156. Zhou ZR, Gao HC, Zhou CJ, Chang YG, Hong J, et al. 2008. Differential ubiquitin binding of the UBA domains from human c-Cbl and Cbl-b: NMR structural and biochemical insights. *Protein Sci.* 17:1805–14
157. Zhu X, Menard R, Sulea T. 2007. High incidence of ubiquitin-like domains in human ubiquitin-specific proteases. *Proteins* 69:1–7
158. Leggett DS, Hanna J, Borodovsky A, Crosas B, Schmidt M, et al. 2002. Multiple associated proteins regulate proteasome structure and function. *Mol. Cell* 10:495–507
159. Luna-Vargas MP, Faesen AC, van Dijk WJ, Rape M, Fish A, Sixma TK. 2011. Ubiquitin-specific protease 4 is inhibited by its ubiquitin-like domain. *EMBO Rep.* 12:365–72
160. Chaugule VK, Burchell L, Barber KR, Sidhu A, Leslie SJ, et al. 2011. Autoregulation of Parkin activity through its ubiquitin-like domain. *EMBO J.* 30:2853–67
161. Polo S, Sigismund S, Faretta M, Guidi M, Capua MR, et al. 2002. A single motif responsible for ubiquitin recognition and monoubiquitination in endocytic proteins. *Nature* 416:451–55
162. Woelk T, Oldrini B, Maspero E, Confalonieri S, Cavallaro E, et al. 2006. Molecular mechanisms of coupled monoubiquitination. *Nat. Cell Biol.* 8:1246–54

163. Hoeller D, Crosetto N, Blagoev B, Raiborg C, Tikkanen R, et al. 2006. Regulation of ubiquitin-binding proteins by monoubiquitination. *Nat. Cell Biol.* 8:163–69
164. Hoeller D, Hecker CM, Wagner S, Rogov V, Dotsch V, Dikic I. 2007. E3-independent monoubiquitination of ubiquitin-binding proteins. *Mol. Cell* 26:891–98
165. Bienko M, Green CM, Sabbioneda S, Crosetto N, Matic I, et al. 2010. Regulation of translesion synthesis DNA polymerase η by monoubiquitination. *Mol. Cell* 37:396–407
166. Isasa M, Katz EJ, Kim W, Yugo V, Gonzalez S, et al. 2010. Monoubiquitination of RPN10 regulates substrate recruitment to the proteasome. *Mol. Cell* 38:733–45
167. Lamark T, Kirkin V, Dikic I, Johansen T. 2009. NBR1 and p62 as cargo receptors for selective autophagy of ubiquitinated targets. *Cell Cycle* 8:1986–90
168. Seibenhener ML, Babu JR, Geetha T, Wong HC, Krishna NR, Wooten MW. 2004. Sequestosome 1/p62 is a polyubiquitin chain binding protein involved in ubiquitin proteasome degradation. *Mol. Cell. Biol.* 24:8055–68
169. Kirkin V, Lamark T, Sou YS, Bjorkoy G, Nunn JL, et al. 2009. A role for NBR1 in autophagosomal degradation of ubiquitinated substrates. *Mol. Cell* 33:505–16
170. Matsumoto M, Wada K, Okuno M, Kurosawa M, Nukina N. 2011. Serine 403 phosphorylation of p62/SQSTM1 regulates selective autophagic clearance of ubiquitinated proteins. *Mol. Cell* 44:279–89
171. Wild P, Farhan H, McEwan DG, Wagner S, Rogov VV, et al. 2011. Phosphorylation of the autophagy receptor optineurin restricts *Salmonella* growth. *Science* 333:228–33
172. Stehmeier P, Muller S. 2009. Phospho-regulated SUMO interaction modules connect the SUMO system to CK2 signaling. *Mol. Cell* 33:400–9
173. Hoeller D, Dikic I. 2009. Targeting the ubiquitin system in cancer therapy. *Nature* 458:438–44
174. Kang Y, Vossler RA, Diaz-Martinez LA, Winter NS, Clarke DJ, Walters KJ. 2006. UBL/UBA ubiquitin receptor proteins bind a common tetraubiquitin chain. *J. Mol. Biol.* 356:1027–35
175. Sims JJ, Haririnia A, Dickinson BC, Fushman D, Cohen RE. 2009. Avid interactions underlie the Lys63-linked polyubiquitin binding specificities observed for UBA domains. *Nat. Struct. Mol. Biol.* 16:883–89
176. Scortegagna M, Subtil T, Qi J, Kim H, Zhao W, et al. 2011. USP13 enzyme regulates Siah2 ligase stability and activity via noncatalytic ubiquitin-binding domains. *J. Biol. Chem.* 286:27333–41
177. Denuc A, Bosch-Comas A, Gonzalez-Duarte R, Marfany G. 2009. The UBA-UIM domains of the USP25 regulate the enzyme ubiquitination state and modulate substrate recognition. *PLoS One* 4:e5571
178. Swanson KA, Kang RS, Stamenova SD, Hicke L, Radhakrishnan I. 2003. Solution structure of Vps27 UIM-ubiquitin complex important for endosomal sorting and receptor downregulation. *EMBO J.* 22:4597–606
179. Westhoff B, Chapple JP, van der Spuy J, Hohfeld J, Cheetham ME. 2005. HSJ1 is a neuronal shuttling factor for the sorting of chaperone clients to the proteasome. *Curr. Biol.* 15:1058–64
180. Hawryluk MJ, Keyel PA, Mishra SK, Watkins SC, Heuser JE, Traub LM. 2006. Epsin 1 is a polyubiquitin-selective clathrin-associated sorting protein. *Traffic* 7:262–81

Ubiquitin-Like Proteins

Annemarthe G. van der Veen[1] and Hidde L. Ploegh[1,2]

[1]Whitehead Institute for Biomedical Research, Cambridge, Massachusetts 02142; email: ploegh@wi.mit.edu

[2]Department of Biology, Massachusetts Institute of Technology, Cambridge, Massachusetts 02142

Keywords

SUMO, Nedd8, ISG15, autophagy, Urm1, Hub1

Abstract

The eukaryotic ubiquitin family encompasses nearly 20 proteins that are involved in the posttranslational modification of various macromolecules. The ubiquitin-like proteins (UBLs) that are part of this family adopt the β-grasp fold that is characteristic of its founding member ubiquitin (Ub). Although structurally related, UBLs regulate a strikingly diverse set of cellular processes, including nuclear transport, proteolysis, translation, autophagy, and antiviral pathways. New UBL substrates continue to be identified and further expand the functional diversity of UBL pathways in cellular homeostasis and physiology. Here, we review recent findings on such novel substrates, mechanisms, and functions of UBLs.

Contents

INTRODUCTION 324
SUMOYLATION: ONE GAME,
    FOUR PLAYERS 325
    SUMO Paralogs 325
    Mechanisms of Paralog Specificity:
        SUMO-Interaction Motifs 326
    Mechanisms of Paralog Specificity:
        SUMO Enzymes 328
    Sumoylation and Cellular Stress ... 329
NEDD8: FROM REGULATOR
    OF UBIQUITYLATION TO
    THERAPEUTIC TARGET 329
    Nedd8 Targets Cullin RING
        E3 Ligases and Promotes
        Ubiquitylation 330
    Nedd8 E3 Ligases 331
    Cell- and Pathogen-Derived
        Inhibitors of Neddylation 332
    Pharmaceutical Inhibition of
        Neddylation as Anticancer
        Therapy 332
UBIQUITIN-LIKE PROTEINS
    FIGHT BACK: ISG15 IN
    HOST DEFENSE 332
    The Interplay Between ISGylation
        and Viruses 334
    ISG15 Targets Host and Viral
        Proteins 334
    Antiviral Effects of ISG15 and
        UBP43: Not All About
        ISGylation 336
FAT10 AND MNSFβ:
    CONNECTIONS TO
    PROTEASOMAL
    DEGRADATION AND
    IMMUNITY 336
UFM1 IN ERYTHROID AND
    MEGAKARYOCYTE
    DEVELOPMENT 337
ATG8 AND ATG12:
    REGULATORS OF
    AUTOPHAGY 338
    Atg8 in Autophagosome
        Biogenesis 338
    Atg8 in Selective Autophagy 340
    Atg8 Paralogs and Autophagy
        Receptors 341
    A Family of Atg8-Directed
        Proteases 342
URM1: BRIDGING THE GAP
    BETWEEN PROKARYOTIC
    SULFUR CARRIERS AND
    EUKARYOTIC PROTEIN
    MODIFIERS 342
    Urm1 in tRNA Thiolation 343
    Urm1 in Protein Modification 344
HUB1 IN RNA SPLICING 344

INTRODUCTION

Cellular homeostasis requires tight control of protein activity, function, stability, and localization. A widely used strategy to achieve this involves posttranslational modifications of proteins. These often consist of a small chemical substituent, as is the case for phosphorylation, acetylation, and methylation. A substrate may also be modified by the covalent attachment of another polypeptide. In 1975, researchers discovered ubiquitin (Ub), which serves as a molecular tag to target proteins to the proteasome for degradation (1). Although best known for its role in proteasomal degradation, ubiquitylation affects a wide variety of processes, including endocytosis, membrane-protein trafficking, cell signaling, and DNA repair (1). Ubiquitylation is the covalent attachment of Ub via its C-terminal glycine to lysine residues (or, less frequently, to other amino acids such as threonine, cysteine, or serine) in target proteins via a series of enzymes, referred to as E1 activating enzymes, E2 conjugating enzymes, and E3 ligases (1–3). Additional flexibility and control is provided by the action of deubiquitylating enzymes (DUBs), which hydrolyze the bond between Ub and its substrates (1, 2). Ubiquitylation is extensively reviewed

E1 activating enzyme: mediates the ATP-dependent activation of the C-terminal glycine of ubiquitin, followed by formation of a thioester-linked Ub-E1 intermediate

in this volume and elsewhere and, therefore, is not further discussed in this article (1–3).

Proteins related in sequence to Ub and of a similar three-dimensional structure are referred to as ubiquitin-like proteins (UBLs) (1, 3). Most UBLs are conjugated to proteins via an enzymatic cascade that resembles ubiquitylation, and many of their components are related to those involved in ubiquitylation. As most UBLs have far fewer known substrates than are known for Ub, they require a more limited number of E2-type conjugating enzymes and E3-type ligases (**Supplemental Table 1**: Follow the **Supplemental Material link** on the Annual Reviews home page at **http://www.annualreviews.org**). Several poorly characterized UBLs are now assigned a function and participate in various cellular processes, e.g., Ufm1 in development and Hub1 in RNA splicing (4, 5). Host UBL pathways are also important in interactions with viral and bacterial pathogens, as best exemplified by the role of ISG15 in antiviral immune defense (6). Other UBLs, such as Urm1 and the archaeal SAMPs, function not only in protein modification, but also in sulfur transfer (1). The dual role of these latter UBLs illustrates how the eukaryotic Ub/UBL protein modifiers evolved from prokaryotic sulfurtransferases. The UBLs SUMO and Atg8 exist as multiple paralogs in metazoans (7–9). How is isoform specificity determined by the conjugation machinery? Are individual paralogs functionally distinct? We address these questions as a recurrent theme in UBL biology and review how various motifs in proteins that interact with UBLs dictate the isoform specificity and function of UBL paralogs. Given the breadth of this review, we focus predominantly on work done in metazoan model organisms, with occasional reference to work in yeast.

SUMOYLATION: ONE GAME, FOUR PLAYERS

The small-ubiquitin-related modifier (SUMO) regulates various cellular processes, including nuclear transport and organization, transcription, chromatin remodeling, DNA repair, and ribosomal biogenesis (7, 8). Although many SUMO substrates are found in or near the nucleus, sumoylation may occur in other compartments as well. At a molecular level, SUMO modification frequently creates or disrupts an interface for intermolecular interactions, or it induces a conformational change (7, 8). Sumoylation of RanGAP1 (Ran GTPase-activating protein), the first SUMO substrate identified, increases its affinity for the nuclear pore component RanBP2, causing it to redistribute from the cytosol to nuclear pore complexes (10, 11). The functions of sumoylation have been expertly reviewed elsewhere and are not further discussed here (7, 8).

SUMO Paralogs

Yeast and invertebrates have only a single SUMO-encoding gene, but vertebrate genomes contain at least four SUMO genes, designated *SUMO1–4*. SUMO2 and SUMO3 are nearly identical (∼97%) and differ by only three N-terminal residues (7, 8). As antibodies cannot distinguish between these two paralogs, they are collectively referred to as SUMO2/3. Although SUMO2/3 are only ∼50% identical to SUMO1, they share their E1 (SAE1/SAE2 heterodimer, SUMO-activating enzyme 1 and 2, also known as AOS1 and Uba2, respectively) and E2 (Ubc9) enzymes (**Supplemental Table 1**) (12). SUMO4 is ∼87% similar to SUMO2, but it is unclear whether this paralog is expressed, processed and conjugated (8).

In yeast, deletion of the SUMO gene *SMT3* (*Saccharomyces cerevisiae*) or *PMT3* (*Schizosaccharomyces pombe*) causes lethality or severe growth impairment, respectively (8). Likewise, Ubc9-deficient mice die at an early embryonic stage (8). Conversely, although SUMO1 haploinsufficiency has been associated with cleft-lip and -palate formation and reduced viability in mice (8), SUMO1-deficient mice appear healthy, suggesting that SUMO2/3 may compensate for loss of SUMO1 (13, 14). Zebrafish embryos that lack all three paralogs

E2 conjugating enzyme: accepts Ub from the E1 enzyme following a *trans*-thioesterification reaction between Ub-E1 and the active-site cysteine of the E2

E3 ligase: mediates the transfer of Ub to the ε-amino group of a lysine residue in the substrate

Deubiquitylating enzyme (DUB): a protease involved in C-terminal processing of Ub precursors and/or that cleaves the isopeptide bond between Ub and its substrate

SUMO consensus motif: the classical sumoylation site (found in ~70% of substrates) is ΨKx[DE], where Ψ is a large hydrophobic residue and K the acceptor lysine

Polysumoylation: polymerization of SUMO following isopeptide bond formation between the C-terminal glycine of SUMO and an internal lysine of another SUMO moiety

SUMO-interaction motif (SIM): a designated region or domain in a protein that mediates its noncovalent binding to SUMO

die, whereas loss of individual paralogs does not affect development (15). In SUMO1-deficient mice, RanGAP1 was increasingly modified with SUMO2/3, but no such effect was observed for promyelocytic leukemia protein (PML), indicating that compensation by SUMO paralogs may be specific to a subset of targets (14).

SUMO1 and SUMO2/3 differ in their intracellular distribution; SUMO1 localizes to the nuclear envelope, nucleoli, and cytoplasm, whereas SUMO2/3 is diffusely present in the nucleoplasm during interphase (16). SUMO1 and SUMO2/3 substrates show considerable (~15%) overlap (e.g., PML, p53, SART1, and TDG), but other proteins are uniquely modified with either paralog (17–19). For example, the microtubule motor protein CENP-E is modified with SUMO2/3; RanGAP1, NEMO (NFκB essential modulator), and Blimp-1 (B lymphocyte–induced maturation protein-1) are preferentially modified with SUMO1, and the nuclear receptors LXRα and LXRβ are conjugated to SUMO2/3 and SUMO1, respectively, in IFNγ-stimulated brain astrocytes (10, 11, 20–23). Finally, SUMO1 and SUMO2/3 differ in extent of conjugation: The majority of SUMO1 occurs in conjugates, and SUMO2/3 exists in free form (24). Application of oxidative or heat stress causes a dramatic increase in SUMO2/3 conjugation but leaves SUMO1 modification unaffected (8, 24).

Hundreds of SUMO substrates have been identified, and many of these contain a SUMO acceptor site or consensus motif that is directly recognized by Ubc9 (7, 8). SUMO2 and SUMO3 both contain an internal sumoylation consensus site (VKTE) and can form polymeric K11-linked chains (7, 8). Whereas SUMO1 may form K7-, K16-, or K17-linked chains in vitro, SUMO1 occurs exclusively in its monomeric form or as chain terminator on SUMO2/3 polymers in vivo (25–28). Yeast Smt3 may polymerize as well, but mutation of its N-terminal lysine residues (K11, K15, K19), which abolishes chain formation, does not affect viability, growth, or stress sensitivity in yeast, indicating that polysumoylation is not essential in *S. cerevisiae* (29).

Mechanisms of Paralog Specificity: SUMO-Interaction Motifs

As SUMO1 and SUMO2/3 require SAE1/SAE2 and Ubc9 for conjugation (12) paralog selection must take place further down the sumoylation pathway. SUMO-interaction motifs (SIMs) mediate noncovalent interactions between SUMO and target proteins and are present in several SUMO substrates, such as PML, the transcription factor Daxx, and the base excision repair enzyme TDG (thymine DNA glycosylase), possibly to recruit SUMO-loaded Ubc9 (7, 8). SIMs are also found in several SUMO enzymes, e.g., PIAS (protein inhibitor of activated STAT) and RanBP2 E3 ligases, and in regulatory and effector proteins that are recruited to interaction platforms via noncovalent SUMO-SIM interactions.

SIMs consist of a hydrophobic core (I/V-X-I/V-I/V or the reverse) flanked by N- or C-terminal acidic or phosphorylatable serine residues (30). The SIM forms a β-strand that can be inserted between the α-helix and β-strand of SUMO, either in a parallel or antiparallel orientation (31, 32). Many proteins contain noncovalent SUMO binding sites that conform to this consensus motif. Of these, several show preferential recruitment of either SUMO1 or SUMO2/3, and—if it concerns a SUMO substrate—this preferential recruitment subsequently dictates selective conjugation to either paralog (**Figure 1a**). The acidic stretch may favor SUMO1 binding (31, 33). Deletion of the acidic residues in the SUMO E3 PIAS proteins or nuclear autoantigen Sp100 abolishes binding to SUMO1 but does not affect interaction with SUMO2/3 (31). Acidic residues adjacent to the SIM of RanBP2 complemented the hydrophobic core residues in promoting binding to SUMO1-modified RanGAP1 (30). Addition of acidic residues to the SIM of TTRAP (TRAP and TNF-receptor-associated protein) alters its preference for SUMO2/3 in favor of SUMO1 (31). Conversely, the preference of ubiquitin-specific peptidase (USP25) for SUMO2/3 lies within seven residues that constitute the SIM hydrophobic core, although the

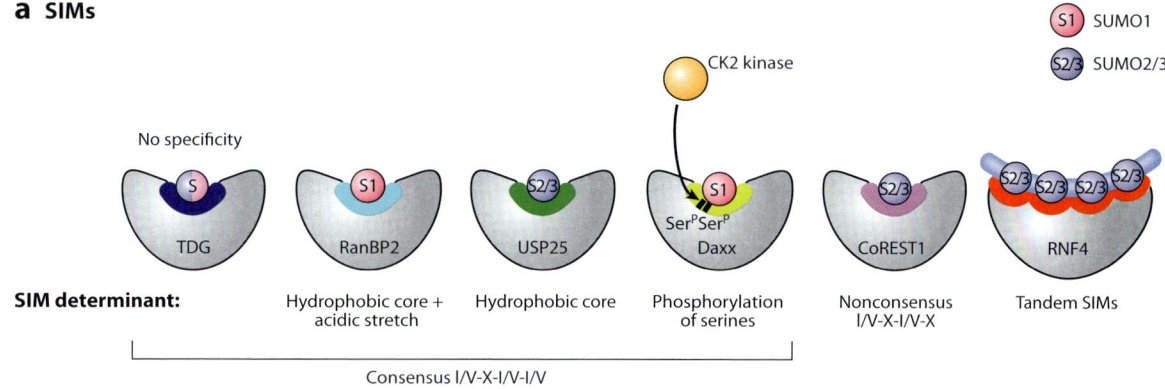

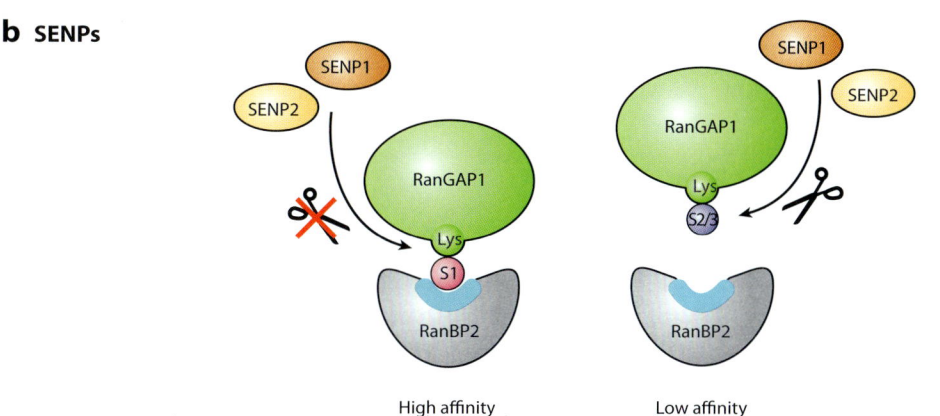

Figure 1

(*a*) Paralog specificity of small ubiquitin-related modifier (SUMO) interaction motifs. Although certain SUMO-interacting motifs (SIMs) do not exhibit selectivity for either SUMO1 or SUMO2/3, the presence of particular residues, posttranslational modifications, and tandem motifs may dictate the preferential binding of a SIM to either paralog. If the SIM is contained within a SUMO substrate, such preferential binding will lead to the exclusive conjugation of the respective paralog. (*b*) Although RanGAP1 may be modified with either SUMO1 or SUMO2/3, only SUMO1 conjugation results in a high-affinity interaction between SUMO-modified RanGAP1 and the SIM of RanBP2, thereby precluding SENPs (sentrin-specific proteases) from accessing SUMO1-RanGAP1. In contrast, the loose association between RanBP2 and SUMO2/3-modified RanGAP1 renders the latter prone to rapid desumoylation by SENPs.

presence of the acidic stretch further enhances binding to SUMO2/3 (34). USP25 is modified at a nonconsensus site, likely facilitated by the SIM-mediated recruitment and stabilization of SUMO2/3-charged Ubc9 (34). Similarly, selective sumoylation of BLM (Bloom syndrome, RecQ helicase like) at nonconsensus sites requires binding between SUMO2/3 and the SIM of the BLM (35).

Phosphorylation of serine residues in the SIM further contributes to paralog selectivity, as shown for modification of the transcriptional repressor Daxx with SUMO1 (36). Phosphorylation of the Daxx SIM by the serine/threonine kinase CK2 enhances its affinity for SUMO1 over SUMO2/3 via electrostatic interactions, and it favors Daxx sumoylation with SUMO1 (36). A phospho-modified SIM in the E3 ligase PIASxα influences its interaction with SUMO1 but not SUMO2/3 (31). Furthermore, CK2-mediated modification of an extended phospho-regulated SIM module in PIAS1, PML, and the

HECT (homologous with E6-associated protein C terminus) domain: E3 ligases form a thioester intermediate with Ub before transfer to a bound substrate

RING (really interesting new gene) domain: E3 ligases act as scaffold to bring a Ub-E2 intermediate near a substrate and subsequently catalyze Ub transfer

exosome component PMSCL1 regulates their association with SUMO, although paralog preference was not investigated in this study (37). Potential CK2 consensus sites are also found adjacent to SIMs in BLM, the SUMO E3 ligase PC2, and the viral SUMO E3 enzyme K-bZIP (36).

In addition to the "classical" hydrophobic SIM motif (I/V-X-I/V-I/V or the reverse), a nonconsensus SIM motif that lacks a hydrophobic residue in the fourth position was found in the transcriptional repressor CoREST1, the histone binding protein RBBP4, and the mRNA processing factor FIP1L1, and it mediates SUMO2-specific binding (38).

Proteins may contain multiple SIMs to facilitate recognition of polymeric SUMO2/3 chains. Tandem SIMs are found in the Ub E3 ligase RNF4 (RING finger protein 4 or Snurf) and its yeast homologs Slx5/Slx8 and Rfp1/Rfp2/Slx8 (39). Interaction between the four SIMs of RNF4 and SUMO2/3 is essential for in vitro and in vivo ubiquitylation and degradation of sumoylated PML and poly(ADP-ribose) polymerase (PARP-1) (40, 41). The Ub polymer may be directly conjugated either to the substrate or to the poly-SUMO2/3 chain (40, 42). Thus, SIM-SUMO2/3 interactions mediate recruitment of the Ub machinery and target polysumoylated proteins for degradation (39). Alternatively, RNF4 may regulate intracellular localization, as demonstrated for the human T cell leukemia virus (HTLV) oncoprotein Tax (43). Whereas RNF4 was the first identified SUMO-targeted Ub ligase (STUbL), the Zn-dependent metalloprotease Wss1 was identified as a SUMO-directed DUB in yeast (44). Its two predicted SIMs likely facilitate substrate recognition (44). Wss1 has isopeptidase activity toward both SUMO chains and SUMO chains capped with additional Ub moieties, but not Ub chains (44). Thus, Wss1 may hydrolyze a mixed SUMO-Ub polymer.

Recently developed SUMO1-specific monobodies (binding modules) inhibit SUMO1-SIM interactions as well as the formation of a thioester-linked complex between SAE2 and SUMO1, but not SUMO3 (45). These SUMO1 inhibitors may help clarify the structural and sequence determinants of SUMO1 and SUMO2/3 binding motifs.

Mechanisms of Paralog Specificity: SUMO Enzymes

The preference of certain SUMO proteases [sentrin-specific proteases (SENPs)] for SUMO1 versus SUMO2/3 may aid in paralog selection. SENP1 and -2 broadly target both SUMO1 and SUMO2/3, SENP3 and -5 favor processing and deconjugation of SUMO2/3 over SUMO1, and SENP6 and -7 are involved in chain editing (i.e., trimming of polymers by cleaving SUMO-SUMO conjugates), but not deconjugation, of SUMO2/3 polymers (8). Deconjugation by SENPs is dynamic and varies with cell type and subcellular localization (8). Sumoylation of RanGAP1 targets it to the nuclear pore complex via association with RanBP2 (10, 11). Although RanGAP1 is equally well modified by SUMO1 and SUMO2/3 in vitro, it is preferentially targeted by SUMO1 in vivo (10, 11). The interaction between sumoylated RanGAP1 and RanBP2 is mediated by two ∼50-residue internal repeats in RanBP2, one of which contains an SIM (30, 46). SUMO1-modified RanGAP1 interacts more tightly with RanBP2, thus shielding it from isopeptidase-mediated deconjugation by SENP1 and SENP2 (**Figure 1***b*) (47). SUMO-SIM interactions and protease susceptibility together favor RanGAP1 modification by SUMO1.

SUMO E3 ligases also contribute to selective paralog conjugation. RanBP2 contains neither a HECT nor a RING domain (48). RanBP2 does not interact with its substrates but instead binds both Ubc9 and SUMO via its two internal repeats (IR1 and IR2) separated by a short linker sequence (M) (46). Whereas IR1 interacts directly with Ubc9 and equally mediates SUMO1 and SUMO2/3 modifications, M-IR2 binds SUMO1 but not SUMO2/3 and may therefore recruit SUMO1-loaded Ubc9 in the absence of a stable E2-E3 interaction (49). As a consequence, autosumoylation of RanBP2 proceeds more rapidly with SUMO1 than it does with SUMO2/3 (49). How this affects RanBP2-mediated modification of other substrates is unclear.

In conclusion, paralog specificity is not controlled by a single determinant but results from the combined effects of SIMs, SENPs, and E3 ligases and perhaps other mechanisms. SUMO1 and SUMO3, but not SUMO2, are phosphorylated in vivo at an N-terminal serine (27). Likewise, several SUMO E3 ligases are subject to phosphorylation (8). Posttranslational modifications of SUMO isoforms or enzymes may further specify paralog selection. Furthermore, whereas in mammals SAE1/SAE2 and Ubc9 do not discriminate between SUMO paralogs, the equivalent enzymes in *Arabidopsis thaliana* may contribute to paralog selection (50).

Sumoylation and Cellular Stress

Severe oxidative, hypoxic, osmotic, genotoxic, or heat stress all result in increased SUMO2/3 conjugation (24, 51). Although SUMO1 modification is not generally increased in stressed cells (24), hypoxic conditions increase SUMO1 conjugation to hypoxia-inducible factor 1α (HIF-1α) via the E3 ligases PIASγ and RSUME (52, 53). In yeast, both ethanol exposure and oxidative stress enhance conjugation of the single Smt3 protein (54). Conversely, mild oxidative stress links the catalytic cysteines of E1 subunits Uba2 and Ubc9 via a disulphide bond and consequently inhibits sumoylation (55). Mild oxidative stress also inhibits Ub-mediated degradation of SENP3 and leads to desumoylation and stabilization of SUMO2/3-modified PML and p300 (56, 57). Serum starvation may induce maturation of SUMO4, suggesting that SUMO4 modification may be a stress-dependent process (58). SUMO4 expression may also be IL-1β inducible, as an IL-1β-responsive NFκB element was identified in the SUMO4 promotor (59).

Pathogens interact with the host sumoylation pathway (60, 61). The adenoviral Gam1 induces ubiquitylation and degradation of the SAE1 subunit of the SUMO E1 (60). Likewise, listeriolysin O causes proteasome-independent degradation of Ubc9 (60). In both cases host sumoylation is suppressed, indicating that SUMO modification is detrimental to bacterial or viral infection. In other cases, sumoylation of viral proteins can stimulate viral growth, replication, and persistence (61). The presence of SIMs in viral proteins further illustrates the coevolution and interplay between viruses and host. KSHV K-bZIP is a SUMO E3 ligase with a SUMO2/3-specific SIM that catalyzes the modification of p53 and Rb (62). The Ub ligase ICP0 from HSV-1 has properties characteristic of a STUbL and targets polysumoylated host proteins for degradation in a SIM-dependent manner (63). The HCMV transactivator IE2 and varicella-zoster-virus open reading frame 61 contain a SIM that is required for its preferential conjugation to SUMO1, whereas the KSHV-latent protein LANA2 and the vaccinia E3 protein have a nonspecific SIM required for modification with either paralog (64–67).

NEDD8: FROM REGULATOR OF UBIQUITYLATION TO THERAPEUTIC TARGET

Of all UBLs identified to date, Nedd8 is closest in sequence to Ub (58% identity) (**Supplemental Table 1**). *Nedd8* was originally identified as one of the ten Nedd (neural precursor cell-expressed, developmentally downregulated) genes abundantly expressed in embryonic mouse brain (68, 69). With the notable exception of *S. cerevisiae*, a defect in the Nedd8 pathway is lethal in *A. thaliana*, *S. pombe*, *Drosophila melanogaster*, *Caenorhabditis elegans*, and mice, (68, 69). The C terminus of Nedd8 is processed by NEDP1 (DEN1, SENP8) or UCHL3 (which also acts on Ub) and conjugated to target proteins by the sequential action of the Nedd8-activating enzyme (NAE), the E2 enzymes Ubc12 (Ube2M) or Ube2F, and a handful of E3 ligases (68, 69). NAE is a heterodimer composed of NAE1 (APP-BP1) and Uba3, which are homologous to the N and C terminus, respectively, of Ube1. Despite the high homology between Ub and Nedd8,

NAE and Ubc12 uniquely mediate conjugation of Nedd8. A single residue in Nedd8, Ala72, dictates its specificity for Uba3 of NAE (69a and the references therein). By comparison, the affinity of Ub for Ube1 is dictated by an Arg residue in the identical position (69a). Mutation of Nedd8 Ala72 to Arg72, abrogates its interaction with Uba3 and allows binding to Ube1 (69a). In the case of Ubc12, specificity for Nedd8 is determined by Ubc12 surface elements that selectively bind Nedd8-loaded NAE, while repelling Ube1 (70). Removal of these surface elements confers specificity for Ube1, rather than NAE (70). Importantly, although under normal conditions Nedd8 does not employ the ubiquitylation machinery, an elevation of the free Nedd8 to Ub ratio (e.g., by overexpression of Nedd8 or depletion of free Ub following proteasome inhibition) results in atypical neddylation that is dependent on Ub enzymes but not Nedd8 enzymes (71).

The best-studied targets of neddylation are members of the Cullin (Cul) family, (68, 69). Cullins are scaffold proteins for the assembly of multicomponent RING E3 ligases (CRLs) that participate in ubiquitylation and proteasomal degradation. The activity of CRLs is regulated by mononeddylation of a conserved lysine residue. Nedd8 may also exist in chains of different topology, although the function of polyneddylation is unknown (72, 73). Removal of Nedd8, or deneddylation, is mediated mostly by zinc-dependent metalloenzyme CSN5 (Jab1), a subunit of the eight-subunit COP9 signalosome (CSN) (74). NEDP1 hydrolyzes Nedd8 from a specific subset of non-cullin adducts (75, 76). UCHL3, UCHL1, Otubain-1, and Ataxin3 may also have Nedd8 isopeptidase activity (68, 77).

Nedd8 Targets Cullin RING E3 Ligases and Promotes Ubiquitylation

Nedd8 modifies nearly all members of the Cul family (Cul1, -2, -3, -4A, -4B, -5, and PARC) with the exception of APC/C component Apc2, which lacks the conserved lysine residue targeted by Nedd8, and Cul7 (69). CRLs share a common architecture in which a Cullin scaffold interacts at its C terminus with the RING-domain protein Rbx1 or Rbx2 (also named ROC1/Hrt1 or ROC2/Hrt2 respectively) and at its N terminus with a substrate-recognition subunit, either directly or via an adaptor protein (69, 74). The availability of hundreds of substrate receptors allows the assembly of a multitude of CRLs that, in turn, ubiquitylate a wide spectrum of targets, many of which are involved in cell-cycle regulation, e.g., p27kip1, Cyclin E, and CDT1 (69, 74). The composition and specificity of CRLs has been expertly reviewed and are not discussed further here (69, 74).

How does Nedd8 modification regulate CRL activity? Neddylation of cullins at a conserved lysine near the Rbx1/2 binding site stimulates the ubiquitylation activity of CRLs. Consequently, mutation of this lysine results in the accumulation of CRL substrates (69, 74). Unmodified Cul5 restrains the E2-binding RING domain of Rbx1 in a binding pocket at its C terminus (78). Neddylation of Cul5 liberates the RING domain from the Cul5 binding pocket, allowing Rbx1 to extend from the Cul5 through a flexible linker, much like a "balloon on a string" (78). The release of Rbx1 upon neddylation facilitates E2 recruitment and, most importantly, bridges the 50-Å gap that exists between the substrate lysine and Ub-loaded E2, thereby enhancing Ub transfer (78, 79). The flexibility of the linker further allows Rbx1 to adopt multiple orientations that promote Ub chain extension (78, 79). A stretch of 50 residues at the extreme C-terminal domain (ECTD) of Cul1 that contacts the RING domain of Rbx1 contains intrinsic autoinhibitory activity. Mutation of this ECTD increases CRL activity and mimics Nedd8 modification of Cul1 (80). Thus, neddylation relieves the autoinhibitory activity of the ECTD and allows the CRL to adopt an open conformation that favors ubiquitylation. The budding yeast homolog of Rbx1 contains an unusually flexible linker that possibly bypasses the need for Nedd8 modification, which may explain why the Nedd8 pathway is not essential in *S. cerevisiae* (81).

Optimal CRL activity requires cycling between an active (neddylated) and inactive (unmodified) state. In *Drosophila* and *Neurospora* (but not in yeast and higher eukaryotes), neddylation renders cullins unstable, and regular deneddylation by CSN5 is required to prevent their degradation (74, 82, 83). Similarly, the CSN-associated DUB Ubp12 (USP15) cooperates with the deneddylase CSN5 to maintain the stability of a subset of CRL adaptors, substrate-recognition subunits, and the Ub E2 Ubc3 (Cdc34) in yeast (84–86).

CRL activity is further regulated by the binding of the inhibitor CAND1 (cullin-associated and neddylation-disassociated) (87, 88). This inhibitor binds to unmodified cullins and prevents both neddylation and the association of adaptor proteins, thereby keeping the CRL in an inactive state (87, 88).

Nedd8 E3 Ligases

The RING E3 ligases Rbx1 and Rbx2 not only catalyze ubiquitylation of CRL substrates, but also bind the Nedd8 E2s Ubc12 and Ube2F, respectively, and enhance neddylation of cullins (89, 90). Previous structural models of Ubc12-Rbx1-Cul1 were incompatible with efficient Nedd8 transfer, as the catalytic cysteine of Ubc12 and the Nedd8 acceptor lysine on Cul1 are separated by a 30-Å gap. More recent analysis shows that the RING domain of Rbx1 may undergo a striking rotation that bridges this gap (91). Rbx1/Rbx2 functions in conjunction with the highly conserved DCN1 (defective in cullin neddylation), a scaffold-type E3 ligase that lacks a RING domain, HECT domain, and catalytically active cysteines, yet stimulates neddylation both in vitro and in vivo (92, 93). DCN1 also promotes the proper positioning of the active site of Ubc12 and Cul1 by restricting the motion of the flexible Rbx1-Ubc12-Nedd8 subcomplex (92–94). Additional nonessential factors may further enhance the ligase activity of DCN1/Rbx1 (92). N-terminal acetylation of Ubc12 promotes its interaction with DCN1 by allowing the complete burial of the *N*-acetyl-methionine of Ubc12 in a hydrophobic pocket of DCN1 (95). Humans express 5 DCN1-like protein (DCNL1-5, also termed DCUN1D1-5) of which DCNL1-3 have nonredundant roles in cullin neddylation (96). Human DCNLs have distinct N-terminal domains, which dictate their intracellular localization. DCNL3 localizes to the plasma membrane via a palmitoylation motif at its N terminus and consequently recruits Cul3 to membranes (96). Mouse embryonic fibroblasts lacking DCNL1 (SCCRO) not only exhibit reduced neddylation but also display reduced nuclear localization of Cul1 (97). Thus, the availability of an expanded set of DCN1-like proteins in higher eukaryotes, each with distinct subcellular localization, may regulate CRL activity beyond promoting cullin neddylation. Unlike Cul1 (Cdc53 in *S. cerevisiae*), the yeast Cul4-type cullin Rtt101 and Cul3 are neddylated by the RING-domain protein Tfb3, a subunit of the transcription factor TFIIH (98). Neddylation of Rtt101 does not require DCN1 or the catalytic activity of Rbx1/Hrt1, although it physically interacts with the latter (98). Thus, various RING-domain-containing proteins may regulate the activity of individual cullins.

The dual activity of Rbx1/2 as both a Ub and a Nedd8 E3 ligase is not unique to Rbx1/2; it is also found in other RING-domain ligases. Most of these additional ligases act on a specific subset of non-cullin Nedd8 substrates. *Drosophila* and mammalian inhibitor of apoptosis proteins (IAPs) are RING E3 ligases that catalyze ubiquitylation as well as neddylation of effector caspases involved in apoptosis (75). Neddylation of caspases by IAPs inhibits their proteolytic activity and prevents apoptosis (75). The Ub RING E3 ligase Mdm2 not only regulates ubiquitylation and proteasomal degradation of p53, but also mediates neddylation of p53, p73, and several ribosomal proteins (e.g., L11). Other non-cullin substrates of Nedd8 have been reported (including VHL, EGFR, BCA3, and APP) (68). However, an increased Nedd8 to Ub ratio triggers atypical neddylation, suggesting that many non-cullin Nedd8 substrates may in fact be Ub substrates that are abnormally modified with Nedd8 upon Nedd8 overexpression and engagement of the Ub machinery (71), as predicted by Rabut & Peter (68). Therefore, the identification of non-cullin proteins as authentic Nedd8 substrates requires careful analysis (68).

Cell- and Pathogen-Derived Inhibitors of Neddylation

We know of several negative regulators of the Nedd8 pathway. The interferon-inducible Nedd8 ultimate buster 1 (NUB1) binds Nedd8 and neddylated substrates, including p53, and sends them to the proteasome for degradation (99, 100). Tip160 relocalizes Mdm2, thereby inhibiting neddylation of p53 (101). Furthermore, Tbata (thymus, brain, and testes-associated gene, also termed Spatial) interacts with the E1 subunit Uba3, thereby inhibiting neddylation and cell proliferation (102).

Pathogens have also evolved strategies to counteract Nedd8 modification. Epstein-Barr virus, *Chlamydia trachomatis*, and the parasite *Plasmodium falciparum* all encode a protease with dual activity toward both Ub and Nedd8 (103–107). The bacterial effectors Cif from enterpathogenic *E. coli* (EPEC) and the Cif homolog from *Burkholderia pseudomallei* are translocated into host cells where they deamidate Gln40 in Nedd8 (108, 109). Nedd8 deamidation abrogates cullin neddylation and inhibits CRL activity, resulting in an accumulation of CRL substrates. Cif-induced Nedd8 deamidation is linked to cytopathic effects associated with bacterial infection (108, 109). Finally, bacteria-induced reactive oxygen species oxidize the catalytic cysteine of Ubc12, resulting in suppression of cullin neddylation (110).

Pharmaceutical Inhibition of Neddylation as Anticancer Therapy

Many CRL substrates are involved in the cell cycle. By controlling cullin function, Nedd8 regulates proper cell-cycle progression (69, 111). Defects in neddylation stabilize cell-cycle regulators (e.g., p27kip1, Cyclin E, CDT1) and cause uncontrolled G1-S transition (69, 111, 112). For example, a temperature-sensitive mutation in NAE1 inactivates the Nedd8 pathway and causes multiple rounds of DNA replication (S-phase) without intervening G2, M, and G1 phases (111). Thus, therapeutic intervention in the Nedd8 pathway may be a means to inhibit growth of cancer cells. A selective inhibitor of the NAE (MLN4924; Millennium Pharmaceuticals) causes accumulation of the CRL substrate CDT1 that regulates entry in S-phase, which in turn leads to DNA rereplication, DNA damage, and eventually cell death (112). NAE-bound MLN4924 attacks thioester-linked Nedd8, resulting in the formation of a covalent MLN4924-Nedd8 adduct (113). MLN4924 inhibits the growth of several tumors and is currently being tested in phase I clinical trials (112).

UBIQUITIN-LIKE PROTEINS FIGHT BACK: ISG15 IN HOST DEFENSE

The first UBL modifier identified is the product encoded by *interferon-stimulated gene 15* (*ISG15*), also called ubiquitin cross-reactive protein (UCRP) (6). ISG15 contains two domains with high sequence homology to Ub, fused by a short hinge (**Supplemental Table 1**) (6). Expression of ISG15 is induced by type I interferons (IFN), including IFNα and IFNβ, which are secreted from virus-infected cells and mediate antiviral immunity (6). Exposure to bacterial lipopolysaccharide (LPS) and viral double-stranded RNA (dsRNA) also induce expression of ISG15 (6). Yeast, *C. elegans*, and *D. melanogaster* lack ISG15. Expression of ISG15, though with a low degree of conservation, is limited to higher eukaryotes that participate in IFN signaling (6). Conjugation of ISG15, or ISGylation, involves an enzymatic cascade that resembles ubiquitylation, and the required E1 (Ube1L), E2 (UbcH6 and UbcH8), and E3 enzymes are also IFN inducible (**Supplemental Table 1** and **Figure 2**) (6). The HECT-type E3 enzyme Herc5 (Ceb1) is the predominant ligase for ISG15, as reduction of Herc5 protein levels results in a significant inhibition in overall IFN-induced ISGylation (6). Two RING E3 ligases, EFP (estrogen-responsive finger

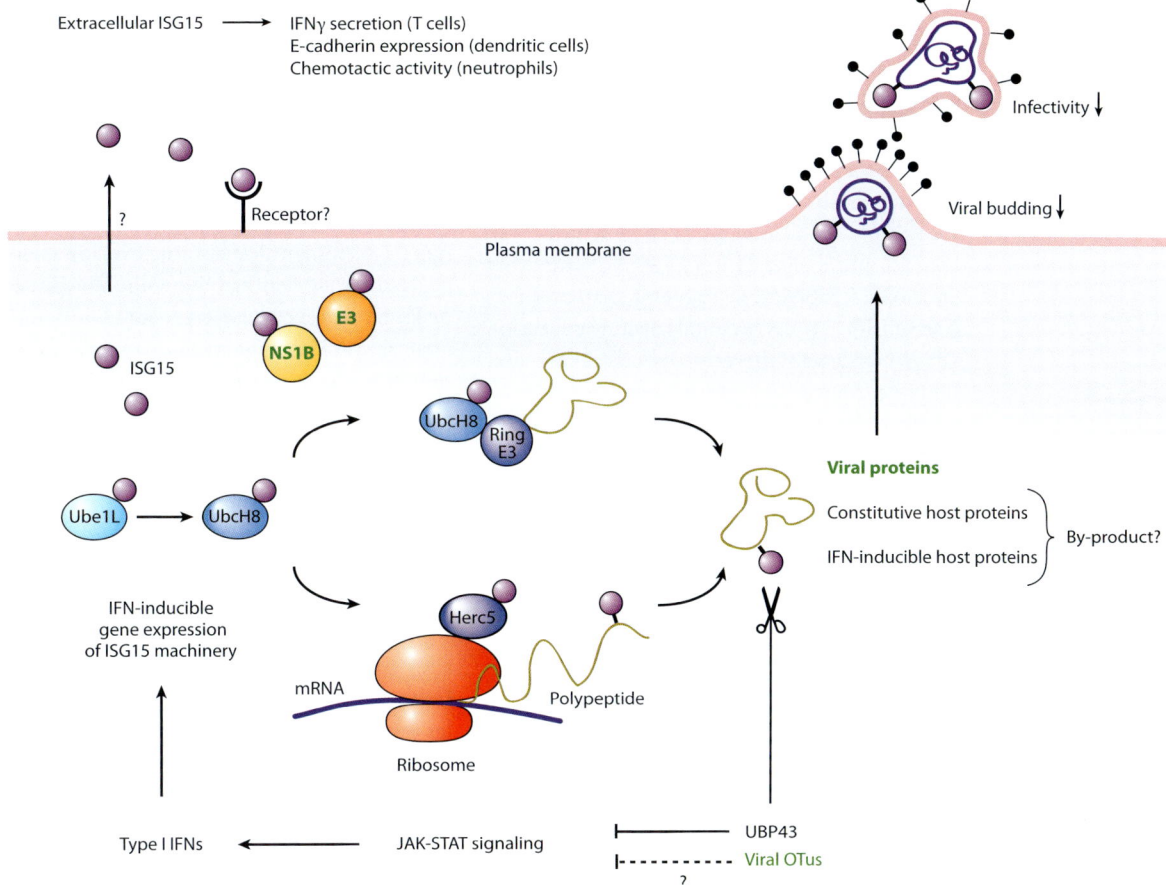

Figure 2
Schematic overview of the ISGylation pathway. Exposure to type I interferons (IFNs), lipopolysaccharide, or double-stranded RNA induces the expression of ISG15 and the ISGylation machinery. ISG15 is conjugated to intracellular substrates, either in a posttranslational or cotranslational manner. The latter is mediated by the HECT E3 ligase Herc5 that is, at least partially, associated with the 60S ribosomal subunit, whereas the RING E3 ligases HHARI and EFP may further mediate ISGylation in the cytosol. ISG15 targets both host and viral proteins, and ISGylation of the latter may negatively affect virion structure and infectivity. In addition, ISG15 may be secreted extracellularly where it has an ill-defined function. The protease UBP43 negatively regulates ISGylation both directly, by catalyzing hydrolysis of the bond between ISG15 and its substrates, and indirectly, by inhibition of IFN signaling via the JAK/STAT pathway. Viral proteins (*green*) employ several strategies to counteract ISG15-mediated antiviral immunity. NS1B (influenza B) and E3 (vaccinia) sequester ISG15 and prevent its association with the E1 Ube1L. Viral proteases of the ovarian tumor (OTU) family function as deISGylating enzymes and, possibly, inhibit IFN signaling in a similar manner to that of UBP43.

protein) and HHARI (human homolog of *Drosophila* ariadne), catalyze ISG15 modification of specific substrates, i.e., 14-3-3σ and 4EHP, respectively (114, 115). ISG15 targets both viral and host proteins, and ISGylated adducts appear ~12 h after IFN treatment (6). PolyISGylation has not been reported, although multimonoISGylation is frequently observed. The IFN-inducible UBP43 (USP18) is the major deISGylase, though not required for C-terminal processing of ISG15 (116). Several deubiquitylases, including USP2, USP5, USP13, USP14, and USP21, show reactivity toward ISG15 in vitro and may either mediate maturation of ISG15 precursors or perform deconjugation (117, 118).

The Interplay Between ISGylation and Viruses

The IFN responsiveness of the ISG15 pathway suggests a role in host defense against viral infections. Two sets of observations support this: First, mice with a deficiency in the IFN-response or ISGylation pathway are more susceptible to various types of viral infection. Second, the identification of viral factors that antagonize ISGylation suggests that conjugation of ISG15 is detrimental to viral replication or infectivity and, therefore, is a target for viral immune evasion.

Mice that lack the IFN-$\alpha\beta$ receptor (IFN-$\alpha\beta$R) are unable to upregulate IFN-stimulated genes, including ISG15, and readily succumb to viral infection. However, infection with a chimeric ISG15-expressing Sindbis virus reduced viral load and limited virus-induced lethality (119). ISG15-deficient mice are more susceptible to both DNA and RNA viruses, including influenza A, influenza B, herpes simplex virus 1 (HSV-1), murine gammaherpesvirus 68 (γHV68), and Sindbis virus (120). The antiviral activity of ISG15 requires its conjugation to substrates, as Ube1L-deficient mice were as susceptible to infection with Sindbis virus or influenza B as were ISG15-deficient mice (121, 122). Chimeric Sindbis virus that expresses an ISG15 mutant that lacks its C-terminal GG motif or that is unable to bind Ube1L fails to rescue IFN-$\alpha\beta$R$^{-/-}$ mice from virus-induced lethality (119, 121). Deletion of the deISGylase UBP43 gene conferred resistance to Sindbis virus, lymphocytic choriomeningitis virus (LCMV), and vesicular stomatitis virus (VSV) (123). ISG15$^{-/-}$ or Ube1L$^{-/-}$ mice did not display increased sensitivity to LCMV or VSV (124, 125), suggesting that UBP43 has a role independent of its activity as deISGylase (discussed below). ISG15 may enhance viral activity during hepatitis C virus (HCV) infection (126), although in separate work HCV replication was suppressed in the presence of a functional ISGylation machinery (127). Thus, ISG15 confers a protective effect during some viral infections, although minimally affecting or even exacerbating other infections.

Immune escape through targeting of the ISGylation pathway was first described for influenza B virus. ISG15 is upregulated upon infection with influenza B virus, but ISG15 conjugation to substrates is inhibited by the viral NS1B protein (128). NS1B binds the N-terminal domain of ISG15, thereby preventing its interaction with Ube1L (128) or E3 ligases (129). NS1B sequesters ISG15 in nuclear speckles, which may further contribute to inhibition of its function (130). The NS1B-mediated inhibition of ISGylation is species specific; although NS1B readily associates with human or primate ISG15, it fails to bind mouse or canine ISG15 (130, 131). The species specificity of NS1B may explain why influenza B viral infection is restricted to humans (130). The E3 protein of vaccinia virus, which has a domain similar to NS1B, also binds ISG15 and suppresses its antiviral activity (132). Other means of immune escape are conferred by viral deISGylases. Ovarian tumor (OTU) domain-containing proteases in nairoviruses and arteriviruses hydrolyze both Ub and ISG15 adducts, unlike mammalian OTU proteases, which uniquely target Ub (133). Similarly, papain-like proteases in coronaviruses and hepatitis E virus exhibit broad specificity toward both Ub and ISG15 (134–136). DeISGylation by viral OTU proteases inhibits ISG15-mediated antiviral immunity against Sindbis virus in mice (133). Structural studies allowed the engineering of viral OTU variants that hydrolyze either Ub or ISG15 (137, 138). The use of such variants will aid our understanding of the relative contribution of ubiquitylation and ISGylation in virus infections.

ISG15 Targets Host and Viral Proteins

ISG15 targets both cellular and viral proteins, and many candidate substrates have been identified by proteomic analysis (6, 139). Although some of the identified substrates are IFN inducible and have a known role in antiviral immunity (e.g., RIG-I, PKR, and MxA), ISGylation affects many constitutively expressed proteins that are localized in different compartments and participate in a wide variety of cellular processes (139). ISGylation of the translational repressor 4EHP via the HHARI RING E3 ligase enhances its affinity for the mRNA 5′-cap structure and suggests that translation of specific, perhaps viral, mRNAs is further reduced by ISGylation (114). ISG15 conjugation to the scaffold protein filamin B causes the dissociation of the signaling molecules RAC1, MEKK1, and MKK4 and consequently prevents JNK activation and apoptosis (140). ISGylation of interferon regulatory factor 3 (IRF3) inhibits its interaction with the Ub E3 ligase Pin1; consequently, Ub-mediated degradation of this factor is reduced upon Sendai virus infection, implicating ISG15 as a positive regulator in IFN signaling (141).

ISG15 also targets a number of viral substrates. ISGylation of the NS1 protein of influenza A (NS1A) protects against infection by preventing association of NS1 with importin-α, consequently preventing its nuclear import (142), or by inhibiting NS1 homodimerization and binding to PKR and/or dsRNA (143). ISG15 modification of NS5A of HCV reduces its stability and suppresses HCV replication (127). In the case of retroviruses and the Ebola virus, ISG15 inhibits virion release by interfering with the budding process. Noncovalent or covalent associations between ISG15 and the Ub E3 ligase Nedd4 inhibit the interaction between Nedd4 and E2 enzymes and consequently suppress ubiquitylation of the Ebola structural protein VP40 and budding of VP40-containing virus-like particles (144, 145). Similarly, ISG15 attenuates ubiquitylation of the ESCRT-I protein Tsg101 and HIV-1 Gag, thus impairing the interaction between Tsg101 and Gag and early stages of virus-like-particle budding (146). ISG15 also interferes with late-stage budding of both HIV-1 and avian sarcoma leukosis virus by inhibiting the association between the ESCRT-III protein Vps4 and Gag or Vps4 and its coactivator Lip5 through ISGylation of the ESCRT-III proteins CHMP2A, CHMP4B, CHMP5, or CHMP6 (147). As ISG15 generally modifies only a few percent of any given substrate (whether of viral or cellular origin), it is unclear how ISGylation of such a small fraction can severely impact viral infectivity.

The above examples illustrate how ISG15 may exert at least some of its antiviral effects through attenuation of ubiquitylation. Indeed, ISGylation of the Ub E2 enzyme Ubc13 suppresses its ability to form its Ub thioester (148, 149). Expression and conjugation of ISG15 is also increased in tumor cells of different origin, concomitant with a decrease in polyubiquitylation (150). Depletion of ISG15 or UbcH8 restored polyubiquitylation, suggestive of a causal relationship between increased ISGylation and decreased ubiquitylation in tumor cells (150). Elevated levels of free and conjugated ISG15 in biopsies from patients with various solid tumors may result from increased IFN release in physical proximity of the tumor and could serve as cancer biomarker (6, 151 and references therein). The relationship between ISGylation, ubiquitylation, and cancer development remains unclear and deserves further attention.

To explain the broad substrate specificity of ISGylation, mediated predominantly by a single E3 ligase, ISG15 was proposed to target only newly synthesized proteins and to do so with limited selectivity. Several human and bacterial proteins not previously identified as targets for ISG15 were ISGylated when exogenously expressed in the presence of the ISGylation machinery with few apparent structural requirements for ISG15 substrate recognition (152). Herc5 cofractionated with polysomes in IFN-treated cells, suggesting that ISGylation takes place cotranslationally (152). ISGylation of as little as 10% of HPV16 L1 capsid protein reduces infectivity of an HPV pseudovirus by 70% (152). Modification of viral structural proteins may explain how ISGylation of a minor fraction can impair the infectivity of a virus, although this needs to be confirmed using additional viral models. ISGylation of newly synthesized host proteins may be considered as collateral damage and kept to a minimum by modification of such a small fraction of any individual target (152). Although this is an attractive hypothesis, several issues require clarification. First, only a few viral proteins, including NS1A and M1 (influenza A) (142, 143), NS5A (HCV) (127), and L1 (HPV16) (152), are known to be ISGylated. One could expect that a wide range of newly synthesized viral proteins would be conjugated to ISG15 if ISGylation targeted nascent chains indiscriminately. Type I IFNs stimulate translation of hundreds of IFN-inducible proteins, many of which have critical roles in antiviral immunity. Such newly synthesized IFN-inducible proteins would be heavily ISGylated, possibly with detrimental effects on their function in antiviral host defense. Likewise, although ISGylation of a small fraction of long-lived host proteins is unlikely to affect cellular processes, this may not be the case for short-lived host proteins, the modification of which may have dramatic effects especially during persistent viral infections.

Antiviral Effects of ISG15 and UBP43: Not All About ISGylation

A unique feature of ISG15 is its secretion from cultured IFN-treated T cells, monocytes, B cells, and epithelial cells, despite the lack of an obvious secretory signal sequence (153). ISG15 was also detected in the serum of IFNβ-treated healthy volunteers (153), and serum levels were elevated in wild-type and Ube1L-deficient mice upon LPS treatment, dsRNA treatment, or infection with influenza B or γHV68 (120, 122). ISG15 is secreted in its unconjugated, mature form. It stimulates IFNγ release from T cells (153, 154) as well as E-cadherin surface expression on dendritic cells (155) and has chemotactic activity toward neutrophils (156). Thus, secreted ISG15 may function as a cytokine, but its secretory mechanism, the identity of a possible receptor, and its precise role in modulating the immune response remain unknown. The susceptibility of ISG15$^{-/-}$, but not Ube1L$^{-/-}$, mice to Chikungunya virus (CHIKV) further points to a conjugation-independent role for ISG15, perhaps involving its extracellular release (157). Levels of proinflammatory cytokines and chemokines are significantly increased in ISG15$^{-/-}$ mice following CHIKV infection, suggesting that ISG15 may be required to mount an effective immune response (157).

UBP43 does not function solely as an isopeptidase in deISGylation; it also has an ISG15-independent role in the negative regulation of IFN signaling. UBP43 competes with the signaling molecule JAK1 for binding to the IFNAR2 subunit of the type I IFN receptor, thereby inhibiting activation of the JAK-STAT pathway and transcription of IFN-stimulated genes (158). Consequently, UBP43-deficient cells are hypersensitive to type I IFN, resulting in prolonged STAT1 tyrosine phosphorylation, sustained IFN-induced gene activation, and, eventually, apoptosis (159). Competition between UBP43 and JAK1 for IFNAR2 binding is independent of the catalytic activity of UBP43. Therefore, this phenomenon does not involve ISGylation (158). Whether OTUs of viral origin manipulate the JAK-STAT pathway as an immune escape mechanism is unknown.

An ISG15-independent role for UBP43 is further underscored by the findings that UBP43-deficient mice suffer from brain abnormalities and are resistant to infection by LCMV and VSV, whereas the combined deletion of UBP43 and Ube1L or UBP43 and ISG15 in mice does not revert these phenotypic alterations, nor does it rescue the type I IFN hypersensitivity of UBP43-deficient mice (123, 124, 160, 161). In addition, deletion of UBP43, but not Ube1L, inhibits replication of hepatitis B virus in mice (162).

FAT10 AND MNSFβ: CONNECTIONS TO PROTEASOMAL DEGRADATION AND IMMUNITY

FAT10 (HLA-F adjacent transcript 10) and MNSFβ (monoclonal nonspecific suppressor factor β) are two UBLs of unknown function that, similar to ISG15, may function in immune defense. Similar to ISG15, FAT10 contains two tandem Ub-like domains; thus, it is also referred to as diubiquitin or ubiquitin D (163). Whereas lower eukaryotes do not express FAT10, mammalian FAT10 is expressed in the cytoplasm of mature dendritic cells and B cells and is induced in other cell types by IFNγ and TNFα (163). Activation and conjugation of FAT10 is mediated by Uba6 (also termed E1-L2 and Ube1L2) and USE1 (Uba6-specific E2 enzyme), respectively, which have dual specificity for both Ub and FAT10 (**Supplemental Table 1**) (164, 165). It is unclear how Ub and FAT10 compete for binding to Uba6 and USE1. When present in equimolar concentrations, Uba6 preferentially binds and activates Ub over FAT10 in vitro (165). Cytokine stimulation may alter this equilibrium in favor of FAT10 or Ub, and FAT10 may form mixed chains.

FAT10 conjugates accumulate upon proteasomal inhibition (166). Does FAT10 provide an alternative route to protein degradation? FAT10 noncovalently interacts with NUB1L (Nedd8-ultimate buster 1 long), which is associated with the 26S proteasome and accelerates degradation of free and conjugated FAT10 both in vitro and in vivo (166, 167). FAT10-mediated degradation does not require an interaction between NUB1L and FAT10, but instead depends on binding of NUB1L to the proteasome (166, 168). Because FAT10 can also directly interact with the 26S proteasome, it has been proposed that NUB1L induces a conformational change in the proteasome that allows FAT10 binding and degradation of the modified substrate (168). Earlier reports suggested that FAT10-mediated degradation is independent of ubiquitylation, as lysine-less FAT10 rapidly disappears. However, a recent study demonstrated that this mutant protein is not degraded by the proteasome, but instead aggregates in an insoluble fraction (168, 169). In contrast, wild-type FAT10 is polyubiquitylated prior to its proteasomal degradation (169). It is unclear how proteasomal degradation via FAT10 complements ubiquitylation and whether FAT10 performs additional functions. FAT10 has been positively correlated with induction of caspase-dependent apoptosis, NF-κB activation, cell-cycle defects, and chromosomal instability (163, 170). p53 is a substrate of FAT10, and fatylation of p53 increases its transcriptional activity (171).

The abundance of FAT10 in lymphoid organs combined with its cytokine-inducible expression implicate FAT10 in the immune response. Indeed, FAT10 as well as Uba6 and NUB1L mRNA levels are upregulated upon maturation of dendritic cells with a variety of stimuli, including the Toll-like-receptor ligands LPS and poly(I:C) (172). Although FAT10-deficient mice seem healthy and normal, they display increased lymphocyte cell death and enhanced sensitivity toward endotoxin (LPS) challenge (173). Furthermore, FAT10 is one of the most upregulated genes in renal cells of HIV-1-infected patients. FAT10 interacts with the viral Vpr protein at the mitochondria, suggesting that FAT10 may have a role in immunity (174).

MNSFβ (also known as Fau ubiquitin-like 1 or Fau) is synthesized as a fusion with the ribosomal protein S30 (175). Posttranslational processing of the fusion products yields free MSNFβ and the mature ribosomal protein. MNSFβ is conjugated by an unidentified mechanism to the proapoptotic protein Bcl-G and endophillin II, thereby inhibiting LPS-induced ERK-MAPK signaling and phagocytosis, respectively, in macrophages (176, 177). Similar to ISG15, MNSFβ may be secreted and regulate proliferation of T cells (175, 178).

UFM1 IN ERYTHROID AND MEGAKARYOCYTE DEVELOPMENT

The ubiquitin-fold modifier 1 (Ufm1) is one of the newest members of the Ub family and is conserved in metazoa and plants (179). Ufm1 was unexpectedly identified in complex with the E1 enzyme Uba5 (179). Indeed, Ufm1 conjugation is greatly impaired in Uba5-deficient MEFs (180). Although the function of Ufm1 remains unclear, Uba5 is required for erythroid differentiation in mice (5). Uba5-deficient mice die in utero from severe anemia caused by impaired differentiation and increased apoptosis of megakaryocyte and erythroid progenitors (5). Transgenic expression of Uba5 in the erythroid and megakaryocyte lineages rescues erythropoiesis and megakaryopoiesis resulting in reduced anemia and prolonged survival of Uba5$^{-/-}$ embryos from E13.5 to E18.5 (5). Rescued embryos eventually die, despite improvement in erythroid and megakaryocyte development; thus, Uba5 is likely involved in later stages of development or in other cell and tissue types. The uncharacterized, ER-resident protein C20orf116 is the only Ufm1 substrate identified so far (180).

ATG8 AND ATG12: REGULATORS OF AUTOPHAGY

Autophagy is a catabolic process by which cells break down their own components. To this end, cytoplasm, macromolecules, and organelles are sequestered in double-membrane vesicles (autophagosomes), and their components are recycled via the lysosomal pathway (9, 181). Autophagy is an essential process for growth and survival during conditions of low nutrient availability and is involved in stress responses, development, immunity, and inflammation (9, 181). Of the different types of autophagy, macroautophagy is the best described. Autophagosome biogenesis is regulated by approximately 30 autophagy-related genes (Atg) that mediate vesicle nucleation, expansion, and fusion with lysosomes (**Figure 3a**) (9, 181). Among the core components of the autophagy machinery are two UBLs, Atg8 and Atg12, that are essential for autophagosomal membrane growth and expansion (182). Atg12 contains a single C-terminal glycine residue that is attached to a lysine in Atg5 via the sequential action of Atg7 (E1) and Atg10 (E2) in the absence of a ligase activity (**Supplemental Table 1**). Atg5 further interacts with Atg16L1, and the Atg12-Atg5-Atg16L1 complex forms a multimer through homo-oligomerization of Atg16L1 (182). The Atg12-Atg5-Atg16L1 oligomer is found on the outer side of the autophagosomal membrane where it functions as an E3 ligase in the conjugation of the second autophagy-related UBL, Atg8 (183, 184). Although the Atg12-Atg5-Atg16L1 complex dissociates from the membrane upon completion of the autophagosome, no Atg12-specific hydrolase has been identified. Although Atg5 was long considered the sole substrate of Atg12, the E2 enzyme Atg3 is an additional Atg12 substrate. Atg12-Atg3 conjugation does not affect starvation-induced autophagy, but rather regulates mitochondrial homeostasis and cell death via an undefined mechanism (185).

Atg8 in Autophagosome Biogenesis

Atg8 is an unusual lipid modifier that it is conjugated to phosphatidylethanolamine (PE) via the Atg7 (E1), Atg3 (E2), and Atg12-Atg5-Atg16L1 (E3) cascade, and it is processed and delipidated by the cysteine protease Atg4 (**Supplemental Table 1**) (9, 181, 182). The Atg8 conjugation cascade involves distinct mechanisms in which homodimeric Atg7 transfers thioester-linked Atg8 from the catalytic cysteine of one protomer to that of Atg3 bound to the opposite Atg7 protomer (186–188). Lipidated Atg8 is membrane associated and required for the expansion of autophagosomal precursors (189, 190). The amount of Atg8 determines the size of the autophagosomes (190). In vitro reconstitution of Atg8 conjugation showed that lipidation of yeast Atg8 may mediate the tethering and hemifusion of liposomal membranes (189). However, repetition of these in vitro experiments in the presence of physiological concentrations of PE demonstrated that hemifusion does not require Atg8, but instead proceeds via a SNARE-dependent mechanism (191, 192).

Unlike yeast, which contains a single Atg8 gene, mammals contain six Atg8 paralogs that, based on homology, are divided into the LC3 (microtubule-associated protein light chain 3) and GABARAP (γ-aminobutyric acid receptor-associated protein) subfamilies

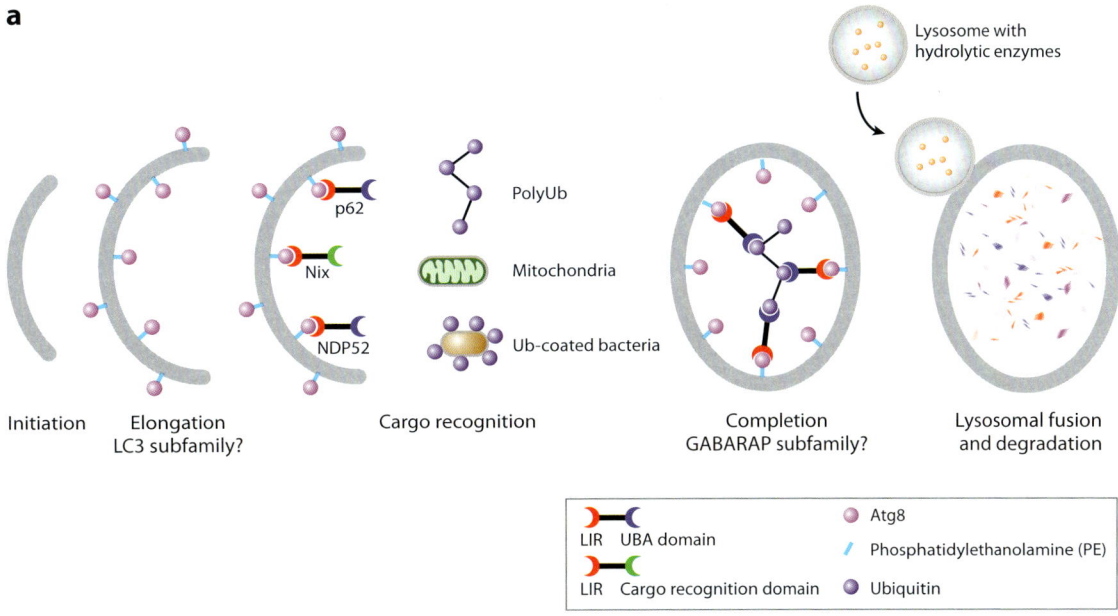

Figure 3

(*a*) The Atg8 family is required for autophagosome biogenesis and cargo recognition. Following initiation of autophagosome formation, PE-modified Atg8 is inserted in the autophagosomal membrane, thereby mediating expansion (LC3 subfamily) and sealing (GABARAP subfamily) of the membrane. In addition, Atg8 family members are required for recognition and recruitment of cargo to the autophagosome. To this end, Atg8 associates with the LIR of selective autophagy receptors, which in turn recruit polyubiquitylated proteins or ubiquitin-coated bacteria (via a UBA domain) or damaged mitochondria (via an undefined mechanism). (*b*) LIRs mediate noncovalent association with Atg8s and have been identified in autophagy cargo receptors, enzymes of the Atg8 (de)lipidation machinery, the microtubule-binding protein FYCO1, the Rab GAP OATL1, calreticulin, and clathrin heavy chain (clathrin hc) (213, 257). The LIR includes a conserved aromatic residue (*red*) in the first position and a hydrophobic (*blue*) core. The interaction between Atg8s and LIRs may be further modulated by acidic residues (*green*) or phosphorylatable serines (*orange*) in or adjacent to the LIR. Abbreviations: LC3, light chain 3; LIR, LC3-interacting region; PE, phosphatidylethanolamine; Ub, ubiquitin; UBA, ubiquitin associated.

Ubiquitin-associated (UBA) domain: mediates noncovalent binding to Ub

(**Supplemental Table 1**). All Atg8 paralogs interact with the conjugation machinery and associate with the autophagosomal/lysosomal membrane upon lipidation (182, 193–195). The existence of multiple Atg8 paralogs is analogous to the expression of at least three SUMO paralogs and raises similar questions about conjugation selectivity and function. Atg8 paralogs interact in vivo with a total of 67 proteins, with significant overlap between the interactomes of individual Atg8 paralogs (193). One-third of all 67 binding partners was shared between the LC3 and GABARAP subfamilies in vivo, and nearly all binding partners associated with both subfamilies in vitro (193). The lack of specificity of interacting proteins for the LC3 or GABARAP subfamilies suggests at least partially redundant roles for Atg8 paralogs. Alternatively, specificity may be controlled by other factors, such as intracellular localization, tissue expression, and abundance of Atg8 paralogs and their binding partners. Consistent with an overlapping role for both subfamilies in autophagosome biogenesis, both LC3B and GABARAPL2 mediate membrane tethering and hemifusion in a cell-free system (196). In both cases, the distinct N-terminal 10 residues were sufficient for fusion activity but involved ionic (LC3B) or hydrophobic (GABARAPL2) interactions with the membrane (196). Lipidation of Atg8 and GABARAP induces a conformational change in the N-terminal α-helix of both proteins. The PE-induced shift from a closed conformation, in which the N-terminal helix points toward the C-terminal Ub-like domain, to an open conformation, in which the helix forms an extended structure that points away from the Ub-like core, promotes its availability for intermolecular interactions and multimerization (197, 198). Although LC3 and GABARAP subfamilies are both indispensible for autophagosome biogenesis, they are required at different stages of maturation. Whereas both LC3B and GABARAPL2 colocalize at autophagosomal membranes, LC3B is involved in elongation of the membrane, and GABARAP is required at a later stage, possibly in the sealing of vesicles (195). Whether the different N termini of both subfamilies mediate these alternative functions remains to be determined. Atg8 paralogs may also promote the fusion between autophagosomes and lysosomes by recruiting the Rab GAP OATL1 (199).

Atg8 in Selective Autophagy

Members of the Atg8 family selectively target cargo to autophagosomal vesicles. Although autophagy is generally a nonselective process that engulfs organelles or cytosolic material nonspecifically, the identification of autophagy receptors that recognize specified cellular structures, such as protein aggregates, mitochondria, and bacteria, suggests that autophagy may also function as a targeted degradation machinery (200). The majority of selective autophagy receptors simultaneously bind Atg8 and Ub via an LC3-interacting region (LIR) and UBA domain, respectively (200). The LIR domain of these receptors mediates the recruitment of the autophagosomal machinery to select cargo (**Figure 3**). Both p62 (also known as Sequestosome 1/SQSTM1) and NBR1 (neighbor of BRCA1 gene) preferentially target K63-linked polyubiquitylated proteins and mediate their aggregation and autophagic clearance in an LIR- and UBA-dependent manner (201, 202). NBR1 and p62 oligomerize but can function independently, explaining the mild phenotype of p62-deficient mice (201, 202). Selective autophagy receptors also function in the removal of damaged organelles. Yeast Atg32 and mammalian Nix (NIP3-like protein X, also termed BNIP3L) target damaged or excess mitochondria for autophagosomal degradation (mitophagy) in an LIR- and Atg8-dependent manner (203, 204). Because the mitochondrial Atg32 and Nix lack a UBA domain, it is unclear what triggers the recruitment of the autophagic machinery upon mitochondrial damage. In mammals, the PINK1-Parkin-p62 pathway may also mediate mitophagy. Upon mitochondrial damage, PINK1 accumulates and recruits the Ub E3 ligase Parkin to the mitochondria, followed by K63-linked polyubiquitylation of mitochondrial substrates and p62 accumulation (204). Autophagy helps clear many damaged organelles, such as ribosomes, peroxisomes, and the ER. The study of these selective degradation pathways will inevitably lead to the identification of additional organelle-specific autophagy receptors.

Selective autophagy targets cytosolic pathogens as part of the immune defense mechanisms (xenophagy). Although intracellular bacteria such as *Salmonella enterica* serotype *typhimurium* and *Streptococcus pyogenes* reside within vacuoles, these can occasionally rupture, leading to rapid Ub coating of the bacterial surface and autophagy-dependent clearance of the bacterium (181). These two events have recently been linked by the identification of autophagy receptors such as p62, NDP52 (nuclear dot protein 52), and OPTN (Optineurin) that recruit the autophagy machinery to Ub-coated cytosolic bacteria via their UBA and LIR domains (205–207). Silencing of these receptors or mutation of their LIR domain leads to inefficient autophagy-mediated clearance of *S. typhimurium* (205–207). *Shigella flexneri*, a professional cytosol-dwelling organism, is not targeted by Ub, NDP52, or the autophagy machinery and may escape this cellular defense pathway through an unknown mechanism (205). P62 also indirectly mediates clearance of *Mycobacterium tuberculosis* by enhancing the delivery of ribosomal and ubiquitylated proteins to autophagosomes, where they are subsequently processed into antimicrobial peptides that are active against *M. tuberculosis* (208). The identity of the ubiquitylated substrates on bacterial surfaces and the E3 ligase(s) that mediate(s) their modification remain(s) unknown. Whereas p62 and OPTN accumulate at both Ub-coated cytosolic bacteria and starvation-induced autophagosomes, it is unclear whether NDP52 has uniquely evolved to recognize bacterial determinants, or targets host cargo as well. The existence of multiple autophagy receptors with seemingly overlapping roles and modes of recognition suggests other, as yet unidentified, functions.

Atg8 Paralogs and Autophagy Receptors

The discovery of autophagy receptors with dual specificity for both Ub and Atg8s illustrates how Ub and Ub-like modifiers may collaborate to eliminate cellular threats. But how do different Atg8 paralogs contribute to selective autophagy? Most known autophagy receptors do not appear to have an exclusive preference for a particular Atg8 paralog. For example, Nix interacts with all Atg8 paralogs with a preference for GABARAPL1 and with the exception of LC3B (203). The LIR motif is also found in other proteins that interact with Atg8 variants in a nonpreferential manner, such as the yeast E2 enzyme Atg3 and the mammalian protease Atg4B (209, 210). By contrast, although p62 equally interacts with cytosolic LC3s and GABARAPs, only membrane-bound LC3-PE, but not GABARAPL2-PE, can incorporate p62 into autophagosomes (211). Given that the N-terminal ten residues of LC3 are required for its interaction with p62, this further illustrates that the unrelated N termini of LC3 and GABARAPL2 are functionally distinct (211).

The LIR motif is composed of a hydrophobic core and a conserved aromatic residue (**Figure 3b**) (210) and interacts with a hydrophobic patch on Atg8 paralogs. Structural analysis of the p62 LIR in association with LC3B shows that these domains form an intermolecular parallel β-sheet (200, 209, 212). The LIR of p62 (WTHL) and NBR1 (YIII) interact in a similar manner with LC3B and GABARAPL1, respectively (213). Whereas the aromatic tryptophan (W) residue in p62 provides a strong interaction with LC3B, its replacement with the weaker binding tyrosine (Y) in NBR1 is compensated by the presence of three residues with increased hydrophobicity (isoleucines, III) (213). The LIR motif in p62 and NBR1 is preceded by an acidic stretch, which further promotes their interaction with Atg8 paralogs (210). Finally, TBK1-dependent phosphorylation of a series of conserved serines near the LIR of OPTN increases its affinity

Molybdopterin (MPT): a component of molybdenum cofactor (MoCo). MoCo is used as a cofactor by several enzymes

Thiamine: a vitamin (called B1) used as cofactor by several enzymes

β-grasp fold: a structural motif commonly found in Ub and UBLs that is characterized by a β-sheet with five antiparallel β-strands and a helical segment

for LC3 (206). Because NDP52 also associates with TBK1 and both Nix and NBR1 contain conserved serine residues adjacent to their LIRs, phosphorylation is probably a versatile mechanism to control Atg8 recruitment and affinity (205, 206). Whether phosphorylation differentially affects the binding of OPTN with Atg8 paralogs remains to be determined.

A Family of Atg8-Directed Proteases

Similar to the expansion of the mammalian Atg8 family, the Atg8-specific protease Atg4 has multiplied, and four Atg4 homologs (Atg4A-D) are expressed in mammals. These proteases are involved in C-terminal processing of Atg precursors as well as delipidation of Atg8 paralogs. Atg4B (autophagin-1) is a promiscuous protease that acts on all Atg8 paralogs (214). Unlike Atg3-, Atg5-, Atg7- or Atg16L1-deficient mice, genetic disruption of Atg4B does not lead to perinatal death. However, it does result in a significant, but not complete, impairment in autophagy, suggesting that Atg4 proteases are at least partially redundant (215). Deletion of Atg4C (autophagin-3) only minimally affects autophagy, suggesting that it has a minor role in the autophagic response (215). Atg4A (autophagin-2) preferentially targets the GABARAP subfamily (195). Likewise, caspase-processed Atg4D (autophagin-4) cleaves the C terminus of GABARAPL1, and to a lesser extent GABARAPL2, more efficiently than that of LC3 (216). Thus, Atg4 proteases may preferentially target a subset of Atg8 paralogs, analogous to SUMO-directed SENPs.

Successful autophagosome formation requires tight regulation of Atg4 protease activity to prevent premature delipidation of Atg8-PE. Atg4 activity may be partially controlled by autophagy-induced reactive oxygen species, which inactivate Atg4A and Atg4B by oxidation of a cysteine residue near the catalytic site of the protease (217). Alternatively, the autophagy proteins Atg18 and Atg21, together with Atg12-Atg5-Atg16L1, may protect Atg8-PE from premature cleavage by blocking access of Atg4 to its substrate (218).

URM1: BRIDGING THE GAP BETWEEN PROKARYOTIC SULFUR CARRIERS AND EUKARYOTIC PROTEIN MODIFIERS

Ub and most UBLs are highly conserved across all eukaryotic species (**Supplemental Table 1**). Although the apparent absence of Ub ancestors in prokaryotes remained puzzling for years, the recent identification of an expanding set of prokaryotic proteins with features characteristic of eukaryotic UBLs, conjugating enzymes, and peptidases suggests that these prokaryotic cognates may have spawned the eukaryotic Ub/UBL modification system (219, 220). The molybdopterin (MPT) and thiamine (vitamin B1) biosynthetic pathways in prokaryotes show a remarkable similarity with Ub modification. A key step in the synthesis of MPT and thiamine involves the transfer of a sulfur atom from a sulfur donor to MPT precursor Z or thiazole, respectively (1). The sulfur donors required for these reactions, MoaD and ThiS, contain a C-terminal diglycine motif derivatized by the enzymatic activity of MoeB or ThiF, respectively, to introduce a sulfur atom in the form of a thiocarboxylate (**Supplemental Figure 1**) (1). Both MoaD and ThiS have low sequence homology with Ub and adopt a β-grasp fold common to Ub and UBLs (221, 222). MoeB and ThiF are evolutionarily related to Ube1, and there are several mechanistic and structural similarities between the ATP-dependent activation of Ub, MoaD, and ThiS by their respective enzymes (**Supplemental Figure 1**) (223–225). Thus, MoaD and ThiS are likely antecedents of the eukaryotic Ub/UBL modifiers. This proposal received support from the identification of the Ub-related modifier Urm1 (226). Urm1 is conserved from *S. cerevisiae* to *Homo sapiens*, and it adopts a β-grasp fold. Furthermore, of all Ub family members, yeast and murine Urm1 have best conserved the structural features of the ancient sulfur carriers (226–228). Analogous to MoaD and ThiS, the C-terminal glycine of Urm1 is derivatized by the E1-like enzyme Uba4 (molybdenum cofactor synthesis 3, MOCS3 in *H. sapiens*) to contain a thiocarboxylate (229–232). The N-terminal domain of Uba4/MOCS3 is homologous to Ube1 and *E. coli* MoeB/ThiF and contains a nucleotide-binding domain required for activation of Urm1 by forming an acyl-adenylate between the C terminus of Urm1 and AMP (226, 230, 232). The C-terminal domain of Uba4/MOCS3 contains a rhodanese-like domain. Rhodaneses are sulfurtransferases, and the rhodanese-like domain of Uba4/MOCS3 transfers sulfur to the activated C-terminal glycine of Urm1 to form a thiocarboxylate (232).

Thiocarboxylation of Urm1 is initiated by the L-cysteine desulfurase Nfs1, which binds the rhodanese-like domain of Uba4/MOCS3 and transfers sulfur from L-cysteine, either directly or with the help of additional sulfurtransferases, onto the active-site cysteine of Uba4 (Cys397) or MOCS3 (Cys412) in the form of a persulfide (**Supplemental Figure 1**) (233, 234). This persulfide forms an acyl-disulfide intermediate with activated Urm1 (232). Although not formallly demonstrated, a cysteine in the N-terminal domain of Uba4 (Cys225) or MOCS3 (Cys239) may release thiocarboxylated Urm1 by reductive cleavage of the acyl-disulfide linkage (232). Dimeric forms of both Urm1 and Uba4 have been reported, and together they may form a heterotetrameric (Uba4-Urm1)$_2$ complex (232, 235).

Urm1 in tRNA Thiolation

The identification of a thiocarboxylate at the C terminus of Urm1 suggested that this unusual UBL functions as a sulfur carrier, reminiscent of the prokaryotic MoaD and ThiS. Indeed, Urm1 functions in the sulfur modification of several tRNA species (229–231, 236). A chemical modification is commonly found on uridine molecules in the wobble position (U_{34}) of the anticodon of tRNA$^{Lys(UUU)}$, tRNA$^{Glu(UUC)}$, and tRNA$^{Gln(UUG)}$. Most frequent at U_{34} are a methoxy-carbonyl-methyl modification at the 5' position (mcm^5), and a thiocarbonyl group at the 2' position (s^2) of the uracil ring (**Supplemental Figure 2**) (237). Sulfur modification, or thiolation, of wobble uridines in eukaryotic tRNAs requires the Urm1 pathway. Thiocarboxylated Urm1 interacts with the thiouridylase Ncs6 (needs Cla4 to survive 6; also known as cytosolic thiouridylase 1 or Ctu1) or its human homolog ATPBD3 (ATP binding domain 3), which in turn binds and adenylates the tRNA moiety, and it likely mediates the transfer of sulfur from the C terminus of Urm1 to the tRNA (229, 231, 238). Genomic loss of yeast Urm1, Uba4, Ncs6, or Nfs1 as well as shRNA-mediated depletion of human Urm1 or ATPBD3 cause a complete loss or strong reduction in thiolation of tRNA$^{Lys(UUU)}$, tRNA$^{Glu(UUC)}$, and tRNA$^{Gln(UUG)}$ (229–231, 236, 238–240). Loss of the cytosolic thiouridylase Ncs2, which forms a complex with Ncs6, also abrogates tRNA thio-modification (229, 230, 236, 240). Analysis of *urm1*Δ, *uba4*Δ, *ncs6*Δ, and *ncs2*Δ yeast strains revealed their resistance to treatment with the *Kluyveromyces lactis* γ-toxin, which cleaves mcm^{5}s^2-modified, but not unmodified, tRNAs (241). Thus, thiocarboxylated Urm1 functions as sulfur donor in a conserved eukaryotic pathway that mediates thio-modification of certain cytsolic tRNA species.

Modification of U_{34} improves base stacking at the codon-anticodon interface of the ribosome and promotes base pairing to purines but not pyrimidines (237). Few studies have addressed the net effect of the mcm^{5}s^2U_{34} modification on translational fidelity and efficiency. Translation of a reporter construct was reduced in *S. pombe* or *C. elegans* strains with a deficiency in the mcm^5 or s^2 pathway, but consequences for translation of endogenous proteins remain undetermined (240, 242). The tRNA thiolation pathway is not essential, although lack of any of its components affects viability during stress conditions such as starvation, oxidative stress, and temperature shifts (226, 243, 244). Likewise, deletion of components of the mcm^5 modification pathway restricts growth but does not affect viability (245, 246). However, simultaneous inactivation of both the mcm^5 and the s^2 pathway causes lethality in both

Wobble position: the first position of an anticodon or the third position of a codon that may base pair with a noncognate nucleotide

S. cerevisiae and in *C. elegans* (239, 242). Finally, in *C. elegans*, deletion of the worm counterpart of Ncs6 (Tuc1) results in failed embryogenesis at high temperatures (242).

Urm1 in Protein Modification

Does Urm1 participate in protein modification in addition to its function in tRNA thiolation? For a time, the peroxiredoxin Ahp1 was the only known substrate identified, and the chemistry of its conjugation to Urm1 remained unexplored (243). Researchers then discovered that an oxidative environment significantly enhances Urm1 protein modification, or urmylation, in both *S. cerevisiae* and *H. sapiens* (247). Urmylation involves its C-terminal thiocarboxylate but resembles ubiquitylation in the formation of a covalent, lysine-linked adduct (**Supplemental Figure 1**) (247). Urmylation requires the activity of MOCS3, but it is unknown whether MOCS3 is directly involved in adduct formation or simply in thiocarboxylation of the C terminus of Urm1 (247). Urm1 is targeted to a specific, but limited, set of substrates upon oxidant exposure (247). These include components of the Urm1 pathway as well as the nucleocytoplasmic shuttling factor CAS (cellular apoptosis susceptibility protein, also known as CSE1L) (247). However, the function of an Urm1-CAS adduct is presently unknown.

The dual role of thiocarboxylated Urm1 in tRNA modification and protein urmylation suggests that Urm1 is an evolutionary intermediate that integrates the functions of prokaryotic sulfur carriers with those of the eukaryotic protein modifiers. The combined chemistry of both the MoaD/ThiS and the Ub/UBL pathway allows Urm1 to exhibit two, seemingly unrelated, roles and sheds light on the evolution of the eukaryotic Ub/UBL protein-modification pathway.

Are protein modifiers unique to eukaryotes or present in other kingdoms of life as well? The prokaryotic Ub-like protein (Pup) in Actinobacteria and *Nitrospira* is covalently attached to target proteins, thereby targeting them to prokaryotic proteasomes (248). Pup does not adopt a β-grasp fold, and it is conjugated to substrates via a C-terminal glutamic acid (248). Two small archaeal modifier proteins (SAMPs) in *Haloferax volcanii* and *Methanosarcina acetivorans* have much in common with eukaryotic UBLs (249, 250). Archaeal SAMPs adopt a β-grasp fold, terminate in a diglycine motif, and are conjugated by the E1-like enzyme UbaA to lysine residues in substrates following nitrogen limitation (249–251). Sampylated substrates include several components of the Urm1 pathway as well as proteins involved in MPT synthesis (249). The stress-induced sampylation of members of the Urm1 pathway, together with conjugation in the absence of an obvious E2 or E3 enzyme, resembles many aspects of urmylation, thus suggesting that SAMPs may function similarly to Urm1. Indeed, SAMP1 and SAMP2 are involved in MPT biosynthesis and tRNA thiolation, respectively, further suggesting that SAMPs are thiocarboxylated in a manner analogous to Urm1 (251). A Ub-like modification system consisting of Ub, E1, E2, and a small RING-finger family protein with structural motifs specific to their eukaryotic counterparts occurs in the archaeon *Caldiarchaeum subterraneum* (252). This eukaryotic-type modification system is distinct from the SAMP pathway but similarly suggests that protein modification is an ancient process not restricted to eukaryotes.

HUB1 IN RNA SPLICING

Similar to Atg8 and Urm1, Hub1 (homologous to Ub, also termed Ubl5 or beacon in mammals) differs from canonical modifiers. Hub1 lacks a terminal glycine and instead harbors a conserved dityrosine motif followed by a non-conserved residue. Although a Hub1 variant with an exposed YY motif is conjugated to the polarity factors Sph1 and Hbt1 in *S. cerevisiae*, no evidence was found for C-terminal processing of Hub1, and SDS-resistant Hub1-protein complexes were formed independent of ATP (253–255). Therefore, endogenous Hub1 is not likely to function as a covalent protein modifier. Instead, Hub1 interacts noncovalently with the spliceosome component Snu66, and loss of

Hub1 in *S. pombe* results in premRNA splicing defects and lethality (254, 256). The Hub1-Snu66 interaction is crucial for the recognition of certain noncanonical 5′ splice sites by the spliceosome, and loss of this interaction results in defective splicing of a subset of targets in yeast (4). How Hub1 promotes the recognition of noncanonical splice sites is unclear. Although alternative splicing is uncommon in *S. cerevisiae*, it frequently occurs in *S. pombe* and metazoans. In addition, 5′ splice sites are much more variable in these organisms, explaining why Hub1 is an essential gene in *S. pombe* but not in *S. cerevisiae* (4). The Hub1-Snu66 interaction depends on an 18–19-amino acid long helical Hub1 interaction domain (HIND) in Snu66 (4). Although loss of Snu66 or Hub1 is lethal in *S. pombe*, mutant strains in which the Snu66-Hub1 interaction is abrogated are viable, suggesting that Hub1 may have additional functions (4). Perhaps HIND(-like) domains are found in other proteins, much like the existence of a variety of SIMs and LIRs in SUMO- and Atg8-interacting proteins, respectively.

SUMMARY POINTS

1. SUMO-interaction motifs mediate noncovalent interactions with SUMO1 and SUMO2/3 and may be major determinants of paralog specificity.

2. The Nedd8 pathway regulates CRL-mediated ubiquitylation and degradation of several cell-cycle regulators, and pharmacological inhibition of neddylation holds promise for the treatment of cancer.

3. The antiviral function of ISG15 involves modification of substrates of both viral and host origin, possibly in a cotranslational manner.

4. The six mammalian Atg8 paralogs not only are required for autophagosomal membrane biogenesis, but also mediate selective autophagy via the recruitment of LIR-containing autophagy receptors that recognize select cargo.

5. The dual role of thiocarboxylated Urm1 in tRNA thiolation and protein modification underscores the notion that Urm1 is an evolutionary intermediate that integrates the functions of prokaryotic sulfur carriers with those of the eukaryotic protein modifiers.

6. The hitherto poorly characterized UBLs Ufm1 and Hub1 have recently been ascribed a function in erythroid and megakaryocyte development and noncanonical splicing, respectively.

FUTURE ISSUES

1. What other determinants specify SUMO paralog specificity? Besides phosphorylation, what additional posttranslational modifications enhance SIM-mediated recognition of SUMO1 or SUMO2/3?

2. What is the function of neddylation of non-cullin substrates? How does MLN4924-mediated inhibition of the Nedd8 pathway affect these functions?

3. What is the precise mechanism of the antiviral activity of ISG15? How does ISGylation affect the function of host proteins? What other viral proteins are targeted by ISG15, and how does this affect virion assembly and infectivity?

4. What is the function of FAT10 and MNSFβ? Are (additional) enzymes involved in their conjugation?
5. What is the function of protein urmylation? Does Urm1 modify other types of RNA species? How does tRNA thiolation affect translation?
6. How do Atg8 paralogs differ in function? Do LIRs differentially recognize Atg8 isoforms? What additional autophagy receptors interact with members of the Atg8 family?
7. Do pathogens hijack or modulate all UBL pathways? How do they affect the function of Urm1, FAT10, and Hub1?

DISCLOSURE STATEMENT

The authors are not aware of any affiliations, memberships, funding, or financial holding that might be perceived as affecting the objectivity of this review.

ACKNOWLEDGMENTS

We apologize to all colleagues whose papers could not be cited owing to space limitations. We thank Christian Schlieker and Ludovico Buti for critical review of this manuscript and Tom DiCesare for assistance with the artwork. A.G.V. is now a member of the Immunobiology Laboratory, Cancer Research UK London Research Institute.

LITERATURE CITED

1. Hochstrasser M. 2009. Origin and function of ubiquitin-like proteins. *Nature* 458:422–29
2. Komander D. 2009. The emerging complexity of protein ubiquitination. *Biochem. Soc. Trans.* 37:937–53
3. Schulman BA, Harper JW. 2009. Ubiquitin-like protein activation by E1 enzymes: the apex for downstream signalling pathways. *Nat. Rev. Mol. Cell Biol.* 10:319–31
4. Mishra SK, Ammon T, Popowicz GM, Krajewski M, Nagel RJ, et al. 2011. Role of the ubiquitin-like protein Hub1 in splice-site usage and alternative splicing. *Nature* 474:173–78
5. Tatsumi K, Yamamoto-Mukai H, Shimizu R, Waguri S, Sou YS, et al. 2011. The Ufm1-activating enzyme Uba5 is indispensable for erythroid differentiation in mice. *Nat. Commun.* 2:181
6. Jeon YJ, Yoo HM, Chung CH. 2010. ISG15 and immune diseases. *Biochim. Biophys. Acta* 1802:485–96
7. Gareau JR, Lima CD. 2010. The SUMO pathway: emerging mechanisms that shape specificity, conjugation and recognition. *Nat. Rev. Mol. Cell Biol.* 11:861–71
8. Wilkinson KA, Henley JM. 2010. Mechanisms, regulation and consequences of protein SUMOylation. *Biochem. J.* 428:133–45
9. Chen Y, Klionsky DJ. 2011. The regulation of autophagy: unanswered questions. *J. Cell Sci.* 124:161–70
10. Matunis MJ, Coutavas E, Blobel G. 1996. A novel ubiquitin-like modification modulates the partitioning of the Ran-GTPase-activating protein RanGAP1 between the cytosol and the nuclear pore complex. *J. Cell Biol.* 135:1457–70
11. Mahajan R, Delphin C, Guan T, Gerace L, Melchior F. 1997. A small ubiquitin-related polypeptide involved in targeting RanGAP1 to nuclear pore complex protein RanBP2. *Cell* 88:97–107
12. Tatham MH, Jaffray E, Vaughan OA, Desterro JM, Botting CH, et al. 2001. Polymeric chains of SUMO-2 and SUMO-3 are conjugated to protein substrates by SAE1/SAE2 and Ubc9. *J. Biol. Chem.* 276:35368–74
13. Zhang FP, Mikkonen L, Toppari J, Palvimo JJ, Thesleff I, Janne OA. 2008. Sumo-1 function is dispensable in normal mouse development. *Mol. Cell. Biol.* 28:5381–90

14. Evdokimov E, Sharma P, Lockett SJ, Lualdi M, Kuehn MR. 2008. Loss of SUMO1 in mice affects RanGAP1 localization and formation of PML nuclear bodies, but is not lethal as it can be compensated by SUMO2 or SUMO3. *J. Cell Sci.* 121:4106–13
15. Yuan H, Zhou J, Deng M, Liu X, Le Bras M, et al. 2010. Small ubiquitin-related modifier paralogs are indispensable but functionally redundant during early development of zebrafish. *Cell Res.* 20:185–96
16. Ayaydin F, Dasso M. 2004. Distinct in vivo dynamics of vertebrate SUMO paralogues. *Mol. Biol. Cell* 15:5208–18
17. Vertegaal AC, Andersen JS, Ogg SC, Hay RT, Mann M, Lamond AI. 2006. Distinct and overlapping sets of SUMO-1 and SUMO-2 target proteins revealed by quantitative proteomics. *Mol. Cell Proteomics* 5:2298–310
18. Rosas-Acosta G, Russell WK, Deyrieux A, Russell DH, Wilson VG. 2005. A universal strategy for proteomic studies of SUMO and other ubiquitin-like modifiers. *Mol. Cell Proteomics* 4:56–72
19. Manza LL, Codreanu SG, Stamer SL, Smith DL, Wells KS, et al. 2004. Global shifts in protein sumoylation in response to electrophile and oxidative stress. *Chem. Res. Toxicol.* 17:1706–15
20. Zhang XD, Goeres J, Zhang H, Yen TJ, Porter AC, Matunis MJ. 2008. SUMO-2/3 modification and binding regulate the association of CENP-E with kinetochores and progression through mitosis. *Mol. Cell* 29:729–41
21. Mabb AM, Wuerzberger-Davis SM, Miyamoto S. 2006. PIASy mediates NEMO sumoylation and NF-kappaB activation in response to genotoxic stress. *Nat. Cell Biol.* 8:986–93
22. Ghisletti S, Huang W, Ogawa S, Pascual G, Lin ME, et al. 2007. Parallel SUMOylation-dependent pathways mediate gene- and signal-specific transrepression by LXRs and PPARgamma. *Mol. Cell* 25:57–70
23. Shimshon L, Michaeli A, Hadar R, Nutt SL, David Y, et al. 2011. SUMOylation of Blimp-1 promotes its proteasomal degradation. *FEBS Lett.* 585:2405–9
24. Saitoh H, Hinchey J. 2000. Functional heterogeneity of small ubiquitin-related protein modifiers SUMO-1 versus SUMO-2/3. *J. Biol. Chem.* 275:6252–58
25. Yang M, Hsu CT, Ting CY, Liu LF, Hwang J. 2006. Assembly of a polymeric chain of SUMO1 on human topoisomerase I in vitro. *J. Biol. Chem.* 281:8264–74
26. Pichler A, Gast A, Seeler JS, Dejean A, Melchior F. 2002. The nucleoporin RanBP2 has SUMO1 E3 ligase activity. *Cell* 108:109–20
27. Matic I, van Hagen M, Schimmel J, Macek B, Ogg SC, et al. 2008. In vivo identification of human small ubiquitin-like modifier polymerization sites by high accuracy mass spectrometry and an in vitro to in vivo strategy. *Mol. Cell Proteomics* 7:132–44
28. Pedrioli PG, Raught B, Zhang XD, Rogers R, Aitchison J, et al. 2006. Automated identification of SUMOylation sites using mass spectrometry and SUMmOn pattern recognition software. *Nat. Methods* 3:533–39
29. Bylebyl GR, Belichenko I, Johnson ES. 2003. The SUMO isopeptidase Ulp2 prevents accumulation of SUMO chains in yeast. *J. Biol. Chem.* 278:44113–20
30. Song J, Durrin LK, Wilkinson TA, Krontiris TG, Chen Y. 2004. Identification of a SUMO-binding motif that recognizes SUMO-modified proteins. *Proc. Natl. Acad. Sci. USA* 101:14373–78
31. Hecker CM, Rabiller M, Haglund K, Bayer P, Dikic I. 2006. Specification of SUMO1- and SUMO2-interacting motifs. *J. Biol. Chem.* 281:16117–27
32. Song J, Zhang Z, Hu W, Chen Y. 2005. Small ubiquitin-like modifier (SUMO) recognition of a SUMO binding motif: a reversal of the bound orientation. *J. Biol. Chem.* 280:40122–29
33. Kerscher O. 2007. SUMO junction-what's your function? New insights through SUMO-interacting motifs. *EMBO Rep.* 8:550–55
34. Meulmeester E, Kunze M, Hsiao HH, Urlaub H, Melchior F. 2008. Mechanism and consequences for paralog-specific sumoylation of ubiquitin-specific protease 25. *Mol. Cell* 30:610–19
35. Zhu J, Zhu S, Guzzo CM, Ellis NA, Sung KS, et al. 2008. Small ubiquitin-related modifier (SUMO) binding determines substrate recognition and paralog-selective SUMO modification. *J. Biol. Chem.* 283:29405–15
36. Chang CC, Naik MT, Huang YS, Jeng JC, Liao PH, et al. 2011. Structural and functional roles of Daxx SIM phosphorylation in SUMO paralog-selective binding and apoptosis modulation. *Mol. Cell* 42:62–74

37. Stehmeier P, Muller S. 2009. Phospho-regulated SUMO interaction modules connect the SUMO system to CK2 signaling. *Mol. Cell* 33:400–9
38. Ouyang J, Shi Y, Valin A, Xuan Y, Gill G. 2009. Direct binding of CoREST1 to SUMO-2/3 contributes to gene-specific repression by the LSD1/CoREST1/HDAC complex. *Mol. Cell* 34:145–54
39. Denuc A, Marfany G. 2010. SUMO and ubiquitin paths converge. *Biochem. Soc. Trans.* 38:34–9
40. Tatham MH, Geoffroy MC, Shen L, Plechanovova A, Hattersley N, et al. 2008. RNF4 is a poly-SUMO-specific E3 ubiquitin ligase required for arsenic-induced PML degradation. *Nat. Cell Biol.* 10:538–46
41. Martin N, Schwamborn K, Schreiber V, Werner A, Guillier C, et al. 2009. PARP-1 transcriptional activity is regulated by sumoylation upon heat shock. *EMBO J.* 28:3534–48
42. Mullen JR, Brill SJ. 2008. Activation of the Slx5-Slx8 ubiquitin ligase by poly-small ubiquitin-like modifier conjugates. *J. Biol. Chem.* 283:19912–21
43. Fryrear KA, Guo X, Kerscher O, Semmes OJ. 2011. The SUMO-targeted ubiquitin ligase RNF4 regulates the localization and function of the HTLV-1 oncoprotein Tax. *Blood* 119:1173–81
44. Mullen JR, Chen CF, Brill SJ. 2010. Wss1 is a SUMO-dependent isopeptidase that interacts genetically with the Slx5-Slx8 SUMO-targeted ubiquitin ligase. *Mol. Cell. Biol.* 30:3737–48
45. Gilbreth RN, Truong K, Madu I, Koide A, Wojcik JB, et al. 2011. Isoform-specific monobody inhibitors of small ubiquitin-related modifiers engineered using structure-guided library design. *Proc. Natl. Acad. Sci. USA* 108:7751–56
46. Reverter D, Lima CD. 2005. Insights into E3 ligase activity revealed by a SUMO-RanGAP1-Ubc9-Nup358 complex. *Nature* 435:687–92
47. Zhu S, Goeres J, Sixt KM, Bekes M, Zhang XD, et al. 2009. Protection from isopeptidase-mediated deconjugation regulates paralog-selective sumoylation of RanGAP1. *Mol. Cell* 33:570–80
48. Pichler A, Knipscheer P, Saitoh H, Sixma TK, Melchior F. 2004. The RanBP2 SUMO E3 ligase is neither HECT- nor RING-type. *Nat. Struct. Mol. Biol.* 11:984–91
49. Tatham MH, Kim S, Jaffray E, Song J, Chen Y, Hay RT. 2005. Unique binding interactions among Ubc9, SUMO and RanBP2 reveal a mechanism for SUMO paralog selection. *Nat. Struct. Mol. Biol.* 12:67–74
50. Castano-Miquel L, Segui J, Lois LM. 2011. Distinctive properties of *Arabidopsis* SUMO paralogs support the in vivo predominant role of AtSUMO1/2 isoforms. *Biochem. J.* 436:581–90
51. Golebiowski F, Matic I, Tatham MH, Cole C, Yin Y, et al. 2009. System-wide changes to SUMO modifications in response to heat shock. *Sci. Signal.* 2:ra24
52. Carbia-Nagashima A, Gerez J, Perez-Castro C, Paez-Pereda M, Silberstein S, et al. 2007. RSUME, a small RWD-containing protein, enhances SUMO conjugation and stabilizes HIF-1alpha during hypoxia. *Cell* 131:309–23
53. Kang X, Li J, Zou Y, Yi J, Zhang H, et al. 2010. PIASy stimulates HIF1alpha SUMOylation and negatively regulates HIF1alpha activity in response to hypoxia. *Oncogene* 29:5568–78
54. Zhou W, Ryan JJ, Zhou H. 2004. Global analyses of sumoylated proteins in *Saccharomyces cerevisiae*. Induction of protein sumoylation by cellular stresses. *J. Biol. Chem.* 279:32262–68
55. Bossis G, Melchior F. 2006. Regulation of SUMOylation by reversible oxidation of SUMO conjugating enzymes. *Mol. Cell* 21:349–57
56. Han Y, Huang C, Sun X, Xiang B, Wang M, et al. 2010. SENP3-mediated de-conjugation of SUMO2/3 from promyelocytic leukemia is correlated with accelerated cell proliferation under mild oxidative stress. *J. Biol. Chem.* 285:12906–15
57. Huang C, Han Y, Wang Y, Sun X, Yan S, et al. 2009. SENP3 is responsible for HIF-1 transactivation under mild oxidative stress via p300 de-SUMOylation. *EMBO J.* 28:2748–62
58. Wei W, Yang P, Pang J, Zhang S, Wang Y, et al. 2008. A stress-dependent SUMO4 sumoylation of its substrate proteins. *Biochem. Biophys. Res. Commun.* 375:454–59
59. Wang CY, Yang P, Li M, Gong F. 2009. Characterization of a negative feedback network between SUMO4 expression and NFkappaB transcriptional activity. *Biochem. Biophys. Res. Commun.* 381:477–81
60. Ribet D, Cossart P. 2010. Pathogen-mediated posttranslational modifications: a re-emerging field. *Cell* 143:694–702
61. Isaacson MK, Ploegh HL. 2009. Ubiquitination, ubiquitin-like modifiers, and deubiquitination in viral infection. *Cell Host Microbe* 5:559–70

62. Chang PC, Izumiya Y, Wu CY, Fitzgerald LD, Campbell M, et al. 2010. Kaposi's sarcoma-associated herpesvirus (KSHV) encodes a SUMO E3 ligase that is SIM-dependent and SUMO-2/3-specific. *J. Biol. Chem.* 285:5266–73
63. Boutell C, Cuchet-Lourenco D, Vanni E, Orr A, Glass M, et al. 2011. A viral ubiquitin ligase has substrate preferential SUMO targeted ubiquitin ligase activity that counteracts intrinsic antiviral defence. *PLoS Pathog.* 7:e1002245
64. Marcos-Villar L, Campagna M, Lopitz-Otsoa F, Gallego P, Gonzalez-Santamaria J, et al. 2011. Covalent modification by SUMO is required for efficient disruption of PML oncogenic domains by Kaposi's sarcoma-associated herpesvirus latent protein LANA2. *J. Gen. Virol.* 92:188–94
65. Kim ET, Kim YE, Huh YH, Ahn JH. 2010. Role of noncovalent SUMO binding by the human cytomegalovirus IE2 transactivator in lytic growth. *J. Virol.* 84:8111–23
66. Gonzalez-Santamaria J, Campagna M, Garcia MA, Marcos-Villar L, Gonzalez D, et al. 2011. Regulation of vaccinia virus e3 protein by small ubiquitin-like modifier proteins. *J. Virol.* 85:12890–900
67. Wang L, Oliver SL, Sommer M, Rajamani J, Reichelt M, Arvin AM. 2011. Disruption of PML nuclear bodies is mediated by ORF61 SUMO-interacting motifs and required for varicella-zoster virus pathogenesis in skin. *PLoS Pathog.* 7:e1002157
68. Rabut G, Peter M. 2008. Function and regulation of protein neddylation. "Protein modifications: beyond the usual suspects" review series. *EMBO Rep.* 9:969–76
69. Watson IR, Irwin MS, Ohh M. 2011. NEDD8 pathways in cancer, sine quibus non. *Cancer Cell* 19:168–76
69a. Souphron J, Waddell MB, Paydar A, Tokgoz-Gromley Z, Roussel MF, Schulman BA. 2008. Structural dissection of a gating mechanism preventing misactivation of ubiquitin by NEDD8's E1. *Biochemistry* 47:8961–69
70. Huang DT, Zhuang M, Ayrault O, Schulman BA. 2008. Identification of conjugation specificity determinants unmasks vestigial preference for ubiquitin within the NEDD8 E2. *Nat. Struct. Mol. Biol.* 15:280–87
71. Hjerpe R, Thomas Y, Chen J, Zemla A, Curran S, et al. 2011. Changes in the ratio of free NEDD8 to ubiquitin triggers NEDDylation by ubiquitin enzymes. *Biochem. J.* 441(Part 3):927–36
72. Jeram SM, Srikumar T, Zhang XD, Anne Eisenhauer H, Rogers R, et al. 2010. An improved SUMmOn-based methodology for the identification of ubiquitin and ubiquitin-like protein conjugation sites identifies novel ubiquitin-like protein chain linkages. *Proteomics* 10:254–65
73. Jones J, Wu K, Yang Y, Guerrero C, Nillegoda N, et al. 2008. A targeted proteomic analysis of the ubiquitin-like modifier Nedd8 and associated proteins. *J. Proteome Res.* 7:1274–87
74. Deshaies RJ, Emberley ED, Saha A. 2010. Control of cullin-ring ubiquitin ligase activity by Nedd8. *Subcell. Biochem.* 54:41–56
75. Broemer M, Tenev T, Rigbolt KT, Hempel S, Blagoev B, et al. 2010. Systematic in vivo RNAi analysis identifies IAPs as NEDD8-E3 ligases. *Mol. Cell* 40:810–22
76. Chan Y, Yoon J, Wu JT, Kim HJ, Pan KT, et al. 2008. DEN1 deneddylates non-cullin proteins in vivo. *J. Cell Sci.* 121:3218–23
77. Edelmann MJ, Iphofer A, Akutsu M, Altun M, di Gleria K, et al. 2009. Structural basis and specificity of human otubain 1-mediated deubiquitination. *Biochem. J.* 418:379–90
78. Duda DM, Borg LA, Scott DC, Hunt HW, Hammel M, Schulman BA. 2008. Structural insights into NEDD8 activation of cullin-RING ligases: conformational control of conjugation. *Cell* 134:995–1006
79. Saha A, Deshaies RJ. 2008. Multimodal activation of the ubiquitin ligase SCF by Nedd8 conjugation. *Mol. Cell* 32:21–31
80. Yamoah K, Oashi T, Sarikas A, Gazdoiu S, Osman R, Pan ZQ. 2008. Autoinhibitory regulation of SCF-mediated ubiquitination by human cullin 1's C-terminal tail. *Proc. Natl. Acad. Sci. USA* 105:12230–35
81. Merlet J, Burger J, Gomes JE, Pintard L. 2009. Regulation of cullin-RING E3 ubiquitin-ligases by neddylation and dimerization. *Cell Mol. Life Sci.* 66:1924–38
82. Wu JT, Lin HC, Hu YC, Chien CT. 2005. Neddylation and deneddylation regulate Cul1 and Cul3 protein accumulation. *Nat. Cell Biol.* 7:1014–20
83. He Q, Cheng P, Liu Y. 2005. The COP9 signalosome regulates the *Neurospora* circadian clock by controlling the stability of the SCFFWD-1 complex. *Genes Dev.* 19:1518–31

84. Schmidt MW, McQuary PR, Wee S, Hofmann K, Wolf DA. 2009. F-box-directed CRL complex assembly and regulation by the CSN and CAND1. *Mol. Cell* 35:586–97
85. Wee S, Geyer RK, Toda T, Wolf DA. 2005. CSN facilitates cullin-RING ubiquitin ligase function by counteracting autocatalytic adapter instability. *Nat. Cell Biol.* 7:387–91
86. Zhou C, Wee S, Rhee E, Naumann M, Dubiel W, Wolf DA. 2003. Fission yeast COP9/signalosome suppresses cullin activity through recruitment of the deubiquitylating enzyme Ubp12p. *Mol. Cell* 11:927–38
87. Liu J, Furukawa M, Matsumoto T, Xiong Y. 2002. NEDD8 modification of CUL1 dissociates p120(CAND1), an inhibitor of CUL1-SKP1 binding and SCF ligases. *Mol. Cell* 10:1511–18
88. Zheng J, Yang X, Harrell JM, Ryzhikov S, Shim EH, et al. 2002. CAND1 binds to unneddylated CUL1 and regulates the formation of SCF ubiquitin E3 ligase complex. *Mol. Cell* 10:1519–26
89. Huang DT, Ayrault O, Hunt HW, Taherbhoy AM, Duda DM, et al. 2009. E2-RING expansion of the NEDD8 cascade confers specificity to cullin modification. *Mol. Cell* 33:483–95
90. Kamura T, Conrad MN, Yan Q, Conaway RC, Conaway JW. 1999. The Rbx1 subunit of SCF and VHL E3 ubiquitin ligase activates Rub1 modification of cullins Cdc53 and Cul2. *Genes Dev.* 13:2928–33
91. Calabrese MF, Scott DC, Duda DM, Grace CR, Kurinov I, et al. 2011. A RING E3-substrate complex poised for ubiquitin-like protein transfer: structural insights into cullin-RING ligases. *Nat. Struct. Mol. Biol.* 18:947–49
92. Kurz T, Chou YC, Willems AR, Meyer-Schaller N, Hecht ML, et al. 2008. Dcn1 functions as a scaffold-type E3 ligase for cullin neddylation. *Mol. Cell* 29:23–35
93. Kurz T, Ozlu N, Rudolf F, O'Rourke SM, Luke B, et al. 2005. The conserved protein DCN-1/Dcn1p is required for cullin neddylation in *C. elegans* and *S. cerevisiae*. *Nature* 435:1257–61
94. Scott DC, Monda JK, Grace CR, Duda DM, Kriwacki RW, et al. 2010. A dual E3 mechanism for Rub1 ligation to Cdc53. *Mol. Cell* 39:784–96
95. Scott DC, Monda JK, Bennett EJ, Harper JW, Schulman BA. 2011. N-terminal acetylation acts as an avidity enhancer within an interconnected multiprotein complex. *Science* 334:674–78
96. Meyer-Schaller N, Chou YC, Sumara I, Martin DD, Kurz T, et al. 2009. The human Dcn1-like protein DCNL3 promotes Cul3 neddylation at membranes. *Proc. Natl. Acad. Sci. USA* 106:12365–70
97. Huang G, Kaufman AJ, Ramanathan Y, Singh B. 2011. SCCRO (DCUN1D1) promotes nuclear translocation and assembly of the neddylation E3 complex. *J. Biol. Chem.* 286:10297–304
98. Rabut G, Le Dez G, Verma R, Makhnevych T, Knebel A, et al. 2011. The TFIIH subunit Tfb3 regulates cullin neddylation. *Mol. Cell* 43:488–95
99. Kamitani T, Kito K, Fukuda-Kamitani T, Yeh ET. 2001. Targeting of NEDD8 and its conjugates for proteasomal degradation by NUB1. *J. Biol. Chem.* 276:46655–60
100. Liu G, Xirodimas DP. 2010. NUB1 promotes cytoplasmic localization of p53 through cooperation of the NEDD8 and ubiquitin pathways. *Oncogene* 29:2252–61
101. Dohmesen C, Koeppel M, Dobbelstein M. 2008. Specific inhibition of Mdm2-mediated neddylation by Tip60. *Cell Cycle* 7:222–31
102. Flomerfelt FA, El Kassar N, Gurunathan C, Chua KS, League SC, et al. 2010. Tbata modulates thymic stromal cell proliferation and thymus function. *J. Exp. Med.* 207:2521–32
103. Gastaldello S, Hildebrand S, Faridani O, Callegari S, Palmkvist M, et al. 2010. A deneddylase encoded by Epstein-Barr virus promotes viral DNA replication by regulating the activity of cullin-RING ligases. *Nat. Cell Biol.* 12:351–61
104. Schlieker C, Korbel GA, Kattenhorn LM, Ploegh HL. 2005. A deubiquitinating activity is conserved in the large tegument protein of the *Herpesviridae*. *J. Virol.* 79:15582–5
105. Misaghi S, Balsara ZR, Catic A, Spooner E, Ploegh HL, Starnbach MN. 2006. *Chlamydia trachomatis*-derived deubiquitinating enzymes in mammalian cells during infection. *Mol. Microbiol.* 61:142–50
106. Artavanis-Tsakonas K, Weihofen WA, Antos JM, Coleman BI, Comeaux CA, et al. 2010. Characterization and structural studies of the *Plasmodium falciparum* ubiquitin and Nedd8 hydrolase UCHL3. *J. Biol. Chem.* 285:6857–66
107. Frickel EM, Quesada V, Muething L, Gubbels MJ, Spooner E, et al. 2007. Apicomplexan UCHL3 retains dual specificity for ubiquitin and Nedd8 throughout evolution. *Cell Microbiol.* 9:1601–10

108. Cui J, Yao Q, Li S, Ding X, Lu Q, et al. 2010. Glutamine deamidation and dysfunction of ubiquitin/NEDD8 induced by a bacterial effector family. *Science* 329:1215–18
109. Jubelin G, Taieb F, Duda DM, Hsu Y, Samba-Louaka A, et al. 2010. Pathogenic bacteria target NEDD8-conjugated cullins to hijack host-cell signaling pathways. *PLoS Pathog.* 6:e1001128
110. Kumar A, Wu H, Collier-Hyams LS, Hansen JM, Li T, et al. 2007. Commensal bacteria modulate cullin-dependent signaling via generation of reactive oxygen species. *EMBO J.* 26:4457–66
111. Soucy TA, Dick LR, Smith PG, Milhollen MA, Brownell JE. 2010. The NEDD8 conjugation pathway and its relevance in cancer biology and therapy. *Genes Cancer* 1:708–16
112. Soucy TA, Smith PG, Milhollen MA, Berger AJ, Gavin JM, et al. 2009. An inhibitor of NEDD8-activating enzyme as a new approach to treat cancer. *Nature* 458:732–36
113. Brownell JE, Sintchak MD, Gavin JM, Liao H, Bruzzese FJ, et al. 2010. Substrate-assisted inhibition of ubiquitin-like protein-activating enzymes: the NEDD8 E1 inhibitor MLN4924 forms a NEDD8-AMP mimetic in situ. *Mol. Cell* 37:102–11
114. Okumura F, Zou W, Zhang DE. 2007. ISG15 modification of the eIF4E cognate 4EHP enhances cap structure-binding activity of 4EHP. *Genes Dev.* 21:255–60
115. Zou W, Zhang DE. 2006. The interferon-inducible ubiquitin-protein isopeptide ligase (E3) EFP also functions as an ISG15 E3 ligase. *J. Biol. Chem.* 281:3989–94
116. Malakhov MP, Malakhova OA, Kim KI, Ritchie KJ, Zhang DE. 2002. UBP43 (USP18) specifically removes ISG15 from conjugated proteins. *J. Biol. Chem.* 277:9976–81
117. Catic A, Fiebiger E, Korbel GA, Blom D, Galardy PJ, Ploegh HL. 2007. Screen for ISG15-crossreactive deubiquitinases. *PLoS One* 2:e679
118. Ye Y, Akutsu M, Reyes-Turcu F, Enchev RI, Wilkinson KD, Komander D. 2011. Polyubiquitin binding and cross-reactivity in the USP domain deubiquitinase USP21. *EMBO Rep.* 12:350–57
119. Lenschow DJ, Giannakopoulos NV, Gunn LJ, Johnston C, O'Guin AK, et al. 2005. Identification of interferon-stimulated gene 15 as an antiviral molecule during Sindbis virus infection in vivo. *J. Virol.* 79:13974–83
120. Lenschow DJ, Lai C, Frias-Staheli N, Giannakopoulos NV, Lutz A, et al. 2007. IFN-stimulated gene 15 functions as a critical antiviral molecule against influenza, herpes, and Sindbis viruses. *Proc. Natl. Acad. Sci. USA* 104:1371–76
121. Giannakopoulos NV, Arutyunova E, Lai C, Lenschow DJ, Haas AL, Virgin HW. 2009. ISG15 Arg151 and the ISG15-conjugating enzyme UbE1L are important for innate immune control of Sindbis virus. *J. Virol.* 83:1602–10
122. Lai C, Struckhoff JJ, Schneider J, Martinez-Sobrido L, Wolff T, et al. 2009. Mice lacking the ISG15 E1 enzyme UbE1L demonstrate increased susceptibility to both mouse-adapted and non-mouse-adapted influenza B virus infection. *J. Virol.* 83:1147–51
123. Ritchie KJ, Hahn CS, Kim KI, Yan M, Rosario D, et al. 2004. Role of ISG15 protease UBP43 (USP18) in innate immunity to viral infection. *Nat. Med.* 10:1374–78
124. Kim KI, Yan M, Malakhova O, Luo JK, Shen MF, et al. 2006. Ube1L and protein ISGylation are not essential for alpha/beta interferon signaling. *Mol. Cell. Biol.* 26:472–79
125. Osiak A, Utermohlen O, Niendorf S, Horak I, Knobeloch KP. 2005. ISG15, an interferon-stimulated ubiquitin-like protein, is not essential for STAT1 signaling and responses against vesicular stomatitis and lymphocytic choriomeningitis virus. *Mol. Cell. Biol.* 25:6338–45
126. Broering R, Zhang X, Kottilil S, Trippler M, Jiang M, et al. 2010. The interferon stimulated gene 15 functions as a proviral factor for the hepatitis C virus and as a regulator of the IFN response. *Gut* 59:1111–19
127. Kim MJ, Yoo JY. 2010. Inhibition of hepatitis C virus replication by IFN-mediated ISGylation of HCV-NS5A. *J. Immunol.* 185:4311–18
128. Yuan W, Krug RM. 2001. Influenza B virus NS1 protein inhibits conjugation of the interferon (IFN)-induced ubiquitin-like ISG15 protein. *EMBO J.* 20:362–71
129. Chang YG, Yan XZ, Xie YY, Gao XC, Song AX, et al. 2008. Different roles for two ubiquitin-like domains of ISG15 in protein modification. *J. Biol. Chem.* 283:13370–77
130. Sridharan H, Zhao C, Krug RM. 2010. Species specificity of the NS1 protein of influenza B virus: NS1 binds only human and non-human primate ubiquitin-like ISG15 proteins. *J. Biol. Chem.* 285:7852–56

131. Versteeg GA, Hale BG, van Boheemen S, Wolff T, Lenschow DJ, Garcia-Sastre A. 2010. Species-specific antagonism of host ISGylation by the influenza B virus NS1 protein. *J. Virol.* 84:5423–30
132. Guerra S, Caceres A, Knobeloch KP, Horak I, Esteban M. 2008. Vaccinia virus E3 protein prevents the antiviral action of ISG15. *PLoS Pathog.* 4:e1000096
133. Frias-Staheli N, Giannakopoulos NV, Kikkert M, Taylor SL, Bridgen A, et al. 2007. Ovarian tumor domain-containing viral proteases evade ubiquitin- and ISG15-dependent innate immune responses. *Cell Host Microbe* 2:404–16
134. Clementz MA, Chen Z, Banach BS, Wang Y, Sun L, et al. 2010. Deubiquitinating and interferon antagonism activities of coronavirus papain-like proteases. *J. Virol.* 84:4619–29
135. Lindner HA, Lytvyn V, Qi H, Lachance P, Ziomek E, Menard R. 2007. Selectivity in ISG15 and ubiquitin recognition by the SARS coronavirus papain-like protease. *Arch. Biochem. Biophys.* 466:8–14
136. Karpe YA, Lole KS. 2011. Deubiquitination activity associated with hepatitis E virus putative papain-like cysteine protease. *J. Gen. Virol.* 92:2088–92
137. Akutsu M, Ye Y, Virdee S, Chin JW, Komander D. 2011. Molecular basis for ubiquitin and ISG15 cross-reactivity in viral ovarian tumor domains. *Proc. Natl. Acad. Sci. USA* 108:2228–33
138. James TW, Frias-Staheli N, Bacik JP, Levingston Macleod JM, Khajehpour M, et al. 2011. Structural basis for the removal of ubiquitin and interferon-stimulated gene 15 by a viral ovarian tumor domain-containing protease. *Proc. Natl. Acad. Sci. USA* 108:2222–27
139. Zhao C, Denison C, Huibregtse JM, Gygi S, Krug RM. 2005. Human ISG15 conjugation targets both IFN-induced and constitutively expressed proteins functioning in diverse cellular pathways. *Proc. Natl. Acad. Sci. USA* 102:10200–5
140. Jeon YJ, Choi JS, Lee JY, Yu KR, Kim SM, et al. 2009. ISG15 modification of filamin B negatively regulates the type I interferon-induced JNK signalling pathway. *EMBO Rep.* 10:374–80
141. Shi HX, Yang K, Liu X, Liu XY, Wei B, et al. 2010. Positive regulation of interferon regulatory factor 3 activation by Herc5 via ISG15 modification. *Mol. Cell. Biol.* 30:2424–36
142. Zhao C, Hsiang TY, Kuo RL, Krug RM. 2010. ISG15 conjugation system targets the viral NS1 protein in influenza A virus-infected cells. *Proc. Natl. Acad. Sci. USA* 107:2253–58
143. Tang Y, Zhong G, Zhu L, Liu X, Shan Y, et al. 2010. Herc5 attenuates influenza A virus by catalyzing ISGylation of viral NS1 protein. *J. Immunol.* 184:5777–90
144. Malakhova OA, Zhang DE. 2008. ISG15 inhibits Nedd4 ubiquitin E3 activity and enhances the innate antiviral response. *J. Biol. Chem.* 283:8783–87
145. Okumura A, Pitha PM, Harty RN. 2008. ISG15 inhibits Ebola VP40 VLP budding in an L-domain-dependent manner by blocking Nedd4 ligase activity. *Proc. Natl. Acad. Sci. USA* 105:3974–79
146. Okumura A, Lu G, Pitha-Rowe I, Pitha PM. 2006. Innate antiviral response targets HIV-1 release by the induction of ubiquitin-like protein ISG15. *Proc. Natl. Acad. Sci. USA* 103:1440–45
147. Kuang Z, Seo EJ, Leis J. 2011. Mechanism of inhibition of retrovirus release from cells by interferon-induced gene ISG15. *J. Virol.* 85:7153–61
148. Takeuchi T, Yokosawa H. 2005. ISG15 modification of Ubc13 suppresses its ubiquitin-conjugating activity. *Biochem. Biophys. Res. Commun.* 336:9–13
149. Zou W, Papov V, Malakhova O, Kim KI, Dao C, et al. 2005. ISG15 modification of ubiquitin E2 Ubc13 disrupts its ability to form thioester bond with ubiquitin. *Biochem. Biophys. Res. Commun.* 336:61–68
150. Desai SD, Haas AL, Wood LM, Tsai YC, Pestka S, et al. 2006. Elevated expression of ISG15 in tumor cells interferes with the ubiquitin/26S proteasome pathway. *Cancer Res.* 66:921–28
151. Zhang D, Zhang DE. 2011. Interferon-stimulated gene 15 and the protein ISGylation system. *J. Interferon Cytokine Res.* 31:119–30
152. Durfee LA, Lyon N, Seo K, Huibregtse JM. 2010. The ISG15 conjugation system broadly targets newly synthesized proteins: implications for the antiviral function of ISG15. *Mol. Cell* 38:722–32
153. D'Cunha J, Ramanujam S, Wagner RJ, Witt PL, Knight E Jr, Borden EC. 1996. In vitro and in vivo secretion of human ISG15, an IFN-induced immunomodulatory cytokine. *J. Immunol.* 157:4100–8
154. Recht M, Borden EC, Knight E Jr. 1991. A human 15-kDa IFN-induced protein induces the secretion of IFN-gamma. *J. Immunol.* 147:2617–23

155. Padovan E, Terracciano L, Certa U, Jacobs B, Reschner A, et al. 2002. Interferon stimulated gene 15 constitutively produced by melanoma cells induces e-cadherin expression on human dendritic cells. *Cancer Res.* 62:3453–58
156. Owhashi M, Taoka Y, Ishii K, Nakazawa S, Uemura H, Kambara H. 2003. Identification of a ubiquitin family protein as a novel neutrophil chemotactic factor. *Biochem. Biophys. Res. Commun.* 309:533–39
157. Werneke SW, Schilte C, Rohatgi A, Monte KJ, Michault A, et al. 2011. ISG15 is critical in the control of chikungunya virus infection independent of UbE1L-mediated conjugation. *PLoS Pathog.* 7:e1002322
158. Malakhova OA, Kim KI, Luo JK, Zou W, Kumar KG, et al. 2006. UBP43 is a novel regulator of interferon signaling independent of its ISG15 isopeptidase activity. *EMBO J.* 25:2358–67
159. Malakhova OA, Yan M, Malakhov MP, Yuan Y, Ritchie KJ, et al. 2003. Protein ISGylation modulates the JAK-STAT signaling pathway. *Genes Dev.* 17:455–60
160. Ritchie KJ, Malakhov MP, Hetherington CJ, Zhou L, Little MT, et al. 2002. Dysregulation of protein modification by ISG15 results in brain cell injury. *Genes Dev.* 16:2207–12
161. Knobeloch KP, Utermohlen O, Kisser A, Prinz M, Horak I. 2005. Reexamination of the role of ubiquitin-like modifier ISG15 in the phenotype of UBP43-deficient mice. *Mol. Cell. Biol.* 25:11030–34
162. Kim JH, Luo JK, Zhang DE. 2008. The level of hepatitis B virus replication is not affected by protein ISG15 modification but is reduced by inhibition of UBP43 (USP18) expression. *J. Immunol.* 181:6467–72
163. Pelzer C, Groettrup M. 2010. FAT10: activated by UBA6 and functioning in protein degradation. *Subcell. Biochem.* 54:238–46
164. Aichem A, Pelzer C, Lukasiak S, Kalveram B, Sheppard PW, et al. 2010. USE1 is a bispecific conjugating enzyme for ubiquitin and FAT10, which FAT10ylates itself in cis. *Nat. Commun.* 1:13
165. Chiu YH, Sun Q, Chen ZJ. 2007. E1-L2 activates both ubiquitin and FAT10. *Mol. Cell* 27:1014–23
166. Hipp MS, Kalveram B, Raasi S, Groettrup M, Schmidtke G. 2005. FAT10, a ubiquitin-independent signal for proteasomal degradation. *Mol. Cell. Biol.* 25:3483–91
167. Schmidtke G, Kalveram B, Groettrup M. 2009. Degradation of FAT10 by the 26S proteasome is independent of ubiquitylation but relies on NUB1L. *FEBS Lett.* 583:591–94
168. Schmidtke G, Kalveram B, Weber E, Bochtler P, Lukasiak S, et al. 2006. The UBA domains of NUB1L are required for binding but not for accelerated degradation of the ubiquitin-like modifier FAT10. *J. Biol. Chem.* 281:20045–54
169. Buchsbaum S, Bercovich B, Ciechanover A. 2011. FAT10 is a proteasomal degradation signal which is itself regulated by ubiquitination. *Mol. Biol. Cell* 23:225–32
170. Gong P, Canaan A, Wang B, Leventhal J, Snyder A, et al. 2010. The ubiquitin-like protein FAT10 mediates NF-kappaB activation. *J. Am. Soc. Nephrol.* 21:316–26
171. Li T, Santockyte R, Yu S, Shen RF, Tekle E, et al. 2011. FAT10 modifies p53 and upregulates its transcriptional activity. *Arch. Biochem. Biophys.* 509:164–69
172. Ebstein F, Lange N, Urban S, Seifert U, Kruger E, Kloetzel PM. 2009. Maturation of human dendritic cells is accompanied by functional remodelling of the ubiquitin-proteasome system. *Int. J. Biochem. Cell Biol.* 41:1205–15
173. Canaan A, Yu X, Booth CJ, Lian J, Lazar I, et al. 2006. FAT10/diubiquitin-like protein-deficient mice exhibit minimal phenotypic differences. *Mol. Cell. Biol.* 26:5180–89
174. Snyder A, Alsauskas Z, Gong P, Rosenstiel PE, Klotman ME, et al. 2009. FAT10: a novel mediator of Vpr-induced apoptosis in human immunodeficiency virus-associated nephropathy. *J. Virol.* 83:11983–88
175. Jeram SM, Srikumar T, Pedrioli PG, Raught B. 2009. Using mass spectrometry to identify ubiquitin and ubiquitin-like protein conjugation sites. *Proteomics* 9:922–34
176. Nakamura M, Shimosaki S. 2009. The ubiquitin-like protein monoclonal nonspecific suppressor factor beta conjugates to endophilin II and regulates phagocytosis. *FEBS J.* 276:6355–63
177. Nakamura M, Yamaguchi S. 2006. The ubiquitin-like protein MNSFbeta regulates ERK-MAPK cascade. *J. Biol. Chem.* 281:16861–69
178. Kondoh T, Nakamura M, Nabika T, Yoshimura Y, Tanigawa Y. 1999. Ubiquitin-like polypeptide inhibits the proliferative response of T cells in vivo. *Immunobiology* 200:140–49
179. Komatsu M, Chiba T, Tatsumi K, Iemura S, Tanida I, et al. 2004. A novel protein-conjugating system for Ufm1, a ubiquitin-fold modifier. *EMBO J.* 23:1977–86

180. Tatsumi K, Sou YS, Tada N, Nakamura E, Iemura S, et al. 2010. A novel type of E3 ligase for the Ufm1 conjugation system. *J. Biol. Chem.* 285:5417–27
181. Levine B, Mizushima N, Virgin HW. 2011. Autophagy in immunity and inflammation. *Nature* 469:323–35
182. Geng J, Klionsky DJ. 2008. The Atg8 and Atg12 ubiquitin-like conjugation systems in macroautophagy. "Protein modifications: beyond the usual suspects" review series. *EMBO Rep.* 9:859–64
183. Fujita N, Itoh T, Omori H, Fukuda M, Noda T, Yoshimori T. 2008. The Atg16L complex specifies the site of LC3 lipidation for membrane biogenesis in autophagy. *Mol. Biol. Cell* 19:2092–100
184. Hanada T, Noda NN, Satomi Y, Ichimura Y, Fujioka Y, et al. 2007. The Atg12-Atg5 conjugate has a novel E3-like activity for protein lipidation in autophagy. *J. Biol. Chem.* 282:37298–302
185. Radoshevich L, Murrow L, Chen N, Fernandez E, Roy S, et al. 2010. ATG12 conjugation to ATG3 regulates mitochondrial homeostasis and cell death. *Cell* 142:590–600
186. Taherbhoy AM, Tait SW, Kaiser SE, Williams AH, Deng A, et al. 2011. Atg8 transfer from Atg7 to Atg3: a distinctive E1-E2 architecture and mechanism in the autophagy pathway. *Mol. Cell* 44:451–61
187. Noda NN, Satoo K, Fujioka Y, Kumeta H, Ogura K, et al. 2011. Structural basis of Atg8 activation by a homodimeric E1, Atg7. *Mol. Cell* 44:462–75
188. Hong SB, Kim BW, Lee KE, Kim SW, Jeon H, et al. 2011. Insights into noncanonical E1 enzyme activation from the structure of autophagic E1 Atg7 with Atg8. *Nat. Struct. Mol. Biol.* 18:1323–30
189. Nakatogawa H, Ichimura Y, Ohsumi Y. 2007. Atg8, a ubiquitin-like protein required for autophagosome formation, mediates membrane tethering and hemifusion. *Cell* 130:165–78
190. Xie Z, Nair U, Klionsky DJ. 2008. Atg8 controls phagophore expansion during autophagosome formation. *Mol. Biol. Cell* 19:3290–98
191. Nair U, Jotwani A, Geng J, Gammoh N, Richerson D, et al. 2011. SNARE proteins are required for macroautophagy. *Cell* 146:290–302
192. Moreau K, Ravikumar B, Renna M, Puri C, Rubinsztein DC. 2011. Autophagosome precursor maturation requires homotypic fusion. *Cell* 146:303–17
193. Behrends C, Sowa ME, Gygi SP, Harper JW. 2010. Network organization of the human autophagy system. *Nature* 466:68–76
194. Chakrama FZ, Seguin-Py S, Le Grand JN, Fraichard A, Delage-Mourroux R, et al. 2010. GABARAPL1 (GEC1) associates with autophagic vesicles. *Autophagy* 6:495–505
195. Weidberg H, Shvets E, Shpilka T, Shimron F, Shinder V, Elazar Z. 2010. LC3 and GATE-16/GABARAP subfamilies are both essential yet act differently in autophagosome biogenesis. *EMBO J.* 29:1792–802
196. Weidberg H, Shpilka T, Shvets E, Abada A, Shimron F, Elazar Z. 2011. LC3 and GATE-16 N termini mediate membrane fusion processes required for autophagosome biogenesis. *Dev. Cell* 20:444–54
197. Coyle JE, Qamar S, Rajashankar KR, Nikolov DB. 2002. Structure of GABARAP in two conformations: implications for GABA(A) receptor localization and tubulin binding. *Neuron* 33:63–74
198. Ichimura Y, Imamura Y, Emoto K, Umeda M, Noda T, Ohsumi Y. 2004. In vivo and in vitro reconstitution of Atg8 conjugation essential for autophagy. *J. Biol. Chem.* 279:40584–92
199. Itoh T, Kanno E, Uemura T, Waguri S, Fukuda M. 2011. OATL1, a novel autophagosome-resident Rab33B-GAP, regulates autophagosomal maturation. *J. Cell Biol.* 192:839–53
200. Kirkin V, McEwan DG, Novak I, Dikic I. 2009. A role for ubiquitin in selective autophagy. *Mol. Cell* 34:259–69
201. Kirkin V, Lamark T, Sou YS, Bjorkoy G, Nunn JL, et al. 2009. A role for NBR1 in autophagosomal degradation of ubiquitinated substrates. *Mol. Cell* 33:505–16
202. Komatsu M, Waguri S, Koike M, Sou YS, Ueno T, et al. 2007. Homeostatic levels of p62 control cytoplasmic inclusion body formation in autophagy-deficient mice. *Cell* 131:1149–63
203. Novak I, Kirkin V, McEwan DG, Zhang J, Wild P, et al. 2010. Nix is a selective autophagy receptor for mitochondrial clearance. *EMBO Rep.* 11:45–51
204. Youle RJ, Narendra DP. 2011. Mechanisms of mitophagy. *Nat. Rev. Mol. Cell Biol.* 12:9–14
205. Thurston TL, Ryzhakov G, Bloor S, von Muhlinen N, Randow F. 2009. The TBK1 adaptor and autophagy receptor NDP52 restricts the proliferation of ubiquitin-coated bacteria. *Nat. Immunol.* 10:1215–21

206. Wild P, Farhan H, McEwan DG, Wagner S, Rogov VV, et al. 2011. Phosphorylation of the autophagy receptor optineurin restricts *Salmonella* growth. *Science* 333:228–33
207. Zheng YT, Shahnazari S, Brech A, Lamark T, Johansen T, Brumell JH. 2009. The adaptor protein p62/SQSTM1 targets invading bacteria to the autophagy pathway. *J. Immunol.* 183:5909–16
208. Ponpuak M, Davis AS, Roberts EA, Delgado MA, Dinkins C, et al. 2010. Delivery of cytosolic components by autophagic adaptor protein p62 endows autophagosomes with unique antimicrobial properties. *Immunity* 32:329–41
209. Noda NN, Kumeta H, Nakatogawa H, Satoo K, Adachi W, et al. 2008. Structural basis of target recognition by Atg8/LC3 during selective autophagy. *Genes Cells* 13:1211–18
210. Noda NN, Ohsumi Y, Inagaki F. 2010. Atg8-family interacting motif crucial for selective autophagy. *FEBS Lett.* 584:1379–85
211. Shvets E, Abada A, Weidberg H, Elazar Z. 2011. Dissecting the involvement of LC3B and GATE-16 in p62 recruitment into autophagosomes. *Autophagy* 7:683–88
212. Ichimura Y, Kumanomidou T, Sou YS, Mizushima T, Ezaki J, et al. 2008. Structural basis for sorting mechanism of p62 in selective autophagy. *J. Biol. Chem.* 283:22847–57
213. Rozenknop A, Rogov VV, Rogova NY, Lohr F, Guntert P, et al. 2011. Characterization of the interaction of GABARAPL-1 with the LIR motif of NBR1. *J. Mol. Biol.* 410:477–87
214. Hemelaar J, Lelyveld VS, Kessler BM, Ploegh HL. 2003. A single protease, Apg4B, is specific for the autophagy-related ubiquitin-like proteins GATE-16, MAP1-LC3, GABARAP, and Apg8L. *J. Biol. Chem.* 278:51841–50
215. Cabrera S, Marino G, Fernandez AF, Lopez-Otin C. 2010. Autophagy, proteases and the sense of balance. *Autophagy* 6:961–63
216. Betin VM, Lane JD. 2009. Caspase cleavage of Atg4D stimulates GABARAP-L1 processing and triggers mitochondrial targeting and apoptosis. *J. Cell Sci.* 122:2554–66
217. Scherz-Shouval R, Shvets E, Fass E, Shorer H, Gil L, Elazar Z. 2007. Reactive oxygen species are essential for autophagy and specifically regulate the activity of Atg4. *EMBO J.* 26:1749–60
218. Nair U, Cao Y, Xie Z, Klionsky DJ. 2010. Roles of the lipid-binding motifs of Atg18 and Atg21 in the cytoplasm to vacuole targeting pathway and autophagy. *J. Biol. Chem.* 285:11476–88
219. Burroughs AM, Iyer LM, Aravind L. 2011. Functional diversification of the RING finger and other binuclear treble clef domains in prokaryotes and the early evolution of the ubiquitin system. *Mol. Biosyst.* 7:2261–77
220. Iyer LM, Burroughs AM, Aravind L. 2006. The prokaryotic antecedents of the ubiquitin-signaling system and the early evolution of ubiquitin-like beta-grasp domains. *Genome Biol.* 7:R60
221. Rudolph MJ, Wuebbens MM, Rajagopalan KV, Schindelin H. 2001. Crystal structure of molybdopterin synthase and its evolutionary relationship to ubiquitin activation. *Nat. Struct. Biol.* 8:42–46
222. Wang C, Xi J, Begley TP, Nicholson LK. 2001. Solution structure of ThiS and implications for the evolutionary roots of ubiquitin. *Nat. Struct. Biol.* 8:47–51
223. Burroughs AM, Iyer LM, Aravind L. 2009. Natural history of the E1-like superfamily: implication for adenylation, sulfur transfer, and ubiquitin conjugation. *Proteins* 75:895–910
224. Lake MW, Wuebbens MM, Rajagopalan KV, Schindelin H. 2001. Mechanism of ubiquitin activation revealed by the structure of a bacterial MoeB-MoaD complex. *Nature* 414:325–29
225. Xi J, Ge Y, Kinsland C, McLafferty FW, Begley TP. 2001. Biosynthesis of the thiazole moiety of thiamin in *Escherichia coli*: identification of an acyldisulfide-linked protein-protein conjugate that is functionally analogous to the ubiquitin/E1 complex. *Proc. Natl. Acad. Sci. USA* 98:8513–18
226. Furukawa K, Mizushima N, Noda T, Ohsumi Y. 2000. A protein conjugation system in yeast with homology to biosynthetic enzyme reaction of prokaryotes. *J. Biol. Chem.* 275:7462–65
227. Xu J, Zhang J, Wang L, Zhou J, Huang H, et al. 2006. Solution structure of Urm1 and its implications for the origin of protein modifiers. *Proc. Natl. Acad. Sci. USA* 103:11625–30
228. Singh S, Tonelli M, Tyler RC, Bahrami A, Lee MS, Markley JL. 2005. Three-dimensional structure of the AAH26994.1 protein from Mus musculus, a putative eukaryotic Urm1. *Protein Sci.* 14:2095–102
229. Leidel S, Pedrioli PG, Bucher T, Brost R, Costanzo M, et al. 2009. Ubiquitin-related modifier Urm1 acts as a sulphur carrier in thiolation of eukaryotic transfer RNA. *Nature* 458:228–32

230. Noma A, Sakaguchi Y, Suzuki T. 2009. Mechanistic characterization of the sulfur-relay system for eukaryotic 2-thiouridine biogenesis at tRNA wobble positions. *Nucleic Acids Res.* 37:1335–52
231. Schlieker CD, Van der Veen AG, Damon JR, Spooner E, Ploegh HL. 2008. A functional proteomics approach links the ubiquitin-related modifier Urm1 to a tRNA modification pathway. *Proc. Natl. Acad. Sci. USA* 105:18255–60
232. Schmitz J, Chowdhury MM, Hanzelmann P, Nimtz M, Lee EY, et al. 2008. The sulfurtransferase activity of Uba4 presents a link between ubiquitin-like protein conjugation and activation of sulfur carrier proteins. *Biochemistry* 47:6479–89
233. Marelja Z, Stocklein W, Nimtz M, Leimkuhler S. 2008. A novel role for human Nfs1 in the cytoplasm: Nfs1 acts as a sulfur donor for MOCS3, a protein involved in molybdenum cofactor biosynthesis. *J. Biol. Chem.* 283:25178–85
234. Matthies A, Rajagopalan KV, Mendel RR, Leimkuhler S. 2004. Evidence for the physiological role of a rhodanese-like protein for the biosynthesis of the molybdenum cofactor in humans. *Proc. Natl. Acad. Sci. USA* 101:5946–51
235. Yu J, Zhou CZ. 2008. Crystal structure of the dimeric Urm1 from the yeast *Saccharomyces cerevisiae*. *Proteins* 71:1050–55
236. Nakai Y, Nakai M, Hayashi H. 2008. Thio-modification of yeast cytosolic tRNA requires a ubiquitin-related system that resembles bacterial sulfur transfer systems. *J. Biol. Chem.* 283:27469–76
237. Agris PF. 2008. Bringing order to translation: the contributions of transfer RNA anticodon-domain modifications. *EMBO Rep.* 9:629–35
238. Nakai Y, Umeda N, Suzuki T, Nakai M, Hayashi H, et al. 2004. Yeast Nfs1p is involved in thio-modification of both mitochondrial and cytoplasmic tRNAs. *J. Biol. Chem.* 279:12363–68
239. Bjork GR, Huang B, Persson OP, Bystrom AS. 2007. A conserved modified wobble nucleoside (mcm5s2U) in lysyl-tRNA is required for viability in yeast. *RNA* 13:1245–55
240. Dewez M, Bauer F, Dieu M, Raes M, Vandenhaute J, Hermand D. 2008. The conserved wobble uridine tRNA thiolase Ctu1-Ctu2 is required to maintain genome integrity. *Proc. Natl. Acad. Sci. USA* 105:5459–64
241. Huang B, Lu J, Bystrom AS. 2008. A genome-wide screen identifies genes required for formation of the wobble nucleoside 5-methoxycarbonylmethyl-2-thiouridine in *Saccharomyces cerevisiae*. *RNA* 14:2183–94
242. Chen C, Tuck S, Bystrom AS. 2009. Defects in tRNA modification associated with neurological and developmental dysfunctions in *Caenorhabditis elegans* elongator mutants. *PLoS Genet.* 5:e1000561
243. Goehring AS, Rivers DM, Sprague GF Jr. 2003. Attachment of the ubiquitin-related protein Urm1p to the antioxidant protein Ahp1p. *Eukaryot. Cell* 2:930–36
244. Goehring AS, Rivers DM, Sprague GF Jr. 2003. Urmylation: a ubiquitin-like pathway that functions during invasive growth and budding in yeast. *Mol. Biol. Cell* 14:4329–41
245. Huang B, Johansson MJ, Bystrom AS. 2005. An early step in wobble uridine tRNA modification requires the Elongator complex. *RNA* 11:424–36
246. Frohloff F, Fichtner L, Jablonowski D, Breunig KD, Schaffrath R. 2001. *Saccharomyces cerevisiae* Elongator mutations confer resistance to the *Kluyveromyces lactis* zymocin. *EMBO J.* 20:1993–2003
247. Van der Veen AG, Schorpp K, Schlieker C, Buti L, Damon JR, et al. 2011. Role of the ubiquitin-like protein Urm1 as a noncanonical lysine-directed protein modifier. *Proc. Natl. Acad. Sci. USA* 108:1763–70
248. Burns KE, Darwin KH. 2010. Pupylation versus ubiquitylation: tagging for proteasome-dependent degradation. *Cell Microbiol.* 12:424–31
249. Humbard MA, Miranda HV, Lim JM, Krause DJ, Pritz JR, et al. 2010. Ubiquitin-like small archaeal modifier proteins (SAMPs) in *Haloferax volcanii*. *Nature* 463:54–60
250. Ranjan N, Damberger FF, Sutter M, Allain FH, Weber-Ban E. 2010. Solution structure and activation mechanism of ubiquitin-like small archaeal modifier proteins. *J. Mol. Biol.* 405:1040–55
251. Miranda HV, Nembhard N, Su D, Hepowit N, Krause DJ, et al. 2011. E1- and ubiquitin-like proteins provide a direct link between protein conjugation and sulfur transfer in archaea. *Proc. Natl. Acad. Sci. USA* 108:4417–22
252. Nunoura T, Takaki Y, Kakuta J, Nishi S, Sugahara J, et al. 2010. Insights into the evolution of Archaea and eukaryotic protein modifier systems revealed by the genome of a novel archaeal group. *Nucleic Acids Res.* 39:3204–23

253. Dittmar GA, Wilkinson CR, Jedrzejewski PT, Finley D. 2002. Role of a ubiquitin-like modification in polarized morphogenesis. *Science* 295:2442–46
254. Yashiroda H, Tanaka K. 2004. Hub1 is an essential ubiquitin-like protein without functioning as a typical modifier in fission yeast. *Genes Cells* 9:1189–97
255. Luders J, Pyrowolakis G, Jentsch S. 2003. The ubiquitin-like protein HUB1 forms SDS-resistant complexes with cellular proteins in the absence of ATP. *EMBO Rep.* 4:1169–74
256. Wilkinson CR, Dittmar GA, Ohi MD, Uetz P, Jones N, Finley D. 2004. Ubiquitin-like protein Hub1 is required for pre-mRNA splicing and localization of an essential splicing factor in fission yeast. *Curr. Biol. CB* 14:2283–88
257. Pankiv S, Alemu EA, Brech A, Bruun JA, Lamark T, et al. 2010. FYCO1 is a Rab7 effector that binds to LC3 and PI3P to mediate microtubule plus end-directed vesicle transport. *J. Cell Biol.* 188:253–69

Toward the Single-Hour High-Quality Genome

Patrik L. Ståhl[1] and Joakim Lundeberg[2]

[1]Department of Cell and Molecular Biology, Karolinska Institutet, SE-171 77, Stockholm, Sweden; email: patrik.stahl@ki.se

[2]Science for Life Laboratory, School of Biotechnology, Division of Gene Technology, KTH Royal Institute of Technology, SE-171 65, Solna, Sweden; email: joakim.lundeberg@scilifelab.se

Keywords

DNA sequencing, second-generation sequencing, sequencing by synthesis, real-time sequencing, nanopore sequencing, sequencing quality

Abstract

Today, resequencing of a human genome can be performed in approximately a week using a single instrument. Thanks to a steady logarithmic rate of increase in performance for DNA sequencing platforms over the past seven years, DNA sequencing is one of the fastest developing technology fields. As the process becomes faster, it opens up possibilities within health care, diagnostics, and entirely new fields of research. Immediate genetic characterization of contagious outbreaks has been exemplified, and with such applications for the direct benefit of human health, expectations of future sensitive, rapid, high-throughput, and cost-effective technologies are steadily growing. Simultaneously, some of the limitations of a rapidly growing field have become apparent, and questions regarding the quality of some of the data deposited into databases have been raised. A human genome sequenced in only an hour is likely to become a reality in the future, but its definition may not be as certain.

Contents

INTRODUCTION 360
SINGLE-GENOME
    SEQUENCING 361
    Massively Parallel Pyrosequencing .. 362
    Four-Color Reversible Chain
        Termination 362
    Single-Color Reversible Chain
        Termination 363
    Sequencing by Chained Ligation ... 363
    Sequencing by Unchained
        Ligation 364
SINGLE-HOUR SEQUENCING 364
    Real-Time Single-Molecule
        Sequencing 366
    Sequencing by Sequential
        Electrical Detection 366
FUTURE EXPECTATIONS:
    SINGLE-HOUR GENOME
    SEQUENCING 367
    Nanopore Nucleotide
        Discrimination 368
DISCUSSION 369

INTRODUCTION

Since 2005, the rate of development and of increases in performance for DNA sequencing technology has outperformed most available references. A popular comparison can be made to the computer hardware industry. In 1965, although adjusted somewhat in the following years, Gordon Moore predicted that approximately every two years there would be a twofold increase in processing capacity and in the number of transistors that could fit into an integrated circuit (1, 2). DNA sequencing once seemed to follow this dogma rather well. The cost for every base sequenced started to readily decrease during the end of the Human Genome Project at the beginning of the twenty-first century (3, 4). With adaption from viral and bacterial genome studies of whole-genome shotgun sequencing into the various genome projects culminating in the Human Genome Project (5, 6), biology and information science were for the first time merged on a broad scale. A few years after that, in 2006, the first commercial next-generation, or second-generation, sequencing technologies started appearing. The new platforms were massively parallelized with impressive throughput but far from able to sequence a human genome in a day (7–9). Today, Moore's law is no longer applicable to DNA sequencing; the once linear graph has acquired a hockey stick shape instead, with the first years after the turn of the century serving as its blade, and recent years and the future making up an ever steeper logarithmically growing shaft (**Figure 1**).

There is steadily progressing development of current sequencing platforms, and the launch and promise of new, future platforms are expected. The chemistry linked to the sequencing process in each particular platform has traditionally been the focus for development; however, the strength and novelty of many of the current technologies rely on the hardware manufacturing side. Our competence in manufacturing micro- and nanoscale structures has allowed us to reach entirely new levels of sensitivity in DNA sequencing, setting the path for sequencing of single DNA molecules in an extremely efficient and inexpensive fashion.

Parallel to an impressive increase in data throughput lies a different reality. The limits of modern DNA sequencing are perhaps best understood when considering the fact that the DNA sequencing industry is outperforming the computer hardware industry in some aspects. As the instrument throughput has increased, so has the demand on data handling, as well as its proper organization and deposition. This has proven a mighty challenge (10–12), and common databases used by the biomedical community are at risk of being overwhelmed by improperly annotated data or false-positive findings (13, 14).

The challenges that lie ahead are thus related not only to making the technology more efficient, but also to making it produce a higher-quality and more reliable output. Then, we must find ways to collect and deliver this output

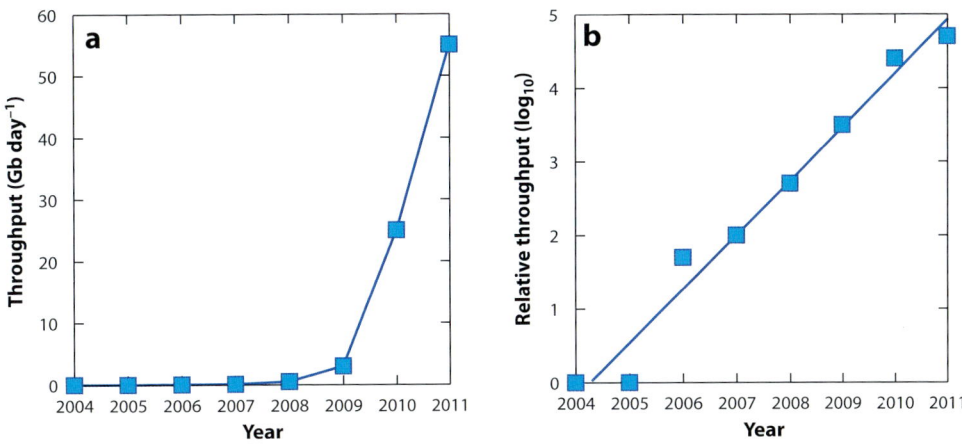

Figure 1
DNA sequencing throughput as a function of time. (*a*) The steep shaft of the curve is indicative of the current evolution in DNA sequencing. (*b*) Throughput for the leading commercial platform has increased more than 40,000-fold since 2004. (Data representing the given time points have been acquired from the platform manufacturers.)

to the rapidly growing community making use of DNA sequencing data.

SINGLE-GENOME SEQUENCING

Today, we can sequence a single human genome in approximately a week on a single instrument and at a cost of $5,000–$10,000. This is a result of the continued development of the second-generation sequencing platforms, first presented in 2005 (7, 9), and of the conceptual high-throughput forerunner technologies, such as massively parallel signature sequencing dating back to before the turn of the century (15). Most of the platforms are in themselves not advanced or sensitive enough to analyze single strands of DNA. Rather the development and combination of a number of supporting methods with the actual sequencing chemistry utilized in each instrument platform have been fundamental to the second-generation sequencing that has revolutionized DNA science for the past six years. Separated massively parallel amplification of clonal DNA fragments is key to most of these systems. It is achieved through powerful methods for DNA library preparation, fragmenting genomic DNA, and introducing universal amplification handles to each fragment (7, 16, 17). After that, clonal amplification of the fragments is achieved through a water-in-oil emulsion polymerase chain reaction on top of beads (7, 9, 18–21), bridge amplification on activated glass surfaces (22–24), or rolling-circle amplification in solution (25). The use of chemically modified nucleotides or oligos carrying fluorescent dyes (24–28) and reversibly terminating moieties (24, 26, 27) is also central to several systems.

Another common denominator for the second-generation sequencing platforms is that their chemistries are all based on a form of sequencing by synthesis either by polymerization or by ligation. The concept of sequencing by synthesis through polymerization was first introduced by Melamede in 1985 (29), although its first commercial application was in pyrosequencing (30). Ligation-based sequencing was first introduced in 1993 (31) and was employed for large-scale polony sequencing by Shendure et al. in 2005 (9). Both pyrosequencing and polony sequencing have been successfully adopted into commercial second-generation sequencing platforms.

Massively Parallel Pyrosequencing

The massively parallel pyrosequencing method originally developed by the company 454 Life Sciences, later acquired by Roche, is powered by a specially built fluidics platform, called the genome sequencer, which sequentially flows each of the four nucleotides across a fiber-optic plate packed with ~1.2 million clonally amplified DNA templates. The templates have been immobilized on the surface of beads, which have then been packed into the picoliter wells in the plate. The fiber-optic construction of the plate is essential because the pyrosequencing chemistry depends on detecting the light generated in each well.

As opposed to the original pyrosequencing protocol, in the genome sequencer, the polymerase, sulfurylase, and luciferase enzymes are all immobilized together with the DNA inside the picoliter reactor wells. The sulfurylase and luciferase are attached to a separate set of beads packed into the well together with a DNA bead and packing beads. The polymerase is incubated together with the DNA beads before the resulting complexes are deposited into the wells. The fourth enzyme, apyrase, is flowed as a part of a washing solution during the washing step in between every nucleotide flow step.

Because of the long read length, which facilitates the assembly of several reads into larger contigs, the genome sequencer quickly became a popular choice for de novo sequencing or resequencing of genomes. Among others, the genome of James Watson has been sequenced using massively parallel pyrosequencing (32, 33). Watson's genome was sequenced at a cost of $2 million (34), which can be compared to the human reference sequence that cost about $3 billion to produce (3, 4).

Today the genome sequencer is primarily applicable in lower-throughput sequencing projects, in sequencing of long amplicons (35), in combination with sequence capture (13), or in de novo sequencing of large genomes as a long-read complement to facilitate interpretation of the massive amounts of shorter-read data generated by other platforms.

Four-Color Reversible Chain Termination

The Solexa sequencing technology is based on fluorescence-labeled chain-terminating nucleotides, a concept clearly inherited from Sanger sequencing, but here the resemblance stops. Today, strategies exist for reversing termination and for cleaving off fluorescent labels by means of light or chemicals (36–40). This is the basis of the sequencing platform developed by the small British company Solexa, today owned by Illumina (41, 42); additionally, the Max-Seq platform, using a set of similarly reversible nucleotides, has also been launched by Intelligent BioSystems (43, 44).

These sequencing methods were outlined for the first time in 1991 and 2001 (40, 41, 45, 46). Early on, the Solexa technology was envisioned as being implemented into single-molecule sequencing platforms, much resembling what is today the Helicos sequencing technology (see below) (26). Today, however, the DNA sample library, prepared by fragmenting DNA and attaching adaptors, is subjected to a clonal amplification step, just like the other second-generation sequencing platforms (47, 48). Instead of emulsion polymerase chain reactions, a bridge amplification step is carried out. Briefly, the DNA fragment library is deposited onto hundreds of millions of primer clusters spread across a chip surface. Amplification is carried out in a flow cell in an automated robot. After clusters have formed, each containing roughly a thousand copies of the original DNA fragment, the chip is transferred into the Illumina fluidics platform, the genome analyzer. Inside, the sequences of the massive amount of clusters are determined.

All four nucleotides are flowed at the same time, each labeled with a different fluorescent dye, making it possible to separate them during imaging. The growing DNA strands are synthesized by polymerases, starting with a primer, and can currently reach lengths of up to 150 bases. However, just like in the massive pyrosequencing method, the theoretical read length is unlimited as long as the background signaling is kept low and the enzymes involved are viable. Furthermore, the introduction of a paired read concept has made it possible to read each DNA fragment from two ends, making the total read length 2 × 150 bp (49).

When the first Solexa 1G Analyzer was launched in 2006, the read length of the system was a mere 30 to 50 bp (41), making it mostly applicable for resequencing of genomic regions, such as in chromatin immunoprecipitation followed by deep sequencing (50, 51). With today's increased read lengths up to 150 bp and with improved sequencing protocols for paired-end and mate-pair sequencing, the latest Illumina platform is probably closest of the second-generation platforms to achieve efficient single-platform de novo sequencing of large genomes (52). So far, the technology has, for instance, been successfully applied to finding structural and sequence variants in human cancer genomes, for resequencing human individuals (53–55), and for the sequencing of de novo genomes, such as that of the panda (56).

Single-Color Reversible Chain Termination

When Helicos BioSciences launched their HeliScope sequencing system in 2008 (57), it was the first single-molecule sequencing system to reach the market. Sensitive single-molecule detection, proven years earlier (58), removes the need for tedious clonal amplification, an important benefit of the platform. It uses reversibly chain-terminating nucleotides, termed virtual terminator nucleotides (59), labeled with cleavable fluorophores, added sequentially to growing strands of DNA immobilized on a chip surface (57). The throughput parameters of the platform, with 35-bp average read lengths and up to 35 gigabases of data per eight-day run, fit in with second-generation specifications (60). Studies undertaken with the HeliScope thus far have been related to direct RNA sequencing (61) and human genome sequencing (62).

Sequencing by Chained Ligation

Like other second-generation platforms, the SOLiD™ from Life Technologies, previously Applied Biosystems, requires the sequenced DNA sample to be fragmented and prepared using either a standard shotgun or a mate-pair protocol. Additionally, Life Technologies has put a lot of effort into optimizing protocols for transcriptome analysis on the SOLiD™. Lately, this type of analysis, termed RNA-Seq, has largely outcompeted the microarray for expression analysis (63, 64). The library is prepared with universal adaptor handles and is subjected to clonal amplification of the fragments on the surface of beads in an emulsion polymerase chain reaction. After enrichment of amplified beads, the beads are deposited onto a chip surface, covalently immobilized by the modified 3′ ends. The beads can be packed with high density, and as in the Illumina system, hundreds of millions of them are fitted onto a single microscopic glass slide (27).

Sequencing is carried out in the SOLiD™ fluidics system through hybridizing a primer to the interrogated fragments and sequentially extending and characterizing the growing strand though as series of ligation steps, using a set of fluorescently labeled octamer oligonucleotides. Each primer is extended through several ligation cycles, where every cycle includes probe ligation, washing, imaging of fluorescent signals, cleaving of the last three generic bases of the probe (thus removing the fluorescent label), and washing. After completion of the series of ligation cycles, the synthesized DNA strand is washed off, and a subsequent primer offset by one base from the previous primer is hybridized (27, 65).

Every ligated octamer is interpreted on the basis of its fluorescent label and the label of overlapping octamers in subsequent primer extensions. Through a clever two-base encoding scheme and with knowledge of the sequencing start site for the respective primers, a complete sequence for the targeted region can be decoded (65).

A seven-ligation cycle scheme results in 35 sequenced bases per fragment; however, the most recent version (5500xl) of the SOLiD™ system can sequence up to 2 × 60 bases per fragment using mate pairs (66).

The short read length of the SOLiD™ system makes it optimal for a variety of resequencing applications. It has successfully been employed for whole-genome sequencing of a cell line (67) and for resequencing of domesticated and wild chickens, but it is perhaps most notable as a tool in deep transcriptome sequencing (64) and for applications like discovery of small RNAs (68).

Sequencing by Unchained Ligation

In a 2010 *Science* paper (69), the in-house method used by the company Complete Genomics to sequence customer samples was outlined in detail. In a proof-of-concept process, the company sequenced three human genomes through a combined sequencing-by-hybridization and sequencing-by-ligation approach (69). This company's chemistry has since also been used to sequence hundreds of human genomes and has been used in studies of genetic inheritance (70).

The library preparation process is key to the Complete Genomics' sequencing technology, where a number of probe anchor sites are introduced into sample DNA fragments through a process of repeated circularization and cutting with restriction enzymes. After the final anchor site has been introduced, the library fragments are circularized and amplified by rolling circle. A proprietary technology keeps the amplifying fragments separated from each other. The generated DNA nanoball clusters are then deposited onto a chip surface with a spotted pattern of sticky probes that make the nanoballs stay immobilized. The balls are sequenced through hybridizing and ligating fluorescently labeled nonamer probes to each of the eight anchor sites, introduced during library preparation. Every hybridization and ligation step allows the decoding of an additional base adjacent to the targeted anchor site, as the query base is offset by one for every probe set hybridized. After five cycles, a degenerate probe is introduced to move the ligation junction another five bases away from the anchor, allowing up to ten bases to be read from each anchor site. The 10 mers are then combined into two 31- to 35-mer mate-pair reads for each nanoball sequenced (25, 71).

The combinatorial probe anchor ligation chemistry of Complete Genomics is in essence the same as was employed by Church's laboratory (9) for polony sequencing by ligation. One of the most important characteristics of the approach, compared to the other second-generation chemistries, is that it is unchained, meaning that, after every detection step, the entire anchor-probe complex is stripped off, and the system is reset. This minimizes the risk of background signals, which is the main bottleneck for long read lengths in the sequencing-by-synthesis-based technologies of Roche/454 Life Sciences, Illumina, and Life Technologies.

SINGLE-HOUR SEQUENCING

The impressively high, albeit slow, throughput of second-generation sequencing systems has enabled studies of genomes and transcriptomes of organisms never before touched by human hands (see the Metagenomics sidebar) as well as studies of plants and animals previously not prioritized owing to ongoing human genome sequencing efforts (56, 72–74). Yet, the implications of faster, close to single-hour sequencing are vast.

With increased speed, modern DNA sequencing methodology is bound to grow way past its fundamental application in basic research efforts targeting static genomes, transcriptomes, and epigenomes. The broad commercial introduction of the sequencing platforms from Ion Torrent (now Life Technologies) and Pacific Biosciences in 2011 opens up new areas of application of rapid DNA sequencing. In health care and in diagnostics, the expectations of fast, accessible sequencing are great. Additionally, old areas of research are being transformed, and new areas of research are already beginning to form.

Genomic epidemiology is perhaps the most rapidly growing field of research in this sense. The analysis of microbes in this area has traditionally been carried out by the means of multilocus sequence typing (75), requiring tedious protocols for obtaining incomplete information about the investigated genome (76). With the help of second-generation sequencing, several important studies concerning outbreaks of *Staphylococcus aureus* (76), *Mycobacterium tuberculosis* (77), and *Salmonella enterica* (78, 79) have been published with phylogenetic information of isolated strains that would have been indistinguishable using traditional genotyping technologies (76). The ability to trace the origin of the outbreak, as in these studies, is fundamental to the field of genomic epidemiology. Additionally, the information obtained may be extremely useful in characterizing future outbreaks or epidemics. With the aid of slow second-generation sequencing platforms, this investigation was done within a limited time frame but not fast enough to enable immediate measures to stop a rapidly spreading outbreak. This brings us to really consider the potential of single-hour sequencing.

In December 2010, a team made up from researchers from Harvard Medical School and Pacific Biosciences published a paper in the *New England Journal of Medicine* describing an Asian origin of the outbreak of cholera in Haiti, which occurred a few months earlier in mid-October (80). From the arrival of samples at Harvard on November 8th, conclusions on the origin of the infectious strains were available on the evening of November 12th (81). For the first time, single-hour DNA sequencing technology was used to characterize an outbreak. This was an important benchmark for both the field of genomic epidemiology and the field of DNA sequencing technology.

In May 2011, another outbreak was suddenly active in Europe. Enterohemorrhagic *Escherichia coli*, a particularly dreadful strain of the common *E. coli* bacterium, spread its way through most of Europe killing at least 27 and causing some 2,800 cases of illness. This time researchers from University Hospital Muenster's Institute of Hygiene managed to isolate the toxin-producing *E. coli* strain and teamed up with members of the German laboratories of Life Technologies. Using the Ion Torrent personal genome machine (PGM^{TM}) at Life Technologies, from the isolation of the strain on May 26th, the sequencing and assembly of the genome was completed by lunchtime on June 2nd (82, 83). In parallel, similar efforts were undertaken by researchers at Beijing Genomics Institute in Shenzhen in collaboration with Germany's University Medical Center Hamburg-Eppendorf (83).

Entering the market for rapid sequencing in late 2011 was the MiSeq^{TM} platform, which is based on the workflow and chemistry of previous sequencing systems from Illumina but with a more rapid cycling of the sequencing steps (84). Like the PGM^{TM} system, the MiSeq^{TM} system is a benchtop sequencer, hinting at other trends for future sequencing platforms—in addition to being rapid, they are likely to be easily accessible as well. A similar yet lower-throughput platform based on the 454 Life

METAGENOMICS

The introduction of second-generation high-throughput sequencing platforms has enabled a great expansion of the field of metagenomics. Without the need to clonally culture organisms before sequencing, any kind of organism can be captured in its natural environment, and its DNA can be subjected to sequencing. Perhaps the most well-known example of current metagenomic efforts is the Sorcerer II Expedition initiated by J. Craig Venter in 2003 to study the microbes in the oceans around the world (133). Numerous discoveries of novel genes have since then been made as well as proof of the great biodiversity among the microbes (72, 134–136). The novel gene content has, as expected, already proven useful for the growth of the synthetic biology field (137, 138). Other earlier metagenomic efforts have included studies on as diverse habitats as soil (139), human gut (140, 141), and the Baltic Sea (142). Future points of sample collection will likely include environments such as clouds and outer space!

Sciences genome sequencer was also introduced by Roche in 2010 (85).

Real-Time Single-Molecule Sequencing

With the Pacific Biosciences' single-molecule real-time sequencing technology, detection is, for the first time, carried out without interruption in real time as a polymerase is monitored while it continuously incorporates nucleotides into a growing DNA strand (86).

The platform, the PacBio *RS*, is based on a nanoscale structure called a zero-mode waveguide (ZMW), an engineered hardware component aiding in the diffraction and detection of single fluorophores (87–89). In the first commercial version, the single-molecule real-time sequencer uses nanofabricated cells containing 150,000 closely packed ZMWs (90). Optimally, each waveguide is fitted with a polymerase enzyme immobilized at the bottom (91), and if a DNA template, a primer, and the four nucleotides are added to the reactor, DNA polymerization can take place (92). To follow the growth of the DNA strand, thus determining the sequence of the template, Pacific Biosciences has fitted all the nucleotides in the reaction with a fluorescent label (92); their investigators have also achieved a chemically engineered φ29 DNA polymerase that incorporates nucleotides at a slower rate than the native version of the polymerase would (93). The nucleotides have been designed so that the fluorescent dye is attached to the terminal phosphate of the nucleotides (92, 94), such that the phosphate is cleaved off when the nucleotide is incorporated by the polymerase. For the actual detection step, the ZMW helps focus the detection point exactly to the active site of the polymerase, avoiding the background noise from fluorophores on nonincorporated nucleotides. The longer processing time of the engineered polymerase also helps the correct signal to be distinguished above noise through the diffracting surface of the ZMW and into a detector. The rather low raw system

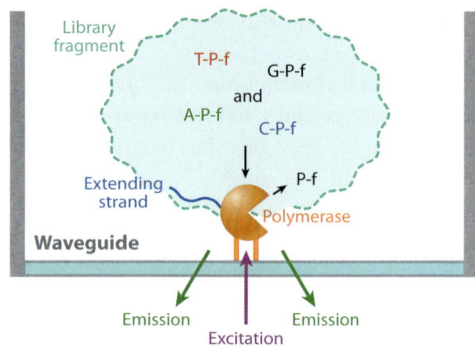

Figure 2

Real-time optical detection. Single-nucleotide incorporation events can be detected thanks to the immobilization of the polymerase at the bottom of the zero-mode waveguide structure. Upon incorporation of a nucleotide, the fluorophore is cleaved off. A circular library fragment enables each base to be read several times for increased accuracy in base calling. Abbreviations: A/C/G/T-P-f denotes a nucleotide coupled via the terminal phosphate to a fluorophore; P-f denotes pyrophosphate coupled to a fluorophore.

accuracy (primarily because of indels) can be improved by the ability to create libraries of circular DNA molecules. Thanks to the strand-displacing feature of the polymerase, these can be sequenced for several laps to achieve an accurate consensus sequence (**Figure 2**) (86).

Although the first commercial systems are just now being distributed, the company has earlier reported achievable read lengths of up to 10 kb, but this is not on average (93). Ongoing platform updates are expected to steadily produce average read lengths of 2,700 bases (95).

Sequencing by Sequential Electrical Detection

The sequencing method of the PGM™ presents something new in DNA-sequencing technology. There is no light detection step and no imaging, making it fast, cheap, and compact.

The platform uses computer hardware industry–type integrated circuits as a consumable. Nucleotide incorporation is detected through measuring the level of free protons released into the reaction compartment. Functioning like a miniature pH meter, a rise in concentration of free protons generates a change in electrical current (96), which is picked up by an underlying electrical circuit and sent off for interpretation and storage. Because nucleotides are flowed one at a time, a change in electric current can be interpreted as an incorporation of the flowed nucleotide (**Figure 3**). Similar to pyrosequencing, homopolymer stretches in the template DNA sequence give rise to larger shifts in current and are interpreted as a multiple of the flowed nucleotide, which is proportional to the magnitude of the shift (97).

The semiconductor chips originally employed in the PGM™ platform were 9 by 9 mm with 1.2 million separate electronic sensors fitted inside. In a layer sitting on top of the sensors, 3.5-μm wells were etched (98). In 2011, a chip containing 6.2 million wells and sensors was launched (with an 11-million-well chip planned for 2012). Additionally, read lengths of 100 bp are to be increased to 200 bp (99). In addition to sequencing of enterohemorrhagic *E. coli* (82), the platform has also been used to decipher the genome of Gordon Moore (100).

The importance of the first nonlight-detecting sequencing system should not be underestimated. The simplicity, use of theoretically low-cost reagents (only native nucleotides and enzymes are needed), speed of detection, and ability to build very compact systems are likely to have impact on where routine DNA sequencing can be performed and by whom. At the moment, however, higher throughput will be required for routine genome sequencing using this technology.

FUTURE EXPECTATIONS: SINGLE-HOUR GENOME SEQUENCING

The evolution of sequencing technology today is as dependent on hardware engineering as it is on the sequencing chemistries them-

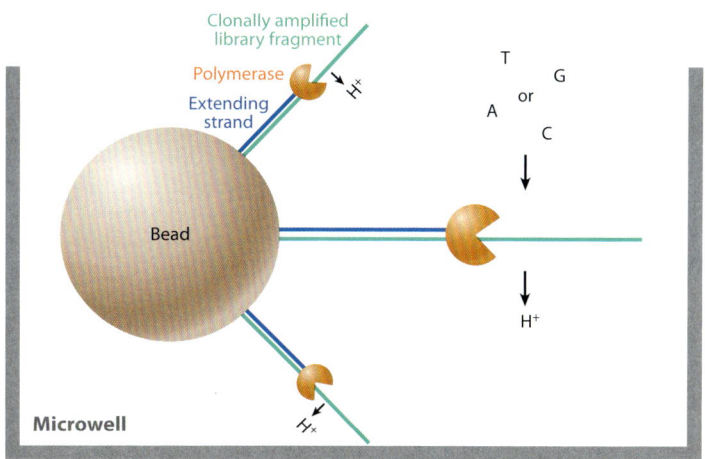

Figure 3

Sequential electrical detection. Nucleotides flow sequentially into the microwell holding a library fragment that has been clonally amplified onto the surface of a bead. Upon incorporation of a nucleotide by the polymerase, protons are released. This causes a shift in current in the well and enables electrical registration of the event. Abbreviations: A, C, G, and T denote their respective nucleotides; H^+ denotes a hydrogen ion.

selves. The second-generation single-genome sequencing platforms have evolved in two different ways into single-hour platforms: one with an inexpensive electrical detection step (PGM™), and the other with single-molecule sensitivity and real-time detection (PacBio RS). The common denominator for the two is that a great deal of effort has been put into creating miniature, highly functionalized, and massively parallelized surface compartments where the sequencing reaction is carried out.

Efforts to combine the best features of the single-hour platforms, single-molecule sensitivity, uninterrupted real-time sequencing, and low-cost detection are currently under way. Most notable in the scientific debate are the expectations regarding nanopore technology.

Nanopore sequencing is based on the concept of utilizing a nanometer-sized pore created through a solid support, typically made out of silicon, to form a reaction chamber, where individual DNA fragments, or even nucleotides, can be threaded or flowed through. Upon

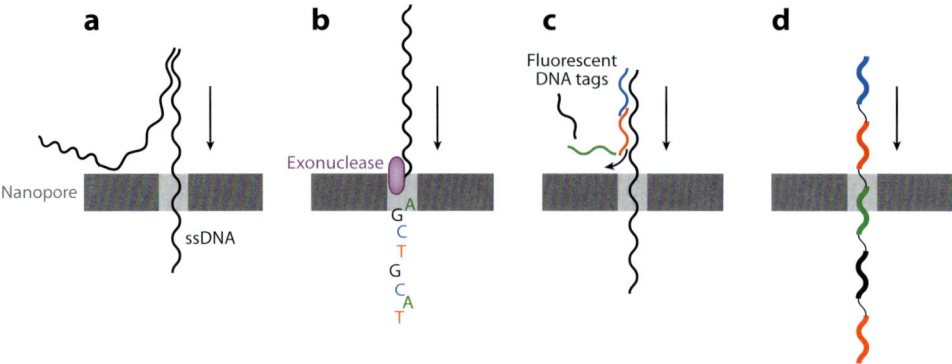

Figure 4

Various principles for nanopore nucleotide discrimination. (*a*) The standard principle of letting a single-stranded DNA (ssDNA) molecule translocate across a nanopore at the same time as its contents are being determined through molecular sensors across the pore. (*b*) The exonuclease principle whereby the ssDNA is chewed into single bases that can more easily be discriminated while they pass through the pore. (*c*) The peel-off principle whereby fluorescent moieties hybridized to the ssDNA are peeled off and detected while the interrogated strand passes through the pore. (*d*) The replacement principle is based on exchanging the small nucleotide features in the original DNA with larger features that can be more easily detected while passing through the pore.

passing through, the molecules or their components can be discriminated from each other through either physical or functional means by utilizing highly sensitive features built into the pore. Examples of such features may be specific proteins or other molecules whose interaction with the DNA bases is measured through shifts in an electrical current applied across the pore (101, 102). Other examples include fluorescent detection principles, where one can benefit from upstream supporting technologies to enhance the signal from each base in the single-stranded DNA, as in the binary conversion developed by LingVitae (103), the conversion and peel-off strategy utilized by NobleGen Biosciences (104), or the expandamers developed by Stratos Genomics Inc. (**Figure 4**) (105). A method utilizing a form of surface-enhanced Raman spectroscopy is also under development by Base 4 Innovation Ltd. (106, 107).

By placing an α-hemolysin protein in the nanopore, Oxford Nanopore Technologies Ltd. has developed a way to create a stable and uniform pore through biological means (108), potentially enabling more consistent conditions for manufacturing arrays of massively parallelized pores.

Nanopore Nucleotide Discrimination

One of the sequencing approaches currently under development at Oxford Nanopore is the first sequencing by degradation approach since the highly toxic method of Maxam & Gilbert (109) and some of the existing sequencing by mass spectrometry approaches (110).

Oxford Nanopore's goal is a system that can carry out single-molecule sequencing, in real time, in a massively parallel fashion, through a nanopore fitted in an electrically charged silicon membrane (108, 111, 112). This makes it the second technology, discussed in depth in this review, that utilizes sequencing-by-electrical signaling, and it has the same implications for cost, speed, and compactness.

The nanopore sequencing technology comes with a couple of additional benefits compared to the sequential electrical detection of the PGM™ technology. Nanopore sequencing works at rates that are only limited by ion stochastics and continuous processing of enzymes, and it does not require constant addition of reagents. On the surface of the silicon membrane, just next to the nanopore, an exonuclease enzyme is immobilized. It is this enzyme that degrades the single-stranded DNA fragments added to the reaction, nucleotide by nucleotide, and spits the nucleotides into the nanopore. Once inside, every kind of nucleotide gives rise to a different change in current across the pore, making it possible to detect which type of nucleotide was cleaved off (113, 114), and enabling the sequence of the degraded DNA strand to be rebuilt.

Because the Oxford Nanopore approach relies on a silicon chip engineered specifically for the application of sequencing, the route toward creating an array of functional nanopores is more difficult compared to using off-the-shelf semiconductor technology as in the PGM™. If successful, however, the varied efforts related to nanopore sequencing have the potential to be groundbreaking (115–117).

DISCUSSION

Not too long ago, a human genome took 13 years to sequence, and the cost for doing it amounted to several billion dollars (3, 4, 32). When the race for the $1,000 genome was started about the time of the introduction of the first second-generation sequencing platform in 2006, its end point, at which routine analysis of human genomes would be possible from a cost perspective (118), could not be foreseen. Today, we can see that the $1,000 diploid genome will most likely be a reality in the near future, and we are already working on the next phase, sequencing massive amounts of healthy and disease-linked human genomes within initiatives like the 1000 Genomes Project (119, 120), the Personal Genome Project (121, 122), the Cancer Genome Atlas (123, 124), the Cancer Genome Project (125, 126), and the International Cancer Genome Consortium (127, 128). With the current rate of development of platform throughput (**Table 1**), it has been estimated that more than 30,000 genomes will have been sequenced by the end of 2011 (129),

Table 1 Platform features and specifications overview 2011

DNA sequencing platforms and companies[a]	Single molecules	Uninterrupted real time	Electrical signaling	Read length (bp)	Throughput per instrument start (gigabases)	Run time (days)	Instrument throughput (gigabases/day)	Estimated relative sequencing cost per base
3730xl × 1, Life Technologies	–	–	–	900	0.002	1	0.002	>>+++++
GS FLX, 454/Roche	–	–	–	700	0.7	1	0.7	+++++
HiSeq 2000, Illumina	–	–	–	2 × 100	600	11	55	+
HeliScope, Helicos	x	–	–	35	35	8	4.4	+++
SOLiD 5500xl × 1, Life Technologies	–	–	–	2 × 60	105	7	15	++
Complete Genomics**	–	–	–	2 × 35	720	12	60	+
MiSeq™, Illumina	–	–	–	2 × 150	1	1	1	+++
PacBio RS, Pacific Biosciences	x	x	–	2,000*	0.08*	0.04*	2*	++++
PGM™, Ion Torrent/Life Technologies	–	–	x	100	0.1	0.08	1.3	+++++
Oxford Nanopore Technologies	x	x	x	10,000*	ND	0.04*	ND	ND

[a]Meaning of symbols: ND, no data; –, not featured in platform; x, featured in platform; >>, much more than; +, relative cost scale (from +, low, typically 10^{-2} scale U.S. dollars per megabase, to +++++, high, typically 10^1 scale U.S. dollars cost per megabase); *, estimated values for the sake of comparison; **, commercially available service (not hardware).

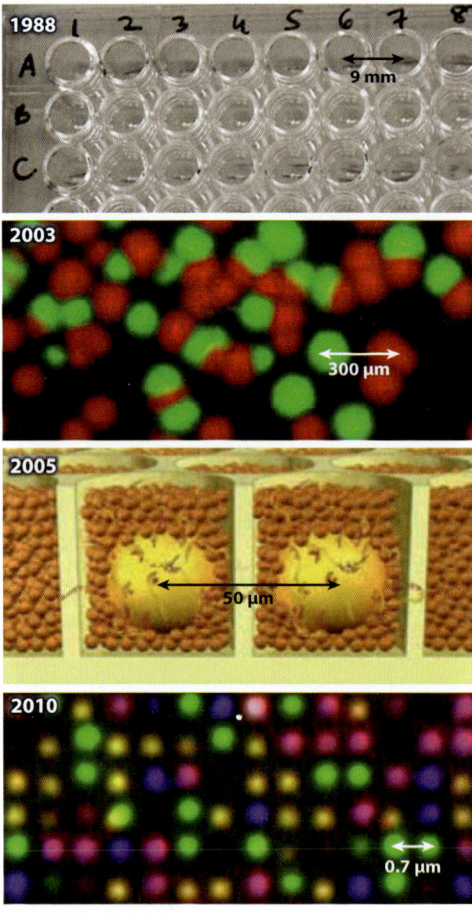

Figure 5

The evolution of surfaces and compartments used for sequencing: from the 1988 microtiter plate (144) used in Sanger sequencing, through the 2003 microscopic glass slide (145) used in several second-generation sequencing technologies, to the 2005 picotiter plate (7) used in massively parallel pyrosequencing, and to 2010 nanostructures etched in a silicon membrane (25) used in sequencing of DNA nanoballs and real-time single-molecule sequencing. The application of increasingly sensitive structures to the sequencing technologies has enabled a reduction in the number of clonally amplified DNA fragments required to generate a detectable signal. Further development of all surface technologies has enabled continuously smaller features to be produced since their original uses in sequencing.

which will be increased to more than a 100,000 at the end of 2012. However, the amount of actual high-quality and high-coverage data relating to these numerous genomes remains to be determined.

Important parameters in the development of the platforms discussed have been the constant evolution of surfaces and the ability to create specialized miniature features applicable for sensitive sequencing. As the surface structures have evolved, so have their sensitivity, and the present trend is toward an increasing fraction of single-molecule platforms (**Figure 5**; **Table 1**).

However, highly sensitive single-molecule detection, just as multiple-molecule detection, still requires a lot from the rest of the connected sequencing process. Although error rates in base calling have decreased since the first introduction of the second-generation platforms, it is still hard to state with absolute certainty if an encountered genomic variant, for instance an SNP, is indeed real or a false positive, miscalled due to low coverage or inherent sequencing artifacts (130–132). Questions have been raised as to whether we are currently populating the common variant databases, such as dbSNP, with low-quality data from massively parallel sequencing, and whether we are thereby at risk of limiting the usefulness of the truly rare variants present in the databases (13, 14). Is the amount of data we are generating simply too big for us to properly verify? Or will the throughput, read length, and sequence quality of future-generation platforms simply enable us to avoid and circumvent subsequent verifications? Currently, Sanger sequencing still remains the state-of-the-art reference for verification of new genomic variants.

This leads us back to the question of what we can expect from a single-hour genome. Much like the $1,000 genome, mentioned earlier, this is likely to happen at some point. For 2012 manufacturers have claimed to be able to provide instrumentation giving a $1,000 human genome in a few hours. But from a quality point of view, we cannot be satisfied with generating a massive amount of small pieces of data in a short time,

Table 2 Overview of examples of upcoming sequencing technologies[a]

Company name	Expected basis for their technology
GnuBIO system	Microfluidics-aided fluorescence-based sequencing
GenapSys	Nanosensor-aided electrical sequencing
Halcyon	Electron microscope-aided sequencing
NABsys	Nanopore-aided electrical sequencing
BioNano Genomics	Nanofluidics-aided fluorescence-based sequencing
NobleGen	Nanopore-aided fluorescence-based sequencing
Genia	Nanopore-aided electrical sequencing
Base 4 Innovation	Raman-spectroscopy-aided sequencing

[a]Adapted and extended from Reference 143.

as is currently the case. Nanopore technologies and other upcoming advanced sequencing technologies (**Table 2**) have the potential to dramatically improve read length and throughput. Ultimately, however, these improvements need to be achieved without compromising quality. Additionally, we must make our methods for interpreting the data more efficient and reliable. Perhaps, when discussing future genome sequencing achievements, is it the single-hour high-quality genome that we should be looking for?

SUMMARY POINTS

1. DNA sequencing is one of the most rapidly developing technology fields in the world.
2. Most modern day sequencing technologies are based on a principle of sequencing by synthesis.
3. Second-generation sequencing technologies currently allow us to sequence a human genome in approximately a week using a single instrument.
4. High-throughput and rapid sequencing have enabled us to study the genomes of animals and plants around us and have created expectations for the future in the areas of health care and diagnostics.
5. Single-hour sequencing enables fast characterization of small organisms, expanding the possibilities for various areas of research, such as genetic epidemiology.
6. Single-molecule sensitivity, combined with real-time electrical detection, seems to be a recipe for fast, cheap, and compact sequencing systems in the future.
7. Nanopore sequencing is a potentially groundbreaking technology, but alternative solutions are under development.
8. Fast, high-throughput future sequencing platforms depend on the continued successful development of advanced micro- and nanostructures.

> **FUTURE ISSUES**
>
> 1. A massive throughput of sequencing data is only useful at a high level of quality with efficient means in place to handle it. This underlines the need for the progression of all aspects of DNA sequencing technology in parallel. A new mode of measuring sequencing platform performance based on quality is needed.
> 2. Currently, more data than can be annotated are deposited into databases, suggesting that the grounds for data deposition needs to be reevaluated to avoid filling up important common resources with erroneous information.
> 3. As the cost of genome sequencing decreases further, a threshold will be reached where sequencing will be useful on a routine basis in health care. For instance, monitoring the effects of a therapy over time could incur significant cost reductions in treatment as a result of routine personal genomics in health care.
> 4. The future likely single-hour genome will enable routine sequencing of human individuals, increasing the need for proper public information about the implications of genome sequencing.

DISCLOSURE STATEMENT

The authors are not aware of any affiliations, memberships, funding, or financial holdings that might be perceived as affecting the objectivity of this review.

ACKNOWLEDGMENTS

Thank you to Sverker Lundin for fruitful discussions. We thank the reviewer for insightful comments and the suggested illustrations for **Figure 5**.

LITERATURE CITED

1. Moore GE. 1965. Cramming more components onto integrated circuits. *Electronics* 38:4
2. Intel Corp. 2005. *Excerpts from A Conversation with Gordon Moore: Moore's Law.* http://download.intel.com/museum/Moores_law/Video-transcripts/excepts_a_Conversation_with_gordon_Moore.pdf
3. Lander ES, Linton LM, Birren B, Nusbaum C, Zody MC, et al. 2001. Initial sequencing and analysis of the human genome. *Nature* 409:860–921
4. Venter JC, Adams MD, Myers EW, Li PW, Mural RJ, et al. 2001. The sequence of the human genome. *Science* 291:1304–51
5. Venter JC, Adams MD, Sutton GG, Kerlavage AR, Smith HO, Hunkapiller M. 1998. Shotgun sequencing of the human genome. *Science* 280:1540–42
6. Int. Hum. Genome Seq. Consort. (IHGS). 2004. Finishing the euchromatic sequence of the human genome. *Nature* 431:931–45
7. Margulies M, Egholm M, Altman WE, Attiya S, Bader JS, et al. 2005. Genome sequencing in microfabricated high-density picolitre reactors. *Nature* 437:376–80
8. Bentley DR. 2006. Whole-genome re-sequencing. *Curr. Opin. Genet. Dev.* 16:545–52
9. Shendure J, Porreca GJ, Reppas NB, Lin X, McCutcheon JP, et al. 2005. Accurate multiplex polony sequencing of an evolved bacterial genome. *Science* 309:1728–32
10. Pop M, Salzberg SL. 2008. Bioinformatics challenges of new sequencing technology. *Trends Genet.* 24:142–49

11. Robinson P, Krawitz P, Mundlos S. 2011. Strategies for exome and genome sequence data analysis in disease-gene discovery projects. *Clin. Genet.* 80:127–32
12. Brown S. 2011. *Involve bioinformatics in design of every experiment...please*. Next-Gen Sequencing blog. http://nextgenseq.blogspot.com/2011/06/involve-bioinformatics-in-design-of.html
13. Ståhl PL, Bjursell MK, Mahdessian H, Hober S, Jirström K, Lundeberg J. 2011. Translational database selection and multiplexed sequence capture for up front filtering of reliable breast cancer biomarker candidates. *PLoS ONE* 6(6):e20794
14. Cirulli ET, Singh A, Shianna KV, Ge D, Smith JP, et al. 2010. Screening the human exome: a comparison of whole genome and whole transcriptome sequencing. *Genome Biol.* 11:R57
15. Brenner S, Johnson M, Bridgham J, Golda G, Lloyd DH, et al. 2000. Gene expression analysis by massively parallel signature sequencing (MPSS) on microbead arrays. *Nat. Biotechnol.* 18:630–34
16. Roach JC, Boysen C, Wang K, Hood L. 1995. Pairwise end sequencing: a unified approach to genomic mapping and sequencing. *Genomics* 26:345–53
17. Ng P, Tan JJS, Ooi HS, Lee YL, Chiu KP, et al. 2006. Multiplex sequencing of paired-end ditags (MS-PET): a strategy for the ultra-high-throughput analysis of transcriptomes and genomes. *Nucleic Acids Res.* 34:e84
18. Dressman D, Yan H, Traverso G, Kinzler KW, Vogelstein B. 2003. Transforming single DNA molecules into fluorescent magnetic particles for detection and enumeration of genetic variations. *Proc. Natl. Acad. Sci. USA* 100:8817–22
19. Kojima T, Takei Y, Ohtsuka M, Kawarasaki Y, Yamane T, Nakano H. 2005. PCR amplification from single DNA molecules on magnetic beads in emulsion: application for high-throughput screening of transcription factor targets. *Nucleic Acids Res.* 33:e150
20. Diehl F, Li M, He Y, Kinzler KW, Vogelstein B, Dressman D. 2006. BEAMing: single-molecule PCR on microparticles in water-in-oil emulsions. *Nat. Methods* 3:551–59
21. Williams R, Peisajovich SG, Miller OJ, Magdassi S, Tawfik DS, Griffiths AD. 2006. Amplification of complex gene libraries by emulsion PCR. *Nat. Methods* 3:545–50
22. Adessi C, Matton G, Ayala G, Turcatti G, Mermod JJ, et al. 2000. Solid phase DNA amplification: characterisation of primer attachment and amplification mechanisms. *Nucleic Acids Res.* 28:e87
23. Mitra RD, Church GM. 1999. In situ localized amplification and contact replication of many individual DNA molecules. *Nucleic Acids Res.* 27:e34
24. Bentley DR, Balasubramanian S, Swerdlow HP, Smith GP, Milton J, et al. 2008. Accurate whole human genome sequencing using reversible terminator chemistry. *Nature* 456:53–59
25. Drmanac R, Sparks AB, Callow MJ, Halpern AL, Burns NL, et al. 2010. Human genome sequencing using unchained base reads on self-assembling DNA nanoarrays. *Science* 327:78–81
26. Harris TD, Buzby PR, Babcock H, Beer E, Bowers J, et al. 2008. Single-molecule DNA sequencing of a viral genome. *Science* 320:106–9
27. Life Technol. 2011. *Overview of SOLiD^{TM} Sequencing Chemistry*. Carlsbad, CA: Life Technol. http://www.appliedbiosystems.com/absite/us/en/home/applications-technologies/solid-next-generation-sequencing/next-generation-systems/solid-sequencing-chemistry.html
28. Eid J, Fehr A, Gray J, Luong K, Lyle J, et al. 2008. Real-time DNA sequencing from single polymerase molecules. *Science* 323:133–38
29. Melamede RJ. 1985. Automatable process for sequencing nucleotide. U.S. Patent No. 4863849
30. Ronaghi M, Uhlén M, Nyrén P. 1998. A sequencing method based on real-time pyrophosphate. *Science* 281:363–365
31. Drmanac R, Drmanac S, Strezoska Z, Paunesku T, Labat I, et al. 1993. DNA sequence determination by hybridization: a strategy for efficient large-scale sequencing. *Science* 260:1649–52
32. Wheeler DA, Srinivasan M, Egholm M, Shen Y, Chen L, et al. 2008. The complete genome of an individual by massively parallel DNA sequencing. *Nature* 452:872–76
33. Olson MV. 2008. Human genetics: Dr Watson's base pairs. *Nature* 452:819–20
34. Pushkarev D, Neff NF, Quake SR. 2009. Single-molecule sequencing of an individual human genome. *Nat. Biotechnol.* 27:847–50

35. Ståhl PL, Stranneheim H, Asplund A, Berglund L, Pontén F, Lundeberg J. 2010. Sun-induced non-synonymous p53 mutations are extensively accumulated and tolerated in normal appearing human skin. *J. Invest. Dermatol.* 131:504–8
36. Ruparel H, Bi L, Li Z, Bai X, Kim DH, et al. 2005. Design and synthesis of a 3′-O-allyl photocleavable fluorescent nucleotide as a reversible terminator for DNA sequencing by synthesis. *Proc. Natl. Acad. Sci. USA* 102:5932–37
37. Bi L, Kim DH, Ju J. 2006. Design and synthesis of a chemically cleavable fluorescent nucleotide, 3′-O-allyl-dGTP-allyl-Bodipy-FL-510, as a reversible terminator for DNA sequencing by synthesis. *J. Am. Chem. Soc.* 128:2542–43
38. Meng Q, Kim DH, Bai X, Bi L, Turro NJ, Ju J. 2006. Design and synthesis of a photocleavable fluorescent nucleotide 3′-O-allyl-dGTP-PC-Bodipy-FL-510 as a reversible terminator for DNA sequencing by synthesis. *J. Org. Chem.* 71:3248–52
39. Turcatti G, Romieu A, Fedurco M, Tairi A-P. 2008. A new class of cleavable fluorescent nucleotides: synthesis and optimization as reversible terminators for DNA sequencing by synthesis. *Nucleic Acids Res.* 36:e25
40. Ansorge WJ. 1991. Process for sequencing nucleic acids without gel sieving media on solid support and DNA chips. *Ger. Patent Appl. DE 41 41 178 A1*
41. Bentley DR. 2006. Whole-genome re-sequencing. *Curr. Opin. Genet. Dev.* 16:545–52
42. Bentley DR, Balasubramanian S, Swerdlow HP, Smith GP, Milton J, et al. 2008. Accurate whole human genome sequencing using reversible terminator chemistry. *Nature* 456:53–59
43. Ju J, Kim DH, Bi L, Meng Q, Bai X, et al. 2006. Four-color DNA sequencing by synthesis using cleavable fluorescent nucleotide reversible terminators. *Proc. Natl. Acad. Sci. USA* 103:19635–40
44. Intell. Bio-Syst. 2011. *Intelligent Bio-Systems Technology*. Waltham, MA: Intell. Bio-Syst. Inc. http://intelligentbiosystems.com/index-1%20mod%201.html
45. Balasubramanian S, Bentley DR. 2001. Polynucleotide arrays and their use in sequencing. *Patent WO 01/157248*
46. Ansorge WJ. 2009. Next-generation DNA sequencing techniques. *New Biotechnol.* 25:195–203
47. Mardis E. 2008. Next-generation DNA sequencing methods. *Annu. Rev. Genomics Hum. Genet.* 9:387–402
48. Pettersson E, Lundeberg J, Ahmadian A. 2009. Generations of sequencing technologies. *Genomics* 93:105–11
49. Illumina. 2011. *System/Genome Analyzer IIx*. http://www.illumina.com/systems/genome_analyzer_iix.ilmn
50. Mardis E. 2007. ChIP-seq: Welcome to the new frontier. *Nat. Methods* 4:613–14
51. Robertson G, Hirst M, Bainbridge M, Bilenky M, Zhao Y, et al. 2007. Genome-wide profiles of STAT1 DNA association using chromatin immunoprecipitation and massively parallel sequencing. *Nat. Methods* 4:651–57
52. Illumina. 2011. *System/HiSeq 2000*. http://www.illumina.com/systems/hiseq_2000.ilmn
53. Pleasance ED, Cheetham RK, Stephens PJ, McBride DJ, Humphray SJ, et al. 2009. A comprehensive catalogue of somatic mutations from a human cancer genome. *Nature* 463:191–96
54. Stephens PJ, McBride DJ, Lin ML, Varela I, Pleasance ED, et al. 2009. Complex landscapes of somatic rearrangement in human breast cancer genomes. *Nature* 462:1005–10
55. Kim JI, Ju YS, Park H, Kim S, Lee S, et al. 2009. A highly annotated whole-genome sequence of a Korean individual. *Nature* 460:1011–15
56. Li R, Fan W, Tian G, Zhu H, He L, et al. 2010. The sequence and de novo assembly of the giant panda genome. *Nature* 463:311–17
57. Harris TD, Buzby PR, Babcock H, Beer E, Bowers J, et al. 2008. Single-molecule DNA sequencing of a viral genome. *Science* 320:106–9
58. Braslavsky I, Hebert B, Kartalov E, Quake SR. 2003. Sequence information can be obtained from single DNA molecules. *Proc. Natl. Acad. Sci. USA* 100:3960–64
59. Bowers J, Mitchell J, Beer E, Buzby PR, Causey M, et al. 2009. Virtual terminator nucleotides for next-generation DNA sequencing. *Nat. Methods* 6:593–95

60. Helicos BioSci. 2011. *True Single Molecule Sequencing (tSMS^{TM}) Performance*. Cambridge, MA: Helicos BioSci. http://www.helicosbio.com/Technology/TrueSingleMoleculeSequencing/tSMStradePerformance/tabid/151/Default.aspx
61. Ozsolak F, Platt AR, Jones DR, Reifenberger JG, Sass LE, et al. 2009. Direct RNA sequencing. *Nature* 461:814–18
62. Pushkarev D, Neff NF, Quake SR. 2009. Single-molecule sequencing of an individual human genome. *Nat. Biotechnol.* 27:847–50
63. Shendure J. 2008. The beginning of the end for microarrays? *Nat. Methods* 5:585–87
64. Cloonan N, Forrest A, Kolle G, Gardiner B, Faulkner G, et al. 2008. Stem cell transcriptome profiling via massive-scale mRNA sequencing. *Nat. Methods* 5:613–19
65. Life Technol. 2011. *SOLiD^{TM} System Sequencing and 2-Base Encoding*. Carlsbad, CA: Life Technol. http://www3.appliedbiosystems.com/cms/groups/mcb_marketing/documents/generaldocuments/cms_057810.pdf
66. Life Technol. 2011. *5500 Series Genetic Analysis Systems*. Carlsbad, CA: Life Technol. http://www.appliedbiosystems.com/absite/us/en/home/applications-technologies/solid-next-generation-sequencing/next-generation-systems.html
67. Clark MJ, Homer N, O'Connor BD, Chen Z, Eskin A, et al. 2010. U87MG decoded: the genomic sequence of a cytogenetically aberrant human cancer cell line. *PLoS Genet.* 6:e1000832
68. Goff LA, Davila J, Swerdel MR, Moore JC, Cohen RI, et al. 2009. Ago2 immunoprecipitation identifies predicted microRNAs in human embryonic stem cells and neural precursors. *PLoS ONE* 4:e7192
69. Drmanac R, Sparks AB, Callow MJ, Halpern A, Burns NL, et al. 2010. Human genome sequencing using unchained base reads on self-assembling DNA nanoarrays. *Science* 327:78–81
70. Roach JC, Glusman G, Smit AF, Huff CD, Hubley R, et al. 2010. Analysis of genetic inheritance in a family quartet by whole-genome sequencing. *Science* 328:636–39
71. Complete Genomics. 2011. *Complete Genomics Analysis Platform*. http://www.completegenomics.com/services/technology/details/
72. Rusch DB, Halpern AL, Sutton G, Heidelberg KB, Williamson S, et al. 2007. The Sorcerer II Global Ocean Sampling expedition: northwest Atlantic through eastern tropical Pacific. *PLoS Biol.* 5(3):e77
73. Klevebring D, Street NR, Fahlgren N, Kasschau KD, Carrington JC, et al. 2009. Genome-wide profiling of populus small RNAs. *BMC Genomics* 10:620
74. Genome 10K Community Sci. 2009. Genome 10K: a proposal to obtain whole-genome sequence for 10,000 vertebrate species. *J. Hered.* 100:659–74
75. Urwin R, Maiden MCJ. 2003. Multi-locus sequence typing: a tool for global epidemiology. *Trends Microbiol.* 11:479–87
76. Harris SR, Feil EJ, Holden MTG, Quail MA, Nickerson EK, et al. 2010. Evolution of MRSA during hospital transmission and intercontinental spread. *Science* 327:469–74
77. Gardy JL, Johnston JC, Ho Sui SJ, Cook VJ, Shah L, et al. 2011. Whole-genome sequencing and social-network analysis of a tuberculosis outbreak. *N. Engl. J. Med.* 364:730–39
78. Lienau EK, Strain E, Wang C, Zheng J, Ottesen AR, et al. 2011. Identification of a salmonellosis outbreak by means of molecular sequencing. *N. Engl. J. Med.* 364:981–82
79. Pflumm M. 2011. Speedy sequencing technologies help track food-borne illness. *Nat. Med.* 17:395
80. Chin C-S, Sorenson J, Harris JB, Robins WP, Charles RC, et al. 2011. The origin of the Haitian cholera outbreak strain. *N. Engl. J. Med.* 364:33–42
81. Davies K. 2010. *Break Out: Pacific Biosciences Team Identifies Asian Origin for Haitian Cholera Bug*. Bio-IT World.com. Dec. 9. http://www.bio-itworld.com/news/12/09/10/PacBio-identifies-Haiti-cholera-outbreak.html
82. Mellmann A, Harmsen D, Cummings CA, Zentz EB. 2011. Prospective genomic characterization of the German enterohemorrhagic *Escherichia coli* O104:H4 outbreak by rapid next generation sequencing. *PLoS ONE* 6:e22751
83. Vence T. 2011. *On Top of Outbreaks*. Genomeweb.com. July/Aug. http://www.genomeweb.com/sequencing/top-outbreaks
84. Illumina. 2011. *Systems/MiSeq Personal Sequencer*. http://www.illumina.com/systems/miseq.ilmn

85. Roche/454. 2011. *The New GS Junior Instrument & Workflow.* http://www.gsjunior.com/instrument-workflow.php
86. Eid J, Fehr A, Gray J, Luong K, Lyle J, et al. 2008. Real-time DNA sequencing from single polymerase molecules. *Science* 323:133–38
87. Levene MJ, Korlach J, Turner SW, Foquet M, Craighead HG, Webb WW. 2003. Zero-mode waveguides for single-molecule analysis at high concentrations. *Science* 299:682–86
88. Foquet M, Samiee KT, Kong X, Chauduri BP, Lundquist PM, et al. 2008. Improved fabrication of zero-mode waveguides for single-molecule detection. *J. Appl. Phys.* 103:034301
89. Lundquist PM, Zhong CF, Zhao P, Tomaney AB, Peluso PS, et al. 2008. Parallel confocal detection of single molecules in real time. *Opt. Lett.* 33:1026–28
90. Pacific Biosci. 2011. $SMRT^{TM}$ *cells.* http://www.pacificbiosciences.com/products/consumables/SMRT-cells
91. Korlach J, Marks PJ, Cicero RL, Gray JJ, Murphy DL, et al. 2008. Selective aluminum passivation for targeted immobilization of single DNA polymerase molecules in zero-mode waveguide nanostructures. *Proc. Natl. Acad. Sci. USA* 105:1176–81
92. Korlach J, Bibillo A, Wegener J, Peluso P, Pham TT, et al. 2008. Long, processive enzymatic DNA synthesis using 100% dye-labeled terminal phosphate-linked nucleotides. *Nucleosides Nucleotides Nucleic Acids* 27:1072–83
93. Karow J. 2010. *At AGBT, Pacific Biosciences Unveils Commercial Instrument; Shows 10 kb Read, Variety of Applications.* Genomeweb.com. March 2. http://www.genomeweb.com/agbt-pacific-biosciences-unveils-commercial-instrument-shows-10-kb-read-variety-
94. Metzker ML. 2010. Sequencing technologies—the next generation. *Nat. Rev. Genet.* 11:31–46
95. Pacific Biosci. 2011. *Pacific Biosciences Contributes Whole Genome Sequence Data for German* E. coli *Outbreak Strain and 11 Related Strains for Comparative Analysis.* July 6. http://www.pacificbiosciences.com/sites/default/files/press_release_assets/Ecoli_data_release_announcement_Final_07062011.pdf
96. Pourmand N, Karhanek M, Persson HH, Webb CD, Lee TH, et al. 2006. Direct electrical detection of DNA synthesis. *Proc. Natl. Acad. Sci. USA* 103:6466–70
97. Life Technol. 2011. *Technology: How does it work?* http://www.iontorrent.com/the-simplest-sequencing-chemistry/
98. Karow J. 2010. *Ion Torrent Systems Presents $50,000 Electronic Sequencer at AGBT.* Genomeweb.com. March 2. http://www.genomeweb.com/sequencing/ion-torrent-systems-presents-50000-electronic-sequencer-agbt
99. Life Technol. 2011. *Ion Torrent Ion Personal Genome Machine^{TM} Performance Overview.* Appl. Note Perform. Spring 2011. http://www.iontorrent.com/lib/images/PDFs/performance_overview_application_note_041211.pdf
100. Rothberg JM, Hinz W, Rearick TM, Schultz J, Mileski W, et al. 2011. An integrated semiconductor device enabling non-optical genome sequencing. *Nature* 475:348–52
101. Branton D, Deamer DW, Marziali A, Bayley H, Benner SA, et al. 2008. The potential and challenges of nanopore sequencing. *Nat. Biotechnol.* 26:1146–53
102. Church GM, Deamer DW, Branton D, Baldarelli R, Kasianowicz J. 1998. Characterization of individual polymer molecules based on monomer-interface interactions. *U.S. Patent* 5,795,782
103. LingVitae. 2011. *Technology.* http://www.lingvitae.com/technology.html
104. McNally B, Singer A, Yu Z, Sun Y, Weng Z, Meller A. 2010. Optical recognition of converted DNA nucleotides for single-molecule DNA sequencing using nanopore arrays. *Nano Lett.* 10:2237–44
105. Stratos Genomics. 2011. *Stratos Genomics Technology.* http://www.stratosgenomics.com/technology/
106. Base 4 Innovation. 2011. *Base 4 Innovation.* http://www.base4.co.uk
107. Sciencebiz. 2010. *Base4 Innovation—Genome Sequencing with Nanostructures—Sciencebiz.* Mar. 18. http://tweetmeme.com/story/842950149/base4-innovation-genome-sequencing-with-nanostructures-sciencebiz-sciencebiz
108. Clarke J, Wu H-C, Jayasinghe L, Patel A, Reid S, Bayley H. 2009. Continuous base identification for single-molecule nanopore DNA sequencing. *Nat. Nanotechnol.* 4:265–70
109. Maxam AM, Gilbert W. 1977. A new method for sequencing DNA. *Proc. Natl. Acad. Sci. USA* 74:560–64

110. Mauger F, Bauer K, Calloway CD, Semhoun J, Nishimoto T, et al. 2007. DNA sequencing by MALDI-TOF MS using alkali cleavage of RNA/DNA chimeras. *Nucleic Acids Res.* 35:e62
111. Oxford Nanopore Technol. 2011. *Platform Technology: The Gridion System.* http://www.nanoporetech.com/sections/index/53
112. Oxford Nanopore Technol. 2011. *Nanopores for DNA Sequencing.* http://www.nanoporetech.com/sections/index/82
113. Astier Y, Braha O, Bayley H. 2006. Toward single molecule DNA sequencing: direct identification of ribonucleoside and deoxyribonucleoside 5'-monophosphates by using an engineered protein nanopore equipped with a molecular adapter. *J. Am. Chem. Soc.* 128:1705–10
114. Clarke J, Wu HC, Jayasinghe L, Patel A, Reid S, Bayley H. 2009. Continuous base identification for single-molecule nanopore DNA sequencing. *Nat. Nanotechnol.* 4:265–70
115. Deamer DW, Akeson M. 2000. Nanopores and nucleic acids: prospects for ultrarapid sequencing. *Trends Biotechnol.* 18:147–51
116. Shendure J, Mitra RD, Varma C, Church G. 2004. Advanced sequencing technologies: methods and goals. *Nat. Rev. Genet.* 5:335–44
117. Branton D, Deamer DW, Marziali A, Bayley H, Benner SA, et al. 2008. The potential and challenges of nanopore sequencing. *Nat. Biotechnol.* 26:1146–53
118. Bennett ST, Barnes C, Cox A, Davies L, Brown C. 2005. Toward the $1,000 human genome. *Pharmacogenomics* 6:373–82
119. 1000 Genomes. 2010. *1000 Genomes: A Deep Catalog of Human Genetic Variation.* http://www.1000genomes.org/home
120. 1000 Genomes Proj. Consort. 2010. A map of human genome variation from population scale sequencing. *Nature* 467:1061–73
121. Lunshof JE, Bobe J, Aach J, Angrist M, Thakuria JV, et al. 2010. Personal genomes in progress: from the human genome project to the personal genome project. *Dialogues Clin. Neurosci.* 12:47–60
122. Pers. Genome Proj. 2011. *Personal Genome Project.* http://www.personalgenomes.org
123. Natl. Cancer Inst. 2011. *The Cancer Genome Atlas.* http://cancergenome.nih.gov/
124. Cancer Genome Atlas Res. Network. 2011. Integrated genomic analyses of ovarian carcinoma. *Nature* 474:609–15
125. Wellcome Trust Sanger Inst. 2011. *The Cancer Genome Project.* http://www.sanger.ac.uk/genetics/CGP/
126. van Haaften G, Dalgliesh GL, Davies H, Chen L, Bignell G, et al. 2009. Somatic mutations of the histone H3K27 demethylase gene *UTX* in human cancer. *Nat. Genet.* 41:521–23
127. Int. Cancer Genome Consort. 2011. *ICGC Cancer Genome Projects.* http://www.icgc.org/
128. Int. Cancer Genome Consort., Hudson TJ, Anderson W, Artez A, Barker AD, et al. 2010. International network of cancer genome projects. *Nature* 464:993–98
129. *Nature.* 2010. Human genome: genomes by the thousand. *Nature* 467:1026–27
130. DePristo MA, Banks E, Poplin R, Garimella KV, Maguire JR, et al. 2011. A framework for variation discovery and genotyping using next-generation DNA sequencing data. *Nat. Genet.* 43:491–98
131. Shen Y, Wan Z, Coarfa C, Drabek R, Chen L, et al. 2010. A SNP discovery method to assess variant allele probability from next-generation resequencing data. *Genome Res.* 20:273–80
132. Hoberman R, Dias J, Ge B, Harmsen E, Mayhew M, et al. 2009. A probabilistic approach for SNP discovery in high-throughput human resequencing data. *Genome Res.* 19:1542–52
133. Sorcerer II Exped. 2011. *Welcome to Sorcerer II Expedition.* http://www.sorcerer2expedition.org/version1/HTML/main.htm
134. Nicholls H. 2007. Sorcerer II: the search for microbial diversity roils the waters. *PLoS Biol.* 5:e74
135. Yooseph S, Sutton G, Rusch DB, Halpern AL, Williamson SJ, et al. 2007. The Sorcerer II Global Ocean Sampling expedition: expanding the universe of protein families. *PLoS Biol.* 5:e16
136. Williamson SJ, Rusch DB, Yooseph S, Halpern AL, Heidelberg KB, et al. 2008. The Sorcerer II Global Ocean Sampling Expedition: metagenomic characterization of viruses within aquatic microbial samples. *PLoS ONE* 3:e1456
137. Legato MJ. 2010. Sailing the sea of synthetic biology: Dr. Venter and the Sorcerer II. *Gend. Med.* 7:276–77

138. Schirmer A, Rude MA, Li X, Popova E, del Cardayre SB. 2010. Microbial biosynthesis of alkanes. *Science* 329:559–62
139. Handelsman J, Rondon MR, Brady SF, Clardy J, Goodman RM. 1998. Molecular biological access to the chemistry of unknown soil microbes: a new frontier for natural products. *Chem. Biol.* 5:R245–49
140. Qin J, Li R, Raes J, Arumugam M, Burgdorf KS, et al. 2010. A human gut microbial gene catalogue established by metagenomic sequencing. *Nature* 464:59–65
141. Andersson AF, Lindberg M, Jakobsson H, Bäckhed F, Nyrén P, Engstrand L. 2008. Comparative analysis of human gut microbiota by barcoded pyrosequencing. *PLoS ONE* 3:e2836
142. Andersson AF, Riemann L, Bertilsson S. 2010. Pyrosequencing reveals contrasting seasonal dynamics of taxa within Baltic Sea bacterioplankton communities. *ISME J.* 4:171–81
143. Church GM. 2011. *What Are All of the Next-Generation DNA Sequencing Technologies?* **http://arep.med.harvard.edu/gmc/nexgen.html**
144. Church GM, Kieffer-Higgins S. 1988. Multiplex DNA sequencing. *Science* 240:185–88
145. Mitra RD, Shendure J, Olejnik J, Edyta-Krzymanska-Olejnik, Church GM. 2003. Fluorescent in situ sequencing on polymerase colonies. *Anal. Biochem.* 320:55–65

Mass Spectrometry–Based Proteomics and Network Biology

Ariel Bensimon,[1,2] Albert J.R. Heck,[1,3] and Ruedi Aebersold[1,2,4]

[1]Department of Biology, Institute of Molecular Systems Biology, and [2]Competence Center for Systems Physiology and Metabolic Diseases, ETH Zurich, CH 8093, Switzerland; email: bensimon@imsb.biol.ethz.ch, aebersold@imsb.biol.ethz.ch

[3]Biomolecular Mass Spectrometry and Proteomics, University of Utrecht, and Netherlands Proteomics Center, 3584 CH Utrecht, The Netherlands; email: a.j.r.heck@uu.nl

[4]Faculty of Science, University of Zurich, CH 8006 Switzerland

Keywords

systems biology, quantitative proteomics, protein interaction network, protein-signaling network, posttranslational modification, data-driven modeling

Abstract

In the life sciences, a new paradigm is emerging that places networks of interacting molecules between genotype and phenotype. These networks are dynamically modulated by a multitude of factors, and the properties emerging from the network as a whole determine observable phenotypes. This paradigm is usually referred to as systems biology, network biology, or integrative biology. Mass spectrometry (MS)–based proteomics is a central life science technology that has realized great progress toward the identification, quantification, and characterization of the proteins that constitute a proteome. Here, we review how MS-based proteomics has been applied to network biology to identify the nodes and edges of biological networks, to detect and quantify perturbation-induced network changes, and to correlate dynamic network rewiring with the cellular phenotype. We discuss future directions for MS-based proteomics within the network biology paradigm.

Contents

MOLECULES TO NETWORKS	380
MASS SPECTROMETRY–BASED PROTEOMICS	381
APPLYING MASS SPECTROMETRY–BASED PROTEOMICS TO NETWORK BIOLOGY	384
DESCRIBING NETWORK WIRING	385
Describing Protein Interaction Networks	385
Describing Protein-Signaling Networks	388
DYNAMIC NETWORK REWIRING	391
Dynamic Rewiring of Protein Interaction Networks	391
Dynamic Rewiring of Protein-Signaling Networks	392
CORRELATING NETWORK WIRING WITH PHENOTYPES	393
Correlating Network Wiring with Phenotypes for Protein Interaction Networks	393
Correlating Network Wiring with Phenotypes for Protein-Signaling Networks	394
OUTLOOK	396

MOLECULES TO NETWORKS

Much of life science research has been focused on understanding the complex relationship between genotype and phenotype. Specifically, research has addressed the fundamental questions of how, when, and where the information encoded in the genome of an organism is expressed and modulated by external (e.g., environmental) or internal (e.g., genomic) factors to generate a specific phenotype.

Over the past decades, such studies have been carried out within the "one gene-one protein-one function" paradigm referred to as the "molecular biology paradigm," which arose from the classical work of Beadle & Tatum (1) on amino acid metabolism in *Neurospora*. This paradigm makes two important assumptions that have dominated the thinking of generations of experimental biologists and guided the development of the techniques of molecular biology. First, it postulates a direct link between gene and protein function, implying that knowledge of all the genes and their translation products can explain biological function. Second, it orders individual proteins and their associated functions in linear pathways, implying that every function "downstream" is affected by an upstream block, whereas every function "upstream" is unaffected by a downstream block (**Figure 1a**, *left*). In the genomic age, powerful technologies have been developed to support research at a global scale within the molecular biology paradigm. These include genome sequencing to identify all protein-coding genes of a genome (2, 3); proteomic methods to identify and quantify the proteins in a biological sample (4); genomic engineering (5); and gene knockout, RNAi technologies, and small-molecule inhibitor screens to inhibit or manipulate specific functions and to identify upstream and downstream events (6). Genome-wide RNAi screens that essentially search the whole genome space have been particularly popular. They are often applied to link genes to phenotypic readouts on a global level (for example, References 7–9). Overall, the technologies to identify, quantify, mutate, and interfere with the expression levels of any conceivable gene or protein of a species have reached a very high level of maturity.

In spite of these impressive technical advances and their wide and successful application, it has generally remained challenging to establish genotype-phenotype links. For example, with the exception of relatively few single gene defects with high penetrance, the molecular basis of most disease phenotypes turned out to be more complex and remain to be determined (10, 11). The reasons for these difficulties are likely conceptual rather than merely technical. The molecule-centric, single

RNAi: small RNA molecules that interfere with messenger RNA and thus prevent the translation of a protein

directional pathway-based paradigm, focusing on the properties of molecules, has turned out to be limited because it has neglected contextual relationships, such as cross talk between linear pathways that were considered to operate in isolation of one another.

Recently, a new paradigm has been emerging, typically referred to as systems biology, network biology, or integrated biology, that takes into account contextual relationships (12–15). In this network view, each node represents a molecule of interest, such as a gene, any of its products, or smaller molecules, such as cofactors, messenger molecules, and metabolites. The edge between two nodes represents a relationship, such as a physical interaction, an enzymatic reaction, or a functional connection. Although the molecule-centric paradigm is concerned with the network nodes, the new paradigm is concerned with the network nodes and edges, placing networks of interacting molecules between genotype and phenotype (**Figure 1a**, *right*). It assumes that the structure and topology of such networks are an expression of the genomic information, that the networks are dynamically modulated at different timescales by external (e.g., environmental factors) or internal (e.g., genomic alterations) perturbations, and that the properties of the entire network determine the phenotype.

The network biology paradigm has several important implications. First, it requires a different, more integrative view of biological processes, as the contextual relationships between molecules move to the forefront. Second, network biology provides new opportunities for and critically depends on new experimental and computational approaches, including methods to visualize networks, methods to infer network topology and structure, and methods to simulate and model the dynamic behavior of networks and their phenotypic consequences. Thus, network biology has been stirring novel technologies that focus on the measurement of contextual relationships of molecules, rather than on simply enumerating molecules in a catalog format. In this review, we discuss the current state of mass spectrometry (MS)–based proteomic approaches that support network biology.

MASS SPECTROMETRY–BASED PROTEOMICS

The main goal of proteomics is the detailed characterization of the proteome. In the molecular biology paradigm, the focus has been on the comprehensive identification and characterization of protein sequences, including their posttranslational modifications (PTMs), and on the comprehensive quantification of the protein components of a biological sample.

Most proteomic studies rely on tandem mass spectrometry as the core technology, specifically on a method referred to as bottom-up proteomics. In bottom-up proteomics, protein samples extracted from cells or tissues are digested into peptides. Peptides in the sample are then separated, typically by liquid chromatography, ionized, and transferred into the mass spectrometer, where peptide fragment ion spectra are recorded. Fragment ion spectra are the currency of information in bottom-up proteomics, as they can be assigned to peptide sequences from which the corresponding proteins are inferred. Fragment ion spectra are also used to detect modified amino acid residues and to identify and locate modifications within the peptide sequence. Peptide ion signals can also be used to infer the quantity of a sample peptide or protein (16, 17). For every step of the process, including sample preparation and fractionation, MS data acquisition, quantification, and data analysis, multiple methods and tools have been developed and reviewed extensively (4, 16–19). This also applies to the MS instrumentation, which enjoys a continued increase in performance in regard to mass accuracy, sensitivity, and analytical robustness (16, 17).

From an extensive menu of available options for each procedure step, individual choices have been combined into different workflows and MS strategies, each addressing different types of biological inquiries (16, 17, 19). These can be applied to various samples, such as whole proteomes, or enriched fractions, such as

MS: mass spectrometry

Posttranslational modifications (PTMs): the enzymatic covalent addition of a molecular entity to a protein, or removal of a molecular entity, or the irreversible change in protein sequence

Liquid chromatography: physical separation of analytes by the differential partition of each analyte between a mobile and a stationary phase in a column

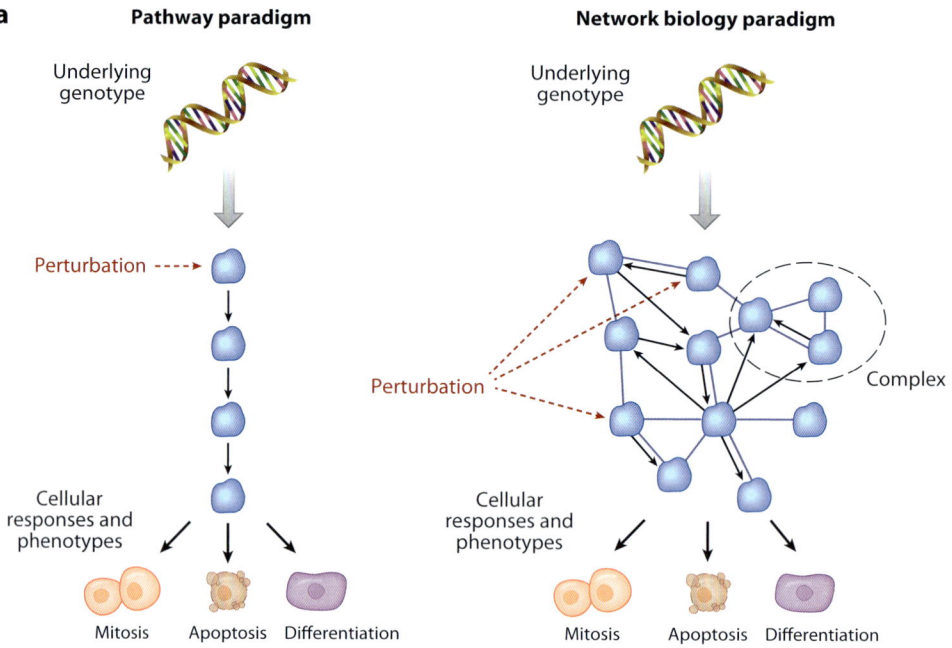
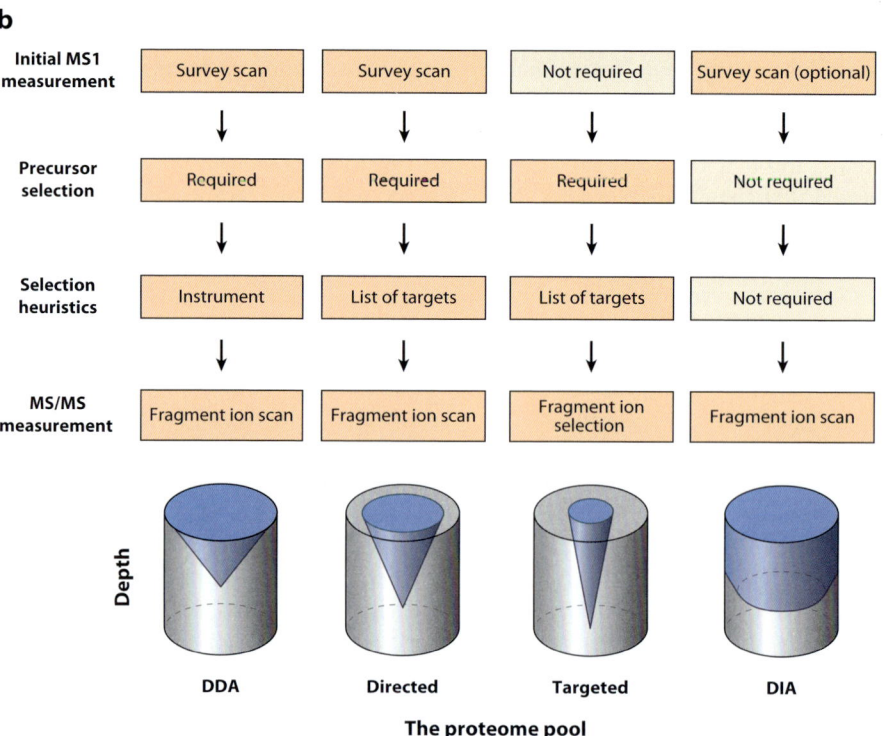

phosphoproteomes. The most frequently used strategy is referred to as shotgun or discovery proteomics. There, precursor ions are detected in a survey scan and selected automatically using a simple heuristic via a process referred to as data-dependent analysis. This strategy results in datasets that can identify vast numbers of proteins and enables quantitative comparison between samples, either with stable isotope labeling or without labeling, in an approach known as label-free quantification (4, 20–22). Shotgun proteomics does not require any prior knowledge of the composition of the sample, and thus each protein in every sample analyzed is newly discovered. In directed proteomics, precursors are only selected for fragmentation if they are detectable in a survey scan and present on a list of predetermined precursor ions, i.e., an "inclusion list" (23, 24). This strategy results in datasets that identify and quantify specific, predetermined segments of a proteome at a higher level of reproducibility, compared to discovery proteomics. In targeted proteomics, only predetermined peptides are selected for detection and quantification in a sample. The main mass spectrometric method supporting targeted proteomics is selected reaction monitoring (SRM) also referred to as multiple reaction monitoring (25, 26). In SRM, specific mass spectrometric assays are generated a priori for each targeted peptide, and these assays are then used to selectively detect and quantify analytes in multiple biological samples (27). This method can generate highly reproducible and accurate datasets of a small, preselected fraction of a proteome (typically one to a few hundred peptides) at a wide dynamic range (17, 28). Finally, with recent advances in instrumentation, a fourth strategy referred to as data-independent analysis (29–32) is emerging in which no selection of precursor ions occurs, i.e., the fragmentation of all precursors is attempted for each sample, the analysis of which can benefit from the availability of extensive spectral libraries (27, 33). Each of these strategies captures a different subset of the "total proteome space" (**Figure 1*b***), balancing trade-offs in comprehensiveness, reproducibility and selectivity, sensitivity, accuracy, and dynamic range (17, 34).

Proteomic studies in the molecular biology paradigm have, for the most part, strived to increase coverage of the discovered, characterized, and quantified proteome (34). This has been technically challenging to accomplish, but it is conceptually simple and largely achievable using the discovery proteomics strategy.

Stable isotope labeling: a technique in which samples are metabolically or enzymatically labeled with stable isotopes of different masses

Figure 1

(*a*) Within the one gene-one protein-one function paradigm, referred to as the molecular biology paradigm, individual proteins and their associated functions are ordered in linear pathways, implying that every function downstream is affected by an upstream block (*left*). In the network biology paradigm, networks of interacting molecules are placed between genotype and phenotype. Each node (represented in *blue*) represents a molecule of interest, such as a gene, any of its products, or smaller molecules, such as metabolites. The edge between two nodes represents any type of relationship, such as protein-protein interactions (*blue lines*), enzyme-substrate relationships (*black arrows*). Both nodes and edges may be subject to perturbation (such as stimulus, inhibition, knockout, etc.). (*b*) Several prominent mass spectrometry (MS)–based strategies are available. The most frequently used strategy is referred to as shotgun or discovery proteomics. This strategy results in datasets that can identify vast numbers of proteins contained in biological samples but is more likely to reproducibly detect the most abundant proteins. In directed proteomics, specific, predetermined segments of a proteome are identified and quantified at a higher level of reproducibility, compared to discovery proteomics. Targeted proteomics generates highly reproducible and accurate datasets of a small, preselected fraction of a proteome (typically one to a few hundred peptides) with high detection sensitivity and dynamic range. With recent advances in instrumentation, a fourth strategy referred to as data-independent analysis (DIA) is emerging in which the identification of all proteins is attempted for each sample. Abbreviation: DDA, data-dependent analysis.

PPI: protein-protein interaction

Yeast two-hybrid (Y2H) screening: a genetic technique used to screen for physical interactions between pairs of proteins

Notable projects have achieved a high degree of coverage for proteomes (35–40) and selected PTMs (41–43), have measured quantitative changes in protein and PTM abundances (44–48), and have estimated the absolute cellular concentrations of significant fractions of proteomes (36, 40, 49, 50). In many of these studies, the main result is a list of abundance-modulated proteins or modified peptides. Typically, one or a few of the list's elements are followed up by classical biochemical or cell biology methods, a strategy that may be successful but ultimately not satisfactory because most of the information collected in the proteomic screen remains unused. Alternatively, the generated lists are subjected to analysis by gene ontology enrichment tools, pathway and signal transduction databases, or protein-protein interaction (PPI) databases (51, 52). The aim of these analyses is (*a*) to relate the contents of the list to prior knowledge in the form of protein functional classes or pathways or (*b*) to provide a visual concept of cellular processes, e.g., in the form of a network (53). We see these analyses as postmeasurement network approaches in a molecular biology paradigm, rather than studies motivated a priori by network biology. As such, although they may provide valuable knowledge, this review does not further focus on such studies.

APPLYING MASS SPECTROMETRY–BASED PROTEOMICS TO NETWORK BIOLOGY

In the network biology paradigm, the technologies available to identify and quantify molecules need to be applied to also determine or infer and quantify the edges of these networks, i.e., the wiring underlying cellular networks. Measuring network edges by large-scale "omics" studies has been addressed primarily by two approaches. The first direct approach uses the affinity between nodes to capture the interacting molecules and to thus directly measure an edge. This is exemplified by technologies such as chromatin immunoprecipitation followed by deep sequencing (ChIP-seq) (54) or by affinity purification (AP)-MS (see below). Second, the indirect approach uses an assay to probe a relationship between two nodes and to thus infer an edge. This is exemplified by yeast two-hybrid (Y2H) screening or genetic interaction networks (55). The limited toolbox of experimental methods to detect and quantify network edges raises the important question of how dynamic networks can be best studied. In a seminal study, Ideker et al. (56) have addressed this question. They described a generic road map for network biology, consisting of the following steps, which can be applied iteratively: (*a*) defining an initial network of nodes and edges from prior information, (*b*) perturbing network components and integrating the experimental data obtained with the network model, and (*c*) refining the network model to better predict experimental data and phenotypes arising from the network. During the past decade, these steps have been explored by a multitude of experimental and computational strategies (12–15).

Although the network biology paradigm has progressed significantly at the conceptual level, it is still substantially bound by molecular biology data collection techniques. To support the road map outlined above, MS-based proteomics, like other data collection technologies, needs to be able to generate datasets that minimally fulfill the following criteria: (*a*) the data have to be complete, i.e., all the network nodes and edges should be measurable; (*b*) the data need to be reproducible, i.e., identical results should be obtained in each repeat measurement of a network; (*c*) the data have to be quantitative to detect dynamic changes of network components; and (*d*) the data need to be measurable at a reasonable throughput to allow iterations within a study. Clearly, in meeting some of these criteria, proteomics has lagged behind other genomic technologies.

In the following sections, we describe steps that have been taken in MS-based proteomics to advance our ability to investigate and compare networks via measurement of molecules.

We also briefly review several network-driven experimental and computational approaches that show promise for use with proteomic data. We structure our review to describe the ways MS-based proteomics has been applied to the generic road map of multidirectional network biology (**Figure 2a**): The first tasks are to identify the network components (nodes and edges) as well as to describe network wiring and the principles of the network organization, the second task is to relate perturbations to quantitative measurements to describe dynamic network rewiring, and the third task is to correlate dynamic network wiring with the cellular phenotype. Finally, we discuss what lies ahead for MS-based proteomics in the path toward network models and their mechanistic analysis and refinement.

For clarity of presentation, we distinguish between two types of networks investigated by MS-based proteomics: protein interaction networks (PINs) and protein-signaling networks (PSNs). PINs are undirected networks, i.e., the edges show no preferred direction, whereas PSNs are directed networks, i.e., the edges have a preferred direction. Clearly, a cellular network is the sum of several such networks (**Figure 2b**).

Importantly, the investigation of network wiring by MS-based proteomics can benefit from network data collected by other technologies. For example, genetic interaction networks examine the dependency in function between two genes, reported by a growth phenotype, and represent a unique case in which the edges are measured (as phenotypes) and nodes are not explicitly measured. Genetic interaction maps are often carried out under basal growth conditions (55, 57) but have also been used to capture context-specific interactions, such as a differential interaction map (58). Because of limited throughput at present, MS-based proteomics cannot be applied as a direct readout for genetic interaction studies, but it can use the complementary information such screens provide to highlight edges, which have functional significance within the context of describing network wiring.

DESCRIBING NETWORK WIRING

Here, we review how qualitative and quantitative MS-based proteomic techniques have been applied toward defining a qualitative network, i.e., to support statements such as "protein X interacts with protein Y," in the case of PINs, or "protein X is a target of protein Y," in the case of PSNs. Such a network is a prerequisite to provide a comprehensive and reproducible description of the underlying cellular network and to serve as an initial map for quantitative/dynamic analyses.

Describing Protein Interaction Networks

PINs, exemplified by protein-protein interaction networks (PPINs), have been a major focus of interest in MS-based proteomics mainly because in these networks, in principle, both nodes and the edges linking them are directly measurable. The prototypical experimental approach to study such networks has been the use of a protein or other biomolecule as a "bait molecule" to capture and isolate "prey" proteins interacting with the bait, and to then identify these proteins (bait and preys) by MS. Although the following discussion mainly focuses on PPINs, proteins evidently also interact with other molecules that represent vital parts of the network (15). Studies focusing on interactions between proteins and other (cellular) molecules include those using RNA or drug molecules as bait to capture and identify interacting proteins (59–61). Conversely, proteins have been used as bait to capture copurified metabolites (62). Although these types of networks have not been investigated as routinely as PPINs, they provide essential information to understand the cellular network dynamics.

Prior to MS-based proteomics, the majority of the PPIN data were collected by Y2H screens, a proteomic technology with a high throughput. Though the quality of the Y2H PPI data has increased substantially over time (55), the method interrogates only binary interactions capturing only a subspace of the whole interactome (57). Therefore, AP of tagged

PIN: protein interaction network

PSN: protein-signaling network

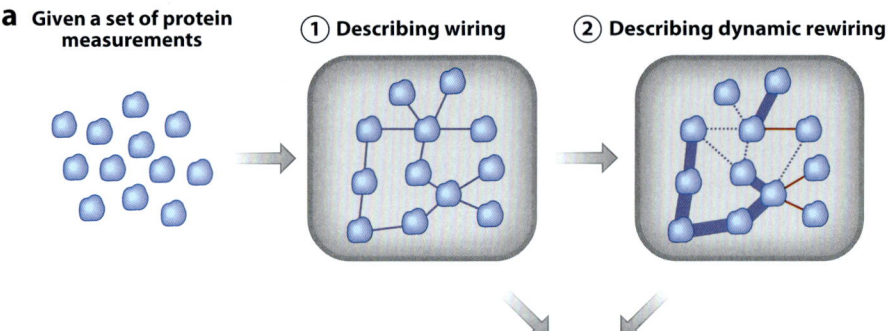
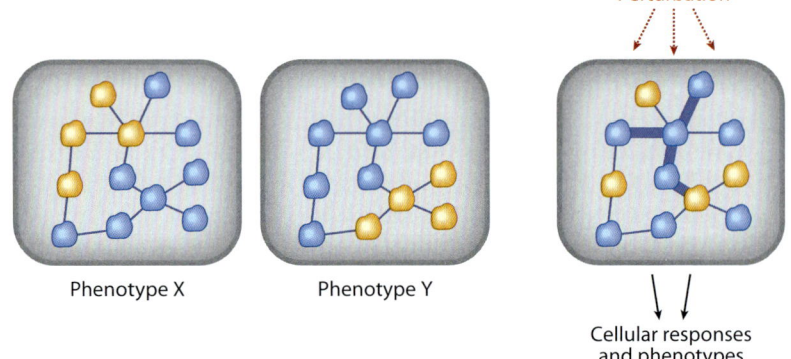
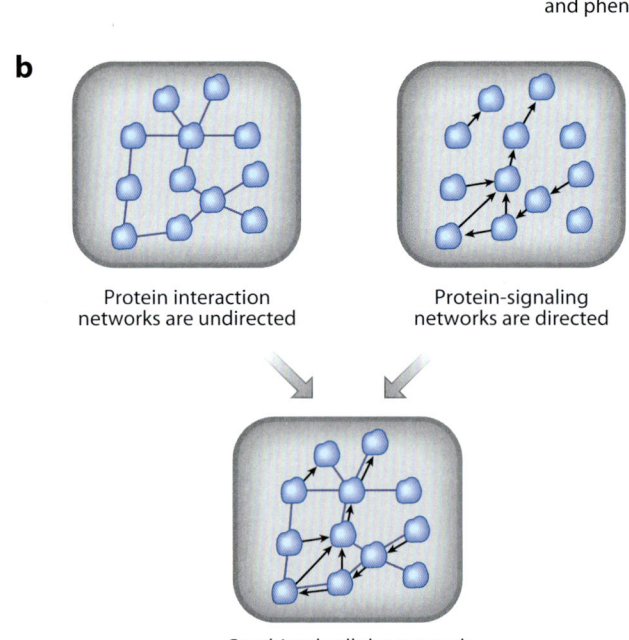

proteins of interest (POI) and identification of the copurified protein components by MS (AP MS) (63) have become a preferred method for the analysis of PPINs because the data it produces reflect more closely the actual multidirectional complexity of a PPIN in the cell. AP-MS data are then interpreted as a network of PPIs, whereby two or more copurified nodes are connected by edges. It should be noted that the majority of these experiments are performed in various cell lines and thus may not reflect correctly the cellular networks in tissues. Moreover, copurification does not necessarily indicate a direct physical interaction, and detailed information about the composition and structural topology of protein complexes is not directly apparent from AP-MS data. These issues can be addressed by cross-linking techniques and mass spectrometric analysis of intact complexes and have been reviewed elsewhere (63–66).

Inevitably as with every technology, AP MS comes with concerns regarding the quality of the obtained data. A first concern relates to the possibility of false-positive PPIs because in most species genetic manipulation is inefficient, and therefore, the tagged protein is often overexpressed from an expression vector and not expressed from the endogenous locus. This will remain an issue until techniques for tagging proteins by genetic manipulation of human cell lines (67) or tagging proteins in whole animals (68) is as feasible as tagging those for yeast cells. One method, suggested to circumvent this issue, is by expression of the protein from a bacterial artificial chromosome (69).

Another way to avoid the problem is by immunoprecipitation of the endogenous protein whenever an antibody is available. For example, Malovannaya et al. (70) performed more than 3,000 immunoprecipitation experiments of endogenous proteins, including reciprocal immunoprecipitations, and uncovered in their vast dataset core complex modules, unique core "isoforms," and complex-complex interactions.

A second concern in this type of experiments is the identification of true and specific PPIs as opposed to nonspecifically copurified proteins. A simple approach to overcome this problem is the identification of common contaminants that copurify nonspecifically owing to their "stickiness" to the matrix or the tag affinity reagent (71, 72). Another way of addressing contaminants is by using quantitative interaction proteomics to compare an AP of a tagged POI to an AP that represents the unspecific background, for example, an untagged POI. A similar abundance of a protein in both purifications would indicate the protein is a contaminant (69, 71, 73, 74). On the basis of this concept, several scoring algorithms were developed to estimate the likelihood of an interaction by different metrics derived from a protein's spectral counts, thus extracting contaminants and identifying "true" interactors (75–77).

Thus far, much effort has been put into providing a description of the cellular PPIN. Several studies have addressed the landscape of PPINs through the analysis by AP MS of large groups of POIs: 75 human deubiquitinating enzymes (76); 32 human proteins linked to autophagy and vesicle trafficking (78); a

POI: protein of interest

AP MS: affinity purification coupled to mass spectrometry

Spectral count: counting the number of times a peptide is identified in a dataset as a proxy to its abundance

Figure 2

(*a*) Given a set of protein measurements by mass spectrometry (MS)-based proteomics, three tasks are faced with the generic road map of network biology: The first tasks are to identify the network edges and to describe network wiring, the second task is to relate perturbations to quantitative measurements to describe dynamic network rewiring (depicted as a variance in the weight of edges), and the third task is to correlate dynamic network rewiring with the cellular phenotype (depicted as *yellow nodes*). (*b*) Two types of networks are investigated by MS-based proteomics: Protein interaction networks are undirected networks, i.e., the edges show no preferred direction. Protein-signaling networks are directed networks, i.e., the edges have a preferred direction. The cellular network integrates a variety of those networks, which in proteomics are often investigated separately.

genome-scale analysis of protein complexes in the bacterium *Mycoplasma pneumonia* (79); and 276 kinases, phosphatases, regulatory subunits, and their scaffolds in yeast (80).

Frequently, PPIs are modulated by PTMs. Vermeulen et al. (81) identified histone mark "readers," proteins that interact with histone tail peptides trimethylated at various lysine residues. For selected readers, they inspected their interactomes by AP MS, the corresponding genomic binding sites by ChIP-seq, and their binding strength in the context of combinatorial histone modifications. This study certainly suggests that describing networks in the future will also have to account for the involvement of PTMs in forming these interactions.

Organizing networks into complexes. AP MS provides a description of the network wiring in the form of edges between nodes identified by MS, but it does not explicitly identify the composition of protein complexes. Therefore, after refining the PPIN for true interactions, the next challenge is to infer true protein complexes from AP-MS data. Experimentally, for any type of AP, a reciprocal AP of the identified copurifying proteins can greatly improve the confidence in the annotation of these relationships and refine the identification of putative complexes (70). Another approach, applied in yeast, is the AP MS of the same bait protein from various strains that were deleted for known interactors. This enabled capturing information about the association between every bait and prey, as well as every prey and the deleted known interactor (82), providing information about how the complex is assembled. In essence, both reciprocal AP and analysis by deletion tackle the identification of complexes by attempting to address the undirected edges in the PPIN from both ends (nodes). Computationally, PPI data have been used to infer complexes from the network topology either by various graph-based approaches (83) or by using MS quantitative data. This has been attempted by nested clustering (84), a biclustering approach, in which baits are first clustered by normalized spectral counts, and then preys are clustered within their respective bait clusters.

In sum, describing the wiring of PPI networks has seen significant advances in throughput and reproducibility of data collection. However, the technique is still far from being comprehensive, as with a few exceptions, such as yeast, for most species only small subsets of their respective proteomes have been thus far covered. Furthermore, PINs, other than PPINs, have been even more challenging to analyze.

Describing Protein-Signaling Networks

PSNs, exemplified by enzyme-substrate relationships, have been a second focus of interest in MS-based proteomics, particularly in the case of enzymes, which modify their substrate by a PTM. The addition or removal of a PTM to or from a peptide results in a change in mass detectable by MS. In PSNs, the interaction between pairs of an enzyme and a substrate tends to be very transient, and its significance lies in the directed transduction of a signal, represented as a directed edge. To describe the wiring of a PSN, we would require reproducible and comprehensive measurements of all proteins involved, thus determining all edges. In particular, the molecular context of these signaling events, in space and time, is also provided by a framework of other proteins, such as adaptor and anchor proteins (85). However, although, in a PPIN, an edge is measured by the concurrent measurement of the two connected nodes, in a PSN, this is not always feasible owing to the transient nature of interaction. Consequently, often only one node (a substrate, for example) of two connected nodes can be measured, and edges need to be inferred. Therefore, much more effort has been put in PSNs to first describe their wiring. As these edges are functional in nature, one could either attempt to identify the edges from qualitative data or attempt to infer edges from quantitative data (**Figure 3**). In particular, perturbation experiments in which defined stimuli

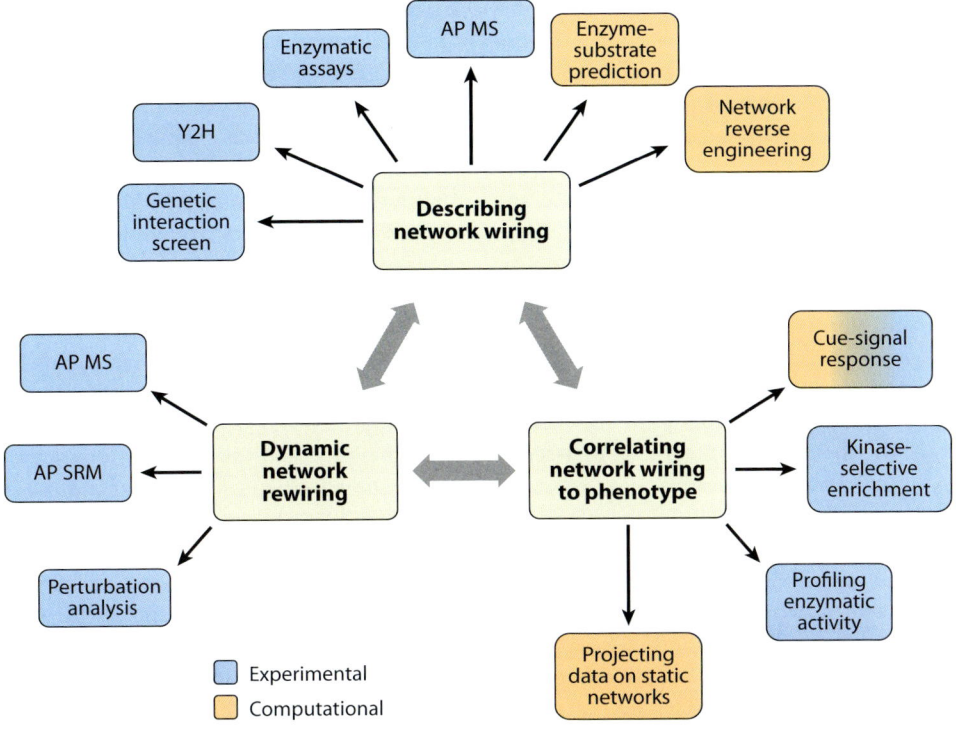

Figure 3

Describing protein-signaling network (PSN) wiring can be performed by either attempting to identify the edges from qualitative data or attempting to infer edges from quantitative data. The yeast two-hybrid (Y2H) system and affinity purification mass spectrometry (AP MS) have been applied to describe protein interaction networks, and various approaches have been applied to assay the relationship between enzymes and substrates. Genetic interaction screens provide complementary information. Dynamic network rewiring in which defined stimuli are used to induce rewiring of the network may highlight the structure of an underlying PSN. It is performed by quantitative analysis of AP-MS or AP-SRM (selected reaction monitoring) experiments and by perturbation experiments using small molecules or gene deletions. Correlating network wiring with phenotypes can use the underlying network of static protein-protein interactions to identify subsets of nodes that have certain network properties and correspond to a given phenotype. Alternatively, the activity of proteins, such as kinases, can be correlated with a given phenotype. In addition, correlating network rewiring with cellular response can be achieved using cue-signal-response compendiums in which samples are perturbed with various cues, preselected network nodes, and cellular phenotypes measured in these samples, and the resulting data are subjected to statistical and modeling frameworks to examine/investigate the signals and responses and to identify those of significance.

are used to induce rewiring of the network may highlight the structure of an underlying PSN. This is discussed further in the next section.

In reviewing PSNs, we refer mainly to protein phosphorylation as a prototypical example of a PSN. Protein phosphorylation is a PTM that has received much attention in proteomics, fueled by rapid improvements in recent years in the enrichment of phosphopeptides from proteome extracts (86–88). More than 160,000 phosphorylation sites are now documented in publicly accessible databases, such as PhosphoSitePlus (89). This vast amount of sites represents the accumulation of data from various studies, and the identification of sites of phosphorylation has become a success story for the application of MS-based proteomic methods used within the molecular biology paradigm.

However, delineating the PSN, i.e., defining which kinases phosphorylate these phosphorylation sites, has lagged behind. Even more challenging have been the questions of which phosphatases dephosphorylate these phosphorylation sites and which other regulatory proteins are involved. Several experimental and computational approaches to provide a description of the kinase-substrate relationships by proteomics have been undertaken (**Figure 3**). They mainly include in vitro kinase assays, prediction of putative kinase-substrate relationships, and perturbation experiments (90, 91).

Methods that are based on in vitro kinase assays provide the most direct data to describe kinase-substrate relationships. For example, Mok et al. (92) conducted protein microarray kinase assays in which a yeast proteome was immobilized on microarrays and incubated with radiolabeled ATP and a kinase of interest. In a conceptually similar approach, peptides were spotted on an array format and phosphorylated by purified kinases to determine consensus recognition motifs or kinases (93). However, the substrates identified in vitro may not represent the bona fide in vivo set of substrates. MS-based proteomics applied to analyze the products of in vitro kinase reactions would obviate the need for radiolabeled ATP and, more importantly, would enable the identification of the phosphorylated residue(s). One successful method relies on engineering a kinase to accept unnatural ATP analogs by modification of the ATP-binding pocket, resulting in the specific thiophosphorylation of a substrate. Substrates of such an engineered kinase are then identified by enrichment of thiophosphopeptides, which can be readily identified by MS (94, 95).

Another approach to capture enzyme-substrate relationships is to stabilize the transient interaction between the two molecules. Bloom et al. (96) expressed in yeast catalytically altered Cdc14 phosphatase mutants that trapped their substrates. Substrates can therefore be isolated while bound to the enzyme and identified by AP MS. "Substrate trapping" has also been applied to a kinase by cross-linking a kinase and a mutated substrate with a specific cross-linker (97).

Finally, a variety of tools have emerged in recent years to predict which kinases putatively phosphorylate a given phosphorylation site, relying on amino acid sequence motifs and a variety of computational methods (91). Of these, NetworKIN (98) should be specifically mentioned because it also uses probabilistic protein association networks. A shortcoming of kinase-substrate predictions is the slower rate at which refinement of kinase consensus motifs occurs, compared to the higher rate at which phosphoproteomics data are acquired. Moreover, these methods do not take into account information from quantitative measurements and rely mostly only on sequence information.

A different computational strategy to infer the wiring of a signaling network is referred to as reconstruction or reverse engineering. Variants of this strategy have been mainly applied to microarray data (reviewed in References 99 and 100). These methods attempt to identify causal relationships between network nodes from quantitative omics data, such as the transcription factors that control modules of coregulated transcripts in the case of microarray data. To the best of our knowledge, MS-based proteomics data have not been used in such approaches, but there is a clear analogy between transcription factors and transcripts measured by microarrays and kinases and the phosphorylation sites measured by phosphoproteomics. However, network inference is not a trivial computational task, and it might not be easily generalized and transferable to MS-based proteomics data.

Other modifications have also been subject to large-scale MS analyses, such as N-terminal and lysine acetylation (41, 101) and N-glycosylation (43). In these studies, thousands of the modified sites have been identified, attesting to the breadth of the identifiable PTM landscape. As in the case of phosphorylation, proteins modified by these PTMs and others are most successfully identified after enrichment. Several studies in recent years have also addressed the identification of proteins modified by ubiquitin and ubiquitin-like modifications, which are small conserved proteins themselves that covalently modify other proteins in a reversible manner (102, 103). In contrast to the rapid increase in studies focused on detecting dynamic change of the phosphoproteome in perturbed cells, the analysis of ubiquitome thus far has lagged behind and has mainly focused on the identification of modified proteins and sites in steady state. Argenzio et al. (104) have explored the epidermal growth factor (EGF)-regulated ubiquitome by a quantitative analysis of cells untreated or treated with EGF. Kim et al. (105) and Wagner et al. (106) used a novel antibody to monitor changes in the ubiquitome in response to proteasomal inhibition. The cleavage of proteins by proteases has also been investigated as a modification in itself, by probing the activity and specificity of proteases (107, 108), as well as by identification of protease cleavage products (109, 110). As there are hundreds of proteins putatively involved in these various protein modifications, clearly the analysis of the corresponding signaling networks is even further in the future than for protein phosphorylation.

DYNAMIC NETWORK REWIRING

MS-based proteomic techniques have been applied toward identifying and quantifying changes in network wiring in response to stimuli or under different physiological conditions. For the most part, these techniques have been used in the molecular biology paradigm to detect abundance changes in network nodes. In the following section, we review studies that focus on either quantifying changes in known edges or identifying novel perturbation-specific edges, thus describing in detail novel network wiring (mentioned in the section above).

Dynamic Rewiring of Protein Interaction Networks

To date most PIN, particularly PPIN, studies have been performed under basal conditions. They do not, therefore, include information about dynamics or reorganization of these PPINs. Understanding the dynamic nature of PPINs as a function of time and/or condition requires quantification of the proteins that are retained, released, or recruited into a complex in response to a perturbation. Few studies so far have attempted time- or perturbation-resolved dynamics. Examples of these studies include the FoxO3A interactomes in response to phosphatidylinositol 3 kinase inhibition (73), the dynamics of the extracellular signal-regulated kinase 1 interactome in response to the EGF and nerve growth factor (111), and the dynamics of the circadian rhythm gene *frequency* (FRQ) interactome during the course of a day (112). In work closing the gap to translational use of PPIN knowledge, Aye et al. (113) applied a chemical proteomics approach to tissue samples of human patients to demonstrate that the PPIN profile of the regulatory subunit of protein kinase A with several of its scaffold proteins became severely altered in the failing heart.

Several techniques have been applied to resolve changes in complex composition: Wepf et al. (114) used a reference peptide, dubbed SH-quant, as part of a protein's affinity tag to calculate the absolute amounts of bait proteins and, by correlational quantification, the amounts of the prey proteins in various AP samples. This approach was then used for the analysis of perturbation-induced quantitative changes in the composition of the protein phosphatase 2A complex, thus resolving protein-complex abundance and dynamic changes in complex components by combining the labeled peptide with label-free quantification. Bennett et al. (115) mapped the basal cullin-RING ubiquitin ligase network and then investigated the changes in the network in response to deneddylation. Moreover, they used AQUA peptides (116) and MS quantification to determine changes in the subunit occupancy of the cullin-RING ubiquitin ligase network. The use of targeted proteomics can facilitate the validation of larger numbers of interactors at high sensitivity and throughput. For example, Bisson et al. (117) used a method they termed AP-SRM

AQUA: absolute quantification using synthesized peptides, incorporated with stable isotopes and spiked in a known concentration into samples as internal standards to study the dynamics of PPINs. They first identified novel interactors of growth factor receptor-bound protein 2 (GRB2), a scaffold protein involved in tyrosine kinase receptor signaling. They then generated SRM assays for 90 interactors and used them to provide a quantitative temporal analysis of the GRB2 PPIN following stimulation with EGF, as well as analysis of growth factor-specific GRB2 PPIs. The success of these studies suggests that the PPINs will increasingly be studied as dynamic networks using SRM-based targeted-MS techniques.

Dynamic Rewiring of Protein-Signaling Networks

Perturbation experiments try to capture the effects of modulating an enzyme activity (inhibition or activation) in a cellular context. The dynamic rewiring of the network could then lead to a context-specific description of the network wiring, which is expected to provide new physiological insights compared to in vitro experiments (**Figure 3**). Two conceptual types of perturbations can be pointed out in the context of network biology: those that affect mainly the edge(s) versus those that affect mainly node(s) and inevitably also their edge(s) (118). Perturbation information comes at the cost of increased complexity and treads a fine line between dependency, causality, and the off-target effects of the molecules used.

In one perturbation setup to study phosphorylation networks, a comparison is made between untreated cells, cells treated with a stimulus, and cells treated with a stimulus in the presence of a kinase inhibitor (for example, References 119 and 120). This experiment can be combined with phosphoproteomics, and from such perturbations, phosphopeptides downregulated in the presence of the inhibitor are considered as kinase dependent, although not necessarily directly kinase mediated. In the context of kinase-substrate networks in which kinase specific inhibitors are employed, it is important to remember that inhibitors actually cover a wider range of kinases at different selectivities (121). This is important not only for clinical investigation of these molecules, but also for biologists interested in the interpretation of experimental perturbation data. To circumvent the issue of inhibitor specificity, several groups have adopted the approach of using analog-sensitive kinases, which have been mutated at the ATP-binding pocket to be specifically and rapidly inhibited by the pyrimidine-based inhibitor, 1-NM-PP1 (122). Holt et al. (123) used this approach to identify novel putative substrates of cyclin-dependent kinase 1 (Cdk1) by identifying those phosphorylation sites that conform to Cdk1's known consensus motif and were downregulated upon Cdk1 inhibition. In a few cases, a phosphoantibody is available for the known consensus motif of a kinase and can be used as an affinity reagent to enrich phosphopeptides from samples in a quantitative experimental setup in which the kinase is activated/inhibited (124, 125). Such samples are then compared by quantitative proteomics to suggest substrates for this known kinase.

One should note that dependence does not necessarily indicate that these are true substrates because phosphorylation sites may presumably be regulated by other kinases, which are themselves regulated. Thus, as phosphoproteomics applied in such setups is an indirect measurement of edges that have to be inferred, it may give only a limited scope of the kinase-substrate relationships embedded in the data. However, it can provide insights to the functional organization of a stimulus-dependent phosphoprotein network, e.g., proteins identified in phosphoproteomic screens of perturbed cells are clustered to identify groups of coregulated proteins, and the resulting clusters are projected onto static PPIN data (46, 119, 126).

From a network perspective, inhibition of a kinase by a small and specific molecule is an "edgetic" approach, removing those edges that represent kinase-substrate relationships. Perturbation can also be carried out by removing a node completely, for example, by deletion of the gene. This would affect not only kinase-substrate relationships, but also kinase interactions with other proteins. Detecting the effects of deleting kinase nodes may be problematic as these effects might be compensated for over time. Bodenmiller et al. (47) analyzed phosphoproteome samples from a large selection of yeast strains in which kinase and phosphatase genes were deleted. The loss of most of these has perturbed significant parts of the phosphoproteome, but only about half of them in the expected direction (e.g., downregulation of a phosphopeptide when a kinase is deleted). These data indicate that the overall wiring of PSNs and their perturbation-induced rewiring is currently beyond the range of MS-based proteomics.

Although these studies tend to be global in approach, two recent studies focused on a specific subset of kinase-substrate relationships and limited their scope of interest in network rewiring to a well-defined biological question. Jorgensen et al. (127) examined bidirectional signaling initiated by cell-cell contacts, studied by the coculture of two cell lines and analysis of their tyrosine phosphoproteomes. The data were used along with results from an siRNA and predicted kinase-substrate networks and PPINs to derive a network model of cell-specific information. Coba et al. (128) examined changes in the phosphoproteome of the neurons' postsynaptic density in response to N-methyl-D-aspartate. Combining these data with information on activated kinases, identified by the Western blot method and in vitro kinase assays on peptide arrays, enabled them to formulate a putative kinase-substrate network of the synapse, benefiting from the lower complexity of postsynaptic density compared to a whole cell. Both of these studies pooled descriptive and perturbation-related quantitative data and suggest an inferred wiring of the kinase-substrate relationships. Although successful, both also highlight that MS-based proteomics applied in perturbation setups is still resulting in the description of network wiring and is not rigorously applied to quantify dynamic rewiring per se.

CORRELATING NETWORK WIRING WITH PHENOTYPES

In a network paradigm, we wish to know how networks capture and process information to induce specific cellular responses or phenotypes. Ideally, to correlate specific network structures with phenotypes, subtle rewiring of specific networks would need to be detected and quantified and related to well-defined, quantifiable phenotypes. Both the detection of network changes and the definition of quantitative phenotypes are challenging. It is therefore not surprising that studies in the field are sparse. A significant fraction of the published studies to date attempt to relate clinical phenotypes to changes in cellular networks, particularly in the field of cancer biology.

Correlating Network Wiring with Phenotypes for Protein Interaction Networks

The actual measurement of PPINs and their dynamic change remains technically challenging. This is in contrast to the description of static PPINs, where significant progress has been achieved. Several studies have therefore attempted to use static PPINs as a priori knowledge to investigate correlations in experimental data generated by various technologies. An underlying network, e.g., a static PPIN, is used to identify subsets of genes that have certain network properties and correlate with a phenotype (**Figure 3**). These have been successfully applied in combination with large-scale nucleic acid data. For example, Taylor et al. (129) constructed a large-scale PPIN and then used the average Pearson corrclation coefficient to quantify the extent to which a hub and its interacting partners were coexpressed. They investigated a cohort of sporadic, nonfamilial breast cancer patients and detected 256 hubs that displayed altered Pearson correlation coefficients of expression between groups with different clinical outcomes. Mani et al. (130) constructed a network by combining PPI data and protein-DNA interaction data measured by microarray technology and searched for genes with unusual numbers of edges, which show a change in correlation in a phenotypic subset of the samples. The ranked list can then be presented in the context of the constructed network. Rather than using a global static PPIN, Chang et al. (131) focused on the analysis of confined networks. First, an initial set of genes in a local network was defined, e.g., the interactors of an oncogene, which was then expanded to identify sets of genes whose signatures reflect modular aspects of expression variation. The activity of selected signatures was investigated in a panel of cancer cell lines and tumor samples. In this study, focused on local networks, the novel signatures are anchored to well-known network modules or local networks and may point to novel mechanisms within the initial context. Of note, Nibbe et al. (132) used targets from a proteomic screen as seeds for searching significant subnetworks in colorectal cancer, which included direct interactors as well as "crosstalkers," proteins in the neighborhood of seeds. Subnetworks were then scored, using mRNA expression data, to estimate the significance in differentiating tumor from control samples, allowing data integration regarding dysregulation at both mRNA and protein levels. Finally, in a unique approach, Lage et al. (133) integrated phenotypic data from targeted mutations in mice with PPIN data to describe a functional network underlying cardiac development. This is perhaps a simple example of a switch from the one gene–one protein-one function paradigm toward the network biology paradigm, where molecular networks are viewed as the determinants of phenotypes.

The above summary indicates that, to date, most studies that attempt to correlate molecular networks with phenotypes have combined large-scale, typically static proteomic data with other data resources. The validity of the conclusions derived from correlating quantitative data and network knowledge depends on the coverage and correctness of the network, the quality of the data, and the method of correlation calculation. Furthermore, the results do not necessarily suggest causality.

Correlating Network Wiring with Phenotypes for Protein-Signaling Networks

As for the use of MS-based proteomic data to describe the wiring and rewiring of PSNs (discussed above), most of the literature investigating the connection of PSNs with phenotypes has been focused on protein kinase-substrate networks. The following discussion is therefore also focused on protein phosphorylation but exemplifies other types of PSNs as well.

Protein kinases are attractive drug targets (121). For this reason, it is of interest to correlate the activity of a kinase with a given phenotype. Studies that attempted to unravel kinase-substrate relationships have indicated extensive indirect and compensatory effects, suggesting a complex and yet incompletely understood kinase-substrate network (47; see the Dynamic Network Rewiring section above). These data also suggest that discerning the rationale for pharmacologic inhibition of protein kinases with the aim of affecting disease phenotypes will be most successful in a network biology paradigm.

Several studies have generated activity profiles of kinases in biological samples. Rivoka et al. (134) examined the extent of tyrosine phosphorylation across many carcinoma cell lines and tumors, assuming that the degree of phosphorylation of the kinase correlated with its state of activity. In a more direct approach, Cutillas et al. (135) used MS-based proteomics to quantify a specific phosphopeptide as a surrogate for kinase activity. The peptide in its unphosphorylated form was incubated with a cell lysate and ATP, and after the reaction was quenched, the abundance of the phosphorylated form of the peptide was quantified using a spiked-in internal standard. Kubota et al. (136) applied a similar approach, but with 90 peptides, in a multiplexed manner. Although not every peptide could be assigned to a unique kinase, profiles of activity could be inferred and were shown to differ between different cell lines and in response to stimuli. In these methods, the signal indicating kinase activity is amplified by the kinase reaction, making even low-abundance kinases detectable, provided they are active. Extending such approaches to a complete kinome level will remain challenging as long as the wiring of the basic kinase-substrate network remains incomplete. Daub et al. (137) therefore profiled the phosphoproteome of kinases themselves by kinase-selective enrichment via affinity chromatography, followed by phosphopeptide enrichment. Monitoring regulatory phosphorylation sites on the enriched kinases, such as phosphorylation events on the activation loop, can then be used as a surrogate marker for changes in kinase activities. The use of kinase inhibitors as affinity reagents is attractive as it can allow capturing dozens of kinases at a time (137, 138). However, the analysis may become complicated in the case of some inhibitors for which the binding affinity of the kinase depends on its activation state and is therefore influenced by the phosphorylation of the activation loop (139).

Given the difficulties of relating complete or extensive PSNs to phenotypes, it is no surprise that studies focused on smaller subnetworks, and selected network nodes have progressed faster (reviewed in References 13, 14, and 140). Such studies are exemplified by the "cue-signal-response" (CSR) compendiums, a type of data acquisition where multiple, typically several dozen, samples are perturbed with various cues—molecules affecting a certain signaling network. Preselected network nodes and cellular phenotypes are then measured in these samples, and the resulting data are subjected to statistical and modeling frameworks to correlate signals with responses. For example, Janes et al. (141) stimulated colon adenocarcinoma cells with nine combinations of tumor necrosis factor, EGF, and insulin (cues), and collected 19 intracellular measurements of the known underlying signaling network (signals) and four apoptotic outputs (responses) along 13 time points. The compendium collected was then analyzed using a data-driven modeling approach to map relationships between the phosphorylation signals measured and cellular death responses. The model captured the two canonical axes of the cellular response, e.g., apoptosis versus survival, and identified previously unknown components of the signaling network, such as autocrine feedback. Saez-Rodriguez et al. (142) applied a different computational approach and used such a compendium to calibrate Boolean logic models of a literature-derived signaling network. They found that a Boolean model, even though it captures only two activation states (on and off), can still fit experimental data and therefore can be used as an approach to harness knowledge already available. The method applied by Janes et al. does not necessarily require a priori mechanistic knowledge (143), but it does entail choosing the informative combinations of CSR, which means a study based on prior knowledge is more likely to provide informative insights. Importantly, although it provides a condensed set of the most informative measurements that fit the data to a model, at the same time the method may overlook key condition-specific modulators, if these conditions are not properly addressed (144). Nelander et al. (145) suggested a method to derive a network structure without any a priori knowledge by applying a CSR setup that includes multiple inputs of drug combinations and measuring multiple outputs of phosphoproteins and phenotypes. Although nodes in the final model are defined as only those precisely perturbed or measured, the model could nevertheless recapitulate known relationships. Finally, Mitsos et al. (146) suggested an approach to identify alterations in a pathway/network of interest in response to drug treatment, regardless of the cellular phenotype. A cell-type-specific network was constructed using integer linear programming and a compendium of phosphorylation site measurements in cells treated with cytokines and known specific inhibitors. The effects of a given drug were detected by reapplying this procedure with the drug instead of the known inhibitor and by identifying altered edges in the model, inferred as drug specific. This approach does not require the measurement of responses, only cues and signals are measured, but it does require a priori knowledge for the network construction and does not offer a uniquely optimal solution.

What the above studies have in common are datasets of relatively few (typically a dozen or so) selected network nodes that were quantified in multiple perturbed samples and that provided insights to the dynamic rewiring correlating to the phenotype. Interestingly, the majority of these data were generated by quantitative Western blotting, a technique that is limited by the availability of antibodies specific for the selected network nodes. It can be expected that the use of suitable MS-based proteomic techniques for CSR compendiums will derive significant advantages. First, the selection of measured network nodes will no longer be constrained by the availability of antibodies: In principle, every protein or phosphorylation site should be measurable. Second, the number of quantified nodes can be extended with a modest increase in experimental cost. Third, the quantitative accuracy of MS-based proteomics should exceed that of Western blotting, and fourth, the sample throughput should be dramatically increased. However, the advantages MS-based proteomics can offer might cause a dramatic increase in computational cost.

As discussed above (see the Mass Spectrometry–Based Proteomics section), different mass spectrometric strategies differ in their performance profiles. Targeted proteomics by SRM is the method of choice if a limited number of analytes needs to be quantified in multiple samples at a high level of reproducibility, sensitivity, and quantitative accuracy, as is the case in the generation of CSR compendiums. The performance of this approach to study the dynamics of biological networks and the resulting phenotype was demonstrated in two recent studies. Costenoble et al. (45) developed SRM assays for 228 proteins that constitute the central carbon and amino acid metabolic network of yeast. This set of proteins was then quantified at five metabolic states. The data uncovered nutritional environment-dependent changes in protein abundances and identified isoenzymes that are preferentially expressed at specific states. The data also suggested that metabolic proteins, which are, according to current network models, not required for growth under certain nutritional environments, are still expressed, presumably to allow adaptation to rapid changes in environment. In a related study, Picotti et al. (44) quantified a more limited set of metabolic proteins in a series of samples collected across different metabolic shifts. Clustering of the resulting quantitative profiles indicated sets of proteins in the metabolic network that are subject to comparable transcriptional/translational control. These studies demonstrated that the nodes of a known network if selected and targeted can be quantified over the whole dynamic range of protein expression in yeast cells. One of the bottlenecks of SRM is the need to design optimal assays for each protein by using a unique set of peptides and their respective transitions. Recently, experimental and computational approaches have been developed to allow faster and more accurate development of these assays (27, 147–149), and for selected species, databases containing assays for each protein of the respective proteome are being developed. We therefore expect a significant increase in the use of targeted-MS proteomics for CSR compendiums and for the analysis of network-phenotype relationships in general.

OUTLOOK

The network biology paradigm places networks of interacting molecules between genotype and phenotype. It makes the assumptions that the network wiring is dynamically and multidirectionally modulated in response to external and internal stimuli and that such changes determine phenotypic changes. The abilities of MS-based proteomics to describe network wiring, to capture dynamic rewiring of networks in response to stimuli, and to correlate network wiring to phenotypes are constantly advancing.

MS-based proteomics faces major technological challenges in achieving the goals of comprehensive, reproducible, and quantitative description of proteomes at reasonable throughput. As these challenges are addressed, new ones arise. For example, with the rapid increase in throughput, estimating the level of confidence in assignment of MS spectra to peptide sequences is often performed by statistical analysis at a fixed false discovery rate. Over time, this can result in the accumulation of false data if large datasets are combined (150, 151). With the introduction of modeling into proteomics, we also hope to see application of experimental design optimization (152) not only with respect to statistical design issues (153), such as replication and blocking, but also with respect to choosing the informative time points, observables, and measurements.

There is no doubt that the rigorous implementation of network biology will require the development of a range of new (proteomic) technologies that are focused on identifying and quantifying the edges of dynamic networks. Although the description of PPINs has seen much progress, the description of interactions between proteins and other molecules, and the description of the PSN, has lagged behind. For the latter, we expect new experimental setups to address the relationships between enzymes and their substrates for various types of PTMs. Techniques to describe network rewiring in response to stimuli have been only recently emerging, and we still largely lack their application to measuring dynamics of known signaling edges. In this respect, datasets that attempt to measure not only nodes or edges, but also cellular responses to correlate them with, would advance the ability to harness proteomics to network biology.

Ideally, the PPIN and the PSN are evaluated in their genuine biological context. At present, the tools are not yet fully developed to chart the PPIN and the PSN in primary cells, tissues, or whole mammalian organisms. Although the basic wiring of PPINs and PSNs may be highly conserved (even from yeast to human), an extra dimension of complexity will be provided by the tissue-, cell-, and organelle-specific features of PPINs and PSNs. Optimistically, many of the proteomics strategies described here may be transferable to analysis of tissue and even of whole mammals.

Finally, using information already gathered to generate new knowledge is a task performed less often in MS-based proteomics. We would like to see more data being used as the generation of informative datasets is refined. At the moment, we deposit large datasets for wet lab biologists to use, with no knowledge if it is actually employed by them. We thus believe that the transition to network biology approaches and applying modeling will not only provide a new wealth of knowledge but also improve proteomic measurements, moving from "what can we measure" to "what should we measure." An iterative approach in which network-driven biological questions are addressed by MS-based proteomics, combined with computational modeling and used to guide the selection of the next set of experiments, will drive our understanding of biological complexity.

SUMMARY POINTS

1. The new paradigm emerging is typically referred to as systems biology, network biology, or integrated biology. Although molecular biology is concerned with the network nodes involving one-dimensional pathways, the new paradigm entails network nodes and edges and places multidimensional and multidirectional networks of interacting molecules between genotypes and phenotypes.

2. Until recently, mass spectrometry (MS)-based proteomics has been used predominantly to identify, characterize, and quantify network nodes without consideration of their context. While the results from such analyses have increased in volume (longer lists) and confidence, the amount of new biology learned has been moderate.

3. Recently, MS-based proteomics has been used to identify network edges, e.g., describe network wiring. Affinity purification followed by MS has been used to interrogate protein-interaction networks. In vitro assays and perturbation experiments have been used to describe or infer protein-signaling networks (PSNs).

4. Describing the dynamic multidirectional rewiring of networks has been emerging recently in MS-based proteomics. Applications to PPINs include the measurement of dynamic changes in protein complexes, whereas applications to PSNs have mainly focused on perturbation experiments used to infer network wiring.

5. Correlating network wiring to phenotypes using proteomics data has been mostly applied by combining static PPINs with dynamic microarray data. Correlating dynamic proteomic PPIN and PSN data awaits application, and cue-signal-response (CSR) may be an experimental setup for this task.

6. Overall, it is apparent that the MS-based proteomic technologies developed within the molecular biology paradigm are useful but not sufficient for network biology. Conceptually, new technologies, rather than incremental advances of current technologies, will therefore be required.

FUTURE ISSUES

1. Can a comprehensive coverage of a complex proteome be obtained in a reproducible manner, with high quantitative accuracy, and at moderate to high throughput?

2. Given that most data collected to date are under basal conditions, how can AP-MS workflows be improved to allow quantitative dynamics of network rewiring?

3. How can the identification or inference of edges in PSNs and their dynamic changes be improved to be quantitative, reproducible, and comprehensive? Can large-scale dynamic rewiring be captured by MS-based proteomics?

4. How can MS proteomics be applied to collect CSR compendiums? Considering their complexity, how can these be analyzed computationally?

5. How can computational analysis of prior data be used in experimental design to direct optimal MS proteomic measurements?

6. Will MS-based proteomics generate data for mechanistic models of signaling networks?

7. How can emerging proteomics technologies that probe network biology be efficiently transferred to primary cells, tissue, and whole animals?

DISCLOSURE STATEMENT

The authors are not aware of any affiliations, memberships, funding, or financial holdings that might be perceived as affecting the objectivity of this review.

ACKNOWLEDGMENTS

Work in our laboratory is supported by research grants from the European Community's Seventh Framework Program [the TRIREME Project (grant 223575), the PROSPECTS Project (grant 201648), SYBILLA (grant 201106), SYSTEM TB (grant 241587), and PRIME-XS (grant 262067)], Swiss National Science Foundation (grant 3100A0-107679), SystemsX.ch, and the Swiss Initiative for Systems Biology, as well as by funds from the ERC advanced grant, Proteomics v 3.0 (grant 233226). A.J.R.H. kindly acknowledges support of the Department of Biology of the ETH Zurich, enabling his sabbatical residence at the Institute for Molecular Systems Biology, which was also supported by a Distinguished Visiting Scientist Stipend of the Netherlands Genomics Initiative.

LITERATURE CITED

1. Beadle GW, Tatum EL. 1941. Genetic control of biochemical reactions in *Neurospora*. *Proc. Natl. Acad. Sci. USA* 27:499–506
2. Hawkins RD, Hon GC, Ren B. 2010. Next-generation genomics: an integrative approach. *Nat. Rev. Genet.* 11:476–86
3. Zhou X, Ren L, Meng Q, Li Y, Yu Y, Yu J. 2010. The next-generation sequencing technology and application. *Protein Cell* 1:520–36
4. Cox J, Mann M. 2011. Quantitative, high-resolution proteomics for data-driven systems biology. *Annu. Rev. Biochem.* 80:273–99
5. Le Provost F, Lillico S, Passet B, Young R, Whitelaw B, Vilotte JL. 2010. Zinc finger nuclease technology heralds a new era in mammalian transgenesis. *Trends Biotechnol.* 28:134–41
6. Boutros M, Ahringer J. 2008. The art and design of genetic screens: RNA interference. *Nat. Rev. Genet.* 9:554–66
7. Zhang EE, Liu AC, Hirota T, Miraglia LJ, Welch G, et al. 2009. A genome-wide RNAi screen for modifiers of the circadian clock in human cells. *Cell* 139:199–210
8. Kondo S, Perrimon N. 2011. A genome-wide RNAi screen identifies core components of the G-M DNA damage checkpoint. *Sci. Signal.* 4:rs1
9. Shapira SD, Gat-Viks I, Shum BO, Dricot A, de Grace MM, et al. 2009. A physical and regulatory map of host-influenza interactions reveals pathways in H1N1 infection. *Cell* 139:1255–67
10. Cazier JB, Tomlinson I. 2010. General lessons from large-scale studies to identify human cancer predisposition genes. *J. Pathol.* 220:255–62
11. Amberger J, Bocchini C, Hamosh A. 2011. A new face and new challenges for Online Mendelian Inheritance in Man (OMIM®). *Hum. Mutat.* 32:564–67
12. Arkin AP, Schaffer DV. 2011. Network news: innovations in 21st century systems biology. *Cell* 144:844–49
13. Kreeger PK, Lauffenburger DA. 2010. Cancer systems biology: a network modeling perspective. *Carcinogenesis* 31:2–8
14. Pe'er D, Hacohen N. 2011. Principles and strategies for developing network models in cancer. *Cell* 144:864–73
15. Vidal M, Cusick ME, Barabasi AL. 2011. Interactome networks and human disease. *Cell* 144:986–98
16. Yates JR, Ruse CI, Nakorchevsky A. 2009. Proteomics by mass spectrometry: approaches, advances, and applications. *Annu. Rev. Biomed. Eng.* 11:49–79
17. Domon B, Aebersold R. 2010. Options and considerations when selecting a quantitative proteomics strategy. *Nat. Biotechnol.* 28:710–21
18. Nesvizhskii AI, Vitek O, Aebersold R. 2007. Analysis and validation of proteomic data generated by tandem mass spectrometry. *Nat. Methods* 4:787–97
19. Mallick P, Kuster B. 2010. Proteomics: a pragmatic perspective. *Nat. Biotechnol.* 28:695–709
20. Ong SE, Blagoev B, Kratchmarova I, Kristensen DB, Steen H, et al. 2002. Stable isotope labeling by amino acids in cell culture, SILAC, as a simple and accurate approach to expression proteomics. *Mol. Cell. Proteomics* 1:376–86

21. Mueller LN, Rinner O, Schmidt A, Letarte S, Bodenmiller B, et al. 2007. *SuperHirn*—a novel tool for high resolution LC-MS-based peptide/protein profiling. *Proteomics* 7:3470–80
22. Neilson KA, Ali NA, Muralidharan S, Mirzaei M, Mariani M, et al. 2011. Less label, more free: approaches in label-free quantitative mass spectrometry. *Proteomics* 11:535–53
23. Jaffe JD, Keshishian H, Chang B, Addona TA, Gillette MA, Carr SA. 2008. Accurate inclusion mass screening: a bridge from unbiased discovery to targeted assay development for biomarker verification. *Mol. Cell. Proteomics* 7:1952–62
24. Schmidt A, Claassen M, Aebersold R. 2009. Directed mass spectrometry: towards hypothesis-driven proteomics. *Curr. Opin. Chem. Biol.* 13:510–17
25. Lange V, Picotti P, Domon B, Aebersold R. 2008. Selected reaction monitoring for quantitative proteomics: a tutorial. *Mol. Syst. Biol.* 4:222
26. Gallien S, Duriez E, Domon B. 2011. Selected reaction monitoring applied to proteomics. *J. Mass Spectrom.* 46:298–312
27. Picotti P, Rinner O, Stallmach R, Dautel F, Farrah T, et al. 2010. High-throughput generation of selected reaction-monitoring assays for proteins and proteomes. *Nat. Methods* 7:43–46
28. Addona TA, Abbatiello SE, Schilling B, Skates SJ, Mani DR, et al. 2009. Multi-site assessment of the precision and reproducibility of multiple reaction monitoring-based measurements of proteins in plasma. *Nat. Biotechnol.* 27:633–41
29. Geromanos SJ, Vissers JP, Silva JC, Dorschel CA, Li GZ, et al. 2009. The detection, correlation, and comparison of peptide precursor and product ions from data independent LC-MS with data dependant LC-MS/MS. *Proteomics* 9:1683–95
30. Geiger T, Cox J, Mann M. 2010. Proteomics on an Orbitrap benchtop mass spectrometer using all-ion fragmentation. *Mol. Cell. Proteomics* 9:2252–61
31. Venable JD, Dong MQ, Wohlschlegel J, Dillin A, Yates JR. 2004. Automated approach for quantitative analysis of complex peptide mixtures from tandem mass spectra. *Nat. Methods* 1:39–45
32. Panchaud A, Scherl A, Shaffer SA, von Haller PD, Kulasekara HD, et al. 2009. Precursor acquisition independent from ion count: how to dive deeper into the proteomics ocean. *Anal. Chem.* 81:6481–88
33. Gillet LC, Navarro P, Tate S, Röst H, Selevsek N, et al. 2012. Targeted data extraction of the MS/MS spectra generated by data independent acquisition: a new concept for consistent and accurate proteome analysis. *Mol. Cell. Proteomics.* In press
34. Ahrens CH, Brunner E, Qeli E, Basler K, Aebersold R. 2010. Generating and navigating proteome maps using mass spectrometry. *Nat. Rev. Mol. Cell Biol.* 11:789–801
35. de Godoy LM, Olsen JV, Cox J, Nielsen ML, Hubner NC, et al. 2008. Comprehensive mass-spectrometry-based proteome quantification of haploid versus diploid yeast. *Nature* 455:1251–54
36. Malmstrom J, Beck M, Schmidt A, Lange V, Deutsch EW, Aebersold R. 2009. Proteome-wide cellular protein concentrations of the human pathogen *Leptospira interrogans*. *Nature* 460:762–65
37. Brunner E, Ahrens CH, Mohanty S, Baetschmann H, Loevenich S, et al. 2007. A high-quality catalog of the *Drosophila melanogaster* proteome. *Nat. Biotechnol.* 25:576–83
38. Munoz J, Low TY, Kok YJ, Chin A, Frese CK, et al. 2011. The quantitative proteomes of human-induced pluripotent stem cells and embryonic stem cells. *Mol. Syst. Biol.* 7:550
39. Nagaraj N, Wisniewski JR, Geiger T, Cox J, Kircher M, et al. 2011. Deep proteome and transcriptome mapping of a human cancer cell line. *Mol. Syst. Biol.* 7:548
40. Beck M, Schmidt A, Malmstroem J, Claassen M, Ori A, et al. 2011. The quantitative proteome of a human cell line. *Mol. Syst. Biol.* 7:549
41. Choudhary C, Kumar C, Gnad F, Nielsen ML, Rehman M, et al. 2009. Lysine acetylation targets protein complexes and co-regulates major cellular functions. *Science* 325:834–40
42. Huttlin EL, Jedrychowski MP, Elias JE, Goswami T, Rad R, et al. 2010. A tissue-specific atlas of mouse protein phosphorylation and expression. *Cell* 143:1174–89
43. Zielinska DF, Gnad F, Wisniewski JR, Mann M. 2010. Precision mapping of an in vivo N-glycoproteome reveals rigid topological and sequence constraints. *Cell* 141:897–907
44. Picotti P, Bodenmiller B, Mueller LN, Domon B, Aebersold R. 2009. Full dynamic range proteome analysis of *S. cerevisiae* by targeted proteomics. *Cell* 138:795–806

45. Costenoble R, Picotti P, Reiter L, Stallmach R, Heinemann M, et al. 2011. Comprehensive quantitative analysis of central carbon and amino-acid metabolism in *Saccharomyces cerevisiae* under multiple conditions by targeted proteomics. *Mol. Syst. Biol.* 7:464
46. Olsen JV, Vermeulen M, Santamaria A, Kumar C, Miller ML, et al. 2010. Quantitative phosphoproteomics reveals widespread full phosphorylation site occupancy during mitosis. *Sci. Signal.* 3:ra3
47. Bodenmiller B, Wanka S, Kraft C, Urban J, Campbell D, et al. 2010. Phosphoproteomic analysis reveals interconnected system-wide responses to perturbations of kinases and phosphatases in yeast. *Sci. Signal.* 3:rs4
48. Van Hoof D, Muñoz J, Braam SR, Pinkse MW, Linding R, et al. 2009. Phosphorylation dynamics during early differentiation of human embryonic stem cells. *Cell Stem Cell* 5:214–26
49. Schwanhausser B, Busse D, Li N, Dittmar G, Schuchhardt J, et al. 2011. Global quantification of mammalian gene expression control. *Nature* 473:337–42
50. Schmidt A, Beck M, Malmström J, Lam H, Claassen M, et al. 2011. Absolute quantification of microbial proteomes at different states by directed mass spectrometry. *Mol. Syst. Biol.* 7:510
51. Huang da W, Sherman BT, Lempicki RA. 2009. Systematic and integrative analysis of large gene lists using DAVID bioinformatics resources. *Nat. Protoc.* 4:44–57
52. Ma'ayan A. 2008. Network integration and graph analysis in mammalian molecular systems biology. *IET Syst. Biol.* 2:206–21
53. Gehlenborg N, O'Donoghue SI, Baliga NS, Goesmann A, Hibbs MA, et al. 2010. Visualization of omics data for systems biology. *Nat. Methods* 7:S56–68
54. Park PJ. 2009. ChIP-seq: advantages and challenges of a maturing technology. *Nat. Rev. Genet.* 10:669–80
55. Suter B, Kittanakom S, Stagljar I. 2008. Two-hybrid technologies in proteomics research. *Curr. Opin. Biotechnol.* 19:316–23
56. Ideker T, Thorsson V, Ranish JA, Christmas R, Buhler J, et al. 2001. Integrated genomic and proteomic analyses of a systematically perturbed metabolic network. *Science* 292:929–34
57. Yu H, Braun P, Yildirim MA, Lemmens I, Venkatesan K, et al. 2008. High-quality binary protein interaction map of the yeast interactome network. *Science* 322:104–10
58. Bandyopadhyay S, Mehta M, Kuo D, Sung MK, Chuang R, et al. 2010. Rewiring of genetic networks in response to DNA damage. *Science* 330:1385–89
59. Butter F, Scheibe M, Morl M, Mann M. 2009. Unbiased RNA-protein interaction screen by quantitative proteomics. *Proc. Natl. Acad. Sci. USA* 106:10626–31
60. Bantscheff M, Scholten A, Heck AJ. 2009. Revealing promiscuous drug-target interactions by chemical proteomics. *Drug Discov. Today* 14:1021–29
61. Rix U, Superti-Furga G. 2009. Target profiling of small molecules by chemical proteomics. *Nat. Chem. Biol.* 5:616–24
62. Li X, Gianoulis TA, Yip KY, Gerstein M, Snyder M. 2010. Extensive in vivo metabolite-protein interactions revealed by large-scale systematic analyses. *Cell* 143:639–50
63. Gingras AC, Gstaiger M, Raught B, Aebersold R. 2007. Analysis of protein complexes using mass spectrometry. *Nat. Rev. Mol. Cell Biol.* 8:645–54
64. Sharon M, Robinson CV. 2007. The role of mass spectrometry in structure elucidation of dynamic protein complexes. *Annu. Rev. Biochem.* 76:167–93
65. Heck AJ. 2008. Native mass spectrometry: a bridge between interactomics and structural biology. *Nat. Methods* 5:927–33
66. Leitner A, Walzthoeni T, Kahraman A, Herzog F, Rinner O, et al. 2010. Probing native protein structures by chemical cross-linking, mass spectrometry, and bioinformatics. *Mol. Cell. Proteomics* 9:1634–49
67. Sigal A, Danon T, Cohen A, Milo R, Geva-Zatorsky N, et al. 2007. Generation of a fluorescently labeled endogenous protein library in living human cells. *Nat. Protoc.* 2:1515–27
68. de Boer E, Rodriguez P, Bonte E, Krijgsveld J, Katsantoni E, et al. 2003. Efficient biotinylation and single-step purification of tagged transcription factors in mammalian cells and transgenic mice. *Proc. Natl. Acad. Sci. USA* 100:7480–85
69. Hubner NC, Bird AW, Cox J, Splettstoesser B, Bandilla P, et al. 2010. Quantitative proteomics combined with BAC TransgeneOmics reveals in vivo protein interactions. *J. Cell Biol.* 189:739–54

70. Malovannaya A, Lanz RB, Jung SY, Bulynko Y, Le NT, et al. 2011. Analysis of the human endogenous coregulator complexome. *Cell* 145:787–99
71. Boulon S, Ahmad Y, Trinkle-Mulcahy L, Verheggen C, Cobley A, et al. 2010. Establishment of a protein frequency library and its application in the reliable identification of specific protein interaction partners. *Mol. Cell. Proteomics* 9:861–79
72. Trinkle-Mulcahy L, Boulon S, Lam YW, Urcia R, Boisvert FM, et al. 2008. Identifying specific protein interaction partners using quantitative mass spectrometry and bead proteomes. *J. Cell Biol.* 183:223–39
73. Rinner O, Mueller LN, Hubálek M, Müller M, Gstaiger M, Aebersold R. 2007. An integrated mass spectrometric and computational framework for the analysis of protein interaction networks. *Nat. Biotechnol.* 25:345–52
74. Glatter T, Wepf A, Aebersold R, Gstaiger M. 2009. An integrated workflow for charting the human interaction proteome: insights into the PP2A system. *Mol. Syst. Biol.* 5:237
75. Sardiu ME, Cai Y, Jin J, Swanson SK, Conaway RC, et al. 2008. Probabilistic assembly of human protein interaction networks from label-free quantitative proteomics. *Proc. Natl. Acad. Sci. USA* 105:1454–59
76. Sowa ME, Bennett EJ, Gygi SP, Harper JW. 2009. Defining the human deubiquitinating enzyme interaction landscape. *Cell* 138:389–403
77. Choi H, Larsen B, Lin ZY, Breitkreutz A, Mellacheruvu D, et al. 2011. SAINT: probabilistic scoring of affinity purification-mass spectrometry data. *Nat. Methods* 8:70–73
78. Behrends C, Sowa ME, Gygi SP, Harper JW. 2010. Network organization of the human autophagy system. *Nature* 466:68–76
79. Kuhner S, van Noort V, Betts MJ, Leo-Macias A, Batisse C, et al. 2009. Proteome organization in a genome-reduced bacterium. *Science* 326:1235–40
80. Breitkreutz A, Choi H, Sharom JR, Boucher L, Neduva V, et al. 2010. A global protein kinase and phosphatase interaction network in yeast. *Science* 328:1043–46
81. Vermeulen M, Eberl HC, Matarese F, Marks H, Denissov S, et al. 2010. Quantitative interaction proteomics and genome-wide profiling of epigenetic histone marks and their readers. *Cell* 142:967–80
82. Lee KK, Sardiu ME, Swanson SK, Gilmore JM, Torok M, et al. 2011. Combinatorial depletion analysis to assemble the network architecture of the SAGA and ADA chromatin remodeling complexes. *Mol. Syst. Biol.* 7:503
83. Wang J, Li M, Deng Y, Pan Y. 2010. Recent advances in clustering methods for protein interaction networks. *BMC Genomics* 11(Suppl. 3):S10
84. Choi H, Kim S, Gingras AC, Nesvizhskii AI. 2010. Analysis of protein complexes through model-based biclustering of label-free quantitative AP-MS data. *Mol. Syst. Biol.* 6:385
85. Scott JD, Pawson T. 2009. Cell signaling in space and time: where proteins come together and when they're apart. *Science* 326:1220–24
86. Bodenmiller B, Mueller LN, Mueller M, Domon B, Aebersold R. 2007. Reproducible isolation of distinct, overlapping segments of the phosphoproteome. *Nat. Methods* 4:231–37
87. Lemeer S, Heck AJ. 2009. The phosphoproteomics data explosion. *Curr. Opin. Chem. Biol.* 13:414–20
88. Thingholm TE, Jensen ON, Larsen MR. 2009. Analytical strategies for phosphoproteomics. *Proteomics* 9:1451–68
89. Hornbeck PV, Chabra I, Kornhauser JM, Skrzypek E, Zhang B. 2004. PhosphoSite: a bioinformatics resource dedicated to physiological protein phosphorylation. *Proteomics* 4:1551–61
90. Cutillas PR, Jørgensen C. 2011. Biological signalling activity measurements using mass spectrometry. *Biochem. J.* 434:189–99
91. Tan CS, Linding R. 2009. Experimental and computational tools useful for (re)construction of dynamic kinase-substrate networks. *Proteomics* 9:5233–42
92. Mok J, Im H, Snyder M. 2009. Global identification of protein kinase substrates by protein microarray analysis. *Nat. Protoc.* 4:1820–27
93. Mok J, Kim PM, Lam HY, Piccirillo S, Zhou X, et al. 2010. Deciphering protein kinase specificity through large-scale analysis of yeast phosphorylation site motifs. *Sci. Signal.* 3:ra12
94. Allen JJ, Li M, Brinkworth CS, Paulson JL, Wang D, et al. 2007. A semisynthetic epitope for kinase substrates. *Nat. Methods* 4:511–16

95. Blethrow JD, Glavy JS, Morgan DO, Shokat KM. 2008. Covalent capture of kinase-specific phosphopeptides reveals Cdk1-cyclin B substrates. *Proc. Natl. Acad. Sci. USA* 105:1442–47
96. Bloom J, Cristea IM, Procko AL, Lubkov V, Chait BT, et al. 2011. Global analysis of Cdc14 phosphatase reveals diverse roles in mitotic processes. *J. Biol. Chem.* 286:5434–45
97. Maly DJ, Allen JA, Shokat KM. 2004. A mechanism-based cross-linker for the identification of kinase-substrate pairs. *J. Am. Chem. Soc.* 126:9160–61
98. Linding R, Jensen LJ, Pasculescu A, Olhovsky M, Colwill K, et al. 2008. NetworKIN: a resource for exploring cellular phosphorylation networks. *Nucleic Acids Res.* 36:D695–99
99. Karlebach G, Shamir R. 2008. Modelling and analysis of gene regulatory networks. *Nat. Rev. Mol. Cell Biol.* 9:770–80
100. De Smet R, Marchal K. 2010. Advantages and limitations of current network inference methods. *Nat. Rev. Microbiol.* 8:717–29
101. Helbig AO, Gauci S, Raijmakers R, van Breukelen B, Slijper M, et al. 2010. Profiling of N-acetylated protein termini provides in-depth insights into the N-terminal nature of the proteome. *Mol. Cell. Proteomics* 9:928–39
102. Kerscher O, Felberbaum R, Hochstrasser M. 2006. Modification of proteins by ubiquitin and ubiquitin-like proteins. *Annu. Rev. Cell Dev. Biol.* 22:159–80
103. Shi Y, Xu P, Qin J. 2011. Ubiquitinated proteome: ready for global? *Mol. Cell. Proteomics* 10:R110.006882
104. Argenzio E, Bange T, Oldrini B, Bianchi F, Peesari R, et al. 2011. Proteomic snapshot of the EGF-induced ubiquitin network. *Mol. Syst. Biol.* 7:462
105. Kim W, Bennett EJ, Huttlin EL, Guo A, Li J, et al. 2011. Systematic and quantitative assessment of the ubiquitin-modified proteome. *Mol. Cell* 44:325–40
106. Wagner SA, Beli P, Weinert BT, Nielsen ML, Cox J, et al. 2011. A proteome-wide, quantitative survey of in vivo ubiquitylation sites reveals widespread regulatory roles. *Mol. Cell. Proteomics* 10:M111 013284
107. Schilling O, auf dem Keller U, Overall CM. 2011. Protease specificity profiling by tandem mass spectrometry using proteome-derived peptide libraries. *Methods Mol. Biol.* 753:257–72
108. Nomura DK, Dix MM, Cravatt BF. 2010. Activity-based protein profiling for biochemical pathway discovery in cancer. *Nat. Rev. Cancer* 10:630–38
109. Mahrus S, Trinidad JC, Barkan DT, Sali A, Burlingame AL, Wells JA. 2008. Global sequencing of proteolytic cleavage sites in apoptosis by specific labeling of protein N termini. *Cell* 134:866–76
110. Kleifeld O, Doucet A, auf dem Keller U, Prudova A, Schilling O, et al. 2010. Isotopic labeling of terminal amines in complex samples identifies protein N-termini and protease cleavage products. *Nat. Biotechnol.* 28:281–88
111. von Kriegsheim A, Baiocchi D, Birtwistle M, Sumpton D, Bienvenut W, et al. 2009. Cell fate decisions are specified by the dynamic ERK interactome. *Nat. Cell Biol.* 11:1458–64
112. Baker CL, Kettenbach AN, Loros JJ, Gerber SA, Dunlap JC. 2009. Quantitative proteomics reveals a dynamic interactome and phase-specific phosphorylation in the *Neurospora* circadian clock. *Mol. Cell* 34:354–63
113. Aye TT, Soni S, van Veen TA, van der Heyden MA, Cappadona S, et al. 2012. Reorganized PKA-AKAP associations in the failing human heart. *J. Mol. Cell. Cardiol.* 52:511–18
114. Wepf A, Glatter T, Schmidt A, Aebersold R, Gstaiger M. 2009. Quantitative interaction proteomics using mass spectrometry. *Nat. Methods* 6:203–5
115. Bennett EJ, Rush J, Gygi SP, Harper JW. 2010. Dynamics of cullin-RING ubiquitin ligase network revealed by systematic quantitative proteomics. *Cell* 143:951–65
116. Gerber SA, Rush J, Stemman O, Kirschner MW, Gygi SP. 2003. Absolute quantification of proteins and phosphoproteins from cell lysates by tandem MS. *Proc. Natl. Acad. Sci. USA* 100:6940–45
117. Bisson N, James DA, Ivosev G, Tate SA, Bonner R, et al. 2011. Selected reaction monitoring mass spectrometry reveals the dynamics of signaling through the GRB2 adaptor. *Nat. Biotechnol.* 29:653–58
118. Zhong Q, Simonis N, Li QR, Charloteaux B, Heuze F, et al. 2009. Edgetic perturbation models of human inherited disorders. *Mol. Syst. Biol.* 5:321
119. Bensimon A, Schmidt A, Ziv Y, Elkon R, Wang SY, et al. 2010. ATM-dependent and -independent dynamics of the nuclear phosphoproteome after DNA damage. *Sci. Signal.* 3:rs3

120. Yu Y, Yoon SO, Poulogiannis G, Yang Q, Ma XM, et al. 2011. Phosphoproteomic analysis identifies Grb10 as an mTORC1 substrate that negatively regulates insulin signaling. *Science* 332:1322–26
121. Karaman MW, Herrgard S, Treiber DK, Gallant P, Atteridge CE, et al. 2008. A quantitative analysis of kinase inhibitor selectivity. *Nat. Biotechnol.* 26:127–32
122. Koch A, Hauf S. 2010. Strategies for the identification of kinase substrates using analog-sensitive kinases. *Eur. J. Cell Biol.* 89:184–93
123. Holt LJ, Tuch BB, Villen J, Johnson AD, Gygi SP, Morgan DO. 2009. Global analysis of Cdk1 substrate phosphorylation sites provides insights into evolution. *Science* 325:1682–86
124. Matsuoka S, Ballif BA, Smogorzewska A, McDonald ER 3rd, Hurov KE, et al. 2007. ATM and ATR substrate analysis reveals extensive protein networks responsive to DNA damage. *Science* 316:1160–66
125. Moritz A, Li Y, Guo A, Villen J, Wang Y, et al. 2010. Akt-RSK-S6 kinase signaling networks activated by oncogenic receptor tyrosine kinases. *Sci. Signal.* 3:ra64
126. Mayya V, Lundgren DH, Hwang SI, Rezaul K, Wu L, et al. 2009. Quantitative phosphoproteomic analysis of T cell receptor signaling reveals system-wide modulation of protein-protein interactions. *Sci. Signal.* 2:ra46
127. Jørgensen C, Sherman A, Chen GI, Pasculescu A, Poliakov A, et al. 2009. Cell-specific information processing in segregating populations of Eph receptor ephrin-expressing cells. *Science* 326:1502–9
128. Coba MP, Pocklington AJ, Collins MO, Kopanitsa MV, Uren RT, et al. 2009. Neurotransmitters drive combinatorial multistate postsynaptic density networks. *Sci. Signal.* 2:ra19
129. Taylor IW, Linding R, Warde-Farley D, Liu Y, Pesquita C, et al. 2009. Dynamic modularity in protein interaction networks predicts breast cancer outcome. *Nat. Biotechnol.* 27:199–204
130. Mani KM, Lefebvre C, Wang K, Lim WK, Basso K, et al. 2008. A systems biology approach to prediction of oncogenes and molecular perturbation targets in B-cell lymphomas. *Mol. Syst. Biol.* 4:169
131. Chang JT, Carvalho C, Mori S, Bild AH, Gatza ML, et al. 2009. A genomic strategy to elucidate modules of oncogenic pathway signaling networks. *Mol. Cell* 34:104–14
132. Nibbe RK, Koyutürk M, Chance MR. 2010. An integrative -omics approach to identify functional subnetworks in human colorectal cancer. *PLoS Comput. Biol.* 6:e1000639
133. Lage K, Mollgard K, Greenway S, Wakimoto H, Gorham JM, et al. 2010. Dissecting spatio-temporal protein networks driving human heart development and related disorders. *Mol. Syst. Biol.* 6:381
134. Rikova K, Guo A, Zeng Q, Possemato A, Yu J, et al. 2007. Global survey of phosphotyrosine signaling identifies oncogenic kinases in lung cancer. *Cell* 131:1190–203
135. Cutillas PR, Khwaja A, Graupera M, Pearce W, Gharbi S, et al. 2006. Ultrasensitive and absolute quantification of the phosphoinositide 3-kinase/Akt signal transduction pathway by mass spectrometry. *Proc. Natl. Acad. Sci. USA* 103:8959–64
136. Kubota K, Anjum R, Yu Y, Kunz RC, Andersen JN, et al. 2009. Sensitive multiplexed analysis of kinase activities and activity-based kinase identification. *Nat. Biotechnol.* 27:933–40
137. Daub H, Olsen JV, Bairlein M, Gnad F, Oppermann FS, et al. 2008. Kinase-selective enrichment enables quantitative phosphoproteomics of the kinome across the cell cycle. *Mol. Cell* 31:438–48
138. Bantscheff M, Eberhard D, Abraham Y, Bastuck S, Boesche M, et al. 2007. Quantitative chemical proteomics reveals mechanisms of action of clinical ABL kinase inhibitors. *Nat. Biotechnol.* 25:1035–44
139. Wodicka LM, Ciceri P, Davis MI, Hunt JP, Floyd M, et al. 2010. Activation state-dependent binding of small molecule kinase inhibitors: structural insights from biochemistry. *Chem. Biol.* 17:1241–49
140. Janes KA, Yaffe MB. 2006. Data-driven modelling of signal-transduction networks. *Nat. Rev. Mol. Cell Biol.* 7:820–28
141. Janes KA, Albeck JG, Gaudet S, Sorger PK, Lauffenburger DA, Yaffe MB. 2005. A systems model of signaling identifies a molecular basis set for cytokine-induced apoptosis. *Science* 310:1646–53
142. Saez-Rodriguez J, Alexopoulos LG, Epperlein J, Samaga R, Lauffenburger DA, et al. 2009. Discrete logic modelling as a means to link protein signalling networks with functional analysis of mammalian signal transduction. *Mol. Syst. Biol.* 5:331
143. Cosgrove BD, Alexopoulos LG, Hang TC, Hendriks BS, Sorger PK, et al. 2010. Cytokine-associated drug toxicity in human hepatocytes is associated with signaling network dysregulation. *Mol. BioSyst.* 6:1195–206

144. Wu Y, Johnson GL, Gomez SM. 2008. Data-driven modeling of cellular stimulation, signaling and output response in RAW 264.7 cells. *J. Mol. Signal.* 3:11
145. Nelander S, Wang W, Nilsson B, She QB, Pratilas C, et al. 2008. Models from experiments: combinatorial drug perturbations of cancer cells. *Mol. Syst. Biol.* 4:216
146. Mitsos A, Melas IN, Siminelakis P, Chairakaki AD, Saez-Rodriguez J, Alexopoulos LG. 2009. Identifying drug effects via pathway alterations using an integer linear programming optimization formulation on phosphoproteomic data. *PLoS Comput. Biol.* 5:e1000591
147. MacLean B, Tomazela DM, Shulman N, Chambers M, Finney GL, et al. 2010. Skyline: an open source document editor for creating and analyzing targeted proteomics experiments. *Bioinformatics* 26:966–68
148. Brusniak MY, Kwok ST, Christiansen M, Campbell D, Reiter L, et al. 2011. ATAQS: a computational software tool for high throughput transition optimization and validation for selected reaction monitoring mass spectrometry. *BMC Bioinform.* 12:78
149. Reiter L, Rinner O, Picotti P, Huttenhain R, Beck M, et al. 2011. mProphet: automated data processing and statistical validation for large-scale SRM experiments. *Nat. Methods* 8:430–35
150. White FM. 2011. The potential cost of high-throughput proteomics. *Sci. Signal.* 4:pe8
151. Reiter L, Claassen M, Schrimpf SP, Jovanovic M, Schmidt A, et al. 2009. Protein identification false discovery rates for very large proteomics data sets generated by tandem mass spectrometry. *Mol. Cell. Proteomics* 8:2405–17
152. Kreutz C, Timmer J. 2009. Systems biology: experimental design. *FEBS J.* 276:923–42
153. Oberg AL, Vitek O. 2009. Statistical design of quantitative mass spectrometry–based proteomic experiments. *J. Proteome Res.* 8:2144–56

Membrane Fission: The Biogenesis of Transport Carriers

Felix Campelo[1] and Vivek Malhotra[1,2]

[1]Department of Cell and Developmental Biology, Center for Genomic Regulation (CRG) and UPF, 08003 Barcelona, Spain; email: vivek.malhotra@crg.eu

[2]ICREA, 08010 Barcelona, Spain

Keywords

dynamin, BAR domain, actin, Sar1, Arf, protein kinase D

Abstract

Membrane-bound transport carriers are used to transfer cargo between membranes of the secretory and the endocytic pathways. The generation of these carriers can be classified into three steps: segregation of cargo away from the residents of a donor compartment (cargo sorting), generation of membrane curvature commensurate with the size of the cargo (membrane budding or tubulation), and finally separation of the nascent carrier from the donor membrane by a scission or membrane fission event. This review summarizes advances in our understanding of some of the best-characterized proteins required for the membrane fission that separates a transport carrier from its progenitor compartment: the large GTPase dynamin, the small guanine nucleotide-binding (G) proteins of the Arf family, BAR (Bin-amphiphysin-Rvs) domain proteins, and protein kinase D. These proteins share their ability to insert into membranes and oligomerize to create the large curvatures; however, the overall process of fission that involves these proteins appears to be quite different.

Contents

INTRODUCTION	408
FISSION PREREQUISITES	408
Membrane Bending and Change of Curvature	409
Energy Requirements for Membrane Fission	411
DYNAMIN: A LARGE GTPase	411
Is Dynamin the Ultimate Membrane Cutter?	412
Key Players in Membrane Fission	413
Sar1 AND Arf1: SMALL G PROTEINS	414
Sar1	414
Arf1	415
PROTEIN KINASE D–DEPENDENT MEMBRANE FISSION	416
A Model for Protein Kinase D–Dependent Fission at the trans-Golgi Network	418
KINETICS OF MEMBRANE FISSION TO REGULATE THE SIZE OF TRANSPORT CARRIERS AT THE ENDOPLASMIC RETICULUM	419
SUMMARY AND FUTURE PERSPECTIVES	419

Fusion: the merger of two topologically separated membranes into one

Fission: the separation by scission of a part of a membrane to form an independent membrane

Clathrin-coated vesicles: coated vesicular transport carriers involved in different transport routes from the TGN, endosomes, and the plasma membrane

INTRODUCTION

More than 35 years ago, George Palade (1) proposed that the best way to maintain compartmental identity during protein secretion and endocytosis is to form cargo-containing carriers (vesicles, tubules, and megavesicles) at the donor compartment and then fuse them with the next compartment along the pathway. Today, the existence of transport carriers is fully accepted, first proven by the accumulation of transport vesicle intermediates in cells harboring mutations in specific genes needed for vesicle consumption (2).

Fusion of transport carriers with their targets is mediated by soluble N-ethylmaleimide-sensitive factor attachment protein receptor (SNARE) and Sec1/Munc18-like proteins (3). In certain cases, additional factors add a layer of regulation, for example, fusion of synaptic vesicles with the presynaptic plasma membrane is triggered by an influx of calcium, which necessitates the involvement of additional regulatory proteins (4). Overall, the process of membrane fusion is relatively well understood with respect to the core proteins that participate: Sec1/Munc18-like proteins regulate SNARE complex assembly at the site of membrane fusion and cooperate with SNAREs to mediate fusion. By contrast, the process of membrane fission is much less well understood. Discussions of membrane fission began with the discovery of dynamin, a large GTPase required for the formation of clathrin-coated vesicles at the plasma membrane. Surprisingly, formation of clathrin-coated vesicles at the yeast plasma membrane is independent of dynamin; thus, other factors must play this role. The formation of coat protein complex II (COPII) vesicles at the endoplasmic reticulum (ER) and COPI vesicles at the Golgi complex is also independent of dynamin or a dynamin-like protein. Thus, not all membrane fission events are mediated by dynamin. In this review, we summarize the physical requirements for cutting a lipid bilayer and discuss the relative contributions of large GTPases and small guanine nucleotide-binding (G) proteins in membrane deformation and facilitation of cutting. Finally, we highlight the role of protein kinase D (PKD) in the regulation of membrane fission at the trans-Golgi network (TGN) and the role of the ER protein TANGO1 in controlling membrane fission to regulate the size of transport carriers to fit large cargoes.

FISSION PREREQUISITES

Membrane fission requires local distortion and remodeling of the lipid bilayer to create a separate membrane-bound compartment without compromising the integrity of the maternal bilayer. Indeed, scission of a nascent transport carrier must occur without permitting any direct exposure of the vesicle lumen with the

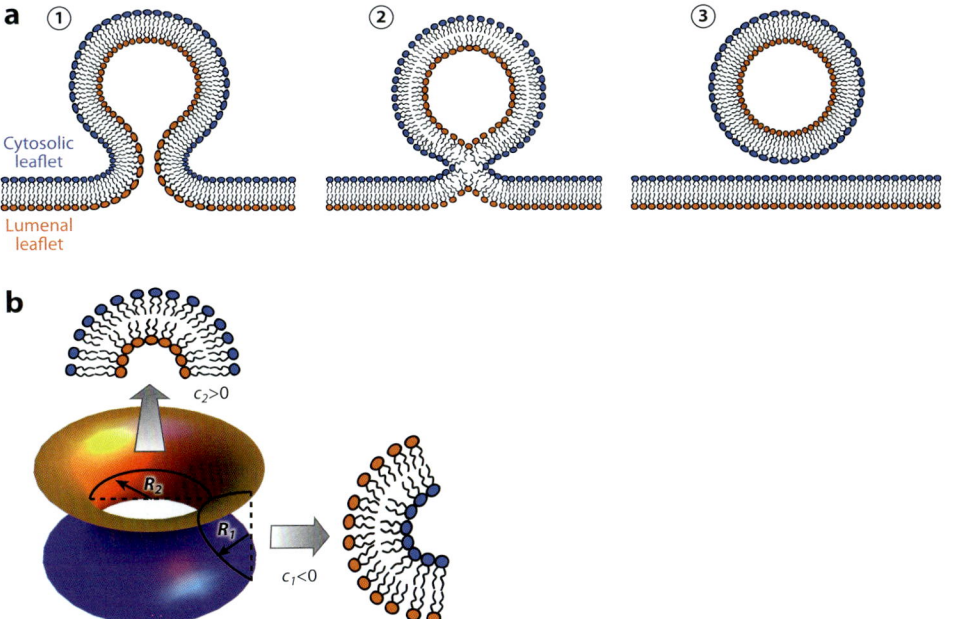

Figure 1

The pathway of membrane fission. (*a*) The steps showing the fission of a vesicular transport carrier, including membrane budding ①, the formation of a hemifission intermediate ②, and the eventual separation of the carrier ③. (*b*) Detailed view of the different curvatures at the neck of a membrane bud. The two principal curvatures, c_1 and c_2, are the inverse of the two radii of curvature, R_1 and R_2, respectively, and they have opposite signs: c_2 is positive (the lipids in the external leaflet bulge in the direction of their polar heads), and c_1 is negative (the lipids in the external leaflet bulge in the direction of their acyl chains). The cytosolic and lumenal leaflets of the membrane are depicted in blue and orange, respectively.

external milieu (5, 6), as the formation of even a small pore would result in the leakage of the carrier's contents (7).

By analogy to leakage free fusion events, membrane fission has been proposed to proceed via a hemifusion-like pathway (8–11). The nonleaky fission pathway through the formation of a hemifission state can be summarized as follows:

1. A constricted neck is formed, and membranes of the fission neck come into close proximity (**Figure 1***a*) (1). This intermembrane distance has been theoretically estimated on the order of less than 3 nm for clathrin-coated buds (8).
2. Contacting monolayers merge in a fission stalk intermediate (hemifission), similar to a fusion stalk (**Figure 1***a*) (2).
3. Decay of the fission stalk and completion of the fission reaction (**Figure 1***a*) (3).

Membrane Bending and Change of Curvature

Membrane fission requires local deformation of the lipid bilayer. Bilayers are elastic sheets that are subject to different kinds of strains including stretching/compression, curvature, and the tilt of the lipid tails. Tension regulates the stretching/compression of the membrane as a whole in the lateral direction and can be regulated in cells, for example, by the actin cytoskeleton. Curvature plays a key role in generating the highly bent vesicle and tubule structures of the transport carriers. Lipid tilt, or the deformation of the lipid hydrocarbon chains to

Coat protein complex II (COPII) vesicles: coated vesicular transport carriers that transport proteins from the ER to the Golgi complex

Coat protein complex I (COPI) vesicles: coated vesicular transport carriers that transport proteins from the Golgi complex back to the ER

Transport and Golgi organization 1 (TANGO1): a protein of 1907 amino acids, which is involved in the exit of collagens from the ER

Hemifission: an intermediate fission state where the proximal monolayer of the bilayer forming the neck self-fuses, forming a stalk

Giant unilamellar vesicles: liposomes of a size larger than a micron

accommodate packing defects in the membrane core, is essential for the events leading to membrane fusion or fission. Each of these modes of deformation is associated with different energy changes and characterized by elastic coefficients that depend on the physical-chemical properties of the lipids forming the bilayer.

The hypothesis that lipid bilayers oppose mechanical resistance against curvature is formulated by the elastic theory of membrane curvature developed by Helfrich (12). In this mathematical formulation, the membrane is characterized by the two principal curvatures at every point of each monolayer's neutral surfaces (13). The principal curvatures, c_1 and c_2, are defined as the inverse of the radii of curvature along each one of the two principal directions, R_1 and R_2 (**Figure 1b**) (14). The total curvature of the membrane, J, is the sum of the two principal curvatures, and the Gaussian curvature, K, is their product. Simply, the Helfrich theory states that the elastic energy of the membrane varies as these two curvatures change.

The bending energy associated with any deviation of the bilayer total curvature, J, from a preferred, or spontaneous, curvature, J_s, is:

$$E_{bend} = \int \kappa/2 (J - J_s)^2 \, dA,$$

where κ is the bending rigidity of the membrane—the elastic coefficient that describes how resistant to bending the membrane is. Measurements of the bending rigidity for different lipid membranes have been achieved using different techniques (15), e.g., by micropipette pressurization of giant unilamellar vesicles (16), giving typical values for lipid bilayers on the order of $\kappa \approx 20 \, k_B T$ (where $k_B T = 4.11 \times 10^{-21} J$ is the thermal energy). The second coefficient in the equation, the spontaneous curvature of the bilayer, is the curvature the membrane would adopt if there were no further constraints. It depends on the lipid composition and asymmetry of each monolayer (17), and can also be tuned by membrane-associated proteins (18, 19).

The energy of the Gaussian curvature is

$$E_G = \int \kappa_G K \, dA,$$

where κ_G is the Gaussian rigidity of the membrane, the value of which has been suggested to be negative (20). Direct measurements of this coefficient remain elusive, but indirect estimations on the basis of the energies of different structural phase transitions of lipid assemblies have been made, giving a value of $\kappa_G \approx -15 \, k_B T$ (21). In most circumstances, the energy of the Gaussian curvature does not depend on the actual shape of the membrane but only on the topology of the membrane. Mathematically, this is described by a global quantity, named the Euler characteristic, χ, defined as twice the number of independent compartments (N) minus the number of handles or holes in the membranes (g): $\chi = 2 \times (N - g)$. The Gaussian energy is then:

$$E_G = 2\pi \kappa_G \chi.$$

This equation shows that the Gaussian energy remains constant during membrane bending but varies upon membrane fission; after this event, the Euler characteristic changes from two to four, and the change on the energy of the Gaussian curvature is $4\pi\kappa_G$. The fact that the value of the Gaussian rigidity is negative implies that the Gaussian energy favors spontaneous fission. Moreover, this elastic coefficient depends on the spontaneous curvature of the bilayer, being more negative when the spontaneous curvature of the bilayer is positive. This means that lipids and/or protein insertions that positively bend the bilayer help promote membrane fission (11, 21a).

The lipids forming the bilayer are also subject to transverse shear deformation, or tilt of the hydrocarbon chains with respect to the surface of the bilayer (22, 23). This type of deformation is especially relevant for the generation of packing defects in the membrane core owing to curvature generation by hydrophobic insertions (24, 25) or to the formation of nonbilayer intermediates during membrane fusion and fission (26, 27).

Energy Requirements for Membrane Fission

How do proteins and lipids recruited to the site of membrane budding make fission an energetically and mechanically feasible process? As described above, lipid rearrangements leading to membrane deformation generate elastic stresses; therefore, elastic energy is stored in the bilayer. To compress or expand a piece of membrane, an external force has to be generated to perform the work of membrane deformation. The same logic holds for bilayer bending and lipid tilting. For membrane fission to occur, the total free energy of the system (the elastic energy stored in the membrane plus the work of the external energy sources) has to decrease: The total free energy of the system after fission completion has to be lower than the energy before the onset of fission.

Additionally, even if the postfission state is energetically favorable with respect to the prefission state, a large energy barrier could arrest the system kinetically in a bud/tube configuration. From a physical point of view, it is difficult to predict the energy landscape of the bilayer transitions during the fission process. It has, however, been possible to calculate the energy of the hemifission intermediate (26), which is the energy barrier that has to be surpassed by thermal fluctuations of the membrane. From these calculations, it is possible to estimate a typical transition time: the timescale by which thermal fluctuations can supply enough energy to the membrane to overcome the energy barrier and create the hemifission intermediate. Although not directly measured, it is thought, on the basis of experimental and theoretical arguments, that energy barriers up to the order of $40 k_B T$ would lead to kinetically feasible fission states (28, 29). Once the hemifission state is formed, and because the energy of the final condition is lower than the hemifission energy, the completion of the fission process should proceed spontaneously (26).

These are some of the key features that can be used to understand the biophysics of the process of membrane fission. But the more important issues are how these events are mediated in a cell and how the timing of these events is controlled to produce transport carriers of the size, shape, and number commensurate with the cargo.

DYNAMIN: A LARGE GTPase

Dynamin was first observed in 1987 (30) and further characterized as a microtubule-associated protein that bridges microtubules and generates force between them (31). In the same year, it was reported that a temperature-sensitive *Drosophila melanogaster* strain harboring a mutation in the *shibire* gene accumulated invaginated pits at the plasma membrane (32); the necks of these pits appeared to be wrapped in electron-dense material. The *shibire* gene was later found to encode an ortholog of dynamin (33–35). Since then, the focus has been to address the role of dynamin assembly at the neck of a budding vesicle and investigate its role in membrane fission during endocytosis. There is now a large literature on dynamin, and although no one challenges its participation in the overall membrane fission process and the fact that it can generate membrane fission in vitro, whether it alone can cut membranes in vivo is not yet certain.

Dynamin is a GTPase of 96 kDa with three homologs in mammals as well as a number of splicing variants (36). There is only one gene encoding dynamin in *D. melanogaster* and in *Caenorhabditis elegans* (36, 37). In mammalian cells, dynamin 1 and dynamin 3 are strongly expressed in neurons, whereas dynamin 2 is ubiquitously expressed (36, 38, 39). In addition to these classical dynamins, the human dynamin GTPase superfamily includes dynamin-like proteins, Mx proteins, optic atrophy 1, mitofusins, and GBP/atlastin-related proteins (40). Classical dynamins are involved in the biogenesis of clathrin-coated vesicles during endocytosis; mitofusins participate in mitochondrial fusion; and atlastins are needed for ER fusion.

GAP: GTPase-activating protein

Phosphatidylinositol-4,5-bisphosphate (PIP$_2$): a signaling lipid that is enriched at the cell surface

Structurally, dynamin consists of a GTPase domain that binds and hydrolyzes GTP; a middle domain; a GTPase effector domain that has been suggested to act as the GTPase-activating protein (GAP) to activate GTP hydrolysis in the GTPase domain; a pleckstrin homology domain that drives membrane anchoring by phosphatidylinositol-4,5-bisphosphate (PIP$_2$) binding and the insertion of a hydrophobic variable loop 1; and a proline-rich domain mediating protein-protein interactions by binding Src homology 3 (SH3) domain-containing proteins, such as the BAR (Bin-amphiphysin-Rvs) domain-containing proteins, amphiphysin, endophilin, syndapin, and sortin nexin 9 (see **Figure 2a**) (40). Dynamin can assemble into higher-order oligomers (41), and this polymerization is thought to depend on the membrane curvature (42). Tetrameric dynamin is reported to bind GTP cooperatively and undergo nucleotide-dependent conformational changes both on membranes and in solution (43–45). But how these conformational changes catalyze fission remains unclear. Numerous recent reviews discuss the involvement of dynamin in membrane fission (40, 46–48). The review by Ramachandran (47) is especially noteworthy.

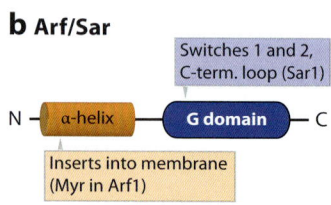
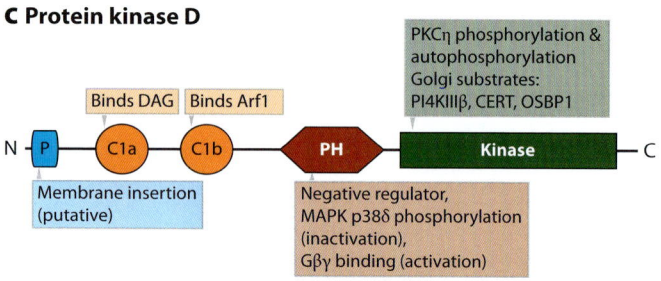

Figure 2
Domain structure of the proteins involved in membrane fission. (*a*) Classical dynamins (dynamin 1, dynamin 2, and dynamin 3) share a common GTPase domain, a middle domain, a pleckstrin homology (PH) domain, a GTPase effector domain (GED), and a proline-rich domain (PRD). (*b*) The small G proteins Arf1 and Sar1 contain an amphipathic α-helix at their N terminus, which is myristoylated in Arf1, and the G domain, which in the case of Sar1 contains the C-terminal loop thought to participate in Sar1 oligomerization. (*c*) Protein kinase D1 (PKD1) and PKD2 contain a hydrophobic stretch at their N termini (P) unlike PKD3, which does not. Otherwise, all PKDs contain two C1 domains, a PH domain, and the catalytic kinase domain. Abbreviations: C-term., C-terminal; C1a and C1b, the two C1 domains of PKD; DAG, diacylglycerol; Gβγ, βγ subunits of heterotrimeric G proteins; MAPK p38δ, mitogen-activated protein kinase p38δ; Myr, myristoylated.

Is Dynamin the Ultimate Membrane Cutter?

Early findings led to the proposal that dynamin assembled into short collars at the neck of budding vesicles and constricted the neck in a GTP-dependent manner (49). But there were problems with this model. For example, *shibire*-like protein collars are only seen in neurons and not in other cells, where dynamin also participates in endocytosis (50). Dynamin assembles into long helices on negatively charged liposomes, which undergo extensive constriction to form long tubules and generate small vesicles in a GTP-dependent manner (49, 51). However, this happens only in tubules anchored to a surface. In suspension, dynamin disassembles and dissociates from the tubules upon GTP addition without generating vesicles (52). Thus, the model in which dynamin represents a choker of the vesicle neck lost its appeal.

Another model was then proposed suggesting that dynamin helices not only constrict but also twist around the tubule upon GTP hydrolysis (53). The torsional forces from GTP binding and hydrolysis were suggested to induce constriction and twisting, thereby breaking the underlying tubule into vesicles. On very short timescales, membranes behave as an elastic medium, and a fast dynamin-induced torsion twist could drive membrane fission (54, 55).

More recent studies suggest an alternative mechanism for dynamin-induced membrane fission (56, 57). Recalling that addition of GTP to dynamin-constricted tubules in solution does not cause fission but instead causes dissociation of dynamin from the membrane and tube retraction upon GTP hydrolysis (52), these studies report that simultaneous addition of dynamin and GTP to tethered membrane tubules or to membranes in suspension generates vesicles (56, 57). Short dynamin necks form when both dynamin and GTP are present, and long dynamin necks form when GTP is absent. Membrane curvature generated by dynamin insertion and polymerization on the membrane relaxes upon severing of the tube neck. Theoretical calculations show that upon dynamin binding, the lipid bilayer rearranges by bending and tilting the lipids. The energy barrier calculated for membrane fission of short necks is on the order of $35k_BT$, making it a kinetically feasible process (57). In other words, different cycles of dynamin binding and dissociation, upon GTP hydrolysis, corresponding to the generation of short collars (*shibire* phenotype) are proposed to catalyze membrane fission (56, 57). However, these experiments do not rule out the possibility that a conformational change in dynamin induced by GTP hydrolysis is needed for proper membrane scission (58).

Here lies the conundrum: Long helices do not occur in intact cells, and the evidence for short collars that could be responsible for dynamin-induced fission is lacking in vivo. More importantly, knockout of dynamin 1 in mouse is not essential for synaptic vesicle recycling. Interestingly, however, the ability to control the size of clathrin-coated pits is lost (39). Ever since the very first description of the *shibire* mutant, the focus has been on the hypothesis that dynamin alone can cut a membrane on the basis of its GTP-dependent recruitment to the membrane neck and GTP hydrolysis-dependent change in conformation (45). More recent data from mouse knockouts of the different dynamin homologs indicate that dynamin might not be essential for membrane cutting and suggest a more supporting role in controlling the efficiency and the timing of the cutting process depending upon the endocytic need (59).

Key Players in Membrane Fission

Another proposal to explain clathrin-coated vesicle fission involves a combination of PIP_2-dependent phase separation and actin-mediated force generation (60, 61). In this scheme, dynamin would control the amount of PIP_2 that is generated and concentrated at the site of a clathrin-coated patch; the patch would then recruit SH3-BAR domain-containing proteins such as SNX9, amphiphysin, syndapin, and endophilin (**Figure 2a**). The BAR domain is a triple helical, coiled-coil element that dimerizes into a banana-shaped structure (62, 63). The concave face of BAR domains has positive charges that electrostatically interact with negatively charged membranes, and the BAR domain acts as a sensor and/or stabilizer of the membrane curvature. Additionally, N-BAR domain-containing proteins possess one or more amphipathic helices that insert into the external leaflet and generate a positive curvature by means of the hydrophobic insertion mechanism (18, 24). Thus, BAR domain-containing proteins can generate curvature. Furthermore, it was recently reported that synaptojanin 1, the major PIP_2 phosphatase in neurons and an endophilin-binding partner (64, 65), is a curvature sensor, meaning that its activity in converting PIP_2 to phosphatidylinositol-4-phosphate [PI(4)P] is enhanced in regions of large curvature. This metabolic reaction could be an important contributor to dynamin-dependent membrane fission (66).

PIP_2 and BAR/N-BAR domain-containing proteins also recruit actin-binding proteins to membrane surfaces, linking newly forming vesicles with the actin cytoskeleton. There are numerous reports suggesting that actin and associated proteins play a critical role in these processes (67–70). Drubin and colleagues (69)

Bin-amphiphysin-Rvs (BAR) domain: dimerized banana-like shaped protein domains that can generate, stabilize, or sense membrane curvature

N-BAR domain: a BAR domain containing an amphipathic helix at the N terminal

Hydrophobic insertion mechanism: membrane curvature generation by proteins based on the shallow insertion of amphipathic helices or other protein domains

Curvature sensor: a protein that binds membranes differently according to their curvature

have visualized the recruitment and dynamics of actin and how it associates with clathrin patches in yeast; they reported that after coat assembly, an actin module composed of four distinct proteins assembles into a branched actin filament meshwork. Interestingly, the BAR domain-containing proteins Rvs161p and Rvs167p arrive after a protein named abp1 (of the actin module) but accumulate simultaneously, which might correspond to the timing of membrane fission (68). More recent direct visualization of actin in clathrin-coated structures in detergent-extracted mammalian cells has revealed the presence of highly branched actin filament barbed ends, the clathrin-coated structures (71). Some of the images reveal an actin network, arranged around the neck of the clathrin-coated buds, generating the force required for further neck constriction and fission. These data support a model in which actin is a central player in the events leading to clathrin-coated vesicle formation. Some caution is warranted, however, because detergent extraction alters the structure of the underlying membranes. It would be useful to carry out this analysis in cells depleted of specific proteins using RNAi. Such an approach might help pinpoint events required for the recruitment of the actin network to clathrin-coated structures. For example, does depletion of dynamin or BAR domain-containing proteins influence the organization and the dynamics of actin network? Dynamin is recruited to membranes by BAR domain-containing proteins. such as amphiphysin, when needed to increase scission efficiency. This scheme could explain the lack of dynamin involvement in the formation of clathrin-coated vesicles at the TGN and during endocytosis in yeast. It would also explain the recent data from mice lacking dynamin 1, where little or no effect is seen in the formation of clathrin-coated vesicles in neurons (39).

Sar1 AND Arf1: SMALL G PROTEINS

Sar1 and Arf1, two low-molecular-weight members of the ADP-ribosylation factor (ARF) family of G proteins, are required for the biogenesis of COPII at the ER, COPI, and clathrin-coated vesicles at the Golgi membranes, respectively (72–74). There is a high level of functional similarity between these two proteins, even though they are recruited to different membranes and bind different effectors. Both Sar1 and Arf1 translocate to the membrane and change conformation upon exchange of bound GDP for GTP, a process that exposes an N-terminal amphipathic helix to permit insertion into the adjacent membrane (**Figure 2b**). Arf1 also contains an N-terminal myristoyl group (75). At ER exit sites, Sar1 recruits the heterodimeric Sec23/24 (inner COPII coat layer) followed by the heterotetrameric Sec13/31 (outer layer of the coat) (see Reference 76). By contrast, Arf1 acts at the Golgi membranes and recruits a complex of seven polypeptides that comprise the COPI coat (77). Arf1 also recruits clathrin coats via tetrameric adaptor proteins or the monomeric, Golgi complex-associated, γ-adaptin homologous, ARF-interacting proteins at endosomes and at the TGN to form clathrin-coated vesicles.

The assembly of these small G proteins bearing GTP with their respective coat proteins parallels a change in membrane curvature to generate small coated buds, and upon GAP catalysis of the hydrolysis of GTP to GDP, these proteins detach from the membrane, and the vesicle detaches from the maternal membrane compartment.

Sar1

Sar1 contains a 23-residue amphipathic helix at its N terminus that is essential for its membrane binding and for protein export from the ER (78, 79). This N-terminal helix inserts into the cytosolic leaflet of the ER membrane and can induce positive curvature into small liposomes. Giant unilamellar vesicles incubated with a constitutively active mutant of Sar1 that is not able to hydrolyze GTP generate tubules of two distinct types: flexible tubes and rigid tubes (80). Electron microscopic analysis of these

30–60 nm diameter tubules has revealed that the flexible tubes have periodic constrictions (beads on a string-like phenotype), whereas the rigid tubules are decorated by an ordered assembly of Sar1, suggesting oligomerization of Sar1 into higher-order structures. A protein bearing a point mutation in the C-terminal Ω loop (Sar1^{T158A}, the Ω loop corresponding to the amino acids 156–171 of the hamster Sar1a) fails to drive COPII vesicle biogenesis from microsomes incubated with purified COPII coats (34). Incubation of Sar1^{QTTG} (residues 156–159 mutated to alanines) with microsomes and cytosol revealed no defect in the recruitment of Sec23/24 or Sec13/31 (80). However, when tested in semi-intact NRK cells in the absence of COPII components, Sar1^{QTTG} GTP, unlike Sar1 GTP, failed to generate rigid tubules. Similar experiments with liposomes revealed that the production of rigid tubules was inhibited by the mutation in the QTTG region of Sar1 GTP. These findings suggest that the C-terminal loop containing the QTTG sequence participates in the organized oligomerization of Sar1 GTP and, therefore, in membrane tubulation.

Unlike BAR domain-containing proteins, the inner coat layer of COPII, containing the Sec23/24 complex, appears to lack the rigidity required for membrane shaping. The outer, Sec13/31 layer has a flexible architecture; thus, it is unlikely that COPII assembly alone generates adequate force to result in membrane fission. Sar1 GTP-induced tubulation requires its insertion into the membrane, and the C-terminal loop may drive higher-order oligomer assembly such that Sar1 could control the fission reaction at the bud neck. Whether COPII-coated vesicle fission occurs by GTP hydrolysis triggered release of Sar1 from the membrane, or through depolymerization of oligomerized Sar1 followed by uncontrolled curvature generation by these proteins, or by some other means, is not yet known.

Arf1

COPI-coated vesicles form in vitro with Arf GTP and a purified coatomer in incubations carried out with isolated Golgi membranes (81). These experiments, however, do not reveal the identity of membrane-associated components that might also be required for the generation of this vesicle type. Arf1 possesses an N-terminal, myristoylated, amphipathic helix that inserts into the outer leaflet of the membrane in a GTP-dependent manner (75). Arf1 GTP appears to form a dimer at the membrane; this dimeric form is proposed to be essential for membrane curvature generation and, in cooperation with the COPI coats, to break the neck of a COPI-coated bud (82). Recent results suggest that Arf1 dimerization is essential for COPI vesicle fission (83). Others have reported that COPI vesicle fission also requires the activity of lyosophosphatidic acid acyltransferases, phospholipase D, and brefeldin-A ADP-ribosylated substrate (84–87). Although these liposome-based analyses have been useful in revealing the involvement of these proteins in membrane fission, none of these findings provide the definite proof that this is all it takes to cut the neck of a COPI-coated vesicle in intact cells. Moreover, these studies do not reveal the order of the events, or how the timing of these events is regulated to control fission, so that it occurs only when and where it is required.

Another suggested mechanism for the regulation of COPI vesicle fission is through kinetic regulation of the levels of GDP- versus GTP-bound Arf1. Membrane-associated Arf1 GTP is inactivated upon GTP hydrolysis catalyzed by ArfGAPs. This series of events must be precisely regulated for a proper vesicle biogenesis because too-rapid GTP hydrolysis by Arf1 would prevent the coatomer recruitment and, thus, vesicle formation. The actual geometry of the bud has been proposed to control the delay of GTP hydrolysis triggered by ArfGAP until the forming coated bud is mature and large enough to be separated from the Golgi membrane (88). ArfGAP1 contains an amphipathic lipid-packing sensor motif that binds with a higher affinity to curved than to flat membranes (see Reference 89 for a review). According to this view, Arf1 GTP recruits coatomers and triggers the initiation of

Ilimaquinone (IQ): a marine sponge metabolite that causes Golgi fragmentation

Diacylglycerol (DAG): a mostly hydrophobic, conically shaped lipid

the budding process. Initially, the curvature of the bud is moderate, and ArfGAP1 does not remove Arf1 from the membrane. When the bud reaches the right size, ArfGAP1 activates GTP hydrolysis, thereby removing Arf1 molecules from the spherical part of the bud, where the curvature is large, but not Arf1 molecules at the neck, where the total curvature would be smaller. Eventual removal of Arf1 from the neck upon sufficient constriction would drive COPI vesicle fission in a mechanism similar to the two-stage model of constriction and stress release suggested for dynamin (57). However, many unanswered questions remain regarding these models (89, 90). Is ArfGAP1 an active component of the coat that binds its own functional effectors? How does it regulate its GAP activity to release Arf1 from the membrane and trigger efficient membrane fission? What is the role of lipids in this process—do they recruit specific proteins to membranes or do they play a more structural role by modulating the biophysical properties of the membranes?

PROTEIN KINASE D–DEPENDENT MEMBRANE FISSION

Treatment of NRK cells with the sponge metabolite ilimaquinone (IQ) vesiculated Golgi membranes (91). IQ was therefore suggested to activate components of the membrane fission machinery at the Golgi complex. The serine/threonine PKD is required for IQ-dependent Golgi vesiculation (92, 93). Thus, PKD is likely to also be essential for membrane fission at the TGN. PKD is mostly cytosolic, and less than 30% of the protein localizes to the TGN. However, a kinase-inactive form of PKD is predominantly TGN associated, and in cells expressing the mutant kinase, cargoes that are ordinarily transported from the TGN to the basolateral surface of cells are found in long tubules attached to the TGN (94, 95). A failure in the separation of the cargo-containing tubules from the TGN thus prevents the delivery of cargo to the cell surface. By contrast, expression of a constitutively active PKD causes extensive vesiculation of the TGN (96). These data strongly suggest that PKD activity is needed for the membrane fission that generates a class of TGN-to-cell surface transport carriers (97, 98). Other transport steps, such as traffic from the ER to the Golgi complex, from the Golgi complex to endosomes, from the Golgi complex to the apical surface, or from the Golgi complex to the ER, are unaffected in cells expressing a kinase-inactive PKD (94, 99). It is still not clear why IQ vesiculates the entire Golgi membranes, whereas PKD activity is required only for the formation of a specific class of transport carriers at the TGN. PKD has been reported to be required for the export of a number of distinct cargoes from the Golgi membranes, including viral budding at the TGN (100–102). Depletion of PKD by siRNA or chemical inactivation also inhibits Golgi complex to cell surface transport but, surprisingly, rarely causes the formation of cargo-containing tubules at the TGN (96). This means that attachment of PKD to the membrane is necessary for the event leading to the tubulation and membrane fission.

The recruitment of PKD to the TGN is mediated by the binding of its first C1 domain to diacylglycerol (DAG) (103) and the second C1 domain to Arf1 (**Figure 3a**) (104). The pleckstrin homology domain of PKD does not bind lipids and is a negative regulator of its activity (105). The pleckstrin homology domain is important for the trimeric G protein βγ-subunit-dependent activation of PKD (92). The activation of PKD is mediated by G protein–coupled receptors and requires Golgi complex-associated PKCη, but the signaling events that trigger PKD translocation to the TGN are not known (106, 107).

There are three homologs of PKD in mammals: PKD1, PKD2, and PKD3. HeLa cells express predominantly PKD2 and PKD3, and at the TGN, PKD2 and PKD3 bind directly (most likely through the cysteine-rich domain) to form homo- and heterodimers (96). Inactivation of either kinase in the dimer inhibits

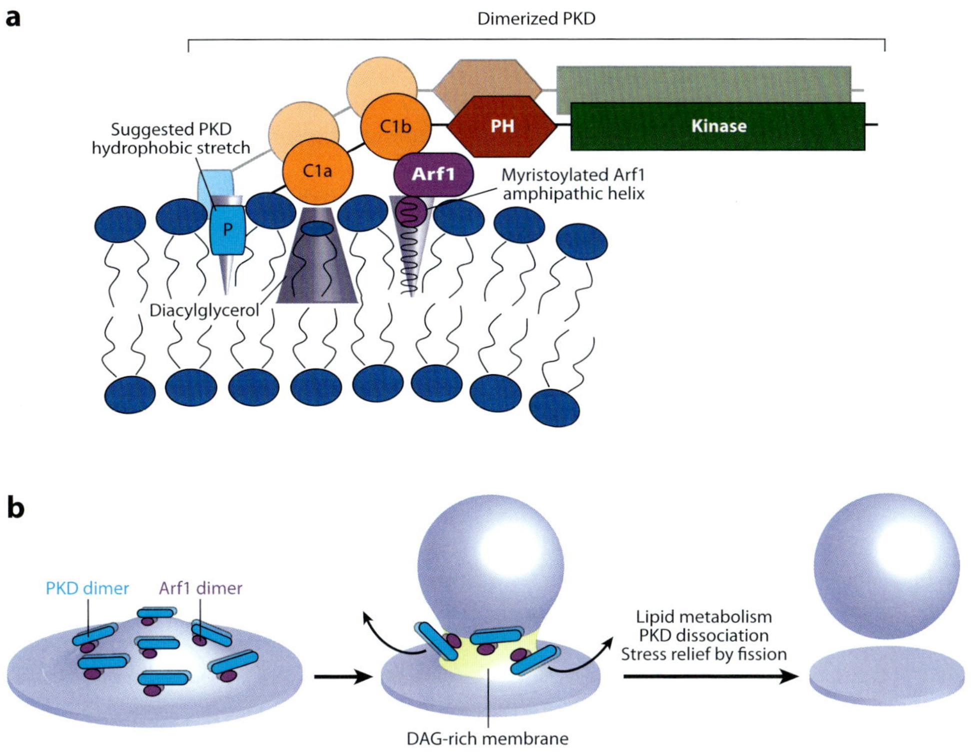

Figure 3

A model for protein kinase D (PKD)-dependent membrane fission. (*a*) PKD inserts into the *trans*-Golgi network (TGN) membrane by binding to diacylglycerol (DAG) by its C1a domain, and to Arf1 by its C1b domain. In addition, PKD1 and PKD2 contain a hydrophobic stretch of amino acids at their N termini that could potentially insert into the membrane core. DAG is a lipid with a negative spontaneous curvature and a fast flip-flop rate, and once bound to PKD could generate negative membrane curvature. The N-terminal myristoylated amphipathic helix of Arf1 inserts into the membrane and can lead to the generation of positive spontaneous curvature by means of the hydrophobic insertion mechanism. In addition, dimeric PKD controls the density of these curving agents on the membrane and could generate positive membrane curvature. (*b*) A model for PKD-dependent fission. Dimeric PKD binds to the TGN by its affinity for DAG and Arf1. This creates positive curvature and therefore the initiation of the budding process, followed by an increase in the concentration of the PKD-DAG-Arf1 complex in the neck region. Release of PKD from the membrane triggers membrane fission by lipid modifications at the neck region or relieving the stress generated by PKD in the neck, or a combination of both.

secretion and generates cargo-containing tubules at the TGN. This suggests that the activity of both kinases is required for transport out of the TGN by regulating the membrane fission reaction.

Activated PKD has three identified TGN-associated substrates: phosphatidylinositol-4-kinase type III β enzyme, ceramide transfer protein (CERT), and oxysterol-binding protein (OSBP). PKD-dependent phosphorylation increases the lipid kinase activity of phosphatidylinositol-4-kinase type III β and therefore increases the production of PI(4)P at the TGN (108). Phosphorylation of CERT (109) and OSBP (110) by the PKD causes their dissociation from the TGN. CERT is required for the nonvesicular transport of ceramide from the ER to the TGN (111). Its PKD-dependent dissociation thus suggests a role in controlling ceramide transport to the TGN. OSBP is considered a sterol sensor, and its PKD-dependent dissociation could represent a means to control sterol levels at the TGN. Thus, PKD has a role in controlling the homeostasis of PI(4)P, ceramide, and sterols at the TGN.

A Model for Protein Kinase D–Dependent Fission at the *trans*-Golgi Network

Dimeric PKD is recruited to the TGN by binding to DAG and Arf1. The known sources of DAG at the TGN include: the sphingomyelin-mediated conversion of phosphatidylcholine and ceramide into DAG and sphingomyelin; dephosphorylation of phosphatidic acid by the lipid phosphate phosphatase enzyme; and a phospholipase C–mediated conversion of PI(4)P (97, 111a). Although the exact source of DAG at the TGN for the generation of transport carriers is unclear, a recent report lends support to the role of sphingomyelin in DAG generation and recruitment of PKD to the TGN (111b). In some species, the N-terminal hydrophobic stretch of PKD (only in PKD1 and PKD2) may insert into the outer leaflet of the TGN (**Figure 3a**). This, and the fact that PKD binds Arf1, which itself inserts into the outer leaflet, changes membrane curvature, analogous to the Sar1- and Arf1-mediated changes in membrane curvature at the ER and the Golgi membranes, respectively. Insertion into the membrane can create a positive curvature that leads to bud growth. DAG is a conical lipid that becomes concentrated at the neck as a result of bud growth (112). This can stabilize the neck, preventing fission by increasing the hemifission energy barrier (26). At a specific time, depending on the packaging of cargo into the bud, PKD bound to DAG dissociates. The DAG is then converted into more destabilizing lipids, such as phosphatidic acid and lyso-phosphatidic acid, by specific phospholipases, triggering membrane fission. Alternatively, the membrane stresses created by PKD at the neck of the bud are released by membrane fission upon dissociation from the membrane (**Figure 3b**). In this model, the mechanism by which PKD regulates membrane fission is analogous to that discussed for dynamin, Sar1, and Arf1. However, in these latter cases, GTP hydrolysis drives the association and the dissociation of the respective protein to the membrane. Although Arf1 is required for PKD binding to the TGN, there is no direct evidence regarding the involvement of GTP hydrolysis in this overall reaction. Interestingly, kinase-dead PKDs tubulate the TGN and do not dissociate from the membrane. PKDs also *trans*-phosphorylate in their homo- and heterodimeric forms. One possibility is that *trans*-phosphorylation is the signal for PKD dissociation and for the events leading to membrane fission. In other words, phosphorylation replaces GTP hydrolysis for the fission of this class of carriers. There are, however, differences. There is no specific coat protein thus far implicated in PKD-dependent cutting of carriers at the TGN. There is certainly no requirement for dynamin in this reaction. But what are the roles of PI(4)P, ceramide, and sterols in PKD-dependent processes? PI(4)P recruits a number of proteins to the TGN, including CERT, OSBP, FAPP1, and FAPP2 (113–116). We suggest that these PI(4)P-binding proteins are crucial for fine-tuning the lipid homeostasis at the TGN and that PKD plays a crucial role in controlling their levels at the TGN. An imbalance in the levels of PI(4)P, ceramide, or sterols affects not only the physical properties of the membrane, thereby affecting membrane rigidity and the propensity to break, but also the location and the activity of proteins required for the overall biogenesis of the PKD-dependent Golgi complex to cell surface transport carriers. For example, in yeast, PI(4)P recruits Sec2p to the Golgi membranes, where it binds the Ypt32p Rab GTPase to generate transport carriers and subsequent trafficking events (117). These findings highlight the role of a number of distinct proteins that are required for the events leading to membrane fission. It is not known how these proteins are recruited and organized spatially at the site of vesicle biogenesis.

KINETICS OF MEMBRANE FISSION TO REGULATE THE SIZE OF TRANSPORT CARRIERS AT THE ENDOPLASMIC RETICULUM

Metazoans secrete bulky cargoes, such as collagens, which raises the obvious question of how these cargoes are secreted from the cells. At the level of the ER, COPII vesicles export both the small and the large cargoes, but how is the size of the containers regulated to accommodate big cargoes? Is there a way to control the kinetics of membrane fission to generate carriers commensurate with cargo size? Recent studies suggest the involvement of the TANGO1 protein, which is localized to ER exit sites and binds collagen VII in the lumen (118). TANGO1 interacts with Sec23/Sec24 of the COPII coats on the cytoplasmic side of the ER. This interaction is suggested to prevent the recruitment of Sec13/31 to Sec23/Sec24 and delays Sar1-GTP hydrolysis. More recent data indicate that TANGO1 forms a dimer with a protein called cTAGE5, and both are required for collagen VII export at the ER (119). In this scheme, TANGO1 binds collagen VII in the lumen of the ER via its SH3-like domain. The second coiled-coil domain of TANGO1 binds the second coiled-coil domain of cTAGE5, and the proline-rich domains of the TANGO1-cTAGE5 dimer bind Sec23/24. This retards the recruitment of Sec13/31 to the nascent Sec23/24-coated carrier. Although not tested directly, it is suggested that the delay or inhibition in the recruitment of Sec13/31 to Sec23/24 by TANGO1-cTAGE5 prevents GTP hydrolysis on Sar1, and as a result, the carrier grows in size. During this growth phase, collagen VII is packed into the carrier, and once the carrier is large enough to encompass collagen VII, collagen VII dissociates from the SH3 domain of TANGO1. This would trigger the release of TANGO1-cTAGE5 proline-rich domains from Sec23/24. Sec13/31 is then recruited by Sec23/24, and the Sec24 and Sec31 GAP complex activates GTP hydrolysis on Sar1, thus terminating the process of COPII carrier biogenesis. In other words, the TANGO1-cTAGE5 dimer thus acts as a transient coat (middle layer) or a kinetic timer for the growth of COPII carriers only when bound to its specific cargo, collagen VII, to delay a Sar1-dependent membrane fission event.

Physiological support of TANGO1's role in collagen secretion was recently obtained in a mouse knockout model (120). Mice lacking TANGO1 were deficient for the secretion of a number of collagens. In *D. melanogaster*, TANGO1 is also required for the export of collagen IV from the ER (120a). Taken together, these results confirm the role of TANGO1 in the secretion of a number of collagens. Proteins like TANGO1 can delay the cutting reaction and therefore help regulate the size of transport carrier so it is commensurate with the size of the cargo.

Collagen VII: a rigid rod-like cargo protein of about 400 nm in size, which travels through the secretory pathway

Line energy: an energy penalty equal to the length of the domain boundary times the line tension

SUMMARY AND FUTURE PERSPECTIVES

The first common feature of all membrane fission events that produce transport carriers is the insertion of an amphipathic helix or a hydrophobic loop into the outer leaflet by specific proteins; this generates a positive curvature and elastic stresses in the membrane. N-BAR domain-containing proteins, dynamin, Sar1, Arf1, and PKD are some of the key candidates that may be inserted into specific membranes. A second common feature is the oligomerization and/or polymerization of these proteins, abating or enhancing curvature generation. After these steps, the events leading to membrane fission take two different routes depending on the budding membrane.

The formation of clathrin-coated vesicles at the plasma membrane involves the concentration of PIP_2 at the neck, which then recruits actin and associated proteins. It has been suggested that this phase-separated PIP_2 domain might create a line energy that promotes membrane fission (60, 61, 121–124). However, it is not clear whether under physiological conditions the line energy is large enough to drive membrane bulging and scission

(124, 125). More importantly, knockout of all three endophilins in mice, as in the case of synaptojanin knockout, results in an increase in the number of clathrin-coated vesicles but not in the number of clathrin-coated pits (126). In other words, these two proteins, which control the fate of PIP_2 at the neck, are functional, after membrane fission, in regulating the uncoating of clathrin-coated vesicles. The significance of line tension in membrane fission therefore remains unclear at present. Both membrane bending created by the insertion of amphipathic helices of N-BAR domain-containing proteins and the force generated by actin-mediated pulling play key roles in this fission process. Dynamin increases the efficiency of cutting by polymerization and constriction.

There is no clear evidence that phosphoinositides and actin play a role in the biogenesis of COPII and COPI carriers. A BAR domain-containing protein is also not required for the events leading to the formation of this class of vesicles.

PKD, as mentioned above, is also recruited by binding DAG and Arf1 at the TGN. The N-terminal amphipathic helix of Arf1, by insertion into the outer leaflet of the TGN, initiates changes in membrane curvature, which is stabilized by PKD bound to DAG. In the absence of the kinase activity of PKD, PKD remains attached to DAG and Arf1, and the bud grows into a tubule owing to a fission defect. PKD has TGN-associated substrates, which result in an increase in the levels of PI(4)P and control the levels of ceramide and sterols. PI(4)P and its effectors, we suggest, could create the phase separation akin to the PIP_2-mediated events suggested for the genesis of clathrin-coated vesicles at the plasma membrane. Although the ARF GTP effector arfaptin, which is a BAR domain-containing protein, localizes to the TGN (127) and there is some evidence that it might have a role in secretion (128, 129), a connection between PKD-dependent membrane fission and arfaptin has not been reported. Although the downstream components in the PKD pathway that cause membrane fission are not known, we suggest that DAG, bound to PKD at the neck, stabilizes the neck and prevents membrane fission. Timely conversion of DAG into PA and LPA might be necessary to break the membrane. There is currently no evidence for the involvement of actin and an actin-dependent force generation in this reaction.

In sum, since the identification of dynamin as a key player of membrane fission, this field has seen tremendous growth. A large number of new proteins have been identified that act at the plasma membrane, ER, and the Golgi complex to cut membranes for the production of specific transport carriers. It would be unfair for us to propose experiments that have not already been considered by experts in the field, and the most obvious issues are probably being addressed by the finest minds. But the swing to the more artificial system of liposomes might in fact be misleading in the long run, as exemplified by the more recent studies on dynamin that have led some to propose that it is by itself capable of cutting membranes. We hope that a hard look at some of the issues raised above will attract more effort to identify the missing components and address the mechanisms of the interactions that result in cutting membranes to produce transport carriers.

SUMMARY POINTS

1. A number of proteins insert into the outer leaflet of membranes of the endocytic and exocytic compartments via an amphipathic helix. This insertion leads to the changes in membrane curvature that are essential for the biogenesis of transport carriers during endocytosis and protein secretion. The downstream events in the membrane fission reaction differ depending on the membrane, and this might reflect the differences in the lipid and protein composition of the budding membranes and the cargoes exported.

2. Dynamin regulates the efficiency and the timing of the membrane fission reaction.

3. Force generation by regulated assembly of actin at the neck of the budding vesicle plays an important role in the biogenesis of clathrin-coated vesicles. However, there is no clear evidence that COPII and COPI vesicles require actin-dependent force generation for their separation by membrane fission from the ER and the Golgi membranes, respectively.

4. BAR domain-containing proteins are essential for curvature sensing and generation and therefore in the events leading to the biogenesis of clathrin-coated vesicles at the plasma membrane. A BAR domain-containing protein has not been identified in the events leading to the biogenesis of COPII and COPI vesicles or to the PKD-dependent vesicle biogenesis at the TGN.

5. PIP_2 is essential for the biogenesis of clathrin-coated vesicles at the plasma membrane. Although PI(4)P is required for the biogenesis of TGN to cell surface carriers, there is no clear evidence for the involvement of phosphoinositides in the biogenesis of COPII and COPI vesicles.

6. The lipid DAG is required for the biogenesis of PKD-dependent transport carriers. PKD bound to DAG at the neck of a budding carrier is proposed to restrict membrane fission, and therefore, this is a key step in the events leading to the separation of this class of transport carrier at the TGN. There is no clear evidence of a similar step in the biogenesis of other classes of transport carriers. PKD regulates the lipid homeostasis and along with Arf1 controls the events leading to membrane fission of a specific class of TGN to cell surface carriers.

7. TANGO1 and cTAGE5 help load big cargoes into COPII carriers. They are proposed to regulate the kinetics of membrane fission by affecting Sar1 GTP hydrolysis.

FUTURE ISSUES

1. The biophysics of the lipid bilayer has revealed theoretical numbers for the forces and energies required for membrane fission. Does the assembly of fission components match these values? In other words, is the assembly of fission components sufficient to generate forces necessary to cut membranes?

2. Are there specific lipids and proteins that concentrate at the necks of the budding carriers?

3. What are the temporal and spatial arrangements of the components of the membrane fission complex at the respective membranes?

4. Does cargo play a role in starting the events that lead to membrane fission?

5. Yeast lack PKD. Does PKD behave like dynamin to control the efficiency of the membrane fission event? Or, is PKD involved in membrane fission that produces a specific class of transport carriers that do not form in yeast? What are the signals that activate the recruitment of PKD to the TGN to generate transport carriers?

6. The bulk of endocytosis is mediated by a dynamin-independent process. How are such transport carriers severed from the plasma membrane?

7. Secretory granules form at the TGN for the secretion of a large number of essential proteins, such as mucins and insulin. Virtually nothing is known about the mechanism by which such granules are separated by fission from the Golgi complex. In vitro reconstitution and a genome-wide screen could help reveal the components of this pathway.

8. Does the TANGO1-cTAGE5 complex affect the kinetics of Sar1-dependent membrane fission? Are similar components involved in regulating the size, shape, and number of other transport carriers by affecting membrane fission?

DISCLOSURE STATEMENT

The authors are not aware of any affiliations, memberships, funding, or financial holdings that might be perceived as affecting the objectivity of this review.

ACKNOWLEDGMENTS

We thank Suzanne Pfeffer, Rajesh Ramachandran, Meir Aridor, Pietro De Camilli, and Aurélien Roux for valuable discussions and critical reading of the manuscript. Felix Campelo is funded by a postdoctoral fellowship from the Juan de la Cierva program from the Spanish government. Vivek Malhotra is an ICREA professor at the Center for Genomic Regulation, and the work in his lab is funded by grants from Plan Nacional (BFU2008-00414), Consolider (CSD2009-00016), and AGAUR SGR2009-1488 Grups de Recerca Emergents (AGAUR-Catalan government), and a European Research Council advanced grant.

LITERATURE CITED

1. Palade G. 1975. Intracellular aspects of the process of protein synthesis. *Science* 189:347–58
2. Novick P, Field C, Schekman R. 1980. Identification of 23 complementation groups required for post-translational events in the yeast secretory pathway. *Cell* 21:205–15
3. Sudhof TC, Rothman JE. 2009. Membrane fusion: grappling with SNARE and SM proteins. *Science* 323:474–77
4. McMahon HT, Kozlov MM, Martens S. 2010. Membrane curvature in synaptic vesicle fusion and beyond. *Cell* 140:601–5
5. Matsuoka K, Orci L, Amherdt M, Bednarek SY, Hamamoto S, et al. 1998. COPII-coated vesicle formation reconstituted with purified coat proteins and chemically defined liposomes. *Cell* 93:263–75
6. Takahashi N, Kishimoto T, Nemoto T, Kadowaki T, Kasai H. 2002. Fusion pore dynamics and insulin granule exocytosis in the pancreatic islet. *Science* 297:1349–52
7. Frolov VA, Dunina-Barkovskaya AY, Samsonov AV, Zimmerberg J. 2003. Membrane permeability changes at early stages of influenza hemagglutinin-mediated fusion. *Biophys. J.* 85:1725–33
8. Kozlovsky Y, Kozlov MM. 2003. Membrane fission: model for intermediate structures. *Biophys. J.* 85:85–96
9. Chernomordik LV, Kozlov MM. 2003. Protein-lipid interplay in fusion and fission of biological membranes. *Annu. Rev. Biochem.* 72:175–207
10. Chernomordik LV, Kozlov MM. 2005. Membrane hemifusion: crossing a chasm in two leaps. *Cell* 123:375–82
11. Kozlov MM, McMahon HT, Chernomordik LV. 2010. Protein-driven membrane stresses in fusion and fission. *Trends Biochem. Sci.* 35:699–706
12. Helfrich W. 1973. Elastic properties of lipid bilayers: theory and possible experiments. *Z. Naturforsch. C* 28:693–703

13. Kozlov MM, Winterhalter M. 1991. Elastic moduli for strongly curved monolayers. Position of the neutral surface. *J. Phys. II* 1:1077–84
14. Do Carmo MP. 1976. *Differential Geometry of Curves and Surfaces*. Englewood Cliffs, NJ: Prentice Hall. 503 pp.
15. Niggemann G, Kummrow M, Helfrich W. 1995. The bending rigidity of phosphatidylcholine bilayers: dependences on experimental method, sample cell sealing and temperature. *J. Phys. II* 5:413–25
16. Rawicz W, Olbrich KC, McIntosh T, Needham D, Evans E. 2000. Effect of chain length and unsaturation on elasticity of lipid bilayers. *Biophys. J.* 79:328–39
17. Gruner SM. 1985. Intrinsic curvature hypothesis for biomembrane lipid composition: a role for non-bilayer lipids. *Proc. Natl. Acad. Sci. USA* 82:3665–69
18. McMahon HT, Gallop JL. 2005. Membrane curvature and mechanisms of dynamic cell membrane remodelling. *Nature* 438:590–96
19. Zimmerberg J, Kozlov MM. 2006. How proteins produce cellular membrane curvature. *Nat. Rev. Mol. Cell Biol.* 7:9–19
20. Templer R, Seddon J, Warrender N. 1994. Measuring the elastic parameters for inverse bicontinuous cubic phases. *Biophys. Chem.* 49:1–12
21. Siegel DP, Kozlov MM. 2004. The Gaussian curvature elastic modulus of N-monomethylated dioleoylphosphatidylethanolamine: relevance to membrane fusion and lipid phase behavior. *Biophys. J.* 87:366–74
21a. Boucrot E, Pick A, Camdere G, Liska N, Evergren E, et al. 2012. Membrane fission is promoted by insertion of amphipathic helices and is restricted by crescent bar domains. *Cell* 149:124–36
22. Hamm M, Kozlov MM. 2000. Elastic energy of tilt and bending of fluid membranes. *Eur. Phys. J. E Soft Matter Biol. Phys.* 3:323–35
23. Hamm M, Kozlov MM. 1998. Tilt model of inverted amphiphilic mesophases. *Eur. Phys. J. B Condens. Matter Complex Syst.* 6:519–28
24. Campelo F, McMahon HT, Kozlov MM. 2008. The hydrophobic insertion mechanism of membrane curvature generation by proteins. *Biophys. J.* 95:2325–39
25. Campelo F, Fabrikant G, McMahon HT, Kozlov MM. 2010. Modeling membrane shaping by proteins: focus on EHD2 and N-BAR domains. *FEBS Lett.* 584:1830–39
26. Kozlovsky Y, Kozlov MM. 2003. Membrane fission: model for intermediate structures. *Biophys. J.* 85:85–96
27. Kozlovsky Y, Kozlov MM. 2002. Stalk model of membrane fusion: solution of energy crisis. *Biophys. J.* 82:882–95
28. Kuzmin PI, Zimmerberg J, Chizmadzhev YA, Cohen FS. 2001. A quantitative model for membrane fusion based on low-energy intermediates. *Proc. Natl. Acad. Sci. USA* 98:7235–40
29. Abidor IG, Arakelyan VB, Chernomordik LV, Chizmadzhev YA, Pastushenko VF, Tarasevich MP. 1979. Electric breakdown of bilayer lipid membranes: I. The main experimental facts and their qualitative discussion. *J. Electroanal. Chem. Interfacial Electrochem.* 104:37–52
30. Paschal BM, Shpetner HS, Vallee RB. 1987. MAP 1C is a microtubule-activated ATPase which translocates microtubules in vitro and has dynein-like properties. *J. Cell Biol.* 105:1273–82
31. Shpetner HS, Vallee RB. 1989. Identification of dynamin, a novel mechanochemical enzyme that mediates interactions between microtubules. *Cell* 59:421–32
32. Koenig JH, Ikeda K. 1989. Disappearance and reformation of synaptic vesicle membrane upon transmitter release observed under reversible blockage of membrane retrieval. *J. Neurosci.* 9:3844–60
33. van der Bliek AM, Meyerowitz EM. 1991. Dynamin-like protein encoded by the *Drosophila shibire* gene associated with vesicular traffic. *Nature* 351:411–14
34. Huang M, Weissman JT, Beraud-Dufour S, Luan P, Wang C, et al. 2001. Crystal structure of Sar1-GDP at 1.7 A resolution and the role of the NH2 terminus in ER export. *J. Cell Biol.* 155:937–48
35. Chen MS, Obar RA, Schroeder CC, Austin TW, Poodry CA, et al. 1991. Multiple forms of dynamin are encoded by *shibire*, a *Drosophila* gene involved in endocytosis. *Nature* 351:583–86
36. Cao H, Garcia F, McNiven MA. 1998. Differential distribution of dynamin isoforms in mammalian cells. *Mol. Biol. Cell* 9:2595–609

37. Urrutia R, Henley JR, Cook T, McNiven MA. 1997. The dynamins: redundant or distinct functions for an expanding family of related GTPases? *Proc. Natl. Acad. Sci. USA* 94:377–84
38. Gray NW, Fourgeaud L, Huang B, Chen J, Cao H, et al. 2003. Dynamin 3 is a component of the postsynapse, where it interacts with mGluR5 and Homer. *Curr. Biol.* 13:510–15
39. Ferguson SM, Brasnjo G, Hayashi M, Wolfel M, Collesi C, et al. 2007. A selective activity-dependent requirement for dynamin 1 in synaptic vesicle endocytosis. *Science* 316:570–74
40. Praefcke GJ, McMahon HT. 2004. The dynamin superfamily: universal membrane tubulation and fission molecules? *Nat. Rev. Mol. Cell Biol.* 5:133–47
41. Muhlberg AB, Warnock DE, Schmid SL. 1997. Domain structure and intramolecular regulation of dynamin GTPase. *EMBO J.* 16:6676–83
42. Roux A, Koster G, Lenz M, Sorre B, Manneville JB, et al. 2010. Membrane curvature controls dynamin polymerization. *Proc. Natl. Acad. Sci. USA* 107:4141–46
43. van der Bliek AM, Redelmeier TE, Damke H, Tisdale EJ, Meyerowitz EM, Schmid SL. 1993. Mutations in human dynamin block an intermediate stage in coated vesicle formation. *J. Cell Biol.* 122:553–63
44. Herskovits JS, Burgess CC, Obar RA, Vallee RB. 1993. Effects of mutant rat dynamin on endocytosis. *J. Cell Biol.* 122:565–78
45. Damke H, Baba T, Warnock DE, Schmid SL. 1994. Induction of mutant dynamin specifically blocks endocytic coated vesicle formation. *J. Cell Biol.* 127:915–34
46. Hinshaw JE. 2000. Dynamin and its role in membrane fission. *Annu. Rev. Cell Dev. Biol.* 16:483–519
47. Ramachandran R. 2011. Vesicle scission: dynamin. *Semin. Cell Dev. Biol.* 22:10–17
48. Schmid SL, Frolov VA. 2011. Dynamin: functional design of a membrane fission catalyst. *Annu. Rev. Cell Dev. Biol.* 27:79–105
49. Takei K, McPherson PS, Schmid SL, De Camilli P. 1995. Tubular membrane invaginations coated by dynamin rings are induced by GTP-gamma S in nerve terminals. *Nature* 374:186–90
50. Koenig JH, Ikeda K. 1990. Transformational process of the endosomal compartment in nephrocytes of *Drosophila melanogaster*. *Cell Tissue Res.* 262:233–44
51. Sweitzer SM, Hinshaw JE. 1998. Dynamin undergoes a GTP-dependent conformational change causing vesiculation. *Cell* 93:1021–29
52. Danino D, Moon KH, Hinshaw JE. 2004. Rapid constriction of lipid bilayers by the mechanochemical enzyme dynamin. *J. Struct. Biol.* 147:259–67
53. Roux A, Uyhazi K, Frost A, De Camilli P. 2006. GTP-dependent twisting of dynamin implicates constriction and tension in membrane fission. *Nature* 441:528–31
54. Morlot S, Lenz M, Prost J, Joanny JF, Roux A. 2010. Deformation of dynamin helices damped by membrane friction. *Biophys. J.* 99:3580–88
55. Lenz M, Prost J, Joanny JF. 2008. Mechanochemical action of the dynamin protein. *Phys. Rev. E* 78:011911
56. Pucadyil TJ, Schmid SL. 2008. Real-time visualization of dynamin-catalyzed membrane fission and vesicle release. *Cell* 135:1263–75
57. Bashkirov PV, Akimov SA, Evseev AI, Schmid SL, Zimmerberg J, Frolov VA. 2008. GTPase cycle of dynamin is coupled to membrane squeeze and release, leading to spontaneous fission. *Cell* 135:1276–86
58. Roux A, Antonny B. 2008. The long and short of membrane fission. *Cell* 135:1163–65
59. Raimondi A, Ferguson SM, Lou X, Armbruster M, Paradise S, et al. 2011. Overlapping role of dynamin isoforms in synaptic vesicle endocytosis. *Neuron* 70:1100–14
60. Liu J, Sun Y, Oster GF, Drubin DG. 2010. Mechanochemical crosstalk during endocytic vesicle formation. *Curr. Opin. Cell Biol.* 22:36–43
61. Liu J, Sun Y, Drubin DG, Oster GF. 2009. The mechanochemistry of endocytosis. *PLoS Biol.* 7:e1000204
62. Peter BJ, Kent HM, Mills IG, Vallis Y, Butler PJ, et al. 2004. BAR domains as sensors of membrane curvature: the amphiphysin BAR structure. *Science* 303:495–99
63. Frost A, Unger VM, De Camilli P. 2009. The BAR domain superfamily: membrane-molding macromolecules. *Cell* 137:191–96
64. McPherson PS, Garcia EP, Slepnev VI, David C, Zhang X, et al. 1996. A presynaptic inositol-5-phosphatase. *Nature* 379:353–57

65. Gad H, Ringstad N, Low P, Kjaerulff O, Gustafsson J, et al. 2000. Fission and uncoating of synaptic clathrin-coated vesicles are perturbed by disruption of interactions with the SH3 domain of endophilin. *Neuron* 27:301–12
66. Chang-Ileto B, Frere SG, Chan RB, Voronov SV, Roux A, Di Paolo G. 2011. Synaptojanin 1-mediated PI(4,5)P$_2$ hydrolysis is modulated by membrane curvature and facilitates membrane fission. *Dev. Cell* 20:206–18
67. Kaksonen M, Toret CP, Drubin DG. 2006. Harnessing actin dynamics for clathrin-mediated endocytosis. *Nat. Rev. Mol. Cell Biol.* 7:404–14
68. Kaksonen M, Toret CP, Drubin DG. 2005. A modular design for the clathrin- and actin-mediated endocytosis machinery. *Cell* 123:305–20
69. Kaksonen M, Sun Y, Drubin DG. 2003. A pathway for association of receptors, adaptors, and actin during endocytic internalization. *Cell* 115:475–87
70. Sun Y, Martin AC, Drubin DG. 2006. Endocytic internalization in budding yeast requires coordinated actin nucleation and myosin motor activity. *Dev. Cell* 11:33–46
71. Collins A, Warrington A, Taylor KA, Svitkina T. 2011. Structural organization of the actin cytoskeleton at sites of clathrin-mediated endocytosis. *Curr. Biol.* 21:1167–75
72. D'Souza-Schorey C, Chavrier P. 2006. ARF proteins: roles in membrane traffic and beyond. *Nat. Rev. Mol. Cell Biol.* 7:347–58
73. Gillingham AK, Munro S. 2007. The small G proteins of the Arf family and their regulators. *Annu. Rev. Cell Dev. Biol.* 23:579–611
74. Donaldson JG, Jackson CL. 2011. ARF family G proteins and their regulators: roles in membrane transport, development and disease. *Nat. Rev. Mol. Cell Biol.* 12:362–75
75. Antonny B, Beraud-Dufour S, Chardin P, Chabre M. 1997. N-terminal hydrophobic residues of the G-protein ADP-ribosylation factor-1 insert into membrane phospholipids upon GDP to GTP exchange. *Biochemistry* 36:4675–84
76. Jensen D, Schekman R. 2011. COPII-mediated vesicle formation at a glance. *J. Cell Sci.* 124:1–4
77. Kirchhausen T. 2000. Three ways to make a vesicle. *Nat. Rev. Mol. Cell Biol.* 1:187–98
78. Bielli A, Haney CJ, Gabreski G, Watkins SC, Bannykh SI, Aridor M. 2005. Regulation of Sar1 NH2 terminus by GTP binding and hydrolysis promotes membrane deformation to control COPII vesicle fission. *J. Cell Biol.* 171:919–24
79. Lee MC, Orci L, Hamamoto S, Futai E, Ravazzola M, Schekman R. 2005. Sar1p N-terminal helix initiates membrane curvature and completes the fission of a COPII vesicle. *Cell* 122:605–17
80. Long KR, Yamamoto Y, Baker AL, Watkins SC, Coyne CB, et al. 2010. Sar1 assembly regulates membrane constriction and ER export. *J. Cell Biol.* 190:115–28
81. Orci L, Palmer DJ, Ravazzola M, Perrelet A, Amherdt M, Rothman JE. 1993. Budding from Golgi membranes requires the coatomer complex of non-clathrin coat proteins. *Nature* 362:648–52
82. Beck R, Sun Z, Adolf F, Rutz C, Bassler J, et al. 2008. Membrane curvature induced by Arf1-GTP is essential for vesicle formation. *Proc. Natl. Acad. Sci. USA* 105:11731–36
83. Beck R, Prinz S, Diestelkotter-Bachert P, Rohling S, Adolf F, et al. 2011. Coatomer and dimeric ADP ribosylation factor 1 promote distinct steps in membrane scission. *J. Cell Biol.* 194:765–77
84. Yang JS, Valente C, Polishchuk RS, Turacchio G, Layre E, et al. 2011. COPI acts in both vesicular and tubular transport. *Nat. Cell Biol.* 13:996–1003
85. Yang JS, Gad H, Lee SY, Mironov A, Zhang L, et al. 2008. A role for phosphatidic acid in COPI vesicle fission yields insights into Golgi maintenance. *Nat. Cell Biol.* 10:1146–53
86. Bonazzi M, Spano S, Turacchio G, Cericola C, Valente C, et al. 2005. CtBP3/BARS drives membrane fission in dynamin-independent transport pathways. *Nat. Cell Biol.* 7:570–80
87. Yang JS, Lee SY, Spano S, Gad H, Zhang L, et al. 2005. A role for BARS at the fission step of COPI vesicle formation from Golgi membrane. *EMBO J.* 24:4133–43
88. Bigay J, Gounon P, Robineau S, Antonny B. 2003. Lipid packing sensed by ArfGAP1 couples COPI coat disassembly to membrane bilayer curvature. *Nature* 426:563–66
89. Antonny B. 2011. Mechanisms of membrane curvature sensing. *Annu. Rev. Biochem.* 80:101–23
90. Hsu VW, Lee SY, Yang JS. 2009. The evolving understanding of COPI vesicle formation. *Nat. Rev. Mol. Cell Biol.* 10:360–64

91. Takizawa PA, Yucel JK, Veit B, Faulkner DJ, Deerinck T, et al. 1993. Complete vesiculation of Golgi membranes and inhibition of protein transport by a novel sea sponge metabolite, ilimaquinone. *Cell* 73:1079–90
92. Jamora C, Yamanouye N, Van Lint J, Laudenslager J, Vandenheede JR, et al. 1999. Gbetagamma-mediated regulation of Golgi organization is through the direct activation of protein kinase D. *Cell* 98:59–68
93. Jamora C, Takizawa PA, Zaarour RF, Denesvre C, Faulkner DJ, Malhotra V. 1997. Regulation of Golgi structure through heterotrimeric G proteins. *Cell* 91:617–26
94. Liljedahl M, Maeda Y, Colanzi A, Ayala I, Van Lint J, Malhotra V. 2001. Protein kinase D regulates the fission of cell surface destined transport carriers from the trans-Golgi network. *Cell* 104:409–20
95. Maeda Y, Beznoussenko GV, Van Lint J, Mironov AA, Malhotra V. 2001. Recruitment of protein kinase D to the *trans*-Golgi network via the first cysteine-rich domain. *EMBO J.* 20:5982–90
96. Bossard C, Bresson D, Polishchuk RS, Malhotra V. 2007. Dimeric PKD regulates membrane fission to form transport carriers at the TGN. *J. Cell Biol.* 179:1123–31
97. Bard F, Malhotra V. 2006. The formation of TGN-to-plasma-membrane transport carriers. *Annu. Rev. Cell Dev. Biol.* 22:439–55
98. Malhotra V, Campelo F. 2011. PKD regulates membrane fission to generate TGN to cell surface transport carriers. *Cold Spring Harb. Perspect. Biol.* 3:a005280
99. Yeaman C, Ayala MI, Wright JR, Bard F, Bossard C, et al. 2004. Protein kinase D regulates basolateral membrane protein exit from *trans*-Golgi network. *Nat. Cell Biol.* 6:106–12
100. Remillard-Labrosse G, Mihai C, Duron J, Guay G, Lippe R. 2009. Protein kinase D-dependent trafficking of the large Herpes simplex virus type 1 capsids from the TGN to plasma membrane. *Traffic* 10:1074–83
101. Irannejad R, Wedegaertner PB. 2010. Regulation of constitutive cargo transport from the *trans*-Golgi network to plasma membrane by Golgi-localized G protein βγ subunits. *J. Biol. Chem.* 285:32393–404
102. Sumara G, Formentini I, Collins S, Sumara I, Windak R, et al. 2009. Regulation of PKD by the MAPK p38delta in insulin secretion and glucose homeostasis. *Cell* 136:235–48
103. Baron CL, Malhotra V. 2002. Role of diacylglycerol in PKD recruitment to the TGN and protein transport to the plasma membrane. *Science* 295:325–28
104. Pusapati GV, Krndija D, Armacki M, von Wichert G, von Blume J, et al. 2010. Role of the second cysteine-rich domain and Pro275 in protein kinase D2 interaction with ADP-ribosylation factor 1, *trans*-Golgi network recruitment, and protein transport. *Mol. Biol. Cell* 21:1011–22
105. Iglesias T, Rozengurt E. 1998. Protein kinase D activation by mutations within its pleckstrin homology domain. *J. Biol. Chem.* 273:410–16
106. Diaz Anel AM, Malhotra V. 2005. PKCη is required for β1γ2/β3γ2- and PKD-mediated transport to the cell surface and the organization of the Golgi apparatus. *J. Cell Biol.* 169:83–91
107. Rozengurt E. 2011. Protein kinase D signaling: multiple biological functions in health and disease. *Physiology* 26:23–33
108. Hausser A, Storz P, Martens S, Link G, Toker A, Pfizenmaier K. 2005. Protein kinase D regulates vesicular transport by phosphorylating and activating phosphatidylinositol-4 kinase IIIbeta at the Golgi complex. *Nat. Cell Biol.* 7:880–86
109. Fugmann T, Hausser A, Schoffler P, Schmid S, Pfizenmaier K, Olayioye MA. 2007. Regulation of secretory transport by protein kinase D-mediated phosphorylation of the ceramide transfer protein. *J. Cell Biol.* 178:15–22
110. Nhek S, Ngo M, Yang X, Ng MM, Field SJ, et al. 2010. Regulation of oxysterol-binding protein Golgi localization through protein kinase D-mediated phosphorylation. *Mol. Biol. Cell* 21:2327–37
111. Hanada K, Kumagai K, Yasuda S, Miura Y, Kawano M, et al. 2003. Molecular machinery for non-vesicular trafficking of ceramide. *Nature* 426:803–9
111a. Carrasco S, Merida I. 2007. Diacylglycerol, when simplicity becomes complex. *Trends Biochem. Sci.* 32:27–36
111b. Subathra M, Qureshi A, Luberto C. 2011. Sphingomyelin synthases regulate protein trafficking and secretion. *PLoS ONE* 6:e23644

112. Shemesh T, Luini A, Malhotra V, Burger KN, Kozlov MM. 2003. Prefission constriction of Golgi tubular carriers driven by local lipid metabolism: a theoretical model. *Biophys. J.* 85:3813–27
113. Levine TP, Munro S. 1998. The pleckstrin homology domain of oxysterol-binding protein recognises a determinant specific to Golgi membranes. *Curr. Biol.* 8:729–39
114. Nishikawa K, Toker A, Wong K, Marignani PA, Johannes FJ, Cantley LC. 1998. Association of protein kinase Cmu with type II phosphatidylinositol 4-kinase and type I phosphatidylinositol-4-phosphate 5-kinase. *J. Biol. Chem.* 273:23126–33
115. Levine TP, Munro S. 2002. Targeting of Golgi-specific pleckstrin homology domains involves both PtdIns 4-kinase-dependent and -independent components. *Curr. Biol.* 12:695–704
116. Godi A, Di Campli A, Konstantakopoulos A, Di Tullio G, Alessi DR, et al. 2004. FAPPs control Golgi-to-cell-surface membrane traffic by binding to ARF and PtdIns(4)P. *Nat. Cell Biol.* 6:393–404
117. Mizuno-Yamasaki E, Medkova M, Coleman J, Novick P. 2010. Phosphatidylinositol 4-phosphate controls both membrane recruitment and a regulatory switch of the Rab GEF Sec2p. *Dev. Cell* 18:828–40
118. Saito K, Chen M, Bard F, Chen S, Zhou H, et al. 2009. TANGO1 facilitates cargo loading at endoplasmic reticulum exit sites. *Cell* 136:891–902
119. Saito K, Yamashiro K, Ichikawa Y, Erlmann P, Kontani K, et al. 2011. cTAGE5 mediates collagen secretion through interaction with TANGO1 at endoplasmic reticulum exit sites. *Mol. Biol. Cell* 22:2301–8
120. Wilson DG, Phamluong K, Li L, Sun M, Cao TC, et al. 2011. Global defects in collagen secretion in a Mia3/TANGO1 knockout mouse. *J. Cell Biol.* 193:935–51
120a. Pastor-Pareja JC, Xu T. 2011. Shaping cells and organs in *Drosophila* by opposing roles of fat body-secreted collagen IV and perlecan. *Dev. Cell* 21:245–56
121. Helfrich W. 1974. The size of bilayer vesicles generated by sonication. *Phys. Lett. A* 50:115–16
122. Lipowsky R. 1993. Domain-induced budding of fluid membranes. *Biophys. J.* 64:1133–38
123. Allain JM, Storm C, Roux A, Ben Amar M, Joanny JF. 2004. Fission of a multiphase membrane tube. *Phys. Rev. Lett.* 93:158104
124. Roux A, Cuvelier D, Nassoy P, Prost J, Bassereau P, Goud B. 2005. Role of curvature and phase transition in lipid sorting and fission of membrane tubules. *EMBO J.* 24:1537–45
125. Sorre B, Callan-Jones A, Manneville JB, Nassoy P, Joanny JF, et al. 2009. Curvature-driven lipid sorting needs proximity to a demixing point and is aided by proteins. *Proc. Natl. Acad. Sci. USA* 106:5622–26
126. Milosevic I, Giovedi S, Lou X, Raimondi A, Collesi C, et al. 2011. Recruitment of endophilin to clathrin coated pit necks is required for efficient vesicle uncoating after fission. *Neuron* 72:587–601
127. Kanoh H, Williger BT, Exton JH. 1997. Arfaptin 1, a putative cytosolic target protein of ADP-ribosylation factor, is recruited to Golgi membranes. *J. Biol. Chem.* 272:5421–29
128. Ho WT, Exton JH, Williger BT. 2003. Arfaptin 1 inhibits ADP-ribosylation factor-dependent matrix metalloproteinase-9 secretion induced by phorbol ester in HT 1080 fibrosarcoma cells. *FEBS Lett.* 537:91–95
129. Man Z, Kondo Y, Koga H, Umino H, Nakayama K, Shin HW. 2011. Arfaptins are localized to the *trans*-Golgi by interaction with Arl1, but not Arfs. *J. Biol. Chem.* 286:11569–78

Emerging Paradigms for Complex Iron-Sulfur Cofactor Assembly and Insertion

John W. Peters and Joan B. Broderick

Department of Chemistry and Biochemistry and the Astrobiology Biogeocatalysis Research Center, Montana State University, Bozeman, Montana 59717; email: john.peters@chemistry.montana.edu, jbroderick@chemistry.montana.edu

Keywords

hydrogenase, nitrogenase, H cluster, FeMo cofactor, radical SAM, biogenesis

Abstract

[FeFe]-hydrogenses and molybdenum (Mo)-nitrogenase are evolutionarily unrelated enzymes with unique complex iron-sulfur cofactors at their active sites. The H cluster of [FeFe]-hydrogenases and the FeMo cofactor of Mo-nitrogenase require specific maturation machinery for their proper synthesis and insertion into the structural enzymes. Recent insights reveal striking similarities in the biosynthetic pathways of these complex cofactors. For both systems, simple iron-sulfur cluster precursors are modified on assembly scaffolds by the activity of radical S-adenosylmethionine (SAM) enzymes.

Radical SAM enzymes are responsible for the synthesis and insertion of the unique nonprotein ligands presumed to be key structural determinants for their respective catalytic activities. Maturation culminates in the transfer of the intact cluster assemblies to a cofactor-less structural protein recipient. Required roles for nucleotide binding and hydrolysis have been implicated in both systems, but the specific role for these requirements remain unclear. In this review, we highlight the progress on [FeFe]-hydrogenase H cluster and nitrogenase FeMo-cofactor assembly in the context of these emerging paradigms.

Contents

- INTRODUCTION 430
  - The Active-Site Iron-Sulfur Clusters of Nitrogenase and Hydrogenase 430
  - Nitrogenase and Nitrogen Fixation 431
  - Hydrogenase and Hydrogen Metabolism 432
- IRON-MOLYBDENUM-COFACTOR BIOSYNTHESIS AND NITROGENASE MATURATION 432
  - Synthesis of Simple Iron-Sulfur Clusters as Precursors 433
  - The NifEN Complex, an Assembly Scaffold Resembling the Molybdenum-Iron Protein 433
  - The Dual Roles of the Magnesium-ATP-Dependent Iron Protein 436
  - NifB and Radical S-Adenosylmethionine Chemistry 436
- BIOSYNTHESIS OF THE H CLUSTER OF THE [FeFe]-HYDROGENASE 437
  - Requirements for [FeFe]-Hydrogenase Maturation . 437
  - HydF: GTPase and Scaffold/Carrier 439
  - Radical S-Adenosylmethionine Enzymes in H-Cluster Maturation 440
  - A Model for [FeFe]-Hydrogenase Activation 441
- FUTURE CHALLENGES AND DIRECTIONS 443

INTRODUCTION

The Active-Site Iron-Sulfur Clusters of Nitrogenase and Hydrogenase

Iron-sulfur clusters are among the most ubiquitous metal cofactors in biology, mediating processes as diverse as electron transfer, gene expression, and catalysis, and playing critical roles in such central metabolic processes as respiration, photosynthesis, and the catalytic interconversions of small molecules (1–6). In their most common forms, these clusters exist as [2Fe-2S], [4Fe-4S], and [3Fe-4S] assemblages of iron ions (+2 or +3 formal oxidation states) and inorganic sulfide, coordinated to the protein most typically by cysteinal ligation at each iron of the cluster (**Figure 1**). Two cysteine thiolates typically coordinate each iron of the [2Fe-2S] cluster while one cysteine thiolate coordinates each iron in typical [4Fe-4S] and [3Fe-4S] clusters, such that in all cases the iron ions are in tetrahedral coordination environments. The protein environment effectively tunes the redox potentials of these clusters to span a range greater than 1 V, such that they are poised in the biologically accessible realm and can undergo facile 1 e^- redox reactions to mediate electron transfer processes and diverse redox reactions in biology.

The active-site cofactors of nitrogenases and [FeFe]-hydrogenases are examples of complicated variations and/or elaborations on these simpler assemblies (**Figure 1**). The iron-molybdenum (FeMo) cofactor at the active site of nitrogenase, as its name implies, is a heterometallic cluster that can be best described as a [4Fe-3S] partial cubane fused to a [Mo-3Fe-4S] partial cubane through three cubane bridging sulfides. Each partial cubane is coordinated to the protein at its distal metal ions, with an Fe ion coordinated by a Cys thiolate, and the single Mo ion of the second partial cubane by the δ-amine nitrogen of histidine. The FeMo cofactor also has nonprotein ligands, including a homocitrate that coordinates the Mo via hydroxyl and carboxylate moieties and what has been reported to be an interstitial carbon coordinating all six Fe ions of the cofactor core (7–10). The assignment of the central carbon is a result of combined computational (10), structural (8), and spectroscopic analysis (9), and its identification provides resolution of the last real unknown with respect to the FeMo-cofactor structure and composition.

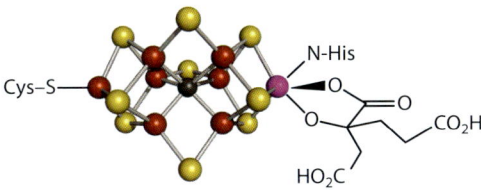
Nitrogenase FeMo cofactor

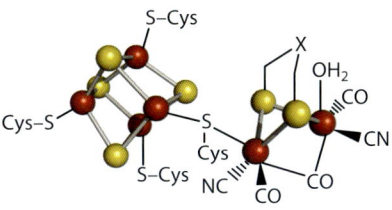
Hydrogenase H cluster

Figure 1
Nitrogenase iron-molybdenum (FeMo) cofactor and [FeFe]-hydrogenase H cluster. For each cluster, the covalent coordinating ligands supplied by the protein are indicated by the three-letter amino acid abbreviation (Cys, His) and show the specific group involved in coordination (S for Cys thiolates, and N for His). The salient features of the inorganic constituents of the iron-sulfur clusters are shown as ball-and-stick representations with sulfur shown in yellow, iron in red, and molybdenum in magenta. Abbreviation: X, the bridgehead atom (most likely N) of the dithiolate ligand of the H cluster.

The H cluster at the active site of [FeFe]-hydrogenase comprises a diiron cluster with several unusual nonprotein ligands bridged to a [4Fe-4S] cluster (11, 12). The [4Fe-4S] subcluster of the H cluster is analogous to aforementioned simple [4Fe-4S] clusters, with the exception that one of the coordinating cysteine thiolates serves as a bridge between this [4Fe-4S] subcluster and the 2Fe subcluster. The 2Fe subcluster of the H cluster has a unique ligand environment, more commonly found in an organometallic compound than a biological assembly. Each Fe ion in the 2Fe subcluster is coordinated by terminal carbon monoxide and cyanide ligands with an additional Fe-bridging carbon monoxide ligand. A nonprotein dithiolate, which has been proposed to exist as dithiodimethylamine (13, 14), also bridges the two Fe ions. The terminal carbon monoxide and cyanide ligands, together with the bridging carbon monoxide and dithiolate ligands, supplant the sulfides and additional protein coordination found in typical iron clusters in biology. Thus, like the FeMo cofactor, the 2Fe subcluster of the H cluster has only minimal protein coordination and, in place of protein ligands, possesses unusual nonprotein ligands that tune the reactivity of the cluster.

The FeMo cofactor and the H cluster thus have in common key features that are not found in other complex iron-sulfur clusters in biology; they are, in essence, modular inorganic/organometallic nanocrystals that have minimal protein ligation but rather are "nested" in the protein environment at the active site, and are tuned by considerable nonprotein ligation. This modularity is reflected in the biosynthetic pathways for these two clusters (15–19), which show striking similarities as discussed below.

Nitrogenase and Nitrogen Fixation

Nitrogen fixation is a key component of global nitrogen cycling, and the availablility of fixed nitrogen is a limiting factor for global nutrition. Biological nitrogen fixation is catalyzed by nitrogenase, an enzyme expressed in a variety of bacterial species and methanogenic archaea but not expressed by any known eukaryotes (20). Although there are alternative forms that either contain vanadium or are heterometal independent, Mo-nitrogenase is by far the most common and most well studied. The FeMo cofactor, discussed above, forms the core of the active site of Mo-nitrogenase and plays a central role in catalyzing the reduction of dinitrogen to ammonia.

The reduction of N_2 to NH_3 has an extremely high activation barrier related to the energy required to break the strong nitrogen-nitrogen triple bond, and thus, it is not surprising that biology has evolved a fairly complicated apparatus to facilitate this reaction. Nitrogenase was one of the first complex iron-sulfur enzymes to be structurally characterized, with the first high-resolution structures published in the early 1990s (21, 22). Mo-nitrogenase is composed of two separable components, the Fe protein and the MoFe protein (20). The Fe protein is a dimeric protein with a single [4Fe-4S] cluster and sites for nucleotide binding and hydrolysis. The MoFe protein is a heterotetrameric protein containing two complex iron-sulfur (FeS) clusters: the P cluster, an [8Fe-7S] cluster reminiscent of two fused [4Fe-4S] clusters, and the aforementioned FeMo cofactor. During nitrogenase catalysis, the Fe protein and MoFe protein associate and dissociate in a manner that couples nucleotide binding and hydrolysis to intermolecular electron transfer. The P cluster functions as an intermediate in electron transfer from the [4Fe-4S] cluster of the Fe protein to the FeMo-cofactor active site of the MoFe protein (23).

Hydrogenase and Hydrogen Metabolism

Although there are a number of enzymes that catalyze biochemical reactions in which hydrogen is a cosubstrate or product, the enzymes that formally catalyze the reversible reduction of protons to produce molecular hydrogen can be divided into two major classes, the [NiFe]-hydrogenases and [FeFe]-hydrogenases (24–27). The first structures of these enzymes appeared in the mid- to late 1990s (28), and the presence of the aforementioned diatomic carbon monoxide and cyanide ligands came as a surprise; these ligands were brought to light not from the structural characterization but from parallel studies examining the enzymes using Fourier transform infrared spectroscopy (28). Although the two classes are not evolutionarily related, they share these unusual features of their active-site metal coordination environments, arguably illustrating one of the more profound cases of convergent evolution. The presence of these diatomic ligands is presumably critical for stabilizing lower-oxidation states of Fe (26), and thus they are key structural determinants for reversible hydrogen oxidation in biology.

Hydrogen metabolism is widespread in biology, where [NiFe]-hydrogenases are commonly associated with bacteria and archaea and [FeFe]-hydrogenases are found in bacteria, some protists, and green algae (28). Many organisms harbor multiple hydrogenases and often a combination of [NiFe] and [FeFe] forms. Their physiological roles are generally biased toward proton reduction for the [FeFe]-hydrogenase and toward hydrogen oxidation for the [NiFe]-hydrogenase. In biological systems, these hydrogenases function either in hydrogen oxidation coupled to energy-yielding reactions or in recycling reduced-electron carriers accumulated during fermentation through proton reduction. The interplay between hydrogen-producing and hydrogen-consuming microorganisms has been well characterized and has even been suggested to be a primary driver for the evolution of the eukaryotic cell (29). Members of each class typically have a specific active-site domain or subunit that contains the H cluster ([FeFe]-hydrogenases) or heterobimetallic NiFe centers ([NiFe]-hydrogenases) and accessory domains containing assemblages of simple FeS clusters that function in electron transfer. The arrangement and composition of these accessory clusters and associated protein domains are responsible for facilitating interactions with specific cellular electron donors and acceptors and for defining their individual physiological roles.

IRON-MOLYBDENUM-COFACTOR BIOSYNTHESIS AND NITROGENASE MATURATION

FeMo-cofactor biosynthesis and nitrogenase maturation are the most well-studied examples of complex cofactor biosynthesis, and many seminal discoveries in this area have led to a powerful paradigm for complex metal center assembly (17–19, 30, 31). Studies initially identified a suite of genes under the nitrogen fixation transcriptional control required for

FeMo-cofactor biosynthesis and nitrogenase maturation. Subsequent biochemical studies revealed that these genes encoded enzymes, specific cluster intermediate carrier proteins, and scaffold proteins that work in concert to synthesize the FeMo cofactor independently of the MoFe protein (**Figure 2**). In a final step of nitrogenase MoFe protein maturation, the synthesized FeMo cofactor is inserted into the MoFe protein through a cationically charged channel.

Synthesis of Simple Iron-Sulfur Clusters as Precursors

One of the more seminal discoveries along the path of delineating the mechanism of FeMo-cofactor biosynthesis was the involvement of simple FeS cluster biosynthetic machinery and simple FeS clusters as precursors to the complex clusters. Deletion mutant analysis of *Azotobacter vinelandii* initially revealed that *nifS* and *nifU* were among the suite of genes required for nitrogen fixation (32). The biochemical characterization of NifS and NifU revealed they work in concert to bring S and Fe together in a scaffolded process to produce simple [2Fe-2S] and [4Fe-4S] clusters (33–35), and it was later revealed that NifU and NifS participate specifically in FeMo-cofactor biosynthesis (36). Subsequent work revealed that an analogous mechanism was employed for the biosynthesis of all simple FeS clusters in biology (5, 6, 37). Homologs of both *nifS* and *nifU* are found in many microorganisms and all domains of life, supporting the thought that this general mechanism for FeS cluster biogenesis is ubiquitous (38).

The primary systems involved in general simple iron-sulfur cluster assembly are the Isc and Suf pathways, which function under normal and oxidative stress conditions, respectively (**Figure 3**) (5, 6, 37). Although the proteins and/or small molecules involved in iron delivery to the Isc and Suf machinery remain a matter of debate, it is known that sulfide is generated through the activity of cysteine desulfurase enzymes (NifS, IscS, or SufS) and pyridoxal phosphate-containing enzymes, which utilize cysteine as a substrate and also liberate and deliver inorganic sulfide via a protein-bound persulfide to the iron-sulfur cluster assembly scaffolds (NifU, IscU, or SufU). The Isc machinery includes IscU, which serves as an assembly scaffold where iron and sulfide are delivered, and the assembly of [2Fe-2S] and [4Fe-4S] clusters occurs. The stepwise details of this process, including the order of delivery of the iron and sulfide components, are still a matter of debate; however, it is clear that IscU facilitates assembly of these simple iron-sulfur clusters and that it can ultimately deliver these clusters to target proteins (39, 40). The IscA protein may serve as an alternate scaffold or cluster carrier (5, 41–43). In the Suf machinery, proteins SufU and SufA, as well as SufB (in a SufBCD complex), can serve as assembly scaffolds and/or carriers that deliver assembled clusters to target proteins (44–46).

In addition to iron and sulfide, the assembly of these simple clusters requires a source of electrons, supplied by ferredoxin in the case of the Isc system, which may be involved in reducing S^0 to S^{2-} or in providing the electrons required to fuse two [2Fe2S] to form a [4Fe-4S] on IscU (39, 40, 47). Furthermore, the Isc machinery includes a dedicated chaperone system consisting of the HscA ATPase and its cochaperone HscB. HscA interacts with a conserved motif on IscU in an ATP hydrolysis–dependent fashion and appears to facilitate cluster transfer from IscU to the target protein (48, 49). For the Suf system, recent results indicate the specific involvement of SufC, an ATPase with similarity to the ABC-type ATPases (50, 51) within a Suf-BCD complex (44, 52) to promote iron-sulfur cluster biosynthesis on SufB (45, 46).

The NifEN Complex, an Assembly Scaffold Resembling the Molybdenum-Iron Protein

Synthesis of the FeMo cofactor occurs on the NifEN heterotetramer, a protein that has primary sequence and structural similarity to the nitrogenase structural protein, NifDK

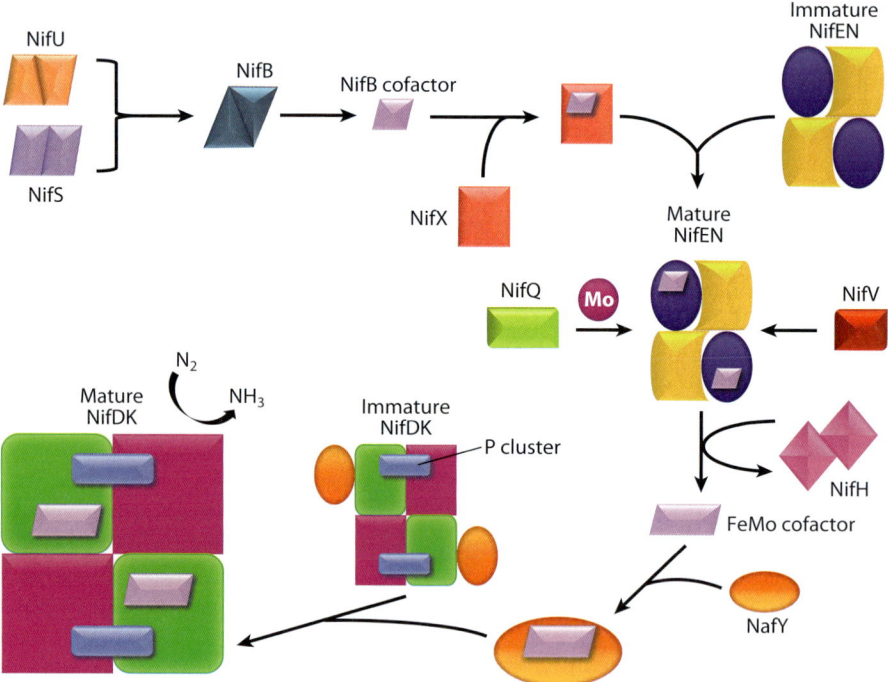

Figure 2

The overall steps in the assembly of the iron-molybdenum (FeMo) cofactor of Mo-nitrogenase. Nitrogenase FeMo-cofactor biosynthesis through the *S*-adenosylmethionine (SAM)-dependent modification of simple FeS clusters by the activity of NifB to form the core of the FeMo cofactor, which is transferred via NifX to the assembly scaffold NifEN. NifEN coordinates the addition of Mo, involving NifH and NifQ as well as homocitrate, produced by the activity of NifV. Final maturation of the MoFe protein (NifDK) occurs by the transfer of the FeMo cofactor from NifEN to a FeMo-cofactor-deficient NifDK in a process facilitated by NafY.

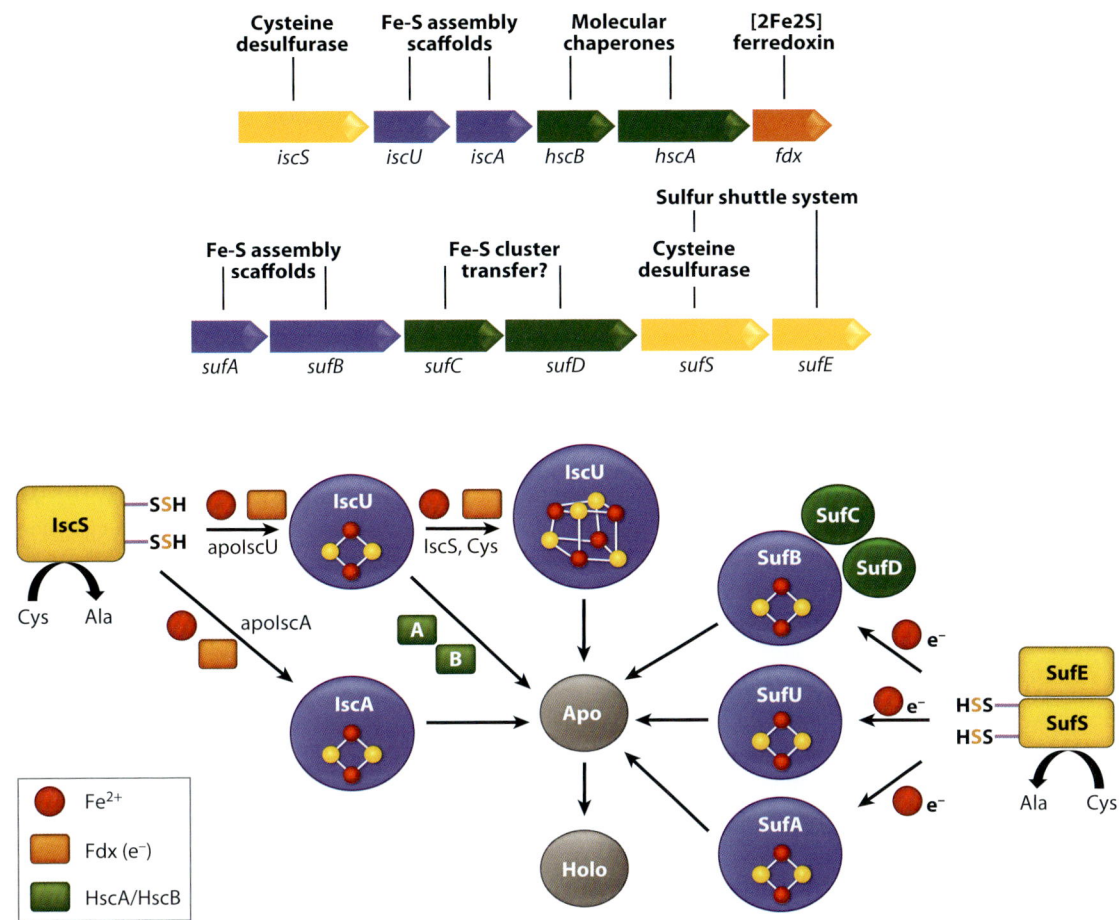

Figure 3

Overview of the synthesis of [2Fe-2S] and [4Fe-4S] clusters in bacteria. The Isc and Suf systems are responsible for simple iron-sulfur cluster biosynthesis. Each system generates sulfide from cysteine through the activity of pyridoxal phosphate-dependent cysteine desulfurase enzymes (IscS or SufS). Cysteine is cleaved to form alanine, and the sulfur is transferred to a cysteine residue at the active site to form an enzyme-bound persulfide. Sulfide is delivered via this persulfide intermediate to scaffold proteins (IscU, IscA, SufA, SufB) and assembled together with iron delivered by specific chaperone proteins (not shown here). Once assembled, the iron-sulfur clusters are transferred intact from the scaffold proteins to target apoproteins via a series of ligand exchange reactions with cysteine residues on the target apoprotein. Abbreviations: Apo, proteins lacking their iron-sulfur cluster cofactor; e−, electron; Fdx, ferredoxin; Holo, proteins containing their requisite iron-sulfur cluster; HscA/HscB, heat shock cognate proteins; SSH or HSS, cysteine persulfide.

(53–55). NifEN is an essential component of the FeMo-cofactor biosynthetic pathway and carries out several key functions in nitrogenase maturation, including the incorporation of additional iron (19, 56–58), molybdenum (57–59), and homocitrate (60) into the FeMo-cofactor precursor. Integral to the function of the NifEN scaffold are its interactions with two other key proteins in the FeMo-cofactor biosynthetic pathway: (*a*) NifB, a radical *S*-adenosylmethionine (SAM) enzyme that appears to scaffold and assemble, using radical SAM chemistry, the NifB-cofactor that is transferred to NifEN (56, 61); and (*b*) NifH, which is required for molybdenum and/or homocitrate incorporation into the

FeMo-cofactor precursor. In fact, it has been proposed that the radical SAM enzyme NifB, the NifEN scaffold, and the iron protein NifH, together comprising the minimal machinery for FeMo-cofactor biosynthesis, form a "biosynthetic factory," a complex of proteins that function in concert to synthesize the FeMo cofactor (19).

The Dual Roles of the Magnesium-ATP-Dependent Iron Protein

Interestingly, NifH, the Fe protein of nitrogenase, which is the unique electron donor to the nitrogenase MoFe protein, is required for biosynthesis of the FeMo cofactor of nitrogenase, as *nifH* deletion analysis revealed that the MoFe protein expressed in a genetic background devoid of NifH was FeMo cofactor deficient (62). Studies involving in vitro FeMo-cofactor biosynthesis have implicated a role for both NifH and MgATP for insertion of molybdenum into a FeMo-cofactor precursor, although there is contradictory evidence as to whether ATPase activity is required for this step (63, 64).

Proteins like NifH that bind and hydrolyze NTPs are central to nearly all major biochemical processes; the most abundant among these are the P-loop NTPases, which have been estimated to comprise 10% or more of the cellular proteins in most organisms (65). These enzymes are broadly conserved (66–68) and carry out critical roles, yet in the vast majority of cases, identifying specific functions has proved elusive. Perhaps the most widely recognized roles for the P-loop NTPases are the critical functions they play in protein translation, where molecular switches, such as EF-Tu, cycle between "on" and "off" states in conjunction with the hydrolysis of GTP to GDP (69, 70). It has now become clear that NTPases, several of them P-loop NTPases, play central roles in the assembly of numerous metal centers in biology, including the FeMo cofactor, common [2Fe2S] and [4Fe4S] clusters, the active-site metal center of the [NiFe]-hydrogenase, the Ni active site of the urease, and the H cluster of hydrogenase.

In the case of the biogenesis of [2Fe2S] and [4Fe4S] clusters, a molecular chaperone (HscA in bacteria, Ssq1 in eukaryotes) hydrolyzes ATP while assisting in the cluster transfer from the scaffold IscU to the target protein (48, 49, 71). Although HscA is of the Hsp70/RNAse H fold and thus is not a P-loop NTPase (70), it appears to play a function similar to the P-loop NTPases discussed herein. SufC, part of the Suf iron-sulfur assembly machinery that is particularly important under oxidative stress conditions, is an ABC-type ATPase, which appears to play a role in iron acquisition during iron-sulfur cluster assembly (45, 50). Furthermore, two proteins central to cytosolic iron-sulfur cluster assembly (the CIA machinery) in yeast, Cfd1 and Nbp35, are P-loop NTPases that form a complex and together act as a scaffold for cytosolic iron-sulfur cluster assembly (72). HypB is a P-loop GTPase that plays a role in nickel insertion during maturation of the active-site metal cluster of [NiFe]-hydrogenase (73–76), and the related P-loop GTPase UreG is involved in nickel insertion into the active site of urease (76, 77). CooC1 is yet another NTPase that is involved in nickel insertion into a target enzyme, in this case the [NiFe] cluster of carbon monoxide dehydrogenase (78).

NifB and Radical *S*-Adenosylmethionine Chemistry

NifB is a radical SAM enzyme thought to be responsible for making the first modifications to simple FeS clusters to produce a core intermediate that is subsequently transferred through the activity of the NifX carrier protein to the NifEN scaffold. Radical SAM enzymes utilize a [4Fe-4S] cluster and SAM to intiate radical reactions in a wide diversity of biological processes (79, 80). The initial mechanistic steps of these enzymes are generally accepted to involve a reduced site-differentiated [4Fe-4S]$^+$ cluster with SAM coordinated to one iron of the cluster via the amino and carboxylate groups of the methionine moiety of SAM. The reduced cluster then transfers an electron to the sulfonium of the coordinated SAM, which results in homolytic S-C bond cleavage to produce a 5′-deoxyadenosyl radical intermediate. This latter intermediate is believed to be the common intermediate in the wide variety of reactions catalyzed by radical SAM enzymes, which include H-atom abstraction from unactivated C-H bonds, C-C bond cleavage, complex multistep rearrangements of organic species, and S-C bond formation. Radical SAM enzymes are found in all kingdoms of life and take part in central biological processes, including ribonucleotide reduction, glucose metabolism, amino acid metabolism, cofactor biosynthesis, and DNA repair.

The radical SAM enzyme NifB is absolutely required for the biogenesis of the FeMo cofactor and appears to play a role not only in catalyzing a SAM-dependent radical reaction, but also in scaffolding the NifB cofactor, a precursor of the FeMo cofactor, which is transferred from NifB to NifEN (56). Biochemical and spectroscopic studies have been interpreted to indicate that the recently assigned central carbon of the FeMo cofactor is also present in the NifB cofactor, supporting the hypothesis that the activity of NifB is directed at generating a six Fe carbon- and sulfide-linked core of the FeMo-cofactor biosynthetic intermediate (81). Interestingly, it has been shown that when NifB cofactor alone is inserted into the FeMo-cofactor site of the MoFe protein, proton reduction activity can be observed, indicating that NifB cofactor is an intact entity that has some of the key structural determinants of catalysis (82).

BIOSYNTHESIS OF THE H CLUSTER OF THE [FeFe]-HYDROGENASE

Major advances in understanding H-cluster maturation (**Figure 4**) have come about only in the past several years, and in many ways, these have followed in the footsteps of the pioneering work on nitrogenase maturation. It was not until the required genes for hydrogenase maturation were revealed in 2004 that solid hypotheses could be generated and substantive experiments could be designed. By analyzing knockout mutants of the green alga *Chlamydomonas reinhardtii* deficient in hydrogen production, two putative genes, in addition to the hydrogen structural gene (*hydA*), were identified as being required for hydrogen production (83, 84). Subsequent coexpression of these genes, designated *hydEF* and *hydG*, with the *hydA* structural gene revealed that these gene products alone were capable of generating an active [FeFe]-hydrogenase (83, 84). Homologs of these genes were found to be associated with all other organisms expressing an active [FeFe]-hydrogenase; however, in bacterial systems, *hydE* and *hydF* were separate genes. The annotation of these genes, indicating that the *hydE* and *hydG* gene products were members of the radical SAM family of enzymes and that *hydF* had a clear nucleotide-binding motif, provided a major breakthrough, allowing the development of hypothetical models for H-cluster biosynthesis (85). In addition, the implication that radical SAM chemistry was a key component of H cluster biosynthesis brought to light the potential for common features between the mechanisms of nitrogenase FeMo-cofactor biosynthesis and H-cluster biosynthesis. The basic tenets that were discussed previously for FeMo-cofactor biosynthesis, including roles for (*a*) simple FeS cluster biosynthetic machinery, (*b*) scaffold proteins, (*c*) NTPase activity, and (*d*) radical SAM chemistry, served as the basis for generating a hypothetical scheme for H-cluster biosynthesis.

Requirements for [FeFe]-Hydrogenase Maturation

Although the active-site H cluster of the [FeFe]-hydrogenase, like the FeMo cofactor, is complex and unique in biology, its specific maturation machinery appears to be remarkably simple in comparison, with only three hydrogenase-specific accessory gene products (HydE, HydF, and HydG) required to biosynthesize the active-site cluster. Because there were a limited number of gene products

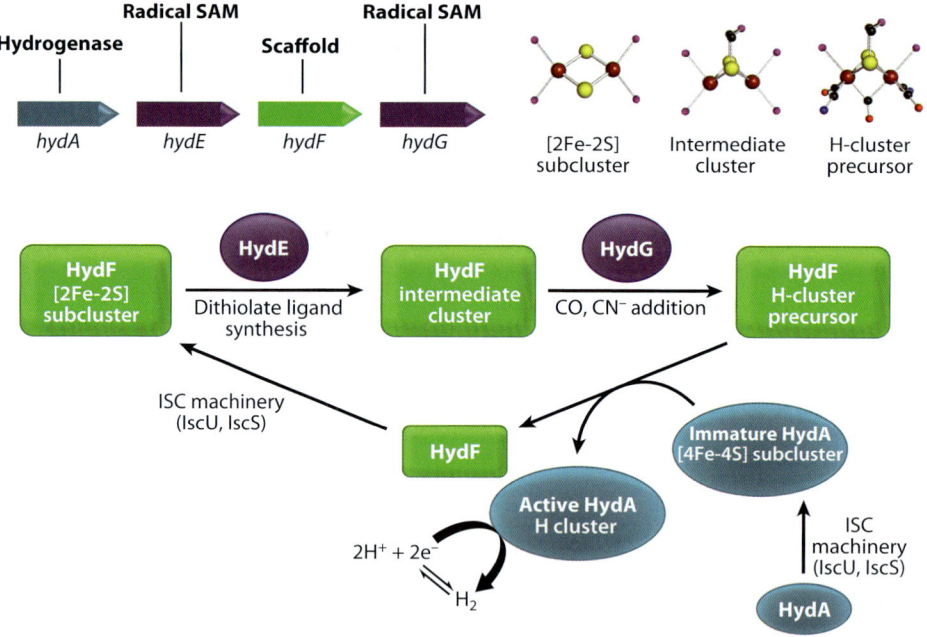

Figure 4

Biogenesis of the H cluster of the [FeFe]-hydrogenase. The current model for H-cluster biogenesis involves the HydF protein acting as a scaffold on which the radical S-adenosylmethionine (SAM) enzymes HydE and HydG modify a [2Fe-2S] cluster to a dithiolate-, CO-, and CN^--ligated 2Fe precursor of the H cluster. HydE is proposed to synthesize the dithiolate ligand of the H cluster, although the substrate and product of HydE catalysis have yet to be identified. HydG utilizes tyrosine as a substrate in the synthesis of CO and CN^-, which are presumably delivered directly to the nascent 2Fe subcluster on HydF. There is no experimental evidence for the order in which HydE and HydG might act in ligand formation; however, a chemical rationale, which may favor the order illustrated, has been presented. The SAM-dependent modifications of the cluster on HydF may require the GTPase activity of HydF as well. In the final step of biogenesis, the assembled 2Fe precursor on HydF is transferred to HydA, which already contains the [4Fe-4S] subcluster of the H cluster in its active site. The bridging of the 2Fe subcluster to the 4Fe subcluster via a cysteine ligand completes the biogenesis of the H cluster to produce the active hydrogenase. Abbreviations: CN^-, cyanide; $2e^-$, two electrons; IscS, cysteine desulfurase; IscU, scaffold protein.

involved in the process, and because the [4Fe-4S] subcluster of the H cluster so closely resembled those clusters known to be synthesized by the aforementioned "housekeeping" iron-sulfur cluster assembly machinery, our initial hypothesis was that specific *hyd*-encoded maturation machinery was directed at the synthesis and insertion of only the unique 2Fe subcluster. We explored this experimentally by expressing the [FeFe]-hydrogenase structural protein, HydA, in *Escherichia coli*, which lacks the accessory Hyd proteins. In the absence of specific maturation enzymes, *E. coli* expresses a hydrogenase containing a [4Fe-4S] cluster, and the presence of this [4Fe-4S] cluster has been shown to be absolutely essential for the subsequent maturation of the active-site cluster by HydE, HydF, and HydG (86). This requirement for the housekeeping machinery, together with the requirement for a preformed [4Fe-4S] cluster on HydA, supports the hypothesis that the hydrogenase-specific maturation machinery is not involved in [4Fe-4S] subcluster synthesis but rather is involved only in the assembly of the 2Fe subcluster of the H cluster.

Early experiments on [FeFe]-hydrogenase maturation demonstrated that when the three accessory proteins are expressed together in

E. coli, they generate a protein-associated activating component that can be transferred in vitro to the structural protein HydA to generate an active hydrogenase (87, 88). Purification of the individual accessory proteins from *E. coli* in which all three had been expressed together revealed that the protein HydF alone harbors the activating element responsible for activating HydA (88). HydF was thus proposed to serve as a scaffold or carrier in H-cluster maturation and to carry out the final delivery of an H-cluster precursor to HydA.

HydF: GTPase and Scaffold/Carrier

The sequence of HydF revealed the presence of Walker A and B motifs, suggesting that HydF was a P-loop NTPase (84). In addition, a series of cysteine and histidine residues in the C-terminal domain of HydF suggested the potential for iron-sulfur cluster binding. Subsequent biochemical experiments confirmed the ability of HydF to bind iron-sulfur clusters and to hydrolyze GTP (89). More detailed analysis of GTP hydrolysis by HydF demonstrated that the activity was gated by monovalent cations, with potassium and rubidium providing the most significant enhancement of GTP hydrolysis rates (90). These results place HydF among the small number of P-loop NTPases that are activated by monovalent cation binding (91). It was further shown that, although the presence or absence of iron-sulfur clusters on HydF had no effect on rates of GTP hydrolysis, the binding of GTP dramatically increased the intensity of the electron paramagnetic resonance (EPR) signals for the reduced clusters on HydF, suggesting communication between GTP- and cluster-binding domains (90).

The iron-sulfur cluster content of HydF has also been examined in detail, with the aim to understand whether HydE and HydG induce changes in the cluster content of HydF that could be correlated with the generation of an activating element on HydF (88, 90). These experiments involved expressing HydF in *E. coli* in either the absence or presence of the other maturation proteins; these HydF proteins were then purified and subjected to spectroscopic analysis. UV-visible spectroscopy provided the first clues that coexpression of HydF with HydE and HydG resulted in dramatic changes in the cluster content on HydF; difference spectra provided evidence that one key difference was the loss of features characteristic of a [2Fe-2S] cluster from HydF when it was expressed in the presence of the other maturases, i.e., HydE and HydG (88). Low-temperature EPR spectroscopy of reduced HydF subsequently showed that, when purified from *E. coli* expressing only HydF, the protein has the characteristic EPR spectral features for both $[2Fe-2S]^+$ and $[4Fe-4S]^+$ clusters. When purified from *E. coli* in which HydE and HydG are also expressed, reduced HydF exhibits only a $[4Fe-4S]^+$ EPR signal; as with the UV-visible experiments, the [2Fe-2S] cluster appears to have been lost upon expression of HydE and HydG with HydF (90). Together, these results suggested that the radical SAM enzymes HydE and HydG modify HydF in vivo so as to convert a [2Fe-2S] cluster to some other form that has neither the characteristic UV-visible nor EPR spectral features of a typical [2Fe-2S] cluster.

Given the evidence that HydA requires the presence of a preformed [4Fe-4S] cluster prior to maturation (86), and the resulting implication that the maturation enzymes are responsible for synthesis of the 2Fe subcluster of the H cluster, it seemed reasonable to suggest that the [2Fe-2S] cluster of HydF was being converted to a 2Fe H-cluster precursor. Fourier transform infrared spectroscopic characterization of HydF supported this supposition by demonstrating that when HydF was expressed in the presence of HydE and HydG in *E. coli*, it contained an H-cluster-like species with CO and CN^- ligands (90). Thus these spectroscopic studies have revealed further insights into the role of HydF, providing evidence that it does indeed serve as a scaffold for biosynthesis of an H-cluster precursor. Specifically, the results suggest that the radical SAM enzymes HydE and HydG modify a standard [2Fe-2S] cluster on HydF such that it becomes the CO- and CN⁻- ligated activating element that can be transferred to HydA to generate an active hydrogenase.

The role of NTP hydrolysis in H-cluster assembly has proved more elusive. As discussed above, NTPases are prevalent in the assembly of metal cofactors, specifically in the biosynthesis of iron-sulfur clusters and the FeMo cofactor. The most commonly implicated function for these NTPases appears to be either metal delivery to the active site or cluster transfer to the target protein. On the basis of these precedents, the involvement of GTP hydrolysis in the transfer of an H-cluster precursor to HydA might be considered the most likely scenario. Experiments have shown, however, that "loaded" HydF could activate HydA equally well regardless of the presence or absence of GTP, GDP, or nonhydrolyzable analogs of GTP in the activation reaction, suggesting that GTP binding and hydrolysis are not essential for transfer of a cluster precursor from HydF to HydA (88). The effect of the presence of the maturation proteins on the rate of GTP hydrolysis by HydF was also examined, and it was shown that, although the presence of HydA had no effect on GTP hydrolysis kinetics, both HydE and HydG stimulated HydF-catalyzed GTP hydrolysis by 50% (88). The demonstrated requirement for the Walker motifs of HydF for in vivo hydrogenase maturation (92), together with the in vitro evidence against a role for GTP binding and hydrolysis in cluster transfer and the stimulation of GTP hydrolysis in the presence of HydE and HydG (88), suggests that GTP hydrolysis plays a role in the interaction of the radical SAM enzymes HydE and HydG with HydF.

Radical S-Adenosylmethionine Enzymes in H-Cluster Maturation

With the evidence that HydF serves as a scaffold or carrier for the 2Fe precursor of the H cluster, the roles for the radical SAM enzymes HydE and HydG seemed likely to be in the synthesis of the unique nonprotein ligands of the H cluster (85), perhaps by interacting directly with HydF in a GTP-dependent fashion. As shown in **Figure 1**, the H cluster contains a bridging dithiolate ligand, widely accepted to be dithiomethylamine (14), as well as two cyanide and three carbon monoxide ligands. Established radical SAM chemistry could conceivably lead to such biologically unusual ligands (85). Specifically, sulfide insertion into C-H bonds to generate alkane thiols has precedents in radical SAM chemistry (93), as such reactions are catalyzed by LipA (synthesis of the lipoyl cofactor) (94) and BioB (synthesis of biotin). Likewise, generation of the diatomic ligands could conceivably occur via radical-mediated degradation of an amino acid (85).

Early studies of HydG demonstrated that it could bind a [4Fe-4S] cluster with properties typical of a radical SAM enzyme, including the SAM-dependent alteration of EPR spectral properties (95). Furthermore, HydG catalyzes uncoupled reductive cleavage of SAM. The significant sequence homology between HydG and ThiH, a radical SAM enzyme that catalyzes the cleavage of tyrosine to generate a precursor for thiamine biosynthesis, led to an examination of tyrosine as a potential substrate of HydG (96). Tyrosine stimulates the rate of SAM cleavage by HydG, and subsequently, HydG was shown to cleave tyrosine to produce p-cresol and presumably dehydroglycine. In the first study demonstrating tyrosine as a substrate for HydG, it was proposed that multiple dehydroglycine molecules produced by HydG could combine to form a dithiomethylamine ligand (96). Although dehydroglycine could conceivably serve as a precursor for the dithiomethylamine ligand of the H cluster, careful analysis of the HydG-catalyzed turnover of tyrosine revealed that both cyanide and carbon monoxide were additional products (97, 98). Cyanide was detected by formation of a fluorescent adduct after denaturation of protein, and both the kinetics and the overall stoichiometry of cyanide production correlated directly with the production of p-cresol (97). The production of 5′-deoxyadenosine (5′-dAdo) from the cleavage of SAM was in slight excess over that of p-cresol and cyanide, indicative of a small amount of uncoupled SAM cleavage during turnover. Production of carbon monoxide during HydG-catalyzed turnover of tyrosine was detected by utilizing deoxyhemoglobin as a carbon monoxide sensor, and monitoring the characteristic UV-visible spectroscopic changes that have been well characterized for the conversion of deoxy- to carbonmonoxy-hemoglobin (98). CO detection was carried out continuously during HydG turnover, with simultaneous sampling to monitor for 5′-dAdo and p-cresol production; the results revealed kinetics comparable to those observed for the cyanide assay. For both the cyanide and carbon monoxide assays, U-^{13}C-tyrosine in the assays was used to demonstrate unequivocally that these diatomic ligands derive from tyrosine (97, 98). Fourier transform infrared characterization of HydA that was activated in a cell-free system containing HydE, HydG, HydF, and ^{13}C-tyrosine further demonstrated that all the diatomic ligands of the H cluster derive from tyrosine (99).

With the demonstration that HydG synthesizes the diatomic ligands known to be present in the H cluster, this leaves HydE to presumably synthesize the dithiolate ligand. Like HydG, early studies of HydE demonstrated its ability to bind a [4Fe-4S] cluster characteristic of radical SAM enzymes and to catalyze the uncoupled cleavage of SAM (95). A high-resolution crystal structure of HydE reveals the presence of a SAM-bound [4Fe-4S] cluster as well as a [2Fe-2S] cluster, although the latter may be the result of reconstitution and thus may not be relevant to the function of HydE (100, 101). The large active-site binding cavity of HydE is somewhat ill defined; however, there are three putative anion-binding sites, potentially providing clues into the chemical nature of the HydE substrate. Computational docking experiments indicate a strong affinity for thiocyanate, a compound that has analogous features of the H-cluster nonprotein ligand set (101). In work involving a creative approach to in vitro activation of [FeFe]-hydrogenases, it was shown that the additional Tyr and Cys stimulate the synthesis of active [FeFe]-hydrogenases (102). The observation that Tyr stimulates this synthesis is directly in line with the biochemical work showing the assignment of Tyr as the substrate for HydG and the source of the carbon monoxide and cyanide in the H cluster (96–99). The stimulation by Cys could be a result of enhanced synthesis of the simple iron-sulfur precursor of the H cluster, as Cys is the substrate for cysteine desulfurase; however, it is also attractive to consider the potential of Cys as a substrate for HydE and as a source of the nonprotein dithiolate ligand. To date, however, the stimulation by Cys and the reaction catalyzed by HydE remain unresolved aspects of H-cluster biosynthesis.

A Model for [FeFe]-Hydrogenase Activation

Identification of the substrate and product for HydG, and a possible substrate for HydE, represents a major step forward in the understanding of H-cluster biosynthesis; however, it represents only part of the story. The evidence for HydF acting as the scaffold for synthesis of the 2Fe H-cluster precursor from a standard [2Fe-2S] cluster, and the evidence that HydE and HydG are the enzymes that catalyze these modifications, suggests that significant protein-protein interactions are central to this maturation process. In the working model, HydE catalyzes a BioB-type sulfur insertion reaction into an unknown substrate, using the [2Fe-2S] cluster on HydF as the sulfur source, with the resulting thiolate and dithiolate ligands bound to the two irons on HydF (**Figure 4**). Subsequently, HydG utilizes tyrosine to synthesize the CO and CN$^-$ ligands, which presumably must be carefully delivered or channeled to the nascent 2Fe H-cluster precursor on HydF. Indeed, qualitative evidence for protein-protein interactions has been reported for some of these proteins (88). The roles of GTP binding and hydrolysis in H-cluster biosynthesis remain to be elucidated, but evidence suggests that these are associated with the interactions between HydF and the radical SAM enzymes HydE and HydG (98).

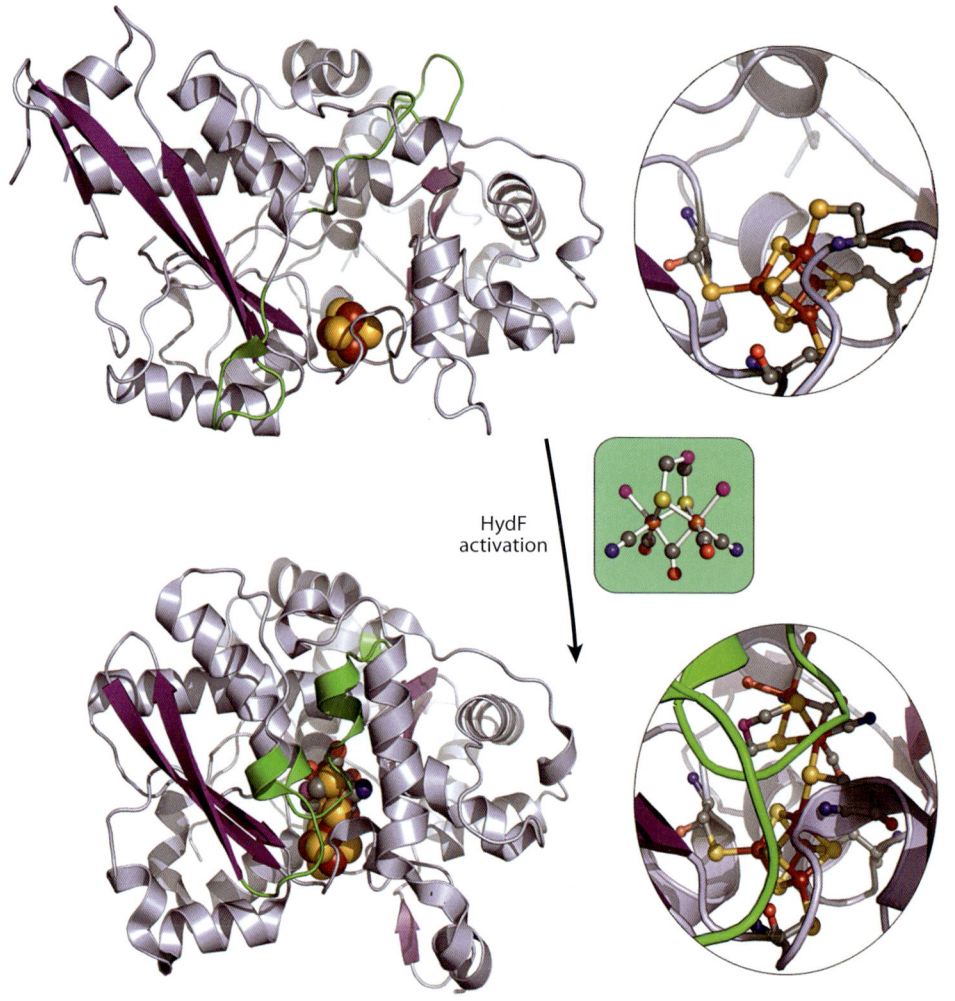

Figure 5
The final step of H-cluster maturation through 2Fe subcluster insertion. A model for H-cluster 2Fe subcluster insertion is shown using ribbon diagrams of HydA structural states with clusters shown as ball-and-stick (2Fe subcluster) or space-filling ([4Fe-4S] subcluster) representations. HydA prior to 2Fe subcluster insertion (*top*) has an open channel leading to a 2Fe subcluster recipient cavity adjacent to the [4Fe-4S] cluster. Upon insertion of the 2Fe subcluster, facilitated by HydF, two conserved loop regions (*green*) undergo a conformational rearrangement leading to the active [FeFe]-hydrogenase (*bottom*).

Once the dithiolate-, CO-, and CN⁻-ligated 2Fe precursors are assembled on HydF, they are delivered to the HydA structural protein, which already contains a preformed [4Fe-4S] cluster. Structural characterization of an intermediate of HydA prior to 2Fe subcluster insertion (**Figure 5**) reveals that the 2Fe subcluster is inserted via a positively charged channel that leads from the protein surface to the cluster-binding site (103). The channel terminates with a binding cavity adjacent to the [4Fe-4S] subcluster, which requires very limited reorientation of amino acid side chains for insertion of the 2Fe subcluster. In the structure of the HydA holoprotein containing the entire 6Fe H cluster, the channel is closed, presumably as a result of a conformational change in a conserved protein loop, burying the catalytic cluster in the interior of the protein. Because there is very little rearrangement of the active site required for insertion of the 2Fe subcluster, the nature or driving force for the conformational rearrangement and channel closure is not clear, perhaps suggesting a role for HydF and protein-protein interaction in this process. Interestingly, cofactor-deficient nitrogenase also contains a cationic channel leading from the protein surface to the cofactor-binding site, and this channel is also closed in the holoenzyme (104). The similar structural rearrangements in HydA and NifDK upon cluster insertion are suggestive of a conserved process for insertion of the complex iron-sulfur cluster into target apoproteins (103). The results also support the idea of a stepwise evolution of an active [FeFe]-hydrogenase (103), consistent with the proposed evolution of nitrogenase (105) and possibly other Fe-S enzymes.

FUTURE CHALLENGES AND DIRECTIONS

[FeFe]-hydrogenase and Mo-nitrogenase are two enzymes that are evolutionarily unrelated yet have evolved strikingly common biochemical strategies for cofactor biosynthesis and maturation (**Figure 6**) (15, 19, 103). The key steps utilize radical SAM chemistry and NTPases to modify simple iron-sulfur cluster precursors on a scaffold and culminate in the synthesis of a preformed cluster or subcluster that, in the final step of maturation, is inserted into a stable cofactor-less form of the enzyme through well-defined channels, which collapse upon final insertion, to generate the mature active enzyme (15, 103, 104). Although parallels can be drawn to the biosynthetic pathways of other complex iron-sulfur clusters as well, the recently elucidated details of the biochemistry of enzyme maturation reveal outstanding similarities between the FeMo-cofactor and H-cluster assembly pathways.

The general features of the biosynthesis of both the FeMo cofactor and the H cluster and the maturation of Mo-nitrogenase and [FeFe]-hydrogenase are rapidly becoming better understood, but there are still numerous aspects that remain to be elucidated. Chief among these outstanding questions are the substrates and products for two essential radical SAM enzymes, the HydE, which presumably synthesizes the bridging dithiolate ligand of the H cluster, and NifB, which has been speculated to provide the central atom of the FeMo cofactor (15, 81, 82). Furthermore, there are key unanswered questions for both pathways regarding the precise role for NTP binding and hydrolysis. Finally, the details of the protein-protein interactions that must be central to these biosynthetic processes remain to be elucidated, particularly in the case of [FeFe]-hydrogenase maturation.

The advances made in the past several years have provided a deeper understanding of the overall complexity or these biosynthetic mechanisms and allow us to revisit aspects of the evolution of these unique and interesting enzymes. The reactions catalyzed by hydrogenases and nitrogenases as well as carbon monoxide dehydrogenases are all reactions that can be placed as having an importance in prebiotic chemistry or Hadean chemistry. This, together with the prevalence of iron-sulfur minerals on early Earth, has prompted the suggestion that iron-sulfur clusters may be among the most ancient of biological cofactors (106). The radical SAM enzymes have also been proposed to be evolutionarily ancient, and the unprecedented reactivity demonstrated for HydG (96–98, 107), generating diatomic ligands derived from the radical-based decomposition of an amino acid, could be considered as an example of "ligand-accelerated catalysis" in modern biology (108, 109). Ligand-accelerated catalysis is defined as the reaction of simple organics (for example, organic acids or amino acids) with metal surfaces, metal clusters, or metals to provide alternative ligand sets that tune reactivity toward accelerating different reactions, and this has been invoked to explain the potential for diversification of reactivity in a prebiotic Earth leading up to the origin of life. The final

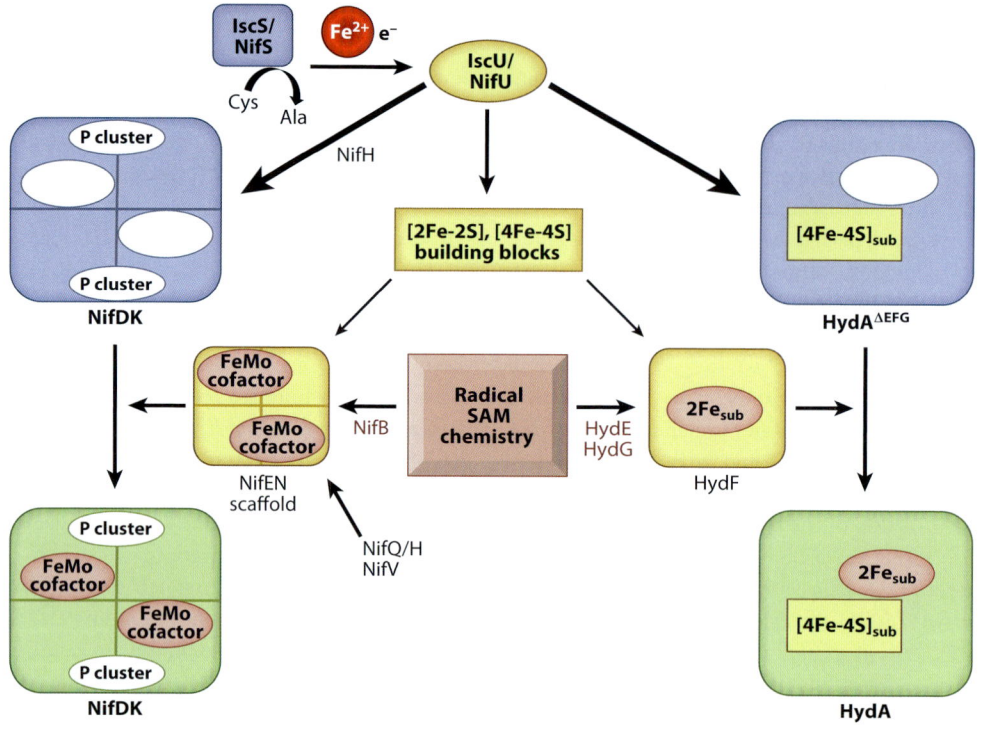

Figure 6

Pathways for maturation of the Mo-nitrogenase and the [FeFe]-hydrogenase. The biogenesis of both the FeMo cofactor and the H cluster begins with the utilization of the housekeeping iron-sulfur cluster assembly machinery to generate simple iron-sulfur cluster building blocks that are subsequently modified. In both cases, these clusters are transferred to the structural proteins, which already have a complement of iron-sulfur clusters (P clusters for nitrogenase and a [4Fe-4S] subcluster ([4Fe-4S]$_{sub}$) of the H cluster for [FeFe] hydrogenase). Final maturation and generation of an active nitrogenase or [FeFe]-hydrogenase occur by the insertion of the modified cluster (FeMo cofactor or H-cluster 2Fe subcluster) into a structural protein through cationically charged channels. Abbreviations: HydA, hydrogenase structural protein; HydA$^{\Delta EFG}$, hydrogenase structural protein expressed in the absence of the Hyd maturases; HydE/HydG, radical S-adenosylmethionine (SAM) maturases for the [FeFe]-hydrogenase; HydF, scaffold or carrier for H-cluster biosynthesis; IscS/NifS, cysteine desulfurases; IscU/NifU, scaffolds in biosynthesis of simple iron-sulfur clusters; NifB, radical SAM enzyme in FeMo-cofactor biosynthesis; NifDK, nitrogenase MoFe structural protein; NifEN, scaffold for assembly of the FeMo-cofactor; NifH, Fe protein of nitrogenase; NifQ/NifV, other proteins required for maturation of Mo-nitrogenase.

steps in maturation of Mo-nitrogenase and [FeFe]-hydrogenase are the type of phenomena thought to be important for exploiting and tuning the reactivity of inorganic moieties in prebiotic chemistry. The cluster insertion observed in these systems is a highly evolved example of inorganic nesting (110), whereby the solubility, accessibility, and reactivity of metals and metal clusters in an aqueous solution are modulated by the binding by organic ligands, much in the same way as described for the role of siderophores in iron acquisition (111).

The presence of iron-sulfur cofactors, the tuning and nesting of these cofactors to elicit specific reactivity, and the catalysis of conceivably prebiotic reactions relate these highly evolved biosynthetic processes as well as the hydrogenase and nitrogenase enzymes themselves to aspects of prebiotic chemistry, but what then about the idea that these enzymes are ancient or even primordial (106)? Certainly the reactions catalyzed by these enzymes, reversible hydrogen oxidation and nitrogen fixation, are reactions that can be easily argued to have had importance in early life on Earth. However, the multiple layers involved in the biosynthesis and maturation of the enzymes suggests that a great deal of time and evolutionary refinement were needed to produce the modern enzymes.

It is an ideal time to examine the merits of the traditional paradigms for the evolution of these enzymes and their biosynthetic pathways and to challenge ourselves to produce rational models for the evolutionary origin of these enzymes by which they might have emerged and been refined in response to specific selective pressures (105).

SUMMARY POINTS

1. The [FeFe]-hydrogenase H cluster and the Mo-nitrogenase FeMo-cofactor are synthesized through specific modifications of simple, biologically ubiquitous FeS cluster precursors.
2. Modifications are imposed on simple FeS cluster precursors through the activity of radical SAM enzymes.
3. The radical SAM enzyme HydG involved in [FeFe]-hydrogenase H cluster biosynthesis and ThiH involved in thiamine biosynthesis are related and use the same substrate, the amino acid tyrosine.
4. The diatomic ligands of the hydrogenase H cluster are derived from the backbone of tyrosine during HydG catalysis.
5. Nitrogenase FeMo-cofactor biosynthesis and [FeFe]-hydrogenase H-cluster biosynthesis both require an NTPase, but the roles of nucleotide binding and hydrolysis in these processes have not yet been definitively determined.
6. The final step in maturation of nitrogenase and [FeFe]-hydrogenase involves the respective transfer of the FeMo-cofactor and the H cluster from specific assembly scaffolds.
7. The unifying paradigms for nitrogenase and [FeFe]-hydrogenase maturation have exciting implications on models for complex iron-sulfur enzyme origin and evolution.

FUTURE ISSUES

1. The reaction catalyzed by the radical SAM enzyme NifB in nitrogenase FeMo-cofactor biosynthesis and also the source and mechanism of insertion of the central carbon of the cofactor remain the central unresolved questions in nitrogenase maturation.
2. The reaction catalyzed by the radical SAM enzyme HydE and the associated mechanism of H cluster dithiolate ligand formation and insertion have yet to be determined.
3. The specific roles played by nucleotide binding and hydrolysis in these and other metalloenzyme maturation systems remain unresolved, and must be elucidated to better understand biological metal center assembly.
4. Future work will allow us to better define paradigms for complex biological metal cluster assembly. Furthermore, tracing the evolutionary history of nitrogenases, hydrogenases, and their biosynthetic accessory proteins will provide insights into the evolutionary origin of these key biological reactivities.

DISCLOSURE STATEMENT

The authors are not aware of any affiliations, memberships, funding, or financial holdings that might be perceived as affecting the objectivity of this review.

ACKNOWLEDGMENTS

The authors thank Kaitlin Duschene and Rachel Hutcheson for assistance in the preparation of figures for this article. Research in the laboratory of J.W. Peters is supported by AFOSR grant FA9550-05-1-0365, Department of Energy (DOE) grant DE-FG02-10ER16194, and NASA Astrobiology Institute grant NNA08CN85A. Research in the laboratory of J.B. Broderick is supported by National Insitutes of Health grant GM54608, DOE grant DE-FG02-10ER16194, and NASA Astrobiology Institute grant NNA08CN85A.

LITERATURE CITED

1. Beinert H, Holm RH, Münck E. 1997. Iron-sulfur clusters: nature's modular, multipurpose structures. *Science* 277:653–59
2. Beinert H. 2000. Iron-sulfur proteins: ancient structures, still full of surprises. *J. Biol. Inorg. Chem.* 5:2–15
3. Noodleman L, Lovell T, Liu T, Himo F, Torres RA. 2002. Insights into properties and energetics of iron-sulfur proteins from simple clusters to nitrogenase. *Curr. Opin. Chem. Biol.* 6:259–73
4. Holm RH, Kennepohl P, Solomon EI. 1996. Structural and functional aspects of metal sites in biology. *Chem. Rev.* 96:2239–314
5. Johnson DC, Dean DR, Smith AD, Johnson MK. 2005. Structure, function, and formation of biological iron-sulfur clusters. *Annu. Rev. Biochem.* 74:247–81
6. Johnson MK. 1998. Iron-sulfur proteins: new roles for old clusters. *Curr. Opin. Chem. Biol.* 2:173–81
7. Einsle O, Tezcan FA, Andrade SLA, Schmid B, Yoshida M, et al. 2002. Nitrogenase MoFe-protein at 1.16 Å resolution: a central ligand in the FeMo-cofactor. *Science* 297:1696–700
8. Spatzal T, Aksoyoglu M, Zhang L, Andrade SL, Schleicher E, et al. 2011. Evidence for interstitial carbon in nitrogenase FeMo cofactor. *Science* 334:940
9. Lancaster KM, Roemelt M, Ettenhuber P, Hu Y, Ribbe MW, et al. 2011. X-ray emission spectroscopy evidences a central carbon in the nitrogenase iron-molybdenum cofactor. *Science* 334:974–77
10. Harris TV, Szilagyi RK. 2011. Comparative assessment of the composition and charge state of nitrogenase FeMo-cofactor. *Inorg. Chem.* 50:4811–24
11. Peters JW, Lanzilotta WN, Lemon BJ, Seefeldt LC. 1998. X-ray crystal structure of the Fe-only hydrogenase (CpI) from *Clostridium pasteurianum* to 1.8 angstrom resolution. *Science* 282:1853–58
12. Nicolet Y, Lemon BJ, Fontecilla-Camps JC, Peters JW. 2000. A novel FeS cluster in Fe-only hydrogenases. *Trends Biochem. Sci.* 25:138–43
13. Nicolet Y, de Lacey AL, Vernede X, Fernandez VM, Hatchikian EC, Fontecilla-Camps JC. 2001. Crystallographic and FTIR spectroscopic evidence of changes in Fe coordination upon reduction of the active site of the Fe-only hydrogenase from *Desulfovibrio desulfuricans*. *J. Am. Chem. Soc.* 123:1596–601
14. Silakov A, Wenk B, Reijerse E, Lubitz W. 2009. (14)N HYSCORE investigation of the H-cluster of [FeFe] hydrogenase: evidence for a nitrogen in the dithiol bridge. *Phys. Chem. Chem. Phys.* 11:6592–99
15. Shepard EM, Boyd ES, Broderick JB, Peters JW. 2011. Biosynthesis of complex iron-sulfur enzymes. *Curr. Opin. Chem. Biol.* 15:319–27
16. Dos Santos PC, Dean DR, Hu Y, Ribbe MW. 2004. Formation and insertion of the nitrogenase iron-molybdenum cofactor. *Chem. Rev.* 104:1159–73
17. Hu Y, Fay AW, Lee CC, Yoshizawa J, Ribbe MW. 2008. Assembly of nitrogenase MoFe protein. *Biochemistry* 47:3973–81
18. Rubio LM, Ludden PW. 2005. Maturation of nitrogenase: a biochemical puzzle. *J. Bacteriol.* 187:405–14
19. Rubio LM, Ludden PW. 2008. Biosynthesis of the iron-molybdenum cofactor of nitrogenase. *Annu. Rev. Microbiol.* 62:93–111

20. Peters JW, Boyd ES, Hamilton TL, Rubio LM. 2011. Biochemistry of Mo-nitrogenase. In *Nitrogen Cycling in Bacteria: Molecular Analysis*, ed. JWB Moir, pp. 59–99. Norwich: Caister Acad.
21. Chan MK, Kim J, Rees DC. 1993. The nitrogenase FeMo-cofactor and P-cluster pair: 2.2 A resolution structures. *Science* 260:792–94
22. Kim J, Rees DC. 1992. Structural models for the metal centers in the nitrogenase molybdenum-iron protein. *Science* 257:1677–82
23. Peters JW, Fisher K, Newton WE, Dean DR. 1995. Involvement of the P cluster in intramolecular electron transfer within the nitrogenase MoFe protein. *J. Biol. Chem.* 270:27007–13
24. Fontecilla-Camps JC. 2009. Structure and function of [NiFe]-hydrogenases. *Metal Ions Life Sci.* 6:151–78
25. Fontecilla-Camps JC, Amara P, Cavazza C, Nicolet Y, Volbeda A. 2009. Structure-function relationships of anaerobic gas-processing metalloenzymes. *Nature* 460:814–22
26. Peters JW. 2009. Carbon monoxide and cyanide ligands in the active site of [FeFe]-hydrogenases. *Metal Ions Life Sci.* 6:179–218
27. Mulder DW, Shepard EM, Meuser JE, Joshi N, King PW, et al. 2011. Insights into [FeFe]-hydrogenase structure, mechanism, and maturation. *Structure* 19:1038–52
28. Volbeda A, Charon MH, Piras C, Hatchikian EC, Frey M, Fontecilla-Camps JC. 1995. Crystal structure of the nickel-iron hydrogenase from *Desulfovibrio gigas*. *Nature* 373:580–87
29. Martin W, Müller M. 1998. The hydrogen hypothesis for the first eukaryote. *Nature* 392:37–41
30. Hu Y, Ribbe MW. 2011. Biosynthesis of the metalloclusters of molybdenum nitrogenase. *Microbiol. Mol. Biol. Rev.* 75:664–77
31. Hu Y, Ribbe MW. 2011. Biosynthesis of nitrogenase FeMoco. *Coord. Chem. Rev.* 255:1218–24
32. Jacobson MR, Cash VL, Weiss MC, Laird NF, Newton WE, Dean DR. 1989. Biochemical and genetic analysis of the nifUSVWZM cluster from *Azotobacter vinelandii*. *Mol. Gen. Genet.* 219:49–57
33. Agar JN, Yuvaniyama P, Jack RF, Cash VL, Smith AD, et al. 2000. Modular organization and identification of a mononuclear iron-binding site within the NifU protein. *J. Biol. Inorg. Chem.* 5:167–77
34. Fu W, Jack RF, Morgan TV, Dean DR, Johnson MK. 1994. nifU gene product from *Azotobacter vinelandii* is a homodimer that contains two identical [2Fe-2S] clusters. *Biochemistry* 33:13455–63
35. Zheng L, White RH, Cash VL, Jack RF, Dean DR. 1993. Cysteine desulfurase activity indicates a role for NIFS in metallocluster biosynthesis. *Proc. Natl. Acad. Sci. USA* 90:2754–58
36. Zhao D, Curatti L, Rubio LM. 2007. Evidence for nifU and nifS participation in the biosynthesis of the iron-molybdenum cofactor of nitrogenase. *J. Biol. Chem.* 282:37016–25
37. Frazzon J, Dean DR. 2003. Formation of iron-sulfur clusters in bacteria: an emerging field in bioinorganic chemistry. *Curr. Opin. Chem. Biol.* 7:166–73
38. Py B, Barras F. 2010. Building Fe-S proteins: bacterial strategies. *Nat. Rev. Microbiol.* 8:436–46
39. Chandramouli K, Unciuleac M-C, Naik S, Dean DR, Huynh BH, Johnson MK. 2007. Formation and properties of [4Fe-4S] clusters on the IscU scaffold protein. *Biochemistry* 46:6804–11
40. Unciuleac M-C, Chandramouli K, Naik S, Mayer S, Huynh BH, et al. 2007. In vitro activation of apoaconitase using a [4Fe-4S] cluster-loaded form of the IscU [Fe-S] cluster scaffolding protein. *Biochemistry* 46:6812–21
41. Ayala-Castro C, Saini A, Outten FW. 2008. Fe-S cluster assembly pathways in bacteria. *Microbiol. Mol. Biol. Rev.* 72:110–25
42. Fontecave M, Ollagnier-de-Choudens S. 2008. Iron-sulfur cluster biosynthesis in bacteria: mechanisms of cluster assembly and transfer. *Arch. Biochem. Biophys.* 474:226–37
43. Bandyopadhyay S, Chandramouli K, Johnson MK. 2008. Iron-sulfur cluster biosynthesis. *Biochem. Soc. Trans.* 36:1112–19
44. Chahal HK, Dai Y, Saini A, Ayala-Castro C, Outten FW. 2009. The SufBCD Fe-S scaffold complex interacts with SufA for Fe-S cluster transfer. *Biochemistry* 48:10644–53
45. Saini A, Mapolelo DT, Chahal HK, Johnson MK, Outten FW. 2010. SufD and SufC ATPase activity are required for iron acquisition during in vivo Fe-S cluster formation on SufB. *Biochemistry* 49:9402–12
46. Wollers S, Layer G, Garcia-Serres R, Signor L, Clemancey M, et al. 2010. Iron-sulfur (Fe-S) cluster assembly: The SufBCD complex is a new type of Fe-S scaffold with a flavin redox cofactor. *J. Biol. Chem.* 285:23331–41

47. Lill R. 2009. Function and biogenesis of iron-sulphur proteins. *Nature* 460:831–38
48. Chandramouli K, Johnson MK. 2006. HscA and HscB stimulate [2Fe-2S] cluster transfer from IscU to apoferredoxin in an ATP-dependent reaction. *Biochemistry* 45:11087–95
49. Bonomi F, Iametti S, Morleo A, Ta D, Vickery LE. 2008. Studies on the mechanism of catalysis of iron-sulfur cluster transfer from IscU[2Fe2S] by HscA/HscB chaperones. *Biochemistry* 47:12795–801
50. Nachin L, Loiseau L, Expert D, Barras F. 2003. SufC: an unorthodox cytoplasmic ABC/ATPase required for [Fe-S] biogenesis under oxidative stress. *EMBO J.* 22:427–37
51. Kitaoka S, Wada K, Hasegawa Y, Minami Y, Fukuyama K, Takahashi Y. 2006. Crystal structure of *Escherichia coli* SufC, an ABC-type ATPase component of the SUF iron-sulfur cluster assembly machinery. *FEBS Lett.* 580:137–43
52. Outten FW, Wood MJ, Muñoz M, Storz G. 2003. The SufE protein and the SufBCD complex enhance SufS cysteine desulfurase activity as part of a sulfur transfer pathway for Fe-S cluster assembly in *Escherichia coli*. *J. Biol. Chem.* 278:45713–19
53. Ugalde RA, Imperial J, Shah VK, Brill WJ. 1984. Biosynthesis of iron-molybdenum cofactor in the absence of nitrogenase. *J. Bacteriol.* 159:888–93
54. Brigle KE, Weiss MC, Newton WE, Dean DR. 1987. Products of the iron-molybdenum cofactor-specific biosynthetic genes, nifE and nifN, are structurally homologous to the products of the nitrogenase molybdenum-iron protein genes, nifD and nifK. *J. Bacteriol.* 169:1547–53
55. Kaiser JT, Hu Y, Wiig JA, Rees DC, Ribbe MW. 2011. Structure of a precursor-bound NifEN: a nitrogenase FeMo cofactor maturase/insertase. *Science* 331:91–94
56. Curatti L, Ludden PW, Rubio LM. 2006. NifB-dependent in vitro synthesis of the iron-molybdenum cofactor of nitrogenase. *Proc. Natl. Acad. Sci. USA* 103:5297–301
57. George SJ, Igarashi RY, Piamonteze C, Soboh B, Cramer SP, Rubio LM. 2007. Identification of a Mo-Fe-S cluster on NifEN by Mo K-edge extended X-ray absorption fine structure. *J. Am. Chem. Soc.* 129:3060–61
58. Corbett MC, Hu Y, Fay AW, Ribbe MW, Hedman B, Hodgson KO. 2006. Structural insights into a protein-bound iron-molybdenum cofactor precursor. *Proc. Natl. Acad. Sci. USA* 103:1238–43
59. Hernandez JA, Curatti L, Aznar CP, Perova Z, Britt RD, Rubio LM. 2008. Metal trafficking for nitrogen fixation: NifQ donates molybdenum to NifEN/NifH for the biosynthesis of the nitrogenase FeMo-cofactor. *Proc. Natl. Acad. Sci. USA* 105:11679–84
60. Rangaraj P, Ludden PW. 2002. Accumulation of (99)Mo-containing iron-molybdenum cofactor precursors of nitrogenase on NifNE, NifH, and NifX of *Azotobacter vinelandii*. *J. Biol. Chem.* 277:40106–11
61. Paustian TD, Shah VK, Roberts GP. 1989. Purification and characterization of the nifN and nifE gene products from *Azotobacter vinelandii* mutant UW45. *Proc. Natl. Acad. Sci. USA* 86:6082–86
62. Robinson AC, Dean DR, Burgess BK. 1987. Iron-molybdenum cofactor biosynthesis in *Azotobacter vinelandii* requires the iron protein of nitrogenase. *J. Biol. Chem.* 262:14327–32
63. Curatti L, Hernandez JA, Igarashi RY, Soboh B, Zhao D, Rubio LM. 2007. In vitro synthesis of the iron-molybdenum cofactor of nitrogenase from iron, sulfur, molybdenum, and homocitrate using purified proteins. *Proc. Natl. Acad. Sci. USA* 104:17626–31
64. Hu Y, Fay AW, Ribbe MW. 2005. Identification of a nitrogenase FeMo cofactor precursor on NifEN complex. *Proc. Natl. Acad. Sci. USA* 102:3236–41
65. Koonin EV, Wolf YI, Avarind L. 2000. Protein fold recognition using sequence profiles and its application in structural genomics. *Adv. Protein Chem.* 54:245–75
66. Walker JE, Saraste M, Runswick MJ, Gay NJ. 1982. Distantly related sequences in the alpha- and beta-subunits of ATP synthase, myosin, kinases and other ATP-requiring enzymes and a common nucleotide binding fold. *EMBO J.* 1:945–51
67. Saraste M, Sibbald PR, Wittinghofer A. 1990. The P-loop—a common motif in ATP- and GTP- binding proteins. *Trends Biochem. Sci.* 15:430–34
68. Milner-White EJ, Coggins JR, Anton IA. 1991. Evidence for an ancestral core structure in nucleotide-binding proteins with the type A motif. *J. Mol. Biol.* 221:751–54
69. Bourne HR, Sanders DA, McCormick F. 1991. The GTPase superfamily: conserved structure and molecular mechanism. *Nature* 349:117–27

70. Leipe D, Wolf YI, Koonin EV, Aravind L. 2002. Classification and evlolution of P-loop GTPases and related ATPases. *J. Mol. Biol.* 317:41–72
71. Vickery LE, Cupp-Vickery JR. 2007. Molecular chaperones HscA/Ssq1 and HscB/Jac1 and their roles in iron-sulfur protein maturation. *Crit. Rev. Biochem. Mol. Biol.* 42:95–111
72. Netz DJ, Pierik AJ, Stümpfig M, Muhlenhoff U, Lill R. 2007. The Cfd1-Nbp35 complex acts as a scaffold for iron-sulfur protein assembly in the yeast cytosol. *Nat. Chem. Biol.* 3:278–86
73. Maier T, Jacobi A, Sauter M, Böck A. 1993. The product of the hypB gene, which is required for nickel incorporation into hydrogenases, is a novel guanine nucleotide-binding protein. *J. Bacteriol.* 175:630–35
74. Maier T, Lottspeich F, Böck A. 1995. GTP hydrolysis by HypB is essential for nickel insertion into hydrogenases of *Escherichia coli*. *Eur. J. Biochem.* 230:133–38
75. Cai F, Ngu TT, Kaluarachchi H, Zamble DB. 2011. Relationship between the GTPase, metal-binding, and dimerization activities of *E. coli* HypB. *J. Biol. Inorg. Chem.* 16:857–68
76. Kaluarachchi H, Chung KCC, Zamble DB. 2010. Microbial nickel proteins. *Nat. Prod. Rep.* 27:681–94
77. Zambelli B, Stola M, Musiani F, De Vriendt K, Samyn B, et al. 2005. UreG, a chaperone in the urease assembly process, is an intrinsically unstructured GTPase that specifically binds Zn^{2+}. *J. Biol. Chem.* 280:4684–95
78. Jeoung J-H, Giese T, Grünwald M, Dobbek H. 2009. CooC1 from *Carboxydothermus hydrogenoformans* is a nickel-binding ATPase. *Biochemistry* 48:11505–13
79. Shepard EM, Broderick JB. 2010. *S*-adenosylmethionine and iron-sulfur clusters in biological radical reactions: the radical SAM superfamily. In *Comprehensive Natural Products II: Chemistry and Biology*, Vol. 8, ed. L Mander, H-W Liu, pp. 625–61. Oxford, UK: Elsevier
80. Frey PA, Hegeman AD, Ruzicka FJ. 2008. The radical SAM superfamily. *Crit. Rev. Biochem. Mol. Biol.* 43:63–88
81. George SJ, Igarashi RY, Xiao Y, Hernandez JA, Demuez M, et al. 2008. Extended X-ray absorption fine structure and nuclear resonance vibrational spectroscopy reveal that NifB-co, a FeMo-co precursor, comprises a 6Fe core with an interstitial light atom. *J. Am. Chem. Soc.* 130:5673–80
82. Soboh B, Boyd ES, Zhao D, Peters JW, Rubio LM. 2010. Substrate specificity and evolutionary implications of a NifDK enzyme carrying NifB-co at its active site. *FEBS Lett.* 584:1487–92
83. Posewitz MC, King PW, Smolinski SL, Smith RD, Ginley AR, et al. 2005. Identification of genes required for hydrogenase activity in *Chlamydomonas reinhardtii*. *Biochem. Soc. Trans.* 33:102–4
84. Posewitz MC, King PW, Smolinski SL, Zhang L, Seibert M, Ghirardi ML. 2004. Discovery of two novel radical S-adenosylmethionine proteins required for the assembly of an active [Fe] hydrogenase. *J. Biol. Chem.* 279:25711–20
85. Peters JW, Szilagyi RK, Naumov A, Douglas T. 2006. A radical solution for the biosynthesis of the H-cluster of hydrogenase. *FEBS Lett.* 580:363–67
86. Mulder DW, Ortillo DO, Gardenghi DJ, Naumov AV, Ruebush SS, et al. 2009. Activation of HydA(DeltaEFG) requires a preformed [4Fe-4S] cluster. *Biochemistry* 48:6240–48
87. McGlynn SE, Ruebush SS, Naumov A, Nagy LE, Dubini A, et al. 2007. In vitro activation of [FeFe] hydrogenase: new insights into hydrogenase maturation. *J. Biol. Inorg. Chem.* 12:443–47
88. McGlynn SE, Shepard EM, Winslow MA, Naumov AV, Duschene KS, et al. 2008. HydF as a scaffold protein in [FeFe] hydrogenase H-cluster biosynthesis. *FEBS Lett.* 582:2183–87
89. Brazzolotto X, Rubach JK, Gaillard J, Gambarelli S, Atta M, Fontecave M. 2006. The [Fe-Fe]-hydrogenase maturation protein HydF from *Thermotoga maritima* is a GTPase with an iron-sulfur cluster. *J. Biol. Chem.* 281:769–74
90. Shepard EM, McGlynn SE, Bueling AL, Grady-Smith C, George SJ, et al. 2010. Synthesis of the 2Fe-subcluster of the [FeFe]-hydrogenase H-cluster on the HydF scaffold. *Proc. Natl. Acad. Sci. USA* 107:10448–53
91. Scrima A, Wittinghofer A. 2006. Dimerisation-dependent GTPase reaction of MnmE: how potassium acts as GTPase-activating element. *EMBO J.* 25:2940–51
92. King PW, Posewitz MC, Ghirardi ML, Seibert M. 2006. Functional studies of [FeFe] hydrogenase maturation in an *Escherichia coli* biosynthetic system. *J. Bacteriol.* 188:2163–72
93. Fontecave M, Ollagnier-de-Choudens S, Mulliez E. 2003. Biological radical sulfur insertion reactions. *Chem. Rev.* 103:2149–66

94. Miller JR, Busby RW, Jordan SW, Cheek J, Henshaw TF, et al. 2000. *Escherichia coli* LipA is a lipoyl synthase: in vitro biosynthesis of lipoylated pyruvate dehydrogenase complex from octanoyl-acyl carrier protein. *Biochemistry* 39:15166–78
95. Rubach JK, Brazzolotto X, Gaillard J, Fontecave M. 2005. Biochemical characterization of the HydE and HydG iron-only hydrogenase maturation enzymes from Thermatoga maritima. *FEBS Lett.* 579:5055–60
96. Pilet E, Nicolet Y, Mathevon C, Douki T, Fontecilla-Camps JC, Fontecave M. 2009. The role of the maturase HydG in [FeFe]-hydrogenase active site synthesis and assembly. *FEBS Lett.* 583:506–11
97. Driesener RC, Challand MR, McGlynn SE, Shepard EM, Boyd ES, et al. 2010. [FeFe]-hydrogenase cyanide ligands derived from S-adenosylmethionine dependent cleavage of tyrosine. *Angew. Chem.* 49:1687–90
98. Shepard EM, Duffus BR, George SJ, McGlynn SE, Challand MR, et al. 2010. [FeFe]-hydrogenase maturation: HydG-catalyzed synthesis of carbon monoxide. *J. Am. Chem. Soc.* 132:9247–49
99. Kuchenreuther JM, George SJ, Grady-Smith CS, Cramer SP, Swartz JR. 2011. Cell-free H-cluster synthesis and [FeFe] hydrogenase activation: All five CO and CN- ligands derive from tyrosine. *PLoS ONE* 6:e20346
100. Nicolet Y, Amara P, Mouesca J-M, Fontecilla-Camps JC. 2009. Unexpected electron transfer mechanism upon AdoMet cleavage in radical SAM proteins. *Proc. Natl. Acad. Sci. USA* 106:14867–71
101. Nicolet Y, Rubach JK, Posewitz MC, Amara P, Mathevon C, et al. 2008. X-ray structure of the [FeFe]-hydrogenase maturase HydE from *Thermotoga maritima. J. Biol. Chem.* 283:18861–72
102. Kuchenreuther JM, Stapleton JA, Swartz JR. 2009. Tyrosine, cysteine, and S-adenosyl methionine stimulate in vitro [FeFe] hydrogenase activation. *PLoS ONE* 4:e7565
103. Mulder DM, Boyd ES, Sarma R, Lange RK, Endrizzi JA, et al. 2010. Stepwise [FeFe]-hydrogenase H-cluster assembly revealed in the structure of HydA$^{\Delta EFG}$. *Nature* 465:248–51
104. Schmid B, Ribbe MW, Einsle O, Yoshida M, Thomas LM, et al. 2002. Structure of a cofactor-deficient nitrogenase MoFe protein. *Science* 296:352–56
105. Boyd ES, Hamilton TL, Peters JW. 2011. An alternative path for the evolution of biological nitrogen fixation. *Front. Microbiol.* 2:205
106. Rees DC, Howard JB. 2003. The interface between the biological and inorganic worlds: iron-sulfur metalloclusters. *Science* 300:929–31
107. Nicolet Y, Martin L, Tron C, Fontecilla-Camps JC. 2010. A glycyl free radical as the precursor in the synthesis of carbon monoxide and cyanide by the [FeFe]-hydrogenase maturase HydG. *FEBS Lett.* 584:4197–202
108. McGlynn SE, Mulder DW, Shepard EM, Broderick JB, Peters JW. 2009. Hydrogenase cluster biosynthesis: organometallic chemistry nature's way. *Dalton Trans.* 2009:4274–85
109. Blackmond DG. 2009. An examination of the role of autocatalytic cycles in the chemistry of proposed primordial reactions. *Angew. Chem.* 121:392–96
110. Milner-White EJ, Russell MJ. 2005. Sites for phosphates and iron-sulfur thiolates in the first membranes: 3 to 6 residue anion-binding motifs (nests). *Orig. Life Evol. Biosph.* 35:19–27
111. Sandy M, Butler A. 2009. Microbial iron acquisition: marine and terrestrial siderophores. *Chem. Rev.* 109:4580–95

Structural Perspective of Peptidoglycan Biosynthesis and Assembly

Andrew L. Lovering,[1,]* Susan S. Safadi,[2,]* and Natalie C.J. Strynadka[2]

[1]School of Biosciences, University of Birmingham, Birmingham B15 2TT, United Kingdom
[2]Department of Biochemistry and Molecular Biology, and the Center for Blood Research, University of British Columbia, Vancouver V6T 1Z3, Canada; email: ncjs@mail.ubc.ca

*Authors contributed equally to this work.

Keywords

murein, cell wall, glycosyltransferase, transpeptidation, antibiotic

Abstract

The peptidoglycan biosynthetic pathway is a critical process in the bacterial cell and is exploited as a target for the design of antibiotics. This pathway culminates in the production of the peptidoglycan layer, which is composed of polymerized glycan chains with cross-linked peptide substituents. This layer forms the major structural component of the protective barrier known as the cell wall. Disruption in the assembly of the peptidoglycan layer causes a weakened cell wall and subsequent bacterial lysis. With bacteria responsible for both properly functioning human health (probiotic strains) and potentially serious illness (pathogenic strains), a delicate balance is necessary during clinical intervention. Recent research has furthered our understanding of the precise molecular structures, mechanisms of action, and functional interactions involved in peptidoglycan biosynthesis. This research is helping guide our understanding of how to capitalize on peptidoglycan-based therapeutics and, at a more fundamental level, of the complex machinery that creates this critical barrier for bacterial survival.

Contents

INTRODUCTION	452
UDP-N-ACETYLMURAMYL PENTAPEPTIDE SYNTHESIS	452
ASSEMBLY OF LIPID II ON THE INNER CYTOPLASMIC MEMBRANE	456
TRANSPORT OF LIPID II ACROSS THE CYTOPLASMIC MEMBRANE	457
POLYMERIZATION OF LIPID II BY PEPTIDOGLYCAN GLYCOSYLTRANSFERASES	459
TRANSPEPTIDATION OF NEWLY POLYMERIZED GLYCAN STRANDS	462
REMOVAL/RECYCLING OF UNDECAPRENYL PYROPHOSPHATE	464
PEPTIDOGLYCAN IN THE CONTEXT OF BACTERIAL CELL ULTRASTRUCTURE	465
CONCLUSIONS	468

INTRODUCTION

Bacterial pathogens cause a wide range of severe illness, including infections of the skin and blood, pneumonia, tuberculosis, and meningitis. One of the most clinically and economically successful sets of antibacterial targets to date has been the biosynthetic pathway for the peptidoglycan layer, the major structural component of the cell wall common to the majority of these pathogenic bacteria. The peptidoglycan layer is an elaborate polymeric mesh of alternating N-acetylglucosamine (GlcNAc) and N-acetylmuramic acid (MurNAc) glycan units cross-linked via peptidyl "bridges." These features presumably allow for the structural rigidity required to preserve cellular integrity against osmotic forces within the infected host or environmental niche but, at the same time, allow the necessary fluidity to adapt to changes in bacterial cell shape during various stages of growth, division, and infection. The biosynthesis of peptidoglycan is a complex process involving approximately 20 enzyme reactions, which likely act in a concerted, dynamic but, as yet, largely uncharacterized assembly. These reactions take place both in the cytoplasm (synthesis of the nucleotide precursors) and on the inner and outer sides of the cytoplasmic membrane (synthesis of lipid-linked intermediates and polymerization reactions, respectively) (**Figure 1**). The enzymes involved in the biosynthetic pathway of the bacterial cell wall are essential, and most do not have any mammalian homologs. Their essential nature and presence in both gram-negative and gram-positive pathogens make them excellent targets for broad-spectrum antibiotics. Indeed, much success has been achieved in this respect (reviewed in Reference 1), but the issues of antibiotic resistance [β-lactam and vancomycin resistance both involve cell wall targets and are particularly well studied (2, 3)] and the need to maintain the clinical "upper hand" against superbugs, such as methicillin-resistant *Staphylococcus aureus* (MRSA), are driving efforts to characterize the less-studied members of the pathway and develop novel antibacterial agents. Aside from this antibiotic-centric view, details of peptidoglycan synthesis, deposition, and ultrastructure are of immense interest to the field of bacterial physiology. Many advances have been made in this area recently, as reviewed here, pushing us closer to the ultimate goal of understanding the complete molecular features defining synthesis, dynamic interactions with host cytoskeleton, and high-resolution three-dimensional molecular features of mature peptidoglycans.

UDP-N-ACETYLMURAMYL PENTAPEPTIDE SYNTHESIS

The first committed step in peptidoglycan synthesis proceeds within the bacterial cytoplasm with the MurA-catalyzed transfer of the enolpyruvyl moiety from phosphoenolpyruvate (PEP) to uridine diphosphate

---

Peptidoglycan:
a polymer of repeating N-acetylmuramic acid and N-acetylglucosamine sugar units linked into a mesh via peptidyl substituents

MurNAc:
N-acetylmuramic acid

UDP: uridine diphosphate

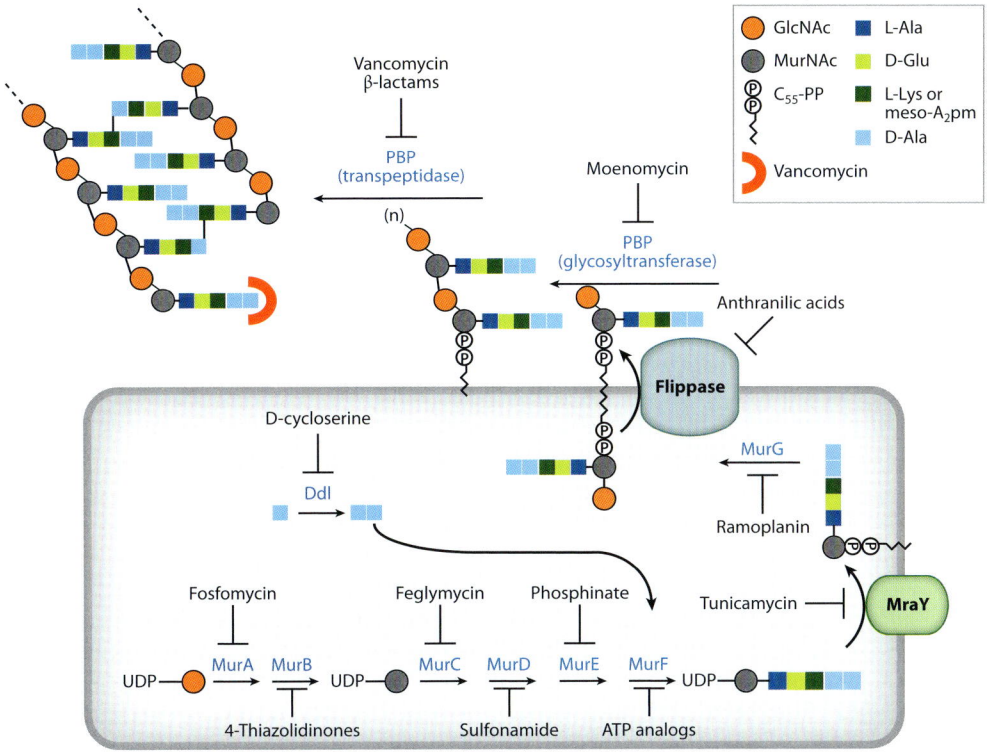

Figure 1

Summary of the peptidoglycan biosynthesis pathway. Protein names (*blue*) and inhibitors (where appropriate) (*black*) are shown. Integral membrane proteins (MraY and flippase) are present at the transition between cytoplasmic synthesis of the uridine diphosphate (UDP)-linked precursor and outer leaflet utilization of the lipid II monomer. Abbreviations: C_{55}-PP, undecaprenyl diphosphate; D-Ala, D-alanine; Ddl, D-alanyl-D-alanine ligase; D-Glu, D-glutamic acid; GlcNAc, *N*-acetylglucosamine; L-Ala, L-alanine; L-Lys, L-lysine; meso-A_2pm, mesodiaminopimelic acid; MurA, MurB, MurC, MurD, MurE, MurF, MurG, enzymes involved in the cytoplasmic biosynthesis steps of peptidoglycan; MurNAc, *N*-acetylmuramic acid; PBP, penicillin-binding protein.

(UDP)-*N*-acetylglucosamine (GlcNAc), generating enolpyruvyl UDP-GlcNAc. Gram-negative bacteria have only one copy of the *murA* gene, and its deletion is lethal (4). Gram-positive bacteria have two *murA* genes, both demonstrated to encode active enzymes (5). The crystal structure of MurA has been determined from several species including *Escherichia coli*, *Enterobacter cloacae* and *Haemophilus influenzae*, both in apoenzyme form and in complex with different ligands (**Supplemental Table 1**; follow the **Supplemental Material link** from the Annual Reviews home page at **http://www.annualreviews.org**) (6–8). The structures contain two globular domains, each of which possess a threefold repeat of a single folding unit made up of a four-stranded β-sheet and two parallel helices. The active site is located between the two domains and upon binding of the substrates (UDP-GlcNAc and PEP) the active site is rearranged by conformational changes in the enzyme, which brings the two domains toward each other to encompass the substrates, thus forming a more closed conformation in the active site. Fosfomycin is a naturally occurring antibiotic [produced by several species of *Pseudomonas* and *Streptomyces* from phosphoenolpyruvate (9)] that inactivates MurA by mimicking PEP and irreversibly modifying the critical active-site nucleophile Cys115 (*E. coli* sequence numbering) through a covalent linkage (10, 11). With an increasing number of pathogenic bacterial strains developing resistance to fosfomycin and other antibiotics, there is a growing demand for new drugs targeting MurA. In addition, as UDP-GlcNAc is also required for assembly of fungal walls, MurA offers a potential target for antifungal as well as antibacterial action. However, despite our knowledge of several liganded and unliganded complex structures in tandem with biochemical and kinetic data on MurA, it has taken almost 50 years since the discovery of fosfomycin to identify a second MurA-specific antibiotic agent, terreic acid; other attempts were hindered by nonspecific interactions and poor antibacterial activity (12, 13). Terreic acid is produced by the fungus *Aspergillus terrus*, and although its antibiotic properties were identified many years ago, MurA was not identified as its target until recently (12, 14). This mode of inhibition is similar to that of fosfomycin, occurring through chemical modification of Cys115; however, terreic acid is ∼50-fold less potent than fosfomycin. The superior potency of fosfomycin appears to arise, in part, from greater complementarity between the bound inhibitor and the active site. Although in vitro studies have shown directly that MurA is inhibited by terreic acid, whether MurA is the only target for the compound in the cell remains unclear. It has been suggested that in *E. coli*, for example, MurA is not the primary target of terreic acid, but rather proteins on the outside of the bacterial cell membrane likely play a more critical role in its antibacterial activity (12).

The next step in peptidoglycan synthesis utilizes the reductase MurB to catalyze the NADPH-dependent conversion of enolpyruvyl UDP-GlcNAc to UDP-MurNAc. Several structures of MurB have been determined from various species including *E. coli*, *S. aureus*, and *Vibrio cholerae*, as well as complexes with MurB's substrate, enolpyruvyl-EDP-GlcNAc and an inhibitor naphthyltetronic acid (15–17). In species of particular pathogenic interest, where a structure has not been determined, homology models may be of use (18). A *Mycobacterium tuberculosis* MurB model (with 33% sequence identity and 50% sequence similarity to existing *E. coli* structures) was used for molecular dynamics and docking analysis, with 28 inhibitors subsequently identified having reasonable correlation between the binding energies generated from Autodock and the experimentally obtained IC_{50} (half maximal inhibitory concentration) values measured in vitro (18).

Following the production of UDP-MurNAc in the cytoplasm, a series of ATP-dependent amino acid ligases proceed to catalyze the stepwise addition of the pentapeptide side chain onto UDP-MurNAc. The four Mur ligases (MurC-F) are highly conserved among various bacterial species. Representative structures for all four enzymes have been determined, and despite limited sequence identity (15%–22%), common structural motifs can be discerned (19, 20). All enzymes contain three structural domains that function in the nonribosomal peptide bond formation, a process dependent on ATP hydrolysis. The three structural domains include an N-terminal Rossmann-type α/β fold domain for binding the growing nucleotide-activated substrate, a central ATPase domain, and a C-terminal domain to bind the appropriate amino acid. Significant progress has been made in the purification and biochemical characterization of these enzymes, and although there is an extensive amount of knowledge on the *E. coli* variants, recent work has advanced our knowledge into Mur ligase action in other highly pathogenic and antibiotic-resistant strains. For example, Patin et al. (21) have characterized biochemical properties (optimal pH and magnesium concentration) of the purified *S. aureus* enzymes, and provided an extensive comparison of kinetic parameters with those of the *E. coli* orthologs.

MurC catalyzes the first addition of L-Ala onto the nucleotide precursor UDP-MurNAc.

MurD then catalyzes the addition of D-Glu to the MurC product UDP-MurNAc-L-Ala. Not surprisingly both enzymes show significant structural similarity (22). Structures and accompanying biochemical analysis from several species, including *E. coli*, *Thermotoga maritima*, and *H. influenzae*, suggest MurC functions by binding three substrates (ATP, UDP-MurNAc, and L-Ala) in an ordered and sequential manner. In this reaction, the terminal carboxyl group of the UDP-MurNAc substrate is activated by an ATP-dependent phosphorylation, generating an acyl phosphate intermediate, which is then attacked by the amino group of the incoming L-Ala. This high-energy intermediate breaks down with elimination of inorganic phosphate, forming the peptide bond (23, 24). MurE functions in the addition of the third peptidyl residue, mesodiaminopimelic (meso-A_2pm) acid in *E. coli* or L-Lys in *S. aureus*, to form UDP-*N*-acetylmuramyl-tripeptide. MurE structures from *E. coli*, and more recently from *M. tuberculosis*, have been determined for both unliganded and liganded proteins, suggesting MurE functions catalytically via an acyl phosphate intermediate mechanism as described for the other Mur enzymes (25, 26). These previous studies have also detailed the structural determinants responsible for meso-A_2pm/L-Lys discrimination by MurE. For all A_2pm-specific enzymes, position 416 is occupied by an Arg to bind the free end of A_2pm substrate; however, in L-Lys-specific enzymes, this position is replaced by an Ala or Asn providing the main structural determinant for L-Lys selection (25). MurF catalyzes the next cytoplasmic step in peptidoglycan synthesis with the addition of D-Ala-D-Ala to the UDP-MurNAc-L-Ala-D-Glu-meso-A_2pm acid product of MurE. MurF is less specific for meso-A_2pm acid/L-Lys in the tripeptide moiety, and its structure, although similar for the most part to the other MurC-MurE ligases in the pathway, shows an entirely unique N-terminal domain that lacks the characteristic Rossmann nucleotide-binding fold. Accordingly, this MurF domain is not involved in fixation of the substrate pyrophosphate group, but rather an extension from the central domain plays the analogous role (27).

The D-Ala-D-Ala dipeptide acceptor substrate of MurF is preformed in the cell by an ATP-dependent D-Ala:D-Ala ligase (28). D-Ala is made by an alanine racemase, which uses a covalently bound pyridoxal 5'-phosphate cofactor to catalyze the racemization of L-Ala. A condensation reaction involving two molecules of D-Ala leads to the final D-Ala-D-Ala product. Once incorporated into the stem peptide, this product is the target of the glycopeptide antibiotic vancomycin. Vancomycin exerts its antibacterial effects through binding to the C-terminal D-Ala-D-Ala in the peptidoglycan peptide moiety, sequestering these substrates from transpeptidases (TPs) and inhibiting the cross-linking of the cell wall. Vancomycin-resistant strains evade this inhibition through alterations to the D-Ala-D-Ala termini of the stem peptide (for example, use of D-Ala-D-Lac or D-Ala-D-Ser provides resistance to vancomycin). Recent developments have been made in redesigning vancomycin to provide dual binding to D-Ala-D-Ala and to D-Ala-D-Lac in vancomycin-resistant infections (29). Inhibition of terminal peptide formation is achieved through the action of D-cycloserine, which is a structural analog of D-Ala and inhibits both the alanine racemase and D-Ala ligase.

With few inhibitors of MurF available, recent work has focused on the identification of new compounds. Screens employing thermal stability assays have been used to determine the binding of inhibitor molecules to MurF; however, several identified compounds lacked measurable antibacterial activity (30). A pharmacophore model was used to develop compounds on the basis of a 4-phenylpiperidine derivative; this is the first compound to inhibit both MurF in vitro and show significant antibacterial activity (30). Additional inhibitors have since been identified from compound library screening, with antibacterial effects on both gram-negative and gram-positive bacteria (31).

> **Transpeptidase (TP):** an enzyme that catalyzes the reaction involving the transfer and cross-linking of one peptidyl substituent to another

In a variety of bacteria, variations in the peptidyl moieties of peptidoglycan structure are also observed (reviewed in Reference 32), including the addition of extra residues onto the stem pentapeptides (resulting in a branched peptidoglycan, common to gram-positive organisms). In the latter, pentaglycine cross-bridges are preformed in the cytoplasm by the FemX, FemA, and FemB proteins. The X-ray structures of FemA from *S. aureus* and FemX from *Weissella viridescens* have been determined (33, 34). However, despite crystallographic structures, the catalytic mechanism of peptide bond formation by this family of proteins remains obscure (35).

ASSEMBLY OF LIPID II ON THE INNER CYTOPLASMIC MEMBRANE

The first membrane-associated step in peptidoglycan synthesis is catalyzed by the integral membrane protein, MraY. This reaction involves a transfer step of the MurNAc-pentapeptide from the cytoplasm onto an undecaprenyl phosphate carrier (C_{55}-P, often commonly referred to as bactoprenol or UP) on the cytoplasmic side of the membrane, resulting in the generation of the product known commonly as lipid I. A two-step catalytic mechanism has been proposed involving attack by a nucleophilic residue within the as yet uncharacterized MraY active site on the β-phosphate of UDP-MurNAc-pentapeptide, resulting in the formation of a covalent enzyme-phospho-MurNAc-pentapeptide intermediate and the release of UMP (36). The second step is proposed to involve an attack by a C_{55}-P oxyanion on the phosphate of the covalent intermediate, resulting in the formation of lipid I. An alternative mechanism has also been suggested, involving the direct attack of the phosphate oxyanion of C_{55}-P onto the β-phosphate of UDP-MurNAc-pentapeptide, leading to the formation of lipid I and the release of UMP in a single step (37).

Sequence analysis and membrane topology experiments on the *S. aureus* and *E. coli* orthologs indicate the presence of ten transmembrane (TM) segments, five cytoplasmic loops (relatively long), and six periplasmic loops (relatively short). Regions of high sequence conservation are located on the cytoplasmic side of the membrane, consistent with the active site being oriented toward the cytoplasm. MraY is a member of a superfamily that includes eukaryotic and prokaryotic prenyl sugar transferases. There are currently no structures of any members of this superfamily available; however, in recent work, the successful purification and characterization of wild-type MraY has been demonstrated (38). A translocase assay has been developed that monitors the addition of the phospho-MurNAc-[^{14}C]pentapeptide moiety of UDP-MurNAc-[^{14}C]pentapeptide to C_{55}-P (38). Following separation by thin-layer chromatography, radioactive spots corresponding to the nucleotide substrate and the lipid I product could be analyzed. The biochemical properties (pH, detergent, salt, and cation preference) of the purified wild-type MraY were determined, identifying an absolute catalytic requirement for a monovalent or divalent metal ion. Nineteen MraY mutants (from various conserved motifs present in cytoplasmic loop regions and TM helices) (37) have also been analyzed via a functional complementation assay in an *E. coli* MraY temperature-sensitive strain. Of these mutants, 14 showed an impact on the catalytic function of MraY and were purified and further characterized, identifying residues most likely to interact with the essential metal ion (37). Despite the existence of several known classes of natural product inhibitors, including amphomycin, tunicamycins, and bacteriophage protein E, MraY is currently not a target for any antibiotics in clinical use (primarily owing to poor bioavailability of the known natural product compounds) (39, 40). However, ongoing efforts to develop novel inhibitors show much promise (41–43).

The pool of lipid I molecules has been estimated at 700 molecules/cell (44). This relatively low abundance is thought to be the result of reaction coupling, which suggests an interaction between MraY and MurG, the final cytoplasmic component of the pathway, which acts to convert lipid I to lipid II via glycosyl transfer of a UDP-activated N-acetyl-D-glucosamine donor (see below). Coupling of the MraY reaction to the subsequent MurG reaction step was confirmed utilizing UDP-[^{14}C]GlcNAc and purified preparations of each of the proteins (38).

High-resolution X-ray structures of a MurG construct from *E. coli* encompassing the full length protein, including the proposed glycosyltransferase (GT) region (residues 164–340), show two distinct domains that each adopt an α/β open sheet motif (characterizing MurG as a member of the EC 2.4.1.227, GT$_{28}$ CAZy family of GT enzymes). These domains are separated by an active-site cleft that is approximately 20 Å deep and 18 Å wide, and the structure showed an extensive hydrophobic surface for potential interaction with the phospholipid bilayer (45). Typical of many GTs, MurG follows an ordered Bi-Bi mechanism in which a UDP-GlcNAc donor binds first, followed by an acceptor (lipid I) with a conserved GGS motif in the loop region, mediating a sequentially ordered catalytic mechanism involving an oxocarbenium-ion-like transition state (46, 47). The structure of MurG captured in the proposed oxocarbenium ion transition state bound to its substrate UDP-GlcNAc has provided the molecular details for structure-based drug design. Synthetic transition state analog variants are linked through an uncharged spacer to a uridine nucleoside possessing IC$_{50}$ values of 400 μM (48). Recent advances in assay development also show promise for the high-throughput screening of potential MurG and MraY inhibitors for antibacterial discovery. Activity of MurG has typically been assayed using the lipid I intermediate generated in situ by membranes containing MraY, which is then converted in the presence of ^{3}H-UDPGlcNAc and MurG to give radiolabeled lipid II (49). Potentially more high-throughput assay methods make use of wild-type membranes, supplemented with decaprenol phosphate, phosphatidylglycerol, and UDP-MurNAc-^{14}C pentapeptide and with the lipid I product immobilized using hydrophobic HP20ss beads (50). An alternative assay has also been described using *E. coli* membranes (from strains lacking PBP1b, the primary polymerizing GT, which would normally utilize the lipid II substrate product for incorporation into peptidoglycan under these conditions) and radiolabeled UDP-GlcNAc to measure the lipid II product via bead-based scintillation proximity (51).

TRANSPORT OF LIPID II ACROSS THE CYTOPLASMIC MEMBRANE

The final cytoplasmic/inner leaflet step in peptidoglycan synthesis functions to flip the lipid II building block produced by MurG to the outer leaflet of the cytoplasmic membrane. Experimental data have shown that this transfer is not a spontaneous process and support the hypothesis that flipping is protein mediated (52). To date, several candidate proteins have been proposed to function as the lipid II flippase. In one study, a reductionist bioinformatics approach was applied by Ruiz (53), sorting 963 possible integral membrane proteins in *E. coli* using several levels of criteria, such as conservation among gram-negative endosymbiotic bacteria that contain peptidoglycan, absence of the protein in bacteria that lack peptidoglycan, and essentiality in *E. coli* (unless redundant). Through this process, the number of possible candidates was reduced to a single protein, identifying MurJ (formerly MviN) as the potential flippase with further support from the observed lethality of *E. coli* MurJ deletion strains (53, 54). Furthermore, following six to seven generations, MurJ-depleted strains incorporated ~70% less ^{3}H-diaminopimelic acid (DAP) into mature peptidoglycan than wild type. Analysis using temperature-sensitive *murJ* mutant strains, ^{14}C-isopentenyl diphosphate to label the polyprenyl moiety of the lipid intermediates, and ^{3}H-DAP resulted in an accumulation of lipid intermediates in mutant cells (55). Taken together, these studies suggested a role for MurJ in peptidoglycan synthesis in *E. coli*. However, the unambiguous confirmation that MurJ plays a direct rather

Lipid II: the lipid-activated substrate building block of peptidoglycan polymerization

Glycosyltransferase (GT): an enzyme that catalyzes the transfer of a glycosyl group to an acceptor compound

than indirect role in lipid II transport requires verification, as do insights into a potential energy source or coupling factor that would drive flippase activity.

Although MurJ is highly conserved among gram-negative bacteria, it is absent in 21 genomes representative of the gram-positive phylum Firmicutes (including *Bacillus*, *Listeria*, *Staphylococcus*, *Enterococcus*, and *Streptococcus*). In a bioinformatics analysis by Fay & Dworkin (56), investigation into potential flippase conservation in gram-positive bacteria did not yield any significant hits when searching with *E. coli* MurJ. Using both BLAST and topology analysis, it was shown that all enzymes in the pathway (including MurA to MraY) had e values of $>10^{-20}$, whereas the top hit for MurJ was YtgP with an e value of 10^{-7}. Using YtgP for a BLASTp query against the *Bacillus subtilis* genome, three additional proteins were identified (SpoVB, YkvU, and YabM). However, chromosomal deletions of these genes in various combinations did not result in defects in vegetative growth. At the same time, confirmation that these proteins are true homologs of MurJ was validated by MurJ's ability to complement SpoVB function in spore formation and by YtgP's and SpoVB's abilities to complement a *murJ* deletion strain in *E. coli*. It should be noted that these results do not exclude further unidentified genes that may lack sequence identity to MurJ and that are functionally redundant. In that regard, additional studies have identified a YtgP ortholog from *Streptococcus pyogenes* (Spy_0390), which complements MurJ-depleted *E. coli* cells.

Decades prior to the proposal of MurJ's role in peptidoglycan synthesis, FtsW and RodA were also identified as potential flippase candidates (57). The aforementioned bioinformatics work (53) had eliminated these proteins because they are present in bacteria that do not possess peptidoglycan. However, it has been shown that several enzymes involved in peptidoglycan synthesis can be found in cell wall–less bacterial species (58, 59). FtsW is an essential division protein, with 10 predicted TM segments, a large periplasmic loop between TMs 7 and 8 (loop 7/8), and both N- and C-terminal ends localized to the cytoplasm (60, 61). RodA is an integral membrane protein that plays an important role in elongation and rod shape maintenance (62). Both proteins belong to the SEDS (shape, elongation, division, and sporulation) family, which also includes SpoVE. RodA, FtsW, and SpoVE have been suggested to participate in separate peptidoglycan synthesis complexes functioning during elongation, division, and sporulation, respectively (62).

Significant advances have recently been made in the reconstitution of the transport of lipid II in vitro (52). Several experiments monitoring this activity have been described using *E. coli* inner membrane vesicles and a fluorescent 7-nitro-2,1,3-benzoxadiazol-4-yl analog of lipid II, as well as a fluorescence resonance energy transfer–based assay (52, 59). Building on their reconstitution work, Breukink and colleagues (59) have recently directly demonstrated the translocation of 7-nitro-2,1,3-benzoxadiazol-4-yl lipid II in *Bacillus subtilis* membranes by FtsW, providing the first direct biochemical evidence of a specific protein functioning as a lipid II flippase. In this work, MurJ was also tested with the outcome that no lipid II flipping could be detected (at least in a *B. subtilis* context) (59). Furthermore, overexpression of MurJ in the membrane did not result in any enhanced translocation, in contrast to that observed with the overexpression of FtsW. Importantly, these fluorescence-based experiments also demonstrated a link between lipid II transport and GT activity, suggesting that the translocation of lipid II is coupled to the GT activity (52). Interaction between FtsW and the TP PBP3 in *E. coli* has been shown previously by in vivo fluorescence resonance energy transfer experiments and in vitro coimmunoprecipitation experiments (63), with the conserved sequence in the periplasmic loop between TM 9 and 10 of FtsW involved in PBP3 recruitment at the division site (64). Although this work is significant because it provides the first biochemical evidence of FtsW's direct involvement in lipid II transport, the underlying mechanism and governing molecular features as well as the possibility of the recruitment of distinct flippase molecules at different points in the biogenesis of the cell wall remain unknown.

POLYMERIZATION OF LIPID II BY PEPTIDOGLYCAN GLYCOSYLTRANSFERASES

The formation of lipid II (and its delivery to the outer leaflet of the inner membrane by flippases) completes the cellular stages of peptidoglycan synthesis; the subsequent extracellular stages utilize the energy of bonds present in the lipid II monomer to drive reactions forward (e.g., phosphodiester-muramic acid bond for polymerization, D-Ala-D-Ala peptide bond for cross-linking). Lipid II polymerization into a nascent peptidoglycan strand occurs via a GT peptidoglycan polymerization (GT_{PGP}) reaction, catalyzed by monotopic membrane proteins that can be either monofunctional or bifunctional enzymes, with the latter possessing an additional TP domain (65). The monofunctional (mGT_{PGP}) and bifunctional enzymes (known historically as PBPs, penicillin-binding proteins) have been shown to functionally interact in *S. aureus*, although the mGT_{PGP} exhibits an essential phenotype only in the absence of bifunctional enzymes (66). Lipid II polymerization (EC 2.4.1.129, CAZy family GT_{51}) involves an initiation stage (reaction of two lipid II monomer units to form a lipid IV product) and an elongation stage (successive addition of further lipid II units). These are essentially identical GT_{PGP} reactions, differing only in the length of substrate used, and involve the formation of a new β-1,4 NAM-NAG linkage with loss of the undecaprenyl-pyrophosphate donor leaving group in each cycle. The first structures of the GT_{PGP} active site were revelatory in terms of understanding this reaction [the bifunctional PBP2 from *S. aureus* (67) and a GT_{PGP} domain from *Aquifex aeolicus* PBP1a (**Figure 2**) (68)] and added significantly to preexisting models of catalysis (69, 70). In a wider context, the GT_{51} structures remained the only high-resolution data available for a non-nucleotide-utilizing GT active site until the structure of an oligosaccharyltransferase involved in protein glycosylation was recently identified (71). The structures of *E. coli* PBP1b (including the single-span TM segment) (72) and a *S. aureus* mGT_{PGP}-soluble mutant (73) have also subsequently been determined indicating several structural features common to members of the GT_{PGP} family.

The GT_{PGP} fold can be divided into a head subdomain that shares homology with λL lysozyme, and a novel jaw subdomain of high hydrophobicity that is inferred to interact with the lipid bilayer (**Figure 2b**); the juxtaposition of the subdomains results in a cleft lined with residues that display strong sequence conservation (74). The conserved residues can be clustered into five separate motifs, which have been shown to be essential in several species including *E. coli* PBP1b (75), *A. aeolicus* PBP1a (76), and *S. aureus* mGT_{PGP} (73). The homology with λL lysozyme extends partly to the catalytic apparatus, where the essential glutamate of conserved motif I (E114 in *S. aureus* PBP2) is orientated similarly to the λL nucleophile E19, with the reaction products of lysozyme situated analogously to the reaction substrates of GT_{PGP} enzymes (in a somewhat "loose" sense given the differences inherent in soluble and lipid-linked sugar units, and catabolism versus anabolism). A cocrystallization of PBP2 with the *Streptomyces* natural product antibiotic moenomycin revealed that this lipid-phosphoglycerate-sugar compound indeed mimicked lipid IV/the extending nascent chain (67), which was corroborated by additional GT_{PGP}:moenomycin structures (72, 73, 77). Thus, it could be inferred that lipid II was the acceptor in the GT_{PGP} reaction (retaining its lipid-pyrophosphate group as the donor nascent chain loses its own) and that catalysis occurs through deprotonation of the acceptor NAG 4-OH by the motif I glutamate, resulting in an S_N2-like reaction with the NAM C1 of the donor. The identification of the donor/acceptor has since been confirmed by elegant experiments performed by the Walker group (78), who have also discovered that initiation is rate limiting (79), that different

PBPs: penicillin-binding proteins

GT$_{PGP}$s produce polymers varying in length (80), and that the donor site is more specific in terms of substrate lipid group length than the acceptor site (81). This latter finding may have implications for inhibitor design, especially in the context of modified moenomycins, which is possible now that the biosynthetic genes have been identified and characterized (82, 83).

The structure of an *E. coli* PBP1b construct containing the TM-spanning helix revealed that this helix (residues 66–96) made close contacts with the GT$_{PGP}$ domain but did not appear to influence the structure of the active site (72). The PBP1b structure provides the best model to date for estimating membrane orientation, an important consideration given the

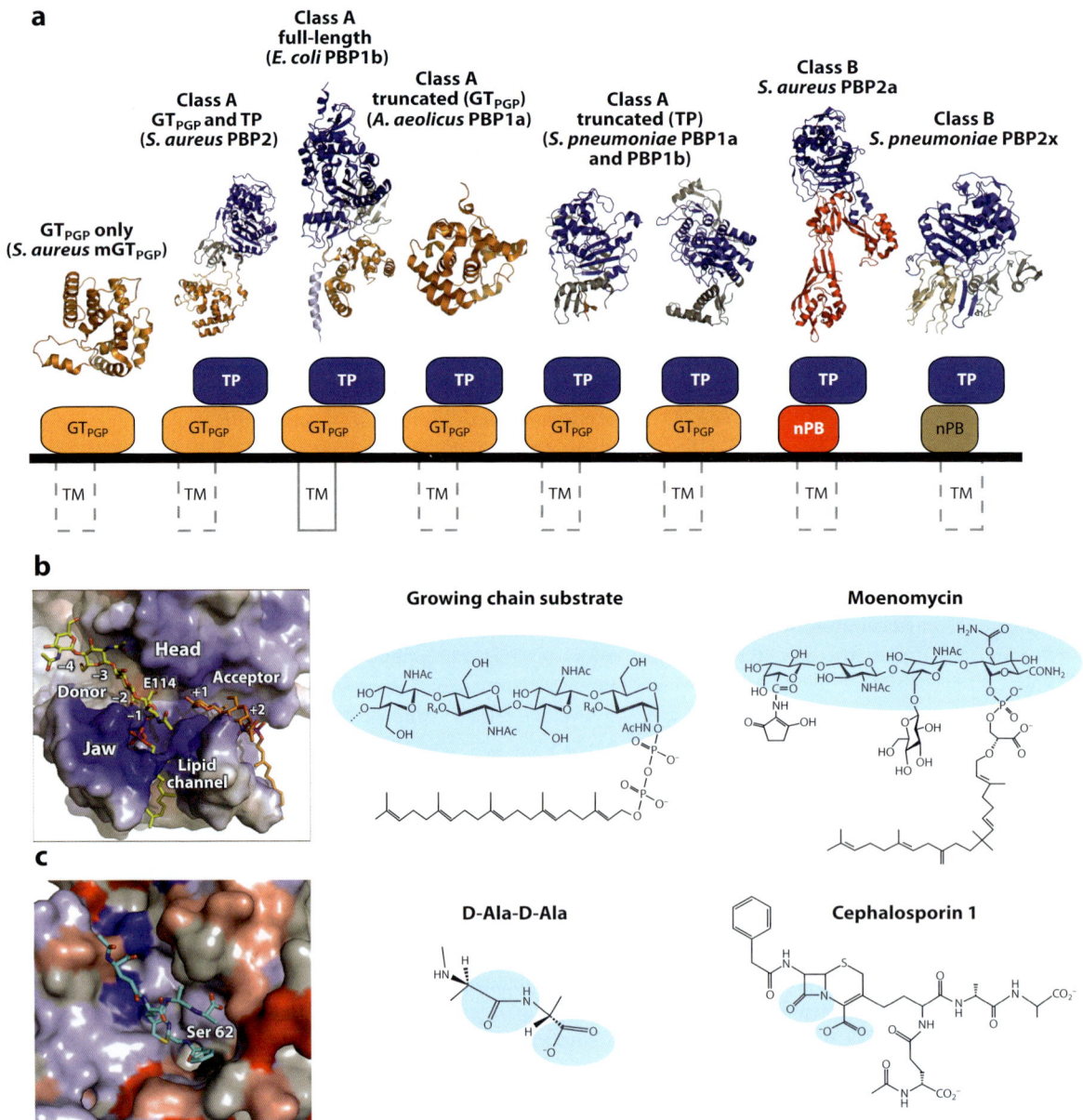

amphipathic nature of the GT_{PGP} substrates and their interaction with integral membrane protein partners. A different crystal form obtained from a partially truncated *S. aureus* PBP2 (PBP2Δ) revealed the conformation of two features within the GT_{PGP} active site (presumably disordered in the other crystal structures) (84): a π-bulge in the outermost α-helix of the jaw subdomain and a β-hairpin that straddles the donor and acceptor binding pockets. These features may be of interest for catalysis as the π-bulge would putatively assist local disorder/unfolding and remove steric barriers to allow transit of the completed reaction product from the acceptor site back into the donor site for another round of catalysis. This movement of product to ensure enzyme processivity would also be aided by GT_{PGP} active-site electrostatics (67). The β-hairpin has gross similarity to a second type of lysozyme (SLT70) and provides more detail on the region of the protein that is likely to make specific contacts with the lipid II acceptor. The *S. aureus* mGT_{PGP} structure was published at approximately the same time as the PBP2Δ study and shows this region (between motifs I and II; PBP2Δ H121-S147, mGT_{PGP} H107-S132) in a completely different conformation; if both orientations represent distinct physiological states, it would make this region of the GT_{PGP} fold an excellent candidate for regulation by accessory proteins.

The interdomain arrangement of the GT_{PGP} and TP domains in the bifunctional enzymes shows varying "flexation," and a small-angle X-ray scattering experiment on *S. pneumoniae* PBP1b indicates that in solution such movement is likely restricted (85). Such interest in the relative distance between the bifunctional active sites (and also PBP oligomerization state) is grounded in the observation that transpeptidation is only seen to occur on glycan-polymerized substrate (69, 86); hence, the GT_{PGP} product likely has an influence on the TP active site. This interplay is likely to play a role in the observed variation in both glycan length and percentage of peptidyl cross-linking found between species (e.g., average chain lengths of 20–35 nm in the pepetidoglycan sacculi and >500 nm in *E. coli* and *B. subtilis*, respectively; cross-linking can range from 25% in *Thermus thermophilus* to 92% in *S. aureus*; for a full description, see Reference 87). Recently, in γ-proteobacteria (with the caveat that a similar approach in other bacteria would be possible using nonhomologous proteins), lipoproteins were observed to influence both GT_{PGP} initiation (88) and TP cross-linking (89). This was remarkable in the sense that these were the first outer membrane regulators of peptidoglycan synthesis, and all previous studies on regulation involved intracellular/TM partners. The lipoproteins were named lpoA and -B (lipoprotein activator of PBP from outer membrane) and achieve specificity by targeting the unique ODD domain of PBP1a (lpoA, formerly YraM) or the UB2H domain of PBP1b (lpoB, formerly YcfM). From the structure of PBP1b, it is apparent that the inner membrane:UB2H domain distance would require lpoB to stretch through the

---

Figure 2

Structural summary of high-molecular-weight penicillin-binding proteins (PBPs). (*a*) Representative members from the different PBPs are shown in ribbon format. From left to right, the items from the Protein Data Bank (PDB) shown are *S. aureus* monofunctional glycosyltransferase peptidoglycan polymerization (mGT_{PGP}) (PDB code 3HZS), *S. aureus* PBP2 (PDB code 2OLU), *E. coli* PBP1b (PDB code 3FWM), *A. aeolicus* PBP1a (PDB code 2OQO), *Streptococcus pneumoniae* PBP1a (PDB code 2C6W), *S. pneumoniae* PBP1b (PDB code 2BG1), *S. aureus* PBP2a (PDB code 1VQQ), and *S. pneumoniae* PBP2x (PDB code 1QME), and they are represented by a domain-based schematic below their structures. (*b*) Model for a productive complex of the GT peptidoglycan polymerization (GT_{PGP}) domain from *S. aureus* PBP2 with the acceptor lipid II and donor glycan chain. Shown next to the structure (*left*) are the molecular structures of lipid IV (*middle*; E-Z stereochemical designation of *cis/trans* bonds of isoprenyl groups not shown) and the donor strand analog moenomycin (*right*). (*c*) Complex structure of the *Streptomyces* R61 D,D carboxypeptidase/transpeptidase (TP) acylated by cephalosporin 1 (PDB code 1HVB) (*left*). The molecular structure of the natural TP substrate D-Ala-D-Ala is shown (*middle*) along with the mimicking β-lactam substrate analog cephalosporin 1, bound in the active site (*right*). The blue highlighted regions represent the functional groups on the substrate molecule and their corresponding groups on the antibiotic that are used to mimic these groups.

peptidoglycan layer to form a complex; indeed, both lipoproteins interacted with peptidoglycan sacculi in pull-down experiments (89). The UB2H/ODD domains may have other roles in regulating cell wall metabolism, and the UB2H region of PBP1b has been shown to interact with the lytic transglycosylase MltA (72).

The structures of the GT_{PGP} domains have been followed by both virtual and experimental inhibitor compound screening (90, 91). Additionally, fluorescence-based assays have been developed for both product detection (92) and moenomycin displacement (93). Aside from inhibitors that directly target the enzyme active site, the GT_{PGP} reaction is sensitive to agents that limit the availability of lipid II—either via chelation or hydrolysis. Interestingly, lipid II is large enough that different antibiotics can be developed that show no cross-resistance, as exemplified in a comparison of vancomycin and the newly identified fungal defensin, plectasin (94). With respect to inhibiting glycan chain formation, it is pertinent to note that other mechanisms for polymerizing lipid II may potentially occur: A *B. subtilis* mutant lacking all of the classical GT_{PGP} enzymes retained some viability and appeared to incorporate radiolabeled UDP-NAG into polymeric chains (95).

TRANSPEPTIDATION OF NEWLY POLYMERIZED GLYCAN STRANDS

Cross-linking of peptidoglycan occurs via transpeptidation of peptide units in adjacent glycan strands (polymerized as above from lipid II precursors). TPs are the important natural target of β-lactam antibiotics, including penicillins and cephalosporins. Thus, enzymes encoding this activity in various species have been identified in the literature as PBPs.

PBPs are divided into two main categories: high-molecular-mass (HMM) PBPs and low-molecular-mass (LMM) PBPs (65). HMM PBPs are multimodular, membrane-anchored enzymes that have been subdivided into one of two nonoverlapping classes, A and B, on the basis of the structure and function of their distinct N-terminal domains. As mentioned above, class A enzymes are bifunctional and contain an N-terminal polymerizing GT activity and a C-terminal domain that catalyzes (likely in a coordinated way with the GT_{PGP} domain) the TP-mediated cross-linking of peptides from two adjacent glycan chains. Class B HMM enzymes are monofunctional and possess a catalytic C-terminal TP domain with an N-terminal domain often associated with other nonpenicillin-binding functions (65). An example of this modularity is seen in PBP2a from *S. aureus*, which contains an N-terminal extension that is not believed to have a catalytic role. This region contains sequence identity (25% identity, 47% similarity) with YoeB, a cell wall–associated protein that moderates the activity of autolysins in cells (96). The HMM class B PBPs can be further differentiated into subclasses, e.g., B1 contains drug-resistant PBPs found in *B. subtilis*, *Staphylococci*, and *Enterococci* with no structural equivalent in gram-negative species. Additional subclasses, such as B2, contain an elongase complex–specific PBP, whereas subclass B3 contains a divisome-specific PBP. Subclass B4 contains PBPs involved in division of gram-positive bacteria, and subclass B5 enzymes are indirectly involved in septation (PBP classes are reviewed in References 97 and 98). The LMM PBPs are primarily soluble enzymes involved in carboxypeptidase "trimming" reactions that cleave the peptide bond between the two terminal D-alanines of the muramyl peptide. This modification moderates the degree of potential cross-linking of the cell wall by TPs (99).

The TP domains from both the HMM and LMM families show overall similar architectures in the catalytic domain (despite often disparate sequence identities) and are typically made up of two subdomains, a five stranded β-sheet bounded by three α-helices and an all-helical domain; the active site lies at the subdomain interface (**Figure 2**). The TP active site contains three conserved sequence motifs: SXXK (which includes the catalytic serine and general base lysine), (S/Y)X(N/C), and (K/H)(S/T)G (65).

Transpeptidation and carboxypeptidation reactions occur in three steps, beginning with the noncovalent binding of a donor strand (muramyl peptide) to the TP. The active-site serine then attacks the C-terminal D-Ala-D-Ala peptide bond of the peptidoglycan stem pentapeptide, forming an acyl-enzyme intermediate and releasing the terminal D-Ala residue. This acyl-enzyme intermediate can then be hydrolyzed by water with release of a shortened peptide (carboxypeptidation) or cross-linked with an amino group on a second peptidoglycan stem peptide (acceptor strand) to form a new peptide bond (transpeptidation), releasing the free enzyme (100). The amino acid residue on the reactive amine varies depending on the bacterial species (for example, glycine in *S. aureus*, lysine in *S. pneumoniae*, or a diaminopimelate group in *E. coli*); however, the overall mechanism is conserved. Recent developments have been made into probing the individual steps of the TP reaction by reconstituting the TP activity using defined substrates and purified proteins (101). Characterizing the TP reaction with natural peptidoglycan substrates has been impeded by difficulties in obtaining homogeneous samples of these complex molecules (101), although success in the synthesis of small analogs encompassing various elements of these substrates is becoming increasingly evident (102).

Given its perhaps unprecedented clinical importance in the treatment of bacterial infections, it is not surprising that β-lactam antibiotic inhibition of peptidoglycan TP activity has been extremely well studied over several decades (103). The β-lactams mimic the D-Ala-D-Ala dipeptide substrate, as predicted in the Tipper-Strominger hypothesis (104). Following attack by the TP active-site serine on the β-lactam ring, a long-lived covalent acyl-enzyme complex is formed, with the β-lactam adduct impairing PBP activity as well as peptidoglycan cross-linking. Several structures of PBPs have been solved in both the apo form and in complex with β-lactams, including several PBPs associated with β-lactam resistance, e.g., PBP2x from penicillin-resistant *S. pneumoniae* (105–107), PBP2a from MRSA (108), and PBP5fm from the naturally resistant *Enterococcus faecium* (**Supplemental Table 2**) (109). Typically, structural differences reported between resistant and sensitive forms are due to mutations located in the active-site region surrounding the SXXK motif, although more far-reaching secondary mutations likely also play a role in the overall observed resistance. The high level of broad-spectrum β-lactam resistance in the MRSA superbug, for example, is attributed to the acquisition of PBP2a. When challenged with β-lactams, MRSA utilizes the GT_{PGP} activity of PBP2 (the only class A enzyme in *S. aureus*) and the TP activity from PBP2a to synthesize the cell wall (the TP activity of PBP2 rendered inactive). The structure of a soluble form of PBP2a was solved to a resolution of 1.8 Å (108); a comparison with the structure of the susceptible PBP2 reveals that the mode of resistance is via a reorganization of the active site of PBP2a, placing the active-site serine in a poor position for nucleophilic attack. Recently, several novel β-lactams have been developed to specifically target MRSA PBP2a. Of these, ceftobiprole, televancin, and ceftaroline show promising broad-spectrum activity against several pathogens (110–113).

Recently, the complexity of the stereochemistry and regulation of transpeptidation activity in cell wall structure has increasingly come into evidence. Although D,D-transpeptidation catalyzing 4-3 cross-links (cross-linking the peptidyl substrate terminus D-Ala^4→A$_2$pm^3) between adjacent peptidoglycan strands is typical, 3-3 cross-links (A$_2$pm^3→A$_2$pm^3) are also found in some bacterial strains, which are generated by distinct L,D-TP enzymes. Interestingly from a clinical perspective, it has been shown in *Clostridium difficile* that cross-links formed by L,D-TPs increase in the presence of ampicillin, indicating that ampicillin does not inhibit the L,D-transpeptidation pathway. However, *C. difficile* remains susceptible to ampicillin, although minimum inhibitory concentration values are higher than other *Clostridium* species (114). Similarly, in *M. tuberculosis*, loss of the L,D-TPs resulted in increased susceptibility to amoxicillin therapy both in vitro and in the mouse model of tuberculosis (115). In addition to the function of L,D-TPs in the cross-linking of peptidoglycan, this family of proteins is also responsible for the attachment of the Braun lipoprotein (one of the most abundant proteins in gram-negative membranes) to peptidoglycan in *E. coli* (116).

The redox state of the cysteine residues in PBPs has also recently been shown to play a role in modulating their structure and function in various species. The structure of PBPA from *M. tuberculosis* displayed a typical TPase fold but with distinct structural features, including the displacement of the cysteine containing (S/Y)X(N/C$_{282}$) motif (117). This alteration, which is thought to be a result of disulfide bridge formation, places the (S/Y)X(N/C$_{282}$) motif of PBPA farther from the active site than in other PBPs. Because the SxN motif plays a central role in deacylation of PBPs, differences in conformations may influence catalysis. Another PBP modulated by redox state is the sporulation-specific PBP, SpoVD, from *B. subtilis*. Therein, Cys332 and Cys351 are surface exposed and are believed to form an intramolecular disulfide bond in the oxidized form. It was shown that both the oxidized and reduced forms of SpoVD react with penicillin, although the reduced PBP bound 1.4-fold more tightly to penicillin. The Cys351 thiol group is positioned within the active-site region, and inactivation of SpoVD results from a direct effect of the intramolecular disulfide. Interestingly, disulfide-mediated SpoVD functionality would be in keeping with the change in redox state known to occur during sporulation (118).

In addition to the effects of cysteine disulfides, select PBPs contain a number of proline residues that can affect enzyme function depending on the state of isomerization. The colocalization of these functional domains in the same cell compartment as the PrsA foldase (a peptidyl-prolyl *cis-trans* isomerase) suggests that the folding and activity of specific PBPs could be dependent on or regulated by the presence of PrsA. Indeed, several *B. subtilis* PBPs (PBP2a, PBP2b, PBP3, and PBP4) showed decreased activity in cells depleted of PrsA as compared with nondepleted cells (119).

Finally, speaking to the historically often-ignored potential importance of the interplay between peptidoglycan and cell wall teichoic acid (WTA, the predominant peptidoglycan modifying anionic polymer of various gram-positive cell walls), it was recently shown that inhibition of TarO, a key enzymatic component in the synthesis of WTA, sensitized MRSA strains to β-lactams. Upon use of low concentrations of the natural product inhibitor tunicamycin to specifically inhibit TarO, methicillin became bactericidal at concentrations that normally would not affect growth. This work outlines a new strategy for treating MRSA infections, through a combination of antimicrobials such as a β-lactam and a TarO (WTA) inhibitor (120).

REMOVAL/RECYCLING OF UNDECAPRENYL PYROPHOSPHATE

The GT reaction generates two products—a nascent polymer and undecaprenyl pyrophosphate (UPP). Levels of UPP are tightly controlled in the outer leaflet of the inner membrane, and free UPP forms the target of the antibiotic bacitracin (121). Bacitracin is a nonribosomal peptide product (from *Bacillus*) that acts by binding and sequestering UPP, thereby preventing dephosphorylation of UPP to undecaprenyl phosphate (UP). For this reason, enzymes catalyzing the dephosphorylation of UPP to UP have historically been linked to bacitracin resistance; hence, the *E. coli* protein BacA has been renamed UppP following confirmation of such an activity [this situation is further complicated in that BacA was originally proposed to phosphorylate UP (122)]. Interestingly, a member of the predatory bacteria genus *Lysobacter* was recently shown to produce an agent that also chelates UPP (a lipopeptide named tripropeptin C), which inhibits growth of both MRSA and vancomycin-resistant enterococci (123). The bacterial UPP:UP balance is critical, not just for recycling of UP(P) units for further lipid II and peptidoglycan biosynthesis, but also for other enzymes that share the prenyl lipid pool, e.g., teichoic acid biosynthesis in gram positives, lipid A phosphorylation, O-antigen biosynthesis, and protein glycosylation in gram negatives (71, 124–126). The Mengin-Lecreulx group (122, 127, 128) has been at the forefront of understanding UPP turnover, discovering that multiple UPPase activities exist—a homologous family (PgpB, YeiU, YbjG, YwoA, and BcrC) with a common catalytic motif and the unrelated BacA/UppP. YeiU is also known as LpxT and is the enzyme implicated in the aforementioned lipid A phosphorylation, catalyzing both UPP dephosphorylation and transfer of the phosphate group to lipid A (125). Further advances have been made in mapping the membrane topology of PgpB and in modeling its active site on phosphatase motifs from a soluble protein (129). The Valvano laboratory members (130) have confirmed a periplasmic-facing active site for YbjG and YeiU, and we await the first structure of a protein from either of the two UPPase families.

PEPTIDOGLYCAN IN THE CONTEXT OF BACTERIAL CELL ULTRASTRUCTURE

The terminal enzymes of the peptidoglycan biosynthesis pathway (that is both monofunctional and bifunctional GT_{PGPs} and TPs) have been shown to interact with a wide variety of proteins—for functional, regulatory, and localization purposes (e.g., interactions of GT_{PGP} domain detailed in Reference 74). The requirement to couple modification of the existing cell wall with insertion of new material is central to one of the original proposals for a multienzyme peptidoglycan "synthase" that is associated with the "three-for-one" growth model whereby an old strand is replaced by three nascent strands (131). We now recognize that specific complexes form for the purposes of elongation and septation; as expected, this process is best understood in the rod-shaped model organism *E. coli* [reviewed by Vollmer et al. (32)]. These systems can be thought of in a simplified manner whereby MreB cytoplasmic filaments "organize" elongation complexes along the long axis of the cell and FtsZ (filamentation temperature-sensitive) septal rings direct complexes involved in division/septum formation. An interesting sidenote to this functional differentiation is that the gram-positive secondary cell wall polymers, lipoteichoic acid and WTA, also display division-specific and elongation-specific roles, respectively (132). Alongside specialized morphogenic roles for PBPs [e.g., PBP2 in elongation and PBP3 in division in *E. coli* (32)], individual PBPs can also be "shuttled" between elongation and division complexes; in *B. subtilis*, it was revealed that PBP1 differential localization is controlled by the EzrA and GpsB proteins (133).

MreB and FtsZ are bacterial homologs of actin and tubulin, respectively, exhibiting nucleotide-dependent polymerization (134). A series of reports in 2005 established that the organization of rod-shaped wall synthesis was achieved by cytoplasmic and periplasmic filaments that exhibited a mutual interdependence (135–138), with the cytoplasmic MreB filament contacting a periplasmic MreC filament through the integral membrane protein MreD. Aside from interaction with MreC, a further interaction brings MreB into contact with the inner membrane—the bitopic rod shape-determining protein RodZ (139, 140) forms a complex with MreB via its cytoplasmic helix-turn-helix domain (141). Indeed, recent experiments have shown that MreB may even bind the membrane directly (142). An example of the comprehensive interactions between the MreBCD proteins and members of the peptidoglycan biosynthesis pathway is provided by White et al. (143) using the bacterial two hybrid approach in *Caulobacter crescentus*. The importance of the *mreBCD* operon is emphasized by its location, often in close proximity to TPs and putative flippases. The structures of MreC from *Listeria monocytogenes* and *S. pneumoniae* revealed an L-shaped protein with extensive interfaces for protein:protein interactions (144, 145). Recent experiments utilizing small-angle

X-ray scattering suggest that *Helicobacter pylori* MreC forms a 1:1 complex with the class B PBP2 enzyme. The region for this interaction mapped to the surface of MreC that opposes the filamentous, coiled-coil interface (146).

The MreB and MreC filaments (out of phase with one another) (135) were previously presumed to be helical in nature through the interpretation of images obtained by fluorescent light microscopy; however, newer experiments utilizing cryofluorescent light microscopy and electron cryotomography suggest these helical structures are artifactual, resulting from relatively slow exposure times and high depth of field (147). Complementary studies have confirmed a nonhelical arrangement of MreB, revealing instead a more patch-like arrangement, where bands of MreB move processively along peripheral tracks orientated perpendicularly to the long axis of the cell (148, 149). The processive movement derives from peptidoglycan polymerization itself, and the studies converge on a model where, instead of providing a rigid filamentous template to guide peptidoglycan formation, the Mre proteins restrict diffusion of the polymerases and provide tension and direction to the emergent new strand. Two previous studies complement such a hypothesis. Peptidoglycan polymerization has also been shown to provide the energetic force for engulfment (the process by which one cell becomes enclosed in the cytoplasm of another cell) during spore formation (150), and experimentally constrained mathematical modeling has suggested that the processivity of synthesis intrinsically guides cell straightening and/or rod shape (151). A variety of filamentous proteins have been shown to influence peptidoglycan deposition, including bactofillins in *C. crescentus* (152), DivIVA in *Streptomyces coelicolor* (153), and the DivIVA-like protein Wag31 in mycobacteria (154).

Control of peptidoglycan insertion and degradation has a proven role in the mechanical processes that govern cell shape (151, 155, 156), and recent studies have provided specific examples of hydrolytic enzymes directly implicated in morphogenesis. These studies include both peptidoglycan amidases [AmiC2, septum modification for multicellularity in cyanobacteria (157); and AmiA, -B, and -C for cytokinesis in *E. coli* (158)] and peptidases [DipM for cytokinesis in *C. crescentus* (159–161); Csd1, -2, and -3 for the helical curvature of *H. pylori* (162); and SwlC for branching in *S. coelicolor* (163)]. Remarkably, peptidoglycan shape and strength is regulated by D-amino acids (distinct from the usual D-Ala and D-Glu of the stem pentapeptide) (164), which can be incorporated by cellular ligases or periplasmic DD- and LD-TPs (101, 165). Incorporation of the nonstandard D-amino acids by nonproducing

Figure 3

Peptidoglycan structural organization. Conceptual models of peptidoglycan organization include the (*a*) layered, (*b*) scaffold, and (*c*) disorganized layered arrangements. Panels *a*–*c* depict the disaccharides as green pills and the stretched peptide cross-links as blue sticks (images from Reference 169). It is contested whether the glycans run parallel or antiparallel to each other, although see Sharif et al. (173) for an examination of the *S. aureus* peptidoglycan. (*d*) Electron cryotomography of sacculi isolated from *C. crescentus* strain CB15N, with a putative glycan 9-mer atomic model placed in density (*inset*) (image from Reference 169). (*e*) Atomic force microscopy (AFM; left side of image is height, right side is phase components) image of *S. aureus* division planes, showing the inherited rib and junction peptidoglycan "piecrust" features characteristic of division in successive 90° planes (images from Reference 170). (*f*) AFM of the inner-facing surface of a *B. subtilis* peptidoglycan layer with increasing magnification and a schematic (*far right side*), revealing the "twisted cable" architecture (*feature* I, *background B*) in all four panels (images from Reference 168). Panels *g* and *h* show the use of AFM and secondary cell wall polymer mutants to probe the nanoscale architecture of cell wall peptidoglycans in living gram-positive bacteria, using a topographic imaging peptidoglycan localized as parallel lines [visible in both *g* and *h* panels, deflection image and adhesion force map (from the square area shown in *g*), respectively, on the surface of these mutants (images from Reference 171)]. (*i*) AFM image of a single *Bacillus atrophaeus* spore germinating under native conditions; the peptidoglycan cell wall structure is evident in the center of the image (image from Reference 167).

organisms suggests that they may perform a signaling role in microbial communities (165).

The three-dimensional structure of peptidoglycan itself remains elusive [a quest reviewed in depth by Vollmer & Seligman (87)], and when considering all possible strain variability, modifications, and state (turnover, new versus old material, elongation, and division), there may be no "core" repeat arrangement. Attempts to address this gap in our knowledge have been made—ranging from small-scale analysis [NMR characterization of isolated, synthetic tetrasaccharide-pentapeptide$_2$ (166)] to sacculi/whole-cell analysis [atomic force microscopy (AFM) on peptidoglycan spores (167); AFM on gently broken *B. subtilis* cells (168); electron cryotomography on isolated *E. coli* and *C. crescentus* sacculi (169); AFM of the division planes of *S. aureus* (170); and AFM on living, secondary cell wall polysaccharide-free *Lactococcus lactis* cells (**Figure 3**) (171)]. These techniques have been used to answer the first

AFM: atomic force microscopy

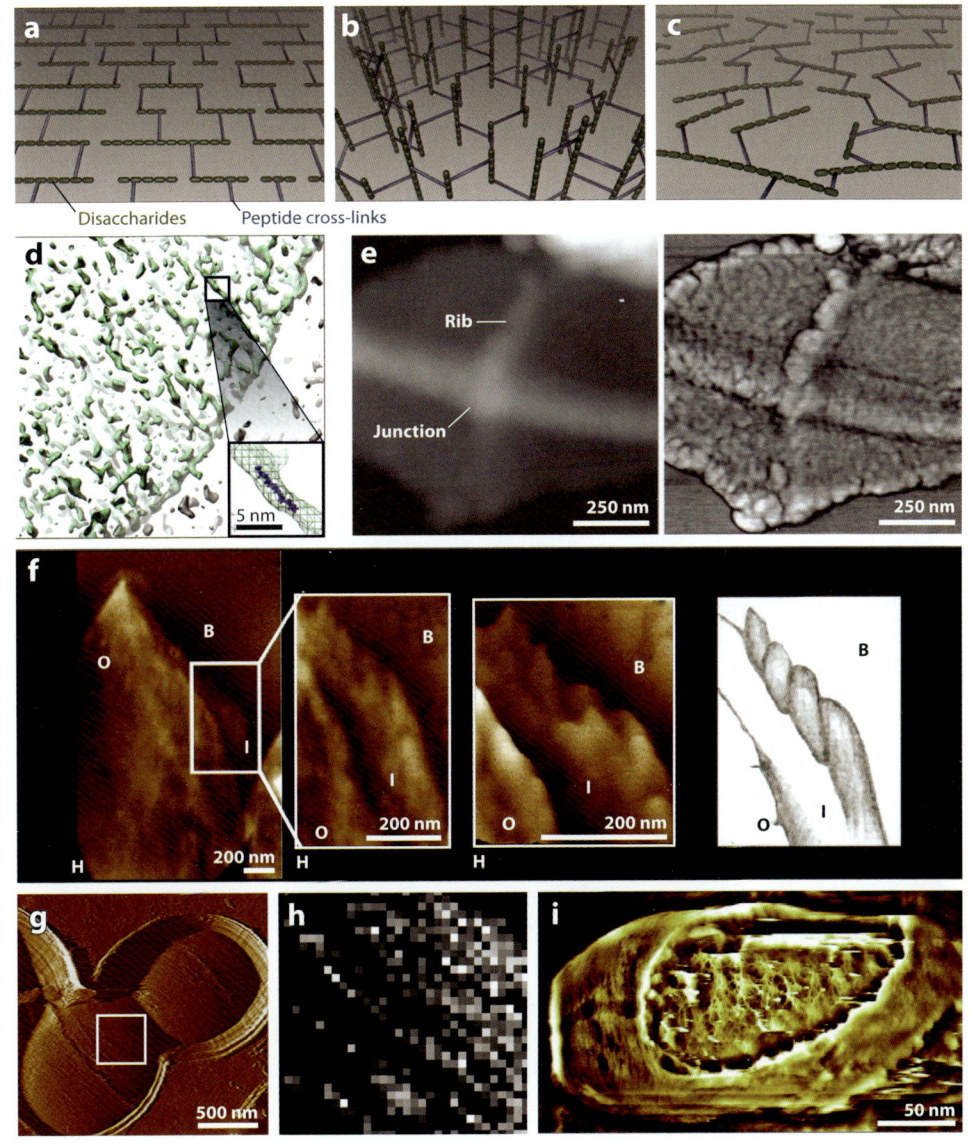

Elongase-specific and divisome complexes: multiprotein complexes responsible for cell wall synthesis/metabolism during elongation or cell division; they assemble around different cellular filaments

topological question about peptidoglycan organization: Do the glycan strands run parallel to the inner membrane or do they radiate outwardly from it in a perpendicular manner? The referenced AFM and cryotomography reports converge on the parallel model, although as mentioned above, one model need not fit all possibilities. Despite their large size ($\sim 3 \times 10^6$ kDa), intact *E. coli* sacculi give high-quality solid-state NMR spectra (172) and can provide details on the interaction with peptidoglycan-binding proteins. Measurement of the intercrossbridge distances of *S. aureus* peptidoglycan using NMR indicates that the glycan chains likely run parallel to one another (173). Solid-state NMR was also used to investigate the peptidoglycan:WTA interaction, resulting in a structural model where divalent cations mediate interactions between teichoic acid phosphate groups and peptidoglycan glutamate/DAP residues (174). The gram-positive covalent modification of peptidoglycan by teichoic acids at NAM-O6 plays a large role in regulation of cross-linking (175), and the synthetic lethality observed when combining mutants of nonessential peptidoglycan and WTA biosynthetic enzymes further highlights the interaction between the two wall polymers (120).

CONCLUSIONS

Research into peptidoglycan biosynthesis is of immense interest, relevant to the fields of both antimicrobial development and bacterial physiology. Recent publications have answered several of the long-standing questions in the field: How are the lipid II repeating units polymerized? How are peptidoglycan metabolism and cell shape linked? Do the polymeric chains run parallel, or perpendicular, to the membrane? Yet, we are still unable to define the exact molecular features of the mature peptidoglycan sacculus. The past few years have brought insight into both pathway commonalities (e.g., the molecular details of the ubiquitous GT step and the quest for the lipid II flippase) and species-specific deviations from the norm (e.g., modification for purposes of vancomycin resistance and the intricate use of endopeptidases in generating the curvature of *Helicobacter*). The vast range of this field can be gauged because it is possible to ask some peptidoglycan-related questions that are relatively abstract: Why does lipid II appear to be involved in division in peptidoglycan-deficient bacteria (and moss!)? What can we learn from analysis of the archaeal polymer pseudomurein that is functionally/structurally similar yet biosynthetically distinct from bacteria? It is tempting to suggest that we may be asking even more questions when summarizing this area in the future.

The challenge ahead is to annotate all of the biosynthetic steps (to obtain a full structure:function understanding of all the enzymatic reactions in the pathway) and develop our knowledge as to how the individual components interact when faced with the two cellular requirements of growth and division, essentially characterizing the elongase-specific and divisome protein complexes in more detail. Given the variability between gram-negative and -positive and differing bacterial morphology, this goal may require systematic investigation in singular organisms, and even then under precisely controlled conditions. Furthermore, the rich history of peptidoglycan biosynthetic enzymes as drug targets is likely to continue, with the additional prospect of potentially developing agents to disrupt these newly described multiprotein complexes.

SUMMARY POINTS

1. High-resolution structures of the soluble Mur enzymes are nearly complete. These structures have delineated the active-site composition and catalytic mechanisms governing the functions for these enzymes, facilitating high-throughput screening of drug libraries that target their essential action in generating peptidoglycan precursor molecules.

2. The covalent attachment of MurNAc-pentapeptide to an activating undecaprenyl phosphate carrier, catalyzed by the membrane-spanning MraY, and the molecular details of lipid II transport across the cytoplasmic membrane by peptidoglycan flippases have remained elusive. However, recent progress has been made in the isolation of candidate enzymes and potential binding partners as well as in the development of appropriate assays to facilitate future structure/function analysis.

3. The polymerization of lipid II precursors into extended glycan strands by the peptidoglycan GT (GT_{PBP}) family has provided fascinating details on the membrane-associated nature of the protein, substrate-binding motifs, catalytic mechanisms, processive polymerization, and determinants of inhibition by potent natural products, such as moenomycin.

4. The TP domain of penicillin-binding proteins (PBPs), and the essential cross-linking activity they provide for cell wall strength, represents one of the most successful targets for antibiotic development. Despite the development of drug resistance through alteration of TP function (e.g., as observed for the superbug MRSA), it still is a prime target for new drug development as evidenced by the recent development of several TP-based therapeutics.

5. The challenging goal to attain high-resolution molecular details of peptidoglycan ultrastructure is being investigated by newly developed methods in AFM and electron cryotomography with recent work answering some of the first topological questions, such as the likely parallel disposition of glycan chains with respect to the cytoplasmic membrane.

6. There has been strong recent research activity into proteins that modify peptidoglycan deposition and tailoring (e.g., cellular filaments and periplasmic peptidases). This activity is providing new antibacterial targets and helping to uncover the mechanistic details underlying bacterial shape and form.

FUTURE ISSUES

1. Is it possible to obtain an atomic resolution understanding of mature peptidoglycan arrangement? It would also be of interest to learn more about the physical interactions between peptidoglycan and other wall components (e.g., lipoproteins, secondary cell wall polymers, S-layers).

2. With the above question, there is a "chicken and egg" scenario where we need to know more about the mechanism/regulation of nascent strand insertion—especially because it appears that the polymerization process itself influences cell shape.

3. Can we define the elongase and divisome complexes in more detail? With the discovery of lpoA and lpoB, and uncertainty over the nature of the lipid II flippase, are there any other proteins that play vital roles in the peptidoglycan biosynthetic pathway that are currently unannotated?

4. There are a few challenging proteins in the pathway remaining to be structurally characterized (e.g., MraY, BacA, FtsW). Will any forthcoming information on these (and substrate complexes of those already determined, e.g., MurG and GT_{PGP}) help design much-needed novel antibacterials?

DISCLOSURE STATEMENT

The authors are not aware of any affiliations, memberships, funding, or financial holdings that might be perceived as affecting the objectivity of this review.

ACKNOWLEDGMENTS

We gratefully acknowledge funding by the Canadian Institute of Health Research and Howard Hughes Medical Institute (to N.C.J.S.). A.L.L. is supported by a New Investigator Award from the Royal Society. N.C.J.S. is a Canada Research Tier 1 Chair in Antibiotic Discovery.

LITERATURE CITED

1. Bugg TD, Braddick D, Dowson CG, Roper DI. 2011. Bacterial cell wall assembly: still an attractive antibacterial target. *Trends Biotechnol.* 29:167–73
2. Pantosti A, Sanchini A, Monaco M. 2007. Mechanisms of antibiotic resistance in *Staphylococcus aureus*. *Future Microbiol.* 2:323–34
3. Wilke MS, Lovering AL, Strynadka NC. 2005. Beta-lactam antibiotic resistance: a current structural perspective. *Curr. Opin. Microbiol.* 8:525–33
4. Brown ED, Vivas EI, Walsh CT, Kolter R. 1995. MurA (MurZ), the enzyme that catalyzes the first committed step in peptidoglycan biosynthesis, is essential in *Escherichia coli*. *J. Bacteriol.* 177:4194–97
5. Du W, Brown JR, Sylvester DR, Huang J, Chalker AF, et al. 2000. Two active forms of UDP-*N*-acetylglucosamine enolpyruvyl transferase in gram-positive bacteria. *J. Bacteriol.* 182:4146–52
6. Skarzynski T, Mistry A, Wonacott A, Hutchinson SE, Kelly VA, Duncan K. 1996. Structure of UDP-*N*-acetylglucosamine enolpyruvyl transferase, an enzyme essential for the synthesis of bacterial peptidoglycan, complexed with substrate UDP-*N*-acetylglucosamine and the drug fosfomycin. *Structure* 4:1465–74
7. Schönbrunn E, Sack S, Eschenburg S, Perrakis A, Krekel F, et al. 1996. Crystal structure of UDP-*N*-acetylglucosamine enolpyruvyltransferase, the target of the antibiotic fosfomycin. *Structure* 4:1065–75
8. Yoon HJ, Lee SJ, Mikami B, Park HJ, Yoo J, Suh SW. 2008. Crystal structure of UDP-*N*-acetylglucosamine enolpyruvyl transferase from *Haemophilus influenzae* in complex with UDP-*N*-acetylglucosamine and fosfomycin. *Proteins* 71:1032–37
9. Rogers TO, Birnbaum J. 1974. Biosynthesis of fosfomycin by *Streptomyces fradiae*. *Antimicrob. Agents Chemother.* 5:121–32
10. Kahan FM, Kahan JS, Cassidy PJ, Kropp H. 1974. The mechanism of action of fosfomycin (phosphonomycin). *Ann. N. Y. Acad. Sci.* 235:364–86
11. Marquardt JL, Brown ED, Lane WS, Haley TM, Ichikawa Y, et al. 1994. Kinetics, stoichiometry, and identification of the reactive thiolate in the inactivation of UDP-GlcNAc enolpyruvoyl transferase by the antibiotic fosfomycin. *Biochemistry* 33:10646–51
12. Han H, Yang Y, Olesen SH, Becker A, Betzi S, Schönbrunn E. 2010. The fungal product terreic acid is a covalent inhibitor of the bacterial cell wall biosynthetic enzyme UDP-*N*-acetylglucosamine 1-carboxyvinyltransferase (MurA). *Biochemistry* 49:4276–82
13. Silver LL. 2006. Does the cell wall of bacteria remain a viable source of targets for novel antibiotics? *Biochem. Pharmacol.* 71:996–1005
14. Sheehan JC, Lawson WB, Gaul RJ. 1958. The structure of terreic acid. *J. Am. Chem. Soc.* 80:5536–38

15. Benson TE, Filman DJ, Walsh CT, Hogle JM. 1995. An enzyme-substrate complex involved in bacterial cell wall biosynthesis. *Nat. Struct. Biol.* 2:644–53
16. Benson TE, Walsh CT, Hogle JM. 1996. The structure of the substrate-free form of MurB, an essential enzyme for the synthesis of bacterial cell walls. *Structure* 4:47–54
17. Benson TE, Walsh CT, Hogle JM. 1997. X-ray crystal structures of the S229A mutant and wild-type MurB in the presence of the substrate enolpyruvyl-UDP-*N*-acetylglucosamine at 1.8-A resolution. *Biochemistry* 36:806–11
18. Kumar V, Saravanan P, Arvind A, Mohan CG. 2011. Identification of hotspot regions of MurB oxidoreductase enzyme using homology modeling, molecular dynamics and molecular docking techniques. *J. Mol. Model.* 17:939–53
19. Bouhss A, Mengin-Lecreulx D, Blanot D, van Heijenoort J, Parquet C. 1997. Invariant amino acids in the Mur peptide synthetases of bacterial peptidoglycan synthesis and their modification by site-directed mutagenesis in the UDP-MurNAc:L-alanine ligase from *Escherichia coli*. *Biochemistry* 36:11556–63
20. Eveland SS, Pompliano DL, Anderson MS. 1997. Conditionally lethal *Escherichia coli* murein mutants contain point defects that map to regions conserved among murein and folyl poly-gamma-glutamate ligases: identification of a ligase superfamily. *Biochemistry* 36:6223–29
21. Patin D, Boniface A, Kovac A, Herve M, Dementin S, et al. 2010. Purification and biochemical characterization of Mur ligases from *Staphylococcus aureus*. *Biochimie* 92:1793–800
22. Deva T, Baker EN, Squire CJ, Smith CA. 2006. Structure of *Escherichia coli* UDP-*N*-acetylmuramoyl:L-alanine ligase (MurC). *Acta Crystallogr. D* 62:1466–74
23. Tomasic T, Zidar N, Sink R, Kovac A, Blanot D, et al. 2011. Structure-based design of a new series of D-glutamic acid based inhibitors of bacterial UDP-*N*-acetylmuramoyl-L-alanine: D-glutamate ligase (MurD). *J. Med. Chem.* 54:4600–10
24. Emanuele JJ Jr, Jin H, Yanchunas J Jr, Villafranca JJ. 1997. Evaluation of the kinetic mechanism of *Escherichia coli* uridine diphosphate-*N*-acetylmuramate:L-alanine ligase. *Biochemistry* 36:7264–71
25. Gordon E, Flouret B, Chantalat L, van Heijenoort J, Mengin-Lecreulx D, Dideberg O. 2001. Crystal structure of UDP-*N*-acetylmuramoyl-L-alanyl-D-glutamate: meso-diaminopimelate ligase from *Escherichia coli*. *J. Biol. Chem.* 276:10999–1006
26. Basavannacharya C, Robertson G, Munshi T, Keep NH, Bhakta S. 2010. ATP-dependent MurE ligase in *Mycobacterium tuberculosis*: biochemical and structural characterisation. *Tuberculosis* 90:16–24
27. Yan Y, Munshi S, Leiting B, Anderson MS, Chrzas J, Chen Z. 2000. Crystal structure of *Escherichia coli* UDPMurNAc-tripeptide D-alanyl-D-alanine-adding enzyme (MurF) at 2.3 A resolution. *J. Mol. Biol.* 304:435–45
28. Fan C, Moews PC, Walsh CT, Knox JR. 1994. Vancomycin resistance: structure of D-alanine: D-alanine ligase at 2.3 A resolution. *Science* 266:439–43
29. Xie J, Pierce JG, James RC, Okano A, Boger DL. 2011. A redesigned vancomycin engineered for dual D-Ala-D-Ala and D-Ala-D-Lac binding exhibits potent antimicrobial activity against vancomycin-resistant bacteria. *J. Am. Chem. Soc.* 133:13946–49
30. Baum EZ, Crespo-Carbone SM, Klinger A, Foleno BD, Turchi I, et al. 2007. A MurF inhibitor that disrupts cell wall biosynthesis in *Escherichia coli*. *Antimicrob. Agents Chemother.* 51:4420–26
31. Baum EZ, Crespo-Carbone SM, Foleno BD, Simon LD, Guillemont J, et al. 2009. MurF inhibitors with antibacterial activity: effect on muropeptide levels. *Antimicrob. Agents Chemother.* 53:3240–47
32. Vollmer W, Blanot D, de Pedro MA. 2008. Peptidoglycan structure and architecture. *FEMS Microbiol. Rev.* 32:149–67
33. Benson TE, Prince DB, Mutchler VT, Curry KA, Ho AM, et al. 2002. X-ray crystal structure of *Staphylococcus aureus* FemA. *Structure* 10:1107–15
34. Biarrotte-Sorin S, Maillard AP, Delettré J, Sougakoff W, Arthur M, Mayer C. 2004. Crystal structures of *Weissella viridescens* FemX and its complex with UDP-MurNAc-pentapeptide: insights into FemABX family substrates recognition. *Structure* 12:257–67
35. Watanabe K, Toh Y, Suto K, Shimizu Y, Oka N, et al. 2007. Protein-based peptide-bond formation by aminoacyl-tRNA protein transferase. *Nature* 449:867–71

36. Heydanek MG Jr, Struve WG, Neuhaus FC. 1969. On the initial stage in peptidoglycan synthesis. 3. Kinetics and uncoupling of phospho-N-acetylmuramyl-pentapeptide translocase (uridine 5′-phosphate). *Biochemistry* 8:1214–21
37. Al-Dabbagh B, Henry X, El Ghachi M, Auger G, Blanot D, et al. 2008. Active site mapping of MraY, a member of the polyprenyl-phosphate N-acetylhexosamine 1-phosphate transferase superfamily, catalyzing the first membrane step of peptidoglycan biosynthesis. *Biochemistry* 47:8919–28
38. Bouhss A, Crouvoisier M, Blanot D, Mengin-Lecreulx D. 2004. Purification and characterization of the bacterial MraY translocase catalyzing the first membrane step of peptidoglycan biosynthesis. *J. Biol. Chem.* 279:29974–80
39. Kotnik M, Anderluh PS, Prezelj A. 2007. Development of novel inhibitors targeting intracellular steps of peptidoglycan biosynthesis. *Curr. Pharm. Des.* 13:2283–309
40. Dini C. 2005. MraY inhibitors as novel antibacterial agents. *Curr. Top. Med. Chem.* 5:1221–36
41. Mravljak J, Monasson O, Al-Dabbagh B, Crouvoisier M, Bouhss A, et al. 2011. Synthesis and biological evaluation of a diazepanone-based library of liposidomycins analogs as MraY inhibitors. *Eur. J. Med. Chem.* 46:1582–92
42. Lecercle D, Clouet A, Al-Dabbagh B, Crouvoisier M, Bouhss A, et al. 2010. Bacterial transferase MraY inhibitors: synthesis and biological evaluation. *Bioorg. Med. Chem.* 18:4560–69
43. Zheng Y, Struck DK, Young R. 2009. Purification and functional characterization of phiX174 lysis protein E. *Biochemistry* 48:4999–5006
44. van Heijenoort Y, Gomez M, Derrien M, Ayala J, van Heijenoort J. 1992. Membrane intermediates in the peptidoglycan metabolism of *Escherichia coli:* possible roles of PBP 1b and PBP 3. *J. Bacteriol.* 174:3549–57
45. Ha S, Walker D, Shi Y, Walker S. 2000. The 1.9 Å crystal structure of *Escherichia coli* MurG, a membrane-associated glycosyltransferase involved in peptidoglycan biosynthesis. *Protein Sci.* 9:1045–52
46. Hu Y, Chen L, Ha S, Gross B, Falcone B, et al. 2003. Crystal structure of the MurG:UDP-GlcNAc complex reveals common structural principles of a superfamily of glycosyltransferases. *Proc. Natl. Acad. Sci. USA* 100:845–49
47. Unligil UM, Rini JM. 2000. Glycosyltransferase structure and mechanism. *Curr. Opin. Struct. Biol.* 10:510–17
48. Trunkfield AE, Gurcha SS, Besra GS, Bugg TD. 2010. Inhibition of *Escherichia coli* glycosyltransferase MurG and *Mycobacterium tuberculosis* Gal transferase by uridine-linked transition state mimics. *Bioorg. Med. Chem.* 18:2651–63
49. Zawadzke LE, Wu P, Cook L, Fan L, Casperson M, et al. 2003. Targeting the MraY and MurG bacterial enzymes for antimicrobial therapeutic intervention. *Anal. Biochem.* 314:243–52
50. Hyland SA, Anderson MS. 2003. A high-throughput solid-phase extraction assay capable of measuring diverse polyprenyl phosphate: sugar-1-phosphate transferases as exemplified by the WecA, MraY, and MurG proteins. *Anal. Biochem.* 317:156–65
51. Ravishankar S, Kumar VP, Chandrakala B, Jha RK, Solapure SM, de Sousa SM. 2005. Scintillation proximity assay for inhibitors of *Escherichia coli* MurG and, optionally, MraY. *Antimicrob. Agents Chemother.* 49:1410–18
52. van Dam V, Sijbrandi R, Kol M, Swiezewska E, de Kruijff B, Breukink E. 2007. Transmembrane transport of peptidoglycan precursors across model and bacterial membranes. *Mol. Microbiol.* 64:1105–14
53. Ruiz N. 2008. Bioinformatics identification of MurJ (MviN) as the peptidoglycan lipid II flippase in *Escherichia coli*. *Proc. Natl. Acad. Sci. USA* 105:15553–57
54. Baba T, Ara T, Hasegawa M, Takai Y, Okumura Y, et al. 2006. Construction of *Escherichia coli* K-12 in-frame, single-gene knockout mutants: the Keio collection. *Mol. Syst. Biol.* 2:2006.0008
55. Inoue A, Murata Y, Takahashi H, Tsuji N, Fujisaki S, Kato J. 2008. Involvement of an essential gene, *mviN*, in murein synthesis in *Escherichia coli*. *J. Bacteriol.* 190:7298–301
56. Fay A, Dworkin J. 2009. *Bacillus subtilis* homologs of MviN (MurJ), the putative *Escherichia coli* lipid II flippase, are not essential for growth. *J. Bacteriol.* 191:6020–28
57. Ikeda M, Sato T, Wachi M, Jung HK, Ishino F, et al. 1989. Structural similarity among *Escherichia coli* FtsW and RodA proteins and *Bacillus subtilis* SpoVE protein, which function in cell division, cell elongation, and spore formation, respectively. *J. Bacteriol.* 171:6375–78

58. Henrichfreise B, Schiefer A, Schneider T, Nzukou E, Poellinger C, et al. 2009. Functional conservation of the lipid II biosynthesis pathway in the cell wall–less bacteria *Chlamydia* and *Wolbachia*: Why is lipid II needed? *Mol. Microbiol.* 73:913–23
59. Mohammadi T, van Dam V, Sijbrandi R, Vernet T, Zapun A, et al. 2011. Identification of FtsW as a transporter of lipid-linked cell wall precursors across the membrane. *EMBO J.* 30:1425–32
60. Gerard P, Vernet T, Zapun A. 2002. Membrane topology of the *Streptococcus pneumoniae* FtsW division protein. *J. Bacteriol.* 184:1925–31
61. Lara B, Ayala JA. 2002. Topological characterization of the essential *Escherichia coli* cell division protein FtsW. *FEMS Microbiol. Lett.* 216:23–32
62. Henriques AO, Glaser P, Piggot PJ, Moran CP Jr. 1998. Control of cell shape and elongation by the *rodA* gene in *Bacillus subtilis*. *Mol. Microbiol.* 28:235–47
63. Fraipont C, Alexeeva S, Wolf B, van der Ploeg R, Schloesser M, et al. 2011. The integral membrane FtsW protein and peptidoglycan synthase PBP3 form a subcomplex in *Escherichia coli*. *Microbiology* 157:251–59
64. Pastoret S, Fraipont C, den Blaauwen T, Wolf B, Aarsman ME, et al. 2004. Functional analysis of the cell division protein FtsW of *Escherichia coli*. *J. Bacteriol.* 186:8370–79
65. Goffin C, Ghuysen JM. 1998. Multimodular penicillin-binding proteins: an enigmatic family of orthologs and paralogs. *Microbiol. Mol. Biol. Rev.* 62:1079–93
66. Reed P, Veiga H, Jorge AM, Terrak M, Pinho MG. 2011. Monofunctional transglycosylases are not essential for *Staphylococcus aureus* cell wall synthesis. *J. Bacteriol.* 193:2549–56
67. Lovering AL, de Castro LH, Lim D, Strynadka NC. 2007. Structural insight into the transglycosylation step of bacterial cell-wall biosynthesis. *Science* 315:1402–5
68. Yuan Y, Barrett D, Zhang Y, Kahne D, Sliz P, Walker S. 2007. Crystal structure of a peptidoglycan glycosyltransferase suggests a model for processive glycan chain synthesis. *Proc. Natl. Acad. Sci. USA* 104:5348–53
69. Terrak M, Ghosh TK, van Heijenoort J, Van Beeumen J, Lampilas M, et al. 1999. The catalytic, glycosyl transferase and acyl transferase modules of the cell wall peptidoglycan-polymerizing penicillin-binding protein 1b of *Escherichia coli*. *Mol. Microbiol.* 34:350–64
70. Welzel P. 2005. Syntheses around the transglycosylation step in peptidoglycan biosynthesis. *Chem. Rev.* 105:4610–60
71. Lizak C, Gerber S, Numao S, Aebi M, Locher KP. 2011. X-ray structure of a bacterial oligosaccharyl-transferase. *Nature* 474:350–55
72. Sung MT, Lai YT, Huang CY, Chou LY, Shih HW, et al. 2009. Crystal structure of the membrane-bound bifunctional transglycosylase PBP1b from *Escherichia coli*. *Proc. Natl. Acad. Sci. USA* 106:8824–29
73. Heaslet H, Shaw B, Mistry A, Miller AA. 2009. Characterization of the active site of *S. aureus* monofunctional glycosyltransferase (Mtg) by site-directed mutation and structural analysis of the protein complexed with moenomycin. *J. Struct. Biol.* 167:129–35
74. Lovering AL, Gretes M, Strynadka NC. 2008. Structural details of the glycosyltransferase step of peptidoglycan assembly. *Curr. Opin. Struct. Biol.* 18:534–43
75. Terrak M, Sauvage E, Derouaux A, Dehareng D, Bouhss A, et al. 2008. Importance of the conserved residues in the peptidoglycan glycosyltransferase module of the class A penicillin-binding protein 1b of *Escherichia coli*. *J. Biol. Chem.* 283:28464–70
76. Barrett D, Wang TS, Yuan Y, Zhang Y, Kahne D, Walker S. 2007. Analysis of glycan polymers produced by peptidoglycan glycosyltransferases. *J. Biol. Chem.* 282:31964–71
77. Yuan Y, Fuse S, Ostash B, Sliz P, Kahne D, Walker S. 2008. Structural analysis of the contacts anchoring moenomycin to peptidoglycan glycosyltransferases and implications for antibiotic design. *ACS Chem. Biol.* 3:429–36
78. Perlstein DL, Zhang Y, Wang TS, Kahne DE, Walker S. 2007. The direction of glycan chain elongation by peptidoglycan glycosyltransferases. *J. Am. Chem. Soc.* 129:12674–75
79. Wang TS, Lupoli TJ, Sumida Y, Tsukamoto H, Wu Y, et al. 2011. Primer preactivation of peptidoglycan polymerases. *J. Am. Chem. Soc.* 133:8528–30
80. Wang TS, Manning SA, Walker S, Kahne D. 2008. Isolated peptidoglycan glycosyltransferases from different organisms produce different glycan chain lengths. *J. Am. Chem. Soc.* 130:14068–69

81. Perlstein DL, Wang TS, Doud EH, Kahne D, Walker S. 2010. The role of the substrate lipid in processive glycan polymerization by the peptidoglycan glycosyltransferases. *J. Am. Chem. Soc.* 132:48–49
82. Ostash B, Doud EH, Lin C, Ostash I, Perlstein DL, et al. 2009. Complete characterization of the seventeen step moenomycin biosynthetic pathway. *Biochemistry* 48:8830–41
83. Ostash B, Saghatelian A, Walker S. 2007. A streamlined metabolic pathway for the biosynthesis of moenomycin A. *Chem. Biol.* 14:257–67
84. Lovering AL, De Castro L, Strynadka NC. 2008. Identification of dynamic structural motifs involved in peptidoglycan glycosyltransfer. *J. Mol. Biol.* 383:167–77
85. Macheboeuf P, Piuzzi M, Finet S, Bontems F, Pérez J, et al. 2011. Solution X-ray scattering study of a full-length class A penicillin-binding protein. *Biochem. Biophys. Res. Commun.* 405:107–11
86. Born P, Breukink E, Vollmer W. 2006. In vitro synthesis of cross-linked murein and its attachment to sacculi by PBP1A from *Escherichia coli*. *J. Biol. Chem.* 281:26985–93
87. Vollmer W, Seligman SJ. 2010. Architecture of peptidoglycan: more data and more models. *Trends Microbiol.* 18:59–66
88. Paradis-Bleau C, Markovski M, Uehara T, Lupoli TJ, Walker S, et al. 2010. Lipoprotein cofactors located in the outer membrane activate bacterial cell wall polymerases. *Cell* 143:1110–20
89. Typas A, Banzhaf M, van den Berg van Saparoea B, Verheul J, Biboy J, et al. 2010. Regulation of peptidoglycan synthesis by outer-membrane proteins. *Cell* 143:1097–109
90. Cheng TJ, Wu YT, Yang ST, Lo KH, Chen SK, et al. 2010. High-throughput identification of antibacterials against methicillin-resistant *Staphylococcus aureus* (MRSA) and the transglycosylase. *Bioorg. Med. Chem.* 18:8512–29
91. Derouaux A, Turk S, Olrichs NK, Gobec S, Breukink E, et al. 2011. Small molecule inhibitors of peptidoglycan synthesis targeting the lipid II precursor. *Biochem. Pharmacol.* 81:1098–105
92. Liu CY, Guo CW, Chang YF, Wang JT, Shih HW, et al. 2010. Synthesis and evaluation of a new fluorescent transglycosylase substrate: lipid II-based molecule possessing a dansyl-C20 polyprenyl moiety. *Org. Lett.* 12:1608–11
93. Cheng TJ, Sung MT, Liao HY, Chang YF, Chen CW, et al. 2008. Domain requirement of moenomycin binding to bifunctional transglycosylases and development of high-throughput discovery of antibiotics. *Proc. Natl. Acad. Sci. USA* 105:431–36
94. Schneider T, Kruse T, Wimmer R, Wiedemann I, Sass V, et al. 2010. Plectasin, a fungal defensin, targets the bacterial cell wall precursor lipid II. *Science* 328:1168–72
95. McPherson DC, Popham DL. 2003. Peptidoglycan synthesis in the absence of class A penicillin-binding proteins in *Bacillus subtilis*. *J. Bacteriol.* 185:1423–31
96. Salzberg LI, Helmann JD. 2007. An antibiotic-inducible cell wall–associated protein that protects *Bacillus subtilis* from autolysis. *J. Bacteriol.* 189:4671–80
97. Sauvage E, Kerff F, Terrak M, Ayala JA, Charlier P. 2008. The penicillin-binding proteins: structure and role in peptidoglycan biosynthesis. *FEMS Microbiol. Rev.* 32:234–58
98. Mattei PJ, Neves D, Dessen A. 2010. Bridging cell wall biosynthesis and bacterial morphogenesis. *Curr. Opin. Struct. Biol.* 20:749–55
99. Ghosh AS, Chowdhury C, Nelson DE. 2008. Physiological functions of D-alanine carboxypeptidases in *Escherichia coli*. *Trends Microbiol.* 16:309–17
100. McDonough MA, Anderson JW, Silvaggi NR, Pratt RF, Knox JR, Kelly JA. 2002. Structures of two kinetic intermediates reveal species specificity of penicillin-binding proteins. *J. Mol. Biol.* 322:111–22
101. Lupoli TJ, Tsukamoto H, Doud EH, Wang TS, Walker S, Kahne D. 2011. Transpeptidase-mediated incorporation of D-amino acids into bacterial peptidoglycan. *J. Am. Chem. Soc.* 133:10748–51
102. Lee M, Hesek D, Shah IM, Oliver AG, Dworkin J, Mobashery S. 2010. Synthetic peptidoglycan motifs for germination of bacterial spores. *ChemBioChem* 11:2525–29
103. Kong KF, Schneper L, Mathee K. 2010. Beta-lactam antibiotics: from antibiosis to resistance and bacteriology. *APMIS* 118:1–36
104. Tipper DJ, Strominger JL. 1965. Mechanism of action of penicillins: a proposal based on their structural similarity to acyl-D-alanyl-D-alanine. *Proc. Natl. Acad. Sci. USA* 54:1133–41

105. Pernot L, Chesnel L, Le Gouellec A, Croize J, Vernet T, et al. 2004. A PBP2x from a clinical isolate of *Streptococcus pneumoniae* exhibits an alternative mechanism for reduction of susceptibility to beta-lactam antibiotics. *J. Biol. Chem.* 279:16463–70
106. Chesnel L, Pernot L, Lemaire D, Champelovier D, Croize J, et al. 2003. The structural modifications induced by the M339F substitution in PBP2x from *Streptococcus pneumoniae* further decreases the susceptibility to beta-lactams of resistant strains. *J. Biol. Chem.* 278:44448–56
107. Dessen A, Mouz N, Gordon E, Hopkins J, Dideberg O. 2001. Crystal structure of PBP2x from a highly penicillin-resistant *Streptococcus pneumoniae* clinical isolate: a mosaic framework containing 83 mutations. *J. Biol. Chem.* 276:45106–12
108. Lim D, Strynadka NC. 2002. Structural basis for the beta lactam resistance of PBP2a from methicillin-resistant *Staphylococcus aureus*. *Nat. Struct. Biol.* 9:870–76
109. Sauvage E, Kerff F, Fonze E, Herman R, Schoot B, et al. 2002. The 2.4-A crystal structure of the penicillin-resistant penicillin-binding protein PBP5fm from *Enterococcus faecium* in complex with benzylpenicillin. *Cell. Mol. Life Sci.* 59:1223–32
110. Dauner DG, Nelson RE, Taketa DC. 2010. Ceftobiprole: a novel, broad-spectrum cephalosporin with activity against methicillin-resistant *Staphylococcus aureus*. *Am. J. Health Syst. Pharm.* 67:983–93
111. Page MG. 2006. Anti-MRSA beta-lactams in development. *Curr. Opin. Pharmacol.* 6:480–85
112. Lim L, Sutton E, Brown J. 2011. Ceftaroline: a new broad-spectrum cephalosporin. *Am. J. Health Syst. Pharm.* 68:491–98
113. Charneski L, Patel PN, Sym D. 2009. Telavancin: a novel lipoglycopeptide antibiotic. *Ann. Pharmacother.* 43:928–38
114. Peltier J, Courtin P, El Meouche I, Lemée L, Chapot-Chartier MP, Pons JL. 2011. *Clostridium difficile* has an original peptidoglycan structure with high level of *N*-acetylglucosamine deacetylation and mainly 3-3 cross-links. *J. Biol. Chem.* 286:29053–62
115. Gupta R, Lavollay M, Mainardi JL, Arthur M, Bishai WR, Lamichhane G. 2010. The *Mycobacterium tuberculosis* protein LdtMt2 is a nonclassical transpeptidase required for virulence and resistance to amoxicillin. *Nat. Med.* 16:466–69
116. Magnet S, Bellais S, Dubost L, Fourgeaud M, Mainardi JL, et al. 2007. Identification of the L,D-transpeptidases responsible for attachment of the Braun lipoprotein to *Escherichia coli* peptidoglycan. *J. Bacteriol.* 189:3927–31
117. Fedarovich A, Nicholas RA, Davies C. 2010. Unusual conformation of the SxN motif in the crystal structure of penicillin-binding protein A from *Mycobacterium tuberculosis*. *J. Mol. Biol.* 398:54–65
118. Liu Y, Carlsson Möller M, Petersen L, Söderberg CA, Hederstedt L. 2010. Penicillin-binding protein SpoVD disulphide is a target for StoA in *Bacillus subtilis* forespores. *Mol. Microbiol.* 75:46–60
119. Hyyrylainen HL, Marciniak BC, Dahncke K, Pietiainen M, Courtin P, et al. 2010. Penicillin-binding protein folding is dependent on the PrsA peptidyl-prolyl *cis-trans* isomerase in *Bacillus subtilis*. *Mol. Microbiol.* 77:108–27
120. Campbell J, Singh AK, Santa Maria JP Jr, Kim Y, Brown S, et al. 2011. Synthetic lethal compound combinations reveal a fundamental connection between wall teichoic acid and peptidoglycan biosyntheses in *Staphylococcus aureus*. *ACS Chem. Biol.* 6:106–16
121. Siewert G, Strominger JL. 1967. Bacitracin: an inhibitor of the dephosphorylation of lipid pyrophosphate, an intermediate in the biosynthesis of the peptidoglycan of bacterial cell walls. *Proc. Natl. Acad. Sci. USA* 57:767–73
122. El Ghachi M, Bouhss A, Blanot D, Mengin-Lecreulx D. 2004. The *bacA* gene of *Escherichia coli* encodes an undecaprenyl pyrophosphate phosphatase activity. *J. Biol. Chem.* 279:30106–13
123. Hashizume H, Sawa R, Harada S, Igarashi M, Adachi H, et al. 2011. Tripropeptin C blocks the lipid cycle of cell wall biosynthesis by complex formation with undecaprenyl pyrophosphate. *Antimicrob. Agents Chemother.* 55:3821–28
124. D'Elia MA, Millar KE, Bhavsar AP, Tomljenovic AM, Hutter B, et al. 2009. Probing teichoic acid genetics with bioactive molecules reveals new interactions among diverse processes in bacterial cell wall biogenesis. *Chem. Biol.* 16:548–56
125. Touze T, Tran AX, Hankins JV, Mengin-Lecreulx D, Trent MS. 2008. Periplasmic phosphorylation of lipid A is linked to the synthesis of undecaprenyl phosphate. *Mol. Microbiol.* 67:264–77

126. Harkness RE, Braun V. 1989. Colicin M inhibits peptidoglycan biosynthesis by interfering with lipid carrier recycling. *J. Biol. Chem.* 264:6177–82
127. El Ghachi M, Derbise A, Bouhss A, Mengin-Lecreulx D. 2005. Identification of multiple genes encoding membrane proteins with undecaprenyl pyrophosphate phosphatase (UppP) activity in *Escherichia coli*. *J. Biol. Chem.* 280:18689–95
128. Bernard R, El Ghachi M, Mengin-Lecreulx D, Chippaux M, Denizot F. 2005. BcrC from *Bacillus subtilis* acts as an undecaprenyl pyrophosphate phosphatase in bacitracin resistance. *J. Biol. Chem.* 280:28852–57
129. Touze T, Blanot D, Mengin-Lecreulx D. 2008. Substrate specificity and membrane topology of *Escherichia coli* PgpB, an undecaprenyl pyrophosphate phosphatase. *J. Biol. Chem.* 283:16573–83
130. Tatar LD, Marolda CL, Polischuk AN, van Leeuwen D, Valvano MA. 2007. An *Escherichia coli* undecaprenyl-pyrophosphate phosphatase implicated in undecaprenyl phosphate recycling. *Microbiology* 153:2518–29
131. Holtje JV. 1998. Growth of the stress-bearing and shape-maintaining murein sacculus of *Escherichia coli*. *Microbiol. Mol. Biol. Rev.* 62:181–203
132. Schirner K, Marles-Wright J, Lewis RJ, Errington J. 2009. Distinct and essential morphogenic functions for wall- and lipo-teichoic acids in *Bacillus subtilis*. *EMBO J.* 28:830–42
133. Claessen D, Emmins R, Hamoen LW, Daniel RA, Errington J, Edwards DH. 2008. Control of the cell elongation-division cycle by shuttling of PBP1 protein in *Bacillus subtilis*. *Mol. Microbiol.* 68:1029–46
134. van den Ent F, Amos L, Lowe J. 2001. Bacterial ancestry of actin and tubulin. *Curr. Opin. Microbiol.* 4:634–38
135. Divakaruni AV, Loo RR, Xie Y, Loo JA, Gober JW. 2005. The cell-shape protein MreC interacts with extracytoplasmic proteins including cell wall assembly complexes in *Caulobacter crescentus*. *Proc. Natl. Acad. Sci. USA* 102:18602–7
136. Dye NA, Pincus Z, Theriot JA, Shapiro L, Gitai Z. 2005. Two independent spiral structures control cell shape in *Caulobacter*. *Proc. Natl. Acad. Sci. USA* 102:18608–13
137. Leaver M, Errington J. 2005. Roles for MreC and MreD proteins in helical growth of the cylindrical cell wall in *Bacillus subtilis*. *Mol. Microbiol.* 57:1196–209
138. Kruse T, Bork-Jensen J, Gerdes K. 2005. The morphogenetic MreBCD proteins of *Escherichia coli* form an essential membrane-bound complex. *Mol. Microbiol.* 55:78–89
139. Shiomi D, Sakai M, Niki H. 2008. Determination of bacterial rod shape by a novel cytoskeletal membrane protein. *EMBO J.* 27:3081–91
140. Alyahya SA, Alexander R, Costa T, Henriques AO, Emonet T, Jacobs-Wagner C. 2009. RodZ, a component of the bacterial core morphogenic apparatus. *Proc. Natl. Acad. Sci. USA* 106:1239–44
141. van den Ent F, Johnson CM, Persons L, de Boer P, Löwe J. 2010. Bacterial actin MreB assembles in complex with cell shape protein RodZ. *EMBO J.* 29:1081–90
142. Salje J, van den Ent F, de Boer P, Löwe J. 2011. Direct membrane binding by bacterial actin MreB. *Mol. Cell* 43:478–87
143. White CL, Kitich A, Gober JW. 2010. Positioning cell wall synthetic complexes by the bacterial morphogenetic proteins MreB and MreD. *Mol. Microbiol.* 76:616–33
144. van den Ent F, Leaver M, Bendezu F, Errington J, de Boer P, Lowe J. 2006. Dimeric structure of the cell shape protein MreC and its functional implications. *Mol. Microbiol.* 62:1631–42
145. Lovering AL, Strynadka NC. 2007. High-resolution structure of the major periplasmic domain from the cell shape-determining filament MreC. *J. Mol. Biol.* 372:1034–44
146. El Ghachi M, Mattei PJ, Ecobichon C, Martins A, Hoos S, et al. Characterization of the elongasome core PBP2:MreC complex of *Helicobacter pylori*. *Mol. Microbiol.* 82:68–86
147. Swulius MT, Chen S, Ding HJ, Li Z, Briegel A, et al. 2011. Long helical filaments are not seen encircling cells in electron cryotomograms of rod-shaped bacteria. *Biochem. Biophys. Res. Commun.* 407:650–55
148. Dominguez-Escobar J, Chastanet A, Crevenna AH, Fromion V, Wedlich-Söldner R, Carballido-López R. 2011. Processive movement of MreB-associated cell wall biosynthetic complexes in bacteria. *Science* 333:225–28
149. Garner EC, Bernard R, Wang W, Zhuang X, Rudner DZ, Mitchison T. 2011. Coupled, circumferential motions of the cell wall synthesis machinery and MreB filaments in *B. subtilis*. *Science* 333:222–25

150. Meyer P, Gutierrez J, Pogliano K, Dworkin J. 2010. Cell wall synthesis is necessary for membrane dynamics during sporulation of *Bacillus subtilis*. *Mol. Microbiol.* 76:956–70
151. Sliusarenko O, Cabeen MT, Wolgemuth CW, Jacobs-Wagner C, Emonet T. 2010. Processivity of peptidoglycan synthesis provides a built-in mechanism for the robustness of straight-rod cell morphology. *Proc. Natl. Acad. Sci. USA* 107:10086–91
152. Kühn J, Briegel A, Mörschel E, Kahnt J, Leser K, et al. 2010. Bactofilins, a ubiquitous class of cytoskeletal proteins mediating polar localization of a cell wall synthase in *Caulobacter crescentus*. *EMBO J.* 29:327–39
153. Hempel AM, Wang SB, Letek M, Gil JA, Flardh K. 2008. Assemblies of DivIVA mark sites for hyphal branching and can establish new zones of cell wall growth in *Streptomyces coelicolor*. *J. Bacteriol.* 190:7579–83
154. Jani C, Eoh H, Lee JJ, Hamasha K, Sahana MB, et al. 2010. Regulation of polar peptidoglycan biosynthesis by Wag31 phosphorylation in mycobacteria. *BMC Microbiol.* 10:327
155. Huang KC, Mukhopadhyay R, Wen B, Gitai Z, Wingreen NS. 2008. Cell shape and cell-wall organization in gram-negative bacteria. *Proc. Natl. Acad. Sci. USA* 105:19282–87
156. Furchtgott L, Wingreen NS, Huang KC. 2011. Mechanisms for maintaining cell shape in rod-shaped gram-negative bacteria. *Mol. Microbiol.* 81:340–53
157. Lehner J, Zhang Y, Berendt S, Rasse TM, Forchhammer K, Maldener I. 2011. The morphogene AmiC2 is pivotal for multicellular development in the cyanobacterium *Nostoc punctiforme*. *Mol. Microbiol.* 79:1655–69
158. Uehara T, Parzych KR, Dinh T, Bernhardt TG. 2010. Daughter cell separation is controlled by cytokinetic ring-activated cell wall hydrolysis. *EMBO J.* 29:1412–22
159. Möll A, Schlimpert S, Briegel A, Jensen GJ, Thanbichler M. 2010. DipM, a new factor required for peptidoglycan remodelling during cell division in *Caulobacter crescentus*. *Mol. Microbiol.* 77:90–107
160. Goley ED, Comolli LR, Fero KE, Downing KH, Shapiro L. 2010. DipM links peptidoglycan remodelling to outer membrane organization in *Caulobacter*. *Mol. Microbiol.* 77:56–73
161. Poggio S, Takacs CN, Vollmer W, Jacobs-Wagner C. 2010. A protein critical for cell constriction in the gram-negative bacterium *Caulobacter crescentus* localizes at the division site through its peptidoglycan-binding LysM domains. *Mol. Microbiol.* 77:74–89
162. Sycuro LK, Pincus Z, Gutierrez KD, Biboy J, Stern CA, et al. 2010. Peptidoglycan crosslinking relaxation promotes *Helicobacter pylori's* helical shape and stomach colonization. *Cell* 141:822–33
163. Haiser HJ, Yousef MR, Elliot MA. 2009. Cell wall hydrolases affect germination, vegetative growth, and sporulation in *Streptomyces coelicolor*. *J. Bacteriol.* 191:6501–12
164. Lam H, Oh DC, Cava F, Takacs CN, Clardy J, et al. 2009. D-amino acids govern stationary phase cell wall remodeling in bacteria. *Science* 325:1552–55
165. Cava F, de Pedro MA, Lam H, Davis BM, Waldor MK. 2011. Distinct pathways for modification of the bacterial cell wall by non-canonical D-amino acids. *EMBO J.* 30:3442–53
166. Meroueh SO, Bencze KZ, Hesek D, Lee M, Fisher JF, et al. 2006. Three-dimensional structure of the bacterial cell wall peptidoglycan. *Proc. Natl. Acad. Sci. USA* 103:4404–9
167. Plomp M, Leighton TJ, Wheeler KE, Hill HD, Malkin AJ. 2007. In vitro high-resolution structural dynamics of single germinating bacterial spores. *Proc. Natl. Acad. Sci. USA* 104:9644–49
168. Hayhurst EJ, Kailas L, Hobbs JK, Foster SJ. 2008. Cell wall peptidoglycan architecture in *Bacillus subtilis*. *Proc. Natl. Acad. Sci. USA* 105:14603–8
169. Gan L, Chen S, Jensen GJ. 2008. Molecular organization of gram-negative peptidoglycan. *Proc. Natl. Acad. Sci. USA* 105:18953–57
170. Turner RD, Ratcliffe EC, Wheeler R, Golestanian R, Hobbs JK, Foster SJ. 2010. Peptidoglycan architecture can specify division planes in *Staphylococcus aureus*. *Nat. Commun.* 1:26
171. Andre G, Kulakauskas S, Chapot-Chartier MP, Navet B, Deghorain M, et al. 2010. Imaging the nanoscale organization of peptidoglycan in living *Lactococcus lactis* cells. *Nat. Commun.* 1:27
172. Kern T, Hediger S, Muller P, Giustini C, Joris B, et al. 2008. Toward the characterization of peptidoglycan structure and protein-peptidoglycan interactions by solid-state NMR spectroscopy. *J. Am. Chem. Soc.* 130:5618–19
173. Sharif S, Singh M, Kim SJ, Schaefer J. 2009. *Staphylococcus aureus* peptidoglycan tertiary structure from carbon-13 spin diffusion. *J. Am. Chem. Soc.* 131:7023–30

174. Kern T, Giffard M, Hediger S, Amoroso A, Giustini C, et al. 2010. Dynamics characterization of fully hydrated bacterial cell walls by solid-state NMR: evidence for cooperative binding of metal ions. *J. Am. Chem. Soc.* 132:10911–19
175. Atilano ML, Pereira PM, Yates J, Reed P, Veiga H, et al. 2010. Teichoic acids are temporal and spatial regulators of peptidoglycan cross-linking in *Staphylococcus aureus*. *Proc. Natl. Acad. Sci. USA* 107:18991–96

Discovery, Biosynthesis, and Engineering of Lantipeptides

Patrick J. Knerr[1] and Wilfred A. van der Donk[1,2]

[1]Department of Chemistry and [2]Howard Hughes Medical Institute, University of Illinois at Urbana-Champaign, Urbana, Illinois 61801; email: knerr1@illinois.edu, vddonk@illinois.edu

Keywords

lantibiotic, ribosomally synthesized natural product, leader peptide, posttranslational modification, genome mining

Abstract

Aided by genome-mining strategies, knowledge of the prevalence and diversity of ribosomally synthesized natural products (RNPs) is rapidly increasing. Among these are the lantipeptides, posttranslationally modified peptides containing characteristic thioether cross-links imperative for bioactivity and stability. Though this family was once thought to be a limited class of antimicrobial compounds produced by gram-positive bacteria, new insights have revealed a much larger diversity of activity, structure, biosynthetic machinery, and producing organisms than previously appreciated. Detailed investigation of the enzymes responsible for installing the posttranslational modifications has resulted in improved in vivo and in vitro engineering systems focusing on enhancement of the therapeutic potential of these compounds. Although dozens of new lantipeptides have been isolated in recent years, bioinformatic analyses indicate that many hundreds more await discovery owing to the widespread frequency of lantipeptide biosynthetic machinery in bacterial genomes.

Contents

1. INTRODUCTION 480
2. LANTIPEPTIDE ACTIVITY, STRUCTURE, AND DIVERSITY 480
3. CLASS I LANTIPEPTIDES: DEDICATED DEHYDRATASE AND CYCLASE ENZYMES 484
   3.1. Biosynthetic Machinery 484
   3.2. Selected New Class I Lantipeptides 486
4. CLASS II LANTIPEPTIDES: BIFUNCTIONAL LANTHIONINE SYNTHETASES 486
   4.1. Biosynthetic Machinery 486
   4.2. Selected New Class II Lantipeptides 488
5. CLASS III LANTIPEPTIDES: TRIFUNCTIONAL SYNTHETASES AND CARBOCYCLIC RINGS 490
6. CLASS IV LANTIPEPTIDES: TRIFUNCTIONAL LANTHIONINE SYNTHETASES 490
7. ROLE OF THE LEADER PEPTIDE 491
8. OTHER POSTTRANSLATIONAL TAILORING REACTIONS 492
9. RECLASSIFICATION OF FORMER LANTIPEPTIDES 493
10. RECENT DEVELOPMENTS IN LANTIPEPTIDE ENGINEERING 493
    10.1. In Vivo Engineering in Producing Strains and Heterologous Hosts 494
    10.2. In Vivo Engineering in *Escherichia coli* 494
    10.3. In Vitro Engineering 495
    10.4. Chemical Synthesis 496
11. OUTLOOK 496

1. INTRODUCTION

Aided by the progress in genome sequencing, the past decade has revealed that ribosomally synthesized natural products (RNPs) possess much greater diversity with respect to structure and biological activity than previously anticipated (1–3). Initially limited to the 20 canonical amino acids during translation, the diversity of these compounds is greatly enhanced by extensive posttranslational modifications (PTMs), such as hetero- or macrocyclization, dehydration, acylation, glycosylation, halogenation, prenylation, and epimerization (3, 4). Owing to the direct connection between the final compound and the gene encoding its precursor, RNPs are particularly suited for genome-mining strategies, which have demonstrated that these pathways are remarkably widespread (1). In addition, the relative brevity of RNP biosynthetic pathways (**Figure 1a**), as well as the substrate promiscuity of the biosynthetic machinery, renders ribosome-derived compounds attractive for bioengineering efforts. Among the best-studied compounds to date are the lantipeptides; sustained interest in this class of RNPs has resulted in comprehensive reviews (5–7), as well as specialized reviews concerning their antimicrobial activity and potential applications (8, 9, 10), mode of action (11–13), and self-resistance (14). This review focuses on the expanding knowledge of lantipeptide biosynthetic pathways, covering the period 2007 to 2011, and how this understanding has yielded breakthroughs in the discovery and bioengineering of new compounds.

2. LANTIPEPTIDE ACTIVITY, STRUCTURE, AND DIVERSITY

Lantipeptides are polycyclic peptides characterized by the presence of the thioether-cross-linked amino acids *meso*-lanthionine (Lan) and (2S,3S,6R)-3-methyllanthionine (MeLan) (**Figure 1b,c**). Formerly known as lantibiotics for lanthionine-containing antibiotics, this family name has been broadened to lantipeptides to encompass the discovery of

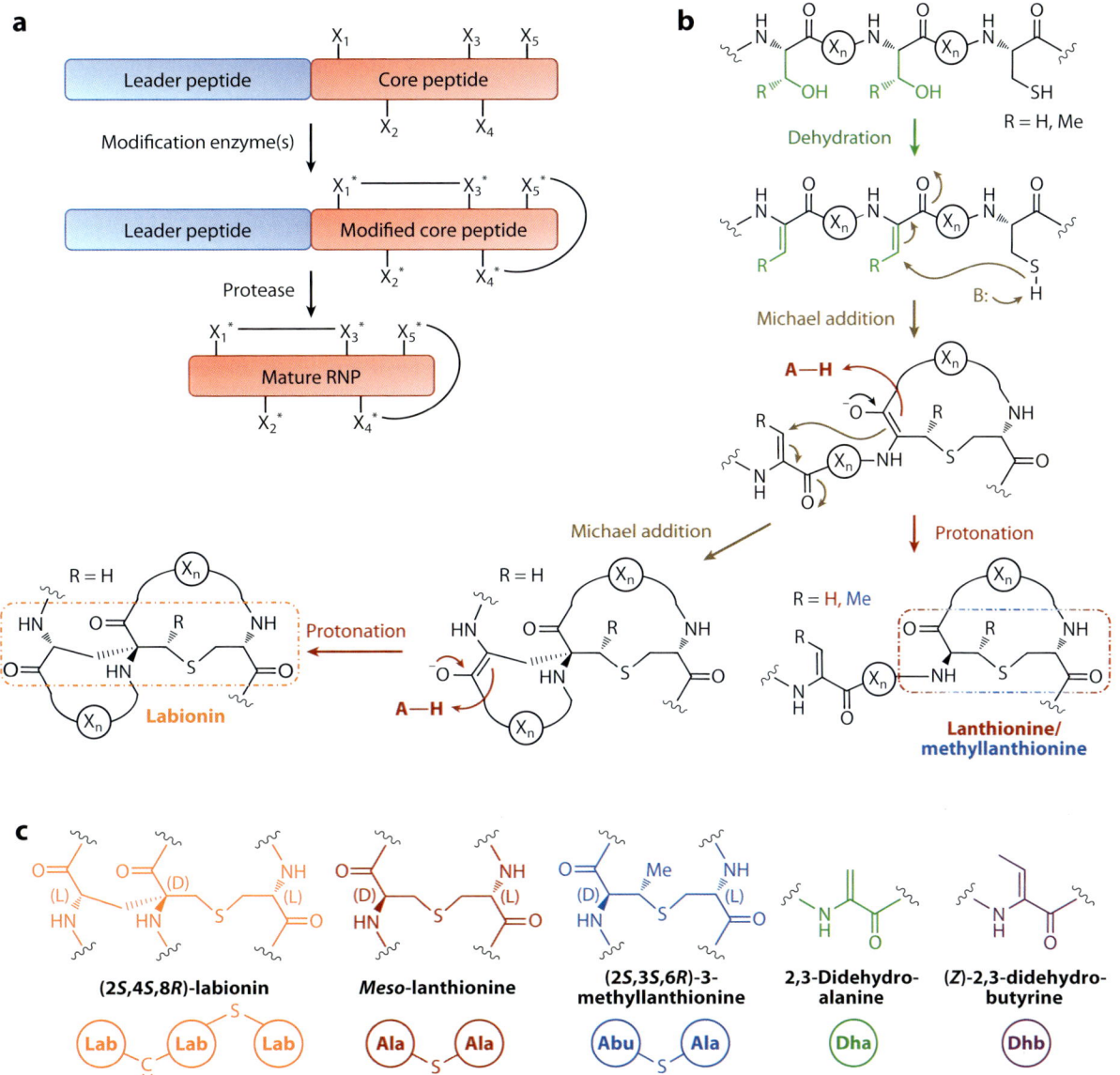

Figure 1

(*a*) General scheme of ribosomally synthesized natural product (RNP) biosynthesis; X_n^* indicates a modified residue. (*b*) Mechanism of thioether cross-link formation in lantipeptides. (*c*) Characteristic lantipeptide posttranslational modifications, with shorthand notation used throughout this review. Abbreviations: Abu, 2-aminobutyric acid; Dha, 2,3-didehydroalanine; Dhb, (*Z*)-2,3-didehydrobutyrine; Lab, labionin.

compounds with nonantimicrobial activities (15). Genome-mining efforts have identified many new members of this family, which now includes over 90 isolated compounds, most with distinct ring topologies (**Figure 2**). Interest in these RNPs as chemotherapeutics stems from the use of the prototypical lantipeptide nisin as a food preservative for more than 50 years without substantial incidence of microbial resistance (16). Lantibiotics possess potent activity against

RNPs: ribosomally synthesized natural products

PTMs: posttranslational modifications

Lantipeptide: a ribosomal natural product containing lanthionine and/or methyllanthionine residues

Lan: *meso*-lanthionine

MeLan: (2*S*,3*S*,6*R*)-3-methyllanthionine

Lantibiotic: a lantipeptide with antimicrobial activity

Dha: 2,3-didehydroalanine

Dhb: (*Z*)-2,3-didehydrobutyrine

LanA: a linear lantipeptide precursor peptide and substrate for lantipeptide modification enzymes

Core peptide: the C-terminal portion of the precursor peptide that becomes the final product after modification and leader peptide removal

Leader peptide: the N-terminal portion of the precursor peptide important for production, but unmodified and cleaved from the final product

many clinically relevant gram-positive bacteria, including drug-resistant strains of *Staphylococcus*, *Streptococcus*, *Enterococcus*, and *Clostridium*, as well as against select gram-negative pathogens, such as *Neisseria* and *Helicobacter* (8). Highlighting the potential of this family are several members currently in clinical development, including duramycin for the treatment of cystic fibrosis (17) and a derivative of actagardine for the treatment of *Clostridium difficile* infections (18, 19). Additionally, mutacin 1140 is currently in preclinical development for the treatment of gram-positive bacterial infections (20). Other applications of this family, including in agriculture and veterinary medicine (9) and in molecular imaging (21), are also emerging.

The mechanisms by which lantipeptides exert biological activity have been studied extensively in only a few instances, but the majority of antibacterial lantipeptides are believed to inhibit cell wall biosynthesis and/or disrupt membrane integrity through pore formation. In several cases, the essential cell wall precursor lipid II serves as the target of the lantibiotic; binding to lipid II inhibits the transglycosylation reaction necessary to synthesize peptidoglycan and also sequesters lipid II into nonfunctional locations (22). Details of the binding interactions with lipid II have been elucidated for nisin (23), which forms contacts with the pyrophosphate moiety of lipid II through the A and B rings conserved in several other class I lantipeptides, including microbisporicin (**Figure 2**) and mutacin 1140. Once bound to lipid II, nisin is able to insert into the membrane and form stable pores consisting of eight nisin and four lipid II molecules (24). The class II lantipeptide mersacidin is known to bind lipid II and inhibit transglycosylation in a Ca^{2+}-dependent manner (25) but does not form pores; this interaction is thought to be mediated through its C ring (26), a structure conserved in other class II lantibiotics (**Figure 2**) (27, 28). For the two-component lantibiotics lacticin 3147 and haloduracin (see Section 4.2), the α-peptide shares the mersacidin-binding motif and interacts with lipid II, which then recruits the β-peptide to form pores (29, 30).

In the case of haloduracin, the complex has been shown to consist of a 1:2:2 stoichiometry of lipid II to α-peptide to β-peptide (31). This mode of action explains the observation that the individual peptides of two-component lantibiotics display little or no activity by themselves, but potent and synergistic activity in combination. Interestingly, some lantibiotics, such as Pep5 and epilancin K7, form pores but do not use lipid II as a docking molecule (32); the high potency of these compounds suggests a specific but alternative molecular target. Further investigations of lantipeptide's mode of action, including compounds such as cinnamycin that do not target lipid II, have been reviewed elsewhere (11–13) and are beyond the scope of this work.

The distinguishing thioether cross-links of lantipeptides (**Figure 1***b*) are installed posttranslationally by one or more biosynthetic enzymes via dehydration of serine or threonine to the α,β-unsaturated residues 2,3-didehydroalanine (Dha) or (*Z*)-2,3-didehydrobutyrine (Dhb), respectively, followed by Michael-type addition of a cysteinyl thiol to give a thioether bridge. In all cases investigated to date, an *S*-configuration is installed at the newly formed Cα-stereocenter of the thioether cross-link (6). These cross-links are critical not only for the activity of the mature compound but also for stability against proteolysis and heat denaturation, as demonstrated with mutants of Pep5 (33), nisin (34), and lacticin 3147 (35), and with artificial lanthionine-containing peptides (36, 37). Other less-common PTMs, such as aminovinylcysteine and N-terminal acyl moieties, are also thought to play roles in stability and/or activity (see Section 8).

In all known cases, lantipeptides are biosynthesized on the ribosome as a precursor LanA peptide; this linear precursor contains both a leader peptide and a core peptide, and the latter of these becomes the mature compound (**Figure 1***a*). The leader peptide is thought to play multiple roles in PTM installation, export, and immunity (see Section 7). The mature lantipeptide is produced from LanA by one or more enzymes that install the thioether

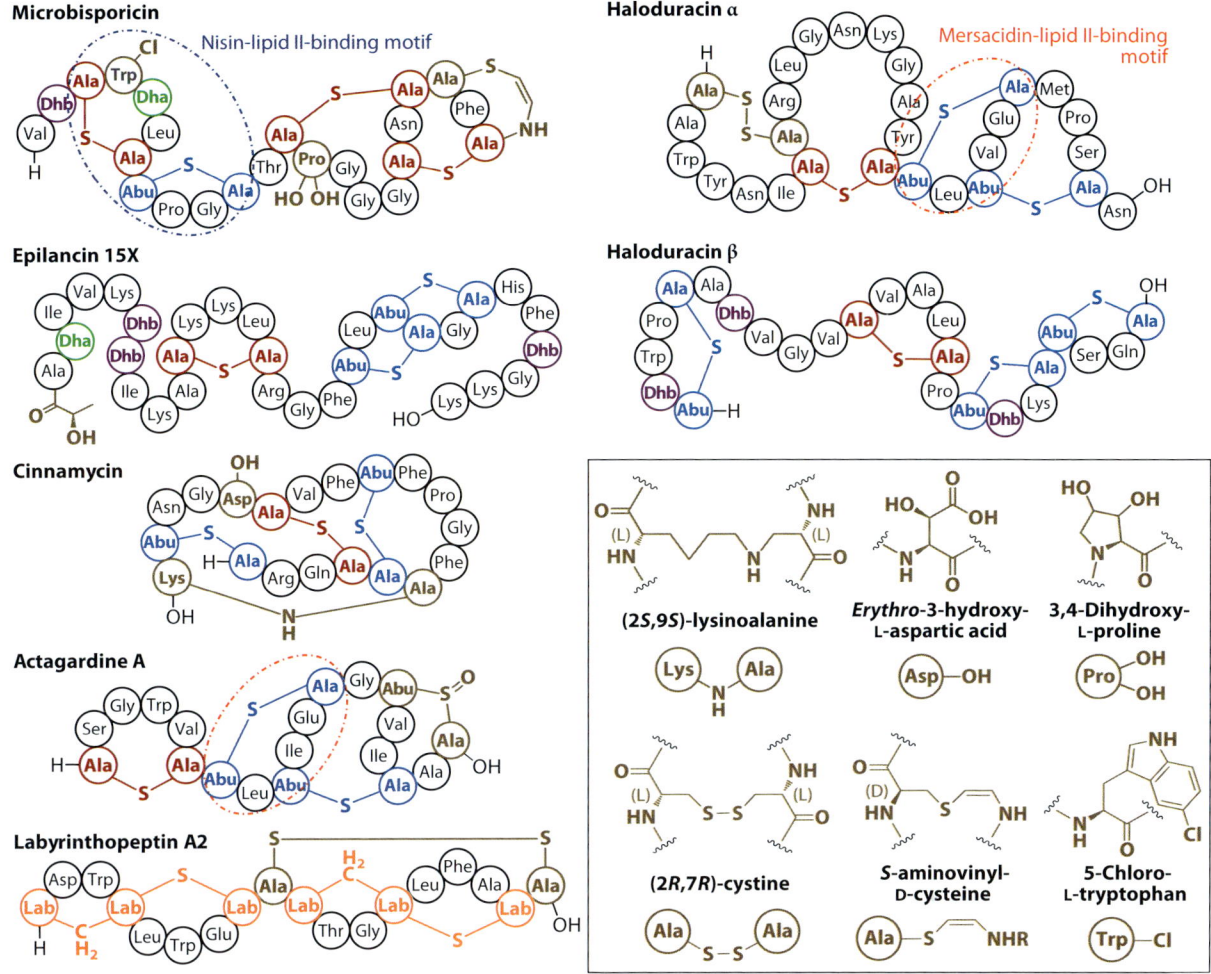

Figure 2

Representative examples of lantipeptide structures, highlighting the canonical nisin- and mersacidin-lipid II–binding motifs (*blue* and *red dashed circles*, respectively) and tailoring posttranslational modifications (*brown*). Shorthand notations for tailoring modifications are defined in the box. For the structure of nisin, see **Figure 4**.

cross-links, followed by removal of the leader peptide and export from the producing cell. Biochemical and genetic analyses have revealed four distinct classes of lantipeptides on the basis of their biosynthetic machinery (**Figure 3**). For class I, serine/threonine dehydration and thioether cyclization are performed by the dehydratase LanB and the cyclase LanC, respectively. For class II, these reactions are carried out by a single bifunctional lantipeptide synthetase, LanM, containing N-terminal dehydratase and C-terminal LanC-like cyclase domains. Class III lantipeptides are modified by a trifunctional synthetase bearing an N-terminal lyase domain, a central kinase domain, and a putative C-terminal cyclase domain, which lacks many of the conserved active-site residues found in LanC/LanM (38, 39). Finally, the recently identified class IV synthetase, LanL, contains N-terminal lyase and kinase domains as in class III, but its C-terminal cyclase domain is analogous to LanC (15, 40). The multiplicity of distinct pathways toward these compounds demonstrates the convergent evolution

LanB: a class I lantipeptide dehydratase

LanC: a class I lantipeptide cyclase

LanM: a bifunctional class II lantipeptide synthetase containing dehydratase and LanC-like cyclase domains

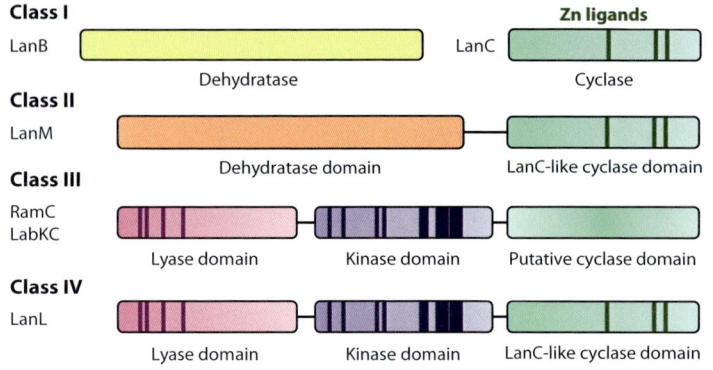

Figure 3

Schematic representation of the four classes of lanthionine-generating enzymes, highlighting conserved motifs. Abbreviations: LabKC, labionin synthetase; LanB, lantipeptide dehydratase; LanC, lantipeptide cyclase; LanL, class IV lantipeptide synthetase; LanM, class II lantipeptide synthetase; RamC, SapB-modifying synthetase.

LanL: a trifunctional class IV lantipeptide synthetase containing lyase, kinase, and cyclase domains

of efficient, ribosome-based biosynthetic strategies and highlights the potential evolutionary advantage of accessing high chemical diversity at low genetic cost (1, 41). In addition, natural combinatorial biosynthesis within one pathway can be achieved using diverse paralogous precursor peptides modified by a single, highly promiscuous synthetase, as demonstrated for the prochlorosins (42). These various means of facile structural diversification through precursor mutagenesis provide the producing organisms with mechanisms for relatively rapid adaptation to changes in their environments.

It was long believed that lantipeptides were produced only by a select group of gram-positive Firmicutes, predominantly species of *Lactococcus*, *Bacillus*, *Staphylococcus*, and *Streptococcus*, and by select strains of *Streptomyces* (1, 7). However, both genome analysis and isolation studies have recently demonstrated that this family is much more widely disseminated. For instance, the number of actinomycetes shown to produce lantipeptides has expanded dramatically. Furthermore, *in silico* searches demonstrate that lantipeptide biosynthetic genes are distributed well beyond firmicutes and actinomycetes: LanB- and LanC-like genes are present in bacteroidetes and chlamydiae (43), and LanM- and LanL-like genes are found in proteobacteria and cyanobacteria (15, 42, 44).

Additionally, a particular lantipeptide topology can be produced by different species, as nisin U from *Streptococcus uberis* represents the first nisin analog not produced by *Lactococcus lactis* (45). Homologs of lantipeptide-modifying enzymes are also involved in the biosynthesis of nonlantipeptide RNPs; for instance, LanB-like dehydratases are implicated in the formation of Dha/Dhb in the thiopeptides (46–49) and potentially in goadsporin (50). Given the rapid pace of genome sequencing, the discovery of novel lantipeptides is expected to continue its acceleration with the implementation of freely available, automated *in silico* genome-mining and annotation software, such as antiSMASH (51), BAGEL (52), and BACTIBASE (53). Indeed, a recent *in silico* study identified lantipeptide biosynthetic genes in 478 of 1,466 reference bacterial genomes (54), demonstrating how prevalent these pathways may be.

3. CLASS I LANTIPEPTIDES: DEDICATED DEHYDRATASE AND CYCLASE ENZYMES

3.1. Biosynthetic Machinery

In class I lantipeptide biosynthesis, two separate enzymes carry out the reactions forming the characteristic thioether cross-links: the dehydratase LanB and the cyclase LanC (**Figure 4**). The *lanB* genes encode proteins of about 1,000 residues, sharing no homology to any other known enzymes; only about 30% sequence identity exists across the family. As no direct experimental evidence has yet been reported, the molecular details of LanB catalysis are currently unknown; however, phosphorylation of the substrate serine/threonine residues is believed to be involved (55). The *lanC* genes encode proteins of about 400 residues and also have low overall sequence identity, 20%–30% across the family. In vitro reconstitution of the nisin cyclase NisC and solution of its X-ray crystal structure shed light on its mechanism, which is believed to involve an active-site zinc ion bound by a cysteine-cysteine-histidine triad, which activates a cysteinyl thiol for nucleophilic attack on Dha/Dhb (56). Mutagenesis of NisC and the subtilin cyclase SpaC revealed additional active-site residues critical to catalysis (57, 58), which may serve in general acid/base catalysis or electrophilic activation of the Michael acceptor.

LanT, a transmembrane ATP-binding cassette (ABC) transporter about 600 residues in length, serves to export the modified peptide from the producing cell, often with relaxed specificity. For instance, the nisin transporter, NisT, is able to export the unmodified precursor peptide NisA as well as unrelated peptide sequences fused to the NisA leader peptide (59); this substrate promiscuity has enabled the use of the nisin biosynthetic machinery for in vivo engineering of diverse peptides containing Lan/MeLan (see Section 10.1) (60). A dedicated subtilisin-like serine protease, LanP, is often, but not universally, encoded in class I biosynthetic gene clusters for removal of the leader peptide; these may be membrane anchored or cytoplasmic (5). The first in vitro reconstitution of a cytoplasmic LanP was recently reported for ElxP, the epilancin 15X protease (55).

Engineered nisin expression systems have greatly enhanced our understanding of class I biosynthesis. Previous coimmunoprecipitation and yeast two-hybrid studies had determined that nisin biosynthetic enzymes NisB, NisC, and NisT each localize predominantly to the membrane, that they each interact with NisA, and that they each maintain activity in the absence of the other enzymes in the complex (16). However, individual contributions of each enzyme and the order of reactions had not been rigorously assessed. A recent study reported that export of modified NisA by NisT occurs prior to proteolysis by the dedicated nisin protease NisP and that deletion of NisP has no effect on the production of modified NisA (61). In contrast, deletion of *nisB*, *nisC*, or *nisT* greatly decreases the overall kinetics of modified peptide production, suggesting synergistic activities. Additionally, the presence of a catalytically incompetent NisC significantly improves production levels of dehydrated NisA compared to levels observed in the absence of NisC (62).

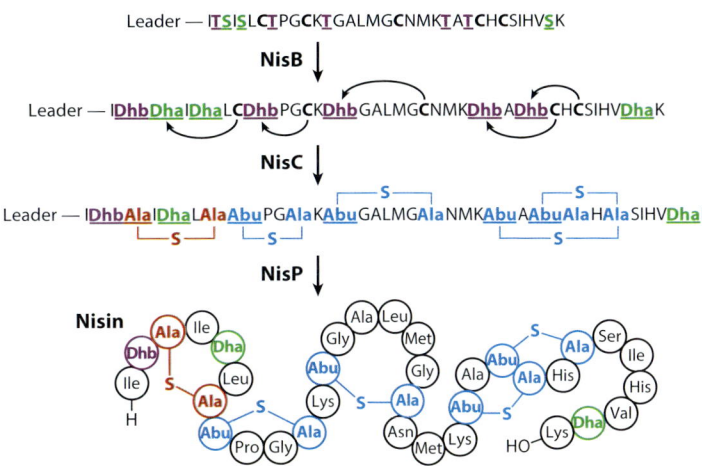

Figure 4

Posttranslational maturation of the class I lantipeptide nisin. Although not explicitly shown, current data suggest that dehydration and cyclization are alternating processes. Abbreviations: Abu, 2-aminobutyric acid; Dha, 2,3-didehydroalanine; Dhb, (Z)-2,3-didehydrobutyrine; NisB, nisin dehydratase; NisC, nisin cyclase; NisP, nisin protease.

Mutagenesis studies on NisA demonstrated that the dehydration of one residue has no influence on other dehydrations, but cross-link formation can protect serine/threonine residues from dehydration (62, 63).

Additional investigations have aided in understanding the interactions between NisA and NisB/C. Biophysical analysis of purified NisA and NisB revealed that NisB is dimeric in solution, that the leader peptide is necessary for interaction between NisA and NisB, and that NisB binds dehydrated NisA more strongly than either unmodified or fully modified (i.e., dehydrated and cyclized) NisA (64). These observations were supported by a more recent study in which NisB and NisC were co-purified from *L. lactis* using C-terminally His-tagged NisA (65). This pull-down system also revealed that NisB binds more strongly than NisC to unmodified NisA and that the NisA leader peptide alone was not sufficient to isolate the NisB/C complex, suggesting that these enzymes make important binding contacts with the core peptide as well as with the leader peptide.

From these data, a model has been constructed in which the NisA modification is

LanT:
a transmembrane ATP-binding cassette lantipeptide transporter; in class II systems, this protein contains an additional cysteine protease domain

ABC: ATP-binding cassette

directional and moves from the N to the C terminus, dehydration and cyclization are cooperative and alternating, and the presence but not necessarily activity of all three enzymes is required for efficient NisA modification (62). The proposed model further suggests that a membrane-localized, dimeric NisB/C/T complex recognizes NisA, which is channeled through the active sites of NisB/C by NisT with the substrate alternating between the two to install the PTMs. Finally, transport of the modified NisA across the membrane by NisT is followed by extracellular cleavage of the leader peptide by NisP.

3.2. Selected New Class I Lantipeptides

A noteworthy addition to class I is microbisporicin (**Figure 2**), a highly modified peptide from the actinomycete *Microbispora corallina*. This compound demonstrates high activity against drug-resistant gram-positive pathogens and also against several clinically relevant gram-negative species (66). Structural elucidation revealed one MeLan and three Lan residues, an aminovinylcysteine, and two unprecedented PTMs for lantipeptides: 5-chlorotryptophan and 3,4-dihydroxyproline. Its A and B rings strongly resemble the corresponding rings of nisin (**Figures 2** and **4**), which mediate antibacterial activity though binding of the essential peptidoglycan precursor lipid II (23). The high activity of this compound against drug-resistant pathogens was also demonstrated in vivo using animal models of severe infection (67).

The microbisporicin biosynthetic gene cluster has been identified (68) and contains both common class I machinery and enzymes involved in the unusual halogenation and hydroxylation (see Section 8). No dedicated LanP was found in the cluster, suggesting leader proteolysis may be performed by endogenous proteases. In addition, an extracytoplasmic function σ factor, MibX, and its anti-σ factor, MibW, were identified as part of an autoinducing, feed-forward pathway, which regulates microbisporicin biosynthesis and represents the first such regulatory system found for lantipeptides (69).

Another recently isolated lantipeptide, planosporicin, from a strain of the uncommon actinomycete *Planomonospora*, was discovered from a screen for peptidoglycan biosynthesis inhibitors (70). A unique ring topology involving five intertwined Lan/MeLan rings was initially presented, but subsequent structural revision (71) yielded a topology reminiscent of microbisporicin, including the N-terminal lipid II–binding motif. Given this revised structure, this peptide is tentatively assigned to class I, although further biosynthetic information is needed to confirm this assignment. The NMR structures of both planosporicin and microbisporicin have recently been solved (72), revealing similar but distinct solution conformations that may account for the higher bioactivity of microbisporicin.

4. CLASS II LANTIPEPTIDES: BIFUNCTIONAL LANTHIONINE SYNTHETASES

4.1. Biosynthetic Machinery

The dehydration and cyclization reactions in class II lantipeptide biosynthesis are carried out by a bifunctional synthetase, LanM (**Figure 5a**). These proteins are typically 900–1,200 residues in length and contain two domains: an N-terminal dehydratase domain, bearing no homology to LanB; and a C-terminal cyclase domain, having ~25% sequence identity to LanC, including conservation of the zinc-binding residues critical for NisC catalysis (57). Similar to LanB and LanC, LanM enzymes do not share high sequence identity with each other.

The successful in vitro reconstitution of the lacticin 481 synthetase, LctM, from *L. lactis* (73) has enabled detailed studies of Lan/MeLan formation during class II biosynthesis. LctM uses ATP and Mg^{2+} to phosphorylate serine/threonine residues in its substrate and subsequently eliminates the resulting phosphate ester to yield Dha/Dhb (**Figure 5b**) (74); cyclization

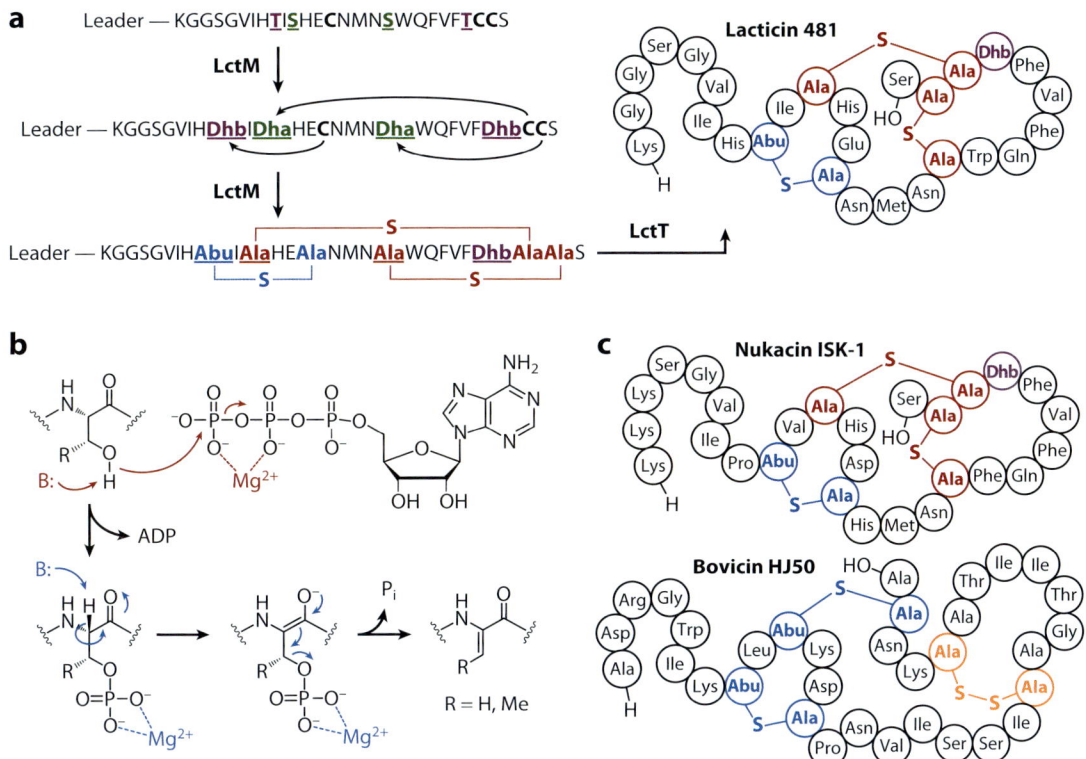

Figure 5

Features of class II lantipeptide biosynthesis. (*a*) Posttranslational maturation of lacticin 481. Although not explicitly shown, current data suggest that dehydration and cyclization are alternating processes. (*b*) Proposed mechanism of serine/threonine dehydration by the lacticin 481 synthetase, LctM. (*c*) Structures of related class II lantipeptides, nukacin ISK-1 and bovicin HJ50. Abbreviations: Abu, 2-aminobutyric acid; Dha, 2,3-didehydroalanine; Dhb, (Z)-2,3-didehydrobutyrine; LctT, lacticin 481 protease/transporter.

activity does not require ATP. LctM does not appear to contain any known ATP-binding domain, but conserved residues involved in phosphorylation have been identified through mutagenesis and resemble critical catalytic residues in canonical serine/threonine kinases (75). Of note, mutation of either arginine 399 or threonine 405 results in an enzyme with retained kinase but prohibited elimination activity, yielding buildup of phosphorylated intermediates. This observation has been exploited to engineer an LctM mutant as a general serine/threonine kinase, able to phosphorylate a variety of peptide substrates attached to the lacticin 481 leader peptide (76). The cyclization activity of the LanC-like domain of LctM was also investigated, and as with NisC, mutation of the conserved zinc-binding ligands abolished cyclase activity (77). However, these mutations had no effect on substrate dehydration. Tandem mass spectrometry showed that phosphorylation of LctA by the elimination-deficient mutant LctM-R399M is distributive, with the product of each phosphorylation being released back into solution, and largely directional, moving from the N to the C terminus (78). Such directionality was also shown for both dehydration and cyclization catalyzed by HalM2, one of the two synthetases involved in haloduracin biosynthesis (see Section 4.2) (78), which mirrors the directionality proposed for nisin biosynthesis (62). Additionally, LctM was shown to overcome the intrinsic cyclization preference for Dha over Dhb, illustrating the impressive regioselectivity of lantipeptide cyclases (79).

Heterologous expression of the structurally related nukacin ISK-1, a bacteriostatic lantibiotic (**Figure 5c**) from *Staphylococcus warneri* (80), has been used to examine in vivo features of class II biosynthesis. As for class I, yeast two-hybrid experiments support a multimeric complex of nukacin synthetase, NukM, and transporter NukT localized at the cell membrane, which may spatially connect processing and export of the nukacin precursor, NukA (81). Mutants of NukA coexpressed in vivo with NukM revealed that, as observed with LctM, the dehydration and cyclization activities of NukM are not coupled and that NukM first phosphorylates its substrate using ATP (82).

Transport and proteolysis in class II lantipeptides are also typically combined into a single LanT enzyme, about 100 residues longer than the class I LanT proteins owing to the presence of a conserved N-terminal, papain-like cysteine protease domain (5). The ABC transporter domain is not required for protease activity, as the isolated protease domain of the lacticin 481 transporter, LctT (residues 1–150), has been reconstituted in vitro (83). It shows relaxed specificity with respect to both the leader and core portions of its substrate, with the exception of the double-glycine-type cleavage site conserved in class II lantipeptides (83). The protease domain cleaves both unmodified and LctM-modified LctA, similar to the in vitro substrate selectivity of ElxP (55), but in contrast to the high in vivo substrate specificity of NisP, which cleaves only modified NisA (84). High selectivity was also demonstrated in vivo and in vitro with the full-length, bifunctional protease/transporter NukT involved in nukacin ISK-1 biosynthesis (85, 86), suggesting the ABC domain may play a role in the more restricted substrate recognition. Leader cleavage by NukT is dependent on ATP hydrolysis, and mutation of conserved residues in either domain of NukT leads to intracellular accumulation of unproteolyzed NukA, suggesting that the protease and ABC domains are cooperative. Proteolysis by NukT occurs on the cytoplasmic side of the membrane, prior to or concurrent with transport (85).

Substrate recognition by LanT enzymes has been proposed to involve recognition of an amphipathic α-helix (83, 87). This hypothesis has been supported by studies on ComA, a LanT-like transporter involved in cleavage and export of the quorum-sensing signal precursor ComC in *Streptococcus pneumoniae*. The leader peptide of ComC appears to transition from a random coil to a helical conformation upon binding to ComA (88). A crystal structure of ComC revealed the protease active site at the end of a narrow cleft, explaining the selectivity for the double-glycine cleavage site (89). Furthermore, this structure displayed a shallow hydrophobic surface proposed to interact with the α-helical ComC leader peptide. The residues making up this proposed binding site are conserved in many class II LanT proteins (89).

4.2. Selected New Class II Lantipeptides

In recent years, a variety of new class II lantipeptides have been introduced, which greatly expands the known spectrum of structures and producing organisms. Among these are two-component lantibiotics, a growing subgroup currently limited to class II, that possess potent and synergistic antimicrobial activity (90). Bioinformatic analyses by two independent groups led to the discovery of haloduracin from *Bacillus halodurans* C-125 (**Figure 2**), the first lantipeptide from an alkaliphilic species (91, 92). Its sequence, ring topology, and biosynthetic machinery bear similarity to other two-component systems, including lacticin 3147 and plantaricin W. The biosynthetic gene cluster contains separate LanM and immunity proteins for each lantipeptide, termed Halα and Halβ, but only a single LanT. Halα contains a disulfide linkage, but this bond is not critical for activity (93). Haloduracin possesses activity comparable to nisin but is more stable at physiological pH, improving its potential for clinical use (30).

Two-component lantibiotic: two synergistic peptides (termed the α- and β-peptides) are required for full activity

Additional genome searches for *lanM*-like genes revealed another two-component lantibiotic, lichenicidin, which was isolated from strains of *Bacillus licheniformis* and partially characterized in 2009 (44, 94). During the following year, purification and full structural characterization of both peptides were completed (95), which revealed that both peptides resemble their counterparts in lacticin 3147 and haloduracin. Each lichenicidin peptide undergoes a large number of dehydrations (7 for the α-peptide and 12 for the β-peptide), leading to four rings and an N-terminal 2-oxobutyryl cap in each peptide. The α-peptides of all characterized two-peptide systems contain the lipid II–binding motif found in mersacidin (**Figure 2**) (26, 90), but two different topologies have been reported for the A ring of lichenicidin-α (95, 96). In addition to a canonical class II protease/transporter LicT, the gene cluster also contains a serine protease LicP, necessary for removing an additional six residues from the N terminus of the β-peptide after leader cleavage by LicT to give the active species (96). Additional post-LanT proteolytic processing is also known for cytolysin (97), plantaricin W (98), and haloduracin (93).

Genome-mining efforts also identified a putative *lanM*-like gene, *procM*, in the genome of the marine cyanobacterium *Prochlorococcus marinus* MIT9313 (42). Intriguingly, 29 different *lanA*-like genes with highly conserved leader sequences, termed *procAs*, were found in the genome despite the presence of only a single *procM* gene. Heterologous expression of the prochlorosin synthetase, ProcM, and a panel of 17 precursor ProcAs revealed that all precursors tested were accepted by ProcM in vitro. The ring topologies of six mature lantipeptides, termed prochlorosins, were determined (**Figure 6**); all are unique compared to known lantipeptides. It is currently unknown how ProcM is able to catalyze generation of such a diversity of structures, or whether some of these thioether rings may be formed nonenzymatically. Although no biological role has yet been established for the prochlorosins, this work gives a striking example of efficient combinatorial biosynthesis via highly evolvable, ribosomally encoded substrates and an extremely promiscuous enzyme. The potential evolutionary benefit of such a system was demonstrated by bioinformatic analysis (42, 99), revealing large numbers of *procM*-like and *procA*-like genes in *Prochlorococcus*, *Synechococcus*, and other organisms.

Strains of the oral probiotic *Streptococcus salivarius* exert their antimicrobial effect through the production of several class II lantipeptides, which are predicted to share the ring topology of lacticin 481 and possess similar spectra of activity. Five different one- to three-residue variants of the first discovered *S. salivarius* lantipeptide, salivaricin A, have been isolated and detected in human saliva (100). Two more diverse sequences, salivaricin

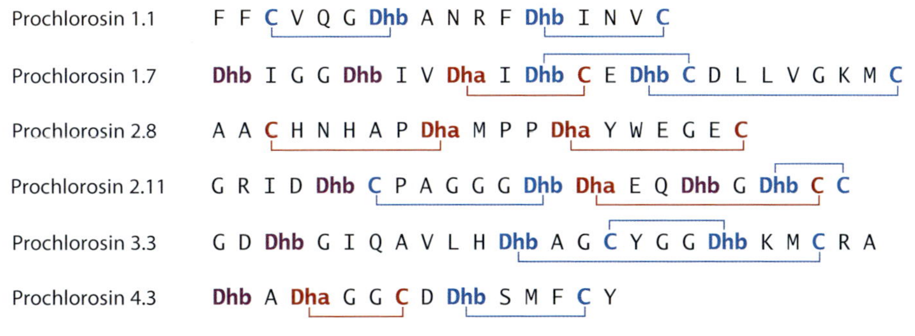

Prochlorosin 1.1	F F C V Q G Dhb A N R F Dhb I N V C
Prochlorosin 1.7	Dhb I G G Dhb I V Dha I Dhb C E Dhb C D L L V G K M C
Prochlorosin 2.8	A A C H N H A P Dha M P P Dha Y W E G E C
Prochlorosin 2.11	G R I D Dhb C P A G G Dhb Dha E Q Dhb G Dhb C C
Prochlorosin 3.3	G D Dhb G I Q A V L H Dhb A G C Y G G Dhb K M C R A
Prochlorosin 4.3	Dhb A Dha G G C D Dhb S M F C Y

Figure 6

Schematic representation of selected prochlorosin ring topologies. Abbreviations: Dha, 2,3-didehydroalanine; Dhb, (Z)-2,3-didehydrobutyrine.

LabKC:
a trifunctional class III lantipeptide synthetase involved in labyrinthopeptin biosynthesis containing lyase, kinase, and putative cyclase domains

B (101) and salivaricin 9 (102), have also been reported. Coproduction of salivaricin B and salivaricin A2 by *S. salivarius* K12 marks the first report of multiple lantipeptides produced by a single *Streptococcus* strain (101). Unlike the prochlorosins, however, a different LanM appears necessary for processing each salivaricin variant. The class II members bovicin HJ50 (**Figure 5c**) from *Streptococcus bovis* HJ50 (103) and thermophilin 1277 from the dairy bacterium *Streptococcus thermophilus* (104) also possess ring topologies resembling lacticin 481 but, interestingly, with a disulfide bond replacing a lanthionine (see Section 8).

5. CLASS III LANTIPEPTIDES: TRIFUNCTIONAL SYNTHETASES AND CARBOCYCLIC RINGS

In 2004, the morphogenetic peptide SapB from *Streptomyces coelicolor* was determined to contain two lanthionines (105). Unlike other lantipeptides known at the time, this peptide did not display antibiotic activity but instead promoted the growth of vegetative hyphae involved in streptomycete sporulation. Analysis of the biosynthetic gene cluster revealed a putative modifying enzyme, RamC, containing an N-terminal domain resembling serine/threonine protein kinases and a C-terminal domain with limited homology to the cyclase domain of LanM, but lacking the zinc-binding active-site residues (**Figure 3**). Because of these striking differences from class II synthetases, a third class of lantipeptide biosynthesis was established (6). More recently, a distinct domain within the proposed RamC kinase domain has been identified with homology to the OspF family of phosphothreonine lyases, suggesting that the enzyme forms Dha/Dhb via a stepwise phosphorylation/β-elimination pathway spanning two different domains (15, 40). No dedicated protease was found in the gene cluster. Subsequently, a second morphogenetic lantipeptide, SapT from *Streptomyces tendae*, was identified, as well as SapB-like gene clusters from other streptomycetes (38).

No further insights into class III lantipeptide biosynthesis were obtained until 2010 with the discovery of the labyrinthopeptins from the actinomycete *Actinomadura namibiensis* (39). A single trifunctional lantipeptide synthetase, LabKC, was identified, which bears significant homology to RamC and processes three similar precursor peptides found in the gene cluster. One of the modified products, labyrinthopeptin A2 (**Figure 2**), demonstrated efficacy against a mouse model of neuropathic pain, a first for lantipeptides. Despite the presence of a disulfide bond, no thiol-disulfide oxidoreductase was found in the gene cluster, which also lacked a dedicated protease but possessed two LanT-like transporters. Also unique to the labyrinthopeptins is the unprecedented carbocyclic residue labionin (**Figure 1c**), the structure of which was confirmed by X-ray crystallography (39). The in vitro reconstitution of LabKC demonstrated that GTP, not ATP, is required for kinase activity (106). A model of labionin formation has been presented in which the enolate resulting from the initial Lan-forming Michael addition is not protonated but instead performs a second Michael addition to yield the carbocycle (**Figure 1b**). It remains to be seen how LabKC catalyzes this double Michael addition without the conserved zinc-binding residues of LanM or whether any of these unique features of labyrinthopeptin biosynthesis are observed in related streptomycetes. Efforts toward the analytical detection of labionin from peptide hydrolysates via gas chromatography–mass spectrometry (107) should facilitate future discovery efforts.

6. CLASS IV LANTIPEPTIDES: TRIFUNCTIONAL LANTHIONINE SYNTHETASES

A fourth class of lantipeptides was established in 2010 with the discovery by genome analysis of a cryptic lantipeptide biosynthetic gene cluster in *Streptomyces venezuelae* (15). The identified synthetase, VenL, contains an N-terminal OspF-like lyase domain and a central serine/threonine kinase domain, both of which align well with the class III synthetase RamC; but unlike RamC, the C-terminal cyclase domain of VenL possesses the zinc-binding motif found in LanC and LanM (**Figure 3**). Similar to class III lantipeptides, a LanT-like transporter, but no dedicated protease or protease domain, was found in the gene cluster. The evolution of both class III and IV synthetases may have resulted from a fusion of serine/threonine kinases and phosphoserine/threonine lyases, which later adopted different cyclization strategies to enrich biologically active ring topologies. Interestingly, LanB and LanM do not contain any discernible kinase or lyase domains, suggesting entirely separate evolutionary histories from the class III and IV synthetases. A search for putative *lanB*, *lanM*, and *lanL* genes identified not only many firmicutes and actinomycetes, but also species from the phyla Bacteroidetes, Cyanobacteria, and Proteobacteria. Additionally, many species were found to contain more than one pathway, with *S. pneumoniae* containing all three (15).

Further insights into class IV biosynthesis were provided by heterologous expression of *venA* and *venL* in *Escherichia coli* (15). VenL was shown to require ATP and MgCl$_2$ for dehydration of VenA. Substrate mutagenesis allowed construction of a proposed ring topology of the putative lantipeptide, venezuelin, which was reminiscent of, but distinct from, the globular structure of cinnamycin. To test the proposed roles of each of the three domains of VenL, truncated enzymes missing one or more domains were expressed and purified. In vitro assays with VenA clearly confirmed phosphorylation activity of the kinase domain, β-elimination activity of the lyase domain, and cyclization activity of the cyclase domain. Further mutagenesis of the lyase domain revealed that some, but not all, of the residues conserved across the LanL, RamC/LabKC, and OspF families are important for catalysis (40).

7. ROLE OF THE LEADER PEPTIDE

The precise mechanism by which the leader peptide facilitates lantipeptide maturation is still enigmatic (2). Mutagenesis of conserved residues in the leader peptides of both class I and II lantipeptides has shown a high degree of plasticity (84, 87, 108–110). The conserved phenylalanine-asparagine-leucine-aspartate box of class I leader peptides is clearly important for efficient in vivo production but appears forgiving to many single substitutions with respect to LanB/C modification (84, 110). Similar observations were made in regard to the mutation of conserved residues of class II (87, 108, 109) and class III (111) leader peptides. A distinction between leader peptides of different classes is that the N-terminal residues are critical for modification in class I (84) and class III (111), whereas the C terminus appears more important for class II owing to the observation that the N terminus could be truncated from class II precursor peptides without loss of modification activity (73). Studies involving classes II and III have suggested that the modification enzymes may recognize an amphipathic, α-helical confirmation of the leader peptide (87, 88, 108, 111), but delineating the exact factors essential for recognition has been difficult. Crystallographic information on leader peptide binding is still lacking, possibly because these enzymes may be inherently flexible in order to permit their substantial substrate promiscuity. Additionally, leader peptide binding was shown not to be very tight in the few investigated cases (57, 64), which is likely to prevent product inhibition.

Although identification of exact binding interactions has been challenging, recognition of the leader peptide is clearly different for the synthetase and protease in both class I and II (83, 84, 108, 109), and thus different parts of the leader peptide may be involved in various steps of the overall modification process. In addition, studies of NisA (84), LctA (112), and the labyrinthopeptin precursor, LabA2 (111), suggest that a certain minimum distance is required between the serine/threonine residues to be dehydrated and the recognition site for binding to NisB, LctM, and LabKC, respectively. This model is similar to that proposed for the PTMs of the thiazole-containing

Tailoring PTM: a PTM other than the characteristic thioether cross-links Lan/MeLan and unsaturated residues Dha/Dhb

nonlantibiotic peptide, microcin B17 (113), where the first modified residue is separated from the end of the leader peptide by an essential spacer region proposed to allow the core region to reach the active site(s) for modification. Indeed, RNPs from different classes may utilize similar recognition mechanisms, as sequence homology observed between LanA leader peptides and those of several other RNP families may suggest a common evolutionary predecessor (2, 38, 99, 108). The first example of homology between LanA leader peptides and other proteins was recently reported with the identification of long leader peptides that have sequence homology with nitrile hydratase and the Nif-11 proteins, thus providing the first glimpses into a potential evolutionary origin (99).

The purported absolute requirement of the leader peptide for lantipeptide processing has been refuted in lacticin 481. LctM retains dehydration activity when the LctA leader and core portions are presented in *trans* or even in complete absence of the leader peptide (114). Interestingly, in *trans* assays resulted in nondirectional substrate dehydration, in contrast to the directional nature of LctM with its natural LctA substrate (78). These observations have led to a proposed model in which leader peptide binding is not strictly required for enzyme activation but instead shifts the equilibrium population from a low to a high concentration of the active enzyme. In contrast, recent work with the labyrinthopeptin synthetase, LabKC, has shown that no substrate modification occurs if the leader peptide is not directly attached to the core peptide (111). Therefore, the absolute necessity of the leader peptide for modification may differ between biosynthetic classes.

8. OTHER POSTTRANSLATIONAL TAILORING REACTIONS

Although Lan/MeLan residues define the lantipeptide family, many other tailoring PTMs are known (**Figure 2**), including β-hydroxylated aspartate and lysinoalanine (in cinnamycin), aminovinylcysteine (in mersacidin and others), D-alanine (in lacticin 3147 and lactocin S), and several α-ketoamide and α-hydroxyamide N-terminal caps (in Pep5, lactocin S, and others). Presumably, the genes responsible for these PTMs were recruited to the gene clusters of these molecules and imparted favorable properties; indeed, many of these functionalities are thought to improve peptide stability through protection from proteolysis. Oxidative decarboxylation of the C terminus by LanD enzymes during aminovinylcysteine formation protects the peptide from carboxypeptidases (115, 116), and masking the N terminus with acyl substituents (55) or disulfide formation (93) protects the peptide from aminopeptidases. The leader peptide appears less critical in these tailoring PTMs than for Lan/MeLan formation, as some enzymes are able to process leaderless precursor peptides, such as the aspartate hydroxylase CinX of cinnamycin (117) and the decarboxylase EpiD of epidermin (118). The importance of the leader peptide has not yet been determined for Cinorf7, a protein critical for forming the unique lysinoalanine cross-link of cinnamycin (117).

In recent years, new tailoring PTMs have been identified, which have further expanded the functional diversity of lantipeptide biosynthetic machinery. The class I lantipeptide microbisporicin (**Figure 2**) exhibits two new tailoring PTMs: tryptophan chlorination and proline hydroxylation. Gene cluster analysis (68) revealed a flavin-dependent tryptophan halogenase MibH and a hypothetical protein MibV; deletion of either resulted in production of deschloromicrobisporicin. A flavin reductase MibS was also identified, which likely acts to regenerate the MibH cofactor. The 3,4-dihydroxylation of proline is likely carried out by a cytochrome P450 homolog, MibO, encoded in the cluster.

Although α-ketoamide functionalities resulting from spontaneous enamine hydrolysis of N-terminal Dha/Dhb had been known, acetylation of the N terminus of mature lantipeptides had not been observed until the structural characterization of paenibacillin (119), isolated from

Paenibacillus polymyxa OSY-DF (120). Acetylation is postulated to be an extracellular tailoring step, as leader peptide cleavage necessary to reveal the modified amine occurs only during or after export from the cell. Similar to other N-terminal capping functionalities, this PTM likely protects the compound from aminopeptidases. Biosynthesis of the unusual D-lactate cap in epilancin 15X (**Figure 2**) has also been investigated (55). The NADPH-dependent oxidoreductase ElxO encoded in the biosynthetic gene cluster was reconstituted in vitro, demonstrating stereospecific reduction of N-terminal pyruvate in model substrates.

Another PTM uncommon among lantipeptides is thioether oxidation, found in the class II member actagardine A (**Figure 2**) from the actinomycete *Actinoplanes garbadinensis*. Even though sulfoxide formation had long been presumed to be a spontaneous side reaction during isolation, this moiety was recently demonstrated to be produced by a luciferase-like monooxygenase, GarO (121). Deletion of *garO* from an engineered producing strain resulted in the exclusive isolation of the nonoxidized product. Interestingly, deoxyactagardine B from *Actinoplanes liguriae* differs from actagardine A by only two residues, and its gene cluster encodes a putative GarO-like monooxygenase, but no sulfoxide formation has been observed in this compound (122).

Bovicin HJ50 contains a disulfide and two MeLan residues in a topology reminiscent of lacticin 481 (**Figure 5a**). A gene encoding a thiol/disulfide oxidoreductase was identified in the biosynthetic gene cluster, but as knockout did not affect the mass of the isolated product, its role remains unverified (123). Only two other known class II lantipeptides contain a disulfide, the α-peptides of haloduracin and plantaricin W, and in neither case does reduction of the bond affect activity. Instead, the disulfide in haloduracin α appears to protect the peptide from proteolysis (93). Although reduction of bovicin HJ50 with dithiothreitol also appears inconsequential for activity (124), mutation of one of the cysteines to alanine in BovA significantly reduces antimicrobial activity (103). Furthermore, heterologous expression of the bovicin gene cluster lacking the oxidoreductase yielded a peptide with a mass corresponding to the reduced species but activity comparable to the disulfide-containing natural product (123). Further studies are needed to better elucidate the role of the disulfide.

9. RECLASSIFICATION OF FORMER LANTIPEPTIDES

Two antimicrobial natural products previously included in the lantipeptide family have been recently reclassified. The structure of sublancin, a 37-residue peptide produced by *Bacillus subtilis* 168, was recently revised from a lantipeptide to a rare S-linked glycopeptide, featuring a cysteine modified with a β-linked glucose (125). Another S-linked glycopeptide has been independently reported, glycocin F from *Lactobacillus plantarum* (126).

A second reclassified peptide is cypemycin, an antibacterial product of *Streptomyces* sp. OH-4156. Owing to the presence of four Dhb residues and an aminovinylcysteine, this peptide had been considered a lantibiotic. However, the cypemycin biosynthetic gene cluster is not evolutionarily related to lantipeptides but instead marks the establishment of a new family of RNPs, the linaridins (127). Dhb residues as well as two other PTMs, N-terminal dimethylation and isoleucine isomerization, are installed by enzymes bearing no similarity to any known lantipeptide-modifying enzymes. In addition, the aminovinylcysteine is formed from two cysteine residues, as opposed to one serine and one cysteine as in lantipeptides. A second member of this family, grisemycin from *Streptomyces griseus*, has been recently isolated (128).

10. RECENT DEVELOPMENTS IN LANTIPEPTIDE ENGINEERING

Improvement in the pharmacological properties of lantipeptides, including potency, stability, and solubility, is needed to advance their therapeutic utility. A deeper understanding of their biosynthesis has accelerated the development of new in vivo and in vitro engineering strategies toward lantipeptide analogs. Several recent reviews have discussed the structure-activity relationships studied for particular lantipeptides (129–131); here, we seek to highlight recently developed methods that expand the scope and impact of lantipeptide engineering.

10.1. In Vivo Engineering in Producing Strains and Heterologous Hosts

Although *lanA* mutagenesis or complementation strategies to produce lantipeptide mutants in homologous and closely related heterologous hosts are known (132), only recently have these approaches yielded analogs with improved activities. A complementation approach combining random and site-saturation mutagenesis of *nisA* yielded nisin mutants in the "hinge region" between the C and D rings, and several of these mutants demonstrated improved activities against clinically relevant pathogens (133). Additionally, random mutagenesis of the A and B rings of nisin in *L. lactis* NZ9000 revealed several A-ring mutants with improved activity and also some that were active against the native producing strain (134). This latter observation illustrates a potential pitfall of in vivo engineering, the possibility that mutants with improved activity may be able to circumvent endogenous immunity mechanisms. One means to overcome potential toxicity is to prevent leader peptide removal within the producer (103, 135, 136).

A panel of studies has demonstrated the utility of in vivo expression of the nisin biosynthetic machinery to install Dha/Dhb and Lan/MeLan into nonlantipeptide "designer" peptides (60). In this system, two plasmids are expressed in *L. lactis* NZ9000, one containing *nisB/C/T* (or variants thereof) and the other containing a gene encoding the NisA leader peptide fused to a peptide of interest, separated by a commercial protease recognition site (84); the modified precursor can thus be isolated from the cell supernatant and cleaved in vitro. Both NisB (137–139) and NisT (59) have demonstrated broad substrate specificity, dehydrating and exporting a wide range of peptides fused to the NisA leader. Similarly, NisC catalyzes the cyclization of a variety of Dha/Dhb-containing, leader-fused peptides, including the regioselective cyclization of overlapping rings (140). This system was used to introduce Lan/MeLan residues in a variety of medicinally relevant peptides (139), including angiotensin-(1–7) (36) and lutenizing hormone release hormone (37), both of which demonstrated significantly improved proteolytic stability and thus enhanced therapeutic potential.

Engineering studies in class II lantipeptides have also been fruitful. Site-directed mutagenesis in the native lacticin 3147 producer *L. lactis* allowed production of a large panel of analogs (35, 141, 142), and one mutant was identified that displayed enhanced activity against certain indicator strains. Similarly, 250 nukacin ISK-1 variants were generated via site-saturation mutagenesis in a heterologous *L. lactis* expression system (143); several mutants with comparable activity to wild-type nukacin were produced in higher yields, and two possessed twofold higher specific activity than the wild type against *Lactobacillus sakei*. Variants of mersacidin were generated via complementation with an *mrsA* saturation mutagenesis library into the native *Bacillus* producer (144) and more recently via gene transfer into the closely related, naturally competent organism *Bacillus amyloliquefaciens* (145). Additionally, a large number of variants of actagardine was obtained via heterologous expression in *Streptomyces lividans* (121).

10.2. In Vivo Engineering in *Escherichia coli*

A promising approach to in vivo lantipeptide engineering is production in *E. coli*, which simplifies and generalizes genetic manipulation. In 2005, the first report of lantipeptide biosynthesis in *E. coli* described production of a truncated nukacin ISK-1 through coexpression of His-tagged *nukA* and *nukM* on a single vector, followed by purification and leader peptide removal with LysC (146, 147). This coexpression system has since been used to produce NukA mutants, one of which contained an additional MeLan residue (82). A 2011 study demonstrated the versatility of *E. coli* production of modified LanAs (135). Four prochlorosin and both haloduracin precursor peptides were fully modified by coexpression of His-tagged *lanA* with the corresponding *lanM*. This study also represents the only report to date of class I lantipeptide production in *E. coli*, as coexpression of His-tagged *nisA* with *nisB* on one vector and *nisC* on a second vector produced the fully modified nisin precursor peptide. Insertion of commercial protease cleavage sites between the LanA leader and core portions allowed production of fully active haloduracin and nisin after in vitro cleavage. This approach also permitted the use of amber stop-codon suppression technology to introduce nonproteinogenic amino acids in LanA peptides. Cinnamycin biosynthesis has also been reconstituted in *E. coli* through the heterologous expression of *cinA*, *cinM*, and *cinX* involved in aspartase hydroxylation, and *cinorf7* involved in lysinoalanine formation (117).

Two other *E. coli* expression systems reported in 2011 have utilized dedicated lantipeptide proteases to produce the mature lantipeptide. The bovicin HJ50 studies described in Section 8 were made possible by coexpression of His-tagged *bovA* and *bovM* in *E. coli*, followed by leader peptide cleavage with an in vitro reconstituted BovT protease domain (103). This system produced a fourfold improvement in yield compared to the producing strain and allowed for deduction of the bovicin HJ50 ring topology via mutation of the four cysteine residues. A second study reported the first expression of an entire lantipeptide biosynthetic pathway in *E. coli*, the two-component lantibiotic lichenicidin, which resulted in extraction of the active species from cell supernatant (96). The transporter LicT was critical for leader peptide removal from both precursor peptides and may interact with the endogenous type I secretion system to export lichenicidin across both membranes of the gram-negative host. Further study revealed that five genes involved in lichenicidin producer immunity are not essential to production in *E. coli* (148). Collectively, these studies validate coexpression in *E. coli* as a general and versatile method to produce lantipeptides and analogs.

10.3. In Vitro Engineering

The use of in vitro reconstituted biosynthetic enzymes to synthesize lantipeptides avoids several of the potential drawbacks of in vivo engineering, such as complications with export or immunity. Haloduracin mutants, generated by in vitro reconstitution of the synthetases HalM1 and HalM2, enabled structure-activity analysis (93) and mode-of-action studies (30). The recent in vitro reconstitution of the cinnamycin synthetase, CinM, and hydroxylase CinX may permit similar investigations for cinnamycin (117).

In vitro approaches using synthetic substrates have allowed inclusion of nonproteinogenic amino acids in lantipeptides. LctM-modified LctA mutants have been constructed that contain peptoids, D- or β-amino acids (112, 149–151), and synthetic threonine analogs (152). Recently, synthetic lacticin 481 analogs were produced by 1,3-dipolar cycloaddition of an alkyne-functionalized LctA leader peptide and azide-functionalized core peptide mutants constructed by solid-phase peptide synthesis (SPPS); then, these substrates were treated with reconstituted LctM and subsequently a commercial protease to remove the leader peptide (**Figure 7a**). Of a panel of 11 lacticin 481 analogs containing nonproteinogenic residues produced in this way, two were shown to possess improved activity (149). This methodology was also used to prepare nonlantipeptide sequences containing thioether cross-links or Dha (151). A drawback of this approach is the proteolysis step, which can be plagued by low yields and requires optimization for each peptide (84, 135). To address these issues, a photocleavable linker incorporated into substrate analogs via SPPS has been developed that enables a general, clean cleavage process (153).

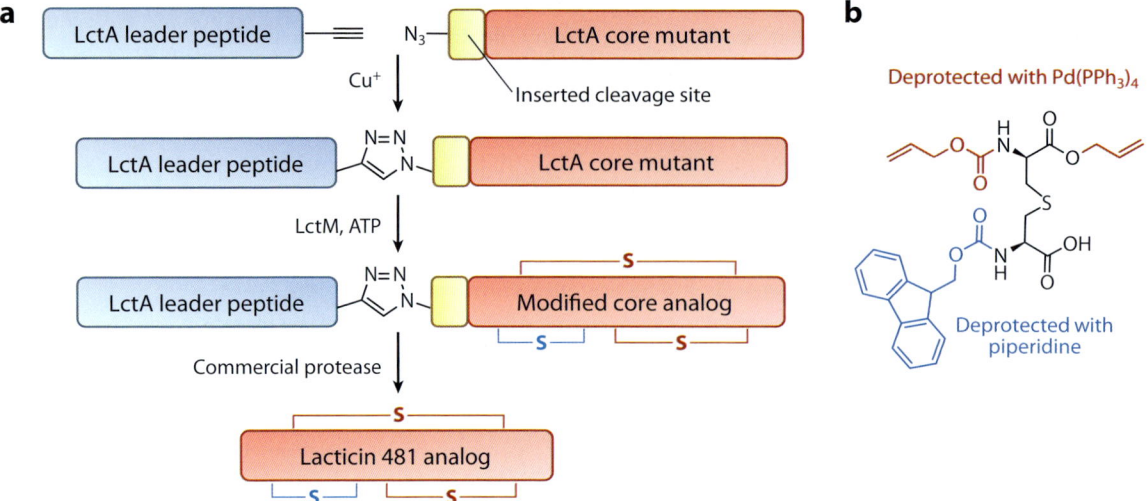

Figure 7

Approaches to in vitro lantipeptide engineering. (*a*) Preparation of lacticin 481 analogs from ligation of synthetic fragments, modification with lacticin 481 synthetase, LctM, and proteolysis. (*b*) Protected Lan amino acid used in the chemical synthesis of lantipeptides, highlighting orthogonal deprotection conditions. Abbreviations: LctA, lacticin 481 precursor peptide; Pd(PPh$_3$)$_4$, tetrakis(triphenylphosphine)palladium(0).

An alternative in vitro biosynthetic scheme was reported in 2007 using a commercial, cell-free rapid translation system derived from *E. coli* lysate (154). Linear constructs of *nisA*, *nisB*, and *nisC* were added to the rapid translation system mixture, resulting in active nisin after cleavage of the leader peptide with trypsin. The presence of authentic nisin was supported with a nisin A antibody blot and a nisin-inducible green fluorescent protein assay, although no purification or characterization of the mature lantibiotic was carried out. Cell-free translation systems that incorporate Dha/Dhb and Lan/MeLan into peptides through chemical transformations of vinylglycine (155) or 4-selenoamino acids (156) have also been reported, although regio- and stereochemical control of cyclization is limited in the absence of a cyclase enzyme.

10.4. Chemical Synthesis

In addition to bioengineering, chemical synthesis of lantipeptides and analogs has advanced greatly in recent years (157). Prior to 2008, the only successful total chemical synthesis of a lantipeptide was the impressive solution-phase synthesis of nisin (158). The use of an orthogonally protected Lan amino acid (**Figure 7b**) together with general advances in SPPS has allowed the solid-supported total synthesis of lacticin S (159) as well as analogs of the lacticin 3147 β-peptide, including MeLan replaced with Lan (160) and thioethers replaced with ethers (161) or olefins (162). This SPPS-amenable methodology has been expanded to the synthesis of cystathionine cross-links as mimics of the disulfide bond in bioactive peptides (163). Recently, the first solid-supported total synthesis of a lantipeptide containing overlapping cross-links, the α-peptide of lacticin 3147, has been reported (164), which expands the complexity of lantipeptide structures amenable to chemical synthesis.

11. OUTLOOK

Research on lantipeptide biosynthesis has advanced at a remarkable pace in recent years, resulting in the discovery of new classes of biosynthetic machinery, new PTMs, and large numbers of lantipeptide-encoding gene clusters. Although some of these clusters have been investigated, the vast majority awaits exploration and may hold valuable lead structures for therapeutic development. Understanding of lantipeptide biosynthesis has also progressed, but detailed structural insights regarding substrate binding and enzyme complex formation are still needed. Nevertheless, engineering strategies continue to increase in sophistication and scale, capitalizing on the remarkable substrate tolerance of the cross-link-forming enzymes. Engineering efforts using the enzymes that install tailoring PTMs have thus far not been conducted and provide a potential area for further expansion. In addition, some of the biosynthetic enzymes found in other classes of RNPs may be used to install further structural diversity or fashion hybrid molecules. Given the growing interest in stabilizing peptides for pharmaceutical use, and the potential for low-cost production of such peptides by fermentation, the recent progress in this area of research holds tantalizing promise.

SUMMARY POINTS

1. Genome-mining and isolation studies have revealed a much wider diversity of lantipeptide structures, PTMs, biological activities and producer organisms than previously known. Producers now include cyanobacteria and a wide variety of actinomycetes.

2. Four distinct classes of biosynthetic enzymes that install Lan/MeLan are currently recognized, highlighting the convergent evolution of a favorable and adaptable strategy to produce high chemical diversity at low genetic cost.

3. Investigation of enzymes involved in Lan/MeLan installation has revealed (*a*) directionality and coordination of the dehydration and cyclization reactions, (*b*) the requirements of the leader peptide in modifications, and (*c*) possible models of membrane-associated, multimeric biosynthetic complexes. Additionally, in vitro reconstitution of LanC, LanM, LanL, and LabKC enzymes has given some insight into their mechanisms of catalysis, including necessary cofactors and critical catalytic residues.

4. Advances in bioengineering and chemical synthesis have led to production of panels of lantipeptide analogs; some of these compounds possess improved therapeutic properties compared to the parent peptide. Heterologous expression in *E. coli* holds particular promise for furthering these endeavors.

FUTURE ISSUES

1. Genome-mining strategies have identified hundreds of bacterial strains containing putative lantipeptide biosynthetic enzymes but that are not yet known to produce lantipeptides. Isolation efforts from these organisms or heterologous expression of the identified gene clusters may yield novel structures and activities.

2. Detailed structural and biophysical studies are needed to refine current models of biosynthetic enzyme complexes, enzyme-substrate interactions, and the role of the leader peptide. Additionally, mechanisms of LanB dehydration and LabKC labionin formation are currently unknown and warrant investigation.

3. Mechanistic study and engineering of the enzymes involved in tailoring reactions would shed light on the roles of these PTMs in activity and stability.

4. The clinical use of lantipeptides depends on improvements in pharmacological properties. Further advances of in vivo and in vitro engineering systems may enable production of such analogs, as well as other cyclic bioactive peptides. Additionally, structure-activity relationship studies made possible by engineering efforts may shed light on the modes of action of these compounds.

DISCLOSURE STATEMENT

The authors are not aware of any affiliations, memberships, funding, or financial holdings that might be perceived as affecting the objectivity of this review.

ACKNOWLEDGMENTS

This work was supported by the Howard Hughes Medical Institute (to W.A.V.), the National Institutes of Health grant GM58822 (to W.A.V.), the Robert C. and Carolyn J. Springborn Endowment (to P.J.K.), and the American Heart Association Midwest Affiliate Predoctoral Fellowship (to P.J.K.).

LITERATURE CITED

1. Velásquez JE, van der Donk WA. 2011. Genome mining for ribosomally synthesized natural products. *Curr. Opin. Chem. Biol.* 15:11–21
2. Oman TJ, van der Donk WA. 2010. Follow the leader: the use of leader peptides to guide natural product biosynthesis. *Nat. Chem. Biol.* 6:9–18
3. McIntosh JA, Donia MS, Schmidt EW. 2009. Ribosomal peptide natural products: bridging the ribosomal and nonribosomal worlds. *Nat. Prod. Rep.* 26:537–59
4. Nolan EM, Walsh CT. 2009. How nature morphs peptide scaffolds into antibiotics. *ChemBioChem* 10:34–53
5. Chatterjee C, Paul M, Xie L, van der Donk WA. 2005. Biosynthesis and mode of action of lantibiotics. *Chem. Rev.* 105:633–84
6. Willey JM, van der Donk WA. 2007. Lantibiotics: peptides of diverse structure and function. *Annu. Rev. Microbiol.* 61:477–501
7. Bierbaum G, Sahl HG. 2009. Lantibiotics: mode of action, biosynthesis and bioengineering. *Curr. Pharm. Biotechnol.* 10:2–18
8. Piper C, Cotter PD, Ross PR, Hill C. 2009. Discovery of medically significant lantibiotics. *Curr. Drug Discov. Technol.* 6:1–18
9. Cotter PD, Hill C, Ross RP. 2005. Bacteriocins: developing innate immunity for food. *Nat. Rev. Microbiol.* 3:777–88
10. Smith L, Hillman JD. 2008. Therapeutic potential of type A (I) lantibiotics, a group of cationic peptide antibiotics. *Curr. Opin. Microbiol.* 11:401–8
11. Asaduzzaman SM, Sonomoto K. 2009. Lantibiotics: diverse activities and unique modes of action. *J. Biosci. Bioeng.* 107:475–87
12. Breukink E, de Kruijff B. 2006. Lipid II as a target for antibiotics. *Nat. Rev. Drug Discov.* 5:321–23
13. Cooper LE, Li B, van der Donk WA. 2009. Biosynthesis and mode of action of lantibiotics. In *Comprehensive Natural Products II: Chemistry and Biology*, ed. L Mander, H-w Liu, 5:217–56. Oxford: Elsevier
14. Draper LA, Ross RP, Hill C, Cotter PD. 2008. Lantibiotic immunity. *Curr. Protein Pept. Sci.* 9:39–49
15. Goto Y, Li B, Claesen J, Shi Y, Bibb MJ, van der Donk WA. 2010. Discovery of unique lanthionine synthetases reveals new mechanistic and evolutionary insights. *PLoS Biol.* 8:e1000339

16. Lubelski J, Rink R, Khusainov R, Moll G, Kuipers O. 2008. Biosynthesis, immunity, regulation, mode of action and engineering of the model lantibiotic nisin. *Cell. Mol. Life Sci.* 65:455–76
17. Grasemann H, Stehling F, Brunar H, Widmann R, Laliberte TW, et al. 2007. Inhalation of Moli1901 in patients with cystic fibrosis. *Chest* 131:1461–66
18. Sedgwick T, Dawson MJ. 2009. Novacta targets *C. difficile* the natural way. *MedNous* May 6:8–9
19. Donadio S, Maffioli S, Monciardini P, Sosio M, Jabés D. 2010. Antibiotic discovery in the twenty-first century: current trends and future perspectives. *J. Antibiot.* 63:423–30
20. Ghobrial O, Derendorf H, Hillman JD. 2010. Pharmacokinetic and pharmacodynamic evaluation of the lantibiotic MU1140. *J. Pharm. Sci.* 99:2521–28
21. Zhao M, Li Z, Bugenhagen S. 2008. ^{99m}Tc-labeled duramycin as a novel phosphatidylethanolamine-binding molecular probe. *J. Nucl. Med.* 49:1345–52
22. Hasper HE, Kramer NE, Smith JL, Hillman JD, Zachariah C, et al. 2006. An alternative bactericidal mechanism of action for lantibiotic peptides that target lipid II. *Science* 313:636–37
23. Hsu S-TD, Breukink E, Tischenko E, Lutters MAG, de Kruijff B, et al. 2004. The nisin-lipid II complex reveals a pyrophosphate cage that provides a blueprint for novel antibiotics. *Nat. Struct. Mol. Biol.* 11:963–67
24. Hasper HE, de Kruijff B, Breukink E. 2004. Assembly and stability of nisin-lipid II pores. *Biochemistry* 43:11567–75
25. Böttiger T, Schneider T, Martinez B, Sahl HG, Wiedemann I. 2009. Influence of Ca^{2+} ions on the activity of lantibiotics containing a mersacidin-like lipid II binding motif. *Appl. Environ. Microbiol.* 75:4427–34
26. Hsu S-TD, Breukink E, Bierbaum G, Sahl HG, de Kruijff B, et al. 2003. NMR study of mersacidin and lipid II interaction in dodecylphosphocholine micelles: Conformational changes are a key to antimicrobial activity. *J. Biol. Chem.* 278:13110–17
27. Dufour A, Hindré T, Haras D, Le Pennec J-P. 2007. The biology of lantibiotics from the lacticin 481 group is coming of age. *FEMS Microbiol. Rev.* 31:134–67
28. Hsu S-TD, Breukink E, Kaptein R. 2006. Structural motifs of lipid II-binding lantibiotics as a blueprint for novel antibiotics. *Anti-Infect. Agents Med. Chem.* 5:245–54
29. Wiedemann I, Böttiger T, Bonelli RR, Wiese A, Hagge SO, et al. 2006. The mode of action of the lantibiotic lacticin 3147—a complex mechanism involving specific interaction of two peptides and the cell wall precursor lipid II. *Mol. Microbiol.* 61:285–96
30. Oman TJ, van der Donk WA. 2009. Insights into the mode of action of the two-peptide lantibiotic haloduracin. *ACS Chem. Biol.* 4:865–74
31. Oman TJ, Lupoli TJ, Wang T-SA, Kahne D, Walker S, van der Donk WA. 2011. Haloduracin α binds the peptidoglycan precursor lipid II with 2:1 stoichiometry. *J. Am. Chem. Soc.* 133:17544–47
32. Brötz H, Josten M, Wiedemann I, Schneider U, Götz F, et al. 1998. Role of lipid-bound peptidoglycan precursors in the formation of pores by nisin, epidermin and other lantibiotics. *Mol. Microbiol.* 30:317–27
33. Bierbaum G, Szekat C, Josten M, Heidrich C, Kempter C, et al. 1996. Engineering of a novel thioether bridge and role of modified residues in the lantibiotic Pep5. *Appl. Environ. Microbiol.* 62:385–92
34. Lubelski J, Overkamp W, Kluskens LD, Moll GN, Kuipers OP. 2008. Influence of shifting positions of Ser, Thr, and Cys residues in prenisin on the efficiency of modification reactions and on the antimicrobial activities of the modified prepeptides. *Appl. Environ. Microbiol.* 74:4680–85
35. Suda S, Westerbeek A, O'Connor PM, Ross RP, Hill C, Cotter PD. 2010. Effect of bioengineering lacticin 3147 lanthionine bridges on specific activity and resistance to heat and proteases. *Chem. Biol.* 17:1151–60
36. Kluskens LD, Nelemans SA, Rink R, de Vries L, Meter-Arkema A, et al. 2009. Angiotensin-(1–7) with thioether bridge: an angiotensin-converting enzyme-resistant, potent angiotensin-(1–7) analog. *J. Pharmacol. Exp. Ther.* 328:849–54
37. Rink R, Arkema-Meter A, Baudoin I, Post E, Kuipers A, et al. 2010. To protect peptide pharmaceuticals against peptidases. *J. Pharmacol. Toxicol. Methods* 61:210–18
38. Willey JM, Gaskell AA. 2011. Morphogenetic signaling molecules of the streptomycetes. *Chem. Rev.* 111:174–87
39. Meindl K, Schmiederer T, Schneider K, Reicke A, Butz D, et al. 2010. Labyrinthopeptins: a new class of carbacyclic lantibiotics. *Angew. Chem. Int. Ed.* 49:1151–54

40. Goto Y, Ökesli A, van der Donk WA. 2011. Mechanistic studies of Ser/Thr dehydration catalyzed by a member of the LanL lanthionine synthetase family. *Biochemistry* 50:891–98
41. Fischbach MA, Clardy J. 2007. One pathway, many products. *Nat. Chem. Biol.* 3:353–55
42. Li B, Sher D, Kelly L, Shi Y, Huang K, et al. 2010. Catalytic promiscuity in the biosynthesis of cyclic peptide secondary metabolites in planktonic marine cyanobacteria. *Proc. Natl. Acad. Sci. USA* 107:10430–35
43. Marsh A, O'Sullivan O, Ross RP, Cotter P, Hill C. 2010. *In silico* analysis highlights the frequency and diversity of type 1 lantibiotic gene clusters in genome sequenced bacteria. *BMC Genomics* 11:679
44. Begley M, Cotter PD, Hill C, Ross RP. 2009. Identification of a novel two-peptide lantibiotic, lichenicidin, following rational genome mining for LanM proteins. *Appl. Environ. Microbiol.* 75:5451–60
45. Wirawan RE, Klesse NA, Jack RW, Tagg JR. 2006. Molecular and genetic characterization of a novel nisin variant produced by *Streptococcus uberis*. *Appl. Environ. Microbiol.* 72:1148–56
46. Wieland Brown LC, Acker MG, Clardy J, Walsh CT, Fischbach MA. 2009. Thirteen posttranslational modifications convert a 14-residue peptide into the antibiotic thiocillin. *Proc. Natl. Acad. Sci. USA* 106:2549–53
47. Kelly WL, Pan L, Li C. 2009. Thiostrepton biosynthesis: prototype for a new family of bacteriocins. *J. Am. Chem. Soc.* 131:4327–34
48. Liao R, Duan L, Lei C, Pan H, Ding Y, et al. 2009. Thiopeptide biosynthesis featuring ribosomally synthesized precursor peptides and conserved posttranslational modifications. *Chem. Biol.* 16:141–47
49. Morris RP, Leeds JA, Naegeli HU, Oberer L, Memmert K, et al. 2009. Ribosomally synthesized thiopeptide antibiotics targeting elongation factor Tu. *J. Am. Chem. Soc.* 131:5946–55
50. Onaka H, Nakaho M, Hayashi K, Igarashi Y, Furumai T. 2005. Cloning and characterization of the goadsporin biosynthetic gene cluster from *Streptomyces* sp. TP-A0584. *Microbiology* 151:3923–33
51. Medema MH, Blin K, Cimermancic P, de Jager V, Zakrzewski P, et al. 2011. antiSMASH: rapid identification, annotation and analysis of secondary metabolite biosynthesis gene clusters in bacterial and fungal genome sequences. *Nucleic Acids Res.* 39:W339–46
52. de Jong A, van Heel AJ, Kok J, Kuipers OP. 2010. BAGEL2: mining for bacteriocins in genomic data. *Nucleic Acids Res.* 38:W647–51
53. Hammami R, Zouhir A, Le Lay C, Ben Hamida J, Fliss I. 2010. BACTIBASE second release: a database and tool platform for bacteriocin characterization. *BMC Microbiol.* 10:22
54. Haft DH, Basu MK. 2011. Biological systems discovery *in silico*: radical *S*-adenosylmethionine protein families and their target peptides for posttranslational modification. *J. Bacteriol.* 193:2745–55
55. Velásquez JE, Zhang X, van der Donk WA. 2011. Biosynthesis of the antimicrobial peptide epilancin 15X and its N-terminal lactate. *Chem. Biol.* 18:857–67
56. Li B, Yu JPJ, Brunzelle JS, Moll GN, van der Donk WA, Nair SK. 2006. Structure and mechanism of the lantibiotic cyclase involved in nisin biosynthesis. *Science* 311:1464–67
57. Li B, van der Donk WA. 2007. Identification of essential catalytic residues of the cyclase NisC involved in the biosynthesis of nisin. *J. Biol. Chem.* 282:21169–75
58. Helfrich M, Entian K-D, Stein T. 2007. Structure-function relationships of the lanthionine cyclase SpaC involved in biosynthesis of the *Bacillus subtilis* peptide antibiotic subtilin. *Biochemistry* 46:3224–33
59. Kuipers A, de Boef E, Rink R, Fekken S, Kluskens LD, et al. 2004. NisT, the transporter of the lantibiotic nisin, can transport fully modified, dehydrated, and unmodified prenisin and fusions of the leader peptide with non-lantibiotic peptides. *J. Biol. Chem.* 279:22176–82
60. Moll GN, Kuipers A, Rink R. 2010. Microbial engineering of dehydro-amino acids and lanthionines in non-lantibiotic peptides. *Antonie van Leeuwenhoek* 97:319–33
61. van den Berg van Saparoea HB, Bakkes PJ, Moll GN, Driessen AJM. 2008. Distinct contributions of the nisin biosynthesis enzymes NisB and NisC and transporter NisT to prenisin production by *Lactococcus lactis*. *Appl. Environ. Microbiol.* 74:5541–48
62. Lubelski J, Khusainov R, Kuipers OP. 2009. Directionality and coordination of dehydration and ring formation during biosynthesis of the lantibiotic nisin. *J. Biol. Chem.* 284:25962–72
63. Kuipers A, Meijer-Wierenga J, Rink R, Kluskens LD, Moll GN. 2008. Mechanistic dissection of the enzyme complexes involved in biosynthesis of lacticin 3147 and nisin. *Appl. Environ. Microbiol.* 74:6591–97

64. Mavaro A, Abts A, Bakkes PJ, Moll GN, Driessen AJM, et al. 2011. Substrate recognition and specificity of NisB, the lantibiotic dehydratase involved in nisin biosynthesis. *J. Biol. Chem.* 286:30552–60
65. Khusainov R, Heils R, Lubelski J, Moll GN, Kuipers OP. 2011. Determining sites of interaction between prenisin and its modification enzymes NisB and NisC. *Mol. Microbiol.* 82:706–18
66. Castiglione F, Lazzarini A, Carrano L, Corti E, Ciciliato I, et al. 2008. Determining the structure and mode of action of microbisporicin, a potent lantibiotic active against multiresistant pathogens. *Chem. Biol.* 15:22–31
67. Jabés D, Brunati C, Candiani G, Riva S, Romanó G, Donadio S. 2011. Efficacy of the new lantibiotic NAI-107 in experimental infections induced by multidrug-resistant gram-positive pathogens. *Antimicrob. Agents Chemother.* 55:1671–76
68. Foulston LC, Bibb MJ. 2010. Microbisporicin gene cluster reveals unusual features of lantibiotic biosynthesis in actinomycetes. *Proc. Natl. Acad. Sci. USA* 107:13461–66
69. Foulston L, Bibb M. 2011. Feed-forward regulation of microbisporicin biosynthesis in *Microbispora corallina*. *J. Bacteriol.* 193:3064–71
70. Castiglione F, Cavaletti L, Losi D, Lazzarini A, Carrano L, et al. 2007. A novel lantibiotic acting on bacterial cell wall synthesis produced by the uncommon actinomycete *Planomonospora* sp. *Biochemistry* 46:5884–95
71. Maffioli SI, Potenza D, Vasile F, De Matteo M, Sosio M, et al. 2009. Structure revision of the lantibiotic 97518. *J. Nat. Prod.* 72:605–7
72. Vasile F, Potenza D, Marsiglia B, Maffioli S, Donadio S. 2011. Solution structure by nuclear magnetic resonance of the two lantibiotics 97518 and NAI-107. *J. Pept. Sci.* 18:129–34
73. Xie L, Miller LM, Chatterjee C, Averin O, Kelleher NL, van der Donk WA. 2004. Lacticin 481: In vitro reconstitution of lantibiotic synthetase activity. *Science* 303:679–81
74. Chatterjee C, Miller LM, Leung YL, Xie L, Yi M, et al. 2005. Lacticin 481 synthetase phosphorylates its substrate during lantibiotic production. *J. Am. Chem. Soc.* 127:15332–33
75. You YO, van der Donk WA. 2007. Mechanistic investigations of the dehydration reaction of lacticin 481 synthetase using site-directed mutagenesis. *Biochemistry* 46:5991–6000
76. You YO, Levengood MR, Furgerson Ihnken LA, Knowlton AK, van der Donk WA. 2009. Lacticin 481 synthetase as a general serine/threonine kinase. *ACS Chem. Biol.* 4:379–85
77. Paul M, Patton GC, van der Donk WA. 2007. Mutants of the zinc ligands of lacticin 481 synthetase retain dehydration activity but have impaired cyclization activity. *Biochemistry* 46:6268–76
78. Lee MV, Furgerson Ihnken LA, You YO, McClerren AL, van der Donk WA, Kelleher NL. 2009. Distributive and directional behavior of lantibiotic synthetases revealed by high-resolution tandem mass spectrometry. *J. Am. Chem. Soc.* 131:12258–64
79. Zhang X, Ni W, van der Donk WA. 2007. On the regioselectivity of thioether formation by lacticin 481 synthetase. *Org. Lett.* 9:3343–46
80. Asaduzzaman SM, Nagao J, Iida H, Zendo T, Nakayama J, Sonomoto K. 2009. Nukacin ISK-1, a bacteriostatic lantibiotic. *Antimicrob. Agents Chemother.* 53:3595–98
81. Nagao J, Aso Y, Sashihara T, Shioya K, Adachi A, et al. 2005. Localization and interaction of the biosynthetic proteins for the lantibiotic, nukacin ISK-1. *Biosci. Biotechnol. Biochem.* 69:1341–47
82. Shioya K, Harada Y, Nagao J, Nakayama J, Sonomoto K. 2010. Characterization of modification enzyme NukM and engineering of a novel thioether bridge in lantibiotic nukacin ISK-1. *Appl. Microbiol. Biotechnol.* 86:891–99
83. Furgerson Ihnken LA, Chatterjee C, van der Donk WA. 2008. In vitro reconstitution and substrate specificity of a lantibiotic protease. *Biochemistry* 47:7352–63
84. Plat A, Kluskens LD, Kuipers A, Rink R, Moll GN. 2011. Requirements of the engineered leader peptide of nisin for inducing modification, export, and cleavage. *Appl. Environ. Microbiol.* 77:604–11
85. Nishie M, Sasaki M, Nagao J, Zendo T, Nakayama J, Sonomoto K. 2011. Lantibiotic transporter requires cooperative functioning of the peptidase domain and the ATP binding domain. *J. Biol. Chem.* 286:11163–69
86. Nishie M, Shioya K, Nagao J, Jikuya H, Sonomoto K. 2009. ATP-dependent leader peptide cleavage by NukT, a bifunctional ABC transporter, during lantibiotic biosynthesis. *J. Biosci. Bioeng.* 108:460–64

87. Nagao J, Morinaga Y, Islam MR, Asaduzzaman SM, Aso Y, et al. 2009. Mapping and identification of the region and secondary structure required for the maturation of the nukacin ISK-1 prepeptide. *Peptides* 30:1412–20
88. Kotake Y, Ishii S, Yano T, Katsuoka Y, Hayashi H. 2008. Substrate recognition mechanism of the peptidase domain of the quorum-sensing-signal-producing ABC transporter ComA from *Streptococcus*. *Biochemistry* 47:2531–38
89. Ishii S, Yano T, Ebihara A, Okamoto A, Manzoku M, Hayashi H. 2010. Crystal structure of the peptidase domain of *Streptococcus* ComA, a bifunctional ATP-binding cassette transporter involved in the quorum-sensing pathway. *J. Biol. Chem.* 285:10777–85
90. Lawton EM, Ross RP, Hill C, Cotter PD. 2007. Two-peptide lantibiotics: a medical perspective. *Mini Rev. Med. Chem.* 7:1236–47
91. McClerren AL, Cooper LE, Quan C, Thomas PM, Kelleher NL, van der Donk WA. 2006. Discovery and in vitro biosynthesis of haloduracin, a two-component lantibiotic. *Proc. Natl. Acad. Sci. USA* 103:17243–48
92. Lawton EM, Cotter PD, Hill C, Ross RP. 2007. Identification of a novel two-peptide lantibiotic, haloduracin, produced by the alkaliphile *Bacillus halodurans* C-125. *FEMS Microbiol. Lett.* 267:64–71
93. Cooper LE, McClerren AL, Chary A, van der Donk WA. 2008. Structure-activity relationship studies of the two-component lantibiotic haloduracin. *Chem. Biol.* 15:1035–45
94. Dischinger J, Josten M, Szekat C, Sahl HG, Bierbaum G. 2009. Production of the novel two-peptide lantibiotic lichenicidin by *Bacillus licheniformis* DSM 13. *PLoS ONE* 4:e6788
95. Shenkarev ZO, Finkina EI, Nurmukhamedova EK, Balandin SV, Mineev KS, et al. 2010. Isolation, structure elucidation, and synergistic antibacterial activity of a novel two-component lantibiotic lichenicidin from *Bacillus licheniformis* VK21. *Biochemistry* 49:6462–72
96. Caetano T, Krawczyk JM, Mösker E, Süssmuth RD, Mendo S. 2011. Heterologous expression, biosynthesis, and mutagenesis of type II lantibiotics from *Bacillus licheniformis* in *Escherichia coli*. *Chem. Biol.* 18:90–100
97. Booth MC, Bogie CP, Sahl HG, Siezen RJ, Hatter KL, Gilmore MS. 1996. Structural analysis and proteolytic activation of *Enterococcus faecalis* cytolysin, a novel lantibiotic. *Mol. Microbiol.* 21:1175–84
98. Holo H, Jeknic Z, Daeschel M, Stevanovic S, Nes IF. 2001. Plantaricin W from *Lactobacillus plantarum* belongs to a new family of two-peptide lantibiotics. *Microbiology* 147:643–51
99. Haft DH, Basu MK, Mitchell DA. 2010. Expansion of ribosomally produced natural products: a nitrile hydratase- and Nif11-related precursor family. *BMC Biol.* 8:70
100. Wescombe PA, Upton M, Dierksen KP, Ragland NL, Sivabalan S, et al. 2006. Production of the lantibiotic salivaricin A and its variants by oral streptococci and use of a specific induction assay to detect their presence in human saliva. *Appl. Environ. Microbiol.* 72:1459–66
101. Hyink O, Wescombe PA, Upton M, Ragland N, Burton JP, Tagg JR. 2007. Salivaricin A2 and the novel lantibiotic salivaricin B are encoded at adjacent loci on a 190-kilobase transmissible megaplasmid in the oral probiotic strain *Streptococcus salivarius* K12. *Appl. Environ. Microbiol.* 73:1107–13
102. Wescombe PA, Upton M, Renault P, Wirawan RE, Power D, et al. 2011. Salivaricin 9, a new lantibiotic produced by *Streptococcus salivarius*. *Microbiology* 157:1290–99
103. Lin Y, Teng K, Huan L, Zhong J. 2011. Dissection of the bridging pattern of bovicin HJ50, a lantibiotic containing a characteristic disulfide bridge. *Microbiol. Res.* 166:146–54
104. Kabuki T, Uenishi H, Seto Y, Yoshioka T, Nakajima H. 2009. A unique lantibiotic, thermophilin 1277, containing a disulfide bridge and two thioether bridges. *J. Appl. Microbiol.* 106:853–62
105. Kodani S, Hudson ME, Durrant MC, Buttner MJ, Nodwell JR, Willey JM. 2004. The SapB morphogen is a lantibiotic-like peptide derived from the product of the developmental gene *ramS* in *Streptomyces coelicolor*. *Proc. Natl. Acad. Sci. USA* 101:11448–53
106. Müller WM, Schmiederer T, Ensle P, Süssmuth RD. 2010. In vitro biosynthesis of the prepeptide of type-III lantibiotic labyrinthopeptin A2 including formation of a C-C bond as a post-translational modification. *Angew. Chem. Int. Ed.* 49:2436–40
107. Pesic A, Henkel M, Süssmuth RD. 2011. Identification of the amino acid labionin and its desulfurised derivative in the type-III lantibiotic LabA2 by means of GC/MS. *Chem. Commun.* 47:7401–3

108. Patton GC, Paul M, Cooper LE, Chatterjee C, van der Donk WA. 2008. The importance of the leader sequence for directing lanthionine formation in lacticin 481. *Biochemistry* 47:7342–51
109. Chen P, Qi FX, Novak J, Krull RE, Caufield PW. 2001. Effect of amino acid substitutions in conserved residues in the leader peptide on biosynthesis of the lantibiotic mutacin II. *FEMS Microbiol. Lett.* 195:139–44
110. Neis S, Bierbaum G, Josten M, Pag U, Kempter C, et al. 1997. Effect of leader peptide mutations on biosynthesis of the lantibiotic Pep5. *FEMS Microbiol. Lett.* 149:249–55
111. Müller WM, Ensle P, Krawczyk B, Süssmuth RD. 2011. Leader peptide-directed processing of labyrinthopeptin A2 precursor peptide by the modifying enzyme LabKC. *Biochemistry* 50:8362–73
112. Chatterjee C, Patton GC, Cooper L, Paul M, van der Donk WA. 2006. Engineering dehydro amino acids and thioethers into peptides using lacticin 481 synthetase. *Chem. Biol.* 13:1109–17
113. Sinha Roy R, Belshaw PJ, Walsh CT. 1998. Mutational analysis of posttranslational heterocycle biosynthesis in the gyrase inhibitor microcin B17: distance dependence from propeptide and tolerance for substitution in a GSCG cyclizable sequence. *Biochemistry* 37:4125–36
114. Levengood MR, Patton GC, van der Donk WA. 2007. The leader peptide is not required for posttranslational modification by lacticin 481 synthetase. *J. Am. Chem. Soc.* 129:10314–15
115. Sit CS, Yoganathan S, Vederas JC. 2011. Biosynthesis of aminovinyl-cysteine-containing peptides and its application in the production of potential drug candidates. *Acc. Chem. Res.* 44:261–68
116. Kupke T, Götz F. 1997. In vivo reaction of affinity-tag-labelled epidermin precursor peptide with flavoenzyme EpiD. *FEMS Microbiol. Lett.* 153:25–32
117. Ökesli A, Cooper LE, Fogle EJ, van der Donk WA. 2011. Nine post-translational modifications during the biosynthesis of cinnamycin. *J. Am. Chem. Soc.* 133:13753–60
118. Kupke T, Kempter C, Jung G, Götz F. 1995. Oxidative decarboxylation of peptides catalyzed by flavoprotein EpiD. Determination of substrate specificity using peptide libraries and neutral loss mass spectrometry. *J. Biol. Chem.* 270:11282–89
119. He Z, Yuan C, Zhang L, Yousef AE. 2008. N-terminal acetylation in paenibacillin, a novel lantibiotic. *FEBS Lett.* 582:2787–92
120. He Z, Kisla D, Zhang L, Yuan C, Green-Church KB, Yousef AE. 2007. Isolation and identification of a *Paenibacillus polymyxa* strain that coproduces a novel lantibiotic and polymyxin. *Appl. Environ. Microbiol.* 73:168–78
121. Boakes S, Cortés J, Appleyard AN, Rudd BAM, Dawson MJ. 2009. Organization of the genes encoding the biosynthesis of actagardine and engineering of a variant generation system. *Mol. Microbiol.* 72:1126–36
122. Boakes S, Appleyard AN, Cortés J, Dawson MJ. 2010. Organization of the biosynthetic genes encoding deoxyactagardine B (DAB), a new lantibiotic produced by *Actinoplanes liguriae* NCIMB41362. *J. Antibiot.* 63:351–58
123. Liu G, Zhong J, Ni J, Chen M, Xiao H, Huan L. 2009. Characteristics of the bovicin HJ50 gene cluster in *Streptococcus bovis* HJ50. *Microbiology* 155:584–93
124. Xiao H, Chen X, Chen M, Tang S, Zhao X, Huan L. 2004. Bovicin HJ50, a novel lantibiotic produced by *Streptococcus bovis* HJ50. *Microbiology* 150:103–8
125. Oman TJ, Boettcher JM, Wang H, Okalibe XN, van der Donk WA. 2011. Sublancin is not a lantibiotic but an S-linked glycopeptide. *Nat. Chem. Biol.* 7:78–80
126. Stepper J, Shastri S, Loo TS, Preston JC, Novak P, et al. 2011. Cysteine S-glycosylation, a new posttranslational modification found in glycopeptide bacteriocins. *FEBS Lett.* 585:645–50
127. Claesen J, Bibb M. 2010. Genome mining and genetic analysis of cypemycin biosynthesis reveal an unusual class of posttranslationally modified peptides. *Proc. Natl. Acad. Sci. USA* 107:16297–302
128. Claesen J, Bibb MJ. 2011. Grisemycin, a new member of the linaridin family of ribosomally synthesized peptides produced by *Streptomyces griseus* IFO 13350: biosynthesis and regulation. *J. Bacteriol.* 193:2510–16
129. Field D, Hill C, Cotter PD, Ross RP. 2010. The dawning of a 'Golden era' in lantibiotic bioengineering. *Mol. Microbiol.* 78:1077–87
130. Ross AC, Vederas JC. 2011. Fundamental functionality: recent developments in understanding the structure-activity relationships of lantibiotic peptides. *J. Antibiot.* 64:27–34

131. Nagao J, Nishie M, Sonomoto K. 2011. Methodologies and strategies for the bioengineering of lantibiotics. *Curr. Pharm. Biotechnol.* 12:1221–30
132. Cortés J, Appleyard AN, Dawson MJ, David AH. 2009. Whole-cell generation of lantibiotic variants. *Methods Enzymol.* 458:559–74
133. Field D, Connor PMO, Cotter PD, Hill C, Ross RP. 2008. The generation of nisin variants with enhanced activity against specific gram-positive pathogens. *Mol. Microbiol.* 69:218–30
134. Rink R, Wierenga J, Kuipers A, Kluskens LD, Driessen AJM, et al. 2007. Dissection and modulation of the four distinct activities of nisin by mutagenesis of rings A and B and by C-terminal truncation. *Appl. Environ. Microbiol.* 73:5809–16
135. Shi Y, Yang X, Garg N, van der Donk WA. 2011. Production of lantipeptides in *Escherichia coli*. *J. Am. Chem. Soc.* 133:2338–41
136. Valsesia G, Medaglia G, Held M, Minas W, Panke S. 2007. Circumventing the effect of product toxicity: development of a novel two-stage production process for the lantibiotic gallidermin. *Appl. Environ. Microbiol.* 73:1635–45
137. Rink R, Wierenga J, Kuipers A, Kluskens LD, Driessen AJM, et al. 2007. Production of dehydroamino acid-containing peptides by *Lactococcus lactis*. *Appl. Environ. Microbiol.* 73:1792–96
138. Rink R, Kuipers A, de Boef E, Leenhouts KJ, Driessen AJM, et al. 2005. Lantibiotic structures as guidelines for the design of peptides that can be modified by lantibiotic enzymes. *Biochemistry* 44:8873–82
139. Kluskens LD, Kuipers A, Rink R, de Boef E, Fekken S, et al. 2005. Post-translational modification of therapeutic peptides by NisB, the dehydratase of the lantibiotic nisin. *Biochemistry* 44:12827–34
140. Rink R, Kluskens LD, Kuipers A, Driessen AJM, Kuipers OP, Moll GN. 2007. NisC, the cyclase of the lantibiotic nisin, can catalyze cyclization of designed nonlantibiotic peptides. *Biochemistry* 46:13179–89
141. Deegan LH, Suda S, Lawton EM, Draper LA, Hugenholtz F, et al. 2010. Manipulation of charged residues within the two-peptide lantibiotic lacticin 3147. *Microb. Biotechnol.* 3:222–34
142. Cotter PD, Deegan LH, Lawton EM, Draper LA, O'Connor PM, et al. 2006. Complete alanine scanning of the two-component lantibiotic lacticin 3147: generating a blueprint for rational drug design. *Mol. Microbiol.* 62:735–47
143. Islam MR, Shioya K, Nagao J, Nishie M, Jikuya H, et al. 2009. Evaluation of essential and variable residues of nukacin ISK-1 by NNK scanning. *Mol. Microbiol.* 72:1438–47
144. Appleyard AN, Choi S, Read DM, Lightfoot A, Boakes S, et al. 2009. Dissecting structural and functional diversity of the lantibiotic mersacidin. *Chem. Biol.* 16:490–98
145. Herzner AM, Dischinger J, Szekat C, Josten M, Schmitz S, et al. 2011. Expression of the lantibiotic mersacidin in *Bacillus amyloliquefaciens* FZB42. *PLoS ONE* 6:e22389
146. Nagao J, Harada Y, Shioya K, Aso Y, Zendo T, et al. 2005. Lanthionine introduction into nukacin ISK-1 prepeptide by co-expression with modification enzyme NukM in *Escherichia coli*. *Biochem. Biophys. Res. Commun.* 336:507–13
147. Nagao J, Shioya K, Harada Y, Okuda K, Zendo T, et al. 2011. Engineering unusual amino acids into peptides using lantibiotic synthetase. *Methods Mol. Biol.* 705:225–36
148. Caetano T, Krawczyk JM, Mosker E, Sussmuth RD, Mendo S. 2011. Lichenicidin biosynthesis in *Escherichia coli*: *licFGEHI* immunity genes are not essential for lantibiotic production or self-protection. *Appl. Environ. Microbiol.* 77:5023–26
149. Levengood MR, Knerr PJ, Oman TJ, van der Donk WA. 2009. In vitro mutasynthesis of lantibiotic analogues containing nonproteinogenic amino acids. *J. Am. Chem. Soc.* 131:12024–25
150. Levengood MR, Kerwood CC, Chatterjee C, van der Donk WA. 2009. Investigation of the substrate specificity of lacticin 481 synthetase by using nonproteinogenic amino acids. *ChemBioChem* 10:911–19
151. Levengood MR, van der Donk WA. 2008. Use of lantibiotic synthetases for the preparation of bioactive constrained peptides. *Bioorg. Med. Chem. Lett.* 18:3025–28
152. Zhang X, van der Donk WA. 2007. On the substrate specificity of dehydration by lacticin 481 synthetase. *J. Am. Chem. Soc.* 129:2212–13
153. Bindman N, Merkx R, Koehler R, Herrman N, van der Donk WA. 2010. Photochemical cleavage of leader peptides. *Chem. Commun.* 46:8935–37

154. Cheng F, Takala TM, Saris PEJ. 2007. Nisin biosynthesis in vitro. *J. Mol. Microbiol. Biotechnol.* 13:248–54
155. Goto Y, Iwasaki K, Torikai K, Murakami H, Suga H. 2009. Ribosomal synthesis of dehydrobutyrine- and methyllanthionine-containing peptides. *Chem. Commun.* 45:3419–21
156. Seebeck FP, Ricardo A, Szostak JW. 2011. Artificial lantipeptides from in vitro translations. *Chem. Commun.* 47:6141–43
157. Tabor AB. 2011. The challenge of the lantibiotics: synthetic approaches to thioether-bridged peptides. *Org. Biomol. Chem.* 9:7606–28
158. Fukase K, Kitazawa M, Sano A, Shimbo K, Horimoto S, et al. 1992. Synthetic study on peptide antibiotic nisin. V. Total synthesis of nisin. *Bull. Chem. Soc. Jpn.* 65:2227–40
159. Ross AC, Liu H, Pattabiraman VR, Vederas JC. 2010. Synthesis of the lantibiotic lactocin S using peptide cyclizations on solid phase. *J. Am. Chem. Soc.* 132:462–63
160. Pattabiraman VR, McKinnie SMK, Vederas JC. 2008. Solid-supported synthesis and biological evaluation of the lantibiotic peptide bis(desmethyl) lacticin 3147 A2. *Angew. Chem. Int. Ed.* 47:9472–75
161. Liu H, Pattabiraman VR, Vederas JC. 2009. Synthesis and biological activity of oxa-lacticin A2, a lantibiotic analogue with sulfur replaced by oxygen. *Org. Lett.* 11:5574–77
162. Pattabiraman VR, Stymiest JL, Derksen DJ, Martin NI, Vederas JC. 2007. Multiple on-resin olefin metathesis to form ring-expanded analogues of the lantibiotic peptide, lacticin 3147 A2. *Org. Lett.* 9:699–702
163. Knerr PJ, Tzekou A, Ricklin D, Qu H, Chen H, et al. 2011. Synthesis and activity of thioether-containing analogues of the complement inhibitor compstatin. *ACS Chem. Biol.* 6:753–60
164. Liu W, Chan ASH, Liu H, Cochrane SA, Vederas JC. 2011. Solid supported chemical syntheses of both components of the lantibiotic lacticin 3147. *J. Am. Chem. Soc.* 133:14216–19

Regulation of Glucose Transporter Translocation in Health and Diabetes

Jonathan S. Bogan

Section of Endocrinology and Metabolism, Department of Internal Medicine, and Department of Cell Biology, Yale University School of Medicine, New Haven, Connecticut 06520-8020; email: jonathan.bogan@yale.edu

Keywords

GLUT4, insulin signaling, vesicle trafficking, metabolic syndrome, Golgi matrix

Abstract

To enhance glucose uptake into muscle and fat cells, insulin stimulates the translocation of GLUT4 glucose transporters from intracellular membranes to the cell surface. This response requires the intersection of insulin signaling and vesicle trafficking pathways, and it is compromised in the setting of overnutrition to cause insulin resistance. Insulin signals through AS160/Tbc1D4 and Tbc1D1 to modulate Rab GTPases and through the Rho GTPase TC10α to act on other targets. In unstimulated cells, GLUT4 is incorporated into specialized storage vesicles containing IRAP, LRP1, sortilin, and VAMP2, which are sequestered by TUG, Ubc9, and other proteins. Insulin mobilizes these vesicles directly to the plasma membrane, and it modulates the trafficking itinerary so that cargo recycles from endosomes during ongoing insulin exposure. Knowledge of how signaling and trafficking pathways are coordinated will be essential to understanding the pathogenesis of diabetes and the metabolic syndrome and may also inform a wide range of other physiologies.

Contents

- INTRODUCTION 508
- GLUCOSE TRANSPORTERS AND GLUCOSE UPTAKE 509
- MULTIPLE INSULIN SIGNALING PATHWAYS CONTROL GLUT4 TRAFFICKING 510
  - Inactivation of Rab GTPase-Activating Proteins: AS160/Tbc1D4 and Tbc1D1 510
  - Signals for Vesicle Fusion at the Plasma Membrane 511
  - The Rho GTPase: TC10α 512
- FORMATION AND REGULATION OF GLUT4 STORAGE VESICLES 513
  - "Static" and "Dynamic" Models of GLUT4 Intracellular Retention 513
  - Two Classes of Exocytic Vesicles ... 515
  - GLUT4 Storage Vesicle Formation by Mass Action 515
  - IRAP, LRP1, and Intracellular Retention of GLUT4 Storage Vesicles 516
- A TETHERING COMPLEX FOR GLUT4 STORAGE VESICLE RETENTION AND RELEASE ... 517
  - GLUT4 Sequestration Requires the TUG Protein 518
  - An Unconventional Secretion Pathway? 519
  - Possible Coordination of Rabs and Tethers 520
- TRANSLOCATION OF GLUT4 STORAGE VESICLES MAY COORDINATELY REGULATE MULTIPLE PHYSIOLOGIC END POINTS 521
  - Vesicles in Other Cell Types That Are Similar to GLUT4 Storage Vesicles 522
- IMPLICATIONS FOR INSULIN RESISTANCE AND DIABETES 522

INTRODUCTION

Glucose is central to metabolism and is used both as a primary substrate for energy production and as a precursor to other carbon-containing molecules. In mammals, the brain requires a constant supply of glucose, and a highly regulated system has evolved to cope with limited or unpredictable food availability. Glucose absorbed from the gut stimulates the release of insulin from pancreatic β-cells, which both suppresses hepatic glucose production and promotes glucose uptake by muscle and adipose tissues. Absorbed glucose is stored as glycogen in liver and skeletal muscle and as triglyceride in adipose. These stores are mobilized during periods of fasting or increased energy demand and, together with gluconeogenesis, maintain blood glucose concentrations at approximately 4–5 mM. Insulin is the primary hormone regulating this physiology and normally causes the prompt return of blood glucose concentrations to this range even after large caloric loads are absorbed.

The concept that insulin accelerates the transmembrane transport of glucose into extrahepatic tissues was proposed by Levine et al. in 1949 (1) and expanded on by Park and coworkers (2). In 1980, Cushman and others showed that insulin translocates a glucose-transporting activity to the plasma membrane from an intracellular storage site (3, 4). In 1985, the first facilitative glucose transporter was cloned (5). Now termed GLUT1, this protein is widely expressed and not markedly translocated by insulin. The main isoform present in insulin-responsive tissues, GLUT4, was cloned in several laboratories in 1989 (6, 7). These groups showed that GLUT4 proteins are mobilized from internal membranes to the plasma membrane by insulin, similar to the previously described activity in adipocytes. Thus, the physiological problem of how insulin accelerates glucose uptake was transformed into the cell-biological problem of how insulin stimulates the translocation of GLUT4 glucose transporters.

Because of its importance for type 2 diabetes, GLUT4 regulation has been the topic of extensive research since 1989. The evolutionary factors that favor energy storage and metabolic efficiency can be maladaptive in the setting of readily available, calorie-dense foods. Thus, the combination of overnutrition and genetic factors results in insulin resistance, which is critical to the pathogenesis of type 2 diabetes. The term insulin resistance encompasses both a defect in the ability of insulin to suppress hepatic glucose production and a defect in insulin-responsive glucose uptake into muscle (8). The latter results from dysregulated GLUT4 trafficking. Together with insufficient insulin secretion, these processes lead to hyperglycemia. Insulin resistance frequently occurs together with hypertension, dyslipidemia, and other factors, referred to as the metabolic syndrome. The components of this syndrome have been difficult to separate in epidemiologic studies, suggesting that they may share some common pathophysiology.

This article focuses on aspects of GLUT4 translocation and glucose uptake. Several reviews have discussed GLUT4 regulation in recent years, but there remains much controversy regarding precisely where and how insulin acts (9–14). Most data support the view that, in unstimulated cells, endocytosed GLUT4 accumulates selectively in intracellular vesicles that are distinct from endosomes. These GLUT4 storage vesicles (GSVs) are mobilized within minutes of insulin addition, so that GLUT4 is inserted into the plasma membrane and glucose uptake is increased. Among other terms, GSVs are also called insulin-responsive vesicles; these terms are synonymous and lack a widely accepted molecular definition. The formation and regulation of this compartment is cell-type specific, and the targeting of glucose transporters to these vesicles is GLUT-isoform specific. How GSVs originate and how they are regulated have been enduring mysteries.

Here, I propose that GLUT4 participates in a fundamental trafficking pathway that is adapted in a cell-type-specific manner to permit regulation by extracellular signals. Similar pathways may act in various cell types. After reviewing signaling pathways that control GLUT4, I consider the hypothesis that a tethering complex, comprising TUG and other proteins, sequesters GSVs intracellularly within unstimulated cells. Insulin acts on this complex and modulates Rab proteins to mobilize these vesicles to the cell surface. As well as GLUT4, GSVs contain proteins that may contribute to the control of vascular tone and lipoprotein metabolism, such as IRAP, LRP1, and sortilin. Thus, vesicle translocation is a mechanism that may coordinately regulate multiple physiologic effects. Finally, an understanding of how GLUT4 translocation may be compromised to cause insulin resistance is beginning to emerge. Further studies in this area will have broad significance as well as direct importance for understanding metabolic disease.

GLUCOSE TRANSPORTERS AND GLUCOSE UPTAKE

The mammalian facilitative glucose transporter (GLUT) family comprises 14 proteins, which typically transport sugars down concentration gradients (15). These proteins have 12 transmembrane domains, with cytosolic amino and carboxyl termini. Each GLUT isoform is expressed in a distinct tissue distribution. In muscle and adipose, GLUT4 is predominant. In most cells, GLUTs mediate the import of glucose, because hexokinase activity maintains low intracellular glucose concentrations. Transport requires two steps: First, glucose binds at the surface of the transporter; second, a conformational change moves it across the membrane and releases it. The transporter toggles between two states, and the sugar associates sequentially with several sites in the transporter channel (16). For most GLUT isoforms, the overall K_m for transport is near or below physiologic glucose concentrations, and the number of transporters in the membrane is a primary determinant of the overall transport rate.

Data from transgenic mice highlight the importance of GLUT4 regulation for overall glucose homeostasis (17). In addition to its

> **GLUT4 storage vesicles (GSVs):** cell-type-specific, regulated exocytic vesicles that contain GLUT4 and that are distinct from endosomes and *trans*-Golgi compartments

GAP: GTPase activating protein

function in muscle and fat, GLUT4 acts in the brain and other tissues. In skeletal muscle, GLUT4 is translocated to the cell surface in response to contraction as well as insulin stimulation, and ischemia causes GLUT4 translocation in the heart (18). Skeletal muscle accounts for the bulk of glucose removal from the blood. Yet, the ability of insulin to redistribute GLUT4 is typically greater in adipose than in muscle, and adipose is a more facile experimental system. Abnormal GLUT4 regulation in adipose may also contribute disproportionately to systemic insulin resistance by modulating the secretion of adipose-derived hormones.

In type 2 diabetes, the ability of insulin to stimulate glucose transport into muscle is impaired. This was shown most elegantly using NMR spectroscopy (19), which complemented earlier work showing that GLUT4 translocation is defective (20). More recent data show that the control of glucose uptake is distributed (21). To enter muscle, glucose must be delivered by blood flow, transported across the sarcolemma and T-tubule membranes, and trapped intracellularly by hexokinase. Transport is rate controlling in resting muscle, when insulin concentrations are low. During maximal insulin stimulation or exercise, the delivery and hexokinase steps exert greater effects on the overall rate of uptake. Even so, GLUT4 translocation is a critical regulatory site and a primary defect in insulin resistance.

The cell biology of GLUT4 trafficking supports the concept that a main role of regulation is to restrict glucose uptake during low-insulin states. Unlike most proteins that recycle at the cell surface, GLUT4 is efficiently excluded from the plasma membrane of cells not stimulated with insulin. Early studies found only ~1% of GLUT4 at the surface of unstimulated brown adipocytes and cardiac myocytes; this fraction increased to ~40% after insulin stimulation (22, 23). In the commonly used 3T3-L1 adipocyte cell line, the basal proportion of GLUT4 at the cell surface is closer to ~5%–10% and is increased to 30%–50% after insulin stimulation (e.g., Reference 24). The latter percentage is typical of proteins that recycle in endosomes. Thus, how GLUT4 is targeted in the absence of insulin and how it is mobilized at the transition from basal to insulin-stimulated states are the most intriguing aspects of its trafficking.

Because GLUT4 regulation is cell-type dependent, studies have employed fat- or muscle-like cultured cells. 3T3-L1 and L6 cells can differentiate into adipocytes and myotubes, respectively, and upregulate endogenous GLUT4. These cultured cells facilitate the manipulation of genes, and results are extended in more physiologic settings. Yet, to the extent that cultured cell lines are poor models, progress has been hampered. Glucose uptake is stimulated 20- to 30-fold by insulin in rat adipocytes, but only 4- to 8-fold in 3T3-L1 adipocytes. In part, this reflects more abundant GLUT1 in 3T3-L1 cells. However, compared with primary adipocytes, the cultured cells have a reduced capacity to sequester GLUT4 intracellularly in the basal (unstimulated) state and, consequently, a blunted insulin response.

MULTIPLE INSULIN SIGNALING PATHWAYS CONTROL GLUT4 TRAFFICKING

Inactivation of Rab GTPase-Activating Proteins: AS160/Tbc1D4 and Tbc1D1

GLUT4 translocation requires the intersection of insulin signaling and vesicle trafficking pathways. Therefore, much excitement surrounded the discovery of an Akt substrate of 160 kDa (AS160, also termed Tbc1D4), which contains a Rab GTPase-activating protein (GAP) domain (25). Insulin signaling through phosphatidylinositol 3-kinase (PI3K) to Akt2 is required for its metabolic actions (26, 27). Because Rab proteins direct vesicle trafficking, AS160 links signaling and trafficking pathways. Insulin causes AS160 phosphorylation at multiple sites and inactivates its GAP activity. Mutation of four phosphorylation sites creates a dominant negative protein, which impairs GLUT4 translocation (28). RNAi-mediated AS160 depletion increases the fraction of GLUT4 at the surface of unstimulated cells (29, 30). This effect is rescued by wild-type AS160, but not by a GAP-domain mutant, suggesting that most, if not all, of the effect of AS160 in unstimulated cells is due to its Rab GAP activity.

AS160 is present on GLUT4-containing vesicles and may be recruited by binding to IRAP and LRP1, proteins that cotraffic in GSVs (30–32). These proteins likely mediate GSV sequestration (32–34), suggesting that AS160 inactivates particular Rab proteins on GSVs in unstimulated cells. AS160 phosphorylation recruits 14-3-3 proteins, which may be important to reduce GAP activity (35, 36). AS160-interacting proteins, RUVBL2 and RIP140, may also participate; however, their roles are not fully defined (37–39).

A related Rab GAP, Tbc1D1, is more abundant than is AS160/Tbc1D4 in skeletal muscle (36, 40). Both Tbc1D1 and AS160/Tbc1D4 are regulated by insulin; however, other upstream signaling inputs are complementary. Muscle contraction stimulates GLUT4 translocation and acts through AMP-activated protein kinase (AMPK) and calmodulin to phosphorylate Tbc1D1 and AS160 (41, 42). An AMPK-Tbc1D1 pathway may be particularly important in muscle (36, 40, 43). In humans, a rare truncation of AS160/TBC1D4 causes severe postprandial insulin resistance (44). The truncated protein dimerizes with wild-type AS160 and likely acts by a dominant negative effect; whether it similarly inhibits Tbc1D1 is unknown. The *TBC1D1* gene has also been implicated in obesity in humans and mice (45, 46).

Although phosphorylation of AS160/Tbc1D4 and Tbc1D1 is compromised in type 2 diabetes, increased phosphorylation is not clearly correlated with improved insulin sensitivity following drug treatment (47). Exercise increases insulin sensitivity and enhances insulin-stimulated AS160 phosphorylation in some, but not all, studies (48, 49). To some extent, the effect of exercise may reflect enhanced blood flow and increased abundance of GLUT4 and hexokinase (48). Enhanced signaling through Tbc1D1 may also contribute (50). Finally, defects in basal GLUT4 targeting, independent of insulin signaling, likely also contribute to insulin resistance (20, 51–53). The relative importance of each of these pathophysiologic mechanisms is poorly understood.

Which Rab proteins are activated by insulin signaling to AS160 and Tbc1D1? The GAP domains have similar specificities and enhance GTP hydrolysis by several Rabs on GLUT4-containing membranes, including Rab2A, -8A, -8B, -10, -11, and -14 (30, 54, 55). Differences in abundance suggest that Rab8A and Rab14 may predominate in muscle, whereas Rab10 dominates in adipose. In 3T3-L1 adipocytes, Rab10 and its GTP exchange factor (GEF), Dennd4C, are required for GLUT4 translocation; effects of Rab10 RNAi are particularly significant at submaximal insulin concentrations (54, 56). Rab10 likely acts on GSVs, because its depletion affects GLUT4 and IRAP, but not transferrin receptor (TfR), trafficking. Silencing of both AS160 and Rab10 only partially restores GLUT4 intracellular retention, implying additional factors downstream of AS160. Rab10 depletion also blunts GLUT4 targeting in the presence of insulin. This may result secondarily from effects on basal targeting, but it also fits with the idea that Rab10 acts through an effector, possibly a myosin isoform, to facilitate exocytosis (57). Together, these data support the notion that, in 3T3-L1 adipocytes, insulin acts through AS160 and Rab10 to regulate GSVs.

Signals for Vesicle Fusion at the Plasma Membrane

Other insulin signaling pathways are important for GLUT4 regulation. AS160 knockdown increases GLUT4 modestly at the surface of unstimulated cells, but it does not mimic the full effect of insulin (29, 30, 58). Moreover, the effects of IRAP and AS160 depletion are additive (33), implying that IRAP interacts with other proteins to sequester GLUT4 in GSVs. Insulin also regulates a site downstream of AS160-dependent vesicle docking at the plasma membrane, as suggested by total internal reflection fluorescence microscopy (TIRFM) (e.g., 59) and in vitro assays (60). What, then, are the other signaling pathways that act together with

GEF: GTP exchange factor

TIRFM: total internal reflection fluorescence microscopy

PLD: phospholipase D

Akt-AS160 to mediate insulin action? Which are quantitatively most important for the physiologic regulation of glucose uptake? Which are disrupted in insulin resistance?

After insulin binds, the insulin receptor phosphorylates itself and substrate (IRS-1 and IRS-2) proteins on tyrosine residues, activating several pathways. The receptor phosphorylates Munc18c, a Sec1/Munc18-like protein that regulates SNARE-complex formation at the plasma membrane (61). Munc18c then acts through syntaxin-4, VAMP2, and SNAP-23 to mediate fusion of GLUT4-containing vesicles and to control glucose uptake in mice. Reduced Munc18c and syntaxin-4 abundance are observed in insulin resistance, suggesting that this step may have pathophysiologic significance (62). In addition, SNAP-23 incorporation into lipid droplets may limit its availability for GLUT4 trafficking and contribute to insulin resistance (63).

Insulin also stimulates calcium influx in muscle (64). Insulin acts through Ca^{2+}/calmodulin-dependent kinase II (CaMKII) in adipocytes to phosphorylate Myo1c, an unconventional myosin required for GLUT4 translocation (65). In muscle, it is not clear if CaMKII mediates effects of insulin, or only of contraction, to promote glucose uptake; Myo1c is required for both (66). Myo1c associates with the small GTPase, RalA, which couples GLUT4-containing vesicles to the exocyst, a complex that tethers exocytic vesicles prior to fusion at the plasma membrane (67). Whereas the interaction of RalA with the exocyst requires RalA activation, its interaction with Myo1c is independent of its GTP-bound state. Signaling through these proteins may coordinate the final steps before GSV fusion at the plasma membrane.

RalA is present on GLUT4-containing vesicles and is activated by insulin. Similar to the Akt-AS160 pathway for Rab activation, Akt2 phosphorylates and inactivates a GAP, the RGC1/2 complex, to activate RalA (68). RalA then binds exocyst components, Sec5 and Exo70, to tether incoming vesicles at the plasma membrane. Disengagement of RalA from the exocyst is not mediated by GTP hydrolysis, but by phosphorylation of Sec5 (67). Phosphorylation is likely mediated by conventional protein kinase C (PKC) isoforms, presumably in response to locally increased calcium and diacylglycerol. Thus, several downstream effects of insulin converge on steps leading to vesicle fusion at the cell surface.

Phospholipase D (PLD) activity is required for GLUT4 insertion into the plasma membrane (69, 70). Insulin recruits PLD1, which generates phosphatidic acid at the plasma membrane. Phosphatidic acid may play a direct role in fusion and also activates phosphatidylinositol 4-phosphate 5-kinase. The resulting increase in phosphatidylinositol 4,5-bisphosphate is required for exocytosis (71). Neither pharmacologic PLD inhibition nor PLD1 knockdown impairs the recruitment of GLUT4-containing vesicles to the cytosolic face of the plasma membrane. Rather, PLD inhibition increases partial "kiss-and-run" fusion, as assessed using TIRFM (70). Thus, PLD activity may be required for the final steps of exocytic vesicle collapse.

The Rho GTPase: TC10α

Insulin activates the Rho-family GTPase, TC10α, to translocate GLUT4 in 3T3-L1 adipocytes (72). The relevance of this PI3K-independent pathway was initially unclear because of nonspecific effects of TC10 overexpression. More recent RNAi data show definitively that TC10α is required for GLUT4 translocation (73). Precisely how insulin activates this GTPase remains uncertain. Insulin stimulates phosphorylation of c-Cbl, which is recruited to the insulin receptor by the adaptor proteins CAP and APS (72, 74). Upon its phosphorylation, Cbl associates with the lipid raft protein, flotillin, and recruits a Crk-C3G complex. C3G (also called RAPGEF1) is a GEF that activates TC10α. Studies of the roles of APS, c-Cbl, and CrkII in this pathway have yielded discrepant results (75–78), perhaps because these proteins are multifunctional; for example, the ubiquitin ligase activity of c-Cbl may not be required for TC10 activation, but it may instead trigger insulin receptor endocytosis (79). Thus, to elucidate fully the upstream pathway that activates TC10, disrupting selected aspects of protein function may be necessary.

Downstream effects of TC10α activation include actin rearrangement (80) and phosphatidylinositol-3-phosphate production (81, 82). Activated TC10 binds CIP4/2 and is then coupled to a Rab GEF, Gapex-5, that activates Rab31 and Rab5 (82). Thus, signaling through TC10 and AS160 may differentially modulate various Rab proteins. Activated TC10 binds Exo70 and promotes tethering of GLUT4-containing vesicles at the plasma membrane (83). A convergent downstream target of TC10 and Akt signaling is atypical protein kinase C (aPKC-λ/ζ) (84), which may couple to various downstream targets (85, 86). Finally, activated TC10 is coupled to PIST, a PDZ-domain protein that interacts specifically with TC10 (13). PIST (also known as CAL, GOPC, and FIG) controls the cell-surface targeting of several proteins and binds to syntaxin-6 and Golgin-160, both of which participate in basal GLUT4 intracellular retention (87–94). As described below, this pathway may be linked to the release of GSVs retained by TUG and other proteins (13, 14, 70, 95, 96).

Signaling through TC10 is impaired in insulin resistance (97–100) and enhanced after exercise (101, 102). In rodents, the abundance of the upstream CAP protein is decreased by diet-induced insulin resistance, increased by exercise, and induced by insulin-sensitizing drugs (101–103). In humans, polymorphisms at the *RAPGEF1* gene encoding the TC10 activator, C3G, are associated with type 2 diabetes (104, 105). Although these variants were identified in candidate-gene studies, they may act in a pathway together with *SORT1*, *SORLA*, and *ASPSCR1* (encoding TUG), which are also associated with metabolic disease (106–109). As is also discussed below, data support the idea that this pathway converges on GSVs, and detailed knowledge of the biochemical mechanisms involved will be important to understand pathophysiologic defects in GSV regulation.

FORMATION AND REGULATION OF GLUT4 STORAGE VESICLES

Kinetic studies first suggested that GLUT4 traffics through at least three compartments, including the plasma membrane, endosomes, and an "insulin-responsive compartment" (110). Models employ differential equations, with first-order rate constants to approximate the transfer of GLUT4 from one compartment to another. What rate constants are affected by insulin? Although insulin modulates the endocytic mechanism, its main effect is to accelerate GLUT4 exocytosis (111, 112). On an exocytic pathway, insulin may act at one or more steps corresponding to vesicle budding; retention and release; movement on cytoskeletal elements; and tethering, docking, and fusion at the plasma membrane (**Figure 1**). The acquisition of glucose transport activity may not require an independent insulin signal but may follow as a consequence of translocation, perhaps from the association or dissociation of an interacting protein (96, 113).

The biochemical correlate of the kinetically defined, insulin-responsive compartment is the pool of GSVs. These 50–70-nm-diameter vesicles accumulate in unstimulated fat and muscle cells but not in fibroblasts or other cell types (14, 114). Although insulin may promote GSV budding, this effect does not account for GLUT4 trafficking. Insulin enhances vesicle fusion at the cell surface, but this can increase the overall exocytic rate only if vesicles are available for fusion. In fact, GSVs are sequestered deep within cells (9, 22, 23). In 3T3-L1 adipocytes and L6 myotubes, insulin-responsive GSVs reside in a perinuclear location and are scattered throughout the periphery (95, 115, 116). How insulin signaling pathways relate to specific trafficking steps for GSV formation, sequestration, and mobilization is not well understood.

"Static" and "Dynamic" Models of GLUT4 Intracellular Retention

Conflicting data have been reported regarding whether GSVs exchange with endosomes and

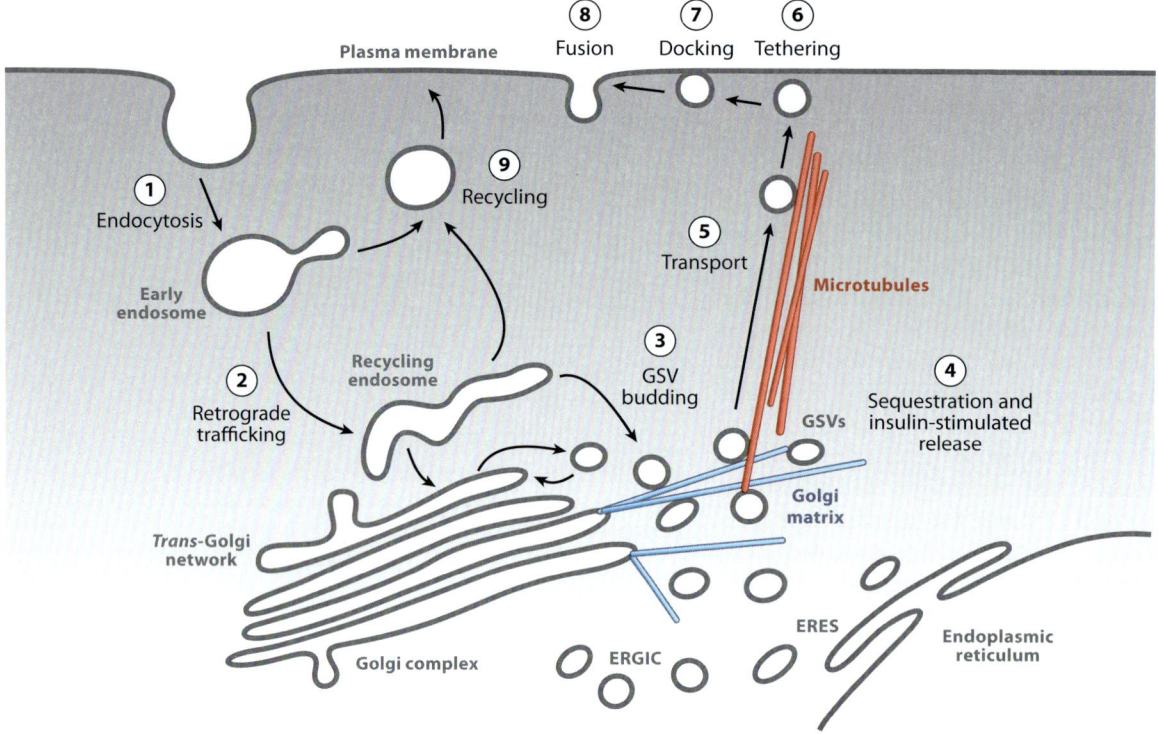

Figure 1
A model of GLUT4 trafficking pathways. After endocytosis at the plasma membrane ①, GLUT4 follows a retrograde pathway to the *trans*-Golgi network ②. In unstimulated cells, GLUT4 storage vesicles (GSVs) bud from a domain of the *trans*-Golgi network and/or recycling endosome ③ and are trapped at the Golgi matrix. The GSVs may then be held in a relatively static configuration or may participate in an intracellular cycle, possibly at the *trans*-Golgi network, to sequester GLUT4 and other cargoes away from the plasma membrane ④. Insulin stimulates the release of GSVs and loads them onto microtubule motors for transport to the cell surface ⑤. The vesicles are tethered to the plasma membrane ⑥, dock ⑦, and fuse ⑧ to insert GLUT4 and facilitate glucose uptake. During ongoing insulin stimulation, GSV components return to the plasma membrane from endosomes ⑨, bypassing the GSV compartment. Upon insulin removal, the internalized cargo is once again directed to accumulate in GSVs. Abbreviations: ERES, endoplasmic reticulum exit sites; ERGIC, endoplasmic reticulum–Golgi intermediate compartment.

the plasma membrane in 3T3-L1 adipocytes not stimulated with insulin. In a dynamic-equilibrium model, the size of the recycling pool of GLUT4 is not altered by insulin. Rather, insulin redistributes GLUT4 by a net effect at several trafficking steps, including flux through GSVs (117, 118). By contrast, in a static-retention model, GSVs are inaccessible to endosomes and the plasma membrane (24, 119). Insulin causes GSVs to enter the recycling pathway by a quantal-release mechanism, in which increasing concentrations of insulin mobilize increasing fractions of total cellular GLUT4, ranging from 10%–20% to ∼70%. In this model, the GSVs are the main site of insulin action. In fact, the main distinction between these models rests on whether flux through GSVs can be detected in unstimulated cells. If this flux is so low as to be undetectable, then the pool is fully sequestered, and the static model applies. If flux is detectable, then the dynamic model better describes the data.

Because GSV formation and GLUT4 intracellular retention are differentiation dependent

(120, 121), the discrepant results may reflect the degree to which 3T3-L1 cells mimic true adipocytes. Experimental manipulations, such as replating 3T3-L1 cells after differentiation, reduce the ability of the cells to sequester GLUT4 (122). Single-molecule, live-cell imaging of GLUT4 shows that approximately two-thirds of GLUT4 is immobile in the absence of insulin, and insulin both reduces this fraction and increases the speed at which the mobile vesicles move (116). Thus, the static model likely better describes GSV targeting in well-differentiated 3T3-L1 or primary adipocytes. An observed component of the dynamic model is an insulin-stimulated increase in the exocytic rate constant (122). Thus, aspects of both the static and dynamic models apply.

Two Classes of Exocytic Vesicles

Recent data demonstrate that GLUT4 is present in two distinct classes of exocytic vesicles in 3T3-L1 adipocytes, further supporting the idea that insulin releases sequestered GSVs into a cell-surface recycling pathway (70). The size of individual exocytic vesicles can be calculated using TIRFM of GLUT4-green fluorescent protein (GFP). In undifferentiated 3T3-L1 fibroblasts, GLUT4 is present in ∼150-nm-diameter vesicles, consistent with recycling from endosomes. After adipocyte differentiation, a new class of ∼60-nm-diameter vesicles is observed. These are presumably GSVs, on the basis of their size and differentiation dependence. Exocytosis of these small vesicles is difficult to detect using TIRFM of GLUT4-GFP but is facilitated using VAMP2-pHluorin. The fluorescence of this reporter increases upon formation of the fusion pore, which neutralizes the mildly acidic pH in the vesicle lumen. TIRFM can then be used to quantify the rate of exocytosis and to distinguish the two types of exocytic vesicles. Insulin stimulates exocytosis by ∼fourfold and changes the size of the exocytic vesicles from ∼150 nm to ∼60 nm in diameter, consistent with GSV mobilization from a sequestered pool. Remarkably, after more prolonged (>15 min) insulin exposure, the increased exocytic rate is maintained, but the vesicles are again ∼150 nm in diameter. This result is consistent with the idea that the GSV compartment is bypassed during ongoing insulin exposure, so that GSV cargoes return directly to the plasma membrane from endosomes.

GLUT4 Storage Vesicle Formation by Mass Action

After insulin is removed, GLUT4 reaccumulates in GSVs. These vesicles likely form by "mass action," in which the individual protein components interact combinatorially to recruit factors involved in vesicle budding (114). The cargoes enter the nascent vesicles as a unit, rather than separately. In rat adipocytes, proteins that are enriched in GSVs include not only GLUT4, but also insulin-responsive aminopeptidase (IRAP), lipoprotein receptor-related protein 1 (LRP1), sortilin (SORT1) and sortilin-related receptor (SORL1), and VAMP2 (30, 32). Lumenal interactions among sortilin, GLUT4, IRAP, and LRP1 drive vesicle formation and recruit GGA2 clathrin adaptors (121, 123). Sortilin, an ortholog of the yeast Vps10p sorting receptor, is induced early during 3T3-L1 adipocyte differentiation, coincident with the appearance of biochemically detectable GSVs. In undifferentiated 3T3-L1 cells, exogenous expression of sortilin is sufficient to drive formation of GSV-like vesicles; when GLUT4 is coexpressed, endogenous IRAP is incorporated into these vesicles (121, 123). Sortilin recruits GGA adaptors at the *trans*-Golgi network (TGN), which is further promoted by GLUT4 ubiquitination and by the generation of phosphatidylinositol-4-phosphate (124, 125). Thus, interactions among GSV proteins promote vesicle budding and drive the selective incorporation of cargo.

ACAP1 is another clathrin adaptor that interacts directly with GLUT4 and promotes the budding of GLUT4-containing vesicles, possibly GSVs (126). ACAP1 has ARF6 GAP activity, and RNAi-mediated ARF6 depletion blocks GLUT4 translocation. Of note, GGA proteins also bind ARF6, suggesting that ACAP1 and

TGN: *trans*-Golgi network

GGA2 may cooperate to mediate GSV budding in a step that also involves sortilin (127). Alternatively, because GSV cargoes can take two distinct routes to the plasma membrane (70), sortilin may act with GGA2 to promote GSV formation at the TGN, and ACAP1 may act with ARF6 and clathrin to promote recycling from endosomes during sustained insulin exposure. The idea that ACAP1 acts at the recycling endosome is consistent with its role in the recycling of other cargoes and, because ACAP1 is a target of Akt phosphorylation, also suggests how Akt may direct GSV cargo to either of the two routes (128). Further studies are required to determine if sortilin and ACAP1 function at common or distinct budding events.

Whether GSVs originate from the TGN or endosomes is uncertain. In part, this may reflect debate about whether a static or dynamic model applies, as discussed above. In unstimulated cells, endocytosed GLUT4 follows a syntaxin-16-dependent retrograde pathway from endosomes to the TGN, suggesting that the TGN is the GSV "donor compartment" (13). GLUT4 traffic in humans requires CHC22 and syntaxin-10, proteins not present in rodents, which also mediate retrograde trafficking (53, 129). IRAP that is enzymatically deglycosylated at the cell surface is resialylated after endocytosis, implying that it is internalized to the TGN or *trans*-Golgi cisterna (94). Finally, newly synthesized GLUT4 and IRAP are targeted to GSVs directly from the biosynthetic pathway, without first traversing the plasma membrane (130, 131). This observation has made it possible to study GSV regulation without interference from steps involved in GLUT4 internalization and sorting (132, 133). By contrast, data suggesting that GSVs participate in bidirectional exchange with endosomes have been reported (117). This result may reflect incomplete GSV sequestration. Most data suggest that the TGN can serve as an origin for GSVs.

A possibility is that GSVs can originate from either endosomes or the TGN. The Golgi matrix is thought to form a tentacular structure, consisting of long coiled-coil proteins anchored at one end to the Golgi stack (134). These proteins contain binding sites for Rab GTPases along their lengths and, as tentacles radiating out from the Golgi complex, capture vesicles bearing cognate, GTP-bound Rab proteins. The origin of a vesicle does not matter; if it is in the vicinity of the Golgi stack and is decorated by the appropriate Rab, then it will be trapped and, once tethered, directed to fuse at its target membrane. A similar model may apply to GSVs, which are likely sequestered at the Golgi matrix (91, 135). The adaptor that links the GSV to the matrix may consist of TUG and other proteins that bind GSV cargo, as well as Rab proteins (see description below). If so, then the GSVs could arise from endosomes or the TGN. Intriguingly, although newly synthesized IRAP requires GGA adaptors to enter GSVs, IRAP endocytosed from the plasma membrane does not (92, 136). A possible explanation for this result is that GGA adaptors may mediate budding of GSVs from the TGN, and ACAP1 may mediate budding from endosomes, as suggested above. Endocytosed GLUT4/IRAP could take either route, but newly synthesized proteins would rely solely on budding at the TGN. Additional studies will be required to understand if this model is correct.

IRAP, LRP1, and Intracellular Retention of GLUT4 Storage Vesicles

IRAP is a widely expressed protein that contains a cytosolic amino terminus of 109 residues, a transmembrane domain, and a large extracellular aminopeptidase domain. It is probably the most abundant protein in GSVs, at >10 molecules per vesicle, whereas GLUT4 is present at approximately 5–6 molecules per vesicle (137). IRAP is more important than GLUT4 for GSV formation and insulin-responsive trafficking (33, 34). Indeed, regulated GSV trafficking can occur, at least to some degree, in the absence of GLUT4, leading to the proposal that GLUT4 is merely a cargo and not integral to the insulin-responsive mechanism (33). This likely overstates the case, given (*a*) luminal interactions of GLUT4 with sortilin, IRAP, and LRP1 (32, 123);

(b) recruitment of peripherally associated proteins such as ACAP1 and TUG, which may be mediated in part by cytosolic domains of GLUT4 (96, 126); and (c) particular mutations in GLUT4 affect its insulin-responsive trafficking, apparently by altering GSV mobilization rather than sorting to the GSVs (138, 139). Nonetheless, the idea that IRAP is the more critical protein for GSV regulation is likely correct. Because IRAP is present in many cell types, this idea highlights how similar regulated vesicles may be present in many cell types. Other cargoes may also contribute the signals that are found on GLUT4, which may enhance the basic mechanism.

The idea that IRAP tethers GSVs to an intracellular anchoring site was suggested initially by competition experiments (140). Microinjection of a peptide corresponding to the amino terminus of IRAP causes acute translocation of GLUT4 to the cell surface of 3T3-L1 adipocytes. Thus, the peptide may compete with IRAP for binding to a protein(s) required for intracellular GLUT4 retention. The identity of this "retention receptor" remains unknown, although a dileucine motif and acidic cluster in IRAP are critical for insulin-regulated trafficking (136). The dileucine mediates binding of IRAP to acyl-CoA dehydrogenases, but the physiological relevance of this interaction is uncertain (141). Several other IRAP-interacting proteins do not require the dileucine but may nonetheless be important for GSV regulation. These include p115, a component of the Golgi matrix, and vimentin, an intermediate filament protein (135, 142). Disruption of these proteins disperses perinuclear GLUT4, consistent with a potential role in anchoring GSVs intracellularly. p115 can also bind two other proteins enriched in GSVs, sortilin and LRP1 (32). Formin homolog overexpressed in spleen (FHOS) binds IRAP and profilin IIa, suggesting a link to the actin cytoskeleton (143). Other IRAP-interacting proteins include AS160 and sortilin (as noted above). Finally, tankyrase is a poly-ADP-ribose polymerase (PARP) that binds IRAP through its ankyrin repeats (34). Tankyrase is required for insulin-responsive GLUT4 trafficking, and its PARP activity is suggested to facilitate budding or sequestration of GSVs, but how this occurs is unknown. Overall, it is most likely that IRAP interacts with a multiprotein complex to retain GSVs within unstimulated cells. Binary interactions among components of such a complex may be weak and may require stabilization by posttranslational modifications [e.g., possibly poly(ADP-ribosyl)ation]. As detailed below, TUG and Ubc9 may participate in such a complex, which is hypothesized to link GSVs to the Golgi matrix and, possibly, to cytoskeletal elements.

LRP1 is a large type 1 transmembrane protein recently identified as a GSV component (32). Also known as the receptor for α2-macroglobulin, LRP1 translocates to the cell surface in response to insulin, but it follows an endocytic pathway distinct from that of GLUT4 (112). Data indicate that LRP1 is an essential component of the GSV trafficking mechanism, similar to IRAP. LRP1 interacts with GLUT4, IRAP, and sortilin in the vesicle lumen, as assessed using cross-linkers. Its cytosolic terminus binds AS160 and p115. GLUT4 protein stability is secondarily affected by its trafficking, and in IRAP-knockout mice, GLUT4 abundance is greatly reduced (13, 144). Thus, it is significant that GLUT4 and sortilin are depleted in primary adipocytes lacking LRP1 and that GLUT4, sortilin, and IRAP are depleted in 3T3-L1 adipocytes depleted of LRP1 using RNAi. These data are consistent with the idea that LRP1 is essential for GSV sequestration. Similar roles are attributed to TUG and Ubc9 (96, 145, 146). How LRP1 may regulate GSVs is not well understood, but its recruitment of AS160, p115, and perhaps other proteins to these vesicles is likely important.

A TETHERING COMPLEX FOR GLUT4 STORAGE VESICLE RETENTION AND RELEASE

TUG is a putative tether, containing a ubiquitin-like UBX domain, for GLUT4 that was identified in the only genetic screen for

GLUT4 regulators (95). This screen used flow cytometry to enrich for particular cDNAs that, when overexpressed, alter the proportion of GLUT4 at the surface of individual cells. Previous data showed that insulin can cause a biphasic translocation response in 3T3-L1 adipocytes (24, 120). After insulin addition, the initial burst of GLUT4 that appears at the cell surface was hypothesized to result from GSV mobilization, whereas subsequently exocytosed GLUT4 may originate from endosomes. The screen assayed the initial burst so as to enrich preferentially for proteins that regulate GSV trafficking. Both of these ideas—that the initially translocated GLUT4 comes from GSVs and that TUG regulates these vesicles—have been supported by subsequent data (13, 70, 96). The human ortholog of murine TUG, ASPL (gene name, *ASPSCR1*), was identified as part of a fusion protein found in alveolar soft part sarcomas (147). The development of these sarcomas may result, at least in part, from overexpression of the downstream gene product (the TFE3 transcription factor) that is driven at high levels from the ASPL/TUG promotor. Here, the protein is referred to as TUG, which emphasizes its physiologic role to regulate GLUT4.

The proposed model is that TUG traps endocytosed GLUT4 and, together with other proteins, tethers GSVs intracellularly within unstimulated cells. Insulin then mobilizes the GSVs by releasing this tether (13, 14). In 3T3-L1 adipocytes, the initial burst of translocated GLUT4 was enhanced by TUG overexpression and reduced by a dominant negative TUG fragment, suggesting that TUG promotes GLUT4 accumulation in GSVs (95). TUG and GLUT4 dissociate rapidly after insulin addition, and the number of complexes that dissociate correlates with the size of the initial burst of GLUT4 (95, 120, 148). In unstimulated 3T3-L1 adipocytes, TUG colocalizes extensively with GLUT4, and not with TfR, consistent with the idea that it retains GLUT4 in a nonendosomal compartment. TUG overexpression enhances GLUT4 accumulation in TfR-negative membranes, whereas a dominant negative TUG fragment causes nearly complete overlap of GLUT4 and TfR (96). Finally, the interaction of TUG and GLUT4 is GLUT-isoform specific, and distinct regions of TUG are required to bind GLUT4 and to retain GLUT4 intracellularly in transfected, nonadipose cells. The latter region (the carboxyl terminus) was presumed to bind an intracellular anchoring site, which now appears to be the Golgi matrix (see below).

GLUT4 Sequestration Requires the TUG Protein

If GSV mobilization accounts for the bulk of insulin action on GLUT4 and if TUG is an essential part of a GSV sequestration mechanism, then TUG disruption should have effects that are quantitatively similar to those of insulin. This prediction has been realized in 3T3-L1 adipocytes (70, 96). In unstimulated cells in which TUG is disrupted using RNAi or a dominant negative fragment, GLUT4 targeting and glucose uptake are similar to that in insulin-stimulated control cells (96). Insulin causes a modest additional effect that may reflect accelerated endosome recycling, enhanced fusion at the plasma membrane, or reduced endocytosis. More detailed studies using TIRFM reveal striking effects of TUG depletion (70). The rate of exocytic vesicle fusion in unstimulated, TUG-depleted cells is essentially identical to that in insulin-stimulated control cells. Vesicles accumulate just beneath the plasma membrane, and insulin causes a transient wave of exocytosis that is likely due to fusion of these "docked" vesicles. The subsequent exocytic rate is, again, similar to that prior to insulin addition. Similar to acute insulin action, TUG depletion mobilizes ~60-nm-diameter vesicles, which correspond in size to GSVs. These are distinct from the ~150-nm vesicles, likely endosome derived, that fuse in unstimulated cells or during ongoing insulin exposure. Insulin also reduces the frequency of partial "kiss-and-run" fusions, in part through action at the plasma membrane (see above). Yet, the GSVs that are mobilized by TUG depletion have intrinsic properties that make them more likely to fuse fully at the plasma membrane. Together, the data support the view that, in 3T3-L1 adipocytes, insulin's action to mobilize GSVs from a TUG-regulated pathway accounts for the bulk of its overall effect to translocate GLUT4.

GLUT4 trafficking and stability are linked, probably because GLUT4 sequestration restricts its movement to lysosomes as well as to the plasma membrane (13). Similar to insulin, RNAi-mediated TUG depletion reduces GLUT4 protein stability (96). Effects on GLUT4 stability are also observed upon manipulation of Ubc9, which binds GLUT4 and likely cooperates with TUG (145, 146). Depletion of IRAP and LRP1 reduces GLUT4 abundance (noted above). Whether TUG or Ubc9 binds these proteins directly is not known, but there is at least an indirect interaction. Two regions of TUG interact with GLUT4, and at least one does so directly (95, 96). The GLUT4 sequences that mediate this interaction include the large intracellular loop and possibly the amino terminus, which are determinants of GLUT-isoform-specific, insulin-responsive trafficking (132). Ubc9 binds to the carboxyl terminus of both GLUT4 and GLUT1, and it may participate in insulin-responsive trafficking without conferring GLUT-isoform specificity (145). Similar to the clustering of proteins proposed to drive GSV vesicle formation (114), the GSV retention mechanism may require the formation of an oligomeric complex. The binary protein interactions among the components of this complex are likely to be weak, which may facilitate insulin-stimulated disassembly (95, 148). This disassembly releases GSVs, enlarges the pool of GLUT4 that recycles at the cell surface, and, together with other insulin actions, mediates the full effect of insulin to stimulate glucose uptake.

An Unconventional Secretion Pathway?

With which proteins does TUG interact, and how does it function? TUG contains two ubiquitin-like domains at its amino terminus, a low-complexity region, a probable coiled-coil motif, a ubiquitin-like UBX domain, and a carboxyl-terminal region of unknown structure (149). UBX-containing proteins are typically adaptors for p97, a hexameric ATPase also called VCP or CDC48 (150). These cofactors recruit p97 activity to a wide range of substrates, often to disassemble protein complexes. In 293 cells, most TUG is bound to p97 (151, 152). Yet unlike most UBX-containing proteins, TUG may not simply couple p97 activity to a substrate. Rather, TUG disassembles the p97 hexamer into monomers in vitro and in transfected cells (152). A similar function is described for the *Arabidopsis* protein, PUX1, which regulates p97 activity and resembles the carboxyl-terminal half of TUG (153). Unexpectedly, in HeLa cells, TUG localizes to the endoplasmic reticulum–Golgi intermediate compartment (ERGIC), endoplasmic reticulum exit sites (ERES), and, to some degree, *cis*-Golgi compartments (152). Thus, TUG may regulate p97 oligomerization at one or more of these locations. Whether amino-terminal sequences of TUG or extracellular stimuli control localized p97 hexamer disassembly is not known. Regardless, this action of TUG may control a trafficking pathway that is present in a wide range of cell types and that is adapted and upregulated to control GLUT4 in adipocytes and myocytes.

That TUG may reside at the ERGIC/ERES fits well with the idea that p115 and Golgin-160 function to retain GSVs intracellularly in cells not stimulated with insulin (91, 135). These Golgi matrix proteins are enriched at the *cis*-Golgi compartments and ERGIC/ERES, and both IRAP and LRP1 bind p115 (32, 135). GSVs may then be mobilized directly to the cell surface (70) by an unconventional secretion pathway, similar to that used by CFTR, β1-integrin, and other proteins (154–156). Golgin-160 interacts with PIST, an effector of the TC10α GTPase, providing a potential link between insulin signaling and GSV mobilization. Of note, PIST, syntaxin-6, and TC10 control the trafficking and degradation of CFTR (157). We have found that PIST and Golgin-160 interact with TUG, and data support the hypothesis that insulin signaling through TC10α releases GSVs in a manner

ERGIC: endoplasmic reticulum–Golgi intermediate compartment

ERES: endoplasmic reticulum exit site

GERL:
Golgi-endoplasmic reticulum-lysosome–related organelle

requiring site-specific endoproteolytic cleavage of TUG (J.S. Bogan, B.R. Rubin, C. Yu, M.G. Löffler, C.M. Orme, et al., unpublished results). This liberates an amino-terminal fragment that may function as a ubiquitin-like modifier, and it separates regions of TUG that bind GSVs from those that bind Golgi matrix proteins. Cleavage is observed in adipocyte but not fibroblast cell lines and is required for fully insulin-responsive GLUT4 translocation. When transfected in HeLa cells, a fragment corresponding to the TUG carboxyl-terminal product accumulates in the nucleus (152). Another regulator of GLUT4 translocation is RIP140, a nuclear receptor corepressor that also functions in the cytosol to inhibit Akt-mediated AS160 phosphorylation (39). These observations raise the possibility that GSV mobilization and gene expression may somehow be coupled in fat and muscle. Ubc9 and Daxx bind GLUT4 and may participate in TUG processing (145, 146, 158). The mobilized vesicles are likely carried toward the cell surface by kinesin motors (159). The idea that TGN-derived GSVs may be trapped at the ERGIC/ERES recalls Novikoff's GERL (Golgi–endoplasmic reticulum–lysosome) concept, which suggested that particular endoplasmic reticulum domains make direct contact with *trans*-Golgi compartments (160). Remarkably, in 1980, Novikoff et al. noted that 3T3-L1 cells contain "an extensive acid phosphatase (AcPase) positive GERL from which coated vesicles apparently arise (these coated vesicles display AcPase activity and are much smaller and far more numerous than the coated vesicles that seem to arise from the plasmalemmal coated pits)" (161, p. 180). One wonders if these vesicles were GSVs!

The idea that insulin liberates sequestered GLUT4 by a quantal-release mechanism, which increases the number of proteins that participate in recycling at the cell surface, fits well with the hypothesis that TUG is destroyed (14, 24, 122). Yet, during ongoing insulin exposure, GSV cargo recycles directly to the cell surface from endosomes (70). This arrangement obviates the need for ongoing TUG destruction (and new synthesis) during sustained insulin exposure, because the GSV-retention mechanism is bypassed. Precisely how insulin acts on TUG will require further study, and mechanisms controlling synthesis of TUG proteins are not understood. Nonetheless, evidence indicates that this is an important site of action involving a novel biochemical mechanism.

Effects of a dominant negative fragment of TUG (UBX-Cter) are similar to those of RNAi-mediated TUG depletion. This fragment likely acts by binding and occupying an anchoring site at the Golgi matrix, excluding endogenous TUG from this site to prevent the capture of GSVs. In unstimulated 3T3-L1 adipocytes, the vesicles then default to the plasma membrane (70, 96). In transgenic mice, muscle-specific expression of this dominant negative fragment causes increased fasting glucose uptake into muscle, reduced circulating glucose and insulin, enhanced fasting glucose turnover, and hypermetabolism (M.G. Löffler, A. Birkenfeld, K. Mitsopoulos, W.M. Philbrick, G.I. Shulman, V.T. Samuel, and J.S. Bogan, unpublished data). This result is consistent with data showing that, in muscle, insulin stimulates the dissociation of TUG from GLUT4 (148) and that TUG and GLUT4 abundances are highly correlated (38). Intriguingly, heterozygous Ubc9-knockout mice have a small but significant increase in blood glucose concentrations, which may reflect Ubc9 action in this pathway (162). Therefore, this mechanism for GSV retention and release likely pertains to muscle and regulates organism-level glucose homeostasis.

Possible Coordination of Rabs and Tethers

Whether AS160 and TUG act coordinately on GSVs is not known. Simultaneous knockdown of AS160 and IRAP reveals additive effects to increase GLUT4 targeting to the surface of unstimulated cells (33). One possibility is that IRAP knockdown disrupts a TUG-mediated retention mechanism. "Tethering" of GSVs may involve a cycle of fusion and budding at TGN (or other) membranes, which may be directed by Rabs on the GSVs and modulated by AS160. Vesicle capture by Golgi tethers, followed by fusion at an appropriate cisterna, is thought to be a mechanism for the action of Golgi matrix proteins (134). If TUG serves as an adaptor to link GSVs to matrix proteins, then it may cooperate with Rabs to direct a cycle of GSV fusion and budding in unstimulated cells. Knockdown of either AS160 or TUG could then impair GSV sequestration by disrupting this cycle.

TUG depletion does not affect insulin signaling to Akt, and targeting of the released vesicles may be controlled to some degree by AS160. Alternatively, AS160 may help direct GLUT4 that is endocytosed during ongoing insulin stimulation to bypass the GSVs and to return to the plasma membrane from endosomes (70). Insulin also modulates the endocytic mechanism, dephosphorylates syntaxin-16, and may act through Akt to phosphorylate ACAP1 (93, 111, 128). These signals may determine if endocytosed GLUT4 is targeted to GSVs or directed back to the cell surface. Understanding if these signaling pathways regulate the same or different trafficking events will require further study.

TRANSLOCATION OF GLUT4 STORAGE VESICLES MAY COORDINATELY REGULATE MULTIPLE PHYSIOLOGIC END POINTS

IRAP and LRP1 have actions that are significant for organism-level physiology, apart from their roles in controlling GSV trafficking and glucose uptake. When present at the cell surface, IRAP cleaves circulating peptides that have an amino-terminal disulfide-linked cysteine residue (163). A physiologic substrate is arginine vasopressin (AVP), which is inactivated by IRAP action (163). AVP controls vascular tone, renal water resorption, and urea transport, and it acts through the epithelial sodium channel, ENaC, to control sodium retention and blood pressure (164, 165). IRAP-knockout mice have a threefold prolongation of AVP clearance and a twofold increase in serum AVP concentrations (163). AVP is used clinically to treat hypotension in septic shock. Remarkably, a polymorphism in the gene encoding the human homolog of rodent IRAP, P-LAP, predicts both AVP clearance and mortality in individuals with septic shock (166). IRAP/P-LAP also cleaves oxytocin and plays a role in reproduction (163, 167). During pregnancy, the ectodomain of human P-LAP (but not rodent IRAP) is shed from the placenta, and alterations in this process may contribute to hypertension in pre-eclampsia. Thus, GSV translocation is predicted to result not only in increased glucose uptake, but also in enhanced AVP degradation.

Why should AVP inactivation be enhanced by insulin action or muscle contraction? One possibility is that this action helps to increase blood flow to metabolically active muscles. AVP inactivation may enhance local vasodilation so that GSV translocation coordinates glucose delivery and uptake. Because of the mass and extensive T-tubule system of skeletal muscle, the surface area that it presents to the extracellular space is enormous (possibly similar to the absorptive area of the gut). Although diffusion and *trans*-capillary passage of glucose/AVP must be taken into account (21, 168), it is conceivable that, in a pathophysiologic setting, dysregulated IRAP translocation has systemic consequences. If systemic AVP action is prolonged, it may contribute to hypertension in the metabolic syndrome.

LRP1 has pleotropic roles in many tissues (169). A member of the LDL receptor family, LRP1 binds multiple ligands and mediates the uptake of triglyceride-rich lipoproteins bearing ApoE and ApoA5 (170, 171). Sortilin and SorLA are also present in GSVs and contribute to the control of lipid metabolism, though the mechanism for this regulation is debated (172). LRP1 also modulates Wnt signaling, which may further regulate lipid metabolism and mitochondrial activity (173, 174). Thus, GSV translocation may contribute to the control of lipid as well as glucose metabolism, and defects in this process may contribute to the dyslipidemia present in the metabolic syndrome. These speculations deserve further attention.

Vesicles in Other Cell Types That Are Similar to GLUT4 Storage Vesicles

GSV-like vesicles are present in other terminally differentiated cell types and likely control a wide range of physiology (9). In dendritic cells, for example, IRAP is required for cross-presentation of antigens in association with MHC class I molecules (175). GSV-like vesicles are present in regions of the brain (176), and IRAP has been suggested to function in spatial memory (177). Similar translocation mechanisms may regulate AQP2 water channels in the renal collecting duct, H^+/K^+ ATPases in gastric parietal cells, and the Na^+/K^+ pump in muscle and kidney (178–180). Distinct upstream signaling mechanisms may be involved in each of these instances. The extent to which common downstream trafficking mechanisms are used and adapted to particular tissue-specific cargoes will be a topic for future research.

IMPLICATIONS FOR INSULIN RESISTANCE AND DIABETES

Substantial data support the concept that insulin resistance in muscle results from the accumulation of lipids, particularly diacylglycerols and sphingolipids including ceramides (8, 181). Possibly, diacylglycerols may act through protein kinase D to regulate trafficking pathways involved in GSV formation, in addition to acting through novel protein kinase C isoforms to impair insulin signaling (8, 182). Lowering cellular ceramide content reverses insulin resistance. In models of insulin resistance, some defects are independent of IRS-1 and Akt (183). Compromised insulin signaling through TC10α or other pathways, or in vesicle trafficking pathways in unstimulated cells, may be responsible. Sphingolipid depletion enhances the formation of GSVs in 3T3-L1 adipocytes (184). Conversely, palmitate treatment of C2C12 myocytes downregulates sortilin and impairs GLUT4 translocation, possibly acting through ceramides (109). These results fit with data from humans suggesting that intracellular GLUT4 is not properly localized in fasting, insulin-resistant individuals (20, 51–53). Thus, insulin resistance may result not only from impaired insulin signal transduction, but also from insulin-independent alterations in GLUT4 targeting. Lipid raft assembly can take place in pre-Golgi compartments and endosomes as well as at the TGN (154). Thus, disrupted membrane composition may compromise GSV formation or intracellular retention. How such a defect may combine with insulin signaling defects to prevent overall translocation of GLUT4 to the cell surface will require further study. The synthesis of these biochemical insights with data from human genetics and nutritional studies promises a deeper understanding of diabetes pathogenesis with implications for prevention and treatment.

SUMMARY POINTS

1. Insulin signals through multiple pathways that act at distinct vesicle trafficking steps to redistribute GLUT4 glucose transporters and enhance glucose uptake in fat and muscle cells.

2. An important signaling pathway is through Akt to AS160/Tbc1D4 and Tbc1D1, thereby modulating Rab GTPases to direct GLUT4 vesicle traffic.

3. GLUT4 storage vesicles (GSVs) are the main targets of insulin regulation. These are formed in a cell-type-specific manner, efficiently sequester GLUT4 within unstimulated cells, and are mobilized to the cell surface by insulin signaling.

4. GSVs contain a limited number of proteins that assemble together to specify vesicle formation and regulated trafficking; these proteins may also coordinate glucose uptake with other physiologic effects.

5. GSVs are retained at the Golgi matrix by the action of TUG, Ubc9, and other proteins and may be mobilized along an unconventional secretory pathway to the cell surface.

6. During ongoing insulin exposure, GLUT4 and other cargoes likely recycle to the plasma membrane directly from endosomes, so that the GSVs are bypassed.

7. GSV-like vesicles are present in a wide range of cell types and may translocate in response to various extracellular stimuli to control diverse biological effects.

8. During the pathogenesis of type 2 diabetes, impaired insulin signaling and GLUT4 vesicle trafficking may both contribute to insulin resistance, and they may result from accumulation of ceramides and other sphingolipids as well as diacylglycerol.

FUTURE ISSUES

1. Are distinct insulin signaling pathways activated differentially at varying insulin concentrations? How is activation affected by nutritional status and membrane lipid composition?

2. What Rab proteins are controlled by AS160/Tbc1D4 and Tbc1D1? Are distinct Rab isoforms used in fat and muscle? Are there other regulators and effectors of these Rab proteins?

3. How are GSVs trapped within unstimulated cells? From what membranes do these vesicles bud? How is GSV budding coupled to capture and retention? Is GSV sequestration modulated by nutritional cues or cellular energy status?

4. How does insulin mobilize GSVs? Are actions through AS160 and TUG coordinated? Are other signals involved?

5. How is GLUT4 regulated by contraction and by ischemia in skeletal and cardiac muscle? What is the relative importance of GSVs versus other sites of regulation? What cross talk occurs with insulin signaling?

6. What steps in insulin signaling and vesicle trafficking are impaired in insulin resistance? What is the role of membrane lipid composition?

7. Are vascular tone and lipid metabolism coordinately regulated with glucose uptake by IRAP and LRP1? Do sortilin or sortilin-like proteins participate in these physiological processes? What other processes are affected? Are other LRP isoforms involved?

8. Do similar mechanisms regulate GSV-like vesicles in other tissues? What cargoes are regulated? How do different upstream signaling pathways act on common downstream mechanisms for vesicle capture and release?

DISCLOSURE STATEMENT

The author is not aware of any affiliations, memberships, funding, or financial holdings that might be perceived as affecting the objectivity of this review.

ACKNOWLEDGMENTS

I am grateful to members of my laboratory and to Drs. Christopher Burd, Konstantin Kandror, and Derek Toomre for helpful discussions. This work was supported by National Institutes of Health grant DK075772 and by a Distinguished Young Scholar Award from the W.M. Keck Foundation.

LITERATURE CITED

1. Levine R, Goldstein M, Klein S, Huddlestun B. 1949. The action of insulin on the distribution of galactose in eviscerated nephrectomized dogs. *J. Biol. Chem.* 179:985–86
2. Morgan HE, Henderson MJ, Regen DM, Park CR. 1961. Regulation of glucose uptake in muscle. I. The effects of insulin and anoxia on glucose transport and phosphorylation in the isolated, perfused heart of normal rats. *J. Biol. Chem.* 236:253–61
3. Cushman SW, Wardzala LJ. 1980. Potential mechanism of insulin action on glucose transport in the isolated rat adipose cell. Apparent translocation of intracellular transport systems to the plasma membrane. *J. Biol. Chem.* 255:4758–62
4. Suzuki K, Kono T. 1980. Evidence that insulin causes translocation of glucose transport activity to the plasma membrane from an intracellular storage site. *Proc. Natl. Acad. Sci. USA* 77:2542–45
5. Mueckler M, Caruso C, Baldwin SA, Panico M, Blench I, et al. 1985. Sequence and structure of a human glucose transporter. *Science* 229:941–45
6. James DE, Strube M, Mueckler M. 1989. Molecular cloning and characterization of an insulin-regulatable glucose transporter. *Nature* 338:83–87
7. Birnbaum MJ. 1989. Identification of a novel gene encoding an insulin-responsive glucose transporter protein. *Cell* 57:305–15
8. Samuel VT, Petersen KF, Shulman GI. 2010. Lipid-induced insulin resistance: unravelling the mechanism. *Lancet* 375:2267–77
9. Bryant NJ, Govers R, James DE. 2002. Regulated transport of the glucose transporter GLUT4. *Nat. Rev. Mol. Cell Biol.* 3:267–77
10. Huang S, Czech MP. 2007. The GLUT4 glucose transporter. *Cell Metab.* 5:237–52
11. Larance M, Ramm G, James DE. 2008. The GLUT4 code. *Mol. Endocrinol.* 22:226–33
12. Foley K, Boguslavsky S, Klip A. 2011. Endocytosis, recycling, and regulated exocytosis of glucose transporter 4. *Biochemistry* 50:3048–61
13. Rubin BR, Bogan JS. 2009. Intracellular retention and insulin-stimulated mobilization of GLUT4 glucose transporters. *Vitam. Horm.* 80:155–92
14. Bogan JS, Kandror KV. 2010. Biogenesis and regulation of insulin-responsive vesicles containing GLUT4. *Curr. Opin. Cell Biol.* 22:506–12
15. Thorens B, Mueckler M. 2010. Glucose transporters in the 21st century. *Am. J. Physiol. Endocrinol. Metab.* 298:E141–45
16. Naftalin RJ. 2010. Reassessment of models of facilitated transport and cotransport. *J. Membr. Biol.* 234:75–112
17. Minokoshi Y, Kahn CR, Kahn BB. 2003. Tissue-specific ablation of the GLUT4 glucose transporter or the insulin receptor challenges assumptions about insulin action and glucose homeostasis. *J. Biol. Chem.* 278:33609–12
18. Russell RR 3rd, Li J, Coven DL, Pypaert M, Zechner C, et al. 2004. AMP-activated protein kinase mediates ischemic glucose uptake and prevents postischemic cardiac dysfunction, apoptosis, and injury. *J. Clin. Investig.* 114:495–503
19. Cline GW, Petersen KF, Krssak M, Shen J, Hundal RS, et al. 1999. Impaired glucose transport as a cause of decreased insulin-stimulated muscle glycogen synthesis in type 2 diabetes. *N. Engl. J. Med.* 341:240–46
20. Garvey WT, Maianu L, Zhu JH, Brechtel-Hook G, Wallace P, Baron AD. 1998. Evidence for defects in the trafficking and translocation of GLUT4 glucose transporters in skeletal muscle as a cause of human insulin resistance. *J. Clin. Investig.* 101:2377–86

21. Wasserman DH. 2009. Four grams of glucose. *Am. J. Physiol. Endocrinol. Metab.* 296:E11–21
22. Slot JW, Geuze HJ, Gigengack S, James DE, Lienhard GE. 1991. Translocation of the glucose transporter GLUT4 in cardiac myocytes of the rat. *Proc. Natl. Acad. Sci. USA* 88:7815–19
23. Slot JW, Geuze HJ, Gigengack S, Lienhard GE, James DE. 1991. Immuno-localization of the insulin regulatable glucose transporter in brown adipose tissue of the rat. *J. Cell Biol.* 113:123–35
24. Govers R, Coster AC, James DE. 2004. Insulin increases cell surface GLUT4 levels by dose dependently discharging GLUT4 into a cell-surface recycling pathway. *Mol. Cell. Biol.* 24:6456–66
25. Kane S, Sano H, Liu SC, Asara JM, Lane WS, et al. 2002. A method to identify serine kinase substrates. Akt phosphorylates a novel adipocyte protein with a Rab GTPase-activating protein (GAP) domain. *J. Biol. Chem.* 277:22115–18
26. Cho H, Mu J, Kim JK, Thorvaldsen JL, Chu Q, et al. 2001. Insulin resistance and a diabetes mellitus–like syndrome in mice lacking the protein kinase Akt2 (PKB beta). *Science* 292:1728–31
27. Garofalo RS, Orena SJ, Rafidi K, Torchia AJ, Stock JL, et al. 2003. Severe diabetes, age-dependent loss of adipose tissue, and mild growth deficiency in mice lacking Akt2/PKB beta. *J. Clin. Investig.* 112:197–208
28. Sano H, Kane S, Sano E, Miinea CP, Asara JM, et al. 2003. Insulin-stimulated phosphorylation of a Rab GTPase-activating protein regulates GLUT4 translocation. *J. Biol. Chem.* 278:14599–602
29. Eguez L, Lee A, Chavez JA, Miinea CP, Kane S, et al. 2005. Full intracellular retention of GLUT4 requires AS160 Rab GTPase-activating protein. *Cell Metab.* 2:263–72
30. Larance M, Ramm G, Stockli J, van Dam EM, Winata S, et al. 2005. Characterization of the role of the Rab GTPase-activating protein AS160 in insulin-regulated GLUT4 trafficking. *J. Biol. Chem.* 280:37803–13
31. Peck GR, Ye S, Pham V, Fernando RN, Macaulay SL, et al. 2006. Interaction of the Akt substrate, AS160, with the glucose transporter 4 vesicle marker protein, insulin-regulated aminopeptidase. *Mol. Endocrinol.* 20:2576–83
32. Jedrychowski MP, Gartner CA, Gygi SP, Zhou L, Herz J, et al. 2010. Proteomic analysis of GLUT4 storage vesicles reveals LRP1 to be an important vesicle component and target of insulin signaling. *J. Biol. Chem.* 285:104–14
33. Jordens I, Molle D, Xiong W, Keller SR, McGraw TE. 2010. Insulin-regulated aminopeptidase is a key regulator of GLUT4 trafficking by controlling the sorting of GLUT4 from endosomes to specialized insulin-regulated vesicles. *Mol. Biol. Cell* 21:2034–44
34. Yeh TY, Sbodio JI, Tsun ZY, Luo B, Chi NW. 2007. Insulin-stimulated exocytosis of GLUT4 is enhanced by IRAP and its partner tankyrase. *Biochem. J.* 402:279–90
35. Koumanov F, Richardson JD, Murrow BA, Holman GD. 2011. AS160 phosphotyrosine-binding domain constructs inhibit insulin-stimulated GLUT4 vesicle fusion with the plasma membrane. *J. Biol. Chem.* 286:16574–82
36. Chen S, Synowsky S, Tinti M, Mackintosh C. 2011. The capture of phosphoproteins by 14-3-3 proteins mediates actions of insulin. *Trends Endocrinol. Metab.* 22:429–36
37. Ho PC, Lin YW, Tsui YC, Gupta P, Wei LN. 2009. A negative regulatory pathway of GLUT4 trafficking in adipocyte: new function of RIP140 in the cytoplasm via AS160. *Cell Metab.* 10:516–23
38. Castorena CM, Mackrell JG, Bogan JS, Kanzaki M, Cartee GD. 2011. Clustering of GLUT4, TUG and RUVBL2 protein levels correlate with myosin heavy-chain isoform pattern in skeletal muscles, but AS160 and TBC1D1 levels do not. *J. Appl. Physiol.* 111:1106–17
39. Xie X, Chen Y, Xue P, Fan Y, Deng Y, et al. 2009. RUVBL2, a novel AS160-binding protein, regulates insulin-stimulated GLUT4 translocation. *Cell Res.* 19:1090–97
40. Chavez JA, Roach WG, Keller SR, Lane WS, Lienhard GE. 2008. Inhibition of GLUT4 translocation by Tbc1d1, a Rab GTPase-activating protein abundant in skeletal muscle, is partially relieved by AMP-activated protein kinase activation. *J. Biol. Chem.* 283:9187–95
41. Deshmukh A, Coffey VG, Zhong Z, Chibalin AV, Hawley JA, Zierath JR. 2006. Exercise-induced phosphorylation of the novel Akt substrates AS160 and filamin A in human skeletal muscle. *Diabetes* 55:1776–82
42. Bruss MD, Arias EB, Lienhard GE, Cartee GD. 2005. Increased phosphorylation of Akt substrate of 160 kDa (AS160) in rat skeletal muscle in response to insulin or contractile activity. *Diabetes* 54:41–50

43. Taylor EB, An D, Kramer HF, Yu H, Fujii NL, et al. 2008. Discovery of TBC1D1 as an insulin-, AICAR-, and contraction-stimulated signaling nexus in mouse skeletal muscle. *J. Biol. Chem.* 283:9787–96
44. Dash S, Sano H, Rochford JJ, Semple RK, Yeo G, et al. 2009. A truncation mutation in TBC1D4 in a family with acanthosis nigricans and postprandial hyperinsulinemia. *Proc. Natl. Acad. Sci. USA* 106:9350–55
45. Meyre D, Farge M, Lecoeur C, Proenca C, Durand E, et al. 2008. R125W coding variant in TBC1D1 confers risk for familial obesity and contributes to linkage on chromosome 4p14 in the French population. *Hum. Mol. Genet.* 17:1798–802
46. Chadt A, Leicht K, Deshmukh A, Jiang LQ, Scherneck S, et al. 2008. Tbc1d1 mutation in lean mouse strain confers leanness and protects from diet-induced obesity. *Nat. Genet.* 40:1354–59
47. Karlsson HK, Hallsten K, Bjornholm M, Tsuchida H, Chibalin AV, et al. 2005. Effects of metformin and rosiglitazone treatment on insulin signaling and glucose uptake in patients with newly diagnosed type 2 diabetes: a randomized controlled study. *Diabetes* 54:1459–67
48. Frosig C, Rose AJ, Treebak JT, Kiens B, Richter EA, Wojtaszewski JF. 2007. Effects of endurance exercise training on insulin signaling in human skeletal muscle: interactions at the level of phosphatidyl-inositol 3-kinase, Akt, and AS160. *Diabetes* 56:2093–102
49. Howlett KF, Mathews A, Garnham A, Sakamoto K. 2008. The effect of exercise and insulin on AS160 phosphorylation and 14-3-3 binding capacity in human skeletal muscle. *Am. J. Physiol. Endocrinol. Metab.* 294:E401–7
50. Frosig C, Pehmoller C, Birk JB, Richter EA, Wojtaszewski JF. 2010. Exercise-induced TBC1D1 Ser237 phosphorylation and 14-3-3 protein-binding capacity in human skeletal muscle. *J. Physiol.* 588:4539–48
51. Maianu L, Keller SR, Garvey WT. 2001. Adipocytes exhibit abnormal subcellular distribution and translocation of vesicles containing glucose transporter 4 and insulin-regulated aminopeptidase in type 2 diabetes mellitus: implications regarding defects in vesicle trafficking. *J. Clin. Endocrinol. Metab.* 86:5450–56
52. Karlsson HK, Ahlsen M, Zierath JR, Wallberg-Henriksson H, Koistinen HA. 2006. Insulin signaling and glucose transport in skeletal muscle from first-degree relatives of type 2 diabetic patients. *Diabetes* 55:1283–88
53. Vassilopoulos S, Esk C, Hoshino S, Funke BH, Chen CY, et al. 2009. A role for the CHC22 clathrin heavy-chain isoform in human glucose metabolism. *Science* 324:1192–96
54. Sano H, Eguez L, Teruel MN, Fukuda M, Chuang TD, et al. 2007. Rab10, a target of the AS160 Rab GAP, is required for insulin-stimulated translocation of GLUT4 to the adipocyte plasma membrane. *Cell Metab.* 5:293–303
55. Ishikura S, Bilan PJ, Klip A. 2007. Rabs 8A and 14 are targets of the insulin-regulated Rab-GAP AS160 regulating GLUT4 traffic in muscle cells. *Biochem. Biophys. Res. Commun.* 353:1074–79
56. Sano H, Peck GR, Kettenbach AN, Gerber SA, Lienhard GE. 2011. Insulin-stimulated GLUT4 protein translocation in adipocytes requires the Rab10 guanine nucleotide exchange factor Dennd4C. *J. Biol. Chem.* 286:16541–45
57. Roland JT, Lapierre LA, Goldenring JR. 2009. Alternative splicing in class V myosins determines association with Rab10. *J. Biol. Chem.* 284:1213–23
58. Brewer PD, Romenskaia I, Kanow MA, Mastick CC. 2011. Loss of AS160 Akt substrate causes Glut4 protein to accumulate in compartments that are primed for fusion in basal adipocytes. *J. Biol. Chem.* 286:26287–97
59. Gonzalez E, McGraw TE. 2006. Insulin signaling diverges into Akt-dependent and -independent signals to regulate the recruitment/docking and the fusion of GLUT4 vesicles to the plasma membrane. *Mol. Biol. Cell* 17:4484–93
60. Koumanov F, Jin B, Yang J, Holman GD. 2005. Insulin signaling meets vesicle traffic of GLUT4 at a plasma-membrane-activated fusion step. *Cell Metab.* 2:179–89
61. Jewell JL, Oh E, Ramalingam L, Kalwat MA, Tagliabracci VS, et al. 2011. Munc18c phosphorylation by the insulin receptor links cell signaling directly to SNARE exocytosis. *J. Cell Biol.* 193:185–99
62. Bergman BC, Cornier MA, Horton TJ, Bessesen DH, Eckel RH. 2008. Skeletal muscle Munc18c and Syntaxin 4 in human obesity. *Nutr. Metab.* 5:21

63. Bostrom P, Andersson L, Rutberg M, Perman J, Lidberg U, et al. 2007. SNARE proteins mediate fusion between cytosolic lipid droplets and are implicated in insulin sensitivity. *Nat. Cell Biol.* 9:128693
64. Lanner JT, Bruton JD, Katz A, Westerblad H. 2008. Ca(2+) and insulin-mediated glucose uptake. *Curr. Opin. Pharmacol.* 8:339–45
65. Yip MF, Ramm G, Larance M, Hoehn KL, Wagner MC, et al. 2008. CaMKII-mediated phosphorylation of the myosin motor Myo1c is required for insulin-stimulated GLUT4 translocation in adipocytes. *Cell Metab.* 8:384–98
66. Toyoda T, An D, Witczak CA, Koh HJ, Hirshman MF, et al. 2011. Myo1c regulates glucose uptake in mouse skeletal muscle. *J. Biol. Chem.* 286:4133–40
67. Chen XW, Leto D, Xiao J, Goss J, Wang Q, et al. 2011. Exocyst function is regulated by effector phosphorylation. *Nat. Cell Biol.* 13:580–88
68. Chen XW, Leto D, Xiong T, Yu G, Cheng A, et al. 2011. A Ral GAP complex links PI 3-kinase/Akt signaling to RalA activation in insulin action. *Mol. Biol. Cell* 22:141–52
69. Huang P, Altshuller YM, Hou JC, Pessin JE, Frohman MA. 2005. Insulin-stimulated plasma membrane fusion of Glut4 glucose transporter-containing vesicles is regulated by phospholipase D1. *Mol. Biol. Cell* 16:2614–23
70. Xu Y, Rubin BR, Orme CM, Karpikov A, Yu C, et al. 2011. Dual-mode of insulin action controls GLUT4 vesicle exocytosis. *J. Cell Biol.* 193:643–53
71. Di Paolo G, Moskowitz HS, Gipson K, Wenk MR, Voronov S, et al. 2004. Impaired PtdIns(4,5)P2 synthesis in nerve terminals produces defects in synaptic vesicle trafficking. *Nature* 431:415–22
72. Chiang SH, Baumann CA, Kanzaki M, Thurmond DC, Watson RT, et al. 2001. Insulin-stimulated GLUT4 translocation requires the CAP-dependent activation of TC10. *Nature* 410:944–48
73. Chang L, Chiang SH, Saltiel AR. 2007. TC10alpha is required for insulin-stimulated glucose uptake in adipocytes. *Endocrinology* 148:27–33
74. Ahn MY, Katsanakis KD, Bheda F, Pillay TS. 2004. Primary and essential role of the adaptor protein APS for recruitment of both c-Cbl and its associated protein CAP in insulin signaling. *J. Biol. Chem.* 279:21526–32
75. Mitra P, Zheng X, Czech MP. 2004. RNAi-based analysis of CAP, Cbl, and CrkII function in the regulation of GLUT4 by insulin. *J. Biol. Chem.* 279:37431–35
76. Minami A, Iseki M, Kishi K, Wang M, Ogura M, et al. 2003. Increased insulin sensitivity and hypoinsulinemia in APS knockout mice. *Diabetes* 52:2657–65
77. Molero JC, Turner N, Thien CB, Langdon WY, James DE, Cooney GJ. 2006. Genetic ablation of the c-Cbl ubiquitin ligase domain results in increased energy expenditure and improved insulin action. *Diabetes* 55:3411–17
78. JeBailey L, Rudich A, Huang X, Di Ciano-Oliveira C, Kapus A, Klip A. 2004. Skeletal muscle cells and adipocytes differ in their reliance on TC10 and Rac for insulin-induced actin remodeling. *Mol. Endocrinol.* 18:359–72
79. Ahmed Z, Smith BJ, Pillay TS. 2000. The APS adapter protein couples the insulin receptor to the phosphorylation of c-Cbl and facilitates ligand-stimulated ubiquitination of the insulin receptor. *FEBS Lett.* 475:31–34
80. Kanzaki M, Watson RT, Hou JC, Stamnes M, Saltiel AR, Pessin JE. 2002. Small GTP-binding protein TC10 differentially regulates two distinct populations of filamentous actin in 3T3L1 adipocytes. *Mol. Biol. Cell* 13:2334–46
81. Falasca M, Hughes WE, Dominguez V, Sala G, Fostira F, et al. 2007. The role of phosphoinositide 3-kinase C2alpha in insulin signaling. *J. Biol. Chem.* 282:28226–36
82. Lodhi IJ, Bridges D, Chiang SH, Zhang Y, Cheng A, et al. 2008. Insulin stimulates phosphatidylinositol 3-phosphate production via the activation of Rab5. *Mol. Biol. Cell* 19:2718–28
83. Inoue M, Chang L, Hwang J, Chiang SH, Saltiel AR. 2003. The exocyst complex is required for targeting of Glut4 to the plasma membrane by insulin. *Nature* 422:629–33
84. Kanzaki M, Mora S, Hwang JB, Saltiel AR, Pessin JE. 2004. Atypical protein kinase C (PKCzeta/lambda) is a convergent downstream target of the insulin-stimulated phosphatidylinositol 3-kinase and TC10 signaling pathways. *J. Cell Biol.* 164:279–90

85. Farese RV, Sajan MP, Yang H, Li P, Mastorides S, et al. 2007. Muscle-specific knockout of PKC-lambda impairs glucose transport and induces metabolic and diabetic syndromes. *J. Clin. Investig.* 117:2289–301
86. Hodgkinson CP, Mander A, Sale GJ. 2005. Protein kinase-zeta interacts with munc18c: role in GLUT4 trafficking. *Diabetologia* 48:1627–36
87. Charest A, Lane K, McMahon K, Housman DE. 2001. Association of a novel PDZ domain-containing peripheral Golgi protein with the Q-SNARE (Q-soluble N-ethylmaleimide-sensitive fusion protein (NSF) attachment protein receptor) protein syntaxin 6. *J. Biol. Chem.* 276:29456–65
88. Cheng J, Moyer BD, Milewski M, Loffing J, Ikeda M, et al. 2002. A Golgi-associated PDZ domain protein modulates cystic fibrosis transmembrane regulator plasma membrane expression. *J. Biol. Chem.* 277:3520–29
89. Yao R, Maeda T, Takada S, Noda T. 2001. Identification of a PDZ domain-containing Golgi protein, GOPC, as an interaction partner of frizzled. *Biochem. Biophys. Res. Commun.* 286:771–78
90. Hicks SW, Machamer CE. 2005. Isoform-specific interaction of golgin-160 with the Golgi-associated protein PIST. *J. Biol. Chem.* 280:28944–51
91. Williams D, Hicks SW, Machamer CE, Pessin JE. 2006. Golgin-160 is required for the golgi membrane sorting of the insulin-responsive glucose transporter GLUT4 in adipocytes. *Mol. Biol. Cell* 17:5346–55
92. Watson RT, Hou JC, Pessin JE. 2008. Recycling of IRAP from the plasma membrane back to the insulin-responsive compartment requires the Q-SNARE syntaxin 6 but not the GGA clathrin adaptors. *J. Cell Sci.* 121:1243–51
93. Perera HK, Clarke M, Morris NJ, Hong W, Chamberlain LH, Gould GW. 2003. Syntaxin 6 regulates Glut4 trafficking in 3T3-L1 adipocytes. *Mol. Biol. Cell* 14:2946–58
94. Shewan AM, Van Dam EM, Martin S, Luen TB, Hong W, et al. 2003. GLUT4 recycles via a *trans*-Golgi network (TGN) subdomain enriched in syntaxins 6 and 16 but not TGN38: involvement of an acidic targeting motif. *Mol. Biol. Cell* 14:973–86
95. Bogan JS, Hendon N, McKee AE, Tsao TS, Lodish HF. 2003. Functional cloning of TUG as a regulator of GLUT4 glucose transporter trafficking. *Nature* 425:727–33
96. Yu C, Cresswell J, Loffler MG, Bogan JS. 2007. The glucose transporter 4–regulating protein TUG is essential for highly insulin-responsive glucose uptake in 3T3-L1 adipocytes. *J. Biol. Chem.* 282:7710–22
97. Sebastian BM, Nagy LE. 2005. Decreased insulin-dependent glucose transport by chronic ethanol feeding is associated with dysregulation of the Cbl/TC10 pathway in rat adipocytes. *Am. J. Physiol. Endocrinol. Metab.* 289:E1077–84
98. Bernard JR, Reeder DW, Herr HJ, Rivas DA, Yaspelkis BB 3rd. 2006. High-fat feeding effects on components of the CAP/Cbl signaling cascade in Sprague-Dawley rat skeletal muscle. *Metabolism* 55:203–12
99. Gupte A, Mora S. 2006. Activation of the Cbl insulin signaling pathway in cardiac muscle: dysregulation in obesity and diabetes. *Biochem. Biophys. Res. Commun.* 342:751–7
100. Jun HS, Hwang K, Kim Y, Park T. 2008. High-fat diet alters PP2A, TC10, and CIP4 expression in visceral adipose tissue of rats. *Obesity* 16:1226–31
101. Bernard JR, Saito M, Liao YH, Yaspelkis BB 3rd, Ivy JL. 2008. Exercise training increases components of the c-Cbl-associated protein/c-Cbl signaling cascade in muscle of obese Zucker rats. *Metabolism* 57:858–66
102. Saito M, Lessard SJ, Rivas DA, Reeder DW, Hawley JA, Yaspelkis BB 3rd. 2008. Activation of atypical protein kinase Czeta toward TC10 is regulated by high-fat diet and aerobic exercise in skeletal muscle. *Metabolism* 57:1173–80
103. Ribon V, Johnson JH, Camp HS, Saltiel AR. 1998. Thiazolidinediones and insulin resistance: peroxisome proliferatoractivated receptor gamma activation stimulates expression of the *CAP* gene. *Proc. Natl. Acad. Sci. USA* 95:14751–56
104. Gaulton KJ, Willer CJ, Li Y, Scott LJ, Conneely KN, et al. 2008. Comprehensive association study of type 2 diabetes and related quantitative traits with 222 candidate genes. *Diabetes* 57:3136–44
105. Hong KW, Jin HS, Lim JE, Ryu HJ, Go MJ, et al. 2009. RAPGEF1 gene variants associated with type 2 diabetes in the Korean population. *Diabetes Res. Clin. Pract.* 84:117–22

106. Kathiresan S, Melander O, Guiducci C, Surti A, Burtt NP, et al. 2008. Six new loci associated with blood low-density lipoprotein cholesterol, high-density lipoprotein cholesterol or triglycerides in humans. *Nat. Genet.* 40:189–97
107. Willer CJ, Sanna S, Jackson AU, Scuteri A, Bonnycastle LL, et al. 2008. Newly identified loci that influence lipid concentrations and risk of coronary artery disease. *Nat. Genet.* 40:161–69
108. Ban HJ, Heo JY, Oh KS, Park KJ. 2010. Identification of type 2 diabetes–associated combination of SNPs using support vector machine. *BMC Genet.* 11:26
109. Tsuchiya Y, Hatakeyama H, Emoto N, Wagatsuma F, Matsushita S, Kanzaki M. 2010. Palmitate-induced down-regulation of sortilin and impaired GLUT4 trafficking in C2C12 myotubes. *J. Biol. Chem.* 285:34371–81
110. Holman GD, Lo Leggio L, Cushman SW. 1994. Insulin-stimulated GLUT4 glucose transporter recycling. A problem in membrane protein subcellular trafficking through multiple pools. *J. Biol. Chem.* 269:17516–24
111. Blot V, McGraw TE. 2006. GLUT4 is internalized by a cholesterol-dependent nystatin-sensitive mechanism inhibited by insulin. *EMBO J.* 25:5648–58
112. Habtemichael EN, Brewer PD, Romenskaia I, Mastick CC. 2011. Kinetic evidence that Glut4 follows different endocytic pathways than the receptors for transferrin and alpha2-macroglobulin. *J. Biol. Chem.* 286:10115–25
113. Zaid H, Talior-Volodarsky I, Antonescu C, Liu Z, Klip A. 2009. GAPDH binds GLUT4 reciprocally to hexokinase-II and regulates glucose transport activity. *Biochem. J.* 419:475–84
114. Kandror KV, Pilch PF. 2011. The sugar is sIRVed: sorting Glut4 and its fellow travelers. *Traffic* 12:665–71
115. Dugani CB, Randhawa VK, Cheng AW, Patel N, Klip A. 2008. Selective regulation of the perinuclear distribution of glucose transporter 4 (GLUT4) by insulin signals in muscle cells. *Eur. J. Cell Biol.* 87:337–51
116. Fujita H, Hatakeyama H, Watanabe TM, Sato M, Higuchi H, Kanzaki M. 2010. Identification of three distinct functional sites of insulin-mediated GLUT4 trafficking in adipocytes using quantitative single molecule imaging. *Mol. Biol. Cell* 21:2721–31
117. Karylowski O, Zeigerer A, Cohen A, McGraw TE. 2004. GLUT4 is retained by an intracellular cycle of vesicle formation and fusion with endosomes. *Mol. Biol. Cell* 15:870–82
118. Martin OJ, Lee A, McGraw TE. 2006. GLUT4 distribution between the plasma membrane and the intracellular compartments is maintained by an insulin-modulated bipartite dynamic mechanism. *J. Biol. Chem.* 281:484–90
119. Coster AC, Govers R, James DE. 2004. Insulin stimulates the entry of GLUT4 into the endosomal recycling pathway by a quantal mechanism. *Traffic* 5:763–71
120. Bogan JS, McKee AE, Lodish HF. 2001. Insulin-responsive compartments containing GLUT4 in 3T3-L1 and CHO cells: regulation by amino acid concentrations. *Mol. Cell. Biol.* 21:4785–806
121. Shi J, Kandror KV. 2005. Sortilin is essential and sufficient for the formation of Glut4 storage vesicles in 3T3-L1 adipocytes. *Dev. Cell* 9:99–108
122. Muretta JM, Romenskaia I, Mastick CC. 2008. Insulin releases Glut4 from static storage compartments into cycling endosomes and increases the rate constant for Glut4 exocytosis. *J. Biol. Chem.* 283:311–23
123. Shi J, Huang G, Kandror KV. 2008. Self-assembly of Glut4 storage vesicles during differentiation of 3T3-L1 adipocytes. *J. Biol. Chem.* 283:30311–21
124. Wang J, Sun HQ, Macia E, Kirchhausen T, Watson H, et al. 2007. PI4P promotes the recruitment of the GGA adaptor proteins to the *trans*-Golgi network and regulates their recognition of the ubiquitin sorting signal. *Mol. Biol. Cell* 18:2646–55
125. Lamb CA, McCann RK, Stockli J, James DE, Bryant NJ. 2010. Insulin-regulated trafficking of GLUT4 requires ubiquitination. *Traffic* 11:1445–54
126. Li J, Peters PJ, Bai M, Dai J, Bos E, et al. 2007. An ACAP1-containing clathrin coat complex for endocytic recycling. *J. Cell Biol.* 178:453–64
127. Takatsu H, Yoshino K, Toda K, Nakayama K. 2002. GGA proteins associate with Golgi membranes through interaction between their GGAH domains and ADP-ribosylation factors. *Biochem. J.* 365:369–78

128. Li J, Ballif BA, Powelka AM, Dai J, Gygi SP, Hsu VW. 2005. Phosphorylation of ACAP1 by Akt regulates the stimulation-dependent recycling of integrin beta1 to control cell migration. *Dev. Cell* 9:663–73
129. Esk C, Chen CY, Johannes L, Brodsky FM. 2010. The clathrin heavy-chain isoform CHC22 functions in a novel endosomal sorting step. *J. Cell Biol.* 188:131–44
130. Watson RT, Khan AH, Furukawa M, Hou JC, Li L, et al. 2004. Entry of newly synthesized GLUT4 into the insulin-responsive storage compartment is GGA dependent. *EMBO J.* 23:2059–70
131. Liu G, Hou JC, Watson RT, Pessin JE. 2005. Initial entry of IRAP into the insulin-responsive storage compartment occurs prior to basal or insulin-stimulated plasma membrane recycling. *Am. J. Physiol. Endocrinol. Metab.* 289:E746–52
132. Khan AH, Capilla E, Hou JC, Watson RT, Smith JR, Pessin JE. 2004. Entry of newly synthesized GLUT4 into the insulin-responsive storage compartment is dependent upon both the amino terminus and the large cytoplasmic loop. *J. Biol. Chem.* 279:37505–11
133. Li LV, Bakirtzi K, Watson RT, Pessin JE, Kandror KV. 2009. The C-terminus of GLUT4 targets the transporter to the perinuclear compartment but not to the insulin-responsive vesicles. *Biochem. J.* 419:105–12
134. Munro S. 2011. The golgin coiled-coil proteins of the Golgi apparatus. *Cold Spring Harb. Perspect. Biol.* 3:a005256
135. Hosaka T, Brooks CC, Presman E, Kim SK, Zhang Z, et al. 2005. p115 interacts with the GLUT4 vesicle protein, IRAP, and plays a critical role in insulin-stimulated GLUT4 translocation. *Mol. Biol. Cell* 16:2882–90
136. Hou JC, Suzuki N, Pessin JE, Watson RT. 2006. A specific dileucine motif is required for the GGA-dependent entry of newly synthesized insulin-responsive aminopeptidase into the insulin-responsive compartment. *J. Biol. Chem.* 281:33457–66
137. Kupriyanova TA, Kandror V, Kandror KV. 2002. Isolation and characterization of the two major intracellular Glut4 storage compartments. *J. Biol. Chem.* 277:9133–38
138. Capilla E, Suzuki N, Pessin JE, Hou JC. 2007. The glucose transporter 4 FQQI motif is necessary for Akt substrate of 160-kilodalton-dependent plasma membrane translocation but not Golgi-localized (gamma)-ear-containing Arf-binding protein-dependent entry into the insulin-responsive storage compartment. *Mol. Endocrinol.* 21:3087–99
139. Blot V, McGraw TE. 2008. Molecular mechanisms controlling GLUT4 intracellular retention. *Mol. Biol. Cell* 19:3477–87
140. Waters SB, D'Auria M, Martin SS, Nguyen C, Kozma LM, Luskey KL. 1997. The amino terminus of insulin-responsive aminopeptidase causes Glut4 translocation in 3T3-L1 adipocytes. *J. Biol. Chem.* 272:23323–7
141. Katagiri H, Asano T, Yamada T, Aoyama T, Fukushima Y, et al. 2002. Acyl-coenzyme A dehydrogenases are localized on GLUT4-containing vesicles via association with insulin-regulated aminopeptidase in a manner dependent on its dileucine motif. *Mol. Endocrinol.* 16:1049–59
142. Hirata Y, Hosaka T, Iwata T, Le CT, Jambaldorj B, et al. 2011. Vimentin binds IRAP and is involved in GLUT4 vesicle trafficking. *Biochem. Biophys. Res. Commun.* 405:96–101
143. Tojo H, Kaieda I, Hattori H, Katayama N, Yoshimura K, et al. 2003. The Formin family protein, formin homolog overexpressed in spleen, interacts with the insulin-responsive aminopeptidase and profilin IIa. *Mol. Endocrinol.* 17:1216–29
144. Keller SR, Davis AC, Clairmont KB. 2002. Mice deficient in the insulin-regulated membrane aminopeptidase show substantial decreases in glucose transporter GLUT4 levels but maintain normal glucose homeostasis. *J. Biol. Chem.* 277:17677–86
145. Giorgino F, de Robertis O, Laviola L, Montrone C, Perrini S, et al. 2000. The sentrin-conjugating enzyme mUbc9 interacts with GLUT4 and GLUT1 glucose transporters and regulates transporter levels in skeletal muscle cells. *Proc. Natl. Acad. Sci. USA* 97:1125–30
146. Liu LB, Omata W, Kojima I, Shibata H. 2007. The SUMO conjugating enzyme Ubc9 is a regulator of GLUT4 turnover and targeting to the insulin-responsive storage compartment in 3T3-L1 adipocytes. *Diabetes* 56:1977–85

147. Ladanyi M, Lui MY, Antonescu CR, Krause-Boehm A, Meindl A, et al. 2001. The der(17)t(X;17)(p11;q25) of human alveolar soft part sarcoma fuses the *TFE3* transcription factor gene to *ASPL*, a novel gene at 17q25. *Oncogene* 20:48–57
148. Schertzer JD, Antonescu CN, Bilan PJ, Jain S, Huang X, et al. 2009. A transgenic mouse model to study glucose transporter 4myc regulation in skeletal muscle. *Endocrinology* 150:1935–40
149. Tettamanzi MC, Yu C, Bogan JS, Hodsdon ME. 2006. Solution structure and backbone dynamics of an N-terminal ubiquitin-like domain in the GLUT4-regulating protein, TUG. *Protein Sci.* 15:498–508
150. Schuberth C, Buchberger A. 2008. UBX domain proteins: major regulators of the AAA ATPase Cdc48/p97. *Cell Mol. Life Sci.* 65:2360–71
151. Alexandru G, Graumann J, Smith GT, Kolawa NJ, Fang R, Deshaies RJ. 2008. UBXD7 binds multiple ubiquitin ligases and implicates p97 in HIF1alpha turnover. *Cell* 134:804–16
152. Orme CM, Bogan JS. 2011. The ubiquitin regulatory X (UBX) domain-containing protein TUG regulates the p97 ATPase and resides at the endoplasmic reticulum–Golgi intermediate compartment. *J. Biol. Chem.* 287:6679–92
153. Rancour DM, Park S, Knight SD, Bednarek SY. 2004. Plant UBX domain-containing protein 1, PUX1, regulates the oligomeric structure and activity of *Arabidopsis* CDC48. *J. Biol. Chem.* 279:54264–74
154. Saraste J, Dale HA, Bazzocco S, Marie M. 2009. Emerging new roles of the pre-Golgi intermediate compartment in biosynthetic-secretory trafficking. *FEBS Lett.* 583:3804–10
155. Wang C, Yoo Y, Fan H, Kim E, Guan KL, Guan JL. 2010. Regulation of integrin β1 recycling to lipid rafts by Rab1a to promote cell migration. *J. Biol. Chem.* 285:29398–405
156. Yoo JS, Moyer BD, Bannykh S, Yoo HM, Riordan JR, Balch WE. 2002. Non-conventional trafficking of the cystic fibrosis transmembrane conductance regulator through the early secretory pathway. *J. Biol. Chem.* 277:11401–9
157. Cheng J, Cebotaru V, Cebotaru L, Guggino WB. 2010. Syntaxin 6 and CAL mediate the degradation of the cystic fibrosis transmembrane conductance regulator. *Mol. Biol. Cell* 21:1178–87
158. Lalioti VS, Vergarajauregui S, Pulido D, Sandoval IV. 2002. The insulin-sensitive glucose transporter, GLUT4, interacts physically with Daxx. Two proteins with capacity to bind Ubc9 and conjugated to SUMO1. *J. Biol. Chem.* 277:19783–91
159. Semiz S, Park JG, Nicoloro SM, Furcinitti P, Zhang C, et al. 2003. Conventional kinesin KIF5B mediates insulin-stimulated GLUT4 movements on microtubules. *EMBO J.* 22:2387–99
160. Novikoff AB. 1976. The endoplasmic reticulum: a cytochemist's view (a review). *Proc. Natl. Acad. Sci. USA* 73:2781–87
161. Novikoff AB, Novikoff PM, Rosen OM, Rubin CS. 1980. Organelle relationships in cultured 3T3-L1 preadipocytes. *J. Cell Biol.* 87:180–96
162. Nacerddine K, Lehembre F, Bhaumik M, Artus J, Cohen-Tannoudji M, et al. 2005. The SUMO pathway is essential for nuclear integrity and chromosome segregation in mice. *Dev. Cell* 9:769–79
163. Wallis MG, Lankford MF, Keller SR. 2007. Vasopressin is a physiological substrate for the insulin-regulated aminopeptidase IRAP. *Am. J. Physiol. Endocrinol. Metab.* 293:E1092–102
164. Bankir L, Bichet DG, Bouby N. 2010. Vasopressin V2 receptors, ENaC, and sodium reabsorption: a risk factor for hypertension? *Am. J. Physiol. Ren. Physiol* 299:F917–28
165. Fenton RA, Knepper MA. 2007. Mouse models and the urinary concentrating mechanism in the new millennium. *Physiol. Rev.* 87:1083–112
166. Nakada TA, Russell JA, Wellman H, Boyd JH, Nakada E, et al. 2011. Leucyl/cystinyl aminopeptidase gene variants in septic shock. *Chest* 139:1042–49
167. Mizutani S, Wright JW, Kobayashi H. 2011. Placental leucine aminopeptidase- and aminopeptidase A-deficient mice offer insight concerning the mechanisms underlying preterm labor and preeclampsia. *J. Biomed. Biotechnol.* 2011:286947
168. Gudbjornsdottir S, Sjostrand M, Strindberg L, Wahren J, Lonnroth P. 2003. Direct measurements of the permeability surface area for insulin and glucose in human skeletal muscle. *J. Clin. Endocrinol. Metab.* 88:4559–64
169. Lillis AP, Van Duyn LB, Murphy-Ullrich JE, Strickland DK. 2008. LDL receptor-related protein 1: unique tissue-specific functions revealed by selective gene knockout studies. *Physiol. Rev.* 88:887–918

170. Nilsson SK, Christensen S, Raarup MK, Ryan RO, Nielsen MS, Olivecrona G. 2008. Endocytosis of apolipoprotein A-V by members of the low density lipoprotein receptor and the VPS10p domain receptor families. *J. Biol. Chem.* 283:25920–27
171. Hofmann SM, Zhou L, Perez-Tilve D, Greer T, Grant E, et al. 2007. Adipocyte LDL receptor-related protein-1 expression modulates postprandial lipid transport and glucose homeostasis in mice. *J. Clin. Investig.* 117:3271–82
172. Willnow TE, Kjolby M, Nykjaer A. 2011. Sortilins: new players in lipoprotein metabolism. *Curr. Opin. Lipidol.* 22:79–85
173. Terrand J, Bruban V, Zhou L, Gong W, El Asmar Z, et al. 2009. LRP1 controls intracellular cholesterol storage and fatty acid synthesis through modulation of Wnt signaling. *J. Biol. Chem.* 284:381–88
174. Yoon JC, Ng A, Kim BH, Bianco A, Xavier RJ, Elledge SJ. 2010. Wnt signaling regulates mitochondrial physiology and insulin sensitivity. *Genes Dev.* 24:1507–18
175. Saveanu L, Carroll O, Weimershaus M, Guermonprez P, Firat E, et al. 2009. IRAP identifies an endosomal compartment required for MHC class I cross-presentation. *Science* 325:213–17
176. Bakirtzi K, Belfort G, Lopez-Coviella I, Kuruppu D, Cao L, et al. 2009. Cerebellar neurons possess a vesicular compartment structurally and functionally similar to Glut4-storage vesicles from peripheral insulin-sensitive tissues. *J. Neurosci. Off. J. Soc. Neurosci.* 29:5193–201
177. Albiston AL, Fernando RN, Yeatman HR, Burns P, Ng L, et al. 2010. Gene knockout of insulin-regulated aminopeptidase: loss of the specific binding site for angiotensin IV and age-related deficit in spatial memory. *Neurobiol. Learn. Mem.* 93:19–30
178. Kim HY, Choi HJ, Lim JS, Park EJ, Jung HJ, et al. 2011. Emerging role of Akt substrate protein AS160 in the regulation of AQP2 translocation. *Am. J. Physiol. Ren. Physiol.* 301:F151–61
179. Alves DS, Farr GA, Seo-Mayer P, Caplan MJ. 2010. AS160 associates with the Na^+,K^+-ATPase and mediates the adenosine monophosphate-stimulated protein kinase-dependent regulation of sodium pump surface expression. *Mol. Biol. Cell* 21:4400–8
180. Aoyama F, Sawaguchi A. 2011. Functional transformation of gastric parietal cells and intracellular trafficking of ion channels/transporters in the apical canalicular membrane associated with acid secretion. *Biol. Pharm. Bull.* 34:813–16
181. Ussher JR, Koves TR, Cadete VJ, Zhang L, Jaswal JS, et al. 2010. Inhibition of de novo ceramide synthesis reverses diet-induced insulin resistance and enhances whole-body oxygen consumption. *Diabetes* 59:2453–64
182. Malhotra V, Campelo F. 2011. PKD regulates membrane fission to generate TGN to cell surface transport carriers. *Cold Spring Harb. Perspect. Biol.* 3:a005280
183. Hoehn KL, Hohnen-Behrens C, Cederberg A, Wu LE, Turner N, et al. 2008. IRS1-independent defects define major nodes of insulin resistance. *Cell Metab.* 7:421–33
184. Cheng ZJ, Singh RD, Wang TK, Holicky EL, Wheatley CL, et al. 2010. Stimulation of GLUT4 (glucose transporter isoform 4) storage vesicle formation by sphingolipid depletion. *Biochem. J.* 427:143–50

Structure and Regulation of Soluble Guanylate Cyclase

Emily R. Derbyshire[1] and Michael A. Marletta[2]

[1]Department of Biological Chemistry and Molecular Pharmacology, Harvard Medical School, Boston, Massachusetts 02115

[2]Department of Chemistry, The Scripps Research Institute, La Jolla, California 92037; email: marletta@scripps.edu

Keywords

nitric oxide, heme, signaling, desensitization, nitrosation

Abstract

Nitric oxide (NO) is an essential signaling molecule in biological systems. In mammals, the diatomic gas is critical to the cyclic guanosine monophosphate (cGMP) pathway as it functions as the primary activator of soluble guanylate cyclase (sGC). NO is synthesized from L-arginine and oxygen (O_2) by the enzyme nitric oxide synthase (NOS). Once produced, NO rapidly diffuses across cell membranes and binds to the heme cofactor of sGC. sGC forms a stable complex with NO and carbon monoxide (CO), but not with O_2. The binding of NO to sGC leads to significant increases in cGMP levels. The second messenger then directly modulates phosphodiesterases (PDEs), ion-gated channels, or cGMP-dependent protein kinases to regulate physiological functions, including vasodilation, platelet aggregation, and neurotransmission. Many studies are focused on elucidating the molecular mechanism of sGC activation and deactivation with a goal of therapeutic intervention in diseases involving the NO/cGMP-signaling pathway. This review summarizes the current understanding of sGC structure and regulation as well as recent developments in NO signaling.

Contents

- INTRODUCTION 534
  - Enzymes Critical to the Nitric Oxide/Cyclic Guanosine Monophosphate Pathway 534
- ISOLATION OF SOLUBLE GUANYLATE CYCLASE 535
- SOLUBLE GUANYLATE CYCLASE ISOFORMS 537
- ARCHITECTURE OF SOLUBLE GUANYLATE CYCLASE 538
  - Heme-Nitric Oxide and Oxygen Binding Domain 538
  - Per/Arnt/Sim and Coiled-Coil Domains 539
  - Catalytic Domain 539
- SOLUBLE GUANYLATE CYCLASE HOMOLOGS 540
  - Eukaryotic Atypical Soluble Guanylate Cyclases 540
  - Prokaryotic Heme-Nitric Oxide and Oxygen Binding Proteins 541
- STRUCTURAL INSIGHTS FROM STUDIES ON SOLUBLE GUANYLATE CYCLASE AND ITS HOMOLOGS 542
- LIGAND SELECTIVITY 545
- REGULATION OF SOLUBLE GUANYLATE CYCLASE BY NITRIC OXIDE 546
  - Activation and Nitric Oxide Association 546
  - Deactivation and Nitric Oxide Dissociation 547
  - Desensitization 548
- MODULATORS OF SOLUBLE GUANYLATE CYCLASE ACTIVITY 549
  - Soluble Guanylate Cyclase Activators 549
  - Soluble Guanylate Cyclase Inhibitors 550

INTRODUCTION

The nitric oxide/cyclic guanosine monophosphate (NO/cGMP) pathway was discovered in the 1980s, but chemical modulation of the pathway for the treatment of angina pectoris had been unknowingly achieved 100 years earlier. This stimulation of cGMP production occurred with the clinical administration of organic nitrites (isoamyl nitrite) (1) and organic nitrates (glyceryl trinitrate; GTN) (2). These compounds alleviate the pain associated with angina by relaxing the vascular smooth muscle, leading to vasodilation. For years, investigations were focused on the mechanism of smooth muscle relaxation by these molecules, and these efforts led to the discovery that NO is a physiologically relevant signaling molecule. Additionally, these efforts led to the identification of the enzymes that biosynthesize NO and cGMP.

Enzymes Critical to the Nitric Oxide/Cyclic Guanosine Monophosphate Pathway

Early studies showed that both cytosolic and particulate fractions of mammalian tissue exhibit guanylate cyclase activity. Within the insoluble fractions, membrane-bound particulate guanylate cyclases are present, which are activated by natriuretic peptides (reviewed in References 3 and 4), whereas the cytosolic fractions contain soluble guanylate cyclases (sGCs), which are activated by NO. NO-responsive guanylate cyclase activity is also associated with cell membranes in certain tissues, including skeletal muscle and brain, as well as in platelets (5–7). Guanylate cyclases are found in most tissues, and the distribution of these proteins in various cells is isoform specific. This provides an additional means to regulate cGMP-dependent responses because localized pools of the signaling molecule can be generated within specific tissues and in proximity to either soluble or membrane-bound cGMP receptors. Thus, tissues can regulate cGMP levels by expression of specific GC isoforms, and the isoforms have a distinct peptide receptor or ligand activator. Additionally, a reciprocal communication between particulate guanylate cyclase and sGC has been observed in the regulation of human and mouse vascular homeostasis (8), and it remains likely that communication between these pathways occurs in several cGMP-dependent processes.

Generally, in eukaryotic NO signaling, the initial event involves calcium release, followed by binding of a calcium/calmodulin complex to nitric oxide synthase (NOS), which activates the enzyme. NO is synthesized and then diffuses into target cells, where it binds to the heme in sGC (**Figure 1**). sGC is a histidine-ligated hemoprotein that binds NO and carbon monoxide (CO), but not oxygen (O_2). This binding event leads to a several hundredfold increase in cGMP synthesis. Once formed, cGMP targets include phosphodiesterases (PDEs), ion-gated channels, and cGMP-dependent protein kinases in the regulation of several physiological functions, including vasodilation, platelet aggregation, and neurotransmission (9–11).

In 1998, the Nobel Prize in Physiology or Medicine was awarded to Robert F. Furchgott, Louis J. Ignarro, and Ferid Murad, in recognition of their achievements toward the discovery of the NO-signaling pathway. Currently, this pathway is actively studied because drugs modulating NO-dependent processes have the potential to treat several maladies, including cardiovascular and neurodegenerative diseases, as well as various airway diseases.

ISOLATION OF SOLUBLE GUANYLATE CYCLASE

Despite many years of research on sGC, an efficient low-cost purification of the protein has

cGMP: cyclic guanosine monophosphate

GTN: glyceryl trinitrate

Vasodilation: blood vessel widening from smooth muscle relaxation

sGC: soluble guanylate cyclase

NOS: nitric oxide synthase

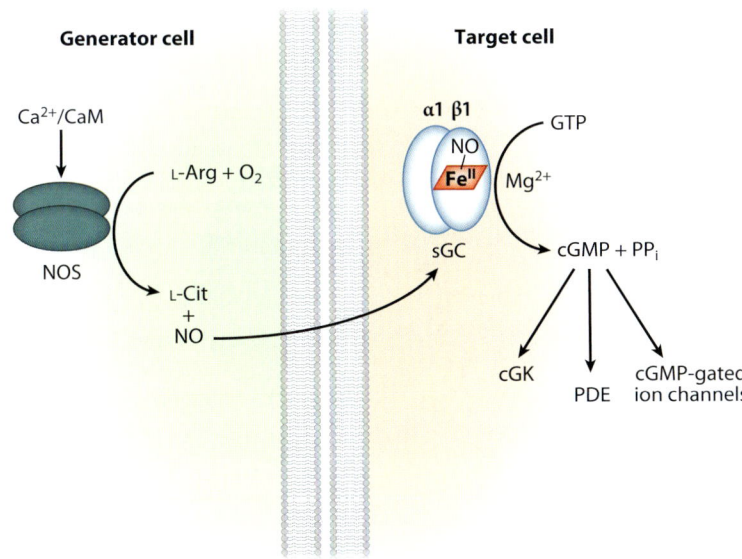

Figure 1
The nitric oxide/cyclic GMP (NO/cGMP)-signaling pathway. A Ca^{2+}/calmodulin (CaM) complex binds nitric oxide synthase (NOS). NOS catalyzes the oxidation of L-arginine (L-Arg) to L-citrulline (L-Cit) and nitric oxide (NO). NO binds to the Fe^{II} heme of α1 β1 soluble guanylate cyclase (sGC) at a diffusion-controlled rate. This binding event leads to significant increases in cGMP and pyrophosphate (PP_i). cGMP then binds to and activates cGMP-dependent protein kinases (cGKs), phosphodiesterases (PDEs) and ion-gated channels. Abbreviations: α1 and β1, soluble guanylate cyclase subunits; CaM, calmodulin.

GTP: guanosine 5′-triphosphate

remained elusive, but several methods have been developed that yield low microgram amounts of homogeneous protein. Initial characterization of sGC was carried out with protein obtained from rat and bovine tissues. By the 1980s, studies were being done with purified sGC from rat lung (12) and liver (13), as well as from bovine lung (14, 15); these studies showed sGC to be a heterodimer. Importantly, it was observed that sGC could be purified with and without the heme cofactor, depending on the purification protocol. In short, the use of solubilizing agents or ammonium sulfate precipitation can lead to misfolded apoprotein. Heme reconstitution of this apoprotein yields an sGC species [later termed sGC^1 by Vogel et al. (16)] with biochemical properties that vary from the native protein [named sGC^2 by Vogel et al. (16)]. To date, the bovine lung sGC prep is the most efficient method of isolating heme-bound protein from source tissue (17, 18). This method typically yields ~1 mg of pure protein per kilogram of lung.

The development of heterologous expression systems for recombinant sGC expression led to significant advances. The first successful heterologous expression system for sGC was accomplished in COS-7 cells (19). Although COS-7 cells do not produce enough sGC for protein purification, the procedure was pivotal to establishing that sGC is an obligate heterodimer composed of α1 and β1 subunits (20). Additionally, COS-7 cells have been used to examine mutants and truncations of sGC via lysate activity assays (21, 22).

The overexpression of rat sGC in insect cells with the Sf9/baculovirus expression system was the first procedure used to isolate pure recombinant protein (21, 23, 24). sGC expression in insect cells was initially accomplished without an affinity tag (21), but current protocols involve a His tag to facilitate the purification process (25–27). Although highly expressed, most (>90%) of the protein is insoluble. This method typically yields 0.2–0.4 mg of pure soluble protein per liter of culture and is now commonly used to obtain purified rat and human sGC. A clear advantage of the Sf9/baculovirus expression system is that it enables in vitro characterization, including the generation of site-directed mutants. However, both the COS-7 and Sf9/baculovirus expression systems facilitated the biochemical characterization of subunit dimerization, allowed for the generation of site-directed mutants, and provided larger quantities of enzyme for in vitro studies.

The full-length mammalian α1β1 heterodimer has not yet been isolated from a bacterial expression system, but several truncations of sGC have been successfully obtained via expression in *Escherichia coli*. These proteins include N-terminal truncations of α1, β1, and β2 (28–30), C-terminal truncations of α1 and β1 (31, 32), and a domain within the central region of β1 (**Figure 2**) (31). These constructs purify with yields ranging from 0.5 to 5 mg

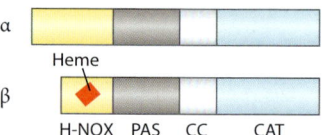

Figure 2

Domain architecture of soluble guanylate cyclase (sGC). (*a*) Heme-nitric oxide/oxygen binding (H-NOX, *yellow*), Per/Arnt/Sim domain (PAS, *gray*), coiled-coil domains (CC, *white*), and catalytic domains (CAT, *blue*) are shown. (*b*) Characterized heme-binding and GTP-binding truncations of rat sGC. Heme is represented by the red parallelogram. α1, β1, and β2 isoforms are shown.

of pure protein per liter of culture. Recently an *E. coli* expression system was used for the full-length *Manduca sexta* α1β1 heterodimer (33). This method provides low amounts (0.5–1 mg/liter) of partially pure full-length protein, but higher yields (1–2 mg/liter) of pure protein were obtained by truncating the C terminus of the α1 and β1 subunits (33). The resulting heterodimeric proteins (msGC-NT1 and msGC-NT2) lack the ability to cyclize GTP but can be purified to homogeneity. Perhaps future work optimizing the expression conditions and/or purification of sGC from *E. coli* will overcome the current limitations in protein yield.

SOLUBLE GUANYLATE CYCLASE ISOFORMS

Heterodimeric sGC consists of two homologous subunits, α and β. The most commonly studied isoform is the α1β1 protein; however, α2 and β2 subunits have also been identified (34, 35). These proteins were first characterized in mammals, but they also exist in insects, such as *Drosophila melanogaster* and *M. sexta*, and in fish. Isoforms of α-subunits are highly homologous with ∼48% sequence identity, and β-subunits have an overall sequence identity of ∼41%.

The localization of each subunit has been studied in mammals, including humans, rats, and cows. Both α1 and β1 subunits are expressed in most tissues, and it is well accepted that these proteins form a physiologically relevant heterodimer (36). By quantitative polymerase chain reaction analysis and Western blotting, the α2 subunit is found in fewer tissues when compared to the α1 and β1 isoforms but is highly expressed in the brain, lung, colon, heart, spleen, uterus, and placenta (36). Studies with purified protein have shown that the α2β1 heterodimer exhibits ligand-binding characteristics identical to the α1β1 heterodimer (37, 38), but a splice variant of the α2 subunit (α2i) forms a dimer with the β1 subunit to form an inactive complex. α2i contains an in-frame insertion of 31 amino acids within the catalytic domain and appears to function as a dominant-negative protein (39). Despite the similar biochemical properties of the two physiologically relevant sGC heterodimers, they have unique roles in cGMP signaling that may be attributed to their varying cellular localization. Specifically, α2β1 has been associated with the membrane in several tissues (5–7), and consequently, α2β1 responds differently than cytosolic α1β1 (37). In rat brain, it was found that this association is mediated by an interaction between the C terminus of the α2 protein and PSD-95 (postsynaptic density-95) protein (6). Most recently, a distinct presynaptic role has been identified for α1β1 in glutamate release in the hippocampus (40).

The β2 isoform is not ubiquitously expressed like the β1 isoform, and analysis of mRNA levels indicates that it is found primarily in the kidney (35). Unlike β1, the C terminus of β2 contains a possible isoprenylation site, but the subcellular localization of this protein is unknown. The β2 isoform has not yet been purified and characterized, but the protein has been studied after transient expression in insect cells. Using this approach, it was found that β2 does not exhibit cyclase activity when expressed with α1 or α2 but that β2 is active in the absence of an α-subunit, suggesting that the β2 protein can function as a homodimer ex vivo (41). This is in contrast to the β1 protein, which has been isolated as an inactive homodimer after overexpression in insect cells (26). In rat kidney, the mRNA levels of the β2 subunit were shown to be developmentally regulated (42); however, it remains unclear what the physiological role of the β2 isoform is in cGMP signaling.

Recently, several genetic studies in mouse models have emphasized the importance of the various sGC isoforms for physiological processes (reviewed in Reference 43). Knockout mice lacking the sGC β1 subunit exhibit elevated blood pressure, reduced heart rate, and dysfunction in gastrointestinal contractility (44). Additionally, deletion of the β1 subunit within only smooth muscle cells implicates the loss of the protein in these cells as the cause of hypertension in the knockout mice (45). Deletion of the β1 subunit is generally viewed

PAS: protein fold named for its association with the Per, ARNT, and Sim proteins

H-NOX: heme-nitric oxide and oxygen binding

as a global sGC knockout because α1 and α2 do not form functional heterodimers with β2. sGC α1 and α2 knockout mice have also been generated. Both proteins were found to be essential for long-term potentiation (46), and vasodilation was mediated primarily by the α1 isoform (47). In addition to clarifying the role of the α1β1 protein in NO-mediated pulmonary vasodilation (48, 49), studies with α1 subunit-deficient mice suggest that both α-subunits are involved in gastric nitrergic relaxation (50) and relaxation of colon tissue (51).

Several invertebrates also contain genes that encode predicted NO-sensitive guanylate cyclases. Likely owing to the genetic tools available in *D. melanogaster*, this organism contains the best-characterized insect guanylate cyclase. In *D. melanogaster*, Gycα-99B and Gycβ-100B are orthologs of the α1 and β1 subunits, respectively. *Drosophila* mutants deficient in the production of these proteins suggest that the NO/cGMP pathway mediates a behavioral phenotype (52) and development of the visual system (53), and this pathway is important for larval foraging locomotion (54). In other organisms, several biochemical studies have been aimed at elucidating the significance of the NO/cGMP-signaling pathway. As mentioned above, the full-length *M. sexta* α1β1 heterodimer has recently been characterized (33). There is evidence that sGC in *M. sexta* is involved in neuronal excitability (55) and odor responsiveness (56). In mollusks, such as the *Limax marginatus* and *Limax maximus* slugs, sGC may modulate the electrical oscillation of interneurons in the central olfactory pathway (57, 58).

ARCHITECTURE OF SOLUBLE GUANYLATE CYCLASE

The rat sGC α1 and β1 subunits are 690 and 619 amino acids in length, respectively. These proteins are part of a large family of sGC subunits that are conserved in eukaryotes. Generally, there is the highest sequence variability at the N terminus of α-subunits and the greatest sequence identity at the C terminus of both the α- and β-proteins. Each sGC subunit consists of four distinct domains. The β1 subunit contains a N-terminal heme-binding domain, a Per/Arnt/Sim (PAS) domain, a coiled-coil domain, and a C-terminal catalytic domain (**Figure 2**) (reviewed in Reference 59).

Heme-Nitric Oxide and Oxygen Binding Domain

Experiments with sGC truncations and site-directed mutants were necessary to localize the minimal heme-binding domain of sGC. These experiments involved the systematic mutation of conserved histidines (60), expression of various truncations in *E. coli* (29), and deletion of the β1 N terminus (61). Taken together, these studies showed that the β1 N terminus constituted the heme-binding domain and suggested that histidine 105 (rat) was the proximal heme ligand. The sGC heme-binding domain has been localized to residues 1 to ∼194 on the β1 subunit (28). Like the full-length sGC, the isolated heme domain binds NO and CO, but not O_2. The N terminus of the α1 subunit has homology to the β1 N terminus and was shown to have affinity for heme despite lacking the proximal histidine ligand (30). The sGC N terminus is part of a conserved family of proteins found in both prokaryotes and eukaryotes (62). This family of proteins has been termed heme-nitric oxide binding (62), sensor of nitric oxide (63), and heme-nitric oxide and oxygen binding (H-NOX) (64). H-NOX is the most used abbreviation and will be used throughout this review. To date, all of the characterized H-NOX proteins bind heme as well as the gaseous heme ligands NO and CO (reviewed in Reference 65). Some H-NOX proteins, including β1 and β2, discriminate against O_2 binding, whereas others form a stable complex with O_2. In eukaryotes, H-NOX proteins have only been found with the known sGC domain architecture. In bacteria, H-NOX domains can be found as proteins of ∼200 amino acids in length with a single predicted function or as a domain within a larger protein on the basis of sequence analysis programs. Additionally, the genes that encode these H-NOX proteins in bacteria are in proximity to genes that encode putative histidine kinases and diguanylate cyclases, or in some cases, a gene encodes the H-NOX as a domain within a larger protein, often fused to a methyl-accepting chemotaxis domain. It is likely that these proteins have an evolutionarily conserved function and serve as gas sensors in prokaryotes and eukaryotes.

Per/Arnt/Sim and Coiled-Coil Domains

The central region of sGC contains two domains of unresolved function. The domain closer to the N terminus is predicted to adopt a PAS-like fold. Typically, PAS domains mediate protein-protein interactions and have often been found to bind heme, a flavin, or a nucleotide (66). The other domain, a coiled-coil domain, appears to be unique to sGC and shares no significant homology with any other protein in the National Center for Biotechnology Information protein database (**http://www.ncbi.nlm.nih.gov/protein**). The coiled-coil domain of the rat $\beta 1$ subunit (residues 348–409) was isolated as a tetramer and structurally elucidated with X-ray crystallography (31). This structural study, in addition to experiments involving site-directed mutagenesis of residues on the $\alpha 1$ subunit (67), a bimolecular fluorescence complementation assay in cells (68), and structural analysis of homologs of the sGC PAS domain (69), suggests that the central regions of both sGC subunits are important for the formation of a functional heterodimer.

Catalytic Domain

The C-terminal regions of the $\alpha 1$ and $\beta 1$ proteins are highly homologous to the particulate guanylate cyclase and adenylate cyclase catalytic domains. In the 1990s, structural insights on the sGC catalytic domains came from homology models on the basis of crystal structures of the adenylate cyclase catalytic domains (70, 71). These models identified key catalytic residues, including two conserved aspartate residues on the $\alpha 1$ subunit (D485 and D529, rat numbering), which are predicted to bind two Mg^{2+} ions (72). These residues are critical to catalysis as the associated metals likely activate both the nucleotide 3′-hydroxyl and the α-phosphate for the reaction, as well as stabilize the charge on the β- and γ-phosphates on both the substrate and product. Additionally, $\beta 1$ N548 is proposed to orient the ribose ring for the reaction. Residues thought to be responsible for base recognition include E473 and C541 on the $\beta 1$ subunit. Other residues on both $\alpha 1$ (R573) and $\beta 1$ (R552) are thought to interact with the nucleotide triphosphate (72). With the identification of these critical residues, predictions can be made about guanylate cyclase activity using sequence analysis. This type of analysis would correctly predict that the $\beta 2$ isoform could function as a homodimer but that $\beta 1$, $\alpha 1$, and $\alpha 2$ need a partner to be active.

The catalytic domains have now been localized to the C-terminal 467–690 and 414–619 residues of the $\alpha 1$ and $\beta 1$ subunits, respectively (32). These catalytic domains must form a heterodimer for cGMP to be synthesized, and in the full-length protein, the catalytic efficiency of the protein is dependent on the heme ligation state of the $\beta 1$ H-NOX domain. sGC is highly selective for GTP as a substrate, but the protein can also cyclize 2′-d-GTP, GTP-γ-S, guanosine 5′-[β,γ-imido]-triphosphate (GMP-PNP), ITP, UTP, and ATP (73–75). The isolated $\alpha 1_{cat}\beta 1_{cat}$ heterodimer is also selective for GTP but can synthesize cAMP from ATP (76). Interestingly, the activity of $\alpha 1_{cat}\beta 1_{cat}$ is inhibited by the presence of the H-NOX domain [$\beta 1$(1–194) or $\beta 1$(1–385)] (32).[1,2] This shows that these domains interact in *trans* and suggests that the NO mechanism of activation involves the relief of an inhibitory interaction between the H-NOX domain and the catalytic domains. In support of this

---

[1] $\alpha 1_{cat}$ and $\beta 1_{cat}$ are the catalytic domains of the rat sGC $\alpha 1$ and sGC $\beta 1$ subunits, respectively.

[2] $\beta 1$(1–194) represents residues 1–194 of the rat sGC $\beta 1$ subunit, the minimum heme-binding domain.

sGC homologs: proteins with high sequence homology to mammalian sGCs

Atypical sGCs: distinct subset of sGCs with reduced sensitivity to NO

proposal, a fluorescence (or Förster) resonance energy transfer-based study showed that the N terminus of both the α1 and α2 subunits interact with the C terminus of the β1 subunit (77). In addition to the heterodimeric rat sGC catalytic domains, the catalytic domain from the *Chlamydomonas reinhardtii* sGC (CYG12) has been biochemically characterized (78). This protein, as well as *Synechocystis PCC6803* Cya2, a particulate guanylate cyclase (79), is discussed in more detail below.

SOLUBLE GUANYLATE CYCLASE HOMOLOGS

After the initial identification in mammals, it was not until the emergence of genome sequencing that the prevalence of sGC and sGC homologs in other organisms was fully realized. Full-length guanylate cyclases with domain architecture similar to the α1 and β1 subunits were found in several eukaryotic organisms. In prokaryotes, sGC-like H-NOX domains were found as stand-alone proteins or fused to other functional domains (62). In some genomes, an sGC-like PAS domain was also identified. To date, the genes for sGC-like PAS domains are always found near genes that encode sGC-like H-NOX domains (62).

Eukaryotic Atypical Soluble Guanylate Cyclases

As mentioned above, an increasing number of eukaryotic organisms are known to contain predicted sGCs. Some sGCs are very similar to the well-characterized rat α1 and β1 subunits, whereas others vary significantly. Collectively, these cyclases have been termed atypical sGCs (80), owing to their distinct activation and dimerization properties. Some atypical sGCs exhibit a very surprising property—the ability to respond to O_2.

The *D. melanogaster* genome contains five genes that code for sGCs. Two of these genes code for subunits with high homology to the α1 and β1 proteins and have been shown to form a highly NO-sensitive heterodimeric sGC (Gycα-99B and Gycβ-100B). The other three genes code for subunits with greater homology to the β2 protein (Gyc-88E, Gyc-89Da, and Gyc-89Db) (81). Like β2, Gyc-88E can function as a homodimer, whereas Gyc-89Da and Gyc-89Db are only active as heterodimers (81). Experiments with cells overexpressing Gyc-88E, Gyc-89Da, and Gyc-89Db indicate that cGMP synthesis in these cyclases is activated in the absence of O_2 (80), and work with purified protein confirms that the Gyc-88E homodimer forms a stable complex with O_2 (82). As expected, Gyc-88E also binds NO and CO, but there are data that support the role of these proteins in mediating behavioral responses in hyperoxic environments in *D. melanogaster* (83). Interestingly, Gyc-88E is inhibited two- to threefold by the binding of NO, CO, and O_2, a property that is quite distinct from the ligand-induced activation of the sGC α1β1 heterodimer.

On the basis of sequence analysis, atypical sGCs exist in several organisms, including *Caenorhabditis elegans* (GCY-31-GCY-37); however, worms do not contain a predicted NOS or NO-sensitive sGC. Seven β-like guanylate cyclases are contained within the *C. elegans* genome, and it is likely that each gene has an important functional role. GCY-35 is involved in social feeding (84), and both GCY-35 and GCY-36 promote aggregation and bordering behaviors (85). Significantly, these cGMP-dependent behavioral responses are mediated by O_2, which suggests the proteins function as O_2 sensors in vivo (84, 86). In support of this proposal, the GCY-35 H-NOX domain was isolated and shown to bind O_2 (84). Analysis of *gcy-31* and *gcy-33* mutants also implicates these genes in O_2-dependent behavioral responses in *C. elegans*, and the proteins encoded by these genes (GCY-31 and GCY-33) likely function in distinct sensory neurons (BAG versus URX) when compared to GCY-35 and GCY-36 (87).

Thus, a distinct class of cyclases exist, which bind O_2 and are inhibited by ligand binding, but there is currently no means to predict if a cyclase is activated or inhibited by gaseous ligand binding on the basis of sequence analysis. However, residues that contribute to gaseous ligand selectivity have been identified, as described below.

Prokaryotic Heme-Nitric Oxide and Oxygen Binding Proteins

Several species of bacteria are known to contain H-NOX proteins, and this number continues to increase as more genomes are sequenced. Interestingly, all of the isolated H-NOX domains from facultative aerobes have ligand-binding properties like sGC, namely they do not bind O_2. In facultative aerobes, the bacterial members of the H-NOX family encode a single domain as a predicted stand-alone protein, and genes that encode either putative histidine kinases or diguanylate cyclases are found in the same predicted operon, suggesting that the domain has a role in two-component signaling in bacteria. In support of this hypothesis, an H-NOX domain and a predicted histidine kinase from *Shewanella oneidensis* were isolated and found to interact in vitro. Additionally, the functional interaction between the H-NOX and kinase was mediated by NO (88). *Vibrio fischeri* also contains an H-NOX protein, termed H-NOX$_{Vf}$, which is proposed to regulate a putative histidine kinase. In *V. fischeri*, H-NOX$_{Vf}$ was shown to regulate genes involved in iron uptake, and these same genes are modulated by the presence of NO (89), suggesting that H-NOX$_{Vf}$ mediates gene expression by sensing NO. *V. fischeri* colonizes the light-emitting organ of the Hawaiian bobtail squid, *Euprymna scolopes*, and it is likely that this mutualistic host-microbe symbiosis is mediated by a NO/H-NOX$_{Vf}$ interaction. Some bacteria, like *Rhodobacter sphaeroides* and *Nostoc punctiforme*, also encode a predicted PAS-like domain upstream of the H-NOX gene (62). The functional significance of this PAS domain to bacterial signaling is unknown, but it has been proposed to mediate protein dimerization (69).

In obligate anaerobes, such as *Thermonanaerobacter tengcongensis*, the N-terminal H-NOX domain is fused to a C-terminal methyl-accepting chemotaxis protein, suggesting a role in a chemotactic/signaling function. This H-NOX domain was found to form a

> ## NITRIC OXIDE SIGNALING IN BACTERIA
>
> A currently expanding topic within biological studies on NO involves the role of the diatomic gas in bacterial signaling. NOS-like proteins have been identified in several prokaryotic organisms, including *Bacillus anthracis*, *Bacillus subtilis*, *Sorangium cellulosum*, and *Streptomyces turgidiscabies*. Additionally, a wide range of bacteria also have nitrite reductases, which generate NO as part of denitrifying, assimilatory, and dissimilatory pathways. This endogenously produced NO is known to regulate several transcription factors via *S*-nitrosation. There are also two potential classes of prokaryotic heme-based NO sensors: globin-like proteins and H-NOX proteins. Microbial globin-like proteins bind O_2, NO, and CO, and are thought to be involved in the nitrosative stress response. Some H-NOX proteins bind O_2, in addition to CO and NO, whereas others exclude O_2 binding. The genes that code for these H-NOX proteins are found in the same operons as predicted histidine kinases or diguanylate cyclases. In vitro studies have shown that the H-NOX protein and histidine kinase from *S. oneidensis* interact in a NO-dependent manner, but the physiological significance of this interaction is unknown. However, NO is known to be important for the symbiosis between the bobtail squid and *V. fischeri*, where the H-NOX from *V. fischeri* regulates colonization of the bacteria within the symbiotic host squid by a mechanism that is dependent on NO. Further experiments in different microbial systems will likely uncover additional NO-dependent signaling processes in bacteria.

very stable heme-O_2 complex (K_d = 90 nM) (90), a molecular distinction from the H-NOX domains from aerobic bacteria. The isolation and characterization of the *T. tengcongensis* H-NOX domain have significantly influenced current understanding of sGC because it was the first H-NOX domain to be structurally determined, and, moreover, it was crystallized bound to the diatomic ligand O_2 (63, 64).

Thus far, these proteins have been used as tools for probing sGC structure and regulation, but functional studies in microbial systems (see the sidebar titled Nitric Oxide Signaling in Bacteria) will be particularly interesting as the variable ligand-binding properties of these H-NOXs may have consequences for their ability to respond to different gases, i.e., some proteins may sense O_2 in addition to NO or CO.

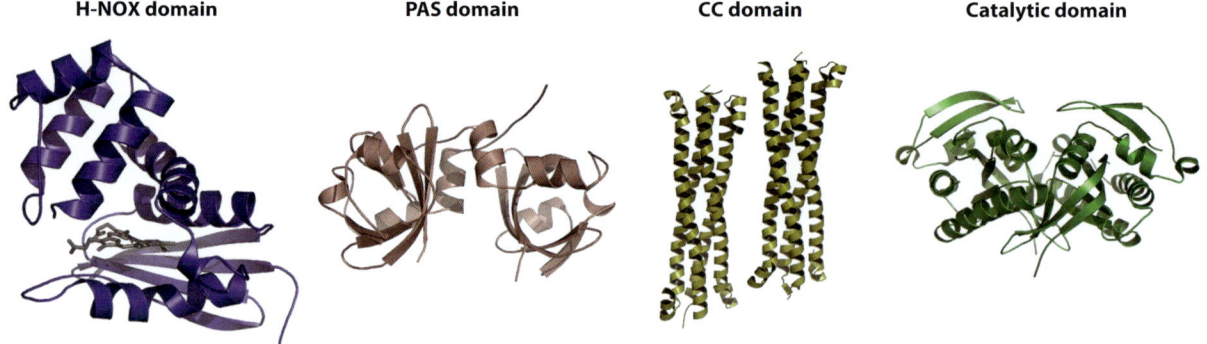

Figure 3

Structures of heme nitric oxide and oxygen binding (H-NOX), PAS, coiled-coil (CC), and catalytic domains. The structures shown are the *T. tengcongensis* H-NOX domain [Protein Data Bank (pdb) code 1U55], the dimerized PAS domain from *N. punctiforme* PCC 73102 (pdb 2p04), the CC domain from the rat β1 subunit (pdb 3hls), and the homodimeric guanylate cyclase domain from *C. reinhardtii* (pdb 3et6).

STRUCTURAL INSIGHTS FROM STUDIES ON SOLUBLE GUANYLATE CYCLASE AND ITS HOMOLOGS

An increasing number of sGC homologs have been isolated and characterized. Significantly, some of these homologs have been amenable to crystallography, thus providing a foundation for structural proposals on the β1 H-NOX, PAS, and catalytic domains. The first crystal structure of an sGC-like domain was that of the O_2-binding H-NOX domain (residues 1–188) from *T. tengcongensis* (**Figure 3**) (63, 64). This protein crystallized in two different six-coordinate heme states, with O_2 bound to the reduced heme iron and in the oxidized heme state. These reports identified several amino acids with critical structural roles that are highly conserved within the H-NOX family. Among the highly conserved amino acids was the *T. tengcongensis* H-NOX heme-coordinating histidine (H102) and three residues that stabilize heme binding (**Figure 4**). Specifically, arginine (R135) is critical to the coordination of the heme propionate groups and forms a hydrogen bond with both carboxyl groups, and tyrosine (Y135) and serine (S133) coordinate to one of the heme carboxyl groups. Together, these residues form a YxSxR motif that is strictly conserved in H-NOX proteins. The proximal histidine, arginine, and tyrosine had been previously identified as critical residues for heme binding in the sGC β1 subunit on the basis of mutagenesis studies (91, 92), and the role of these residues became clear when the *T. tengcongensis* H-NOX structure was solved.

Another striking characteristic of the *T. tengcongensis* H-NOX structure is a highly nonplanar heme conformation; it contains one of the most highly distorted hemes reported in the Protein Data Bank. Interestingly, different molecules of the O_2-bound H-NOX structure exhibit varying degrees of deformation, suggesting that the heme can exist in a range of conformations. This proposal is supported by the observation that the crystal structure of the H-NOX protein from *Nostoc* sp. contains a moderately distorted heme (93). This heme distortion is not an artifact of crystallization as it is also observed in solution on the basis of an NMR study of the *S. oneidensis* H-NOX (94) and resonance Raman experiments with *T. tengcongensis* H-NOX (95) and α1β1 sGC (96–98). In fact, the dynamic range of heme conformations can be accessed in solution by site-directed mutagenesis (95) or, in the case of sGC, by addition of the allosteric activators YC-1 or BAY 41-2272 (96–98). Specifically, the α1β1 sGC heme becomes more planar upon activator binding;

YC-1:
3-(5′-hydroxymethyl-2′-furyl)-1-benzyl indazole

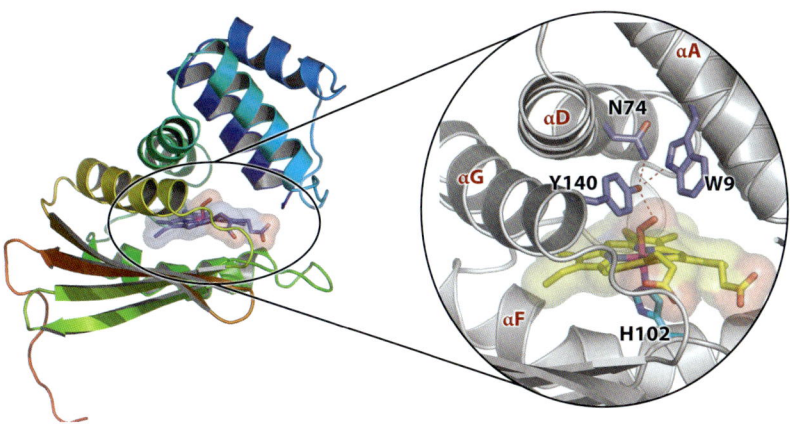

Figure 4

Distal heme pocket of the *T. tengcongensis* H-NOX domain [Protein Data Bank (pdb) 1U55 code]. Residues important to ligand selectivity (W9, N74, and Y140) and heme binding (H102) are shown. Image reproduced with permission of Elsevier (65).

however, it is unclear if this is a cause or consequence of enzyme activation. Thus, although the functional importance of heme distortion in sGC activation remains unknown, there is a potential to utilize changes in distortion in biological responses.

Among the residues thought to be critical to maintaining the nonplanar heme conformation are I5, D45, R135 (in the YxSxR motif), L144, and P115, all of which are highly conserved in H-NOX proteins (64). In *T. tengcongensis* H-NOX, mutation of proline 115 to alanine leads to relaxation of the distorted heme (95, 99). This heme relaxation was not observed after mutation of proline 118 to alanine (P115 in *T. tengcongensis* H-NOX) in α1β1 sGC (96); however, it is not surprising that additional residues or domain interactions are important for maintaining the heme conformation in the mammalian protein.

The crystal structure of the non-O_2-binding H-NOX from *Nostoc* sp. has been solved in the five-coordinate unligated state and as the six-coordinate NO- and CO-heme complexes (93). *Nostoc* sp. H-NOX could function as a redox or NO sensor (100). Comparison of the *T. tengcongensis* H-NOX and *Nostoc* sp. H-NOX structures in different ligation states (Fe^{II}-CO, Fe^{II}-NO, and Fe^{II}-unligated states) led to speculation on a molecular mechanism of sGC activation. Specifically, the differential pivoting and bending in the H-NOX heme upon NO or CO binding was suggested to account for the varying degree of activation induced by the two ligands (200-fold versus fourfold, respectively) (93). In support of this proposal, an UV resonance Raman study found that significant conformational changes occurred at the N terminus of sGC upon activator binding (96).

In sGC, the breaking of the proximal iron (Fe)-His bond is an essential event in the activation by NO, and thus, a structure of a five-coordinate H-NOX protein would be very important. Although a five-coordinate NO complex has not yet been determined, there are two recently solved structures where the net effect is the severing of the proximal Fe-His bond; the solution structure of a *S. oneidensis* H-NOX mutant (94) and the crystal structure of BAY 58-2667-bound *Nostoc* sp. H-NOX (101). Wild-type *S. oneidensis* H-NOX was shown to modulate the activity of a histidine kinase in a ligand-dependent manner: Kinase activity is inhibited by the Fe^{II}-NO H-NOX but not the Fe^{II}-unligated H-NOX (88). The H103G Fe^{II}-CO mutant mimics the kinase-inhibitory activity of the wild-type Fe^{II}-NO H-NOX (94). Because H103 is the proximal iron ligand in

S. oneidensis H-NOX, mutation of this residue to glycine leads to the isolation of apoprotein. Similar to previous reports with other heme proteins (92, 102), heme binding in *S. oneidensis* H-NOX H103G can be rescued by imidazole. H103 is in α-helix F, and in the heme-binding rescued mutant, this helix is free to adopt a position like that in a five-coordinate NO complex. The *S. oneidensis* H103G H-NOX structure, along with the structure of the weakly active wild-type Fe^{II}-CO *S. oneidensis* H-NOX, was solved by NMR. The solution structures of both of these proteins indicate that major changes in heme planarity and H-NOX conformation occur upon cleavage of the proximal histidine heme ligand.

Major structural changes are also observed between the unbound and BAY 58-2667-bound *Nostoc* sp. H-NOX structures (101). BAY 58-2667 is a NO- and heme-independent sGC activator (103) and is a candidate to treat decompensated heart failure. The crystal structure shows that BAY 58-2667 is able to displace the heme and occupy the heme-binding site, thereby leading to a shift in the α-helix F that contains the proximal histidine residue. Additional experiments mutating residues around the proximal histidine on α-helix F suggest that D102 could play a critical role in enzyme activation (104). In addition to providing a structural basis for proposals on sGC activation, these H-NOX structures enabled the development of homology models of the heme-binding domains of both NO-sensitive and atypical cyclases (28, 63, 105, 106), as well as density functional theory analysis and computer simulations to address questions concerning structural dynamics (107, 108).

The crystal structure of a domain from the *N. punctiforme* signal transduction histidine kinase has also been determined (**Figure 3**). This domain has high sequence identity (35%–38%) to the sGC PAS domain, and the crystal structure showed that the domain dimerized and adopted a PAS fold (69). There is no structure of a eukaryotic sGC PAS domain, but the crystal structure of the coiled-coil domain of the β1 subunit has been solved (**Figure 3**) (31). This structure indicates that the coiled-coil domain forms a tetramer composed of a dimer of dimers. In addition to identifying potential residues involved in mediating dimerization, the authors propose that interhelix salt-bridge formation selects for heterodimerization versus homodimerization in sGCs on the basis of their structure (31).

The catalytic domains of two different functional guanylate cyclases (78, 79), and the inactive human β1β1 catalytic domains [Protein Data Bank (pdb) code 2WZ1], have been structurally elucidated (**Figure 3**). The functional cyclase domains include a soluble homodimeric guanylate cyclase from the eukaryotic algae *C. reinhardtii* (78) and a particulate homodimeric guanylate cyclase from the unicellular cyanobacterium *Synechocystis* PCC6803 Cya2 (79). Both proteins cyclize GTP and likely contain two catalytic sites per dimer. On the basis of kinetic analysis of cGMP synthesis, the eukaryotic algae catalytic domains exhibit positive cooperativity with a Hill coefficient of 1.5. This is similar to the cooperativity observed in particulate guanylate cyclases and may be a common feature of homodimeric sGCs (78). The α1β1 heterodimer contains one active site and a proposed pseudosymmetric site. This proposed pseudosymmetric site is thought to constitute an allosteric nucleotide-binding site that communicates with the catalytic nucleotide-binding site. On the basis of the algae sGC structure, this communication may involve residues contained on the β2-β3 loop of each catalytic domain monomer. Specifically, the interaction of D527 or E523 (*C. reinhardtii* sGC numbering) with a nucleotide in one active site could alter the loop conformation and thus lead to a conformational change in the other nucleotide-binding site (78).

Unfortunately, both structures are in an inactive state, but they confirm the guanylate cyclase residues critical to metal binding and nucleotide recognition that were predicted from the adenylate cyclase crystal structures. Together, the reports of the *C. reinhardtii* sGC and Cya2 catalytic domains provided the first structures of a guanylate cyclase, and the structures show that there is high homology between guanylate cyclase and adenylate cyclase domains; however, the elucidation of a mammalian heterodimeric structure remains an important task for understanding sGC regulation. Details about how movement in the β1 H-NOX domain affects the catalytic domain, and the role of the PAS and coiled-coil domain in relaying a signal from the H-NOX domain, may remain open questions until the full-length sGC structure is elucidated.

LIGAND SELECTIVITY

The α1β1 heme environment is unique when compared to the globins because the cofactor efficiently binds NO while having no affinity for O_2. Additionally, the α1β1 heterodimer exhibits an extremely slow rate of oxidation and has among the highest midpoint potential reported for a high-spin heme protein (+187 mV versus +58 mV for myoglobin) (109). This ability of mammalian sGC to select against O_2 binding is important for it to function as a NO sensor, as O_2 is present at much higher levels than NO in vivo, and Fe^{II}-O_2 proteins react rapidly with NO. Hemoproteins in the Fe^{III} oxidation state also form weak NO complexes and could inadvertently serve as a sink for NO. Since the discovery of sGC, several potential mechanisms of ligand discrimination against O_2 have been proposed, including a weak Fe-His bond strength, a negatively charged distal pocket, and a sterically constrained distal pocket (109–111). One proposal on the mechanism by which α1β1 excludes O_2 is based on analysis of the *T. tengcongensis* H-NOX crystal structure (63, 64). *T. tengcongensis* H-NOX stabilizes O_2 binding with a hydrogen-bonding network involving a tyrosine (Y140), a tryptophan (W9), and an asparagine (N74) (**Figure 4**). These residues appear to be absent in the β1 H-NOX protein and other O_2-excluding H-NOXs on the basis of primary sequence alignments and homology modeling. Conversely, several atypical sGCs, including the known O_2-binding cyclase Gyc-88E, encode a tyrosine that aligns with *T. tengcongensis* H-NOX Y140. Therefore, the absence of a hydrogen bond donor in the α1β1 distal heme pocket, and the subsequent increase in the O_2 off rate, likely contributes to the protein's inability to form a stable Fe^{II}-O_2 complex (64, 90). Molecular dynamics simulations of O_2-binding and non-O_2-binding H-NOX proteins suggest that they have varying tunnel systems, which may also contribute to their different ligand-binding properties (112), and site-directed mutagenesis within the proposed tunnel indicates that it is important for diffusion of gaseous ligands (113).

Several biochemical studies aimed at probing ligand selectivity in sGC have been reported. The distal-pocket tyrosine in the O_2-binding *T. tengcongensis* H-NOX was mutated to leucine (Y140L), and this mutation significantly reduced O_2 affinity. Additionally, the introduction of a distal-pocket tyrosine in a non-O_2-binding H-NOX from *Legionella pneumophilia* (L2 H-NOX) enabled the protein to bind O_2 (90). Mutagenesis studies introducing a tyrosine into the distal pocket of the β1 H-NOX-PAS domain β1(1–385) also produced a protein that was capable of binding O_2 (90), but the same mutation in full-length sGC did not facilitate O_2 binding (106, 114). However, the introduction of a tyrosine and glutamine into the β1 heme pocket (I145Y/I149Q) resulted in a full-length α1β1 protein with altered reactivity to O_2 (105). A homology model of the O_2-binding guanylate cyclase Gyc-88E places these amino acids within the predicted distal heme pocket. Thus, some O_2-binding guanylate cyclases may utilize a Tyr/Gln hydrogen-bonding network, similar to several truncated globins (115–118), to stabilize ligand binding. The marked variability in O_2 reactivity between full-length sGC and heme-binding truncations of the β1 protein highlights the potential for other sGC domains to influence ligand selectivity. Thus, despite significant progress in our understanding of ligand discrimination in sGC, including the identification of residues critical for stabilizing O_2 binding in sGC homologs, there remain some unknown variables that may contribute to the ligand specificity of these heme proteins.

REGULATION OF SOLUBLE GUANYLATE CYCLASE BY NITRIC OXIDE

Physiological responses to NO, such as smooth muscle relaxation, are rapidly induced by low levels of the diatomic gas. In cells, cGMP levels rise within milliseconds after exposure to nanomolar concentrations of NO, and this fast response occurs because sGC efficiently binds to and is activated by NO. When NO dissociates from the protein, the amount of cGMP decreases to a basal level. Thus, both the rise and fall of cGMP levels must be tightly regulated for proper function of cGMP-dependent processes. Upon repeated exposure to NO or NO-donors, like GTN, maximal sGC activation decreases, and this process is called sGC desensitization. sGC activation, deactivation, and desensitization have been extensively studied in vivo and in vitro. A major challenge with understanding these processes occurs when reconciling discrepancies in results obtained with purified protein versus sGC examined in the cellular milieu.

Activation and Nitric Oxide Association

As mentioned above, the activation of sGC in vitro and in vivo is rapid (occurs in milliseconds). Using a NO donor or NO gas, the apparent 50% effective concentration for NO acting on purified sGC is between 80 and 250 nM (119, 120). A similar potency of NO has been measured in rat cerebellar cells (apparent 50% effective concentration ~45 nM) (121). On the basis of studies with purified protein, it is known that NO binds to the heme of sGC at a diffusion-controlled rate to form an initial six-coordinate complex, which rapidly converts to a five-coordinate ferrous nitrosyl (Fe^{II}-NO) complex (120). The six-coordinate Fe^{II}-NO complex is not stable and has only been observed with time-resolved spectroscopic methods (120, 122, 123). The presence of Mg^{2+}GTP or Mg^{2+}/cGMP/pyrophosphate (PP_i) accelerates the formation of the five-coordinate sGC-NO complex, and this effect is blocked by the addition of ATP (122, 123).

Breakage of the Fe-His bond is thought to be a critical step in the activation of sGC by NO; however, recent data have shown that NO coordination to the heme is not sufficient for full activation ex vivo (122, 123). A low-activity Fe^{II}-NO complex can be formed in the presence of stoichiometric amounts of NO, whereas a high-activity Fe^{II}-NO complex is formed in the presence of excess NO. These two five-coordinate Fe^{II}-NO species are indistinguishable by electronic absorption spectroscopy, but they exhibit distinct signals by electron paramagnetic spectroscopy (124). Preincubation of sGC with substrate Mg^{2+}GTP or the reaction products Mg^{2+}/cGMP/PP_i produces a high-activity Fe^{II}-NO species in the presence of stoichiometric amounts of NO (123). ATP can compete with the GTP effect, and in the presence of ATP and GTP, a low-activity Fe^{II}-NO species is formed (122). This clear difference in sGC activity indicates that preincubation of the enzyme with the small molecules leads to a conformational change such that sGC is highly activated by low levels of NO. It was also found that the small-molecule YC-1 can activate the low-activity Fe^{II}-NO complex to the high-activity state (122, 125). Thus, both YC-1 and excess NO activate the low-activity five-coordinate Fe^{II}-NO complex that is formed in the absence of substrate or reaction products.

Two mechanisms have been proposed to account for the varying activity of the sGC Fe^{II}-NO complex observed ex vivo. One proposal is that excess NO activates the Fe^{II}-NO complex by binding to nonheme sites on the protein (122). If NO binds to a nonheme site, it is likely that cysteines would comprise this binding site as experiments with the thiol reactive reagent methyl methanethiosulfonate suggest that reduced cysteines are necessary for the mechanism of NO activation (125). The second proposal regarding sGC activation involves excess NO binding to the heme to form a transient dinitrosyl complex, which then converts

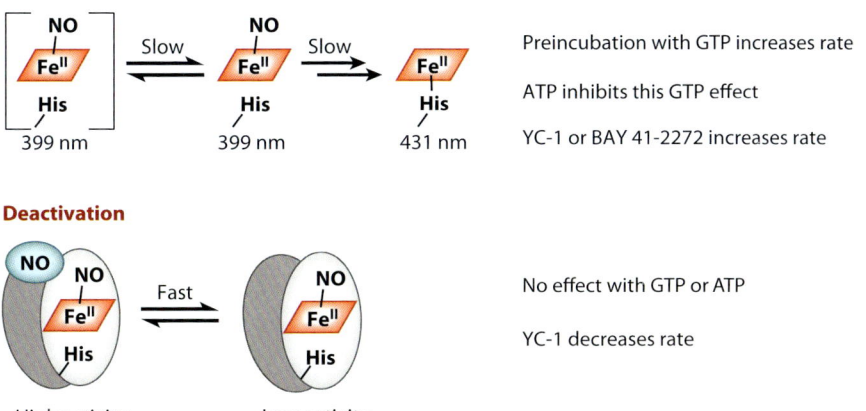

Figure 5

Model of nitric oxide (NO) dissociation and deactivation of NO-stimulated soluble guanylate cyclase (sGC) in vitro. The sGC-NO complex consists of two different five-coordinate species that slowly interconvert. NO dissociation from the sGC heme is slow, whereas NO deactivation is rapid on the basis of in vitro measurements, including spectroscopic methods (Abs$_{max}$ indicated in nanometers) and cyclic GMP analysis. The presence of allosteric modulators (GTP, ATP, YC-1, or BAY 41-2272) influences the deactivation and NO dissociation rates differently. Taken together, these results indicate that the deactivated sGC species is distinct from the unligated sGC species (431 nm) produced from NO dissociation from the heme and suggests that two molecules of NO are able to bind to sGC.

to a five-coordinate complex with NO bound in the proximal heme pocket (123). Such proximal heme pocket NO binding has been observed with other histidine-ligated heme proteins including cytochrome *c'* (126).

The importance of the two different sGC-NO states in vivo remains unclear. One report examining sGC activity in endothelial cells proposed that enzyme activation by excess NO is important for the physiological activation of sGC (125), whereas another report examining sGC activation in rat platelets and cerebellar cells proposed that the observed activation kinetics were consistent with a single NO-binding event (127). Thus, further experiments are necessary to resolve the mechanism of sGC activation in vivo.

Deactivation and Nitric Oxide Dissociation

Deactivation of the sGC Fe^{II}-NO complex has been extensively studied to understand the lifetime of the NO signal in vivo. This process was originally thought to correlate directly with NO dissociation from the sGC heme; however, as mentioned above, a NO-bound protein that is partially activated has been characterized. This indicates that sGC deactivation and NO dissociation from the heme cofactor are not explicitly linked, and thus, deactivation and heme-NO dissociation measurements cannot be used interchangeably (**Figure 5**).

NO-induced relaxation of smooth muscle cells dissipates within seconds upon removal of free NO due to the deactivation of sGC (128). sGC deactivation, the rate of decline in cGMP synthesis after an activator is removed, has been determined in various cells and with purified protein. In cells, sGC deactivation is rapid ($t_{1/2} < 5$ s at 37°C) in the presence of an NO trap (oxyhemoglobin) to limit NO rebinding (121, 129). Activity assays with the cytosol of retina homogenate indicate that the sGC-NO deactivation rate is not influenced by the presence

S-nitrosation: posttranslational modification involving the oxidative addition of NO to a thiol

of Mg^{2+}GTP and is only slightly increased by the presence of reducing agents, such as glutathione and dithiothreitol (129). In agreement with cellular data, the deactivation rate of purified protein is also rapid ($t_{1/2} \sim 4$ s at 37°C) (130). This deactivation rate is unaffected by the presence of a GTP analog and/or ATP (122), but the rate is significantly decreased (140-fold) by the presence of the allosteric activator YC-1 ($t_{1/2} > 10$ min at 37°C) (130).

The rate of NO dissociation from the sGC heme must be spectroscopically determined, and as a consequence, this rate has only been measured in vitro. Early reports found that the sGC heme-NO complex is very stable ($t_{1/2} \sim 87$ min at 37°C) in the absence of NO traps or allosteric effectors (23, 131). Unlike most heme proteins, which rapidly oxidize in the presence of O_2 and NO, a reduced Fe^{II} protein is formed when NO dissociates from the sGC heme. Using a trap to scavenge NO, like CO/dithionite or oxymyoglobin, the dissociation rate can be determined without interference from NO rebinding. Reductants like dithiothreitol or glutathione react with NO and can also be used to prevent rebinding (23). With these NO trapping methods, it was determined that the NO dissociation rate is relatively slow ($t_{1/2} = 3$–8 min at 37°C) (23, 27, 131) compared to previously determined deactivation rates but increases ($\sim$50-fold) in the presence of Mg^{2+}GTP (131). However, if Mg^{2+}GTP is added to sGC after the NO complex is formed, it has no effect on the dissociation rate (23, 123, 131), indicating that the allosteric affect of GTP is dependent on substrate binding in the absence of NO. In contrast to the previously discussed deactivation results, the GTP effect on the dissociation rate is inhibited by the presence of ATP (122). Therefore, in the presence of both GTP and ATP, sGC exhibits a slow NO dissociation rate but rapid deactivation. In the presence of YC-1, the NO dissociation rate increases (27), whereas the deactivation rate decreases (130). A model summarizing deactivation of sGC and NO dissociation is shown in **Figure 5**.

Desensitization

The sGC response to GTN or NO decreases upon repeated exposure to the activators. This desensitization is rapid, such that a single pretreatment of cells with GTN or NO can significantly reduce cGMP stimulation upon the second exposure (132, 133). Purified protein in the absence of reducing agents also exhibits a decrease in NO-stimulated activity after pretreatment with NO. This desensitization has implications for therapies used in the treatment of heart disease as this loss of responsiveness, or tolerance, is also observed when organic nitrates are administered to patients with angina pectoris (134). As a consequence, these drugs cannot be used repeatedly. Over the past 30 years, several proposals have been advanced to explain the phenomenon of tolerance, including increased PDE activity (135), inhibition of mitochondrial aldehyde dehydrogenase (136), and most recently S-nitrosation (137, 138). There is currently significant evidence in support of the proposal that nitrosation, the oxidative addition of NO to a thiol, contributes to sGC desensitization.

It has long been known that thiol oxidation inhibits sGC activity. Cysteines can oxidize sequentially to form a sulfenic acid, sulfinic acid, sulfonic acid, or a disulfide bond in the presence of O_2, or form a nitrosothiol in the presence of NO and O_2. In the 1980s, the formation of sGC-cysteine mixed disulfides was shown to reversibly inhibit cGMP production (139). More recently, the induction of reactive O_2 species in vascular smooth muscle cells was found to inhibit cGMP synthesis and lead to the oxidative modification of sGC cysteines (140). Additionally, the thiol modifying reagent methyl methanethiosulfonate was shown to inhibit sGC activity in vitro and in primary endothelial cells. In vitro sGC inhibition by methyl methanethiosulfonate is reversible, and the molecule was shown to modify several cysteines on both the α1 and β1 subunits (125). The NO-dependent oxidation of cysteines on sGC has also been reported. sGC is S-nitrosated in the presence of low levels of NO, and this modification has been linked to a reduction in NO-stimulated activity (137). In addition, GTN has been shown to induce sGC S-nitrosation and desensitization in a concentration-dependent manner (138). Cysteines on both the α1 (C243) and β1 (C122) subunits have been identified as targets of this oxidative modification.

It has become clear that sGC requires free thiols for proper function. Without a crystal structure, it is difficult to address why these cysteines are necessary for NO-induced sGC activation. Perhaps free thiols are important to the structural integrity of the protein, involved in a conformational change to the activated enzyme state, and/or directly involved in the NO mechanism of activation. If sGC S-nitrosation is the primary molecular mechanism of nitrate tolerance, then molecules developed to protect these thiols from oxidation could be useful for the treatment of diseases related to sGC dysfunction. Furthermore, the apo- and heme-oxidized sGC states have been proposed to be physiologically relevant sGC species in diseased tissue.

MODULATORS OF SOLUBLE GUANYLATE CYCLASE ACTIVITY

Soluble Guanylate Cyclase Activators

sGC is a therapeutic target in the treatment of heart disease. Organic nitrites and organic nitrates (like GTN) are perhaps the earliest agents used to target the NO/cGMP-signaling pathway and have been in clinical use for over 100 years. The first description of GTN as a therapeutic agent for the treatment of angina pectoris appeared in 1879, and it remains the drug of choice to treat the disease. Despite decades of clinical use, the precise mechanism of action of GTN is unknown; however, it is generally considered a nitrovasodilator or a NO-donor. This NO release may occur by spontaneous decomposition or bioconversion to result in NO-dependent sGC activation.

In addition to NO, other known heme ligands, including CO, nitrosoalkanes, and alkyl isocyanides, can bind to the sGC heme but can only weakly activate the protein (17, 141, 142). The binding of these compounds leads to the formation of a six-coordinate complex and to a two- to fourfold increase in the rate of cGMP production, significantly lower than the 100- to 400-fold increase in cGMP synthesis observed with NO. Other compounds that have been reported to activate sGC by targeting the heme-binding pocket include protoporphyrin IX (15) and Co^{2+} protoporphyrin IX (143).

It has become clear that small molecules can modulate the activity of sGC and that new therapeutics might be developed for the treatment of various diseases. This prompted a search for novel sGC activators, and several compounds were screened for the ability to increase cGMP levels in cell lysates. Such a screen led to the identification of YC-1, a benzylindazole derivative that activates sGC without coordinating to the heme (144). YC-1 only activates the Fe^{II}-unligated sGC state two- to fourfold but significantly increases sGC activity when a ligand is bound at the Fe^{II} heme (141, 142, 145, 146). This synergistic activation leads to an Fe^{II}-CO complex that is activated 100- to 400-fold and an Fe^{II}-NO complex that is activated 200- to 800-fold. The molecular mechanism of YC-1 activation is unknown. Experiments with equilibrium dialysis suggest that sGC binds one equivalent of YC-1 per heterodimer (147). This binding site may be contained within the N terminus of the α1 subunit (33, 148) or within the pseudosymmetric substrate site (149–151). What is clear is that YC-1 binding to sGC induces a conformational change that elevates cGMP production. This highly active conformational state has been characterized by spectroscopic methods, including electronic absorption (152), resonance Raman spectroscopy (147, 153), and electron paramagnetic resonance spectroscopy (124, 147). There are also clear kinetic effects of adding YC-1 to CO- or NO-bound sGC (27, 130, 154).

The discovery of the novel sGC stimulator YC-1 led several groups to carry out structure activity relationships to improve both the solubility and efficacy of YC-1. With this work

Nitrate tolerance: loss of sensitivity to nitrates

came the identification of several other compounds, including BAY 41-2272, BAY 41-8543, CMF-1571, and A-350619 (reviewed in Reference 103). Although there is some debate over the possible inhibition of PDE5 by BAY 41-2272 (155, 156), it is commonly accepted that the molecule activates sGC without coordinating to the heme. Collectively, these molecules constitute a novel class of sGC stimulators that require the presence of the heme moiety and have the ability to synergistically stimulate sGC with both NO and CO. These molecules were promising drug candidates but had unfavorable drug metabolism. Medicinal chemistry efforts to reduce the problems associated with these compounds led to the discovery of BAY 63-2521 (riociguat). This compound induces vasodilation by activating sGC and is currently in Phase III clinical trails (reviewed in Reference 157).

There are also small-molecule activators that target heme-oxidized or -deficient sGC; therefore, they are classified as NO and heme independent. One of these is BAY 58-2667 (Cinaciguat) (158). BAY 58-2667 is generally cited as an activator of both heme-deficient and heme-oxidized (Fe^{III}) sGC (158), but there is one report that proposes the molecule exclusively targets the heme-deficient sGC state (159). BAY 58-2667 is proposed to activate sGC by binding within the heme pocket, thereby serving as a mimic of the heme-NO complex (91, 101). In addition to activating the enzyme, BAY 58-2667 also stabilizes sGC and protects the protein from degradation (160). In rats, BAY 58-2667 was shown to lower blood pressure (161) and protect against ischemic injury (162), and the compound has been tested on patients with acute decompensated heart failure (163–165).

Soluble Guanylate Cyclase Inhibitors

There has been significantly more work on identifying sGC activators, but compounds that inhibit sGC activity have also been reported. These compounds are not as selective as the identified activators and include general oxidants, molecules that target hemoproteins, and nucleotide cyclase inhibitors. Compounds proposed to target the sGC H-NOX domain include heme, hematin (15), and 1H-[1,2,4]oxadiazolol[4,3-a]quinoxalin-1-one (ODQ) (166). These compounds inhibit sGC by oxidation of the ferrous iron in the heme cofactor. Molecules such as hydrogen peroxide, superoxide, cystine, and S-nitrosocysteine also reduce cGMP production by oxidizing critical residues on sGC (138–140). Other known sGC inhibitors, such as LY83583 (167) and methylene blue (168), inhibit the enzyme via generation of superoxide anion (169, 170). Inhibition of sGC by these oxidants emphasizes the importance of the redox environment for proper protein function.

Substrate analogs that bind to the sGC catalytic domain inhibit guanylate cyclase activity. To date, several such analogs have been identified as sGC inhibitors, including, but not limited to, ITP, XTP, CTP, ATP, ADP, AMP, 2′-deoxyadenosine 5′-diphosphate (2′-dADP), GDP, guanosine-5′-[(α,β)-methylene]triphosphate (GMP-CPP), and N-methylanthraniloyl (MANT)-nucleotides (75, 151, 171, 172). These compounds inhibit sGC by varying mechanisms depending on whether they target the pseudosymmetric site in addition to the substrate-binding site. For example, AMP-PNP is a competitive inhibitor (151), while ATP is a mixed-type inhibitor and substrate (76).

SUMMARY POINTS

1. Soluble guanylate cyclase (sGC) is the most thoroughly characterized receptor for the signaling molecule nitric oxide (NO). NO activation of sGC is essential to several physiological processes, and thus dysfunction in sGC is linked to several diseases. Compounds that modulate sGC are in clinical trails for the treatment of heart disease.

2. Each sGC subunit consists of an H-NOX, PAS, coiled-coil, and catalytic domain. Functional truncations of sGC include the isolated β1 H-NOX domain, the β1 coiled-coil domain, and the α1β1 catalytic domains.

3. Genome mining has revealed that several predicted proteins exist with high sequence homology to sGC, including prokaryotic H-NOXs and eukaryotic NO-sensitive and atypical sGCs.

4. The structure of full-length sGC is not yet determined, but studies on H-NOXs and sGC homologs have illuminated many important structural features of the enzyme. On the basis of these studies, sGC likely contains a hydrophobic heme-binding pocket, without a distal-pocket hydrogen bond donor, and heterodimerization is mediated by contacts on the PAS, coiled-coil, and catalytic domains.

5. α1β1 sGC does not form a stable complex with O_2, but several sGC homologs bind O_2. Hydrogen bond donors in the heme-distal pocket are often found in the H-NOXs and atypical sGCs that bind O_2. The presence or absence of amino acids capable of forming a hydrogen bond in the heme-distal pocket contributes to selectivity for gaseous ligands.

6. sGC regulation by NO and nucleotides in vitro is complex. Two molecules of NO can bind to the protein, and both GTP and ATP can influence NO binding, dissociation, and activation. In cells, it is likely that NO, GTP, and ATP modulate cGMP production, and both reduced cysteines and a reduced heme iron are important for this activity.

FUTURE ISSUES

Over the past few years, sGC has been implicated in an expanding number of physiological processes and diseases. Continued progress toward elucidating the mechanism of sGC activation is important in the development of therapeutics to treat disorders relating to the NO-signaling pathway. Despite significant advances in our understanding of NO as a signaling agent, many questions remain unanswered. In mammals, the sGC response to NO is complicated, and the precise events that lead to activation remain a topic of debate. Furthermore, sGC is regulated by allosteric interactions with ATP and GTP, and there are likely other important factors, including *S*-nitrosation and phosphorylation, that modulate sGC activity. Future experiments considering these factors will be necessary to determine the influence of each regulatory factor in the activation, deactivation, and desensitization of sGC. How do the α1 and β1 subunits interact when sGC is activated? What is the nature of the allosteric activator-binding site? What conformational changes occur when ligands bind to sGC? All of these questions may be addressed with the structural elucidation of the full-length heterodimeric protein. Additionally, structural analysis may provide insight into the molecular mechanisms that lead to sGC dysfunction.

DISCLOSURE STATEMENT

M.A.M. is a cofounder of Omniox, Inc., a company commercializing oxygen delivery technology for a broad range of clinical indications in cancer, surgical, and cardiovascular markets. The authors are not aware of any other affiliations, memberships, funding, or financial holdings that might be perceived as affecting the objectivity of this review.

ACKNOWLEDGMENTS

This article was written while M.A.M. was employed by the University of California, Berkeley. The authors acknowledge members of the Marletta lab, past and present, who have contributed to studies on sGC, NOS, and H-NOXs. Research on sGC in the authors' laboratory was supported by the National Institutes of Health (NIH grant GM077365), and research on H-NOXs by NIH grant GM070671.

LITERATURE CITED

1. Brunton TL. 1867. On the use of nitrite of amyl in angina pectoris. *Lancet* 90:97–98
2. Murrell W. 1879. Nitro-glycerine as a remedy for angina pectoris. *Lancet* 113:113–15
3. Garbers DL, Chrisman TD, Wiegn P, Katafuchi T, Albanesi JP, et al. 2006. Membrane guanylyl cyclase receptors: an update. *Trends Endocrinol. Metab.* 17:251–58
4. Padayatti PS, Pattanaik P, Ma X, van den Akker F. 2004. Structural insights into the regulation and the activation mechanism of mammalian guanylyl cyclases. *Pharmacol. Ther.* 104:83–99
5. Feussner M, Richter H, Baum O, Gossrau R. 2001. Association of soluble guanylate cyclase with the sarcolemma of mammalian skeletal muscle fibers. *Acta Histochem.* 103:265–77
6. Russwurm M, Wittau N, Koesling D. 2001. Guanylyl cyclase/PSD-95 interaction: targeting of the nitric oxide-sensitive $\alpha 2\beta 1$ guanylyl cyclase to synaptic membranes. *J. Biol. Chem.* 276:44647–52
7. Zabel U, Kleinschnitz C, Oh P, Nedvetsky P, Smolenski A, et al. 2002. Calcium-dependent membrane association sensitizes soluble guanylyl cyclase to nitric oxide. *Nat. Cell Biol.* 4:307–11
8. Madhani M, Okorie M, Hobbs AJ, MacAllister RJ. 2006. Reciprocal regulation of human soluble and particulate guanylate cyclases in vivo. *Br. J. Pharmacol.* 149:797–801
9. Munzel T, Feil R, Mulsch A, Lohmann SM, Hofmann F, Walter U. 2003. Physiology and pathophysiology of vascular signaling controlled by guanosine 3′,5′-cyclic monophosphate-dependent protein kinase. *Circulation* 108:2172–83
10. Sanders KM, Ward SM, Thornbury KD, Dalziel HH, Westfall DP, Carl A. 1992. Nitric oxide as a non-adrenergic, non-cholinergic neurotransmitter in the gastrointestinal tract. *Jap. J. Pharmacol.* 58(Suppl.):P220–25
11. Warner TD, Mitchell JA, Sheng H, Murad F. 1994. Effects of cyclic GMP on smooth muscle relaxation. *Adv. Pharmacol.* 26:171–94
12. Garbers DL. 1979. Purification of soluble guanylate cyclase from rat lung. *J. Biol. Chem.* 254:240–43
13. Braughler JM, Mittal CK, Murad F. 1979. Purification of soluble guanylate cyclase from rat liver. *Proc. Natl. Acad. Sci. USA* 76:219–22
14. Gerzer R, Hofmann F, Schultz G. 1981. Purification of a soluble, sodium-nitroprusside-stimulated guanylate cyclase from bovine lung. *Eur. J. Biochem.* 116:479–86
15. Ignarro LJ, Wood KS, Wolin MS. 1982. Activation of purified soluble guanylate cyclase by protoporphyrin IX. *Proc. Natl. Acad. Sci. USA* 79:2870–73
16. Vogel KM, Hu S, Spiro TG, Dierks EA, Yu AE, Burstyn JN. 1999. Variable forms of soluble guanylyl cyclase: protein-ligand interactions and the issue of activation by carbon monoxide. *J. Biol. Inorg. Chem.* 4:804–13
17. Stone JR, Marletta MA. 1994. Soluble guanylate cyclase from bovine lung: activation with nitric oxide and carbon monoxide and spectral characterization of the ferrous and ferric states. *Biochemistry* 33:5636–40
18. Russwurm M, Koesling D. 2005. Purification and characterization of NO-sensitive guanylyl cyclase. *Methods Enzymol.* 396:492–501
19. Harteneck C, Koesling D, Soling A, Schultz G, Bohme E. 1990. Expression of soluble guanylyl cyclase. Catalytic activity requires two enzyme subunits. *FEBS Lett.* 272:221–23
20. Buechler WA, Nakane M, Murad F. 1991. Expression of soluble guanylate cyclase activity requires both enzyme subunits. *Biochem. Biophys. Res. Commun.* 174:351–57
21. Friebe A, Wedel B, Harteneck C, Foerster J, Schultz G, Koesling D. 1997. Functions of conserved cysteines of soluble guanylyl cyclase. *Biochemistry* 36:1194–98

22. Zhou Z, Gross S, Roussos C, Meurer S, Muller-Esterl W, Papapetropoulos A. 2004. Structural and functional characterization of the dimerization region of soluble guanylyl cyclase. *J. Biol. Chem.* 279:24935–43
23. Brandish PE, Buechler W, Marletta MA. 1998. Regeneration of the ferrous heme of soluble guanylate cyclase from the nitric oxide complex: acceleration by thiols and oxyhemoglobin. *Biochemistry* 37:16898–907
24. Hoenicka M, Becker EM, Apeler H, Sirichoke T, Schroder H, et al. 1999. Purified soluble guanylyl cyclase expressed in a baculovirus/Sf9 system: stimulation by YC-1, nitric oxide, and carbon monoxide. *J. Mol. Med.* 77:14–23
25. Lee YC, Martin E, Murad F. 2000. Human recombinant soluble guanylyl cyclase: expression, purification, and regulation. *Proc. Natl. Acad. Sci. USA* 97:10763–68
26. Wagner C, Russwurm M, Jager R, Friebe A, Koesling D. 2005. Dimerization of nitric oxide-sensitive guanylyl cyclase requires the $\alpha 1$ N terminus. *J. Biol. Chem.* 280:17687–93
27. Winger JA, Derbyshire ER, Marletta MA. 2007. Dissociation of nitric oxide from soluble guanylate cyclase and heme-nitric oxide/oxygen binding domain constructs. *J. Biol. Chem.* 282:897–907
28. Karow DS, Pan D, Davis JH, Behrends S, Mathies RA, Marletta MA. 2005. Characterization of functional heme domains from soluble guanylate cyclase. *Biochemistry* 44:16266–74
29. Zhao Y, Marletta MA. 1997. Localization of the heme binding region in soluble guanylate cyclase. *Biochemistry* 36:15959–64
30. Zhong F, Pan J, Liu X, Wang H, Ying T, et al. 2011. A novel insight into the heme and NO/CO binding mechanism of the alpha subunit of human soluble guanylate cyclase. *J. Biol. Inorg. Chem.* 16:1227–39
31. Ma X, Beuve A, van den Akker F. 2010. Crystal structure of the signaling helix coiled-coil domain of the $\beta 1$ subunit of the soluble guanylyl cyclase. *BMC Struct. Biol.* 10:2
32. Winger JA, Marletta MA. 2005. Expression and characterization of the catalytic domains of soluble guanylate cyclase: interaction with the heme domain. *Biochemistry* 44:4083–90
33. Hu X, Murata LB, Weichsel A, Brailey JL, Roberts SA, et al. 2008. Allostery in recombinant soluble guanylyl cyclase from *Manduca sexta*. *J. Biol. Chem.* 283:20968–77
34. Harteneck C, Wedel B, Koesling D, Malkewitz J, Bohme E, Schultz G. 1991. Molecular cloning and expression of a new α-subunit of soluble guanylyl cyclase. Interchangeability of the α-subunits of the enzyme. *FEBS Lett.* 292:217–22
35. Yuen PS, Potter LR, Garbers DL. 1990. A new form of guanylyl cyclase is preferentially expressed in rat kidney. *Biochemistry* 29:10872–78
36. Budworth J, Meillerais S, Charles I, Powell K. 1999. Tissue distribution of the human soluble guanylate cyclases. *Biochem. Biophys. Res. Commun.* 263:696–701
37. Bellingham M, Evans TJ. 2007. The $\alpha 2 \beta 1$ isoform of guanylyl cyclase mediates plasma membrane localized nitric oxide signalling. *Cell Signal.* 19:2183–93
38. Russwurm M, Behrends S, Harteneck C, Koesling D. 1998. Functional properties of a naturally occurring isoform of soluble guanylyl cyclase. *Biochem J.* 335(Part 1):125–30
39. Behrends S, Harteneck C, Schultz G, Koesling D. 1995. A variant of the α_2 subunit of soluble guanylyl cyclase contains an insert homologous to a region within adenylyl cyclases and functions as a dominant negative protein. *J. Biol. Chem.* 270:21109–13
40. Neitz A, Mergia E, Eysel UT, Koesling D, Mittmann T. 2011. Presynaptic nitric oxide/cGMP facilitates glutamate release via hyperpolarization-activated cyclic nucleotide-gated channels in the hippocampus. *Eur. J. Neurosci.* 33:1611–21
41. Koglin M, Vehse K, Budaeus L, Scholz H, Behrends S. 2001. Nitric oxide activates the β_2 subunit of soluble guanylyl cyclase in the absence of a second subunit. *J. Biol. Chem.* 276:30737–43
42. Behrends S, Budaeus L, Kempfert J, Scholz H, Starbatty J, Vehse K. 2001. The $\beta 2$ subunit of nitric oxide-sensitive guanylyl cyclase is developmentally regulated in rat kidney. *Naunyn-Schmiedebergs Arch. Pharmacol.* 364:573–76
43. Friebe A, Koesling D. 2009. The function of NO-sensitive guanylyl cyclase: What we can learn from genetic mouse models. *Nitric Oxide* 21:149–56
44. Friebe A, Mergia E, Dangel O, Lange A, Koesling D. 2007. Fatal gastrointestinal obstruction and hypertension in mice lacking nitric oxide-sensitive guanylyl cyclase. *Proc. Natl. Acad. Sci. USA* 104:7699–704

45. Groneberg D, Konig P, Wirth A, Offermanns S, Koesling D, Friebe A. 2010. Smooth muscle-specific deletion of nitric oxide-sensitive guanylyl cyclase is sufficient to induce hypertension in mice. *Circulation* 121:401–9
46. Haghikia A, Mergia E, Friebe A, Eysel UT, Koesling D, Mittmann T. 2007. Long-term potentiation in the visual cortex requires both nitric oxide receptor guanylyl cyclases. *J. Neurosci.* 27:818–23
47. Mergia E, Friebe A, Dangel O, Russwurm M, Koesling D. 2006. Spare guanylyl cyclase NO receptors ensure high NO sensitivity in the vascular system. *J. Clin. Investig.* 116:1731–37
48. Nimmegeers S, Sips P, Buys E, Brouckaert P, Van de Voorde J. 2007. Functional role of the soluble guanylyl cyclase α1 subunit in vascular smooth muscle relaxation. *Cardiovasc. Res.* 76:149–59
49. Vermeersch P, Buys E, Pokreisz P, Marsboom G, Ichinose F, et al. 2007. Soluble guanylate cyclase-α1 deficiency selectively inhibits the pulmonary vasodilator response to nitric oxide and increases the pulmonary vascular remodeling response to chronic hypoxia. *Circulation* 116:936–43
50. Vanneste G, Dhaese I, Sips P, Buys E, Brouckaert P, Lefebvre RA. 2007. Gastric motility in soluble guanylate cyclase α1 knock-out mice. *J. Physiol.* 584:907–20
51. Dhaese I, Vanneste G, Sips P, Buys E, Brouckaert P, Lefebvre RA. 2008. Involvement of soluble guanylate cyclase α_1 and α_2, and SK_{Ca} channels in NANC relaxation of mouse distal colon. *Eur. J. Pharmacol.* 589:251–59
52. Tinette S, Zhang L, Garnier A, Engler G, Tares S, Robichon A. 2007. Exploratory behaviour in NO-dependent cyclase mutants of *Drosophila* shows defects in coincident neuronal signalling. *BMC Neurosci.* 8:65
53. Gibbs SM, Becker A, Hardy RW, Truman JW. 2001. Soluble guanylate cyclase is required during development for visual system function in *Drosophila*. *J. Neurosci.* 21:7705–14
54. Riedl CA, Neal SJ, Robichon A, Westwood JT, Sokolowski MB. 2005. *Drosophila* soluble guanylyl cyclase mutants exhibit increased foraging locomotion: behavioral and genomic investigations. *Behav. Genet.* 35:231–44
55. Zayas RM, Trimmer BA. 2007. Characterization of NO/cGMP-mediated responses in identified motoneurons. *Cell Mol. Neurobiol.* 27:191–209
56. Wilson CH, Christensen TA, Nighorn AJ. 2007. Inhibition of nitric oxide and soluble guanylyl cyclase signaling affects olfactory neuron activity in the moth, *Manduca sexta*. *J. Comp. Physiol. A* 193:715–28
57. Fujie S, Yamamoto T, Murakami J, Hatakeyama D, Shiga H, et al. 2005. Nitric oxide synthase and soluble guanylyl cyclase underlying the modulation of electrical oscillations in a central olfactory organ. *J. Neurobiol.* 62:14–30
58. Gelperin A, Flores J, Raccuia-Behling F, Cooke IR. 2000. Nitric oxide and carbon monoxide modulate oscillations of olfactory interneurons in a terrestrial mollusk. *J. Neurophysiol.* 83:116–27
59. Cary SP, Winger JA, Derbyshire ER, Marletta MA. 2006. Nitric oxide signaling: no longer simply on or off. *Trends Biochem. Sci.* 31:231–39
60. Wedel B, Humbert P, Harteneck C, Foerster J, Malkewitz J, et al. 1994. Mutation of His-105 in the β1 subunit yields a nitric oxide-insensitive form of soluble guanylyl cyclase. *Proc. Natl. Acad. Sci. USA* 91:2592–96
61. Wedel B, Harteneck C, Foerster J, Friebe A, Schultz G, Koesling D. 1995. Functional domains of soluble guanylyl cyclase. *J. Biol. Chem.* 270:24871–75
62. Iyer LM, Anantharaman V, Aravind L. 2003. Ancient conserved domains shared by animal soluble guanylyl cyclases and bacterial signaling proteins. *BMC Genomics* 4:5
63. Nioche P, Berka V, Vipond J, Minton N, Tsai AL, Raman CS. 2004. Femtomolar sensitivity of a NO sensor from *Clostridium botulinum*. *Science* 306:1550–53
64. Pellicena P, Karow DS, Boon EM, Marletta MA, Kuriyan J. 2004. Crystal structure of an oxygen-binding heme domain related to soluble guanylate cyclases. *Proc. Natl. Acad. Sci. USA* 101:12854–59
65. Boon EM, Marletta MA. 2005. Ligand discrimination in soluble guanylate cyclase and the H-NOX family of heme sensor proteins. *Curr. Opin. Chem. Biol.* 9:441–46
66. Moglich A, Ayers RA, Moffat K. 2009. Structure and signaling mechanism of Per-ARNT-Sim domains. *Structure* 17:1282–94
67. Shiga T, Suzuki N. 2005. Amphipathic α-helix mediates the heterodimerization of soluble guanylyl cyclase. *Zool. Sci.* 22:735–42

68. Rothkegel C, Schmidt PM, Atkins DJ, Hoffmann LS, Schmidt HH, et al. 2007. Dimerization region of soluble guanylate cyclase characterized by bimolecular fluorescence complementation in vivo. *Mol. Pharmacol.* 72:1181–90
69. Ma X, Sayed N, Baskaran P, Beuve A, van den Akker F. 2008. PAS-mediated dimerization of soluble guanylyl cyclase revealed by signal transduction histidine kinase domain crystal structure. *J. Biol. Chem.* 283:1167–78
70. Tesmer JJ, Sunahara RK, Gilman AG, Sprang SR. 1997. Crystal structure of the catalytic domains of adenylyl cyclase in a complex with $G_{s\alpha}$ • GTPγS. *Science* 278:1907–16
71. Zhang G, Liu Y, Ruoho AE, Hurley JH. 1997. Structure of the adenylyl cyclase catalytic core. *Nature* 386:247–53
72. Sunahara RK, Beuve A, Tesmer JJ, Sprang SR, Garbers DL, Gilman AG. 1998. Exchange of substrate and inhibitor specificities between adenylyl and guanylyl cyclases. *J. Biol. Chem.* 273:16332–38
73. Brandwein HJ, Lewicki JA, Waldman SA, Murad F. 1982. Effect of GTP analogues on purified soluble guanylate cyclase. *J. Biol. Chem.* 257:1309–11
74. Garbers DL, Suddath JL, Hardman JG. 1975. Enzymatic formation of inosine 3′,5′-monophosphate and of 2′-deoxyguanosine 3′,5′-monophosphate. Inosinate and deoxyguanylate cyclase activity. *Biochim. Biophys. Acta* 377:174–85
75. Gille A, Lushington GH, Mou TC, Doughty MB, Johnson RA, Seifert R. 2004. Differential inhibition of adenylyl cyclase isoforms and soluble guanylyl cyclase by purine and pyrimidine nucleotides. *J. Biol. Chem.* 279:19955–69
76. Derbyshire ER, Fernhoff NB, Deng S, Marletta MA. 2009. Nucleotide regulation of soluble guanylate cyclase substrate specificity. *Biochemistry* 48:7519–24
77. Haase T, Haase N, Kraehling JR, Behrends S. 2010. Fluorescent fusion proteins of soluble guanylyl cyclase indicate proximity of the heme nitric oxide domain and catalytic domain. *PLoS ONE* 5:e11617
78. Winger JA, Derbyshire ER, Lamers MH, Marletta MA, Kuriyan J. 2008. The crystal structure of the catalytic domain of a eukaryotic guanylate cyclase. *BMC Struct. Biol.* 8:42
79. Rauch A, Leipelt M, Russwurm M, Steegborn C. 2008. Crystal structure of the guanylyl cyclase Cya2. *Proc. Natl. Acad. Sci. USA* 105:15720–25
80. Morton DB. 2004. Atypical soluble guanylyl cyclases in *Drosophila* can function as molecular oxygen sensors. *J. Biol. Chem.* 279:50651–53
81. Morton DB, Langlais KK, Stewart JA, Vermehren A. 2005. Comparison of the properties of the five soluble guanylyl cyclase subunits in *Drosophila melanogaster*. *J. Insect Sci.* 5:12
82. Huang SH, Rio DC, Marletta MA. 2007. Ligand binding and inhibition of an oxygen-sensitive soluble guanylate cyclase, Gyc-88E, from *Drosophila*. *Biochemistry* 46:15115–22
83. Vermehren-Schmaedick A, Ainsley JA, Johnson WA, Davies SA, Morton DB. 2010. Behavioral responses to hypoxia in *Drosophila* larvae are mediated by atypical soluble guanylyl cyclases. *Genetics* 186:183–96
84. Gray JM, Karow DS, Lu H, Chang AJ, Chang JS, et al. 2004. Oxygen sensation and social feeding mediated by a *C. elegans* guanylate cyclase homologue. *Nature* 430:317–22
85. Cheung BH, Arellano-Carbajal F, Rybicki I, de Bono M. 2004. Soluble guanylate cyclases act in neurons exposed to the body fluid to promote *C. elegans* aggregation behavior. *Curr. Biol.* 14:1105–11
86. Cheung BH, Cohen M, Rogers C, Albayram O, de Bono M. 2005. Experience-dependent modulation of *C. elegans* behavior by ambient oxygen. *Curr. Biol.* 15:905–17
87. Zimmer M, Gray JM, Pokala N, Chang AJ, Karow DS, et al. 2009. Neurons detect increases and decreases in oxygen levels using distinct guanylate cyclases. *Neuron* 61:865–79
88. Price MS, Chao LY, Marletta MA. 2007. *Shewanella oneidensis* MR-1 H-NOX regulation of a histidine kinase by nitric oxide. *Biochemistry* 46:13677–83
89. Wang Y, Dufour YS, Carlson HK, Donohue TJ, Marletta MA, Ruby EG. 2010. H-NOX-mediated nitric oxide sensing modulates symbiotic colonization by *Vibrio fischeri*. *Proc. Natl. Acad. Sci. USA* 107:8375–80
90. Boon EM, Huang SH, Marletta MA. 2005. A molecular basis for NO selectivity in soluble guanylate cyclase. *Nat. Chem. Biol.* 1:53–59
91. Schmidt PM, Schramm M, Schroder H, Wunder F, Stasch JP. 2004. Identification of residues crucially involved in the binding of the heme moiety of soluble guanylate cyclase. *J. Biol. Chem.* 279:3025–32

92. Zhao Y, Schelvis JP, Babcock GT, Marletta MA. 1998. Identification of histidine 105 in the β1 subunit of soluble guanylate cyclase as the heme proximal ligand. *Biochemistry* 37:4502–9
93. Ma X, Sayed N, Beuve A, van den Akker F. 2007. NO and CO differentially activate soluble guanylyl cyclase via a heme pivot-bend mechanism. *EMBO J.* 26:578–88
94. Erbil WK, Price MS, Wemmer DE, Marletta MA. 2009. A structural basis for H-NOX signaling in *Shewanella oneidensis* by trapping a histidine kinase inhibitory conformation. *Proc. Natl. Acad. Sci. USA* 106:19753–60
95. Tran R, Boon EM, Marletta MA, Mathies RA. 2009. Resonance Raman spectra of an O_2-binding H-NOX domain reveal heme relaxation upon mutation. *Biochemistry* 48:8568–77
96. Ibrahim M, Derbyshire ER, Marletta MA, Spiro TG. 2010. Probing soluble guanylate cyclase activation by CO and YC-1 using resonance Raman spectroscopy. *Biochemistry* 49:3815–23
97. Ibrahim M, Derbyshire ER, Soldatova AV, Marletta MA, Spiro TG. 2010. Soluble guanylate cyclase is activated differently by excess NO and by YC-1: resonance Raman spectroscopic evidence. *Biochemistry* 49:4864–71
98. Pal B, Tanaka K, Takenaka S, Kitagawa T. 2010. Resonance Raman spectroscopic investigation of structural changes of CO-heme in soluble guanylate cyclase generated by effectors and substrate. *J. Raman Spectrosc.* 41:1178–84
99. Olea C, Boon EM, Pellicena P, Kuriyan J, Marletta MA. 2008. Probing the function of heme distortion in the H-NOX family. *ACS Chem. Biol.* 3:703–10
100. Tsai AL, Berka V, Martin F, Ma X, van den Akker F, et al. 2010. Is *Nostoc* H-NOX a NO sensor or redox switch? *Biochemistry* 49:6587–99
101. Martin F, Baskaran P, Ma X, Dunten PW, Schaefer M, et al. 2010. Structure of cinaciguat (BAY 58-2667) bound to Nostoc H-NOX domain reveals insights into heme-mimetic activation of the soluble guanylyl cyclase. *J. Biol. Chem.* 285:22651–57
102. Barrick D. 1994. Replacement of the proximal ligand of sperm whale myoglobin with free imidazole in the mutant His-93→Gly. *Biochemistry* 33:6546–54
103. Evgenov OV, Pacher P, Schmidt PM, Hasko G, Schmidt HH, Stasch JP. 2006. NO-independent stimulators and activators of soluble guanylate cyclase: discovery and therapeutic potential. *Nat. Rev. Drug Discov.* 5:755–68
104. Baskaran P, Heckler EJ, van den Akker F, Beuve A. 2011. Aspartate 102 in the heme domain of soluble guanylyl cyclase has a key role in NO activation. *Biochemistry* 50:4291–97
105. Derbyshire ER, Deng S, Marletta MA. 2010. Incorporation of tyrosine and glutamine residues into the soluble guanylate cyclase heme distal pocket alters NO and O_2 binding. *J. Biol. Chem.* 285:17471–78
106. Rothkegel C, Schmidt PM, Stoll F, Schroder H, Schmidt HHHW, Stasch JP. 2006. Identification of residues crucially involved in soluble guanylate cyclase activation. *FEBS Lett.* 580:4205–13
107. Capece L, Estrin DA, Marti MA. 2008. Dynamical characterization of the heme NO oxygen binding (HNOX) domain. Insight into soluble guanylate cyclase allosteric transition. *Biochemistry* 47:9416–27
108. Xu C, Ibrahim M, Spiro TG. 2008. DFT analysis of axial and equatorial effects on heme-CO vibrational modes: applications to CooA and H-NOX heme sensor proteins. *Biochemistry* 47:2379–87
109. Makino R, Park SY, Obayashi E, Iizuka T, Hori H, Shiro Y. 2011. Oxygen binding and redox properties of the heme in soluble guanylate cyclase: implications for the mechanism of ligand discrimination. *J. Biol. Chem.* 286:15678–87
110. Deinum G, Stone JR, Babcock GT, Marletta MA. 1996. Binding of nitric oxide and carbon monoxide to soluble guanylate cyclase as observed with resonance Raman spectroscopy. *Biochemistry* 35:1540–7
111. Jain R, Chan MK. 2003. Mechanisms of ligand discrimination by heme proteins. *J. Biol. Inorg. Chem.* 8:1–11
112. Zhang Y, Lu M, Cheng Y, Li Z. 2010. H-NOX domains display different tunnel systems for ligand migration. *J. Mol. Graph. Model.* 28:814–19
113. Winter MB, Herzik MA, Kuriyan J, Marletta MA. 2011. Tunnels modulate ligand flux in a heme nitric oxide/oxygen binding (H-NOX) domain. *Proc. Natl. Acad. Sci. USA* 108:E881–89
114. Martin E, Berka V, Bogatenkova E, Murad F, Tsai AL. 2006. Ligand selectivity of soluble guanylyl cyclase: effect of the hydrogen-bonding tyrosine in the distal heme pocket on binding of oxygen, nitric oxide, and carbon monoxide. *J. Biol. Chem.* 281:27836–45

115. Das TK, Samuni U, Lin Y, Goldberg DE, Rousseau DL, Friedman JM. 2004. Distal heme pocket conformers of carbonmonoxy derivatives of *Ascaris* hemoglobin: evidence of conformational trapping in porous sol-gel matrices. *J. Biol. Chem.* 279:10433–41
116. Giangiacomo L, Ilari A, Boffi A, Morea V, Chiancone E. 2005. The truncated oxygen-avid hemoglobin from *Bacillus subtilis*: X-ray structure and ligand binding properties. *J. Biol. Chem.* 280:9192–202
117. Marti MA, Capece L, Bikiel DE, Falcone B, Estrin DA. 2007. Oxygen affinity controlled by dynamical distal conformations: the soybean leghemoglobin and the *Paramecium caudatum* hemoglobin cases. *Proteins* 68:480–87
118. Milani M, Pesce A, Ouellet Y, Dewilde S, Friedman J, et al. 2004. Heme-ligand tunneling in group I truncated hemoglobins. *J. Biol. Chem.* 279:21520–25
119. Schrammel A, Behrends S, Schmidt K, Koesling D, Mayer B. 1996. Characterization of 1H-[1,2,4]oxadiazolo[4,3-a]quinoxalin-1-one as a heme-site inhibitor of nitric oxide-sensitive guanylyl cyclase. *Mol. Pharmacol.* 50:1–5
120. Stone JR, Marletta MA. 1996. Spectral and kinetic studies on the activation of soluble guanylate cyclase by nitric oxide. *Biochemistry* 35:1093–99
121. Bellamy TC, Garthwaite J. 2001. Sub-second kinetics of the nitric oxide receptor, soluble guanylyl cyclase, in intact cerebellar cells. *J. Biol. Chem.* 276:4287–92
122. Cary SP, Winger JA, Marletta MA. 2005. Tonic and acute nitric oxide signaling through soluble guanylate cyclase is mediated by nonheme nitric oxide, ATP, and GTP. *Proc. Natl. Acad. Sci. USA* 102:13064–69
123. Russwurm M, Koesling D. 2004. NO activation of guanylyl cyclase. *EMBO J.* 23:4443–50
124. Derbyshire ER, Gunn A, Ibrahim M, Spiro TG, Britt RD, Marletta MA. 2008. Characterization of two different five-coordinate soluble guanylate cyclase ferrous-nitrosyl complexes. *Biochemistry* 47:3892–99
125. Fernhoff NB, Derbyshire ER, Marletta MA. 2009. A nitric oxide/cysteine interaction mediates the activation of soluble guanylate cyclase. *Proc. Natl. Acad. Sci. USA* 106:21602–7
126. Lawson DM, Stevenson CE, Andrew CR, Eady RR. 2000. Unprecedented proximal binding of nitric oxide to heme: implications for guanylate cyclase. *EMBO J.* 19:5661–71
127. Roy B, Garthwaite J. 2006. Nitric oxide activation of guanylyl cyclase in cells revisited. *Proc. Natl. Acad. Sci. USA* 103:12185–90
128. Palmer RM, Ferrige AG, Moncada S. 1987. Nitric oxide release accounts for the biological activity of endothelium-derived relaxing factor. *Nature* 327:524–26
129. Margulis A, Sitaramayya A. 2000. Rate of deactivation of nitric oxide-stimulated soluble guanylate cyclase: influence of nitric oxide scavengers and calcium. *Biochemistry* 39:1034–39
130. Russwurm M, Mergia E, Mullershausen F, Koesling D. 2002. Inhibition of deactivation of NO-sensitive guanylyl cyclase accounts for the sensitizing effect of YC-1. *J. Biol. Chem.* 277:24883–88
131. Kharitonov VG, Russwurm M, Magde D, Sharma VS, Koesling D. 1997. Dissociation of nitric oxide from soluble guanylate cyclase. *Biochem. Biophys. Res. Commun.* 239:284–86
132. Bellamy TC, Wood J, Goodwin DA, Garthwaite J. 2000. Rapid desensitization of the nitric oxide receptor, soluble guanylyl cyclase, underlies diversity of cellular cGMP responses. *Proc. Natl. Acad. Sci. USA* 97:2928–33
133. Schroder H, Leitman DC, Bennett BM, Waldman SA, Murad F. 1988. Glyceryl trinitrate-induced desensitization of guanylate cyclase in cultured rat lung fibroblasts. *J. Pharmacol. Exp. Ther.* 245:413–18
134. Elkayam U, Kulick D, McIntosh N, Roth A, Hsueh W, Rahimtoola SH. 1987. Incidence of early tolerance to hemodynamic effects of continuous infusion of nitroglycerin in patients with coronary artery disease and heart failure. *Circulation* 76:577–84
135. Kim D, Rybalkin SD, Pi X, Wang Y, Zhang C, et al. 2001. Upregulation of phosphodiesterase 1A1 expression is associated with the development of nitrate tolerance. *Circulation* 104:2338–43
136. Sydow K, Daiber A, Oelze M, Chen Z, August M, et al. 2004. Central role of mitochondrial aldehyde dehydrogenase and reactive oxygen species in nitroglycerin tolerance and cross-tolerance. *J. Clin. Investig.* 113:482–89
137. Sayed N, Baskaran P, Ma X, van den Akker F, Beuve A. 2007. Desensitization of soluble guanylyl cyclase, the NO receptor, by S-nitrosylation. *Proc. Natl. Acad. Sci. USA* 104:12312–17

138. Sayed N, Kim DD, Fioramonti X, Iwahashi T, Duran WN, Beuve A. 2008. Nitroglycerin-induced S-nitrosylation and desensitization of soluble guanylyl cyclase contribute to nitrate tolerance. *Circ. Res.* 103:606–14
139. Brandwein HJ, Lewicki JA, Murad F. 1981. Reversible inactivation of guanylate cyclase by mixed disulfide formation. *J. Biol. Chem.* 256:2958–62
140. Maron BA, Zhang YY, Handy DE, Beuve A, Tang SS, et al. 2009. Aldosterone increases oxidant stress to impair guanylyl cyclase activity by cysteinyl thiol oxidation in vascular smooth muscle cells. *J. Biol. Chem.* 284:7665–72
141. Derbyshire ER, Marletta MA. 2007. Butyl isocyanide as a probe of the activation mechanism of soluble guanylate cyclase. Investigating the role of non-heme nitric oxide. *J. Biol. Chem.* 282:35741–48
142. Derbyshire ER, Tran R, Mathies RA, Marletta MA. 2005. Characterization of nitrosoalkane binding and activation of soluble guanylate cyclase. *Biochemistry* 44:16257–65
143. Makino R, Matsuda H, Obayashi E, Shiro Y, Iizuka T, Hori H. 1999. EPR characterization of axial bond in metal center of native and cobalt-substituted guanylate cyclase. *J. Biol. Chem.* 274:7714–23
144. Ko FN, Wu CC, Kuo SC, Lee FY, Teng CM. 1994. YC-1, a novel activator of platelet guanylate cyclase. *Blood* 84:4226–33
145. Friebe A, Schultz G, Koesling D. 1996. Sensitizing soluble guanylyl cyclase to become a highly CO-sensitive enzyme. *EMBO J.* 15:6863–68
146. Stone JR, Marletta MA. 1998. Synergistic activation of soluble guanylate cyclase by YC-1 and carbon monoxide: implications for the role of cleavage of the iron-histidine bond during activation by nitric oxide. *Chem. Biol.* 5:255–61
147. Makino R, Obayashi E, Homma N, Shiro Y, Hori H. 2003. YC-1 facilitates release of the proximal His residue in the NO and CO complexes of soluble guanylate cyclase. *J. Biol. Chem.* 278:11130–37
148. Stasch JP, Becker EM, Alonso-Alija C, Apeler H, Dembowsky K, et al. 2001. NO-independent regulatory site on soluble guanylate cyclase. *Nature* 410:212–15
149. Friebe A, Russwurm M, Mergia E, Koesling D. 1999. A point-mutated guanylyl cyclase with features of the YC-1-stimulated enzyme: implications for the YC-1 binding site? *Biochemistry* 38:15253–57
150. Lamothe M, Chang FJ, Balashova N, Shirokov R, Beuve A. 2004. Functional characterization of nitric oxide and YC-1 activation of soluble guanylyl cyclase: structural implication for the YC-1 binding site? *Biochemistry* 43:3039–48
151. Yazawa S, Tsuchiya H, Hori H, Makino R. 2006. Functional characterization of two nucleotide-binding sites in soluble guanylate cyclase. *J. Biol. Chem.* 281:21763–70
152. Kharitonov VG, Sharma VS, Magde D, Koesling D. 1999. Kinetics and equilibria of soluble guanylate cyclase ligation by CO: effect of YC-1. *Biochemistry* 38:10699–706
153. Denninger JW, Schelvis JP, Brandish PE, Zhao Y, Babcock GT, Marletta MA. 2000. Interaction of soluble guanylate cyclase with YC-1: kinetic and resonance Raman studies. *Biochemistry* 39:4191–98
154. Sharma VS, Magde D, Kharitonov VG, Koesling D. 1999. Soluble guanylate cyclase: effect of YC-1 on ligation kinetics with carbon monoxide. *Biochem. Biophys. Res. Commun.* 254:188–91
155. Bischoff E, Stasch JP. 2004. Effects of the sGC stimulator BAY 41-2272 are not mediated by phosphodiesterase 5 inhibition. *Circulation* 110:e320–21; author reply e20–1
156. Mullershausen F, Russwurm M, Friebe A, Koesling D. 2004. Inhibition of phosphodiesterase type 5 by the activator of nitric oxide-sensitive guanylyl cyclase BAY 41-2272. *Circulation* 109:1711–13
157. Belik J. 2009. Riociguat, an oral soluble guanylate cyclase stimulator for the treatment of pulmonary hypertension. *Curr. Opin. Investig. Drugs* 10:971–79
158. Stasch JP, Schmidt P, Alonso-Alija C, Apeler H, Dembowsky K, et al. 2002. NO- and haem-independent activation of soluble guanylyl cyclase: molecular basis and cardiovascular implications of a new pharmacological principle. *Br. J. Pharmacol.* 136:773–83
159. Roy B, Mo E, Vernon J, Garthwaite J. 2008. Probing the presence of the ligand-binding haem in cellular nitric oxide receptors. *Br. J. Pharmacol.* 153:1495–504
160. Meurer S, Pioch S, Pabst T, Opitz N, Schmidt PM, et al. 2009. Nitric oxide-independent vasodilator rescues heme-oxidized soluble guanylate cyclase from proteasomal degradation. *Circ. Res.* 105:33–41
161. Kalk P, Godes M, Relle K, Rothkegel C, Hucke A, et al. 2006. NO-independent activation of soluble guanylate cyclase prevents disease progression in rats with 5/6 nephrectomy. *Br. J. Pharmacol.* 148:853–59

162. Korkmaz S, Radovits T, Barnucz E, Hirschberg K, Neugebauer P, et al. 2009. Pharmacological activation of soluble guanylate cyclase protects the heart against ischemic injury. *Circulation* 120:677–86
163. Frey R, Muck W, Unger S, Artmeier-Brandt U, Weimann G, Wensing G. 2008. Pharmacokinetics, pharmacodynamics, tolerability, and safety of the soluble guanylate cyclase activator cinaciguat (BAY 58-2667) in healthy male volunteers. *J. Clin. Pharmacol.* 48:1400–10
164. Lapp H, Mitrovic V, Franz N, Heuer H, Buerke M, et al. 2009. Cinaciguat (BAY 58-2667) improves cardiopulmonary hemodynamics in patients with acute decompensated heart failure. *Circulation* 119:2781–88
165. Mueck W, Frey R. 2010. Population pharmacokinetics and pharmacodynamics of cinaciguat, a soluble guanylate cyclase activator, in patients with acute decompensated heart failure. *Clin. Pharmacokinet.* 49:119–29
166. Garthwaite J, Southam E, Boulton CL, Nielsen EB, Schmidt K, Mayer B. 1995. Potent and selective inhibition of nitric oxide-sensitive guanylyl cyclase by 1H-[1,2,4]oxadiazolo[4,3-a]quinoxalin-1-one. *Mol. Pharmacol.* 48:184–88
167. Mulsch A, Busse R, Liebau S, Forstermann U. 1988. LY 83583 interferes with the release of endothelium-derived relaxing factor and inhibits soluble guanylate cyclase. *J. Pharmacol. Exp. Ther.* 247:283–88
168. Gruetter CA, Kadowitz PJ, Ignarro LJ. 1981. Methylene blue inhibits coronary arterial relaxation and guanylate cyclase activation by nitroglycerin, sodium nitrite, and amyl nitrite. *Can. J. Physiol. Pharmacol.* 59:150–56
169. Lee YS, Wurster RD. 1995. Mechanism of potentiation of LY83583-induced growth inhibition by sodium nitroprusside in human brain tumor cells. *Cancer Chemother. Pharmacol.* 36:341–44
170. Wolin MS, Cherry PD, Rodenburg JM, Messina EJ, Kaley G. 1990. Methylene blue inhibits vasodilation of skeletal muscle arterioles to acetylcholine and nitric oxide via the extracellular generation of superoxide anion. *J. Pharmacol. Exp. Ther.* 254:872–76
171. Chang FJ, Lemme S, Sun Q, Sunahara RK, Beuve A. 2005. Nitric oxide-dependent allosteric inhibitory role of a second nucleotide binding site in soluble guanylyl cyclase. *J. Biol. Chem.* 280:11513–19
172. Suzuki T, Suematsu M, Makino R. 2001. Organic phosphates as a new class of soluble guanylate cyclase inhibitors. *FEBS Lett.* 507:49–53

The MPS1 Family of Protein Kinases

Xuedong Liu[1] and Mark Winey[2]

[1]Department of Chemistry and Biochemistry, [2]Department of Molecular, Cellular and Developmental Biology, University of Colorado, Boulder, Colorado 80309; email: xuedong.liu@colorado.edu, mark.winey@colorado.edu

Keywords

TTK, spindle checkpoint, mitosis, kinetochore, cell cycle

Abstract

MPS1 protein kinases are found widely, but not ubiquitously, in eukaryotes. This family of potentially dual-specific protein kinases is among several that regulate a number of steps of mitosis. The most widely conserved MPS1 kinase functions involve activities at the kinetochore in both the chromosome attachment and the spindle checkpoint. MPS1 kinases also function at centrosomes. Beyond mitosis, MPS1 kinases have been implicated in development, cytokinesis, and several different signaling pathways. Family members are identified by virtue of a conserved C-terminal kinase domain, though the N-terminal domain is quite divergent. The kinase domain of the human enzyme has been crystallized, revealing an unusual ATP-binding pocket. The activity, level, and subcellular localization of Mps1 family members are tightly regulated during cell-cycle progression. The mitotic functions of Mps1 kinases and their overexpression in some tumors have prompted the identification of Mps1 inhibitors and their active development as anticancer drugs.

Contents

1. INTRODUCTION 562
   1.1. Discovery and Initial Characterization 562
   1.2. Mps1 Kinase Features 563
   1.3. MPS1 Distribution and Diversity 563
2. MPS1 FUNCTIONS 563
   2.1. Spindle Pole Assembly 563
   2.2. Kinetochores and the Spindle Assembly Checkpoint 564
   2.3. Other Signaling Pathways 566
   2.4. Cytokinesis 567
   2.5. Meiosis 567
3. MPS1 STRUCTURE, ENZYMOLOGY AND INHIBITORS 568
   3.1. Structure of the Mps1 Catalytic Domain 568
   3.2. Regulation of Mps1 Activity by Phosphorylation 568
   3.3. Diversity in Phosphorylation Site Selection and Substrate Recognition 571
   3.4. Dimerization 572
   3.5. Mutant and Analog-Sensitive Alleles 572
   3.6. Small-Molecule Mps1 Inhibitors 573
4. REGULATION OF MPS1 KINASES 573
   4.1. Transcription 573
   4.2. Localization 575
   4.3. Degradation and Inactivation ... 577
   4.4. Misregulation in Tumor Cells 577

1. INTRODUCTION

MPS1: monopolar spindle 1

SPB: spindle pole body

Protein kinases are critical regulators of cell division. Apart from the cyclin-dependent kinases (CDKs), which are considered the master regulators, a suite of additional conserved kinases control progression through mitosis, including Polo, Aurora, Bub, NEK/NimA, and MPS1 kinases. Collectively, these have been called the mitotic kinases because of the widely conserved nature of their functions in mitosis. Here, we review the MPS1 family of protein kinases, which are still being discovered, dissected, and potentially exploited for therapeutics owing to their critical functions in the control of mitosis.

1.1. Discovery and Initial Characterization

The *MPS1* gene (monopolar spindle) was first identified in the budding yeast, *Saccharomyces cerevisiae*. The original mutant allele, *mps1-1*, causes yeast cells to fail at a restrictive temperature in spindle pole body (SPB) assembly (the yeast centrosome) (1), a critical cell-cycle event that is necessary to form a bipolar spindle. The *MPS1* gene was first cloned as an essential gene encoding an apparent protein kinase by Poch et al. (2). It was named *RPK1*, but *MPS1* is used because it was the first published name. Lauze et al. (3) demonstrated that glutathione S-transferase (GST)-tagged Mps1 was indeed a protein kinase. GST-Mps1 exhibited robust autophosphorylation, as well as substrate phosphorylation of several common in vitro kinase substrates. Mps1 was able to phosphorylate serines/threonines and tyrosines, suggesting that it is a dual-specificity protein kinase, but thus far, no biologically relevant substrate is known to be phosphorylated on tyrosine by Mps1.

Phenotypic analysis of the original yeast *MPS1* mutants identified key functions of the kinase. As noted above, *mps1-1* was first discovered because of a defect in yeast SPB duplication, leading to an aberrant monopolar spindle. It was also observed that *mps1-1* mutant cells failed to arrest in mitosis with the monopolar spindle defect unlike other mutants defective in SPB duplication [*kar1* (4), *cdc31* and *mps2* (1)]. A subsequent study demonstrated that this phenotype was because of the role of Mps1 in the spindle checkpoint (5). In addition, Hardwick et al. (6) showed that Mps1 overexpression caused mitotic arrest by triggering the spindle checkpoint and identified the checkpoint protein, Mad1, as the first Mps1 substrate. All of the original *MPS1* alleles caused defects in both SPB duplication and in the spindle assembly checkpoint at a restrictive temperature (7). Interestingly, electron microscopic examination of the SPBs in these various strains revealed that Mps1 is required for multiple steps of SPB assembly.

Prior to cloning of the yeast *MPS1* gene, Mps1 orthologs from humans [phosphotyrosine-picked threonine kinase/threonine and tyrosine kinase (PYT/TTK) aka hMPS1] (8, 9) and mice [esk (EC STY kinase aka MmMps1)] (10) had been discovered. Intriguingly, these kinase genes were identified in screens using antiphosphotyrosine antibodies that tested expression libraries for protein kinases that autophosphorylate on tyrosine residues. Indeed, these kinases phosphorylate serine, threonine, and tyrosine residues in vitro, offering the initial evidence that this is a family of dual-specificity protein kinases. Also, it was observed that both mRNA and protein levels of Mps1/TTK are readily detectable in all proliferating human cells and tissues but are markedly reduced or absent in resting cells and in tissues with a low proliferative index (11).

1.2. Mps1 Kinase Features

MPS1 kinase family members are ~85–95 kDa and have conserved C-terminal kinase domains. The N-terminal domains of the family members appear unrelated, and they lack any unifying motif [such as the Polo box observed in the Polo family of kinases (12)]. Nonetheless, the kinase domain is distinctive enough to identify family members (**http://www.signaling-gateway.org/molecule/query?afcsid=A000882&type=blast&adv=latest**). The crystal structure of the kinase domain reveals interesting features that are discussed below.

1.3. MPS1 Distribution and Diversity

MPS1 kinase genes are found in most eukaryotes. Interestingly, there is no well-documented case of a genome containing paralogs of an *MPS1* gene. However, *MPS1* isoforms generated by alternative splicing have been predicted in humans and observed in mice (10). Orthologs are easily identified in fungi, vertebrates, and invertebrates, like *Drosophila*, as well as in plants, including the ancient plant lineage of lycophytes (*Selaginella moellendorffii*), diatoms (*Phaeodactylum tricornutum*), and alga (*Chlamydomonas*) (13). Although no validated *MPS1* has been identified in the nematode *Caenorhabditis elegans*, there are orthologs in the pathogenic nematode (*Globodera*), as well in flat and round worms.

2. MPS1 FUNCTIONS

MPS1 kinases have multiple roles in mitosis that we briefly survey here. The most widely conserved and prominent function of these kinases is to ensure proper biorientation of sister chromatids on the mitotic spindle at kinetochores, and this function involves the spindle checkpoint. Along with being implicated in other cellular processes, MPS1 kinases also function from the earliest steps of mitosis, including spindle pole duplication, to the latest steps of mitotic exit and cytokinesis.

2.1. Spindle Pole Assembly

As described above, the budding yeast *MPS1* gene was identified by a mutation that is defective in SPB (centrosome) duplication. A collection of *MPS1* alleles revealed that the kinase acts in multiple steps of the duplication pathway (7), and the kinase has been shown to phosphorylate numerous SPB components. These include Spc29 and Spc42, which are fungus specific, whose assembly and stability are controlled by Mps1 phosphorylation (14–16). The more widely conserved SPB components that are Mps1 substrates include centrin (Cdc31) (14), the γ-tubulin complex component Spc98 (17), and the Spc110 tether that holds the γ-tubulin complex (Tub4, Spc98, Spc97) to the SPB (18). Centrin (Cdc31) is a small, EF-hand calcium-binding protein (19), and its phosphorylation by Mps1 influences its interaction with a binding partner (14). Phosphorylation of Spc98 is only found on the nuclear pool of the γ-tubulin

Centrosome: the cellular structure that contains centrioles, nucleates microtubule formation, and organizes mitotic spindles

Autophosphorylation: the action of a kinase adding one or more phosphate groups to itself

Dual-specificity protein kinase: a protein kinase that exhibits Ser/Thr and Tyr phosphorylation activities

Spindle checkpoint: this mechanism ensures proper chromosome attachment to microtubules prior to chromosome segregation (aka spindle assembly checkpoint or mitotic checkpoint)

Kinetochore: this structure, assembled at centromeres, captures spindle microtubules and serves as the signaling platform for the spindle checkpoint

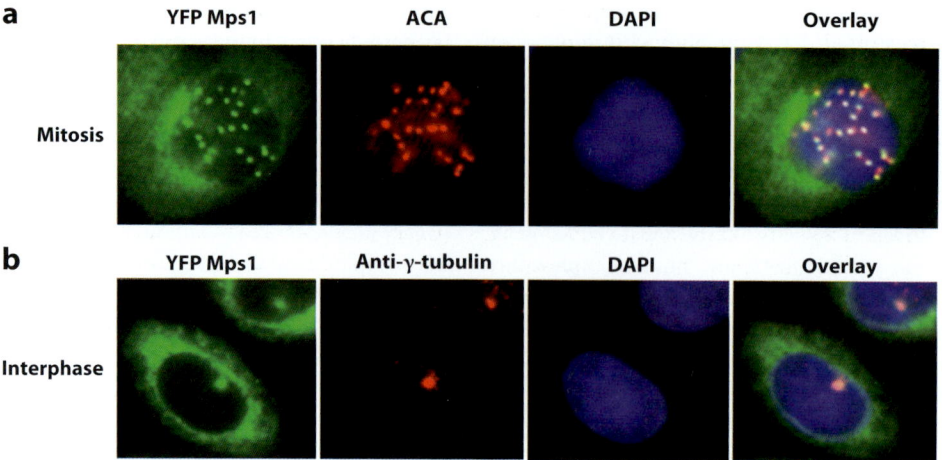

Figure 1

Localization of Mps1 in vertebrate mitotic and interphase cells. (*a*) Kinetochore localization of yellow fluorescent protein (YFP) Mps1 in mitotic SW480 cells. Anti-centromere antibodies (ACA) and 4′,6-diamidino-2-phenylindole (DAPI) were used to stain kinetochores and chromosomes. (*b*) YFP Mps1 is localized to centrosomes and the cytosol during interphase. Centrosomes and nuclei were stained by anti-γ-tubulin and DAPI, respectively.

complex, possibly influencing its interaction with Spc110 (17). Likewise, Mps1 phosphorylation of Spc110 (in conjunction with phosphorylation by Cdc28; the yeast CDK) is required for interaction with Spc97 (18). Interestingly, both Mps1 and Cdk phosphorylate most of these substrates such that combinatorial control appears to be the rule (20, 21).

Mps1 is localized to SPBs in yeast (15), and the mammalian enzymes are localized to centrosomes (**Figure 1**). Overexpression of mammalian Mps1 leads to overduplication of the centrosomes, and overexpression of a kinase-inactive allele blocks centrosome duplication (reviewed in Reference 22). Despite these results, RNAi experiments have produced contradictory results concerning a requirement for Mps1 in centrosome duplication (22). Recently, Mps1 was deleted from human cell lines using cre-lox, and these cells were capable of centrosome duplication (23). Similarly, the fission yeast ortholog, Mph1 (the *Schizosaccharomyces pombe* Mps1 homolog), is not required for SPB duplication (24), nor is the *Drosophila* ortholog, Ald, required for centrosome duplication (25). Finally, *C. elegans* appears to lack an MPS1 ortholog and can execute centriole and centrosome duplication.

Nonetheless, Mps1 influences centrosome duplication in human cells (reviewed by Reference 22). Furthermore, the centrosomal levels of hMps1 are exquisitely controlled, separately from other pools of the kinase (discussed below). Additionally, hMps1 centrosomal substrates, such as mortalin (26) and centrin 2 (27), have been identified. The phosphorylation of centrin 2 is required for its ability to stimulate centriole (the microtubule-based structural core of the centrosome) assembly (27). Remarkably, the major site of centrin 2 phosphorylation by hMps1 is T118 (27), which is the site analogous to the yMps1-phosphorylated T110 on the yeast centrin Cdc31 (14). These results suggest a deeply conserved regulatory event.

2.2. Kinetochores and the Spindle Assembly Checkpoint

MPS1 kinases have universally conserved functions at kinetochores. In yeast, Mps1 was implicated in the spindle checkpoint (5, 6) and was later shown to be localized to kinetochores

(15). The spindle checkpoint monitors the correct bipolar attachment and tension of all chromosomes to the mitotic spindle. The cells are held at metaphase until every chromosome is properly attached; then the cells can proceed into anaphase. The molecular target of the checkpoint is the anaphase-promoting complex (APC), a ubiquitin ligase that targets mitotic cyclins and other proteins for destruction, allowing the cells to segregate their chromosomes. The APC is controlled by the Cdc20 activator, which in turn can be inactivated by the checkpoint protein Mad2. Mad2 is activated in the course of cycling on and off unattached kinetochores. A variety of other checkpoint proteins act with Mad2, both on and off of the kinetochore, to inhibit Cdc20 activity and therefore inhibit APC.

Xenopus Mps1, XlMps1, was the first vertebrate ortholog implicated in the spindle checkpoint and localized to kinetochores (28). *Xenopus* oocyte extracts require XlMps1 for mitotic arrest and for the recruitment of Mad2 and other checkpoint proteins to the kinetochore. Human Mps1 is also found at kinetochores and is required for the activation and maintenance of the spindle checkpoint (29, 30). These results have been repeated in several recent studies using a variety of tools, including depletion of hMps1, conditional deletion of the hMps1 gene, and small-molecule inhibition of the native or engineered forms of the kinase (reviewed in Reference 31). In most systems, kinetochore localization of Mad2 also requires active hMps1, although Mad2 appears not to be a hMps1 substrate.

Kinetochore localization of checkpoint proteins is important for their function. In *Xenopus*, Cenp-E localization to kinetochores requires XlMps1 (28), and CENP-E has been identified as an Mps1 substrate in vitro (32). However, this result, and similar dependencies on Mps1 for kinetochore localization of various checkpoint proteins, has not been universally observed. Lan & Cleveland (31) carefully document the various discrepancies and propose that they arise from the use of various Mps1-inactivating methods and different cell types. Some of these studies, but not all, show the loss of several checkpoint proteins from the kinetochore when Mps1 is inactivated, consistent with work in *Xenopus*. Similarly, the Mps1 overexpression-induced arrest in yeast is dependent on the function of each of the checkpoint proteins. Collectively, these results suggest Mps1 is near the top of the checkpoint-signaling pathway. However, the complexity of the data indicates that a linear pathway may be too simple a model for the checkpoint. An alternative view is that hMps1 is a linchpin in a checkpoint network such that the absence of Mps1 activity disrupts several checkpoint activities, leading to catastrophic failure of the network and other spindle-related functions (31). These predictions are complicated by the fact that Mps1 likely has several substrates in this pathway or network. Already known Mps1 checkpoint substrates are Mad1 (in yeast), Cenp-E, and Mps1 itself (6, 32, 33).

Interestingly, Mps1 can act in the checkpoint without being present at the kinetochore. The overexpression of yMps1 is capable of imposing a mitotic arrest in *ndc10-1* strains (34), which normally do not arrest because the mutation destroys the kinetochore and obviates its ability to act in checkpoint signaling. A similar phenomenon has been observed in human cells using a truncated allele of hMps1 that does not localize to kinetochores but retains the catalytic domain (23). This allele can still activate the mitotic checkpoint complex (35), which inhibits Cdc20 during mitosis. Mps1 also contributes to the formation of an interphase APC inhibitor that shares components with the mitotic checkpoint complex, such as Mad2 and BubR1 (23). Although these proteins can inhibit Cdc20 in vitro without Mps1 (36), their association in vivo is dependent on Mps1 activity (23). Indeed, Mps1 is so critical to controlling normal mitotic progression that cells lacking Mps1 activity transit mitosis faster than cells with the activity (reviewed in Reference 31).

Prior to checkpoint signaling, both the Aurora B and Mps1 kinases are required for the proper bipolar spindle attachment of chromosomes. These kinases are involved in resolving syntelic attachments in which both

APC: anaphase-promoting complex

Syntelic attachments: an aberrant chromosome attachment, where both sister chromosomes are attached to a single spindle pole instead of a bipolar attachment

kinetochores of replicated sister chromatids are attached to the same pole instead of their correct bipolar attachment (23, 37–41). The dependency relationship between Aurora B and Mps1 in vertebrates is a point of contention. In some studies, Aurora B activity is reduced upon reduction of Mps1 activity (42, 43), placing Aurora B downstream of Mps1. One reported mechanism for this dependency is via phosphorylation of the chromosomal passenger protein Borealin, which influences Aurora B activity (42, 44). However, other studies, using various hMps1 inhibitors or inhibitable hMps1 alleles, found that Aurora B activity was unchanged by reducing Mps1 activity (23, 39, 40). Indeed, two of these studies (39, 40) have shown instead that reducing Aurora B activity results in reduced hMps1 at kinetochores, similar to findings in *Xenopus* extracts (45). Saurin et al. (46) report that Aurora B and the kinetochore component Hec1/hNdc80 are both required for Mps1 recruitment to the kinetochore, a precondition that can be circumvented by tethering Mps1 to the kinetochore. Furthermore, the kinetochore recruitment of Mps1 is necessary for timely activation of Mps1 in mitosis. These results place Aurora B upstream of hMps1 in the spindle checkpoint pathway, as seen in yeast (37). However, the yeast kinetochore protein Dam1 is a substrate of both Ipl1 [the yeast Aurora B kinase (47)] and yMps1 (48), suggesting that the kinases collaborate in controlling kinetochore attachment. Similarly, in reviewing recent findings, Lan & Cleveland (31) also suggest that shared substrates may explain some of the complexity in Aurora B and Mps1 interactions. Their candidate substrate for analysis is CENP-E, which is phosphorylated by both Aurora B (49) and Mps1 (32). Ndc80/Hec1 should be considered as well, as the yeast Ndc80 protein is an Mps1 substrate (50), and Hec1/hNdc80 is a substrate of Aurora B (51). Finally, Ipl1 and the chromosomal passenger complex in yeast can act in a pathway distinct from yMps1, Sgo1 (shugoshin), and Bub1 in processing syntelic attachments (52).

The collective Mps1 functions in chromosome segregation are sufficient to make the enzyme essential in most organisms [though *S. pombe* Mps1 (Mph1) is not essential (24)]. The use of inhibitable alleles in budding yeast (53) and in human cells (reviewed in Reference 31) reveals that cells die without Mps1 function, likely because of severe aneuploidy. In zebrafish, Mps1 (called nightcap) has also been found to be especially crucial for the very rapid and extensive cell proliferation during tissue regeneration, presumably because it prevents excess aneuploidy (54–56).

2.3. Other Signaling Pathways

Several lines of evidence implicate Mps1 in the genotoxic stress response. Genotoxic stress, such as DNA damage, causes tumor cells to arrest at G2/M or G1, or to commit apoptosis depending on the status of p53. Mps1 influences these responses through multiple mechanisms. Upon exposure to X-ray or UV irradiation, robust G2/M arrest of HeLa cells requires the activity of Mps1, which has been attributed to direct interaction between Mps1 and CHK2/Rad53/Cds1. Mps1 has been shown to phosphorylate CHK2 at Thr68 (57). CHK2 reciprocates Mps1's action by phosphorylating Mps1 on Thr288 and increasing its stability, thereby creating a positive feedback loop for CHK2 Thr68 phosphorylation (58). Disruption of this positive feedback attenuates the DNA damage checkpoint at G2/M arrest (57). The Bloom syndrome protein (BLM) is another Mps1 target in the DNA damage pathway (59). BLM phosphorylation at Ser144 by Mps1 promotes its association with and phosphorylation by Polo-like kinase. Ser144 phosphorylation is important for sustained mitotic arrest in response to microtubule poisons and for accurate chromosome segregation (59).

Mps1 may also be involved in another facet of genotoxic stress response by regulating phosphorylation and subcellular localization of c-Abl. c-Abl is phosphorylated at Thr735, and pThr735-c-Abl normally localizes in cytosol, but it enters the nucleus upon exposure to oxidative stress (60). Mps1 has been identified as the Thr735 kinase, and phosphorylation of c-Abl at Thr735 enhances its association with 14-3-3 protein and cytoplasmic sequestration (60). It has been proposed that Mps1 phosphorylates c-Abl under normal and oxidative stress conditions.

The function of Mps1 in genotoxic response depends on the status of p53. Mps1 is required for the apoptotic response in p53 null cells exposed to the topoisomerase I inhibitor (CPT-11) (61), and Mps1 suppression partially overrides CPT-11-induced cell death. In the presence of functional p53, however, CPT-11 treatment leads to growth arrest without mitotic entry (61). This result suggests that cancer cells with high levels of Mps1 but defective p53 checkpoint pathways are susceptible to DNA damage–induced cell death. There is also a direct link between Mps1 and p53. In response to microtubule poisons, p53 is stabilized and phosphorylated at Thr18, which has been attributed to Mps1 (62). Phosphorylation of p53 by Mps1 may contribute to the postmitotic checkpoint, which arrests cells in G1, thereby preventing further increase in DNA content and genome polyploidization (62).

Mps1 has also been implicated in modulating other cellular signaling responses. Depolymerization of microtubules by nocodazole results in phosphorylation of Smad2 and Smad3 proteins at the C-terminal SSXS motif, a site normally targeted by TGF-β type I receptor (63, 64). Interestingly, this event requires Mps1 and is independent of TGF-β type I receptor (64). Mps1 interacts with Smad4 and phosphorylates Smad2/3 in vitro. Phosphorylation of Smad2 at the C-terminal SSXS motif and at the linker region increases significantly during mitosis. However, the significance of Smad2 mitotic phosphorylation remains unclear. Also, Mps1 has been shown to be a negative regulator of the MAP kinase (MAPK) pathway in yeast (65), although the specific Mps1 target(s) have not been identified.

2.4. Cytokinesis

Mps1 RNAi in human cells led to the appearance of multinucleated cells and to the proposal that Mps1 is involved in the exit from mitosis and/or cytokinesis (66). Hints of this from budding yeast include localization of Mps1 to the bud neck (67) and interaction with a Mob1 (Mps1 one binder) protein (68). Mob1 binds and activates the Dbf2 protein kinase, and the complex acts in the mitotic exit network (69). Although the interaction of yMps1 and Mob1 is not understood, Mps1 is inactivated as cells exit mitosis (70). A more direct link between Mps1 and cytokinesis comes from the discovery of an hMps1-binding partner and substrate, MIP1 (Mps1 interacting protein 1), which is a component of the actin cytoskeleton (71). MIP1 RNAi led to the accumulation of multinucleate cells and disorganization of the actin cytoskeleton. Live-cell recordings revealed a spindle rocking phenotype indicative of difficulties in organizing the cytokinetic furrow. It is not known how MIP1's interaction with, or phosphorylation by, hMPS1 affects its function.

2.5. Meiosis

Disruption of MPS1 function in meiotic yeast cells (72), during meiosis I in mouse oocytes (73), female meiosis in *Drosophila melanogaster* (74, 75), and germ cell production in zebrafish (76), leads to severe chromosome missegregation and aneuploidy. These defects may all arise from the Mps1 mutants' failing to maintain the spindle checkpoint and/or to properly attach chromosomes on the meiotic spindle, suggesting similar Mps1 functions in meiosis as detailed above in mitosis. In fact, Straight et al. (72) were able to demonstrate defects in both meiosis I and meiosis II segregation. Much of this work was done with hypomorphic alleles (25, 72, 76), suggesting that meiotic chromosome segregation is particularly sensitive to disruption in Mps1 activity. For instance, during meiosis I, hypomorphic alleles of the *Drosophila* Mps1 gene, *Ald*, destroy the metaphase pause, which normally leaves the spindles ample time to segregate the nonexchange chromosomes, leading to their loss in *ald* mutant flies (25).

Separate from spindle-related functions, Mps1 has been implicated in other meiotic and

Ald: *Drosophila* Mps1 homolog

germ cell formation functions. Yeast Mps1 is required for both rounds of meiotic SPB duplication (72). Also in yeast, Mps1 is required for the postmeiotic event of spore formation. This developmental pathway depends on a transcriptional regulatory network and the function of a MAPK cascade. Mutant strains in *MPS1* retain the normal function of the transcription regulatory network, but the *mps1⁻* phenotypes resemble defects in the Ste20-family kinase Sps1, such that Mps1 may function with this kinase to control specific aspects of the spore formation program (72).

XlMps1 has been implicated in CSF (cytostatic factor) function, which causes meiosis II metaphase arrest in eggs by inhibiting APC. Two distinct pathways, Mos dependent and CDK2/cyclin E dependent, contribute to the CSF arrest. XlMps1 is required for the CDK2/cyclin E-mediated arrest of cycling extracts (77). Paradoxically, XlMps1 activity is reduced at CSF arrest and must be restrained for extracts to exit CSF arrest. This regulation, as well as Mps1 synthesis, is dependent on CDK2/cyclin E and is associated with different electrophoretic versions of XlMps1, suggesting control by phosphorylation (77). Finally, *Drosophila Ald* has been implicated in hypoxia-induced arrest and the arrests of polar bodies, both of which may reflect a checkpoint function (74). Interestingly, *Drosophila Ald*, along with Polo kinase, is found in novel filamentous structures in oocytes that appear at the end of prophase and are maintained until egg activation (25, 78).

3. MPS1 STRUCTURE, ENZYMOLOGY AND INHIBITORS

3.1. Structure of the Mps1 Catalytic Domain

Several groups have solved the structure of the hMps1 kinase domain with very good agreement (79–82). The Mps1 kinase domain adopts the typical protein kinase bilobe architecture. The N-terminal small lobe consists of a standard five-stranded β-sheet and an αC helix, a canonical feature seen in many protein kinases (**Figures 2a** and **3**). In addition, Mps1 contains an extra β-strand (β0) at the N terminus of the small lobe, which, together with part of β1, covers the twisted β-sheet (**Figure 2a**). The two lobes are joined by a hinge loop (Glu603-Gly605) at the back of the active-site cleft. The C-terminal large lobe shows a standard kinase structure, composed of a two-stranded β-sheet (β6 and β7) adjacent to the small lobe, seven α-helices, the catalytic loop, and the activation loop. The loop between helices αEF and αF (700–708) and the C-terminal tail are disordered (795–857). All of the reported Mps1 catalytic domain structures adopt an inactive conformation, as indicated by incorrect positioning of the αC helix, which prevents ion pairing between the conserved αC glutamate (Glu571) and the active-site lysine (Lys553), the unstructured activation loop, and the inactive conformation of the P+1 loop (684–688). In two structures, a polyethylene glycol molecule, which is a widely used precipitant in protein crystallization, is present as a ring surrounding the catalytic lysine (Lys553). Even though polyethylene glycol is artificially introduced into Mps1 by the crystallization conditions, its presence created a secondary pocket unseen in other kinases, a feature that could be exploited for inhibitor design. The Mps1 kinase domain has been cocrystalized with ATP (83). However, the ATP did not significantly alter the kinase domain conformation in that the Mps1-ATP structure is indistinguishable from the apo or inhibitor-bound conformations. This result suggests that ATP binding is insufficient for switching the kinase to an active conformation, raising tantalizing questions about the active kinase conformation (83).

3.2. Regulation of Mps1 Activity by Phosphorylation

The Mps1 C-terminal catalytic domain undergoes autophosphorylation and is active toward exogenous substrates (33, 80, 81, 84, 85). Initial structure characterization efforts focused

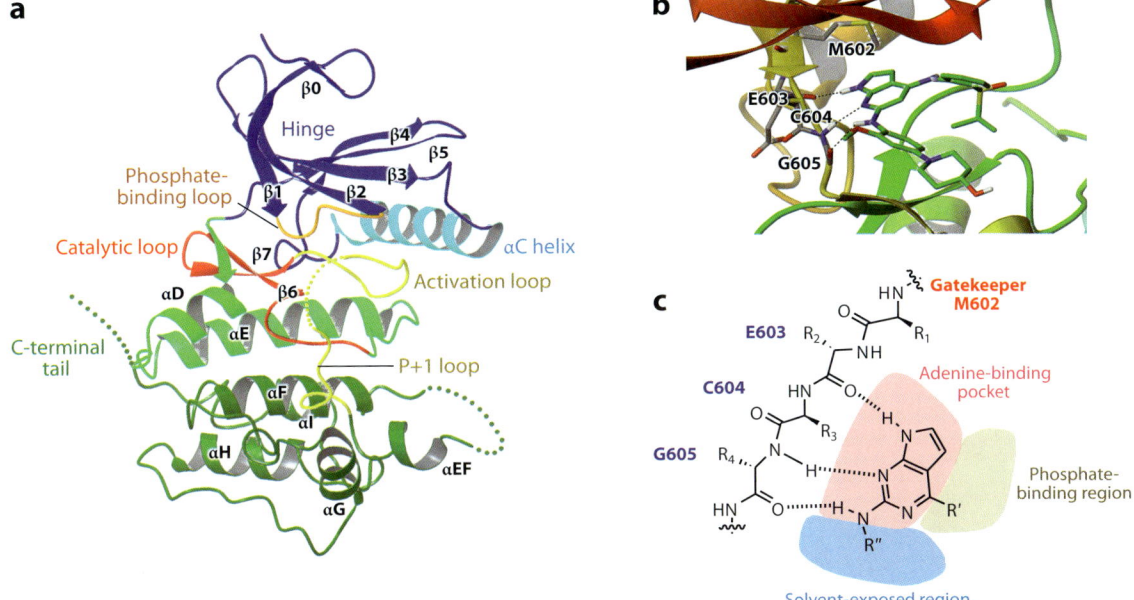

Figure 2

(*a*) Ribbon representation of the structure of the Mps1 catalytic domain. Key structural elements are labeled. The structure has been rendered from the Protein Data Bank (PDB) entry 3DBQ, using the Maestro interface from Schrödinger. The dotted lines represent the disordered regions in the activation loop, the loop between αEF and αF and also at the C-terminal tail. (*b*) Ribbon representation of the structure of Mps1 in complex with a small-molecule inhibitor, Mps1-IN-1. The structure has been rendered from the PDB entry 3GFW using the Maestro interface. The residues in the hinge region are shown. (*c*) Illustration of the inhibitor-binding mode. The gatekeeper residue M602, hinge region residues that interact with the inhibitor and the ATP-binding pocket are shown. The dotted lines represent hydrogen bonds.

on either dephosphorylated, kinase-defective mutants or the normal kinase complexed with small-molecule inhibitors (79, 80). These structures reveal the expected inactive kinase domain conformation, indicated by an unstructured activation, a P+1 loop, and an incorrect αC helix position (**Figures 2a** and **3**) (79, 80). Surprisingly, structures of the wild-type kinase reveal that the catalytic domain still adopts an inactivate conformation, despite extensive autophosphorylation at nine different sites (81). The active catalytic domain may be quite heterogeneous owing to extensive posttranslational modifications; this makes it challenging to obtain crystals of the highly active enzyme. Although dephosphorylated enzymes can be prepared, the enzyme reactivates when ATP is introduced (33, 79). The puzzle remains as to why the active catalytic domain alone or in complex with ATP does not lead to an active enzyme conformation.

Mps1 undergoes extensive autophosphorylation in vitro. Mps1 isolated from mitotic HeLa cells or insect cells were also phosphorylated at numerous sites (84–88). Phosphorylation occurs predominantly at Ser/Thr sites, although Tyr phosphorylation is also observed in vitro (33, 79). Among a myriad of phosphorylation sites, Thr676 and Thr686 are observed to have significant effects on kinase activity (33, 79, 80, 84, 85, 89). Thr676 lies in the activation loop, whereas Thr686 is on the P+1 loop. Mutation of the Thr676 residue to Ala reduces Mps1 transphosphorylation kinase activity by sevenfold. Interestingly, this mutation causes only a 1.4-fold reduction in autophosphorylation (33). Nevertheless, Thr676 phosphorylation is required for Mps1 to function optimally

Transphosphorylation: the action of a kinase mediating transfer of phosphate to its cognate substrates

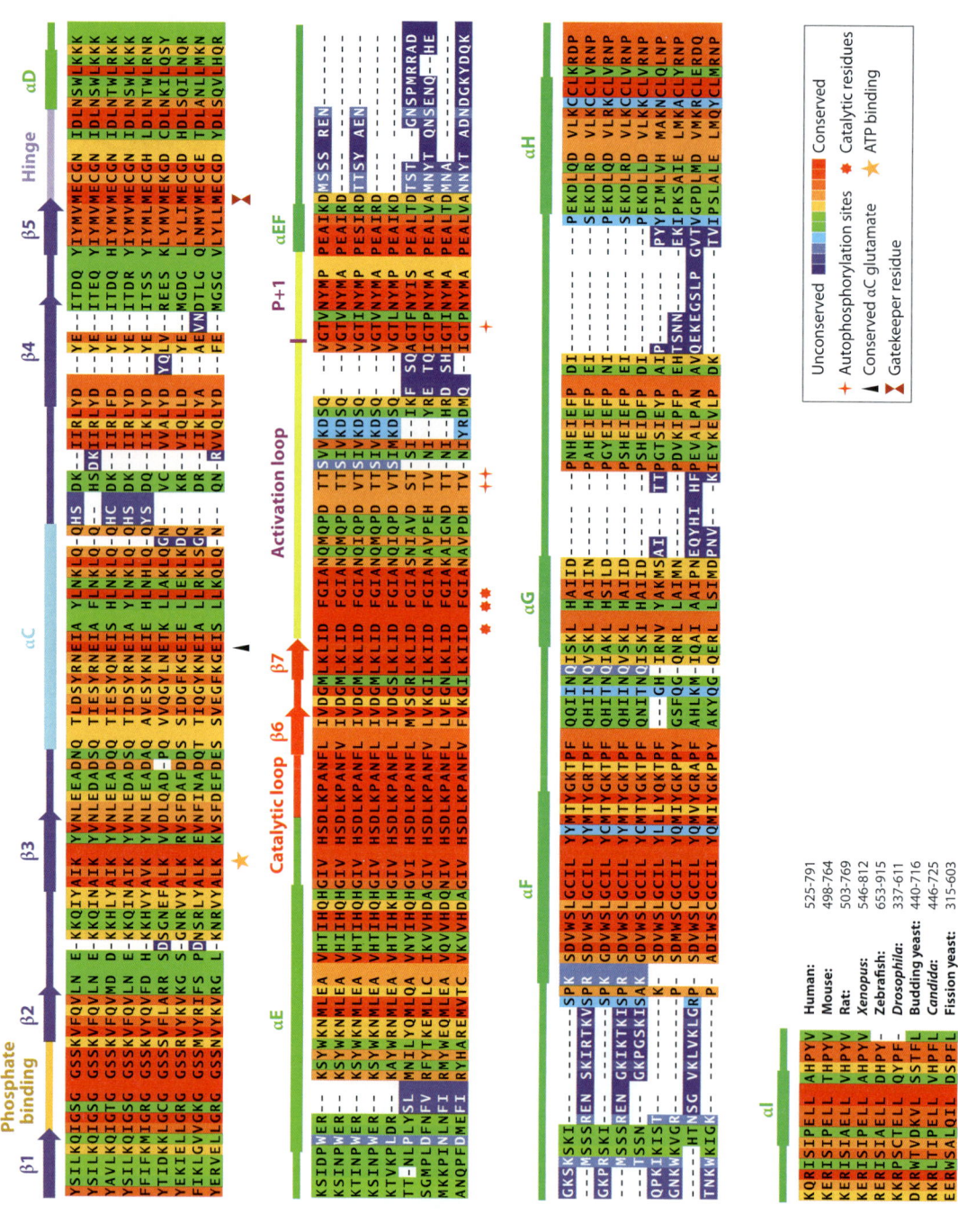

Figure 3

Multiple sequence alignment of the kinase domain of Mps1 from representative species using PRALINE multiple sequence alignment (http://www.ibi.vu.nl/programs/pralineww/). Amino acid conservation is shown by a color-coded heat map from unconserved to conserved. The secondary structure of Mps1 is shown above the Mps1. Key amino acid residues are shown below the alignment.

in yeast and in human cells (33, 84, 85). Supporting the theory that autophosphorylation increases kinase activity, kinetic analysis of Mps1 phosphorylation revealed a lag phase in product formation that is eliminated by preincubation with cold ATP (80, 89). Therefore, Thr676 phosphorylation is likely a priming event for kinase activation. Without an active Mps1 structure, we can only speculate about how phosphorylation stabilizes the activation loop. Mps1 lacks the basic RD pocket, which is referred to as a catalytic loop between the β6 and β7 strands featuring highly conserved Arg (R) and Asp (D) residues. The basic RD pocket, present in many protein kinases regulated through activation loop phosphorylation, directly interacts with the phospho residue in the activation loop, causing a switch to an active conformation. In the place of the RD pocket, it has been proposed that the Mps1 pThr676 phospho group may interact with one of the three lysine residues in the disordered loop between αEF and αF (700–708) (84). Confirmation or repudiation of this hypothesis awaits the availability of an active Mps1 catalytic domain structure.

The Thr residue at the beginning of the P+1 loop is invariant in numerous Ser/Thr kinases, including the AGC, CAMK, and CMGC group of kinases (90). Thr686 is the corresponding residue in Mps1 and, unlike many other Ser/Thr kinases, is autophosphorylated both in vitro and in vivo (33, 79, 80, 84, 85, 87, 88). Mutation of Thr686 reduces the kinase activity by at least 40-fold in vitro and inactivates Mps1 function in vivo (33). Phosphorylated Thr686 is likely a feature of active Mps1 kinase as a phospho-T686 antibody can deplete the active kinase (80, 87). What is unique about Thr686 phosphorylation in Mps1 is that the hydroxyl groups of this residue in other Ser/Thr kinases (e.g., Cdk2) form hydrogen bonds with conserved catalytic residues of the active kinases. The equivalent residue in the P+1 loop of tyrosine kinases is a proline, which is involved in substrate binding. Phosphorylation of the P+1 loop is unique to Mps1, and it is tempting to speculate that modulating the P+1 loop conformation via phosphorylation could be a novel mechanism for kinase activation and that this difference in the P+1 loop is associated with the dual specificity of Mps1 (80).

3.3. Diversity in Phosphorylation Site Selection and Substrate Recognition

Many Mps1 substrates have been identified. Surveying a variety of Mps1 auto- and transphosphorylation sites makes clear that there is no strict consensus phosphorylation sequence associated with the Mps1 kinase. A recent study reported a preference for D/E/N/Q at the −2 position (88), a recognition feature that is also associated with Plk1. This finding suggests the interesting possibilities that Mps1 and Plk1 may share some common physiological substrates and that some of the Mps1 in vitro autophosphorylation sites could be targeted by Plk1 in vivo (88). Despite these notable preferences, the sites targeted by Mps1 are highly diverse, and it is impossible to predict the authentic Mps1 phosphorylation sites in vitro and in vivo.

How Mps1 recognizes diverse substrates remains a mystery, although there are some hints that there may be different requirements for Mps1 autophosphorylation and transphosphorylation. As mentioned above, Thr676 mutation affects transphosphorylation more than autophosphorylation (33). Another unexpected finding came from deletion analysis of the C-terminal region of MPS1, a 65-amino-acid region that is disordered in all known Mps1 structures. This region is susceptible to proteolysis, which may explain the disagreement in abundance measurements using N-terminal or C-terminal antibodies (89). Removal of the 65-amino acid (noncatalytic) tail reduces Mps1 transphosphorylation by about sixfold but has little impact on Mps1 autophosphorylation (89). The most straightforward interpretation of this result is that this region of Mps1 is involved in exogenous substrate recognition. The significance of this region is underscored by the fact that, without it, Mps1 is defective in the spindle assembly checkpoint response, demonstrating that the presence of a kinase domain

> **Analog-sensitive allele:** a variant protein kinase carrying a mutation at the gatekeeper residue, allowing it to accept a bulky analog of ATP for inhibition
>
> **Gatekeeper residue:** the residue located in the ATP-binding pocket of a protein kinase that controls the selectivity and sensitivity to small-molecule inhibitors

alone is insufficient for Mps1 function in vivo (89).

3.4. Dimerization

Kinase autophosphorylation can occur through intermolecular or intramolecular mechanisms. Mps1 transphosphorylation was first shown in vitro (33, 80, 84). Induced dimerization of Mps1 is sufficient to activate its kinase activity in cells (84). Mps1 dimerization and transphosphorylation have also been demonstrated using differentially tagged Mps1 constructs in a coimmunoprecipitation assay (39). Finally, kinetic studies of Mps1 phosphorylation in vitro support the notion that Mps1 undergoes intermolecular autophosphorylation in vitro, as the rate of autophosphorylation increases with increasing concentrations of Mps1 (89). Dimerization of Mps1 may have implications about where Mps1 is initially activated during cell-cycle progression. One proposal is that the kinetochore localization of Mps1 could raise its local concentration, leading to its activation during mitosis via more efficient intermolecular autophosphorylation (84, 91, 92). Although this may be the case for elevated Mps1 activity during spindle assembly checkpoint signaling, Mps1 activity also needs to increase prior to its relocalization to kinetochores to control centrosome duplication. It is equally possible that the initial activation of Mps1 occurs at the centrosome, where Mps1 is also highly concentrated prior to mitotic entry (**Figure 1**).

3.5. Mutant and Analog-Sensitive Alleles

Several mutant *MPS1* alleles have been discovered in yeast, fly, and zebrafish. The original yeast temperature-sensitive-for-growth *mps1-1* allele (C696Y/αH/domain XI), as well as five additional temperature-sensitive alleles all arose from point mutations in the kinase domain (7). The mutations varied in their effect on in vitro kinase activity from very severe (e.g., *mps1-1*) to rather mild (e.g., *mps1-6* C642Y/αF/domain IX). The importance of kinase activity for Mps1 function is reinforced by the finding that the hypomorphic nightcap mutation in zebrafish is an Ile843Lys mutation in subdomain VI of the kinase domain, which is conserved in Mps1 kinases as a hydrophobic Ile or Leu residue (655 in hMps1/β6/catalytic loop/domain VI) (54). Finally, a null allele of the *Drosophila* Mps1 ortholog, *ald^{C3}*, was found to contain a nine-amino acid deletion in the kinase domain (codon 369–377 between the β3 and αC/domain III) (25).

The N-terminal, noncatalytic region of Mps1 kinases is also mutated in some alleles of *MPS1* genes. The original hypomorphic Ald mutation in *Drosophila*, *ald^1*, is an Arg to His substitution at amino acid 7. Similarly, hypomorphic alleles of yeast *MPS1*, *pac8-1* and *pac8-2*, identified as synthetically lethal with the deletion allele of the kinesin Cin8 (93), are point mutations in the N terminus (M. Winey, unpublished observation). These alleles are likely defective in a kinetochore function of Mps1, as is *mps1-7*, which contains a point mutation in the N terminus and is defective in the spindle checkpoint but competent for SPB duplication (94). The *mps1-8* allele is a temperature-sensitive allele that contains several mutations in the N terminus, which do not affect the kinase activity but are defective in SPB duplication (15).

Analog-sensitive alleles have been a valuable tool in probing kinase function and identifying authentic kinase substrates. The gatekeeper residue of Mps1 is a Met at positions 516 and 602 in yeast and human enzymes, respectively. The yeast Mps1-as1 allele was first created by changing the bulky gatekeeper residue Met to a smaller Gly, which makes the kinase specifically sensitive to a cell-permeable ATP analog inhibitor, 1-NM-PP1 (53). The Mps1-as1 allele not only reaffirmed the function of Mps1 in SPB duplication and the spindle checkpoint but was also used to show that Mps1 acts in bipolar chromosome attachment (38, 53) and to show that Mps1 must be inactivated to exit the cell cycle (70). Mps1-as alleles have also been used to resolve some of the controversies surrounding Mad1 kinetochore recruitment and Aurora

B activation by Mps1, which is discussed below (23, 43, 95).

3.6. Small-Molecule Mps1 Inhibitors

The Mps1 kinase has emerged as a novel drug target for cancer therapy. Cincreasin was the first reported small inhibitor of Mps1, though it is not particularly potent with 50% inhibitory concentration (IC_{50}) = 700 μM (96). SP600125, a JNK inhibitor, was found to inhibit Mps1 off target (97). In recent years, a variety of structurally diverse Mps1 inhibitors have been described (**Tables 1** and **2**). Several Mps1 kinase catalytic domain crystal structures, apo or complexed with an inhibitor, were solved recently (79, 81–83). These structures helped shed light on how a small-molecule inhibitor binds with the Mps1 kinase and on how to design selective Mps1 inhibitors. Shown in **Figure 2b** is the crystal structure of the Mps1 kinase domain in complex with Mps1-IN-1 to illustrate kinase inhibitor-binding modes (81). As shown in **Figure 2**, the pyrrolo-pyrimidine scaffold forms the anchor of the inhibitor. It sits in the adenine-binding pocket, making hydrogen bond interactions between the substitutions on the scaffold and the protein inside the adenosine-binding pocket; these substitutions could also extend out into the solvent-exposed region and phosphate-binding region (**Figure 2c**). Cys604, a hinge residue that varies between kinases, has been explored in designing more selective Mps1 inhibitors (82).

Because Mps1 is an essential gene in the pathogenic fungi *Candida albicans* and significant sequence divergence exists between human and *Candida* Mps1, species-specific inhibitors of Mps1 kinase could be employed as antifungal agents. In an effort to identify novel antifungal chemotherapeutics, the guanylate cyclase inhibitor LY83583 was found to inhibit *Candida* Mps1 without affecting hMps1 activity (98). Further advancing the feasibility of species-specific targeting of Mps1, SP600125, which inactivates human Mps1, has only modest inhibitory effects on *Candida* Mps1 and is nontoxic to *Candida* (98). Thus, sequence variations in Mps1 may offer a window of opportunity for new therapeutics in combating human pathogens.

4. REGULATION OF MPS1 KINASES

To accomplish a myriad of functions, MPS1 kinases must be exquisitely regulated. Indeed, experimental changes in the expression levels of MPS1 kinases (active and inactive) are detrimental in a variety of cell types. In general, MPS1 kinases are expressed at low levels, and the most important regulatory mechanisms operate via the posttranslational mechanisms of phosphorylation (discussed above with reference to the catalytic domain structure) and degradation.

4.1. Transcription

Many mammalian proteins that function in mitosis and mitotic checkpoint signaling, including Mps1, are controlled at the transcriptional level by the E2F family of transcription factors (99–103). Mps1 mRNA peaks in mitosis (11), and E2F4 and, to a lesser degree, E2F1 bind the *MPS1* promoter region (103). In mouse embryonic fibroblasts lacking p107 and p130, which are two Retinoblastoma family transcriptional repressors of E2F transcription, *MPS1* transcription is derepressed, and the mRNA transcribed at a higher level than in the wild-type control. These data suggest that the retinoblastoma E2F complex may be directly involved in repressing *MPS1* transcription in interphase cells (103). Whether *MPS1* transcription is directly regulated by E2F4 family transcription factors remains to be investigated. Finally, *MPS1* mRNA levels are elevated in freshly isolated peripheral blood lymphocyte or T cell blasts (104). IL-2 incubation also induces Mps1 expression in proliferating peripheral blood lymphocyte blasts (104). Thus, transcription of Mps1 is upregulated when cells enter the cell cycle and transit through mitosis.

Table 1 Summary of published potent hMps1 inhibitors

Inhibitor	Structure	IC$_{50}$ (nM)	References (patent number)	Protein Data Bank crystal structure
SP600125		692	78, 83, 97	2ZMC
AZ3146		35	39 (WO2009024824)	
Mps1-IN-1		370	81 (WO2009032694)	3GFW
Mps1-IN-2		145	81 (WO2010080712)	3H9F
NMS-P715		8	82 (WO2009156315)	2X9E
MPI-0479605		3.5	106, 132	
Reversine		3/6	40	
Staurosporine		102	83	3HMO
Cpd4		38,000	83	3HMP

Table 2 Additional patented hMps1 inhibitors

Claimed structure	Example	IC$_{50}$(nM)	Patent number (company)
(structure)	(structure)	3	JP2010111624 (Shionogi & Co., Ltd.)
(structure)	(structure)	2.6	WO2010124826 (Bayer Pharma)
(structure)	(structure)	1	WO2011063907 WO2011063908 WO2011064328 (Bayer Pharma)
(structure)	(structure)	3.9	WO2011013729 (OncoTherapy Science, Inc.)
(structure)	(structure)	4	WO2011016472 (OncoTherapy Science, Inc.)

4.2. Localization

Subcellular localization of Mps1 is both spatially and temporally regulated during cell-cycle progression (29, 30, 66, 86). In mammalian cells, Mps1 primarily resides within the cytosol during G1. In late G2, Mps1 accumulates on centrosomes and the nuclear envelope (29, 66, 86, 105). At the G2/M boundary, Mps1 abruptly enters into the nucleus prior to nuclear membrane breakdown (106). Nuclear import of Mps1 requires two LXXLL motifs in the N terminus of Mps1. In interphase cells, Mps1 likely shuttles between nucleus and cytosol constantly, as leptomycin B treatment can lead to redistribution of Mps1 into the nucleus (106). As cells move into prophase, Mps1 preferentially associates with kinetochores and is slowly lost until the onset of anaphase, when Mps1 disassociates from kinetochores (91, 92, 105). The noncatalytic N-terminal domain is necessary and sufficient for localization to kinetochores in isolation, whereas the C-terminal domain by itself cannot locate hMps1 to kinetochores (23, 29, 86, 107). However, the function of the kinetochore-targeting signal in the N terminus could be masked by the sequence in the C-terminal region of Mps1. For example, without phosphorylation of Ser844 of XlMps1 by MAPK, XlMps1 cannot relocate to kinetochores even though the N-terminal-targeting signal is intact (108). Similar observations were made with hMps1 (86, 109). These results imply that the C-terminal region of Mps1 may regulate access to the kinetochore-targeting signal that resides in the N terminus of Mps1.

Besides MAPK, two other kinases, PRP4 (premessenger RNA processing 4) and Aurora B, have been implicated in the regulation of

Mps1 kinetochore localization. PRP4 protein kinase associates with kinetochores during mitosis and is required for efficient Mps1 kinetochore targeting (110). Depletion of PRP4 induces mitotic acceleration, chromosome misalignment, and defects in Mad2 localization, which are phenotypes observed with inactivation of Mps1. The mechanism by which PRP4 regulates Mps1 remains to be determined. Similarly, as mentioned above, inhibition of Aurora B by various methods reduces Mps1 localization to unattached kinetochores throughout mitosis (46). However, Mad2 recruitment to kinetochores, which requires Mps1 activity, is significantly affected only in early mitosis, suggesting that Aurora B may regulate the timing or amplitude of Mps1 activation. Delayed Mps1 activation caused by Aurora B inhibition also causes a delay in establishment of the spindle checkpoint. The defects in Mps1 kinetochore targeting in early mitosis and spindle checkpoint delay can be rescued by tethering Mps1 to the kinetochore (46). This result suggests that Aurora B acts upstream in promoting early mitotic Mps1 kinetochore targeting. The effects of Aurora B on Mps1 have also been investigated in conjunction with Hec1/Ndc80, a core component of the kinetochore essential for organizing microtubule attachment sites (51). Hec1 is required for the recruitment of Mps1 kinase and Mad1/Mad2 complexes to kinetochores (111). Although there has been no report of direct interaction between Hec1 and Mps1, the budding yeast Hec1 ortholog, Ndc80, directly interacts with yeast Mps1 (50). Hec1 may well be the kinetochore-bound acceptor for Mps1 in mammalian cells. Consistent with a direct role of Hec1 on Mps1 targeting, depletion of Hec1 results in more dramatic effects on Mps1 targeting than Aurora B inactivation. However, one intriguing possibility is that Aurora B may act by phosphorylating the N terminus of Hec1 to regulate Mps1 kinetochore localization (51). It will be interesting to investigate whether Hec1 phosphorylation by Aurora B creates a docking site for Mps1 to bind kinetochores.

Mps1 centrosome localization is mediated by its N-terminal domain, but the precise motifs have not been characterized. Nonetheless, distinct regions of the N terminus of the yeast and human enzymes have been implicated in its centrosome function. In yeast, deletion analysis with the clever use of analog-sensitive alleles revealed that amino acids 201–300 are required for SPB duplication and are distinct from amino acids 151–200, which are required for chromosome biorientation (14). Similarly, a region internal to the N terminus of hMps1, amino acids 420–507, called the MDS (Mps1 degradation signal), is critical in controlling centrosomal levels of hMps1 (112). Deletion of this region stabilizes the protein and localizes it to centrosomes, driving excess centrosome production. The MDS is recognized by centrosome-localized OAZ (antizyme), which is responsible for the degradation of centrosomal Mps1 (113). The level of the centrosomal Mps1 is also regulated by its phosphorylation on Thr468 within the MDS by Cdk2 (112), which stabilizes the protein, opposing OAZ-mediated degradation to create a regulatory circuit that controls centrosomal hMps1 levels (22).

Phosphorylation of Mps1 is also important for its localization. Mutating the nine autophosphorylation sites in the N terminus of Mps1 causes a significant decrease in kinetochore targeting of Mps1 without affecting centrosomal localization in SW480 cells (86). This result suggests that the kinetochore-targeting signal is independent of the centrosome-localization signal. Among these sites, T12 and S15 appear to be critical in mediating Mps1 accumulation on kinetochores (86). Consistent with these results, kinase-inactive Mps1, expressed in SW480 cells, exhibits reduced kinetochore relocalization upon depletion of endogenous Mps1 (86). Reduced kinetochore localization of endogenous Mps1 was also observed when US2OS cells were treated with the inhibitor NMS-P715 (82). However, kinetochore accumulation of transiently transfected Mps1 or Mps1KD in HeLa increases when cells are treated with the inhibitor AZ3146, suggesting that kinase activity inhibits kinetochore recruitment in HeLa cells (39). It is interesting to note that the results of AZ3146 on recruitment of spindle checkpoint proteins to kinetochores depend on whether the inhibitor is administrated before or after mitotic entry. To reconcile the timing effect, it was proposed that there are two phases of checkpoint protein recruitment to kinetochores: an initial phase prior to mitotic entry and a subsequent maintenance phase during mitosis (39). Increased Mps1 accumulation on kinetochores in the presence of AZ3146 may be a result of reduced release from kinetochores. In this way, the Mps1 kinase activity may be required for targeting to kinetochores and also for release from kinetochores. The details about these requirements remain unsettled and may, like the role of Mps1 at the centrosome, depend on the cell type and experimental conditions.

4.3. Degradation and Inactivation

The major route of Mps1 inactivation is degradation. The expression and activity of Mps1 are cell cycle dependent in both yeast and mammalian cells (70, 89, 107). Expression peaks in metaphase and declines when cells enter anaphase. Timely inactivation of Mps1 is required for proper cell-cycle progression and termination of spindle checkpoint signaling. During normal cell-cycle progression, Mps1 is partially degraded in anaphase by the ubiquitin E3 ligase APCCdc20. Overexpression of Mps1 in anaphase can activate the checkpoint by inhibiting APCCdc20 and blocks mitotic exit in yeast (70). There are three D-boxes in the N terminus of yeast Mps1, which are required for proteolysis by APCCdc20 (70). Therefore, APCCdc20 and Mps1 are mutually inhibitory, forming a double negative feedback loop. This circuit may enable the metaphase to anaphase transition to be switch like and irreversible. Human Mps1 contains only one canonical D-box, and it is sequentially degraded by APCCdc20 and APCCdh1 in a D-box-dependent manner (114). A D-box-deficient hMps1 perturbs normal mitosis and causes centrosome overreplication in human cells. Efficient degradation of Mps1 is also aided by Ufd2, a U-box-containing ubiquitin-protein ligase, in both yeast and mammalian cells (115). Hence, proteolysis regulates temporal expression and activity of Mps1.

Degradation of Mps1 also occurs spatially. Centrosome accumulation of hMps1 is greatly enhanced by phosphorylation at Thr468 by Cdk2 (112). Phosphomimetic mutations at Thr468 or deletion of the region surrounding Thr468 protects Mps1 from degradation at centrosomes (112). Yeast Mps1 is stabilized by CDK phosphorylation of Thr29, but the mechanism is unknown (20). Kinetochore-associated Mps1 also may be regulated by proteolysis. The retention time for Mps1 on unattached kinetochores in checkpoint-arrested cells is about 10 s (91). Treatment with both MG132 and Mps1 inhibitors enhances its accumulation at kinetochores (39), suggesting a role for proteolysis. Because Mps1 has a distinct subcellular localization during cell-cycle progression, it is possible that different pools of Mps1 are differentially regulated by proteolysis.

Another possible mechanism of Mps1 inactivation is dephosphorylation. Mps1 is hyperphosphorylated in mitosis (29, 30) and rapidly dephosphorylated upon anaphase entry (29). To date, phosphatases that specifically act on MPS1 family kinases have not been identified. Early in vitro studies show that PTP1B can remove the phospho-Tyr epitope produced by mouse Mps1 autophosphorylation (10). PP1γ has also been shown to dephosphorylate Mps1 in vitro (77). Whether any of these phosphatases inactivate Mps1 in vivo remains unknown.

4.4. Misregulation in Tumor Cells

Like many cell-cycle regulators, Mps1 transcription is deregulated in a variety of human tumors. Elevated Mps1 mRNA levels are found in several human cancers, including thyroid papillary carcinoma, breast cancer, gastric cancer tissue, bronchogenic carcinoma, and lung cancers (8, 116–120). Furthermore, high levels of Mps1 correlate with a high histological grade in breast cancers (119). Conversely, Mps1 mRNA is markedly reduced or absent in resting cells and in tissues with a low proliferative index

(11). Thus, there is a correlation between elevated Mps1 levels and cell proliferation as well as with tumor aggressiveness. Consistent with the notion that oncogenic signaling promotes Mps1 expression, the levels and activity of Mps1 are increased by 3- and 10-fold, respectively, in human melanoma cell lines containing the B-Raf^{V600E} mutant (121). Inhibition of B-Raf or MEK1 reduces Mps1 expression (109, 121).

The observation that tumor cells frequently overexpress spindle checkpoint proteins is perplexing as the conventional wisdom would postulate that tumor cells would have a weakened checkpoint, contributing to chromosome missegregation and aneuploidy. Indeed, significant evidence from yeast to mice supports the notion that a weakened checkpoint leads to chromosome instability (122). However, mutations in key checkpoint proteins are rare in human tumors, and correlative evidence showing that compromised checkpoint signaling directly contributes to the development of human tumors has been elusive. *MPS1* missense mutations have been found in the noncatalytic N terminus in bladder (123) and lung cancers (124), as well as in the kinase domain in pancreatic (125) and lung cancers (124). Interestingly, frameshift mutations that truncate the protein arise from microsatellite instability in the hMps1 gene in gastric (126) and colorectal cancers (127). Thus, mutations in hMPS1 have been detected in tumor-derived cells; however, their influence on tumorigenesis is not known.

The prevalence of high levels of checkpoint protein expression, such as Mps1, in human tumors prompts an alternative hypothesis regarding the potential role of checkpoint proteins in cancer cells, i.e., overexpression of these proteins may promote either cancer initiation or survival of aneuploid cancer cells (119, 128). Accordingly, reductions in key checkpoint proteins should severely decrease human cancer cell viability. This prediction is confirmed for several checkpoint proteins, including Mps1 (66, 119), BubRI (129), and Mad2 (130, 131). Suppression of Mps1 expression in Hs578T breast cancer cells also reduces the tumorigenicity of these cells in xenografts. Cancer cell death is likely the result of severe chromosome segregation errors when the checkpoint is disabled. Interestingly, cells that survived reduced Mps1 levels often display lower levels of aneuploidy, suggesting that lower levels of Mps1 potentially inactivating the checkpoint are incompatible with aneuploidy (119). This concept is in excellent agreement with the observation that reduction in checkpoint proteins makes tumor cells more sensitive than untransformed human fibroblasts to low doses of spindle poisons (129). Differential cellular responses to checkpoint inhibition between normal and tumor cells could be key in developing new anticancer drugs targeting hMps1. Recent results from at least one hMps1 inhibitor, NMS-P715, show great promise in preclinical cancer models (82). We anxiously await the determination of whether inhibitors of Mps1 are efficacious and safe, either as single agents or in combination, in clinically relevant settings.

SUMMARY POINTS

1. Mps1 kinases with their conserved, C-terminal kinase domains are widely, but not ubiquitously, distributed among eukaryotes.
2. Mps1 kinases are localized at kinetochores, where they function with Aurora B kinases to ensure proper bipolar attachment.
3. Mps1 kinases act at an early step in the spindle checkpoint, and the functions of most of the checkpoint proteins are dependent, directly or indirectly, on Mps1 activity.
4. Mps1 kinases are found in centrosomes, are required for SPB (centrosome) assembly in yeast, and influence centrosome assembly in mammals.

5. Mps1 kinases exhibit significant levels of autophosphorylation, which is essential for its activation and subcellular localization.
6. Mps1 kinases are inactivated by APC-dependent degradation, which is necessary for cells to exit mitosis correctly.
7. Mps1 kinase genes are misregulated in tumors, supporting the hypothesis that the checkpoint is necessary for the viability of aneuploid tumor cells.
8. Mps1 kinases have become of interest for the development of small-molecule inhibitors. It is anticipated that some of the inhibitors discovered will be tested in clinical trials.

FUTURE ISSUES

Since the discovery of the first Mps1 allele, there has been tremendous progress in understanding the biological function and underlying mechanisms of this protein kinase. However, many important questions regarding Mps1 function remain. For example:

1. What are the molecular mechanisms of Mps1 in its known functions in kinetochore attachment, the spindle checkpoint, and centrosome assembly? Particularly, what are the pertinent Mps1 substrates for these various functions?
2. What are all of the Mps1 kinase functions? Mps1 kinases function in genotoxic stress, the actin cytoskeleton, and likely in other contexts that remain to be identified.
3. Is the lack of Mps1 paralogs functionally significant? Could the myriad and complex functions of these kinases require a single isoform for correct regulation?
4. What protein kinases carry out the various functions of Mps1 kinases in organisms lacking this kinase?
5. What governs Mps1 subcellular localization and its changes during the cell cycle?
6. Are Mps1 kinases inactivated by reversible mechanisms, such as dephosphorylation? A biosensor assay for active Mps1 would be critical for this work, and it would be useful in examining Mps1 at its various cellular locations.
7. What is the active conformation of Mps1, and what can it tell us about the mechanisms of Mps1 activation and substrate recognition?
8. Will Mps1 be found to be a good drug target for antitumor therapy?

Answers to these questions will undoubtedly provide a more lucid and exciting picture of how Mps1 orchestrates normal cell-cycle progression and its deviation in tumorigenesis.

DISCLOSURE STATEMENT

The authors are not aware of any affiliations, memberships, funding, or financial holdings that might be perceived as affecting the objectivity of this review.

ACKNOWLEDGMENTS

We are indebted to Quanbin Xu for the images in **Figure 1**. We thank Harold Fisk and Shelly Jones for critically reading the manuscript. We also thank Gan Zhang and Robert Holton-Burke for preparing the inhibitor tables and structure figures. X.L. is supported by the National Institutes of Health (NIH) grants CA107089 and GM083172. M.W.'s work on Mps1 is supported by NIH grant GM51312.

LITERATURE CITED

1. Winey M, Goetsch L, Baum P, Byers B. 1991. MPS1 and MPS2: novel yeast genes defining distinct steps of spindle pole body duplication. *J. Cell Biol.* 114:745–54
2. Poch O, Schwob E, de Fraipont F, Camasses A, Bordonne R, Martin RP. 1994. RPK1, an essential yeast protein kinase involved in the regulation of the onset of mitosis, shows homology to mammalian dual-specificity kinases. *Mol. Gen. Genet.* 243:641–53
3. Lauze E, Stoelcker B, Luca FC, Weiss E, Schutz AR, Winey M. 1995. Yeast spindle pole body duplication gene *MPS1* encodes an essential dual specificity protein kinase. *EMBO J.* 14:1655–63
4. Rose MD, Fink GR. 1987. *KAR1*, a gene required for function of both intranuclear and extranuclear microtubules in yeast. *Cell* 48:1047–60
5. Weiss E, Winey M. 1996. The *Saccharomyces cerevisiae* spindle pole body duplication gene *MPS1* is part of a mitotic checkpoint. *J. Cell Biol.* 132:111–23
6. Hardwick KG, Weiss E, Luca FC, Winey M, Murray AW. 1996. Activation of the budding yeast spindle assembly checkpoint without mitotic spindle disruption. *Science* 273:953–56
7. Schutz AR, Winey M. 1998. New alleles of the yeast *MPS1* gene reveal multiple requirements in spindle pole body duplication. *Mol. Biol. Cell* 9:759–74
8. Mills GB, Schmandt R, McGill M, Amendola A, Hill M, et al. 1992. Expression of TTK, a novel human protein kinase, is associated with cell proliferation. *J. Biol. Chem.* 267:16000–6
9. Lindberg RA, Fischer WH, Hunter T. 1993. Characterization of a human protein threonine kinase isolated by screening an expression library with antibodies to phosphotyrosine. *Oncogene* 8:351–59
10. Douville EM, Afar DE, Howell BW, Letwin K, Tannock L, et al. 1992. Multiple cDNAs encoding the esk kinase predict transmembrane and intracellular enzyme isoforms. *Mol. Cell. Biol.* 12:2681–89
11. Hogg D, Guidos C, Bailey D, Amendola A, Groves T, et al. 1994. Cell cycle dependent regulation of the protein kinase TTK. *Oncogene* 9:89–96
12. Lowery DM, Mohammad DH, Elia AE, Yaffe MB. 2004. The Polo-box domain: a molecular integrator of mitotic kinase cascades and Polo-like kinase function. *Cell Cycle* 3:128–31
13. Chu M, Eyers P. 2010. *UCSD-Nature Molecule Pages: MPS1*. **http://www.signaling-gateway.org/molecule/query?afcsid=A000882**
14. Araki Y, Gombos L, Migueleti SP, Sivashanmugam L, Antony C, Schiebel E. 2010. N-terminal regions of Mps1 kinase determine functional bifurcation. *J. Cell Biol.* 189:41–56
15. Castillo AR, Meehl JB, Morgan G, Schutz-Geschwender A, Winey M. 2002. The yeast protein kinase Mps1p is required for assembly of the integral spindle pole body component Spc42p. *J. Cell Biol.* 156:453–65
16. Holinger EP, Old WM, Giddings TH, Wong C, Yates JR, Winey M. 2009. Budding yeast centrosome duplication requires stabilization of Spc29 via Mps1-mediated phosphorylation. *J. Biol. Chem.* 284:12949–55
17. Pereira G, Knop M, Schiebel E. 1998. Spc98p directs the yeast gamma-tubulin complex into the nucleus and is subject to cell cycle–dependent phosphorylation on the nuclear side of the spindle pole body. *Mol. Biol. Cell* 9:775–93
18. Friedman DB, Kern JW, Huneycutt BJ, Vinh DB, Crawford DK, et al. 2001. Yeast Mps1p phosphorylates the spindle pole component Spc110p in the N-terminal domain. *J. Biol. Chem.* 276:17958–67
19. Schiebel E, Bornens M. 1995. In search of a function for centrins. *Trends Cell Biol.* 5:197–201
20. Jaspersen SL, Huneycutt BJ, Giddings TH, Resing KA, Ahn NG, Winey M. 2004. Cdc28/Cdk1 regulates spindle pole body duplication through phosphorylation of Spc42 and Mps1. *Dev. Cell* 7:263–74
21. Keck JM, Jones MH, Wong CC, Binkley J, Chen D, et al. 2011. A cell cycle phosphoproteome of the yeast centrosome. *Science* 332:1557–61

22. Pike AN, Fisk HA. 2011. Centriole assembly and the role of Mps1: defensible or dispensable? *Cell Div.* 6:9
23. Maciejowski J, George KA, Terret M-E, Zhang C, Shokat KM, Jallepalli PV. 2010. Mps1 directs the assembly of Cdc20 inhibitory complexes during interphase and mitosis to control M phase timing and spindle checkpoint signaling. *J. Cell Biol.* 190:89–100
24. He X, Jones MH, Winey M, Sazer S. 1998. Mph1, a member of the Mps1-like family of dual specificity protein kinases, is required for the spindle checkpoint in *S. pombe*. *J. Cell Sci.* 111(Part 12):1635–47
25. Gilliland WD, Hughes SE, Cotitta JL, Takeo S, Xiang Y, Hawley RS. 2007. The multiple roles of mps1 in *Drosophila* female meiosis. *PLoS Genet.* 3:e113
26. Kanai M, Ma Z, Izumi H, Kim S-H, Mattison CP, et al. 2007. Physical and functional interaction between mortalin and Mps1 kinase. *Genes Cells* 12:797–810
27. Yang CH, Kasbek C, Majumder S, Yusof AM, Fisk HA. 2010. Mps1 phosphorylation sites regulate the function of centrin 2 in centriole assembly. *Mol. Biol. Cell* 21:4361–72
28. Abrieu A, Magnaghi-Jaulin L, Kahana JA, Peter M, Castro A, et al. 2001. Mps1 is a kinetochore-associated kinase essential for the vertebrate mitotic checkpoint. *Cell* 106:83–93
29. Liu S-T, Chan GKT, Hittle JC, Fujii G, Lees E, Yen TJ. 2003. Human MPS1 kinase is required for mitotic arrest induced by the loss of CENP-E from kinetochores. *Mol. Biol. Cell* 14:1638–51
30. Stucke VM, Sillje HHW, Arnaud L, Nigg EA. 2002. Human Mps1 kinase is required for the spindle assembly checkpoint but not for centrosome duplication. *EMBO J.* 21:1723–32
31. Lan W, Cleveland DW. 2010. A chemical tool box defines mitotic and interphase roles for Mps1 kinase. *J. Cell Biol.* 190:21–24
32. Espeut J, Gaussen A, Bieling P, Morin V, Prieto S, et al. 2008. Phosphorylation relieves autoinhibition of the kinetochore motor Cenp-E. *Mol. Cell* 29:637–43
33. Mattison CP, Old WM, Steiner E, Huneycutt BJ, Resing KA, et al. 2007. Mps1 activation loop autophosphorylation enhances kinase activity. *J. Biol. Chem.* 282:30553–61
34. Fraschini R, Beretta A, Lucchini G, Piatti S. 2001. Role of the kinetochore protein Ndc10 in mitotic checkpoint activation in *Saccharomyces cerevisiae*. *Mol. Genet. Genomics* 266:115–25
35. Chan GK, Yen TJ. 2003. The mitotic checkpoint: a signaling pathway that allows a single unattached kinetochore to inhibit mitotic exit. *Prog. Cell Cycle Res.* 5:431–39
36. Kulukian A, Han JS, Cleveland DW. 2009. Unattached kinetochores catalyze production of an anaphase inhibitor that requires a Mad2 template to prime Cdc20 for BubR1 binding. *Dev. Cell* 16:105–17
37. Biggins S, Murray AW. 2001. The budding yeast protein kinase Ipl1/Aurora allows the absence of tension to activate the spindle checkpoint. *Genes Dev.* 15:3118–29
38. Maure J-F, Kitamura E, Tanaka TU. 2007. Mps1 kinase promotes sister-kinetochore bi-orientation by a tension-dependent mechanism. *Curr. Biol.* 17:2175–82
39. Hewitt L, Tighe A, Santaguida S, White AM, Jones CD, et al. 2010. Sustained Mps1 activity is required in mitosis to recruit O-Mad2 to the Mad1-C-Mad2 core complex. *J. Cell Biol.* 190:25–34
40. Santaguida S, Tighe A, D'Alise AM, Taylor SS, Musacchio A. 2010. Dissecting the role of MPS1 in chromosome biorientation and the spindle checkpoint through the small molecule inhibitor reversine. *J. Cell Biol.* 190:73–87
41. Lampson MA, Renduchitala K, Khodjakov A, Kapoor TM. 2004. Correcting improper chromosome-spindle attachments during cell division. *Nat. Cell Biol.* 6:232–37
42. Jelluma N, Brenkman AB, van den Broek NJF, Cruijsen CWA, van Osch MHJ, et al. 2008. Mps1 phosphorylates Borealin to control Aurora B activity and chromosome alignment. *Cell* 132:233–46
43. Sliedrecht T, Zhang C, Shokat KM, Kops GJ. 2010. Chemical genetic inhibition of Mps1 in stable human cell lines reveals novel aspects of Mps1 function in mitosis. *PLoS ONE* 5:e10251
44. Bourhis E, Lingel A, Phung Q, Fairbrother WJ, Cochran AG. 2009. Phosphorylation of a Borealin dimerization domain is required for proper chromosome segregation. *Biochemistry* 48:6783–93
45. Vigneron S, Prieto S, Bernis C, Labbe JC, Castro A, Lorca T. 2004. Kinetochore localization of spindle checkpoint proteins: Who controls whom? *Mol. Biol. Cell* 15:4584–96
46. Saurin AT, van der Waal MS, Medema RH, Lens SMA, Kops GJPL. 2011. Aurora B potentiates Mps1 activation to ensure rapid checkpoint establishment at the onset of mitosis. *Nat. Commun.* 2:316

47. Cheeseman IM, Anderson S, Jwa M, Green EM, Kang J, et al. 2002. Phospho-regulation of kinetochore-microtubule attachments by the Aurora kinase Ipl1p. *Cell* 111:163–72
48. Shimogawa MM, Graczyk B, Gardner MK, Francis SE, White EA, et al. 2006. Mps1 phosphorylation of Dam1 couples kinetochores to microtubule plus ends at metaphase. *Curr. Biol.* 16:1489–501
49. Kim Y, Holland AJ, Lan W, Cleveland DW. 2010. Aurora kinases and protein phosphatase 1 mediate chromosome congression through regulation of CENP-E. *Cell* 142:444–55
50. Kemmler S, Stach M, Knapp M, Ortiz J, Pfannstiel J, et al. 2009. Mimicking Ndc80 phosphorylation triggers spindle assembly checkpoint signalling. *EMBO J.* 28:1099–110
51. DeLuca JG, Gall WE, Ciferri C, Cimini D, Musacchio A, Salmon ED. 2006. Kinetochore microtubule dynamics and attachment stability are regulated by Hec1. *Cell* 127:969–82
52. Storchova Z, Becker JS, Talarek N, Kogelsberger S, Pellman D. 2011. Bub1, Sgo1, and Mps1 mediate a distinct pathway for chromosome biorientation in budding yeast. *Mol. Biol. Cell* 22:1473–85
53. Jones MH, Huneycutt BJ, Pearson CG, Zhang C, Morgan G, et al. 2005. Chemical genetics reveals a role for Mps1 kinase in kinetochore attachment during mitosis. *Curr. Biol.* 15:160–65
54. Poss KD, Nechiporuk A, Hillam AM, Johnson SL, Keating MT. 2002. Mps1 defines a proximal blastemal proliferative compartment essential for zebrafish fin regeneration. *Development* 129:5141–49
55. Poss KD, Wilson LG, Keating MT. 2002. Heart regeneration in zebrafish. *Science* 298:2188–90
56. Wills AA, Kidd AR, Lepilina A, Poss KD. 2008. Fgfs control homeostatic regeneration in adult zebrafish fins. *Development* 135:3063–70
57. Wei JH, Chou YF, Ou YH, Yeh YH, Tyan SW, et al. 2005. TTK/hMps1 participates in the regulation of DNA damage checkpoint response by phosphorylating CHK2 on threonine 68. *J. Biol. Chem.* 280:7748–57
58. Yeh YH, Huang YF, Lin TY, Shieh SY. 2009. The cell cycle checkpoint kinase CHK2 mediates DNA damage-induced stabilization of TTK/hMps1. *Oncogene* 28:1366–78
59. Leng M, Chan DW, Luo H, Zhu C, Qin J, Wang Y. 2006. MPS1-dependent mitotic BLM phosphorylation is important for chromosome stability. *Proc. Natl. Acad. Sci. USA* 103:11485–90
60. Nihira K, Taira N, Miki Y, Yoshida K. 2008. TTK/Mps1 controls nuclear targeting of c-Abl by 14-3-3-coupled phosphorylation in response to oxidative stress. *Oncogene* 27:7285–95
61. Bhonde MR, Hanski ML, Budczies J, Cao M, Gillissen B, et al. 2006. DNA damage-induced expression of p53 suppresses mitotic checkpoint kinase hMps1: the lack of this suppression in p53MUT cells contributes to apoptosis. *J. Biol. Chem.* 281:8675–85
62. Huang YF, Chang MD, Shieh SY. 2009. TTK/hMps1 mediates the p53-dependent postmitotic checkpoint by phosphorylating p53 at Thr18. *Mol. Cell. Biol.* 29:2935–44
63. Dong C, Li Z, Alvarez R Jr, Feng XH, Goldschmidt-Clermont PJ. 2000. Microtubule binding to Smads may regulate TGF beta activity. *Mol. Cell* 5:27–34
64. Zhu S, Wang W, Clarke DC, Liu X. 2007. Activation of Mps1 promotes transforming growth factor-beta-independent Smad signaling. *J. Biol. Chem.* 282:18327–38
65. Cappell SD, Baker R, Skowyra D, Dohlman HG. 2010. Systematic analysis of essential genes reveals important regulators of G protein signaling. *Mol. Cell* 38:746–57
66. Fisk HA, Mattison CP, Winey M. 2003. Human Mps1 protein kinase is required for centrosome duplication and normal mitotic progression. *Proc. Natl. Acad. Sci. USA* 100:14875–80
67. Huh WK, Falvo JV, Gerke LC, Carroll AS, Howson RW, et al. 2003. Global analysis of protein localization in budding yeast. *Nature* 425:686–91
68. Luca FC, Winey M. 1998. *MOB1*, an essential yeast gene required for completion of mitosis and maintenance of ploidy. *Mol. Biol. Cell* 9:29–46
69. Mohl DA, Huddleston MJ, Collingwood TS, Annan RS, Deshaies RJ. 2009. Dbf2-Mob1 drives relocalization of protein phosphatase Cdc14 to the cytoplasm during exit from mitosis. *J. Cell Biol.* 184:527–39
70. Palframan WJ, Meehl JB, Jaspersen SL, Winey M, Murray AW. 2006. Anaphase inactivation of the spindle checkpoint. *Science* 313:680–84
71. Mattison CP, Stumpff J, Wordeman L, Winey M. 2011. Mip1 associates with both the Mps1 kinase and actin, and is required for cell cortex stability and anaphase spindle positioning. *Cell Cycle* 10:783–93
72. Straight PD, Giddings TH, Winey M. 2000. Mps1p regulates meiotic spindle pole body duplication in addition to having novel roles during sporulation. *Mol. Biol. Cell* 11:3525–37

73. Hached K, Xie SZ, Buffin E, Cladiere D, Rachez C, et al. 2011. Mps1 at kinetochores is essential for female mouse meiosis I. *Development* 138:2261–71
74. Fischer MG, Heeger S, Hacker U, Lehner CF. 2004. The mitotic arrest in response to hypoxia and of polar bodies during early embryogenesis requires *Drosophila* Mps1. *Curr. Biol.* 14:2019–24
75. Gilliland WD, Wayson SM, Hawley RS. 2005. The meiotic defects of mutants in the *Drosophila mps1* gene reveal a critical role of Mps1 in the segregation of achiasmate homologs. *Curr. Biol.* 15:672–77
76. Poss KD, Nechiporuk A, Stringer KF, Lee C, Keating MT. 2004. Germ cell aneuploidy in zebrafish with mutations in the mitotic checkpoint gene *mps1*. *Genes Dev.* 18:1527–32
77. Grimison B, Liu J, Lewellyn AL, Maller JL. 2006. Metaphase arrest by cyclin E-Cdk2 requires the spindle-checkpoint kinase Mps1. *Curr. Biol.* 16:1968–73
78. Gilliland WD, Vietti DL, Schweppe NM, Guo F, Johnson TJ, Hawley RS. 2009. Hypoxia transiently sequesters Mps1 and Polo to collagenase-sensitive filaments in *Drosophila* prometaphase oocytes. *PLoS ONE* 4:e7544
79. Chu MLH, Chavas LMG, Douglas KT, Eyers PA, Tabernero L. 2008. Crystal structure of the catalytic domain of the mitotic checkpoint kinase Mps1 in complex with SP600125. *J. Biol. Chem.* 283:21495–500
80. Wang W, Yang Y, Gao Y, Xu Q, Wang F, et al. 2009. Structural and mechanistic insights into Mps1 kinase activation. *J. Cell Mol. Med.* 13:1679–94
81. Kwiatkowski N, Jelluma N, Filippakopoulos P, Soundararajan M, Manak MS, et al. 2010. Small-molecule kinase inhibitors provide insight into Mps1 cell cycle function. *Nat. Chem. Biol.* 6:359–68
82. Colombo R, Caldarelli M, Mennecozzi M, Giorgini ML, Sola F, et al. 2010. Targeting the mitotic checkpoint for cancer therapy with NMS-P715, an inhibitor of MPS1 kinase. *Cancer Res.* 70:10255–64
83. Chu ML, Lang Z, Chavas LM, Neres J, Fedorova OS, et al. 2010. Biophysical and X-ray crystallographic analysis of Mps1 kinase inhibitor complexes. *Biochemistry* 49:1689–701
84. Kang J, Chen Y, Zhao Y, Yu H. 2007. Autophosphorylation-dependent activation of human Mps1 is required for the spindle checkpoint. *Proc. Natl. Acad. Sci. USA* 104:20232–37
85. Jelluma N, Brenkman AB, McLeod I, Yates JR, Cleveland DW, et al. 2008. Chromosomal instability by inefficient Mps1 auto-activation due to a weakened mitotic checkpoint and lagging chromosomes. *PLoS ONE* 3:e2145
86. Xu Q, Zhu S, Wang W, Zhang X, Old W, et al. 2009. Regulation of kinetochore recruitment of two essential mitotic spindle checkpoint proteins by Mps1 phosphorylation. *Mol. Biol. Cell* 20:10–20
87. Tyler RK, Chu MLH, Johnson H, McKenzie EA, Gaskell SJ, Eyers PA. 2009. Phosphoregulation of human Mps1 kinase. *Biochem. J.* 417:173–81
88. Dou Z, von Schubert C, Korner R, Santamaria A, Elowe S, Nigg EA. 2011. Quantitative mass spectrometry analysis reveals similar substrate consensus motif for human Mps1 kinase and Plk1. *PLoS ONE* 6:e18793
89. Sun T, Yang X, Wang W, Zhang X, Xu Q, et al. 2010. Cellular abundance of Mps1 and the role of its carboxyl terminal tail in substrate recruitment. *J. Biol. Chem.* 285:38730–39
90. Nolen B, Taylor S, Ghosh G. 2004. Regulation of protein kinases: controlling activity through activation segment conformation. *Mol. Cell* 15:661–75
91. Howell BJ, Moree B, Farrar EM, Stewart S, Fang G, Salmon ED. 2004. Spindle checkpoint protein dynamics at kinetochores in living cells. *Curr. Biol.* 14:953–64
92. Jelluma N, Dansen TB, Sliedrecht T, Kwiatkowski NP, Kops GJPL. 2010. Release of Mps1 from kinetochores is crucial for timely anaphase onset. *J. Cell Biol.* 191:281–90
93. Geiser JR, Schott EJ, Kingsbury TJ, Cole NB, Totis LJ, et al. 1997. *Saccharomyces cerevisiae* genes required in the absence of the *CIN8*-encoded spindle motor act in functionally diverse mitotic pathways. *Mol. Biol. Cell* 8:1035–50
94. Castillo A. 2000. *Analysis of MPS1 Separation-of-Function Alleles Reveals a Novel Interaction at the Spindle Pole Bodies in Yeast*. Bolder, CO: Univ. Colorado. 392 pp.
95. Tighe A, Staples O, Taylor S. 2008. Mps1 kinase activity restrains anaphase during an unperturbed mitosis and targets Mad2 to kinetochores. *J. Cell Biol.* 181:893–901
96. Dorer RK, Zhong S, Tallarico JA, Wong WH, Mitchison TJ, Murray AW. 2005. A small-molecule inhibitor of Mps1 blocks the spindle-checkpoint response to a lack of tension on mitotic chromosomes. *Curr. Biol.* 15:1070–76

97. Schmidt M, Budirahardja Y, Klompmaker R, Medema RH. 2005. Ablation of the spindle assembly checkpoint by a compound targeting Mps1. *EMBO Rep.* 6:866–72
98. Tsuda K, Nishiya N, Umeyama T, Uehara Y. 2011. Identification of LY83583 as a specific inhibitor of *Candida albicans* MPS1 protein kinase. *Biochem. Biophys. Res. Commun.* 409:418–23
99. Ishida S, Huang E, Zuzan H, Spang R, Leone G, et al. 2001. Role for E2F in control of both DNA replication and mitotic functions as revealed from DNA microarray analysis. *Mol. Cell. Biol.* 21:4684–99
100. Polager S, Kalma Y, Berkovich E, Ginsberg D. 2002. E2Fs up-regulate expression of genes involved in DNA replication, DNA repair and mitosis. *Oncogene* 21:437–46
101. Zhu W, Giangrande PH, Nevins JR. 2005. Temporal control of cell cycle gene expression mediated by E2F transcription factors. *Cell Cycle* 4:633–36
102. Zhu W, Giangrande PH, Nevins JR. 2004. E2Fs link the control of G1/S and G2/M transcription. *EMBO J.* 23:4615–26
103. Ren B, Cam H, Takahashi Y, Volkert T, Terragni J, et al. 2002. E2F integrates cell cycle progression with DNA repair, replication, and G(2)/M checkpoints. *Genes Dev.* 16:245–56
104. Schmandt R, Hill M, Amendola A, Mills GB, Hogg D. 1994. IL-2-induced expression of TTK, a serine, threonine, tyrosine kinase, correlates with cell cycle progression. *J. Immunol.* 152:96–105
105. Dou Z, Sawagechi A, Zhang J, Luo H, Brako L, Yao XB. 2003. Dynamic distribution of TTK in HeLa cells: insights from an ultrastructural study. *Cell Res.* 13:443–49
106. Zhang X, Yin Q, Ling Y, Zhang Y, Ma R, et al. 2011. Two LXXLL motifs in the N terminus of Mps1 are required for Mps1 nuclear import during G 2/M transition and sustained spindle checkpoint responses. *Cell Cycle* 10:2742–50
107. Stucke VM, Baumann C, Nigg EA. 2004. Kinetochore localization and microtubule interaction of the human spindle checkpoint kinase Mps1. *Chromosoma* 113:1–15
108. Zhao Y, Chen R-H. 2006. Mps1 phosphorylation by MAP kinase is required for kinetochore localization of spindle-checkpoint proteins. *Curr. Biol.* 16:1764–69
109. Borysova MK, Cui Y, Snyder M, Guadagno TM. 2008. Knockdown of B-Raf impairs spindle formation and the mitotic checkpoint in human somatic cells. *Cell Cycle* 7:2894–901
110. Montembault E, Dutertre S, Prigent C, Giet R. 2007. PRP4 is a spindle assembly checkpoint protein required for MPS1, MAD1, and MAD2 localization to the kinetochores. *J. Cell Biol.* 179:601–9
111. Martin-Lluesma S, Stucke VM, Nigg EA. 2002. Role of Hec1 in spindle checkpoint signaling and kinetochore recruitment of Mad1/Mad2. *Science* 297:2267–70
112. Kasbek C, Yang C-H, Yusof AM, Chapman HM, Winey M, Fisk HA. 2007. Preventing the degradation of Mps1 at centrosomes is sufficient to cause centrosome reduplication in human cells. *Mol. Biol. Cell* 18:4457–69
113. Kasbek C, Yang C-H, Fisk HA. 2010. Antizyme restrains centrosome amplification by regulating the accumulation of Mps1 at centrosomes. *Mol. Biol. Cell* 21:3878–89
114. Cui Y, Cheng X, Zhang C, Zhang Y, Li S, et al. 2010. Degradation of the human mitotic checkpoint kinase Mps1 is cell cycle-regulated by APC-c^{Cdc20} and APC-c^{Cdh1} ubiquitin ligases. *J. Biol. Chem.* 285:32988–98
115. Liu C, van Dyk D, Choe V, Yan J, Majumder S, et al. 2011. Ubiquitin ligase Ufd2 is required for efficient degradation of Mps1 kinase. *J. Biol. Chem.* 286:43660–67
116. Salvatore G, Nappi TC, Salerno P, Jiang Y, Garbi C, et al. 2007. A cell proliferation and chromosomal instability signature in anaplastic thyroid carcinoma. *Cancer Res.* 67:10148–58
117. Yuan B, Xu Y, Woo JH, Wang Y, Bae YK, et al. 2006. Increased expression of mitotic checkpoint genes in breast cancer cells with chromosomal instability. *Clin. Cancer Res.* 12:405–10
118. Kilpinen S, Ojala K, Kallioniemi O. 2010. Analysis of kinase gene expression patterns across 5681 human tissue samples reveals functional genomic taxonomy of the kinome. *PLoS ONE* 5:e15068
119. Daniel J, Coulter J, Woo J-H, Wilsbach K, Gabrielson E. 2010. High levels of the Mps1 checkpoint protein are protective of aneuploidy in breast cancer cells. *Proc. Natl. Acad. Sci. USA* 108:5384–89
120. Landi MT, Dracheva T, Rotunno M, Figueroa JD, Liu H, et al. 2008. Gene expression signature of cigarette smoking and its role in lung adenocarcinoma development and survival. *PLoS ONE* 3:e1651
121. Cui Y, Guadagno TM. 2008. B-Raf^{V600E} signaling deregulates the mitotic spindle checkpoint through stabilizing Mps1 levels in melanoma cells. *Oncogene* 27:3122–33

122. Weaver BA, Cleveland DW. 2005. Decoding the links between mitosis, cancer, and chemotherapy: the mitotic checkpoint, adaptation, and cell death. *Cancer Cell* 8:7–12
123. Olesen SH, Thykjaer T, Orntoft TF. 2001. Mitotic checkpoint genes *hBUB1*, *hBUB1B*, *hBUB3* and *TTK* in human bladder cancer, screening for mutations and loss of heterozygosity. *Carcinogenesis* 22:813–15
124. Nakagawa Y, Daigo Y, Nakatsuru S. 2008. Japan Patent No. WO 2008/072777 A2
125. Carter H, Samayoa J, Hruban RH, Karchin R. 2010. Prioritization of driver mutations in pancreatic cancer using cancer-specific high-throughput annotation of somatic mutations (CHASM). *Cancer Biol. Ther.* 10:582–87
126. Ahn CH, Kim YR, Kim SS, Yoo NJ, Lee SH. 2009. Mutational analysis of *TTK* gene in gastric and colorectal cancers with microsatellite instability. *Cancer Res. Treat.* 41:224–28
127. Niittymaki I, Gylfe A, Laine L, Laakso M, Lehtonen HJ, et al. 2011. High frequency of *TTK* mutations in microsatellite-unstable colorectal cancer and evaluation of their effect on spindle assembly checkpoint. *Carcinogenesis* 32:305–11
128. Sotillo R, Hernando E, Díaz-Rodríguez E, Teruya-Feldstein J, Cordón-Cardo C, et al. 2007. Mad2 overexpression promotes aneuploidy and tumorigenesis in mice. *Cancer Cell* 11:9–23
129. Janssen A, Kops GJPL, Medema RH. 2009. Elevating the frequency of chromosome mis-segregation as a strategy to kill tumor cells. *Proc. Natl. Acad. Sci. USA* 106:19108–13
130. Kops GJ, Foltz DR, Cleveland DW. 2004. Lethality to human cancer cells through massive chromosome loss by inhibition of the mitotic checkpoint. *Proc. Natl. Acad. Sci. USA* 101:8699–704
131. Michel L, Diaz-Rodriguez E, Narayan G, Hernando E, Murty VV, Benezra R. 2004. Complete loss of the tumor suppressor MAD2 causes premature cyclin B degradation and mitotic failure in human somatic cells. *Proc. Natl. Acad. Sci. USA* 101:4459–64
132. Tardif KD, Rogers A, Cassiano J, Roth BL, Cimbora DM, et al. 2011. Characterization of the cellular and antitumor effects of MPI-0479605, a small-molecule inhibitor of the mitotic kinase Mps1. *Mol. Cancer Ther.* 10:2267–75

The Structural Basis for Control of Eukaryotic Protein Kinases

Jane A. Endicott,[1] Martin E.M. Noble,[1] and Louise N. Johnson[2]

[1]Northern Institute for Cancer Research, Medical School, Newcastle University, Newcastle upon Tyne NE2 4HH, United Kingdom; email: martin.noble@ncl.ac.uk

[2]Laboratory of Molecular Biophysics, Department of Biochemistry, University of Oxford, Oxford OX1 3QU, United Kingdom; email: louise.johnson@bioch.ox.ac.uk

Keywords

signal transduction, phosphorylation, substrate recognition, dimerization, pseudokinases

Abstract

Eukaryotic protein kinases are key regulators of cell processes. Comparison of the structures of protein kinase domains, both alone and in complexes, allows generalizations to be made about the mechanisms that regulate protein kinase activation. Protein kinases in the active state adopt a catalytically competent conformation upon binding of both the ATP and peptide substrates that has led to an understanding of the catalytic mechanism. Docking sites remote from the catalytic site are a key feature of several substrate recognition complexes. Mechanisms for kinase activation through phosphorylation, additional domains or subunits, by scaffolding proteins and by kinase dimerization are discussed.

Contents

INTRODUCTION	588
THE PROTEIN KINASE FOLD	588
CATALYTIC MECHANISM	591
Substrate Recognition	591
Kinase Catalysis	593
MECHANISMS OF PROTEIN KINASE ACTIVATION	596
Activation by Accessory Proteins or Domains	596
Dimerization and Activation	599
PSEUDOKINASES	604
FUTURE DIRECTIONS	606

INTRODUCTION

Most eukaryotic cellular processes and cell signaling pathways are regulated by protein phosphorylation (1, 2). Protein kinases in turn are regulated by inhibitory or activating protein partners, phosphorylation, cellular localization that limits availability of substrates and activators, protein degradation, and gene transcription. Deregulation of protein kinase activity through mutation to constitutively active forms, loss of negative regulators, and chromosomal rearrangements that lead to the formation of active fusion proteins are associated with a number of disorders. Protein kinases have become major targets for therapy, and protein kinase structures have had a significant impact on the development of selective and specific targeted therapies. An appraisal of these studies is outside the scope of this review, but a number of relevant, recent reviews are recommended (3–5).

Phosphorylation of protein substrates can have profound effects. Phosphorylation can result in enzyme activation, enzyme inhibition, the creation of recognition sites for recruitment of other proteins, and transitions in protein state from order to disorder or disorder to order (6). Reflecting the importance of kinases for eukaryotic cell signal transduction and metabolism, there are more than 518 human protein kinases (7, 8; and **http://kinase.com**) recognized through their conserved sequence motifs. These constitute the third most populous protein family and represent ~1.7% of the human genome. Of the total, 478 protein kinases are typical kinases, and 40 are atypical. The typical kinases are divided into those that phosphorylate serine or threonine residues (388 kinases) and those that phosphorylate tyrosine residues (90 kinases). Atypical kinases are proteins reported to have biochemical kinase activity but lack sequence similarity to the conventional eukaryotic kinases. By April 2011, 170 unique kinase domain structures from humans or closely related orthologs had been determined (9; and **http://www.thesgc.org/resources/kinases**).

A distinguishing feature of the protein kinase family is the different structures that they adopt between the active and inactive states. This family characteristic was first appreciated following the determination of the first protein kinase structures of protein kinase A (PKA) [Protein Data Bank (PDB) code 1ATP] (10) in the active conformation and cyclin-dependent protein kinase 2 (Cdk2) (PDB code 1HCK) (11) in an inactive conformation. Adoption of the active state occurs in response to specific signaling events, which are transduced via kinase-associated regulatory domains in *cis* or in *trans*, and/or by phosphorylation of the kinase domain. Structural details are now emerging on the importance of kinase scaffolding to kinase activation and substrate selection and on the role of pseudokinase domains in regulation. In this review, we summarize the results of structural studies on protein kinases that have provided insights into regulation and into the exquisite substrate specificity shown by protein kinases, which ensures fidelity in cell signaling.

THE PROTEIN KINASE FOLD

Serine/threonine- and tyrosine-specific protein kinases share a catalytic domain of ~290 residues in which the active site is sandwiched between an N-terminal lobe composed of a β-sheet and a single α-helix (the "C helix") and

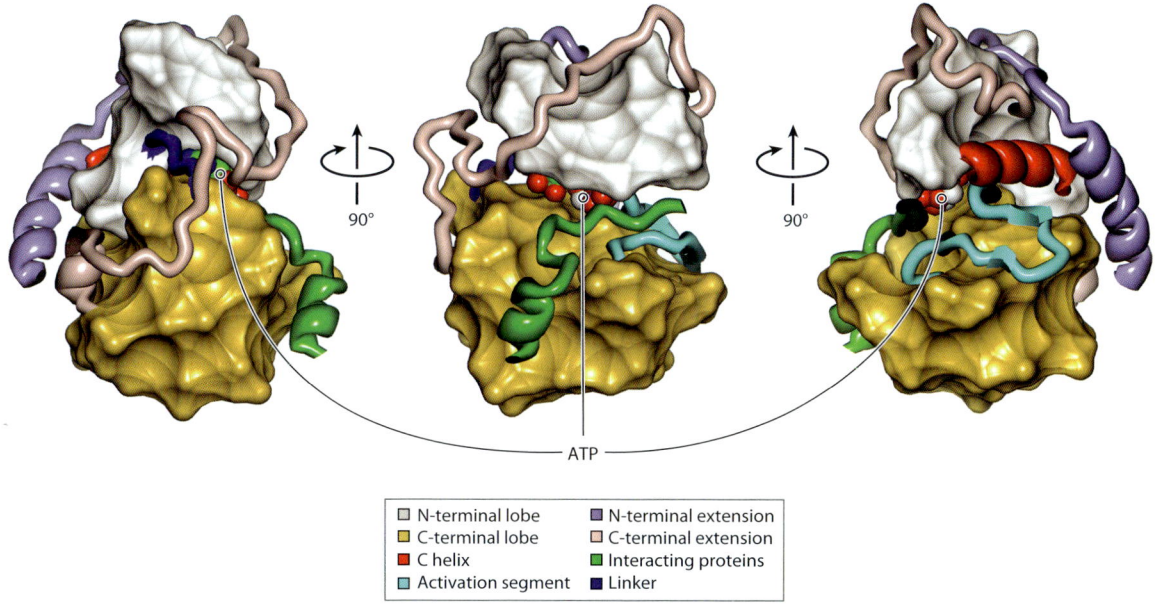

Figure 1

Architecture of a prototypical protein kinase. A highly reduced representation of the key structural and regulatory elements of a protein kinase, as exemplified by protein kinase A, Protein Data Bank (PDB) code 1ATP, illustrating the N-terminal lobe (*main chain surface representation*), C-terminal lobe (*main chain surface representation*), C helix (*ribbon representation*), activation segment (*ribbon representation*), N-terminal extension (*ribbon representation*), C-terminal extension (*ribbon representation*), and interacting proteins (here the proteinaceous inhibitor PKI, ribbon representation). Three consecutive views are related by a 90° rotation about a vertical axis. The same color scheme has been applied throughout the figures in this article. C helix is a secondary structural element named according to the nomenclature in Reference 10.

a larger C-terminal lobe, connected by a linker (**Figure 1**). The C-terminal lobe is predominantly α-helical and includes the activation segment, a region of 20–35 residues located between a conserved DFG (using the single-letter amino acid codes) motif and an APE motif, which is less conserved (**Figures 1** and **2a**) (12, 13). In the active conformation, the C helix packs against the N-terminal lobe, and the aspartate of the DFG chelates an Mg^{2+} ion to orientate the ATP substrate (**Figures 1** and **2b**). In the inactive conformation, this latter interaction is often disrupted, and the phenylalanine of the DFG motif is turned in toward the ATP site (**Figure 2c**). In some kinases, the catalytic domain is flanked by N- and C-terminal extensions that may be involved in regulation (**Figures 1** and **2a**).

Within the conserved ATP-binding pocket, the adenine ring forms specific hydrogen bonds between N1 and N6 and the peptide backbone of the hinge region, and nonpolar aliphatic groups line the pocket and provide van der Waals contacts to the purine moiety. The ribose O2′ and O3′ hydrogen bond to a glutamate side chain (E127) (residue numbers correspond to those of PKA unless otherwise indicated) and the main chain carbonyl oxygen of E170, respectively.

The triphosphate group points out of the adenosine pocket for transfer of the γ-phosphate to the peptide substrate. From the N-terminal lobe, a conserved glutamate residue within the C helix (E91) and a lysine located on β3 (K72) assist to optimally position the α- and β-phosphate groups. A second network of

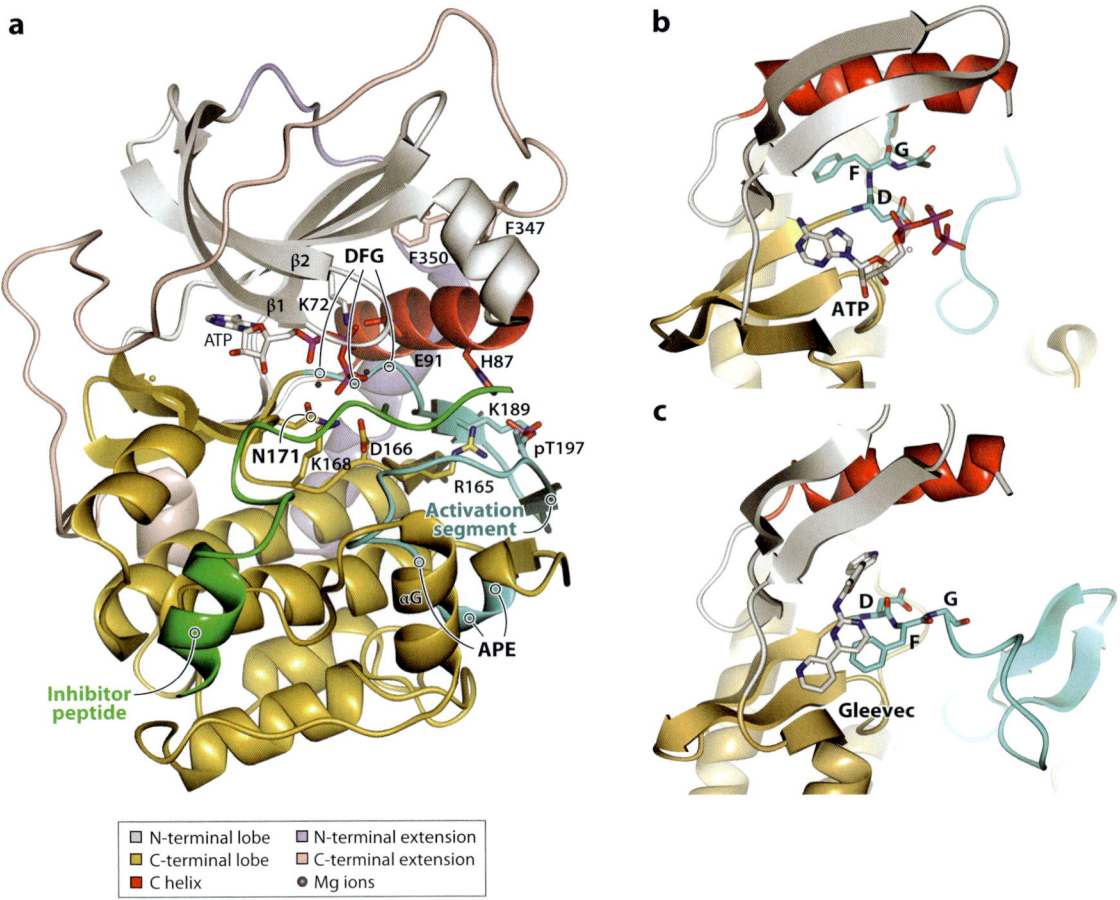

Figure 2

Active and inactive kinases. (*a*) A schematic representation of the conserved residues at the protein kinase A (PKA)-active catalytic site with ATP and substrate. The start of the activation segment is D184, the aspartate that chelates one of the Mg ions (*spheres*). Hydrogen bonds are not shown for simplicity, Protein Data Bank (PDB) code 1ATP. (*b*) Details of the C helix and DFG motif at the start of the activation segment for PKA in the active conformation. (*c*) Details of the C helix and DFG motif at the start of the activation segment for Abl tyrosine kinase complexed with Gleevec in the inactive conformation, PDB code 1IEP. Abbreviations: αG, a secondary structural element named according to the nomenclature in Reference 10; APE, a conserved eukaryotic protein kinase sequence motif at the end of the activation segment; DFG, a conserved eukaryotic protein kinase sequence motif at the start of the activation segment.

interactions to the α- and γ-phosphate groups mediated by a magnesium ion (Mg_2) bound between the aspartate (D184) of the DFG motif and an asparagine (N171) collectively ensure correct positioning required for ATP binding and catalysis. A second magnesium ion (Mg_1) is bound to D184 and the β- and γ-phosphate groups. Additional interactions between the ATP β- and γ-phosphate groups and the glycine-rich loop located between β1 and β2 further stabilize the ATP conformation (**Figure 2***a*).

The activation segment forms a crucial part of the substrate-binding site. In inactive kinases, the activation segment is often partially disordered. Adoption of the catalytically competent conformation to form the peptide-binding platform is triggered in many kinases by phosphorylation. As illustrated for PKA, the phosphothreonine 197 acts as an organizing center and hydrogen bonds to the side chains of H87 from the C helix, R165 located immediately N-terminal to the catalytic aspartate, and K189 from the activation segment (**Figure 2a**). The phosphate group promotes closure of the two lobes of the domain and the correct conformation of the activation segment for substrate binding. In other kinases [e.g., phosphorylase kinase (PhK), PDB code 1PHK; epidermal growth factor receptor (EGFR), PDB code 2GS2; cyclin-dependent kinase 5 (Cdk5), PDB code 1H4L], the activation segment does not require phosphorylation for activity and is able to adopt the correct conformation through other interactions. As more kinase structures emerge, atypical activation segments have been observed that include additional secondary structures (14, 15).

In all kinases, the substrate is oriented so that the hydroxyl is directed toward the catalytic aspartate (D166). In serine/threonine kinases, a lysine residue two residues away (K168) contacts the γ-phosphate and is poised to stabilize the developing local negative charge during catalysis. In tyrosine kinases, the stabilizing residue is four residues away and is an arginine to allow for the larger tyrosine residue.

In addition to the conserved residues that directly interact with the bound substrates, two additional chains or "spines" of conserved hydrophobic residues termed the catalytic and regulatory spines have been defined that traverse the N- and C-terminal lobes (16–18). These spines assemble as a response to changes within the catalytic cleft upon kinase activation and devolve those changes to the rest of the domain. The regulatory spine describes an assembly of interactions that is promoted by the conformation of the activation segment and is responsive to peptide binding, whereas the interactions that generate the catalytic spine include a number between residues of the N- and C-terminal lobes and the ATP adenine ring.

In contrast to the active kinases, which share a common catalytically competent conformation, the inactive kinases are structurally diverse (19). This diversity arises because no catalytic requirements constrain the fold when it is inactive, allowing the proliferation of different conformations that, nevertheless, share a number of common structural themes. These themes were first identified in Cdk2 and Src kinase (PDB code 1FMK) and further elaborated in the EGFR kinase and others. Analysis of these structures has provided insights into mechanisms for kinase activation that depend on the structures of the C helix and the activation segment (20).

CATALYTIC MECHANISM

Substrate Recognition

Protein kinases catalyze the transfer of a phosphoryl group from the γ-phosphate of ATP to the hydroxyl group of serine, threonine, or tyrosine residues in protein substrates, a process that may be summarized by the reaction scheme in Equation 1:

$$\text{Protein-OH} + \text{ATP}^{4-}\cdot\text{Mg}^{2+} \rightarrow$$
$$\text{Protein-O-PO}_3^{2-} + \text{ADP}^{3-}\cdot\text{Mg}^{2+} + \text{H}^+.$$

Most protein kinases show specificity for the local region around the site of phosphorylation where certain residues are required for recognition. Examples are given in **Table 1**. Protein kinases normally phosphorylate sites that are in less well-ordered parts of the protein that are exposed on the surface (21). The preference for disordered regions allows the kinase to mold the region of the protein substrate to an extended conformation that fits the catalytic site and allows the localization of the specificity-determining residues to recognition pockets on the kinase (22). It may also allow those regions to act, upon phosphorylation, as specific interaction motifs capable of partnering with diverse proteins. In many kinases, specificity is also conferred by remote docking sites located either on the kinase at sites separate from the catalytic site [as in mitogen-activated protein kinases (MAPKs), for example, PDB code 2GPH, reviewed in Reference 23] or on separate domains or subunits as in the Cdk2/cyclin complexes (PDB codes 1H24, 1H26, 1H27, and 1H28) (24) or in Polo-like kinase 1 [(Plk1)

Table 1 Some protein kinases and their preferred substrate specificities

Name	Consensus sequence[a,b]
Serine and threonine kinases (abbreviation, if any)	
Cyclic AMP-dependent kinase (PKA)	-R-R-X-**S/T**-Φ
Protein kinase B (PKB) (Akt)	R-X-X-R-X-X-**S/T**-Φ
Phosphorylase kinase (PhK)	-R-X-X-**S/T**-Φ-R
Cyclin-dependent protein kinase 2 (Cdk2)	-**S/T**-P-X-K/R
Extracellular-regulated kinase 2 (ERK2)	-P-X-**S/T**-P
Polo-like kinase 1 (Plk1)[c]	-D/E/N-X-**S/T**-(Φ//not P)
Aurora B[c]	-R-R/K-**S/T**-(not P)
Tyrosine kinases (abbreviation)	
Insulin receptor kinase (Irk)	-D-**Y**-M-M
Cellular form of the Rous sarcoma virus transforming agent (c-Src)	-E-E-I-**Y**-X-X-F
C-terminal Src kinase (Csk)	-I-**Y**-M-F-F
Epidermal growth factor receptor kinase (EGFRK)	-E-E-E-**Y**-F

[a]The phosphorylated serine, threonine, and tyrosine residues are indicated in bold.
[b]Φ is a hydrophobic residue.
[c]Some kinases (e.g., Plk1 or Aurora B) discriminate against proline in the P+1 site (136).

PDB code 1Q4K], which has an N-terminal Polo box domain, which recognizes substrates that have been phosphorylated by Cdk1/cyclin B to dock the kinase on its substrate (25).

There are relatively few structural studies on protein kinases in complex with their protein substrate (**Table 2**). These studies have been achieved by crystallization in the presence of an inactive ATP analog or in PKA with an inhibitor peptide in which the residue to be phosphorylated has been substituted by alanine. The affinity of protein kinases for peptide substrates is weak (typically 2×10^{-4} M), requiring millimolar concentrations to saturate the kinase in crystallization studies, and this may partly explain why so few kinase/peptide substrate complexes have been cocrystallized. A comparison of recognition sites indicates that discrimination for serine/threonine or tyrosine kinases (**Figure 3**) is mostly achieved by a subelement of the activation segment (13). An inward orientation of the activation segment toward the catalytic site is observed for serine/threonine kinases, whereas an outward facing orientation, allowing a larger residue, is observed for tyrosine kinases.

There are three structures where the kinase has been cocrystallized with its intact protein substrate: ROCK-1 kinase with RhoE (PDB code 2V55) (26), RNA-dependent protein kinase (PKR) in complex with the α-subunit of the translation initiation factor eIF2 (PDB code 2A1A) (27), and C-terminal Src kinase in complex with the authentic, endogenous form of the Src kinase, c-Src (PDB code 3D7T) (**Figure 4**) (28). In these structures, the phospho-acceptor region was disordered, and so its position at the kinase catalytic site could not be determined. However, the location of the major part of the protein substrate was evident, and in the complexes for PKR and ROCK-1, despite quite different molecular partners, the protein substrate was located at

Table 2 Protein kinase/peptide substrate complex structures

Protein kinase	Substrate[a]	References
PKA	Inhibitor peptide fragment 5–24	137, 138
PhK	R-Q-M-**S**-F-R-L	22
pCdk2/cyclin A	H-H-A-**S**-P-R-K	139
PKB	G-R-P-R-T-T-**S**-F-A-E	56
Irk	G-D-**Y**-M-N-M	71

[a]The phosphorylated serine, threonine, and tyrosine residues are indicated in bold.

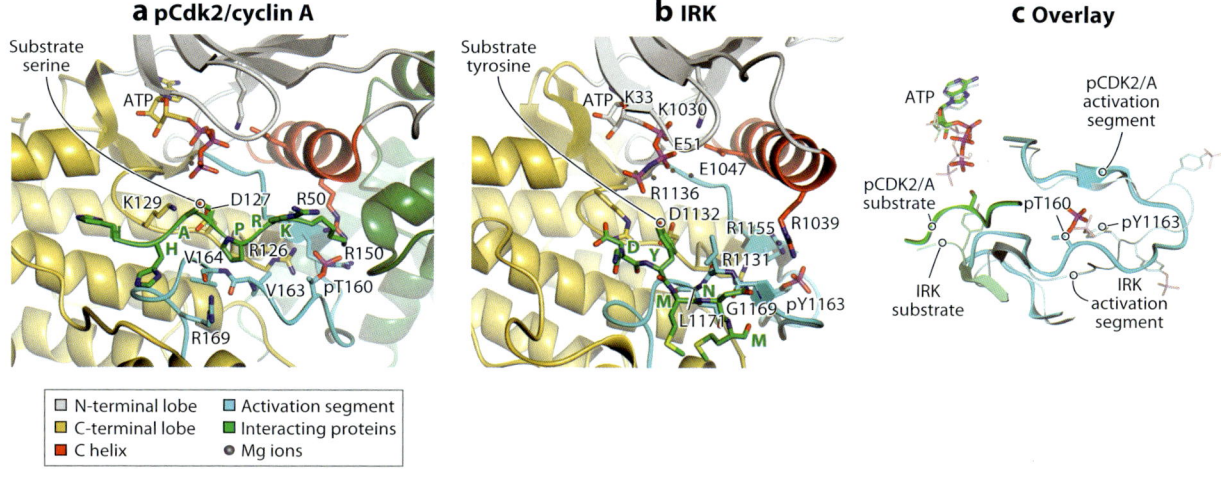

Figure 3

Protein kinase peptide substrates. (*a*) Human phospho-cyclin-dependent protein kinase 2 (pCdk2)/cyclin A in complex with the peptide substrate (HHASRK) in green. The catalytic aspartate, D127, is in hydrogen bond distance of the substrate serine hydroxyl, and K129 is poised to assist catalysis. The substrate is positioned by the docking of the adjacent proline of the serine-proline motif into a hydrophobic pocket created by a left-handed conformation of the activation segment residues V163 and V164, which is stabilized by R169. The residues interacting with phospho-T160 (pT160), which acts as an organizing center, are shown (PDB code 1QMZ). (*b*) Human insulin receptor kinase (IRK) in complex with the substrate peptide (GDYMNM) in green. The catalytic aspartate, D1132, is in hydrogen bond distance of the substrate tyrosine hydroxyl, and R1136 is poised to assist catalysis. The substrate is positioned by a short stretch of β-sheet between residues G1169 and L1171 of the IRK activation segment and the substrate. pY1163 acts as an organizing center. The γ-phosphate of ATP is misaligned for catalysis in this complex (PDB code 1IR3). (*c*) Overlay of the peptide-substrate complexes of pCdk2/cyclin A and IRK. For clarity, only the activation segment, peptide substrate, and ATP molecules are shown. The standard color scheme has been applied but structural elements of IRK are shown with decreased color saturation and with thinner bonds and ribbons.

a similar docking region comprising the kinase αG helix and part of the αF helix (**Figure 4a,b**). Docking at this site placed the likely positions of the phospho-acceptor sites within reach of the kinase catalytic site. This docking region of the kinase is also used to locate the regulatory R subunit of PKA, which allows the R subunit to engage its inhibitory segment at the catalytic site of PKA (PDB code 1U7E) (29), and a similar region on phospho-Cdk2 is used to position the protein phosphatase KAP to allow the KAP catalytic site to reach phospho-T160 of Cdk2 (PDB code 1FQ1) (30).

The location of a substrate through a secondary, remote binding site allows the kinase to phosphorylate sites with suboptimal local sequences, in some cases increasing the apparent affinity for substrate by 1,000-fold (22, 27). The remote docking site allows a stable association, whereas the presence of the substrate at the catalytic site is transient. A recent study of the PKR-eIF2α complex suggests that substrate docking might fulfill a more active role in promoting kinase selectivity and prevent promiscuous substrate phosphorylation (31). Using mutant proteins designed following a comparison of the free and PKR-bound eIF2α structures, a model was proposed in which eIF2α binding to PKR elicits a conformational change in eIF2α that alters the accessibility and mobility of the PKR-phosphorylated residue (S51) enhancing its phosphorylation.

Kinase Catalysis

The phosphoryl transfer step is chemically simple and is dependent on the correct orientation of the two substrates, the γ-phosphate of ATP and the hydroxyl group of the serine, threonine, or tyrosine residue to be phosphorylated. The

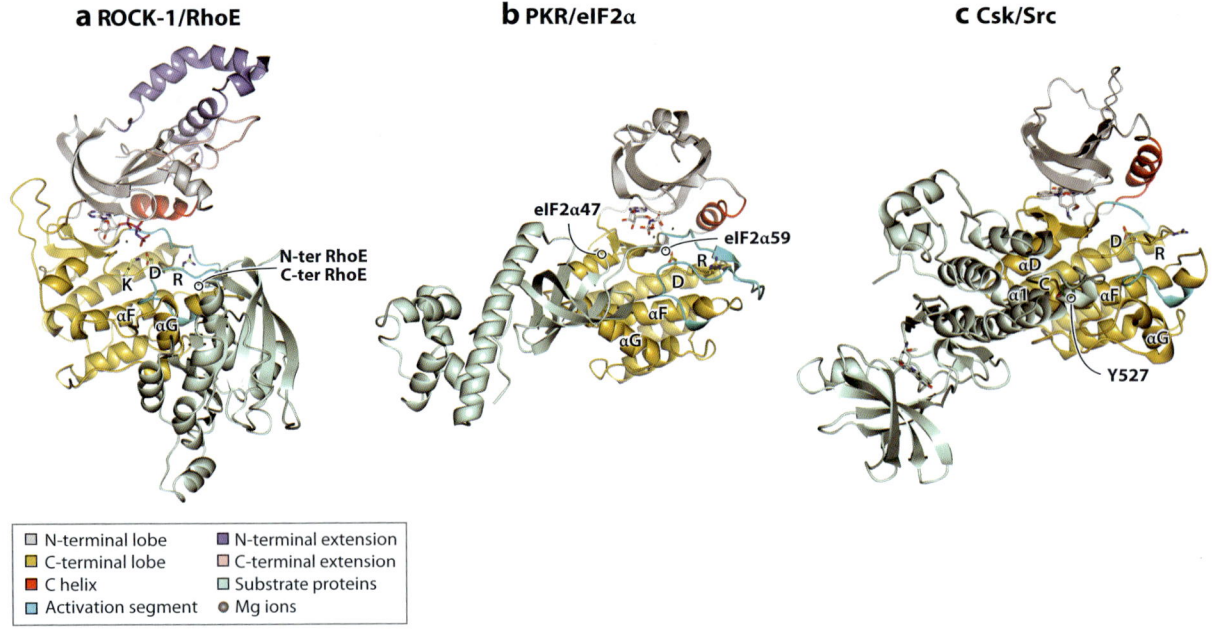

Figure 4

Protein kinase/protein substrate complexes. The ATP molecules and ATP-competitive inhibitors are shown bound, and the residues R and D mark the catalytic site, where the D is the catalytic aspartate. (*a*) ROCK-1 kinase in complex with RhoE substrate. The beginnings of the N-terminal (N-ter) and C-terminal (C-ter) regions of RhoE are marked. These regions contain the two and five ROCK-1 phosphorylation sites, respectively. They are disordered, but it is possible for them to reach the catalytic site some 20 Å away (PDB code 2V55). (*b*) RNA-dependent protein kinase (PKR) in complex with eukaryotic translation initiation factor 2α (eIF2α). The view is rotated ∼15° about the vertical axis from the standard view for clarity and shows the docking of the β-sheet of eIF2α against the PKR G helix. S51, the site of phosphorylation on eIF2α, is in a disordered region between residues 47–59. However, it is positioned close enough to reach the catalytic site (PDB code 2A1A). (*c*) The kinase domains of C-terminal Src kinase (Csk) and Src in complex. Csk uses part of its D helix (αD) to bind Src through the region between helices H and I. The Csk activation segment is disordered. Src Y527, the site of phosphorylation, is 12 Å from the Csk catalytic site but could reach the site through conformational change. Staurosporine, a nonspecific kinase inhibitor, is bound at the ATP-binding sites of Csk and Src (PDB code 3D7T). The helices (αD–αG) within the C-terminal lobe are labeled according to the nomenclature originally used to describe the protein kinase A structure (PDB code 1ATP).

rate (k_{cat}/K_M) of a kinase-catalyzed transfer of phosphate to a serine residue is fast compared with the uncatalyzed reaction. For example, the phosphorylase kinase action on glycogen phosphorylase has $k_{cat} = 28$ s^{-1} and K_M (ATP) $= 7.0 \times 10^{-5}$ M, giving $k_{cat}/K_M = 4 \times 10^5$ s^{-1}M^{-1} (32). The rate of the uncatalyzed methanolysis of ATP^{2-}·Mg to methyl phosphate has $k_{cat}/K_M = 3.8 \times 10^{-9}$ s^{-1}M^{-1} (33). The enzyme catalyzed reaction demonstrates an extraordinary enhancement of ∼10^{14}.

Kinetic and catalytic mechanisms of protein kinases were reviewed in detail in 2001 (34).

Kinetic studies with ^{32}P-labeled ATP or radiolabeled peptide substrate with PKA indicate that both substrates have unrestricted access to the catalytic site and the binding of one does not exclude the other, although at high ATP concentrations, which are typical in the cell, there is a preference for ATP binding first (35). This is consistent with the arrangement of the ATP and substrate sites observed in the crystal structures, where the ATP site is partially shielded by the substrate peptide. With larger natural protein substrates, the ATP site could be even more shielded, and indeed, several kinases have now been shown to demonstrate an ordered catalytic mechanism.

The kinase reaction proceeds with an in-line mechanism in which the attacking group (serine, threonine, or tyrosine OH) comes in opposite to the leaving group (phosphate ester oxygen), leading to inversion of configuration at the phosphorus. This geometry was supported by structural studies with PKA cocrystallized with a putative transition state analog (PDB code 1L3R) (36). One of the roles proposed for the bound magnesium ions is to stabilize the significant amount of negative charge that develops on the bridging oxygen as the reaction proceeds and thereby to aid departure of the leaving group, ADP (34). The transition state for the intermediate could be either dissociative or associative, as reviewed recently for phosphoryl transfer mechanisms for nonenzymatic and enzymatic reactions (37). In the dissociative mechanism, the reaction proceeds through a metaphosphate intermediate in which the bond to the leaving group is broken before the bond by the attacking group is made. In the associative mechanism, the reaction proceeds through a phosphorane pentavalent intermediate in which bond making occurs in advance or at the same time as bond breaking. Structural evidence from pCdk2/cyclin A crystallized with a putative transition state analog [PDB codes 3QHR, 3QHW (38), and 1GY3 (39)] supported the notion of the dissociative mechanism, whereas studies with PKA indicated that there may also be a small percentage (11%) of an associative mechanism (PDB code 1L3R) (36).

We may simplify the steps in the reaction as shown in Scheme 1 below:

where E represents the kinase, S and P are the protein substrate and product, respectively; k_3 is the rate constant for the catalytic step of the phosphoryl transfer; and k_4 is the rate constant for the dissociation of products. Kinetic studies in solvents of different viscosity that allowed binding and dissociation events to be distinguished from chemical catalytic steps showed that the catalytic step was fast ($k_3 \sim$ 300–500 s^{-1}), and the release of products relatively slow ($k_4 \sim$ 20–30 s^{-1}) (32, 35, 40, 41). Such measurements require careful interpretation and need to take account of possible conformational changes. In summary, once the substrates have been correctly oriented, the kinase chemical step is easy, and the rate-limiting step is the release of products, i.e., ADP and phosphorylated proteins.

A simplified representation of a kinase mechanism is shown in **Figure 5**. Kinase-catalyzed phosphoryl transfer can be envisaged as comprising three major steps: orientation of the substrates; nucleophilic attack by the substrate hydroxyl group, followed by general base catalysis from the catalytic aspartate; and subsequent general acid catalysis for the transfer of the proton (32). Support for the notion that deprotonation of the nucleophile by a catalytic base in the early stages of the reaction is not a rate-limiting step came from studies with the tyrosine kinase C-terminal Src kinase (42). A peptide containing the unnatural amino acid trifluorotyrosine showed similar efficiency as a substrate compared with the corresponding tyrosine-containing peptide despite a four-unit change in the phenolic pK_a. However, a residue with the ability to develop a negative charge at the correct separation is

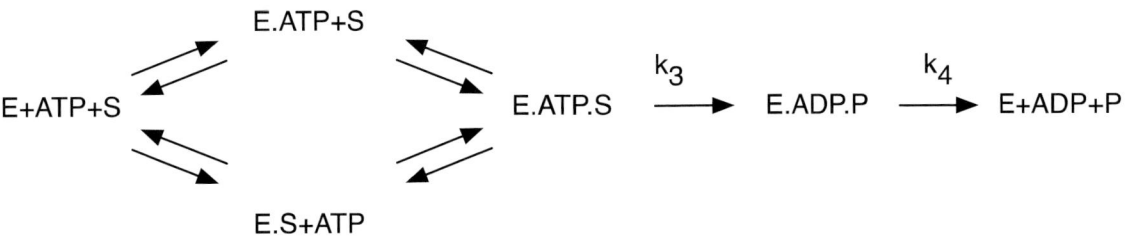

Scheme 1

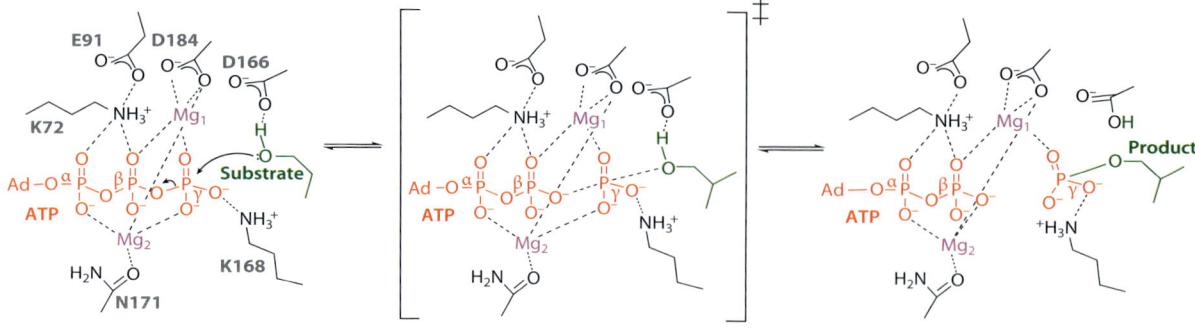

Figure 5
Schematic diagram of protein kinase catalytic mechanism. The reaction proceeds from the enzyme/substrate complex through the transition state (*center*) to the enzyme/product complex. Residue numbers for protein kinase A (PKA) are shown in the left panel. The OH group of the protein substrate is aligned so that the lone pair of electrons on the oxygen are directed in-line through the γ-phosphorus atom to the $\beta\gamma$-bridging oxygen of the bound ATP. The transition state involves a metaphosphate intermediate in which the bond breaking of the $\beta\gamma$-bridging oxygen of ATP is well advanced, while the incoming nucleophile bond making to the phosphorus is only just beginning. The negative charge on the γ-phosphate is compensated by the Mg ions and nearby lysine residue. As the reaction proceeds, the acidity of the substrate hydroxyl group increases, and its pK_a will become lower than the pK_a of the nearby catalytic aspartate, thus allowing transfer of a proton from the hydroxyl (normal pK_a ~12) to the aspartate (normal pK_a ~ 4.5). This proton is probably eventually transferred to the phosphate dianion of the product restoring the catalytic site aspartate to the carboxylate state. Abbreviations: Ad, adenosine; NH_3^+, represents the charged alternative form of the epsilon amino group of the side chain of K72 (labeled in the LHS panel).

important because mutation of the catalytic aspartate to asparagine, alanine, or glutamate results in a reduction of the k_{cat} by $\sim10^4$ but little change in the K_M (32, 42).

MECHANISMS OF PROTEIN KINASE ACTIVATION

Activation by Accessory Proteins or Domains

The protein kinase fold is pliable and may be manipulated to an active or inactive conformation by extra domains or separate subunits. The cell cycle kinase Cdk2 is dependent on the association with a cyclin subunit for activity. The cyclin associates in the region of the C helix and promotes a rotation approximately about the axis of the helix (PDB code 1FIN) (43) so that an isoleucine from the PSTAIRE motif at the start of the C helix is buried in a hydrophobic pocket on the cyclin, and the glutamate of this motif contacts the β3 lysine to create an ion pair that is part of the ATP-binding site. In this conformation, the Cdk C helix hydrophobic residues are shielded by cyclin binding. In parallel with the shift of the C helix, a movement of the activation segment, starting at the DFG motif, takes the activation segment out of the catalytic site so that the threonine becomes accessible for phosphorylation and the aspartate of the DFG shifts to an internal site where it chelates the Mg^{2+} ion for ATP binding (**Figure 6a**). As an example of a constitutively active kinase, the PhK catalytic subunit in isolation possesses a canonical amphipathic C helix that requires no further interactions to adopt the active conformation (22).

In contrast, the kinase domain of Src is held with the C helix in an inactive conformation by restraining interactions with its SH2 and SH3 domains that pack on the opposite side of the kinase and are not in direct contact with the C helix (PDB code 1FMK) (44, 45). When these restraints are removed, either through the SH2 and SH3 domains docking to recognition proteins or by phosphatase-mediated hydrolysis of the phosphorylated tyrosine that forms the SH2-docking site, the kinase is able to relax to its active conformation of the C helix (PDB code 1Y57) (**Figure 6b**) (46).

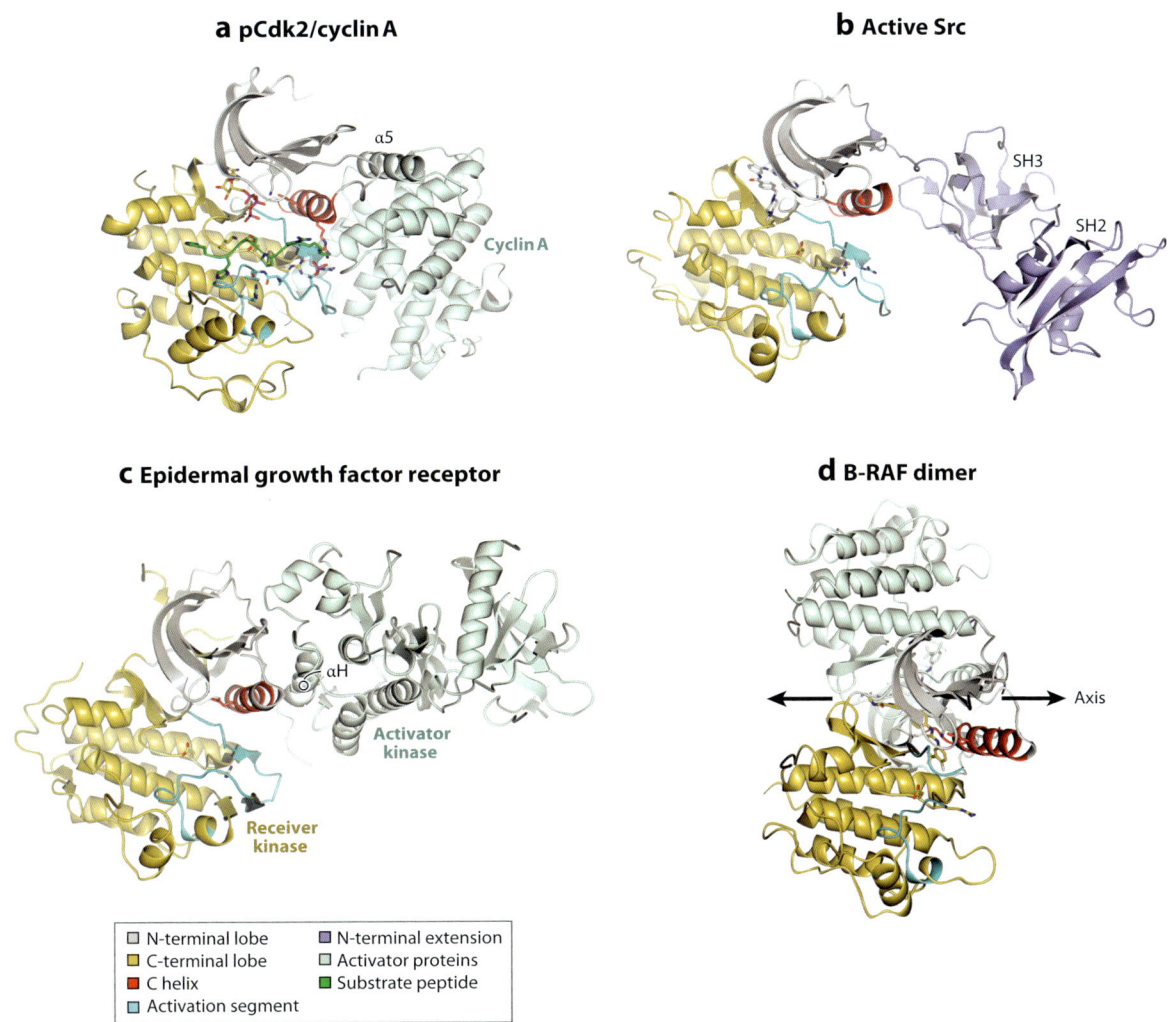

Figure 6

Stabilizing interactions for C helix conformations. (*a*) Human phospho-cyclin-dependent protein kinase 2 (pCdk2)/cyclin A. The cyclin H5 helix docks against the C helix to promote the active conformation (PDB code 1QMZ). (*b*) Human Src kinase in the active conformation. The SH2 and SH3 domains are liberated and no longer restrain the kinase in the inactive conformation (PDB code 1Y57), and may stabilize an active C-helix conformation. (*c*) Human epidermal growth factor receptor kinase. The receiver kinase (left) is activated by interaction between its C helix and the activator kinase H helix (*right*) (PDB code 2GS2). (*d*) Human B-RAF side-by-side dimer where the two monomers interact about a twofold axis (marked by the *double-headed arrow*) to promote an active conformation (PDB code 1UWH). Abbreviation: α5, the fifth α-helix of the N-terminal cyclin box fold of cyclin A2; αH, an α-helix structurally equivalent to the eighth helix of cyclic AMP-dependent protein kinase.

Kuriyan and colleagues (20) have illustrated the recurring theme of regions engaging the C helix hydrophobic patch to activate kinases. In PKA (**Figures 1** and **2***a*) (47) and extracellular signal-regulated kinase 2 (Erk2) (PDB code 1ERK) (48), C-terminal helical extensions wrap around the C helix; while in the p21-activated kinase 1 (PDB codes 1YHV and 1YHW) (49), an N-terminal helix performs this role; and in the protein kinases RET (PDB codes 2IVS and 2IVT) (50) and c-Kit (PDB code 1PKG) (51), an N-terminal region from the end of the juxtamembrane region shields the C helix. In the TGFβ receptor kinase, part of the juxtamembrane sequence, known as the GS region, is held against the C helix in an inactive state by FKBP12, a small immunophilin protein. Inhibition is relieved by phosphorylation of the GS region creating a docking site for the Smad substrate (PDB code 1IAS) (52). The tyrosine kinase Fes utilizes its own SH2 domain to promote the active conformation of the Fes C helix (PDB code 3BKB) (53).

Among examples of stabilization of the C helix hydrophobic surface in *trans*, Aurora A kinase associates with the mitotic spindle assembly protein TPX2 to achieve activation (PDB code 1OL5) (54). TPX2 is located in a groove and interacts with the C helix in a similar manner to the flanking regions of PKA.

AGC kinases. The AGC group of kinases (7) comprises 60 members, including PKA (reviewed in Reference 55). For many of these kinases, activation involves phosphorylation not only on the activation segment but also on a hydrophobic motif (HM) located toward the C terminus with consensus sequence FXXF(S/T/D)Y. Phosphorylation of the HM promotes the intramolecular association of this region with an N-terminal lobe groove, where the extensive hydrophobic interactions from the two phenylalanines promote the active conformation of the C helix and lead to activation. Protein kinase B (PKB/Akt) is activated by phosphorylation on a serine residue (S473) in the HM by the kinase mammalian target of rapamycin complex 2 and by phosphorylation on the activation segment catalyzed by 3-phosphoinositide-dependent kinase-1 (PDK1) when both are colocalized at the plasma membrane (56). In PKA, the chain ends at the second phenylalanine and does not have the residue that is phosphorylated. However, its C-terminal region still folds against the N-terminal lobe to dock the two phenylalanines into a hydrophobic pocket in an identical manner to that achieved by those kinases activated by HM phosphorylation (**Figures 1** and **2a**). Similar interactions occur within protein kinase C, where the HM is phosphorylated (PDB code 2I0E) (57), and within ROCK, where like PKA the HM is not phosphorylated (PDB code 2F2U) (58).

PDK1 is a key member of the AGC family and phosphorylates several AGC kinases on their activation segment. PDK1 autophosphorylates on the activation segment and is constitutively active, but activity for substrates is augmented by recognition through the phosphorylated HM motif. PDK1 does not possess an HM in its C-terminal region, but it does have the hydrophobic pocket in the N-terminal lobe (PDB code 1H1W) (59). Kinases such as S6K, SGK, and RSK are recognized by PDK1 through binding of their phosphorylated HM motifs to the PDK1 hydrophobic pocket.

AMP-activated kinase. AMP-activated kinase (AMPK) plays a homeostatic role in mammalian cells to maintain ATP levels by phosphorylating and inactivating acetyl-coenzyme A (CoA) carboxylase and HMG-CoA reductase, two rate-limiting enzymes in fatty acid and cholesterol biosynthesis, respectively (reviewed in Reference 60). It is a complex of three proteins in which the α-subunit is the catalytic kinase and the β- and γ-subunits play regulatory roles. The α-subunit requires phosphorylation within the activation segment at T172, the residue equivalent to T197 in PKA for activity. The γ-subunit has four cystathionine β-synthase motifs (sites 1–4), and each contains a potential nucleotide-binding site (61–63). Comparative studies suggest that the nucleotide-binding preference (ATP versus ADP versus AMP), affinity, and functional consequences of ligand binding at each site vary both between sites and at any comparable site across species (64).

Significant insights into how nucleotide binding to the γ-subunit allosterically regulates the activity of the α-subunit have been revealed by the determination of the structure of the regulatory core of the trimeric complex (composed of the γ-subunit and C-terminal fragments of the α- and β-subunits) bound to the kinase domain in which the activation segment is ordered as a result of phosphorylation on T172 (PDB code 2Y94) (**Figure 7**)

(65). First, the activation segment interacts with the C-terminal regions of the α- and β-subunits. As a result, the activation segment is constrained and not available to protein phosphatases, whose activity determines to a large extent the amount of phosphorylated AMPK (66). In at least one instance, the structure has shown that phosphatases require a mobile activation segment to dephosphorylate (30). Second, within the α-subunit, a motif known as the "α-hook" from the C-terminal extension regulatory segment docks into site 3 of the γ-subunit. Supported by both mutagenesis data and a comparison of this structure with previously determined structures of the γ-subunit bound to either ADP or Mg.ATP, the authors propose a switch model to explain the regulation of AMPK activity by nucleotides. In this model, the proportion of ADP and AMP bound within the γ-subunit at regulatory site 3 is read out by interactions of this site with the α-hook. Exchange of ADP or AMP by Mg.ATP leads to a steric clash with the α-hook, resulting in the dissociation of the hook and an increase in the flexibility of the linker to the kinase domain. In this form, the activation loop is no longer protected (see also Reference 67 for a supporting electron microscopy study).

AMPK and the AMPK-related protein kinases encode an additional regulatory region within the α-subunit, termed the autoinhibitory domain (AID) sequence C-terminal to the catalytic domain (residues 289–338 in human AMPK), which is largely unstructured in the above structure (**Figure 7**). However, in the structure of a *Schizosaccharomyces pombe* truncated α-subunit encoding the unphosphorylated kinase domain and the AID, the AID sequence is ordered into a three α-helical bundle with structural homology to the ubiquitin-associated (UBA) domain fold (PDB code 3H4J) (68). The AID binds to the kinase domain via helices αC and αE and the hinge sequence. This structure suggests that ordering of the AID sequence in the unphosphorylated AMPK structure could assist in maintaining the enzyme in its inactive state by stabilizing displacement of the C helix and possibly by

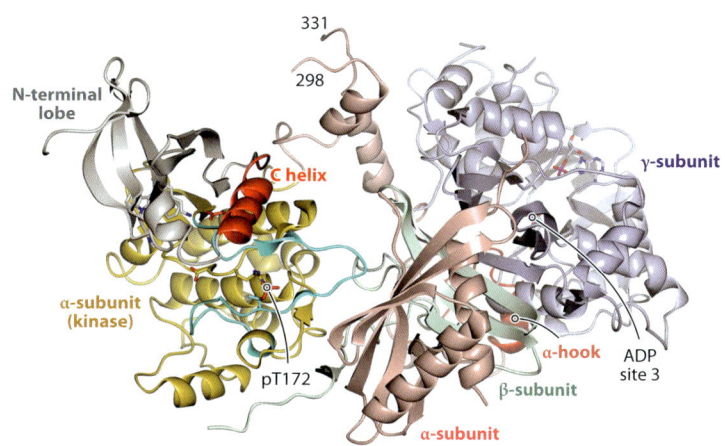

Figure 7

AMP-activated kinase (AMPK) regulation through activation segment contacts. The mammalian AMPK structure with the kinase (α-subunit) in the standard conformation showing phosphothreonine (pT) 172 on the activation segment and the glutamate/lysine pair within the ATP site, which also contains the inhibitor staurosporine. The C-terminal extension of the α-subunit (*pink*) interacts with the β-subunit (*pale green*) and the γ-subunit (*violet*), which contains the AMP/ADP activatory nucleotide-binding sites. The α-hook is in proximity to ADP bound at site 3 of the γ-subunit. The autoinhibitory region between residues 299–330 is disordered (PDB code 2Y94).

eliciting more global changes to the kinase fold (68). How potential structural rearrangements within the AID sequence communicate with those that mediate allosteric regulation of the kinase domain by nucleotide binding to the regulatory core of the trimeric complex remains to be determined.

Dimerization and Activation

Many protein kinases dimerize as part of their activation mechanism, and dimerization can be regarded as a special case of kinase activation by accessory proteins or domains. In such cases, either both partners are activated by reciprocal phosphorylation or one partner (the activator kinase) activates the other (the receiver kinase) through an allosteric mechanism. Though activation subsequent to dimerization was first observed in members of the receptor tyrosine kinase subfamily, a number of examples of serine/threonine kinases that employ this mechanism have also been reported.

Binding of a ligand to the extracellular portion of a receptor tyrosine kinase results in activation of the protein kinase domain and subsequent downstream signaling. Almost exclusively, this requires receptor dimerization to permit phosphorylation and activation in *trans* (reviewed in References 12 and 69). The receptor complexes can be either preexisting, as is the case for the receptors for insulin and insulin-like growth factor 1, or assembled as a result of bivalent ligand binding (for example, the KIT and VEGF receptors). An inhibitory sequence is removed from the active site to permit rearrangement of the activation segment to form the peptide substrate-binding site. The inhibitory sequence can originate from different locations within the protein sequence, and subsequent to its removal, the structure of the rearranged activation segment is frequently secured by phosphorylation on a conserved threonine or tyrosine residue equivalent to T197 in PKA.

The receptor tyrosine kinases. Structures of the insulin receptor tyrosine kinase (IRK) domain were the first to delineate this mechanism of inhibition (PDB codes 1IRK, 1IR3) (70, 71). In this receptor, the inhibitory sequence originates from the activation segment. It contains three tyrosine residues and, in its inactive state, one (Y1162) occludes the ATP-binding site in *cis* by mimicking, in part, the interactions of the adenine ring of the ATP substrate. The resulting conformation of the activation segment also precludes protein substrate binding. Activation promotes autophosphorylation of the three activation segment tyrosine residues in *trans*, disrupting the inhibitory network of interactions within the ATP-binding site and stabilizing the activation segment in its peptide-binding conformation. Similar, but distinct, mechanisms involving the activation segment are also employed to inhibit the kinase domain of FGFR1 (PDB code 1CVS) (72) and the muscle-specific kinase (PDB domain 1LUF) (73). In the latter case, subsequent activation in *trans* is promoted by binding of the adaptor protein Dok7 (PDB code 3ML4) (74).

Subsequent structures of the inhibited forms of the intracellular domains of KIT (PDB code 1PKG) (51), EphB2 (PDB code 1JPA) (75), FLT3 (PDB code 1RJB) (76), and Tie2 (PDB code 1FVR) (77) revealed how their active sites could also be blocked by the binding of peptides derived from sequences outside the catalytic domain, the juxtamembrane region in the cases of KIT, Ephb2, and FLT3, and the C-terminal tail in the Tie2 structure (reviewed in Reference 78). Frequently, as mentioned above, additional sequences engage the C helix to ensure maintenance of the "off" state.

Serine/threonine protein kinases. Within the serine/threonine protein kinase family, dimerization-dependent phosphorylation of the activation segment in *trans* to promote kinase activation has also been reported, and a general molecular model has been proposed (reviewed in Reference 79). Studies of checkpoint kinase 2 (Chk2) generated the first structure and data in support of this model (PDB code 2CN5) (80, 81), but subsequent structure determinations of death-associated protein kinase 3 (PDB code 2J90), lymphocyte-originated kinase (PDB code 2J7T), and STE20-like kinase (PDB code 2J51) provided further support (82). Chk2 dimerizes after phosphorylation of T68 in the N-terminal serine-glutamine/threonine-glutamine cluster domain by the protein kinase Ataxia-telangiectasia mutated, which generates a ligand for the central fork head–associated domain of a second Chk2 molecule (reviewed in Reference 83). The crystal structure of the Chk2 kinase domain revealed an intimate dimer in which the activation segments exchange, forming interactions across the interface such that each sequence completes the active site of the reciprocal molecule by adopting the activation segment structure present in active kinase structures (**Figure 8a**). It was proposed that small rearrangements within the activation segment could bring its two phosphorylated residues (T383 and T387) into a position in which phosphotransfer from the ATP bound to the other Chk2 molecule could occur.

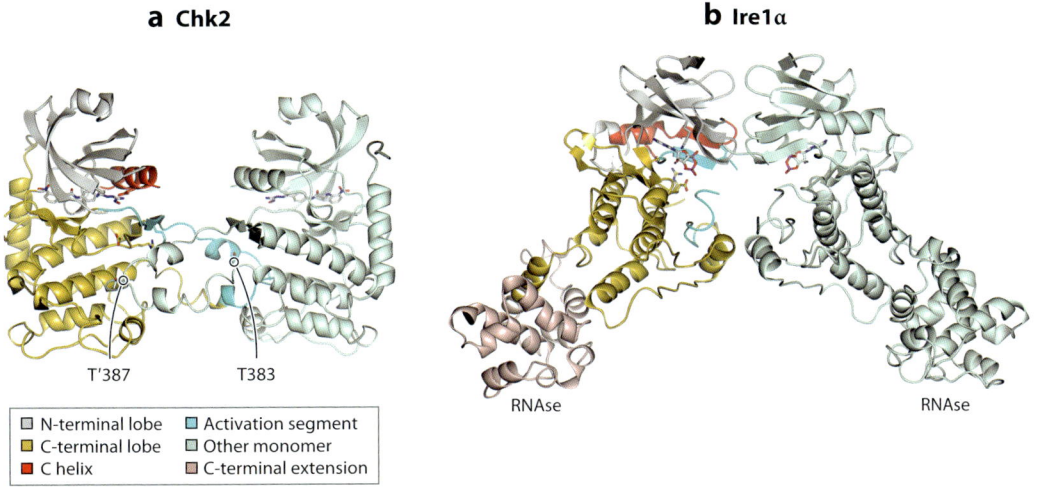

Figure 8

Dimerization through face-to-face activation segment exchange. (*a*) Human checkpoint kinase 2 in the presence of the inhibitor Pv1533. The sites of phosphorylation on the activation segment are T383 and T387. T383 is shown for one subunit and T'387 for the other subunit (PDB code 2XK9). (*b*) Human Ire1α in the presence of Mg.ADP in which the activation segments are disordered but are directed to the catalytic site of the other monomer for phosphorylation of S724 in the activation segment (PDB code 3P23).

Mechanistically, dimerization plays a role that is similar to a remote substrate-docking site in that, by optimally positioning the substrate, kinase-substrate affinity is enhanced, and phosphorylation at noncanonical sites is promoted.

A model for the mechanism of activation of Ire1 through dimerization has also recently been proposed on the basis of a series of Ire1 structures. Ire1 is a transmembrane serine/threonine kinase that is essential for the endoplasmic reticulum (ER) unfolded protein response (reviewed in Reference 84). The association of its luminal domain with the ER Hsp70 protein Bip maintains it in an inactive, monomeric state. However, accumulation of misfolded proteins in the ER activates the unfolded protein response, leading to disengagement of Bip from Ire1 and permitting Ire1 dimerization. A structure of an N-terminally truncated human Ire1α fragment encoding both the unphosphorylated kinase domain bound to Mg.ADP and the ribonuclease domain revealed a face-to-face orientation of the kinase domains (**Figure 8*b***) (PDB code 3P23) (85). This orientation is predicted to promote autophosphorylation and kinase activation in *trans*. Structures of the yeast Ire1 cytoplasmic domain in which the kinase domain is phosphorylated and as a result the RNase domain is proposed to be in its active conformation have revealed an alternative Ire1 dimeric structure in which the kinase domains are arranged back-to-back (PDB codes 2RIO, 3FBV, and 3LJ0) (86–88). Taken together, the structures could be reconciled to an Ire1 activation model in which engagement of the Ire1 ER luminal domains would promote kinase autophosphorylation in *trans*, rearranging the kinase domain into a catalytically active conformation that would then engender substantial rearrangement of the Ire1 dimer to its RNAase-active back-to-back form. Ire1 has been identified as a potential drug target to treat inflammation (84), and a detailed mechanistic understanding of Ire1 activation is required if it is going to be targeted effectively for therapy (3, 89).

The epidermal growth factor receptor kinase family. The EGFR kinase is unusual among the family of receptor tyrosine kinases in that it does not require phosphorylation of

Table 3 Key motifs that are not conserved in selected human pseudokinases[a]

Kinase	Glycine-rich loop[b]	β3 lysine	Catalytic aspartate, lysine, and magnesium-binding asparagine	Magnesium-binding activation segment aspartate-phenylalanine-glycine (DFG) motif	Crystal structure
Conventional active kinases					
PKA (for example)	**GTGSFG**	**YAMK**	**YRDLKPEN**	**DFG**	137, 138
Unconventional active kinases					
WNK1	G<u>R</u>GSFK	VA<u>W</u>C	**HRDLK**C**DN**	D<u>L</u>G	114
Titin	G<u>R</u>GEFG	YMA<u>K</u>	**H**F**DI**R**PEN**	<u>E</u>FG	115
CASK	G<u>K</u>GPFS	FA<u>V</u>K	**HRD**V**K**P**HC**	<u>G</u>FG	116
Kinases that regulate other kinases					
KSR1	G<u>QG</u>RWG	VA<u>I</u>R	**HKDLKSKN**	**DFG**	108
Her3	GSG<u>V</u>FG	V<u>C</u>IK	**H**R**NLA**A**RN**	**DFG**	92
STRADα	GKGF<u>ED</u>	V<u>IV</u>R	**H**R**SVKA**S**H**	<u>G</u>L<u>R</u>	118
JAK JH2	G<u>R</u>GT<u>R</u>I	VI<u>L</u>K	H**GN**V**C**T**KN**	D<u>P</u>G	
Scaffold proteins					
ILK	<u>NENH</u>SG	I<u>VV</u>K	**R**HA**LN**S**R**S	D<u>VK</u>	123
VRK3	<u>TRDNQ</u>G	FS<u>L</u>K	H**GN**V**T**A**EN**	<u>G</u>FG	125

[a]Adapted from Reference 125.
[b]Key residues are in bold when conserved and underlined when not conserved.

the activation segment for full activity and its juxtamembrane sequence is required to activate, rather than to inhibit, the kinase domain. An elegant series of structures have delineated the role played by receptor dimerization, augmented by sequences outside the catalytic domain in regulating EGFR activity (reviewed in References 20 and 90). The EGFR family consists of four members, EGFR (ErbB1), HER2 (ErbB2, HER2/neu), HER3 (ErbB3), and HER4 (ErbB4), that can form both homo- and heterodimers. Notably, HER3 is regarded as pseudokinase as the key catalytic residues equivalent to E91 (in the C helix) and D166 (the proposed catalytic base) in PKA are replaced by a histidine and asparagine, respectively (**Table 3**). However, HER3 has been reported to autophosphorylate (91) and can act as an activator kinase when paired with a catalytically competent receiver kinase (92).

A comparison of the crystal structure of the inactive EGFR kinase domain (PDB code 3GOP) (93, 94) with the active structures of EGFR (PDB code 1M14) (95), HER2 (PDB code 3PP0) (96), and HER4 (PDB code 3BCE) (97) has revealed a conserved mechanism of allosteric activation in which one molecule (the receiver kinase) is remodeled and activated following formation of an asymmetric dimer with a second activator kinase (**Figure 6c**) (98, 99). The conformational changes that accompany activation within the activated monomer and the character and location of the interface within the dimer are both reminiscent of the mechanism of CDK activation by cyclin binding (reviewed in Reference 20). However, the dimer formed between the kinase domains is not stable, and at least in vitro dimer formation requires the cytoplasmic sequence between the membrane and the start of the kinase domain (93, 94). This sequence is required for receptor activation, and its mechanism has been elucidated by structural studies. After the transmembrane helix, the chain forms a helix that dimerizes across a receptor pair, and the sequence originating from the receiver kinase (now termed the juxtamembrane latch) binds to the C-terminal lobe of the activating subunit to stabilize the active asymmetric kinase dimer (93, 94).

RAF kinases. A second example of where kinase dimerization plays a crucial role in kinase activation and in which it now appears one of the monomers plays a scaffolding role and need not have catalytic activity is provided by the cytoplasmic serine/threonine kinase, RAF, and its close relative kinase suppressor of Ras (KSR) (100, 101). RAF, together with mitogen-activated protein kinase kinase (MEK) and the ERKs, comprises one of the evolutionarily conserved MAPK pathways that collectively signal to regulate cell growth, differentiation, and survival (reviewed in Reference 102).

The B-RAF kinase domain bound to the RAF inhibitor BAY43-9006 adopts an inactive structure (PDB code 1UWH) (103). Although some catalytically important residues were properly positioned (most notably the N-terminal glycine-rich loop, the catalytic loop, the lysine equivalent to K72, and the glutamate on the C helix equivalent to E91 in PKA, respectively), the B-RAF structure required movement of the DFG motif and activation segment to adopt the active structure seen in other protein kinases. Subsequent crystal structures of the RAF catalytic domain consistently revealed the same side-to-side dimer structure present in the original B-RAF crystal lattice (PDB codes 3C4C, 3C4D, 3C4E, and 3C4F) (104), (PDB code 3D4Q) (105), and (PDB code 2FB8) (106). It was noted that this dimer is mediated by interactions between the kinase N-terminal lobes, and from this observation, it was predicted that this arrangement retains the C helix in a position to support catalysis (**Figure 6d**) (107). Collectively, these structures suggested that activation of RAF might, like the EGFR kinase, be mediated by an allosteric mechanism in which one kinase subunit acts as a scaffold (the activating kinase) to stabilize the active conformation of the other (the receiving kinase). The functional relevance of the crystallographically observed dimer structure was supported by a mutational study showing that alterations to residues predicted to be at the B-RAF/KSR interface affected the ability of KSR to activate B-RAF. Further experiments demonstrated that activation of B-RAF through formation of a KSR heterodimer was greater than activation subsequent to B-RAF homodimer formation (107). This result suggested that the mechanism of B-RAF activation by KSR might be twofold, firstly by allosterically activating the B-RAF kinase domain and subsequently by acting as a scaffolding protein to promote association of B-RAF with MEK.

The details of this model have subsequently been refined by the determination of the structure of a KSR2 kinase domain [KSR2(KD)]/MEK1 heterodimer (PDB code 2Y4I) (108). Within the crystal lattice, two heterodimers associate to form a tetramer through adjacent KSR2(KD) molecules. The KSR2(KD) and MEK1 face each other in the heterodimer complex, generating an interface that is composed of residues from their respective activation segments and αG helices. As a result, the activation segments are mutually constrained, and the KSR subunit is in an inactive conformation with the C helix displaced. The KSR2(KD) homodimer, however, forms a side-to-side association that is reminiscent of but distinct from that previously observed in the B-RAF homodimer structures. Only a subset of the intersubunit contacts is common to both structures. As a result the quaternary arrangements of the two complexes are different. An analysis of the residue conservation at the observed interfaces and the overall structures predicts that a KSR2(KD)/B-RAF heterodimer would resemble the B-RAF homodimer structure [as proposed (107)] and would be compatible with the KSR2(KD) αC helix only when it is in an active conformation.

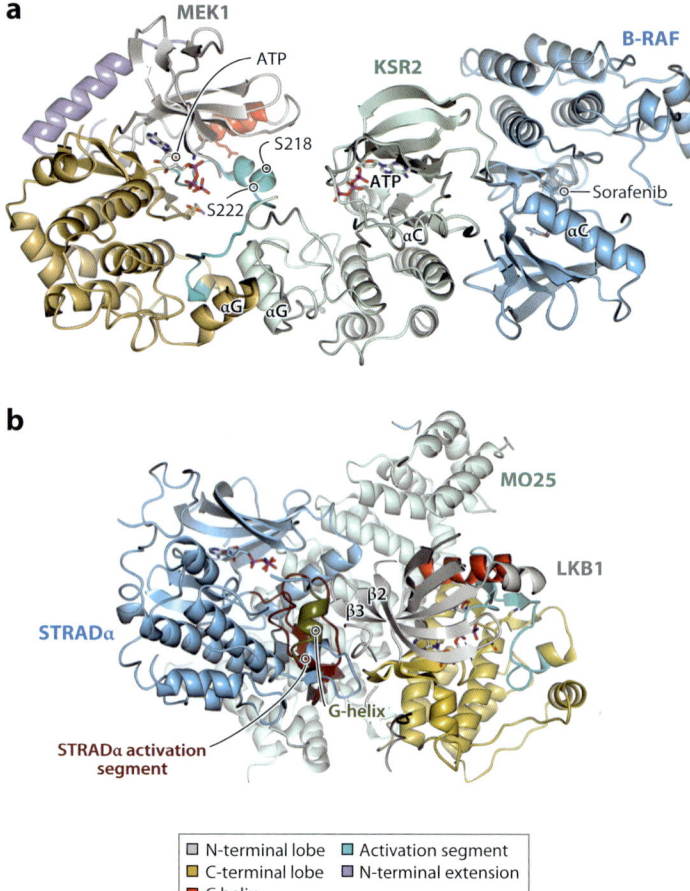

Figure 9

Pseudokinases exhibit different roles in the activation of their partner kinases. (*a*) The proposed trimeric assembly of human MEK1/KSR2/B-RAF kinases based on the structures of MEK1/KSR2 (PDB code 2Y4I) and dimeric B-RAF (PDB code 2UWH) in which MEK1, the target of regulation, is shown. A regulatory RAF is proposed to activate the pseudokinase KSR2 through the side-to-side dimer interface. In turn, KSR2 phosphorylates MEK1 in the N-terminal region through the face-to-face dimer interface. This results in a conformation in which an external catalytic RAF dimer phosphorylates MEK1 on the activation segment S218 and S222. (*b*) The human trimeric assembly of STRADα/MO25/LKB1. The pseudokinase STRADα (STE20-related adaptor kinase, *blue* in *standard orientation*) acts with MO25 (*pale green*) to activate the kinase LKB1 (*nonstandard view*). The STRADα activation segment (*dark red*), and the G helix (*olive green*) contact the β2-β3 loop and the C-terminal region of LKB1 while MO25 contacts the LKB1 C helix. Both STRADα and LKB1 have an active conformation, although STRADα has no catalytic activity (PDB code 2WTK).

Subsequent studies showed that addition of a B-RAF kinase-deficient mutant was able to promote KSR(KD)-dependent phosphorylation of MEK, supporting the model that the KSR(KD) does adopt an active conformation and that its observed low levels of activity are functionally significant within a physiological complex (108, 109). Taken together, the experimental data (107, 108) support a model in which the interactions between B-RAF, KSR(KD), and MEK in *cis* lead to an allosteric activation of KSR by B-RAF that promotes KSR phosphorylation of MEK, leading to a change in the accessibility of the MEK subunit to phosphorylation by an activating B-RAF molecule in *trans* (**Figure 9***a*). This chain of events results in MEK activation and subsequent downstream signaling. The importance of having a detailed mechanistic understanding of this pathway for the development of ATP-competitive inhibitors targeting B-RAF has been highlighted in recent studies (110, 111).

PSEUDOKINASES

Pseudokinases are defined by the lack of conservation of one or more of the catalytic site residues in the kinase core. The human phylogenetic kinome contains 48 pseudokinases distributed throughout the seven families (7, 112). These kinases are expressed, indicating that they are transcribed genes, but their function has been obscure. They are now recognized to be more than passive bystanders; some do exhibit activity, and others participate in signal transduction (reviewed in Reference 113). Some have been discussed above, including the roles of KSR in activation of RAF and HER3 as an activator of EGFR kinase.

The motifs that are changed in the pseudokinases include the glycine-rich loop, and the VAIK (β3 lysine), HRD (catalytic aspartate), and DFG motifs or combinations of these (**Table 3**) (112). The WNK [whose name is a contraction of with-no-K (Lys)] kinase lacks the K of the VAIK motif, but structural studies showed that a lysine residue in an adjacent region substituted for the missing lysine, and the kinase is active (PDB code 3FPQ) (114). Similarly, in the giant protein titin that contains a kinase domain with DFG substituted by EFG, the structure is modulated so that the glutamate can perform a similar role to that of the aspartate (PDB code 1TKI) (115). Ca^{2+}/calmodulin-activated Ser/Thr kinase (CASK) catalyzes kinase activity in the absence of Mg^{2+}. The DFG motif is replaced by GFG, and CASK lacks the asparagine that contributes to Mg^{2+} binding at the catalytic site. Structural analysis showed that a histidine partially performs the role of neutralizing the charge on the phosphates of ATP (PDB code 3C0H) (116).

The LKB1/STRAD/MO25 complex demonstrates an activating role for a pseudokinase. The protein kinase LKB1 phosphorylates and activates AMPK and thereby couples the cell's function to energy supply (reviewed in Reference 117). LKB1 activity is regulated by the pseudokinase STE20-related adaptor kinase (STRADα) and the scaffolding protein MO25α through an allosteric mechanism. STRADα has a serine in place of the catalytic aspartate, the DFG motif is GLR, and the VAIK motif lysine is arginine (**Table 3**). Despite these alterations, STRADα can still bind ATP using the arginine in its GLR motif and a histidine to take the place of Mg^{2+}. STRADα presents an active kinase conformation when bound to ATP and to MO25α, where the scaffolding protein interacts with and orients the C helix, rather like cyclin in Cdk2 (see above). The heterotrimeric LKB1/STRADα/MO25α complex revealed an unusual allosteric mechanism of LKB1 activation (PDB code 2WTK) (118). MO25α contacts STRADα and the LKB1 C helix and activation segment, while STRADα uses its activation segment and G helix to contact LKB1 (**Figure 9b**). The combined interactions result in an activated LKB1 in which the C helix is correctly oriented for ATP binding and the activation segment is ordered to accept the substrate. Neither STRADα nor MO25α alone are able to produce activation of LKB1.

A role for one kinase domain modulating another has also been demonstrated for the Janus tyrosine kinases (JAKs). JAKs together with STATs (signal transducers and activators of transcription) mediate cytokine receptor signaling to the nucleus. JAKs have a functional C-terminal kinase domain (JH1) and an N-terminal region consisting of a FERM domain and a pseudokinase domain (JH2) that binds to JH1 and in which the catalytic aspartate of the HRD motif is asparagine (119, 120). Despite this mutation, the JH2 domain is catalytically active and downregulates JAK2 activity by phosphorylating two negative regulatory sites, S523 and Y570 (121). A mutation in JH2 (V617F) depresses its activity, resulting in disregulated JAK2, and is present in patients with hematopoietic proliferation diseases. Structures of the JH1 kinase domain have been productive in the search for potent inhibitors for treatment of inflammatory diseases (PDB codes 3LXK and 3LXL) (122), but there is no structure as yet for the JH2-JH1 complex.

Several pseudokinases have a scaffolding role. Integrin-linked kinase (ILK) binds to the C-terminal tails of integrin and mediates integrin to actin regulation in cell-adhesion-dependent processes. ILK comprises an ankyrin repeat domain and a pseudokinase domain that has alterations in four critical regions: the glycine-rich loop, the catalytic aspartate, the next-but-one lysine, and the asparagine that chelates Mg_2 (**Table 3**). Structural studies of the kinase domain showed that it bound ATP, despite alterations in the glycine-rich loop, but had a degraded catalytic site that could not support catalysis (PDB codes 3KMU, 3KMW, and 3REP) (123, 124). The ILK kinase domain was demonstrated to bind to the α-parvin CH2 domain, a component of the signaling complex, using recognition sites on the G and EF helices, as observed for several kinase/substrate complexes (discussed above). When bound to parvin, the pseudokinase can still interact with integrin tails and recruit focal adhesion proteins.

VRK3 is a member of the vaccinia-related kinase family, which includes two active paralogs VRK1 and VRK2. VRK3 has alterations in three critical regions, rendering it inactive as a kinase (**Table 3**) (PDB code 2JII) (125). However, the overall fold represents that of an active kinase. Further analysis indicated an "inverted" pattern of conservation in which portions of the molecular surface, but not the catalytic site, showed conservation. This result suggests that VRK3 may form interactions with other proteins that may explain its evolutionary retention.

FUTURE DIRECTIONS

Protein kinases have to be activated not only in response to appropriate signals but also at the right time and in the right place. Many individual protein kinases are dispatched to intracellular compartments by the presence of short localization sequences that can be provided in *cis* or in *trans*. However, the mechanisms that control the spatial and temporal integrity of other protein kinases are complex.

For example, multiple kinases can be tethered together via scaffolding proteins to ensure specificity and enhance activity. In certain cases, emerging evidence suggests that they play more active roles in signal transduction than had previously been anticipated (see above and reviewed in Reference 126). The *Saccharomyces cerevisiae* mating pathway was one of the first for which scaffold proteins were shown to play a dynamic role in modulating the activity of a signaling pathway and provides an indication of the complexity to be unraveled. *S. cerevisiae* Ste5 organizes the MAPK cascade composed of Ste11 (a MAPK kinase kinase), Ste7 (a MAPK kinase), and Fus3 (a MAPK) within the mating-type pathway (reviewed in Reference 127). However, in response to an alternative signal (nitrogen starvation), the filamentous growth pathway is activated in which Ste11 phosphorylates Ste7, which then phosphorylates an alternative MAPK Kss1 in a cascade that does not require Ste5. In an elegant study, Good and colleagues (128) demonstrated that, in addition to having a role in tethering Ste7 to Fus3, Ste5 also acts as a substrate-specific Ste7 cocatalyst, enhancing its activity toward Fus3 but leaving its activity toward Kss1 unchanged.

Many protein kinases exist either as domains within large multidomain proteins or as components of large macromolecular complexes. Both cases provide challenges for structure determination by X-ray crystallography. A combination of electron microscopy and other biophysical methods can generate lower-resolution models that are amenable to interpretation with structures determined by X-ray crystallography and verification by biochemical and cellular assays. As examples, models for the allosteric regulation of protein kinase C βII (129) and for the mechanism of activation of calcium-calmodulin-dependent protein kinase II within a dodecameric holoenzyme assembly have been recently established [PDB codes 3KK8, 3KK9, and 3KL8 (130); PDB codes 2WEL and 2UX0 (131); PDB code 3SOA (132)].

Very few structures are available for intact protein kinase-substrate complexes, reflecting the transient nature of their interactions. In some cases, gradients of kinase activity, determined by their anchoring within specialized regions within the cell, dictate substrate selection. The conversion of cell-based models, such as those that describe the role of the Aurora B gradient for successful execution of mitosis (133–135), to structural molecular models is a significant challenge for the future.

SUMMARY POINTS

1. Eukaryotic protein kinases have a common catalytic fold in which conformational elements, including the C helix and the activation segment, are key to correctly position catalytic residues. Catalysis is promoted by the precise alignment of ATP and the substrate OH group, by participation of a catalytic aspartate residue as an acid/base, and by a basic residue to stabilize the transition state.

2. Substrate recognition at the catalytic site involves specific residues in the region near the site of phosphorylation. Association between kinase and substrate is often low affinity, and greater stability is achieved through docking sites that are remote from the catalytic site.

3. Protein kinases are pliable, and different kinases may adopt distinct conformations in the inactive state. This diversity arises because there are no catalytic requirements to constrain the fold. Although distinct, they do share a number of common structural themes.

4. Mechanisms for achieving kinase activation vary but include phosphorylation on the activation segment, phosphorylation at other sites, removal of inhibitory sequences or subunits, and/or association of other domains or subunits. Engagement of the C helix hydrophobic patch by internal regions or by external subunits has emerged as a common theme for activation of some kinases. Protection of the activation segment by regulatory domains is another theme.

5. Kinase dimerization can be regarded as a special case of regulation by additional subunits. Mechanisms of dimerization include relocation of inhibitory sequences in certain receptor kinases, mutual strand exchange of activation segments to allow active phosphorylation of noncanonical sequences, and, as discovered in EGFR, a mechanism in which one kinase acts as a receiver and is allosterically activated by the other kinase acting as an activator.

6. Scaffolding proteins may play a dynamic and catalytic role in kinase activation and substrate selection as seen in a B-RAF/KSR/MEK complex.

7. Pseudokinases are more than inactive bystanders. Some do exhibit activity, and others participate as scaffolds to activate other kinases.

FUTURE ISSUES

1. The role of scaffolding proteins in determining kinase activity and substrate selection.

2. Integration of structural results from electron microscopy, X-ray crystallography and nuclear magnetic resonance approaches to understand the dynamics of the formation of complexes containing protein kinases.

3. The conversion of cell-based models describing kinase activity to detailed structural mechanistic models.

DISCLOSURE STATEMENT

The authors are not aware of any affiliations, memberships, funding, or financial holdings that might be perceived as affecting the objectivity of this review.

ACKNOWLEDGMENTS

We acknowledge the major contributions from many scientists and regret that we have had to omit many citations, often citing just one work. This article was written while J.A.E. and M.E.M.N. were employees of the Department of Biochemistry, Oxford University.

LITERATURE CITED

1. Brognard J, Hunter T. 2011. Protein kinase signaling networks in cancer. *Curr. Opin. Genet. Dev.* 21:4–11
2. Cohen P. 2002. The origins of protein phosphorylation. *Nat. Cell Biol.* 4:E127–30
3. Dar AC, Shokat KM. 2011. The evolution of protein kinase inhibitors from antagonists to agonists of cellular signaling. *Annu. Rev. Biochem.* 80:769–95
4. Johnson LN. 2009. Protein kinase inhibitors: contributions from structure to clinical compounds. *Q. Rev. Biophys.* 42:1–40
5. Zhang J, Yang PL, Gray NS. 2009. Targeting cancer with small molecule kinase inhibitors. *Nat. Rev. Cancer* 9:28–39
6. Johnson LN, Lewis RJ. 2001. Structural basis for control by phosphorylation. *Chem. Rev.* 101:2209–42
7. Manning G, Whyte DB, Martinez R, Hunter T, Sudarsanam S. 2002. The protein kinase complement of the human genome. *Science* 298:1912–34
8. Martin DM, Miranda-Saavedra D, Barton GJ. 2009. Kinomer v. 1.0: a database of systematically classified eukaryotic protein kinases. *Nucleic Acids Res.* 37:D244–50
9. Eswaran J, Knapp S. 2010. Insights into protein kinase regulation and inhibition by large scale structural comparison. *Biochim. Biophys. Acta* 1804:429–32
10. Knighton DR, Zheng JH, Ten Eyck LF, Ashford VA, Xuong NH, et al. 1991. Crystal structure of the catalytic subunit of cyclic adenosine monophosphate-dependent protein kinase. *Science* 253:407–14
11. De Bondt HL, Rosenblatt J, Jancarik J, Jones HD, Morgan DO, Kim SH. 1993. Crystal structure of cyclin-dependent kinase 2. *Nature* 363:595–602
12. Huse M, Kuriyan J. 2002. The conformational plasticity of protein kinases. *Cell* 109:275–82
13. Nolen B, Taylor S, Ghosh G. 2004. Regulation of protein kinases; controlling activity through activation segment conformation. *Mol. Cell* 15:661–75
14. Eswaran J, Bernad A, Ligos JM, Guinea B, Debreczeni JE, et al. 2008. Structure of the human protein kinase MPSK1 reveals an atypical activation loop architecture. *Structure* 16:115–24
15. Eswaran J, Patnaik D, Filippakopoulos P, Wang F, Stein RL, et al. 2009. Structure and functional characterization of the atypical human kinase haspin. *Proc. Natl. Acad. Sci. USA* 106:20198–203
16. Kornev AP, Haste NM, Taylor SS, Eyck LF. 2006. Surface comparison of active and inactive protein kinases identifies a conserved activation mechanism. *Proc. Natl. Acad. Sci. USA* 103:17783–88
17. Kornev AP, Taylor SS, Ten Eyck LF. 2008. A helix scaffold for the assembly of active protein kinases. *Proc. Natl. Acad. Sci. USA* 105:14377–82
18. Ten Eyck LF, Taylor SS, Kornev AP. 2008. Conserved spatial patterns across the protein kinase family. *Biochim. Biophys. Acta* 1784:238–43
19. Noble ME, Endicott JA, Johnson LN. 2004. Protein kinase inhibitors: insights into drug design from structure. *Science* 303:1800–5
20. Jura N, Zhang X, Endres NF, Seeliger MA, Schindler T, Kuriyan J. 2011. Catalytic control in the EGF receptor and its connection to general kinase regulatory mechanisms. *Mol. Cell* 42:9–22
21. Iakoucheva LM, Radivojac P, Brown CJ, O'Connor TR, Sikes JG, et al. 2004. The importance of intrinsic disorder for protein phosphorylation. *Nucleic Acids Res.* 32:1037–49
22. Lowe ED, Noble ME, Skamnaki VT, Oikonomakos NG, Owen DJ, Johnson LN. 1997. The crystal structure of a phosphorylase kinase peptide substrate complex: kinase substrate recognition. *EMBO J.* 16:6646–58
23. Goldsmith EJ, Akella R, Min X, Zhou T, Humphreys JM. 2007. Substrate and docking interactions in serine/threonine protein kinases. *Chem. Rev.* 107:5065–81
24. Cheng KY, Noble ME, Skamnaki V, Brown NR, Lowe ED, et al. 2006. The role of the phospho-CDK2/cyclin A recruitment site in substrate recognition. *J. Biol. Chem.* 281:23167–79
25. Lowery DM, Lim D, Yaffe MB. 2005. Structure and function of Polo-like kinases. *Oncogene* 24:248–59
26. Komander D, Garg R, Wan PT, Ridley AJ, Barford D. 2008. Mechanism of multi-site phosphorylation from a ROCK-I:RhoE complex structure. *EMBO J.* 27:3175–85
27. Dar AC, Dever TE, Sicheri F. 2005. Higher-order substrate recognition of eIF2alpha by the RNA-dependent protein kinase PKR. *Cell* 122:887–900

28. Levinson NM, Seeliger MA, Cole PA, Kuriyan J. 2008. Structural basis for the recognition of c-Src by its inactivator Csk. *Cell* 134:124–34
29. Kim C, Xuong NH, Taylor SS. 2005. Crystal structure of a complex between the catalytic and regulatory (RIalpha) subunits of PKA. *Science* 307:690–96
30. Song H, Hanlon N, Brown NR, Noble ME, Johnson LN, Barford D. 2001. Phosphoprotein-protein interactions revealed by the crystal structure of kinase-associated phosphatase in complex with phospho-CDK2. *Mol. Cell* 7:615–26
31. Dey M, Velyvis A, Li JJ, Chiu E, Chiovitti D, et al. 2011. Requirement for kinase-induced conformational change in eukaryotic initiation factor 2alpha (eIF2alpha) restricts phosphorylation of Ser51. *Proc. Natl. Acad. Sci. USA* 108:4316–21
32. Skamnaki VT, Owen DJ, Noble ME, Lowe ED, Lowe G, et al. 1999. Catalytic mechanism of phosphorylase kinase probed by mutational studies. *Biochemistry* 38:14718–30
33. Stockbridge RB, Wolfenden R. 2009. The intrinsic reactivity of ATP and the catalytic proficiencies of kinases acting on glucose, N-acetylgalactosamine, and homoserine: a thermodynamic analysis. *J. Biol. Chem.* 284:22747–57
34. Adams JA. 2001. Kinetic and catalytic mechanisms of protein kinases. *Chem. Rev.* 101:2271–90
35. Grant BD, Adams JA. 1996. Pre-steady-state kinetic analysis of cAMP-dependent protein kinase using rapid quench flow techniques. *Biochemistry* 35:2022–29
36. Madhusudan, Akamine P, Xuong NH, Taylor SS. 2002. Crystal structure of a transition state mimic of the catalytic subunit of cAMP-dependent protein kinase. *Nat. Struct. Biol.* 9:273–77
37. Lassila JK, Zalatan JG, Herschlag D. 2011. Biological phosphoryl-transfer reactions: understanding mechanism and catalysis. *Annu. Rev. Biochem.* 80:669–702
38. Bao ZQ, Jacobsen DM, Young MA. 2011. Briefly bound to activate: transient binding of a second catalytic magnesium activates the structure and dynamics of CDK2 kinase for catalysis. *Structure* 19:675–90
39. Cook A, Lowe ED, Chrysina ED, Skamnaki VT, Oikonomakos NG, Johnson LN. 2002. Structural studies on phospho-CDK2/cyclin A bound to nitrate, a transition state analogue: implications for the protein kinase mechanism. *Biochemistry* 41:7301–11
40. Adams JA, Taylor SS. 1992. Energetic limits of phosphotransfer in the catalytic subunit of cAMP-dependent protein kinase as measured by viscosity experiments. *Biochemistry* 31:8516–22
41. Callaway K, Waas WF, Rainey MA, Ren P, Dalby KN. 2010. Phosphorylation of the transcription factor Ets-1 by ERK2: rapid dissociation of ADP and phospho-Ets-1. *Biochemistry* 49:3619–30
42. Cole PA, Grace MR, Phillips RS, Burn P, Walsh CT. 1995. The role of the catalytic base in the protein tyrosine kinase Csk. *J. Biol. Chem.* 270:22105–8
43. Jeffrey PD, Russo AA, Polyak K, Gibbs E, Hurwitz J, et al. 1995. Mechanism of CDK activation revealed by the structure of a cyclinA-CDK2 complex. *Nature* 376:313–20
44. Sicheri F, Moarefi I, Kuriyan J. 1997. Crystal structure of the Src family tyrosine kinase Hck. *Nature* 385:602–9
45. Xu W, Harrison SC, Eck MJ. 1997. Three-dimensional structure of the tyrosine kinase c-Src. *Nature* 385:595–602
46. Cowan-Jacob SW, Fendrich G, Manley PW, Jahnke W, Fabbro D, et al. 2005. The crystal structure of a c-Src complex in an active conformation suggests possible steps in c-Src activation. *Structure* 13:861–71
47. Zheng J, Knighton DR, Ten Eyck LF, Karlsson R, Xuong N, et al. 1993. Crystal structure of the catalytic subunit of cAMP-dependent protein kinase complexed with magnesium-ATP and peptide inhibitor. *Biochemistry* 32:2154–61
48. Zhang F, Strand A, Robbins D, Cobb MH, Goldsmith EJ. 1994. Atomic structure of the MAP kinase ERK2 at 2.3 Å resolution. *Nature* 367:704–11
49. Lei M, Robinson MA, Harrison SC. 2005. The active conformation of the PAK1 kinase domain. *Structure* 13:769–78
50. Knowles PP, Murray-Rust J, Kjaer S, Scott RP, Hanrahan S, et al. 2006. Structure and chemical inhibition of the RET tyrosine kinase domain. *J. Biol. Chem.* 281:33577–87
51. Mol CD, Lim KB, Sridhar V, Zou H, Chien EY, et al. 2003. Structure of a c-Kit product complex reveals the basis for kinase transactivation. *J. Biol. Chem.* 278:31461–64

52. Huse M, Muir TW, Xu L, Chen YG, Kuriyan J, Massague J. 2001. The TGFβ receptor activation process: an inhibitor- to substrate-binding switch. *Mol. Cell* 8:671–82
53. Filippakopoulos P, Kofler M, Hantschel O, Gish GD, Grebien F, et al. 2008. Structural coupling of SH2-kinase domains links Fes and Abl substrate recognition and kinase activation. *Cell* 134:793–803
54. Bayliss R, Sardon T, Vernos I, Conti E. 2003. Structural basis of Aurora-A activation by TPX2 at the mitotic spindle. *Mol. Cell* 12:851–62
55. Pearce LR, Komander D, Alessi DR. 2010. The nuts and bolts of AGC protein kinases. *Nat. Rev. Mol. Cell Biol.* 11:9–22
56. Yang J, Cron P, Good VM, Thompson V, Hemmings BA, Barford D. 2002. Crystal structure of an activated Akt/protein kinase B ternary complex with GSK3-peptide and AMP-PNP. *Nat. Struct. Biol.* 9:940–44
57. Grodsky N, Li Y, Bouzida D, Love R, Jensen J, et al. 2006. Structure of the catalytic domain of human protein kinase C beta II complexed with a bisindolylmaleimide inhibitor. *Biochemistry* 45:13970–81
58. Yamaguchi H, Kasa M, Amano M, Kaibuchi K, Hakoshima T. 2006. Molecular mechanism for the regulation of rho-kinase by dimerization and its inhibition by fasudil. *Structure* 14:589–600
59. Biondi RM, Komander D, Thomas CC, Lizcano JM, Deak M, et al. 2002. High resolution crystal structure of the human PDK1 catalytic domain defines the regulatory phosphopeptide docking site. *EMBO J.* 21:4219–28
60. Steinberg GR, Kemp BE. 2009. AMPK in health and disease. *Physiol. Rev.* 89:1025–78
61. Xiao B, Heath R, Saiu P, Leiper FC, Leone P, et al. 2007. Structural basis for AMP binding to mammalian AMP-activated protein kinase. *Nature* 449:496–500
62. Townley R, Shapiro L. 2007. Crystal structures of the adenylate sensor from fission yeast AMP-activated protein kinase. *Science* 315:1726–29
63. Amodeo GA, Rudolph MJ, Tong L. 2007. Crystal structure of the heterotrimer core of *Saccharomyces cerevisiae* AMPK homologue SNF1. *Nature* 449:492–95
64. Oakhill JS, Steel R, Chen ZP, Scott JW, Ling N, et al. 2011. AMPK is a direct adenylate charge-regulated protein kinase. *Science* 332:1433–35
65. Xiao B, Sanders MJ, Underwood E, Heath R, Mayer FV, et al. 2011. Structure of mammalian AMPK and its regulation by ADP. *Nature* 472:230–33
66. Carling D, Mayer F, Sanders M, Gamblin S. 2011. AMP-activated protein kinase: nature's energy sensor. *Nat. Chem. Biol.* 7:512–18
67. Zhu L, Chen L, Zhou XM, Zhang YY, Zhang YJ, et al. 2011. Structural insights into the architecture and allostery of full-length AMP-activated protein kinase. *Structure* 19:515–22
68. Chen L, Jiao ZH, Zheng LS, Zhang YY, Xie ST, et al. 2009. Structural insight into the autoinhibition mechanism of AMP-activated protein kinase. *Nature* 459:1146–49
69. Lemmon MA, Schlessinger J. 2010. Cell signaling by receptor tyrosine kinases. *Cell* 141:1117–34
70. Hubbard SR, Wei L, Ellis L, Hendrickson WA. 1994. Crystal structure of the tyrosine kinase domain of the human insulin receptor. *Nature* 372:746–54
71. Hubbard SR. 1997. Crystal structure of the activated insulin receptor tyrosine kinase in complex with peptide substrate and ATP analog. *EMBO J.* 16:5572–81
72. Mohammadi M, Schlessinger J, Hubbard SR. 1996. Structure of the FGF receptor tyrosine kinase domain reveals a novel autoinhibitory mechanism. *Cell* 86:577–87
73. Till JH, Becerra M, Watty A, Lu Y, Ma Y, et al. 2002. Crystal structure of the MuSK tyrosine kinase: insights into receptor autoregulation. *Structure* 10:1187–96
74. Bergamin E, Hallock PT, Burden SJ, Hubbard SR. 2010. The cytoplasmic adaptor protein Dok7 activates the receptor tyrosine kinase MuSK via dimerization. *Mol. Cell* 39:100–9
75. Wybenga-Groot LE, Baskin B, Ong SH, Tong J, Pawson T, Sicheri F. 2001. Structural basis for autoinhibition of the EphB2 receptor tyrosine kinase by the unphosphorylated juxtamembrane region. *Cell* 106:745–57
76. Griffith J, Black J, Faerman C, Swenson L, Wynn M, et al. 2004. The structural basis for autoinhibition of FLT3 by the juxtamembrane domain. *Mol. Cell* 13:169–78

77. Shewchuk LM, Hassell AM, Ellis B, Holmes WD, Davis R, et al. 2000. Structure of the Tie2 RTK domain: self-inhibition by the nucleotide binding loop, activation loop, and C-terminal tail. *Structure* 8:1105–13
78. Hubbard SR. 2004. Juxtamembrane autoinhibition in receptor tyrosine kinases. *Nat. Rev. Mol. Cell Biol.* 5:464–71
79. Oliver AW, Knapp S, Pearl LH. 2007. Activation segment exchange: a common mechanism of kinase autophosphorylation? *Trends Biochem. Sci.* 32:351–56
80. Oliver AW, Paul A, Boxall KJ, Barrie SE, Aherne GW, et al. 2006. Trans-activation of the DNA-damage signalling protein kinase Chk2 by T-loop exchange. *EMBO J.* 25:3179–90
81. Xu YJ, Davenport M, Kelly TJ. 2006. Two-stage mechanism for activation of the DNA replication checkpoint kinase Cds1 in fission yeast. *Genes Dev.* 20:990–1003
82. Pike AC, Rellos P, Niesen FH, Turnbull A, Oliver AW, et al. 2008. Activation segment dimerization: a mechanism for kinase autophosphorylation of non-consensus sites. *EMBO J.* 27:704–14
83. Bartek J, Lukas J. 2007. DNA damage checkpoints: from initiation to recovery or adaptation. *Curr. Opin. Cell Biol.* 19:238–45
84. Zhang K, Kaufman RJ. 2008. From endoplasmic-reticulum stress to the inflammatory response. *Nature* 454:455–62
85. Ali MM, Bagratuni T, Davenport EL, Nowak PR, Silva-Santisteban MC, et al. 2011. Structure of the Ire1 autophosphorylation complex and implications for the unfolded protein response. *EMBO J.* 30:894–905
86. Lee KP, Dey M, Neculai D, Cao C, Dever TE, Sicheri F. 2008. Structure of the dual enzyme Ire1 reveals the basis for catalysis and regulation in nonconventional RNA splicing. *Cell* 132:89–100
87. Korennykh AV, Egea PF, Korostelev AA, Finer-Moore J, Zhang C, et al. 2009. The unfolded protein response signals through high-order assembly of Ire1. *Nature* 457:687–93
88. Wiseman RL, Zhang Y, Lee KP, Harding HP, Haynes CM, et al. 2010. Flavonol activation defines an unanticipated ligand-binding site in the kinase-RNase domain of IRE1. *Mol. Cell* 38:291–304
89. Papa FR, Zhang C, Shokat K, Walter P. 2003. Bypassing a kinase activity with an ATP-competitive drug. *Science* 302:1533–37
90. Ferguson KM. 2008. Structure-based view of epidermal growth factor receptor regulation. *Annu. Rev. Biophys.* 37:353–73
91. Shi F, Telesco SE, Liu Y, Radhakrishnan R, Lemmon MA. 2010. ErbB3/HER3 intracellular domain is competent to bind ATP and catalyze autophosphorylation. *Proc. Natl. Acad. Sci. USA* 107:7692–97
92. Jura N, Shan Y, Cao X, Shaw DE, Kuriyan J. 2009. Structural analysis of the catalytically inactive kinase domain of the human EGF receptor 3. *Proc. Natl. Acad. Sci. USA* 106:21608–13
93. Red Brewer M, Choi SH, Alvarado D, Moravcevic K, Pozzi A, et al. 2009. The juxtamembrane region of the EGF receptor functions as an activation domain. *Mol. Cell* 34:641–51
94. Jura N, Endres NF, Engel K, Deindl S, Das R, et al. 2009. Mechanism for activation of the EGF receptor catalytic domain by the juxtamembrane segment. *Cell* 137:1293–307
95. Stamos J, Sliwkowski MX, Eigenbrot C. 2002. Structure of the epidermal growth factor receptor kinase domain alone and in complex with a 4-anilinoquinazoline inhibitor. *J. Biol. Chem.* 277:46265–72
96. Aertgeerts K, Skene R, Yano J, Sang BC, Zou H, et al. 2011. Structural analysis of the mechanism of inhibition and allosteric activation of the kinase domain of HER2 protein. *J. Biol. Chem.* 286:18756–65
97. Qiu C, Tarrant MK, Choi SH, Sathyamurthy A, Bose R, et al. 2008. Mechanism of activation and inhibition of the HER4/ErbB4 kinase. *Structure* 16:460–67
98. Zhang X, Gureasko J, Shen K, Cole PA, Kuriyan J. 2006. An allosteric mechanism for activation of the kinase domain of epidermal growth factor receptor. *Cell* 125:1137–49
99. Monsey J, Shen W, Schlesinger P, Bose R. 2010. Her4 and Her2/neu tyrosine kinase domains dimerize and activate in a reconstituted in vitro system. *J. Biol. Chem.* 285:7035–44
100. Claperon A, Therrien M. 2007. KSR and CNK: two scaffolds regulating RAS-mediated RAF activation. *Oncogene* 26:3143–58
101. Kolch W. 2005. Coordinating ERK/MAPK signalling through scaffolds and inhibitors. *Nat. Rev. Mol. Cell Biol.* 6:827–37
102. Udell CM, Rajakulendran T, Sicheri F, Therrien M. 2011. Mechanistic principles of RAF kinase signaling. *Cell Mol. Life Sci.* 68:553–65

103. Wan PT, Garnett MJ, Roe SM, Lee S, Niculescu-Duvaz D, et al. 2004. Mechanism of activation of the RAF-ERK signaling pathway by oncogenic mutations of B-RAF. *Cell* 116:855–67
104. Tsai J, Lee JT, Wang W, Zhang J, Cho H, et al. 2008. Discovery of a selective inhibitor of oncogenic B-Raf kinase with potent antimelanoma activity. *Proc. Natl. Acad. Sci. USA* 105:3041–46
105. Hansen JD, Grina J, Newhouse B, Welch M, Topalov G, et al. 2008. Potent and selective pyrazole-based inhibitors of B-Raf kinase. *Bioorg. Med. Chem. Lett.* 18:4692–95
106. King AJ, Patrick DR, Batorsky RS, Ho ML, Do HT, et al. 2006. Demonstration of a genetic therapeutic index for tumors expressing oncogenic BRAF by the kinase inhibitor SB-590885. *Cancer Res.* 66:11100–5
107. Rajakulendran T, Sahmi M, Lefrancois M, Sicheri F, Therrien M. 2009. A dimerization-dependent mechanism drives RAF catalytic activation. *Nature* 461:542–45
108. Brennan DF, Dar AC, Hertz NT, Chao WC, Burlingame AL, et al. 2011. A Raf-induced allosteric transition of KSR stimulates phosphorylation of MEK. *Nature* 472:366–69
109. Hu J, Yu H, Kornev AP, Zhao J, Filbert EL, et al. 2011. Mutation that blocks ATP binding creates a pseudokinase stabilizing the scaffolding function of kinase suppressor of Ras, CRAF and BRAF. *Proc. Natl. Acad. Sci. USA* 108:6067–72
110. Poulikakos PI, Zhang C, Bollag G, Shokat KM, Rosen N. 2010. RAF inhibitors transactivate RAF dimers and ERK signalling in cells with wild-type BRAF. *Nature* 464:427–30
111. McKay MM, Ritt DA, Morrison DK. 2011. RAF inhibitor-induced KSR1/B-RAF binding and its effects on ERK cascade signaling. *Curr. Biol.* 21:563–68
112. Boudeau J, Miranda-Saavedra D, Barton GJ, Alessi DR. 2006. Emerging roles of pseudokinases. *Trends Cell Biol.* 16:443–52
113. Zeqiraj E, van Aalten DM. 2010. Pseudokinases—remnants of evolution or key allosteric regulators? *Curr. Opin. Struct. Biol.* 20:772–81
114. Min X, Lee BH, Cobb MH, Goldsmith EJ. 2004. Crystal structure of the kinase domain of WNK1, a kinase that causes a hereditary form of hypertension. *Structure* 12:1303–11
115. Mayans O, van der Ven PF, Wilm M, Mues A, Young P, et al. 1998. Structural basis for activation of the titin kinase domain during myofibrillogenesis. *Nature* 395:863–69
116. Mukherjee K, Sharma M, Urlaub H, Bourenkov GP, Jahn R, et al. 2008. CASK functions as a Mg^{2+}-independent neurexin kinase. *Cell* 133:328–39
117. Alessi DR, Sakamoto K, Bayascas JR. 2006. LKB1-dependent signaling pathways. *Annu. Rev. Biochem.* 75:137–63
118. Zeqiraj E, Filippi BM, Deak M, Alessi DR, van Aalten DM. 2009. Structure of the LKB1-STRAD-MO25 complex reveals an allosteric mechanism of kinase activation. *Science* 326:1707–11
119. Saharinen P, Vihinen M, Silvennoinen O. 2003. Autoinhibition of Jak2 tyrosine kinase is dependent on specific regions in its pseudokinase domain. *Mol. Biol. Cell* 14:1448–59
120. Sanz A, Ungureanu D, Pekkala T, Ruijtenbeek R, Touw IP, et al. 2011. Analysis of Jak2 catalytic function by peptide microarrays: the role of the JH2 domain and V617F mutation. *PLoS ONE* 6:e18522
121. Ungureanu D, Wu J, Pekkala T, Niranjan Y, Young C, et al. 2011. The pseudokinase domain of JAK2 is a dual-specificity protein kinase that negatively regulates cytokine signaling. *Nat. Struct. Mol. Biol.* 18:971–76
122. Chrencik JE, Patny A, Leung IK, Korniski B, Emmons TL, et al. 2010. Structural and thermodynamic characterization of the TYK2 and JAK3 kinase domains in complex with CP-690550 and CMP-6. *J. Mol. Biol.* 400:413–33
123. Fukuda K, Gupta S, Chen K, Wu C, Qin J. 2009. The pseudoactive site of ILK is essential for its binding to alpha-Parvin and localization to focal adhesions. *Mol. Cell* 36:819–30
124. Fukuda K, Knight JD, Piszczek G, Kothary R, Qin J. 2011. Biochemical, proteomic, structural, and thermodynamic characterizations of integrin-linked kinase (ILK): cross-validation of the pseudokinase. *J. Biol. Chem.* 286:21886–95
125. Scheeff ED, Eswaran J, Bunkoczi G, Knapp S, Manning G. 2009. Structure of the pseudokinase VRK3 reveals a degraded catalytic site, a highly conserved kinase fold, and a putative regulatory binding site. *Structure* 17:128–38
126. Good MC, Zalatan JG, Lim WA. 2011. Scaffold proteins: hubs for controlling the flow of cellular information. *Science* 332:680–86

127. Schwartz MA, Madhani HD. 2004. Principles of MAP kinase signaling specificity in *Saccharomyces cerevisiae*. *Annu. Rev. Genet.* 38:725–48
128. Good M, Tang G, Singleton J, Remenyi A, Lim WA. 2009. The Ste5 scaffold directs mating signaling by catalytically unlocking the Fus3 MAP kinase for activation. *Cell* 136:1085–97
129. Leonard TA, Rozycki B, Saidi LF, Hummer G, Hurley JH. 2011. Crystal structure and allosteric activation of protein kinase C betaII. *Cell* 144:55–66
130. Chao LH, Pellicena P, Deindl S, Barclay LA, Schulman H, Kuriyan J. 2010. Intersubunit capture of regulatory segments is a component of cooperative CaMKII activation. *Nat. Struct. Mol. Biol.* 17:264–72
131. Rellos P, Pike AC, Niesen FH, Salah E, Lee WH, et al. 2010. Structure of the CaMKIIdelta/calmodulin complex reveals the molecular mechanism of CaMKII kinase activation. *PLoS Biol.* 8:e1000426
132. Chao LH, Stratton MM, Lee IH, Rosenberg OS, Levitz J, et al. 2011. A mechanism for tunable autoinhibition in the structure of a human Ca^{2+}/calmodulin-dependent kinase II holoenzyme. *Cell* 146:732–45
133. Liu D, Vader G, Vromans MJ, Lampson MA, Lens SM. 2009. Sensing chromosome bi-orientation by spatial separation of Aurora B kinase from kinetochore substrates. *Science* 323:1350–53
134. Welburn JP, Vleugel M, Liu D, Yates JR 3rd, Lampson MA, et al. 2010. Aurora B phosphorylates spatially distinct targets to differentially regulate the kinetochore-microtubule interface. *Mol. Cell* 38:383–92
135. Xu Z, Vagnarelli P, Ogawa H, Samejima K, Earnshaw WC. 2010. Gradient of increasing Aurora B kinase activity is required for cells to execute mitosis. *J. Biol. Chem.* 285:40163–70
136. Alexander J, Lim D, Joughin BA, Hegemann B, Hutchins JR, et al. 2011. Spatial exclusivity combined with positive and negative selection of phosphorylation motifs is the basis for context-dependent mitotic signaling. *Sci. Signal.* 4:ra42
137. Bossemeyer D, Engh RA, Kinzel V, Ponstingl H, Huber R. 1993. Phosphotransferase and substrate binding mechanism of the cAMP-dependent protein kinase catalytic subunit from porcine heart as deduced from the 2.0 Å structure of the complex with Mn^{2+} adenylyl imidodiphosphate and inhibitor peptide PKI(5–24). *EMBO J.* 12:849–59
138. Knighton DR, Zheng JH, Ten Eyck LF, Xuong NH, Taylor SS, Sowadski JM. 1991. Structure of a peptide inhibitor bound to the catalytic subunit of cyclic adenosine monophosphate-dependent protein kinase. *Science* 253:414–20
139. Brown NR, Noble ME, Endicott JA, Johnson LN. 1999. The structural basis for specificity of substrate and recruitment peptides for cyclin-dependent kinases. *Nat. Cell Biol.* 1:438–43

Measurements and Implications of the Membrane Dipole Potential

Liguo Wang

Department of Biological Structure, University of Washington, Seattle, Washington 98195; email: LW32@uw.edu

Keywords

membrane protein, fluorescence, electron microscopy, air electrode, ion transport

Abstract

There are three kinds of membrane potentials: the surface potentials, resulting from the accumulation of charges at the membrane surfaces; the transmembrane potential, determined by imbalance of charge in the aqueous solutions; and the dipole potential, a membrane-internal potential from the dipolar components of the phospholipids and interface water. The absolute value of the dipole potential has been very difficult to measure, although its value has been estimated to be in the range of 200–1,000 mV from ion translocation rates (determined by the planar lipid bilayer method), the surface potential of lipid monolayers (determined by the lipid monolayer method), molecular-dynamics calculations, and electron scattering using cryoelectron microscopy (cryo-EM). Spectroscopy methods have also been used to monitor the dipole potential changes on the basis of the observed fluorescence changes of voltage-sensitive probes. The dipole potential accounts for the much larger permeability of a bare phospholipid membrane to anions than cations and affects the conformation and function of membrane proteins.

Contents

INTRODUCTION 616
MEASUREMENTS OF THE
　　DIPOLE POTENTIAL 618
　Planar Lipid Bilayers 618
　Lipid Monolayer Method 620
　Cryoelectron Microscopy Method .. 620
VOLTAGE-SENSITIVE
　　FLUORESCENT PROBES 621
　Styryl Probes 622
　Hydroxychromone Probes.......... 623
MOLECULAR-DYNAMICS
　　STUDIES OF THE DIPOLE
　　POTENTIAL..................... 624
PHYSIOLOGICAL
　　IMPLICATIONS 627
　Interaction between Peptides and
　　Membranes..................... 627
　Function of Membrane Proteins 627
　Partition and Translocation of
　　Macromolecules 628

INTRODUCTION

The lipid bilayer membrane is a thin sheet, about 4 nm in thickness, consisting of two leaflets of lipid molecules. Lipid bilayers comprise the outer membrane of cells, as well as many intracellular structures. These membranes constitute the barriers that keep ions, proteins, and other molecules where they are needed, as their nonpolar hydrocarbon interior presents an enormous energy barrier to the passage of water-soluble (hydrophilic) molecules. To overcome this barrier, specialized membrane proteins—channels, pumps, and carriers—are embedded in lipid membranes to transport ions, metabolites, and other molecules across the membrane. The selective transport of ions results in a difference in the ion concentrations at the two sides of the lipid membrane, leading to an imbalance of charge and an electrical potential difference across the membrane, termed as the transmembrane potential $\Delta\Psi$ shown in **Figure 1**. The size of the transmembrane potential is on the order of 10–100 mV in biological systems (1) and is important in regulating the function of membrane proteins, most famously members of the voltage-gated ion channel family (Na^+, K^+, and Ca^{2+} channels) (2). In the absence of current flows, the transmembrane potential is constant within the bulk solution on either side of the membrane and can accurately be measured using electrodes (see reviews in References 3 and 4).

The surface potential Ψ_s is another kind of electrical potential, which spans between the membrane surface and the bulk water, associated with lipid membranes. It is generated by the charged head groups of phospholipids and the adsorbed ions at the interface. Almost all cell surfaces are intrinsically negatively charged because of the negatively charged lipids, primarily phosphatidylserine in mammalian membranes (4–6). Biological membranes typically have 10%–20% negatively charged lipids, and the surface potentials are some tens of millivolts in size (1, 7). The surface potential decays exponentially away from the membrane surface with a spatial spread given by the Debye length, about 1 nm in physiological solutions. It therefore controls ion distributions very close to the membrane surface; Ψ_s concentrates cations and depletes anions (8). It is difficult to measure Ψ_s directly, but it can be calculated using the

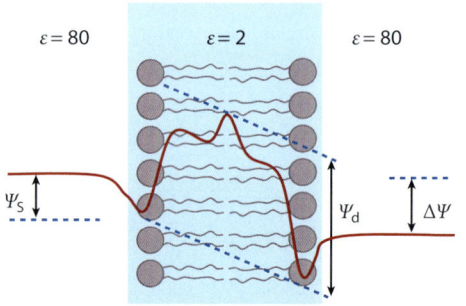

Figure 1

Three kinds of electrostatic potentials associated with membranes. $\Delta\Psi$ is the transmembrane potential, Ψ_s is the surface potential, and Ψ_d is the dipole potential. Values of the dielectric constant ε are given.

Lipid membrane: a thin sheet (∼4 nm in thickness) of two layers of lipid molecules, also called the lipid bilayer

Gouy-Chapman-Stern model (5, 9, 10). A remnant of the surface potential appears as the ζ potential (6, 8, 9, 11), the electrostatic potential at the hydrodynamic plane of shear, which is the boundary that separates the mobile fluid and the immobilized surface layer that moves together with the particle or cell. Although the exact position of this boundary cannot be determined either experimentally or computationally, the ζ potential can be determined by electrophoretic mobility (11) and is much smaller than Ψ_s. The surface potential has a demonstrated role in ion uptake and intoxication (5, 8).

The last, but not the least, kind of electrical potential is the dipole potential Ψ_d. The dipole potential originates from the alignment of dipolar residues of the lipids and water molecules. The membrane dipole potential was first discovered by Liberman & Topaly (12) when they were studying the carrier mechanism of ion transport. They found that fat-soluble (i.e., lipophilic or hydrophobic) ions could cross lipid membranes, and they were able to measure the electrical conductivity changes after the addition of fat-soluble ions to the aqueous phase. In comparing fat-soluble ions of similar size, they were surprised to find that the membrane conductivity was 10^5 times higher with tetraphenylborate (TPB$^-$) than with triphenylmethylphosphonium (TMP$^+$) at the same concentration (**Figure 2**). Assuming the diffusion coefficients of the fat-soluble ions in the membrane were similar, they attributed this conductance difference to the varying partition coefficients between the lipid phase and the aqueous medium. Together with the negative surface charge of membranes, they hypothesized that the membrane had an internal positive charge, which implied a positive potential within the membrane.

The term dipole potential was first used by Hladky & Haydon (13) when they were studying the specific conductance of artificial planar membranes of phosphatidylcholine (PC), glyceryl monooleate, and cholesterol. They attributed the conductance difference of PC and glyceryl monooleate membranes to the difference in their surface potentials. Because neither PC nor glyceryl monooleate carries a net charge at pH 7 and the surface potentials they measured were not affected by the potassium chloride concentration, they concluded that the potentials must arise from layers of oriented molecular dipoles in the membrane interior.

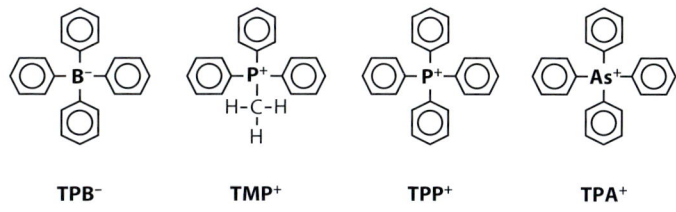

Figure 2
Structures of hydrophobic ions: tetraphenylborate (TPB$^-$), triphenylmethylphosphonium (TMP$^+$), tetraphenylphosphonium (TPP$^+$), and tetraphenylarsonium (TPA$^+$).

The first model accounting for the interactions between the hydrophobic ions and the membranes was proposed by Ketterer et al. (14). In this model, the potential energy consisted of two terms. One accounts for the electrostatic interaction of the ion with dipoles and charges in the membrane, whereas the other contains all other interactions that would be experienced by a hypothetical neutral ion. The total potential energy shows two deep minima near the membrane surfaces and a potential barrier in the interior, similar to the energy curve shown in **Figure 3**. However, this model makes no distinction between differently charged hydrophobic ions. To account for that, Flewelling & Hubbell (15) developed a total potential energy profile to include the dipole potential contribution in addition to the Born self-charging energy, image-charge contributions, and a neutral energy term. Many experiments confirmed this potential energy profile (15–18). To include both the surface potential and the dipole potential, Benz & Cros (19) proposed a total electrostatic potential profile similar to that shown in **Figure 1**. The magnitude of the dipole potential depends on the structure of the lipid, in particular the degree of unsaturation and nature of the linkage between head group and hydrocarbon chain (ester or ether) (20–22). The dipole

Dipole potential: a positive electrostatic potential in the lipid membrane originating from the alignment of dipolar residues of the lipids and water molecules

TPB$^-$: tetraphenylborate

PC: phosphatidylcholine

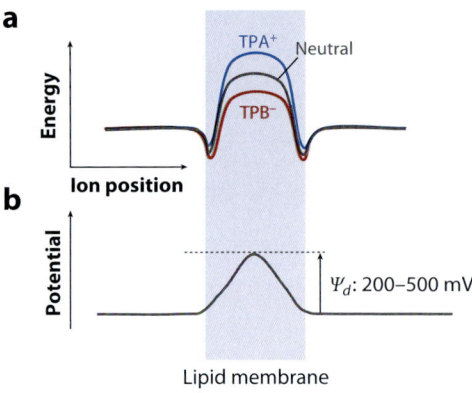

Figure 3
Schematic free-energy profiles across a lipid membrane for the hydrophobic ions tetraphenylborate (TPB$^-$) and tetraphenylarsonium (TPA$^+$). (*a*) Free energy. (*b*) Electrical potential.

potential can also be strongly influenced through intercalation of dipolar molecules. It can be increased by adding 6-ketocholestanol to the membrane (7, 23) or reduced by adding phloretin (24–26). Although there is an inconsistency between the values measured by different methods, the dipole potential appears to be on the order of 300 mV (7).

MEASUREMENTS OF THE DIPOLE POTENTIAL

Two commonly used methods to estimate the value of the dipole potential involve transport measurements in planar lipid bilayers and electrostatic measurements on lipid monolayers. A recently developed method is to use quantitative electron microscopy to measure the potential in a flash-frozen suspension of lipid-membrane vesicles (liposomes). Predictions of the dipole potential also come from molecular-dynamics (MD) simulations of lipid bilayers. Experimental estimates of the dipole potential from various methods range from +200 mV to +500 mV, whereas estimates from MD simulations range as high as +1,000 mV (**Tables 1** and **2**). As the measurements on planar lipid bilayers and lipid monolayers have

been reviewed extensively (1, 4, 27–29), only a brief introduction is given here.

Planar Lipid Bilayers

The planar lipid bilayer method is an indirect electrical measurement, which is based on the kinetics of the hydrophobic ion transport. It measures the specific conductance of a lipid bilayer in the presence of hydrophobic ions (19, 20, 30–33). The planar lipid bilayer can be formed by painting a solution of phospholipid in an organic solvent, such as decane or hexadecane, across a hole in a Teflon® wall between two aqueous chambers. The size of the hole (about 1 mm or smaller) has no influence on the probe concentration in the membrane or on the relaxation time constants (20). After forming the membrane, a small concentration of a salt of a hydrophobic ion, such as TPA$^+$ or TPB$^-$, is added to both chambers. Then a voltage is applied via electrodes inserted in each chamber, and the current is recorded. The conductance of the membrane can be determined by dividing the stationary current by the applied voltage. Based on the Arrhenius equation, the ratio of the conductance g is dependent on the difference in energy barriers experienced by hydrophobic cations and anions when they cross the membrane (**Figure 3**):

$$\ln \frac{g_-}{g_+} = \frac{\Delta G}{RT} = \frac{2F\Psi_d}{RT}. \qquad 1.$$

Here g_- and g_+ are the conductances when hydrophobic anions or cations are present, R is the gas constant, T is the temperature, ΔG is the free energy difference between cations and anions when crossing the lipid membrane, and F is the Faraday constant. Owing to the structural similarity between TPB$^-$ and TPA$^+$, the free-energy terms involved in the transfer of hydrophobic ions from water to the membrane are assumed to be negligible. In this way, only the electrostatic energy difference (15, 28) is considered, and the dipole potential is determined as follows:

$$\Psi_d = \frac{RT}{2F} \ln \frac{g_-}{g_+}. \qquad 2.$$

Table 1 Dipole potential estimates (in millivolts) from experiments

Method used[a]	Phosphatidylcholine (ester)	Phosphatidylcholine (ether)	Lipid used (References)
Planar lipid bilayers	227	109	DPPC (22)
Planar lipid bilayers	240	114	DPPC (135)
Planar lipid bilayers	228	—	DPhPC (135)
Planar lipid bilayers + ΔG_{hydr}[b]	346	228	DPPC (36)
Lipid monolayers	450	360	DMPC (91)
Lipid monolayers	475	—	Egg PC (39)
Lipid monolayers	415	—	Egg PC (40)
Cryo-EM	510	260	DPhPC (44)

[a]Abbreviations: Cryo-EM, cryoelectron microscopy; DMPC, dimyristoylphosphatidylcholine; DPhPC, diphytanoyl phosphatidylcholine; DPPC, dipalmitoyl phosphatidylcholine; PC, phosphatidylcholine; —, no data.
[b]ΔG_{hydr} is the correction made to account for the hydration energy difference between tetraphenylarsonium (TPA$^+$) and tetraphenylborate (TPB$^-$). Phospholipids conventionally have an ester linkage from the head group to the hydrocarbon chain but can exist with an ether linkage instead.

As shown by Pickar & Benz (30), TPB$^-$ has a permeability of about 1.1×10^7 times larger than the identically sized TPA$^+$ in dioleoylphosphatidylcholine membranes. The estimated dipole potential is about 200 mV. In subsequent work, a relaxation of the assumption of identical transfer energies of the ions (34, 35), in particular taking into account the smaller hydration energy of TPB$^-$ (36), yields estimates for the dipole potential that are about 100 mV larger (**Table 1**, bilayer + ΔG_{hydr} values).

ΔG_{hydr}: the correction made to account for the hydration energy difference between TPA$^+$ and TPB$^-$

Table 2 Dipole potentials estimated from molecular-dynamics simulations

Lipid simulated[a]	Ψ_d (V)	Number of lipids	Force field	Long-range force	Water model	Temperature (Kelvin)	Type[b]	Reference
DPPC	0.60	64	Amber UA	NA	TIP3P	323	II	107
DPPC	0.75	128	Gromos UA	PME	NA	323	II	106
DPPC	0.56	256	Gromos UA	NA	SPC	323	II	108
DMPC	0.77	128	Gromos UA	PME	SPC	310	II	109
DMPC	0.90	512	CHARMM[c]	PME	TIP3P	298	I	104
DPhPC	1.00	72	CHARMM	Ewald	TIP3P	298	I	51
Ether-DPhPC	0.57	72	CHARMM	Ewald	TIP3P	298	I	51
DMPC	0.90	72	CHARMM (Cheq)	PME	TIP4PFQ	323	I	136
DPPC	0.35	72	CHARMM (Pol)	PME	NA	320	NA	116
DPPC	0.30	128	MARTINI, CG	PME	BMW	310	III	119

[a]Abbreviations: BMW, big multipole water; DMPC, dimyristoylphosphatidylcholine; DPhPC, diphytanoyl phosphatidylcholine; DPPC, dipalmitoyl phosphatidylcholine; ether-DPhPC: diphytanyl phosphatidylcholine; CG, coarse grained; NA, not applicable; PME, particle mesh Ewald method; UA, united atom.
[b]Type I represents the profile having a maximum peaked at the membrane center. Type II represents the profile having two maxima peaked at the lipid head group regions. Type III represents the profile having a plateau in the membrane interior.
[c]The CHARMM force field variants are the Cheq (charge equilibration) method and Pol (polarizable, Drude oscillator) model.

cryo-EM: cryoelectron microscopy

Lipid Monolayer Method

In an alternative method, lipids are spread on an air-water interface. A monolayer is spontaneously formed, with the hydrophobic chains pointing into the air. An "air electrode" is placed above the lipid monolayer, and a second electrode is placed in the aqueous phase below the lipid monolayer. The potential is measured, and the difference between this potential and that of the bare air-water interface is taken to be a measure of the dipole potential; values of ∼450 mV are obtained (27, 37–40). The air electrode is implemented in two different ways. In the Kelvin method, the air gap between the air electrode and the air-water or air-lipid interface is nonconductive. When the air electrode is vibrated (150–200 Hz), a current is generated. By applying a voltage between the two electrodes, the alternating current can be nulled, and at this time the applied voltage is the measured "air potential." In the ionizing electrode method, the air above the surface is ionized through the use of a radioactive ionizing source, typically the α-particle emitters ^{241}Am or ^{210}Po. The mobile ions in the air allow the potential to be measured; this method is much simpler to implement compared to the Kelvin method. Values obtained with both methods are reproducible within 10 mV.

In the lipid monolayer method, the measured potential is actually the change in surface potential between the air-water and air-lipid-water interfaces. Unfortunately, it is not known what the absolute size of the air-water potential difference is, whether the subtraction of this potential is justified, and whether a monolayer is an adequate model for half a bilayer. As the measured potential is a surface potential, the value is dependent on the surface pressure. The dipole potential in a palmitoyloleoyl phosphatidylcholine monolayer is decreased by about 150 mV when the area per lipid molecule is increased from 50 to 100 Å^2 (27). Also, the choice of spreading solvent for the lipid monolayer affects the measurement. The surface potential of dipalmitoyl lecithin varied up to 100 mV when different spreading solvents were used (41).

Cryoelectron Microscopy Method

The large size of the hydrophobic ions, 4.2 Å and 4.0 Å for TPB$^-$ and tetraphenylphosphonium (42), respectively, make the planar bilayer method susceptible to nontrivial electrostatic and nonelectrostatic effects (43). An ideal and direct measurement of dipole potentials in lipid membranes would use point charge probes instead of large hydrophobic ions or voltage-sensitive dye molecules. Wang et al. (44) developed a cryoelectron microscopy (cryo-EM) method to use electrons as probes for the measurement of the dipole potential in rapidly frozen lipid membranes.

In cryo-EM, the primary mechanism for image contrast is the phase shift in the electron wave function (elastic scattering) as it passes through the specimen. The total phase shift is proportional to the projected potential, that is, the integrated electrostatic potential along the path of the electron. The intensity of the recorded image is expected to vary in proportion to the phase shift when the weak phase object approximation is employed for defocused imaging (45). Thus, the recorded image intensity is a reflection of the projected potential of the specimen. For isolated, neutral atoms, the electrostatic potential of the specimen is the linear superposition of the atomic potential, that is, the coulomb potential experienced by a probe passing through each atom. When bonds are formed between atoms, the outer valence electrons rearrange themselves, and additional electrostatic potentials arise from the resulting charge displacements. Zhong et al. (46) have shown that atomic-resolution cryo-EM data are better described when molecular bonding effects are taken into consideration. However, for the low-resolution data considered in the cryo-EM method, the projected potential is assumed to be the superposition of the atomic potential and any additional electrostatic potentials.

In the cryo-EM method, highly spherical, osmotically swollen liposomes with diameters of 50–100 nm are trapped in a layer of vitreous ice of 100–150 nm. From images of these spherical shells of lipid bilayers, application of the Fourier slice theorem allows the profile of a bilayer to be obtained (47–49). Images of liposomes are corrected for the microscope contrast-transfer function (50), rotationally averaged, and an inverse Abel transform yields the phase-shift profile $\gamma(z)$ as a function of the distance z normal to the membrane plane. The neutral-atom phase-shift γ_n (units of mrad/Å) is computed according to

$$\gamma_n(z) = \sigma_e \sum_i V_i \rho_i(z). \qquad 3.$$

Here, V_i is the spatially integrated atomic potential for the neutral atom i, obtained from published parameters (45), σ_e describes the first-order dependence of electron phase on projected potential (45), $\sum_i$ is the summation over each atom i, and ρ_i is the atom density that can be obtained from MD simulations (51). The magnitude and the profile of the dipole potential along the bilayer normal are obtained by subtracting the neutral-atom phase-shift γ_n from the cryo-EM electron phase-shift profile $\gamma(z)$. The peak dipole potential was estimated to be 510 mV and 260 mV for diphytanoyl phosphatidylcholine (DPhPC) and diphytanyl phosphatidylcholine, respectively. Compared with values obtained from other methods (**Tables 1 and 2**), this value is smaller than that in MD simulations but larger than that in both planer lipid bilayer and monolayer measurements.

The cryo-EM method has substantial advantages: It uses electrons as probes, and it can provide information about the spatial profile of the electrostatic potential. The method has drawbacks in that it has low sensitivity (the random error in a measurement from one membrane vesicle is on the order of 100 mV), and of course, its use rests on a number of assumptions. The first assumption is that the rapidly frozen specimen, imaged at a liquid-nitrogen temperature, preserves the electrostatic features of the native structure. It is to be expected that the lipid bilayer structure remains intact during the rapid freezing, which occurs at $\sim 10^6$ K/s (52) and vitrifies water. The second assumption is that a traditional first-order model for electron elastic scattering, including an added amplitude-contrast term, is the proper model of the cryo-EM image formation. Finally, this method relies heavily on structural data from MD simulations to uncouple the neutral-atom and dipole potential contributions to the image intensity.

VOLTAGE-SENSITIVE FLUORESCENT PROBES

Voltage-sensitive dyes have been used as probes for membrane potentials for more than 30 years (28, 53–68). Three general mechanisms have been exploited to make membrane-potential probes. Electrochromic probes have direct responses of the ground-state and excited-state energies to the local electric field, resulting in the shifts of absorption and emission spectra (electrochromism). The slow potential-sensitive probes, by contrast, make use of potential-dependent redistribution across the membrane or sorption-desorption to provide a sensitive measure of membrane potential, and a sensitivity of 100% fluorescence change per 100 mV is seen (60). These dyes (e.g., carbocyanines, rhodamines, and oxonols) are used mainly to study the transmembrane potential. Because they reside more than 16 Å away from the membrane center (69), where the dipole potential is almost completely offset by water dipoles (51), these probes are not used to study the dipole potential inside the membrane. Finally, pH-sensitive probes located close to membrane surface report potential-dependent responses to proton concentration. These dyes are exposed to the aqueous medium and can be used to estimate the surface potential Ψ_s, but not the dipole potential Ψ_d inside the membrane.

The observations of dipole potentials have made use of electrochromic dyes, including the well-studied styryl dyes and the more recently introduced hydroxychromone (HC) dyes. None of these probes can provide an absolute measure of the dipole potential, but they can report spatial and temporal changes in the potential. In the electrochromic dyes, light absorption causes electron redistribution within the chromophore to produce the excited state.

DPhPC: diphytanoyl phosphatidylcholine

Electrochromism (aka the Stark effect): the shifts of electronic (absorption and fluorescence) spectra under the influence of electric fields

The energy of the electronic transition is affected by local electric field as follows:

$$h\Delta v = -\Delta\vec{\mu} \cdot \vec{E} - \Delta\alpha|E|^2, \quad 4.$$

where h is Planck's constant; Δv is spectral shift induced by electric field; $\vec{E}$ is electric field vector, $|E|$ is the magnitude of electric field; $\Delta\vec{\mu}$, and $\Delta\alpha$ [on the order of 10^{-22} cm^3 (70)] are changes in dipole moment and polarizability upon excitation, respectively. The second term in this equation is not significant when the electric field is smaller than 10^9 volts/meter (V/m) (64, 70). Usually, the charge is located at one end of the chromophore in the ground state and at the other end in the excited state. To maximize the first term in Equation 4, a polar head group and nonpolar side chains are appended to opposite ends of the chromophore to ensure the direction of the electron displacement, and the dipole changes are perpendicular to the membrane.

Styryl Probes

The styryl probes are a subgroup of the cyanine dyes, including aminostyrylpyridinium (64, 66, 67), RH421 (65, 71), di-4-ANEPPS (61, 72), and di-8-ANEPPS (**Figure 4**) (59). The electron-donor and electron-acceptor substituents are at the opposite ends of their rod-shaped aromatic conjugated moieties (**Figure 4**). In early studies, the fluorescence signal change of the dyes was small (66, 67). With the successful synthesis of new probes (73) and the introduction of a rational design of new probes (64, 74), the fluorescence fractional change of the optical signal increased to 14% per 100 mV for RH421 (65) and up to 25% per 100 mV for ANNINE (75). As the shifts in emission spectra are distorted by molecular relaxations, shifts in excitation are commonly used for measurement. The fractional change in fluorescence is affected by the dye or cell concentration, bleaching effects, and uneven staining in cells. To tackle these, a dual-wavelength ratiometric approach has been developed (72, 76, 77). Gross et al. (59) used this dual-wavelength ratiometric method to study the effect of agents on the dipole potential. The ratio of di-8-ANEPPS fluorescence intensity at 620 nm excited at 440 nm to that excited at 530 nm increased 2.5 times as the mole fraction

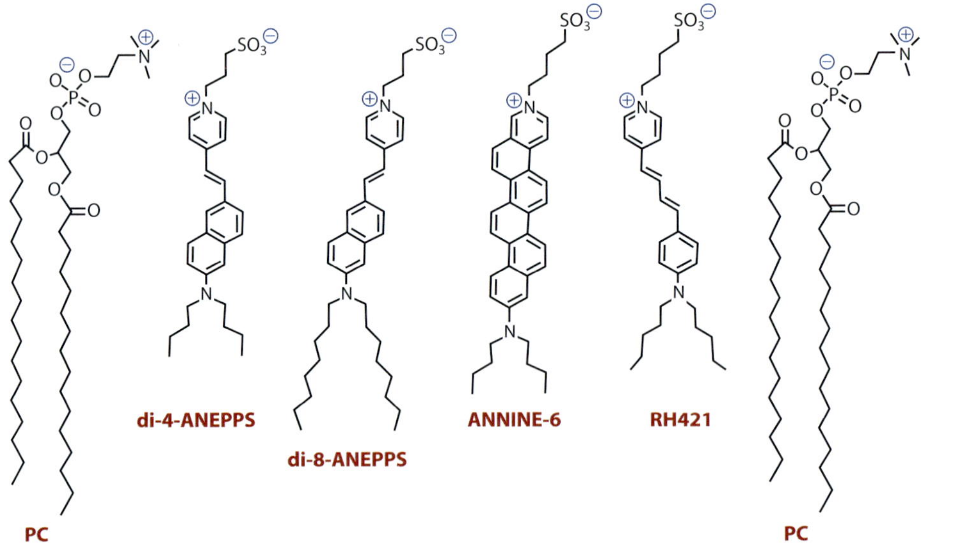

Figure 4
Structures of styryl probes. The probe location is illustrated by showing two phosphatidylcholine (PC) molecules constituting the outer membrane leaflet into which these dyes spontaneously insert.

of 6-ketocholestanol was increased from 0% to 30%, and it decreased 3.0 times as the mole fraction of phloretin increased from 0% to 15% (59). However, the choice of the emission wavelength is critical to exclude viscosity effects. The value of the fluorescence ratio for both RH421 and di-8-ANEPPS is only insensitive to membrane fluidity above the phase transition temperature of the lipid and only if the fluorescence is detected at the red edge of the emission spectrum (78). To avoid a distortion of potentials by the probes themselves, it is important to make sure there is excess lipid around the probes, for example, a 200-fold excess of lipid molecules to RH421 probes. The nature of the monitored potential is usually interpreted on the basis of the position of the chromophore (**Figure 4**). The transmembrane potential, dipole potential, or surface potential are sensed when the chromophore is deep in the membrane, close to the phosphate groups and the glycerol backbone linkage, or near the water-lipid interface, respectively. The sensitivity is high for dipole potential, moderate for transmembrane potential, and low for surface potential (59, 76). The sensitivity also depends on the length of the chromophore. The newly developed long rigid analogs of styryl dyes (ANNINE probes) span from the surface to the middle of the lipid acyl chain, yielding a sensitivity to transmembrane potential up to 25% per 100 mV (75).

The styryl probes can be incorporated into membranes of different types, and the response is on the nanosecond scale as there is no relocation of molecules involved. They can be used in broad areas, including two-photon spectroscopy and microscopy (57). However, there exist some limitations. First, these probes sense the integrated local electric field at the site of its location; thus, the fluorescence signal depends on both the length and position of the chromophore (**Figure 4**). Therefore, the relationship of the signals to the overall potentials is not easily defined. Consequently, the use of these probes is limited to studies of changes in potentials because of various factors, such as phloretin, 6-KC, and cholesterol on membrane potentials. Second, in addition to potentials, these dyes also respond to environmental properties, such as the fluidity of the membrane. For these reasons, experiments need to be carefully designed to exclude undesired factors. For example, the emission ratio does not correlate with the dipole potential of vesicles made from different lipids (79); the excitation spectra are affected by the solvation of highly polar ground states; and both RH421 and di-8-ANEPPS show significant temperature-dependent shifts of their excitation spectra (78).

6-KC: 6-ketocholestanol

Hydroxychromone Probes

The recently developed HC probes are based on the excited-state intermolecular proton transfer (ESIPT) mechanism, which couples the electrochromic spectral shift to a photochemical reaction (56, 80). The effect can be illustrated by studying the 3-hydroxyflavone (3HF) derivatives of HC probes (figure 3 in Reference 56). There exist six forms of 3HF derivatives in lipid membranes: The initially excited normal (N*) form is in equilibrium with the ESIPT product–excited tautomeric (T*) form; the excited H-bond (H-N*) form; and the three corresponding ground states (H-N, N, and T). The N* state exhibits a much larger dipole moment than the T* state, 10.3 D versus 2.7 D for 3HF probe type 3 (81–83). The equilibrium between the N* and T* forms is determined by the relative energies of these states (e.g., electrostatic potentials in membranes), and the timescale to establish the ESIPT equilibrium is below 300 ps (84, 85). A change in emission ratio (I_{N^*}/I_{T^*}) reports changes in environmental polarity and electrostatics. The H-N* form exhibits a single emission band positioned at a wavelength between those of the N* and T* forms at the blue and red edges of the spectrum. The relative contribution of the H-N* form with respect to the N* and T* forms in the emission spectrum provides a measurement of the so-called hydration parameter. This parameter and the emission ratio (I_{N^*}/I_{T^*}) are intrinsically independent and describe

Force field: the combination of mathematical formulas and associated parameters that are used to describe the interactions between atoms in terms of their relative positions

different membrane properties. Thus, the HC probes are also called multiparametric probes.

In contrast to styryl probes, where part of the chromophore includes the charged groups interacting with lipid heads, the positively charged anchor in 3HF derivatives is not part of the chromophore. The connecting spacer can be of variable length; thus, it is possible to move the chromophore in the lipid membrane freely (figure 1 in Reference 86) and even to design probes with opposite orientations with respect to the normal membrane [e.g., F4N1 versus BPPZ (54)]. The probe di-SFA locates near the membrane center, which can sense transmembrane potential (87, 88); probes F8N1S and BPPZ are located close to the glycerol backbone linkage and are suitable to sense dipole potential; and the probe F2N12S stays close to the water-lipid membrane and responds to the changes of surface potential and hydration (80, 89).

The HC probes have been used to study the dipole potential in lipid membranes (86–87). When 6-KC is added to a membrane to increase the dipole potential (7, 23, 90), the emission ratiometric response of BPPZ has a similar magnitude but opposite sign to that of F4N1, as expected because they have opposite orientations. However, the response of BPPZ to the addition of phloretin, which decreases the dipole potential (24–26, 54, 90–92), is significantly smaller than that of F4N1; hence, contributions other than dipole potential may also be involved. This difference has been attributed to the effect of phloretin and 6-KC on the lipid-membrane hydration (90, 93). After the deconvolution of the emission profile into H-N*, N*, and T* bands, the correlation between the emission ratio of F4N1 and F8N1S and the excitation ratio of di-8-ANEPPS with the addition of 6-KC or phloretin improved (87). However, this new method has not yet been used to interpret the discrepancy in the responses of BPPZ and F4N1 to the addition of phloretin.

The key advantage of these HC probes lies in their ratiometric response in emission spectra. This makes measurements much simpler than those with styryl probes, which have excitation-spectral shifts. The HC probes are thus ideally adapted for fluorescence microscopy with the use of single-excitation and two-color fluorescence detection. These probes show selective and strong two-band ratiometric responses in fluorescence spectra to variations in dipole potentials of cell plasma membranes of living cells (86). Another important advantage of HC probes is the capability to monitor both the electrostatic potential and the hydration. By deconvoluting the emission spectra, the effects of the hydration and electrostatic potential can be decoupled (87).

MOLECULAR-DYNAMICS STUDIES OF THE DIPOLE POTENTIAL

MD simulations provide a unique tool to study lipid membranes at the atomic level (57, 94–98). In MD simulations, a force field, which describes the interactions between atoms in terms of their positions, is applied through Newton's equation of motion to update the atom positions. This procedure is iterated with time steps on the order of 1 fs to simulate the temporal evolution of the system. Modern MD simulations are quite accurate in reproducing the structural and thermodynamic properties of proteins, nucleic acids, lipids, and their complexes (57, 94–100).

Any force field contains two kinds of interactions. The bonded term accounts for the stretching, bending, and rotation of bonds, whereas the nonbonded term captures electrostatics (Coulomb interactions) and the dispersion and Pauli exclusion (van der Waals') forces. To include electrostatics in force fields, each atom is assigned a partial point charge, which may vary depending on the force fields. The determination of partial changes depends strongly on the parameterization rules of a particular force field. The widely used biomolecular force fields are AMBER (101), CHARMM (102), and GROMOS (103). Although CHARMM has been the dominant all-atom force field in use, force fields for united-atom (UA) models are included in AMBER and GROMOS.

The electrostatic potentials in lipid membranes can be easily calculated in MD simulations. First, the charge density, ρ, is averaged in a series of narrow slabs parallel to the membrane plane. The electrostatic potential $\Psi(z)$ is related to the charge density $\rho(z)$ along the membrane normal z via the Poisson equation,

$$\frac{d^2\Psi(z)}{dz^2} = -\frac{\rho(z)}{\varepsilon_0}. \qquad 5.$$

The potential can be calculated by integrating the charge density twice,

$$\Psi(z) = -\frac{1}{\varepsilon_0}\int_{z_0}^{z}\int_{z_0}^{z'}\rho(z'')dz''dz', \qquad 6.$$

where ε_0 is the electrostatic permittivity of vacuum, and z_0 is the position where the potential is assumed to be zero (e.g., the center of the water layer, or the center of the lipid bilayer).

The potential profile across a lipid membrane (**Figure 5**) has been obtained in all-atom MD simulations for DMPC (104) and DPhPC (51) bilayers by two independent research groups. The profile indicates that the negative contribution of lipid charges in the phosphate region (z values between 17 and 20 Å) is almost completely offset by the positive contribution owing to the dipole moment of water molecules. In contrast, the field from deeper dipoles resulting from the glycerol backbone linkage (12–17 Å) is not compensated by waters. A further remarkable feature of the profile is the peak of potential at the membrane center, about one-fifth of the total magnitude of the dipole potential and around 10 Å in width. This feature was seen in both DMPC and DPhPC membranes (51, 104, 105) and comes mainly from the partial charges on the terminal methyl group at the lipid tail (**Figure 6**).

The dipole potential profile across lipid membranes has also been computed (106–109) using UA models in which hydrogen atoms are not simulated separately. In these simulations, the dipole potential is constant in the interior of the membrane, increases in the head group region, and then drops to the potential of bulk water. The total width of the plateau of positive potential is about 15 Å greater than that observed in all-atom simulations. The reason for the wider dipole potential profile and offset maximum in UA models may lie in the fact that electrostatic details of the lipid membrane are lost when atoms are grouped into pseudoatoms;

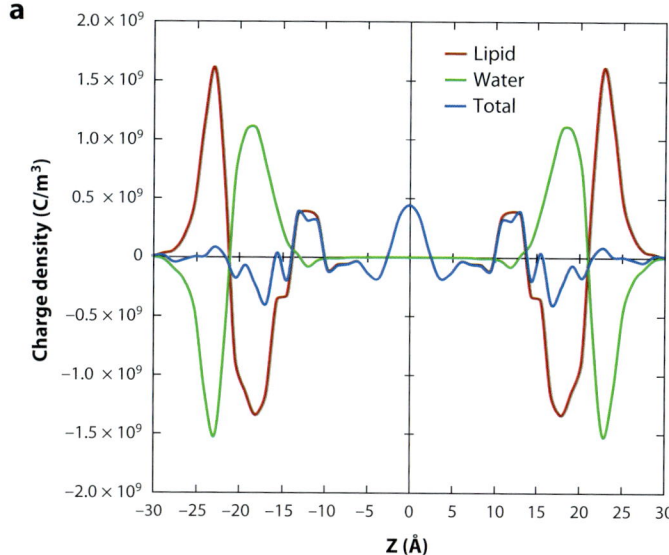
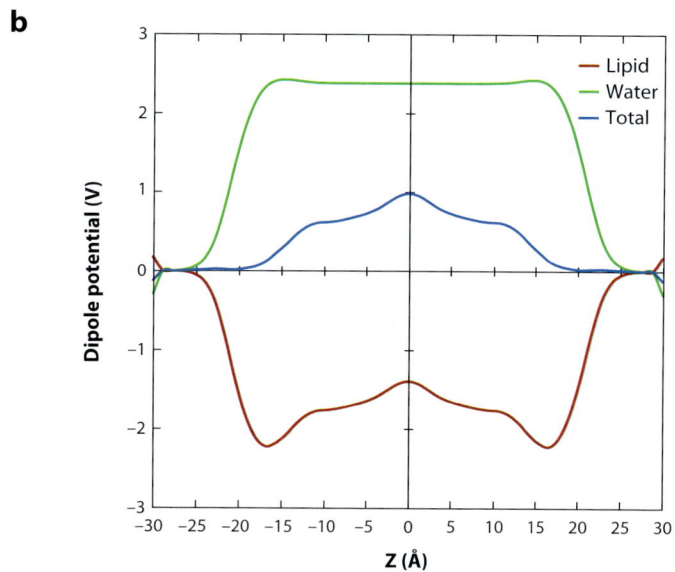

Figure 5
(*a*) Charge density and (*b*) electrostatic potential across a diphytanoyl phosphatidylcholine; membrane from a molecular-dynamics simulation. Atom densities and potentials were kindly provided by Dr. Wataru Shinoda, National Institute for Advanced Science and Technology, Tsukuba, Japan (51). Abbreviation: C/m^3, coulomb/m^3.

DMPC: dimyristoylphosphatidylcholine

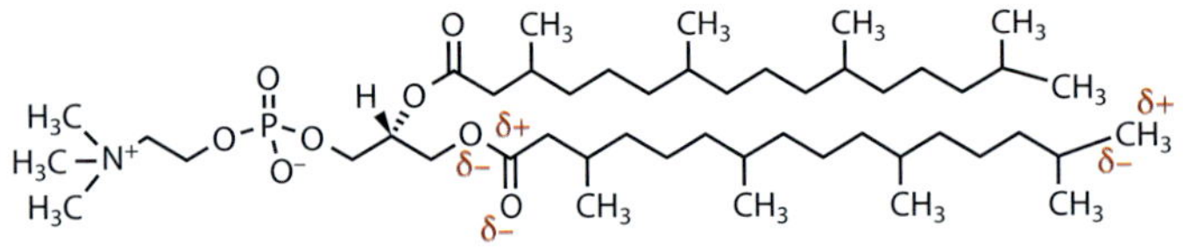

Figure 6

Structure of diphytanoyl phosphatidylcholine. The primary partial charges that contribute to the dipole potential are labeled as $\delta+$ and $\delta-$ on one chain of the lipid molecule.

for example, the four atoms of a methyl group, each having a fractional charge, are grouped into a single pseudoatom.

As shown in **Table 2**, the predicted values of the dipole potential vary widely and are large, ranging up to 1,000 mV. There are several possible reasons the discrepancies among MD simulations and between simulations and experiments. First, not all force fields have been optimized for lipids, so the partial charges of individual atoms and the dipole moments of the lipid head groups differ substantially among the force fields (96). Second, different water models have been used in MD simulations. The water molecule dipole moment in the various models is 2.35 D for TIP3P (110), 2.18 D for TIP4P (111), 2.29 D for TIP5P (112), and 2.27 D for SPC (113), respectively. Third, different treatments of the long-range electrostatics were employed. The predicted dipole potential changed by about 100 mV when the truncation distance varied from 1.8 nm to 2.5 nm (114). The dipole potentials were 620 mV, 720 mV, and 830 mV in the cases of particle-mesh Ewald electrostatics, simple cutoff electrostatics, and the moving-boundary field approaches, respectively (115).

Fourth and most importantly, electronic polarization effects have generally not been modeled in MD simulations. In the low-dielectric environment of the hydrophobic core of a membrane, such effects become highly significant. A very important advance in MD simulation therefore has been the recent development of polarizable force fields. Using the conventional CHARMM force field, Roux and colleagues (116) obtained a dipole potential of 800 mV, but this value fell to 350 mV, more typical of the experimental results when their new force field, which was based on a Drude oscillator model, was employed (117, 118). The contribution to the dipole potential from the hydrocarbon chains in response to the electric field was about 500 mV in the polarizable force field, which simulates a dielectric constant of about two in the hydrophobic region, but this contribution was completely missing in the nonpolarizable force field, which has, in effect, a dielectric constant of one. By contrast, the dipole potential varied only by 100 mV when the polarizable CHARMM force field, which was based on the charge-equilibrium formalism, was employed (105).

MD simulations on very large systems, such as on realistic membrane models or on the long timescales that are necessary to equilibrate membrane models, may require such large computer resources that they cannot easily be studied by traditional all-atom methods. In these cases, one can sometimes tackle the problem by using coarse-grained (CG) models, where 10 to 30 atoms are grouped into one CG particle (e.g., four water molecules into one CG particle). The MD simulations of the CG particles are two orders of magnitude more efficient than atomistic simulations. With a new CG model, big multipole water in which the water molecules cluster into a single unit, a simulation with the MARTINI force field yielded a dipole potential of about 300 mV in a dipalmitoyl phosphatidylcholine membrane, close to that estimated in experiments (119).

MD simulations have been used to study various factors that may affect the dipole potential. Shinoda et al. (51) investigated the effect of the lipid carbonyl group on the dipole potential. The dipole potential of ether-linked lipid (diphytanyl phosphatidylcholine) was about half of that of ester-linked lipid (DPhPC), which is consistent with experimental observations (22). Mojumdar & Lyubartsev (109) used the UA model, which was based on GROMOS force field, to model DMPC lipids and studied the effect of anesthetic articaine. Both charged and uncharged articaine increase the electrostatic potential in the membrane interior by

70–200 mV. In addition to these, MD simulations were used to study cholesterol in dipalmitoyl phosphatidylcholine (120) and a cationic lipid (DMTAP) in DMPC (121), as well as the sodium chloride (NaCl) effect in solution (104, 106).

PHYSIOLOGICAL IMPLICATIONS

It is not surprising that the dipole potential can play a functional role in biomembranes as the electric field generated by a dipole potential that is in the range of 10^8–10^9 V/m, much higher than the field strength generated by a 100 mV of transmembrane potential ($\sim 2.5 \times 10^7$ V/m). The transmembrane potential affects the conformation and function of many membrane proteins (2), so that one expects that the dipole potential, being so large, may have very dramatic effects. It should be kept in mind, however, that the large size of the dipole potential results in part from the very low-dielectric constant of the hydrophobic membrane interior. Where proteins are present in the membrane, especially proteins with aqueous cavities, such as ion channels and transporters, the effective dielectric constant is much higher, and the local dipole potential is reduced proportionately. Nevertheless, the dipole potential affects the translocation rates of ions across lipid membranes as well as the partition and translocation of macromolecules (55, 122, 123). It affects the structure and function of membrane-incorporated proteins, such as peptides (124–127), gramicidin A (128–130), the Na^+-K^+-ATPase (131), and phospholipase A (132). It has also been suggested that the dipole potential may play a role in the function and conformation of proteins in lipid rafts, where the dipole potential is different from surrounding lipids owing to associated sterols within the raft structure (133).

Interaction between Peptides and Membranes

The interaction of the mitochondrial amphipathic signal peptide p25 with lipid vesicles was studied with di-8-ANEPPS (124). This peptide induced a decrease in the dipole potential of PC lipid vesicles, as evaluated by the excitation ratiometric response of di-8-ANEPPS. The decrease caused by the insertion of p25 was larger or smaller when 6-KC or phloretin was incorporated, respectively. Together with attenuated total-reflectance-Fourier transform infrared spectroscopy, the authors found that the presence of 6-KC induced a higher amount of helicoidal structure, whereas the presence of phloretin did not affect the secondary structure of the peptide.

Cladera et al. (125) studied the interaction of membranes and the synthetic peptide corresponding to the 12-residue N-terminal region of the viral envelope glycoprotein gp32 in simian immunodeficiency virus. This region plays an equivalent role to that of the viral envelope glycoprotein gp41 in human immunodeficiency virus type 1 (HIV-1), where it is involved in the fusion between the viral and cellular membranes (125). Upon addition of the peptide, the dipole potential decreased as measured as the excitation ratiometric response of di-8-ANEPPS. The percentage of measured membrane fusion increased or decreased with the addition of 6-KC or phloretin, respectively. Because of the complexity of the interaction between the peptide and the membrane, no correlation between the dipole potential and the peptide structure was made. The addition of sifuvirtide or T-1249, inhibitors of HIV-1 fusion, caused a concentration-dependent decrease in the dipole potential (126, 127) of blood cell membranes. The inhibition might be caused by the decrease of the dipole potential, which in turn decreases membrane fusion, in addition to the effects of binding to gp41 to prevent its fusogenic conformation.

Function of Membrane Proteins

It was established that the dipole potential affects the function of the gramicidin channel formed by peptide gramicidin A (128–130). The gramicidin-mediated cation current through black lipid membranes decreased upon the addition of phloretin, whereas the kinetics was markedly accelerated upon the addition of 6-KC (128). The proton conductance was increased or decreased upon the addition of 6-KC or phloretin, respectively (129), but the channel lifetime decreased or increased upon the addition of 6-KC or phloretin, respectively. It has been suggested that the process of channel dissociation is affected via the interaction between the dipole potential and the dipole moment of gramicidin molecules, as gramicidin A is as potent as phloretin in modifying the membrane dipole potential (130).

The effect of the dipole potential on the activity of Na^+-K^+-ATPase was investigated using the excitation ratiometric method of di-8-ANEPPS (131). The variation of the dipole potential was implemented by the use of different lipids extracted from kidney and brain of 11 different animal species. The dipole potential varied from 236 mV to 334 mV for vesicles prepared from the total membrane lipids. The increase of the molecular activity of Na^+-K^+-ATPase, defined as the maximal activity (i.e., the number of ATP molecules hydrolyzed) divided by the Na^+-K^+-ATPase concentration, correlated well with the increase in dipole potential. However, this correlation does not necessarily imply that the cause of the enzyme activity change was the change in the dipole potential. The change in enzyme activity might instead be the result of changes in other membrane properties.

The activity of phosphohydrolytic enzyme phospholipase A_2 (PLA_2) was also investigated under the effect of the dipole potential (132). PLA_2 was incorporated into lipid monolayers of dilauroylphosphatidic acid or dilauroylphosphatidylcholine, and the activity was measured by the degradation of phospholipids. The effect of the dipole potential on the activity of PLA_2 was carried out differently. A voltage was applied to a silanized semiconductor electrode, which was carefully brought to make direct contact with the lipid hydrocarbon chains on the air side of the monolayer surface. Although high voltages (-25 V to $+50$ V) were applied, the calculated maximum electric field was about 10,000 V/m, which is much lower than the electric field generated by the dipole potential (10^8–10^9 V/m). With the application of a negative voltage (-25 V) to the surface electrode, the enzyme activity against films of pure dilauroylphosphatidic acid increased to about twice the rate observed in the absence of the field, and the changes were reversible when the voltage was switched off. A 74% decrease in the enzyme activity was also observed when a positive voltage ($+25$ V) was applied.

Partition and Translocation of Macromolecules

Pregnanolone is a steroid-based progesterone metabolite used as an anesthetic. It decreases the membrane dipole potential (122), an effect that also has been reported for other anesthetics (134). The penetration of pregnanolone into lipid monolayers (observed as an increase in surface pressure) decreased upon the addition of phloretin. This may be explained by the electrostatic dipole-dipole interactions between pregnanolone and the membrane lipids with their associated water molecules.

In addition to affecting the permeability of the membrane to hydrophobic ions and the binding of such ions to the membrane, the dipole potential might also affect skin permeability. When human skin was treated with liposomes containing phloretin for 24 h, the delivery of lignocaine hydrochloride was enhanced 3–5 times compared with the untreated skin (123). Similar enhancement was observed for the delivery of the peptide antibiotic bacitracin using fluorescein-labeled bacitracin and confocal microscopy (55). It was proposed that the mode of action of phloretin is to reduce the dipole potential.

SUMMARY POINTS

1. There exist three kinds of membrane potentials: surface potential, dipole potential, and transmembrane potential.

2. The dipole potential is a positive potential centered in the lipid bilayer. It underlies a difference of five orders of magnitude in the rate at which hydrophobic anions and cations cross bare lipid bilayers.

3. The magnitude of the dipole potential is in the range of 200–500 mV as estimated using planar bilayer, lipid monolayer, and cryo-EM methods.

4. Voltage-sensitive fluorescent probes are used to monitor changes in dipole potentials. The fluorescence response depends on both the position and the size of the chromophore, and the sensed electrostatic potential is just a portion of the entire potential profile across lipid membranes.

5. Molecular-dynamics (MD) simulations provide a unique tool to study both the amplitude and the shape of dipole potential across the lipid membrane. The recent development of polarizable force fields makes MD simulations of membranes more realistic.

6. In addition to the translocation rates of ions across lipid membranes, the dipole potential also affects the partition and translocation of macromolecules and the structure and function of membrane proteins.

FUTURE ISSUES

1. Further investigation is needed on the energy difference between hydrophobic cations and anions in crossing lipid membranes. This is needed to resolve the discrepancy in dipole potential measurements from the planar lipid membrane and lipid monolayer methods.

2. In MD simulations there exist two types of dipole potential profiles. The type showing a maximum at the membrane center is supported by estimates from the cryo-EM method. However, the resolution of the cryo-EM method is not high and needs further improvement.

3. The polarizable force field shines light on resolving the difference in estimated dipole potentials between experiments and simulations. However, further development is required to confirm the observed effect of a polarizable force field on the dipole potential as that effect was not observed in some polarizable force fields. In addition, systematic parameterization of the widely used force fields is necessary for more accurate and realistic simulations of biological systems.

4. Current voltage-sensitive fluorescent probes are widely used in monitoring the changes of membrane potential. However, the resolution is limited due to the large size of the chromophore, and the interpretation is not obvious because of the dependence of the dipole potential on the location and length of the chromophore in the membrane. Development of new probes with a smaller size and high sensitivity is of great interest.

5. Studies about how the membrane dipole potential affects the functions of membrane proteins have shown some promising evidence. However, in some studies, the dipole potential was not directly changed, which leaves some uncertainty in the interpretations. Further direct investigation is needed.

DISCLOSURE STATEMENT

The author is not aware of any affiliations, memberships, funding, or financial holdings that might be perceived as affecting the objectivity of this review.

ACKNOWLEDGMENTS

I thank Dr. Fred J. Sigworth (Yale University) for helpful discussions and Dr. Wataru Shinoda (Research Institute for Computational Sciences, Ibaraki, Japan) for providing the atom densities and potentials from his MD simulations of ester- and ether-DPhPC membranes. Studies of membrane dipole potentials using cryo-EM were supported by the National Institutes of Health grant NS21501 (to F.J. Sigworth).

LITERATURE CITED

1. Honig BH, Hubbell WL, Flewelling RF. 1986. Electrostatic interactions in membranes and proteins. *Annu. Rev. Biophys. Biophys. Chem.* 15:163–93
2. Hille B. 2001. *Ion Channels of Excitable Membranes*. Sunderland, MA: Sinauer
3. Fraser JA, Huang CLH. 2007. Quantitative techniques for steady-state calculation and dynamic integrated modelling of membrane potential and intracellular ion concentrations. *Prog. Biophys. Mol. Biol.* 94:336–72
4. Poignard C, Silve A, Campion F, Mir LM, Saut O, Schwartz L. 2011. Ion fluxes, transmembrane potential, and osmotic stabilization: a new dynamic electrophysiological model for eukaryotic cells. *Eur. Biophys. J.* 40:235–46
5. Shomer I, Novacky AJ, Pike SM, Yermiyahu U, Kinraide TB. 2003. Electrical potentials of plant cell walls in response to the ionic environment. *Plant Physiol.* 133:411–22
6. Kinraide TB. 2001. Ion fluxes considered in terms of membrane-surface electrical potentials. *Aust. J. Plant Physiol.* 28:605–16
7. Franklin JC, Cafiso DS. 1993. Internal electrostatic potentials in bilayers: measuring and controlling dipole potentials in lipid vesicles. *Biophys. J.* 65:289–99
8. Wang P, Zhou D, Kinraide TB, Luo X, Li L, et al. 2008. Cell membrane surface potential plays a dominant role in the phytotoxicity of copper and arsenate. *Plant Physiol.* 148:2134–43
9. Kinraide TB, Yermiyahu U, Rytwo G. 1998. Computation of surface electrical potentials of plant cell membranes. *Plant Physiol.* 118:505–12
10. Kinraide TB. 2006. Plasma membrane surface potential (ψ_{pm}) as a determinant of ion bioavailability: a critical analysis of new and published toxicological studies and a simplified method for the computation of plant ψpm. *Environ. Toxicol. Chem.* 25:3188–98
11. Delgado AV, González-Caballero F, Hunter RJ, Koopal LK, Lyklema J. 2007. Measurement and interpretation of electrokinetic phenomena. *J. Colloid Interface Sci.* 309:194–224
12. Liberman EA, Topaly VP. 1969. [Permeability of bimolecular phospholipid membranes for fat-soluble ions]. *Biofizika* 14:452–61 (In Russian)
13. Hladky SB, Haydon DA. 1973. Membrane conductance and surface potential. *Biochim. Biophys. Acta* 318:464–68
14. Ketterer B, Neumcke B, Läuger P. 1971. Transport mechanism of hydrophobic ions through lipid bilayer membranes. *J. Membr. Biol.* 5:225–45
15. Flewelling RF, Hubbell WL. 1986. The membrane dipole potential in a total membrane potential model. Applications to hydrophobic ion interactions with membranes. *Biophys. J.* 49:541–52
16. Haydon DA, Hladky SB. 1972. Ion transport across thin lipid membranes: a critical discussion of mechanisms in selected systems. *Q. Rev. Biophys.* 5:187–282
17. Hladky SB. 1979. The carrier mechanism. *Curr. Top. Membr. Transp.* 12:53–164
18. Flewelling RF, Hubbell WL. 1986. Hydrophobic ion interactions with membranes. Thermodynamic analysis of tetraphenylphosphonium binding to vesicles. *Biophys. J.* 49:531–40

19. Benz R, Cros D. 1978. Influence of sterols on ion transport through lipid bilayer membranes. *Biochim. Biophys. Acta* 506:265–80
20. Benz R, Gisin BF. 1978. Influence of membrane structure on ion transport through lipid bilayer membranes. *J. Membr. Biol.* 40:293–314
21. Starke-Peterkovic T, Clarke R. 2009. Effect of headgroup on the dipole potential of phospholipid vesicles. *Eur. Biophys. J.* 39:103–10
22. Gawrisch K, Ruston D, Zimmerberg J, Parsegian VA, Rand RP, Fuller N. 1992. Membrane dipole potentials, hydration forces, and the ordering of water at membrane surfaces. *Biophys. J.* 61:1213–23
23. Simon SA, McIntosh TJ, Magid AD, Needham D. 1992. Modulation of the interbilayer hydration pressure by the addition of dipoles at the hydrocarbon/water interface. *Biophys. J.* 61:786–99
24. Andersen OS, Finkelstein A, Katz I, Cass A. 1976. Effect of phloretin on the permeability of thin lipid membranes. *J. Gen. Physiol.* 67:749–71
25. Melnik E, Latorre R, Hall JE, Tosteson DC. 1977. Phloretin-induced changes in ion transport across lipid bilayer membranes. *J. Gen. Physiol.* 69:243–57
26. Perkins WR, Cafiso DS. 1987. Procedure using voltage-sensitive spin-labels to monitor dipole potential changes in phospholipid vesicles: the estimation of phloretin-induced conductance changes in vesicles. *J. Membr. Biol.* 96:165–73
27. Brockman H. 1994. Dipole potential of lipid-membranes. *Chem. Phys. Lipids* 73:57–79
28. Clarke RJ. 2001. The dipole potential of phospholipid membranes and methods for its detection. *Adv. Colloid Interface Sci.* 89–90:263–81
29. Pohl EE. 2005. Dipole potential of bilayer membranes. In *Advances in Planar Lipid Bilayers and Liposomes*, ed. HT Tien, A Ottova-Leitmannova, 1:77–100. London: Elsevier/Academic
30. Pickar AD, Benz R. 1978. Transport of oppositely charged lipophilic probe ions in lipid bilayer membranes having various structures. *J. Membr. Biol.* 44:353–76
31. Läuger P, Benz R, Stark G, Bamberg E, Jordan PC, et al. 1981. Relaxation studies of ion transport systems in lipid bilayer membranes. *Q. Rev. Biophys.* 14:513–98
32. Benz R, Fröhlich O, Läuger P. 1977. Influence of membrane structure on the kinetics of carrier-mediated ion transport through lipid bilayers. *Biochim. Biophys. Acta* 464:465–81
33. Benz R, Stark G, Janko K, Läuger P. 1973. Valinomycin mediated ion transport through neutral lipid membranes: influence of hydrocarbon chain length and temperature. *J. Membr. Biol.* 14:339–64
34. Stangret J, Kamienska-Piotrowicz E. 1997. Effect of tetraphenylphosphonium and tetraphenylborate ions on the water structure in aqueous solutions; FTIR studies of HDO spectra. *J. Chem. Soc. Faraday Trans.* 93:3463–66
35. Coetzee JF, Sharpe WR. 1971. Solute-solvent interactions. 6. Specific interactions of tetraphenylarsonium, tetraphenylphosphonium, and tetraphenylborate ions with water and other solvents. *J. Phys. Chem.* 75:3141–46
36. Schamberger J, Clarke RJ. 2002. Hydrophobic ion hydration and the magnitude of the dipole potential. *Biophys. J.* 82:3081–88
37. Bangham AD, Mason W. 1979. The effect of some general anaesthetics on the surface potential of lipid monolayers. *Br. J. Pharmacol.* 66:259–65
38. Reyes J, Greco F, Motais R, Latorre R. 1983. Phloretin and phloretin analogs: mode of action in planar lipid bilayers and monolayers. *J. Membr. Biol.* 72:93–103
39. Haydon DA, Elliott JR. 1986. Surface potential changes in lipid monolayers and the 'cut-off' in anaesthetic effects of N-alkanols. *Biochim. Biophys. Acta* 863:337–40
40. Gabev E, Kasianowicz J, Abbott T, McLaughlin S. 1989. Binding of neomycin to phosphatidylinositol 4,5-bisphosphate (PIP2). *Biochim. Biophys. Acta* 979:105–12
41. Cadenhead DA, Kellner BMJ. 1974. Some observations on monolayer spreading solvents with special reference to phospholipid monolayers. *J. Colloid Interface Sci.* 49:143–45
42. Grunwald E, Baughman G, Kohnstam G. 1960. The solvation of electrolytes in dioxane-water mixtures, as deduced from the effect of solvent change on the standard partial molar free energy. *J. Am. Chem. Soc.* 82:5801–11
43. Cevc G. 1990. Membrane electrostatics. *Biochim. Biophys. Acta* 1031:311–82

44. Wang L, Bose PS, Sigworth FJ. 2006. Using cryo-EM to measure the dipole potential of a lipid membrane. *Proc. Natl. Acad. Sci. USA* 103:18528–33
45. Kirkland EJ. 1998. *Advanced Computing in Electron Microscopy*. New York: Plenum
46. Zhong SJ, Dadarlat VM, Glaeser RM, Head-Gordon T, Downing KH. 2002. Modeling chemical bonding effects for protein electron crystallography: the transferable fragmental electrostatic potential (TFESP) method. *Acta Crystallogr. Sect. A* 58:162–70
47. Stallmeyer MJB, Aizawa SI, Macnab RM, DeRosier DJ. 1989. Image reconstruction of the flagellar basal body of *Salmonella typhimurium*. *J. Mol. Biol.* 205:519–28
48. Sosinsky GE, Francis NR, Stallmeyer MJB, DeRosier DJ. 1992. Substructure of the flagellar basal body of *Salmonella typhimurium*. *J. Mol. Biol.* 223:171–84
49. Francis NR, Sosinsky GE, Thomas D, Derosier DJ. 1994. Isolation, characterization and structure of bacterial flagellar motors containing the switch complex. *J. Mol. Biol.* 235:1261–70
50. Frank J. 2006. *Three-Dimensional Electron Microscopy of Macromolecular Assemblies*. New York: Oxford Univ. Press
51. Shinoda K, Shinoda W, Baba T, Mikami M. 2004. Comparative molecular dynamics study of ether- and ester-linked phospholipid bilayers. *J. Chem. Phys.* 121:9648–54
52. Dubochet J, Adrian M, Chang JJ, Homo JC, Lepault J, et al. 1988. Cryo-electron microscopy of vitrified specimens. *Q. Rev. Biophys.* 21:129–228
53. Clarke RJ. 1997. Effect of lipid structure on the dipole potential of phosphatidylcholine bilayers. *Biochim. Biophys. Acta* 1327:269–78
54. Klymchenko AS, Duportail G, Mély Y, Demchenko AP. 2003. Ultrasensitive two-color fluorescence probes for dipole potential in phospholipid membranes. *Proc. Natl. Acad. Sci. USA* 100:11219–24
55. Cladera J, O'Shea P, Hadgraft J, Valenta C. 2003. Influence of molecular dipoles on human skin permeability: use of 6-ketocholestanol to enhance the transdermal delivery of bacitracin. *J. Pharm. Sci.* 92:1018–27
56. Demchenko AP, Mély Y, Duportail G, Klymchenko AS. 2009. Monitoring biophysical properties of lipid membranes by environment-sensitive fluorescent probes. *Biophys. J.* 96:3461–70
57. Demchenko AP, Yesylevskyy SO. 2009. Nanoscopic description of biomembrane electrostatics: results of molecular dynamics simulations and fluorescence probing. *Chem. Phys. Lipids* 160:63–84
58. Zouni A, Clarke RJ, Holzwarth JF. 1994. Kinetics of the solubilization of styryl dye aggregates by lipid vesicles. *J. Phys. Chem.* 98:1732–38
59. Gross E, Bedlack RS Jr, Loew LM. 1994. Dual-wavelength ratiometric fluorescence measurement of the membrane dipole potential. *Biophys. J.* 67:208–16
60. Plášek J, Sigler K. 1996. Slow fluorescent indicators of membrane potential: a survey of different approaches to probe response analysis. *J. Photochem. Photobiol. B* 33:101–24
61. Fluhler E, Burnham VG, Loew LM. 1986. Spectra, membrane binding, and potentiometric responses of new charge shift probes. *Biochemistry* 24:5749–55
62. Cohen LB, Salzberg BM. 1978. Optical measurement of membrane potential. *Rev. Physiol. Biochem. Pharmacol.* 83:35–88
63. Grinvald A, Hildesheim R. 2004. VSDI: a new era in functional imaging of cortical dynamics. *Nat. Rev. Neurosci.* 5:874–85
64. Loew LM. 1982. Design and characterization of electrochromic membrane probes. *J. Biochem. Biophys. Methods* 6:243–60
65. Grinvald A, Hildesheim R, Farber IC, Anglister L. 1982. Improved fluorescent probes for the measurement of rapid changes in membrane potential. *Biophys. J.* 39:301–8
66. Cohen LB, Salzberg BM, Davila HV. 1974. Changes in axon fluorescence during activity: molecular probes of membrane potential. *J. Membr. Biol.* 19:1–36
67. Davila HV, Salzberg BM, Cohen LB. 1973. A large change in axon fluorescence that provides a promising method for measuring membrane potential. *Nat. New Biol.* 241:159–60
68. Bücher H, Wiegand J, Snavely BB, Beck KH, Kuhn H. 1969. Electric field induced changes in the optical absorption of a merocyanine dye. *Chem. Phys. Lett.* 3:508–11
69. Kachel K, Asuncion-Punzalan E, London E. 1998. The location of fluorescence probes with charged groups in model membranes. *Biochim. Biophys. Acta* 1374:63–76

70. Fischer JK, von Brüning DM, Labhart H. 1976. Light modulation by electrochromism. *Appl. Opt.* 25:2812–16
71. Zouni A, Clarke RJ, Visser AJWG, Visser NV, Holzwarth JF. 1993. Static and dynamic studies of the potential-sensitive membrane probe RH421 in dimyristoylphosphatidylcholine vesicles. *Biochim. Biophys. Acta* 1153:203–12
72. Loew LM, Cohen LB, Dix J, Fluhler EN, Montana V, et al. 1992. A naphthyl analog of the aminostyryl pyridinium class of potentiometric membrane dyes shows consistent sensitivity in a variety of tissue, cell, and model membrane preparations. *J. Membr. Biol.* 130:1–10
73. Gupta RK, Salzberg BM, Grinvald A. 1981. Improvements in optical methods for measuring rapid changes in membrane potential. *J. Membr. Biol.* 58:123–37
74. Loew LM. 1978. Charge shift optical probes of membrane potential. *Theory Biochem.* 17:4065–71
75. Fromherz P, Hübener G, Kuhn B, Hinner MJ. 2008. ANNINE-6plus, a voltage-sensitive dye with good solubility, strong membrane binding and high sensitivity. *Eur. Biophys. J.* 37:509–14
76. Montana V, Farkas DL, Loew LM. 1989. Dual-wavelength ratiometric fluorescence measurements of membrane potential. *Biochemistry* 28:4536–39
77. Bedlack RS Jr, Wei MD, Loew LM. 1992. Localized membrane depolarizations and localized calcium influx during electric field-guided neurite growth. *Neuron* 9:393–403
78. Clarke RJ, Kane DJ. 1997. Optical detection of membrane dipole potential: avoidance of fluidity and dye-induced effects. *Biochim. Biophys. Acta* 1323:223–39
79. Vitha MF, Clarke RJ. 2007. Comparison of excitation and emission ratiometric fluorescence methods for quantifying the membrane dipole potential. *Biochim. Biophys. Acta* 1768:107–14
80. Klymchenko AS, Duportail G, Ozturk T, Pivovarenko VG, Mély Y, Demchenko AP. 2002. Novel two-band ratiometric fluorescence probes with different location and orientation in phospholipid membranes. *Chem. Biol.* 9:1199–208
81. Klymchenko AS, Demchenko AP. 2003. Multiparametric probing of intermolecular interactions with fluorescent dye exhibiting excited state intramolecular proton transfer. *Phys. Chem. Chem. Phys.* 5:461–68
82. Chou P-T, Pu S-C, Cheng Y-M, Yu W-S, Yu Y-C, et al. 2005. Femtosecond dynamics on excited-state proton/charge-transfer reaction in 4'-N,N-diethylamino-3-hydroxyflavone. The role of dipolar vectors in constructing a rational mechanism. *J. Phys. Chem. A* 109:3777–87
83. Yesylevskyy SO, Klymchenko AS, Demchenko AP. 2005. Semi-empirical study of two-color fluorescent dyes based on 3-hydroxychromone. *J. Mol. Struct. Theochem.* 755:229–39
84. Das R, Klymchenko AS, Duportail G, Mély Y. 2008. Excited state proton transfer and solvent relaxation of a 3-hydroxyflavone probe in lipid bilayers. *J. Phys. Chem. B* 112:11929–35
85. Das R, Klymchenko AS, Duportail G, Mély Y. 2009. Unusually slow proton transfer dynamics of a 3-hydroxychromone dye in protic solvents. *Photochem. Photobiol. Sci.* 8:1583–89
86. Shynkar VV, Klymchenko AS, Duportail G, Demchenko AP, Mély Y. 2005. Two-color fluorescent probes for imaging the dipole potential of cell plasma membranes. *Biochim. Biophys. Acta* 1712:128–36
87. M'Baye G, Shynkar VV, Klymchenko AS, Mély Y, Duportail G. 2006. Membrane dipole potential as measured by ratiometric 3-hydroxyflavone fluorescence probes: Accounting for hydration effects. *J. Fluoresc.* 16:35–42
88. Klymchenko AS, Stoeckel H, Takeda K, Mély Y. 2006. Fluorescent probe based on intramolecular proton transfer for fast ratiometric measurement of cellular transmembrane potential. *J. Phys. Chem. B* 110:13624–32
89. Shynkar VV, Klymchenko AS, Kunzelmann C, Duportail G, Muller CD, et al. 2007. Fluorescent biomembrane probe for ratiometric detection of apoptosis. *J. Am. Chem. Soc.* 129:2187–93
90. Jendrasiak GL, Smith RL, McIntosh TJ. 1997. The effect of phloretin on the hydration of egg phosphatidylcholine multilayers. *Biochim. Biophys. Acta* 1329:159–68
91. Lairion F, Disalvo EA. 2004. Effect of phloretin on the dipole potential of phosphatidylcholine, phosphatidylethanolamine, and phosphatidylglycerol monolayers. *Langmuir* 20:9151–55
92. Cseh R, Benz R. 1999. Interaction of phloretin with lipid monolayers: relationship between structural changes and dipole potential change. *Biophys. J.* 77:1477–88
93. Klymchenko AS, Mély Y, Demchenko AP, Duportail G. 2004. Simultaneous probing of hydration and polarity of lipid bilayers with 3-hydroxyflavone fluorescent dyes. *Biochim. Biophys. Acta* 1665:6–19

94. van Gunsteren WF, Bakowies D, Baron R, Chandrasekhar I, Christen M, et al. 2006. Biomolecular modeling: goals, problems, perspectives. *Angew. Chem. Int. Ed.* 45:4064–92
95. Berendsen HJC. 1996. Bio-molecular dynamics comes of age. *Science* 271:954–55
96. Guvench O, MacKerell AD Jr. 2008. Comparison of protein force fields for molecular dynamics simulations. *Methods Mol. Biol.* 443:63–88
97. Mackerell AD Jr. 2004. Empirical force fields for biological macromolecules: overview and issues. *J. Comput. Chem.* 25:1584–604
98. Tieleman DP, Marrink SJ, Berendsen HJC. 1997. A computer perspective of membranes: molecular dynamics studies of lipid bilayer systems. *Biochim. Biophys. Acta Rev.* 1331:235–70
99. Sagui C, Darden TA. 1999. Molecular dynamics simulations of biomolecules: long-range electrostatic effects. *Annu. Rev. Biophys. Biomol. Struct.* 28:155–79
100. Anezo C, de Vries AH, Holtje HD, Tieleman DP, Marrink SJ. 2003. Methodological issues in lipid bilayer simulations. *J. Phys. Chem. B* 107:9424–33
101. Cornell WD, Cieplak P, Bayly CI, Gould IR, Merz KM Jr, et al. 1995. A second generation force field for the simulation of proteins, nucleic acids, and organic molecules. *J. Am. Chem. Soc.* 117:5179–97
102. MacKerell AD Jr, Bashford D, Bellott M, Dunbrack RL Jr, Evanseck JD, et al. 1998. All-atom empirical potential for molecular modeling and dynamics studies of proteins. *J. Phys. Chem. B* 102:3586–616
103. Oostenbrink C, Villa A, Mark AE, van Gunsteren WF. 2004. A biomolecular force field based on the free enthalpy of hydration and solvation: the GROMOS force-field parameter sets 53A5 and 53A6. *J. Comput. Chem.* 25:1656–76
104. Sachs JN, Crozier PS, Woolf TB. 2004. Atomistic simulations of biologically realistic transmembrane potential gradients. *J. Chem. Phys.* 121:10847–51
105. Davis JE, Rahaman O, Patel S. 2009. Molecular dynamics simulations of a DMPC bilayer using nonadditive interaction models. *Biophys. J.* 96:385–402
106. Pandit SA, Bostick D, Berkowitz ML. 2003. Molecular dynamics simulation of a dipalmitoylphosphatidylcholine bilayer with NaCl. *Biophys. J.* 84:3743–50
107. Smondyrev AM, Berkowitz ML. 1999. United atom force field for phospholipid membranes: constant pressure molecular dynamics simulation of dipalmitoylphosphatidicholine/water system. *J. Comput. Chem.* 20:531–45
108. Villarreal MA, Diaz SB, Disalvo EA, Montich GG. 2004. Molecular dynamics simulation study of the interaction of trehalose with lipid membranes. *Langmuir* 20:7844–51
109. Mojumdar EH, Lyubartsev AP. 2010. Molecular dynamics simulations of local anesthetic articaine in a lipid bilayer. *Biophys. Chem.* 153:27–35
110. Jorgensen WL, Chandrasekhar J, Madura JD, Impey RW, Klein ML. 1983. Comparison of simple potential functions for simulating liquid water. *J. Chem. Phys.* 79:926–35
111. Abascal JLF, Vega C. 2005. A general purpose model for the condensed phases of water: TIP4P/2005. *J. Chem. Phys.* 123:1–12
112. Mahoney MW, Jorgensen WL. 2000. A five-site model for liquid water and the reproduction of the density anomaly by rigid, non-polarizable potential functions. *J. Chem. Phys.* 112:8910–22
113. Berendsen HJC, Postma JPM, van Gunsteren WF, Hermans J. 1981. Interaction models for water in relation to protein hydration. In *Intermolecular Forces*, ed. B Pullman, pp. 331–42. Dordrecht: Reidel
114. Patra M, Karttunen M, Hyvönen MT, Falck E, Lindqvist P, Vattulainen I. 2003. Molecular dynamics simulations of lipid bilayers: major artifacts due to truncating electrostatic interactions. *Biophys. J.* 84:3636–45
115. Anézo C, de Vries AH, Höltje H-D, Tieleman DP, Marrink S-J. 2003. Methodological issues in lipid bilayer simulations. *J. Phys. Chem. B* 107:9424–33
116. Harder E, MacKerell AD Jr, Roux B. 2009. Many-body polarization effects and the membrane dipole potential. *J. Am. Chem. Soc.* 131:2760–61
117. Lamoureux G, Harder E, Vorobyov IV, Roux B, MacKerell AD Jr. 2006. A polarizable model of water for molecular dynamics simulations of biomolecules. *Chem. Phys. Lett.* 418:245–49
118. Lamoureux G, MacKerell AD Jr, Roux B. 2003. A simple polarizable model of water based on classical drude oscillators. *J. Chem. Phys.* 119:5185–97

119. Wu Z, Cui Q, Yethiraj A. 2010. A new coarse-grained model for water: the importance of electrostatic interactions. *J. Phys. Chem. B* 114:10524–29
120. Hofsäß C, Lindahl E, Edholm O. 2003. Molecular dynamics simulations of phospholipid bilayers with cholesterol. *Biophys. J.* 84:2192–206
121. Gurtovenko AA, Patra M, Karttunen M, Vattulainen I. 2004. Cationic DMPC/DMTAP lipid bilayers: molecular dynamics study. *Biophys. J.* 86:3461–72
122. Alakoskela J-MI, Söderlund T, Holopainen JM, Kinnunen PKJ. 2004. Dipole potential and head-group spacing are determinants for the membrane partitioning of pregnanolone. *Mol. Pharmacol.* 66:161–68
123. Valenta C, Cladera J, O'Shea P, Hadgraft J. 2001. Effect of phloretin on the percutaneous absorption of lignocaine across human skin. *J. Pharm. Sci.* 90:485–92
124. Cladera J, O'Shea P. 1998. Intramembrane molecular dipoles affect the membrane insertion and folding of a model amphiphilic peptide. *Biophys. J.* 74:2434–42
125. Cladera J, Martin I, Ruysschaert J-M, O'Shea P. 1999. Characterization of the sequence of interactions of the fusion domain of the simian immunodeficiency virus with membranes. *J. Biol. Chem.* 274:29951–59
126. Matos PM, Freitas T, Castanho MARB, Santos NC. 2010. The role of blood cell membrane lipids on the mode of action of HIV-1 fusion inhibitor sifuvirtide. *Biochem. Biophys. Res. Commun.* 403:270–74
127. Matos PM, Castanho MARB, Santos NC. 2010. HIV-1 fusion inhibitor peptides enfuvirtide and T-1249 interact with erythrocyte and lymphocyte membranes. *PLoS ONE* 5:e9830
128. Rokitskaya TI, Antonenko YN, Kotova EA. 1997. Effect of the dipole potential of a bilayer lipid membrane on gramicidin channel dissociation kinetics. *Biophys. J.* 73:850–54
129. Rokitskaya TI, Kotova EA, Antonenko YN. 2002. Membrane dipole potential modulates proton conductance through gramicidin channel: movement of negative ionic defects inside the channel. *Biophys. J.* 82:865–73
130. Shapovalov VL, Kotova EA, Rokitskaya TI, Antonenko YN. 1999. Effect of gramicidin A on the dipole potential of phospholipid membranes. *Biophys. J.* 77:299–305
131. Starke-Peterkovic T, Turner N, Else PL, Clarke RJ. 2005. Electric field strength of membrane lipids from vertebrate species: membrane lipid composition and Na^+-K^+-ATPase molecular activity. *Am. J. Physiol. Regul. Integr. Comp. Physiol.* 288:R663–70
132. Maggio B. 1999. Modulation of phospholipase A(2) by electrostatic fields and dipole potential of glycosphingolipids in monolayers. *J. Lipid Res.* 40:930–39
133. O'Shea P. 2005. Physical landscapes in biological membranes: physico-chemical terrains for spatio-temporal control of biomolecular interactions and behaviour. *Philos. Trans. R. Soc. Lond. Ser. A* 363:575–88
134. Cafiso DS. 1998. Dipole potentials and spontaneous curvature: membrane properties that could mediate anesthesia. *Toxicol. Lett.* 100–101:431–39
135. Peterson U, Mannock DA, Lewis R, Pohl P, McElhaney RN, Pohl EE. 2002. Origin of membrane dipole potential: contribution of the phospholipid fatty acid chains. *Chem. Phys. Lipids* 117:19–27
136. Davis JE, Patel S. 2009. Charge equilibration force fields for lipid environments: Applications to fully hydrated DPPC bilayers and DMPC-embedded gramicidin A. *J. Phys. Chem. B* 113:9183–96

GTPase Networks in Membrane Traffic

Emi Mizuno-Yamasaki,[1,*] Felix Rivera-Molina,[2,*] and Peter Novick[3]

[1]Institute for Molecular and Cellular Regulation, Gunma University, Maebashi 371-8512, Japan; email: eyamasak@gunma-u.ac.jp

[2]Department of Cell Biology, Yale University School of Medicine, New Haven, Connecticut 06510; email: felix.rivera-molina@yale.edu

[3]Department of Cellular and Molecular Medicine, University of California, San Diego, La Jolla, California 92093; email: pnovick@ucsd.edu

*These authors made equal contributions.

Keywords

Rab, ARF, guanine nucleotide exchange factor, GTPase-activating protein, effector, vesicular transport

Abstract

Members of the Rab or ARF/Sar branches of the Ras GTPase superfamily regulate almost every step of intracellular membrane traffic. A rapidly growing body of evidence indicates that these GTPases do not act as lone agents but are networked to one another through a variety of mechanisms to coordinate the individual events of one stage of transport and to link together the different stages of an entire transport pathway. These mechanisms include guanine nucleotide exchange factor (GEF) cascades, GTPase-activating protein (GAP) cascades, effectors that bind to multiple GTPases, and positive-feedback loops generated by exchange factor-effector interactions. Together these mechanisms can lead to an ordered series of transitions from one GTPase to the next. As each GTPase recruits a unique set of effectors, these transitions help to define changes in the functionality of the membrane compartments with which they are associated.

Contents

- INTRODUCTION 638
  - Overview of the GTPase Cycle 639
  - GTPases in Membrane Traffic 639
- GUANINE NUCLEOTIDE EXCHANGE FACTOR CASCADES 640
  - Rab/Ypt Cascades in Endosomal Traffic 640
  - Rab/Ypt Cascades in the Secretory Pathway 642
  - ADP Ribosylation Factors and Arf-Like Proteins 644
  - Rho and Cdc42 GTPases 644
  - Cross Talk Between Rab1, GBF1, and Arf1 645
- RAB-GUANINE NUCLEOTIDE EXCHANGE FACTOR POSITIVE-FEEDBACK LOOPS 646
  - Rab5, Rabex-5, and Rabaptin-5 646
  - Sec4p, Sec2p, and Sec15p 646
- GTPase-ACTIVATING PROTEIN CASCADES 647
  - Ypt1p, Gyp1p, and Ypt31p/Ypt32p 647
  - Rab5, TBC-2, and Rab7 648
  - Sec4p, Msb3p, Msb4p, and Cdc42p 649
  - Rab35, Centaurinβ2, and Arf6 649
- RABS WITH COMMON EFFECTORS 649
  - Rab4, Rab14, and Rabip4/RUFY1 .. 650
  - Rab4, Rab11, and D-AKAP2 650
  - Families of Interacting Proteins: Rab11 and Rab14 650
  - Rab6, Rab11, and R6IP1 651
  - Rab1, Rab6, and Giantin 651
  - Rab8, Rab13, and JRAB/MICAL-L2 652
- CONCLUDING REMARKS 652

INTRODUCTION

Eukaryotic cells contain a variety of membrane-bounded organelles, each with its own specific set of resident components that confer to these organelles their unique functions. Organelles of the exocytic and endocytic pathways, including the endoplasmic reticulum (ER), Golgi apparatus, plasma membrane, endosomes, and lysosomes are linked by rapid, bidirectional membrane traffic, primarily mediated by vesicular transport. One of the major challenges facing the membrane traffic field has been to understand how the unique composition of each organelle is maintained despite this high level of interorganelle traffic. Tightly controlled selection of cargo from the donor compartment during vesicle formation and faithful targeting of vesicles to the appropriate recipient compartment prior to fusion are essential to avoid the rapid dispersal and homogenization of all resident components among the interconnected organelles. GTPases from several branches of the Ras superfamily play essential roles in regulating these steps. However, these GTPases must themselves be regulated both spatially and temporally to maintain the orderly transport of cargo through a series of compartments. Over the past decade, evidence has mounted that these GTPases do not act as independent agents but rather are networked to one another by several different mechanisms, both to coordinate the various biochemical steps of each individual stage of membrane traffic and to link together the different stages of an entire transport pathway. In this review, we discuss specific examples of GTPase networks in the regulation of membrane traffic. Although it is still too early to make broad generalizations regarding the mechanisms by which GTPases are networked, these examples offer some paradigms that may help speed the analysis of other stages of membrane traffic in diverse organisms.

The Ras GTPase superfamily includes several major branches that are primarily dedicated to the regulation of membrane traffic. The Arf/Sar branch is best known for the roles that its members play in the regulation of vesicle budding from donor compartments. Sar helps coordinate cargo selection and budding of coat protein II (COPII)-coated vesicles from

ER exit sites (1), whereas Arf proteins play analogous roles in the budding of clathrin-coated and COPI-coated vesicles at the Golgi apparatus and plasma membrane. The Rab branch typically controls subsequent events, such as the active transport of vesicles along cytoskeletal elements, the initial recognition and tethering of vesicles to the target compartment, and the regulation of vesicle fusion with the acceptor compartment. Members of other branches of the Ras superfamily, including Rho and Ral, have important, but more specialized roles in membrane traffic.

Overview of the GTPase Cycle

All members of the Ras superfamily function as nucleotide-dependent switches. They exhibit high affinity for both GDP and GTP, and assume different conformations in two so-called switch regions, depending upon the nucleotide bound. On their own, these proteins can only slowly interconvert from one form to the other owing to low intrinsic rates of GDP dissociation and GTP hydrolysis. The switch mechanism is therefore controlled by accessory proteins: guanine nucleotide exchange factors (GEFs) catalyze the displacement of prebound GDP, allowing its replacement with GTP, while GTPase-activating proteins (GAPs) greatly stimulate the slow intrinsic hydrolysis of GTP to GDP. The nucleotide-binding and hydrolysis cycles of both Arf/Sar and Rab proteins are coupled to cycles of membrane association and dissociation. In the case of Rab proteins, a protein termed Rab GDP dissociation inhibitor (Rab GDI) binds specifically to the GDP-bound form, masking the C-terminal prenyl lipid moieties and thereby extracting the inactive Rab from the membrane (2). Arf and Sar members are released from membranes following GTP hydrolysis by changing conformation in such a way that they mask their own N-terminal membrane-binding domains, without the involvement of a GDI protein (3, 4). Because the nucleotide cycle is coupled to membrane attachment and release, the localization of the GEFs and GAPs can help define the localization of the active form of the regulated GTPase.

In their GTP-bound form, members of the Ras superfamily bind to a specific set of effectors. These interactions recruit the effectors to their site of action and can directly stimulate their activity or promote their assembly into protein complexes. Arf and Sar effectors include the coat proteins needed for cargo selection and vesicle budding (5). Rab effectors represent a very diverse collection of proteins, including motors needed to transport vesicles on the cytoskeleton, tethers that direct the initial recognition of the target compartment, and regulators of SNARE complex assembly (6). By recruiting a new set of effectors, a GTPase can help to change the functional identity of the membrane with which it is associated. Understanding how GTPase activation and inactivation are spatially and temporally controlled is critical toward understanding the mechanism of membrane traffic.

GTPases in Membrane Traffic

The number of GTPases involved in membrane traffic varies over a large range from species to species. For example, yeast express only 10 Rab proteins, whereas humans express more than 60 (7). Much of the expansion appears to have arisen from gene duplication followed by specialization of function. In this way, one ancestral traffic pathway can be subdivided into multiple pathways under regulation by distinct Rabs. For the purposes of this review, we very briefly outline a basic cast of GTPases on the endocytic and exocytic pathways, with the recognition that many more are involved in many cell types (for more comprehensive reviews see References 8 and 9). Formation of the initial endocytic vesicles requires Arf6. Homotypic fusion of these vesicles and fusion to the early endosome require Rab5. Rapid recycling to the cell surface requires Rab4. Conversion of the early endosome to a late endosome requires Rab7. Transport from the endosome to the

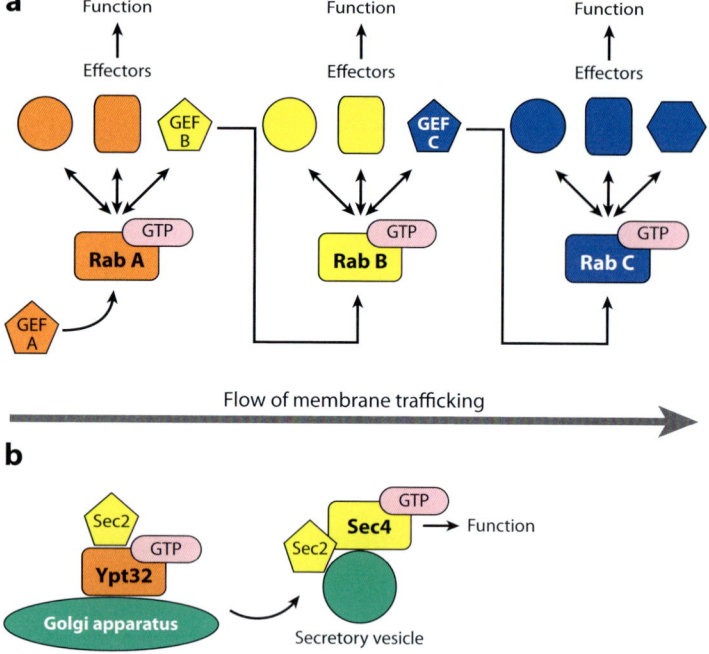

Figure 1

Rab-guanine nucleotide exchange factor (GEF) cascade. (*a*) A Rab GTPase is activated by its own GEF. The GTP-bound active form recruits several effectors to the organelle membrane to fulfill their functions. One of the effectors is a GEF for the downstream Rab: Active Rab A recruits GEF B for the downstream Rab B (*left*). GEF B activates Rab B, and then active Rab B recruits effectors, including GEF C for the downstream Rab C (*middle*). Again, GEF C activates the downstream Rab C (*right*). Thus, several Rab GTPases in the pathway are sequentially recruited to the membrane and efficiently activated. (*b*) The first Rab GEF cascade was found in yeast. Golgi-localized GTP-bound Rab Ypt31p/Ypt32p interacts with Sec2p, a GEF for the next Rab Sec4p (10).

GUANINE NUCLEOTIDE EXCHANGE FACTOR CASCADES

The cytoplasmic surface of every organelle of the endocytic and exocytic pathways is tagged with a specific set of Rab GTPases that play important roles in defining organelle function and identity. Typically, several Rab GTPases work cooperatively on the same traffic pathway and must be coordinately regulated to fulfill their functions in a sequential fashion. Rab activation requires a specific GEF; however, it is still unclear how these GEFs are regulated so that each Rab can be activated at the right time and place. One regulatory mechanism has been termed a Rab GEF cascade (**Figure 1***a*). This mechanism was first documented through analysis of the yeast secretory pathway (10). Ypt31p and Ypt32p are closely related and redundant Rab11 homologs that localize to a late Golgi compartment. Active Ypt31p/Ypt32p recruit Sec2p, a GEF for the Rab GTPase Sec4p, which acts just downstream of Ypt31p/Ypt32p on Golgi-derived secretory vesicles (**Figure 1***b*) (11, 12). Thus, two different Rabs are functionally linked in a regulatory cascade through a GEF. Subsequent studies have revealed the existence of several such cascades in both yeast and mammalian cells and at additional stages of membrane traffic, suggesting that the Rab GEF cascade is a common, conserved regulatory mechanism. In this article, we discuss various Rab GEF cascades.

Rab/Ypt Cascades in Endosomal Traffic

Rab5-HOPS-Rab7 cascade on endosome.

Rab5 plays a key role in the early endocytic pathway (**Figure 2**). Rab5 has various functions: cargo sequestration and budding of endocytic vesicles from the plasma membrane, uncoating of clathrin-coated vesicles, vesicle motility along microtubules, tethering of vesicles to acceptor membranes, and membrane fusion (13–17). After recruitment of Rab5 to the endosomal membrane, Rab5 must be activated by its nucleotide exchange factor, Rabex-5, to fulfill

Golgi apparatus requires Rab9, and export of cargo from the recycling endosome back to the cell surface requires Rab11. On the exocytic pathway, the formation of vesicles from the ER requires Sar1. Fusion of those vesicles with the Golgi apparatus requires Rab1. Arf1 is needed to exit the Golgi apparatus, and Rab8 is needed for fusion of Golgi-derived vesicles with the plasma membrane. Rab6 controls retrograde transport within the Golgi apparatus. In the following sections, we review how some of these GTPases are linked to each other through coordinated regulation of exchange or hydrolysis and how some are linked to each other through a common effector.

its functions. Active Rab5 binds various effectors, including the phosphatidylinositol 3-OH kinase, hVps34, leading to the local production of phosphatidylinositol 3-phosphate. Phosphatidylinositol 3-phosphate, together with Rab5-GTP, then recruits early endosome antigen 1 (EEA1) to promote endosome fusion.

Early endosomes contain not only Rab5, but also Rab4 and Rab11; however, each of these Rabs exhibits a distinct localization pattern, forming its own Rab-specific subdomains on early endosomal membranes. Similarly, late endosomes have both Rab7 and Rab9 subdomains (18). Fast live-cell imaging revealed that the Rab7 domain is derived from Rab5-positive endosomes (19) by the time-dependent disappearance of Rab5 and the acquisition of Rab7 on the same endosomes in human A431 cells (20). This Rab5-to-Rab7 transition was thought to be mediated by the class C vacuolar protein sorting (Vps)/homotypic fusion and vacuole protein sorting (HOPS) complex. The HOPS complex is highly conserved from yeast to human and serves as a tethering complex for vacuole homotypic fusion (21). Four of the six HOPS components bind to the GTP-bound form of Rab5, implying that HOPS is an effector of Rab5. Interestingly, one of the HOPS components, Vps39p, was claimed to have GEF activity toward Ypt7p, a yeast ortholog of Rab7 (21, 22). Thus, it appeared plausible that Rab5-GTP would recruit HOPS, thereby leading to activation of Rab7 and maturation of an early endosome into a late endosome. However, a recent study, discussed in the next section, revealed that the HOPS subunit Vps39p does not possess Rab7 GEF activity and that the Mon1p (monensin sensitivity 1)-Ccz1p (calcium caffeine zinc sensitivity) complex instead has this activity (23).

Rab5-SAND-1/Mon1-Rab7 cascade on endosomes.
Poteryaev et al. (24) showed that Rab5-to-Rab7 conversion is also observed on endosomes in *Caenorhabditis elegans*, suggesting that Rab conversion is a general mechanism in metazoans. Importantly, SAND-1, an ortholog of yeast Mon1p, which was identified as an

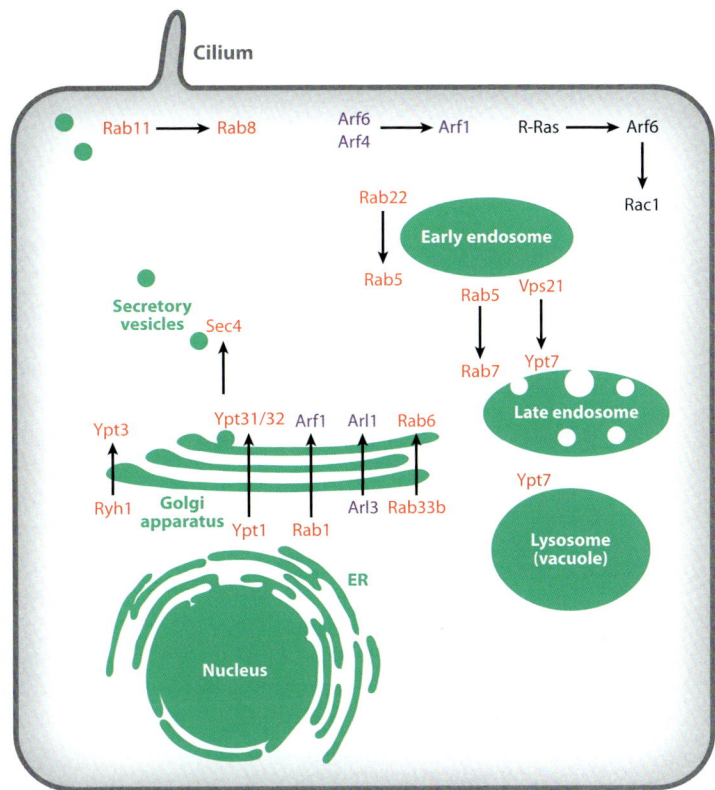

Figure 2

Cellular localization of GTPase cascades in an epithelial cell. On endosomes, the Rab5-to-Rab7 transition is mediated by the Rab7 guanine nucleotide exchange factor (GEF), SAND1/Mon1-Ccz1 complex, to promote the endocytic pathway (24). The Vps21p-Ypt7p transition (yeast orthologs of Rab5 and Rab7) is also mediated by the Mon1-Ccz1 complex (23). Rab22 and Rab5 also form a cascade on endosomes through the Rab5 GEF, Rabex-5, regulating endosome fusion (28). In the endoplasmic reticulum (ER)-Golgi trafficking pathway, yeast Ypt1p (homologous to Rab1) and Ypt31p/Ypt32p (homologous to Rab11) work sequentially, probably through a Ypt32p GEF (37). Ypt32p interacts with Sec2p, a GEF for the next Rab, Sec4p (Rab8 homolog) (10). Sec2p activates Sec4p on the secretory vesicle, which leads to vesicle tethering and fusion with plasma membrane thorough recruitment of the exocyst complex (11). A mammalian Rab11-Rab8 cascade is mediated by Rabin8, a GEF for Rab8; both Rab8 and Rabin8 are required for primary cilium formation (42, 43). A Rab33b-Rab6 cascade is proposed to work in the intra-Golgi apparatus retrograde trafficking (48). Fission yeast Ryh1p-Ypt3p (mammalian Rab6-Rab11) probably forms a cascade through the Ric1p/Rgp1p complex and works on the exocytic/recycling pathway (49). An Arl3-Arl1 cascade localizes to the Golgi apparatus, where it regulates protein sorting (58, 59). Rab1-Arf1 regulates coat protein I vesicle formation for ER-to-Golgi apparatus transport, which is mediated by GBF1, a Rab GEF for Arf1 (76). On the plasma membrane, Arl4 and Arf6 recruit Arno, a GEF for downstream GTPase Arf1, to regulate cytoskeletal organization and endocytosis (55–57). The R-Ras-Arf6-Rac1 pathway is required for cell spreading and migration, which is also mediated by Arno (83, 85, 86). Abbreviations: Arf1–Arf, Rab1–Rab33b, Rac1, members of the GTPase superfamily; R-Ras, a Ras-related protein.

essential regulator of Rab7 function on endosomes, mediates this Rab transition (24, 25). SAND-1/Mon1 is recruited to the membrane by a direct interaction with Rabex-5, a GEF for Rab5, and phosphatidylinositol 3-phosphate. Mon1 forms a stable complex with Ccz1 and works as a GEF for Rab7. SAND-1/Mon1 directly interacts with the HOPS complex and then recruits Rab7 through the HOPS complex. This observation again strongly suggests that SAND-1/Mon1 works as a switch for the Rab5-to-Rab7 transition to promote endosome maturation.

Rab22-Rabex-5-Rab5 cascade on endosomes. A new Rab GEF cascade was found to act at an earlier stage of the endosomal trafficking pathway. In mammalian cells, the Rab5 subfamily includes Rab5, Rab21, Rab22, and Rab31 (26, 27). The Rab5 GEF, Rabex-5, binds to Rab22-GTP yet does not exhibit GEF activity toward Rab22, indicating that Rabex-5 is a novel effector of Rab22. Rab22 recruits Rabex-5 to early endosomes, thus promoting Rab5 activation (28). In the absence of Rabex-5, Rab22 and Rab5 show only partial colocalization, whereas cells overexpressing both Rab22 and Rab5 exhibited enlarged endosomes, suggesting that these two Rab GTPases work cooperatively in the regulation of endosome fusion and the trafficking of epidermal growth factor receptors (28). A recent study demonstrated that RIN (Ras and Rab interactor) family protein Rin-like (Rinl) binds to the nucleotide-free form of Rab5a and Rab22, and Rinl has GEF activity toward both proteins, but it acts with higher efficiency on Rab22 (29). The functional relationship between Rin1 and Rabex-5 is not currently known. Rinl shows higher expression in thymus and spleen; thus, it might contribute to an additional level of regulation in those tissues.

Vps21p-CORVET-Mon1p/Ccz1p-Ypt7p cascade. Vps21p is a yeast ortholog of mammalian Rab5 (**Figure 2**) (30). Vps21p-GTP interacts with an endosomal tethering complex, CORVET (class C core vacuole/endosome tethering), which shares four of the six components of the HOPS complex, and thus, like HOPS, the CORVET complex is an effector of Vps21p (31). The CORVET complex can form an intermediate complex with a HOPS-specific subunit and then finally mature into the HOPS complex. The HOPS complex directly recruits Ypt7p and was believed to catalyze the Ypt7p nucleotide exchange reaction (22). However, the Mon1p-Ccz1p complex was recently shown to be the Ypt7p-specific GEF (23). Mon1p forms a stable complex with Ccz1p and shows GEF activity that is independent of the HOPS complex (23, 32, 33). Taken together, a molecular mechanism for the Rab conversion from Vps21p to Ypt7p can be proposed as follows: (*a*) active Vps21p recruits Mon1p-Ccz1p and the CORVET complex; (*b*) CORVET forms an intermediate complex, and Mon1p-Ccz1p directly associates with Vps39p, one of the components of the intermediate complex; (*c*) Mon1p-Ccz1p recruits and activates Ypt7p and (*d*) active Ypt7p triggers assembly of the final form of the HOPS complex needed for homotypic fusion of the vacuole.

Rab/Ypt Cascades in the Secretory Pathway

Ypt1p-Ypt32p cascade. A Rab GEF cascade mechanism is also utilized on the exocytic pathway. Ypt1p is a yeast ortholog of Rab1 that acts in the ER-to-Golgi and intra-Golgi trafficking (**Figure 2**) (34). Ypt1p is activated by a large protein complex, named TRAPP (transport protein particle) (35, 36). TRAPP has nucleotide exchange activity toward Ypt1p, but not Ypt31p or Ypt32p, Rab GTPases that act downstream of Ypt1p. Genetic experiments have suggested a functional relationship between Ypt1 and Ypt32. Overexpression of Ypt31p or Ypt32p suppresses the growth defect of cells expressing Ypt1p-D124N, an allele that is unable to bind guanine nucleotide or a temperature-sensitive TRAPP mutant in which Ypt1p activation is blocked. Wang & Ferro-Novick (37) showed that beads carrying Ypt1p loaded with the nonhydrolyzable GTP analog GTPγS could retain a factor from a yeast lysate that stimulated nucleotide exchange on Ypt32p, suggesting that the GEF for Ypt32p is an effector of Ypt1p. The identity of the GEF for Ypt31p and/or Ypt32p is still unclear; however, the results suggest that Ypt1p and Ypt32p are linked by a Rab GEF cascade mechanism. The mammalian Ypt1p homolog, Rab1, has several documented effectors, but none to date have been reported to have GEF activity.

Ypt32p-Sec2p-Sec4p cascade. On the yeast secretory pathway, Sec2p acts as a GEF for Sec4p, a Rab8 homolog that plays an essential role in the regulation of the final stage of the exocytic pathway (**Figure 2**) (12). Although Sec2p is normally concentrated on secretory vesicles along with its substrate Sec4p, temperature-sensitive *sec2* mutants (*sec2-59*, *sec2-78*) mislocalize to the cytoplasm, and secretion is blocked. All of these phenotypes are suppressed by overexpression of Ypt32p. Sec2p interacts with the GTP-bound form of Ypt32p and has no GEF activity toward Ypt32p, indicating that Sec2p is an effector of Ypt32p (10). A subsequent study demonstrated that the Sec2p-Ypt32p interaction is required for Sec2p localization to the secretory vesicles (38). Thus, Ypt32p recruits Sec2p to the late Golgi membrane and possibly promotes vesicle budding from the Golgi compartment. Sec2p then catalyzes nucleotide exchange on the downstream Rab, Sec4p, which, in turn, leads to the delivery of vesicles and their fusion with the plasma membrane.

Rab11-Rabin8-Rab8 cascade. The mammalian orthologs of Ypt32p, Sec2p, and Sec4p are Rab11, Rabin8, and Rab8, respectively (**Figure 2**). Rab8 is involved in primary cilium formation, and its activation by Rabin8 is required for this function (39, 40). Rabin8 has specific GEF activity toward Rab8 (41). Rabin8 binds to the GTP-bound form of Rab11, and the interaction promotes activation of Rab8, which is required for ciliogenesis and apical membrane formation, indicating that a Rab GEF cascade directly analogous to the Ypt32-Sec2-Sec4 cascade is conserved in mammals (42, 43). Although Rab8 localizes to the primary cilium, Rabin8 is detected on Rab11-positive vesicles near the cilium base (44). Therefore, Rab11-positive vesicles may be converted into Rab8-positive compartments by a Rab transition mechanism. A recent study showed that Rabin8 binds to the mammalian TRAPP complex, and the localization of Rabin8 to the vesicle is TRAPP dependent; thus TRAPP may act together with Rab11 to recruit Rabin8 (44).

Rab33b-Rab6 cascade. The Golgi apparatus is composed of *cis*-, medial-, and *trans*-Golgi compartments (cisternae), and each cisterna has a distinct set of Golgi enzymes. These compartments may be maintained by a mechanism called "cisternal maturation," in which each compartment gradually acquires, by vesicular traffic, the enzymes typifying the subsequent cisterna and donates its own resident enzymes to a preceding cistern (45, 46). By this model, resident proteins must be rapidly transported by retrograde traffic within the Golgi apparatus via COPI vesicles. Rab33b and Rab6 locate to the medial- and *trans*-Golgi compartments, respectively, and regulate retrograde intra-Golgi trafficking (**Figure 2**). Rab6 genetically interacts with the retrograde Golgi tether complex, conserved oligomeric Golgi (COG)3 and Zeste White 10 (ZW10), and knockdown of Rab6 suppresses COG3- or ZW10-depletion-induced disruption of the Golgi complex (47). Knockdown of Rab33b exhibits a phenotype similar to Rab6 depletion and overexpression of Rab33b displaces Rab6 from Golgi membranes (48). Although the precise molecular mechanism is unknown, Rab6 and Rab33b functionally overlap and might form a cascade for retrograde traffic within the Golgi apparatus.

Ryh1p-Ric1p/Rgp1p-Ypt3p cascade. The fission yeast proteins Ryh1p and Ypt3p are homologs of Rab6 (budding yeast Ypt6p) and Rab11 (budding yeast Ypt31p/Ypt32p), respectively (**Figure 2**). Deletion of *ryh1* is synthetically lethal with a *ypt3* mutation, and overexpression of the GDP-locked form of Ryh1p-T25N inhibits growth of a *ypt3* mutant, indicating a genetic interaction between these two GTPases. Both Ryh1p and Ypt3p localize to the Golgi apparatus and endosome, where they regulate protein secretion and the recycling of the exocytic SNARE from endosomes to Golgi apparatus (49). The GEF for Ryh1p is presumed to be Ric1p, a component of the Ric1p/Rgp1p complex, which in budding yeast serves as a GEF for Ypt6p (50, 51). Although the GEF activity of Ric1p toward Ryh1p is unproven and it is unclear if Ric1p is an effector of Ypt3p, one could speculate that Ryh1p and Ypt3p form a Rab GEF cascade through the Ric1p/Rgp1p complex.

ADP Ribosylation Factors and Arf-Like Proteins

The ADP ribosylation factors (Arf) and Arf-like proteins (Arl) are small GTPases that regulate membrane traffic (8). Arf and Arl require nucleotide exchange for their function, and a number of GEFs for Arf and Arl have been identified to date (52). Arf and Arl also network with Rab proteins through cascade mechanisms.

Arf6-ARNO-Arf1 and Arl4-ARNO-Arf6 cascades. Arf1 localizes predominantly to the Golgi apparatus, where it regulates the formation of coated vesicles, whereas Arf6 localizes to the plasma membrane and is involved in cytoskeletal organization and endocytosis (**Figure 2**) (53). Although the functions of Arf1 and Arf6 are distinct, their GTP-bound conformations are very similar, and they share common effectors (54). The Arf nucleotide-binding site opener (Arno) has GEF activity toward both Arf1 and Arf6 but seems to prefer Arf1 over Arf6 (55). Arno has a pleckstrin homology (PH) domain that binds phosphatidylinositol 4,5-phosphate on the plasma membrane. Interestingly, the PH domain of Arno also binds to the GTP-bound form of Arf6, and both lipid-binding and Arf6-binding abilities are required for recruitment of Arno to the plasma membrane. Arno leads to relocalization of Arf1 from the Golgi apparatus to the plasma membrane, at least in part (55). A recent study showed that Arf6-GTP and liposomes promote the nucleotide exchange activity of Arno on Arf1 (56).

Arno also binds to the GTP-bound form of the Arf-like GTPase Arl4 together with phosphatidylinositol-4,5-bisphosphate and thereby relocates to the plasma membrane (57). Although Arno's target at the plasma membrane is still unclear, these observations suggest that GTPases of the Arf branch can form a cascade through their interactions with a GEF that is analogous to the Rab GEF cascades discussed above.

Arl3p-Imh1p-Arl1p cascade. Two independent studies showed that the Golgi protein Imh1p, a yeast homolog of mammalian Golgin, mislocalizes to the cytoplasm when *arl1* or *arl3* is deleted (58, 59). Arl1p is a Golgi-localized GTPase, implicated in regulating Golgi structure and protein sorting (**Figure 2**) (60). The GRIP domain of Imh1p is able to bind directly to the GTP-bound form of Arl1p but not to Arl3p. Golgi localization of both Arl1p and Imh1p is nevertheless dependent on Arl3p, suggesting that Arl3p might recruit the GEF for Arl1p, forming a cascade.

Rho and Cdc42 GTPases

Rho (Ras homolog) GTPases play numerous cellular roles, including regulation of polarized cell growth, organization of the actin cytoskeleton, and axon signaling (61–63). Their role in regulating cellular polarity has been well studied. Rho proteins regulate both the assembly of the actin cytoskeleton and delivery, docking, and fusion of secretory vesicles with the plasma membrane. Yeast has six Rho GTPases, Rho1–5p, and Cdc42p (cell division cycle 42), yet only Rho1p and Cdc42p are essential. In mammals, 23 Rho member proteins have been identified, but only RhoA, Rac, and Cdc42 have been studied in detail (64). Interestingly, a number of Rho GEFs and Rho GAPs have been identified, and a single Rho protein is often regulated by more than one GEF or GAP. However, it is not known if Rho family proteins form cascades through their GEFs, as have been described for Rab and Arf. Below, we introduce a few examples of Rho networks relevant to membrane traffic.

Both Rho3p and Cdc42p are Exo70p effectors.
Cellular polarization in yeast requires vectorial transport of secretory vesicles along actin cables. Rho3p and Cdc42p are key regulators of polarization of actin, whereas the type V myosin, Myo2p, directs the active delivery of vesicles on actin cables (65, 66). One of the Rho3p effectors is Exo70p, a component of the octameric exocyst complex needed for tethering secretory vesicles to the plasma membrane before fusion (67). Rho3p and Cdc42p mutants show similar trafficking defects, and both of them are required for efficient exocytosis, suggesting overlapping functions. Genetic studies have suggested that Exo70p is a target for both Rho3p and Cdc42p, and Wu et al. (68) showed that the GTP-locked forms of both Rho3p and Cdc42p are able to bind Exo70p, provided that they have undergone C-terminal prenylation. Thus, Exo70p is a common effector for both Rho3p and Cdc42p in exocytosis. Another exocyst component, Sec3p, works as an effector of both Rho1p and Cdc42p (69). The Cdc42p-Exo70p and Rho1p-Sec3p interactions appear to fulfill partially overlapping functions in exocyst localization; thus cross talk between these pathways may provide additional levels of regulation.

Rho1p and Cdc42p in vacuole fusion.
In addition to regulation of cell polarity, both Rho1p and Cdc42p play roles in homotypic vacuole fusion (70, 71). This is supported by the observation that Rho guanine nucleotide dissociation inhibitor, Rdi1p, which extracts Rho1p and Cdc42p from the vacuole membrane, blocks vacuole fusion (70). Vacuole fusion is mediated by actin polymerization, which is initiated by Cdc42p (72). However, the mechanisms that trigger the activation of Cdc42p and Rho1p at the vacuole are unknown. Active Rho pools can be detected with probes containing the Rho-binding domains of their downstream effectors. Using such probes it was recently demonstrated that activation of Cdc42p and Rho1p occurs with different kinetics and conditions (73). Cdc42p is activated in an ATP-independent manner, whereas Rho1p activation requires ATP. Moreover, an in vitro vacuole fusion experiment showed that Cdc42p activation occurs within 5 min, but fusion signals reach a maximum at 60 min and correlate with Rho1p activation. Thus, Cdc42p and Rho1p may be sequentially activated to promote vacuole fusion.

Cross Talk Between Rab1, GBF1, and Arf1

The later stage of ER-to-Golgi transport requires COPI vesicle formation, which is mediated by Arf1 (74, 75). Arf1 activation is catalyzed by its nucleotide exchange factor, GBF1 (Golgi-specific brefeldin A resistance factor 1) (76). Rab1b is localized to the Golgi apparatus and ER-Golgi interface, and Rab1b is also required for ER-to-Golgi transport (77). A Rab1-Arf1 GTPase cascade has been proposed on the basis of genetic studies in yeast (78). Indeed, mammalian Rab1b in its GTP-bound form physically interacts with GBF1 (79). Rab1b is required for membrane association of GBF1 and Arf1 at ER exit sites. Therefore, a Rab1b-GBF1-Arf1 cascade stabilizes activated Arf1 on the membrane and promotes COPI recruitment and vesicle formation.

Ras-RLIP76-Arf6-Rac.
The Ras family GTPase related-Ras (R-Ras) regulates pleiotopic cellular functions, including inhibition of cell proliferation and promotion of cell adhesion, spreading, and migration (80–84). R-Ras mediates these pathways through its effectors and thereby augments the activation of another GTPase, Rac1 (83). RLIP76 (RalBP1) was identified as an R-Ras specific effector (85). The interaction between active R-Ras and RLIP76 is required for cell spreading by promoting Rac1 GTPase activity. By contrast, the Arf6 GTPase regulates endosomal

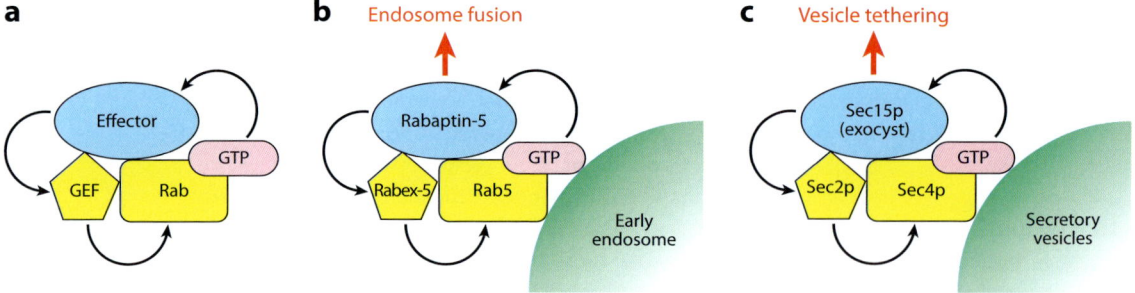

Figure 3

Rab-guanine nucleotide exchange factor (GEF)-effector positive-feedback loop. (*a*) Upon Rab activation by its GEF, a GTP-bound Rab recruits its effector(s). An effector then binds to the GEF and increases the exchange activity on the Rab GTPase. This Rab-GEF-effector positive-feedback loop stabilizes the active Rab region on the membrane. (*b*) Rab5 is activated by its GEF Rabex-5, and then active Rab5 recruits its effector Rabaptin-5. Rabaptin-5, in turn, binds to Rabex-5 directly (87, 88). Association with Rabaptin-5 increases Rabex-5 activity toward Rab5, thereby stabilizing Rab5 in a GTP-bound state on the early endosome, which is needed for endosomal fusion (89). (*c*) In the yeast secretory pathway, Sec2p, which is recruited to the membrane by Ypt32p-GTP, activates Sec4p on the secretory vesicles (12). The GTP-bound form of Sec4p recruits Sec15p, which is a component of the exocyst complex (90). Sec15p binds to Sec2p and displaces Ypt32p (11). Sec2p is able to activate Sec4p, and Sec4p again recruits Sec15p, which leads to vesicle tethering with the plasma membrane through the exocyst complex prior to membrane fusion.

trafficking of Rac1 and mediates Rac1 activation (86). Interestingly, the cell spreading defect in RLIP76-depleted cells can be rescued by activation of Arf through overexpression of the Arf GEF Arno. Indeed, Arno directly binds to RLIP76 in vivo (77, 85). Thus, active R-Ras recruits RLIP76, which, in turn, binds to Arno to activate Arf6 and then consequently activates Rac1. This multiple GTPase cascade promotes cell spreading and migration.

RAB-GUANINE NUCLEOTIDE EXCHANGE FACTOR POSITIVE-FEEDBACK LOOPS

Local activation of Rab GTPases is important for their functions on specific membrane regions. Thus, proper localization of the GEF is the key to the activation of Rab at the right place. GEFs often directly bind to one of the effectors of the substrate Rab. This results in a Rab-GEF-effector complex, potentially generating a positive-feedback loop: (*a*) The GEF activates the Rab, (*b*) the activated Rab recruits its effector, (*c*) the effector binds to GEF, and (*d*) the GEF activates more Rab protein (**Figure 3***a*). This positive-feedback loop could lead to the formation of a metastable platform of highly activated Rab proteins and highly concentrated effectors.

Rab5, Rabex-5, and Rabaptin-5

As mentioned in a previous section, Rab5 is activated by its GEF, Rabex-5. A number of Rab5 effectors have been identified, and each of them seems to regulate distinct downstream events. One of Rab5 effectors is Rabaptin-5 (87). Rabaptin-5 directly binds to Rabex-5, and the interaction is essential for endosomal fusion (88). Later it was shown that Rabaptin-5 stimulates Rabex-5 exchange activity in vitro (89). Although Rabaptin-5 is found to bind Rab5-GTP, recruitment of Rabaptin-5 to the membrane depends on its association with Rabex-5. This observation suggests that Rabex-5 and Rabaptin-5 work as a complex to stabilize an active Rab5 cluster on the endosome. Thus, the Rabex-5-Rabaptin-5 complex is recruited to the membrane, activates Rab5, and then Rabaptin-5 binds to Rab5 to form a positive-feedback loop promoting endosomal fusion (**Figure 3***b*).

Sec4p, Sec2p, and Sec15p

In yeast, Sec2p, Sec4p and Sec15p work together on the secretory pathway. Sec15p, a component of exocyst complex, binds to GTP-bound Sec4p, following activation by its GEF, Sec2p (12, 90). Interestingly, Sec2p and Sec15p directly bind to each other on the surface of secretory vesicles; thus the Rab, its GEF, and effector (Sec4p-Sec2p-Sec15p) form a complex (**Figure 3***c*) (11). This situation is somewhat different from the case of Rab5-Rabex-5-Rabaptin-5 because Sec15p does not appear to modulate the GEF activity of Sec2p.

A recent study (38) has shown that Sec2p is recruited to the membrane by the combined signals of Ypt32p-GTP and phosphatidylinositol-4-phosphate (PI4P). Ypt32p and Sec15p compete for Sec2p binding, and the interaction of these proteins is regulated by PI4P (11, 38). Sec2p-Sec15p binding is inhibited by PI4P, yet the PI4P level on secretory vesicles appears to decrease as the vesicles are delivered to exocytic sites (38). Thus, after Sec2p recruitment to the Golgi membrane by active Ypt32p and PI4P, PI4P levels decrease, allowing Sec15p to displace Ypt32p and form a complex with Sec2p. Thus a switch from a Rab GEF cascade (Ypt32p-Sec2p-Sec4p) to a Sec2p-Sec4p-Sec15p positive-feedback loop transition may be regulated by PI4P levels, facilitating vesicle tethering and fusion with the plasma membrane. In mammals, the Ypt32p ortholog Rab11 directly binds to Sec15, indicating that the exocyst functions as a Rab11 effector, which is somewhat different from the yeast system (91).

GTPase-ACTIVATING PROTEIN CASCADES

The sequential recruitment and activation of Rabs, in an ordered series, may help define the directionality of membrane traffic. The Rab GEF cascade mechanism, discussed above, proposes that an upstream Rab recruits the GEF for the Rab acting just downstream, thereby initiating a Rab conversion along a transport pathway. However, the spatial and temporal regulation of Rab function requires not only GEFs to activate the appropriate Rab at the right time and place but also GAPs to inactivate the Rab after it has fulfilled its function and as it migrates beyond its limited domain on the surface of an organelle. Live-cell imaging has demonstrated that the colocalization of adjacent Rabs is typically only transient (92), suggesting the existence of a mechanism to avoid a prolonged overlap of active Rabs on the same membrane domain. One possibility is a GAP cascade mechanism in which the downstream Rab, in its active form, recruits the GAP that inactivates the upstream Rab, thus limiting overlap. The GEF and GAP cascades working in a countercurrent fashion could serve as self-organizing systems to promote an efficient, spatially and temporally controlled Rab conversion processes. We describe several examples of GAP cascades along the exocytic and endocytic pathways of eukaryotic cells that promote Rab conversions. In addition, we describe examples of GAP cascades involving members from different branches of the Ras superfamily.

Ypt1p, Gyp1p, and Ypt31p/Ypt32p

In *Saccharomyces cerevisiae* cells, Ypt1p and Ypt31p/Ypt32p are essential Rabs that act sequentially. Ypt1p is the homolog of Rab1, and it plays an important role in membrane traffic from the ER to the Golgi apparatus and within the Golgi apparatus (**Figure 2**) (93, 94). Ypt31p and Ypt32p are highly similar and functionally redundant homologs of Rab11 required for the exit of exocytic cargo from a late-Golgi compartment (95, 96). Even though both Ypt1p and Ypt31p/Ypt32p are associated with Golgi compartments, their overlap is low (97). Furthermore, live-cell imaging of individual Golgi compartments in yeast cells shows that they undergo a transition from Ypt1p to Ypt31p/Ypt32p, consistent with a Golgi maturation model (45, 46). Gyp1p, a Tre-2/Bub2/Cdc16 (TBC) domain-containing protein, which localizes to the Golgi apparatus, has been identified as the GAP that downregulates Ypt1p in vivo (98, 99). A GAP cascade model predicts that Gyp1p would be recruited to the Golgi membrane by binding to the active form of Ypt31p/Ypt32p. Yeast two-hybrid analysis and in vitro binding experiments demonstrated that Gyp1p interacts with Ypt31p/Ypt32p (97). The interaction between Gyp1p and Ypt32p depends on the first 200 amino acids of Gyp1p, and not the C-terminal portion, where the Gyp1p TBC GAP catalytic domain resides. Moreover, this interaction is important for the recruitment of Gyp1p to the Golgi compartment because a Gyp1p construct missing the first 200 amino acids remains cytosolic and loss of Ypt31p/Ypt32p results in mislocalization and destabilization of Gyp1p. Further support for a Rab GAP cascade mechanism came from live-cell imaging analysis of Ypt1p and Ypt32p localization in *gyp1*-deficient cells. On the basis of the GAP cascade model, the loss of Gyp1p should prolong the overlap between Ypt1p and Ypt32p on Golgi compartments. Live-cell imaging analysis of Ypt1p and Ypt32p in *gyp1*-deficient cells shows an increase in the colocaliziation between Ypt1p and Ypt32p in comparison with normal cells. In normal cells, Ypt1p compartments undergo a rapid conversion to Ypt32p compartments; however, in *gyp1*-deficient cells, this transition is greatly slowed. In addition, the absence of Gyp1p leads to increased colocalization of the Ypt1p effector, Cog3p, with the late Golgi marker Sec7p, demonstrating the mistargeting of a Ypt1p effector. The mislocalization of Cog3p may explain some of the Golgi complex–associated trafficking defects described in *gyp1*-deficient cells (98, 100). A question that remains to be answered regarding this GAP cascade is how the recruitment of other Ypt32p effectors, such as Sec2p or Rcy1p, might affect the interaction of Gyp1p with Ypt32p (10, 101).

Rab5, TBC-2, and Rab7

Mammalian endosomes undergo a conversion from Rab5 to Rab7 as they mature from early endosomes to late endosomes (**Figure 2**) (92), and this conversion appears to involve a GEF cascade mechanism. However, a mathematical model of Rab5-to-Rab7 conversion suggests the need for a cut-out switch mechanism in which Rab7 activation leads to removal of Rab5 (102). A cut-out switch could be explained by a Rab GAP cascade in which a GAP for Rab5 is recruited once the proper level of active Rab7 is acquired by the endosomal membrane. However, identifying a GAP cascade that would trigger the loss of Rab5 as Rab7 is activated has been hampered owing to the difficulty of identifying the relevant GAP for Rab5. Studies have identified several GAPs that affect Rab5 activity (RabGAP-5, RN-Tre) in mammalian cells; however, the loss of these GAPs does not lead to an increased overlap between Rab5 and Rab7, as expected for a GAP cascade, but rather it affects the functionality of the early endosome (103–105).

Conversion from Rab5 to Rab7 has also been observed during the phagocytosis of apoptotic cells in *C. elegans* (106). *C. elegans* express 21 gene products containing TBC GAP domains, in comparison to the approximately 50 TBC proteins in mammalian cells. The powerful genetic and cell biological tools of *C. elegans* have helped to establish TBC-2 as a GAP for Rab5 that is important for a Rab5-to-Rab7 conversion (107). TBC-2, a homolog of human TBC1D2 and TBC1D2B, contains three conserved domains: an N-terminal PH domain, a coiled-coil region, and a C-terminal TBC domain. Worms deficient for TBC-2 accumulate large intestinal vesicles, which contain Rab7 and other late endosome markers on their membranes (108). These enlarged endosomes also contained Rab5 but not other Rabs, such as Rab10 or Rab11, suggesting an increased colocalization of Rab5 and Rab7 on the same compartment, a phenotype expected upon disruption of a Rab GAP cascade. These effects can be phenocopied by the expression of a Rab5 GTP hydrolysis-deficient mutant (Rab5-Q78L), demonstrating that the accumulation of these large vesicles is associated with the failure to inactivate Rab5. In further support of the Rab GAP cascade model is the observation that TBC-2 requires Rab7 to localize to endosomal compartments (108). The domain of TBC-2 that is important for the interaction with Rab7 remains to be elucidated, and it is important to determine if the PH domain of TBC-2 regulates this interaction.

Sec4p, Msb3p, Msb4p, and Cdc42p

In budding yeast, Sec4p, a homolog of mammalian Rab8, serves as the final Rab on the exocytic pathway (**Figure 2**). Sec4p is highly concentrated on secretory vesicles, where it acts to recruit several effectors, including the exocyst tethering complex, and to ensure the fusion of vesicles at the site of polarized secretion (90, 109). It is difficult to envisage how a Rab GAP cascade mechanism could remove Sec4p following exocytic fusion because there is no Rab that acts downstream of Sec4p on the plasma membrane. However, Rho GTPases are known to concentrate on the plasma membrane, where they control cell polarity and actin polymerization (110). A variant GAP cascade mechanism, which includes a Rho GTPase that recruits the GAP for Sec4p to the site of polarized secretion, may explain this conundrum. Msb3p/Msb4p are redundant GAPs that localize to the site of polarized secretion in yeast and exhibit GAP activity toward Sec4p in vitro (111). The localization and function of these GAPs require interactions with the polarisome subunit, Spa2p, and the Rho GTPase, Cdc42p (112). Moreover, the N-terminal domains of Msb3p/Msb4p are important for their interaction with Spa2p (112). However, the domain important for the interaction with Cdc42p has not yet been determined. The recruitment of Msb3p/Msb4p by these interactions would promote the inactivation of Sec4p at sites of polarized cell growth and would thereby ensure that Sec4p and its effectors are efficiently recycled from the membrane following fusion.

Rab35, Centaurinβ2, and Arf6

Mammalian cells contain more than 60 Rabs, and an important aspect of characterizing their function has been to identify effectors for each Rab. Recent screens have identified novel interacting partners for some of these Rabs (113, 114). Centaurinβ2 was identified as an effector of Rab35. Centaurinβ2 contains five domains: two coiled-coil domains, a PH domain, an Arf GAP domain, and an ANKR domain. In vitro binding experiments demonstrated that the interaction with Rab35 depends on the ANKR domain, and localization of Centaurinβ2 to the plasma membrane of cells depends on the expression of Rab35. The fact that Centaurinβ2 also contains an Arf GAP domain suggests that the interaction with Rab35 is important for the inactivation of an Arf GTPase. Analysis of the GTP-Arf6 levels on cells expressing Rab35 and Centaurinβ2 show a reduced level of GTP-Arf6, implicating Arf6 as a possible target of the Rab35-Centaurinβ2 complex (114).

The same screen also indicated that Rab22A interacts with the Rab GAP mKIAA1055 and that Rab36 interacts with GAPCenA protein (114). The mKIAA1055 interaction with Rab22A occurs through its coiled-coil region, and GAPCenA interacts with Rab36 through its PTB domain. Neither of these Rab GAPs showed significant GAP activity against these Rab-binding partners; however, they colocalized with them when coexpressed in mammalian cells. These results suggest the possibility of a Rab GAP cascade, yet the Rab substrates for mKIAA1055 and GAPCenA remain to be identified. It is important to determine if the loss of these GAPs increases the overlap of their substrate Rabs with Rab22A or Rab36.

RABS WITH COMMON EFFECTORS

The Rab family includes a large number of members, each concentrated on a specific membrane compartment or set of compartments (89). Many organelles, such as the Golgi apparatus and endosomes, are marked by multiple Rabs (**Figure 2**) (115), as would be expected because of the extensive membrane exchange that occurs between subcellular compartments. Still, the main function of Rabs is to direct the recruitment of their specific effectors to these locations. Recent studies have shown that certain effectors can interact with multiple Rabs, sometimes in combinations. Compartments carrying these multivalent effectors may represent crossroads where the sorting of protein and membrane must be regulated.

Rab4, Rab14, and Rabip4/RUFY1

Rab4 and Rab14 play significant roles in membrane traffic at the endosomal level. Rab4 is important in directing cargo between the sorting endosome and the recycling endosome, where it colocalizes with Rab5 and Rab11 (116, 117). Rab14 is important for traffic between the endosomes and a late Golgi compartment (118). In addition, Rab14 is associated with vesicles carrying the glucose transporter 4. These vesicles are generated from endosome and Golgi membranes and are translocated to the plasma membrane in response to insulin (119). Rabip4/RUFY1 was first identified as a Rab4 effector; however, it has also been shown to interact with Rab5 and play a role in trafficking from endosomes (120, 121). Rabip4 is composed of a RUN domain, two coiled-coil domain, and a FYVE motif. A recent study found that Rabip4 is also an effector of Rab14 (122). The interaction of Rabip4 with Rab14 is important for the localization of Rabip4 to the endosomes, where it is also colocalizes with Rab4 and Rab5 (122). Moreover, when Rab4, Rab14, and Rabip4 are coexpressed, endosomes become larger than normal. The study suggests a sequential interaction of Rabip4 with Rab14 and Rab4 to promote endosome tethering and fusion. Interestingly, Rabip4 can also interact with, and be phosphorylated by, the tyrosine kinase Etk, and this modification is important for Rabip4 localization to the endosomal membrane (123). This suggests a regulated mechanism that depends on the phosphorylation of Rabip4 to control its localization to the endosomes. It will be interesting to determine if such a mechanism can regulate the Rabip4 interaction with Rab14.

Rab4, Rab11, and D-AKAP2

Rab4 and Rab11 are important for the proper sorting of cargo and recruitment of effectors to the sorting and recycling endosome, respectively (116, 117). These Rabs also regulate the kinetics of recycling from endosomes; Rab4 controls fast recycling and Rab11 controls slow recycling from a perinuclear compartment (124). D-AKAP2 (dual-specific A-kinase-anchoring protein 2) contains three regulators of G protein–signaling domains, a protein kinase A–interacting domain, and a PDZ-binding motif (125). D-AKAP2 interacts with Rab4 and Rab11 through the region containing a regulator of G protein–signaling domain, and this interaction is important for the localization of D-AKAP2 to compartments containing Rab4 and Rab11 (125). However, yeast two-hybrid analysis and in vitro binding studies demonstrated preferential binding of D-AKAP2 to Rab11wt or Rab11S25N (mimicking the GDP-bound form), as well as to Rab4-Q67L (hydrolysis-deficient, GTP-bound form). These results suggest that D-AKAP2 is an effector of Rab4 and that it might have GEF activity toward Rab11, suggesting a possible GEF cascade. Interestingly, depletion of D-AKAP2 causes Rab11 compartments to accumulate in the cell periphery and causes the endosomal cargos to be released more rapidly from the cells (125). It is important to elucidate whether the protein kinase A–interacting domain of D-AKAP2 and the possible recruitment of PKA play any role in the function of this protein at the endosomal level.

Families of Interacting Proteins: Rab11 and Rab14

The Rab11 family of interacting proteins (FIPs) comprise five Rab11 effectors. These have been divided into two classes on the basis of their domain composition. Class I contains a C2 lipid-binding domain at the N terminus, a coiled-coil region, and a Rab11-binding domain (RBD) domain at the C terminus. Members of this class are Rip11, FIP2, and RCP. Class II contains EF domains at the N terminus and multiple coiled-coil domains with the RBD at the C terminus. Members of this class are FIP3 and FIP4

(126). A yeast two-hybrid screen revealed that FIP2 can interact with Rab14 (127) and that the GTPase-deficient mutant, Rab14-Q70L, interacts with the other two class I FIPs (127). The same study demonstrated that Rab14-Q70L localizes to recycling endosomes, whereas Rab14 S25N localizes to the Golgi apparatus. Rab14 interacts with the RBD domain of FIPs, suggesting either competition or cooperation with Rab11 for this binding site (127). It will be interesting to determine if any of the interactions between Rab14 and class I FIPs are regulated in a cell cycle–dependent manner to ensure the targeting of Rab14-Rab11 containing vesicles to the cleavage furrows of dividing cells.

Rab6, Rab11, and R6IP1

Rab6 plays a variety of roles in membrane traffic to and from the Golgi apparatus (128–131). More recently, it has also been associated with trafficking of post-Golgi vesicles to the plasma membrane and in the fission of vesicles from the Golgi apparatus (132–134). R6IP1A and R6IP1B are isoforms of a Rab6 effector, identified in a yeast two-hybrid screen, that contain two RUN domains (135). RUN domains have been shown to interact with both Rap and Rab GTPases (136). In addition to binding Rab6, R6IP1 can also interact with Rab11 (137). Interestingly, overexpression of Rab11 relocalizes R6IP1 onto endosomes from its normal Golgi membrane association mediated by binding Rab6 (137). The interaction of R6IP1 with both Rab11 and Rab6 might direct the traffic of vesicles from a Rab11 compartment to a Rab6 compartment when active Rab11 levels are high. For example, cells depleted of R6IP1 displayed a defect in cell division, a process that requires high temporal and spatial regulation of Rab11 endosomes (137, 138). More extensive analysis of the interaction between Rab11 and R6IP1 is required to determine which domain of R6IP1 is important for the interaction and to investigate how changes in this domain may change the ability of R6IP1 to localize to Rab11-positive compartments or how this interaction may change the ability of Rab11 to interact with the different FIP proteins.

Rab1, Rab6, and Giantin

There are several long coiled-coil proteins present on Golgi or endosomal membranes in mammalian cells that have been characterized as tethering factors (139) and Rab effectors. Giantin is the largest coiled-coil protein identified (3260 amino acids), and it has a putative *trans*-membrane domain at its C terminus that anchors it to the Golgi membrane. At the *cis*-Golgi level, the coiled-coil proteins p115 and GM130 have been established as effectors of Rab1 (140–142). In addition, these proteins interact with each other to regulate SNARE complex assembly (140, 143–145). Moreover, it has been demonstrated that Giantin can interact with and compete against GM130 for binding to the C terminus of p115 (146–148). The interaction of GM130 and Giantin with p115 is important for the tethering of COPII and COPI containing vesicles at the Golgi level (145). In addition, Giantin also interacts with Rab1 and Rab6, suggesting a possible mechanism of regulation of tethering of different classes of vesicles at the Golgi level, which may depend on the interaction of these Rabs with multiple coiled-coil proteins (149). The interaction of Giantin with Rab1 and Rab6 occurs at a similar region, amino acids 162–357; however, in vitro analysis has shown a better binding of Giantin with Rab6 than with Rab1 (149). Because of the recent findings that Rab6 can play an important role in exocytosis and vesicle fission from the Golgi apparatus (132, 133), it will be interesting to determine if the interaction with Giantin plays any role in these Rab6-dependent pathways. This would suggest a mechanism whereby traffic within the Golgi apparatus depends on the interactions of Giantin with both Rabs, but exit from the Golgi apparatus depends solely on the interaction of Giantin with Rab6.

Rab8, Rab13, and JRAB/MICAL-L2

MICAL represents a family of five proteins with the following domain composition: a FAD domain, a CH domain, a LIM domain, and coiled-coil domains. MICAL members have been identified as effectors for various Rabs (150). JRAB/MICAL-L2 is an effector of Rab13 that is important for the proper targeting of occludin to tight junctions (TJs) of polarized ephitelial cells (151, 152). This role depends on the C-terminal coiled-coil domain because deletion of this domain impaired the recycling of occludin, but not the recycling of the transferrin receptor to the plasma membrane of epithelial cells (152). Rab8 has also been implicated in traffic from endosomes to the basolateral membranes of polarized cells (153, 154). In polarized cells, TJs act as boundaries between the apical and basolateral membranes, whereas adherens junctions (AJs) are needed to stabilize cell-to-cell contacts (155). The TJ and AJ are composed of similar proteins; however, their localization at the plasma membrane is different. Moreover, it was observed that Rab13 depletion affects occludin delivery but not that of E-cadherin, suggesting distinct pathways for the delivery of these proteins to the plasma membrane (155). Nonetheless, JRAB depletion reduces the delivery of both proteins. By analyzing the interaction of JRAB with other Rabs, it was found that JRAB interacts with Rab8 in addition to Rab13. Interestingly, the interaction with Rab8 also depends on the coiled-coil domain of JRAB, and binding experiments demonstrated that Rab8 and Rab13 compete for binding to JRAB (155). Moreover, the interaction of JRAB with Rab8 and Rab13 occurs at different cellular compartments. The location and interaction of JRAB with these Rabs support its role in establishing both a TJ and an AJ and demonstrate how regulating the interaction of an effector with different Rabs can regulate the sorting of different cargos. It is important to determine if the interaction with a specific Rab allows JRAB to interact with specific cargos at the endosome level and if the interaction can promote JRAB to interact with other proteins. Recently, it has been demonstrated that MICAL3, Rab6, and Rab8 play roles in the docking and fusion of vesicles; however, Rab6 was proposed to interact with MICAL3/Rab8 through its effector ELKS, which can also bind MICAL3 (156).

CONCLUDING REMARKS

We have presented some examples of the different mechanisms that link together two or more GTPases into a regulatory circuit that controls membrane traffic. As research continues, other mechanisms, no doubt, may come to light. Over time, we will learn how commonly these mechanisms are used in the regulation of the many stages of membrane traffic and to what extent they are conserved between species. A more complete description will require mathematical modeling of these regulatory networks with quantitative evaluation of all of the key parameters. This would represent an important step toward an understanding at a molecular level of the exocytic and endocytic pathways as stable, self-organizing systems.

SUMMARY POINTS

GEF cascades and GAP cascades, working in concert, can direct a programmed series of GTPase transitions, while GEF-effector interactions can generate positive-feedback loops that lead to the formation of membrane domains marked by highly activated pools of a GTPase. The net effect of these various mechanisms is the establishment of regulatory circuits that control membrane traffic.

FUTURE ISSUES

Additional studies are needed to establish the generality of the mechanisms presented here and to define at a quantitative level the effects of these mechanisms on membrane traffic. It will be interesting and informative to test the effects of "rewiring" the circuitry by mixing and matching the different domains of GEFs, GAPs, and effectors.

DISCLOSURE STATEMENT

The authors are not aware of any affiliations, memberships, funding, or financial holdings that might be perceived as affecting the objectivity of this review.

ACKNOWLEDGMENTS

This study was supported by grants GM35370 and GM082861 from the National Institutes of Health (P.N.). E.M.Y. was supported by a fellowship from the Human Frontier Science Program.

LITERATURE CITED

1. Spang A, Herrmann JM, Hamamoto S, Schekman R. 2001. The ADP ribosylation factor-nucleotide exchange factors Gea1p and Gea2p have overlapping, but not redundant functions in retrograde transport from the Golgi to the endoplasmic reticulum. *Mol. Biol. Cell* 12(4):1035–45
2. Schalk I, Zeng K, Wu SK, Stura EA, Matteson J, et al. 1996. Structure and mutational analysis of Rab GDP-dissociation inhibitor. *Nature* 381(6577):42–48
3. Huang M, Weissman JT, Beraud-Dufour S, Luan P, Wang C, et al. 2001. Crystal structure of Sar1-GDP at 1.7 A resolution and the role of the NH2 terminus in ER export. *J. Cell Biol.* 155(6):937–48
4. Goldberg J. 1998. Structural basis for activation of ARF GTPase: mechanisms of guanine nucleotide exchange and GTP-myristoyl switching. *Cell* 95(2):237–48
5. Gillingham AK, Munro S. 2007. The small G proteins of the Arf family and their regulators. *Annu. Rev. Cell Dev. Biol.* 23:579–611
6. Grosshans BL, Ortiz D, Novick P. 2006. Rabs and their effectors: achieving specificity in membrane traffic. *Proc. Natl. Acad. Sci. USA* 103(32):11821–27
7. Pereira-Leal JB, Seabra MC. 2001. Evolution of the Rab family of small GTP-binding proteins. *J. Mol. Biol.* 313(4):889–901
8. D'Souza-Schorey C, Chavrier P. 2006. ARF proteins: roles in membrane traffic and beyond. *Nat. Rev. Mol. Cell Biol.* 7(5):347–58
9. Stenmark H. 2009. Rab GTPases as coordinators of vesicle traffic. *Nat. Rev. Mol. Cell Biol.* 10(8):513–25
10. Ortiz D, Medkova M, Walch-Solimena C, Novick P. 2002. Ypt32 recruits the Sec4p guanine nucleotide exchange factor, Sec2p, to secretory vesicles; evidence for a Rab cascade in yeast. *J. Cell Biol.* 157(6):1005–15
11. Medkova M, France YE, Coleman J, Novick P. 2006. The rab exchange factor Sec2p reversibly associates with the exocyst. *Mol. Biol. Cell* 17(6):2757–69
12. Walch-Solimena C, Collins RN, Novick PJ. 1997. Sec2p mediates nucleotide exchange on Sec4p and is involved in polarized delivery of post-Golgi vesicles. *J. Cell Biol.* 137(7):1495–509
13. McLauchlan H, Newell J, Morrice N, Osborne A, West M, Smythe E. 1998. A novel role for Rab5-GDI in ligand sequestration into clathrin-coated pits. *Curr. Biol.* 8(1):34–45
14. Semerdjieva S, Shortt B, Maxwell E, Singh S, Fonarev P, et al. 2008. Coordinated regulation of AP2 uncoating from clathrin-coated vesicles by rab5 and hRME-6. *J. Cell Biol.* 183(3):499–511
15. Nielsen E, Severom F, Backer JM, Hyman AA, Zerial M. 1999. Rab5 regulates motility of early endosomes on microtubules. *Nat. Cell Biol.* 1(6):376–82

16. Rubino M, Miaczynska M, Lippé R, Zerial M. 2000. Selective membrane recruitment of EEA1 suggests a role in directional transport of clathrin-coated vesicles to early endosomes. *J. Biol. Chem.* 275(6):3745–48
17. Stenmark H, Parton RG, Steele-Mortimer O, Lütcke A, Gruenberg J, Zerial M. 1994. Inhibition of rab5 GTPase activity stimulates membrane fusion in endocytosis. *EMBO J.* 13(6):1287–96
18. Barbero P, Bittova L, Pfeffer SR. 2002. Visualization of Rab9-mediated vesicle transport from endosomes to the trans-Golgi in living cells. *J. Cell Biol.* 156(3):511–18
19. Vonderheit A, Helenius A. 2005. Rab7 associates with early endosomes to mediate sorting and transport of Semliki forest virus to late endosomes. *PLoS Biol.* 3(7):e233
20. Rink J, Ghigo E, Kalaidzdis Y, Zerial M. 2005. Rab conversion as a mechanism of progression from early to late endosomes. *Cell* 122(5):735–49
21. Seals DF, Eitzen G, Margolis N, Wickner WT, Price A. 2000. A Ypt/Rab effector complex containing the Sec1 homolog Vps33p is required for homotypic vacuole fusion. *Proc. Natl. Acad. Sci. USA* 97(17):9402–7
22. Wurmser AE, Sato TK, Emr SD. 2000. New component of the vacuolar class C-Vps complex couples nucleotide exchange on the Ypt7 GTPase to SNARE-dependent docking and fusion. *J. Cell Biol.* 151(3):551–62
23. Nordmann M, Cabrera M, Perz A, Bröcker C, Ostrowicz C, et al. 2010. The Mon1-Ccz1 complex is the GEF of the late endosomal Rab7 homolog Ypt7. *Curr. Biol.* 20(18):1654–59
24. Poteryaev D, Datta S, Ackema K, Zerial M, Spang A. 2010. Identification of the switch in early-to-late endosome transition. *Cell* 141(3):497–508
25. Poteryaev D, Fares H, Bowerman B, Spang A. 2007. *Caenorhabditis elegans* SAND-1 is essential for RAB-7 function in endosomal traffic. *EMBO J.* 26(2):301–12
26. Pereira-Leal JB, Seabra MC. 2000. The mammalian Rab family of small GTPases: definition of family and subfamily sequence motifs suggests a mechanism for functional specificity in the Ras superfamily. *J. Mol. Biol.* 301(4):1077–87
27. Stenmark H, Olkkonen VM. 2001. The Rab GTPase family. *Genome Biol.* 2(5):REVIEWS3007
28. Zhu H, Liang Z, Li G. 2009. Rabex-5 is a Rab22 effector and mediates a Rab22-Rab5 signaling cascade in endocytosis. *Mol. Biol. Cell* 20:4720–29
29. Woller B, Luiskandl S, Popovic M, Prieter BE, Ikonge G, et al. 2011. Rin-like, a novel regulator of endocytosis, acts as guanine nucleotide exchange factor for Rab5a and Rab22. *Biochim. Biophys. Acta* 1813(6):1198–210
30. Horazdovsky BF, Busch GR, Emr SD. 1994. VPS21 encodes a rab5-like GTP binding protein that is required for the sorting of yeast vacuolar proteins. *EMBO J.* 13(6):1297–309
31. Peplowska K, Markgraf DF, Ostrowicz CW, Bange G, Ungermann C. 2007. The CORVET tethering complex interacts with the yeast Rab5 homolog Vps21 and is involved in endo-lysosomal biogenesis. *Dev. Cell* 12(5):739–50
32. Ostrowicz CW, Bröcker C, Ahnert F, Nordmann M, Lachmann J, et al. 2010. Defined subunit arrangement and Rab interactions are required for functionality of the HOPS tethering complex. *Traffic* 11(10):1334–46
33. Wang CW, Stromhaug PE, Kauffman EJ, Weisman LS, Klionsky DJ. 2003. Yeast homotypic vacuole fusion requires the Ccz1-Mon1 complex during the tethering/docking stage. *J. Cell Biol.* 163(5):973–85
34. Jones S, Litt RJ, Richardson CJ, Segev N. 1995. Requirement of nucleotide exchange factor for Ypt1 GTPase mediated protein transport. *J. Cell Biol.* 130(5):1051–61
35. Wang W, Sacher M, Ferro-Novick S. 2000. TRAPP stimulates guanine nucleotide exchange on Ypt1p. *J. Cell Biol.* 151(2):289–96
36. Sacher M, Barrowman J, Wang W, Horecka J, Zhang Y, et al. 2001. TRAPP I implicated in the specificity of tethering in ER-to-Golgi transport. *Mol. Cell* 7(2):433–42
37. Wang W, Ferro-Novick S. 2002. A Ypt32p exchange factor is a putative effector of Ypt1p. *Mol. Biol. Cell* 13:3336–43
38. Mizuno-Yamasaki E, Medkova M, Coleman J, Novick P. 2010. Phosphatidylinositol 4-phosphate controls both membrane recruitment and a regulatory switch of the Rab GEF Sec2p. *Dev. Cell* 18(5):828–40
39. Yoshimura S, Egerer J, Fuchs E, Haas AK, Barr FA. 2007. Functional dissection of Rab GTPases involved in primary cilium formation. *J. Cell Biol.* 178(3):363–69

40. Nachury MV, Loktev AV, Zhang Q, Westlake CJ, Peränen J, et al. 2007. A core complex of BBS proteins cooperates with the GTPase Rab8 to promote ciliary membrane biogenesis. *Cell* 129(6):1201–13
41. Hattula K, Furuhjelm J, Arffman A, Paränen J. 2002. A Rab8-specific GDP/GTP exchange factor is involved in actin remodeling and polarized membrane transport. *Mol. Biol. Cell* 13:3268–80
42. Knödler A, Feng S, Zhang J, Zhang X, Das A, et al. 2010. Coordination of Rab8 and Rab11 in primary ciliogenesis. *Proc. Natl. Acad. Sci. USA* 107(14):6346–51
43. Bryant DM, Datta A, Rodríguez-Fraticelli AE, Peränen J, Martín-Belmonte F, Mostov KE. 2010. A molecular network for de novo generation of the apical surface and lumen. *Nat. Cell Biol.* 12(11):1035–45
44. Westlake CJ, Baye LM, Nachury MV, Wright KJ, Ervin KE, et al. 2011. Primary cilia membrane assembly is initiated by Rab11 and transport protein particle II (TRAPPII) complex-dependent trafficking of Rabin8 to the centrosome. *Proc. Natl. Acad. Sci. USA* 108(7):2759–64
45. Matsuura-Tokita K, Takeuchi M, Ichihara A, Mikuriya K, Nakano A. 2006. Live imaging of yeast Golgi cisternal maturation. *Nature* 441(7096):1007–10
46. Losev E, Reinke CA, Jellen J, Strongin DE, Bevis BJ, Glick BS. 2006. Golgi maturation visualized in living yeast. *Nature* 441(7096):1002–6
47. Sun Y, Shestakova A, Hunt L, Sehgal S, Lupashin V, Storrie B. 2007. Rab6 regulates both ZW10/RINT-1 and conserved oligomeric Golgi complex-dependent Golgi trafficking and homeostasis. *Mol. Biol. Cell* 18(10):4129–42
48. Starr T, Sun Y, Wilkins N, Storrie B. 2010. Rab33b and Rab6 are functionally overlapping regulators of Golgi homeostasis and trafficking. *Traffic* 11(5):626–36
49. He Y, Sugiura R, Ma Y, Kita A, Deng L, et al. 2006. Genetic and functional interaction between Ryh1 and Ypt3: two Rab GTPases that function in *S. pombe* secretory pathway. *Genes Cells* 11(3):207–21
50. Siniossoglou S, Peak-Chew SY, Pelham HR. 2000. Ric1p and Rgp1p form a complex that catalyses nucleotide exchange on Ypt6p. *EMBO J.* 19(18):4885–94
51. Ma Y, Sugiura R, Zhang L, Zhou X, Takeuchi M, et al. 2010. Isolation of a fission yeast mutant that is sensitive to valproic acid and defective in the gene encoding Ric1, a putative component of Ypt/Rab-specific GEF for Ryh1 GTPase. *Mol. Genet. Genomics* 284(3):161–71
52. Casanova JE. 2007. Regulation of Arf activation: the Sec7 family of guanine nucleotide exchange factors. *Traffic* 8(11):1476–85
53. Gillingham AK, Munro S. 2007. The small G proteins of the Arf family and their regulators. *Annu. Rev. Cell Dev. Biol.* 23:579–611
54. Pasqualato S, Ménétrey J, Franco M, Cherfils J. 2001. The structural GDP/GTP cycle of human Arf6. *EMBO Rep.* 2(3):234–38
55. Cohen LA, Honda A, Varnai P, Brown FD, Balla T, Donaldson JG. 2007. Active Arf6 recruits ARNO/cytohesin GEFs to the PM by binding their PH domains. *Mol. Biol. Cell* 18(6):2244–53
56. Stalder D, Barelli H, Gautier R, Macia E, Jackson CL, Antonny B. 2011. Kinetic studies of the Arf activator Arno on model membranes in the presence of Arf effectors suggest control by a positive feedback loop. *J. Biol. Chem.* 286(5):3873–83
57. Hofmann I, Thompson A, Sanderson CM, Munro S. 2007. The Arl4 family of small G proteins can recruit the cytohesin Arf6 exchange factors to the plasma membrane. *Curr. Biol.* 17(8):711–16
58. Setty S, Shin ME, Yoshino A, Marks MS, Burd CG. 2003. Golgi recruitment of GRIP domain proteins by Arf-like GTPase 1 is regulated by Arf-like GTPase 3. *Curr. Biol.* 13:401–4
59. Panic B, Whyte JRC, Munro S. 2003. The ARF-like GTPases Arl1p and Arl3p act in a pathway that interacts with vesicle-tethering factors at the Golgi apparatus. *Curr. Biol.* 13:405–10
60. Munro S. 2005. The Arf-like GTPase Arl1 and its role in membrane traffic. *Biochem. Soc. Trans.* 33(4):601–5
61. Perez P, Rincón SA. 2010. Rho GTPases: regulation of cell polarity and growth in yeasts. *Biochem. J.* 426(3):243–53
62. Sit ST, Manser E. 2011. Rho GTPases and their role in organizing the actin cytoskeleton. *J. Cell Sci.* 124(Pt. 5):679–83
63. Samuel F, Hynds D. 2010. RHO GTPase signaling for axon extension: Is prenylation important? *Mol. Neurobiol.* 42(2):133–42

64. Wherlock M, Mellor H. 2002. The Rho GTPase family: a Racs to Wrchs story. *J. Cell Sci.* 115(Pt. 2): 239–40
65. Adamo JE, Moskow JJ, Gladfelter AS, Viterbo D, Lew DJ, Brennwald PJ. 2001. Yeast Cdc42 functions at a late step in exocytosis, specifically during polarized growth of the emerging bud. *J. Cell Biol.* 155(4):581–92
66. Adamo JE, Rossi G, Brennwald P. 1999. The Rho GTPase Rho3 has a direct role in exocytosis that is distinct from its role in actin polarity. *Mol. Biol. Cell* 10:4121–33
67. Robinson NGG, Guo L, Imai J, Toh-E A, Matsui Y, Tamanoi F. 1999. Rho3 of *Saccharomyces cerevisiae*, which regulates the actin cytoskeleton and exocytosis, is a GTPase which interacts with Myo2 and Exo70. *Mol. Cell. Biol.* 19(5):3580–87
68. Wu H, Turner C, Gardner J, Temple B, Brennwald P. 2010. The Exo70 subunit of the exocyst is an effector for both Cdc42 and Rho3 function in polarized exocytosis. *Mol. Biol. Cell* 21(3):430–42
69. Roumanie O, Wu H, Molk JN, Rossi G, Bloom K, Brennwald P. 2005. Rho GTPase regulation of exocytosis in yeast is independent of GTP hydrolysis and polarization of the exocyst complex. *J. Cell Biol.* 170(4):583–94
70. Eitzen G, Thorngren N, Wickner W. 2001. Rho1p and Cdc24p act after Ypt7p to regulate vacuole docking. *EMBO J.* 20(20):5650–56
71. Müller O, Johnson DI, Mayer A. 2001. Cdc42p functions at the docking stage of yeast vacuole membrane fusion. *EMBO J.* 20(20):5657–65
72. Isgandarova S, Jones L, Forsberg D, Loncar A, Dawson J, et al. 2007. Stimulation of actin polymerization by vacuoles via Cdc42p-dependent signaling. *J. Biol. Chem.* 282(42):30466–75
73. Logan MR, Jones L, Eitzen G. 2010. Cdc42p and Rho1p are sequentially activated and mechanistically linked to vacuole membrane fusion. *Biochem. Biophys. Res. Commun.* 394(1):64–69
74. Donaldson JG, Cassel D, Kahn RA, Klausner RD. 1992. ADP-ribosylation factor, a small GTP-binding protein, is required for binding of the coatomer protein beta-COP to Golgi membranes. *Proc. Natl. Acad. Sci. USA* 89(14):6408–12
75. Bonifacino JS, Lippincott-Schwartz J. 2003. Coat proteins: shaping membrane transport. *Nat. Rev. Mol. Cell Biol.* 4(5):409–14
76. Garcia-Mata R, Sztul E. 2003. The membrane-tethering protein p115 interacts with GBF1, an ARF guanine-nucleotide-exchange factor. *EMBO Rep.* 4(3):320–25
77. Tisdale EJ, Bourne JR, Khosravi-Far R, Der CJ, Balch WE. 1992. GTP-binding mutants of rab1 and rab2 are potent inhibitors of vesicular transport from the endoplasmic reticulum to the Golgi complex. *J. Cell Biol.* 119(4):749–61
78. Jones S, Jedd G, Kahn RA, Franzusoff A, Bartolini F, Segev N. 1999. Genetic interactions in yeast between Ypt GTPases and Arf guanine nucleotide exchangers. *Genetics* 152(4):1543–56
79. Monetta P, Slavin I, Romero N, Alvarez C. 2007. Rab1b interacts with GBF1 and modulates both ARF1 dynamics and COPI association. *Mol. Biol. Cell* 18(7):2400–10
80. Komatsu M, Ruoslahti E. 2005. R-Ras is a global regulator of vascular regeneration that suppresses intimal hyperplasia and tumor angiogenesis. *Nat. Med.* 11(12):1346–50
81. Zhang Z, Vuori K, Wang H, Reed JC, Ruoslahti E. 1996. Integrin activation by R-ras. *Cell* 85(1):61–69
82. Keely PJ, Rusyn EV, Cox AD, Parise LV. 1999. R-Ras signals through specific integrin alpha cytoplasmic domains to promote migration and invasion of breast epithelial cells. *J. Cell Biol.* 145(5):1077–88
83. Holly SP, Larson MK, Parise LV. 2005. The unique N-terminus of R-ras is required for Rac activation and precise regulation of cell migration. *Mol. Biol. Cell* 16(5):2458–69
84. Ada-Nguema AS, Xenias H, Hofman JM, Wiggins CH, Sheetz MP, Keely PJ. 2006. The small GTPase R-Ras regulates organization of actin and drives membrane protrusions through the activity of PLCepsilon. *J. Cell Sci.* 119(Pt. 7):1307–19
85. Goldfinger LE, Ptak C, Jeffery ED, Shabanowitz J, Hunt DF, Ginsberg MH. 2006. RLIP76 (RalBP1) is an R-Ras effector that mediates adhesion-dependent Rac activation and cell migration. *J. Cell Biol.* 174(6):877–88
86. Santy LC, Ravichandran KS, Casanova JE. 2005. The DOCK180/Elmo complex couples ARNO-mediated Arf6 activation to the downstream activation of Rac1. *Curr. Biol.* 15(19):1749–54

87. Stenmark H, Vitale G, Ullrich O, Zerial M. 1995. Rabaptin-5 is a direct effector of the small GTPase Rab5 in endocytic membrane fusion. *Cell* 83(3):423–32
88. Horiuchi H, Lippé R, McBride HM, Rubino M, Woodman P, et al. 1997. A novel Rab5 GDP/GTP exchange factor complexed to Rabaptin-5 links nucleotide exchange to effector recruitment and function. *Cell* 90(6):1149–59
89. Lippé R, Miaczynska M, Rybin V, Runge A, Zeria M. 2001. Functional synergy between Rab5 effector Rabaptin-5 and exchange factor Rabex-5 when physically associated in a complex. *Mol. Biol. Cell* 12:2219–28
90. Guo W, Roth D, Walch-Solimena C, Novick P. 1999. The exocyst is an effector for Sec4p, targeting secretory vesicles to sites of exocytosis. *EMBO J.* 18(4):1071–80
91. Zhang XM, Ellis S, Sriratana A, Mitchell CA, Rowe T. 2004. Sec15 is an effector for the Rab11 GTPase in mammalian cells. *J. Biol. Chem.* 279(41):43027–34
92. Rink J, Ghigo E, Kalaidzidis Y, Zerial M. 2005. Rab conversion as a mechanism of progression from early to late endosomes. *Cell* 122(5):735–49
93. De Antoni A, Schmitzová J, Trepte HH, Gallwitz D, Albert S. 2002. Significance of GTP hydrolysis in Ypt1p-regulated endoplasmic reticulum to Golgi transport revealed by the analysis of two novel Ypt1-GAPs. *J. Biol. Chem.* 277(43):41023–31
94. Suvorova ES, Duden R, Lupashin VV. 2002. The Sec34/Sec35p complex, a Ypt1p effector required for retrograde intra-Golgi trafficking, interacts with Golgi SNAREs and COPI vesicle coat proteins. *J. Cell Biol.* 157(4):631–43
95. Benli M, Döring F, Robinson DG, Yang X, Gallwitz D. 1996. Two GTPase isoforms, Ypt31p and Ypt32p, are essential for Golgi function in yeast. *EMBO J.* 15(23):6460–75
96. Jedd G, Mulholland J, Segev N. 1997. Two new Ypt GTPases are required for exit from the yeast *trans*-Golgi compartment. *J. Cell Biol.* 137(3):563–80
97. Rivera-Molina FE, Novick PJ. 2009. A Rab GAP cascade defines the boundary between two Rab GTPases on the secretory pathway. *Proc. Natl. Acad. Sci. USA* 106(34):14408–13
98. Du LL, Novick P. 2001. Yeast Rab GTPase-activating protein Gyp1p localizes to the Golgi apparatus and is a negative regulator of Ypt1p. *Mol. Biol. Cell* 12(5):1215–26
99. Pan X, Eathiraj S, Munson M, Lambright DG. 2006. TBC-domain GAPs for Rab GTPases accelerate GTP hydrolysis by a dual-finger mechanism. *Nature* 442(7100):303–6
100. Lafourcade C, Galan M, Gloor Y, Haquenauer-Tsapis R, Peter M. 2004. The GTPase-activating enzyme Gyp1p is required for recycling of internalized membrane material by inactivation of the Rab/Ypt GTPase Ypt1p. *Mol. Cell. Biol.* 24(9):3815–26
101. Chen SH, Chen S, Tokarev AA, Liu F, Jedd G, Segev N. 2005. Ypt31/32 GTPases and their novel F-box effector protein Rcy1 regulate protein recycling. *Mol. Biol. Cell* 16(1):178–92
102. Del Conte-Zerial P, Brusch L, Rink JC, Collinet C, Kalaidzidis Y, et al. 2008. Membrane identity and GTPase cascades regulated by toggle and cut-out switches. *Mol. Syst. Biol.* 4:206
103. Haas AK, Fuchs E, Kopajtich R, Barr FA. 2005. A GTPase-activating protein controls Rab5 function in endocytic trafficking. *Nat. Cell Biol.* 7(9):887–93
104. Lanzetti L, Palamidessi A, Areces L, Scita G, Di Fiore PP. 2004. Rab5 is a signalling GTPase involved in actin remodelling by receptor tyrosine kinases. *Nature* 429(6989):309–14
105. Lanzetti L, Rybin V, Malabarba MG, Christoforidis S, Scita G, et al. 2000. The Eps8 protein coordinates EGF receptor signalling through Rac and trafficking through Rab5. *Nature* 408(6810):374–77
106. Kinchen JM, Doukoumetzidis K, Almendinger J, Stergiou L, Tosello-Trampont A, et al. 2008. A pathway for phagosome maturation during engulfment of apoptotic cells. *Nat. Cell Biol.* 10(5):556–66
107. Li W, Zou W, Zhao D, Yan J, Zhu Z, et al. 2009. *C. elegans* Rab GTPase activating protein TBC-2 promotes cell corpse degradation by regulating the small GTPase RAB-5. *Development* 136(14):2445–55
108. Chotard L, Mishra AK, Sylvain MA, Tuck S, Lambright DG, Rocheleau CE. 2010. TBC-2 regulates RAB-5/RAB-7-mediated endosomal trafficking in *Caenorhabditis elegans*. *Mol. Biol. Cell* 21(13):2285–96
109. TerBush DR, Maurice T, Roth D, Novick P. 1996. The Exocyst is a multiprotein complex required for exocytosis in *Saccharomyces cerevisiae*. *EMBO J.* 15(23):6483–94
110. Pruyne D, Bretscher A. 2000. Polarization of cell growth in yeast. I. Establishment and maintenance of polarity states. *J. Cell Sci.* 113(Pt. 3):365–75

111. Matsukawa J, Nakayama K, Nagao T, Ichijo H, Urushidani T. 2003. Role of ADP-ribosylation factor 6 (ARF6) in gastric acid secretion. *J. Biol. Chem.* 278(38):36470–75
112. Tcheperegine SE, Gao XD, Bi E. 2005. Regulation of cell polarity by interactions of Msb3 and Msb4 with Cdc42 and polarisome components. *Mol. Cell. Biol.* 25(19):8567–80
113. Fukuda M, Kanno E, Ishibashi K, Itoh T. 2008. Large scale screening for novel Rab effectors reveals unexpected broad Rab binding specificity. *Mol. Cell Proteomics* 7(6):1031–42
114. Kanno E, Ishibashi K, Kobayashi H, Matsui T, Ohbayashi N, Fukuda M. 2010. Comprehensive screening for novel rab-binding proteins by GST pull-down assay using 60 different mammalian Rabs. *Traffic* 11(4):491–507
115. Hutagalung AH, Novick PJ. 2011. Role of Rab GTPases in membrane traffic and cell physiology. *Physiol. Rev.* 91(1):119–49
116. Sonnichsen B, De Renzis S, Nielsen E, Rietdorf J, Zerial M. 2000. Distinct membrane domains on endosomes in the recycling pathway visualized by multicolor imaging of Rab4, Rab5, and Rab11. *J. Cell Biol.* 149(4):901–14
117. de Renzis S, Sonnichsen B, Zerial M. 2002. Divalent Rab effectors regulate the sub-compartmental organization and sorting of early endosomes. *Nat. Cell Biol.* 4(2):124–33
118. Junutula JR, De Maziere AM, Peden AA, Ervin KE, Advani RJ, et al. 2004. Rab14 is involved in membrane trafficking between the Golgi complex and endosomes. *Mol. Biol. Cell* 15(5):2218–29
119. Larance M, Ramm G, Stockli J, van Dam EM, Winata S, et al. 2005. Characterization of the role of the Rab GTPase-activating protein AS160 in insulin-regulated GLUT4 trafficking. *J. Biol. Chem.* 280(45):37803–13
120. Cormont M, Mari M, Galmiche A, Hofman P, Le Marchand-Brustel Y. 2001. A FYVE-finger-containing protein, Rabip4, is a Rab4 effector involved in early endosomal traffic. *Proc. Natl. Acad. Sci. USA* 98(4):1637–42
121. Fouraux MA, Deneka M, Ivan V, van der Heijden A, Raymackers J, et al. 2004. rabip4' is an effector of rab5 and rab4 and regulates transport through early endosomes. *Mol. Biol. Cell* 15(2):611–24
122. Yamamoto H, Koga H, Katoh Y, Takahashi S, Nakayama K, Shin HW. 2010. Functional cross-talk between Rab14 and Rab4 through a dual effector, RUFY1/Rabip4. *Mol. Biol. Cell* 21(15):2746–55
123. Yang J, Kim O, Wu J, Qiu Y. 2002. Interaction between tyrosine kinase Etk and a RUN domain- and FYVE domain-containing protein RUFY1. A possible role of ETK in regulation of vesicle trafficking. *J. Biol. Chem.* 277(33):30219–26
124. Maxfield FR, McGraw TE. 2004. Endocytic recycling. *Nat. Rev. Mol. Cell Biol.* 5(2):121–32
125. Eggers CT, Schafer JC, Goldenring JR, Taylor SS. 2009. D-AKAP2 interacts with Rab4 and Rab11 through its RGS domains and regulates transferrin receptor recycling. *J. Biol. Chem.* 284(47):32869–80
126. Horgan CP, McCaffrey MW. 2009. The dynamic Rab11-FIPs. *Biochem. Soc. Trans.* 37(Pt. 5):1032–36
127. Kelly EE, Horgan CP, Adams C, Patzer TM, Ni Shuilleabhain DM, et al. 2010. Class I Rab11-family interacting proteins are binding targets for the Rab14 GTPase. *Biol. Cell* 102(1):51–62
128. Burguete AS, Fenn TD, Brunger AT, Pfeffer SR. 2008. Rab and Arl GTPase family members cooperate in the localization of the Golgin GCC185. *Cell* 132(2):286–98
129. Del Nery E, Miserey-Lenkei S, Falguieres T, Nizak C, Johannes L, et al. 2006. Rab6A and Rab6A' GTPases play non-overlapping roles in membrane trafficking. *Traffic* 7(4):394–407
130. Hayes GL, Brown FC, Haas AK, Nottingham RM, Barr FA, Pfeffer SR. 2009. Multiple Rab GTPase binding sites in GCC185 suggest a model for vesicle tethering at the *trans*-Golgi. *Mol. Biol. Cell* 20(1):209–17
131. Mallard F, Tang BL, Galli T, Tenza D, Saint-Pol A, et al. 2002. Early/recycling endosomes-to-TGN transport involves two SNARE complexes and a Rab6 isoform. *J. Cell Biol.* 156(4):653–64
132. Grigoriev I, Splinter D, Keijzer N, Wulf PS, Demmers J, et al. 2007. Rab6 regulates transport and targeting of exocytotic carriers. *Dev. Cell* 13(2):305–14
133. Miserey-Lenkei S, Chalancon G, Bardin S, Formstecher E, Goud B, Echard A. 2010. Rab and actomyosin-dependent fission of transport vesicles at the Golgi complex. *Nat. Cell Biol.* 12(7):645–54
134. Valente C, Polishchuk R, De Matteis MA. 2010. Rab6 and myosin II at the cutting edge of membrane fission. *Nat. Cell Biol.* 12(7):635–38

135. Monier S, Jollivet F, Janoueix-Lerosey I, Johannes L, Goud B. 2002. Characterization of novel Rab6-interacting proteins involved in endosome-to-TGN transport. *Traffic* 3(4):289–97
136. Callebaut I, de Gunzburg J, Goud B, Mornon JP. 2001. RUN domains: a new family of domains involved in Ras-like GTPase signaling. *Trends Biochem. Sci.* 26(2):79–83
137. Miserey-Lenkei S, Couedel-Courteille A, Del Nery E, Bardin S, Piel M, et al. 2007. Rab6-interacting protein 1 links Rab6 and Rab11 function. *Traffic* 8(10):1385–403
138. Simon GC, Prekeris R. 2008. Mechanisms regulating targeting of recycling endosomes to the cleavage furrow during cytokinesis. *Biochem. Soc. Trans.* 36(Pt. 3):391–94
139. Gillingham AK, Munro S. 2003. Long coiled-coil proteins and membrane traffic. *Biochim. Biophys. Acta* 1641(2–3):71–85
140. Allan BB, Moyer BD, Balch WE. 2000. Rab1 recruitment of p115 into a cis-SNARE complex: programming budding COPII vesicles for fusion. *Science* 289(5478):444–48
141. Moyer BD, Allan BB, Balch WE. 2001. Rab1 interaction with a GM130 effector complex regulates COPII vesicle *cis*-Golgi tethering. *Traffic* 2(4):268–76
142. Weide T, Bayer M, Koster M, Siebrasse JP, Peters R, Barnekow A. 2001. The Golgi matrix protein GM130: a specific interacting partner of the small GTPase rab1b. *EMBO Rep.* 2(4):336–41
143. Diao A, Frost L, Morohashi Y, Lowe M. 2008. Coordination of golgin tethering and SNARE assembly: GM130 binds syntaxin 5 in a p115-regulated manner. *J. Biol. Chem.* 283(11):6957–67
144. Seemann J, Jokitalo EJ, Warren G. 2000. The role of the tethering proteins p115 and GM130 in transport through the Golgi apparatus in vivo. *Mol. Biol. Cell* 11(2):635–45
145. Shorter J, Beard MB, Seemann J, Dirac-Svejstrup AB, Warren G. 2002. Sequential tethering of Golgins and catalysis of SNAREpin assembly by the vesicle-tethering protein p115. *J. Cell Biol.* 157(1):45–62
146. Alvarez C, Garcia-Mata R, Hauri HP, Sztul E. 2001. The p115-interactive proteins GM130 and giantin participate in endoplasmic reticulum-Golgi traffic. *J. Biol. Chem.* 276(4):2693–700
147. Lesa GM, Seemann J, Shorter J, Vandekerckhove J, Warren G, et al. 2000. The amino-terminal domain of the Golgi protein giantin interacts directly with the vesicle-tethering protein p115. *J. Biol. Chem.* 275(4):2831–36
148. Linstedt AD, Jesch SA, Mehta A, Lee TH, Garcia-Mata R, et al. 2000. Binding relationships of membrane tethering components. The giantin N terminus and the GM130 N terminus compete for binding to the p115 C terminus. *J. Biol. Chem.* 275(14):10196–201
149. Rosing M, Ossendorf E, Rak A, Barnekow A. 2007. Giantin interacts with both the small GTPase Rab6 and Rab1. *Exp. Cell Res.* 313(11):2318–25
150. Rahajeng J, Giridharan SS, Cai B, Naslavsky N, Caplan S. 2010. Important relationships between Rab and MICAL proteins in endocytic trafficking. *World J. Biol. Chem.* 1(8):254–64
151. Nokes RL, Fields IC, Collins RN, Folsch H. 2008. Rab13 regulates membrane trafficking between TGN and recycling endosomes in polarized epithelial cells. *J. Cell Biol.* 182(5):845–53
152. Terai T, Nishimura N, Kanda I, Yasui N, Sasaki T. 2006. JRAB/MICAL-L2 is a junctional Rab13-binding protein mediating the endocytic recycling of occludin. *Mol. Biol. Cell* 17(5):2465–75
153. Ang AL, Folsch H, Koivisto UM, Pypaert M, Mellman I. 2003. The Rab8 GTPase selectively regulates AP-1B-dependent basolateral transport in polarized Madin-Darby canine kidney cells. *J. Cell Biol.* 163(2):339–50
154. Henry L, Sheff DR. 2008. Rab8 regulates basolateral secretory, but not recycling, traffic at the recycling endosome. *Mol. Biol. Cell* 19(5):2059–68
155. Yamamura R, Nishimura N, Nakatsuji H, Arase S, Sasaki T. 2008. The interaction of JRAB/MICAL-L2 with Rab8 and Rab13 coordinates the assembly of tight junctions and adherens junctions. *Mol. Biol. Cell* 19(3):971–83
156. Grigoriev I, Yu KL, Martinez-Sanchez E, Serra-Marques A, Smal I, et al. 2011. Rab6, Rab8, and MICAL3 cooperate in controlling docking and fusion of exocytotic carriers. *Curr. Biol.* 21(11):967–74

Roles for Actin Assembly in Endocytosis

Olivia L. Mooren, Brian J. Galletta, and John A. Cooper

Department of Cell Biology and Physiology, Washington University School of Medicine, Saint Louis, Missouri 63110; email: jcooper@wustl.edu

Keywords

clathrin, dynamin, membrane, pinocytosis, filaments, Arp2/3

Abstract

Endocytosis includes a number of processes by which cells internalize segments of their plasma membrane, enclosing a wide variety of material from outside the cell. Endocytosis can contribute to uptake of nutrients, regulation of signaling molecules, control of osmotic pressure, and function of synapses. The actin cytoskeleton plays an essential role in several of these processes. Actin assembly can create protrusions that encompass extracellular materials. Actin can also support the processes of invagination of a membrane segment into the cytoplasm, elongation of the invagination, scission of the new vesicle from the plasma membrane, and movement of the vesicle away from the membrane. We briefly discuss various types of endocytosis, including phagocytosis, macropinocytosis, and clathrin-independent endocytosis. We focus mainly on new findings on the relative importance of actin in clathrin-mediated endocytosis (CME) in yeast versus mammalian cells.

Contents

1. INTRODUCTION.................. 662
   1.1. Types of Endocytosis........... 662
   1.2. Endocytosis in Normal Tissues
        and Disease...................... 663
2. THE ACTIN CYTOSKELETON
   IN CLATHRIN-MEDIATED
   ENDOCYTOSIS.................... 663
   2.1. The Need for Actin in
        Clathrin-Mediated Endocytosis:
        Yeast versus Mammalian
        Systems 663
   2.2. Actin and Clathrin-Mediated
        Endocytosis in Other Cell
        Types 666
   2.3. Regulation of Actin Dynamics at
        Sites of Clathrin-Mediated
        Endocytosis 667
   2.4. Coordination of Membrane
        Curvature Proteins with Actin ... 672
   2.5. Mechanisms of Membrane
        Scission 674
   2.6. Mathematical Modeling of Actin
        Assembly and Clathrin-Mediated
        Endocytosis 675
   2.7. Models of Force Generation
        by Actin Assembly During
        Clathrin-Mediated Endocytosis.. 676
3. CONCLUSIONS AND
   FUTURE DIRECTIONS........... 678

Clathrin: a protein component that associates with itself to form triskelions, which polymerize into the curved coat that surrounds the endocytic vesicle during clathrin-mediated endocytosis

CME: clathrin-mediated endocytosis

1. INTRODUCTION

Endocytosis includes a number of processes by which materials are taken up by cells, and a role for the actin cytoskeleton has been uncovered for a number of these processes. In this article, we review progress in our understanding of how actin is involved in these various processes of endocytosis. A number of reviews addressing various aspects of this subject have been published recently. We refer the reader to them and place our focus on the most recent publications.

1.1. Types of Endocytosis

Cellular uptake is an important process for function in a wide variety of cell types (see Section 1.2 below). Multiple pathways for uptake have been identified including clathrin-mediated endocytosis (CME), caveolae, phagocytosis, macropinocytosis, circular dorsal ruffles, and several clathrin-independent pathways. Doherty & McMahon (1) wrote a remarkably comprehensive review of the molecular mechanisms involved in various types of endocytosis. On the topic of CME, McMahon & Boucrot (2) published a comprehensive review in 2011. For further information on clathrin-independent endocytosis, we recommend a review by Howes and colleagues (3).

The endocytic pathways of phagocytosis, macropinocytosis, and circular dorsal ruffles clearly require actin (4–8). The forces that contribute to deforming the plasma membrane, to bringing the material into the cytoplasm, and to executing fission may come from the polymerization of actin filaments or from the action of myosin motor proteins on actin filaments. In these settings, actin polymerization is likely to depend on the concerted action of filament nucleation proteins, such as Arp2/3, formins, and WH2 domain proteins. Actin-binding proteins, such as ADF/cofilin, VASP, profilin, α-actinin, and capping protein, are likely to play important roles (9–13). In fact, many key elements of the processes remain to be defined, especially with regard to signaling and regulatory controls.

Other endocytic pathways, for which the role of actin is less certain and less well-defined, include the CLIC/GEEC pathway, caveolae, transcytosis, and endocytic pathways that involve Arf6 and flotillin (1). Several studies of clathrin-independent endocytosis have demonstrated a requirement for actin. A recent study of endocytic tubules caused by the Shiga toxin produced by enteropathogenic bacteria revealed a need for actin in the process of scission by which tubules form vesicles (14).

1.2. Endocytosis in Normal Tissues and Disease

In normal cells that either are undergoing development and differentiation or are functioning in their differentiated physiologic state, endocytosis can play a variety of important roles. For example, professional phagocytic cells ingest and degrade apoptotic or necrotic cells. Immune cells that process antigens, such as dendritic cells, use macropinocytosis to ingest relatively large volumes of extracellular fluid and thus take up potential antigens for presentation. Binding of ligands to their cellular receptors induces signaling that is propagated into the cell and can result in endocytosis of the ligand. This is necessary for uptake of vital substances, and the subsequent degradation or recycling of those receptors is a critical element of the control of receptor availability on the cell surface.

Mutations in components of CME and of other types of endocytosis have been implicated in a wide range of human diseases, including various cancers, neurodegenerative diseases, and metabolic syndromes (1, 2). During infections, a number of microorganisms, including viruses, bacteria, and fungi, enter host cells via a wide variety of endocytic processes, including CME, phagocytosis, and macropinocytosis (4, 15–17). The ability to specifically interfere with endocytosis of pathogens, including prions, may be important in the future for treatment of diseases.

2. THE ACTIN CYTOSKELETON IN CLATHRIN-MEDIATED ENDOCYTOSIS

2.1. The Need for Actin in Clathrin-Mediated Endocytosis: Yeast versus Mammalian Systems

CME has been studied most completely in yeasts and cultured mammalian cells. In yeast, it is clear that actin is absolutely required for the successful progression and execution of the process of CME, reviewed in Reference 18. In mammalian cells, actin is not always necessary for CME to occur, but it does appear to become necessary in certain settings, such as the ingestion of larger cargoes or endocytosis in locations that are richly dense with actin filaments.

In yeast, actin-rich foci on the plasma membrane, termed actin patches, are the locations of CME. The evidence for this conclusion comes from the colocalization of endocytic and actin-binding proteins, as well as from mutations in actin-binding proteins resulting in endocytic defects. Advances in live-cell imaging techniques have been crucial in determining the dynamics and recruitment of endocytic and actin regulating proteins to the sites of CME in yeast (see the box Advancing Our Understanding of the Endocytic Process). Also, targeting actin by treatment of cells with toxins, such as latrunculin, has demonstrated a need for actin filament assembly in CME (19–21). The requirement for actin assembly at sites of CME in yeast may be a consequence of the turgor pressure of these cells, which is high relative to that in animal cells (22).

Contradictory evidence of a requirement for actin in metazoan CME has led to debate regarding the role of the actin cytoskeleton. As in yeast, there is convincing biochemical evidence for interactions between several endocytic proteins and actin-binding proteins. The use of various actin toxins has demonstrated that actin may participate in multiple stages of endocytosis, including invagination, elongation of the neck of the endocytic vesicle, fission, and movement away from the plasma membrane (23, 24). However, depending on the cell type used or which surface of the cell was measured, effects were minimal in certain cases (25, 26).

A study by De Camilli and colleagues (27) found similarities between yeast and mammalian CME in the coordination of lipid curvature-inducing proteins with actin. Loss of dynamin 1 and 2 in mouse fibroblasts arrested CME, with long tubules capped by clathrin attached to the plasma membrane. Actin, actin regulators, and curvature-inducing BAR domain proteins were found at the base of the tubule. Latrunculin B treatment led to loss of the actin foci and the tubules, along with

Arp2/3: a seven-subunit protein complex containing the actin-related proteins Arp2 and Arp3; it nucleates actin filament assembly and creates end-to-side branches

Scission: fission of membrane tubule attaching the incipient endocytic vesicle to the plasma membrane

Latrunculin A and B: cell-permeable toxins that bind actin monomers, inhibit actin assembly at low concentrations and short times, and depolymerize filaments at high concentrations and longer times

Dynamin: a protein with GTPase activity capable of binding lipid membranes and facilitating fission

BAR domain: Bin-amphiphysin-Rvs protein domain

ADVANCING OUR UNDERSTANDING OF THE ENDOCYTIC PROCESS

Improvements in technology and methodology associated with light and electron microscopy have been crucial for advancing our understanding of molecular mechanisms. The advent of fluorescent proteins, including green fluorescent protein (GFP) and its relatives, has enabled one to follow the accumulation of individual components over time, to directly compare the time courses of accumulation of two or more proteins with each other, and to follow the movement of endocytic structures. The development of pHluorin as a GFP derivative with fluorescence sensitivity to pH has been a very useful tool to mark the time of scission of the endocytic vesicle from the plasma membrane.

Great improvements in signal-to-noise ratios have been provided by the development of confocal microscopy, both laser scanning and spinning disk, and by the implementation of total internal reflection fluorescence microscopy. These technological advancements have been essential for improving sensitivity and the visualization of relatively faint fluorescent signals. Super-resolution fluorescence microscopy, such as stochastic optical reconstruction microscopy (STORM) or photoactivated localization microscopy (PALM), has also contributed to improved visualization, and further applications of these novel approaches are expected to provide new insights.

Advances in electron microscopy, especially improvements in our ability to visualize clathrin assemblies and actin filaments provided by freeze-etch, platinum replica, and tomographic approaches, have been invaluable. The correlation of fluorescence with high-resolution electron microscopy images has likewise made important contributions.

Quantitative measurements of GFP fluorescence can be converted to absolute numbers of molecules by comparison with standards in cells. Computer-assisted analysis of movie data can track individual entities, such as endocytic vesicles and clathrin-rich sites, providing data on their fluorescence intensity and position over time. The computer programs allow one to analyze data for hundreds to thousands of such entities.

Two key issues related to GFP fusions are the level of expression of the fusion protein and the ability of the fusion to functionally replace the endogenous protein. Ideally, one would like to replace the endogenous protein with a GFP fusion without changing the total expression level of the protein. Gene integration technology, which is well established in yeast and other model organisms and recently available in cultured mammalian cells, allows the fusion protein to be expressed from the appropriate endogenous locus. However, this alone does not guarantee proper protein levels because the presence of the GFP fusion can influence rates of synthesis, folding, and degradation of either RNA or protein.

In addition, one needs to document that the GFP fusion functions identically to the endogenous protein, at least in the processes of interest, namely endocytosis. Experience from model organisms indicates that documenting rescue of null mutant phenotypes is an important criterion, but also that one needs to consider the nature of the assay and the genetic background. In particular, in cells carrying one or more mutations of endocytic components, the ability of a given GFP fusion to function may be compromised more than would be appreciated from otherwise wild-type cells.

widening of the neck of clathrin pits. The authors concluded that actin is important for progression of CME, owing to a role, in coordination with BAR domain proteins, in narrowing and elongating the neck of clathrin pits. These features were reminiscent of CME in yeast. In addition, dynamin appeared to limit actin polymerization and tubule elongation, which also bears some similarities to CME in yeast (discussed in more detail below, in Section 2.4 on scission).

A subsequent study with a cell-free system by the same group also found actin to be required for the formation of plasma membrane tubules (28). Cultured fibroblasts on a poly-L-lysine-coated surface were sonicated, leaving plasma membrane sheets. Addition of brain cytosol with ATP and GTPγS caused the formation of tubules capped by clathrin. Tubules did not form upon treatment with latrunculin B; however, clathrin pits still accumulated. A need for actin in CME was also found in a study of mammalian cells cultured on poly-L-lysine, which leads to a relatively tight attachment of the cell to the surface (24). These observations suggest that actin may be needed in mammalian CME, for invagination or tubulation, only when cells are firmly attached to the substrate or when the membrane is under tension.

Other lines of evidence also indicate that tension may determine the requirement for actin involvement in CME. In yeast, Ayscough and coworkers (22) demonstrated that the high turgor pressure in the cell creates a need for actin to provide enough force for membrane invagination. In addition, treatment of mammalian cells with actin filament disruption toxins (latrunculin B or cytochalasin D), which should decrease cortical membrane tension, allowed BAR and F-BAR domain proteins to induce tubule formation (29). Actin involvement in CME of larger clathrin-coated structures (CCS), termed clathrin plaques, has been observed on the adherent surface of certain cell types (26). In addition, increased adhesion of a cell to a surface inhibited CME and increased the lifetime of clathrin-coated pits (CCPs), as did proximity of a CCP to a cell-surface attachment plaque (30). In this study, latrunculin B treatment of the cells increased the lifetimes of all CCPs, indicating a general need for actin in CME, and the effect of latrunculin was greater on CCPs in adherent regions, confirming that actin is more important when adhesion is high. Together, these results suggest that adhesion of cells to their substrate requires more force for internalization of CCPs and that this increased force is provided by actin.

In striking contrast, in other settings, actin filament networks appear to inhibit CME. Actin filaments are important for generation of cortical tension, and high cortical tension appears to decrease the efficiency of CME. In a study of cultured cells (BSC1) attached to fibronectin-coated micropatterned substrates, CCP lifetimes were heterogeneous. Relaxation of the cortical actin network with inhibitors of actin or myosin led to decreased CCP lifetimes and loss of lifetime heterogeneity, i.e., increased efficiency of CME (31). Most of the CCPs analyzed were located between, not directly coincident with, integrin attachment sites. The authors proposed that stress fibers, anchored at sites of integrin attachment, spanned across the cell and increased cortical tension, which impeded the invagination machinery. In these experiments, the levels of latrunculin A were sufficiently low that most filamentous actin was preserved, in contrast to other studies where high levels of latrunculin A caused actin depolymerization at a gross level. Thus, these results do not negate a possible requirement for actin in CME; instead, they suggest that increased cortical tension may be a means by which actin can inhibit CME.

Different conclusions regarding cortical tension and the role of actin in CME were reached by Kirchhausen and colleagues (32) in a recent study with cultured polarized Madin-Darby canine kidney (MDCK) epithelial cells. CME at the apical surface of these polarized cells was known to depend on actin because of a previous study finding inhibition of CME by cytochalasin D (33). In the new study, disruption of the actin network by latrunculin A treatment resulted in decreased CME, with a large number of U-shaped clathrin pits on the apical surface, apparently arrested in the process of invagination. This dependence of CME on actin correlated with the presence of microvilli on the apical surface. Microvillar structure depends on a bundle of actin filaments as a structural core, so the effect on CME may have resulted from the loss of actin filaments or microvilli. Previously, membrane tension was found to be higher on the apical surface of polarized cells, relative to the basal surface, on the basis of membrane-tether pulling experiments (34). This increased membrane tension was due primarily to cortical actin, revealed by the observation of less tension when tethers were pulled from membrane blebs. Kirchhausen and colleagues (32) found that subjecting cells to swelling (with hypotonic medium) or to mechanical stretching created a requirement for actin in CME on the basolateral surface of the cells, and, again, arrested clathrin pits were seen. They proposed that increased

Cytochalasin D: a cell-permeable toxin that blocks elongation from barbed ends of actin filaments

F-BAR: a BAR protein domain that is an extensions of the FCH (Fes/CIP4 homology) domain

CCS: clathrin-coated structure

CCP: clathrin-coated pit

Jasplakinolide: an actin toxin that is cell permeable and binds to and stabilizes filaments

membrane tension decreased the ability of the clathrin lattice to produce invaginations, creating a need for assistance from the actin cytoskeleton. In this study, latrunculin A and jasplakinolide both caused arrest of CCPs; this seems puzzling because the former inhibits or reverses actin filament polymerization, whereas the latter stabilizes actin filaments. Perhaps in each case, the net effect was to prevent architectural remodeling of the actin filament network needed for CME.

Internalization of large cargo by CME appears to be another setting in which actin becomes necessary for CME. Vesicular stomatitis virus requires actin to enter the cell via CME (35). The large size and shape of the virus would seem to preclude complete encasement of the virus by the relatively smaller clathrin-coated vesicle. However, a shorter form of vesicular stomatitis virus, the defective interfering particle, DI-T, was completely encased into a clathrin-coated vesicle, and the requirement for actin was lost (36). This suggests that additional force was needed for internalization of the whole virus and that actin provided that force.

Finally, a recent electron microscopy (EM) study by Svitkina and coworkers (37) has demonstrated that many CCSs on the ventral surface of unroofed cultured mammalian cells, B16F1s, have networks of short, highly branched actin filaments associated with them. These actin filament networks were not always observed by fluorescence microscopy, raising concern that the detection limitations of fluorescence microscopy might underestimate the number of CCPs that have actin filaments associated with them. In this study, treatment of cells with actin-disrupting drugs (latrunculin B or cytochalasin D) led to dissolution of stress fibers but not complete removal of actin filaments associated with CCSs. This observation raises another important experimental caveat. In studies observing no effect on CME upon addition of such drugs, perhaps the loss of actin filaments at the site of CME is incomplete.

To summarize the different experimental findings, actin appears to be necessary for CME in mammalian cells in certain settings, perhaps because actin provides an increased level of force needed for invagination, scission, or vesicle movement. The role of membrane tension, and the contribution of cortical actin to membrane tension, is somewhat confusing. In some studies, cortical actin appeared to contribute to increased membrane tension, which impeded the process of CME. In other studies, regions of cells with high membrane tension appeared to require actin for CME to occur. The connection between tension and actin in CME is likely to be a physiologically important one, assuming that cells in different tissues modulate their cortical tension in response to their environment. It does seems reasonable to suggest that actin might be constitutively recruited to CCSs, along with other endocytic proteins, because of the abundance of interactions between endocytic and actin-regulatory proteins. Perhaps the amount of actin polymerization needed in a given case is determined by the amount of force required to counteract membrane or cortical tension. The molecular mechanisms of the processes by which the cell senses the tension and regulates the actin assembly should be interesting areas for future investigation.

2.2. Actin and Clathrin-Mediated Endocytosis in Other Cell Types

Although endocytosis in plants has not been investigated extensively, plants appear to use CME as a primary means of internalization (38–40). Plants have homologs of many of the same endocytic proteins conserved in yeast and mammals, such as AP-2, epsin, Arp2/3, and dynamin (41); however, studies of their functional roles have been limited. Several studies have demonstrated a requirement for actin in plant endocytosis by treating cells with actin-depolymerizing drugs (42–46), reviewed in Reference 47. One might expect actin to be required for CME in plants, by analogy to yeast, because both types of cells have high turgor pressure. Interestingly, CCPs in yeast and plants are smaller than those in mammalian cells (2), which may be consistent with the notion that high turgor pressure impedes invagination. Further investigation into the role of actin in plant CME will help test ideas about the role of actin in the face of high membrane or cortical tension.

In *Dictyostelium* cells, actin filaments are clearly essential for a number of endocytic mechanisms, notably phagocytosis and macropinocytosis (reviewed in Reference 48). Relatively less is known about the role of actin in CME in *Dictyostelium*. During the course of CME, a brief burst of actin polymerization can be observed prior to the loss of clathrin from the membrane (49, 50), and several actin-binding proteins localize to CCSs, including WASP (51), Arp2/3, myosin-IB, and coronin (49). Treatment of cells with actin toxins, cytochalasin A or latrunculin A, results in an increased number of clathrin puncta at the plasma membrane, consistent with a requirement for actin in CME progression (50). O'Halloran and coworkers (50) hypothesized that huntingtin-interacting protein 1-related (Hip1r) protein, a component of CCPs, may coordinate clathrin with actin polymerization because Hip1r interacts with both actin and clathrin. Epsin, a $PI(4,5)P_2$-binding protein that promotes early invagination and binds clathrin, recruited Hip1r to sites of CME and promoted its phosphorylation. In cells lacking epsin or Hip1r, clathrin foci persisted at the membrane for longer times than seen in wild-type cells, and actin structures were more diffuse, with a larger and irregular shape and increased lateral motility. In yeast, Hip1r (Sla2) mutants have altered actin dynamics at endocytic patches (20), and clathrin light chain, Clc1, binds to Sla2 and inhibits the interaction of Sla2 with actin (52). In mammalian cells, Hip1R binds cortactin, clathrin, and actin, which may form a cap to negatively regulate actin polymerization (53). Therefore, a conserved function of Hip1r may be to regulate actin polymerization at clathrin pits, thereby promoting efficient CME.

Drosophila also use CME as a means of cellular uptake. CCSs in *Drosophila* hemocytes appeared at the periphery of filopodial membranes and moved centripetally along the membrane before scission occurred (54).

Latrunculin A stopped this lateral movement of CCSs as well as the uptake of maleylated BSA (mBSA), a CME cargo. Arp2/3-mediated actin assembly was implicated by the observations that *wasp* mutants displayed no centripetal movement of CCSs or uptake of mBSA. In dynamin mutants, *shibiri*^{ts2}, elongated tubules formed from the plasma membrane at the restrictive temperature, and the uptake of mBSA was increased because of trapping within the elongated tubules. Latrunculin A treatment prior to the shift to a restrictive temperature decreased the internalization of mBSA, suggesting a role for actin in the formation of the elongated tubules, similar to what has been reported in mammalian systems (27, 29).

A study in *Drosophila* S2R$^+$ cells suggested that actin is involved in the internalization and motility of CCPs via F-BAR domain protein CIP4/Toca-1 (cdc42-interacting protein 4/transducer of cdc42-dependent actin assembly 1) regulation of the Arp2/3 activators WASp (Wiskott-Aldrich syndrome protein) and SCAR/WAVE (WASp family verprolin-homologous protein) (55). Latrunculin A treatment abolished the movement of endocytic vesicles through the cytoplasm following scission. The SH3 domain of CIP4 interacted directly with WASP and indirectly with WAVE via the Abi subunit. Overexpression of CIP4ΔSH3-EGFP in cells caused long tubules to form from the plasma membrane, suggesting that Arp2/3-mediated actin assembly was needed for scission. The loss of Cip4 in flies caused the formation of multiple wing hairs, as did loss of function of Arp2/3, WAVE, or dynamin. Thus, dynamin and CIP4, by regulating actin, may promote endocytic scission and may be important for wing hair development.

2.3. Regulation of Actin Dynamics at Sites of Clathrin-Mediated Endocytosis

Actin networks that form at sites of endocytosis must be tightly regulated for efficient internalization. Indeed, many of the components involved in the dendritic nucleation model localize to CME sites. These include

WASp: Wiskott-Aldrich syndrome protein

SCAR/WAVE: WASp family verprolin-homologous protein

Nucleation-promoting factor (NPF): a protein that binds to and activates Arp2/3

nucleation-promoting factors (NPFs), which increase the ability of Arp2/3 to form a branched actin network; capping protein, which limits the length of filaments; and depolymerization factors, such as cofilin, which turn over older filaments for recycling of actin subunits. The majority of the research on the regulation of actin at sites of CME has been in budding and fission yeast. We discuss that work in the following sections with some comparisons to other systems.

2.3.1. Phases of the endocytic process.
Modern cell biological techniques have solidified the relationship between actin and endocytosis. Fast time resolution, dual-color imaging, in both yeast and mammalian cells, reveals stereotypical temporal relationships between the accumulation of endocytic proteins and regulators of actin assembly. These movies also reveal distinct phases of movement by the endocytic ensemble. Endocytosis can be viewed as a three-phase process. During phase I, endocytic coat proteins, including clathrin, Sla1 (CD2AP/Cin85), Sla2 (Hip1r), and End3, are recruited, and the endocytic structure remains at the cell cortex undergoing relatively minor movements. Regulators of actin assembly are recruited, and near the end of this stage, actin polymerization begins. During phase II, the proteins make a short movement away from the cortex into the cytoplasm. At the end of this phase, endocytic proteins are lost from the vesicle, presumably coinciding with scission of the vesicle from the plasma membrane. Finally, in phase III, the vesicle and its actin machinery execute faster and longer movements about the cytoplasm. For recent reviews on this topic, see References 18 and 56–59.

The polymerization of actin is required for the movements observed during phase I and phase II. When cells are treated with latrunculin A to inhibit actin polymerization, the endocytic site forms, but it never leaves the cortex (20, 21). However, the initial production of membrane curvature may not depend on actin, according to EM studies (60). The architecture of the actin filament network and the protein composition of the endocytic sites strongly suggest that Arp2/3-mediated actin assembly occurs in the manner described by the dendritic nucleation model (10). Actin filaments at sites of endocytosis make end-to-side branches characteristic of Arp2/3 (61). Sites of endocytosis contain Arp2/3 and other actin-regulatory proteins associated with the dendritic nucleation model, including a capping protein and cofilin.

2.3.2. The nucleator Arp2/3 and its regulators.
To date, the only nucleator of actin filaments found at sites of endocytosis is Arp2/3. Conditional mutations affecting Arp2/3 result in a significant disruption of the normal behavior of endocytic sites (62, 63). Other actin nucleators in budding yeast include two formin proteins, and conditional loss of their activity does not affect the movement of endocytic sites (21). To nucleate new actin filaments, mammalian Arp2/3 requires the binding of an activator protein, sometimes called an NPF protein. Binding of an NPF protein to Arp2/3 induces a major conformational change in the complex, bringing the actin-like Arp2 and Arp3 subunits into a configuration that mimics the barbed end of an actin filament (64). In yeast, Arp2/3 has a significant amount of nucleation activity in the absence of an activator, especially when assayed with yeast actin (65). Therefore, the role of Arp2/3 regulatory proteins in yeast may be to localize or fine-tune the activity of Arp2/3. Six proteins are capable of regulating the in vitro activity of Arp2/3 in yeast: five positive regulators, WASp/Las17, two type I myosins (Myo3 and Myo5), Pan1, and Abp1, and one negative regulator, coronin.

The best studied of the Arp2/3 regulators in yeast is the WASp homolog Las17. WASp/Las17 arrives at sites of endocytosis during the nonmotile phase, prior to the arrival of Arp2/3 and actin (20). Complete loss of Las17 leads to a loss of cortical actin patches (66). However, mutations affecting the Arp2/3-binding region of Las17 have far less severe consequences on actin assembly, perturbing the timing and movement of endocytic structures, but not eliminating actin assembly (67, 68).

The differences in phenotype between the *las17* deletion and the *las17*-Arp2/3-binding mutant may result from functional overlap of Las17 with other Arp2/3 regulators, and Las17 may also have a critical role for actin assembly at sites of endocytosis independent of its interaction with Arp2/3.

Las17 interacts with a number of proteins in addition to Arp2/3, some of which appear to regulate the activity of Las17. Because full-length Las17 is not strongly autoinhibited (65), other proteins may inhibit Las17. Indeed, several proteins can negatively regulate Las17, and many of them are recruited early during the assembly of an endocytic site. The first protein recruited to endocytic sites identified to date is Syp1, a muniscin family protein containing an EFC/F-BAR domain, capable of binding and bending membranes (69–71). Beyond its ability to bend membranes, Syp1 can negatively regulate the activity of WASp/Las17. The endocytic proteins Sla1 and Bbc1, also recruited early in the endocytic process, cooperate to inhibit Arp2/3 activation by Las17 (20, 72, 73). An intriguing model is that the early-arriving proteins inhibit actin assembly at sites of endocytosis in the earliest stages, and this inhibition is lifted as Syp1 leaves the patch.

Alternatively, the negative action of Sla1/Bbc1 may be relieved by Bzz1, an endocytic protein that interacts with and appears to positively regulate Las17 (67). Bzz1 arrives at the patch after Las17 and Syp1 just prior to the start of actin polymerization (67, 74). Bzz1 contains an F-BAR domain, so it may help sense the curvature of the membrane to control the timing and localization of actin polymerization.

Arp2/3 activators other than Las17 may also be regulated. The two type I myosins in yeast, Myo3 and Myo5, have acidic/DDW regions that bind Arp2/3. The TH1 domain of Myo5 can interact with its own C-terminal tail, which includes the acidic/DDW Arp2/3-binding region (75). This interaction interferes with Myo5 binding to Vrp1 (verprolin), a mammalian WIP homolog, which is critical for the ability of Myo5 to activate actin filament nucleation by Arp2/3. In addition, inhibition of Myo5 activation of Arp2/3 appears to be, at least in part, mediated by calmodulin (Cmd1). Calcium reduced the association of Cmd1 with Myo5 in vitro and promoted the Myo5/Vrp1 interaction. The dissociation of Cmd1 from Myo5 by calcium addition on beads could also trigger the polymerization of actin in cell extracts. In cells, mutations that disrupt the interaction of Cmd1 and Myo5 led to premature actin assembly at endocytic sites, suggesting a role in regulating the activity of Myo5 toward Arp2/3 (75). In fission yeast, Cam2, a light chain for Myo1, is important for recruiting the type I myosin to the site of endocytosis (76).

The motor activity of the type I myosins in yeast is also important for endocytosis. Eliminating motor activity inhibits fluid-phase endocytosis and inward movement of endocytic patches (67). Motor activity is likely to be regulated; phosphorylation of the TEDS site, which accelerates its ATPase activity, is required for ligand-induced, but not constitutive, endocytosis in yeast (77).

Yeast Pan1 (mammalian intersectin) is another Arp2/3 activator with an acidic/DDW region, and Pan1 also appears to be under negative regulation. Similar to Las17, Pan1 is recruited to actin patches before Arp2/3 and actin, so its ability to activate Arp2/3 may be suppressed at first. In fact, Sla2, an early endocytic protein, can inhibit the ability of Pan1 to activate Arp2/3 in vitro (78). In vivo a *sla2* mutation that impairs the ability of Sla2 to interact with Pan1 displayed endocytic and actin phenotypes that result from excessive activation of Arp2/3 by Pan1.

Another protein that plays a critical role in endocytosis in yeast, likely at the step of regulating actin assembly, is verprolin, or Vrp1, analogous to the mammalian WASp-interacting protein (WIP). Vrp1 can interact with Las17, type I myosins, and actin monomers. Endocytic patches form in *vrp1* null mutant cells, but endocytosis was blocked (reviewed in Reference 79). The patches were immotile, but they did assemble F-actin (67, 72).

Understanding the molecular mechanism by which Vrp1 contributes to the assembly of a functional actin network at sites of endocytosis has been a challenge. An appealing model for Vrp1 function is to provide actin monomers to Arp2/3 or to Arp2/3 activators, such Las17 and type I myosins, to nucleate new daughter filaments. Testing this model has been complicated because Vrp1 interacts genetically and biochemically with Las17 and the type I myosins, which themselves appear to have overlapping functions. However, progress has been made. Unexpectedly, the interaction between Las17 and Vrp1 did not appear to be critical for endocytosis on the basis of fluid-phase end-point assays (80, 81), but more detailed assays may be informative. Other mutations, including ones that disrupt the Las17/Arp2/3 interaction, can affect the dynamics of protein recruitment at sites of endocytosis without affecting uptake of fluid-phase markers. Of course, Las17 and type I myosins may overlap in function for activation of Arp2/3, and the interaction between Vrp1 and type I myosins may be sufficient to mask the loss of the Vrp1/Las17 interaction. Also surprising were observations that mutations affecting actin-binding domains of Vrp1 did not prevent fluid-phase endocytosis (79, 82). Again, overlapping function, in this case between the actin-binding domains of Vrp1 or those of Las17, may mitigate the effects of single mutations.

2.3.3. Origin of the first mother actin filament.
One interesting question for actin assembly at sites of endocytosis is what triggers the beginning of the process. Arp2/3 requires an existing mother filament on which to create a new branch. What is the origin of that filament? Evidence has been provided by a recent study of Dip1, the WASH/DIP ortholog in *Schizosaccharomyces pombe* (83). WASH/DIP proteins regulate Arp2/3 and the formin mDia2 in mammalian cells. In *dip1* mutants, the timing between the arrival of WASp and the recruitment of Arp2/3 was more variable than in wild-type cells. However, once actin polymerization began, the process progressed with normal kinetics. Thus, Dip1 appears to act as a switch to initiate actin polymerization. The interesting observation was that, in wild-type and *dip1* mutant cells, the movement of an existing actin patch near to a site of incipient endocytosis appeared to trigger the assembly of actin at that new site. Perhaps, a filament in the existing actin patch can serve as the first mother actin filament and thereby trigger the switch to turn on actin polymerization.

In mammalian cells, the origin of the first mother actin filament is suggested by recent electron micrographs of unroofed cells. The pointed end of an actin filament associated with a CCS was frequently found to be in contact with the side of an actin filament extending from the surrounding cortical actin network or a nearby stress fiber (37). In general, mammalian cells have a high density of actin filaments, particularly near the cortex, so the first mother actin filament could easily be one that happens to collide with an active Arp2/3 molecule at a CCS. In yeast, the actin filament density in the cytoplasm is substantially less, but small oligomers of actin appear to exist (84). Such an oligomer might serve as the first mother filament.

2.3.4. Termination of growth.
The dendritic nucleation model predicts that proper termination of the growth of actin filaments may be important for regulating actin assembly and the movements produced by actin assembly. Capping barbed ends is the proposed means of terminating actin assembly of individual filaments in the branched network. Deletion of the heterodimeric actin capping protein, Cap1/Cap2 in yeast, affects all phases of endocytosis, delaying the start of movement and affecting both the slow and fast movements away from the membrane. Consistent with a need for a proper balance between filament growth and capping, overexpression of capping proteins also affected the assembly of the actin network at sites of endocytosis (21). Loss of capping proteins had no effect on the movement of presumably endocytic vesicles in the cytoplasm (21); however, the movement appears to depend on

Arp2/3-mediated actin polymerization (85, 86). Thus, this feature of the dendritic nucleation model may not hold in this setting.

2.3.5. Actin turnover and disassembly. An emerging aspect of the regulation of the actin filament network at sites of endocytosis is the role of actin turnover proteins. The dendritic nucleation model predicts that the disassembly of actin filaments is essential for the proper function of an actin network to recycle actin subunits. Cofilin, coronin, and Aip1 potently disassemble actin filaments in vitro, especially in combination (87, 88). In yeast cells, these three proteins are recruited to the endocytic patch at the same time after actin filament assembly is complete, consistent with a role in filament disassembly (89, 90). In mammalian cells, cofilin and coronin are also recruited to endocytic sites after actin assembles (91).

In terms of function in yeast, cofilin is essential for viability, and conditional loss-of-function mutations strongly inhibit movement of endocytic sites (89, 90). However, the phenotypes of coronin and Aip1 mutants in budding yeast indicate that their contribution to cofilin function is relatively minor (68, 90), apparently not consistent with what was expected from biochemical studies. Loss of coronin prolonged the actin assembly time for endocytic patches and increased the movement of endocytic patches late in their life; however, neither of these effects was nearly as large as those resulting from loss of cofilin function. Understanding the precise roles of coronin in regulating actin assembly at endocytic sites may be a significant challenge. In addition to cooperating with cofilin in disassembling actin filaments in vitro, coronin could bind to actin filaments and protect them from cofilin when the actin was in its ATP-bound state (92). Coronin was also able to activate and inhibit nucleation by Arp2/3, depending on their concentrations (93, 94). Point mutations that abolish the interaction of coronin with Arp2/3 had subtle affects on the assembly of endocytic sites (94). Finally, with regard to Aip1, loss of function had a very mild phenotype, far less than those resulting from loss of cofilin or coronin (90). Thus, understanding whether and how this set of three proteins causes filament disassembly in endocytosis requires further study.

2.3.6. Actin cross-linking proteins. Actin cross-linking proteins may also play a role in assembling a productive actin filament network during CME. In biochemical studies, Arp2/3 can connect the side of one filament with the pointed end of another filament, and Sac6, mammalian fimbrin, can bind the sides of two actin filaments, forming a bundle. In fission yeast, fimbrin (Fim1) is important for endocytic actin patch function. Fim1 truncation leads to loss of its ability to cross-link actin filaments; in cells, this mutation increased the lifetime of actin at the patch and decreased the number of actin patches that move inward. In *fim1* null mutants, tropomyosin localizes ectopically to actin patches; however, this did not occur with the truncated Fim1 unable to cross-link, indicating that Fim1's bundling activity is important (95, 96).

Actin filament cross-linking is also important in budding yeast. Fimbrin (Sac6) null mutations caused defects in actin organization, cell polarity, and endocytic uptake of α-factor (97, 98). Loss of Sac6 inhibited inward movement of endocytic structures, completely in one study and partially in another (72, 99). The *sac6* null mutant phenotype was enhanced by deletion of the gene encoding another actin cross-linking protein, Ssp1 (99). The possible role of ectopic tropomyosin, seen in fission yeast, has yet to be examined in these budding yeast mutants. The amount of fimbrin at sites of endocytosis in fission yeast is much greater than that in budding yeast, so fimbrin's function may differ in the two organisms (100; B.J. Galletta & J.A. Cooper, unpublished information).

The end-to-side branch connections between actin filaments created by Arp2/3 are lost over time, and the rate of their loss is likely to be an important parameter for the function of the filament network. In vitro loss of these connections was found to be increased by a cofilin-like protein, glial maturation

N-WASP: neural Wiskott-Aldrich syndrome protein

N-BAR: BAR domain preceded by an N-terminal amphipathic helix

I-BAR: inverse-BAR protein domain

factor (GMF, known as Aim7 in budding yeast) (101). In cells, GMF/Aim7 displayed genetic interactions with cofilin, so its debranching activity may be functionally relevant.

2.3.7. Time profiles of recruitment.
Observations of the time course of recruitment of proteins to the CCS have produced time profiles that allow one to divide the many proteins into groups or modules (72, 91). In yeast, four modules were described: the coat protein module, the actin module, the WASP/Myo module, and the amphiphysin module. A recent comprehensive study in mammalian cells by Merrifield and colleagues (91) found the same four modules and uncovered additional ones. The WASP/Myo module included dynamin and was termed the dynamin module in this study. The coat module was divided into a clathrin submodule and an adaptor/F-BAR submodule, and the actin module was divided into assembly and disassembly submodules. New modules included GAK, Rab5, and FBP17/CIP4, making a total of nine groups. The similarities in these results between yeast and mammalian cells argue for common molecular functions. Future studies of mutations in mammalian cells, similar to those in yeast, can test that hypothesis.

Quantitative analysis of the time profiles for recruitment has led to other novel and important findings. Working in mammalian cells, Merrifield and coworkers (91) employed a pH-sensitive fluorescent transferrin receptor, which revealed the time of scission, allowing them to align the recruitment time profiles of various proteins relative to the time of scission. One unexpected result among the recruitment time profiles was that FBP17, an F-BAR protein, showed biphasic recruitment and peaked after scission. SNX9 (sorting nexin 9), a BAR protein, also peaked after scission. These recruitment profiles were not expected because these proteins interact with dynamin and N-WASP, which are recruited earlier, and BAR domain proteins are predicted to precede and promote scission, not follow it, as described in more detail in the next sections. Both proteins are functionally important in that knockdown of either FBP17 or SNX9 inhibited CME (29, 102, 103).

Another interesting result from Merrifield's group (91) was that ~50% of scission events were categorized as nonterminal, meaning that scission did not result in total clearance of the CCS from the membrane. These nonterminal events may have skewed CCP internalization lifetime measurements in the past based on observing the disappearance of fluorescent puncta. This study also described shorter lifetime events, termed abortive, consistent with previous observations (104).

However, conclusions from the lifetime results are not without controversy. Studies in mammalian cells have used fluorescent fusion proteins, generally expressed at levels somewhat higher than those of the endogenous proteins. This fact raises concern about the possibility of altering the molecular processes under examination. In a recent study (105), the advent of gene-editing techniques in mammalian cells allowed investigators to use fluorescent fusions of endocytic proteins expressed from their endogenous genetic loci. When the levels of the fusion proteins were that of endogenous and likely lower expression than in the past, CCP lifetimes were shorter. Also, almost all clathrin-positive foci led to productive endocytic events because the foci proceeded to a dynamin-positive state (105). The sensitivity of the fluorescence microscopy now becomes an important variable, one that may account for the apparent differences among the various studies. Results from experiments combining high-sensitivity live-cell fluorescence microscopy with gene-editing methodologies should be very interesting. Observations of recruitment time profiles in mammalian cell mutants, analogous to experiments in yeast, will also be exciting.

2.4. Coordination of Membrane Curvature Proteins with Actin

The BAR domain superfamily of proteins induces or senses membrane curvature, and its members are involved in CME in yeast and mammalian cells. The superfamily has three families that modulate curvature in distinct ways: BAR/N-BAR (N-terminal amphipathic helix preceding the BAR domain) proteins, F-BAR (BAR domain is extension of an FCH domain) proteins, and I-BAR (inverse BAR) proteins (106). BAR/N-BAR and F-BAR proteins preferentially bind highly curved and relatively flatter membranes, respectively, which may influence the stage at which they interact with CCSs.

A number of F-BAR proteins have been linked to actin-regulatory proteins and found to be involved in CME, including FBP17, Toca-1/CIP4, syndapin (Bzz1 in yeast), and FCHo1 and -2 (Syp1 in yeast) (107–109). F-BAR proteins are thought to participate in early invagination steps to form the tubules of CCPs, in part because they can bind and deform relatively flat membranes, such as the plasma membrane (106, 110). Overexpression of F-BAR proteins induces tubule formation in mammalian cells (29, 103). Early recruitment of Syp1 in yeast (69–71) or FCHo1 in mammalian cells to sites of CME (37, 111) suggests that these proteins participate in creating the initial membrane curvature and also in recruiting clathrin, thus helping to specify the site of CME (111). Additionally, Syp1 negatively regulates WASp/Las17 in yeast (71), suggesting that actin polymerization is inhibited until an initial level of membrane curvature is achieved. Other F-BAR proteins are recruited later to sites of CME (67, 91), and they stimulate Arp2/3-dependent actin polymerization in mammalian cells via N-WASP/WASp (Las17 in yeast) or the N-WASP/WIP complex (Las17/ Vrp1 in yeast) (55, 67, 74, 103, 109, 112–115). Somewhat surprisingly, the F-BAR proteins CIP4 and FBP17 displayed biphasic recruitment to CCPs in mammalian cells (NIH-3T3), with accumulation phases before and after scission (91), as noted above. Therefore, their functions must differ somewhat from what was predicted.

In mammalian cells, BAR/N-BAR proteins (including amphiphysins, endophilin, Tuba, and SNX9) are involved in CME. CME was inhibited by knockdown of SNX9 (102) or overexpression of dominant-negative mutant forms of BAR domain proteins, such as amphiphysin (116), SNX9 (117) or Tuba (118). BAR/N-BAR proteins prefer to bind to highly curved membranes, and they can induce curvature in cells. Overexpression of amphiphysin or endophilin caused long tubules to form from the plasma membrane (119, 120). Therefore, during CME, BAR/N-BAR proteins are predicted to be recruited to invaginated CCPs to promote tubule elongation or stabilization (110, 121). Several cellular experiments support this prediction (reviewed in References 110 and 122). However, as noted above, the recruitment of SNX9 to sites of CME peaks after scission (91), suggesting a more complex or different mechanism.

BAR/N-BAR proteins may also induce actin assembly at sites of CME. In mammalian cells, many BAR/N-BAR proteins are able to bind and activate N-WASP (Las17 in yeast) and thereby stimulate Arp2/3-mediated actin polymerization (118, 123, 124). Alternatively, the BAR protein PICK1 can bind to and regulate Arp2/3 directly (125).

Together, the number of F-BAR and BAR/N-BAR proteins that sense and/or promote membrane curvature is relatively high in yeast and mammalian cells, as is the number that can interact with actin polymerization machinery. Thus, the field is faced with the challenging and important tasks of understanding how and when each protein is recruited to a site of CME, how each protein contributes to the regulation of actin assembly, and how the various proteins are coordinated with each other.

I-BAR proteins bind to membranes and induce negative curvature, the opposite of what is produced by BAR/N-BAR or F-BAR proteins. Overexpression of I-BAR proteins in mammalian cells promotes filopodia-like protrusions, so the general role for I-BAR proteins may be to cause membrane protrusions, not invaginations. I-BAR proteins may stabilize the neck of an invaginating CCP, where the curvature is negative.

I-BAR proteins interact with several actin regulators, including N-WASp (126), WAVE2 (127), Mena/VASP (128), and mDia (129). The

I-BAR protein missing-in-metastasis (MIM), interacts with the Arp2/3 regulator cortactin (130), and it contains a WH2 domain, which can bind actin monomers (131, 132).

On the basis of recent studies, I-BAR proteins do appear to have a role in CME. Mouse embryo fibroblasts isolated from MIM homozygous knockout mice were impaired in their ability to perform receptor-mediated CME and fluid-phase pinocytosis (133). The MIM$^{-/-}$ cells had an abnormally high number of platelet-derived growth factor (PDGF) receptors remaining on the surface after PDGF stimulation and an increased number of phosphorylated PDGF receptors, also consistent with decreased CME. In addition, MIM overexpression led to increased receptor-mediated CME, suggesting that MIM promotes CME. However, another found that MIM inhibited endocytosis in mouse embryo fibroblasts and *Drosophila* border cells (134). Here, researchers used siRNA-mediated knockdown of MIM in a cultured cell line, a different approach from the first study. They also found that MIM competes with the endophilin/CD2AP complex for cortactin binding, which may relate to the endocytosis and cell migration phenotypes reported. IBARa, a *Dictyostelium* homolog of the I-BAR protein IRSp53, has been implicated in CME because of its localization to clathrin pits (135). Also, BAIAP2L2, a member of the IRSp53 family in mammals, was localized to clathrin-coated plaques in cultured cell lines.

2.5. Mechanisms of Membrane Scission

In mammalian cells, dynamin is a critical component for fission of the CCP from the plasma membrane (27, 136, 137). Dynamin is a GTPase with five functional domains: the N-terminal GTPase domain, a middle domain responsible for oligomerization, a PH domain that binds PIP$_2$, a GED domain also responsible for oligomerization and for stimulating the GTPase activity of the N-terminal domain, and a PRD domain that interacts with many SH3 domain partners. In cells, the recruitment time profile of dynamin shows relatively late accumulation, just prior to scission (91).

Many biochemical interactions between dynamin and actin exist in mammalian cells either by actin-binding proteins, such as cortactin, mAbp1, syndapin, and myosin1E (138, 139), or by direct binding to actin (140). Also, most of the F-BAR or BAR domain proteins involved in CME bind both dynamin and either N-WASP or SCAR/WAVE (121), further linking dynamin to the actin cytoskeleton. Some of dynamin's interactions, including those with cortactin or SNX9, can increase the GTPase activity of dynamin in vitro (102, 141), which is important for dynamin's mechanochemical function. However, how these connections between dynamin and actin contribute to CME is still unclear.

Several studies have indicated that actin may contribute to scission. Cells treated with latrunculin, jasplakinolide, or thymosin β4 contained stalled endocytic pits, as determined by EM or specialized uptake assays, suggesting that actin promotes constriction of the vesicle neck (23, 24, 32, 142). Stalled clathrin pits resulting from actin toxin treatment of MDCK cells did not recruit dynamin (32). Perhaps actin contributes to scission by constricting the vesicle neck, providing force to assist dynamin in scission.

Actin does play a role in fission of membrane tubules in one form of clathrin-independent endocytosis (14). In cells, actin was important for the fission of membrane tubules that were induced by Shiga toxin. Actin was also important for fission in vitro in a synthetic system of liposomes in which membrane tubules were induced to form by Shiga toxin. Actin and actin-binding proteins were components of the synthetic system, but dynamin was absent, suggesting that actin polymerization at the tubule membrane can be sufficient for fission. In this study, fission required the presence of cholesterol in the liposomes, consistent with the notion that boundaries between membrane domains create interfacial force that promotes fission, discussed below in Section 2.6.

In yeast, dynamin also contributes to scission but is less critical. The dynamin homolog

Vps1 was originally found to have a role in CME because there were null mutant phenotypes in receptor internalization and actin organization (143). Subsequent studies found that Vps1 localized transiently to sites of endocytosis, with a time profile similar to that of the amphiphysin Rvs167, a likely marker for vesicle scission (144). Also, $vps1\Delta$ null mutant cells had abnormalities in CME, including the ultrastructure of the membrane invagination and uptake of the membrane dye FM4-64 (144, 145). The $vps1\Delta$ mutant had delayed actin patch movement and decreased recruitment of the amphiphysin Rvs167 (144). In sum, Vps1 appears to promote invagination and trigger scission, with a less critical role than what has been observed for dynamin in other cell types.

In yeast, BAR domain–containing amphiphysins, Rvs161 and Rvs167, also have a role in membrane scission. They are recruited to endocytic patches just before the patch begins to move away from the membrane, they move for a short distance, and then they are lost, consistent with a role in scission (72). Rvs167 was observed at the neck of the membrane invagination by immunoelectron microscopy (60), confirmed by recent correlated light microscopy/EM (146). In vitro purified Rvs161 and Rvs167 proteins formed a heterodimer, and the heterodimer was able to bind liposomes and induce the formation of tubules (147). In $rvs161\Delta$ and $rvs167\Delta$ single and double-null mutant cells, ~20%–30% of endocytic patches began to move away from the plasma membrane and then retracted back, consistent with a failure of scission (72, 147). In another study, $rvs167$ and $vps1$ single mutants showed 30%–40% of endocytic patches retracting back to the membrane, with a greater percentage in an $rvs167\ vsp1$ double mutant, supporting the notion that amphiphysins and dynamin both have roles in scission (144).

In mammalian cells, BAR domain proteins are important for CME, and they appear to coordinate the action of dynamin for scission. Many BAR domain proteins, including SNX9, interact directly with dynamin (102, 122, 148–151). SNX9, endophilin, and amphiphysin are recruited to endocytic sites with temporal profiles similar to that of dynamin (91, 102, 152). BAR domain proteins alone do not seem to cause scission; the long tubules induced by overexpression of amphiphysin or endophilin are stable and do not undergo fission (119, 120).

In mammalian cells, recent evidence suggests a role for phospholipid metabolism in scission. Synaptojanin, a $PI(4,5)P_2$ phosphatase, induced fragmentation of tubules (created by overexpression of the BAR domain protein endophilin) in conjunction with endophilin and dynamin (120). Synaptojanin2β1 was recruited to endocytic sites with a timing profile similar to that of dynamin, peaking at scission (91). These findings support the idea that tension at the interface of membrane lipid domains may contribute to scission (57, 153, 154), which we discuss in more detail below.

2.6. Mathematical Modeling of Actin Assembly and Clathrin-Mediated Endocytosis

The small size of endocytic sites creates a challenge for the analysis of how actin can provide forces driving membrane deformation. Direct measurement of many of the important physical parameters are correspondingly difficult, especially in living cells. The molecular complexity of the endocytic site also provides a challenge, owing to the large number of protein and lipid molecules associated with the membrane, not to mention the polymeric carbohydrates and proteins of the cell wall outside the membrane. In this setting, molecular modeling may provide critical insight into these processes. Two groups have described complementary models of endocytosis aimed at explaining how the actin filament network assembles and how the forces for bending the membrane and causing fission are generated.

Liu and colleagues (57, 153, 154) developed a model to predict curvature formation and scission on the basis of interfacial forces produced by lipid boundaries in the membrane of the forming vesicle. Mechanochemical feedback is a key feature of the model. BAR domain proteins sense curvature in the membrane and bind to the membrane, and this promotes further curvature, which positively feeds back to promote more binding of BAR domain proteins (154). Next, the presence of BAR domain proteins at the neck of the short tubule protects $PI(4,5)P_2$ from the action of phosphatases. PIP_2 hydrolysis continues unabated at the tip of the tubule, providing an interfacial force between the two regions of the membrane owing to the difference in their PIP_2 compositions. Positive feedback occurs again at this point; the interfacial force enhances curvature, and curvature promotes the action of phosphatases. Phosphatase-catalyzed hydrolysis of PIP_2 further increases the difference in PIP_2 composition between the two membrane regions, which increases the interfacial force. The model can be applied to CME in both yeast and mammalian systems. The model also incorporates actin, myosin, and dynamin as components that promote curvature and invagination (154).

In another study, Pollard and colleagues (155) developed a detailed biochemical model for the assembly and disassembly of the actin filament network at sites of endocytosis in fission yeast, using the dendritic nucleation model. The model incorporated experimental values for kinetic rate constants for many of the chemical reactions, which had been determined from in vitro biochemical experiments. The model was further constrained by values for the numbers of molecules at the endocytic sites, which were obtained from quantitative fluorescence microscopy (100, 155). When the model was used to fit experimental data for protein dynamics, some biochemical parameters needed to be changed to provide a good fit, suggesting new experiments to test the model. One prediction was that some of the chemical reactions pertaining to actin needed to occur faster than what had been measured for muscle actin. Confirming that prediction, actin from *S. pombe* was recently found to have a faster rate of nucleotide exchange, dimer formation, and phosphate release when compared to muscle actin (156). This model provides an important mechanism to begin to understand the temporal relationships among the many reactants involved in forming a branched actin filament network. In the future, extending the model with spatial and mechanical information will be informative.

2.7. Models of Force Generation by Actin Assembly During Clathrin-Mediated Endocytosis

As we have discussed above, actin assembly is tightly coupled to and necessary for CME in yeast and, in some situations, mammalian cells. How might actin contribute to the force required for efficient CME? One mechanism includes force productions by myosins, interacting by their motor within actin filaments. Type I myosins localize to sites of CME in yeast and mammals, and they are functionally important for endocytosis (18, 91, 139, 157). In cultured mammalian cells, myosin 1E and myosin VI were recruited to sites of CME as part of the dynamin module (91). Myosin VI, a minus-end-directed motor myosin, was necessary for CME at the apical brush border of epithelial cells in the intestine and the kidney (158, 159). This region of the plasma membrane includes microvilli and the terminal web, both of which are very dense with actin filaments. For most of these actin filaments, their pointed ends are directed away from the membrane, suggesting that myosin VI may walk along these filaments and pull the CME vesicle away from the membrane.

Another possible mechanism involves force produced by Arp2/3-mediated dendritic nucleation at membrane surfaces, resulting from membrane-bound NPFs. In one such model, NPFs are located at the base of the clathrin-coated vesicle on the plasma membrane. The actin filaments created by these NPFs add new actin monomers onto filament barbed ends that are directed toward the plasma membrane. The growing filament ends push on the membrane, which does not move, and the actin filament network flows away from the membrane (20, 160). Attachment of the tip of the invagination to the actin filament network results in the tip being pulled into the cytoplasm. As

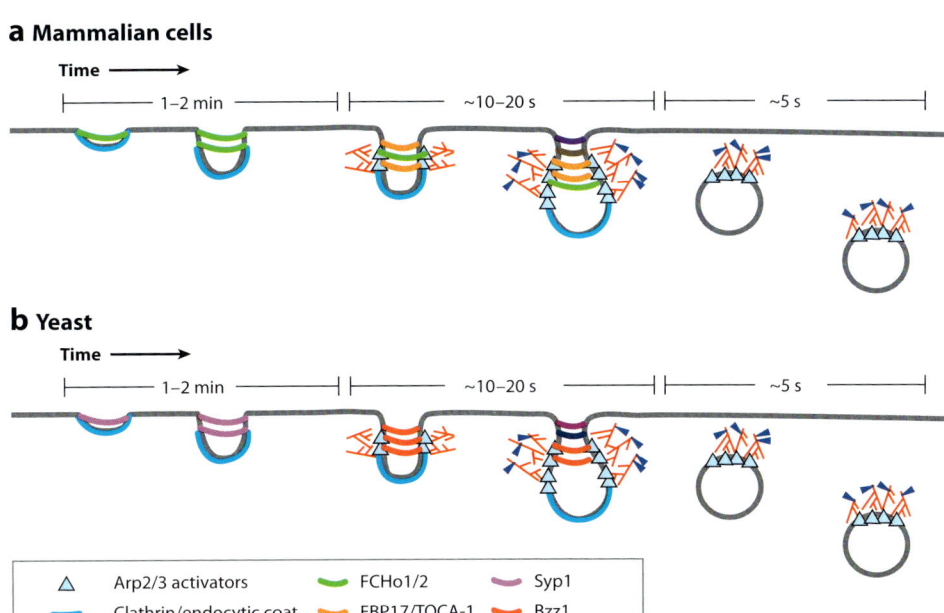

Figure 1

Models for actin assembly during endocytosis in mammalian cells (*a*) and budding yeast (*b*); these models are based on many of the studies discussed above. Many features are found in common. Clathrin and adaptors are recruited at very early stages and provide initial curvature. Regulators of actin nucleation follow. Actin polymerization proceeds, curvature increases, and the invagination elongates. Amphiphysins and dynamin are recruited, and membrane fission occurs. Actin disassembly proteins are recruited late.

invagination proceeds and scission occurs, actin-binding proteins move away from the membrane; however, the actin filament network does not constitute a solid column extending from the membrane to the actin patch. Instead, a clear zone separates the actin patch from the membrane, as found by fluorescence microscopy (18). Furthermore, immunoelectron microscopy images of the invagination reveal two pools of actin and myosin-I (Myo5), one at the base of the invagination and one near the tip (60). Thus, the network appears to assemble at or near the vesicle, as well as at the plasma membrane.

An alternative actin assembly model incorporates these features. Membrane-bound NPFs still nucleate actin filament assembly with barbed ends pointed toward the membrane. In this case, NPFs are located on the plasma membrane, and they remain on the invaginating membrane (161). In this model, illustrated in **Figure 1**, actin filament assembly pushes on the membrane at the initial curvature site to promote invagination, elongation, and fission. This model is supported by new results on the architecture of the actin filament network during CME, revealed by an EM study of CCSs in mammalian cells by Svitkina and colleagues (37). Many CCSs were associated at their edges with short actin filaments in a highly branched network characteristic of Arp2/3-mediated assembly. The barbed ends of the actin filaments were directed toward the CCSs. Shallow CCSs were surrounded by a collar-like network of filaments, and invaginated and constricted CCSs were associated with elongated patches of actin filaments. The results support the hypothesis that the polymerization of free barbed ends of actin filaments may push on the membrane to exert force at the neck of vesicles. This model is consistent with the idea, discussed above, that regulators of actin polymerization can be recruited or activated by BAR domain proteins (108, 121, 162). Actin filaments arranged in this fashion would be poised to drive the movement of the vesicle after it separates from the membrane (37).

3. CONCLUSIONS AND FUTURE DIRECTIONS

In conclusion, actin is essential for many types of endocytosis, including several that do not involve clathrin. For CME, actin is essential in yeast, but in metazoans, it varies in importance, depending on the cell type and experimental design. Comparative molecular studies of CME in metazoans and yeast, including budding and fission yeast, reveal a striking level of similarity despite great phylogenetic distance. Thus, the molecular machinery may be assembled and function in similar ways.

In the future, addressing the issue of how and where actin filaments create force at an endocytic vesicle will be important. In mammalian cells, when actin is important for efficient CME, one would like to define the molecular mechanisms that sense the need for actin and the signaling pathway that connects the sensor information to the appropriate actin assembly machinery. A related issue is understanding how CME is regulated by tension in the plasma membrane and in the underlying actin cortex, a topic currently associated with controversy. Actin polymerization, or myosin motors, or perhaps some unknown mechanism may generate the force during invagination of an endocytic vesicle. Regardless of the source of the force, knowledge of the detailed architecture of the actin filament network over time, with respect to the plasma membrane and the endocytic vesicle, will be critical. Myosins are clearly important for CME, and understanding the molecular nature of their function will be valuable, for myosin-I in yeast and metazoans and for myosin-VI in mammalian cells. Two important issues are (*a*) whether myosins function as motors to move vesicular cargo along actin filaments and (*b*) the molecular nature of the connection between myosin and cargo. CME has been studied intensively, but only in a limited number of organisms, so it will be interesting to explore the commonality of mechanisms across a broader range of organisms. Finally, the types of endocytosis and the underlying molecular mechanisms that occur in cells in the context of tissues and organs will be important areas for future research.

SUMMARY POINTS

1. Actin is essential for clathrin-mediated endocytosis (CME) in yeast.
2. Actin can vary from important to dispensable for CME in metazoans, depending on the circumstances.
3. Actin is essential for many types of endocytosis that do not involve clathrin.
4. Comparative molecular studies of CME in budding and fission yeast reveal a striking level of similarity despite great phylogenetic distance.

FUTURE ISSUES

1. How and where do actin filaments create force?
2. What is the detailed architecture of the actin filament network at sites of endocytosis? How does that architecture change over time?

3. How does myosin function in CME? In particular, does it function as a motor moving along actin filaments? If so, what are the molecular compositions of its cargo?

4. How does CME depend on tension in the plasma membrane and in the actin cortex, which is associated with the plasma membrane?

5. How common are the molecular mechanisms for CME, including the role of actin, across a broad range of organisms?

6. What are the molecular mechanisms that sense the need for actin in a given location for CME? And how does the sensor signal to the actin assembly machinery?

7. What types of endocytosis and molecular mechanisms are found in cells operating in tissues and organs?

DISCLOSURE STATEMENT

The authors are not aware of any affiliations, memberships, funding, or financial holdings that might be perceived as affecting the objectivity of this review.

ACKNOWLEDGMENTS

We are grateful to the many colleagues who shared their insight about the field or provided us with information about their research prior to publication. We apologize to the many researchers whose work was not discussed owing to space constraints. The writing of this article was supported by the following National Institutes of Health grants: R01 GM 38542 to J.A.C., F32 GM083538 to O.L.M., and R01 GM086882 to Anders Carlsson.

LITERATURE CITED

1. Doherty GJ, McMahon HT. 2009. Mechanisms of endocytosis. *Annu. Rev. Biochem.* 78:857–902
2. McMahon HT, Boucrot E. 2011. Molecular mechanism and physiological functions of clathrin-mediated endocytosis. *Nat. Rev. Mol. Cell Biol.* 12:517–33
3. Howes MT, Mayor S, Parton RG. 2010. Molecules, mechanisms, and cellular roles of clathrin-independent endocytosis. *Curr. Opin. Cell Biol.* 22:519–27
4. Groves E, Dart AE, Covarelli V, Caron E. 2008. Molecular mechanisms of phagocytic uptake in mammalian cells. *Cell. Mol. Life Sci.* 65:1957–76
5. Orth JD, McNiven MA. 2006. Get off my back! Rapid receptor internalization through circular dorsal ruffles. *Cancer Res.* 66:11094–96
6. Kerr MC, Teasdale RD. 2009. Defining macropinocytosis. *Traffic* 10:364–71
7. Jaumouille V, Grinstein S. 2011. Receptor mobility, the cytoskeleton, and particle binding during phagocytosis. *Curr. Opin. Cell Biol.* 23:22–29
8. Lim JP, Gleeson PA. 2011. Macropinocytosis: an endocytic pathway for internalising large gulps. *Immunol. Cell Biol.* 89:836–43
9. Michelot A, Drubin DG. 2011. Building distinct actin filament networks in a common cytoplasm. *Curr. Biol.* 21:R560–69
10. Pollard TD, Cooper JA. 2009. Actin, a central player in cell shape and movement. *Science* 326:1208–12
11. Firat-Karalar EN, Welch MD. 2011. New mechanisms and functions of actin nucleation. *Curr. Opin. Cell Biol.* 23:4–13
12. Kovar DR, Sirotkin V, Lord M. 2011. Three's company: the fission yeast actin cytoskeleton. *Trends Cell Biol.* 21:177–87

13. Chesarone MA, Dupage AG, Goode BL. 2010. Unleashing formins to remodel the actin and microtubule cytoskeletons. *Nat. Rev. Mol. Cell Biol.* 11:62–74
14. Romer W, Pontani LL, Sorre B, Rentero C, Berland L, et al. 2010. Actin dynamics drive membrane reorganization and scission in clathrin-independent endocytosis. *Cell* 140:540–53
15. Dunn JD, Valdivia RH. 2010. Uncivil engineers: *Chlamydia*, *Salmonella* and *Shigella* alter cytoskeleton architecture to invade epithelial cells. *Future Microbiol.* 5:1219–32
16. Taylor MP, Koyuncu OO, Enquist LW. 2011. Subversion of the actin cytoskeleton during viral infection. *Nat. Rev. Microbiol.* 9:427–39
17. Cossart P, Veiga E. 2008. Non-classical use of clathrin during bacterial infections. *J. Microsc.* 231:524–28
18. Galletta BJ, Mooren OL, Cooper JA. 2010. Actin dynamics and endocytosis in yeast and mammals. *Curr. Opin. Biotechnol.* 21:604–10
19. Ayscough KR, Stryker J, Pokala N, Sanders M, Crews P, Drubin DG. 1997. High rates of actin filament turnover in budding yeast and roles for actin in establishment and maintenance of cell polarity revealed using the actin inhibitor latrunculin-A. *J. Cell Biol.* 137:399–416
20. Kaksonen M, Sun Y, Drubin DG. 2003. A pathway for association of receptors, adaptors, and actin during endocytic internalization. *Cell* 115:475–87
21. Kim K, Galletta BJ, Schmidt KO, Chang FS, Blumer KJ, Cooper JA. 2006. Actin-based motility during endocytosis in budding yeast. *Mol. Biol. Cell* 17:1354–63
22. Aghamohammadzadeh S, Ayscough KR. 2009. Differential requirements for actin during yeast and mammalian endocytosis. *Nat. Cell Biol.* 11:1039–42
23. Merrifield CJ, Perrais D, Zenisek D. 2005. Coupling between clathrin-coated-pit invagination, cortactin recruitment, and membrane scission observed in live cells. *Cell* 121:593–606
24. Yarar D, Waterman-Storer CM, Schmid SL. 2005. A dynamic actin cytoskeleton functions at multiple stages of clathrin-mediated endocytosis. *Mol. Biol. Cell* 16:964–75
25. Fujimoto LM, Roth R, Heuser JE, Schmid SL. 2000. Actin assembly plays a variable, but not obligatory role in receptor-mediated endocytosis in mammalian cells. *Traffic* 1:161–71
26. Saffarian S, Cocucci E, Kirchhausen T. 2009. Distinct dynamics of endocytic clathrin-coated pits and coated plaques. *PLoS Biol.* 7:e1000191
27. Ferguson SM, Raimondi A, Paradise S, Shen H, Mesaki K, et al. 2009. Coordinated actions of actin and BAR proteins upstream of dynamin at endocytic clathrin-coated pits. *Dev. Cell* 17:811–22
28. Wu M, Huang B, Graham M, Raimondi A, Heuser JE, et al. 2010. Coupling between clathrin-dependent endocytic budding and F-BAR-dependent tubulation in a cell-free system. *Nat. Cell Biol.* 12:902–8
29. Itoh T, Erdmann KS, Roux A, Habermann B, Werner H, De Camilli P. 2005. Dynamin and the actin cytoskeleton cooperatively regulate plasma membrane invagination by BAR and F-BAR proteins. *Dev. Cell* 9:791–804
30. Batchelder EM, Yarar D. 2010. Differential requirements for clathrin-dependent endocytosis at sites of cell-substrate adhesion. *Mol. Biol. Cell* 21:3070–79
31. Liu AP, Loerke D, Schmid SL, Danuser G. 2009. Global and local regulation of clathrin-coated pit dynamics detected on patterned substrates. *Biophys. J.* 97:1038–47
32. Boulant S, Kural C, Zeeh J-C, Ubelmann F, Kirchhausen T. 2011. Actin dynamics counteract membrane tension during clathrin-mediated endocytosis. *Nat. Cell Biol.* 13:1124–31
33. Gottlieb TA, Ivanov IE, Adesnik M, Sabatini DD. 1993. Actin microfilaments play a critical role in endocytosis at the apical but not the basolateral surface of polarized epithelial cells. *J. Cell Biol.* 120:695–710
34. Dai J, Sheetz MP. 1999. Membrane tether formation from blebbing cells. *Biophys. J.* 77:3363–70
35. Cureton DK, Massol RH, Saffarian S, Kirchhausen TL, Whelan SP. 2009. Vesicular stomatitis virus enters cells through vesicles incompletely coated with clathrin that depend upon actin for internalization. *PLoS Pathog.* 5:e1000394
36. Cureton DK, Massol RH, Whelan SP, Kirchhausen T. 2010. The length of vesicular stomatitis virus particles dictates a need for actin assembly during clathrin-dependent endocytosis. *PLoS Pathog.* 6:1001127
37. Collins A, Warrington A, Taylor KA, Svitkina T. 2011. Structural organization of the actin cytoskeleton at sites of clathrin-mediated endocytosis. *Curr. Biol.* 21:1167–75

38. Dhonukshe P, Aniento F, Hwang I, Robinson DG, Mravec J, et al. 2007. Clathrin-mediated constitutive endocytosis of PIN auxin efflux carriers in *Arabidopsis*. *Curr. Biol.* 17:520–27
39. Perez-Gomez J, Moore I. 2007. Plant endocytosis: it is clathrin after all. *Curr. Biol.* 17:R217–19
40. Kitakura S, Vanneste S, Robert S, Löfke C, Teichmann T, et al. 2011. Clathrin mediates endocytosis and polar distribution of PIN auxin transporters in *Arabidopsis*. *Plant Cell* 23:1920–31
41. Holstein SE. 2002. Clathrin and plant endocytosis. *Traffic* 3:614–20
42. Baluska F, Hlavacka A, Samaj J, Palme K, Robinson DG, et al. 2002. F-actin-dependent endocytosis of cell wall pectins in meristematic root cells. Insights from brefeldin A-induced compartments. *Plant Physiol.* 130:422–31
43. Baluska F, Samaj J, Hlavacka A, Kendrick-Jones J, Volkmann D. 2004. Actin-dependent fluid-phase endocytosis in inner cortex cells of maize root apices. *J. Exp. Bot.* 55:463–73
44. Dhonukshe P, Grigoriev I, Fischer R, Tominaga M, Robinson DG, et al. 2008. Auxin transport inhibitors impair vesicle motility and actin cytoskeleton dynamics in diverse eukaryotes. *Proc. Natl. Acad. Sci. USA* 105:4489–94
45. Lisboa S, Scherer GE, Quader H. 2008. Localized endocytosis in tobacco pollen tubes: visualisation and dynamics of membrane retrieval by a fluorescent phospholipid. *Plant Cell Rep.* 27:21–28
46. Eggenberger K, Mink C, Wadhwani P, Ulrich AS, Nick P. 2011. Using the peptide BP100 as a cell-penetrating tool for the chemical engineering of actin filaments within living plant cells. *ChemBioChem* 12:132–37
47. Samaj J, Baluska F, Voigt B, Schlicht M, Volkmann D, Menzel D. 2004. Endocytosis, actin cytoskeleton, and signaling. *Plant Physiol.* 135:1150–61
48. Rivero F. 2008. Endocytosis and the actin cytoskeleton in *Dictyostelium discoideum*. *Int. Rev. Cell Mol. Biol.* 267:343–97
49. Heinrich D, Youssef S, Schroth-Diez B, Engel U, Aydin D, et al. 2008. Actin-cytoskeleton dynamics in non-monotonic cell spreading. *Cell Adhes. Migr.* 2:58–68
50. Brady RJ, Damer CK, Heuser JE, O'Halloran TJ. 2010. Regulation of Hip1r by epsin controls the temporal and spatial coupling of actin filaments to clathrin-coated pits. *J. Cell Sci.* 123:3652–61
51. Veltman DM, Insall RH. 2010. WASP family proteins: their evolution and its physiological implications. *Mol. Biol. Cell* 21:2880–93
52. Boettner DR, Friesen H, Andrews B, Lemmon SK. 2011. Clathrin light chain directs endocytosis by influencing the binding of the yeast Hip1R homologue, Sla2, to F-actin. *Mol. Biol. Cell* 22:3699–714
53. Le Clainche C, Pauly BS, Zhang CX, Engqvist-Goldstein AE, Cunningham K, Drubin DG. 2007. A Hip1R-cortactin complex negatively regulates actin assembly associated with endocytosis. *EMBO J.* 26:1199–210
54. Kochubey O, Majumdar A, Klingauf J. 2006. Imaging clathrin dynamics in *Drosophila melanogaster* hemocytes reveals a role for actin in vesicle fission. *Traffic* 7:1614–27
55. Fricke R, Gohl C, Dharmalingam E, Grevelhörster A, Zahedi B, et al. 2009. *Drosophila* Cip4/Toca-1 integrates membrane trafficking and actin dynamics through WASP and SCAR/WAVE. *Curr. Biol.* 19:1429–37
56. Conibear E. 2010. Converging views of endocytosis in yeast and mammals. *Curr. Opin. Cell Biol.* 22:513–18
57. Liu J, Sun Y, Oster GF, Drubin DG. 2010. Mechanochemical crosstalk during endocytic vesicle formation. *Curr. Opin. Cell Biol.* 22:36–43
58. Robertson AS, Smythe E, Ayscough KR. 2009. Functions of actin in endocytosis. *Cell. Mol. Life Sci.* 66:2049–65
59. Girao H, Geli MI, Idrissi FZ. 2008. Actin in the endocytic pathway: from yeast to mammals. *FEBS Lett.* 582:2112–19
60. Idrissi FZ, Grötsch H, Fernández-Golbano IM, Presciatto-Baschong C, Riezman H, Geli MI. 2008. Distinct acto/myosin-I structures associate with endocytic profiles at the plasma membrane. *J. Cell Biol.* 180:1219–32
61. Young ME, Cooper JA, Bridgman PC. 2004. Yeast actin patches are networks of branched actin filaments. *J. Cell Biol.* 166:629–35

62. Winter D, Podtelejnikov AV, Mann M, Li R. 1997. The complex containing actin-related proteins Arp2 and Arp3 is required for the motility and integrity of yeast actin patches. *Curr. Biol.* 7:519–29
63. Martin AC, Xu XP, Rouiller I, Kaksonen M, Sun Y, et al. 2005. Effects of Arp2 and Arp3 nucleotide-binding pocket mutations on Arp2/3 complex function. *J. Cell Biol.* 168:315–28
64. Rodal AA, Sokolova O, Robins DB, Daugherty KM, Hippenmeyer S, et al. 2005. Conformational changes in the Arp2/3 complex leading to actin nucleation. *Nat. Struct. Mol. Biol.* 12:26–31
65. Wen KK, Rubenstein PA. 2005. Acceleration of yeast actin polymerization by yeast Arp2/3 complex does not require an Arp2/3-activating protein. *J. Biol. Chem.* 280:24168–74
66. Li R. 1997. Bee1, a yeast protein with homology to Wiscott-Aldrich syndrome protein, is critical for the assembly of cortical actin cytoskeleton. *J. Cell Biol.* 136:649–58
67. Sun Y, Martin AC, Drubin DG. 2006. Endocytic internalization in budding yeast requires coordinated actin nucleation and myosin motor activity. *Dev. Cell* 11:33–46
68. Galletta BJ, Chuang DY, Cooper JA. 2008. Distinct roles for Arp2/3 regulators in actin assembly and endocytosis. *PLoS Biol.* 6:e1
69. Reider A, Barker SL, Mishra SK, Im YJ, Maldonado-Báez L, et al. 2009. Syp1 is a conserved endocytic adaptor that contains domains involved in cargo selection and membrane tubulation. *EMBO J.* 28:3103–16
70. Stimpson HE, Toret CP, Cheng AT, Pauly BS, Drubin DG. 2009. Early-arriving Syp1p and Ede1p function in endocytic site placement and formation in budding yeast. *Mol. Biol. Cell* 20:4640–51
71. Boettner DR, D'Agostino JL, Torres OT, Daugherty-Clarke K, Uygur A, et al. 2009. The F-BAR protein Syp1 negatively regulates WASp-Arp2/3 complex activity during endocytic patch formation. *Curr. Biol.* 19:1979–87
72. Kaksonen M, Toret CP, Drubin DG. 2005. A modular design for the clathrin- and actin-mediated endocytosis machinery. *Cell* 123:305–20
73. Rodal AA, Manning AL, Goode BL, Drubin DG. 2003. Negative regulation of yeast WASp by two SH3 domain-containing proteins. *Curr. Biol.* 13:1000–8
74. Soulard A, Lechler T, Spiridonov V, Shevchenko A, Shevchenko A, et al. 2002. *Saccharomyces cerevisiae* Bzz1p is implicated with type I myosins in actin patch polarization and is able to recruit actin-polymerizing machinery in vitro. *Mol. Cell. Biol.* 22:7889–906
75. Grötsch H, Giblin JP, Idrissi FZ, Fernández-Golbano IM, Collette JR, et al. 2010. Calmodulin dissociation regulates Myo5 recruitment and function at endocytic sites. *EMBO J.* 29:2899–914
76. Sammons MR, James ML, Clayton JE, Sladewski TE, Sirotkin V, Lord M. 2011. A calmodulin-related light chain from fission yeast that functions with myosin-I and PI 4-kinase. *J. Cell Sci.* 124:2466–77
77. Grosshans BL, Grötsch H, Mukhopadhyay D, Fernández IM, Pfannstiel J, et al. 2006. TEDS site phosphorylation of the yeast myosins I is required for ligand-induced but not for constitutive endocytosis of the G protein–coupled receptor Ste2p. *J. Biol. Chem.* 281:11104–14
78. Toshima J, Toshima JY, Duncan MC, Cope MJ, Sun Y, et al. 2007. Negative regulation of yeast Eps15-like Arp2/3 complex activator, Pan1p, by the Hip1R-related protein, Sla2p, during endocytosis. *Mol. Biol. Cell* 18:658–68
79. Munn AL, Thanabalu T. 2009. Verprolin: a cool set of actin-binding sites and some very HOT prolines. *IUBMB Life* 61:707–12
80. Rajmohan R, Wong MH, Meng L, Munn AL, Thanabalu T. 2009. Las17p-Vrp1p but not Las17p-Arp2/3 interaction is important for actin patch polarization in yeast. *Biochim. Biophys. Acta* 1793:825–35
81. Wong MH, Meng L, Rajmohan R, Yu S, Thanabalu T. 2010. Vrp1p-Las17p interaction is critical for actin patch polarization but is not essential for growth or fluid phase endocytosis in *S. cerevisiae*. *Biochim. Biophys. Acta* 1803:1332–46
82. Thanabalu T, Rajmohan R, Meng L, Ren G, Vajjhala PR, Munn AL. 2007. Verprolin function in endocytosis and actin organization. Roles of the Las17p (yeast WASP)-binding domain and a novel C-terminal actin-binding domain. *FEBS J.* 274:4103–25
83. Basu R, Chang F. 2011. Characterization of Dip1p reveals a switch in Arp2/3-dependent actin assembly for fission yeast endocytosis. *Curr. Biol.* 21:905–16
84. Okreglak V, Drubin DG. 2010. Loss of Aip1 reveals a role in maintaining the actin monomer pool and an in vivo oligomer assembly pathway. *J. Cell Biol.* 188:769–77

85. Chang FS, Stefan CJ, Blumer KJ. 2003. A WASp homolog powers actin polymerization-dependent motility of endosomes in vivo. *Curr. Biol.* 13:455–63
86. Chang FS, Han GS, Carman GM, Blumer KJ. 2005. A WASp-binding type II phosphatidylinositol 4-kinase required for actin polymerization-driven endosome motility. *J. Cell Biol.* 171:133–42
87. Brieher WM, Kueh HY, Ballif BA, Mitchison TJ. 2006. Rapid actin monomer-insensitive depolymerization of *Listeria* actin comet tails by cofilin, coronin, and Aip1. *J. Cell Biol.* 175:315–24
88. Kueh HY, Charras GT, Mitchison TJ, Brieher WM. 2008. Actin disassembly by cofilin, coronin, and Aip1 occurs in bursts and is inhibited by barbed-end cappers. *J. Cell Biol.* 182:341–53
89. Okreglak V, Drubin DG. 2007. Cofilin recruitment and function during actin-mediated endocytosis dictated by actin nucleotide state. *J. Cell Biol.* 178:1251–64
90. Lin MC, Galletta BJ, Sept D, Cooper JA. 2010. Overlapping and distinct functions for cofilin, coronin and Aip1 in actin dynamics in vivo. *J. Cell Sci.* 123:1329–42
91. Taylor MJ, Perrais D, Merrifield CJ. 2011. A high precision survey of the molecular dynamics of mammalian clathrin-mediated endocytosis. *PLoS Biol.* 9:e1000604
92. Gandhi M, Achard V, Blanchoin L, Goode BL. 2009. Coronin switches roles in actin disassembly depending on the nucleotide state of actin. *Mol. Cell* 34:364–74
93. Humphries CL, Balcer HI, D'Agostino JL, Winsor B, Drubin DG, et al. 2002. Direct regulation of Arp2/3 complex activity and function by the actin binding protein coronin. *J. Cell Biol.* 159:993–1004
94. Liu SL, Needham KM, May JR, Nolen BJ. 2011. Mechanism of a concentration-dependent switch between activation and inhibition of Arp2/3 complex by coronin. *J. Biol. Chem.* 286:17039–46
95. Skau CT, Kovar DR. 2010. Fimbrin and tropomyosin competition regulates endocytosis and cytokinesis kinetics in fission yeast. *Curr. Biol.* 20:1415–22
96. Skau CT, Courson DS, Bestul AJ, Winkelman JD, Rock RS, et al. 2011. Actin filament bundling by fimbrin is important for endocytosis, cytokinesis, and polarization in fission yeast. *J. Biol. Chem.* 286:26964–77
97. Kubler E, Riezman H. 1993. Actin and fimbrin are required for the internalization step of endocytosis in yeast. *EMBO J.* 12:2855–62
98. Adams AEM, Botstein D, Drubin DG. 1991. Requirement of yeast fimbrin for actin organization and morphogenesis in vivo. *Nature* 354:404–8
99. Gheorghe DM, Aghamohammadzadeh S, Smaczynska-de Rooij II, Allwood EG, Winder SJ, Ayscough KR. 2008. Interactions between the yeast SM22 homologue Scp1 and actin demonstrate the importance of actin bundling in endocytosis. *J. Biol. Chem.* 283:15037–46
100. Sirotkin V, Berro J, Macmillan K, Zhao L, Pollard TD. 2010. Quantitative analysis of the mechanism of endocytic actin patch assembly and disassembly in fission yeast. *Mol. Biol. Cell* 21:2894–904
101. Gandhi M, Smith BA, Bovellan M, Paavilainen V, Daugherty-Clarke K, et al. 2010. GMF is a cofilin homolog that binds Arp2/3 complex to stimulate filament debranching and inhibit actin nucleation. *Curr. Biol.* 20:861–67
102. Soulet F, Yarar D, Leonard M, Schmid SL. 2005. SNX9 regulates dynamin assembly and is required for efficient clathrin-mediated endocytosis. *Mol. Biol. Cell* 16:2058–67
103. Tsujita K, Suetsugu S, Sasaki N, Furutani M, Oikawa T, Takenawa T. 2006. Coordination between the actin cytoskeleton and membrane deformation by a novel membrane tubulation domain of PCH proteins is involved in endocytosis. *J. Cell Biol.* 172:269–79
104. Loerke D, Mettlen M, Yarar D, Jaqaman K, Jaqaman H, et al. 2009. Cargo and dynamin regulate clathrin-coated pit maturation. *PLoS Biol.* 7:e57
105. Doyon JB, Zeitler B, Cheng J, Cheng AT, Cherone JM, et al. 2011. Rapid and efficient clathrin-mediated endocytosis revealed in genome-edited mammalian cells. *Nat. Cell Biol.* 13:331–37
106. Saarikangas J, Zhao H, Lappalainen P. 2010. Regulation of the actin cytoskeleton-plasma membrane interplay by phosphoinositides. *Physiol. Rev.* 90:259–89
107. Roberts-Galbraith RH, Gould KL. 2010. Setting the F-BAR: functions and regulation of the F-BAR protein family. *Cell Cycle* 9:4091–97
108. Suetsugu S. 2010. The proposed functions of membrane curvatures mediated by the BAR domain superfamily proteins. *J. Biochem.* 148:1–12

109. Kessels MM, Qualmann B. 2006. Syndapin oligomers interconnect the machineries for endocytic vesicle formation and actin polymerization. *J. Biol. Chem.* 281:13285–99
110. Frost A, Unger VM, De Camilli P. 2009. The BAR domain superfamily: membrane-molding macromolecules. *Cell* 137:191–96
111. Henne WM, Boucrot E, Meinecke M, Evergren E, Vallis Y, et al. 2010. FCHo proteins are nucleators of clathrin-mediated endocytosis. *Science* 328:1281–84
112. Soulard A, Friant S, Fitterer C, Orange C, Kaneva G, et al. 2005. The WASP/Las17p-interacting protein Bzz1p functions with Myo5p in an early stage of endocytosis. *Protoplasma* 226:89–101
113. Takano K, Toyooka K, Suetsugu S. 2008. EFC/F-BAR proteins and the N-WASP-WIP complex induce membrane curvature-dependent actin polymerization. *EMBO J.* 27:2817–28
114. Ho HY, Rohatgi R, Lebensohn AM, Le M, Li J, et al. 2004. Toca-1 mediates Cdc42-dependent actin nucleation by activating the N-WASP-WIP complex. *Cell* 118:203–16
115. Kamioka Y, Fukuhara S, Sawa H, Nagashima K, Masuda M, et al. 2004. A novel dynamin-associating molecule, formin-binding protein 17, induces tubular membrane invaginations and participates in endocytosis. *J. Biol. Chem.* 279:40091–99
116. Slepnev VI, Ochoa GC, Butler MH, De Camilli P. 2000. Tandem arrangement of the clathrin and AP-2 binding domains in amphiphysin 1 and disruption of clathrin coat function by amphiphysin fragments comprising these sites. *J. Biol. Chem.* 275:17583–89
117. Lundmark R, Carlsson SR. 2003. Sorting nexin 9 participates in clathrin-mediated endocytosis through interactions with the core components. *J. Biol. Chem.* 278:46772–81
118. Salazar MA, Kwiatkowski AV, Pellegrini L, Cestra G, Butler MH, et al. 2003. Tuba, a novel protein containing Bin/amphiphysin/Rvs and Dbl homology domains, links dynamin to regulation of the actin cytoskeleton. *J. Biol. Chem.* 278:49031–43
119. Lee E, Marcucci M, Daniell L, Pypaert M, Weisz OA, et al. 2002. Amphiphysin 2 (Bin1) and T-tubule biogenesis in muscle. *Science* 297:1193–96
120. Chang-Ileto B, Frere SG, Chan RB, Voronov SV, Roux A, Di Paolo G. 2011. Synaptojanin 1-mediated PI(4,5)P2 hydrolysis is modulated by membrane curvature and facilitates membrane fission. *Dev. Cell* 20:206–18
121. Suetsugu S, Toyooka K, Senju Y. 2010. Subcellular membrane curvature mediated by the BAR domain superfamily proteins. *Semin. Cell Dev. Biol.* 21:340–49
122. Slepnev VI, De Camilli P. 2000. Accessory factors in clathrin-dependent synaptic vesicle endocytosis. *Nat. Rev. Neurosci.* 1:161–72
123. Colwill K, Field D, Moore L, Friesen J, Andrews B. 1999. In vivo analysis of the domains of yeast Rvs167p suggests Rvs167p function is mediated through multiple protein interactions. *Genetics* 152:881–93
124. Yarar D, Waterman-Storer CM, Schmid SL. 2007. SNX9 couples actin assembly to phosphoinositide signals and is required for membrane remodeling during endocytosis. *Dev. Cell* 13:43–56
125. Rocca DL, Martin S, Jenkins EL, Hanley JG. 2008. Inhibition of Arp2/3-mediated actin polymerization by PICK1 regulates neuronal morphology and AMPA receptor endocytosis. *Nat. Cell Biol.* 10:259–71
126. Lim KB, Bu W, Goh WI, Koh E, Ong SH, et al. 2008. The Cdc42 effector IRSp53 generates filopodia by coupling membrane protrusion with actin dynamics. *J. Biol. Chem.* 283:20454–72
127. Miki H, Yamaguchi H, Suetsugu S, Takenawa T. 2000. IRSp53 is an essential intermediate between Rac and WAVE in the regulation of membrane ruffling. *Nature* 408:732–35
128. Krugmann S, Jordens I, Gevaert K, Driessens M, Vandekerckhove J, Hall A. 2001. Cdc42 induces filopodia by promoting the formation of an IRSp53:Mena complex. *Curr. Biol.* 11:1645–55
129. Fujiwara T, Mammoto A, Kim Y, Takai Y. 2000. Rho small G-protein-dependent binding of mDia to an Src homology 3 domain-containing IRSp53/BAIAP2. *Biochem. Biophys. Res. Commun.* 271:626–29
130. Lin J, Liu J, Wang Y, Zhu J, Zhou K, et al. 2005. Differential regulation of cortactin and N-WASP-mediated actin polymerization by missing in metastasis (MIM) protein. *Oncogene* 24:2059–66
131. Mattila PK, Salminen M, Yamashiro T, Lappalainen P. 2003. Mouse MIM, a tissue-specific regulator of cytoskeletal dynamics, interacts with ATP-actin monomers through its C-terminal WH2 domain. *J. Biol. Chem.* 278:8452–59
132. Woodings JA, Sharp SJ, Machesky LM. 2003. MIM-B, a putative metastasis suppressor protein, binds to actin and to protein tyrosine phosphatase delta. *Biochem. J.* 371:463–71

133. Yu D, Zhan XH, Niu S, Mikhailenko I, Strickland DK, et al. 2011. Murine missing in metastasis (MIM) mediates cell polarity and regulates the motility response to growth factors. *PLoS ONE* 6:e20845
134. Quinones GA, Jin J, Oro AE. 2010. I-BAR protein antagonism of endocytosis mediates directional sensing during guided cell migration. *J. Cell Biol.* 189:353–67
135. Veltman DM, Auciello G, Spence HJ, Machesky LM, Rappoport JZ, Insall RH. 2011. Functional analysis of *Dictyostelium* IBARa reveals a conserved role of the I-BAR domain in endocytosis. *Biochem. J.* 436:45–52
136. Mettlen M, Pucadyil T, Ramachandran R, Schmid SL. 2009. Dissecting dynamin's role in clathrin-mediated endocytosis. *Biochem. Soc. Trans.* 37:1022–26
137. Ramachandran R. 2011. Vesicle scission: dynamin. *Semin. Cell Dev. Biol.* 22:10–17
138. Schafer DA. 2004. Regulating actin dynamics at membranes: a focus on dynamin. *Traffic* 5:463–69
139. Krendel M, Osterweil EK, Mooseker MS. 2007. Myosin 1E interacts with synaptojanin-1 and dynamin and is involved in endocytosis. *FEBS Lett.* 581:644–50
140. Gu C, Yaddanapudi S, Weins A, Osborn T, Reiser J, et al. 2010. Direct dynamin-actin interactions regulate the actin cytoskeleton. *EMBO J.* 29:3593–606
141. Mooren OL, Kotova TI, Moore AJ, Schafer DA. 2009. Dynamin2 GTPase and cortactin remodel actin filaments. *J. Biol. Chem.* 284:23995–4005
142. Lamaze C, Fujimoto LM, Yin HL, Schmid SL. 1997. The actin cytoskeleton is required for receptor-mediated endocytosis in mammalian cells. *J. Biol. Chem.* 272:20332–35
143. Yu X, Cai M. 2004. The yeast dynamin-related GTPase Vps1p functions in the organization of the actin cytoskeleton via interaction with Sla1p. *J. Cell Sci.* 117:3839–53
144. Smaczynska-de Rooij II, Allwood EG, Aghamohammadzadeh S, Hettema EH, Goldberg MW, Ayscough KR. 2010. A role for the dynamin-like protein Vps1 during endocytosis in yeast. *J. Cell Sci.* 123:3496–506
145. Nannapaneni S, Wang D, Jain S, Schroeder B, Highfill C, et al. 2010. The yeast dynamin-like protein Vps1:*vps1* mutations perturb the internalization and the motility of endocytic vesicles and endosomes via disorganization of the actin cytoskeleton. *Eur. J. Cell Biol.* 89:499–508
146. Kukulski W, Schorb M, Welsch S, Picco A, Kaksonen M, Briggs JA. 2011. Correlated fluorescence and 3D electron microscopy with high sensitivity and spatial precision. *J. Cell Biol.* 192:111–19
147. Youn JY, Friesen H, Kishimoto T, Henne WM, Kurat CF, et al. 2010. Dissecting BAR domain function in the yeast amphiphysins Rvs161 and Rvs167 during endocytosis. *Mol. Biol. Cell* 21:3054–69
148. Ringstad N, Nemoto Y, De Camilli P. 1997. The SH3p4/Sh3p8/SH3p13 protein family: binding partners for synaptojanin and dynamin via a Grb2-like Src homology 3 domain. *Proc. Natl. Acad. Sci. USA* 94:8569–74
149. Ringstad N, Gad H, Low P, Di Paolo G, Brodin L, et al. 1999. Endophilin/SH3p4 is required for the transition from early to late stages in clathrin-mediated synaptic vesicle endocytosis. *Neuron* 24:143–54
150. David C, McPherson PS, Mundigl O, de Camilli P. 1996. A role of amphiphysin in synaptic vesicle endocytosis suggested by its binding to dynamin in nerve terminals. *Proc. Natl. Acad. Sci. USA* 93:331–35
151. Ramjaun AR, Micheva KD, Bouchelet I, McPherson PS. 1997. Identification and characterization of a nerve terminal-enriched amphiphysin isoform. *J. Biol. Chem.* 272:16700–6
152. Perera RM, Zoncu R, Lucast L, De Camilli P, Toomre D. 2006. Two synaptojanin 1 isoforms are recruited to clathrin-coated pits at different stages. *Proc. Natl. Acad. Sci. USA* 103:19332–37
153. Liu J, Kaksonen M, Drubin DG, Oster G. 2006. Endocytic vesicle scission by lipid phase boundary forces. *Proc. Natl. Acad. Sci. USA* 103:10277–82
154. Liu J, Sun Y, Drubin DG, Oster GF. 2009. The mechanochemistry of endocytosis. *PLoS Biol.* 7:e1000204
155. Berro J, Sirotkin V, Pollard TD. 2010. Mathematical modeling of endocytic actin patch kinetics in fission yeast: disassembly requires release of actin filament fragments. *Mol. Biol. Cell* 21:2905–15
156. Ti SC, Pollard TD. 2011. Purification of actin from fission yeast *Schizosaccharomyces pombe* and characterization of functional differences from muscle actin. *J. Biol. Chem.* 286:5784–92
157. McConnell RE, Tyska MJ. 2010. Leveraging the membrane-cytoskeleton interface with myosin-1. *Trends Cell Biol.* 20:418–26
158. Collaco A, Jakab R, Hegan P, Mooseker M, Ameen N. 2010. Alpha-AP-2 directs myosin VI-dependent endocytosis of cystic fibrosis transmembrane conductance regulator chloride channels in the intestine. *J. Biol. Chem.* 285:17177–87

159. Gotoh N, Yan Q, Du Z, Biemesderfer D, Kashgarian M, et al. 2010. Altered renal proximal tubular endocytosis and histology in mice lacking myosin-VI. *Cytoskeleton* 67:178–92
160. Kaksonen M, Toret CP, Drubin DG. 2006. Harnessing actin dynamics for clathrin-mediated endocytosis. *Nat. Rev. Mol. Cell Biol.* 6:404–14
161. Merrifield CJ. 2004. Seeing is believing: imaging actin dynamics at single sites of endocytosis. *Trends Cell Biol.* 14:352–58
162. Suetsugu S. 2009. The direction of actin polymerization for vesicle fission suggested from membranes tubulated by the EFC/F-BAR domain protein FBP17. *FEBS Lett.* 583:3401–4

RELATED RESOURCES

Cell Migr. Consort. 2011. *Cell Migration Gateway.* **http://www.cellmigration.org/**

Sci. Signal. 2011. *Database of Cell Signaling.* **http://stke.sciencemag.org/cm/**

H.T. McMahon home page. 2011. *Researching Endocytic Mechanisms.* **http://www.endocytosis.org/**

Lipid Droplets and Cellular Lipid Metabolism

Tobias C. Walther[1] and Robert V. Farese Jr.[2]

[1]Department of Cell Biology, Yale University School of Medicine, New Haven, Connecticut 06520; email: tobias.walther@yale.edu

[2]Gladstone Institute of Cardiovascular Disease, Department of Medicine, and Department of Biochemistry and Biophysics, University of California, San Francisco, California 94158; email: bfarese@gladstone.ucsf.edu

Keywords

organelle, energy metabolism, triacylglycerol, membranes, obesity, oil, fat

Abstract

Among organelles, lipid droplets (LDs) uniquely constitute a hydrophobic phase in the aqueous environment of the cytosol. Their hydrophobic core of neutral lipids stores metabolic energy and membrane components, making LDs hubs for lipid metabolism. In addition, LDs are implicated in a number of other cellular functions, ranging from protein storage and degradation to viral replication. These processes are functionally linked to many physiological and pathological conditions, including obesity and related metabolic diseases. Despite their important functions and nearly ubiquitous presence in cells, many aspects of LD biology are unknown. In the past few years, the pace of LD investigation has increased, providing new insights. Here, we review the current knowledge of LD cell biology and its translation to physiology.

Contents

1. INTRODUCTION 688
2. GENERAL PROPERTIES OF LIPID DROPLETS 689
   2.1. Historical Aspects of Lipid Droplets 689
   2.2. Lipid Droplets Are Found in Most Cells 689
   2.3. Lipid Droplets Separate a Hydrophobic Phase from the Aqueous Cytosol 689
3. LIPID DROPLETS SERVE MULTIPLE FUNCTIONS IN CELLS 691
4. LIPID DROPLETS ARE ANALOGOUS TO LIPOPROTEINS 692
5. LIPID DROPLET FORMATION 692
   5.1. Models of Lipid Droplet Formation 692
   5.2. Identifying Genes in Lipid Droplet Formation 694
6. LIPID DROPLETS GROW BY EXPANSION OR BY COALESCENCE 694
7. LIPID DROPLET PROTEINS .. 695
   7.1. Specific Proteins Localize to the Surfaces of Lipid Droplets 695
   7.2. Protein Signals that Mediate Lipid Droplet Targeting 696
   7.3. Cellular Pathways Involved in Targeting Proteins to Lipid Droplets 698
   7.4. Removal of Lipid Droplet Proteins 698
8. LIPID DROPLETS INTERACT WITH OTHER CELLULAR ORGANELLES 699
9. LIPID DROPLET MOVEMENT WITHIN CELLS 699
10. LIPIDS IN LIPID DROPLETS ARE MOBILIZED BY LIPASES 700
    10.1. Lipid Droplet Catabolism by Lipolysis 700
    10.2. The Fate of Hydrolyzed Lipids 701
    10.3. Lipid Droplet Catabolism by Autophagy 701
11. LIPID DROPLETS FIGURE PROMINENTLY IN PHYSIOLOGY AND DISEASE 701
    11.1. Lipid Droplets and Lipid Storage in Tissues 701
    11.2. Excessive Lipid Droplet Storage and Disease 703
    11.3. Lipodystrophies and Too Few Lipid Droplets 704
    11.4. Lipid Droplets and Cancer 704
    11.5. Lipid Droplets and Lactation 704
12. LIPID DROPLETS ARE TARGETS OF INDUSTRIAL OIL PRODUCTION 705

1. INTRODUCTION

Lipid droplet (LD): the cytoplasmic organelle composed of a hydrophobic core of neutral lipids bounded by a phospholipid monolayer and specific proteins

Lipid droplets (LDs) are dynamic cytoplasmic organelles found nearly ubiquitously in cells. They are linked to many cellular functions, including lipid storage for energy generation and membrane synthesis, viral replication, and protein degradation. LD biology is connected to myriad physiological processes, metabolic diseases, and oil production. Interest in LD biology has surged in recent years, reflecting recognition of the importance of this relatively understudied organelle and opportunities to unravel questions concerning LD biology.

Here, we focus primarily on recent discoveries in LD biology, building on other reviews (1–3), and highlight many areas of LD biology where knowledge is incomplete or controversial.

2. GENERAL PROPERTIES OF LIPID DROPLETS

2.1. Historical Aspects of Lipid Droplets

LDs were identified by light microscopy as cellular organelles in the nineteenth century. For many years, LDs often were called liposomes. In the late 1960s, a method was developed to generate vesicles in vitro, also called liposomes, and these vesicles assumed the name. Since then, LDs have been referred to as LDs, lipid bodies, fat bodies, oil bodies, spherosomes, or adiposomes. In the past decade, the field settled primarily on the name LDs.

For years, LDs were largely ignored in cell biology research, presumably because they were perceived as inert lipid globules with little functional relevance. In 1991, seminal work by Constantine Londos and coworkers (4) identified a protein, perilipin, that specifically localized to LD surfaces, and this opened the door to mechanistically studying the organelle. More recently, research on LD biology has accelerated dramatically. This reflects increased interest in the basic biology of prevalent metabolic diseases linked to LDs (e.g., obesity) and technological advances, which have shed new light on LDs and highlighted many unanswered basic questions.

2.2. Lipid Droplets Are Found in Most Cells

Nearly all cells have LDs or the capacity to form them. Several bacteria store lipids in LDs, including predominantly the actinomycetes group (e.g., *Mycobacteria*, *Rhodococcus*, *Streptomyces*, and *Nocardia*) (5).

Among eukaryotes, LDs are easily detected in budding yeast (*Saccharomyces cerevisiae*), facilitating genetic screens for altered LD morphology (6, 7). In higher eukaryotes, many cells have some LDs, and culturing cells with fatty acids (FAs) stimulates LD formation. Some cells, such as adipocytes or hepatocytes, exhibit many LDs at baseline owing to active lipid synthesis and storage. LD abundance varies dynamically in cells. For example, in *S. cerevisiae*, LDs are found most prominently during the stationary phase, and catabolism of LDs in the exponential phase is coordinated with an increased need for phospholipids during cell division (8). Also, the number of LDs increases up to severalfold in yeast with cellular stress (9, 10) and in cells of some cancers (11). The connections of LDs to these states are poorly understood.

LD number and size in different cell types or between individual cells of a population differ considerably. Many cells have small LDs (100–200-nm diameters), whereas LDs in white adipocytes have diameters up to 100 μm and fill almost the entire cytoplasm. With diameters from 100 nm to 100 μm, the corresponding surface area and volume for LDs vary by 10^6 and 10^9, respectively. Many LDs are visible with light microscopy (**Figure 1** and the sidebar titled Imaging of Lipid Droplets).

LD size can change rapidly. Oleate loading of *Drosophila* S2 cells increases the LD mean diameter nearly threefold within hours, corresponding to an almost 30-fold volume increase (12). LDs also grow rapidly during adipogenesis, when cells increase their capacity to synthesize lipids de novo. In contrast, LDs shrink within hours of culturing cells with limited nutrients.

2.3. Lipid Droplets Separate a Hydrophobic Phase from the Aqueous Cytosol

Among cellular organelles, LDs are uniquely composed of an organic phase of neutral lipids. This hydrophobic core is separated from the aqueous cytosol by a monolayer of surface phospholipids. The cytoplasm contains an emulsion of LDs in the cytosol. The LD phase of the emulsion provides a large interface for interactions with amphipathic molecules.

The primary neutral lipids of the LD core are sterol esters (SEs) and triacylglycerols

---

Phase: a region of a system in which the properties, measured in terms of order parameters, are uniform. In an emulsion, two liquid phases are mixed

Neutral lipid: a nonpolar, hydrophobic lipid

Sterol ester (SE): an ester of cholesterol (or a related sterol) and a fatty acid moiety; a neutral lipid

a Schematic LD architecture **b** LDs visualized by fluorescence microscopy **c** LDs visualized by electron microscopy

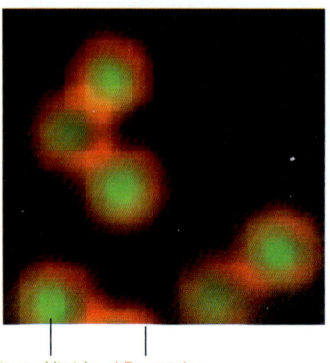
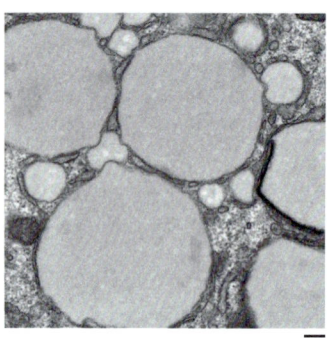

Neutral lipids   LD proteins

200 nm

- Neutral lipids
- Phospholipid monolayer
- LD proteins

Figure 1
Examples of lipid droplets (LDs). Shown are a schematic overview of LD structure, LDs in *Drosophila* S2 cells (*middle panel*), and in murine adipocytes. Images courtesy of Natalie Krahmer and Caroline Mrejen.

Triacylglycerol (TG): a glycerol ester in which each of the hydroxyl groups is combined with a fatty acid. A type of neutral lipid

DGAT: acyl-CoA:diacylglycerol acyltransferase

ACAT: acyl-CoA:cholesterol acyltransferase

Ether lipid: a lipid in which an alcohol is linked to glycerol by an ether (as opposed to ester) bond

PC: phosphatidylcholine

(TGs). Their relative amount varies between cell types. For example, yeast LDs have a mix of SE and TG, possibly arranged in layers (13). LDs of adipocytes contain primarily TG, and those of macrophage foam cells contain mostly SE. Neutral lipid synthesis is catalyzed by various enzymes (14). In mammalian cells, acyl-coenzyme A (acyl-CoA):diacylglycerol acyltransferase (DGAT) enzymes, DGAT1 and DGAT2, synthesize TG, and acyl-CoA:cholesterol acyltransferase (ACAT) enzymes, ACAT1 and ACAT2, generate SEs. In yeast, the corresponding neutral lipid synthesis enzymes are Are1 and Are2, which synthesize primarily SEs, as well as Dga1 and Lro1, which synthesize TGs. Neutral lipid synthesis enzymes reside primarily in the endoplasmic reticulum (ER), with the exception of DGAT2, which also localizes to LDs (15, 16).

Various other hydrophobic lipids are found in LDs. Retinyl esters are found in LDs of hepatic stellate cells (17) and in the retina (18). LDs also contain wax esters and ether lipids, which are derived from peroxisomes and constitute 10%–20% of neutral lipids in some mammalian cells (19). Additionally, long-chain isoprenoids are found within LDs. Natural rubber consists of long-chain isoprene polymers in monolayer-bound particles that appear to be LDs of the rubber tree (20).

The LD surface comprises polar, amphipathic lipids. In mammalian LDs, phosphatidylcholine (PC) is the main surface phospholipid, followed by phosphatidylethanolamine (PE) and phosphatidylinositol (19). Compared with other membranes, LDs are deficient in phosphatidylserine and phosphatidic acid but enriched in lyso-PC and lyso-PE. More than 160 phospholipids species of varying head groups and side chains were detected in CHO cells (19). In *Drosophila*, PE is more abundant than PC, and this is reflected in the LD phospholipids (12). Nevertheless, PC is important in emulsifying LDs (12). LD surfaces contain other polar lipids, such as sterols, and the relative amount of these lipids depends on cell type. Sphingolipids are not present in appreciable quantities (19).

IMAGING OF LIPID DROPLETS

Given their size (0.5 μm to tens of microns), many LDs are readily detected by light microscopy (see **Figure 1**, middle panel). Their refracting nature makes them identifiable with phase-contrast imaging. Several hydrophobic dyes can be used to image LDs. Oil red O is visible by light microscopy and typically used to stain fixed samples. Nile red and BODIPY are concentrated in LDs and fluoresce upon stimulation. They provide a robust signal in living or fixed cells. Care must be taken to avoid artifacts. BODIPY appears to be more specific for staining LDs; Nile red also stains endocytic vesicles. Additionally, BODIPY-tagged FAs can stain LDs, as the tagged FAs in many cases are taken up by cells and esterified and incorporated into neutral lipids.

Because LDs are closely associated with other organelles, it can be difficult to distinguish them from surrounding membranous organelles. This limitation was addressed by the recent development of "super high-resolution" microscopy that overcomes the classical resolution limits of light microscopy, in some cases improving resolution by a factor of 10 (i.e., from ~250 nm to ~25 nm) (166). In one approach, exemplified by photoactivated localization microscopy or stochastic optical reconstruction microscopy, the sequential photoactivation of single fluorescent molecules allows for the precise determination of their position by fitting a Gaussian curve to the resulting signal, leading to high-resolution reconstructed images. In a complementary approach (stimulated emission depletion microscopy), the resolution limit is increased further by depleting the excited state of fluorophores around a narrow focus of light, thus restricting molecules that emit a signal.

Several approaches based on Raman spectroscopy are also used to image LDs. Most prominently, coherent anti-Stokes Raman scattering microscopy uses inelastic scattering of photons interacting with matter to detect molecule-specific vibrations and resulting blue-shifted light (167). These approaches also can quantify lipids.

EM methods are also used to analyze LDs (see **Figure 1**, right panel), including transmission EM and freeze-fracture EM. Difficulties of fixing and staining lipids during sample preparation often cause LDs to appear as empty spheres in conventional EMs. Fixation and sectioning problems may lead to artifacts, but improved protocols can avoid these problems (89, 168).

3. LIPID DROPLETS SERVE MULTIPLE FUNCTIONS IN CELLS

Foremost, LDs are intracellular lipid reservoirs, providing building blocks for membranes or substrates for energy metabolism. Packaging highly reduced, hydrophobic lipids, such as TGs, in a phase without water provides the most efficient form of energy storage. For example, an average non-obese person stores up to 2,500 kJ of metabolic energy in glycogen, but >500,000 kJ in TGs of adipocyte LDs, enough to run ~30 marathons.

LDs also serve as organizing centers for synthesizing specific lipids. TGs, for example, are synthesized in the ER and at LDs (15). Other enzymes of lipid synthesis (e.g., ergosterol) localize to LDs, suggesting links between lipid synthesis pathways and LDs (21).

LDs have been linked to protein storage (22). For example, during embryogenesis in *Drosophila*, histones localize to LDs until

HCV: hepatitis C virus

Lipoprotein: a lipid-protein particle that circulates in the bloodstream and distributes lipids throughout the body

needed for rapid nuclear division associated with embryo segmentation (23). LDs may also temporarily store unfolded membrane proteins before proteasomal degradation.

LDs are involved in hepatitis C virus (HCV) assembly. During viral replication, the HCV core protein is cleaved from the precursor viral polypeptide and binds to LDs (reviewed in Reference 24) via amphipathic helices. New HCV virions are assembled in LDs and ER membranes, where viral RNA is packaged with capsid proteins. Mature viruses are secreted as part of a lipoprotein-virus particle. LDs may provide a location for HCV core proteins until viral assembly or until cells degrade the overexpressed protein. Notably, HCV core protein localization to LDs requires DGAT1 activity, and HCV core proteins bind to DGAT1 in vitro (25). Moreover, blocking DGAT1 activity diminishes viral replication (25).

4. LIPID DROPLETS ARE ANALOGOUS TO LIPOPROTEINS

LDs share structural features with plasma lipoproteins. Both contain neutral lipid cores encased in polar lipid monolayers. Additionally, both particles are decorated with specific surface proteins, often possessing amphipathic α-helices. However, LDs (with diameters of 100 nm up to 100 microns) are generally much larger than lipoproteins [diameters range from <20 nm (e.g., high-density lipoprotein) to ∼500 nm (e.g., chylomicrons)].

Although the formation of both LDs and lipoproteins is linked to neutral lipid synthesis in the ER, their physiological functions differ: LDs primarily store lipids, and lipoproteins distribute lipids in the body. Also, only a few cell types (e.g., hepatocytes, enterocytes, and yolk sac endodermal cells) express the required proteins, such as apolipoprotein (apo) B and the microsomal TG transfer protein, for lipoprotein assembly, whereas most cells make LDs. Thus, lipoproteins may have evolved by adapting LDs to secretion.

How cells regulate the fate of newly synthesized neutral lipids—storage in LDs versus secretion on lipoproteins—is mostly unknown. Secretory cells may store lipids in LDs only after exceeding the capacity to assemble and secrete lipoproteins. Alternatively, secretion may be activated when a storage threshold is achieved. Specific LD proteins [including cell death–inducing DFF45-like effector (CIDE) proteins] may direct the LD pool of TG toward lipoprotein formation (26). Secretion of TG via lipoproteins is thought to involve their hydrolysis at LDs and resynthesis in the ER (27).

5. LIPID DROPLET FORMATION

5.1. Models of Lipid Droplet Formation

Many organelles self-replicate. However, LDs likely form de novo. In bacteria, LDs form by lipid synthesis in the cell-delimiting membrane (28). In eukaryotes, LDs may arise primarily from the ER, where the enzymes that synthesize neutral lipids reside (14). In yeast genetically engineered to lack LDs, induction of LD formation shows they invariably arise from or close to the ER (7, 29). LDs appear to remain in contact with the ER once formed, and proteins that associate with both compartments move between them (29). However, light microscopy, with limited resolution, cannot determine if such proteins reside directly on the LD surface or in ER membranes in close apposition. Many studies employing electron microscopy (EM) suggest a tight assocation of LDs and the ER. In mammalian cells, such studies show membrane cisternae, which could be connected to the ER, in close proximity to LDs (30, 31). Also, LDs in hepatocytes are tightly associated with ER membrane cisternae, marked by luminal apo B100 protein, and these cisternae may be continuous with LDs (31).

Despite these findings, the molecular mechanisms of LD formation are not understood. How does a monolayer-coated LD arise from a bilayer membrane? Because neutral lipid synthesis enzymes reside in the ER, the products of these enzymes might occupy space between the membrane bilayers, forming a lens of neutral

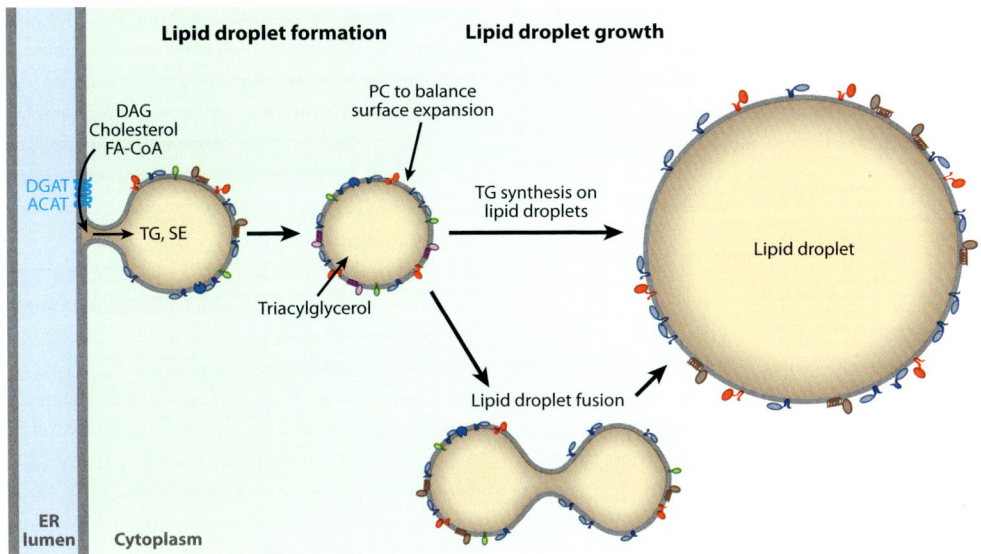

Figure 2

Models of lipid droplet (LD) formation and expansion. For formation, the budding model is shown, with the LD either remaining attached to the endoplasmic reticulum (ER) or detached. The neutral lipid synthesis enzymes, ACAT or DGAT, catalyze the formation of lipids that fill the core. For lipid droplet expansion, the models of local synthesis of triacylglycerol (TG) at the LD and of LD fusion (coalescence) are shown. Abbreviations: DAG, diacylglycerol; FA-CoA, fatty acyl-coenzyme A; SE, sterol ester; PC, phosphatidylcholine.

lipids. In vitro studies with ER microsomes from sunflower (*Helianthus annuus*) found formation of small liquid lenses of 60-nm diameter in the ER bilayer that, similar to LDs, contain freely mobile TG and recruit oleosins (32). It is unclear, however, how neutral lipids are organized into nascent LDs, whether this happens at specific locations in the ER, or whether specific proteins are involved. In mammalian cells, perilipin3/Tip47 is recruited to the ER during lipid storage (33) and was suggested to be required for LD formation (34). However, many organisms (e.g., yeast) have no perilipin 3/Tip47 ortholog, and many cells appear not to express it (e.g., *Drosophila* S2 cells), so this mechanism alone cannot mediate LD formation.

Several models of LD formation have been proposed (1). They include (*a*) ER budding, where LDs grow from the ER bilayer and remain connected or bud off (**Figure 2**); (*b*) bicelle formation in which an entire lipid lens is excised from the ER (35); (*c*) vesicular budding in which a bilayer vesicle forms, followed by filling of the bilayer intramembranous space with neutral lipids (36); and (*d*) an "eggcup" model in which a LD grows within a concave depression of the ER through transport of neutral lipids from the ER (37). The latter transfer process resembles a model for prokaryotes (28). All models posit LDs forming toward the cytosolic face of the ER membrane. However, cells, such as hepatocytes, also secrete neutral lipids into the ER lumen (38), and LDs could be derived from luminal origins. Such a model was recently proposed for yeast (29).

Which model is correct? Several obstacles prevent an easy answer. Most cells have LDs, complicating identification of nascent LDs, and systems of induced LD formation in mammals are lacking. Also, the LD size during the initial stages of formation is below the resolution of light microscopy. Thus, combinations of inducible systems with super high-resolution light microscopy and EM will likely be required

Lipodystrophy: a pathological condition characterized by severely diminished or absent adipose tissue

Lipolysis: the regulated breakdown of complex neutral lipids, such as triacylglycerols or sterol esters, into their components

to gain further insights (see the sidebar titled Imaging of Lipid Droplets). Alternatively, characterizing a system for in vitro LD formation from ER membranes might allow molecular dissection of different steps of formation.

5.2. Identifying Genes in Lipid Droplet Formation

To identify genes involved in LD formation, several studies utilized genome-wide screens with LD morphology as a phenotypic readout. Three screens in S. cerevisiae (6, 7, 39) identified gene deletions leading to different LD morphologies. Notably, deleting the yeast ortholog of seipin (FLD1) caused abnormal LDs. Seipin is of particular interest in LD biogenesis. It is a multimeric ER membrane protein (40) whose deficiency in humans results in lipodystrophy. Surprisingly, there was otherwise little overlap between genes identified in the yeast screens, and no single-gene deletion was found to cause the complete absence of LDs.

Genome-wide screens to find genes regulating LD morphology have been performed by RNAi in Drosophila cells (41, 42). In one screen employing oleate loading of cells, over 200 genes were identified. Interestingly, they binned into five phenotypic classes (41). In another screen, over 800 genes showed an effect on LD accumulation (42). Among overlapping hits of both screens knockdowns of proteasome genes yielded fewer LDs, knockdowns of the ARF/COPI vesicular transport machinery gave large and disperse LDs, and knockdowns of genes related to phospholipid synthesis led to very large LDs. The latter two classes showed defects in lipolysis, indicating functional consequences of these knockdowns. The ARF/COPI machinery functions in retrograde transport of proteins from the Golgi apparatus to the ER, but its role in lipolysis is uncertain. Adipose triglyceride lipase (ATGL), or its Drosophila homolog Brummer, fails to target to LDs in cells depleted of ARF/COPI machinery (30, 43). The link between phospholipid synthesis genes and LD size is discussed below.

Several systematic screens of genes in whole organisms also focused on lipid storage. Screens in Caenorhabditis elegans (44) and Drosophila (45) revealed a plethora of genes involved in fat storage. Although many of these genes participate in feeding behavior and energy expenditure, some may be involved directly with LDs. Other surveys identified proteins that might be important in LD formation. These include fat-inducing transcript (FIT) proteins, which are ER proteins that bind TG and have been implicated LD assembly (46, 47). FIT2 overexpression results in more LDs and knockdown in fewer LDs and TG, but there is no evidence that FIT proteins affect DGAT activity (47). How FIT proteins organize TG into LDs is unclear. Phospholipases might also be required for LD biogenesis (48).

6. LIPID DROPLETS GROW BY EXPANSION OR BY COALESCENCE

To accommodate more TG, cells form new LDs and grow existing ones (**Figure 2**). Adding neutral lipids to existing LDs requires local synthesis or transfer from the ER.

Neutral lipids might be synthesized locally at LDs, particularly during lipid loading. DGAT2, which catalyzes the final step of TG synthesis and is normally found in the ER, localizes to LDs during oleate loading (16, 49), as does its yeast ortholog Dga1 (29). It is unclear whether more proximal steps in TG synthesis also localize to LDs. In contrast to TG synthesis, sterol esterification likely occurs primarily in the ER because ACAT enzymes reside in the ER and do not localize to LDs (50).

The synthesis of neutral lipids in the ER necessitates their transfer to LDs, through membrane connections or via interorganelle transport by transfer proteins. Although TG transfer proteins move TG from the ER bilayer to nascent lipoproteins in the ER lumen, no such mechanism is known for cytosolic LDs.

LDs may also grow by coalescence (**Figure 2**) (51). Coalescence of neutral LDs minimizes the phase boundary area and free energy of the emulsion. Thus, the challenge for cells is to prevent coalescence and maintain LDs as isolated entities. LD emulsions are stabilized by surface proteins (e.g., oleosins) or surfactants (e.g., PC) (12), which prevent coalescence by lowering surface tension. In contrast, accumulation of fusogenic lipids (e.g., phosphatidic acid) might induce LD coalescence (39).

During rapid expansion, LD diameters increase more than threefold within hours—a nearly tenfold increase in surface area. Thus, cells need to synthesize and transport large amounts of phospholipids to expanding LD surfaces. Among phospholipids, PC is key for coating LDs and preventing their coalescence. PC can emulsify artificial LDs in vitro and prevent coalescence (12). During expansion, LDs become PC deficient and CTP:phosphocholine cytidylyltransferase enzymes, which catalyze the rate-limiting step in PC biosynthesis, are activated by binding to LDs (12). The de novo synthesized PC is transported to the expanding LD surface through unknown mechanisms. In this manner, de novo synthesis of PC is activated to maintain adequate PC levels at LD surfaces.

PC to coat the surfaces of LDs is also synthesized from lyso-PC and fatty acyl-CoA precursors by lyso-PC acyltransferase enzymes, which function in the Lands cycle of phospholipid remodeling of FA moieties. Lyso-PC acyltransferase 1 and 2 enzymes localize to and are active on LDs (49). They likely function in remodeling the surface phospholipids rather than providing net PC synthesis, as they cannot yield increased PC unless lyso-PC is provided.

In *Drosophila* cells, LD fusion happens rarely except under specific conditions, such as PC deficiency (12, 41). PC deficiency likely renders LDs prone to coalesce. A recent report found homotypic LD fusion to occur rarely in murine embryonic fibroblasts or NIH-3T3 cells under normal conditions (52), although higher rates of fusion were also reported (51). However, numerous pharmacological agents stimulated LD fusion in different cell types. These included propranolol and other drugs, albeit at supraphysiological concentrations, which may trigger fusion by inserting into and disrupting LD surface monolayers (52). Fusion was relatively slow, occurring over seconds to minutes. The volumes of fusing LDs were conserved, and surface areas were decreased, suggesting excess surface lipids are removed during fusion.

Proteins might be involved in catalyzing LD fusion. One such protein is a fat-specific protein of 27 kDa [FSP27, or cell death-inducing DFF45-like effector C (Cidec)]. In adipocytes, *FSP27* is expressed as a peroxisome proliferator-activated receptor-γ (PPARγ)-regulated gene and promotes the formation of unilocular LDs (53–55). Mice lacking *Fsp27/Cidec* have multilocular white adipocytes, implying the protein forms large LDs (53, 55, 56). If Fsp27 is expressed ectopically in fibrobasts, LD sizes increase. The mechanism is unclear but may include promoting LD fusion or inhibiting lipolysis (53–55, 57, 58).

Because SNARE proteins (soluble *N*-ethylmaleimide-sensitive factor attachment receptor proteins) mediate homotypic fusion of bilayer-bounded vesicles during cellular trafficking, they are attractive candidates for LD fusion. In studies performed in NIH 3T3 cells, knockdowns of genes encoding SNAP23, syntaxin-5, and VAMP4 were reported to decrease the rate of LD fusion (59). However, it is unclear how SNARE proteins would mediate fusion of monolayer-bound vesicles, and more studies to address SNARE involvement are needed.

7. LIPID DROPLET PROTEINS

7.1. Specific Proteins Localize to the Surfaces of Lipid Droplets

LDs are characterized by specific proteins on their surfaces. Analyses of LD proteomes in different organisms yielded diverse lists of LD-associated proteins (21, 60–65). The relatively easy purification of LDs (by centrifugation) facilitates proteomic analyses, but their close association with other organelles, particularly the ER, confounds such analyses. Thus, it is often unclear which proteins are genuine LD proteins. In addition, researchers may overlook LD proteins because they have other well-known functions. For example, histones were unexpectedly found by LD proteomics and subsequently confirmed to transiently target LDs in *Drosophila* embryos (23). LDs may similarly transiently store other proteins that otherwise might aggregate. For example, α-synuclein, a Parkinson's disease–associated protein prone to self-aggregation, localizes to LDs (66).

Some data on LD protein localization can be misleading. By fluorescence microscopy, many proteins reportedly localize to LDs upon oleate loading of cells. These studies assume that a characteristic ring-like appearance of proteins surrounding LDs indicates targeting to the LD-delimiting surface monolayer. In ultrastructural studies, however, membranous structures often accumulate juxtaposed to LDs, so some proteins might localize to LD-associated membranes rather than the LD surface itself (30). Indeed, many reports show no overlap between the LD signal (e.g., from neutral lipid staining) and the LD-encircling protein rings. Because light microscope resolution is ∼250 nm, such proteins might actually localize to membranes at a distance from the LD surface.

7.2. Protein Signals that Mediate Lipid Droplet Targeting

How proteins target LDs is not understood, but some mechanisms are emerging (**Figure 3**), including (*a*) direct binding to the monolayer surface via amphipathic helices, (*b*) embedding short hydrophobic regions localized at the N terminus, (*c*) embedding internal protein domains that penetrate the LD with hydrophobic spans (e.g., hairpin-shaped hydrophobic helices), (*d*) lipid anchors, and (*e*) interaction with other LD-bound proteins. By contrast, proteins with hydrophilic regions on either side of bilayer-spanning helices (e.g., DGAT1) are unlikely to be on the monolayer surface of LDs, as this would be energetically unfavorable.

Examples of proteins containing amphipathic helices as targeting signals include the HCV core protein (67), viperin (endogenous inhibitor of HCV) (68), and CTP: phosphocholine cytidylyltransferase enzymes

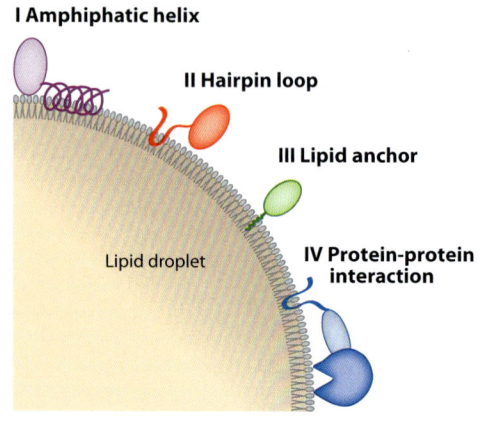
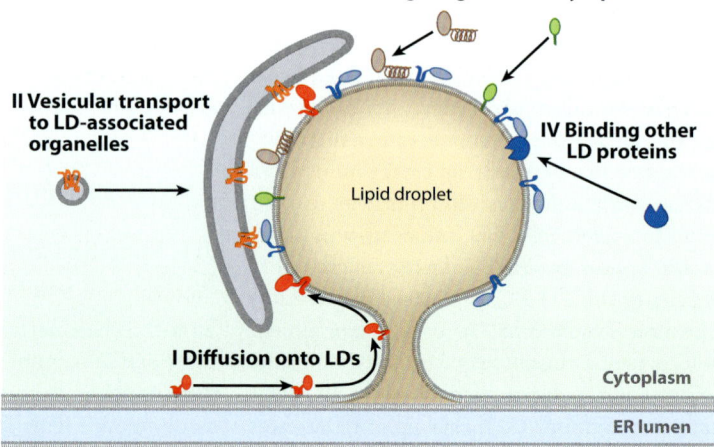

Figure 3
Models of proteins that target lipid droplets (LDs) and the pathways for their targeting. (*a*) Some examples of the types of proteins that target the phospholipid monolayer of a LD surface are shown. (*b*) Several mechanisms by which proteins may target the surfaces of LDs or adjacent membranes are shown. Abbreviation: ER, endoplasmic reticulum.

(12, 41). How amphipathic helices distinguish the LD surface monolayer from other cellular membranes is unclear. CTP:phosphocholine cytidylyltransferase 1 binds to and is activated by PC-deficient LD surfaces (12), one of few cases where a specific lipid composition of the LD surface mediates amphipathic helix recruitment. Another example is recruitment of perilipin3/Tip47 to the ER with diacylglycerol (DAG) accumulation, which was suggested to coordinate DAG buildup with LD protein recruitment and LD formation (33).

For other organelles, specific lipids (e.g., phosphoinositides, phosphatidylserine, and phosphatidic acid) mediate protein targeting to membranes. No lipid signals are known to target proteins to LDs, but it is possible because the lipid composition of the LD surface differs from that of the ER (19, 69).

Several proteins (e.g., AAM-B, UBXD8, or ALDI) (70, 71) possess an N-terminal, hydrophobic, short domain (minimally ~28 amino acids) that is necessary and sufficient for LD targeting. These proteins localize to the ER under some conditions (e.g., in the absence of LDs), suggesting they first are inserted into the ER and subsequently transported to LDs. Their domain structure and the molecular mechanism(s) to target LDs are unknown.

Some proteins (e.g., caveolin, oleosins, and 17-hydroxysteroid dehydrogenases) have an internal hydrophobic domain of variable length that likely forms a hairpin structure that could be integrated into a phospholipid bilayer or LD monolayer. For plant oleosins, members of this class, the topology and some requirements have been elucidated. Oleosins, which account for up to 8% of seed protein, are important for storing oils in seeds. LD targeting of oleosins initiates with their synthesis at the ER (72, 73), followed by targeting to LDs. A hairpin contains a central proline knot (three prolines in 12 amino acids) that is essential for LD targeting and might induce a sharp bend between two adjacent hydrophobic protein segments (74). The hairpin domain might thermodynamically favor LD localization (i.e., the space between the ER membrane bilayer may be limiting). Alternatively, proteins recognizing such hairpins may be in involved in LD localization.

In mammalian cells, caveolins contain a similar hairpin motif. Caveolins primarily localize to caveolae at the plasma membrane but also to LDs. Initially, LD targeting of caveolin was observed in cells treated with brefeldin A, which blocks membrane trafficking, or with caveolin-1 mutants (75–77). However, targeting occurs under physiological conditions (e.g., during lipid loading) (78, 79). At the plasma membrane, caveolin topology includes a central hairpin membrane anchor and adjacent lipid-binding domains. For LDs, the topology of caveolin is unknown, but it may be similar. The hydrophobic sequences of AAM-B, ALDI, or DGAT2 may also form hairpins. For enzymes, such as DGAT2, it has yet to be demonstrated that the protein localizes directly on the LD surface (versus the adjacent ER), and, if so, whether it is active at this location.

Several different LD targeting mechanisms exist for perilipins (perilipin1, perilipin2/ADRP/adipophilin, perlipin3/Tip47, perilipin4/S3-12, and perilipin5/OXPAT), which are the first identified specific LD marker proteins (80). Perilipins are not essential for LD formation (e.g., yeast lack perilipins) (34) but are important for regulating lipid metabolism at LDs. In mammals, several perilipins are expressed ubiquitously (e.g., perilipin2/ADRP and perilipin3/Tip47), whereas others are expressed specifically in certain cell types (e.g., perilipin1 in white adipocytes and perilipin5/OXPAT in highly oxidative cell types).

Adipocytes express high levels of perilipin1, which is involved in the regulation of lipolysis. Perilipin1 contains a combination of domains interacting with LDs. Particularly important are three C-terminal hydrophobic stretches, which may penetrate LDs (81). Additional amphipathic stretches of the protein likely interact with the LD surface. Similarly, both N- and C-terminal regions of perilipin2/ADRP contribute to LD binding (82, 83). Perilipin2/ADRP localization to LDs may also involve ARF1/COPI (84).

Perilipin3/Tip47 may bind to LDs like apolipoprotein E (apo E) binds to lipoproteins. Its C-terminal four-helix bundle and α/β-domain (85) are similar to the N-terminal four-helix bundle of apo E. Apo E binds lipoproteins by opening its four-helix bundle and exposing hydrophobic, amphipathic sequences to the lipid surface (86). Perilipin3/Tip47 might use a similar mechanism. Unlike perilipin1 and perilipin2/ADRP, which are unstable when not bound to LDs, perilipin3/Tip47 is found in the cytoplasm when LDs are absent (87).

As an alternative to protein segments mediating LD interactions, lipid modifications of proteins may serve as anchors to the LD surface. Rab18, a small GTPase, localizes to LDs, where it mediates ER interactions (88, 89). By analogy to other Rabs with C-terminal lipid anchors, Rab18 might target to LDs by a prenylation anchor combined with protein-protein interactions. Unlike other Rabs, Rab18 contains one, rather than two, lipid modification.

Proteins might also localize to LDs by interacting with LD-bound proteins. Regulated recruitment of hormone-sensitive lipase (HSL) to LDs is an example. Under basal conditions, HSL is mostly cytosolic, and access to LDs is restricted by perilipin1. Upon hormonal stimulation, perilipin1 is phosphorylated, which recruits HSL to LDs.

7.3. Cellular Pathways Involved in Targeting Proteins to Lipid Droplets

Pathways targeting proteins to LDs are less understood than the protein signals. Several mechanisms may be involved, including direct recruitment from the cytosol (**Figure 2**). For some membrane proteins, vesicular trafficking may be important. The ARF/COPI machinery, which mediates vesicular trafficking from the Golgi apparatus to the ER, is required for normal LD turnover and localization of the major TG lipase, ATGL, or its *Drosophila* homolog, *brummer*, to LDs (30, 90). A fraction of ATGL colocalized with ER-exit sites (ER membrane domains dedicated to formation of secretory vesicles) and the expression of a dominant-negative Sar1 (GTPase that is required for vesicles to leave the ER exit sites) inhibited ATGL targeting to LDs (30). ATGL requires a C-terminal hydrophobic stretch for LD localization (91) and may behave biochemically as a membrane-associated protein (30). Thus, ATGL may contain a hydrophobic sequence (e.g., hairpin loop) targeted to the ER and trafficked to LDs. From EM studies, cisternal structures around LDs could represent a LD-target compartment for vesicular trafficking (30). It is unclear how ATGL or similar proteins would move from this compartment to the LD surface.

Physical bridges may connect LDs and the ER (29, 31, 92). Such bridges may allow membrane-bound proteins, such as those with hairpins, to diffuse from the bilayer to the LD surface. Data from yeast support this model; fluorescently labeled LD proteins, such as Dga1, rapidly exchange with other membrane pools (29).

7.4. Removal of Lipid Droplet Proteins

Mechanisms to remove LD proteins are likely similar to those for membrane proteins. Endocytosed membrane proteins or autophagic vesicles are degraded in lysosomes. ER membrane proteins can be degraded by proteasomes (ER-associated degradation). Both mechanisms may degrade LD proteins. LDs are delivered to autophagosomes (93), and some LD proteins are modified by ubiquitin for proteasomal degradation (e.g., some perilipins) (94–100). Additionally, a nonbiased screen identified proteasome components as required for normal LD morphology (41).

Molecular links between LDs and proteasomal degradation were recently identified. Among LD proteins, ancient ubiquitous protein 1 (Aup1), located in the ER or directly on LDs, binds ubiquitin E2 ligase, Ube2g2 (101, 102). The ubiquitin E3 ligase, spartin/SPG20, localizes to LDs, and its depletion or overexpression leads to LD accumulation, perhaps from the altered turnover of LD proteins, such as perilipin2/ADRP (97, 100). These findings imply that a specific machinery degrades LD-associated proteins, but the details are unknown.

Some ER-associated degradation substrates may transiently localize to LDs (e.g., HMG-CoA reductase when proteasomal degradation was inhibited) (103). However, at least in yeast, LDs are not required for ER-associated degradation (104).

8. LIPID DROPLETS INTERACT WITH OTHER CELLULAR ORGANELLES

LDs interact with other organelles, including the ER, endosomes, mitochondria, and peroxisomes. Freeze-fracture imaging showed ER membranes in mammalian cells are often wrapped around LDs in a shape resembling an eggcup (37). Because interactions between membrane-bound organelles are implicated in lipid exchange, apposition of LDs with the ER may facilitate lipid transfer between them. Oxysterol-binding proteins localize to LDs (105), where they might be involved in lipid trafficking or neutral lipid metabolism. In steroidogenic cells, ER-LD interactions might function in the synthesis or catabolism of steroid hormones (106). In yeast, ER-LD junctions at least partly reflect intermediates of LD formation (6, 7, 29, 40), and seipin may participate in this process.

Mechanisms mediating ER association with LDs are unknown, but the small GTPase Rab18 may be involved (88, 89). Because Rab18 is recruited to LDs late, its localization likely does not reflect a role during LD formation. Rab18 localization is mutually exclusive with LD localization of perilipin 2/ADRP or a dominant-negative mutant of caveolin (88, 89), suggesting a specialized function. Indeed, Rab18 localization to LDs in cultured adipocytes is related to lipolysis stimulation (89).

Sometimes endosomal structures appear to enwrap LDs (89). These interactions may be important for delivering LDs to lysosomes by autophagy in macrophages to generate cholesterol for ABCA1-dependent efflux (107). Rab5 was detected on purified LDs and implicated in the interaction of early endosomes with LDs (108).

LDs also associate with mitochondria (109) and peroxisomes (110). These interactions may channel FAs liberated from lipolysis to sites of oxidation. Supporting this notion, exercise training increases the number of LDs and their contacts with mitochondria in skeletal muscle (111).

9. LIPID DROPLET MOVEMENT WITHIN CELLS

LDs are usually dispersed and move in small oscillations. However, in mammalian cells, LD movement is sometimes coordinated. Perilipin1 expression leads to LD clustering, which is reversed upon perilipin1 phosphorylation (112). Similarly, when *Drosophila* cells (which lack perilipin1) synthesize TG for a prolonged time (e.g., during 24 h of oleate loading), newly formed LDs eventually cluster into organized superstructures (41). Increased clustering is also observed when genes involved in protein synthesis are depleted from S2 cells (41). What determines this clustering and its purpose are unknown.

LD distribution in the cell may be partially determined by active transport via microtubules. LDs move directionally in axons of *Aplysia* by uncharacterized mechanisms (113). LDs in the *Drosophila* embryonal syncitium move synchronously from the periphery to a central location and back again. Although the function of this movement is unclear, its disruption leads to altered transparency of the embryo, providing a system to screen for mutants of LD transport machinery (114). Such studies revealed that LDs move bidirectionally along microtubules with dynein and kinesin motors, which pull LDs in opposite directions. Net movement depends on the relative activities of the motors. The LD protein Halo interacts with kinesin1 as part of a complex that includes the regulatory Klar protein and the perilipin-like protein Lsd2 (115). The complex appears to link LDs to microtubules and to regulate kinesin1's activity during LD transport. Microtubules have also been implicated in LD clustering in mammalian cells, but supporting evidence is limited. Interestingly, the HCV core protein promotes clustering of LDs around microtubule organizing centers (116).

10. LIPIDS IN LIPID DROPLETS ARE MOBILIZED BY LIPASES

10.1. Lipid Droplet Catabolism by Lipolysis

In lipolysis, lipids are hydrolyzed to liberate FAs. Cells storing TG use lipolysis to generate energy and membrane lipids. In multicellular organisms, energy storage occurs predominantly in adipose tissue, and most of the knowledge concerning lipolysis comes from studies in adipocytes, where FAs are hydrolyzed sequentially by ATGL, HSL, and monoacylglycerol lipase. During lipolysis, LDs shrink as core lipids are catabolized. With decreased volume, the surface area contracts. It is unknown if excess proteins and phospholipids are resorbed into the ER or degraded.

ATGL contains an N-terminal domain with similarity to patatin, a plant acyl-hydrolase expressed highly in the potato tuber, which contains the catalytic activity. It is unknown which of the TG fatty acyl-esters are hydrolyzed by ATGL. Among the potential DAG species produced, only *sn*-1,2-DAG is known to act as a signaling molecule, and *sn*-1,3-DAG is not likely to be an optimal substrate for re-esterification by DGAT enzymes.

The ATGL N terminus also interacts with CGI-58, which has a hydrolase fold and is an ATGL activator. The activation mechanism is unknown, but it requires interaction of both proteins and CGI-58 with LDs (117). G0S2 may inhibit ATGL, although its function is unclear (118). Interaction of ATGL and perilipin1 regulates lipolysis. In mammalian adipocytes, lipolysis is initiated by hormones that trigger β-adrenergic stimulation to activate protein kinase A (PKA). In unstimulated cells, CGI-58 is bound to LD-localized perilipin1. Perilipin1 restricts basal lipolysis by sequestering CGI-58 or by sterically shielding LDs from lipolysis, or by a combination of both (119, 120). Upon activation, perilipin1 is phosphorylated by cAMP-dependent PKA at Ser492 and Ser517, releasing CGI-58 to bind and activate ATGL (121).

ATGL's interaction with CGI-58 also targets the lipase to LDs, where it is activated (122, 123). Similarly, interactions of ATGL and perilipin5/OXPAT (but not other perilipins) recruit ATGL to LDs, but in this case, the interaction limits lipolysis (122, 123). Because perilipin5 is expressed mostly in oxidative tissues, perilipin5's interaction with ATGL might be used in those tissues to regulate ATGL differently than it does in others. Perilipin5/OXPAT interacts with CGI-58, possibly forming high-molecular-weight assemblies (122). These data suggest a complex interplay among ATGL, perilipins, and CGI-58 to mediate LD targeting and lipolysis (122).

HSL catalyzes the second step of lipolysis and is also regulated by hormones. PKA activation leads to HSL phosphorylation at multiple sites, of which Ser650 (human HSL) is particularly important (124). Phosphorylation increases HSL lipolytic activity twofold. This mechanism is combined with regulated recruitment of HSL to LDs (125). Specifically, PKA-phosphorylated perilipin1 (at Ser81, Ser222, and Ser276, particularly) recruits HSL from the cytosol to LDs (125). With HSL phosphorylation, recruitment to LDs increases HSL activity 100-fold (126).

In the final lipolysis step, monoacylglycerol is hydrolyzed by monoacylglycerol lipase and possibly by HSL (127). No evidence indicates the monoacylglycerol lipase–catalyzed reaction occurs at LDs or is regulated by hormones.

Lipolysis in nonadipocyte cells is not as well understood, but similar steps must exist. Unfortunately, sequence homology cannot predict functions in proteins with hydrolase motifs. Adiponurtrin/PNPLA3, another member of the patatin-like family, is expressed in liver, and a human adiponurtrin/PNPLA3 variant (I148M) is associated with nonalcoholic fatty liver disease (128). However, adiponurtrin/PNPLA3 knockout mice do not have LD and TG accumulation in liver (129, 130), so the biochemical and physiological roles of this protein are unknown.

In addition to TG, other lipids are hydrolyzed at LD surfaces by phospholipases and other enzymes. For example, anandamide and other N-acylethanolamines traffic to and are catabolized at LDs (131).

10.2. The Fate of Hydrolyzed Lipids

FAs liberated from TGs in LDs have one of several fates. They can be re-esterified to TG, used for β-oxidation to generate energy, used as building blocks for membrane lipid synthesis, or used as cofactors for cell signaling, or exported.

In adipocytes, many FAs generated by lipolysis are re-esterified by DGAT or MGAT enzymes to TG, although the percentage drops with fasting (132). It is unclear whether re-esterification occurs in the ER or at LD surfaces. Because DGAT enzymes use activated FAs bound to CoA, re-esterification includes generating fatty acyl-CoA esters by long-chain acyl-CoA synthetases (ACSLs) in a reaction that requires the input of cellular energy. Among ACSL proteins, ACSL3 and -4 are recruited to LDs specifically during lipolysis, and ACSL1 constitutively localizes there (62, 65). Specific ACSLs, therefore, may act at LD surfaces as part of the re-esterification cycle. The function of TG turnover in this futile cycle is unknown but may fine-tune lipolysis and preserve stores of essential FAs.

FAs liberated from LD hydrolysis can be utilized directly for β-oxidation. Contact sites between LDs and mitochondria and peroxisomes may directly transfer FAs (discussed above). FAs may also be used to synthesize membrane lipids. Like re-esterification, this reaction requires FA activation by attaching a CoA moiety, possibly at LDs (65, 133). FAs liberated by ATGL also serve as endogenous activators of PPARα, as shown by lack of PPARα signaling in hearts of ATGL-deficient mice (134).

10.3. Lipid Droplet Catabolism by Autophagy

Lipids in LDs can also be degraded by lipases within lysosomes after delivery by macroautophagy. Like lipolysis, macroautophagy is induced during nutrient deprivation and responds to hormonal signals. In macroautophagy of LDs (or macrolipophagy), an LD, whole or in part, is engulfed by a membrane bilayer with the activated PE-modified LC3-II protein. The resulting autophagosome containing an LD is delivered to the lysosome where acid lipases liberate FAs from TG. This last step recapitulates the fate of lipoprotein-derived TG delivered to lysosomes via endocytosis. The relative contributions of LD autophagy to LD catabolism are unclear and may vary among tissues. LD autophagy was first reported in hepatocytes, where expression levels of ATGLs and HSLs are relatively low (93). Hepatic-specific inactivation of autophagy leads to hepatic steatosis, consistent with a function of macroautophagy in hepatic lipid metabolism. The molecular events that trigger LD autophagy are unknown.

11. LIPID DROPLETS FIGURE PROMINENTLY IN PHYSIOLOGY AND DISEASE

We cannot review in detail the metabolic and physiological processes that involve lipid storage and LDs. Instead, we focus on describing LDs in various cell types and present examples of how LDs are linked to physiology and disease. Other reviews address medical aspects of LD biology (135).

11.1. Lipid Droplets and Lipid Storage in Tissues

11.1.1. Adipose tissue. Some insects and most vertebrates have highly specialized white adipocytes and adipose tissue. In most invertebrates, TG is distributed throughout the body, often in connective tissue that fills gaps between other tissues, or in enterocytes of *C. elegans*. In flies, fat is located predominantly in a centrally located fat body. In vertebrates, adipose tissue is distributed to distinct regions and regulated by factors, such as hormones. With the evolutionary advent of homeothermy (in birds and mammals), adipose tissue gained a function as a subcutaneous insulator (136).

Adipocytes are the most highly specialized cell type for storing lipids in LDs, and in white adipocytes, a single large LD frequently occupies most of the cytoplasm. Human fat cells can store vast quantities of energy in LDs, primarily as TG. Adipocyte LDs also store cholesterol esters and fat-soluble vitamins. As expected, the adipocyte gene expression profile reflects the high flux of lipid synthesis, storage, and turnover. Adipocyte fat content is coupled to leptin expression, the main adipocyte-derived hormone that regulates long-term energy homeostasis (e.g., by regulating food intake) (137).

Brown adipocytes in mammals catabolize lipids in mitochondria that are uncoupled from oxidative phosphorylation to generate heat. LDs are prominent but smaller and more numerous than those in white adipocytes. The study of brown adipocytes is likely to shed light on interactions among LDs, mitochondria, and FA oxidation.

11.1.2. Liver. Next to adipose tissue, liver has the greatest capacity to store lipids in LDs. At basal conditions, the fraction of the cytoplasm occupied by LDs in hepatocytes is smaller than in adipoctyes, but they can hold massive amounts of TG, as illustrated by foie gras. In humans, abnormal LD accumulation is called hepatic steatosis.

LDs in hepatocytes are typically multilocular, and their formation and consumption are dynamic in normal physiology. For example, during overnight fasting in mice, large amounts of FAs are mobilized from TG stores in white adipose tissue and taken up by the liver to form LDs, generating a fatty liver. Hepatocytes express DGATs and ACATs, and the lipid composition of LDs partly reflects substrate availability. The liver also has large numbers of stellate cells that store retinol (vitamin A) as retinyl esters in LDs (17).

11.1.3. Small intestine. LDs are dynamic, prominent organelles in intestinal enterocytes (138). They reflect the enormous capacity of small intestine to absorb fat from a meal. Enterocytes have a large surface area and significant ability to synthesize and store TGs. The intestine absorbs over 95% of the TG in a typical meal, leading to rapid formation of LDs in enterocytes until TGs are exported from the cell as part of chylomicrons. This process typically takes minutes to hours after a meal. Many aspects of this process remain unclear.

11.1.4. Yolk sac. During embryonic development in many animals, the yolk sac endoderm functions like the liver and intestine to hold lipids from the mother that are destined for the developing embryo (139). As in the intestinal or liver cells, similar processes of fat uptake, storage in LDs, and export on apoB-containing lipoproteins occur in the yolk sac (139).

11.1.5. Skeletal muscle. Oxidative (type I, slow-twitch) muscles have a high rate of oxidative metabolism and are rich in mitochondria and LDs. In obese individuals, intramyocellular TG accumulation in type I muscles is associated with insulin resistance. However, in highly trained athletes, TG stores in LDs in oxidative muscle are increased without negative consequences. This "athlete's paradox" has been linked to DGAT1 function (140, 141). LDs in oxidative muscle are often close to mitochondria (111, 142). Maintaining coupling of lipid storage with consumption of lipids for fuel appears to be important for efficient energy utilization.

11.1.6. Adrenal cortex. LDs are prominent in the adrenal cortex. They store large amounts of SEs, presumably as cholesterol for steroid hormone synthesis. In fact, the yellow color of adrenal glands is due to lipid storage, and mice lacking ACAT1, and thus cholesterol esters and LDs, have pale-colored adrenal glands.

Surprisingly, adrenocortical hormone synthesis in ACAT1 knockout mice is normal. Like adrenocortical cells, other steroidogenic cells, such as Leydig cells of the testes, often have prominent LDs. Additionally, LDs are found in oocytes.

11.1.7. Macrophages. Macrophages store lipids in LDs (e.g., after phagocytosis of modified lipoproteins in the arterial wall) until they are exported. Macrophages express ACAT and DGAT enzymes, and the content of their LDs reflects their exposure to substrates. Macrophage foam cells store large amounts of cholesterol from lipoproteins, such as low-density lipoproteins. Macrophages utilize LDs to store arachidonic acid, the precursor for bioactive lipids (e.g., prostaglandins and leukotrienes) in inflammation.

11.2. Excessive Lipid Droplet Storage and Disease

Many metabolic diseases are characterized by excessive lipid storage in LDs. In these diseases, lipids may exceed the cellular capacity to store them and to buffer against their toxic effects. Unesterified lipids, such as cholesterol or FAs and their derivatives, can trigger inflammatory responses that result in tissue damage, fibrosis, scarring, and potentially organ failure.

11.2.1. Obesity and diabetes as diseases of excessive lipid droplets. Obesity is a state of excess lipid storage and LDs in the adipose tissue, often accompanied by lipid deposition and excessive LDs in nonadipose tissue, which cause lipotoxicity or tissue dysfunction. Whether obesity results in complications, such as type 2 diabetes mellitus, may relate to the capacity of LDs in adipocytes or macrophages to store lipids. In transgenic mice, increasing adipocyte capacity to store TG results in obesity without diabetes (143, 144), thereby uncoupling the two. Similarly, increasing DGAT1 expression selectively in macrophages avoids many metabolic consequences of obesity (145). The LD capacity of macrophages to store neutral lipids therefore might profoundly influence pathophysiology. Taken together, LD-targeted strategies for therapies for pathologies associated with obesity can be aimed at blocking lipid absorption and LD formation in the small intestine, preventing influx into the system; catabolizing, rather than storing, excess lipids through oxidation; and increasing LD capacity in adipocytes or macrophages to store toxic lipids and prevent associated inflammation and tissue damage.

Obesity is often accompanied by hepatic steatosis or LD accumulation in the liver, affecting millions of people. Many individuals progress to hepatic inflammation (called nonalcoholic steatohepatitis), and some go on to develop fibrosis and liver failure. Possibly, the capacity of hepatocytes to store lipids in LDs is exceeded, promoting inflammation. Hepatic steatosis is also strongly associated with hepatic insulin resistance, although the causative mechanism is unclear. Proposed contributing mechanisms include alterations in insulin signaling pathways mediated by bioactive lipids, ER stress, and activation of inflammation signaling (146). A common theme, however, relates to limited hepatocyte capacity to secrete or store lipids, thereby leading to lipid excess and liver dysfunction.

11.2.2. Atherosclerosis and macrophage lipid droplets. Like obesity, atherosclerosis is associated with excessive deposition of lipids in tissues, in this instance cholesterol, derived from apo B-containing lipoproteins in arterial walls. Excess cholesterol is taken up and esterified by macrophages. Cholesterol esters in turn are stored in LDs until they can be mobilized and effluxed to high-density lipoproteins, which are transported to the liver for clearance of cholesterol in bile. Macrophages full of cholesterol ester–containing LDs are called foam cells. LDs protect macrophage foam cells from toxicity resulting from excess free cholesterol. Deletion of ACAT1, the primary enzyme responsible for esterifying cholesterol in macrophages, in mice does not prevent atherosclerosis and even exacerbates it (147). Additionally, lack of cholesterol esterification in macrophages results in deposition of large amounts of free cholesterol in the skin and brains of ACAT1 knockout mice that were crossed into a hyperlipidemic genetic background (147). These studies highlight the lipid-buffering function of LDs and how storage of potentially toxic lipids prevents lipotoxicity.

11.2.3. The heart and lipid droplets. Although FAs are a fuel for heart muscle, few large LDs are normally present in cardiac myocytes. With lipid overload, however, LDs accumulate and heart dysfunction may occur. In murine models, overexpression of ACSL1 in the heart results in cardiomyopathy and severe heart dysfunction (148). Interestingly, this can be rescued by targeting DGAT1 overexpression to the heart, which esterifies the excess fatty acyl-CoAs to store them as TG in LDs (149). In contrast, ATGL deletion causes lipid accumulation in the heart and severe cardiomyopathy, which results in death of mice (150). Intriguingly, this phenotype can be rescued by treatment with a PPARα agonist (134), which likely increases fat oxidation in the tissue.

11.3. Lipodystrophies and Too Few Lipid Droplets

Absence or deficiency of white adipose tissue, or lipodystrophy (reviewed in Reference 151), results in an LD deficiency. Genetic causes for lipodystrophy include mutations in genes that encode for an acylglycerol-phosphate acyltransferase (*AGPAT2*), seipin (*BCSL2*), caveolin (*CAV1*) for generalized lipodystrophy, and lamin A/C (*LMNA*), peroxisome proliferator-activated receptor-γ (*PPARG*), Akt2/protein kinase B (*AKT2*), and endoprotease Face-1 (*ZMPSTE24*) for partial lipodystrophy. The pathogenesis of lipodystrophies is not understood for all causes but can relate to impairments in the development of white adipose tissue; reduced ability to synthesize glycerolipids and TGs; or an apparent deficiency in LD formation, as in lipodystrophy related to the absence of seipin (*BSCL2*). When LDs of white adipose tissue are insufficient to store TGs, the liver assumes this role, and massive hepatic steatosis often results. Associated leptin deficiency results in many metabolic abnormalities, including lipid deposition in nonadipose tissue and insulin resistance. As a consequence, leptin replacement has emerged as a therapy for some lipodystrohies.

11.4. Lipid Droplets and Cancer

Metabolism in cancer cells, including lipid metabolism (11), has received increased interest for therapeutic interventions. Most cancer cells upregulate FA synthesis, presumably to provide FAs for membrane proliferation. Some tumor cells and associated inflammatory cells have prominent LDs (11). The mechanism for LD accumulation in cancer cells is unclear but could relate to increased FA synthesis or impairments in FA oxidation.

11.5. Lipid Droplets and Lactation

LDs are prominent in mammary epithelial cells of lactating mammary glands, where they essentially serve as a secreted organelle, transferring lipid nutrients and energy from mother to child. In this process, LDs are secreted from the apical side of epithelial cells (reviewed in Reference 152), and at least three proteins are involved. Butyrophilin, a transmembrane protein, localizes to the apical plasma membrane. Xanthine oxidoreductase, an enzyme oxidizing purines and other molecules, has a role in the envelopment of LDs by the plasma membrane (153). Perilipin2/ADRP may act as a LD-specific anchor. The disruptions of butyrophilin and xanthine oxidoreductase genes resulted in impaired lactation (153, 154). Interestingly, the lactation defect in perilipin2/ADRP gene disruption was not severe, most likely because a truncated protein was still produced from the allele (155). Disruption of DGAT1 also impairs lactation in homozygous females (156), which relates to impaired development of the mammary gland and lack of LDs for milk production (157).

12. LIPID DROPLETS ARE TARGETS OF INDUSTRIAL OIL PRODUCTION

Like other organisms, plants store TG. In particular, seeds store oil, and different species store varying amounts of oil in LDs (often referred to as oil bodies) in different parts of the seed. Most plants, including soybean (*Glycine max*), sunflower (*Helianthus annuus*), safflower (*Carthamus tinctorius*), and *Brasicaceae*, store oil in the embryo itself. In other plants, such as castor bean (*Ricinus communis*), oil accumulates mostly in the triploid endosperm surrounding the embryonic tissue. In most of these species, the TG provides metabolic energy for the seedling upon germination. Some species, such as oat grains, fuse large numbers of LDs in the endosperm, and this oil attracts fruit predators that disperse seeds. The FA composition of TG in seeds differs widely with various chain length and levels of desaturation and is genetically controlled. The synthesis of TG is mediated by enzymes analogous to those in mammalian cells, including GPATs, AGPATs, lipin (PA hydrolase), and DGATs. In addition to DGAT1 and DGAT2, an additional DGAT enzyme was recently purified from the cytosol of peanuts (*Arachis hypogaea*) (158). A polymorphism of DGAT in maize (*Zea mays*) increases oil content substantially (159). As an additional complexity in plants, TG generation involves cross talk between four different compartments: the cytosol for glycerol production, plastids for the production of FAs, mitochondria for the production of very-long-chain FAs in some plants, and the ER for LD (oil body) formation.

LD production in plants is part of a highly regulated differentiation program, and many TG synthetic enzymes are under combinatorial control of several transcription factors, including activators and repressors (160). Most likely from this highly integrated process, attempts to increase oil production in seeds by overexpressing TG synthesis enzymes have thus far been disappointing. However, overexpression of *Arabidposis thaliana* DGAT in tobacco plants increased TG content of leaves 20-fold, providing a potential strategy for biofuel production (161).

When seeds germinate, they mobilize TG in oil bodies, using a lipase (Sdp1) similar to ATGL (162). Additional proteins, such as sterolesins, regulate lipolysis and β-oxidation during germination (163). In EMs, plant LDs appear to contact glyoxisomes, where FAs may be transported for β-oxidation (164).

Similar to animal LDs, plant oil bodies are decorated with specific proteins. Most notably, oleosins prevent the coalescence of oil bodies and shield them from lipolysis (165), with the latter function similar to that of mammalian perilipins.

Industries have targeted biodiesel production from crops, such as palm or oilseed rape. However, increasing production of these plants in which only a fraction is used for oil presents challenges because it puts them in competition for arable land with food crops. Thus, TG production in algae, or other microorganisms, for biofuel has received increasing attention. To generate biofuels, algae must be grown and harvested, and the lipids extracted and processed. Considerable effort has gone into identifying suitable species for cultivation in open ponds or photobioreactors and into optimizing other production steps. To be economcially and environmentally viable, biofuel production from algae must have a positive life-cycle analysis, meaning a positive balance of energy yield compared with the energy required for biofuel generation. This is challenging, and thus efforts to optimize growth systems and conditions are underway. Nevertheless, engineering of metabolism in algae toward oil production (e.g., by overexpressing DGAT or acetyl-CoA carboxylase enzymes) is promising.

SUMMARY POINTS

1. LDs are ubiquitous organelles in most cells and most organisms.
2. LDs primarily store lipids for metabolic energy and for membrane components. They are also implicated in protein storage and degradation and in complex processes, such as viral replication.
3. LDs represent an emulsion of a hydrophobic phase of neutral lipids, such as TGs and SEs, within the aqueous cytosol. Individual LDs are stabilized by a monolayer of phospholipids into which specific proteins are embedded. The molecular architectures of LDs and lipoproteins are similar.
4. LDs are believed to be formed at the ER, where neutral lipids are produced, and to bud toward the cytoplasm. They may remain attached to the ER or detach. LDs may originate from other cell membranes. The mechanism for LD biogenesis is unclear.
5. LDs grow by expansion or coalescence (fusion). The molecular details of these processes are being unraveled. PC is important as a surfactant and to prevent coalescence.
6. Specific proteins target LDs through several general targeting motifs. The mechanism of how proteins traffic to LDs remains mostly unknown.
7. Lipids are mobilized from LDs through the sequential actions of lipases. In adipose tissue, this process is highly regulated. LDs may also be catabolized via macroautophagy.
8. LDs in different mammalian tissues have important functions in energy and membrane homeostasis. In adipose tissue, but also in liver and intestine, LDs store TGs for energy generation. Macrophages utilize LDs to store and clear surplus cholesterol from the periphery of the body. Excess or failure to form LDs leads to lipotoxicity and metabolic diseases, including obesity, lipodystrophy, or atherosclerosis.
9. LDs are the cellular organelles that store oils in plants that are used for agricultural production of food oils and biofuels.

FUTURE ISSUES

1. How are LDs formed? Is the ER the only place of LD formation in eukaryotes? What mechanisms are involved in their organization and budding reactions from the ER membrane?
2. How are proteins targeted to LDs? Are there one or multiple targeting mechanisms? What is their relationship to secretory trafficking? How are LD proteins degraded?
3. How do LDs interact with other organelles, such as mitochondria, peroxisomes, and the ER? What are the functions of these interactions?
4. What are the mechanisms of lipolysis? How do lipases access neutral lipids within LDs? What are the fates and functions of lipolysis intermediates and products? What are the relative contributions of lipolysis and macroautophagy to catabolism of LDs?
5. What determines the capacities of cells to store lipids and form LDs? What molecular mechanisms mediate lipotoxicity in different tissues?

6. How can agricultural organisms be engineered to produce large amounts of desired lipids in an easy, accessible, and energy-efficient manner?

DISCLOSURE STATEMENT

The authors are not aware of any affiliations, memberships, funding, or financial holdings that might be perceived as affecting the objectivity of this review.

ACKNOWLEDGMENTS

We acknowledge the many scientists who contributed to the knowledge in the field, some of which we were unable to cite owing to space constraints. In particular, we salute the pioneering contributions of cell biologists Constantine Londos and Richard Anderson, who helped to launch the field of LD biology. We thank current and former members of the Walther and Farese laboratories for their many contributions to the ideas contained in the review, Sylvia Richmond for assistance with manuscript preparation, and Gary Howard for editorial assistance.

LITERATURE CITED

1. Farese RV Jr, Walther TC. 2009. Lipid droplets finally get a little R-E-S-P-E-C-T. *Cell* 139:855–60
2. Fujimoto T, Parton RG. 2011. Not just fat: the structure and function of the lipid droplet. *Cold Spring Harb. Perspect. Biol.* 3:a004838
3. Thiele C, Spandl J. 2008. Cell biology of lipid droplets. *Curr. Opin. Cell Biol.* 20:378–85
4. Greenberg AS, Egan JJ, Wek SA, Garty NB, Blanchette-Mackie EJ, Londos C. 1991. Perilipin, a major hormonally regulated adipocyte-specific phosphoprotein associated with the periphery of lipid storage droplets. *J. Biol. Chem.* 266:11341–46
5. Alvarez HM, Steinbuchel A. 2002. Triacylglycerols in prokaryotic microorganisms. *Appl. Microbiol. Biotechnol.* 60:367–76
6. Fei W, Shui G, Gaeta B, Du X, Kuerschner L, et al. 2008. Fld1p, a functional homologue of human seipin, regulates the size of lipid droplets in yeast. *J. Cell Biol.* 180:473–82
7. Szymanski K, Binns D, Bartz R, Grishin N, Li W, et al. 2007. The lipodystrophy protein seipin is found at endoplasmic reticulum lipid droplet junctions and is important for droplet morphology. *Proc. Natl. Acad. Sci. USA* 104:20890–95
8. Kurat C, Wolinski H, Petschnigg J, Kaluarachchi S, Andrews B, et al. 2009. Cdk1/Cdc28-dependent activation of the major triacylglycerol lipase Tgl4 in yeast links lipolysis to cell-cycle progression. *Mol. Cell.* 33:53–63
9. Yamamoto K, Takahara K, Oyadomari S, Okada T, Sato T, et al. 2010. Induction of liver steatosis and lipid droplet formation in ATF6α-knockout mice burdened with pharmacological endoplasmic reticulum stress. *Mol. Biol. Cell* 21:2975–86
10. Fei W, Wang H, Fu X, Bielby C, Yang H. 2009. Conditions of endoplasmic reticulum stress stimulate lipid droplet formation in *Saccharomyces cerevisiae*. *Biochem. J.* 424:61–67
11. Bozza PT, Viola JP. 2010. Lipid droplets in inflammation and cancer. *Prostaglandins Leukot. Essent. Fatty Acids* 82:243–50
12. Krahmer N, Guo Y, Wilfling F, Hilger M, Lingrell S, et al. 2012. Phosphatidylcholine synthesis for lipid droplet expansion is mediated by localized activation of CTP:phosphocholine cytidylyltransferase. *Cell Metab.* 14:504–15
13. Czabany T, Wagner A, Zweytick D, Lohner K, Leitner E, et al. 2008. Structural and biochemical properties of lipid particles from the yeast *Saccharomyces cerevisiae*. *J. Biol. Chem.* 283:17065–75

14. Buhman KK, Chen HC, Farese RV Jr. 2001. The enzymes of neutral lipid synthesis. *J. Biol. Chem.* 276:40369–72
15. Kuerschner L, Moessinger C, Thiele C. 2008. Imaging of lipid biosynthesis: How a neutral lipid enters lipid droplets. *Traffic* 9:338–52
16. Stone SJ, Levin MC, Zhou P, Han J, Walther TC, Farese RV Jr. 2009. The endoplasmic reticulum enzyme DGAT2 is found in mitochondria-associated membranes and has a mitochondrial targeting signal that promotes its association with mitochondria. *J. Biol. Chem.* 284:5352–61
17. Blaner WS, O'Byrne SM, Wongsiriroj N, Kluwe J, D'Ambrosio DM, et al. 2009. Hepatic stellate cell lipid droplets: a specialized lipid droplet for retinoid storage. *Biochim. Biophys. Acta* 1791:467–73
18. Orban T, Palczewska G, Palczewski K. 2011. Retinyl ester storage particles (retinosomes) from the retinal pigmented epithelium resemble lipid droplets in other tissues. *J. Biol. Chem.* 286:17248–58
19. Bartz R, Li WH, Venables B, Zehmer JK, Roth MR, et al. 2007. Lipidomics reveals that adiposomes store ether lipids and mediate phospholipid traffic. *J. Lipid Res.* 48:837–47
20. Cornish K, Wood DF, Windle JJ. 1999. Rubber particles from four different species, examined by transmission electron microscopy and electron-paramagnetic-resonance spin labeling, are found to consist of a homogeneous rubber core enclosed by a contiguous, monolayer biomembrane. *Planta* 210:85–96
21. Athenstaedt K, Zweytick D, Jandrositz A, Kohlwein S, Daum G. 1999. Identification and characterization of major lipid particle proteins of the yeast *Saccharomyces cerevisiae*. *J. Bacteriol.* 181:6441–48
22. Welte MA. 2007. Proteins under new management: Lipid droplets deliver. *Trends Cell Biol* 17:363–69
23. Cermelli S, Guo Y, Gross SP, Welte MA. 2006. The lipid-droplet proteome reveals that droplets are a protein-storage depot. *Curr. Biol.* 16:1783–95
24. Herker E, Ott M. 2011. Unique ties between hepatitis C virus replication and intracellular lipids. *Trends Endocrinol. Metab.* 22:241–48
25. Herker E, Harris C, Hernandez C, Carpentier A, Kaehlcke K, et al. 2010. Efficient hepatitis C virus particle formation requires diacylglycerol acyltransferase-1. *Nat. Med.* 16:1295–98
26. Ye J, Li JZ, Liu Y, Li X, Yang T, et al. 2009. Cideb, an ER- and lipid droplet-associated protein, mediates VLDL lipidation and maturation by interacting with apolipoprotein B. *Cell Metab.* 9:177–90
27. Dolinsky VW, Gilham D, Alam M, Vance DE, Lehner R. 2004. Triacylglycerol hydrolase: role in intracellular lipid metabolism. *Cell. Mol. Life Sci.* 61:1633–51
28. Wältermann M, Hinz A, Robenek H, Troyer D, Reichelt R, et al. 2005. Mechanism of lipid-body formation in prokaryotes: How bacteria fatten up. *Mol. Microbiol.* 55:750–63
29. Jacquier N, Choudhary V, Mari M, Toulmay A, Reggiori F, Schneiter R. 2011. Lipid droplets are functionally connected to the endoplasmic reticulum in *Saccharomyces cerevisiae*. *J. Cell Sci.* 124:2424–37
30. Soni K, Mardones G, Sougrat R, Smirnova E, Jackson C, Bonifacino J. 2009. Coatomer-dependent protein delivery to lipid droplets. *J. Cell Sci.* 122:1834–41
31. Ohsaki Y, Cheng J, Suzuki M, Fujita A, Fujimoto T. 2008. Lipid droplets are arrested in the ER membrane by tight binding of lipidated apolipoprotein B-100. *J. Cell Sci.* 121:2415–22
32. Lacey DJ, Beaudoin F, Dempsey CE, Shewry PR, Napier JA. 1999. The accumulation of triacylglycerols within the endoplasmic reticulum of developing seeds of *Helianthus annuus*. *Plant J.* 17:397–405
33. Skinner JR, Shew TM, Schwartz DM, Tzekov A, Lepus CM, et al. 2009. Diacylglycerol enrichment of endoplasmic reticulum or lipid droplets recruits perilipin 3/TIP47 during lipid storage and mobilization. *J. Biol. Chem.* 284:30941–48
34. Bulankina AV, Deggerich A, Wenzel D, Mutenda K, Wittmann JG, et al. 2009. TIP47 functions in the biogenesis of lipid droplets. *J. Cell Biol.* 185:641–55
35. Ploegh HL. 2007. A lipid-based model for the creation of an escape hatch from the endoplasmic reticulum. *Nature* 448:435–38
36. Walther TC, Farese RV Jr. 2009. The life of lipid droplets. *Biochim. Biophys. Acta* 1791:459–66
37. Robenek H, Hofnagel O, Buers I, Robenek MJ, Troyer D, Severs NJ. 2006. Adipophilin-enriched domains in the ER membrane are sites of lipid droplet biogenesis. *J. Cell Sci.* 119:4215–24
38. Hamilton RL, Wong JS, Cham CM, Nielsen LB, Young SG. 1998. Chylomicron-sized lipid particles are formed in the setting of apolipoprotein B deficiency. *J. Lipid Res.* 39:1543–57
39. Fei W, Shui G, Zhang Y, Krahmer N, Ferguson C, et al. 2011. A role for phosphatidic acid in the formation "supersized" lipid droplets. *PLoS Genet.* 7:e1002201

40. Binns D, Lee S, Hilton CL, Jiang QX, Goodman JM. 2010. Seipin is a discrete homooligomer. *Biochemistry* 49:10747–55
41. Guo Y, Walther TC, Rao M, Stuurman N, Goshima G, et al. 2008. Functional genomic screen reveals genes involved in lipid-droplet formation and utilization. *Nature* 453:657–61
42. Beller M, Riedel D, Jänsch L, Dieterich G, Wehland J, et al. 2006. Characterization of the *Drosophila* lipid droplet subproteome. *Mol. Cell. Proteomics* 5:1082–94
43. Beller M, Sztalryd C, Southall N, Bell M, Jäckle H, et al. 2008. COPI complex is a regulator of lipid homeostasis. *PLoS Biol.* 6:e292
44. Ashrafi K, Chang FY, Watts JL, Fraser AG, Kamath RS, et al. 2003. Genome-wide RNAi analysis of *Caenorhabditis elegans* fat regulatory genes. *Nature* 421:268–72
45. Pospisilik JA, Schramek D, Schnidar H, Cronin SJ, Nehme NT, et al. 2010. *Drosophila* genome-wide obesity screen reveals hedgehog as a determinant of brown versus white adipose cell fate. *Cell* 140:148–60
46. Gross DA, Snapp EL, Silver DL. 2010. Structural insights into triglyceride storage mediated by fat storage-inducing transmembrane (FIT) protein 2. *PLoS ONE* 5:e10796
47. Kadereit B, Kumar P, Wang WJ, Mirnada D, Snapp EL, et al. 2008. Evolutionarily conserved gene family important for fat storage. *Proc. Natl. Acad. Sci. USA* 105:94–99
48. Gubern A, Barcelo-Torns M, Casas J, Barneda D, Masgrau R, et al. 2009. Lipid droplet biogenesis induced by stress involves triacylglycerol synthesis that depends on group VIA phospholipase A2. *J. Biol. Chem.* 284:5697–708
49. Moessinger C, Kuerschner L, Spandl J, Shevchenko A, Thiele C. 2011. Human lysophosphatidylcholine acyltransferases 1 and 2 are located in lipid droplets where they catalyze the formation of phosphatidylcholine. *J. Biol. Chem.* 286:21330–39
50. Khelef N, Buton X, Beatini N, Wang H, Meiner V, et al. 1998. Immunolocalization of acyl-coenzyme A:cholesterol O-acyltransferase in macrophages. *J. Biol. Chem.* 273:11218–24
51. Bostrom P, Rutberg M, Ericsson J, Holmdahl P, Andersson L, et al. 2005. Cytosolic lipid droplets increase in size by microtubule-dependent complex formation. *Arterioscler. Thromb. Vasc. Biol.* 25:1945–51
52. Murphy S, Martin S, Parton RG. 2010. Quantitative analysis of lipid droplet fusion: inefficient steady state fusion but rapid stimulation by chemical fusogens. *PloS One* 5:e15030
53. Puri V, Konda S, Ranjit S, Aouadi M, Chawla A, et al. 2007. Fat-specific protein 27, a novel lipid droplet protein that enhances triglyceride storage. *J. Biol. Chem.* 282:34213–18
54. Keller P, Petrie JT, De Rose P, Gerin I, Wright WS, et al. 2008. Fat-specific protein 27 regulates storage of triacylglycerol. *J. Biol. Chem.* 283:14355–65
55. Nishino N, Tamori Y, Tateya S, Kawaguchi T, Shibakusa T, et al. 2008. FSP27 contributes to efficient energy storage in murine white adipocytes by promoting the formation of unilocular lipid droplets. *J. Clin. Investig.* 118:2808–21
56. Puri V, Virbasius JV, Guilherme A, Czech MP. 2008. RNAi screens reveal novel metabolic regulators: RIP140, MAP4k4 and the lipid droplet associated fat specific protein (FSP) 27. *Acta Physiol.* 192:103–15
57. Jambunathan S, Yin J, Khan W, Tamori Y, Puri V. 2011. FSP27 promotes lipid droplet clustering and then fusion to regulate triglyceride accumulation. *PLOS One* 6:e28614
58. Gong J, Sun Z, Wu L, Xu W, Schieber N, et al. 2011. Fsp27 promotes lipid droplet growth by lipid exchange and transfer at lipid droplet contact sites. *J. Cell Biol.* 195:953–63
59. Boström P, Andersson L, Rutberg M, Perman J, Lidberg U, et al. 2007. SNARE proteins mediate fusion between cytosolic lipid droplets and are implicated in insulin sensitivity. *Nat. Cell Biol.* 9:1286–93
60. Zweytick D, Athenstaedt K, Daum G. 2000. Intracellular lipid particles of eukaryotic cells. *Biochim. Biophys. Acta* 1469:101–20
61. Liu P, Ying Y, Zhao Y, Mundy DI, Zhu M, Anderson RG. 2004. Chinese hamster ovary K2 cell lipid droplets appear to be metabolic organelles involved in membrane traffic. *J. Biol. Chem.* 279:3787–92
62. Fujimoto Y, Itabe H, Sakai J, Makita M, Noda J, et al. 2004. Identification of major proteins in the lipid droplet-enriched fraction isolated from the human hepatocyte cell line HuH7. *Biochim. Biophys. Acta* 1644:47–59
63. Umlauf E, Csaszar E, Moertelmaier M, Schuetz GJ, Parton RG, Prohaska R. 2004. Association of stomatin with lipid bodies. *J. Biol. Chem.* 279:23699–709

64. Wu CC, Howell KE, Neville MC, Yates JR III, McManaman JL. 2000. Proteomics reveal a link between the endoplasmic reticulum and lipid secretory mechanisms in mammary epithelial cells. *Electrophoresis* 21:3470–82
65. Brasaemle DL, Dolios G, Shapiro L, Wang R. 2004. Proteomic analysis of proteins associated with lipid droplets of basal and lipolytically stimulated 3T3-L1 adipocytes. *J. Biol. Chem.* 279:46835–42
66. Cole NB, Murphy DD, Grider T, Rueter S, Brasaemle D, Nussbaum RL. 2002. Lipid droplet binding and oligomerization properties of the Parkinson's disease protein α-synuclein. *J. Biol. Chem.* 277:6344–52
67. Barba G, Harper F, Harada T, Kohara M, Goulinet S, et al. 1997. Hepatitis C virus core protein shows a cytoplasmic localization and associates to cellular lipid storage droplets. *Proc. Natl. Acad. Sci. USA* 94:1200–5
68. Hinson ER, Cresswell P. 2009. The antiviral protein, viperin, localizes to lipid droplets via its N-terminal amphipathic alpha-helix. *Proc. Natl. Acad. Sci. USA* 106:20452–57
69. Tauchi-Sato K, Ozeki S, Houjou T, Tagushi R, Fujimoto T. 2002. The surface of lipid droplets is a phospholipid monolayer with a unique fatty acid composition. *J. Biol. Chem.* 277:44507–12
70. Zehmer JK, Bartz R, Liu P, Anderson RG. 2008. Identification of a novel N-terminal hydrophobic sequence that targets proteins to lipid droplets. *J. Cell Sci.* 121:1852–60
71. Turro S, Ingelmo-Torres M, Estanyol JM, Tebar F, Fernandez MA, et al. 2006. Identification and characterization of associated with lipid droplet protein 1: A novel membrane-associated protein that resides on hepatic lipid droplets. *Traffic* 7:1254–69
72. Napier JA, Stobart AK, Shewry PR. 1996. The structure and biogenesis of plant oil bodies: the role of the ER membrane and the oleosin class of proteins. *Plant Mol. Biol.* 31:945–56
73. Qu R, Wang S-M, Lin Y-H, Vance VB, Huang AHC. 1986. Characteristics and biosynthesis of membrane proteins of lipid bodies in the scutella of maize (*Zea mays* L.). *Biochem. J.* 235:57–65
74. Abell BM, Holbrook LA, Abenes M, Murphy DJ, Hills MJ, Moloney MM. 1997. Role of the proline knot motif in oleosin endoplasmic reticulum topology and oil body targeting. *Plant Cell* 9:1481–93
75. Pol A, Luetterforst R, Lindsay M, Heinoo S, Ikonen E, Parton RG. 2001. A caveolin dominant negative mutant associates with lipid bodies and induces intracellular cholesterol imbalance. *J. Cell Biol.* 152:1057–70
76. Fujimoto T, Kogo H, Ishiguro K, Tauchi K, Nomura R. 2001. Caveolin-2 is targeted to lipid droplets, a new "membrane domain" in the cell. *J. Cell Biol.* 152:1079–85
77. Ostermeyer A, Paci J, Zeng Y, Lublin D, Munro S, Brown D. 2001. Accumulation of caveolin in the endoplasmic reticulum redirects the protein to lipid storage droplets. *J. Cell Biol.* 152:1071–78
78. Pol A, Martin S, Fernandez MA, Ingelmo-Torres M, Ferguson C, et al. 2005. Cholesterol and fatty acids regulate dynamic caveolin trafficking through the Golgi complex and between the cell surface and lipid bodies. *Mol. Biol. Cell* 16:2091–105
79. Le Lay S, Hajduch E, Lindsay MR, Le Liepvre X, Thiele C, et al. 2006. Cholesterol-induced caveolin targeting to lipid droplets in adipocytes: a role for caveolar endocytosis. *Traffic* 7:549–61
80. Greenberg AS, Egan JJ, Wek SA, Garty NB, Blanchette-Mackie EJ, Londos C. 1991. Perilipin, a major hormonally regulated adipocyte-specific phosphoprotein associated with the periphery of lipid storage droplets. *J. Biol. Chem.* 266:11341–46
81. Subramanian V, Garcia A, Sekowski A, Brasaemle DL. 2004. Hydrophobic sequences target and anchor perilipin A to lipid droplets. *J. Lipid Res.* 45:1983–91
82. McManaman JL, Zabaronick W, Schaack J, Orlicky DJ. 2003. Lipid droplet targeting domains of adipophilin. *J. Lipid Res.* 44:668–73
83. Targett-Adams P, Chambers D, Gledhill S, Hope RG, Coy JF, et al. 2003. Live cell analysis and targeting of the lipid droplet-binding adipocyte differentiation-related protein. *J. Biol. Chem.* 278:15998–6007
84. Nakamura N, Fujimoto T. 2003. Adipose differentiation-related protein has two independent domains for targeting to lipid droplets. *Biochem. Biophys. Res. Commun.* 306:333–38
85. Hickenbottom SJ, Kimmel AR, Londos C, Hurley JH. 2004. Structure of a lipid droplet protein: the PAT family member TIP47. *Structure* 12:1199–207
86. Hatters DM, Peters-Libeu CA, Weisgraber KH. 2006. Apolipoprotein E structure: insights into function. *Trends Biochem. Sci.* 31:445–54

87. Wolins NE, Quaynor BK, Skinner JR, Schoenfish MJ, Tzekov A, Bickel PE. 2005. S3–12, adipophilin, and TIP47 package lipid in adipocytes. *J. Biol. Chem.* 280:19146–55
88. Ozeki S, Cheng J, Tauchi-Sato K, Hatano N, Taniguchi H, Fujimoto T. 2005. Rab18 localizes to lipid droplets and induces their close apposition to the endoplasmic reticulum-derived membrane. *J. Cell Sci.* 118:2601–11
89. Martin S, Driessen K, Nixon SJ, Zerial M, Parton RG. 2005. Regulated localization of Rab18 to lipid droplets: effects of lipolytic stimulation and inhibition of lipid droplet catabolism. *J. Biol. Chem.* 280:42325–35
90. Beller M, Bulankina AV, Hsiao HH, Urlaub H, Jäckle H, Kühnlein RP. 2010. PERILIPIN-dependent control of lipid droplet structure and fat storage in *Drosophila*. *Cell Metab.* 12:521–32
91. Schweiger M, Schoiswohl G, Lass A, Radner FP, Haemmerle G, et al. 2008. The C-terminal region of human adipose triglyceride lipase affects enzyme activity and lipid droplet binding. *J. Biol. Chem.* 283:17211–20
92. Wanner G, Formanet H, Theimer RR. 1981. The ontogeny of lipid bodies (spherosomes) in plant cells. *Planta* 151:109–23
93. Singh R, Kaushik S, Wang Y, Xiang Y, Novak I, et al. 2009. Autophagy regulates lipid metabolism. *Nature* 458:1131–35
94. Xu G, Sztalryd C, Lu X, Tansey JT, Gan J, et al. 2005. Post-translational regulation of adipose differentiation-related protein by the ubiquitin/proteasome pathway. *J. Biol. Chem.* 280:42841–47
95. Masuda Y, Itabe H, Odaki M, Hama K, Fujimoto Y, et al. 2006. ADRP/adipophilin is degraded through the proteasome-dependent pathway during regression of lipid-storing cells. *J. Lipid Res.* 47:87–98
96. Xu G, Sztalryd C, Londos C. 2006. Degradation of perilipin is mediated through ubiquitination-proteasome pathway. *Biochim. Biophys. Acta* 1761:83–90
97. Hooper C, Puttamadappa SS, Loring Z, Shekhtman A, Bakowska JC. 2010. Spartin activates atrophin-1-interacting protein 4 (AIP4) E3 ubiquitin ligase and promotes ubiquitination of adipophilin on lipid droplets. *BMC Biol.* 8:72
98. Nian Z, Sun Z, Yu L, Toh SY, Sang J, Li P. 2010. Fat-specific protein 27 undergoes ubiquitin-dependent degradation regulated by triacylglycerol synthesis and lipid droplet formation. *J. Biol. Chem.* 285:9604–15
99. Edwards TL, Clowes VE, Tsang HT, Connell JW, Sanderson CM, et al. 2009. Endogenous spartin (SPG20) is recruited to endosomes and lipid droplets and interacts with the ubiquitin E3 ligases AIP4 and AIP5. *Biochem. J.* 423:31–39
100. Eastman SW, Yassaee M, Bieniasz PD. 2009. A role for ubiquitin ligases and Spartin/SPG20 in lipid droplet turnover. *J. Cell Biol.* 184:881–94
101. Spandl J, Lohmann D, Kuerschner L, Moessinger C, Thiele C. 2011. Ancient ubiquitous protein 1 (AUP1) localizes to lipid droplets and binds the E2 ubiquitin conjugase G2 (Ube2g2) via its G2 binding region. *J. Biol. Chem.* 286:5599–606
102. Klemm EJ, Spooner E, Ploegh HL. 2011. The dual role of ancient ubiquitous protein 1 (AUP1) in lipid droplet accumulation and ER protein quality control. *J. Biol. Chem.* 286:37602–14
103. Hartman IZ, Liu P, Zehmer JK, Luby-Phelps K, Jo Y, et al. 2010. Sterol-induced dislocation of 3-hydroxy-3-methylglutaryl coenzyme A reductase from endoplasmic reticulum membranes into the cytosol through a subcellular compartment resembling lipid droplets. *J. Biol. Chem.* 285:19288–98
104. Olzmann JA, Kopito RR. 2011. Lipid droplet formation is dispensable for endoplasmic reticulum-associated degradation. *J. Biol. Chem.* 286:27872–74
105. Hynynen R, Suchanek M, Spandl J, Bäck N, Thiele C, Olkkonen VM. 2009. OSBP-related protein 2 is a sterol receptor on lipid droplets that regulates the metabolism of neutral lipids. *J. Lipid Res.* 50:1305–15
106. Yokoi Y, Horiguchi Y, Araki M, Motojima K. 2007. Regulated expression by PPARα and unique localization of 17β-hydroxysteroid dehydrogenase type 11 protein in mouse intestine and liver. *FEBS J.* 274:4837–47
107. Ouimet M, Franklin V, Mak E, Liao X, Tabas I, Marcel YL. 2011. Autophagy regulates cholesterol efflux from macrophage foam cells via lysosomal acid lipase. *Cell Metab.* 13:655–67
108. Liu P, Bartz R, Zehmer JK, Ying YS, Zhu M, et al. 2007. Rab-regulated interaction of early endosomes with lipid droplets. *Biochim. Biophys. Acta* 1773:784–93

109. Blanchette-Mackie EJ, Scow RO. 1983. Movement of lipolytic products to mitochondria in brown adipose tissue of young rats: an electron microscope study. *J. Lipid Res.* 24:229–44
110. Binns D, Januszewski T, Chen Y, Hill J, Markin VS, et al. 2006. An intimate collaboration between peroxisomes and lipid bodies. *J. Cell Biol.* 173:719–31
111. Tarnopolsky MA, Rennie CD, Robertshaw HA, Fedak-Tarnopolsky SN, Devries MC, Hamadeh MJ. 2007. Influence of endurance exercise training and sex on intramyocellular lipid and mitochondrial ultrastructure, substrate use, and mitochondrial enzyme activity. *Am. J. Physiol. Regul. Integr. Comp. Physiol.* 292:R1271–78
112. Marcinkiewicz A, Gauthier D, Garcia A, Brasaemle DL. 2006. The phosphorylation of serine 492 of perilipin A directs lipid droplet fragmentation and dispersion. *J. Biol. Chem.* 281:11901–9
113. Savage MJ, Goldberg DJ, Schacher S. 1987. Absolute specificity for retrograde fast axonal transport displayed by lipid droplets originating in the axon of an identified *Aplysia* neuron in vitro. *Brain Res.* 406:215–23
114. Welte MA, Gross SP, Postner M, Block SM, Wieschaus EF. 1998. Developmental regulation of vesicle transport in *Drosophila* embryos: forces and kinetics. *Cell* 92:547–57
115. Shubeita GT, Tran SL, Xu J, Vershinin M, Cermelli S, et al. 2008. Consequences of motor copy number on the intracellular transport of kinesin-1-driven lipid droplets. *Cell* 135:1098–107
116. Boulant S, Douglas MW, Moody L, Budkowska A, Targett-Adams P, McLauchlan J. 2008. Hepatitis C virus core protein induces lipid droplet redistribution in a microtubule- and dynein-dependent manner. *Traffic* 9:1268–82
117. Gruber A, Cornaciu I, Lass A, Schweiger M, Poeschl M, et al. 2010. The N-terminal region of comparative gene identification-58 (CGI-58) is important for lipid droplet binding and activation of adipose triglyceride lipase. *J. Biol. Chem.* 285:12289–98
118. Yang X, Lu X, Lombès M, Rha GB, Chi Y-I, et al. 2010. The G_0/G_1 switch gene 2 regulates adipose lipolysis through association with adipose triglyceride lipase. *Cell Metab.* 11:194–205
119. Brasaemle DL, Rubin B, Harten IA, Gruia-Gray J, Kimmel AR, Londos C. 2000. Perilipin A increases triacylglycerol storage by decreasing the rate of triacylglycerol hydrolysis. *J. Biol. Chem.* 275:38486–93
120. Souza SC, de Vargas LM, Yamamoto MT, Lien P, Franciosa MD, et al. 1998. Overexpression of perilipin A and B blocks the ability of tumor necrosis factor α to increase lipolysis in 3T3-L1 adipocytes. *J. Biol. Chem.* 273:24665–69
121. Subramanian V, Rothenberg A, Gomez C, Cohen AW, Garcia A, et al. 2004. Perilipin A mediates the reversible binding of CGI-58 to lipid droplets in 3T3-L1 adipocytes. *J. Biol. Chem.* 279:42062–71
122. Granneman JG, Moore HP, Mottillo EP, Zhu Z, Zhou L. 2011. Interactions of perilipin-5 (Plin5) with adipose triglyceride lipase. *J. Biol. Chem.* 286:5126–35
123. Wang H, Bell M, Sreenevasan U, Hu H, Liu J, et al. 2011. Unique regulation of adipose triglyceride lipase (ATGL) by perilipin 5, a lipid droplet-associated protein. *J. Biol. Chem.* 286:15707–15
124. Krintel C, Osmark P, Larsen MR, Resjo S, Logan DT, Holm C. 2008. Ser649 and Ser650 are the major determinants of protein kinase A-mediated activation of human hormone-sensitive lipase against lipid substrates. *PLoS ONE* 3:e3756
125. Wang H, Hu L, Dalen K, Dorward H, Marcinkiewicz A, et al. 2009. Activation of hormone-sensitive lipase requires two steps, protein phosphorylation and binding to the PAT-1 domain of lipid droplet coat proteins. *J. Biol. Chem.* 284:32116–25
126. Schweiger M, Schreiber R, Haemmerle G, Lass A, Fledelius C, et al. 2006. Adipose triglyceride lipase and hormone-sensitive lipase are the major enzymes in adipose tissue triacylglycerol catabolism. *J. Biol. Chem.* 281:40236–41
127. Taschler U, Radner FP, Heier C, Schreiber R, Schweiger M, et al. 2011. Monoglyceride lipase deficiency in mice impairs lipolysis and attenuates diet-induced insulin resistance. *J. Biol. Chem.* 286:17467–77
128. He S, McPhaul C, Li JZ, Garuti R, Kinch L, et al. 2010. A sequence variation (I148M) in PNPLA3 associated with nonalcoholic fatty liver disease disrupts triglyceride hydrolysis. *J. Biol. Chem.* 285:6706–15
129. Chen W, Chang B, Li L, Chan L. 2010. Patatin-like phospholipase domain-containing 3/adiponutrin deficiency in mice is not associated with fatty liver disease. *Hepatology* 52:1134–42
130. Basantani MK, Sitnick MT, Cai L, Brenner DS, Gardner NP, et al. 2011. Pnpla3/adiponutrin deficiency in mice does not contribute to fatty liver disease or metabolic syndrome. *J. Lipid Res.* 52:318–29

131. Kaczocha M, Glaser ST, Chae J, Brown DA, Deutsch DG. 2010. Lipid droplets are novel sites of N-acylethanolamine inactivation by fatty acid amide hydrolase-2. *J. Biol. Chem.* 285:2796–806
132. Leibel RL, Hirsch J, Berry EM, Gruen RK. 1985. Alterations in adipocyte free fatty acid re-esterification associated with obesity and weight reduction in man. *Am. J. Clin. Nutr.* 42:198–206
133. Fujimoto Y, Itabe H, Kinoshita T, Homma KJ, Onoduka J, et al. 2007. Involvement of ACSL in local synthesis of neutral lipids in cytoplasmic lipid droplets in human hepatocyte HuH7. *J. Lipid Res.* 48:1280–92
134. Wölkart G, Schrammel A, Dörffel K, Haemmerle G, Zechner R, Mayer B. 2011. Cardiac dysfunction in adipose triglyceride lipase deficiency: Treatment with a PPARα agonist. *Br. J. Pharmacol.* 165:380–89
135. Greenberg AS, Coleman RA, Kraemer FB, McManaman JL, Obin MS, et al. 2011. The role of lipid droplets in metabolic disease in rodents and humans. *J. Clin. Investig.* 121:2102–10
136. Gesta S, Tseng YH, Kahn CR. 2007. Developmental origin of fat: tracking obesity to its source. *Cell* 131:242–56
137. Friedman JM. 2009. Leptin at 14 y of age: an ongoing story. *Am. J. Clin. Nutr.* 89:S973–79
138. Zhu J, Lee B, Buhman KK, Cheng JX. 2009. A dynamic, cytoplasmic triacylglycerol pool in enterocytes revealed by ex vivo and in vivo coherent anti-Stokes Raman scattering imaging. *J. Lipid Res.* 50:1080–89
139. Farese RV Jr, Cases S, Ruland SL, Kayden HJ, Wong JS, et al. 1996. A novel function for apolipoprotein B: Lipoprotein synthesis in the yolk sac is critical for maternal-fetal lipid transport in mice. *J. Lipid Res.* 37:347–60
140. Liu L, Zhang Y, Chen N, Shi X, Tsang B, Yu YH. 2007. Upregulation of myocellular DGAT1 augments triglyceride synthesis in skeletal muscle and protects against fat-induced insulin resistance. *J. Clin. Investig.* 117:1679–89
141. Schenk S, Horowitz JF. 2007. Acute exercise increases triglyceride synthesis in skeletal muscle and prevents fatty acid-induced insulin resistance. *J. Clin. Investig.* 117:1690–98
142. Shaw CS, Jones DA, Wagenmakers AJM. 2008. Network distribution of mitochondria and lipid droplets in human muscle fibres. *Histochem. Cell Biol.* 129:65–72
143. Chen HC, Stone SJ, Zhou P, Buhman KK, Farese RV Jr. 2002. Dissociation of obesity and impaired glucose disposal in mice overexpressing acyl coenzyme A:diacylglycerol acyltransferase 1 in white adipose tissue. *Diabetes* 51:3189–95
144. Franckhauser S, Muñoz S, Pujol A, Casellas A, Riu E, et al. 2002. Increased fatty acid re-esterification by PEPCK overexpression in adipose tissue leads to obesity without insulin resistance. *Diabetes* 51:624–30
145. Koliwad SK, Streeper RS, Monetti M, Cornelissen I, Chan L, et al. 2010. DGAT1-dependent triacylglycerol storage by macrophages protects mice from diet-induced insulin resistance and inflammation. *J. Clin. Investig.* 120:756–67
146. Nagle CA, Klett EL, Coleman RA. 2009. Hepatic triacylglycerol accumulation and insulin resistance. *J. Lipid Res.* 50(Suppl.):S74–79
147. Accad M, Smith SJ, Newland DL, Sanan DA, King LE Jr, et al. 2000. Massive xanthomatosis and altered composition of atherosclerotic lesions in hyperlipidemic mice lacking acyl CoA:cholesterol acyltransferase 1. *J. Clin. Invest* 105:711–19
148. Chiu H-C, Kovacs A, Ford DA, Hsu F-F, Garcia R, et al. 2001. A novel mouse model of lipotoxic cardiomyopathy. *J. Clin. Investig.* 107:813–22
149. Liu L, Shi X, Bharadwaj KG, Ikeda S, Yamashita H, et al. 2009. DGAT1 expression increases heart triglyceride content but ameliorates lipotoxicity. *J. Biol. Chem.* 284:36312–23
150. Haemmerle G, Lass A, Zimmermann R, Gorkiewicz G, Meyer C, et al. 2006. Defective lipolysis and altered energy metabolism in mice lacking adipose triglyceride lipase. *Science* 312:734–37
151. Garg A, Agarwal AK. 2009. Lipodystrophies: disorders of adipose tissue biology. *Biochim. Biophys. Acta* 1791:507–13
152. McManaman JL, Russell TD, Schaack J, Orlicky DJ, Robenek H. 2007. Molecular determinants of milk lipid secretion. *J. Mammary Gland Biol. Neoplasia* 12:259–68
153. Vorbach C, Scriven A, Capecchi MR. 2002. The housekeeping gene xanthine oxidoreductase is necessary for milk fat droplet enveloping and secretion: gene sharing in the lactating mammary gland. *Genes Dev.* 16:3223–35

154. Ogg SL, Weldon AK, Dobbie L, Smith AJ, Mather IH. 2004. Expression of butyrophilin (Btn1a1) in lactating mammary gland is essential for the regulated secretion of milk-lipid droplets. *Proc. Natl. Acad. Sci. USA* 101:10084–89
155. Russell TD, Palmer CA, Orlicky DJ, Bales ES, Chang BH, et al. 2008. Mammary glands of adipophilin-null mice produce an amino-terminally truncated form of adipophilin that mediates milk lipid droplet formation and secretion. *J. Lipid Res.* 49:206–16
156. Smith SJ, Cases S, Jensen DR, Chen HC, Sande E, et al. 2000. Obesity resistance and multiple mechanisms of triglyceride synthesis in mice lacking DGAT. *Nat. Genet.* 25:87–90
157. Cases S, Zhou P, Schillingford J, Wiseman B, Fish J, et al. 2004. Development of the mammary gland requires DGAT1 expression in stromal and epithelial tissues. *Development* 131:3047–55
158. Saha S, Enugutti B, Rajakumari S, Rajasekharan R. 2006. Cytosolic triacylglycerol biosynthetic pathway in oilseeds. Molecular cloning and expression of peanut cytosolic diacylglycerol acyltransferase. *Plant Physiol.* 141:1533–43
159. Zheng P, Allen WB, Roesler K, Williams ME, Zhang S, et al. 2008. A phenylalanine in DGAT is a key determinant of oil content and composition in maize. *Nat. Genet.* 40:367–72
160. Baud S, Lepiniec L. 2010. Physiological and developmental regulation of seed oil production. *Prog. Lipid Res.* 49:235–49
161. Andrianov V, Borisjuk N, Pogrebnyak N, Brinker A, Dixon J, et al. 2010. Tobacco as a production platform for biofuel: Overexpression of *Arabidopsis DGAT* and *LEC2* genes increases accumulation and shifts the composition of lipids in green biomass. *Plant Biotechnol. J.* 8:277–87
162. Eastmond PJ. 2006. *SUGAR-DEPENDENT1* encodes a patatin domain triacylglycerol lipase that initiates storage oil breakdown in germinating *Arabidopsis* seeds. *Plant Cell* 18:665–75
163. Graham IA. 2008. Seed storage oil mobilization. *Annu. Rev. Plant. Biol.* 59:115–42
164. Hayashi Y, Hayashi M, Hayashi H, Hara-Nishimura I, Nishimura M. 2001. Direct interaction between glyoxysomes and lipid bodies in cotyledons of the *Arabidopsis thaliana* ped1 mutant. *Protoplasma* 218:83–94
165. Schmidt MA, Herman EM. 2008. Suppression of soybean oleosin produces micro-oil bodies that aggregate into oil body/ER complexes. *Mol. Plant* 1:910–24
166. Toomre D, Bewersdorf J. 2010. A new wave of cellular imaging. *Annu. Rev. Cell Dev. Biol.* 26:285–314

Adipogenesis: From Stem Cell to Adipocyte

Qi Qun Tang[1,2] and M. Daniel Lane[1]

[1]Department of Biological Chemistry, The Johns Hopkins University School of Medicine, Baltimore, Maryland 21205; email: dlane@jhmi.edu

[2]Key Laboratory of Molecular Medicine, Ministry of Education, Department of Biochemistry and Molecular Biology, Fudan University Shanghai Medical College, Shanghai 200032, People's Republic of China; email: qqtang@shmu.edu.cn

Keywords

lineage commitment, BMP, C/EBP, PPARγ, Wnt

Abstract

Excessive caloric intake without a rise in energy expenditure promotes adipocyte hyperplasia and adiposity. The rise in adipocyte number is triggered by signaling factors that induce conversion of mesenchymal stem cells (MSCs) to preadipocytes that differentiate into adipocytes. MSCs, which are recruited from the vascular stroma of adipose tissue, provide an unlimited supply of adipocyte precursors. Members of the BMP and Wnt families are key mediators of stem cell commitment to produce preadipocytes. Following commitment, exposure of growth-arrested preadipocytes to differentiation inducers [insulin-like growth factor 1 (IGF1), glucocorticoid, and cyclic AMP (cAMP)] triggers DNA replication and reentry into the cell cycle (mitotic clonal expansion). Mitotic clonal expansion involves a transcription factor cascade, followed by the expression of adipocyte genes. Critical to these events are phosphorylations of the transcription factor CCATT enhancer-binding protein β (C/EBPβ) by MAP kinase and GSK3β to produce a conformational change that gives rise to DNA-binding activity. "Activated" C/EBPβ then triggers transcription of peroxisome proliferator–activated receptor-γ (PPARγ) and C/EBPα, which in turn coordinately activate genes whose expression produces the adipocyte phenotype.

Contents

- INTRODUCTION 716
- THE ADIPOCYTE 716
  - Origin of Adipocytes 717
  - Pluripotent Stem Cell Lines 717
  - Preadipocyte Cell Lines 717
  - Mesenchymal Stem Cells 718
- COMMITMENT: PLURIPOTENT STEM CELL TO PREADIPOCYTE 718
  - Bone Morphogenetic Protein Signaling 718
  - Wnt Signaling 721
  - Hedgehog Signaling 722
  - Retinoblastoma Protein Signaling ... 722
- DIFFERENTIATION: PREADIPOCYTE TO ADIPOCYTE 722
  - Induction of Differentiation 723
  - Mitotic Clonal Expansion 723
  - Transcription Factor Cascade 724
  - Histones and Chromatin Remodeling 729
  - Role of microRNAs in Adipogenesis 729

Mesenchymal stem cell (MSC) lineage commitment: MSCs are pluripotent stem cells, which have the capacity to undergo commitment and differentiation into adipocytes, chondrocytes, myocytes, and osteocytes

INTRODUCTION

Animals possess highly integrated systems to regulate energy storage and expenditure. These systems have evolved to promote energy storage during periods of food surplus and mobilization of these stores when food is scarce (1–4). Such systems are regulated both at the cellular level and in the whole organism by the coordinated actions of circulating hormones and efferent neural signals from the central nervous system both to higher brain centers and to peripheral tissues, including the liver, muscle, and adipose tissue. When food is abundant, carbohydrates are stored as small reserves of glycogen in the liver and as a much larger reserve as fat—primarily in adipocytes. When food is limited, these stores are mobilized to meet energy needs (1, 2). Despite these control mechanisms to survive feast-or-famine situations, lifestyles in developed societies have led to excessive consumption of energy-rich foods and sedentary behavior. As a consequence obesity and its accompanying pathological consequences, most notably insulin resistance, type 2 diabetes, and heart disease, have become serious medical problems (5–8).

Excessive caloric intake relative to expenditure produces a metabolic state that promotes hyperplasia and hypertrophy of adipocytes (5). Hyperplasia involves stem cell recruitment to the adipocyte lineage and a consequent increase in adipocyte number. Here, we review both the mechanisms by which pluripotent mesenchymal stem cells (MSCs) undergo commitment to become preadipocytes and the program by which preadipocytes differentiate into adipocytes. Together, these processes are responsible for the increase in adiposity produced by excessive energy intake, which is accompanied by low energy expenditure. The reader is referred to several recent reviews that deal with related aspects of this topic (2, 6–9).

THE ADIPOCYTE

The mature adipocyte contains a single large fat droplet surrounded by a thin rim of cytoplasm that lies between the droplet and the plasma membrane. When triggered by lipolytic hormones, cytoplasmic hormone-sensitive lipase (10) and adipocyte-triglyceride lipase (11, 12) translocate to the surface of the fat droplet, where lipolysis of triglyceride occurs (13). Also bound to the surface of the fat droplet are accessory proteins, such as perilipin (14), that facilitate this process. Thus, fatty acids derived from triglyceride are released into the blood stream to supply peripheral tissues, especially those of the skeletal and heart muscle and the liver, with an energy-rich fuel. During periods of excess caloric intake, lipogenic enzymes, localized in the cytoplasm and endoplasmic reticulum (ER), synthesize triglyceride, which is incorporated into the fat droplet. As discussed in the section Differentiation: Preadipocyte to Adipocyte, see below, the genes encoding these and other adipocyte proteins are coordinately expressed during preadipocyte differentiation to produce the adipocyte phenotype.

Two technical approaches have been of paramount importance in advancing progress in studies on the metabolism and differentiation with adipocytes and their progenitors. The use of collagenase dissociation to isolate functional adipocytes from adipose tissue (15–17) made possible the biochemical analysis of adipocyte function. Using isolated adipocytes, it was shown that these cells are exquisitely responsive to hormones, such as insulin, that promote glucose uptake and lipogenesis as well as to adrenergic agents, such as epinephrine and cyclic AMP (cAMP), that promote triglyceride mobilization (5, 15, 17). Studies on gene expression/regulation and differentiation, however, had to await the development of clonal preadipocyte cell lines that could be differentiated into adipocytes in cell culture (18–20). Established cell lines, e.g., the 3T3-L1 preadipocyte line, that can be cultured indefinitely have served as faithful model systems for characterizing the differentiation program.

Recent studies have revealed that adipocytes also have endocrine functions and secrete hormones and cytokines that play key roles in global energy metabolism. Certain of these hormones act locally as paracrine factors, and others, such as leptin (21–25) and adiponectin (26–28), have long-range effects and act on the feeding centers of the central nervous system/brain, notably the hypothalamus. Leptin, which is expressed and secreted by adipocytes in proportion to adipose tissue mass, is anorectic and acts to limit energy storage when adipose tissue reserves have been filled (29). Leptin interacts with specific receptors in the hypothalamus (30) to reduce food intake. In the obese state, however, resistance to leptin occurs, limiting its effectiveness. Adipose tissue is also under the control of the central nervous system because it is innervated by neurons of the sympathetic nervous system that secrete adrenergic hormones (epinephrine/norepinephrine) that promote fat mobilization (17).

Origin of Adipocytes

Adipocytes are derived from pluripotent MSCs that have the capacity to develop into several cell types, i.e., adipocytes, myocytes, chondrocytes, and osteocytes (7, 31, 32). These stem cells reside in the vascular stroma of adipose tissue as well as in the bone marrow, and when appropriately stimulated undergo a multistep process of commitment in which the progenitor cells become restricted to the adipocyte lineage. Recruitment to this lineage gives rise to preadipocytes, which, when induced, undergo multiple rounds of mitosis (mitotic clonal expansion) and then differentiate into adipocytes. Model cell culture systems (discussed below) have been indispensable in identifying/characterizing the steps in the commitment and differentiation programs.

Pluripotent Stem Cell Lines

Several cell lines faithfully mimic the functional characteristics of the mesenchymal pluripotent cell type in vivo and can be induced to undergo commitment to the adipose, muscle, cartilage, and bone lineages. The best characterized of these pluripotent lines is the CH310T1/2 line established in 1973 in the Heidelberger laboratory at the University of Wisconsin (33). In addition, embryonic stem cell knockdowns and mouse knockouts have proven useful (32, 34). The adipocyte commitment process can be initiated by factors such as BMP4 or BMP2 (bone morphogenetic proteins) (35) or factors in the downstream signaling pathway (36).

Preadipocyte Cell Lines

Preadipocyte lines that represent the next step (differentiation) in adipocyte development were first established in Green's laboratory (18, 19, 37). Notable among these is the 3T3-L1 preadipocyte line, which has become the "gold standard" for investigating preadipocyte differentiation. This line (along with the 3T3-F442A line) faithfully recapitulates the steps in adipocyte differentiation. Induction of

Mitotic clonal expansion: the process by which G_1-growth-arrested preadipocytes, induced to differentiate, reenter the cell cycle and undergo approximately two rounds of division, a prerequisite for differentiation

Adipogenesis: the processes by which mesenchymal stem cells commit to the adipose lineage and differentiate into adipocytes

differentiation is triggered with a "cocktail" of agents to initiate the cascade of synchronous steps in the differentiation process.

Mesenchymal Stem Cells

Pluripotent MSCs derived from the embryonic mesoderm (7, 31) have the capacity to differentiate into adipocytes and have been used to verify many findings made with the established 3T3-L1 preadipocyte line (38). It was definitively shown that a subpopulation of MSCs is pluripotent and capable of committing to the adipocyte lineage (36, 38).

COMMITMENT: PLURIPOTENT STEM CELL TO PREADIPOCYTE

The vascular stroma of adipose tissue possesses a resident population of pluripotent MSCs that has the potential to undergo commitment and differentiation into adipocytes, chondrocytes, myocytes, and osteocytes (7, 31, 35, 39). Use of these cells for genetic studies is limited, however, because of their short lifetime in culture. Pluripotent C3H10T1/2 cells in culture have served as a faithful MSC model for long-term genetic studies of the adipocyte developmental program. When appropriately induced, C3H10T1/2 stem cells can be prompted to commit and differentiate into cells of the adipocyte, myocyte, chondrocyte or osteocyte lineages.

Recruitment to the adipocyte lineage in vivo is prompted by excessive energy intake and elevated glucose uptake (5) over an extended time period. This metabolic state appears to generate a signal (or signals), yet to be identified, that induces MSCs to enter the commitment pathway leading to hyperplasia and the preadipocyte phenotype. Several factors have been identified that commit or inhibit the conversion of pluripotent stem cells to the adipocyte lineage. These include BMP family members BMP4 and BMP2 (6, 35), Wnt (40, 41), and Hh (hedgehog) (42–44). BMP4 and BMP2 have an activating role, whereas Hh signaling has an inhibitory role, and Wnt appears to have both an activating role in commitment (41) and an inhibitory role in adipocyte differentiation, see below (45).

Mounting evidence indicates that lineage determination is regulated by a network of extracellular signaling factors that ultimately impinge on the promoters of lineage-specific transcription factors. It is the balance of these signaling molecules that determines the developmental pathway, often simultaneously promoting one pathway while inhibiting another. For example, Wnt10b promotes osteogenesis and possibly myogenesis, and inhibits adipogenesis (46); BMP4 promotes adipogenesis while inhibiting myogenesis (45). Conversely, peroxisome proliferator–activated receptor-γ (PPARγ) inhibits chondrogenesis and stimulates adipogenesis, whereas Msx2 stimulates osteogenesis while inhibiting adipogenesis (47, 48) by inhibiting the transcriptional activity of PPARγ.

Bone Morphogenetic Protein Signaling

BMP4 and BMP2 have been implicated in the commitment of pluripotent stem cells to the adipocyte lineage (32, 35, 36, 41, 49–52). Thus, exposure of dividing C3H10T1/2 stem cells to either BMP4 or BMP2 gives rise to preadipocyte-like cells which, when treated at growth arrest with differentiation inducers, enter the adipose development pathway, express adipocyte markers, and acquire the adipocyte phenotype (6, 32, 35).

The role for BMP4 in the commitment process has been validated using another approach, clonal selection after blocking DNA methylation (6). Thus, exposure of proliferating C3H10T1/2 stem cells to 5-azacytidine, an inhibitor of DNA methylation, generated clonal subpopulations of cells, e.g., the A33 line, that converted into adipocytes when exposed to differentiation inducers in the absence of exogenous BMP4 (6). Remarkably, A33 cells express and secrete BMP4 in the same proliferation time window at which exogenous BMP4 must be added to induce adipocyte commitment of "naive" C3H10T1/2 cells.

Moreover, exposure of A33 cells to the naturally occurring BMP4-binding antagonist, noggin, during this critical time window blocked conversion/differentiation into adipocytes (**Figure 1**). The role of BMP4 in adipocyte lineage commitment is further supported by the expression in proliferating C3H10T1/2 stem cells and A33 preadipocytes of genes and proteins that are known to be involved in the BMP signaling pathway (6, 35, 36). These include BMP4; the BMP receptors, BMPr2 and BMPr1a; and Smad-1, -5, -8. BMPs are known to signal through two receptor types, BMPr1 and BMPr2, which form cell-surface complexes with serine/threonine kinase activity (53, 54). Binding of BMP to the BMPr1:BMPr2 complex induces phosphorylation and, thus, activation of the BMPr1 kinase. The BMP receptor phosphorylates Smad-1,-5,-8, which forms a complex with Smad4 that translocates into the nucleus and regulates gene expression (32). Furthermore, overexpression of a constitutively active (CA) BMP receptor, CA-BMPr1A or CA-BMPr1B, induces commitment in the absence of BMP2 or BMP4, whereas overexpression of a dominant-negative receptor, dominant-negative-BMPr1A, suppresses commitment induced by BMP. Also, knockdown of the expression of Smad4 (a coregulator in the BMP/Smad signaling pathway) with RNAi disrupts commitment by the BMPs (see **Figure 2**).

BMP promoter switch during commitment.
The mouse BMP4 gene is known to have two promoters that are differentially regulated during development (55, 56). Analysis of RNA early in the commitment program revealed that promoter 1 is used to drive expression of BMP4 in A33 cells but not in C3H10T1/2 cells (6). On day seven when BMP4 synthesis decreases, use of promoter 1 and expression of BMP4 ceases. In contrast, mRNA from uncommitted C3H10T1/2 cells showed that promoter 2 was used exclusively to drive expression of the BMP4 gene (6). These findings suggest that commitment to the adipocyte lineage induced by 5-azacytidine altered the methylation

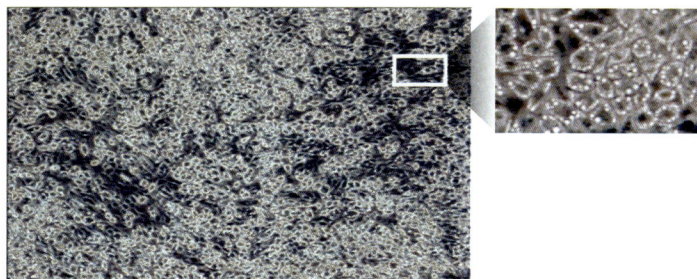

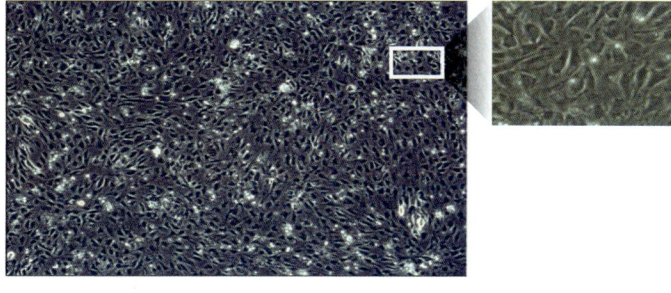

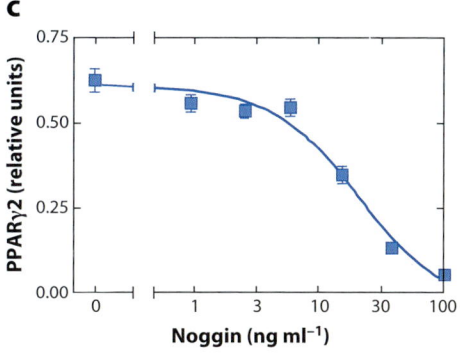

Figure 1

Effect of the BMP4 antagonist, noggin, on the acquisition of adipocyte characteristics by A33 cells [which are cloned CH310T1/2 cell derivatives with preadipocyte characteristics; see the text (33)]. One day after plating, A33 cells were exposed or not to recombinant mouse noggin. Treatment with noggin was continued until growth arrest at confluence when cells were treated with differentiation inducers. Photomicrographs were taken after four days, and RNA was harvested at six days after induction. Panel (*a*) shows control A33 cells, and panel (*b*) shows A33 cells treated with noggin. (*c*) Effect of noggin on the expression of the adipocyte marker PPARγ2 mRNA. Figure is from Reference 6.

status of the BMP4 gene, rendering promoter 1 more accessible for transcriptional activation, and indicate that a region of the BMP4 locus, rich in CpG islands, has fewer methylated cytosines in genomic DNA from A33 cells than

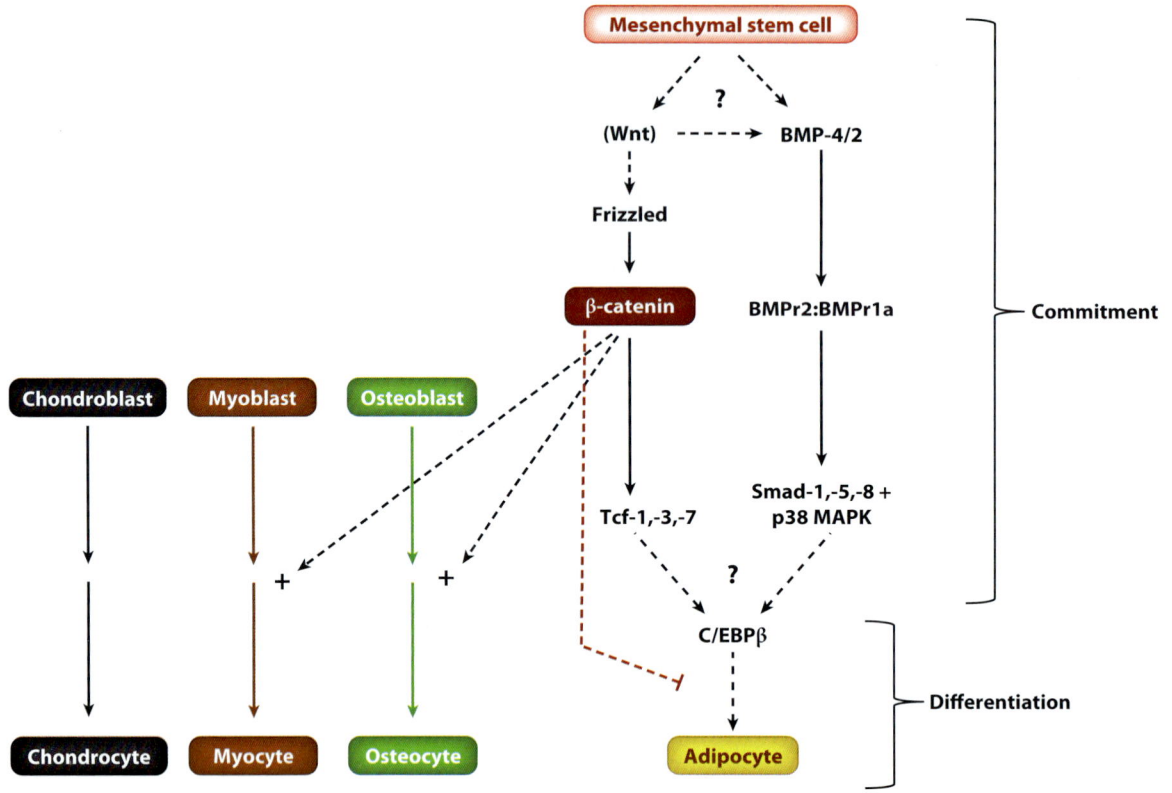

Figure 2

Proposed schema of events for the commitment of mesenchymal stem cells (MSCs) to the adipocyte lineage. BMP refers to bone morphogenetic protein. Frizzled refers to the Wnt cell-surface receptor. The Wnt and BMP pathways are shown as both linear and parallel events since both pathways may be involved. Wnt appears to function both in an activating and an inhibitory capacity in MSC commitment (10, 42, 50) and differentiation (41). Dashed lines indicate some uncertainty.

from 10T1/2 cells. The BMP4 gene locus in naive C3H10T1/2 stem cells possesses a highly methylated region of CpG islands, whereas this region in committed 10T1/2 cells possesses fewer methylated cytosines (6). This difference correlates with the switch in BMP promoter usage (promoter 2 to promoter 1) during commitment that is induced by the methylation inhibitor, suggesting that a change in methylation status renders promoter 1 more accessible for transcriptional activation.

Downstream targets of BMP signaling. Proteomic analysis revealed that three cytoskeleton-associated proteins [i.e., lysyl oxidase (Lox), translationally controlled tumor protein 1 (Tpt1), and αB crystallin] are downstream target genes in the BMP signaling pathway and play important roles during adipocyte lineage commitment (52). Eight proteins were found to be upregulated by BMP2, and 27 proteins were upregulated by BMP4. Five unique proteins were upregulated ≥10-fold by both BMPs, including three cytoskeleton-associated proteins (i.e., Lox, Tpt1, and αB-crystallin). Commitment was completely blocked by knockdown of Lox, whereas it was partially inhibited by knockdown of Tpt1 and αB-crystallin expression. Dramatic changes in cell shape normally occur during commitment. Knockdown of these cytoskeleton-associated proteins prevented these cell shape changes and restored F-actin organization into stress fibers and inhibited commitment to the adipocyte lineage. These differentially expressed proteins may determine the ability of MSCs to commit to the adipocyte lineage via cell shape regulation.

Wnt Signaling

The Wnts comprise a family of secreted signaling glycoproteins whose effects are mediated through the frizzled receptor and low-density lipoprotein-related protein 5/6 coreceptor (57). The Wnt proteins can act via the more prominent "canonical" Wnt signaling pathway (58–60) or a "noncanonical" Wnt signaling pathway (61). A linkage between Wnt signaling and adipogenesis was first recognized through the finding that expression of Wnt10b decreased dramatically during adipocyte differentiation (40). Consistent with this finding, forced expression of Wnt10b prevented adipocyte differentiation by blocking the expression of the key adipogenic transcription factors, PPARγ and C/EBPα. Surprisingly, the Wnts act at two points in the adipose development program (**Figure 2**), both early in the program as an activator of lineage commitment (41) and late in the program as an inhibitor of adipocyte differentiation (40, 45), perhaps through the actions of different Wnt proteins.

Wnt activates lineage commitment of MSCs.
The canonical Wnt pathway functions early in the lineage commitment process of the canonical pathway. In this pathway cytosolic β-catenin is embedded in a "destruction complex" containing adenomatous polyposis coli, axin, and glycogen synthase kinase-3β (GSK-3β) (60). In the absence of Wnt stimulation, GSK-3β phosphorylates β-catenin, priming it for ubiquitination and proteasomal degradation (60). Conversely, activation of Wnt signaling through binding of Wnts to their cell-surface receptor, frizzled, and the associated low-density lipoprotein receptor-related coactivator 5/6 promotes dissociation of the destruction complex, thereby allowing β-catenin to accumulate and translocate to the nucleus (59, 62). Thus, during Wnt signaling, β-catenin accumulates in the nucleus, where it activates the transcription factors lymphoid enhancer factor and/or T-cell factor (**Figure 2**), triggering the transcription of downstream genes (including c-myc and Cyclin D1) (39).

Microarray studies combined with quantitative RT-PCR analyses identified several genes of the Wnt signaling pathway that are differentially expressed by A33 cells (preadipocytes derived from C3H10T1/2 stem cells, see above) (41). Of particular interest are the newly described R-spondins-2 and -3, which activate the canonical Wnt signaling pathway, and are dramatically upregulated in proliferating A33 preadipocytes compared with their progenitor C3H10T1/2 stem cells. Likewise, Lef1 and Tcf, downstream transcription factors of the Wnt signaling pathway, are expressed differently in A3 preadipocytes relative to their expression in C3H10T1/2 stem cells (41). These events occur concurrently with the accumulation of β-catenin in the nuclei of proliferating A33 cells. Taken together, these findings implicate Wnt signaling as an early event in stem cell commitment to the adipose lineage.

Wnt inhibits terminal differentiation.
Late in the adipogenic program, the canonical Wnt signaling pathway regulates the balance among myogenic, osteoblastogenic, and adipogenetic fates (40, 46, 57) and, by doing so, decreases adipogenesis. Myoblasts in cell culture retain their plasticity in developmental potential because they can be induced to undergo myogenic, adipogenic, or osteoblastogenic differentiation (57). β-catenin-dependent signaling has been reported to promote both myogenesis (40) and osteogenesis (46) while inhibiting the differentiation of preadipocytes into adipocytes (41, 63). Consistent with the inhibitory effect of Wnt on adipogenesis, myoblasts isolated from Wnt10b-null mice exhibit increased adipogenic potential (57). In addition, activation of the Wnt signaling pathway enhances myogenesis and inhibits adipogenesis in cultured MSCs (40, 64).

The Wnt signaling pathway is more active in proliferating A33 cells (a clonal line with preadipocyte characteristics) than in its C3H10T1/2 progenitors, perhaps because

Wnt signaling serves to increase the number of preadipocytes during commitment. It has been speculated that Wnt signaling serves to increase the number of preadipocytes during commitment and mitotic clonal expansion (41) but that this signal must ultimately be terminated as preadipocytes/A33 cells enter growth arrest, before the newly recruited cells can undergo terminal differentiation into adipocytes (see the Mitotic Clonal Expansion section below).

Hedgehog Signaling

Three vertebrate Hh ligands have been identified, Sonic (Shh), Indian (Ihh), and Desert (Dhh), which initiate a signaling cascade mediated by Patched (Ptch-1 and Ptch-2) receptors (65, 66). In the presence of an Hh ligand, the membrane-spanning protein Smoothened (Smo), a homolog of G protein–coupled receptors, is activated, and the signal is transmitted via phosphorylation and nuclear localization of GliA (65, 66).

Hh signaling has an inhibitory effect on adipogenesis in murine cells, e.g., C3H10T1/2, KS483, calvaria MSC lines, and mouse adipose-derived stromal cells (67) as visualized by decreased cytoplasmic fat accumulation and the expression of adipocyte marker genes (42–44). In genetically obese (ob/ob) mice, a negative effect on adipogenesis is accompanied by a reduction in the expression of Smo, Gli1, Gli2, and Gli3. Moreover, reduced total white fat mass and epididymal adipocyte cell size were observed in naturally occurring spontaneous mesenchymal dysplasia (mes) adult mice (Ptch1 mes/mes), which carry a deletion of Ptch1, a negative regulator of Hh signaling, at the C-terminal cytoplasmic region (68). Although it is generally agreed that expression of Hh has an inhibitory effect on preadipocyte differentiation, the mechanisms linking Hh signaling and adipogenesis remain poorly defined.

Retinoblastoma Protein Signaling

The retinoblastoma protein (Rb) inhibits the cell cycle by binding to and repressing the transcriptional activity of E2F (69). Upon hyperphosphorylation of Rb ($\Rightarrow$ pRb) by cyclin-dependent kinases, E2F is released and promotes transcriptional activation of genes that encode cell-cycle regulators required for S-phase entry and progression of the cell cycle (70). These events are critical for mitotic clonal expansion (see the Mitotic Clonal Expansion section below), an obligate step in the adipocyte differentiation program (38).

Additionally, pRb has been shown to regulate several transcription factors that are key differentiation inducers (38, 71). Depending on the differentiation factor and its cellular context, pRb can either suppress or promote transcriptional activity of such factors. Thus, pRb binds to Runx2 and potentiates its ability to promote osteogenic differentiation (72). In contrast, Rb acts with E2F to suppress peroxisome proliferator-activated receptor c (PPARγ2-c) subunit, the master activator of adipogenesis (73, 74). Because osteoblasts and adipocytes can both arise from MSCs, these observations suggest that pRb may play a role in the choice between these two lineage fates. And these findings indicate that Rb status plays a key role in establishing fate choice between bone and brown adipose tissue in vivo (75).

DIFFERENTIATION: PREADIPOCYTE TO ADIPOCYTE

The synchronous events undergone by preadipocytes during differentiation into adipocytes are described below. Cell culture studies with established preadipocyte lines have been indispensable for the identification and characterization of the key steps in the differentiation program. A large body of evidence shows that these preadipocyte models faithfully recapitulate differentiation of mouse embryonic fibroblasts (MEFs) in cell culture (34, 38, 76). Likewise, the authenticity of model systems has been validated in vivo by demonstrating that these preadipocyte lines give rise to normal adipose tissue when implanted subcutaneously into athymic mice without exogenous inducers (40, 77, 78).

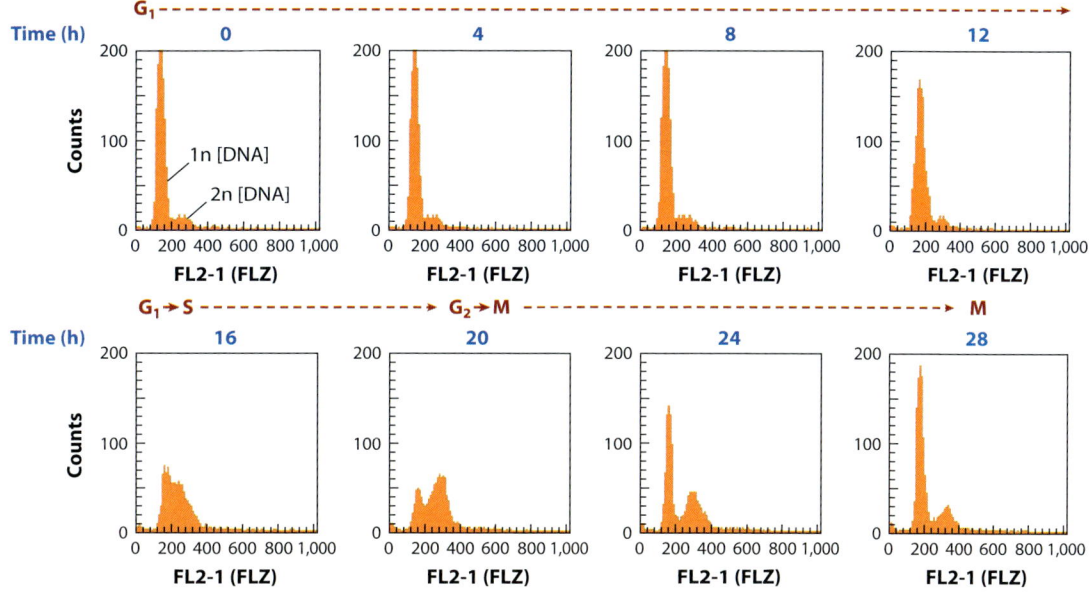

Figure 3

Synchrony of DNA replication during mitotic clonal expansion of the adipocyte differentiation program. After staining with propidium iodide, DNA content was determined by FACS analysis and expressed (y axis) in arbitrary units. Time (h) refers to the elapsed time (in hours) after induction of differentiation of 3T3-L1 preadipocytes with the inducer cocktail. G_1 refers to the growth-arrested phase of the cell cycle. S, M, and G_2 refer to the subsequent phases of the cell cycle. The plots were constructed using data from Reference 38. Abbreviations: 1n and 2n, one- to twofold; FLZ, fluorescence intensity.

Induction of Differentiation

Protocols have been developed to induce differentiation and track the synchronous progression of cells through the program. Preadipocytes, e.g., 3T3-L1 preadipocytes, or MEFs are grown to growth arrest, i.e., at the G_1 phase of the cell cycle. At this point, differentiation is initiated with a cocktail of inducers, including a high level of insulin[1] (or low level of IGF1) (79) and dexamethasone, as well as an agent to elevate cellular cAMP in fetal calf serum-containing medium (80). These inducers activate the IGF1-, glucocorticoid-, and cAMP-signaling pathways, respectively. Induction initiates a series of events that regulate staging of the differentiation program. Following a delay of ~16–20 h after induction, preadipocytes synchronously reenter the cell cycle (**Figure 3**) (39, 80) and undergo several rounds of mitosis, referred to as mitotic clonal expansion (see below). The cells then exit the cell cycle, lose their fibroblastic morphology, accumulate cytoplasmic triglyceride, and acquire the appearance and metabolic features of adipocytes (18, 80). Triglyceride accumulation is closely correlated with an increased rate of de novo lipogenesis and a coordinate rise in expression of the enzymes of fatty acid and triacylglycerol biosynthesis (79–81). Likewise, numerous regulatory proteins, characteristic of adipocytes in situ, are coordinately expressed, including insulin receptors (82, 83), the insulin-responsive glucose transporter GLUT4 (84), leptin (85, 86), and others (79, 86, 87).

Mitotic Clonal Expansion

Following induction of G_1-phase growth-arrested cells, preadipocytes reenter the cell

---

[1] It has been established that the IGF1, rather than insulin, is the true inducer ligand. IGF1 is active as an inducer at much lower concentrations than insulin binding; IGF1 binds ~100-fold more tightly than insulin to the IGF receptor (79).

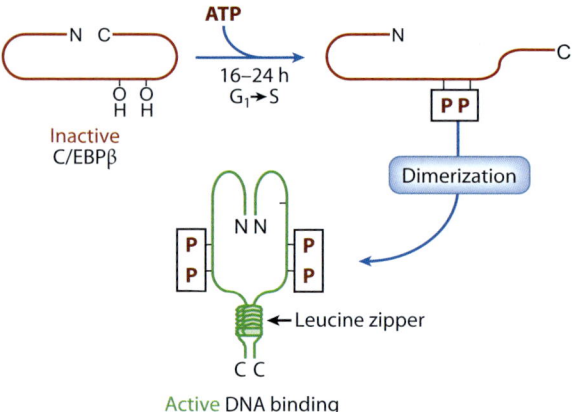

Figure 4

Schema for the dual phosphorylation-induced conformational changes in C/EBPβ, which activates DNA binding. Dual phosphorylation (indicated by P) of C/EBPβ occurs on Thr188 by MAPK and on Ser184 or Thr179 by GSK3β; this induces a conformational change (or changes) that promotes dimerization and gives rise to binding activity. $G_1 \rightarrow S$ refers to the mitotic clonal expansion phase of the cell cycle 16–24 h after the induction of differentiation The data is from References 93 and 114. Abbreviation: C/EBPβ, CCAAT/enhancer-binding protein.

cycle and undergo about two rounds of division, referred to as mitotic clonal expansion. Mitotic clonal expansion is a required step in the adipocyte differentiation program. Thus, blocking DNA replication by various means, e.g., by inhibitors of DNA polymerase or by blocking progression of the cell cycle (38, 88–91), prevents differentiation. For example, the constitutive overexpression of the cell-cycle inhibitor p27 prevents cells from entering the S-phase of the cell cycle and thereby disrupts all subsequent steps of differentiation (88).

Growth arrest. Immediately (≤5 min) after induction of differentiation, cyclic AMP response element-binding protein (CREB) becomes phosphorylated and activates the expression of C/EBPβ (34, 90). However, at this point, C/EBPβ lacks DNA-binding activity.

G_1- to S-phase transition. At 14–16 h after induction, C/EBPβ acquires DNA-binding activity as the preadipocytes reenter the cell cycle (**Figure 3**). Between 16–20 h after induction, the cell-cycle markers of S-phase entry are synchronously expressed (or inactivated) concomitant with the initiation of DNA replication (38, 92). Acquisition of DNA-binding activity is delayed for until ∼16 h when C/EBPβ undergoes phosphorylation by GSK3β (see the CCAAT/enhancer-binding protein-β and -δ section below and **Figure 4**) (93) coincident with reentry into the cell cycle at the G_1/S boundary (**Figure 5**). These events are closely correlated with the expression of histone H4 (94), the coordinated expression of cell-cycle proteins (39), and DNA replication (92). Expression of histone H4 increases dramatically 12–14 h after induction, just prior to entry into the S phase, concurrent with the gain of function by C/EBPβ. It is noteworthy that the histone H4 gene (hist4h4) promoter possesses a functional C/EBP-binding site (94). Moreover, knockdown of C/EBPβ with siRNA prevents the induction of histone H4 expression and arrests the cells in G_1-growth arrest as indicated by bromodeoxyuridine incorporation and FACS analysis of DNA content (94).

Transcription Factor Cascade

The key features of the transcription factor cascade of the differentiation program are shown below (see also **Figure 5**):

Differentiation inducers ⇒ ⇑ [CREB
⇒ P − CREB] ⇒ ⇓ CHOP10
⇒ ⇑ [C/EBPβ ⇒ P_2 − C/EBPβ]
⇒ ⇑ [C/EBPα/PPAR]γ
⇒ ⇑ SREBP1c ⇒ ⇑ adipocyte genes.

Cyclic AMP response element-binding protein.

Because cAMP itself or forskolin or other agents, which elevate the cAMP l level, can substitute for methylisobutylxanthine in the differentiation inducer cocktail, CREB was considered a likely intermediate in the signaling pathway. Many lines of evidence support this view; these include the following:

1. The cellular target of cAMP, protein kinase A, catalyzes the phosphorylation and activation of CREB (95).

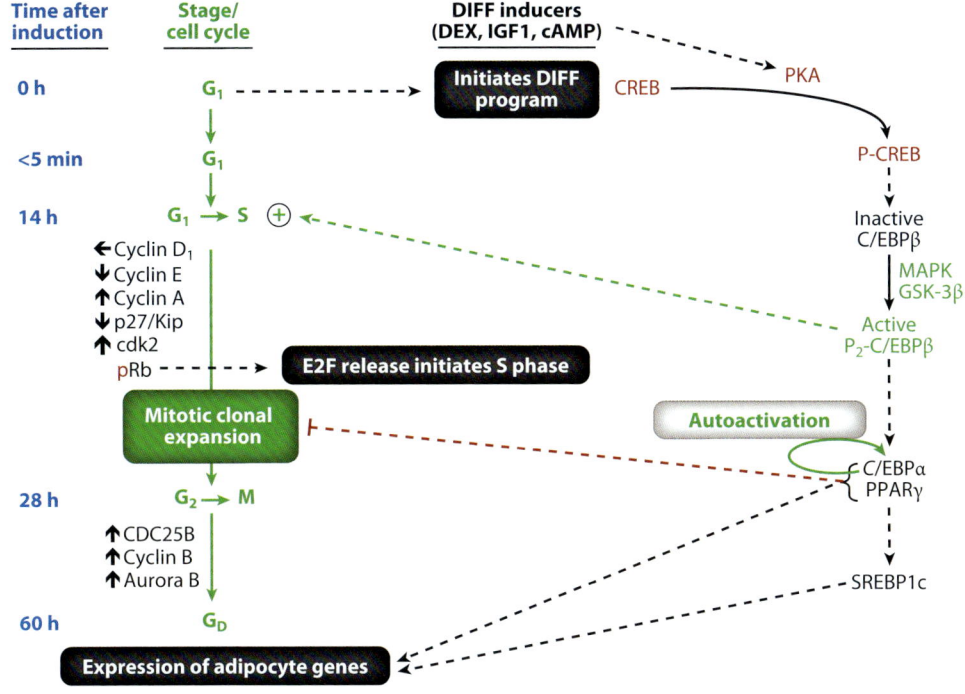

Figure 5

The adipocyte differentiation program. The adipocyte differentiation activation cascade. Abbreviations: CDC25B, cell division cycle 25 homolog B; cdk2, cyclin-dependent kinase 2; C/EBP, CCAAT/enhancer-binding protein; CREB, cyclic AMP regulatory element-binding protein; DEX, dexamethasone; DIFF, differentiation; E2F, eukaryotic transcription factor 2; GSK-3β, glycogen synthase kinase-3β; IGF1, insulin-like growth factor 1; MAPK, mitogen-activated protein kinase; P, phosphoryl group; PPARγ, peroxisome proliferator–activated receptor-γ; pRb, hyperphosphorylated-retinoblastoma binding protein; SREBP1c, sterol regulatory element binding protein 1c; G_1 and S, G_1- and S-phases of the cell cycle, respectively. Dashed lines indicate some uncertainty.

2. Forced expression of CREB in 3T3-L1 preadipocytes promotes differentiation (34).
3. The proximal promoter of the C/EBPβ gene possesses dual *cis*-regulatory elements that contain core CREB-binding sites (34, 96). C/EBPβ promoter-reporter genes with 5′-truncations or site-directed mutations in the TGA regulatory elements revealed that both are required for maximal promoter function.
4. Electromobility shift assay and chromatin immunoprecipitation analyses of wild-type MEFs and 3T3-L1 preadipocytes show that CREB associates with the proximal promoter and that interaction of phospho-CREB, the active form of CREB, with the C/EBPβ gene promoter occurs only after induction of differentiation of 3T3-L1 preadipocytes or MEFs (34, 90).
5. Constitutively active CREB activates C/EBPβ promoter-reporter genes (34, 90), induces expression of endogenous C/EBPβ, and promotes adipogenesis in the absence of hormonal inducers.
6. Conversely, dominant-negative CREB blocks promoter-reporter activity, as well as the expression of C/EBPβ and adipogenesis (90).
7. Subjecting wild-type MEFs to the standard differentiation protocol induces differentiation into adipocytes, whereas

CREB$^{-/-}$ MEFs treated similarly exhibit reduced expression of C/EBPβ and differentiation (90).

Consistent with the role of CREB in the differentiation process, C/EBPβ expression and accumulation of cytoplasmic triacylglycerol are markedly reduced in CREB$^{-/-}$ MEFs. Together these findings show that CREB functions early in the adipocyte differentiation program by transcriptionally activating the C/EBPβ gene.

The CCAAT/enhancer-binding family. Although C/EBPα was the first C/EBP family member to be cloned and to have its function shown in adipogenesis, three other isoforms have also been implicated, notably C/EBPα (97–99) C/EBPβ (100–105); C/EBPδ (104, 105); and CHOP10 (106). Three of the isoforms contain a DNA-binding domain (C/EBPα, C/EBPβ, and C/EBPδ) and an adjacent C-terminal leucine zipper dimerization domain that allows formation of homo- or heterodimers with other C/EBP family members; dimerization is required for DNA binding (107). CHOP10 has a short nonfunctional DNA-binding domain and b-ZIP type C-terminal leucine zipper dimerization domain and can heterodimerize with the other isoforms to produce dominant-negative C/EBP dimers (106). Both C/EBPα and C/EBPβ also occur as N-terminally truncated polypeptides whose functions remain obscure (49, 108, 109).

CCAAT/enhancer-binding protein-β and -δ. C/EBPβ and C/EBPδ are rapidly (in <4 h) expressed following induction of differentiation (38). The importance of C/EBPβ is indicated by its ability to promote adipogenesis when overexpressed in 3T3-L1 preadipocytes or in NIH-3T3 cells in the absence of hormone inducers—an activity not shared by C/EBPδ (110). Although rapidly expressed upon induction of differentiation, C/EBPβ is inactive and unable to bind DNA until later in the differentiation program. Acquisition of DNA-binding activity is achieved at ∼16 h after induction, approximately concomitant with entry into S phase and mitotic clonal expansion (see below). Another indicator of multiple C/EBP consensus-binding sites is the centromeric satellite DNA (38, 111). The role of this binding to centromeres has not been established.

Although C/EBPβ is required for adipocyte differentiation in cell culture (38, 93, 112, 113), a knockout of the C/EBPβ gene in mice has little effect on adipose tissue accumulation (113). Nevertheless, adipose tissue mass in the double knockout [C/EBPβ$^{-/-}$/C/EBPδ$^{-/-}$] is markedly reduced (113). This reduction of adipose tissue mass is because of the decreased cell number, supporting the view that C/EBPβ functions in mitotic clonal expansion. That the double C/EBPβ$^{-/-}$/C/EBPδ$^{-/-}$ knockout produces a significant effect, whereas single knockouts of C/EBPβ$^{-/-}$ or C/EBPδ$^{-/-}$ do not, suggests redundancy of function by members of the C/EBP family (113).

Supporting the dependence of adipogenesis upon C/EBPβ, disruption of the gene prevents mitotic clonal expansion, which is itself required for differentiation. Thus, C/EBPβ$^{-/-}$ MEFs fail to undergo mitotic clonal expansion and do not differentiate into adipocytes, while C/EBPβ$^{-/-}$ MEFs undergo mitotic clonal expansion and differentiate normally (89, 92). Further support for the role of C/EBPβ in clonal expansion is derived from the use of A-C/EBP, which contains a leucine zipper but lacks the functional DNA-binding and transactivation domains and thus acts as a dominant-negative of C/EBP proteins. Forced expression of A-C/EBP disrupts both mitotic clonal expansion and differentiation in 3T3-L1 cells (90). Moreover, the turnover of p27—a requirement for progression from G$_1$-growth arrest to S phase—is blocked by the forced expression of A-C/EBP (88, 114). The mechanism for this effect is that formation of a tight C/EBPβ:A-C/EBP heterodimer prevents the translocation of C/EBPβ into the nucleus, a process that is necessary for mitotic clonal expansion (90). Thus, the primary site of A-C/EBP—the first C/EBP family member to be expressed in the transcription factor cascade—is most likely to block C/EBPβ function. Moreover, when A-C/EBP, under the control of the aP2 adipocyte-specific promoter, is expressed in transgenic mice, the animals become fatless, i.e., devoid of white adipose tissue (115). This finding provides further compelling proof that C/EBPβ is required for adipogenesis in vivo.

Phosphorylation is an important posttranslational modification of C/EBPβ and leads to the acquisition of DNA-binding function as preadipocytes traverse to the G_1-S checkpoint at the onset of mitotic clonal expansion (93). C/EBPβ is phosphorylated sequentially, first by MAP kinase and then much later by GSK-3β (93). Phosphorylation on Thr188 by MAP kinase occurs ∼4 h after induction of differentiation and is required for mitotic clonal expansion, C/EBPβ DNA-binding activity, and terminal differentiation; however, Thr188 phosphorylation is insufficient on its own to produce DNA-binding activity by C/EBPβ (93). Phosphorylation on Thr179 or Ser184 by GSK3β occurs between 12 and 16 h after induction. Dual phosphorylation of C/EBPβ at two of these sites (Thr188 and Thr179 or Thr188 and Ser184) leads to acquisition of DNA-binding activity. Hence, it appears that a phosphorylation-induced conformational change is involved. On the basis of these and other findings (116), it is thought that, following induction of differentiation, C/EBPβ is first rapidly (in ∼2 h) phosphorylated by MAPK, and then phosphorylated at a second site ∼14 h later by GSK3β (**Figure 4**). A conformational change is induced that renders the leucine zipper of monomeric C/EBPβ accessible for dimerization (117). It appears that dimerization brings the basic regions of the two monomers of C/EBPβ together to create a "scissors-like" DNA-binding pocket just above the coiled-coil leucine zipper as suggested by Vinson et al. (118).

CCAAT/enhancer-binding protein-α. C/EBPα and PPARγ function together as pleiotropic transcriptional activators of the large group of genes that produce the adipocyte phenotype (79, 86, 119). Within their proximal promoters, both the C/EBPα and PPARγ genes possess C/EBP regulatory elements at which C/EBPβ binds to coordinately activate transcription (111, 120–122). Once expressed, C/EBPα is thought to maintain expression of both the C/EBPα and PPARγ genes via transactivation mediated by their respective C/EBP regulatory elements (111, 120–122).

The promoters of many adipocyte genes contain C/EBP and PPAR regulatory elements and are *trans*-activated by C/EBPα and PPARγ (79). Forced expression of C/EBPα or PPARγ in 3T3-L1 preadipocytes induces adipogenesis in the absence of hormonal induction (123–125). Furthermore, blocking expression of C/EBPα with antisense RNA suppressed adipogenesis (126), and knocking out the C/EBPα gene in mice led to decreased lipid accumulation (127, 128). These and other findings indicate the requirement for C/EBPα in the adipocyte differentiation program. Beginning 18–24 h after the induction of differentiation, the C/EBPα and PPARγ genes are transcriptionally activated by C/EBPβ through C/EBP regulatory elements in their proximal promoters.

Once C/EBPα is expressed, its expression is maintained through autoactivation (120). The following question is raised: Why is there such redundancy in the expression of the C/EBP genes during adipogenesis? One possible explanation is that C/EBPα is antimitotic, and thus, its premature expression would prevent preadipocytes from entering mitotic clonal expansion, a required step for subsequent differentiation. Therefore, it appears that C/EBPα must remain repressed until the opportune time window. Several mechanisms cause a delay in the expression of C/EBPα. Binding of AP-2α (129) and Sp1 (130) to the C/EBPα promoter repress promoter activity, and delay acquisition of DNA binding by C/EBPβ and C/EBPδ (see the section titled CCAAT/enhancer-binding protein-β and -δ, above). Both AP-2α and Sp1 are downregulated concurrent with the upregulation of C/EBPβ.

CHOP10. CHOP10 was cloned from a 3T3-L1 adipocyte cDNA library on the basis of its ability to interact with the C/EBPβ C-terminal leucine zipper domain (106). CHOP10 contains proline and glycine residues in the DNA-binding region, which abolishes its ability to bind DNA but not its ability to form heterodimers. Thus, CHOP10 acts as a dominant-negative isoform (106) and when expressed ectopically in 3T3-L1 cells blocks adipogenesis (131). CHOP10 is normally expressed by G_1-growth-arrested preadipocytes, but is downregulated as the expression of C/EBPβ declines. Because heterodimerization of CHOP10 with C/EBPβ prevents it from acquiring DNA-binding activity (132), CHOP10 provides a "fail-safe" mechanism to prevent the acquisition of DNA-binding activity by C/EBPβ until preadipocytes have entered mitotic clonal expansion.

Peroxisome proliferator–activated receptor-γ. PPARγ exists as three isoforms (PPARγ1, PPARγ2, and PPARγ3) that are transcribed from the same gene through alternative splicing and promoter usage (133, 134); PPARγ2 is the primary adipocyte-specific isoform (9). All isoforms possess transactivation, DNA-binding, and dimerization domains (9). To bind at peroxisome proliferator response elements in target genes, PPARs must first form heterodimers with the retinoid X receptor (9). Although naturally occurring ligands for PPARγ have not yet been identified, several potent synthetic thiazolidinediones and prostanoids, e.g., 15-deoxy-Δ12,14 prostaglandin J2 (37, 38), bind with high affinity. It is unlikely, however, that the 15-deoxy-Δ12,14 prostaglandin J2 concentration in vivo is of biological importance (9).

Although PPARγ2 appears to act as a "master" regulator of the adipogenesis program, like C/EBPα, it also participates in other diverse systems, including hepatic lipogenesis (9). The ectopic expression of PPARγ2 in naive preadipocytes and nonadipogenic fibroblasts activates expression of adipocyte genes and differentiation (9, 10, 135). Moreover, dominant-negative mutants (49, 136) and knockouts of the mouse gene block these functions.

As indicated above, C/EBPβ (and C/EBPδ) are rapidly (in <4 h) expressed after induction of differentiation. Following a delay of 16–20 h, C/EBPβ coordinately activates the expression of PPARγ2 and C/EBPα through C/EBP regulatory elements in the proximal promoters of their respective genes (38, 89, 137). Following their expression, PPARγ2 and C/EBPα coordinately transactivate a large group of genes that produce the adipocyte phenotype. Once expressed, PPARγ and C/EBPα positively cross activate each other through their respective C/EBP regulatory elements (136, 138, 139). Presumably, this action perpetuates the adipocyte phenotype in the mature adipocyte.

Sterol regulatory element-binding protein.
The sterol regulatory element-binding proteins (SREBPs) are basic helix-loop-helix-leucine zipper proteins. SREBP2 regulates transcription of the genes of cholesterol metabolism, while SREBP1c and its mouse homolog, ADD1, regulate lipogenesis (140, 141). In the adipocyte differentiation program, the expression of SREBP1c/ADD1 mRNA is activated after the expression of C/EBPα and PPARγ, i.e., at ~20 h after induction of differentiation (142). Current evidence implicates SREBP1c/ADD1 in the following terminal events of adipocyte differentiation. (*a*) SREBP1c/ADD1 is expressed as a membrane-bound precursor bound to SCAP (SREBP-cleavage-activating protein) and tethered in the ER by Insig-2a. (*b*) Upon its release from the ER (stimulated by insulin), SCAP·SREBP1c/ADD1 moves to the Golgi apparatus, where proteolytic cleavage frees its basic helix-loop-helix component for translocation to the nucleus. (*c*) Once in the nucleus, transcription of genes encoding lipogenic enzymes occurs—events that produce adipocyte characteristics.

Recent evidence (143, 144) indicates that insulin regulates the release of SCAP·SREBP1c/ADD1 from the ER by downregulating the level of Insig-2a through increased turnover of its mRNA. Reducing the level of Insig-2a facilitates export of SCAP·SREBP1c/ADD1 from the ER to the Golgi for proteolytic cleavage and subsequent translocation of the basic helix-loop-helix component to the nucleus. As a result, the transcription of lipogenic genes is activated, including those that support fatty acid synthesis, desaturation, and uptake, as well as triacylglycerol synthesis. In a metabolic sense, this is consistent with the recent finding that SREBP1c/ADD1 is phosphorylated by AMP kinase (145), which promotes energy mobilization. Insulin, on the other hand, promotes energy storage by activating lipogenic enzyme expression.

Histones and Chromatin Remodeling

Genome-wide mapping studies have revealed changes in chromatin structure that occur during the differentiation of 3T3-L1 preadipocytes into adipocytes (145, 146). DNase I-hypersensitive site analysis revealed that alteration of chromatin structure occurs early in the differentiation program, coinciding with the cooperative binding of early transcription factors, such as C/EBPβ and C/EBPδ, to regulatory element hot spots. Of particular interest, C/EBPβ hot spots were observed prior to induction of differentiation and chromatin remodeling. Furthermore, a subset of the early remodeled C/EBPβ hot spots persisted throughout differentiation and was later occupied by PPARγ, suggesting that early C/EBP family members may act as initiating factors for subsequent binding of PPARγ. These findings are, however, inconsistent with those obtained in cell culture studies (93). These studies showed that, in the G_1-growth-arrested state prior to induction, C/EBPβ is in its dephosphorylated inactive state and unable to bind DNA (i.e., to consensus C/EBP promoter-binding sites); synchronous entry into S phase occurs. Synchronous entry into S phase occurs only after completion of dual phosphorylation of C/EBPβ and acquisition of DNA-binding activity (**Figure 4**) (93). A possible explanation for this discrepancy is that the preadipocytes in this study were not in the G_1-growth-arrested state (147). Dividing preadipocytes are known to express high levels of active phosphorylated C/EBPβ. Thus, C/EBPβ hot spots, attributed to "uninduced" preadipocytes, may well have been due to cells that had traversed the G_1-S cell-cycle checkpoint. See the sections above (i.e., Mitotic Clonal Expansion and CCAAT/enhancer-binding protein-β and -δ, above). C/EBPβ must be phosphorylated multiple times to bind DNA. Thus, it will be necessary to definitively prove by cell-cycle analysis that all preadipocytes prior to induction were in the G_1-growth-arrested state as illustrated in **Figure 3**.

Role of microRNAs in Adipogenesis

A number of microRNAs have been identified that appear to play a role in adipogenesis. Some of these microRNAs seem to accelerate adipogenesis (148, 149), while others negatively regulate adipogenesis (150, 151). Mammalian homologs of miR-8 promote adipogenesis by inhibiting Wnt signaling (148). Ectopic introduction of let-7 into 3T3-L1 cells inhibited clonal expansion as well as terminal differentiation (151). However, the mechanisms by which these microRNAs act have not been definitively linked to specific aspects of the differentiation program. This should be a fertile area for future research.

SUMMARY POINTS

1. Excessive caloric intake leads to adipocyte hyperplasia and adiposity.
2. Adipocyte hyperplasia is caused by recruitment of pluripotent mesenchymal stem cells (MSCs) from the vascular stroma of adipose tissue.

3. The BMP and Wnt families are key mediators of MSC commitment to produce preadipocytes.

4. Exposure of growth-arrested preadipocytes to differentiation inducers (IGF1, glucocorticoid, and cAMP) triggers DNA replication, reentry of the cell cycle (a process known as mitotic clonal expansion), and a transcription factor cascade, which leads to expression of adipocyte genes.

5. The transcription factor cascade includes the following: Induction → ⇑[CREB → **P-CREB**] → ⇓CHOP10 → ⇑[C/EBPβ → **P₂**-C/EBPβ] → ⇑[C/EBPα / PPAR]γ → ⇑SREBP1c → ⇑adipocyte genes.

FUTURE ISSUES

1. A more detailed understanding is needed of the mechanisms by which lineage commitment and differentiation occur.

2. Are the Wnt and BMP pathways, involved in MSC commitment, linear/sequential or parallel?

3. Characterization is lacking of the conformational changes, induced by phosphorylation of C/EBPβ, that promote dimerization and acquisition of DNA-binding activity. The three-dimensional structures of phospho- and dephospho-C/EBPβ by X-ray crystallography/magnetic resonance imaging are needed.

4. The role of microRNAs in lineage commitment and differentiation needs to be determined.

DISCLOSURE STATEMENT

The authors have nothing to disclose regarding potential bias and are not aware of any affiliations, memberships, funding, or financial holdings that might be perceived as affecting the objectivity of this review.

LITERATURE CITED

1. Cahill GF Jr. 2006. Fuel metabolism in starvation. *Annu. Rev. Nutr.* 26:1–22
2. Otto TC, Lane MD. 2005. Adipose development: from stem cell to adipocyte. *Crit. Rev. Biochem. Mol. Biol.* 40:229–42
3. Wolfgang MJ, Lane MD. 2006. The role of hypothalamic malonyl-CoA in energy homeostasis. *J. Biol. Chem.* 281:37265–69
4. Wahren J, Ekberg K. 2007. Splanchnic regulation of glucose production. *Annu. Rev. Nutr.* 27:329–45
5. Shepherd PR, Gnudi L, Tozzo E, Yang H, Leach F, Kahn BB. 1993. Adipose cell hyperplasia and enhanced glucose disposal in transgenic mice overexpressing GLUT4 selectively in adipose tissue. *J. Biol. Chem.* 268:22243–46
6. Bowers RR, Kim JW, Otto TC, Lane MD. 2006. Stable stem cell commitment to the adipocyte lineage by inhibition of DNA methylation: role of the BMP-4 gene. *Proc. Natl. Acad. Sci. USA* 103:13022–27
7. Covas DT, Panepucci RA, Fontes AM, Silva WA, Orellana MD, et al. 2008. Multipotent mesenchymal stromal cells obtained from diverse human tissues share functional properties and gene-expression profile with CD146+ perivascular cells and fibroblasts. *Exp. Hematol.* 36:642–54

8. Wolfgang MJ, Lane MD. 2006. Control of energy homeostasis: role of enzymes and intermediates of fatty acid metabolism in the central nervous system. *Annu. Rev. Nutr.* 26:23–44
9. Tontonoz P, Spiegelman BM. 2008. Fat and beyond: the diverse biology of PPARgamma. *Annu. Rev. Biochem.* 77:289–312
10. Egan JJ, Greenberg AS, Chang MK, Wek SA, Moos MC Jr, Londos C. 1992. Mechanism of hormone-stimulated lipolysis in adipocytes: translocation of hormone-sensitive lipase to the lipid storage droplet. *Proc. Natl. Acad. Sci. USA* 89:8537–41
11. Ahmadian M, Abbott MJ, Tang T, Hudak CS, Kim Y, et al. 2011. Desnutrin/ATGL is regulated by AMPK and is required for a brown adipose phenotype. *Cell Metab.* 13:739–48
12. Ahmadian M, Wang Y, Sul HS. 2010. Lipolysis in adipocytes. *Int. J. Biochem. Cell Biol.* 42:555–59
13. Londos C, Brasaemle DL, Schultz CJ, Adler-Wailes DC, Levin DM, et al. 1999. On the control of lipolysis in adipocytes. *Ann. N. Y. Acad. Sci.* 892:155–68
14. Londos C, Brasaemle DL, Gruia-Gray J, Servetnick DA, Schultz CJ, et al. 1995. Perilipin: unique proteins associated with intracellular neutral lipid droplets in adipocytes and steroidogenic cells. *Biochem. Soc. Trans.* 23:611–15
15. Rodbell M. 1964. Metabolism of isolated fat cells I. Effects of hormones on glucose metabolism and lipolysis. *J. Biol. Chem.* 239:375–80
16. Rodbell M. 1965. Modulation of lipolysis in adipose tissue by fatty acid concentration in fat cell. *Ann. N. Y. Acad. Sci.* 131:302–14
17. Rodbell M. 1966. The metabolism of isolated fat cells. IV. Regulation of release of protein by lipolytic hormones and insulin. *J. Biol. Chem.* 241:3909–17
18. Green H, Kehinde O. 1974. Sublines of mouse 3T3 cells that accumulate lipid. *Cell* 1:113–16
19. Green H, Kehinde O. 1975. An established preadipose cell line and its differentiation in culture II. Factors affecting the adipose conversion. *Cell* 5:19–27
20. Green H, Kehinde O. 1976. Spontaneous heritable changes leading to increased adipose conversion in 3T3 cells. *Cell* 7:105–13
21. Friedman JM. 2002. The function of leptin in nutrition, weight, and physiology. *Nutr. Rev.* 60:S1–14; discussion S68–84, 85–87
22. Friedman JM. 2009. Leptin at 14 y of age: an ongoing story. *Am. J. Clin. Nutr.* 89:S973–79
23. Batra SK, Castelino-Prabhu S, Wikstrand CJ, Zhu X, Humphrey PA, et al. 1995. Epidermal growth factor ligand-independent, unregulated, cell-transforming potential of a naturally occurring human mutant *EGFRvIII* gene. *Cell Growth Differ.* 6:1251–59
24. Zhang Y, Proenca R, Maffei M, Barone M, Leopold L, Friedman JM. 1994. Positional cloning of the mouse *obese* gene and its human homologue. *Nature* 372:425–32
25. Leroy P, Dessolin S, Villageois P, Moon BC, Friedman JM, et al. 1996. Expression of *ob* gene in adipose cells. *J. Biol. Chem.* 271:2365–68
26. Hug C, Lodish HF. 2005. The role of the adipocyte hormone adiponectin in cardiovascular disease. *Curr. Opin. Pharmacol.* 5:129–34
27. Tsao TS, Murrey HE, Hug C, Lee DH, Lodish HF. 2002. Oligomerization state-dependent activation of NF-kappa B signaling pathway by adipocyte complement-related protein of 30 kDa (Acrp30). *J. Biol. Chem.* 277:29359–62
28. Kim MJ, Yoo KH, Kim JH, Seo YT, Ha BW, et al. 2007. Effect of pinitol on glucose metabolism and adipocytokines in uncontrolled type 2 diabetes. *Diabetes Res. Clin. Pract.* 77(Suppl.):S247–51
29. Friedman JM. 1997. Leptin, leptin receptors and the control of body weight. *Eur. J. Med. Res.* 2:7–13
30. Sternson SM, Shepherd GM, Friedman JM. 2005. Topographic mapping of VMH → arcuate nucleus microcircuits and their reorganization by fasting. *Nat. Neurosci.* 8:1356–63
31. Lin CS, Xin ZC, Deng CH, Ning H, Lin G, Lue TF. 2010. Defining adipose tissue-derived stem cells in tissue and in culture. *Histol. Histopathol.* 25:807–15
32. Huang HY, Song TJ, Li X, Hu LL, He Q, et al. 2009. BMP signaling pathway is required for commitment of C3H10T1/2 pluripotent stem cells to the adipocyte lineage. *Proc. Natl. Acad. Sci. USA* 106:12670–75
33. Reznikoff KA, Brankow DW, Heidelberger C. 1973. Establishment and characterization of a cloned line of C3H mouse embryo cells sensitive to postconfluence inhibition of division. *Cancer Res.* 33:3231–38

34. Zhang JW, Klemm DJ, Vinson C, Lane MD. 2004. Role of CREB in transcriptional regulation of CCAAT/enhancer-binding protein beta gene during adipogenesis. *J. Biol. Chem.* 279:4471–78
35. Tang QQ, Otto TC, Lane MD. 2004. Commitment of C3H10T1/2 pluripotent stem cells to the adipocyte lineage. *Proc. Natl. Acad. Sci. USA* 101:9607–11
36. Bowers RR, Lane MD. 2007. A role for bone morphogenetic protein-4 in adipocyte development. *Cell Cycle* 6:385–89
37. Green H, Meuth M. 1974. An established pre-adipose cell line and its differentiation in culture. *Cell* 3:127–33
38. Tang Q-Q, Lane MD. 1999. Activation and centromeric localization of CCAAT/enhancer binding-proteins during the mitotic clonal expansion of adipocyte differentiation. *Genes Dev.* 13:2231–41
39. Davis LA, Zur Nieden NI. 2008. Mesodermal fate decisions of a stem cell: the Wnt switch. *Cell Mol. Life Sci.* 65:2658–74
40. Ross SE, Hemati N, Longo KA, Bennett CN, Lucas PC, et al. 2000. Inhibition of adipogenesis by Wnt signaling. *Science* 289:950–53
41. Bowers RR, Lane MD. 2008. Wnt signaling and adipocyte lineage commitment. *Cell Cycle* 7:1191–96
42. Zehentner BK, Leser U, Burtscher H. 2000. BMP-2 and sonic hedgehog have contrary effects on adipocyte-like differentiation of C3H10T1/2 cells. *DNA Cell Biol.* 19:275–81
43. Spinella-Jaegle S, Rawadi G, Kawai S, Gallea S, Faucheu C, et al. 2001. Sonic hedgehog increases the commitment of pluripotent mesenchymal cells into the osteoblastic lineage and abolishes adipocytic differentiation. *J. Cell Sci.* 114:2085–94
44. van der Horst G, Farih-Sips H, Lowik CWGM, Karperien M. 2003. Hedgehog stimulates only osteoblastic differentiation of undifferentiated KS483 cells. *Bone* 33:899–910
45. Kang S, Bennett CN, Gerin I, Rapp LA, Hankenson KD, Macdougald OA. 2007. Wnt signaling stimulates osteoblastogenesis of mesenchymal precursors by suppressing CCAAT/enhancer-binding protein alpha and peroxisome proliferator-activated receptor gamma. *J. Biol. Chem.* 282:14515–24
46. Bennett CN, Longo KA, Wright WS, Suva LJ, Lane TF, et al. 2005. Regulation of osteoblastogenesis and bone mass by Wnt10b. *Proc. Natl. Acad. Sci. USA* 102:3324–29
47. Isenmann S, Arthur A, Zannettino ACW, Turner JL, Shi ST, et al. 2009. TWIST family of basic helix-loop-helix transcription factors mediate human mesenchymal stem cell growth and commitment. *Stem Cells* 27:2457–68
48. Xu Y, Zhou YL, Erickson RL, Macdougald OA, Snead ML. 2007. Physical dissection of the CCAAT/enhancer-binding protein alpha in regulating the mouse amelogenin gene. *Biochem. Biophys. Res. Commun.* 354:56–61
49. Ahrens M, Ankenbauer T, Schroder D, Hollnagel A, Mayer H, Gross G. 1993. Expression of human bone morphogenetic proteins-2 or -4 in murine mesenchymal progenitor C3H10T1/2 cells induces differentiation into distinct mesenchymal cell lineages. *DNA Cell Biol.* 12:871–80
50. Wang EA, Israel DI, Kelly S, Luxenberg DP. 1993. Bone morphogenetic protein-2 causes commitment and differentiation in C3H10T1/2 and 3T3 cells. *Growth Factors* 9:57–71
51. Bowers RR, Kim JW, Otto TC, Lane MD. 2006. Stable stem cell commitment to the adipocyte lineage by inhibition of DNA methylation: role of the BMP-4 gene. *Proc. Natl. Acad. Sci. USA* 103:13022–27
52. Huang HY, Hu LL, Song TJ, Li X, He Q, et al. 2011. Involvement of cytoskeleton-associated proteins in the commitment of C3H10T1/2 pluripotent stem cells to adipocyte lineage induced by BMP2/4. *Mol. Cell. Proteomics* 10:1–8
53. Heldin C, Miyazono K, ten Dijke P. 1997. TGF-beta signalling from cell membrane to nucleus through SMAD proteins. *Nature* 390:465–71
54. Massague J, Wotton K. 2000. Transcriptional control by the TGF-beta/Smad signaling system. *EMBO J.* 19:1745–54
55. Feng JQ, Harris MA, Ghosh-Choudhury N, Feng M, Mundy GR, Harris SE. 1994. Structure and sequence of mouse bone morphogenetic protein-2 gene (BMP-2): comparison of the structures and promoter regions of BMP-2 and BMP-4 genes. *Biochim. Biophys. Acta* 1218:221–24
56. Helvering LM, Sharp RL, Ou X, Geiser AG. 2000. Regulation of the promoters for the human bone morphogenetic protein 2 and 4 genes. *Gene* 256:123–38

57. Komiya Y, Habas R. 2008. Wnt signal transduction pathways. *Organogenesis* 4:68–75
58. Kikuchi A, Yamamoto H, Sato A. 2009. Selective activation mechanisms of Wnt signaling pathways. *Trends Cell Biol.* 19:119–29
59. Kikuchi A, Yamamoto H, Kishida S. 2007. Multiplicity of the interactions of Wnt proteins and their receptors. *Cell. Signal.* 19:659–71
60. Reya T, Clevers H. 2005. Wnt signalling in stem cells and cancer. *Nature* 434:843–50
61. Christodoulides C, Lagathu C, Sethi JK, Vidal-Puig A. 2009. Adipogenesis and WNT signalling. *Trends Endocrinol. Metab.* 20:16–24
62. Cadigan KM, Liu YI. 2006. Wnt signaling: complexity at the surface. *J. Cell Sci.* 119:395–402
63. Kennell JA, O'Leary EE, Gummow BM, Hammer GD, MacDougald OA. 2003. T-cell factor 4N (TCF-4N), a novel isoform of mouse TCF-4, synergizes with beta-catenin to coactivate C/EBPalpha and steroidogenic factor 1 transcription factors. *Mol. Cell. Biol.* 23:5366–75
64. Kohn AD, Moon RT. 2005. Wnt and calcium signaling: beta-catenin-independent pathways. *Cell Calcium* 38:439–46
65. Cohen MM Jr. 2003. The hedgehog signaling network. *Am. J. Med. Genet. A* 123A:5–28
66. Varjosalo M, Taipale J. 2008. Hedgehog: functions and mechanisms. *Genes Dev.* 22:2454–72
67. James AW, Leucht P, Levi B, Carre AL, Xu Y, et al. 2010. Sonic Hedgehog influences the balance of osteogenesis and adipogenesis in mouse adipose-derived stromal cells. *Tissue Eng. Part A* 16:2605–16
68. Li Z, Zhang H, Denhard LA, Liu LH, Zhou HX, Lan ZJ. 2008. Reduced white fat mass in adult mice bearing a truncated Patched 1. *Int. J. Biol. Sci.* 4:29–36
69. Bamshad M, Song CK, Bartness TJ. 1999. CNS origins of the sympathetic nervous system outflow to brown adipose tissue. *Am. J. Physiol.* 276:R1569–78
70. Burkhart DL, Sage J. 2008. Cellular mechanisms of tumour suppression by the retinoblastoma gene. *Nat. Rev. Cancer* 8:671–82
71. Korenjak M, Brehm A. 2005. E2F-Rb complexes regulating transcription of genes important for differentiation and development. *Curr. Opin. Genet. Dev.* 15:520–27
72. Thomas DM, Carty SA, Piscopo DM, Lee JS, Wang WF, et al. 2001. The retinoblastoma protein acts as a transcriptional coactivator required for osteogenic differentiation. *Mol. Cell* 8:303–16
73. Fajas L, Egler V, Reiter R, Hansen J, Kristiansen K, et al. 2002. The retinoblastoma-histone deacetylase 3 complex inhibits PPAR gamma and adipocyte differentiation. *Dev. Cell* 3:903–10
74. Fajas L, Landsberg RL, Huss-Garcia Y, Sardet C, Lees JA, Auwerx J. 2002. E2Fs regulate adipocyte differentiation. *Dev. Cell* 3:39–49
75. Calo E, Quintero-Estades JA, Danielian PS, Nedelcu S, Berman SD, Lees JA. 2010. Rb regulates fate choice and lineage commitment in vivo. *Nature* 466:1110–14
76. Vertino AM, Taylor-Jones JM, Longo KA, Bearden ED, Lane TF, et al. 2005. Wnt10b deficiency promotes coexpression of myogenic and adipogenic programs in myoblasts. *Mol. Biol. Cell* 16:2039–48
77. Green H, Kehinde O. 1979. Formation of normally differentiated subcutaneous fat pads by an established preadipose cell line. *J. Cell. Physiol.* 101:169–72
78. Mandrup S, Loftus TM, MacDougald OA, Kuhajda FP, Lane MD. 1997. Obese gene expression at in vivo levels by fat pads derived from s.c. implanted 3T3-F442A preadipocytes. *Proc. Natl. Acad. Sci. USA* 94:4300–5
79. MacDougald OA, Lane MD. 1995. Transcriptional regulation of gene expression during adipocyte differentiation. *Annu. Rev. Biochem.* 64:345–73
80. Student AK, Hsu RY, Lane MD. 1980. Induction of fatty acid synthetase synthesis in differentiating 3T3-L1 preadipocytes. *J. Biol. Chem.* 255:4745–50
81. Coleman RA, Reed BC, Mackall JC, Student AK, Lane MD, Bell RM. 1978. Selective changes in microsomal enzymes of triacylglycerol phosphatidylcholine, and phosphatidylethanolamine biosynthesis during differentiation of 3T3-L1 preadipocytes. *J. Biol. Chem.* 253:7256–61
82. Reed BC, Lane MD. 1980. Insulin receptor synthesis and turnover in differentiating 3T3-L1 preadipocytes. *Proc. Natl. Acad. Sci. USA* 77:285–89
83. Reed BC, Lane MD. 1980. Expression of insulin receptors during preadipocyte differentiation. *Adv. Enzym. Regul.* 18:97–117

84. Kaestner KH, Christy RJ, McLenithan JC, Braiterman LT, Cornelius P, et al. 1989. Sequence, tissue distribution, and differential expression of mRNA for a putative insulin-responsive glucose transporter in mouse 3T3-L1 adipocytes. *Proc. Natl Acad. Sci. USA* 86:3150–54

85. Hwang C-S, Mandrup S, MacDougald OA, Geiman DE, Lane MD. 1996. Transcriptional activation of the mouse obese (*ob*) gene by CCAAT/enhancer binding protein α. *Proc. Natl. Acad. Sci. USA* 93:873–77

86. Hwang CS, Loftus TM, Mandrup S, Lane MD. 1997. Adipocyte differentiation and leptin expression. *Annu. Rev. Cell Dev. Biol.* 13:231–59

87. Cornelius P, MacDougald OA, Lane MD. 1994. Regulation of adipocyte development. *Annu. Rev. Nutr.* 14:99–129

88. Patel YM, Lane MD. 2000. Mitotic clonal expansion during preadipocyte differentiation: calpain-mediated turnover of p27. *J. Biol. Chem.* 275:17653–60

89. Tang Q-Q, Otto TC, Lane MD. 2003. Mitotic clonal expansion: asynchronous process required for adipogenesis. *Proc. Natl. Acad. Sci. USA* 100:44–49

90. Zhang JW, Tang Q-Q, Vinson C, Lane MD. 2004. Dominant-negative C/EBP disrupts mitotic clonal expansion and differentiation of 3T3-L1 preadipocytes. *Proc. Natl. Acad. Sci. USA* 101:43–47

91. Chao LC, Bensinger SJ, Villanueva CJ, Wroblewski K, Tontonoz P. 2008. Inhibition of adipocyte differentiation by Nur77, Nurr1, and Nor1. *Mol. Endocrinol.* 22:2596–608

92. Tang Q-Q, Otto TC, Lane MD. 2003. CCAAT/enhancer-binding protein beta is required for mitotic clonal expansion during adipogenesis. *Proc. Natl. Acad. Sci. USA* 100:850–55

93. Tang QQ, Gronborg M, Huang H, Kim JW, Otto TC, et al. 2005. Sequential phosphorylation of CCAAT enhancer-binding protein beta by MAPK and glycogen synthase kinase 3beta is required for adipogenesis. *Proc. Natl. Acad. Sci. USA* 102:9766–71

94. Zhang YY, Li X, Qian SW, Guo L, Huang HY, et al. 2011. Transcriptional activation of histone H4 by C/EBPbeta during the mitotic clonal expansion of 3T3-L1 adipocyte differentiation. *Mol. Biol. Cell* 22:2165–74

95. Fox KE, Colton LA, Erickson PF, Friedman JE, Cha HC, et al. 2008. Regulation of cyclin D1 and Wnt10b gene expression by cAMP-responsive element-binding protein during early adipogenesis involves differential promoter methylation. *J. Biol. Chem.* 283:35096–105

96. Niehof M, Manns MP, Trautwein C. 1997. CREB controls LAP/C/EBP beta transcription. *Mol. Cell. Biol.* 17:3600–13

97. Johnson PF, Landschulz WH, Graves BJ, McKnight SL. 1987. Identification of a rat liver nuclear protein that binds to the enhancer core element of three animal viruses. *Genes Dev.* 1:133–46

98. Landschulz WH, Johnson PF, Adashi EY, Graves BJ, McKnight SL. 1988. Isolation of a recombinant copy of the gene encoding C/EBP. *Genes Dev.* 2:786–800

99. Landschulz WH, Johnson PF, McKnight SL. 1988. The leucine zipper: a hypothetical structure common to a new class of DNA binding proteins. *Science* 240:1759–64

100. Akira S, Isshiki H, Sugita T, Tanabe O, Kinoshita S, et al. 1990. A nuclear factor for IL-6 expression (NF-IL6) is a member of a C/EBP family. *EMBO J.* 6:1897–906

101. Chang C-J, Chen T-T, Lei H-Y, Chen D-S, Lee S-C. 1990. Molecular cloning of a transcription factor, AGP/EBP, that belongs to members of the C/EBP family. *Mol. Cell. Biol.* 10:6642–53

102. Descombes P, Chojkier M, Lichtsteiner S, Falvey E, Schibler U. 1990. LAP, a novel member of the C/EBP gene family, encodes a liver-enriched transcriptional activator protein. *Genes Dev.* 4:1541–51

103. Poli V, Mancini FP, Cortese R. 1990. IL-6DBP, a nuclear protein involved in interleukin-6 signal transduction, defines a new family of leucine zipper proteins related to C/EBP. *Cell* 63:643–53

104. Cao Z, Umek RM, McKnight SL. 1991. Regulated expression of three C/EBP isoforms during adipose conversion of 3T30L1 cells. *Genes Dev.* 5:1538–52

105. Williams SC, Cantwell CA, Johnson PF. 1991. A family of C/EBP-related proteins capable of forming covalently linked leucine zipper dimers in vitro. *Genes Dev.* 5:1553–67

106. Ron D, Habener JF. 1992. CHOP, a novel developmentally regulated nuclear protein that dimerizes with transcription factors C/EBP and LAP and functions as a dominant-negative inhibitor of gene transcription. *Genes Dev.* 6:439–53

107. Landschulz WH, Johnson PF, McKnight SL. 1989. The DNA binding domain of the rat liver nuclear protein C/EBP is bipartite. *Science* 243:1681–88

108. Descombes P, Schibler U. 1991. A liver-enriched transcriptional activator protein, LAP, and a transcriptional inhibitory protein, LIP, are translated from the same mRNA. *Cell* 67:569–79
109. Ossipow V, Descombes P, Schibler U. 1993. CCAAT/enhancer-binding protein mRNA is translated into multiple proteins with different transcription activation potentials. *Proc. Natl. Acad. Sci. USA* 90:8219–23
110. Yeh W-C, Cao Z, Classon M, McKnight SL. 1995. Cascade regulation of terminal adipocyte differentiation by three members of the C/EBP family of leucine zipper proteins. *Genes Dev.* 9:168–81
111. Lane MD, Jiang M-S, Tang. 1999. Role of C/EBPα in adipocyte differentiation. In *Nutrition Genetics Obesity: Pennington Center Nutrition Series*, ed. GA Bray, DH Ryan, 9:459–76. Baton Rouge, LA: La. State Univ. Press
112. Tang QQ, Otto TC, Lane MD. 2003. CCAAT/enhancer-binding protein beta is required for mitotic clonal expansion during adipogenesis. *Proc. Natl. Acad. Sci. USA* 100:850–55
113. Tanaka T, Yoshida N, Kishimoto T, Akira S. 1997. Defective adipocyte differentiation in mice lacking the C/EBPbeta and/or C/EBPdelta gene. *EMBO J.* 16:7432–43
114. Patel YM, Lane MD. 1999. Role of calpain in adipocyte differentiation. *Proc. Natl. Acad. Sci. USA* 96:1279–84
115. Moitra J, Mason MM, Olive M, Krylov D, Gavrilova O, et al. 1998. Life without white fat: a transgenic mouse. *Genes Dev.* 12:3168–81
116. Ahn J, Lee H, Kim S, Ha T. 2007. Resveratrol inhibits TNF-alpha-induced changes of adipokines in 3T3-L1 adipocytes. *Biochem. Biophys. Res. Commun.* 364:972–77
117. Kim JW, Tang QQ, Li X, Lane MD. 2007. Effect of phosphorylation and S-S bond-induced dimerization on DNA binding and transcriptional activation by C/EBPbeta. *Proc. Natl. Acad. Sci. USA* 104:1800–4
118. Vinson CR, Sigler PB, McKnight SL. 1989. Scissor-grip model for DNA recognition by a family of leucine zipper proteins. *Science* 246:911–16
119. Rosen ED, Spiegelman BM. 2000. Molecular regulation of adipogenesis. *Annu. Rev. Cell Dev. Biol.* 16:145–71
120. Christy RJ, Kaestner KH, Geiman DE, Lane MD. 1991. CCAAT/enhancer binding protein gene promoter: binding of nuclear factors during differentiation of 3T3-L1 preadipocytes. *Proc. Natl. Acad. Sci. USA* 88:2593–97
121. Abboud TK, Zhu J, Richardson M, Peres da Silva E, Donovan M. 1995. Desflurane: a new volatile anesthetic for cesarean section. Maternal and neonatal effects. *Acta Anaesthesiol. Scand.* 39:723–26
122. Clarke SL, Robinson CE, Gimble JM. 1997. CAAT/enhancer binding proteins directly modulate transcription from the peroxisome proliferator-activated receptor gamma 2 promoter. *Biochem. Biophys. Res. Commun.* 240:99–103
123. Freytag SO, Geddes TJ. 1992. Reciprocal regulation of adipogenesis by myc and C/EBPα. *Science* 256:379–82
124. Freytag SO, Paielli DL, Gilbert JD. 1994. Ectopic expression of the CCAAT/enhancer-binding protein a promotes the adipogenic program in a variety of mouse fibroblastic cells. *Genes Dev.* 8:1654–63
125. Lin F-T, Lane MD. 1994. CCAAT/enhancer binding protein α is sufficient to initiate the 3T3-L1 adipocyte differentiation program. *Proc. Natl. Acad. Sci. USA* 91:8757–61
126. Lin F-T, Lane MD. 1992. Antisense CCAAT/enhancer-binding protein RNA suppresses coordinate gene expression and triglyceride accumulation during differentiation of 3T3-L1 preadipocytes. *Genes Dev.* 6:533–44
127. Wang ND, Finegold MJ, Bradley A, Ou CN, Abdelsayed SV, et al. 1995. Impaired energy homeostasis in C/EBP alpha knockout mice. *Science.* 26:1108–12
128. Flodby P, Barlow C, Kylefjord H, Ahrlund-Richter L, Xanthopoulos KG. 1996. Increased hepatic cell proliferation and lung abnormalities in mice deficient in CCAAT/enhancer binding protein α. *J. Biol. Chem.* 271:24753–60
129. Jiang MS, Tang QQ, McLenithan J, Geiman D, Shillinglaw W, et al. 1998. Derepression of the C/EBPalpha gene during adipogenesis: identification of AP-2alpha as a repressor. *Proc. Natl. Acad. Sci. USA* 95:3467–71
130. Tang QQ, Jiang MS, Lane MD. 1997. Repression of transcription mediated by dual elements in the CCAAT/enhancer binding protein alpha gene. *Proc. Natl. Acad. Sci. USA* 94:13571–75

131. Batchvarova N, Wang XZ, Ron D. 1995. Inhibition of adipogenesis by the stress-induced protein CHOP (Gadd153). *EMBO J.* 14:4654–61
132. Tang Q-Q, Lane MD. 2000. Role of C/EBP homologous protein (CHOP-10) in the programmed activated activation of CCAAT/enhancer-binding protein-β during adipogenesis. *Proc. Natl. Acad. Sci. USA* 97:12446–50
133. Zhu Y, Qi C, Korenberg JR, Chen XN, Noya D, et al. 1995. Structural organization of mouse peroxisome proliferator-activated receptor gamma (mPPAR gamma) gene: alternative promoter use and different splicing yield two mPPAR gamma isoforms. *Proc. Natl. Acad. Sci. USA* 92:7921–25
134. Fajas L, Fruchart JC, Auwerx J. 1998. PPARgamma3 mRNA: a distinct PPARgamma mRNA subtype transcribed from an independent promoter. *FEBS Lett.* 438:55–60
135. Tontonoz P, Hu E, Spiegelman BM. 1994. Stimulation of adipogenesis in fibroblasts by PPAR gamma 2, a lipid-activated transcription factor. *Cell* 79:1147–56
136. Date T, Doiguchi Y, Nobuta M, Shindo H. 2004. Bone morphogenetic protein-2 induces differentiation of multipotent C3H10T1/2 cells into osteoblasts, chondrocytes, and adipocytes in vivo and in vitro. *J. Orthop. Sci.* 9:503–8
137. Morrison PF, Farmer SR. 1999. Role of PPARγ in regulating a cascade expression of cyclin-dependent kinase inhibitors, p18 (INK4c), and p21 (Waf1/Cip1), during adipogenesis. *J. Biol. Chem.* 274:17088–97
138. Schwarz EJ, Reginato MJ, Shao D, Krakow SL, Lazar MA. 1997. Retinoic acid blocks adipogenesis by inhibiting C/EBPbeta-mediated transcription. *Mol. Cell. Biol.* 17:1552–61
139. Elberg G, Gimble JM, Tsai SY. 2000. Modulation of the murine peroxisome proliferator-activated receptor gamma 2 promoter activity by CCAAT/enhancer-binding proteins. *J. Biol. Chem.* 275:27815–22
140. Horton JD. 2002. Sterol regulatory element-binding proteins: transcriptional activators of lipid synthesis. *Biochem. Soc. Trans.* 30:1091–95
141. Tontonoz P, Kim JB, Graves RA, Spiegelman BM. 1993. ADD1: a novel helix-loop-helix transcription factor associated with adipocyte determination and differentiation. *Mol. Cell. Biol.* 13:4753–59
142. Kim JB, Spiegelman BM. 1996. ADD1/SREBP1 promotes adipocyte differentiation and gene expression linked to fatty acid metabolism. *Genes Dev.* 10:1096–107
143. Yellaturu CR, Deng X, Cagen LM, Wilcox HG, Mansbach CM 2nd, et al. 2009. Insulin enhances post-translational processing of nascent SREBP-1c by promoting its phosphorylation and association with COPII vesicles. *J. Biol. Chem.* 284:7518–32
144. Yellaturu CR, Deng X, Park EA, Raghow R, Elam MB. 2009. Insulin enhances the biogenesis of nuclear sterol regulatory element-binding protein (SREBP)-1c by posttranscriptional down-regulation of Insig-2A and its dissociation from SREBP cleavage-activating protein (SCAP)·SREBP-1c complex. *J. Biol. Chem.* 284:31726–34
145. Aagaard MM, Siersbaek R, Mandrup S. 2011. Molecular basis for gene-specific transactivation by nuclear receptors. *Biochim. Biophys. Acta* 1812:824–35
146. Steger DJ, Lazar MA. 2011. Adipogenic hotspots: where the action is. *EMBO J.* 30:1418–19
147. Siersbaek R, Nielsen R, John S, Sung MH, Baek S, et al. 2011. Extensive chromatin remodelling and establishment of transcription factor 'hotspots' during early adipogenesis. *EMBO J.* 30:1459–72
148. Kennell JA, Gerin I, MacDougald OA, Cadigan KM. 2008. The microRNA miR-8 is a conserved negative regulator of Wnt signaling. *Proc. Natl. Acad. Sci. USA* 105:15417–22
149. Wang Q, Li YC, Wang J, Kong J, Qi Y, et al. 2008. miR-17-92 cluster accelerates adipocyte differentiation by negatively regulating tumor-suppressor Rb2/p130. *Proc. Natl. Acad. Sci. USA* 105:2889–94
150. Kim SY, Kim AY, Lee HW, Son YH, Lee GY, et al. 2010. miR-27a is a negative regulator of adipocyte differentiation via suppressing PPARgamma expression. *Biochem. Biophys. Res. Commun.* 392:323–28
151. Sun T, Fu M, Bookout AL, Kliewer SA, Mangelsdorf DJ. 2009. MicroRNA let-7 regulates 3T3-L1 adipogenesis. *Mol. Endocrinol.* 23:925–31

Pluripotency and Nuclear Reprogramming

Marion Dejosez[1] and Thomas P. Zwaka[2]

[1]Department of Molecular and Human Genetics and [2]Department of Molecular and Cellular Biology, Baylor College of Medicine, Houston, Texas 77030; email: dejosez@bcm.edu, tpzwaka@bcm.edu

Keywords

epigenetics, transcription factors, embryogenesis, stem cells

Abstract

Pluripotency is a "blank" cellular state characteristic of specific cells within the early embryo (e.g., epiblast cells) and of certain cells propagated in vitro (e.g., embryonic stem cells, ESCs). The terms pluripotent cell and stem cell are often used interchangeably to describe cells capable of differentiating into multiple cell types. In this review, we discuss the prevailing molecular and functional definitions of pluripotency and the working parameters employed to describe this state, both in the context of cells residing within the early embryo and cells propagated in vitro.

Contents

1. INTRODUCTION 738
2. CELLULAR ORIGINS OF PLURIPOTENTIALITY 738
   - 2.1. Totipotency and the Pre-Embryonic Stage 738
   - 2.2. Primordial Pluripotency 739
   - 2.3. Refined (Late) Pluripotency 739
   - 2.4. Trapping Pluripotency 739
   - 2.5. Artificial Acquisition of Pluripotency 744
3. MOLECULAR CONTROL OF PLURIPOTENCY 744
   - 3.1. The Developmental Core Module 745
   - 3.2. The Cell Growth Module 747
   - 3.3. Signaling and the Signaling Module 747
   - 3.4. Transcriptional Control in Pluripotent Stem Cells 750
   - 3.5. Pluripotency-Associated Chromatin Structure 751
   - 3.6. Noncoding RNA and Pluripotency 752
4. CONTROLLED SHUTDOWN OF PLURIPOTENCY DURING DIFFERENTIATION 753
   - 4.1. Feed-Forward Generated Destabilization Signals 753
   - 4.2. Variability in the Expression of Pluripotency Factors 753
   - 4.3. Metastable States in Stem Cell Culture 754
   - 4.4. Rapid Proteolytic Removal of Pluripotency Factors 754
   - 4.5. Transcriptionally Mediated Downregulation of Pluripotency Factors 755
5. CONCLUDING REMARKS 755

Pluripotency: the unique ability of a cell to differentiate into all somatic and germ line cells of the developing embryo

1. INTRODUCTION

Mammalian development is characterized by an extremely well-orchestrated transition from an initial single cell, or zygote, to the entire spectrum of cells found in the body. This process is governed by an intrinsic regulatory program that employs transcription factors to decipher both sequence-specific and epigenetic information. Additional layers of control are provided by extrinsic information incorporating the spatial and temporal distributions of cells, cell shape, properties of the extracellular matrix, the actions of morphogens, the functions of other effectors intricately intertwined with those of transcription factors, and the activities of enzymes that modify epigenetic information.

Importantly, studies on pluripotent cells have identified intrinsic autonomous programs (defined by their ability to differentiate into precursor cells representing the founding populations of all three embryonic germ layers) that are of paramount importance in maintenance of the inherent pluripotency of these cells. Here, we review both the cellular and molecular design principles of pluripotency.

2. CELLULAR ORIGINS OF PLURIPOTENTIALITY

2.1. Totipotency and the Pre-Embryonic Stage

Totipotency is defined as the ability of a particular cell to give rise to all cell types of an organism, including the extraembryonic cell lineages that are required for appropriate development of the embryo (1). These extraembryonic lineages are crucial for proper implantation and maintenance of the embryo in the uterus and also provide inductive signals for pattern formation and axis development (2). In vivo totipotent cells exist only transiently in the early embryo and are generated via a natural reprogramming process initiated by fertilization of the oocyte. Although neither the oocyte nor the zygote is totipotent, the fertilized egg undergoes a transition from a specialized cell type with a restricted fate into equipotent blastomeres. At the two- and four-cell stages, murine blastomeres are totipotent, and each can develop into an entire animal (3).

A series of transplantation experiments demonstrated that maternal proteins and RNAs that had been deposited in the oocyte could reset the epigenetic state of somatic chromatin and transform the oocyte into totipotent blastomeres following fertilization (reviewed by Reference 4). Nevertheless, it has not been possible to establish totipotent cell lines in vitro. Indeed, this may be impossible because of the transient nature of the totipotent blastomeres in vivo; these blastomeres are not actually self-renewing or dividing but rather are generated via cleavage during the so-called pre-embryonic stage of embryonic development.

2.2. Primordial Pluripotency

As cleavage of the early embryo proceeds to the 16-cell stage, a gradual restriction in developmental potency is evident, and the cells are committed to development into two distinct lineages: the trophoblast lineage and the inner cell mass (ICM). Differentiation commences with flattening of the blastomeres and strengthening of cell-to-cell contact. This process has been termed compaction and is mediated by changes in cell-cell adhesion and expression of extracellular matrix proteins (5, 6). During compaction, cadherins, especially E-cadherin, play central roles (7). Cells located on the inside of the embryo at the compaction stage give rise to the ICM, whereas the outer cells form the trophectoderm, eventually leading to generation of the placenta. The ICM next gives rise to a second extraembryonic lineage, the hypoblast (or primitive endoderm), which will form the yolk sac. Simultaneously, the remaining cells transition into the epiblast (the primitive ectoderm), originating the embryo. In contrast to the totipotent precursors, cells of the epiblast (at this stage) are termed pluripotent because they can give rise to all somatic and germ line cells of the developing embryo but do not contribute to extraembryonic lineages. Hence, the epiblast cells exhibit "primordial" pluripotency, which has been termed the ground state of pluripotency (8); see **Figure 1** and **Table 1**.

2.3. Refined (Late) Pluripotency

Developmentally, upon implantation into the uterine wall, epiblast cells become rearranged into an epithelial structure that lines the central proamniotic cavity. During this early postimplantation period, termed the egg-cylinder stage, epiblast cells are alkaline phosphatase (AP) (9) and stage-specific embryonic antigen 1 (SSEA1) positive (10) and express high levels of Oct4 (for a detailed discussion of Oct4 see Section 3.1). At this stage, the epiblast cells exhibit "refined" pluripotency [also termed primed pluripotency (8)]. This definition refers to the restricted developmental potential of the late epiblast. Eventually, the primitive streak forms in a localized region of the epiblast adjacent to the embryonic/extraembryonic junction, and gastrulation commences at 6.5 days postcoitum (dpc).

In the interval from implantation to the onset of gastrulation, epiblast cells lose the ability to colonize the blastocyst, X chromosome inactivation takes place, the length of the cell cycle is reduced (10), and the extent of genome methylation increases. In the early embryo, the genome is demethylated, but the DNA next becomes progressively methylated, and by 6.5 dpc, the methylation pattern characteristic of adult tissues is established (11). During the subsequent steps of gastrulation, the embryo becomes transformed into a multilayered structure wherein all primordia are arranged according to the body plan, and the primitive streak marks the future posterior of the embryo.

2.4. Trapping Pluripotency

Pluripotent cell lines are of enormous interest especially in the field of regenerative medicine. Stem cells have been "trapped" in vitro by isolation from tissues at various stages of embryonic and postnatal development as well as from adult and tumor tissues. Such cells share certain features, but usually reflect aspects of their in vivo counterparts at the molecular level. Independent of origin, the gold standard used

Totipotency: the ability of the zygote and blastomeres to differentiate into all embryonic and extraembryonic lineages of the developing embryo

Epigenetic state: a summary of epigenetic modifications that collectively determine cellular identity in the lineage hierarchy

Trophoblast (trophectoderm): the outer layer of the blastocyst, generated during the first differentiation event when mammalian morula cells segregate into two lineages

Inner cell mass (ICM): this second lineage of the early embryo inside the blastocyst is surrounded by the trophectoderm and originates all embryonic tissues

Hypoblast (primitive endoderm): an epithelial layer derived from inner cell mass cells that are in contact with the blastocyst cavity forming the yolk sac

Epiblast (primitive/embryonic ectoderm): the second tissue derived from inner cell mass cells during the second differentiation event of embryonic development forming the embryo

Table 1 Properties of various pluripotent cell types in vitro[a]

	mESC	mEGC	mECC	miPSC	mEpiSC	hESC	hEGC	hiPSC
Origin	Blastocyst	Embryonic gonad	Teratoma	Somatic cells	Late epiblast	Blastocyst	Embryonic gonad	Somatic cells
Blastocyst chimera contribution	Somatic and germ line	Somatic and germ line	Somatic, low frequency of germ line	Somatic and germ line contribution	No	Not determined	Not determined	Not determined
Teratomas	Yes	Yes	Yes	Yes	Yes	Yes	Yes	Yes
Spontaneous trophoblast differentiation	No	Not determined	Not determined	Not determined	Yes	Yes	Not determined	Not determined
Growth factor conditions	Lif, Bmp4	Derivation: Fgf2, Lif, SCF Maintenance: Lif and FBS	Lif, FBS	Lif	Fgf2, Activin	Fgf2, Activin, MEF conditioned medium	Lif, Fgf2, Forksolin	Fgf2, Activin, MEF conditioned medium
Morphology in culture	Domed	Domed	Domed	Domed	Flat	Flat	Domed	Flat
XX status	XaXa	Not determined	XaXa	XaXa	XaXi	XaXi	Not determined	XaXi
Reference	17	26, 27	12	21, 56, 63	25, 24	38	28	58
Pluripotent state		Primordial			Refined			Not determined
Positive regulators	Lif/Stat3, Bmp4, Wnt, Igf				Tgf-β, Activin, Fgf2, Erk1 and -2, Wnt, Igf			
Negative regulators	Tgf-β, Activin, Fgf2, Erk1 and -2				Bmp4			
Pluripotency factors	Oct4, Nanog, Sox2, Klf2, Klf4				Oct4, Sox2, Nanog			
Response to Lif/Stat3	Self-renewal				None			
Response to Fgf/Erk	Differentiation				Self-renewal			
Response to 2i	Self-renewal				Differentiation/death			
Clonogenicity	High				Low			

[a]Abbreviations: Fgf2, fibroblast growth factor 2; Bmp4, bone morphogenic protein 4; Erk, extracellular signal regulated kinase; FBS, fetal bovine serum; Fgf, fibroblast growth factor; hEGC, human embryonic germ cell; hiPSC, human induced pluripotent stem cell; hESC, human embryonic stem cell; Igf, insulin growth factor; Lif, leukemia inhibitory factor; mECC, mouse embryonic carcinoma cell; MEF, mouse embryonic fibroblast; mEGC, mouse embryonic germ cell; mEpiSC, mouse epiblast stem cell; miPSC, mouse induced pluripotent stem cell; mESC, mouse embryonic stem cell; SCF, stem cell factor; Stat3, signal transducer and activator of transcription 3; Wnt, wingless-type MMTV integration site family; Xa, activated X chromosome; Xi, inactivated X chromosome; XX status, X chromosome status; 2i, two-inhibitor cocktail targeting Erk and Gsk3.

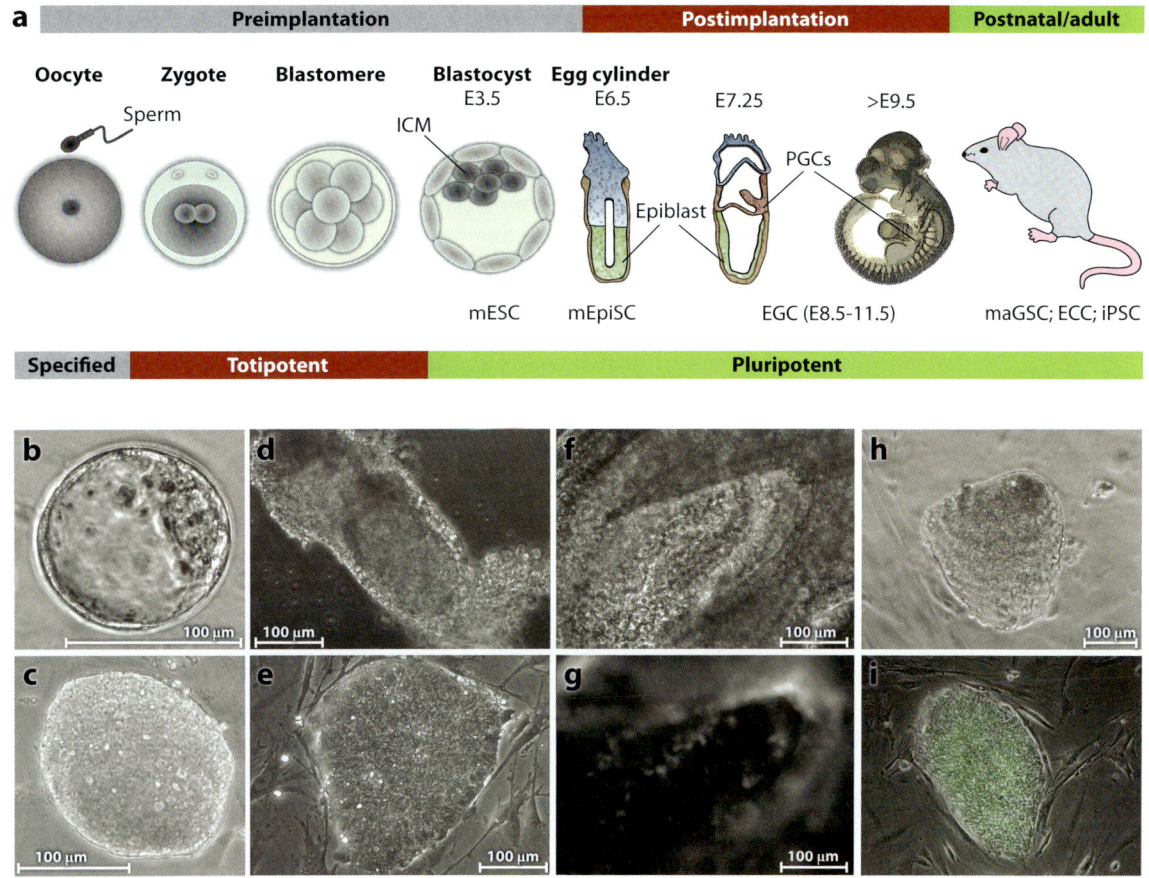

Figure 1

Origins of pluripotent stem cells and the morphology thereof in vitro. (*a*) Pluripotent stem cells can be trapped at various stages of embryonic development. (*b*) A phase-contrast micrograph of a blastocyst at embryonic day (E) 3.5 when the originating embryonic stem cells (ESCs) with domed morphology are evident (*c*). (*d*) A phase-contrast image of a mouse egg-cylinder stage embryo, which can give rise to epiblast stem cells (EpiSCs), at E6.5. (*e*) An EpiSC colony growing as a flat layer of cells. (*f*) A phase-contrast micrograph of a region from the hindgut of a mouse embryo at E8.5, wherein Blimp1-positive primordial germ cells (PGCs) may be noted. (*g*) Immunofluorescence from the region shown in (*f*) in which Blimp1-positive cells in a Blimp1-Cre reporter mouse are shown. (*h*) A phase-contrast image of a mouse induced pluripotent stem cell (iPSC) colony after transfection with DNAs encoding the transcription factors Oct4, Sox2, Klf4, and c-Myc. (*i*) Overlay of a phase-contrast image of a human embryonic stem cell (hESC) colony and an immunofluorescence image of the same colony after immunostaining using an anti-Oct4 antibody. Abbreviations: ECC, embryonic carcinoma cell; EGC, embryonic germ cell; ICM, inner cell mass; maGSC, multipotent adult germ line stem cell; mESC, mouse embryonic stem cell.

to test pluripotentiality is the ability to contribute to the formation of chimeric animals, including colonization of the germ line after injection into host blastocysts or via evaluation of the ability to form multilineage tumors, so-called teratomas, after injection into immunocompromised mice.

2.4.1. Embryonic carcinoma cells. Pluripotent stem cells were first derived from teratocarcinomas, germ line tumors that arise spontaneously from the adult testis or ovary of mice and humans (12). Studies in the early 1960s demonstrated the multilineage potential of single teratocarcinoma cells, and the first pluripotent embryonic carcinoma cell (ECC) lines were derived from such tumors about a decade later (12). Although ECCs can contribute to formation of chimeric animals, this contribution is often of low level, and germ line transmission occurs only sporadically (13). Additionally, many chimeric animals develop tumors derived from the injected cells, attributable to mutations and abnormal karyotypes that accumulated during the development of the teratocarcinoma from which the ECCs were initially derived (14).

ESC: embryonic stem cell

2.4.2. Blastocyst-derived stem cells and mouse embryonic stem cells.
At the time of implantation (~4.5 dpc), the blastocyst contains three distinct lineages of which the epiblast population is the smallest, with only 20–25 cells being embedded between the trophectoderm and the primitive endoderm. Under certain cell culture conditions, three stem cell types can be derived from this preimplantation stage, reflecting the three distinct cell lineages of the blastocyst. These are trophoblast stem (TS) cells from the trophectoderm, extraembryonic endoderm (XEN) cells from the primitive endoderm, and mouse embryonic stem cells (mESCs) from the epiblast (see **Figure 1**). TS cells require fibroblast growth factor (Fgf) and Activin/Nodal signaling to self-renew (see Section 3.3), and rely on the activity of TS-specific transcription factors, including Cdx2 and Eomes (15). In contrast, XEN cells are characterized by Gata6 and Sox7 expression, requiring Fgf signaling only during derivation (16).

mESCs were derived in 1981 from preimplantation blastocysts of the mouse strain 129 upon coculture on a feeder layer of irradiated mouse embryonic fibroblasts, in the presence of fetal bovine serum (17, 18). In this environment, the cytokine leukemia inhibitory factor (Lif) provides the principal self-renewal signal by activating the signal transducer and activator of transcription 3 (Stat3) pathway (19, 20). Lif acts together with bone morphogenic protein 4 (Bmp4) to support the pluripotent state of mESCs via induction of inhibitor of differentiation (Id) genes.

mESCs have been termed naive pluripotent cells, reflecting the primordial pluripotentiality of the early epiblast (blastocyst), but it remains possible that embryonic stem cells (ESCs) are in fact an artifact of tissue culture. mESCs resemble their in vivo counterparts in terms of expression of canonical pluripotency genes, such as *Oct4*, *Sox2*, and *Nanog*, and in expression of *AP* and *Ssea1*. Furthermore, ESC lines isolated from female embryos harbor two active X chromosomes. However, the cells are characterized by the ability to self-renew indefinitely and by possession of a hypermethylated genome (21).

It has been suggested, however, that mESCs are the counterparts of naive epiblasts, rather than being created in cell culture. This suggestion is based on the observation that inhibition of extracellular-regulated MAP kinase (Erk) signaling in early embryos suppresses hypoblast formation and causes the entire ICM to develop into an epiblast (22). Inhibition of glycogen synthase kinase-3 (Gsk3) leads to expansion of the nascent epiblast in situ and to generation of ESCs when such cells are explanted. Hence, the early epiblast may resemble ESCs in this regard (23).

2.4.3. Mouse epiblast stem cells.
Epiblast stem cells (EpiSCs) have been derived from early postimplantation epiblasts between 5.5 and 7.5 dpc (**Figure 1**). Representing the egg-cylinder stage, they have already undergone X chromosome inactivation and exhibit a gene expression profile characteristic of the postimplantation epiblast rather than the ICM. This profile is characterized by low-level expression of the pluripotency-related genes *Nanog*, *Rex2*, and *Klf4*, but also by elevated levels of differentiation markers, such as *Fgf5* (24). Self-renewal of EpiSCs is maintained by Activin, Fgf2, Erk1 and -2, and transforming growth factor-β (Tgf-β) but not by Lif. Furthermore, growth is in fact inhibited by Bmp4 (24, 25). In contrast to the three-dimensional morphology of ESC colonies, EpiSC colonies are flat and do not expand well when diluted to single cells after

dissociation with trypsin. EpiSCs can form teratocarcinomas but contribute only minimally to chimeric animals. Therefore, EpiSCs may reflect the refined (late) pluripotency of the postimplantation epiblast.

2.4.4. Embryonic germ cells.

The first ESC-like cells not derived from blastocysts were mouse and human embryonic germ cells (EGCs), originating from primordial germ cells (PGCs) derived from the developing gonad (26–28). During the specification process, such cells successively express *Fragilis*, *Blimp1*, and *Stella*. PGCs are large and round in shape, and these cells exhibit high-level expression of tissue-nonspecific alkaline phosphatase and *Oct4*. PGCs can be isolated in the presence of steel factor and Lif at 8.5 dpc or from genital ridges between 10.5 and 12.5 dpc. However, under such conditions, PGCs exhibit a finite proliferative capacity (26, 29). Addition of Fgf2 to the culture cocktail during the initial period leads to continuous division and establishment of EGC colonies. EGCs are similar to ESCs in terms of morphology, are able to form teratocarcinomas, and contribute to formation of chimeric animals, with apparent germ line transmission (26, 27). Interestingly, when EGCs are fused with somatic cells, they can induce demethylation of somatically imprinted genes, reflecting expression of an enzymatic activity that resets imprinting during development (30). Even though EGCs are derived from a later developmental stage than EpiSCs, they exhibit primordial rather than defined pluripotency. Study of the molecular events leading to PGC specification, and the molecular characterization of EGCs, has shown that EGCs are remarkably similar to ESCs. On the basis of this similarity, an alternative hypothesis on the true origin of ESCs has been advanced (31). Accordingly, the cells might transition through a PGC-like phase in vitro rather than arising directly from the epiblast. This route of ESC derivation is suggested by the presence of early PGC markers in ESCs, whereas late markers characteristic of mature germ cells are not expressed (32–34). In male mouse embryos, PGCs eventually develop into spermatogonial stem cells, which have been isolated from both newborn and adult male gonads and can give rise to embryonic stem (ES)-like cells in vitro (35–37).

2.4.5. Human embryonic stem cells.

Almost two decades after the isolation of the first mESC lines from mouse blastocysts, the first human embryonic stem cells (hESCs) were derived from human blastomeres in the presence of Fgf2 (38). hESCs express high levels of telomerase and cell surface markers that are characteristic of undifferentiated nonhuman primate ES and human EC cells, including SSEA3, SSEA4, TRA1-60, TRA1-81, and AP. hESCs maintain the potential to form derivatives of all three embryonic germ layers and to produce teratocarcinomas (38). Although both X chromosomes are active in mESCs, hESCs show a propensity toward X chromosome inactivation. Several pathways have been implicated in hESC self-renewal, including those involving Fgf2 (39), Tgf/Activin-A/Nodal (40), sphingosine-1-phosphate/platelet-derived growth factor (S1P/PDGF) (41), and insulin growth factor (IGF)/insulin (42), as reviewed in Reference 43.

It is argued that hESCs are in fact the counterpart of mouse EpiSCs because of the similarities between the cell types. Both types of cells grow as flat colonies and do not grow well after dissociation into single cells. Both rely on Activin/Nodal signaling to maintain pluripotency, and in contrast to mESCs, neither cell type can be maintained in medium supplemented with Lif. Additionally, female cells of both cell types harbor one active and one inactive X chromosome (44). Nevertheless, crucial differences in gene expression profiles are evident. hESCs express SSEA3 and -4 on the surface and are AP positive, whereas mouse EpiSCs express Ssea1 and lack Ap. Additionally, hESCs express *DPPA3* (45) and *KLF4*, as do mESCs; these factors are not expressed by EpiSCs. Both Dppa3 and Klf4 play critical roles in conversion of EpiSCs into mESCs. Klf4 alone can in fact drive this conversion (46), and Dppa3 is activated during this process (47). Furthermore, hESCs express *REX1* (48, 49),

Induced pluripotent stem cells (iPSCs): cells generated by artificial induction of pluripotency in vitro

which is expressed in mESCs but not EpiSCs, and hESCs do not express the EpiSC marker *Fgf5* (50). Finally, hESCs absolutely require Fgf2, whereas mouse EpiSCs do not (51).

2.5. Artificial Acquisition of Pluripotency

2.5.1. Cell fusion.
Cell fusion experiments generating pluripotent cells have been successfully performed using somatic cells and EGCs or ESCs (52–54). Tada et al. (52, 55) demonstrated that EGCs or mESCs could reprogram the nuclei of somatic cells after fusion with thymic lymphocytes. This procedure yields tetraploid cells resembling EGCs in which the originally inactive somatic X chromosome in female cells becomes reactivated and the differential DNA methylation of imprinted loci is erased. After introduction into diploid host blastocysts, those tetraploid cells are able to contribute to chimeric embryos. However, any such contribution is modest and is attributable to the tetraploid nature of the cells (52, 55). On the basis of these studies, it was shown that both bone marrow cells (54) and brain cells isolated from the central nervous system (53) could be fused with mESCs in culture, giving rise to cells that were stem cell–like in terms of morphology and other characteristics (53, 54).

2.5.2. Direct reprogramming of somatic cells into induced pluripotent stem cells.
In an unprecedented study that appeared in 2006, Yamanaka's group (56) first described reprogramming of somatic cells into pluripotent cells via exogenous expression of only four transcription factors. Introduction of the reprogramming factors *Oct4*, *Sox2*, *Klf4*, and *c-Myc* into mouse fibroblasts via retroviral infection, and subsequent selection in mESC medium, resulted in establishment of induced pluripotent stem cells (iPSCs). These cells exhibited all characteristic features of ESCs, including endogenous expression of pluripotency markers, reactivation of both X chromosomes in female cells, and (most importantly) the ability to generate chimeric animals, including contributions to the germ line (56).

On the basis of the initial protocol, different combinations of transcription factors, including *Nanog*, *Lin28*, or *Nr5a2*, and small molecules have been used in subsequent reprogramming strategies employing mouse and human cells (57, 58). Technically, the reprogramming process has been advanced via development of inducible (59) virus-free systems (60), whereby all transcription factors are encoded on a single DNA cassette (61) and integrated at a specific site in the genome. This cassette may even be excised after reprogramming, thus generating unmodified iPSCs (62, 63).

The simplicity of reprogramming is remarkable, but the molecular events involved in this process remain only poorly understood (for a review of current models, see Reference 64). Reprogramming of somatic cells is accompanied by extensive remodeling of epigenetic marks, including DNA demethylation of key pluripotency genes, such as *Oct4* and *Nanog*. A fundamental question is whether iPSCs are equivalent to ESCs. The major concern is that genetic or epigenetic aberrations will alter the differentiation and transplantation properties of iPSCs. Eventually, such abnormalities might lead to malignancy if such cells are used in regenerative medicine. Various studies have shown that iPSC clones appear to be indistinguishable from ESCs. Some studies demonstrated that small differences between iPSCs and ESCs are not attributable to distinct expression signatures but rather to experimental variation (65, 66). Furthermore, the notion of epigenetic memory of the donor cell has been raised (67), describing the reminiscence of functionally relevant epigenetic marks of the somatic donor cell after reprogramming. These may confer different properties to the reprogrammed cells under specific circumstances.

3. MOLECULAR CONTROL OF PLURIPOTENCY

In stem cells, transcription factors can be categorized into three core groups (see **Figure 2**): (*a*) those entrenching a position in the developmental hierarchy (factors of the

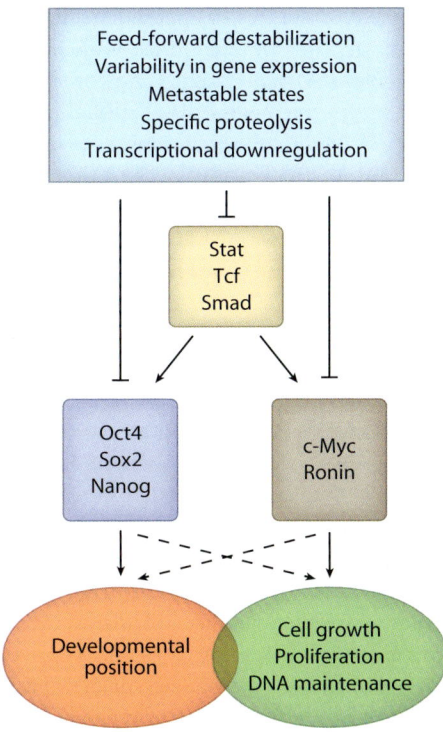

Figure 2

The transcription factor network in pluripotent stem cells. Abbreviations: c-Myc, myelocytomatosis oncogene; Nanog, Nanog homeobox; Oct4, POU domain class 5 transcription factor; Ronin, THAP domain containing 11; Smad, mad homology family; Sox2, SRY-box containing gene 2; Stat, signal transducer and activator of transcription; Tcf, transcription factor.

developmental module); (*b*) those controlling aspects of cell growth and homeostasis (factors of the cell growth module); and (*c*) factors functioning at the interface between complex cell context-specific signaling pathways of the former two modules (those of the cell signaling module). We consider the three genetic modules as separate entities. However, although these modules operate independently to a certain extent, in reality, significant interconnectivity is evident.

3.1. The Developmental Core Module

The best-understood pluripotency factor is Oct4. This protein belongs to the octamer class of transcription factors and is characterized by an ability to recognize the eight-base pair DNA sequence ATGCAAAT (68, 69). Oct4 contains a low-affinity DNA-binding domain (termed POU_S) and a homeotype domain with higher DNA affinity (termed POU_{HD}) (70). Although detailed structural studies on Oct4 have yet to be performed, analysis of Oct1 (Pou2f1) in a complex with target DNA has revealed that two principal configurations exist: Either POU_{HD} alone interacts with DNA, or both POU_S and POU_{HD} do so (71–74). Structural studies of Oct1 in conjunction with another core pluripotency factor of the developmental module, Sox2, have revealed that the latter protein reinforces binding of both forms of Oct1 to DNA (72). This synergy may well explain the peculiar stereotypic configuration found at Oct4 and Oct4/Sox2-recognition sites in the genome.

Oct4 is unique among pluripotency-associated factors in exhibiting an exceptionally confined expression pattern (exclusive to totipotent, pluripotent, and germ cells). Oct4 is also unique because the protein is absolutely required for pluripotency both in vivo (75) and in vitro (76) and for epigenetic reprogramming (64). Upon silencing of Oct4, ESCs spontaneously exit the self-renewal process and differentiate into trophoblast-like cells. This finding was totally unexpected because, in the developing embryo, lineage commitment to the trophoblast line precludes induction of pluripotency (see Section 2.1). It is perhaps even more remarkable that Oct4, in the appropriate molecular context, seems to promote differentiation of ESCs into cells of mesodermal and endodermal lineages (76–78).

Detailed studies have shown that Oct4 interacts with many supplementary transcription factors. Often, these cofactors interact directly with DNA. Specifically, Oct4 appears to interact with Sall4, Hdac2, Sp1, Nanog, Dax1, Nac1, Tcfp2l1, and Essrb (79–81). It is apparent that these various interaction partners do not form a single protein nexus but rather engage (with Oct4) in the formation of multiple complexes differing in composition. This phenomenon may explain, at least in part, the existence of different pluripotent states (see Section 4.3).

The conspicuous interplay between Oct4 and Sox2 is of particular interest. Sox2 exhibits, to some extent, an expression pattern similar to that of Oct4 (82). However, genetic ablation studies indicate that silencing of *Sox2* affects a somewhat later stage of embryogenesis, possibly because of a stronger maternal contribution of Sox2 protein from the oocyte/zygote (82). Interestingly, although acute loss of Sox2 mRNA results in unscheduled differentiation of ESCs into trophoblast-like cells, only a very small subset of Sox2 target genes appear to be affected by *Sox2* loss (83). Thus, the key feature of acute *Sox2* loss appears to be an inability to sustain appropriate Oct4 levels. The observation that forced Oct4 expression can desensitize ESCs to effects to which such cells are normally responsive after Sox2 loss supports this suggestion (83).

The third principal member of the core developmental module is Nanog. This is a homeodomain-containing DNA-binding factor that is not markedly affiliated with any homeobox-containing protein family. Protein expression was originally found to be mandatory for maintenance of pluripotency and, indeed, *Nanog* expression alone is sufficient, under certain circumstances, to perpetuate pluripotency under adverse conditions both in vitro and in vivo (84, 85). Recently, however, functional studies on Nanog have shifted toward exploration of the role played by the protein in inauguration of pluripotency in vitro and in vivo rather than maintenance of the pluripotent state per se (86, 87). As is true of Oct4, Nanog interacts extensively with a number of protein partners including Smad1, Small3, Nr0b1, Nac1, Essrb, Zfp281, Hdac2, and Sp1 (79, 81, 88).

One of the most peculiar facets of the developmental module is that, whereas core pluripotency factors cooperate extensively at the protein level, they also collectively bind to a near-identical repertoire of target genes (**Figure 3**). This has given rise to the hypothesis that the factors form a regulatory circuit. Accordingly, expression of core transcription is self-modulated, and interconnected autoregulatory loops are formed. The core factors firmly control expression of an extensive set of genes required for maintenance of the pluripotent state (89–91). A direct correlation is evident between the number of pluripotency factors bound to each target gene and the level of gene expression. Furthermore, a hierarchical arrangement is apparent. For example, binding of Oct4 increases the likelihood that other factors will be recruited to the same site; such factors include Smad1, Stat3, Dax1, Essrb, and/or Xist (80, 90, 92). Although many genes to which core pluripotency factors bind are transcriptionally stimulated, a particular set of genes, associated with additional cofactors, is repressed by the core developmental module. Such genes are typically involved in lineage-specific regulation, and their expression is essential if a productive physiological differentiation process is to proceed (91, 93–97).

In summary, the core "engine" driving pluripotency consists of a set of highly specialized transcription factors that form a regulatory web. This web is based on extensive interactions and the formation of protein complexes that can vary in stoichiometry and size. Collectively, the core factors transcriptionally control an arsenal of developmental regulators that together

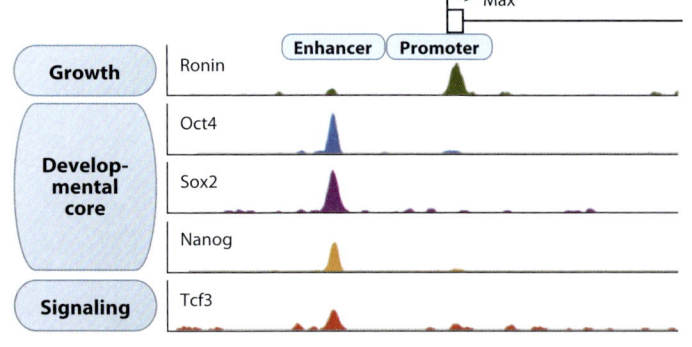

Figure 3
Simultaneous binding of multiple pluripotency-related transcription factors to target genes. Abbreviations: Max, myc-associated factor X; Nanog, Nanog homeobox; Oct4, octamer-binding protein 4; Ronin, THAP domain containing 11; Sox2, SRY-box containing gene 2; Tcf3, transcription factor 3.

regulate development in a lineage hierarchy equivalent to that of peri- or preimplantation epiblast.

3.2. The Cell Growth Module

The transcription factor consortium discussed above firmly controls decisions relevant to developmental fate. However, a second independent gene network can be distinguished. This latter network enables pluripotent stem cells to grow and proliferate, see **Figure 2**.

An important member of this second regulatory module is the proto-oncogene c-Myc. It regulates ESC proliferation and is associated with activation of transcription and opening of chromatin. c-Myc controls the transcription of genes of various functional categories, including cellular metabolism, protein production, and cell cycle regulation (98). c-Myc expression enables ESCs to self-renew even under conditions that do not normally favor such activity (99). Genetic ablation of *c-Myc* causes premature exit from self-renewal (100). One role played by c-Myc is inhibition of expression of the prodifferentiation factor *Gata6* (101). Functional overlap with genes of the developmental module is evident (102, 103). At the protein level, c-Myc has been associated with both Tip60 (an acetyltransferase complex) and the chromatin-remodeling complex p400 (102). Although the Myc-controlled gene set supports ESC self-renewal, the genes also control specific aspects of cancer growth (102).

c-Myc expression is regulated by the transcription factor Ronin (Thap11) (104–106); this is an unorthodox zinc finger–like transcription factor (107, 108) that is a member of a small family of proteins carrying an N-terminal THAP (Thanatos-associated protein) domain. Genes encoding THAP domain-containing proteins evolved from the P element transposon, likely via a process termed molecular domestication (109). In contrast to Oct4 or Nanog, Ronin is exceptionally well conserved between mouse and human in terms of both amino acid sequence (over 95% when the polyQ tract is excluded) (108) and DNA target sequence (108, 110). Loss of Ronin expression is associated with cell death, whereas ectopic overexpression of the protein assists the undifferentiated growth of mESCs (108).

Ronin binds to the promoter regions of numerous metabolism-associated genes, and a significant target overlap with genes bound by c-Myc is evident (106). Genes that regulate development are conspicuously absent from the Ronin-binding set. As with c-Myc, *Ronin* expression is not strictly confined to pluripotent cells; an influence of Ronin activity is suspected in other cellular contexts, including cancer development (104, 105, 111–113). Ronin interacts extensively with other proteins, including the transcriptional coregulator Hcf-1. The antidifferentiation property of Ronin (after Lif removal) is entirely dependent on interaction with Hcf-1 (106); expression of the latter protein is regulated in a complex manner, and Hcf-1 can tether both histone deacetylase and histone lysine-4 methyltransferase at chromatin sites (114). This may explain how Ronin affects transcription of target genes. It appears relevant that Hcf-1 requires association with and proteolytic activation by O-linked *N*-acetylglucosamine transferase, which has been connected to cellular metabolic activity (115). Although core developmental factors interact extensively at the protein level, very few such interactions occur between proteins associated with c-Myc and Ronin. This suggests that, although the systems cooperate functionally, different protein classes are targeted.

3.3. Signaling and the Signaling Module

Several key signaling pathways, and the relevant target transcription factors, converge in the developmental core, permitting extrinsic modulation of pluripotency by developmental cues. These pathways directly reinforce self-renewal and/or maintain pluripotency by blocking differentiation cues. As outlined in **Figure 4**, important pathways in pluripotent cells include those involved in Lif/Stat, Bmp, Fgf2, Tgf-β, and Wnt signaling.

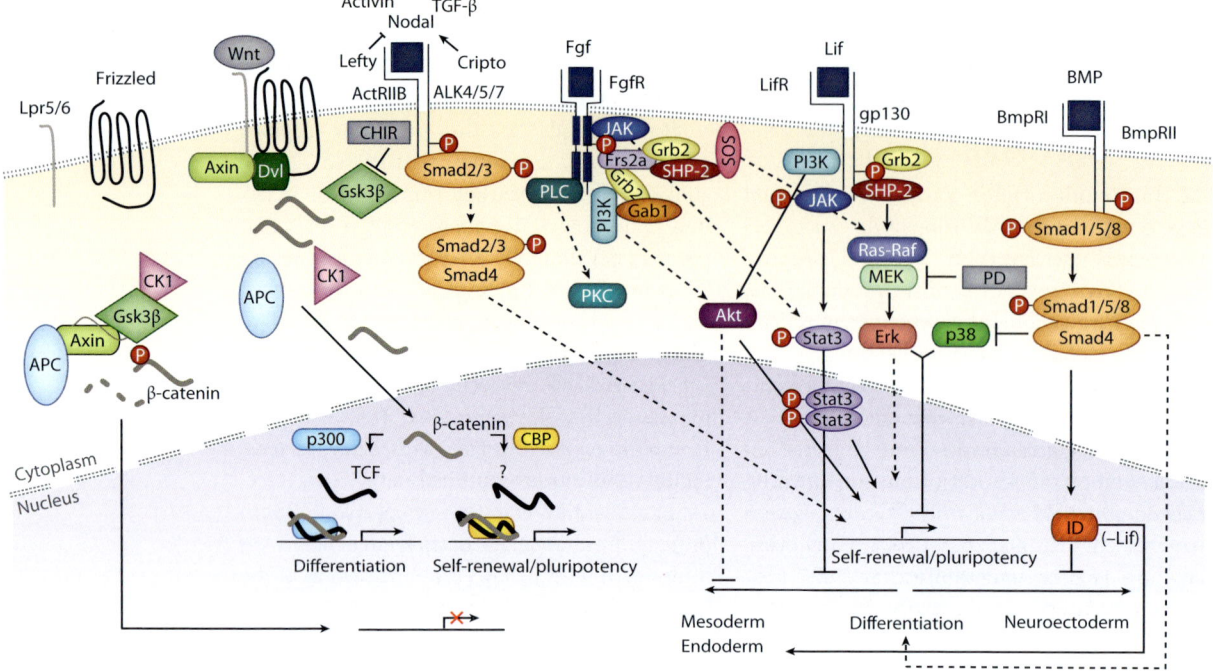

Figure 4

Signaling pathways involved in self-renewal and differentiation of pluripotent stem cells. Black lines and arrows indicate pathways common in primordial and refined pluripotent cells, whereas dashed lines and arrows depict events specific for human embryonic stem cells and cells with refined pluripotency. Abbreviations: ActRIIB, activin receptor type IIB; Akt, thymoma viral proto-oncogene; APC, adenomatosis polyposis coli; ALK4, ALK5, ALK7, activin receptor-like kinase 4, 5, and 7; β-catenin, cadherin-associated protein β1; BMP, bone morphogenic protein; BmpRI and -RII, bone morphogenic protein receptor I and -II; CBP, Creb-binding protein; CHIR, GSK3 inhibitor CHIR99021; CK1, casein kinase 1; Cripto, teratocarcinoma-derived growth factor 1; Dvl, Disheveled protein; Erk, extracellular regulated kinase; Fgf, fibroblast growth factor; Gab1, GRB2-associated binder 1; gp130, glycoprotein 130; Grb2, growth factor receptor-bound protein 2; Gsk3β, glycogen synthase kinase 3β; ID, inhibitor of differentiation; JAK, janus kinase; Lefty, left-right determination factor 1; Lpr5 and -6, low-density lipoprotein receptor-related protein 5 and -6; Lif, leukemia inhibitory factor; LifR, leukemia inhibitory factor receptor; MEK, Mapk/Erk kinase; P, phosphorylated site; PD, Erk inhibitor PD184352; PI3K, phosphatidylinositol 3-OH kinase; PKC, protein kinase C; PLC, phosphoinositide phospholipase C; p38, protein kinase 38; p300, histone acetyltransferase p300; Ras-Raf, rat sarcoma-raf proto-oncogene; SHP-2, SH2 domain-containing protein tyrosine phosphatase-2; SOS, son of sevenless homolog; Smad1, -2, -3, -4, -5, and -8, Smad family members; Stat3, signal transducer and activator of transcription 3; TCF, transcription factor; TGF-β, transforming growth factor-β; Wnt, wingless-type MMTV integration site family.

3.3.1. Leukemia inhibitory factor/signal transducer and activator of transcription 3 signaling.

Lif performs important functions in the early-stage mouse embryo (116) and can promote self-renewal of mESCs even in the absence of a feeder layer. Although Lif expression does not appear to increase the growth rate of mESCs, it enhances the probability that such cells will undergo self-renewal rather than differentiation (117). Lif is a member of the interleukin-6 family of cytokines, and the Lif-signaling effect is mediated via a heterodimeric complex composed of a specific low-affinity Lif receptor (LifR) chain and the gp130 protein. This complex has been shown to activate three pathways: (*a*) the Jak (Janus-associated tyrosine kinase)/Stat3 pathway (118, 119), (*b*) the phosphatidylinositol 3-OH kinase (PI3K)-Akt pathway, and (*c*) the mitogen-activated protein kinase (Mapk) pathway (120). Although the PI3K and Mapk pathways of ESCs are activated via multiple signals, the

Jak/Stat pathway is regulated exclusively by Lif. The Lif receptor complex activates Jak, which in turn phosphorylates specific tyrosine residues in the intracellular domain of gp130. These residues subsequently act as docking sites for proteins containing Src homology 2 (SH2) domains; Stat3 is one such protein. Phosphorylation of Stat3 promotes its dimerization and translocation into the nucleus, wherein the dimer acts as a transcriptional activator of pluripotency-related genes (119).

At the transcriptional level, Lif-mediated signaling via Stat3 and PI3K is integrated into the ESC developmental core module by means of enhanced expression of the transcription factors Klf4 and Tcf3; these proteins in turn activate transcription of *Sox2* and *Nanog*, whereas the Mapk/Erk pathway activity results in diminished amounts of nuclear Tbx3 and therefore loss of its transcriptional activity (121).

3.3.2. Fibroblast growth factor signaling.

FgfR1 is the most abundant receptor of ESCs (39). Importantly, stable interactions between Fgfs and their receptors require the presence of heparin or heparin sulfate. Fgf signaling is activated by ligand-receptor interactions at the cell surface, resulting in autophosphorylation of tyrosine residues in the intracellular region of an FgfR. These altered sites then recruit diverse proteins containing SH2 or phosphotyrosine-binding domains, facilitating assembly of signaling complexes, and in turn allocating signaling to any of four distinct pathways: (*a*) the Jak/Stat pathway, (*b*) the phosphoinositide phospholipase C-γ pathway, (*c*) the PI3K pathway, or (*d*) the Erk pathway (122).

Fgf2 was the first factor identified as being crucial for hESC maintenance, and it is widely accepted that a serum-free culture of hESCs on mouse feeder cells requires soluble Fgf2 if proliferation in an undifferentiated state is to continue (123). The precise mode by which Fgf2 influences cell fate decisions remains to be established.

3.3.3. Tgf-β superfamily.

The Tgf-β superfamily contains structurally related signaling proteins, including Tgf-β, Activin, Nodal, growth differentiation factors (GDFs) and Bmps. All family members are important in maintaining pluripotency. Recent data suggest the existence of a subtle balance between the outcomes of Activin/Nodal- and Bmp-mediated signaling in hESCs; GDF3 may function at the intersection of these pathways, driving self-renewal and blocking differentiation (124, 125).

Canonical Bmp signaling is mediated by phosphorylation of Smad1, -5, and -8 by a BmpRI and -II receptor complex, which sequentially bind to Smad4 and mediate translocation of the complex into the nucleus, wherein specific promoter sequences are bound and expression of Bmp target genes results. Such target genes include those encoding the Id proteins (126), which are helix-loop-helix (HLH) proteins that lack the basic DNA-binding domain, resulting in the formation of inactive dimers with bHLH transcription factors. Such proteins have been shown to inhibit neural, but not mesodermal or endodermal, differentiation (127) and can substitute for Bmp to maintain the pluripotency of mESCs. Lif is still required under such circumstances. These findings were surprising because it is well-established that, in addition to inhibition of neural cell fate via Id activity, Bmp can induce mesodermal differentiation of mESCs in the absence of Lif (128) and plays a conserved role in mesoderm and endoderm patterning during embryogenesis. This suggests that factors downstream of Lif and Bmp must interact to allow maintenance of pluripotency. It has been suggested that these interactions could involve binding of Lif to Smad1 or Stat3 (129), or to the core pluripotency factor Nanog (130). Such binding could negatively regulate expression of Bmp target genes, the transcription of which drives differentiation. Additionally, it has been shown that Bmp4 can support self-renewal by inhibiting Mapk pathways (131). Interestingly, addition of Bmp to hESC medium can trigger differentiation toward a variety of cell types. Antagonization of Bmp signaling enhances ESC self-renewal and neural differentiation (132, 133).

Nodal and Activin are two distinct growth factors that share the same type I receptors (Alk4 and Alk7) and the same type II receptor (ActRIIB), whereas Tgf-β1 generally employs a different receptor (Alk5) to activate Smad2 and -3 signaling. Although the downstream effectors are common to both pathways, the receptors control distinct biological events and are therefore regulated by binding of numerous cofactors. For example, association with teratocarcinoma-derived growth factor 1 (Cripto) is necessary if Nodal is to activate Alk4/ActRIIb receptors, whereas the left-right determination factor 1 (Lefty) can block Nodal receptor binding (134). An additional layer of complexity is added by the involvement of a large number of Smad partners, including transcription factors (such as Runx1 and Gata3), transcriptional repressors (such as E2f4), a calcium-binding protein (calmodulin), and trafficking proteins (such as Importin) (reviewed in Reference 135). Smad signaling is important in control of pluripotency and influences the transcriptional regulation of key transcription factors of which Nanog appears to be the most responsive. In hESCs, Smad2 and -3 bind to the *Nanog* promoter and activate gene expression, whereas Smad1, -5, and -8 (activated by Bmps) bind to the promoter but rather inhibit *Nanog* expression. Additionally, Smad2 or -3 or Fgf2 signaling suppresses *Bmp4* expression, thus preventing spontaneous differentiation (51).

3.3.4. Wnt signaling.
Wnt signaling plays a crucial role during embryonic development. Wnt signaling induces expression of mesodermal and endodermal markers, including *Brachyury*, *Flk1*, *Foxa2*, *Lxh1*, and *Afp* in mESCs; it has additionally been suggested that Wnt signaling probably maintains pluripotency via modulation of *Oct4*, *Sox2*, and *Nanog* expression (136).

The canonical Wnt pathway involves a series of reactions eventually resulting in translocation of β-catenin into the nucleus, to transiently activate target genes that are otherwise repressed by proteins of the Lef/Tcf protein family (137). Cytoplasmic β-catenin is normally phosphorylated at several N-terminal sites (Ser33, Ser37, Ser45, and Thr41) (138), resulting in ubiquitin-mediated proteolysis. Phosphorylation is catalyzed by the so-called destruction complex composed of Axin, Gsk3-β, casein kinase-1 (Ck1), and adenomatous polyposis coli (Apc). Binding of Wnt to the cognate Fz receptor and to the coreceptors low-density lipoprotein receptor-related proteins 5 and 6 (Lrp5 and -6) recruits the protein Dishevelled (Dvl) to the receptor complex. In turn, Dvl binds the scaffold protein Axin and phosphorylated Lrp, as well as Wnt-activated trimeric G proteins participating in degradation of the destruction complex (139). β-catenin remains unphosphorylated and can be translocated into the nucleus to activate its target genes. The protein cannot bind directly to DNA, interacting instead with a number of transcription factors, including Tcf/Lefs, Smads, and nuclear receptors, as well as transcriptional coactivators, such as p300 and Creb-binding protein (CBP) (140). Recent studies suggest that binding to p300 mediates differentiation, whereas formation of the β-catenin/CBP complex maintains the option of mESC self-renewal via interaction with binding partners, including Oct4 and Lrh1, thus activating expression of key pluripotency factors (141, 142). The upstream mechanisms controlling formation of one or the other complex remain under investigation (for a detailed review see Reference 143).

3.4. Transcriptional Control in Pluripotent Stem Cells

The transcriptional circuits considered above are distinguished by the presence of discrete target gene repertoires. In general, the core developmental regulators primarily recruit RNA polymerase II and associated factors and promote transcription initiation. Frequently, however, the RNA polymerase II transcription complex stops on the DNA template shortly after initiation (this is termed stalling or pausing) (144). At least one growth module member (c-Myc) appears, in some instances, to be capable of overcoming this transcriptional impediment by enlisting specific antipausing factors, such as p-TEFb.

Accordingly, Oct4, Sox2, and Nanog typically bind to canonical enhancer regions of target genes. These DNA sequences are associated with nucleosomes bearing histone marks typically found at developmental enhancers, including H3K27me1 and H3K4me1, but are also associated with the acetyltransferase p300 and the mediator complex (90, 145–148).

The functional difference between Oct4 and analogous transcription factors, on the one hand, and Myc, on the other hand, is further emphasized by the presence of different histone marks and chromatin-associated proteins at the binding sites. In contrast to Oct4/Sox2 and Nanog, Myc gene targets are enriched in H3K4me3 and are almost completely devoid of H3K27me3 (90). Another feature characteristic of members of the developmental module is the presence of "bivalent" chromatin domains (see Section 3.5).

3.5. Pluripotency-Associated Chromatin Structure

Each cell type has a unique chromatin architecture, but several features render pluripotency-associated chromatin fundamentally different from that of any other cellular state (149). ESC's chromatin is thought to be more "open" and "dynamic." In pluripotent cells, heterochromatin-associated foci appear to be more diffusely distributed, and the associated histones are generally hyperacetylated (150).

In pluripotent stem cells, chromatin-modifying enzymes interact with the transcriptional modules, discussed above (those controlling development, growth, and cell signaling), by virtue of their control of genes encoding key enzymes capable of covalently modifying proteins. For example, Oct4 directly controls transcription of the H3K9 methyltransferase SetDB1. When histone marks characteristic of this enzyme are established, Polycomb group proteins are recruited, triggering subsequent events, including methylation of H3K27 residues, further enabling additional chromatin-associated modifications (151). Another example features the K9 demethylases of the Jumonji family. Transcription of *Jmjd1a* (*Kdm2a*) and *Jmjd2c* (*Kdm4b*) is directly controlled by Oct4. Loss-of-function studies have revealed prominent roles for these enzymes in terms of ESC self-renewal (152). Another instance of extensive cross connectivity between the developmental core and chromatin-modifying enzymes involves Jarid1b (Kdm5b), a K4 demethylase primarily targeting regions enriched in H3K36me3. This enzyme promotes cell self-renewal, at least in part, by inhibiting cryptic intragenetic transcription (153).

Finally, the Polycomb protein system plays an essential role in control of cell fate decisions in pluripotent stem cells (154). In general, gene regions occupied by these complexes, or by nucleosomes featuring the H3K27me3 modification, are transcriptionally silent. It has been suggested that many Polycomb group -targeted genes are locked in a "poised" state prior to productive expression (94, 95). In many instances, the K27me3 domain, which can span large regions (more than 100 kilobases) of the genome, is coaligned with smaller H3K4me-enriched domains. This configuration has been termed bivalent and has been proposed to provide the molecular basis for developmental gene priming (155). Recently, a Jmj protein (Jarid2) has been shown to be directly associated with the Polycomb complex, although the functional consequences of this linkage are not yet fully resolved (90, 156–158). Polycomb proteins are direct targets of several kinases that mediate cell signaling (159) and are thus intimately connected to cellular signaling processes (see Section 3.3).

The flip side of the "Polycomb coin" is canonical K4 methyltransferase activity. This activity is associated with proteins of the Trx group and therefore is linked to genes that are actively transcribed. Trx proteins have been firmly linked to the effective functioning of pluripotency-specific transcription factors and include Set1a and -b; Mll1-4; the histone mark "reader" Wdr5; and the associated proteins Ash2, Rbp5, and Dpy-30 (160, 161).

Covalent histone modifications provide relatively "hard-wired" substrates affecting transcriptional control at the epigenetic level. A "softer" (thus more dynamic) higher-level nucleosome reconfiguration also plays a critical role in supporting the transcriptional framework necessary for pluripotency. Of particular relevance in this context is ATP-dependent nucleosome remodeling associated with ESC pluripotency and epigenetic reprogramming (93, 162–164). ESCs express a specialized form of the Swi/Snf-related protein complex, termed esBAF. The pluripotency-associated specificity is derived not only from its unique structure but also from the close associations that exist with the transcriptional networks described above. esBAF binds to many Oct4, Sox2, Nanog, Sall4, and c-Myc target genes (165). Functional studies have shown that overexpression of esBAF components enhances epigenetic reprogramming (164). Yet another critical remodeling factor linked to ESC pluripotency is Chd1, a member of the Snf helicase family. Chd1 can recognize H3K4me2 and H3K4me3 marks via an intrinsic chromodomain and is therefore associated with active genes. Chd1 knockdown causes differentiation, increases the presence of heterochromatin-associated features (including H3K9me3 and Hp1), and reduces the level of the exchange linker H1.

In addition to covalent and noncovalent mechanisms promoting the pluripotency-specific transcriptional program, other chromatin/epigenetic regulatory mechanisms have been characterized. Even though 60%–80% of all CpG sites in the ESC genome are methylated (166), ESC lines can be readily isolated and propagated from embryos deficient in DNA methyltransferase activity (167). However, such cells cannot properly differentiate (168); this has been partly attributed to the inability of the cells to appropriately silence pluripotency-associated factors, such as the genes encoding *Oct4* and *Nanog*, which are typically permanently modified in somatic tissues via DNA methylation. Hence, although loss of DNA methylation capacity can be compensated, or is irrelevant, in terms of ESC propagation, it is detrimental in terms of the ability of ESCs to appropriately differentiate.

Ultimately, one of the best-understood epigenetic phenomena (in the context of pluripotency) is X chromosome inactivation and reactivation. In fact, X chromosome status is frequently used as a marker of change in the cellular state of pluripotent cells because, in most somatic cells, one X chromosome is inactive (169). The activation state of the X chromosome is considered to be directly linked to the core developmental regulatory network because several pluripotency factors, including Oct4, Sox2, and Nanog, bind to a regulatory element in the first exon of *Xist* (92). In particular, reactivation of an inactive chromosome is associated with *Nanog* expression, and blastomeres lacking the *Nanog* gene fail to reactivate an inactive X chromosome (86).

3.6. Noncoding RNA and Pluripotency

An emerging major category of genes controlled by the core module is transcribed into noncoding RNAs. Among these, the best-understood class in terms of function is the family of microRNAs (miRNAs), short noncoding RNAs capable of destabilizing and repressing specific target RNAs. Generally, miRNAs are processed by the enzymes Dicer and Dcgr8, and genetic ablation of the genes encoding these enzymes affects the cell cycle and differentiation of ESCs (170–173). As noted above, when other pluripotency-controlling mechanisms were discussed, the key developmental regulatory core firmly controls expression of the miRNAs found in pluripotent cells. Oct4, Sox2, and Nanog bind directly to the miRNAs mir302 and mir290–295 (96). Conceptually, the dominant idea is that a major function of miRNAs in pluripotent stem cells involves control of the cell cycle.

Another very important member of the miRNA family is let7, which targets some pluripotency-associated genes (174, 175). This in turn provides the basis for establishment of negative feedback loops in which let7 expression is negatively regulated by the RNA-binding protein Lin28. Upon differentiation of pluripotent cells, Lin28 is downregulated, resulting in stabilization of and an increase in the level of let7, which has differentiation-promoting activities itself (probably partly via downregulation of Myc) (176). In this context, it is relevant to note that mir290–295 enhances reprogramming in a Myc-dependent fashion (177, 178).

It has been reported that the mir302/367 cluster alone is capable of reprogramming somatic cells to the pluripotent state, possibly via targeting of specific epigenetic factors, such as Aof1, Aof2, Mecp1, and Mecp2 (179–181). This finding, if confirmed, indicates that the pluripotent state is fundamentally distinct from other cellular states in the sense that pluripotency is the common cellular form into which virtually all cell types convert. Recently, Oct4 has been shown to control and activate the expression of specific large intergenic noncoding RNAs (lincRNAs). Importantly, in some instances, knockdown of the expression of such RNAs caused growth defects and increased the level of apoptosis (182); some lincRNAs have been implicated in self-renewal and reprogramming (182, 183).

4. CONTROLLED SHUTDOWN OF PLURIPOTENCY DURING DIFFERENTIATION

Given that the pluripotent state is well-protected from the actions of differentiation stimuli via use of a series of autoregulatory feed-forward and feedback loops, it is logical to ask how robustly self-renewing pluripotent cells can eschew self-renewal.

4.1. Feed-Forward Generated Destabilization Signals

Although developmental regulators such as Oct4 and Sox2 firmly reinforce maintenance of the pluripotent state, by blocking expression of core developmental regulators of hypoblast and trophoblast differentiation, Oct4 and Sox2 also activate a set of genes that, paradoxically, promote differentiation and hence destabilize self-renewal. The first such factor characterized was the growth factor Fgf4. Oct4 and Sox2 transcriptionally activate Fgf4 (184). Fgf4 is a forceful activator of the Erk-signaling pathway, which promotes differentiation in cell types exhibiting primordial pluripotency (see Section 3.3). Hence, ESCs create, when self-renewing, an environment destructive of pluripotency via autoactivation of Erk signaling. However, elevation of Erk activity does not necessarily induce differentiation but renders cells more susceptible to the actions of additional differentiation-promoting signals (78). Indeed, the situation is even more complex. Oct4 not only activates Fgf4 signaling but also triggers expression of a genetic program responsible for mesendoderm differentiation, whereas Sox2 inhibits the pathway (77). Accordingly, quantitative differences in stoichiometry, posttranslational modifications, and other means permit substates of pluripotency, some of which are more prone to differentiation.

4.2. Variability in the Expression of Pluripotency Factors

Examination of a cross section of cells in ES culture at any given moment reveals a broad spectrum of Nanog expression among individual ESCs. This heterogeneity affects the propensity of ESCs to follow particular developmental pathways; the ESCs expressing Nanog at the highest level are relatively protected from the effects of differentiation stimuli and Erk signaling, but cells expressing low levels of or no Nanog exhibit a greatly increased tendency toward lineage commitment (86).

Thus, Nanog insulates ESCs from prodifferentiation signals, in particular those caused by Erk signaling. Nanog fluctuation provides an example in which heterogeneity in ESC culture defines subsets of cells that are more prone to differentiation stimuli. This could open a window of opportunity whereby such cells could exit the self-renewal mode. The mechanisms underlying this fluctuation are currently unknown but may involve gene oscillation and/or stochastic (noisy) genetic expression (185).

4.3. Metastable States in Stem Cell Culture

4.3.1. Intrinsic factors that can stimulate transitions. Although Nanog expression is probably the best-characterized example of transcription factor heterogeneity in tissue culture, in line with phenotypic differences in stem cell behavior, other transcription factors exhibiting apparently similar behavior have been described (47, 186). For example, Stella expression can be used to divide the stem cell population into two subpopulations: Stella-positive cells that more closely resemble cells of the ICM, and Stella-negative cells that are more similar to epiblast cells. Conceptually, the underlying idea is that several separate transcriptional states exist. These individual states are frequently described as metastable states and would mathematically most closely resemble "attractor" states.

Although the idea that stable distinct states exist is attractive, it is equally possible that cell states vary over a much wider spectrum, with no cell permanently locked into any given state (187, 188). Finally, it is possible, although speculative, that the intrinsic heterogeneity of pluripotent stem cells, and the observed interconvertability among states, allow establishment of a lineage hierarchy within which only a small subset of cells is truly pluripotent. These cells spin off derivatives, all of which are to some extent compromised in terms of developmental and differentiation potential.

4.3.2. Extrinsic factors that stimulate transitions. If culture conditions are varied, or if the expression levels of transcription factors are changed, cellular properties diverge, leading to conversion of one pluripotent cell type into another. For example, ESCs can change into EpiSCs in the presence of Fgf2 and Activin. This process is accompanied by silencing of one X chromosome in female cells. The reverse conversion, including reactivation of one X chromosome, has been accomplished via artificial expression of Klf4 (46) and upon culture with Lif, Bmp4, and "2i" (a cocktail of small-molecule inhibitors targeting Erk and Gsk) in the presence of a feeder layer (189). Furthermore, EpiSCs can be converted into EGCs via PGC intermediates upon culture with Bmp4, followed by addition of Noggin, Chordin, Activin, and Fgf2. Once PGCs become established, addition of Lif, FCS, and Fgf2 triggers development of EGCs (47). Additionally, ESCs can give rise to TS cells upon conditional deletion of Oct4 or upon forced expression of *Cdx2* when cultured in MEF-conditioned medium in the presence of Fgf4 (190).

4.4. Rapid Proteolytic Removal of Pluripotency Factors

The pluripotency transcription factor network is a complex genetic system that includes network hubs, that is, proteins that interact with many other proteins. Oct4, Nanog, and Myc serve as such hubs. In general, cells are very sensitive to changes in the levels of these three proteins. A very effective approach toward a change in cellular state is obstruction of protein hub action. This can be achieved, in part, via site-specific proteolysis catalyzed by caspases. These enzymes are typically discussed in the context of programmed cell death (apoptosis) (191) but have recently been investigated in a different context, namely cell specialization (192, 193). In particular, caspase-3, a key "executor" caspase, is involved in cell fate decisions by specifically cleaving Nanog (194). Nanog depletion reduces self-renewal capacity and propels ESCs toward differentiation. In line with this notion, genetic ablation or blocking of caspase activity causes a significant delay in differentiation. The effect is tissue culture specific as caspase-3 knockout embryos survive beyond the interval during which pluripotent cells are present (195), likely owing to compensatory mechanisms (196). It is also likely that other critical transcription factors at hubs of the pluripotency network are recognized and cleaved by caspases; these factors may include Ronin, a recently characterized member of the growth module (108). Finally, caspase activity seems to promote not only differentiation but also epigenetic reprogramming (197) by depleting the retinoblastoma (Rb) protein and thus promoting transition toward an ESC-like state.

4.5. Transcriptionally Mediated Downregulation of Pluripotency Factors

Direct downregulation of transcription factor synthesis is probably as important as the above-mentioned mechanisms in terms of deconstructing the pluripotent state. One critical factor destabilizing pluripotency is Gcnf (198), which binds directly to the Oct4 promoter and subsequently "locks" pluripotent stem cells into a differentiated state via recruitment of newly expressed DNA methyltransferases (199). Another enzyme that may silence Oct4 expression (either transiently after initiation of differentiation or more permanently later) is the histone H3K9 methyltransferase G9a (200).

Similar chromatin-modifying activities are likely to act in other pluripotency-associated gene regions. Furthermore, direct binding of certain transcriptional repressors (including ARP-1/COUP-TFI and EAR3-COUP-TFII) to the Oct4 promoter has been reported (199). In general, it appears that some transcription factors, including Tcf3, antagonize the action of the core developmental module; Tcf3 and related proteins can rapidly shut down the pluripotency network (200).

5. CONCLUDING REMARKS

Although much information is available on the molecular regulation of pluripotency, a unifying view is lacking. At present we cannot with any degree of certainty predict whether a particular cell is totipotent, pluripotent, or multipotent, or whether it can contribute to the germ line. Similarly, we do not understand the major differences that certainly exist between pluripotent stem cells in vitro and in vivo. Only when this is possible will we be able to state that a major advance toward an understanding of pluripotency and nuclear reprogramming has been made.

SUMMARY POINTS

1. Pluripotency is acquired by a subset of cells during early embryonic development and is distinct from the totipotent state of the zygote and blastomeres. Totipotent cells can differentiate into all cell types of the embryo and all extraembryonic lineages, whereas pluripotent cells cannot contribute to the development of extraembryonic tissue.

2. Pluripotent stem cells are characterized by the ability to self-renew and to differentiate into any cell type of the developing embryo. They have been trapped in vitro, at various stages of embryonic development, and from postnatal and adult tissues.

3. Primordial and refined pluripotency are apparent during embryonic development, and in different forms of pluripotent stem cells in culture. Primordial pluripotency, equivalent to that exhibited by the preimplantation epiblast, is characteristic of cells that can differentiate into any embryonic cell lineage (including the germ line). Cells harboring a refined pluripotency, equivalent to that of the late epiblast, can differentiate into cells of all germ layers in teratoma formation assays but cannot contribute to chimeric animals, and therefore have restricted developmental potential.

4. Important pathways in pluripotent cells include the Lif/Stat, Bmp, Fgf2-, Tgfβ-, and Wnt-signaling cascades. It is not surprising that the activities of such pathways are also critical during early-stage embryogenesis.
5. The core factors transcriptionally control an arsenal of developmental regulators that together regulate development in a lineage hierarchy equivalent to that of a pre- or peri-implantation epiblast.
6. The cell growth network enables pluripotent stem cells to grow and proliferate. This function is highly adaptive and serves the very specific context-dependent needs of pluripotent stem cells.
7. Pluripotent stem cells have a unique chromatin architecture, differing fundamentally from that of any other cell type. ESC chromatin is thought to be more open and dynamic.
8. Key signaling pathways, and the relevant DNA-targeting transcription factors, converge in the developmental core, permitting extrinsic modulation by developmental cues.
9. Direct downregulation of transcription factor synthesis and protein degradation are probably the most important mechanisms operative in terms of deconstructing the pluripotent state.

FUTURE ISSUES

1. What is the true origin of mouse embryonic stem cells?
2. How can hESCs with primordial pluripotency be isolated?
3. What are the controls mediated by all transcriptional and epigenetic modules in pluripotent stem cells?
4. How is the pluripotent state deconstructed?

DISCLOSURE STATEMENT

The authors are not aware of any affiliations, memberships, funding, or financial holdings that might be perceived as affecting the objectivity of this review.

ACKNOWLEDGMENTS

We thank Li-Fang Chu for providing us with microphotographs. We apologize to authors whose work has not been cited owing to space limitations. T.P.Z. is supported by the National Institutes of Health (GM077442, GM81627, and 1RC2AR058919) and the Huffington Foundation.

LITERATURE CITED

1. Selwood L, Johnson MH. 2006. Trophoblast and hypoblast in the monotreme, marsupial and eutherian mammal: evolution and origins. *BioEssays* 28(2):128–45
2. Beddington RS, Robertson EJ. 1999. Axis development and early asymmetry in mammals. *Cell* 96(2):195–209

3. Kelly SJ. 1977. Studies of the developmental potential of 4- and 8-cell stage mouse blastomeres. *J. Exp. Zool.* 200(3):365–76
4. Egli D, Birkhoff G, Eggan K. 2008. Mediators of reprogramming: transcription factors and transitions through mitosis. *Nat. Rev. Mol. Cell Biol.* 9(7):505–16
5. Johnson MH, Ziomek CA. 1981. The foundation of two distinct cell lineages within the mouse morula. *Cell* 24(1):71–80
6. Pratt HP, Bolton VN, Gudgeon KA. 1983. The legacy from the oocyte and its role in controlling early development of the mouse embryo. *Ciba Found. Symp.* 98:197–227
7. Ozawa M, Ringwald M, Kemler R. 1990. Uvomorulin-catenin complex formation is regulated by a specific domain in the cytoplasmic region of the cell adhesion molecule. *Proc. Natl. Acad. Sci. USA* 87(11):4246–50
8. Nichols J, Smith A. 2009. Naive and primed pluripotent states. *Cell Stem Cell* 4(6):487–92
9. Hahnel AC, Rappolee DA, Millan JL, Manes T, Ziomek CA, et al. 1990. Two alkaline phosphatase genes are expressed during early development in the mouse embryo. *Development* 110(2):555–64
10. Gardner RL, Beddington RS. 1988. Multi-lineage 'stem' cells in the mammalian embryo. *J. Cell Sci. Suppl.* 10:11–27
11. Monk M, Boubelik M, Lehnert S. 1987. Temporal and regional changes in DNA methylation in the embryonic, extraembryonic and germ cell lineages during mouse embryo development. *Development* 99(3):371–82
12. Hogan BL. 1976. Changes in the behaviour of teratocarcinoma cells cultivated in vitro. *Nature* 263(5573):136–37
13. Brinster RL. 1974. The effect of cells transferred into the mouse blastocyst on subsequent development. *J. Exp. Med.* 140(4):1049–56
14. Papaioannou VE, McBurney MW, Gardner RL, Evans MJ. 1975. Fate of teratocarcinoma cells injected into early mouse embryos. *Nature* 258(5530):70–73
15. Tanaka S, Kunath T, Hadjantonakis AK, Nagy A, Rossant J. 1998. Promotion of trophoblast stem cell proliferation by FGF4. *Science* 282(5396):2072–75
16. Kunath T, Arnaud D, Uy GD, Okamoto I, Chureau C, et al. 2005. Imprinted X-inactivation in extra-embryonic endoderm cell lines from mouse blastocysts. *Development* 132(7):1649–61
17. Evans MJ, Kaufman MH. 1981. Establishment in culture of pluripotential cells from mouse embryos. *Nature* 292(5819):154–56
18. Martin GR. 1981. Isolation of a pluripotent cell line from early mouse embryos cultured in medium conditioned by teratocarcinoma stem cells. *Proc. Natl. Acad. Sci. USA* 78(12):7634–38
19. Smith AG, Heath JK, Donaldson DD, Wong GG, Moreau J, et al. 1988. Inhibition of pluripotential embryonic stem cell differentiation by purified polypeptides. *Nature* 336(6200):688–90
20. Williams RL, Hilton DJ, Pease S, Willson TA, Stewart CL, et al. 1988. Myeloid leukaemia inhibitory factor maintains the developmental potential of embryonic stem cells. *Nature* 336(6200):684–87
21. Meissner A, Wernig M, Jaenisch R. 2007. Direct reprogramming of genetically unmodified fibroblasts into pluripotent stem cells. *Nat. Biotechnol.* 25(10):1177–81
22. Chazaud C, Yamanaka Y, Pawson T, Rossant J. 2006. Early lineage segregation between epiblast and primitive endoderm in mouse blastocysts through the Grb2-Mapk pathway. *Dev. Cell* 10(5):615–24
23. Silva J, Smith A. 2008. Capturing pluripotency. *Cell* 132(4):532–36
24. Tesar PJ, Chenoweth JG, Brook FA, Davies TJ, Evans EP, et al. 2007. New cell lines from mouse epiblast share defining features with human embryonic stem cells. *Nature* 448(7150):196–99
25. Brons IGM, Smithers LE, Trotter MWB, Rugg-Gunn P, Sun B, et al. 2007. Derivation of pluripotent epiblast stem cells from mammalian embryos. *Nature* 448(7150):191–95
26. Matsui Y, Zsebo K, Hogan BL. 1992. Derivation of pluripotential embryonic stem cells from murine primordial germ cells in culture. *Cell* 70(5):841–47
27. Resnick JL, Bixler LS, Cheng L, Donovan PJ. 1992. Long-term proliferation of mouse primordial germ cells in culture. *Nature* 359(6395):550–51
28. Shamblott MJ, Axelman J, Wang S, Bugg EM, Littlefield JW, et al. 1998. Derivation of pluripotent stem cells from cultured human primordial germ cells. *Proc. Natl. Acad. Sci. USA* 95(23):13726–31

29. Godin I, Deed R, Cooke J, Zsebo K, Dexter M, Wylie CC. 1991. Effects of the steel gene product on mouse primordial germ cells in culture. *Nature* 352(6338):807–9
30. Surani MA. 1999. Reprogramming a somatic nucleus by trans-modification activity in germ cells. *Semin. Cell Dev. Biol.* 10(3):273–77
31. Zwaka TP, Thomson JA. 2005. A germ cell origin of embryonic stem cells? *Development* 132(2):227–33
32. Clark AT, Bodnar MS, Fox M, Rodriquez RT, Abeyta MJ, et al. 2004. Spontaneous differentiation of germ cells from human embryonic stem cells in vitro. *Hum. Mol. Genet.* 13(7):727–39
33. Hübner K, Fuhrmann G, Christenson LK, Kehler J, Reinbold R, et al. 2003. Derivation of oocytes from mouse embryonic stem cells. *Science* 300(5623):1251–56
34. Geijsen N, Horoschak M, Kim K, Gribnau J, Eggan K, Daley GQ. 2004. Derivation of embryonic germ cells and male gametes from embryonic stem cells. *Nature* 427(6970):148–54
35. Kanatsu-Shinohara M, Inoue K, Lee J, Yoshimoto M, Ogonuki N, et al. 2004. Generation of pluripotent stem cells from neonatal mouse testis. *Cell* 119(7):1001–12
36. Guan K, Nayernia K, Maier LS, Wagner S, Dressel R, et al. 2006. Pluripotency of spermatogonial stem cells from adult mouse testis. *Nature* 440(7088):1199–203
37. Seandel M, James D, Shmelkov SV, Falciatori I, Kim J, et al. 2007. Generation of functional multipotent adult stem cells from GPR125[+] germline progenitors. *Nature* 449(7160):346–50
38. Thomson JA, Itskovitz-Eldor J, Shapiro SS, Waknitz MA, Swiergiel JJ, et al. 1998. Embryonic stem cell lines derived from human blastocysts. *Science* 282(5391):1145–47
39. Dvorak P, Dvorakova D, Koskova S, Vodinska M, Najvirtova M, et al. 2005. Expression and potential role of fibroblast growth factor 2 and its receptors in human embryonic stem cells. *Stem Cells* 23(8):1200–11
40. Vallier L, Alexander M, Pedersen RA. 2005. Activin/Nodal and FGF pathways cooperate to maintain pluripotency of human embryonic stem cells. *J. Cell Sci.* 118(Pt. 19):4495–509
41. Pébay A, Wong RCB, Pitson SM, Wolvetang EJ, Peh GS-L, et al. 2005. Essential roles of sphingosine-1-phosphate and platelet-derived growth factor in the maintenance of human embryonic stem cells. *Stem Cells* 23(10):1541–48
42. Bendall SC, Stewart MH, Menendez P, George D, Vijayaragavan K, et al. 2007. IGF and FGF cooperatively establish the regulatory stem cell niche of pluripotent human cells in vitro. *Nature* 448(7157):1015–21
43. Avery S, Inniss K, Moore H. 2006. The regulation of self-renewal in human embryonic stem cells. *Stem Cells Dev.* 15(5):729–40
44. Vallier L, Mendjan S, Brown S, Chng Z, Teo A, et al. 2009. Activin/Nodal signalling maintains pluripotency by controlling Nanog expression. *Development* 136(8):1339–49
45. Clark AT, Rodriguez RT, Bodnar MS, Abeyta MJ, Cedars MI, et al. 2004. Human *STELLAR, NANOG*, and *GDF3* genes are expressed in pluripotent cells and map to chromosome 12p13, a hotspot for teratocarcinoma. *Stem Cells* 22(2):169–79
46. Guo G, Yang J, Nichols J, Hall JS, Eyres I, et al. 2009. Klf4 reverts developmentally programmed restriction of ground state pluripotency. *Development* 136(7):1063–69
47. Hayashi K, Lopes SMCdS, Tang F, Surani MA. 2008. Dynamic equilibrium and heterogeneity of mouse pluripotent stem cells with distinct functional and epigenetic states. *Cell Stem Cell* 3(4):391–401
48. Int. Stem Cell Initiat., Adewumi O, Aflatoonian B, Ahrlund-Richter L, Amit M, et al. 2007. Characterization of human embryonic stem cell lines by the International Stem Cell Initiative. *Nat. Biotechnol.* 25(7):803–16
49. Babaie Y, Herwig R, Greber B, Brink TC, Wruck W, et al. 2007. Analysis of Oct4-dependent transcriptional networks regulating self-renewal and pluripotency in human embryonic stem cells. *Stem Cells* 25(2):500–10
50. Assou S, Le Carrour T, Tondeur S, Ström S, Gabelle A, et al. 2007. A meta-analysis of human embryonic stem cells transcriptome integrated into a Web-based expression atlas. *Stem Cells* 25(4):961–73
51. Greber B, Wu G, Bernemann C, Joo JY, Han DW, et al. 2010. Conserved and divergent roles of FGF signaling in mouse epiblast stem cells and human embryonic stem cells. *Cell Stem Cell* 6(3):215–26
52. Tada M, Tada T, Lefebvre L, Barton SC, Surani MA. 1997. Embryonic germ cells induce epigenetic reprogramming of somatic nucleus in hybrid cells. *EMBO J.* 16(21):6510–20

53. Ying Q-L, Nichols J, Evans EP, Smith AG. 2002. Changing potency by spontaneous fusion. *Nature* 416(6880):545–48
54. Terada N, Hamazaki T, Oka M, Hoki M, Mastalerz DM, et al. 2002. Bone marrow cells adopt the phenotype of other cells by spontaneous cell fusion. *Nature* 416(6880):542–45
55. Tada M, Takahama Y, Abe K, Nakatsuji N, Tada T. 2001. Nuclear reprogramming of somatic cells by in vitro hybridization with ES cells. *Curr. Biol.* 11(19):1553–58
56. Takahashi K, Yamanaka S. 2006. Induction of pluripotent stem cells from mouse embryonic and adult fibroblast cultures by defined factors. *Cell* 126(4):663–76
57. Ichida JK, Blanchard J, Lam K, Son EY, Chung JE, et al. 2009. A small-molecule inhibitor of Tgf-β signaling replaces *Sox2* in reprogramming by inducing *Nanog*. *Cell Stem Cell* 5(5):491–503
58. Yu J, Vodyanik MA, Smuga-Otto K, Antosiewicz-Bourget J, Frane JL, et al. 2007. Induced pluripotent stem cell lines derived from human somatic cells. *Science* 318(5858):1917–20
59. Hanna J, Markoulaki S, Schorderet P, Carey BW, Beard C, et al. 2008. Direct reprogramming of terminally differentiated mature B lymphocytes to pluripotency. *Cell* 133(2):250–64
60. Woltjen K, Michael IP, Mohseni P, Desai R, Mileikovsky M, et al. 2009. *piggyBac* transposition reprograms fibroblasts to induced pluripotent stem cells. *Nature* 458(7239):766–70
61. Carey BW, Markoulaki S, Hanna J, Saha K, Gao Q, et al. 2009. Reprogramming of murine and human somatic cells using a single polycistronic vector. *Proc. Natl. Acad. Sci. USA* 106(1):157–62
62. Stadtfeld M, Nagaya M, Utikal J, Weir G, Hochedlinger K. 2008. Induced pluripotent stem cells generated without viral integration. *Science* 322(5903):945–49
63. Okita K, Nakagawa M, Hyenjong H, Ichisaka T, Yamanaka S. 2008. Generation of mouse induced pluripotent stem cells without viral vectors. *Science* 322(5903):949–53
64. Hanna JH, Saha K, Jaenisch R. 2010. Pluripotency and cellular reprogramming: facts, hypotheses, unresolved issues. *Cell* 143(4):508–25
65. Guenther MG, Frampton GM, Soldner F, Hockemeyer D, Mitalipova M, et al. 2010. Chromatin structure and gene expression programs of human embryonic and induced pluripotent stem cells. *Cell Stem Cell* 7(2):249–57
66. Newman AM, Cooper JB. 2010. Lab-specific gene expression signatures in pluripotent stem cells. *Cell Stem Cell* 7(2):258–62
67. Kim K, Doi A, Wen B, Ng K, Zhao R, et al. 2010. Epigenetic memory in induced pluripotent stem cells. *Nature* 467(7313):285–90
68. Falkner FG, Zachau HG. 1984. Correct transcription of an immunoglobulin kappa gene requires an upstream fragment containing conserved sequence elements. *Nature* 310(5972):71–74
69. Parslow TG, Blair DL, Murphy WJ, Granner DK. 1984. Structure of the 5′ ends of immunoglobulin genes: a novel conserved sequence. *Proc. Natl. Acad. Sci. USA* 81(9):2650–54
70. Herr W, Cleary MA. 1995. The POU domain: versatility in transcriptional regulation by a flexible two-in-one DNA-binding domain. *Genes Dev.* 9(14):1679–93
71. Williams DC Jr, Cai M, Clore GM. 2004. Molecular basis for synergistic transcriptional activation by Oct1 and Sox2 revealed from the solution structure of the 42-kDa Oct1·Sox2·*Hoxb1*-DNA ternary transcription factor complex. *J. Biol. Chem.* 279(2):1449–57
72. Lundbäck T, Chang JF, Phillips K, Luisi B, Ladbury JE. 2000. Characterization of sequence-specific DNA binding by the transcription factor Oct-1. *Biochemistry* 39(25):7570–79
73. Phillips K, Luisi B. 2000. The virtuoso of versatility: POU proteins that flex to fit. *J. Mol. Biol.* 302(5):1023–39
74. Nichols J, Zevnik B, Anastassiadis K, Niwa H, Klewe-Nebenius D, et al. 1998. Formation of pluripotent stem cells in the mammalian embryo depends on the POU transcription factor Oct4. *Cell* 95(3):379–91
75. Niwa H, Miyazaki J, Smith AG. 2000. Quantitative expression of Oct-3/4 defines differentiation, dedifferentiation or self-renewal of ES cells. *Nat. Genet.* 24(4):372–76
76. Thomson M, Liu SJ, Zou L-N, Smith Z, Meissner A, Ramanathan S. 2011. Pluripotency factors in embryonic stem cells regulate differentiation into germ layers. *Cell* 145(6):875–89
77. Kunath T, Saba-El-Leil MK, Almousailleakh M, Wray J, Meloche S, Smith A. 2007. FGF stimulation of the Erk1/2 signalling cascade triggers transition of pluripotent embryonic stem cells from self-renewal to lineage commitment. *Development* 134(16):2895–902

78. Wang J, Rao S, Chu J, Shen X, Levasseur DN, et al. 2006. A protein interaction network for pluripotency of embryonic stem cells. *Nature* 444(7117):364–68
79. van den Berg DLC, Snoek T, Mullin NP, Yates A, Bezstarosti K, et al. 2010. An Oct4-centered protein interaction network in embryonic stem cells. *Cell Stem Cell* 6(4):369–81
80. Liang J, Wan M, Zhang Y, Gu P, Xin H, et al. 2008. Nanog and Oct4 associate with unique transcriptional repression complexes in embryonic stem cells. *Nat. Cell Biol.* 10(6):731–79
81. Avilion AA, Nicolis SK, Pevny LH, Perez L, Vivian N, Lovell-Badge R. 2003. Multipotent cell lineages in early mouse development depend on SOX2 function. *Genes Dev.* 17(1):126–40
82. Masui S, Nakatake Y, Toyooka Y, Shimosato D, Yagi R, et al. 2007. Pluripotency governed by SOX2 via regulation of Oct3/4 expression in mouse embryonic stem cells. *Nat. Cell Biol.* 9(6):625–35
83. Mitsui K, Tokuzawa Y, Itoh H, Segawa K, Murakami M, et al. 2003. The homeoprotein Nanog is required for maintenance of pluripotency in mouse epiblast and ES cells. *Cell* 113(5):631–42
84. Chambers I, Colby D, Robertson M, Nichols J, Lee S, et al. 2003. Functional expression cloning of Nanog, a pluripotency sustaining factor in embryonic stem cells. *Cell* 113(5):643–55
85. Chambers I, Silva J, Colby D, Nichols J, Nijmeijer B, et al. 2007. Nanog safeguards pluripotency and mediates germline development. *Nature* 450(7173):1230–34
86. Ivanova N, Dobrin R, Lu R, Kotenko I, Levorse J, et al. 2006. Dissecting self-renewal in stem cells with RNA interference. *Nature* 442(7102):533–38
87. Wu Q, Chen X, Zhang J, Loh Y-H, Low T-Y, et al. 2006. Sall4 interacts with Nanog and co-occupies Nanog genomic sites in embryonic stem cells. *J. Biol. Chem.* 281(34):24090–94
88. Boyer LA, Lee TI, Cole MF, Johnstone SE, Levine SS, et al. 2005. Core transcriptional regulatory circuitry in human embryonic stem cells. *Cell* 122(6):947–56
89. Chen X, Xu H, Yuan P, Fang F, Huss M, et al. 2008. Integration of external signaling pathways with the core transcriptional network in embryonic stem cells. *Cell* 133(6):1106–17
90. Loh Y-H, Wu Q, Chew J-L, Vega VB, Zhang W, et al. 2006. The Oct4 and Nanog transcription network regulates pluripotency in mouse embryonic stem cells. *Nat. Genet.* 38(4):431–40
91. Navarro P, Chambers I, Karwacki-Neisius V, Chureau C, Morey C, et al. 2008. Molecular coupling of *Xist* regulation and pluripotency. *Science* 321(5896):1693–95
92. Bilodeau S, Kagey MH, Frampton GM, Rahl PB, Young RA. 2009. SetDB1 contributes to repression of genes encoding developmental regulators and maintenance of ES cell state. *Genes Dev.* 23(21):2484–89
93. Boyer LA, Plath K, Zeitlinger J, Brambrink T, Medeiros LA, et al. 2006. Polycomb complexes repress developmental regulators in murine embryonic stem cells. *Nature* 441(7091):349–53
94. Lee TI, Jenner RG, Boyer LA, Guenther MG, Levine SS, et al. 2006. Control of developmental regulators by polycomb in human embryonic stem cells. *Cell* 125(2):301–13
95. Marson A, Levine SS, Cole MF, Frampton GM, Brambrink T, et al. 2008. Connecting microRNA genes to the core transcriptional regulatory circuitry of embryonic stem cells. *Cell* 134(3):521–33
96. Pasini D, Bracken AP, Jensen MR, Lazzerini Denchi E, Helin K. 2004. Suz12 is essential for mouse development and for EZH2 histone methyltransferase activity. *EMBO J.* 23(20):4061–71
97. Eilers M, Eisenman RN. 2008. Myc's broad reach. *Genes Dev.* 22(20):2755–66
98. Cartwright P, McLean C, Sheppard A, Rivett D, Jones K, Dalton S. 2005. LTF/STAT3 controls ES cell self-renewal and pluripotency by a Myc-dependent mechanism. *Development* 132(5):885–96
99. Smith KN, Lim J-M, Wells L, Dalton S. 2011. Myc orchestrates a regulatory network required for the establishment and maintenance of pluripotency. *Cell Cycle* 10(4):592–97
100. Smith KN, Singh AM, Dalton S. 2010. Myc represses primitive endoderm differentiation in pluripotent stem cells. *Cell Stem Cell* 7(3):343–54
101. Kim J, Woo AJ, Chu J, Snow JW, Fujiwara Y, et al. 2010. A Myc network accounts for similarities between embryonic stem and cancer cell transcription programs. *Cell* 143(2):313–24
102. Lin C-H, Jackson AL, Guo J, Linsley PS, Eisenman RN. 2009. Myc-regulated microRNAs attenuate embryonic stem cell differentiation. *EMBO J.* 28(20):3157–70
103. Nakamura S, Yokota D, Tan L, Nagata Y, Takemura T, et al. 2012. Down-regulation of Thanatos-associated protein 11 by BCR-ABL promotes CML cell proliferation through c-Myc expression. *Int. J. Cancer* 130(5):1046–59

104. Zhu C-Y, Li C-Y, Li Y, Zhan Y-Q, Li Y-H, et al. 2009. Cell growth suppression by thanatos-associated protein 11(THAP11) is mediated by transcriptional downregulation of c-Myc. *Cell Death Differ.* 16(3):395–405
105. Dejosez M, Levine SS, Frampton GM, Whyte WA, Stratton SA, et al. 2010. Ronin/Hcf-1 binds to a hyperconserved enhancer element and regulates genes involved in the growth of embryonic stem cells. *Genes Dev.* 24(14):1479–84
106. Zwaka TP. 2008. Ronin and caspases in embryonic stem cells: a new perspective on regulation of the pluripotent state. *Cold Spring Harb. Symp. Quant. Biol.* 73:163–69
107. Dejosez M, Krumenacker JS, Zitur LJ, Passeri M, Chu L-F, et al. 2008. Ronin is essential for embryogenesis and the pluripotency of mouse embryonic stem cells. *Cell* 133(7):1162–74
108. Roussigne M, Kossida S, Lavigne A-C, Clouaire T, Ecochard V, et al. 2003. The THAP domain: a novel protein motif with similarity to the DNA-binding domain of P element transposase. *Trends Biochem. Sci.* 28(2):66–69
109. Xie X, Lu J, Kulbokas EJ, Golub TR, Mootha V, et al. 2005. Systematic discovery of regulatory motifs in human promoters and 3′ UTRs by comparison of several mammals. *Nature* 434(7031):338–45
110. Wang D, Manali D, Wang T, Bhat N, Hong N, et al. 2011. Identification of pluripotency genes in the fish medaka. *Int. J. Biol. Sci.* 7:440–51
111. Walker BA, Leone PE, Chiecchio L, Dickens NJ, Jenner MW, et al. 2010. A compendium of myeloma-associated chromosomal copy number abnormalities and their prognostic value. *Blood* 116(15):e56–65
112. Johnson RA, Wright KD, Poppleton H, Mohankumar KM, Finkelstein D, et al. 2010. Cross-species genomics matches driver mutations and cell compartments to model ependymoma. *Nature* 466(7306):632–36
113. Wysocka J, Myers MP, Laherty CD, Eisenman RN, Herr W. 2003. Human Sin3 deacetylase and trithorax-related Set1/Ash2 histone H3-K4 methyltransferase are tethered together selectively by the cell-proliferation factor HCF-1. *Genes Dev.* 17(7):896–911
114. Capotosti F, Guernier S, Lammers F, Waridel P, Cai Y, et al. 2011. O-GlcNAc transferase catalyzes site-specific proteolysis of HCF-1. *Cell* 144(3):376–88
115. Nichols J, Chambers I, Taga T, Smith A. 2001. Physiological rationale for responsiveness of mouse embryonic stem cells to gp130 cytokines. *Development* 128(12):2333–39
116. Zandstra PW, Le HV, Daley GQ, Griffith LG, Lauffenburger DA. 2000. Leukemia inhibitory factor (LIF) concentration modulates embryonic stem cell self-renewal and differentiation independently of proliferation. *Biotechnol. Bioeng.* 69(6):607–17
117. Dani C, Chambers I, Johnstone S, Robertson M, Ebrahimi B, et al. 1998. Paracrine induction of stem cell renewal by LIF-deficient cells: a new ES cell regulatory pathway. *Dev. Biol.* 203(1):149–62
118. Niwa H, Burdon T, Chambers I, Smith A. 1998. Self-renewal of pluripotent embryonic stem cells is mediated via activation of STAT3. *Genes Dev.* 12(13):2048–60
119. Paling NRD, Wheadon H, Bone HK, Welham MJ. 2004. Regulation of embryonic stem cell self-renewal by phosphoinositide 3-kinase-dependent signaling. *J. Biol. Chem.* 279(46):48063–70
120. Niwa H, Ogawa K, Shimosato D, Adachi K. 2009. A parallel circuit of LIF signalling pathways maintains pluripotency of mouse ES cells. *Nature* 460(7251):118–22
121. Eiselleova L, Matulka K, Kriz V, Kunova M, Schmidtova Z, et al. 2009. A complex role for FGF-2 in self-renewal, survival, and adhesion of human embryonic stem cells. *Stem Cells* 27(8):1847–57
122. Amit M, Carpenter MK, Inokuma MS, Chiu CP, Harris CP, et al. 2000. Clonally derived human embryonic stem cell lines maintain pluripotency and proliferative potential for prolonged periods of culture. *Dev. Biol.* 227(2):271–78
123. Xu R-H, Sampsell-Barron TL, Gu F, Root S, Peck RM, et al. 2008. NANOG is a direct target of TGFbeta/activin-mediated SMAD signaling in human ESCs. *Cell Stem Cell* 3(2):196–206
124. Levine AJ, Brivanlou AH. 2006. GDF3 at the crossroads of TGF-beta signaling. *Cell Cycle* 5(10):1069–73
125. Benezra R, Davis RL, Lockshon D, Turner DL, Weintraub H. 1990. The protein Id: a negative regulator of helix-loop-helix DNA binding proteins. *Cell* 61(1):49–59

126. Tropepe V, Hitoshi S, Sirard C, Mak TW, Rossant J, van der Kooy D. 2001. Direct neural fate specification from embryonic stem cells: a primitive mammalian neural stem cell stage acquired through a default mechanism. *Neuron* 30(1):65–78
127. Ying Y, Qi X, Zhao GQ. 2001. Induction of primordial germ cells from murine epiblasts by synergistic action of BMP4 and BMP8B signaling pathways. *Proc. Natl. Acad. Sci. USA* 98(14):7858–62
128. Ying QL, Nichols J, Chambers I, Smith A. 2003. BMP induction of Id proteins suppresses differentiation and sustains embryonic stem cell self-renewal in collaboration with STAT3. *Cell* 115(3):281–92
129. Suzuki A, Raya A, Kawakami Y, Morita M, Matsui T, et al. 2006. Nanog binds to Smad1 and blocks bone morphogenetic protein-induced differentiation of embryonic stem cells. *Proc. Natl. Acad. Sci. USA* 103(27):10294–99
130. Qi X, Li T-G, Hao J, Hu J, Wang J, et al. 2004. BMP4 supports self-renewal of embryonic stem cells by inhibiting mitogen-activated protein kinase pathways. *Proc. Natl. Acad. Sci. USA* 101(16):6027–32
131. Itsykson P, Ilouz N, Turetsky T, Goldstein RS, Pera MF, et al. 2005. Derivation of neural precursors from human embryonic stem cells in the presence of noggin. *Mol. Cell Neurosci.* 30(1):24–36
132. Pera MF, Andrade J, Houssami S, Reubinoff B, Trounson A, et al. 2004. Regulation of human embryonic stem cell differentiation by BMP-2 and its antagonist noggin. *J. Cell Sci.* 117(Pt. 7):1269–80
133. Strizzi L, Bianco C, Normanno N, Salomon D. 2005. Cripto-1: a multifunctional modulator during embryogenesis and oncogenesis. *Oncogene* 24(37):5731–41
134. Reynolds D, Vallier L, Chng Z, Pedersen R. 2009. Signaling pathways in embryonic stem cells. In *Regulatory Networks in Stem Cells*, ed. VK Rajasekhar, MC Vemuri, pp. 293–308. New York: Humana
135. Sato N, Meijer L, Skaltsounis L, Greengard P, Brivanlou AH. 2004. Maintenance of pluripotency in human and mouse embryonic stem cells through activation of Wnt signaling by a pharmacological GSK-3-specific inhibitor. *Nat. Med.* 10(1):55–63
136. Hagen T, Vidal-Puig A. 2002. Characterisation of the phosphorylation of β-catenin at the GSK-3 priming site Ser45. *Biochem. Biophys. Res. Commun.* 294(2):324–28
137. Mao J, Wang J, Liu B, Pan W, Farr GH 3rd, et al. 2001. Low-density lipoprotein receptor-related protein-5 binds to Axin and regulates the canonical Wnt signaling pathway. *Mol. Cell* 7(4):801–9
138. Hecht A, Kemler R. 2000. Curbing the nuclear activities of β-catenin. Control over Wnt target gene expression. *EMBO Rep.* 1(1):24–28
139. Wagner RT, Xu X, Yi F, Merrill BJ, Cooney AJ. 2010. Canonical Wnt/β-catenin regulation of liver receptor homolog-1 mediates pluripotency gene expression. *Stem Cells* 28(10):1794–804
140. Kelly KF, Ng DY, Jayakumaran G, Wood GA, Koide H, Doble BW. 2011. β-catenin enhances Oct-4 activity and reinforces pluripotency through a TCF-independent mechanism. *Cell Stem Cell* 8(2):214–27
141. Miki T, Yasuda S-Y, Kahn M. 2011. Wnt/β-catenin signaling in embryonic stem cell self-renewal and somatic cell reprogramming. *Stem Cell Rev.* 7(4):836–46
142. Rahl PB, Lin CY, Seila AC, Flynn RA, McCuine S, et al. 2010. c-Myc regulates transcriptional pause release. *Cell* 141(3):432–45
143. Creyghton MP, Cheng AW, Welstead GG, Kooistra T, Carey BW, et al. 2010. Histone H3K27ac separates active from poised enhancers and predicts developmental state. *Proc. Natl. Acad. Sci. USA* 107(50):21931–36
144. Pan G, Tian S, Nie J, Yang C, Ruotti V, et al. 2007. Whole-genome analysis of histone H3 lysine 4 and lysine 27 methylation in human embryonic stem cells. *Cell Stem Cell* 1(3):299–312
145. Rada-Iglesias A, Bajpai R, Swigut T, Brugmann SA, Flynn RA, Wysocka J. 2011. A unique chromatin signature uncovers early developmental enhancers in humans. *Nature* 470(7333):279–83
146. Chen X, Vega VB, Ng H-H. 2008. Transcriptional regulatory networks in embryonic stem cells. *Cold Spring Harb. Symp. Quant. Biol.* 73:203–9
147. Gaspar-Maia A, Alajem A, Meshorer E, Ramalho-Santos M. 2011. Open chromatin in pluripotency and reprogramming. *Nat. Rev. Mol. Cell Biol.* 12(1):36–47
148. Meshorer E, Misteli T. 2006. Chromatin in pluripotent embryonic stem cells and differentiation. *Nat. Rev. Mol. Cell Biol.* 7(7):540–46
149. Young RA. 2011. Control of the embryonic stem cell state. *Cell* 144(6):940–54

150. Loh Y-H, Zhang W, Chen X, George J, Ng H-H. 2007. Jmjd1a and Jmjd2c histone H3 Lys 9 demethylases regulate self-renewal in embryonic stem cells. *Genes Dev.* 21(20):2545–57
151. Xie L, Pelz C, Wang W, Bashar A, Varlamova O, et al. 2011. KDM5B regulates embryonic stem cell self-renewal and represses cryptic intragenic transcription. *EMBO J.* 30(8):1473–84
152. Margueron R, Reinberg D. 2011. The Polycomb complex PRC2 and its mark in life. *Nature* 469(7330):343–49
153. Bernstein BE, Mikkelsen TS, Xie X, Kamal M, Huebert DJ, et al. 2006. A bivalent chromatin structure marks key developmental genes in embryonic stem cells. *Cell* 125(2):315–26
154. Li G, Margueron R, Ku M, Chambon P, Bernstein BE, Reinberg D. 2010. Jarid2 and PRC2, partners in regulating gene expression. *Genes Dev.* 24(4):368–80
155. Pasini D, Cloos PAC, Walfridsson J, Olsson L, Bukowski J-P, et al. 2010. JARID2 regulates binding of the Polycomb repressive complex 2 to target genes in ES cells. *Nature* 464(7286):306–10
156. Peng JC, Valouev A, Swigut T, Zhang J, Zhao Y, et al. 2009. Jarid2/Jumonji coordinates control of PRC2 enzymatic activity and target gene occupancy in pluripotent cells. *Cell* 139(7):1290–302
157. Kaneko S, Li G, Son J, Xu C-F, Margueron R, et al. 2010. Phosphorylation of the PRC2 component Ezh2 is cell cycle-regulated and up-regulates its binding to ncRNA. *Genes Dev.* 24(23):2615–20
158. Ang Y-S, Tsai S-Y, Lee D-F, Monk J, Su J, et al. 2011. Wdr5 mediates self-renewal and reprogramming via the embryonic stem cell core transcriptional network. *Cell* 145(2):183–97
159. Jiang H, Shukla A, Wang X, Chen W-y, Bernstein BE, Roeder RG. 2011. Role for Dpy-30 in ES cell-fate specification by regulation of H3K4 methylation within bivalent domains. *Cell* 144(4):513–25
160. Ho L, Crabtree GR. 2010. Chromatin remodelling during development. *Nature* 463(7280):474–84
161. Schultz DC, Ayyanathan K, Negorev D, Maul GG, Rauscher FJ 3rd. 2002. SETDB1: a novel KAP-1-associated histone H3, lysine 9-specific methyltransferase that contributes to HP1-mediated silencing of euchromatic genes by KRAB zinc-finger proteins. *Genes Dev.* 16(8):919–32
162. Singhal N, Graumann J, Wu G, Araúzo-Bravo MJ, Han DW, et al. 2010. Chromatin-remodeling components of the BAF complex facilitate reprogramming. *Cell* 141(6):943–55
163. Lessard JA, Crabtree GR. 2010. Chromatin regulatory mechanisms in pluripotency. *Annu. Rev. Cell Dev. Biol.* 26:503–32
164. Meissner A. 2010. Epigenetic modifications in pluripotent and differentiated cells. *Nat. Biotechnol.* 28(10):1079–88
165. Tsumura A, Hayakawa T, Kumaki Y, Takebayashi S, Sakaue M, et al. 2006. Maintenance of self-renewal ability of mouse embryonic stem cells in the absence of DNA methyltransferases Dnmt1, Dnmt3a and Dnmt3b. *Genes Cells* 11(7):805–14
166. Jackson-Grusby L, Jaenisch R. 1996. Experimental manipulation of genomic methylation. *Semin. Cancer Biol.* 7(5):261–68
167. Stadtfeld M, Maherali N, Breault DT, Hochedlinger K. 2008. Defining molecular cornerstones during fibroblast to iPS cell reprogramming in mouse. *Cell Stem Cell* 2(3):230–40
168. Pauli A, Rinn JL, Schier AF. 2011. Non-coding RNAs as regulators of embryogenesis. *Nat. Rev. Genet.* 12(2):136–49
169. Kanellopoulou C, Muljo SA, Kung AL, Ganesan S, Drapkin R, et al. 2005. Dicer-deficient mouse embryonic stem cells are defective in differentiation and centromeric silencing. *Genes Dev.* 19(4):489–501
170. Murchison EP, Partridge JF, Tam OH, Cheloufi S, Hannon GJ. 2005. Characterization of dicer-deficient murine embryonic stem cells. *Proc. Natl. Acad. Sci. USA* 102(34):12135–40
171. Wang Y, Medvid R, Melton C, Jaenisch R, Blelloch R. 2007. DGC8 is essential for microRNA biogenesis and silencing of embryonic stem cell self-renewal. *Nat. Genet.* 39(3):380–85
172. Melton C, Judson RL, Blelloch R. 2010. Opposing microRNA families regulate self-renewal in mouse embryonic stem cells. *Nature* 463(7281):621–26
173. Xu N, Papagiannakopoulos T, Pan G, Thomson JA, Kosik KS. 2009. MicroRNA-145 regulates OCT4, SOX2, and KLF4 and represses pluripotency in human embryonic stem cells. *Cell* 137(4):647–58
174. Helland Å, Anglesio MS, George J, Cowin PA, Johnstone CN, et al. 2011. Deregulation of *MYCN*, *LIN28B* and *LET7* in a molecular subtype of aggressive high-grade serous ovarian cancers. *PLoS ONE* 6(4):e18064

175. Judson RL, Babiarz JE, Venere M, Blelloch R. 2009. Embryonic stem cell-specific microRNAs promote induced pluripotency. *Nat. Biotechnol.* 27(5):459–61
176. Subramanyam D, Lamouille S, Judson RL, Liu JY, Bucay N, et al. 2011. Multiple targets of miR-302 and miR-372 promote reprogramming of human fibroblasts to induced pluripotent stem cells. *Nat. Biotechnol.* 29(5):443–48
177. Anokye-Danso F, Trivedi CM, Juhr D, Gupta M, Cui Z, et al. 2011. Highly efficient miRNA-mediated reprogramming of mouse and human somatic cells to pluripotency. *Cell Stem Cell* 8(4):376–88
178. Lin S-L, Chang DC, Lin C-H, Ying S-Y, Leu D, Wu DTS. 2011. Regulation of somatic cell reprogramming through inducible mir-302 expression. *Nucleic Acids Res.* 39(3):1054–65
179. Lin S-L, Chang DC, Ying S-Y, Leu D, Wu DTS. 2010. MicroRNA miR-302 inhibits the tumorigenecity of human pluripotent stem cells by coordinate suppression of the CDK2 and CDK4/6 cell cycle pathways. *Cancer Res.* 70(22):9473–82
180. Loewer S, Cabili MN, Guttman M, Loh Y-H, Thomas K, et al. 2010. Large intergenic non-coding RNA-RoR modulates reprogramming of human induced pluripotent stem cells. *Nat. Genet.* 42(12):1113–17
181. Sheik Mohamed J, Gaughwin PM, Lim B, Robson P, Lipovich L. 2010. Conserved long noncoding RNAs transcriptionally regulated by Oct4 and Nanog modulate pluripotency in mouse embryonic stem cells. *RNA* 16(2):324–37
182. Dailey L, Yuan H, Basilico C. 1994. Interaction between a novel F9-specific factor and octamer-binding proteins is required for cell-type-restricted activity of the fibroblast growth factor 4 enhancer. *Mol. Cell. Biol.* 14(12):7758–69
183. Balázsi G, van Oudenaarden A, Collins JJ. 2011. Cellular decision making and biological noise: from microbes to mammals. *Cell* 144(6):910–25
184. Han DW, Tapia N, Joo JY, Greber B, Araúzo-Bravo MJ, et al. 2010. Epiblast stem cell subpopulations represent mouse embryos of distinct pregastrulation stages. *Cell* 143(4):617–27
185. Hough SR, Laslett AL, Grimmond SB, Kolle G, Pera MF. 2009. A continuum of cell states spans pluripotency and lineage commitment in human embryonic stem cells. *PLoS ONE* 4(11):e7708
186. Laslett AL, Grimmond S, Gardiner B, Stamp L, Lin A, et al. 2007. Transcriptional analysis of early lineage commitment in human embryonic stem cells. *BMC Dev. Biol.* 7:12
187. Bao S, Tang F, Li X, Hayashi K, Gillich A, et al. 2009. Epigenetic reversion of post-implantation epiblast to pluripotent embryonic stem cells. *Nature* 461(7268):1292–95
188. Niwa H. 2007. How is pluripotency determined and maintained? *Development* 134(4):635–46
189. Thornberry NA, Lazebnik Y. 1998. Caspases: enemies within. *Science* 281(5381):1312–16
190. De Maria R, Zeuner A, Eramo A, Domenichelli C, Bonci D, et al. 1999. Negative regulation of erythropoiesis by caspase-mediated cleavage of GATA-1. *Nature* 401(6752):489–93
191. Janzen V, Fleming HE, Riedt T, Karlsson G, Riese MJ, et al. 2008. Hematopoietic stem cell responsiveness to exogenous signals is limited by caspase-3. *Cell Stem Cell* 2(6):584–94
192. Fujita J, Crane AM, Souza MK, Dejosez M, Kyba M, et al. 2008. Caspase activity mediates the differentiation of embryonic stem cells. *Cell Stem Cell* 2(6):595–601
193. Kuida K, Zheng TS, Na S, Kuan C, Yang D, et al. 1996. Decreased apoptosis in the brain and premature lethality in CPP32-deficient mice. *Nature* 384(6607):368–72
194. Zheng TS, Hunot S, Kuida K, Momoi T, Srinivasan A, et al. 2000. Deficiency in caspase-9 or caspase-3 induces compensatory caspase activation. *Nat. Med.* 6(11):1241–47
195. Li F, He Z, Shen J, Huang Q, Li W, et al. 2010. Apoptotic caspases regulate induction of iPSCs from human fibroblasts. *Cell Stem Cell* 7(4):508–20
196. Fuhrmann G, Chung AC, Jackson KJ, Hummelke G, Baniahmad A, et al. 2001. Mouse germline restriction of Oct4 expression by germ cell nuclear factor. *Dev. Cell* 1(3):377–87
197. Gu P, Xu X, Le Menuet D, Chung AC-K, Cooney AJ. 2011. Differential recruitment of methyl CpG-binding domain factors and DNA methyltransferases by the orphan receptor germ cell nuclear factor initiates the repression and silencing of Oct4. *Stem Cells* 29(7):1041–51
198. Feldman N, Gerson A, Fang J, Li E, Zhang Y, et al. 2006. G9a-mediated irreversible epigenetic inactivation of *Oct-3/4* during early embryogenesis. *Nat. Cell Biol.* 8(2):188–94

199. Ben-Shushan E, Sharir H, Pikarsky E, Bergman Y. 1995. A dynamic balance between ARP-1/COUP-TFII, EAR-3/COUP-TFI, and retinoic acid receptor:retinoid X receptor heterodimers regulates Oct-3/4 expression in embryonal carcinoma cells. *Mol. Cell. Biol.* 15(2):1034–48
200. Wray J, Kalkan T, Gomez-Lopez S, Eckardt D, Cook A, et al. 2011. Inhibition of glycogen synthase kinase-3 alleviates Tcf3 repression of the pluripotency network and increases embryonic stem cell resistance to differentiation. *Nat. Cell Biol.* 13(7):838–45

Endoplasmic Reticulum Stress and Type 2 Diabetes

Sung Hoon Back[1] and Randal J. Kaufman[2]

[1] School of Biological Sciences, University of Ulsan, Ulsan, Republic of Korea 680-749; email: shback@ulsan.ac.kr

[2] Degenerative Disease Research Program, Neuroscience, Aging, and Stem Cell Research Center, Sanford Burnham Medical Research Institute, La Jolla, California 92037; email: rkaufman@sanfordburnham.org

Keywords

unfolded protein response, ER stress, free fatty acid, glucose, pancreatic β-cell

Abstract

Given the functional importance of the endoplasmic reticulum (ER), an organelle that performs folding, modification, and trafficking of secretory and membrane proteins to the Golgi compartment, the maintenance of ER homeostasis in insulin-secreting β-cells is very important. When ER homeostasis is disrupted, the ER generates adaptive signaling pathways, called the unfolded protein response (UPR), to maintain homeostasis of this organelle. However, if homeostasis fails to be restored, the ER initiates death signaling pathways. New observations suggest that both chronic hyperglycemia and hyperlipidemia, known as important causative factors of type 2 diabetes (T2D), disrupt ER homeostasis to induce unresolvable UPR activation and β-cell death. This review examines how the UPR pathways, induced by high glucose and free fatty acids (FFAs), interact to disrupt ER function and cause β-cell dysfunction and death.

Contents

- INTRODUCTION 768
- PANCREATIC β-CELL AND THE ENDOPLASMIC RETICULUM .. 768
- SIGNALING FROM TRANSMEMBRANE SENSORS OF THE UNFOLDED PROTEIN RESPONSE 769
  - Inositol-Requiring Protein 1–Mediated Signaling Pathways 769
  - PKR-Like Endoplasmic Reticulum Kinase-Mediated Signaling Pathways 771
  - Activating Transcription Factor 6–Mediated Signaling Pathways 772
- UNFOLDED PROTEIN RESPONSE-MEDIATED CELL DEATH 773
  - PERK-Mediated Cell Death Pathways 773
  - IRE1α-Mediated Cell Death Pathways 777
- ENDOPLASMIC RETICULUM STRESS STIMULI AND β-CELL DEATH IN TYPE 2 DIABETES .. 777
  - Endoplasmic Reticulum Stress by Lipotoxicity 777
  - Endoplasmic Reticulum Stress by Glucotoxicity 782
  - Endoplasmic Reticulum Stress by Islet Amyloid 784
- CONCLUSIONS 785

INTRODUCTION

Modern lifestyles, characterized by the overconsumption of energy-rich foods and reduced physical activity, have dramatically increased the frequency of type 2 diabetes (T2D). T2D is a major cause of morbidity and mortality, decreasing both the quality of life and life expectancy. In 2010, diabetes affected about 285 million people worldwide and is expected to increase by 1.5 fold (439 million people) by 2030 (1). Between 2010 and 2030, there will be a 69% increase in the number of adults with diabetes in developing countries and a 20% increase in developed countries. T2D is a complex heterogeneous group of metabolic conditions characterized by increased levels of blood glucose owing to insulin resistance in adipose tissue, muscle, and liver and/or to impaired insulin secretion from pancreatic β-cells (2). Obesity is mechanistically linked to insulin resistance and T2D. Although both obesity and insulin resistance are associated with T2D, the disease only develops in insulin-resistant subjects with the onset of β-cell dysfunction (3). To adapt to the increased metabolic load caused by obesity and insulin resistance, normal pancreatic islets usually respond by increasing β-cell mass through an increase in β-cell proliferation and neogenesis (4) as well as by enhancing β-cell function (5). However, β-cell adaptation eventually fails as a consequence of genetic and environmental factors that cause a progressive decline in β-cell function and survival. As a consequence, individuals progress from normal glucose tolerance to impaired glucose tolerance and then to established T2D with reduced β-cell mass (3). Although the molecular mechanisms underlying β-cell failure/death remain to be clarified, recent studies suggest that β-cell loss in T2D results from endoplasmic reticulum (ER) stress responses induced by gluco/lipotoxicity (6) and amyloid accumulation (7). In this review, we examine the mechanisms behind UPR-mediated pancreatic cell death with specific attention to the connection between ER stress and T2D.

PANCREATIC β-CELL AND THE ENDOPLASMIC RETICULUM

The islets of Langerhans constitute ~2% of the total pancreatic mass, and each islet is composed of ~2,000 cells of which insulin-secreting β-cells constitute ~60% of the total (8). Insulin, a blood glucose-lowering hormone, is only secreted by the pancreatic β-cells of the islets of Langerhans. After proinsulin is synthesized in the ER, it is processed into its biologically active form and stored in secretory granules. When blood glucose levels increase, insulin is released from storage granules to maintain a normal blood glucose level (9). The β-cell contains a large pool of cytoplasmic proinsulin mRNA (∼20% of the total mRNA), one of the most abundant mRNA species (10). One profound example of physiological fluctuation in the protein-folding load in the ER is the unique translational response of pancreatic β-cells to variations in blood glucose (11). Blood glucose level changes lead to a maximum 25-fold increase in proinsulin synthesis, and ∼1 × 10^6 proinsulin molecules are synthesized per minute upon glucose stimulation (12). Therefore, proinsulin synthesis imposes a heavy biosynthetic burden upon the β-cell.

The ER is the first station of the secretory pathway and the site of synthesis for proteins resident in the ER or destined for the Golgi compartment, endosomes, lysosomes, the plasma membrane, or the extracellular milieu (13). It also serves as a site of biosynthesis for steroids, cholesterol, other lipids, and Ca^{2+} storage, and also as a signaling platform between the nucleus and cytosol (14). Efficient ER function relies on numerous resident quality control factors, such as molecular chaperones and folding enzymes, a high level (several hundred micromolars) of Ca^{2+}, and an oxidative environment. Nascent polypeptides that are translocated into the ER lumen undergo posttranslational modifications and rounds of folding interactions required for optimal function. Properly folded proteins exit the ER and progress through the secretory pathway, whereas irreparably unfolded or misfolded proteins are retro-translocated from the ER and degraded by cytoplasmic proteasomes (15). The flux of nascent polypeptides into the ER is variable because it changes rapidly in response to the physiological state and environmental conditions of the cell. To handle this dynamic situation, cells must adjust their ER protein-folding capacity according to environmental and physiological contexts, thereby ensuring that the quality of membranous and secreted proteins is maintained with high fidelity. Given the importance of ER function for normal cellular function, it is not surprising that altered ER homeostasis affects a diverse number of cellular processes, including transcription, translation, cell cycle control, intracellular signaling, and programmed cell death. When ER homeostasis is altered, signaling pathways are activated, eliciting an adaptive response that is collectively called the unfolded protein response (UPR). The UPR includes expansion of ER size, enhanced folding capacity, reduced protein synthesis through transcriptional and translational controls, and increased clearance of unfolded or misfolded proteins (16). When these mechanisms fail to restore ER homeostasis, cell death signaling pathways are activated (17). There are three transmembrane ER-proximal sensors of unfolded proteins that initiate UPR signaling in mammals: inositol-requiring protein 1 (IRE1), activating transcription factor 6 (ATF6), and PKR-like ER kinase (PERK); all three are present in all cell types (**Figure 1**). These three main UPR sensors are integral membrane proteins that sense the unfolded protein loaded in the ER lumen and transmit this information across the ER membrane to the cytosol. Considerable progress has been made in our understanding of the signaling pathways and the pathophysiological significance of the UPR. The goal of this review is to summarize our current insight by considering the role of the UPR in T2D as a disease in which there is a structural and pathological change in β-cells.

UPR: unfolded protein response

IRE1: inositol-requiring protein 1

ATF6: activating transcription factor 6

PERK: PKR (double-stranded RNA-dependent kinase)-like endoplasmic reticulum kinase

SIGNALING FROM TRANSMEMBRANE SENSORS OF THE UNFOLDED PROTEIN RESPONSE

Inositol-Requiring Protein 1–Mediated Signaling Pathways

There are two mammalian homologs of yeast Ire1p: IRE1α, expressed in most cells and tissues with high levels of mRNA expression in the pancreas (18), and IRE1β, mainly expressed in the epithelium of the gastrointestinal tract (19).

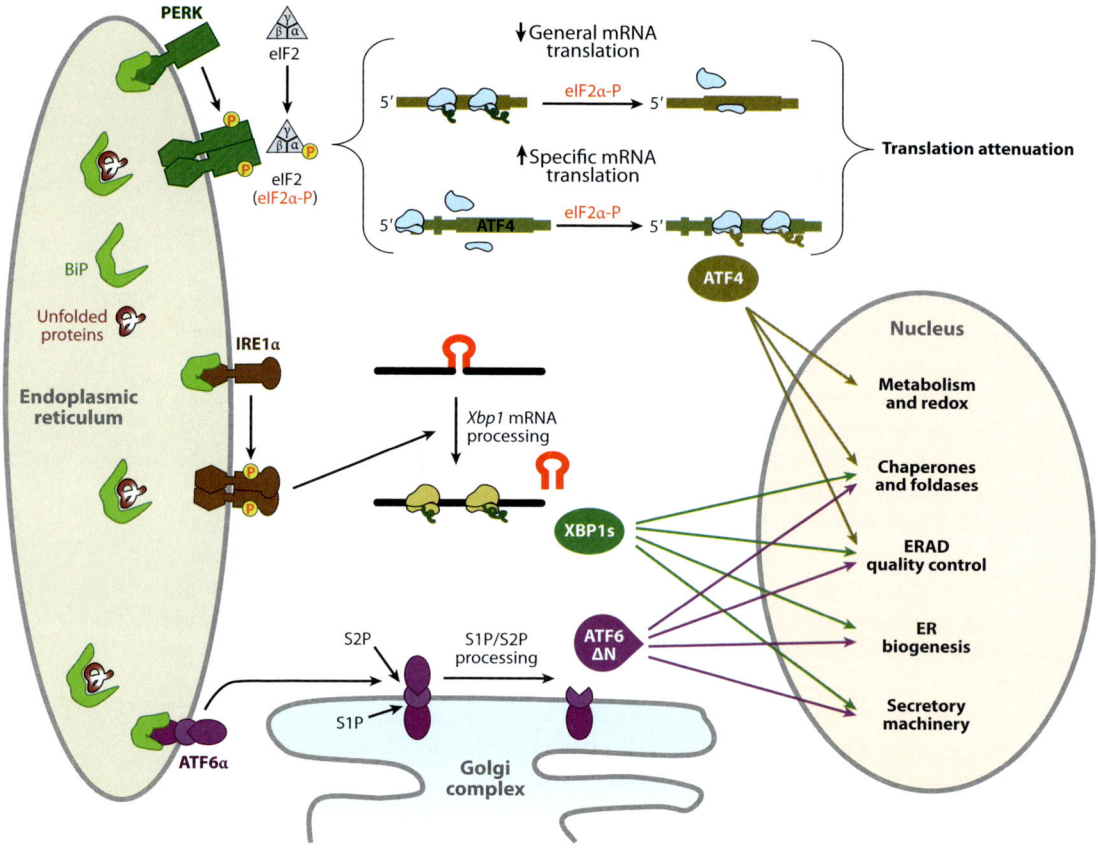

Figure 1

The adaptive unfolded protein response (UPR). Activation of three UPR pathways initiates the adaptive endoplasmic reticulum (ER) stress response. During activation of the UPR in mammals, BiP (immunoglobulin heavy chain binding protein, also known as GRP78) is sequestered through binding to unfolded or misfolded polypeptide chains, thereby leading to BiP release from the ER stress sensors for their activation. Unconventional cytoplasmic splicing, mediated by IRE1α, removes a 26-nucleotide intron from unspliced *X-box-binding protein 1* (*Xbp1*) mRNA (encoding 267 amino acids) to produce a translational frameshift, yielding a fusion protein encoded from two evolutionarily conserved open reading frames (16). The fusion protein, XBP1s, acts as a potent transcription factor for expression of UPR target genes involved in protein folding and export from the ER, export and degradation of misfolded proteins, and lipid biosynthesis, to resolve ER stress (16). Upon accumulation of unfolded protein in the ER lumen, oligomerization of the PKR-like ER kinase (PERK) in ER membranes induces its autophosphorylation and kinase domain activation (141, 142). Activated PERK phosphorylates serine 51 on the α-subunit of heterotrimeric eIF2 (143). When eukaryotic translation initiation factor 2α (eIF2α) is phosphorylated, the eIF2 complex shows increased affinity for its guanine nucleotide exchange factor eIF2B and sequesters all available eIF2B. Because the cellular level of eIF2B is 10- to 20-fold lower than the level of eIF2, very small changes in eIF2α phosphorylation can dramatically change the rate of translation initiation (144). Inhibition of general mRNA translation by the phosphorylation of eIF2α reduces accumulation of misfolded protein in the ER lumen (22), thereby protecting the cell from diverse stimuli that perturb the ER homeostasis. In contrast to inhibition of general mRNA translation, the PERK/eIF2α pathway stimulates the translation of several specific mRNAs containing multiple 5′-upstream open reading frames, such as *Atf4* and *Atf5*, *Chop*, *Gadd34*, and the cationic amino acid transporter 1 (*Cat-1*, an Na+-independent transporter of L-arginine and L-lysine) (28, 145). Among them, ATF4 activates transcription of the adaptive genes that encode functions in ER protein folding, endoplasmic reticulum–associated degradation (ERAD), amino acid biosynthesis and transportation, and the antioxidative stress response (24). Under ER stress, ATF6α and ATF6β are released from BiP and translocate to the Golgi complex, where they are cleaved by Golgi-resident proteases, first by S1P (site 1 protease) and then in the intramembrane region by S2P (site 2 protease), to release the N-terminal basic leucine zipper protein (bZIP) transcription factor domain (16). The bZIP domain of ATF6α then translocates into the nucleus, where it activates the transcription of genes encoding ER-localized molecular chaperones and folding enzymes, ERAD, protein secretion machineries, and ER biogenesis (146), in some cases in cooperation with XBP1s (42).

IRE1α, the most fundamental ER stress sensor, is highly conserved in all eukaryotic cells and is well studied (16). The luminal domain of the IRE1α protein responds to the accumulation of unfolded proteins in the ER and undergoes kinase activation, which triggers a specific endoribonuclease activity in its cytoplasmic domain, where initiation of splicing of the mRNA encoding X-box binding protein 1 (XBP1) occurs (20). β-Cell-specific deletion of *Xbp1* causes β-cell failure in a mouse model, suggesting that IRE1α-XBP1 signaling is essential for β-cell function (21). **Figure 1** provides a schematic view of the IRE1α-mediated signaling pathway, including the other UPR pathways.

PKR-Like Endoplasmic Reticulum Kinase-Mediated Signaling Pathways

PERK, a type I transmembrane protein located in the ER, has serine/threonine kinase activity in its cytoplasmic domain (**Figure 1**) (22, 23). The catalytic domain of PERK shares substantial homology with other eIF2α family kinases (GCN2, HRI, and PKR) (24). Approximately half of *Perk*-null mice, a mouse model of the human genetic disease Wolcott-Rallison syndrome (25), die pre- or postnatally. The surviving half of *Perk*-null mice gradually develop a multitude of metabolic and growth abnormalities, including hyperglycemia, growth retardation, skeletal defects, and atrophy of the exocrine and endocrine pancreas (26, 27). Different from *Perk*-deficient mice, all eIF2α A/A homozygous neonates die within 24 h after birth and suffer from hypoglycemia, possibly owing to defective gluconeogenesis [low phosphoenolpyruvate carboxykinase (PEPCK) activity], reduced glycogen storage in the liver, and a deficiency of pancreatic β-cells (28). For β-cell function, PERK activity is required to prevent an abnormal increase in insulin translation in islet cells responding to a high-glucose load (26), but eIF2α phosphorylation is required to control insulin translation in response to both low- and high-glucose conditions (29). Furthermore, it seems that PERK through eIF2α phosphorylation is a positive regulator of ER chaperone and ER-associated degradation (ERAD) function and thereby contributes to ER and Golgi anterograde trafficking, retrotranslocation from the ER to the cytoplasm, and proteasomal degradation (29–31). Therefore, *Perk* and/or eIF2α phosphorylation-deficient β-cells show retention of misfolded proinsulin in the ER lumen and defective trafficking of proinsulin, and thereby a reduced number of insulin granules in β-cells, indicating that the mutant β-cells experience ER stress, accompanied by increased cell death, leading to progressive diabetes.

In pancreatic β-cells, the extracellular glucose level modulates the activity of the UPR sensors. PERK phosphorylation is differentially regulated by glucose in the β-cell. In β-cell, eIF2α phosphorylation is gradually decreased with the increase of glucose levels. Its phosphorylation inversely correlates with the rate of proinsulin synthesis (32). However, both low blood glucose and chronic high blood glucose activate eIF2α phosphorylation. Chronically high-glucose concentrations stimulate proinsulin transcription and translation. As a consequence, it is believed that proinsulin synthesis overcomes the ER folding machinery, leading to PERK activation to reduce protein influx into the ER (33). However, there remains some controversy whether chronically high-glucose exposure (more than 18 h) actually causes severe ER stress, activating PERK (33, 34). Yet glucose stimulation of β-cells growing in acute high glucose causes eIF2α dephosphorylation, likely through a protein phosphatase 1 (PP1)-like phosphatase (32) that dephosphorylates eIF2α. Although this kinase/phosphatase model can easily explain the changes in eIF2α phosphorylation in response to glucose, it is not known how PP1 is regulated under these conditions. The kinase responsible for low-glucose eIF2α phosphorylation has not been identified (31). It is most likely that PERK is the kinase that phosphorylates eIF2α in low glucose. This is supported by studies from several groups, including Gomez and colleagues (35). Moreover, Gomez et al. (35) propose that PERK may sense levels of cellular ATP/energy in

Eukaryotic translation initiation factor 2α (eIF2α): a subunit of heterotrimeric eIF2 complex, which mediates the binding of methionyl-tRNA to ribosome

Wolcott-Rallison syndrome: a rare autosomal recessive disease, characterized by neonatal/early-onset nonautoimmune insulin-requiring diabetes associated with skeletal dysplasia and growth retardation

eIF2α A/A: a homozygous mutant mouse model harboring an alanine mutation at the phosphorylation site (serine 51 amino acid) in eIF2α

SERCA: sarco/endoplasmic reticulum Ca^{2+}-ATPase

pancreatic β-cells. It has been shown that PERK, but not IRE1α, is activated by a decrease in glucose concentration or intracellular energy level induced by mitochondrial inhibitors (35). Therefore, it is possible that PERK in pancreatic β-cells is also activated by a mechanism independent of IRE1α activation or by the unfolded protein accumulation. It was also reported that a decrease in glucose concentration leads to a concentration-dependent reduction in ER Ca^{2+} that parallels the activation of PERK and the phosphorylation of eIF2α. It was proposed that an ER Ca^{2+} decrease is caused by a decrease in SERCA activity, mediated by a reduction in its cell energy status (154). However, this study did not suggest a precise mechanism that described why IRE1α is not activated by an ER Ca^{2+} decrease, which is induced by low glucose, although it is possible that PERK and IRE1α may have different thresholds for activation in response to a decrease in ER Ca^{2+}. Clearly, further studies are required to elucidate the precise molecular mechanisms involved in energy/glucose-dependent regulation of eIF2α phosphorylation and its biological meaning.

It has been suggested that the cytosolic function of PERK is also controlled by P58^{IPK}, first identified as a PKR inhibitor (36, 37). A more recent study (38), however, suggested that P58^{IPK} localizes mainly to the ER lumen and functions as a molecular cochaperone for BiP in the ER lumen. Therefore, if P58^{IPK} is a major regulator of PERK function, it is likely through some chaperone function (38). Thus, the precise inactivation mechanism of PERK remains to be clarified.

Activating Transcription Factor 6–Mediated Signaling Pathways

ATF6 encodes a basic leucine zipper protein (bZIP)-containing transcription factor localized to the ER membrane (39). Upon accumulation of unfolded protein in the ER, ATF6 traffics to the Golgi complex, where it is cleaved by site 1 and site 2 proteases, S1P and S2P, in a process called regulated intramembrane proteolysis. In mammals, there are two *Atf6* genes, *Atf6α* and *Atf6β/creb-rp/g13* (40). **Figure 1** provides a schematic view of the ATF6-mediated signaling pathway. The deletion of *Atf6α* resulted in increased sensitivity to chronic stress in mice challenged with chemical ER stressers, and it was suggested that ATF6α is required to optimize protein folding, secretion, and degradation during ER stress and thus facilitates recovery from acute stress and confers tolerance to chronic stress (41, 42). Unlike ATF6α, the role(s) for ATF6β is still unknown. Moreover, ATF6β (like ATF6α) is dispensable in embryonic and postnatal development, and a deficiency of ATF6β does not alter UPR gene induction (42). However, the double knockout of *Atf6α* and *Atf6β* causes early embryonic lethality, suggesting that ATF6α and ATF6β possess at least overlapping functions that are essential for mouse development. ATF6α and ATF6β are ubiquitously expressed (40). However, in recent years, several tissue-specific bZIP transcription factors located in the ER membrane and regulated by regulated intramembrane proteolysis have been identified: CREBH in hepatocytes, the pyloric stomach, and small intestine (43); OASIS in astrocytes (44); BBF2H7/CREB3L2 in most tissues (45); Tisp40 (transcript induced in spermatogenesis) in testis (46); Luman/LZIP/CREB3 in most tissues (47, 48); and CREB4 possibly in most tissues (49). The discovery of these numerous regulated intramembrane proteolysis-activated transcription factors raises a question: Why have cells evolved multiple ATF6-like molecules in the ER? These ATF6-like molecules might be evolutionarily chosen to respond to specific conditions of ER stress that can occur in different tissues to activate tissue-specific expression of genes to resolve ER stress. Further studies are required to answer these issues.

Wolfram syndrome (which is caused by mutations in the *Wfs1* gene, encoding wolframin) is a disorder of progressive neurodegeneration and β-cell failure leading to diabetes. Wolframin is a transmembrane ER protein involved in preventing ER stress signaling. Recently, Fonseca et al. (50) reported that WFS1 negatively regulates ATF6α through the ubiquitin-proteasome pathway during ER stress. In the absence of ER stress or in late stages of ER stress, WFS1 may limit ER stress by recruiting ATF6α to the E3 ubiquitin ligase, HRD1, for ubiquitination-mediated degradation of ATF6α. Therefore, *Wfs1*-null murine pancreata display higher levels of ATF6α protein and lower levels of Hrd1 protein compared with those in the control littermates. Additionally, another study suggested that the ectopic overexpression of an active form of ATF6α in β-cells causes apoptosis (51). These results imply that dysregulated ER stress signaling through hyperactivation of ATF6α is one pathogenic pathway for diseases involving chronic, unresolvable ER stress, such as pancreatic β-cell death in diabetes. Therefore, understanding the role of ATF6α in diabetic mouse models is an important step toward understanding the physiological roles of ATF6α in pancreatic β-cells.

UNFOLDED PROTEIN RESPONSE-MEDIATED CELL DEATH

Activation of each arm of the UPR initiates adaptive mechanisms to relieve the stresses accumulating in the ER. The adaptive UPR responses, signaled through activating transcription factor 4 (ATF4), cleaved ATF6α, and spliced XBP1, include inhibition of general mRNA translation by rapid PERK-mediated phosphorylation of eIF2α as well as induction of genes to increase the ER protein folding capacity and remove misfolded proteins from the ER. In 2006, it was recognized that preemptive quality control may also reduce the burden of misfolded substrates entering the ER by inhibiting translocation of many, but not all, polypeptides into the ER lumen (52). Thus, the adaptive pathways maintain cellular function and avoid apoptosis during ER stress. However, if ER stress is severe and chronic, the UPR-mediated efforts to correct the protein-folding defect fail, and then several apoptotic pathways are activated (53). Although both mitochondrial-dependent and -independent cell death pathways (17) execute apoptosis in response to ER stress, it is proposed that the ER serves as an important compartment where apoptotic signals are generated and integrated to elicit cell death in response to the unresolvable accumulation of unfolded proteins. Below, we briefly describe two UPR sensor-mediated cell death pathways.

PERK-Mediated Cell Death Pathways

During unresolvable ER stress, sustained activation of the PERK pathway also induces the proapoptotic pathway, similar to IRE1α (**Figure 2**). There are no reports that show PERK association with adaptors or modulators involves apoptosis. The persistent phosphorylation of eIF2α by PERK increases proapoptotic CHOP/GADD153 (C/EBP homology protein/growth arrest and DNA damage 153) expression through the transcription factor ATF4. CHOP is a member of the C/EBP family of bZIP transcription factors (54). The major inducers of the UPR (ATF4, ATF6, and XBP1) regulate *Chop* through both an ER stress response element and a C/EBP-ATF composite site (55–57). *Perk*- and *Atf4*-null cells, as well as eIF2α (*S51A*) knock-in cells, fail to induce CHOP (28, 58, 59), whereas *Ire1α*-, *Xbp1*-, and *Atf6α*-null cells or mouse liver tissues show increased and/or persistent *Chop* expression during ER stress (41, 42, 60, 61), suggesting that the PERK-eIF2α-ATF4 pathway is the main contributor toward ER stress-dependent CHOP expression. The deletion of the murine *Chop* gene attenuates ER stress-induced cell death in cultured fibroblasts and partially protects mice from renal toxicity owing to pharmacological induction of ER stress by tunicamycin (62). Furthermore, *Chop* deletion (*a*) prevents neuronal apoptosis induced by ischemia (63) and neuronal oxidative injury in a model of Parkinson's disease (64) and (*b*) protects the murine liver from ER stress and

Preemptive quality control: a cotranslational rerouting pathway reducing the burden of misfolded substrates entering the ER by cytosolic degradation

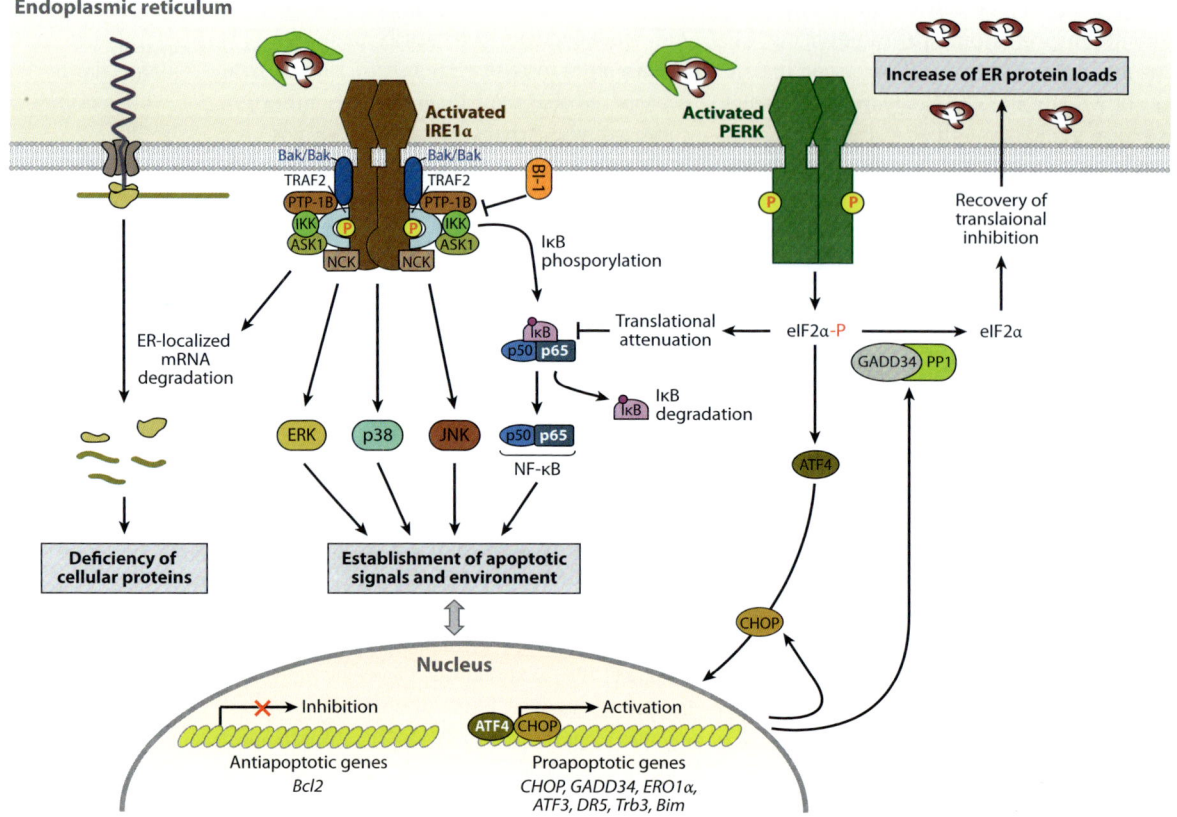

Figure 2

IRE1α- and PERK-mediated cell death pathways. During endoplasmic reticulum (ER) stress, inositol-requiring protein 1α (IRE1α) forms a hetero-oligomeric complex with TNF receptor-associated factor 2 (TRAF2) (76) and apoptosis signal-regulating kinase 1 (ASK1) (77) and then recruits the protein kinase JNK, leading to the activation of JNK (76). The IRE1α-TRAF2 complex recruits IκB kinase (IKK), which phosphorylates inhibitor of κB (IκB), leading to the degradation of IκB and the nuclear translocation of nuclear factor κ-light-chain-enhancer of activated B cells (NF-κB) (147). IRE1α also modulates the activation of other "alarm genes," such as p38 and ERK (80), possibly by the binding of the Src homology (SH) 2/3 -containing adaptor proteins Nck and TRAF2, respectively. Furthermore, several proapoptotic (i.e., BAX/BAK, AIP1, and maybe PTP-1B) or antiapoptotic proteins (i.e., BI-1) interact with IRE1α, regulating its activation state (81–84). Thus, the formation of a macromolecular signaling complex of IRE1α with several proapoptotic proteins can generate apoptotic signals and establish an apoptotic environment. In addition, the endoribonuclease activity of IRE1α, aside from specific cleavage of *Xbp1* mRNA, degrades ER-targeted mRNAs that can decrease cellular functions, such as proinsulin synthesis in β-cells. In contrast to the adaptive response by the PERK-phosphorylated eIF2α-ATF4 pathway, this pathway also contributes to stress-induced cell death by ATF4-mediated induction of proapoptotic genes, including CHOP, ATF3, and GADD34 (16). The induced transcription factor CHOP contributes to increased expression of the proapoptotic factors, such as death receptor 5 (DR5) (148), *tribbles*-related protein 3 (Trb3) (149), and binding to microtubule (Bim) (150), and it can suppress B cell lymphoma 2 (Bcl2) expression. Bim is also activated through protein phosphatase 2A-mediated dephosphorylation, which prevents its ubiquitination and proteasomal degradation (150). PERK-mediated translational attenuation upregulates NF-κB-dependent transcription because IκB has a shorter half-life than NF-κB, so NF-κB is released to translocate to the nucleus (151). Recovery from translational repression is mediated by eIF2α dephosphorylation by the two regulatory subunits of protein phosphatase 1 (PP1), GADD34 and CReP (constitutive repressor of eIF2α phosphorylation) (16). GADD34 is induced transcriptionally during ER stress by ATF4, whereas *CreP* is a constitutive activator of PP1. The premature dephosphorylation of eIF2α by the GADD34-PP1 complex restores translation of general mRNAs, which may be detrimental if the ER protein-folding defect is not resolved.

oxidative damage induced from clotting factor VIII expression (65), which is prone to misfolding and pancreatic β-cell death resulting from either accumulation of misfolded mutant proinsulin or exposure to nitric oxide (66, 67). *Chop* deletion even promotes β-cell survival in multiple diabetic mouse models induced by a leptin receptor deficiency or a high-fat diet treatment with haplo-insufficiency of eIF2α phosphorylation (68). The mechanism by which CHOP-mediated apoptosis occurs in response to ER stress is not well established. It was suggested that overexpression of CHOP decreases expression of the antiapoptotic Bcl-2 protein, depletes cellular glutathione, and exaggerates production of reactive oxygen species (ROS) (69, 70). However, a study of *Chop*-deleted cells revealed a different view of CHOP-mediated cell death. CHOP directly activates *Gadd34*, which promotes ER client protein biosynthesis by dephosphorylating eIF2α in stressed cells, and causes further accumulation of high-molecular-weight detergent-resistant stress-associated ER complexes in the ER (**Figure 2**) (70). Therefore, impaired GADD34 expression reduces client protein load and ER stress in *Chop*-null cells exposed to ER stress. The *Chop*- and *Gadd34*-null mutant cells accumulate fewer high-molecular-weight protein complexes in their stressed ER than wild-type cells. Furthermore, ERO1α (ER oxidoreductase α), a direct CHOP target gene, causes hyperoxidation of the ER to increase abnormal high-molecular-weight protein complexes. Thus, CHOP may disrupt ER function by promoting protein synthesis and oxidation.

Increasing evidence suggests that an accumulation of misfolded proteins in the ER lumen generates ROS (65, 71, 72). Formation of incorrect intermolecular and/or intramolecular disulfide bonds depletes glutathione that is required for their reduction. Disulfide bond formation in the ER is ushered in by the oxidative folding pathway, catalyzed by protein disulfide isomerase (PDI) and ERO1-mediated oxidation of substrate polypeptides, that produces ROS (**Figure 3**). For example, overexpression of a misfolded protein CPY (yeast vacuolar protein carboxypeptidase Y) activates the UPR, causes oxidative stress, and induces apoptosis. However, removal of all cysteine residues in CPY reduced the UPR, oxidative stress, and cell death (73). Furthermore, in macrophages, cholesterol and ER stress inducers oxidize inositol-1,4,5-trisphosphate (IP_3) receptors in the ER membrane to release Ca^{2+} and activate Ca^{2+}/calmodulin-dependent protein kinase IIγ (CaMKIIγ) (74). This study suggested also that there are three CaMKIIγ-mediated apoptotic pathways: (*a*) CaMKIIγ-mediated c-Jun N-terminal kinase (JNK) activation; (*b*) CaMKIIγ-increased mitochondrial Ca^{2+}; and (*c*) CaMKIIγ activation of STAT1, a proapoptotic signal transducer (74). These hypotheses are supported by the finding that either *Chop* or *Ero1α* deletion in yeast and mice reduces ROS accumulation and protects from ER stress-mediated cell death.

Recent studies demonstrated significant increases in ER stress marker proteins $P58^{IPK}$, BiP, and CHOP in islets in tissue sections from obese diabetic individuals (T2D patients) (75). Additionally, analysis of leptin receptor-deficient ($Lepr^{db/db}$) mice, as well as other murine models of T2D, revealed that insulin resistance increases proinsulin synthesis in β-cells beyond the capacity for folding of nascent polypeptides within the ER lumen, thereby disrupting ER homeostasis and triggering the UPR (68). The pancreatic β-cells in the $Lepr^{db/db}$ mice displayed slightly increased expression of several UPR genes encoding adaptive functions to improve ER folding capacity, such as *BiP*, *Grp94*, *Fkbp11*, and $P58^{IPK}$, as well as accumulated oxidative damage. By contrast, the deletion of *Chop* gene increased expression of UPR and oxidative stress response genes and reduced levels of oxidative damage, such as products of protein oxidation (carbonyls) and lipid peroxidation (hydroxyoctadecadienoic acid). These findings suggest that CHOP can be a fundamental factor linking protein misfolding in the ER to oxidative stress and apoptosis in β-cells under conditions of increased insulin demand, such as in T2D. Therefore, small molecules that modulate the expression or

ROS: reactive oxygen species

$Lepr^{db/db}$ **mouse:** a mouse model of obesity, diabetes, and dyslipdemia wherein leptin receptor activity is deficient

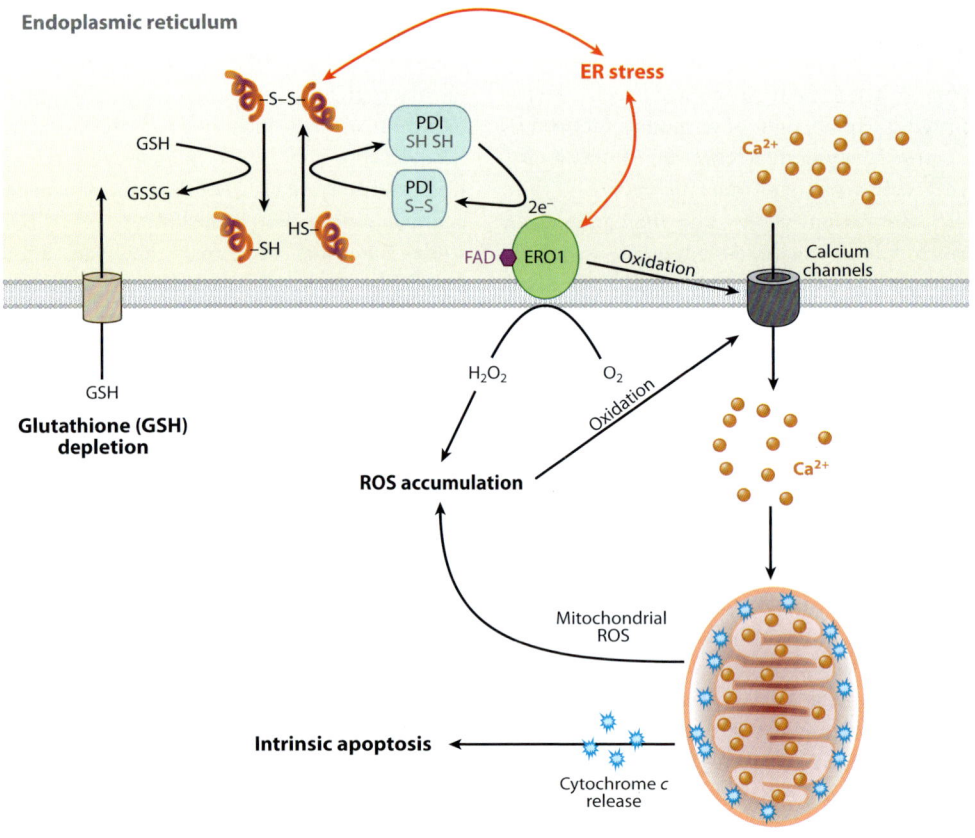

Figure 3

The role of calcium and reactive oxygen species (ROS) in endoplasmic reticulum (ER) stress-mediated cell death. In the ER lumen, oxidative protein folding is catalyzed by protein disulfide isomerase (PDI) and ER oxidoreductase (ERO1). In this reaction, an oxidant flavin adenine dinucleotide (FAD)-bound ERO1 oxidizes PDI, which subsequently oxidizes folding proteins directly. FAD-bound ERO1 then passes two electrons to molecular oxygen, resulting in the production of hydrogen peroxide (73). During unfolded protein response activation, CHOP-mediated induction of ERO1α hyperoxidizes the ER lumen and causes oxidation-induced activation of the ER Ca^{2+} release channel inositol 1,4,5-trisphosphate receptor (152), causing a large and transient release of Ca^{2+} from the ER. Increased cytosolic Ca^{2+} is taken up into the mitochondrial matrix, and this stimulates mitochondrial ROS production through disruption of mitochondrial electron transport (153). High levels of ROS generated from mitochondria, in turn, further increase Ca^{2+} release from the ER. The increase in mitochondrial Ca^{2+} eventually dissociates cytochrome c from the inner membrane cardiolipin, which triggers permeability transition pore opening and cytochrome c release across the outer membrane. Now the vicious cycle of ER calcium release and mitochondrial ROS production activates cytochrome c-mediated apoptosis. In addition, ER stress may cause consumption of excessive cellular glutathione (GSH) because reduced GSH may also assist in reducing nonnative disulfide bonds in misfolded proteins, resulting in the production of oxidized glutathione (GSSG).

transcriptional activity of CHOP may be useful to alleviate ER stress-mediated cell death.

IRE1α-Mediated Cell Death Pathways

Studies have suggested that the IRE1α/TRAF2/ASK1 complex promotes apoptosis (**Figure 2**) through JNK phosphorylation (76). Although polyglutamine aggregates are typically cytoplasmic, it was suggested that polyglutamine inhibits the proteasome and induces ER stress. It was demonstrated that *Ask1*-null primary neurons are resistant to ER stress-induced cell death (77). Activated ASK1 leads to JNK-mediated phosphorylation and activation of the proapoptotic protein Bim (78), but inhibits the antiapoptotic protein Bcl-2 (79). Furthermore, a cell-based chemical library screen identified compounds that enhance phosphorylation of serine 967 of ASK1, promoting 14-3-3 protein binding and thereby suppressing ASK1 function (80). The deficiency of ER-localized proapoptotic Bcl-2 family members BAX and BAK (81), ASK1-interacting protein 1 (AIP1) (82), or protein tyrosine phosphatase-1B (PTP-1B) (83) in cells and mice impairs IRE1α activation, thereby attenuating *Xbp1* splicing, JNK phosphorylation, expression of XBP1 target genes, and ER stress-induced apoptosis. Conversely, BAX inhibitor-1-deficient cells exhibit hyperactivation of IRE1α associated with increased *Xbp1* mRNA splicing and upregulation of XBP1s-dependent genes, activation of JNK, and increased cell death (84). On the basis of these compelling data, we speculate that the IRE1α UPRosome initiates multiple signaling responses through interaction with adaptors and modulators in a highly regulated manner. However, more studies are needed to define the physiological significance of these findings.

Recent reports suggest that during unresolvable ER stress, hyperactivation of IRE1α's RNase causes endonucleolytic decay of many ER-localized mRNAs, including those encoding chaperones, thereby culminating in cellular dysfunction and the death of β-cells as well as other cell types (85, 86). This process was termed regulated IRE1-dependent RNA degradation. It was proposed that the RNase of IRE1α can yield different outputs in RNA cleavage depending on the conditions and/or the intensity of ER stress (85). Under conditions of low-level ER stress or artificial dimerization of IRE1α without ER stress, IRE1α mainly cleaves unspliced *Xbp1* mRNA to promote an adaptive response, whereas persistent and/or strong activation causes IRE1α to cleave both unspliced *Xbp1* mRNA and ER-targeted mRNAs, including insulin. These findings suggest (*a*) that chronic exposure of β-cells to high glucose causes ER stress and hyperactivation of IRE1α, leading to degradation of insulin mRNA; and (*b*) that IRE1α kinase/endoribonuclease can function as an apoptotic switch in response to persistent ER stress.

UPRosome: a complex of IRE1α and multiple signaling molecules, which can modulate the amplitude and duration of IRE1α signaling

ENDOPLASMIC RETICULUM STRESS STIMULI AND β-CELL DEATH IN TYPE 2 DIABETES

Obesity is nearly invariably associated with insulin resistance, but T2D only develops in genetically predisposed and insulin-resistant subjects with the onset of β-cell dysfunction (3). Pancreatic β-cell failure and loss of islet mass are the primary determinants in the pathogenesis of T2D. Many mechanisms for β-cell dysfunction and death in T2D have been proposed, including lipotoxicity, glucotoxicity, oxidative stress, amyloid deposition, and others. There is growing evidence that β-cell failure and death are caused by unresolvable ER stress, leading to chronic and/or strong activation of IRE1α and/or PERK. Indeed, ER stress alone can initiate and propagate all the characteristics of β-cell failure and death observed in T2D (28). Here, we review how insulin resistance and obesity may cause ER stress in the β-cell leading to T2D.

Endoplasmic Reticulum Stress by Lipotoxicity

Many studies suggest that high-fat diets and obesity are associated with elevated levels of plasma free fatty acids (FFAs) (2). FFAs are now

FFAs: free fatty acids

considered as important mediators of β-cell dysfunction and apoptosis in T2D (3). Long saturated FFAs, such as palmitate, mediate apoptotic β-cell death in vivo and in vitro (87, 88), although both unsaturated and saturated FFAs eventually inhibit proinsulin synthesis and glucose-stimulated insulin secretion in β-cells (89). However, the precise mechanisms causing β-cell dysfunction and apoptosis by saturated FFAs (called lipotoxicity) are not fully understood. Accumulating evidence suggests that saturated long-chain FFAs induce ER stress and thereby cause β-cell failure and cell death, whereas unsaturated long-chain FFAs induce it to a lesser extent (6) and may even protect against these processes in some instances. The mechanisms by which saturated FFAs activate ER stress signaling pathways are also unclear. Several reports demonstrated that palmitate treatment of β-cells and/or islets activates the PERK pathway, including expression of ATF4 and CHOP by eIF2α phosphorylation (**Figure 4**) (6). However, detection of IRE1α activation and spliced *Xbp1* mRNAs' or XBP1s' protein was dependent on the palmitate preparation and/or the β-cell lines used in different laboratories (75, 90, 91). Further studies are required under physiologically relevant conditions. There is controversy about whether ATF6α is activated by palmitate. Although some reports suggest that the ATF6α pathway is activated by palmitate treatment (75, 92), other studies did not observe an increase in BiP expression induced by ATF6α activation in palmitate-treated β-cells (90, 93). It is also unknown whether palmitate is a specific inducer for the ATF6α branch because both palmitate and oleate treatment increased expression of known ATF6α-target genes, total *Xbp1*, and *BiP/Grp78* (91). The protective effect of BiP overexpression is also controversial in palmitate-treated β-cells (90). These inconsistencies need reconciliation through direct evidence of ATF6α cleavage and BiP expression in β-cells of palmitate-fed animals. The saturated fatty acid palmitate has multiple deleterious effects on pancreatic β-cells: (*a*) activation of PKC-δ; (*b*) accumulation of long-chain acyl-coenzyme As (CoAs) or lipid derivatives, such as diacylglycerol, lysophosphatic acid, and sphingolipids (ceramide and others); and (*c*) perturbation of ER Ca^{2+} to increase cytosolic Ca^{2+}. Below, we discuss to what extent each of these mechanisms contributes to ER stress and β-cell apoptosis.

First, blocking of PKCδ translocation by the phospholipase C inhibitor, U-73122 (94), or direct inhibition of PKCδ by rotterlin (95) substantially reduced palmitate-induced apoptosis. Moreover, overexpression of dominant-negative PKCδ in pancreatic β-cells protected against high-fat diet-induced glucose

Figure 4

Apoptotic unfolded protein response (UPR) pathways induced by free fatty acids (FFAs) and chronically high glucose in β-cells. In contrast to unsaturated FFAs, saturated FFAs serve as poor substrates for mitochondrial fatty acid oxidation and de novo triglyceride synthesis. However, saturated FFAs serve as intermediates in ceramide biosynthesis. The saturated FFAs activate UPR pathways (primarily PKR-like ER kinase, PERK) by perturbation of ER Ca^{2+} mobilization through inhibition of sarco/endoplasmic reticulum Ca^{2+}-ATPase (SERCA), or activation of inositol-1,4,5-trisphosphate (IP_3) receptors, and/or direct impairment of endoplasmic reticulum (ER) homeostasis. In addition, chronically high glucose increases biosynthesis of proinsulin and islet amyloid polypeptide in β-cells, which increases accumulation of misfolded proteins [insulin and islet amyloid polypeptide (IAPP)] and oxidative protein folding-mediated reactive oxygen species (ROS) production. The oxidative stress created by ROS and toxic IAPP oligomers perturb ER Ca^{2+} mobilization through activation of IP_3 receptors to release ER Ca^{2+}. Perturbation of ER Ca^{2+} causes protein misfolding in the ER and activates the UPR pathways (primarily inositol-requiring protein 1α, IRE1α) that induce proapoptotic signals, including proinsulin mRNA degradation as described in **Figure 2**. Abbreviations: ATF4: activating transcription factor 4, ATF6α: activating transcription factor 6α, CHOP: CCAAT-enhancer-binding protein (C/EBP) homology protein, eIF2α-P: phosphorylated form of eukaryotic translation initiation factor 2α.

intolerance and β-cell dysfunction in mice (96). Therefore, it is possible that FFA-induced activation of PKCδ may contribute to β-cell loss in T2D (97). It was recently shown that overexpression of dominant-negative PKCδ inhibited palmitate-induced nuclear accumulation of forkhead box protein O1 (FoxO1) in cultured β-cells and islets (96), although it remains to be confirmed whether apoptosis by activated PKCδ is caused by changes in FoxO1 nuclear localization. In pancreatic β-cells, the FoxO1 transcription factor is implicated in regulating differentiation, proliferation, and apoptosis (98). Furthermore, inhibition of FoxO1, which requires JNK inhibition, protects pancreatic β-cells against FFAs (99). In addition, FoxO1 activity was increased by an ER stress inducer, thapsigargin, and dominant-negative FoxO1 expression protected β-cells from thapsigargin-induced

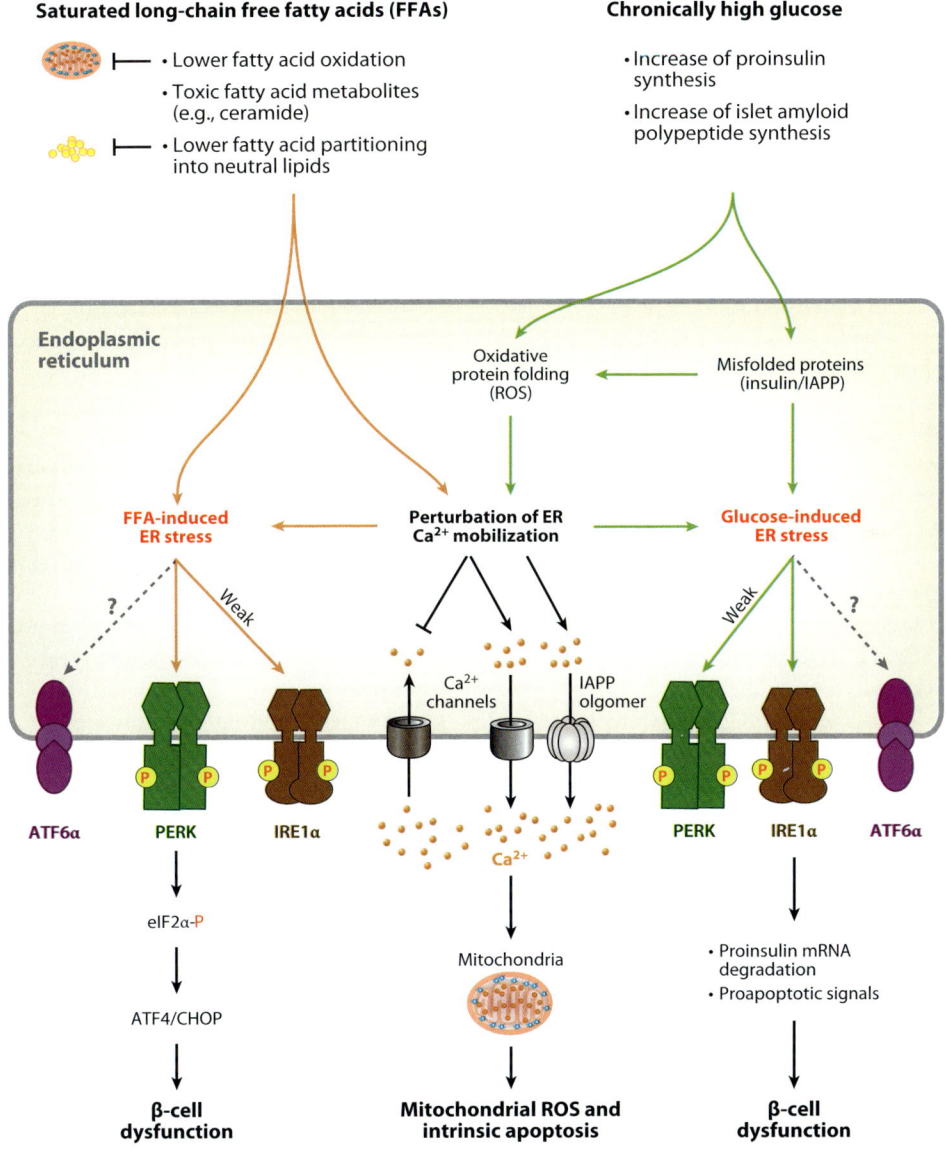

cell death (99). Recently, Qi & Mochly-Rosen (100) reported that during ER stress, PKCδ complexes with c-Abl (a protein tyrosine kinase involved in genotoxic and oxidative stresss) to cause JNK activation (101). Moreover, PKCδ inhibition reduced JNK activation and inhibited ER stress-mediated apoptosis. It is possible that FFAs or ER stress may activate the PKCδ-JNK pathway to activate FoxO1 directly or through inhibition of AKT-mediated insulin signaling. Therefore, the implication of the PKCδ-JNK-FoxO1 pathway in FFA-induced β-cell apoptosis warrants further investigation. It is also important to address how PKCδ is activated in FFA-treated or ER stress-induced β-cells. It is interesting that hepatic XBP1s, or even DNA-binding-defective mutant XBP1s, bind FoxO1 and promote its degradation, and thereby improve glucose homeostasis in leptin-deficient *ob/ob* T2D model mice (102). This suggests that increased expression of XBP1s in FFA- or ER stress-exposed β-cells may provide a new therapeutic approach to modulate the activity of FoxO1 for the treatment of T2D. However, prolonged XBP1s overexpression interferes with β-cell function via inhibition of insulin, pancreatic duodonal homeobox 1 (PDX1), and v-maf musculoaponeurotic fibrosarcoma oncogene homolog A (MafA) expression, eventually leading to β-cell apoptosis (103). Therefore, this issue requires further investigation.

Second, several studies suggest that the differential toxicity between saturated FFAs and unsaturated FFAs is directly related to their ability to promote triglyceride accumulation and fatty acid oxidation (**Figure 4**) (104, 105). In β-cells, oleate treatment leads to triglyceride accumulation and is well tolerated, whereas palmitate is poorly incorporated into triglyceride and causes ER stress and apoptosis (105). The lipotoxicity caused by excess oleate in acyl-CoA:diacylglycerol transferase 1 (DGAT1)-deficient fibroblasts (DGAT catalyzes the formation of triglycerides from diacylglycerol and acyl-CoA) (104) emphasizes the importance of a harmonious match between cellular lipid influx and lipid utilization.

Moreover, other studies have revealed that overexpression of stearoyl-CoA desaturase 1 (SCD1) to desaturate excess saturated FFAs is sufficient to prevent lipotoxicity of palmitate (104, 106). By contrast, knockdown of SCD in INS-1 β-cells decreased desaturation of palmitate to monounsaturated fatty acid, lowered FFA partitioning into complex neutral lipids, and augmented palmitate-induced ER stress and apoptosis (107). The importance of lipid content in ER function was further suggested in diabetic murine models. Loss of SCD1 worsened diabetes in leptin-deficient obese mice (108). In addition, *Scd1* and *Scd2* mRNA expression was induced in islets from prediabetic hyperinsulinemic Zucker diabetic fatty (ZDF) rats, whereas several fatty acid desaturases, including *Scd1* mRNA levels, were markedly reduced in diabetic ZDF rat islets (107).

Studies using inhibitors (e.g., etomoxir or bromopalmitate) or activators (e.g., T090217) of fatty acid mitochondrial β-oxidation suggest that increased fatty acid oxidation is important to prevent lipotoxicity and ER stress (92, 109). For example, overexpression of the mitochondrial fatty acid transporter carnitine palmitoyltransferase 1 alleviated palmitate-induced apoptosis and decreased expression of ER stress markers, eIF2α−P and CHOP, in β-cells (110). Fatty acid oxidation and triglyceride formation were most pronounced in oleate-treated β-cells when compared to palmitate (92, 105), although the responsible genes are not known.

Several saturated lipid intermediates (e.g., lysophosphatidic acid, phosphatidic acid, and diacylglycerols) of the saturated FFA esterification pathway (3) and sphingolipids (such as ceramide) de novo synthesized from saturated FFAs (111, 112) were proposed as toxic molecules that can induce β-cell dysfunction and apoptosis. Among these toxic lipid intermediates, studies using proximal inhibitors (such as myriocin or fumonisin B1) of de novo ceramide synthesis suggested that ceramide induces ER stress and cell death in several cell lines, including β-cells (92, 111, 112). Recently, Boslem et al. (113) employed mass spectrometry in a comprehensive lipidomic screen of MIN6 β-cells treated with palmitate. They observed that the major alterations following palmitate exposure were in the sphingolipid class [glucosylceramide (GlcCer), lactosylceramide, trihexosylceramide] without any significant alteration in the amounts of either ceramide or sphingomyelin. Among the sphingolipid class, the amount of GlcCer changed most significantly, and increased conversion of ceramide to GlcCer by overexpression of GlcCer synthase reduced ER stress and apoptosis and also ameliorated the palmitate-mediated ER-to-Golgi protein trafficking defect. Although these results support ceramide as a causative factor in palmitate-induced β-cell death and possibly in β-cell dysfunction in T2D, it remains unknown how ceramide causes ER stress and cell death without an increase in the steady-state level of ceramide, as other accumulated sphingolipids are not toxic to β-cells.

Third, studies have revealed that ER Ca^{2+} homeostasis is important for β-cell function. ER Ca^{2+} measurement using an ER-targeted cameleon showed that the MIN6 β-cell has about 250 μM resting ER Ca^{2+} (56). The high intraluminal Ca^{2+} concentration in the ER is important to maintain Ca^{2+}-dependent ER chaperone functions, and Ca^{2+} released into the cytosol is an important signaling molecule (114, 115). In response to a variety of external stimuli, Ca^{2+} is released from the lumen of the ER via two Ca^{2+} channels; the IP_3 receptor and the ryanodine receptor. The concentration of Ca^{2+} in the cytosol is maintained at a low level by active Ca^{2+} transport into the ER via SERCA, creating a large [Ca^{2+}] gradient between the cytosol and the ER lumen (0.1 μM versus 400 μM). Ca^{2+} is buffered in the ER by proteins (e.g., calnexin, calreticulin, ORP 150, ERp57, and others) that bear multiple low-affinity Ca^{2+}-binding sites. More importantly, Ca^{2+} binding to molecular chaperones regulates their activity, and therefore ER Ca^{2+} directly affects posttranslational protein folding, modification, and trafficking. For example, the interaction between calreticulin and other ER chaperones (PDI and ERp57) depends on the ER Ca^{2+} concentration. Therefore, blocking the SERCA pump by thapsigargin, a selective inhibitor of SERCA, depletes ER Ca^{2+}, inhibits ER functions, and thereby causes cell death through hyperactivation of the IRE1α and PERK pathways (116). ER stress and cell death could occur if saturated FFAs perturb ER Ca^{2+} homeostasis through reduction of SERCA activity or by activation of IP_3 or ryanodine receptors. A previous study demonstrated that islets from *db/db* mice, an animal model of typical T2D, lack the initial reduction of intracellular Ca^{2+} and subsequent intracellular Ca^{2+} oscillations following stimulation with high glucose, possibly caused by a defect in cytosolic Ca^{2+} sequestration secondary to a reduction in SERCA activity (117). Several recent studies have indicated that palmitate triggers ER stress in pancreatic β-cells through perturbation of ER Ca^{2+} levels (**Figure 4**) (6). Long-term palmitate treatment depleted ER Ca^{2+} (by 40%) to a greater extent than oleate (by 24%) (91). The reduced ER Ca^{2+} content is explained by impaired ER Ca^{2+} uptake, which may reflect inhibition of the SERCA pump in palmitate-treated β-cells. The inhibition of ER Ca^{2+} uptake was observed at an early time (3 h) after palmitate treatment and was followed by UPR activation. Furthermore, CHOP depletion by siRNA partially protected against palmitate-induced β-cell apoptosis, suggesting that ER stress is involved in palmitate-induced apoptosis through ER Ca^{2+} depletion. However, this model is challenged by evidence described below. In the absence of extracellular Ca^{2+}, the palmitate-induced Ca^{2+} signal was mostly inhibited, except for the initiating Ca^{2+} signal from the ER (118). Several plasma membrane L-type Ca^{2+} channel blockers (nifedipine, nimodipine, and verapamil) and the K_{ATP} channel opener (diazoxide) efficiently inhibited the palmitate-induced Ca^{2+} signals and significantly reduced CHOP expression and cell death in pancreatic β-cells and islets, whereas the IP_3 receptor Ca^{2+} channel blocker (xestospongin *c*) and ryanodine receptor Ca^{2+} channel blocker (dantrolin) did not significantly alter the palmitate-induced Ca^{2+} signal and did not protect from palmitate-induced apoptosis

(95). This suggests that a Ca^{2+} influx through the voltage-sensitive Ca^{2+} channel of the L-type coupled to membrane depolarization through closure of the K_{ATP} channel is crucial for cytosolic Ca^{2+} increase in response to palmitate. Moreover, several mono-/polyunsaturated FFAs (oleate, linoleic acid, and α-linolenic acid) in addition to palmitate also elicited intracellular Ca^{2+} signals (119) and reduced ER Ca^{2+} content (118). This suggests that the increase in cytosolic Ca^{2+} alone is not sufficient to induce cell death in palmitate-treated pancreatic β-cells. However, the unsaturated FFAs were less toxic than palmitate to β-cells. Futhermore, palmitate immediately induced PERK activation in 5 min and subsequently caused a significant increase in both XBP1s and CHOP in β-cells. Therefore, it is possible that reduced ER Ca^{2+} content is a prerequisite for activation of palmitate-mediated UPR signaling or apoptosis in pancreatic β-cells. Clearly, further studies are required to unravel the relative contribution of reduced ER Ca^{2+} content in palmitate-induced ER stress and apoptosis in β-cells.

Endoplasmic Reticulum Stress by Glucotoxicity

In T2D, absolute or relative insulin deficiency associated with insulin resistance causes blood glucose levels to remain high, called hyperglycemia (120). In chronic hyperglycemia, consistently overstimulated β-cells show a gradual decrease of glucose-induced insulin secretion and insulin gene expression and eventually impaired β-cell function and survival, a process called glucotoxicity. As diabetic hyperglycemia becomes chronic, the glucose that normally serves as fuel or substrate is used to generate detrimental metabolites for β-cells. Glucotoxicity is mediated at least in part by accumulation of excess ROS generated by several metabolic pathways, including mitochondrial oxidative phosphorylation and other alternative metabolic pathways, such as glucose autoxidation, hexosamine metabolism, sorbitol metabolism, and increased protein glycation (120, 121). The pancreatic β-cell is vulnerable to ROS because it has a low antioxidative stress response (122). β-Cells do not express catalase and only low levels of glutathione peroxidase. Furthermore, oxidative stress by elevated ROS reduced proinsulin synthesis by decreasing mRNA expression through inactivation of the β-cell-specific transcription factors, PDX1 and MafA, that regulate expression of proinsulin genes and multiple downstream genes required for β-cell differentiation, proliferation, and survival (123). Therefore, it is thought that oxidative stress is an important factor in β-cell failure.

Another possible mechanism of ROS generation from hyperglycemia-exposed pancreatic β-cells has emerged in recent years. Protein folding pathways in the ER and ROS production are closely linked events (**Figure 3**) (65, 71, 73). Prolonged UPR activation leads to the accumulation of ROS via two sources: the UPR-regulated oxidative protein folding machinery in the ER and oxidative phosphorylation in mitochondria. First, in hyperglycemia-exposed β-cells, the increased demand for insulin requires increased disulfide bond formation, which generates ROS during the process (73). Moreover, the increased amount of proinsulin synthesis further depletes glutathione, used for reducing nonnative disulfide bonds in misfolded proinsulin molecules. It has been estimated that approximately 25% of the ROS generated in a cell may result from formation of disulfide bonds in the ER during oxidative protein folding (124). Second, ER Ca^{2+} uptake to mitochondria mediates mitochondrial ROS production through disruption of mitochondrial electron transport and eventually induces mitochondrial apoptotic pathways (**Figure 3**). Specifically, more than 50% of the protein synthesized in the ER during glucose stimulation is proinsulin that requires delicate intermolecular disulfide bond formation and exchange (9). If chronic hyperglycemia leads to an increased rate of proinsulin synthesis, it would overcrowd the ER and increase the rate of oxidative protein folding. This would place an enormous burden on the ER and would enhance the probability of protein misfolding. Thus, during hyperglycemia, the accumulation of misfolded proinsulin in the ER lumen can generate ROS by the UPR-regulated oxidative protein folding machinery and by functional perturbation of mitochondria by a Ca^{2+} leak from the ER. In this manner, glucotoxicity may cause ER stress-regulated ROS accumulation, leading to diminished insulin gene expression, β-cell failure, and apoptosis. However, additional studies need to test the notion whether the hyperglycemia-increased proinsulin synthesis-ER stress-ROS pathway is a main cause for the onset and/or the progression of T2D.

Under conditions of chronic hyperglycemia, increased proinsulin biosynthesis may overwhelm the ER protein folding capacity, leading to UPR activation. Chronic high-glucose exposure (24 h) causes hyperactivation of IRE1α to splice *Xbp1* mRNA, whereas acute exposure (1–3 h) to high glucose activates IRE1α without *Xbp1* mRNA splicing (**Figure 4**) (86). During chronically high glucose exposure over several days, hyperactivated IRE1α, which may have a different activation states (85), degrades ER proinsulin mRNA, possibly including ER-localized mRNAs, contributing to the reduction of proinsulin biosynthesis and further β-cell demise (85, 125). Further studies under physiological conditions are necessary to elucidate the validity of these in vitro systems that use prolonged states of high glucose exposure, which are rarely observed in vivo.

It appears that β-cell death in T2D is not associated with hyperglycemia or hyperlipidemia alone but is a combination of hyperglycemia and hyperlipidemia, which is called glucolipotoxicity (3). FFA-mediated UPR signaling pathways are potentiated by high-glucose cosupplementation of β-cells as high glucose exacerbates β-cell lipotoxicity (**Figure 5**) (126, 127). Although high glucose activates PERK and IRE1α phosphorylation in palmitate-treated β-cells, it remains to be determined how high glucose synergizes with palmitate to alter ER stress signaling, especially the PERK pathway. In general, acute high glucose, leading to increased proinsulin synthesis and folding, represses PERK activation, which limits excessive proinsulin synthesis and causes expression of the integrated stress response genes (such as *Atf4*, *Chop*, *Gadd34*, and others) (33, 128). The potential role of high glucose as an enhancer of FFA-mediated ER stress now warrants further investigation to fully understand glucolipotoxicity in T2D.

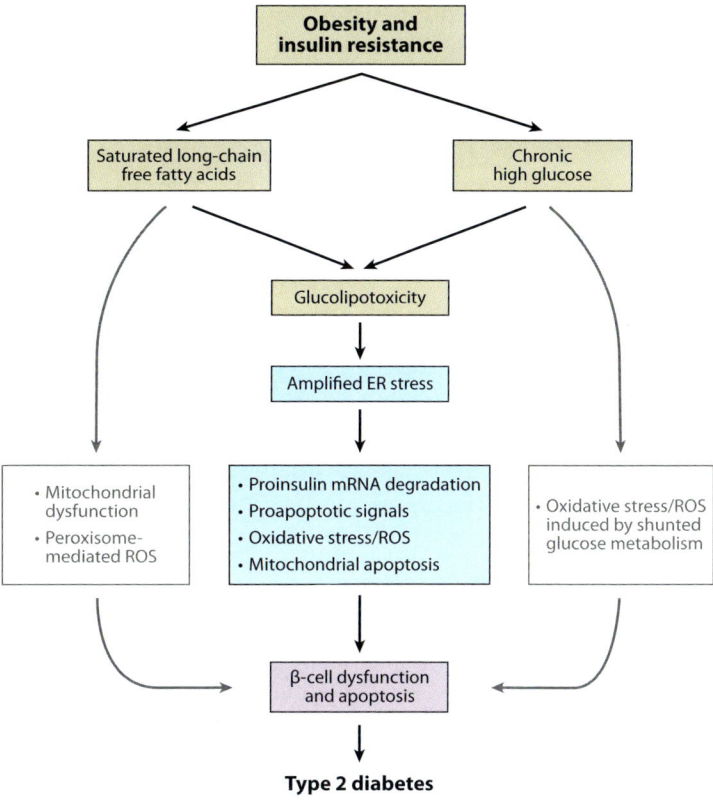

Figure 5

Amplified endoplasmic reticulum (ER) stress and β-cell death in type 2 diabetes (T2D). Conditions of insulin resistance and obesity cause hyperglycemia and hyperlipidemia, which result in glucolipotoxicity for the β-cell. Studies suggest that free fatty acid-mediated unfolded protein response signaling pathways are potentiated by high-glucose cosupplementation to β-cells as high glucose exacerbates β-cell lipotoxicity (126, 127). The amplified ER stress response leads to β-cell dysfunction and apoptosis through proinsulin mRNA degradation, oxidative stress, proapoptotic signals, and mitochondrial apoptosis, eventually culminating in T2D. Abbreviation: ROS, reactive oxygen species.

Integrated stress response: eIF2α phosphorylation-dependent, stress-inducible signaling pathways that can be activated by eIF2 kinases (PERK, PKR, HRI, GCN2)

Endoplasmic Reticulum Stress by Islet Amyloid

In addition to β-cell failure and insulin resistance, islet hyalinosis (hyaline deposits in β-cells), reported for the first time by Opie in 1901 (129), also plays a role in T2D. The hyaline deposits in pancreatic β-cells occur in approximately 90% of individuals with T2D and are associated with reduced β-cell volume (130). The hyaline deposits are known as amyloids (131), insoluble fibrous protein aggregates sharing specific structural traits. A number of human neurodegenerative diseases (such as Alzheimer's disease, Parkinson's disease, Huntington's disease) are now thought to be associated with the formation of amyloids or amyloid-like fibrils (132). In 1987, the islet amyloid in T2D was shown to be an islet amyloid polypeptide (IAPP) (133). It is coexpressed with insulin in pancreatic β-cells, traffics through the insulin secretory pathway and is secreted with insulin after food ingestion (134). The mature IAPP is a 37-amino acid polypeptide derived from an 89-amino acid precursor by proteolytic processing (135). The primary sequence of processed IAPP in humans, nonhuman primates, and cats is highly conserved, but rodent IAPPs show sequence differences between the twentieth and twenty-ninth amino acids. In vitro IAPP in humans, nonhuman primates, and cats, but not rodents, forms amyloid fibrils and conveys cellular toxicity to pancreatic β-cells (136). It is believed that the sequence differences in IAPP conveys its propensity to form a fibrillar amyloid in aqueous environments and confers toxicity to β-cells (7). This amyloid hypothesis for T2D was further tested by β-cell-specific overexpression of human IAPP (hIAPP) in transgenic mice. The hIAPP-transgenic mice recapitulated the metabolic characteristics of T2D, i.e., hyperglycemia, impaired insulin secretion, and insulin resistance (7). Recent examination of amyloid toxicity found that the toxic form of amyloidogenic proteins appears not to be the extracellular or intracellular large amyloid deposits detected by Congo red dye under light microscopy but rather smaller intracellular nonfibrillar oligomers that can be detected by a specific antibody against amyloid β protein toxic oligomers (7). Through use of the toxic oligomer-specific antibody, Lin et al. (137) were able to show that toxic oligomers of hIAPP in hIAPP-expressing transgenic mice accumulate in β-cells, although there are no data from the human pancreata from patients with T2D to confirm the existence of these toxic hIAPP oligomers.

How and why IAPP amyloids are formed are important questions to be answered in the future. Although the exact mechanism linking hIAPP oligomer formation and apoptotic β-cell death is unknown, several in vitro studies suggest the notion that the toxicity of IAPP oligomers is related to the formation of a membrane channel inducing an unregulated Ca^{2+} influx or membrane disruption (**Figure 5**) (138, 139). Therefore, ER accumulation of the toxic IAPP oligomers might reasonably be expected to induce Ca^{2+} leakage into the cytoplasm, which can induce ER stress. As predicted, Huang and coworkers (7) observed ER stress in β-cell lines as well as in islets of transgenic mice and rats expressing human IAPP. Several ER stress markers, including CHOP and XBP1s, were observed in both the islets and β-cells expressing hIAPP, whereas overexpression of rat IAPP did not provoke ER stress. Furthermore, siRNA-mediated inhibition of CHOP reduced apoptotic cell death in hIAPP-expressing β-cells. The β-cells of hIAPP transgenic mice have increased polyubiquitinated proteins, which are not present in transgenic mice that express rat IAPP, implying that β-cells in hIAPP transgenic mice may experience ER stress induced by the accumulation of misfolded proteins. However, there are several challenges to the hIAPP oligomer-mediated ER stress model. Recently, Hull et al. (140) did not detect ER stress in hIAPP-expressing islets, although amyloid formation in these murine islets was associated with reduced β-cell mass. TUNEL-positive β-cells in hIAPP transgenic mice were not always positive for CHOP expression. Moreover, it was curious why the induced CHOP was localized in the perinuclear region instead of in the nucleus. Therefore, experimental designs using chemical chaperones or *Chop* knockout in hIAPP transgenic mice should provide additional insight into the involvement of other apoptotic pathways. Nevertheless, Huang and coworkers' data suggest that, in patients with obesity and T2D, hIAPP oligomer-mediated ER stress is an important mechanism, leading to increased β-cell apoptosis.

CONCLUSIONS

T2D develops only when β-cell dysfunction appears. It is thought that β-cell failure is mainly caused by increases in blood glucose and FFAs that cannot be properly disposed or stored. Excess glucose and fatty acids can overload the cell and disrupt ER and mitochondrial functions. Indeed, both chronically high glucose and saturated fatty acids cause β-cell failure and loss in islet mass. It was suggested that chronically high glucose may cause oxidative stress through its metabolism, and saturated FFAs produce toxic lipid metabolites. However, recent studies suggest that high glucose and saturated FFAs interfere with ER function, which subsequently disrupts proinsulin synthesis, folding, and processing in β-cells. The chronically stressed ER generates several proapoptotic signals, induces oxidative stress, and initiates mitochondrial apoptosis. Consequently, β-cell function deteriorates, and cells eventually die.

SUMMARY POINTS

1. When ER homeostasis is altered, signaling pathways mediated by three ER-proximal sensors: IRE1, PERK and ATF6 are activated, eliciting an adaptive response that is collectively called the unfolded protein response (UPR). However, when the adaptive mechanisms fail to restore ER homeostasis, several death signaling pathways are activated.

2. During the death response, the PERK-mediated cell death pathway is mainly initiated by CHOP expression, which causes an increase in misfolded protein accumulation and oxidative stress, activates the mitochondrial death signal through ER Ca^{2+} release, and regulates expression of pro- and antiapoptotic genes.

3. During the death response, IRE1α may propagate death by signaling activation of JNK and by regulating IRE1-dependent RNA degradation.

4. The UPR pathway is activated by chronic exposure to high glucose or saturated long-chain FFAs, and activation of the UPR pathways is amplified by cosupplementation of both high glucose and saturated long-chain FFAs.

5. The activation of the UPR in T2D β-cells is connected with saturated long-chain fatty acid-mediated proapoptotic mechanisms that include PKCδ activation, accumulation of lipid derivatives, and ER Ca^{2+} release.

6. Under chronic hyperglycemia in T2D, increased proinsulin synthesis induces oxidative stress by ROS production and ER Ca^{2+} release, leading to mitochondria-dependent cell death.

7. Under chronic hyperglycemia conditions, hyperactivated IRE1α contributes to β-cell failure by degrading proinsulin mRNA, possibly including ER-localized mRNAs.

8. Activation of the UPR induced by ER accumulation of the toxic hIAPP oligomer may be an important mechanism in β-cell death in humans.

FUTURE ISSUES

1. Is hyperglycemia and/or hyperlipidemia essential for β-cell failure in T2D?
2. What is the molecular mechanism of glucose level-dependent differential activation of PERK and IREα in pancreatic β-cells?
3. How does ceramide activate the UPR? What other lipid intermediates activate or inhibit the UPR?
4. How does the UPR activate PKCδ in palmitate-treated β-cells?
5. How does palmitate induce ER calcium release?
6. How does high glucose activate the UPR in β-cells? Is there any direct evidence of ER stress?
7. What are the molecular mechanisms involved in the formation of the toxic hIAPP oligomers and UPR induction by hIAPP oligomers?
8. Further investigation is needed to identify effective chemical chaperones that can improve ER function to prevent hyperactivation of the UPR.

DISCLOSURE STATEMENT

Dr. Kaufman is an Scientific Advisory Board member of Alnylam Pharmaceuticals, Inc., Cambridge, MA; and Proteostasis Therapeutics, Inc., Cambridge, MA.

ACKNOWLEDGMENTS

We apologize to our colleagues whose work was not cited owing to space limitations. We especially thank Hyun Ju Yoo at the Asan Insitutute of Life Science for helpful comments. This work was funded by the Basic Science Research Program through the National Research Foundation of Korea (2011-0011433) to S.H.B. This work was partially supported by National Institute of Health grants DK042394, DK088227, DK093074, HL052173, and HL057346 to R.J.K.

LITERATURE CITED

1. Shaw JE, Sicree RA, Zimmet PZ. 2010. Global estimates of the prevalence of diabetes for 2010 and 2030. *Diabetes Res. Clin. Pract.* 87:4–14
2. Kahn SE, Hull RL, Utzschneider KM. 2006. Mechanisms linking obesity to insulin resistance and type 2 diabetes. *Nature* 444:840–46
3. Prentki M, Nolan CJ. 2006. Islet beta cell failure in type 2 diabetes. *J. Clin. Investig.* 116:1802–12
4. Jetton TL, Lausier J, LaRock K, Trotman WE, Larmie B, et al. 2005. Mechanisms of compensatory beta-cell growth in insulin-resistant rats: roles of Akt kinase. *Diabetes* 54:2294–304
5. Liu YQ, Jetton TL, Leahy JL. 2002. β-cell adaptation to insulin resistance. Increased pyruvate carboxylase and malate-pyruvate shuttle activity in islets of nondiabetic Zucker fatty rats. *J. Biol. Chem.* 277:39163–68
6. Cnop M, Ladriere L, Igoillo-Esteve M, Moura RF, Cunha DA. 2010. Causes and cures for endoplasmic reticulum stress in lipotoxic beta-cell dysfunction. *Diabetes Obes. Metab.* 12(Suppl. 2):76–82
7. Haataja L, Gurlo T, Huang CJ, Butler PC. 2008. Islet amyloid in type 2 diabetes, and the toxic oligomer hypothesis. *Endocr. Rev.* 29:303–16

8. Rutter GA. 2001. Nutrient-secretion coupling in the pancreatic islet beta-cell: recent advances. *Mol. Aspects Med.* 22:247–84
9. Dodson G, Steiner D. 1998. The role of assembly in insulin's biosynthesis. *Curr. Opin. Struct. Biol.* 8:189–94
10. Van Lommel L, Janssens K, Quintens R, Tsukamoto K, Vander Mierde D, et al. 2006. Probe-independent and direct quantification of insulin mRNA and growth hormone mRNA in enriched cell preparations. *Diabetes* 55:3214–20
11. Goodge KA, Hutton JC. 2000. Translational regulation of proinsulin biosynthesis and proinsulin conversion in the pancreatic beta-cell. *Semin. Cell Dev. Biol.* 11:235–42
12. Schuit FC, In't Veld PA, Pipeleers DG. 1988. Glucose stimulates proinsulin biosynthesis by a dose-dependent recruitment of pancreatic beta cells. *Proc. Natl. Acad. Sci. USA* 85:3865–69
13. Ellgaard L, Molinari M, Helenius A. 1999. Setting the standards: quality control in the secretory pathway. *Science* 286:1882–88
14. McMaster CR. 2001. Lipid metabolism and vesicle trafficking: more than just greasing the transport machinery. *Biochem. Cell Biol.* 79:681–92
15. Meusser B, Hirsch C, Jarosch E, Sommer T. 2005. ERAD: the long road to destruction. *Nat. Cell Biol.* 7:766–72
16. Ron D, Walter P. 2007. Signal integration in the endoplasmic reticulum unfolded protein response. *Nat. Rev. Mol. Cell Biol.* 8:519–29
17. Kim I, Xu W, Reed JC. 2008. Cell death and endoplasmic reticulum stress: disease relevance and therapeutic opportunities. *Nat. Rev. Drug Discov.* 7:1013–30
18. Tirasophon W, Welihinda AA, Kaufman RJ. 1998. A stress response pathway from the endoplasmic reticulum to the nucleus requires a novel bifunctional protein kinase/endoribonuclease (Ire1p) in mammalian cells. *Genes Dev.* 12:1812–24
19. Bertolotti A, Wang X, Novoa I, Jungreis R, Schlessinger K, et al. 2001. Increased sensitivity to dextran sodium sulfate colitis in IRE1β-deficient mice. *J. Clin. Investig.* 107:585–93
20. Tirasophon W, Lee K, Callaghan B, Welihinda A, Kaufman RJ. 2000. The endoribonuclease activity of mammalian IRE1 autoregulates its mRNA and is required for the unfolded protein response. *Genes Dev.* 14:2725–36
21. Lee AH, Heidtman K, Hotamisligil GS, Glimcher LH. 2011. Dual and opposing roles of the unfolded protein response regulated by IRE1alpha and XBP1 in proinsulin processing and insulin secretion. *Proc. Natl. Acad. Sci. USA* 108:8885–90
22. Harding HP, Zhang Y, Ron D. 1999. Protein translation and folding are coupled by an endoplasmic-reticulum-resident kinase. *Nature* 397:271–74
23. Shi Y, Vattem KM, Sood R, An J, Liang J, et al. 1998. Identification and characterization of pancreatic eukaryotic initiation factor 2 α-subunit kinase, PEK, involved in translational control. *Mol. Cell. Biol.* 18:7499–509
24. Wek RC, Cavener DR. 2007. Translational control and the unfolded protein response. *Antioxid. Redox Signal.* 9:2357–71
25. Delepine M, Nicolino M, Barrett T, Golamaully M, Lathrop GM, Julier C. 2000. EIF2AK3, encoding translation initiation factor 2-alpha kinase 3, is mutated in patients with Wolcott-Rallison syndrome. *Nat. Genet.* 25:406–9
26. Harding HP, Zeng H, Zhang Y, Jungries R, Chung P, et al. 2001. Diabetes mellitus and exocrine pancreatic dysfunction in *Perk−/−* mice reveals a role for translational control in secretory cell survival. *Mol. Cell* 7:1153–63
27. Zhang P, McGrath B, Li S, Frank A, Zambito F, et al. 2002. The PERK eukaryotic initiation factor 2 alpha kinase is required for the development of the skeletal system, postnatal growth, and the function and viability of the pancreas. *Mol. Cell. Biol.* 22:3864–74
28. Scheuner D, Song B, McEwen E, Liu C, Laybutt R, et al. 2001. Translational control is required for the unfolded protein response and in vivo glucose homeostasis. *Mol. Cell* 7:1165–76
29. Back SH, Scheuner D, Han J, Song B, Ribick M, et al. 2009. Translation attenuation through eIF2alpha phosphorylation prevents oxidative stress and maintains the differentiated state in beta cells. *Cell Metab.* 10:13–26

30. Gupta S, McGrath B, Cavener DR. 2010. PERK (EIF2AK3) regulates proinsulin trafficking and quality control in the secretory pathway. *Diabetes* 59:1937–47
31. Zhang W, Feng D, Li Y, Iida K, McGrath B, Cavener DR. 2006. PERK EIF2AK3 control of pancreatic beta cell differentiation and proliferation is required for postnatal glucose homeostasis. *Cell Metab.* 4:491–97
32. Vander Mierde D, Scheuner D, Quintens R, Patel R, Song B, et al. 2007. Glucose activates a protein phosphatase-1-mediated signaling pathway to enhance overall translation in pancreatic beta-cells. *Endocrinology* 148:609–17
33. Hou ZQ, Li HL, Gao L, Pan L, Zhao JJ, Li GW. 2008. Involvement of chronic stresses in rat islet and INS-1 cell glucotoxicity induced by intermittent high glucose. *Mol. Cell Endocrinol.* 291:71–78
34. Elouil H, Bensellam M, Guiot Y, Vander Mierde D, Pascal SM, et al. 2007. Acute nutrient regulation of the unfolded protein response and integrated stress response in cultured rat pancreatic islets. *Diabetologia* 50:1442–52
35. Gomez E, Powell ML, Bevington A, Herbert TP. 2008. A decrease in cellular energy status stimulates PERK-dependent eIF2alpha phosphorylation and regulates protein synthesis in pancreatic beta-cells. *Biochem. J.* 410:485–93
36. Yan W, Frank CL, Korth MJ, Sopher BL, Novoa I, et al. 2002. Control of PERK eIF2α kinase activity by the endoplasmic reticulum stress-induced molecular chaperone P58^{IPK}. *Proc. Natl. Acad. Sci. USA* 99:15920–25
37. Lee TG, Tang N, Thompson S, Miller J, Katze MG. 1994. The 58,000-dalton cellular inhibitor of the interferon-induced double-stranded RNA-activated protein kinase (PKR) is a member of the tetratricopeptide repeat family of proteins. *Mol. Cell. Biol.* 14:2331–42
38. Rutkowski DT, Kang SW, Goodman AG, Garrison JL, Taunton J, et al. 2007. The role of p58IPK in protecting the stressed endoplasmic reticulum. *Mol. Biol. Cell* 18:3681–91
39. Haze K, Yoshida H, Yanagi H, Yura T, Mori K. 1999. Mammalian transcription factor ATF6 is synthesized as a transmembrane protein and activated by proteolysis in response to endoplasmic reticulum stress. *Mol. Biol. Cell* 10:3787–99
40. Haze K, Okada T, Yoshida H, Yanagi H, Yura T, et al. 2001. Identification of the G13 (cAMP-response-element-binding protein-related protein) gene product related to activating transcription factor 6 as a transcriptional activator of the mammalian unfolded protein response. *Biochem. J.* 355:19–28
41. Wu J, Rutkowski DT, Dubois M, Swathirajan J, Saunders T, et al. 2007. ATF6alpha optimizes long-term endoplasmic reticulum function to protect cells from chronic stress. *Dev. Cell* 13:351–64
42. Yamamoto K, Sato T, Matsui T, Sato M, Okada T, et al. 2007. Transcriptional induction of mammalian ER quality control proteins is mediated by single or combined action of ATF6alpha and XBP1. *Dev. Cell* 13:365–76
43. Luebke-Wheeler J, Zhang K, Battle M, Si-Tayeb K, Garrison W, et al. 2008. Hepatocyte nuclear factor 4alpha is implicated in endoplasmic reticulum stress-induced acute phase response by regulating expression of cyclic adenosine monophosphate responsive element binding protein H. *Hepatology* 48:1242–50
44. Kondo S, Murakami T, Tatsumi K, Ogata M, Kanemoto S, et al. 2005. OASIS, a CREB/ATF-family member, modulates UPR signalling in astrocytes. *Nat. Cell Biol.* 7:186–94
45. Kondo S, Saito A, Hino S, Murakami T, Ogata M, et al. 2007. BBF2H7, a novel transmembrane bZIP transcription factor, is a new type of endoplasmic reticulum stress transducer. *Mol. Cell. Biol.* 27:1716–29
46. Nagamori I, Yabuta N, Fujii T, Tanaka H, Yomogida K, et al. 2005. Tisp40, a spermatid specific bZip transcription factor, functions by binding to the unfolded protein response element via the Rip pathway. *Genes Cells* 10:575–94
47. Liang G, Audas TE, Li Y, Cockram GP, Dean JD, et al. 2006. Luman/CREB3 induces transcription of the endoplasmic reticulum (ER) stress response protein Herp through an ER stress response element. *Mol. Cell. Biol.* 26:7999–8010
48. Raggo C, Rapin N, Stirling J, Gobeil P, Smith-Windsor E, et al. 2002. Luman, the cellular counterpart of herpes simplex virus VP16, is processed by regulated intramembrane proteolysis. *Mol. Cell. Biol.* 22:5639–49
49. Stirling J, O'Hare P. 2006. CREB4, a transmembrane bZip transcription factor and potential new substrate for regulation and cleavage by S1P. *Mol. Biol. Cell* 17:413–26

50. Fonseca SG, Ishigaki S, Oslowski CM, Lu S, Lipson KL, et al. 2010. Wolfram syndrome 1 gene negatively regulates ER stress signaling in rodent and human cells. *J. Clin. Investig.* 120:744–55
51. Seo HY, Kim YD, Lee KM, Min AK, Kim MK, et al. 2008. Endoplasmic reticulum stress-induced activation of activating transcription factor 6 decreases insulin gene expression via up-regulation of orphan nuclear receptor small heterodimer partner. *Endocrinology* 149:3832–41
52. Kang SW, Rane NS, Kim SJ, Garrison JL, Taunton J, Hegde RS. 2006. Substrate-specific translocational attenuation during ER stress defines a pre-emptive quality control pathway. *Cell* 127:999–1013
53. Rutkowski DT, Kaufman RJ. 2007. That which does not kill me makes me stronger: adapting to chronic ER stress. *Trends Biochem. Sci.* 32:469–76
54. Oyadomari S, Mori M. 2004. Roles of CHOP/GADD153 in endoplasmic reticulum stress. *Cell Death Differ.* 11:381–89
55. Ma Y, Brewer JW, Diehl JA, Hendershot LM. 2002. Two distinct stress signaling pathways converge upon the CHOP promoter during the mammalian unfolded protein response. *J. Mol. Biol.* 318:1351–65
56. Yoshida H, Okada T, Haze K, Yanagi H, Yura T, et al. 2000. ATF6 activated by proteolysis binds in the presence of NF-Y (CBF) directly to the *cis*-acting element responsible for the mammalian unfolded protein response. *Mol. Cell. Biol.* 20:6755–67
57. Yoshida H, Okada T, Haze K, Yanagi H, Yura T, et al. 2001. Endoplasmic reticulum stress-induced formation of transcription factor complex ERSF including NF-Y (CBF) and activating transcription factors 6alpha and 6beta that activates the mammalian unfolded protein response. *Mol. Cell. Biol.* 21:1239–48
58. Lu PD, Jousse C, Marciniak SJ, Zhang Y, Novoa I, et al. 2004. Cytoprotection by pre-emptive conditional phosphorylation of translation initiation factor 2. *EMBO J.* 23:169–79
59. Ma Y, Hendershot LM. 2004. Herp is dually regulated by both the endoplasmic reticulum stress-specific branch of the unfolded protein response and a branch that is shared with other cellular stress pathways. *J. Biol. Chem.* 279:13792–99
60. Lee AH, Iwakoshi NN, Glimcher LH. 2003. XBP-1 regulates a subset of endoplasmic reticulum resident chaperone genes in the unfolded protein response. *Mol. Cell. Biol.* 23:7448–59
61. Rutkowski DT, Wu J, Back SH, Callaghan MU, Ferris SP, et al. 2008. UPR pathways combine to prevent hepatic steatosis caused by ER stress-mediated suppression of transcriptional master regulators. *Dev. Cell* 15:829–40
62. Zinszner H, Kuroda M, Wang X, Batchvarova N, Lightfoot RT, et al. 1998. CHOP is implicated in programmed cell death in response to impaired function of the endoplasmic reticulum. *Genes Dev.* 12:982–95
63. Tajiri S, Oyadomari S, Yano S, Morioka M, Gotoh T, et al. 2004. Ischemia-induced neuronal cell death is mediated by the endoplasmic reticulum stress pathway involving CHOP. *Cell Death Differ.* 11:403–15
64. Silva RM, Ries V, Oo TF, Yarygina O, Jackson-Lewis V, et al. 2005. CHOP/GADD153 is a mediator of apoptotic death in substantia nigra dopamine neurons in an in vivo neurotoxin model of parkinsonism. *J. Neurochem.* 95:974–86
65. Malhotra JD, Miao H, Zhang K, Wolfson A, Pennathur S, et al. 2008. Antioxidants reduce endoplasmic reticulum stress and improve protein secretion. *Proc. Natl. Acad. Sci. USA* 105:18525–30
66. Oyadomari S, Koizumi A, Takeda K, Gotoh T, Akira S, et al. 2002. Targeted disruption of the *Chop* gene delays endoplasmic reticulum stress-mediated diabetes. *J. Clin. Investig.* 109:525–32
67. Oyadomari S, Takeda K, Takiguchi M, Gotoh T, Matsumoto M, et al. 2001. Nitric oxide-induced apoptosis in pancreatic beta cells is mediated by the endoplasmic reticulum stress pathway. *Proc. Natl. Acad. Sci. USA* 98:10845–50
68. Song B, Scheuner D, Ron D, Pennathur S, Kaufman RJ. 2008. *Chop* deletion reduces oxidative stress, improves β cell function, and promotes cell survival in multiple mouse models of diabetes. *J. Clin. Investig.* 118:3378–89
69. McCullough KD, Martindale JL, Klotz LO, Aw TY, Holbrook NJ. 2001. Gadd153 sensitizes cells to endoplasmic reticulum stress by down-regulating Bcl2 and perturbing the cellular redox state. *Mol. Cell. Biol.* 21:1249–59
70. Marciniak SJ, Yun CY, Oyadomari S, Novoa I, Zhang Y, et al. 2004. CHOP induces death by promoting protein synthesis and oxidation in the stressed endoplasmic reticulum. *Genes Dev.* 18:3066–77

71. Haynes CM, Titus EA, Cooper AA. 2004. Degradation of misfolded proteins prevents ER-derived oxidative stress and cell death. *Mol. Cell* 15:767–76
72. Harding HP, Zhang Y, Zeng H, Novoa I, Lu PD, et al. 2003. An integrated stress response regulates amino acid metabolism and resistance to oxidative stress. *Mol. Cell* 11:619–33
73. Tu BP, Weissman JS. 2004. Oxidative protein folding in eukaryotes: mechanisms and consequences. *J. Cell Biol.* 164:341–46
74. Timmins JM, Ozcan L, Seimon TA, Li G, Malagelada C, et al. 2009. Calcium/calmodulin-dependent protein kinase II links ER stress with Fas and mitochondrial apoptosis pathways. *J. Clin. Investig.* 119:2925–41
75. Laybutt DR, Preston AM, Akerfeldt MC, Kench JG, Busch AK, et al. 2007. Endoplasmic reticulum stress contributes to beta cell apoptosis in type 2 diabetes. *Diabetologia* 50:752–63
76. Urano F, Wang X, Bertolotti A, Zhang Y, Chung P, et al. 2000. Coupling of stress in the ER to activation of JNK protein kinases by transmembrane protein kinase IRE1. *Science* 287:664–66
77. Nishitoh H, Matsuzawa A, Tobiume K, Saegusa K, Takeda K, et al. 2002. ASK1 is essential for endoplasmic reticulum stress-induced neuronal cell death triggered by expanded polyglutamine repeats. *Genes Dev.* 16:1345–55
78. Lei K, Davis RJ. 2003. JNK phosphorylation of Bim-related members of the Bcl2 family induces Bax-dependent apoptosis. *Proc. Natl. Acad. Sci. USA* 100:2432–37
79. Yamamoto K, Ichijo H, Korsmeyer SJ. 1999. BCL-2 is phosphorylated and inactivated by an ASK1/Jun N-terminal protein kinase pathway normally activated at G_2/M. *Mol. Cell. Biol.* 19:8469–78
80. Kim I, Shu CW, Xu W, Shiau CW, Grant D, et al. 2009. Chemical biology investigation of cell death pathways activated by endoplasmic reticulum stress reveals cytoprotective modulators of ASK1. *J. Biol. Chem.* 284:1593–603
81. Hetz C, Bernasconi P, Fisher J, Lee AH, Bassik MC, et al. 2006. Proapoptotic BAX and BAK modulate the unfolded protein response by a direct interaction with IRE1alpha. *Science* 312:572–76
82. Luo D, He Y, Zhang H, Yu L, Chen H, et al. 2008. AIP1 is critical in transducing IRE1-mediated endoplasmic reticulum stress response. *J. Biol. Chem.* 283:11905–12
83. Gu F, Nguyen DT, Stuible M, Dube N, Tremblay ML, Chevet E. 2004. Protein-tyrosine phosphatase 1B potentiates IRE1 signaling during endoplasmic reticulum stress. *J. Biol. Chem.* 279:49689–93
84. Lisbona F, Rojas-Rivera D, Thielen P, Zamorano S, Todd D, et al. 2009. BAX inhibitor-1 is a negative regulator of the ER stress sensor IRE1alpha. *Mol. Cell* 33:679–91
85. Han D, Lerner AG, Vande Walle L, Upton JP, Xu W, et al. 2009. IRE1alpha kinase activation modes control alternate endoribonuclease outputs to determine divergent cell fates. *Cell* 138:562–75
86. Lipson KL, Fonseca SG, Ishigaki S, Nguyen LX, Foss E, et al. 2006. Regulation of insulin biosynthesis in pancreatic beta cells by an endoplasmic reticulum-resident protein kinase IRE1. *Cell Metab.* 4:245–54
87. McGarry JD, Dobbins RL. 1999. Fatty acids, lipotoxicity and insulin secretion. *Diabetologia* 42:128–38
88. Kharroubi I, Ladriere L, Cardozo AK, Dogusan Z, Cnop M, Eizirik DL. 2004. Free fatty acids and cytokines induce pancreatic beta-cell apoptosis by different mechanisms: role of nuclear factor-kappaB and endoplasmic reticulum stress. *Endocrinology* 145:5087–96
89. Bollheimer LC, Skelly RH, Chester MW, McGarry JD, Rhodes CJ. 1998. Chronic exposure to free fatty acid reduces pancreatic beta cell insulin content by increasing basal insulin secretion that is not compensated for by a corresponding increase in proinsulin biosynthesis translation. *J. Clin. Investig.* 101:1094–101
90. Lai E, Bikopoulos G, Wheeler MB, Rozakis-Adcock M, Volchuk A. 2008. Differential activation of ER stress and apoptosis in response to chronically elevated free fatty acids in pancreatic beta-cells. *Am. J. Physiol. Endocrinol. Metab.* 294:E540–50
91. Cunha DA, Hekerman P, Ladriere L, Bazarra-Castro A, Ortis F, et al. 2008. Initiation and execution of lipotoxic ER stress in pancreatic beta-cells. *J. Cell Sci.* 121:2308–18
92. Choi SE, Jung IR, Lee YJ, Lee SJ, Lee JH, et al. 2011. Stimulation of lipogenesis as well as fatty acid oxidation protects against palmitate-induced INS-1 beta-cell death. *Endocrinology* 152:816–27
93. Karaskov E, Scott C, Zhang L, Teodoro T, Ravazzola M, Volchuk A. 2006. Chronic palmitate but not oleate exposure induces endoplasmic reticulum stress, which may contribute to INS-1 pancreatic beta-cell apoptosis. *Endocrinology* 147:3398–407

94. Eitel K, Staiger H, Rieger J, Mischak H, Brandhorst H, et al. 2003. Protein kinase C delta activation and translocation to the nucleus are required for fatty acid-induced apoptosis of insulin-secreting cells. *Diabetes* 52:991–97
95. Choi SE, Kim HE, Shin HC, Jang HJ, Lee KW, et al. 2007. Involvement of Ca^{2+}-mediated apoptotic signals in palmitate-induced MIN6N8a beta cell death. *Mol. Cell Endocrinol.* 272:50–62
96. Hennige AM, Ranta F, Heinzelmann I, Dufer M, Michael D, et al. 2010. Overexpression of kinase-negative protein kinase Cdelta in pancreatic beta-cells protects mice from diet-induced glucose intolerance and beta-cell dysfunction. *Diabetes* 59:119–27
97. Welters HJ, Smith SA, Tadayyon M, Scarpello JH, Morgan NG. 2004. Evidence that protein kinase Cdelta is not required for palmitate-induced cytotoxicity in BRIN-BD11 beta-cells. *J. Mol. Endocrinol.* 32:227–35
98. Cheng Z, White MF. 2011. Targeting Forkhead box O1 from the concept to metabolic diseases: lessons from mouse models. *Antioxid. Redox Signal.* 14:649–61
99. Martinez SC, Tanabe K, Cras-Meneur C, Abumrad NA, Bernal-Mizrachi E, Permutt MA. 2008. Inhibition of Foxo1 protects pancreatic islet beta-cells against fatty acid and endoplasmic reticulum stress-induced apoptosis. *Diabetes* 57:846–59
100. Qi X, Mochly-Rosen D. 2008. The PKCdelta -Abl complex communicates ER stress to the mitochondria—an essential step in subsequent apoptosis. *J. Cell Sci.* 121:804–13
101. Sun X, Majumder P, Shioya H, Wu F, Kumar S, et al. 2000. Activation of the cytoplasmic c-Abl tyrosine kinase by reactive oxygen species. *J. Biol. Chem.* 275:17237–40
102. Zhou Y, Lee J, Reno CM, Sun C, Park SW, et al. 2011. Regulation of glucose homeostasis through a XBP-1-FoxO1 interaction. *Nat. Med.* 17:356–65
103. Allagnat F, Christulia F, Ortis F, Pirot P, Lortz S, et al. 2010. Sustained production of spliced X-box binding protein 1 (XBP1) induces pancreatic beta cell dysfunction and apoptosis. *Diabetologia* 53:1120–30
104. Listenberger LL, Han X, Lewis SE, Cases S, Farese RV Jr, et al. 2003. Triglyceride accumulation protects against fatty acid-induced lipotoxicity. *Proc. Natl. Acad. Sci. USA* 100:3077–82
105. Thorn K, Bergsten P. 2010. Fatty acid-induced oxidation and triglyceride formation is higher in insulin-producing MIN6 cells exposed to oleate compared to palmitate. *J. Cell. Biochem.* 111:497–507
106. Busch AK, Gurisik E, Cordery DV, Sudlow M, Denyer GS, et al. 2005. Increased fatty acid desaturation and enhanced expression of stearoyl coenzyme A desaturase protects pancreatic beta-cells from lipoapoptosis. *Diabetes* 54:2917–24
107. Green CD, Olson LK. 2011. Modulation of palmitate-induced endoplasmic reticulum stress and apoptosis in pancreatic beta-cells by stearoyl-CoA desaturase and Elovl6. *Am. J. Physiol. Endocrinol. Metab.* 300:E640–49
108. Flowers JB, Rabaglia ME, Schueler KL, Flowers MT, Lan H, et al. 2007. Loss of stearoyl-CoA desaturase-1 improves insulin sensitivity in lean mice but worsens diabetes in leptin-deficient obese mice. *Diabetes* 56:1228–39
109. Briaud I, Harmon JS, Kelpe CL, Segu VB, Poitout V. 2001. Lipotoxicity of the pancreatic beta-cell is associated with glucose-dependent esterification of fatty acids into neutral lipids. *Diabetes* 50:315–21
110. Sol EM, Sargsyan E, Akusjarvi G, Bergsten P. 2008. Glucolipotoxicity in INS-1E cells is counteracted by carnitine palmitoyltransferase 1 over-expression. *Biochem. Biophys. Res. Commun.* 375:517–21
111. Maedler K, Spinas GA, Dyntar D, Moritz W, Kaiser N, Donath MY. 2001. Distinct effects of saturated and monounsaturated fatty acids on beta-cell turnover and function. *Diabetes* 50:69–76
112. Lupi R, Dotta F, Marselli L, Del Guerra S, Masini M, et al. 2002. Prolonged exposure to free fatty acids has cytostatic and pro-apoptotic effects on human pancreatic islets: evidence that beta-cell death is caspase mediated, partially dependent on ceramide pathway, and Bcl-2 regulated. *Diabetes* 51:1437–42
113. Boslem E, MacIntosh G, Preston AM, Bartley C, Busch AK, et al. 2011. A lipidomic screen of palmitate-treated MIN6 beta-cells links sphingolipid metabolites with endoplasmic reticulum (ER) stress and impaired protein trafficking. *Biochem. J.* 435:267–76
114. Bygrave FL, Benedetti A. 1996. What is the concentration of calcium ions in the endoplasmic reticulum? *Cell Calcium* 19:547–51
115. Corbett EF, Michalak M. 2000. Calcium, a signaling molecule in the endoplasmic reticulum? *Trends Biochem. Sci.* 25:307–11

116. Luciani DS, Gwiazda KS, Yang TL, Kalynyak TB, Bychkivska Y, et al. 2009. Roles of IP3R and RyR Ca^{2+} channels in endoplasmic reticulum stress and beta-cell death. *Diabetes* 58:422–32

117. Roe MW, Philipson LH, Frangakis CJ, Kuznetsov A, Mertz RJ, et al. 1994. Defective glucose-dependent endoplasmic reticulum Ca^{2+} sequestration in diabetic mouse islets of Langerhans. *J. Biol. Chem.* 269:18279–82

118. Gwiazda KS, Yang TL, Lin Y, Johnson JD. 2009. Effects of palmitate on ER and cytosolic Ca^{2+} homeostasis in beta-cells. *Am. J. Physiol. Endocrinol. Metab.* 296:E690–701

119. Schnell S, Schaefer M, Schofl C. 2007. Free fatty acids increase cytosolic free calcium and stimulate insulin secretion from beta-cells through activation of GPR40. *Mol. Cell Endocrinol.* 263:173–80

120. Poitout V, Robertson RP. 2008. Glucolipotoxicity: fuel excess and beta-cell dysfunction. *Endocr. Rev.* 29:351–66

121. Robertson RP. 2004. Chronic oxidative stress as a central mechanism for glucose toxicity in pancreatic islet beta cells in diabetes. *J. Biol. Chem.* 279:42351–54

122. Tiedge M, Lortz S, Drinkgern J, Lenzen S. 1997. Relation between antioxidant enzyme gene expression and antioxidative defense status of insulin-producing cells. *Diabetes* 46:1733–42

123. Robertson RP. 2006. Oxidative stress and impaired insulin secretion in type 2 diabetes. *Curr. Opin. Pharmacol.* 6:615–19

124. Kaufman RJ, Back SH, Song B, Han J, Hassler J. 2010. The unfolded protein response is required to maintain the integrity of the endoplasmic reticulum, prevent oxidative stress and preserve differentiation in beta-cells. *Diabetes Obes. Metab.* 12(Suppl. 2):99–107

125. Lipson KL, Ghosh R, Urano F. 2008. The role of IRE1alpha in the degradation of insulin mRNA in pancreatic beta-cells. *PLoS ONE* 3:e1648

126. Bachar E, Ariav Y, Ketzinel-Gilad M, Cerasi E, Kaiser N, Leibowitz G. 2009. Glucose amplifies fatty acid–induced endoplasmic reticulum stress in pancreatic beta-cells via activation of mTORC1. *PLoS ONE* 4:e4954

127. Tanabe K, Liu Y, Hasan SD, Martinez SC, Cras-Méneur C, et al. 2011. Glucose and fatty acids synergize to promote B-cell apoptosis through activation of glycogen synthase kinase 3beta independent of JNK activation. *PLoS ONE* 6:e18146

128. Jonas JC, Bensellam M, Duprez J, Elouil H, Guiot Y, Pascal SM. 2009. Glucose regulation of islet stress responses and beta-cell failure in type 2 diabetes. *Diabetes Obes. Metab.* 11(Suppl. 4):65–81

129. Opie EL. 1901. The relation of diabetes mellitus to lesions of the pancreas: hyaline degeneration of the islands of Langerhans. *J. Exp. Med.* 5:527–40

130. Cnop M, Welsh N, Jonas JC, Jorns A, Lenzen S, Eizirik DL. 2005. Mechanisms of pancreatic beta-cell death in type 1 and type 2 diabetes: many differences, few similarities. *Diabetes* 54(Suppl. 2):S97–107

131. Ehrlich JC, Ratner IM. 1961. Amyloidosis of the islets of Langerhans. A restudy of islet hyalin in diabetic and non-diabetic individuals. *Am. J. Pathol.* 38:49–59

132. Chiti F, Dobson CM. 2006. Protein misfolding, functional amyloid, and human disease. *Annu. Rev. Biochem.* 75:333–66

133. Cooper GJ, Willis AC, Clark A, Turner RC, Sim RB, Reid KB. 1987. Purification and characterization of a peptide from amyloid-rich pancreases of type 2 diabetic patients. *Proc. Natl. Acad. Sci. USA* 84:8628–32

134. Butler PC, Chou J, Carter WB, Wang YN, Bu BH, et al. 1990. Effects of meal ingestion on plasma amylin concentration in NIDDM and nondiabetic humans. *Diabetes* 39:752–56

135. Sanke T, Bell GI, Sample C, Rubenstein AH, Steiner DF. 1988. An islet amyloid peptide is derived from an 89-amino acid precursor by proteolytic processing. *J. Biol. Chem.* 263:17243–46

136. Lorenzo A, Razzaboni B, Weir GC, Yankner BA. 1994. Pancreatic islet cell toxicity of amylin associated with type-2 diabetes mellitus. *Nature* 368:756–60

137. Lin CY, Gurlo T, Kayed R, Butler AE, Haataja L, et al. 2007. Toxic human islet amyloid polypeptide (h-IAPP) oligomers are intracellular, and vaccination to induce anti-toxic oligomer antibodies does not prevent h-IAPP-induced beta-cell apoptosis in h-IAPP transgenic mice. *Diabetes* 56:1324–32

138. Janson J, Ashley RH, Harrison D, McIntyre S, Butler PC. 1999. The mechanism of islet amyloid polypeptide toxicity is membrane disruption by intermediate-sized toxic amyloid particles. *Diabetes* 48:491–98

139. Kawahara M, Kuroda Y, Arispe N, Rojas E. 2000. Alzheimer's beta-amyloid, human islet amylin, and prion protein fragment evoke intracellular free calcium elevations by a common mechanism in a hypothalamic GnRH neuronal cell line. *J. Biol. Chem.* 275:14077–83
140. Hull RL, Zraika S, Udayasankar J, Aston-Mourney K, Subramanian SL, Kahn SE. 2009. Amyloid formation in human IAPP transgenic mouse islets and pancreas, and human pancreas, is not associated with endoplasmic reticulum stress. *Diabetologia* 52:1102–11
141. Bertolotti A, Zhang Y, Hendershot LM, Harding HP, Ron D. 2000. Dynamic interaction of BiP and ER stress transducers in the unfolded-protein response. *Nat. Cell Biol.* 2:326–32
142. Liu CY, Schroder M, Kaufman RJ. 2000. Ligand-independent dimerization activates the stress response kinases IRE1 and PERK in the lumen of the endoplasmic reticulum. *J. Biol. Chem.* 275:24881–85
143. Proud CG. 2005. eIF2 and the control of cell physiology. *Semin. Cell Dev. Biol.* 16:3–12
144. Hershey JWB, Merrick WC. 2000. The pathway and mechanism of initiation of protein synthesis. In *Translational Control of Gene Expression*, ed. N Sonenberg, JWB Hershey, MB Mathews, pp. 33–88. Cold Spring Harbor, NY: Cold Spring Harb. Lab. Press
145. Harding HP, Novoa I, Zhang Y, Zeng H, Wek R, et al. 2000. Regulated translation initiation controls stress-induced gene expression in mammalian cells. *Mol. Cell* 6:1099–108
146. Bommiasamy H, Back SH, Fagone P, Lee K, Meshinchi S, et al. 2009. ATF6alpha induces XBP1-independent expansion of the endoplasmic reticulum. *J. Cell Sci.* 122:1626–36
147. Hu P, Han Z, Couvillon AD, Kaufman RJ, Exton JH. 2006. Autocrine tumor necrosis factor alpha links endoplasmic reticulum stress to the membrane death receptor pathway through IRE1alpha-mediated NF-kappaB activation and down-regulation of TRAF2 expression. *Mol. Cell. Biol.* 26:3071–84
148. Yamaguchi H, Wang HG. 2004. CHOP is involved in endoplasmic reticulum stress-induced apoptosis by enhancing DR5 expression in human carcinoma cells. *J. Biol. Chem.* 279:45495–502
149. Ohoka N, Yoshii S, Hattori T, Onozaki K, Hayashi H. 2005. *TRB3*, a novel ER stress-inducible gene, is induced via ATF4-CHOP pathway and is involved in cell death. *EMBO J.* 24:1243–55
150. Puthalakath H, O'Reilly LA, Gunn P, Lee L, Kelly PN, et al. 2007. ER stress triggers apoptosis by activating BH3-only protein Bim. *Cell* 129:1337–49
151. Deng J, Lu PD, Zhang Y, Scheuner D, Kaufman RJ, et al. 2004. Translational repression mediates activation of nuclear factor kappa B by phosphorylated translation initiation factor 2. *Mol. Cell. Biol.* 24:10161–68
152. Li G, Mongillo M, Chin KT, Harding H, Ron D, et al. 2009. Role of ERO1-alpha-mediated stimulation of inositol 1,4,5-triphosphate receptor activity in endoplasmic reticulum stress-induced apoptosis. *J. Cell Biol.* 186:783–92
153. Brookes PS, Yoon Y, Robotham JL, Anders MW, Sheu SS. 2004. Calcium, ATP, and ROS: a mitochondrial love-hate triangle. *Am. J. Physiol. Cell Physiol.* 287:C817–33
154. Moore CE, Omikorede O, Gomez E, Willars GB, Herbert TP. 2011. PERK activation at low glucose concentration is mediated by SERCA pump inhibition and confers preemptive cytoprotection to pancreatic beta-cells. *Mol. Endocrinol.* 25:315–26

Structure Unifies the Viral Universe

Nicola G.A. Abrescia,[1,2] Dennis H. Bamford,[3] Jonathan M. Grimes,[4,5] and David I. Stuart[4,5]

[1] Structural Biology Unit, CIC bioGUNE, CIBERehd, 48160 Derio, Spain; email: nabrescia@cicbiogune.es

[2] Ikerbasque, Basque Foundation for Science, 48011 Bilbao, Spain

[3] Institute of Biotechnology and Department of Biosciences, Viikki Biocenter, University of Helsinki, Viikinkari 5, FI-00014, Finland; email: dennis.bamford@helsinki.fi

[4] Division of Structural Biology, The Wellcome Trust Center for Human Genetics, University of Oxford, Headington, Oxford, OX3 7BN, United Kingdom; email: dave@strubi.ox.ac.uk, jonathan@strubi.ox.ac.uk

[5] Diamond Light Source Limited, Harwell Science and Innovation Campus, Didcot, OX11 0DE, United Kingdom

Keywords

viruses, evolution, three-dimensional structure, classification

Abstract

Is it possible to meaningfully comprehend the diversity of the viral world? We propose that it is. This is based on the observation that, although there is immense genomic variation, every infective virion is restricted by strict constraints in structure space (i.e., there are a limited number of ways to fold a protein chain, and only a small subset of these have the potential to construct a virion, the hallmark of a virus). We have previously suggested the use of structure for the higher-order classification of viruses, where genomic similarities are no longer observable. Here, we summarize the arguments behind this proposal, describe the current status of structural work, highlighting its power to infer common ancestry, and discuss the limitations and obstacles ahead of us. We also reflect on the future opportunities for a more concerted effort to provide high-throughput methods to facilitate the large-scale sampling of the virosphere.

Contents

INTRODUCTING THE VIRAL UNIVERSE	796
VIRUS CLASSIFICATION	798
LESSONS FROM EARLY STRUCTURES	800
PROBLEMS OF COINCIDENCE AND ANALOGY	801
TOOLS FOR STRUCTURE-BASED PHYLOGENY	802
VIRION ARCHITECTURE COUPLES TO GENOME PACKAGING	803
CURRENT STATUS: DETECTING VIRAL LINEAGES	804
Major Lineages	804
Icosahedral dsRNA Viruses	805
Enveloped Viruses: Structural Links with Host Machines?	808
Helical Viruses	811
IMPLICATIONS AND FUTURE PERSPECTIVES	812

INTRODUCING THE VIRAL UNIVERSE

Viruses were discovered approximately a century ago, and much research has focused on the prevention and control of viral infections in humans, animals, and plants. This work has seen notable triumphs; we can now hold many diseases in check through vaccination and antiviral drugs, and we have shown that pathogens can be eradicated (1, 2). Nevertheless, great challenges remain, with increasing population densities and global traffic producing the constant threat of emerging pandemics (3). In addition, environmental virologists have demonstrated the fundamental importance of viruses to the entire biosphere, extending to geochemical cycles and even climate change (4–6). Viruses also lurk, integrated, in the genomes of cellular organisms or as individually replicating genetic elements; in the human genome, the amount of DNA of viral origin is roughly equal to the entire coding region (7). To learn how viruses operate, where they come from, and what the relationships between different viruses are, we need to consider the entire virosphere. However, the genomic diversity and astronomical population of viruses ($>10^{31}$ viral particles in the biosphere, the vast majority infecting microbes) create a massive selective pressure on cellular organisms (4–6, 8). Although huge challenges lie ahead of us in understanding the virosphere, it will not be a one-way street. Viruses were a central driver in the revolution in molecular biology when bacterial genetics, phage biology, and X-ray diffraction studies pushed forward our understanding of central biological components and processes, including the nature and structure of genetic material, replication, transcription, and translation (9, 10). Since then, viruses have continued to be central to many new discoveries and will doubtless remain so.

The use of symmetry, almost invariably icosahedral, helical, or a combination of the two, underlies the structure of most virus capsids, although some are pleomorphic (with a lipid envelope) and others, especially those infecting archaea, are bottle, lemon, and spindle shaped (11). As we explore below, the array of viruses whose structure we know are built according to a modest number of architectural templates, the majority of which are found across viruses infecting host cells residing in all three domains of cellular life (bacteria, archaea, and eukarya) (12), although some may infect only a single domain (13, 14). The simplest virions such as porcine circoviruses (15) contain only two genes, one for replicating the viral genome and the other coding for the protein making the icosahedral shell of the virion. In contrast, the most complex viruses are larger than the smallest cells; for example, mimiviruses have a virion diameter of about 0.65 μm housing a genome encoding >900 genes (16–18).

Within this zoo of structures, can we find a guiding principle to distinguish viruses from other self-replicating genetic elements, such as plasmids and transposons? The answer is simple: Once a replicon incorporates a gene(s) that allows it to make a capsid to enclose the

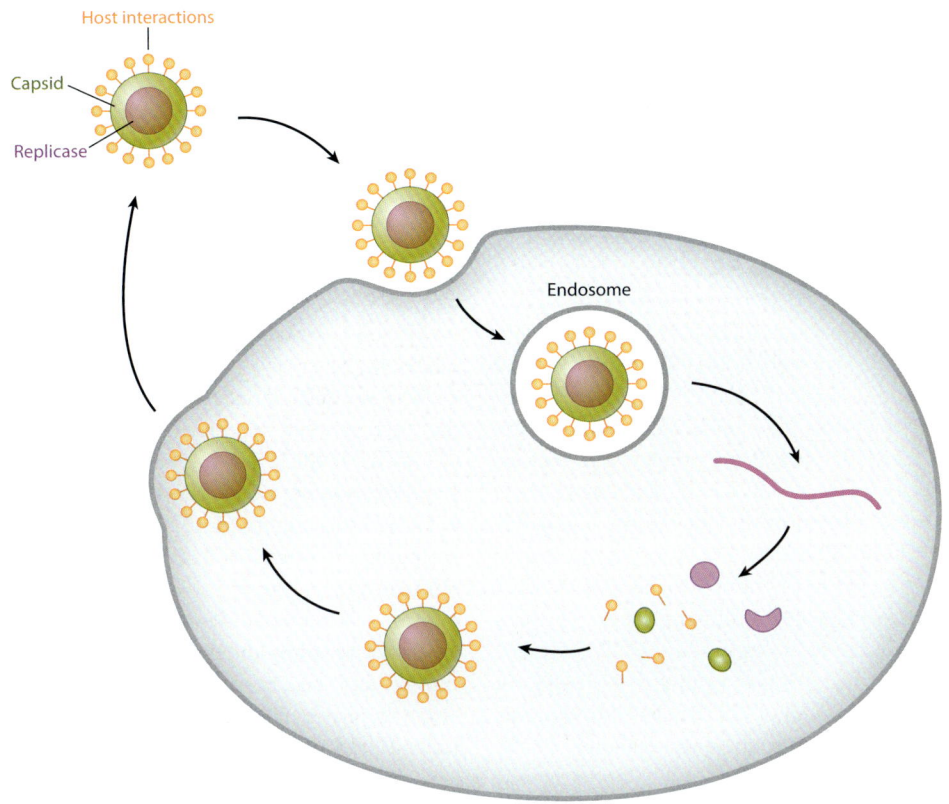

Figure 1
What makes a virus? Components of a virus. Throughout the life cycle these include the capsid- and genome-packaging machinery (*green*), the portion responsible for host interactions (*orange*), and the replicative machinery and genome (*purple*).

replicon, then a functional and structural entity called a virion is produced, and a virus is born. At a minimum, the virion delivers the viral genome to the next host cell to initiate a new round of virus production. Thus, although the genome is an essential element of a virus, the hallmark of a functional virus is the presence of a virion (**Figure 1**). Viral genomes integrated within a cell have frequently lost the capacity to form an infectious virion and, according to our nomenclature, are no longer viruses. Our capsid-centered approach immediately simplifies the extreme complexity of the viral sequence space by mapping it to the reduced complexity of viral structure space. Obviously, where sequence similarity can be discerned, it will remain the major comparison criterion (i.e., for closely related viruses), but three-dimensional structures penetrate much more deeply in time. This is because the requirement to maintain a functional capsid imposes strict constraints directly on that structure; redundancy in sequence space means that a given structure can be produced from an array of quite different amino acid chains.

There has been much speculation on the origins of viruses (for summaries, see References 19–22). Three major hypotheses have been put forward: (*a*) Viruses originated from times before cellular life was invented, (*b*) viruses originated from cells by reduction, and (*c*) viruses escaped from cells by utilizing cellular replicative elements removed from cellular control. The first idea has been rejected mostly because all known viruses utilize cellular hosts. The second proposal would require intermediates between cells and viruses. These have not been found, although there are examples of horizontal gene transfer of the coat protein between virus and host in archaea (23) and bacteria (24), and recently bacterial proteins involved in ethanolamine utilization have been shown to self-assemble, creating bacterial microcompartments resembling virus capsids in architecture but built from a protein fold not seen in viruses (25, 26). The third proposal struggles to explain how it is possible to build complex virion structures, such as genome packaging devices, from cellular components where there are no known cellular counterparts for these structures and functions. Comparative structural analyses of virion architectures [summarized in previous reviews (12, 27–30)] have shown protein fold similarities between viruses infecting hosts residing in all three domains of life, strongly arguing that viruses are ancient and are not reduced cells, although this argument was revived by the isolation of the mimivirus (16–18). The escape model is also less consistent with the ancient origin of viruses. At this stage, we need more viral structures and sequences to reliably evaluate the hypotheses.

VIRUS CLASSIFICATION

Two classification schemes have had an enduring influence on the way we envision the virosphere. The Baltimore scheme (31) divides viruses into seven groups depending on the nature of the genome (chemical type, number of strands, and, for a single-strand genome, whether the genome strand is directly used as a template for translation). The groups are dsDNA, ssDNA, dsRNA, plus (sense) ssRNA, and minus (antisense) ssRNA viruses. Two additional virus categories utilize reverse transcriptase, ssRNA viruses with a DNA intermediate in the life cycle and dsDNA viruses with an RNA intermediate.

The second classification, carried out by the International Committee on Taxonomy of Viruses (ICTV, http://ictvonline.org/virusTaxonomy.asp?version=2009&bhcp=1) of the International Union of Microbiological Societies, aims to develop an internationally agreed upon taxonomy for viruses, including defining the names for virus taxa in communication with the virology community. The agreed taxonomic levels are order, family, genus, and species. The organizational structure of the ICTV reflects the host organism a particular virus (or virus group) infects (e.g., plant virus), with each having a subcommittee. Thus, their classification tends to map the Baltimore scheme into the host type. Obviously, the rapid accumulation of genetic information plays a central role in formulating the relationships between viruses, and the work of the ICTV is never done. The ICTV classification, taken from the last published report, is shown in **Figure 2a**.

Figure 2

The virosphere. (*a*) The International Committee on Taxonomy of Viruses (ICTV) scheme (adapted from the Universal Virus Database 2005) is shown with the currently defined orders colored, and the virus families within the order defined by letters, colored to correspond to the Virus Order. The key defines the coloring for each order and those families that in 2005 were not assigned to an order are drawn in black (*Adenoviridae, Arenaviridae, Ascoviridae, Asfarviridae, Astroviridae, Baculoviridae, Barnaviridae, Birnaviridae, Bromoviridae, Bunyaviridae, Caliciviridae, Caulimoviridae, Chrysoviridae, Circoviridae, Closteroviridae, Corticoviridae, Cystoviridae, Flaviviridae, Fuselloviridae, Geminiviridae, Guttaviridae, Hepadnaviridae, Hypoviridae, Inoviridae, Iridoviridae, Leviviridae, Lipothrixviridae, Luteoviridae, Metaviridae, Microviridae, Nanoviridae, Narnaviridae, Nimaviridae, Nodaviridae, Orthomyxoviridae, Papillomaviridae, Partitiviridae, Parvoviridae, Phycodnaviridae, Plasmaviridae, Polydnaviridae, Polyomaviridae, Potyviridae, Poxviridae, Pseudoviridae, Reoviridae, Retroviridae, Rudiviridae, Tectiviridae, Tetraviridae, Togaviridae, Tombusviridae, Totiviridae*). (*b*) Structure-based viral lineages mapped onto current ICTV taxonomy with each lineage colored separately. Individual viral families within each lineage are labeled and colored according to the key. Abbreviations: BTV, bluetongue virus; HK97, bacteriophage Hong Kong 97; PRD1, prototype member of the *Tectiviridae* double-stranded DNA bacteriophages.

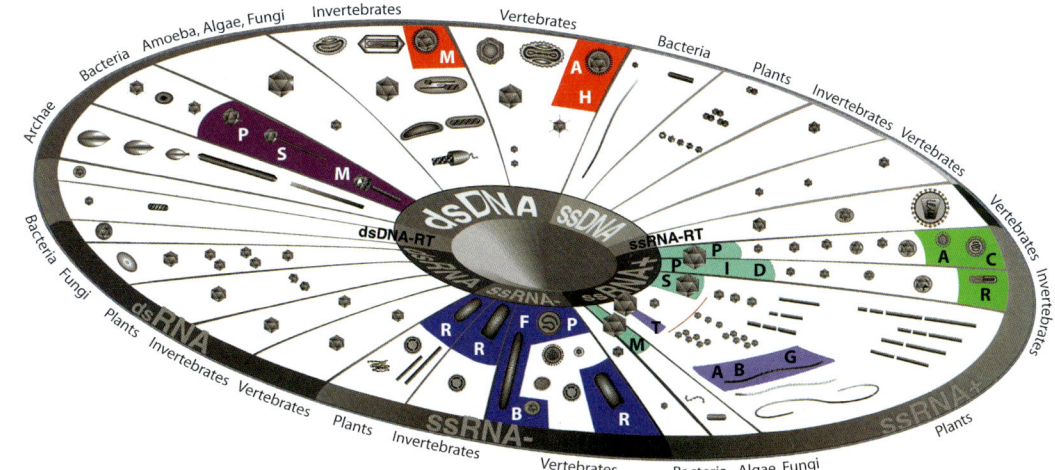

Order: Caudovirales
M: *Myoviridae*
P: *Podoviridae*
S: *Siphoviridae*

Order: Herpesvirales
A: *Alloherpesviridae*
H: *Herpesviridae*
M: *Malacoherpesviridae*

Order: Mononegavirales
B: *Bornaviridae*
F: *Filoviridae*
P: *Paramyxoviridae*
R: *Rhabdoviridae*

Order: Picornavirales
D: *Dicistroviridae*
I: *Iflaviridae*
M: *Marnaviridae*
P: *Picornaviridae*
S: *Secoviridae*

Order: Tymovirales
A: *Alphaflexiviridae*
B: *Betaflexiviridae*
G: *Gammaflexiviridae*
T: *Tymoviridae*

Order: Nidovirales
A: *Arteriviridae*
C: *Coronaviridae*
R: *Roniviridae*

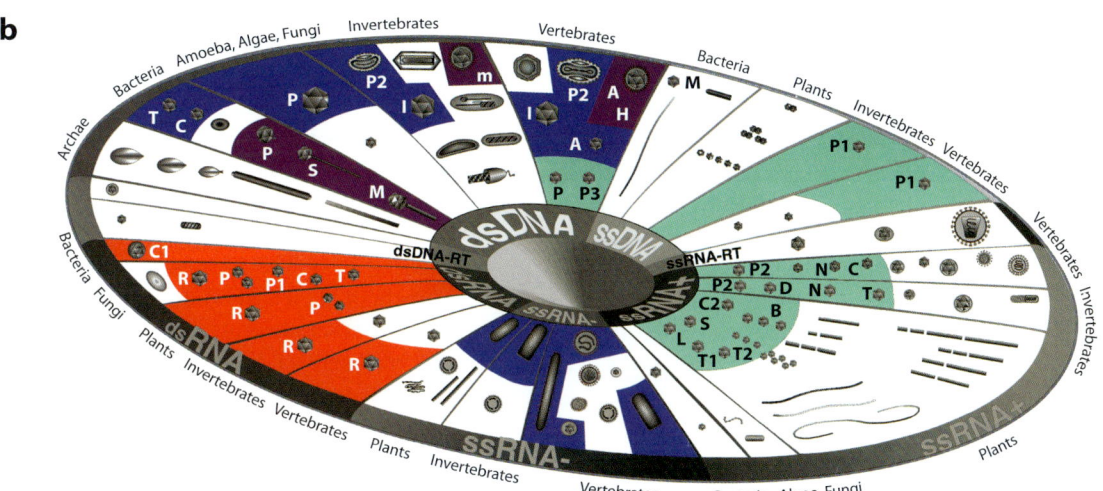

Lineage: BTV like
C: *Chrysoviridae*
C1: *Cystoviridae*
P: *Partitiviridae*
P1: *Picobirnaviridae*
R: *Reoviridae*
T: *Totiviridae*

Lineage: PRD1/Adeno
A: *Adenoviridae*
C: *Corticoviridae*
I: *Iridoviridae*
P: *Phycodnaviridae*
P2: *Poxviridae*
T: *Tectiviridae*

Lineage: HK97 like
M: *Myoviridae*
P: *Podoviridae*
S: *Siphoviridae*
A: *Alloherpesviridae*
H: *Herpesviridae*
m: *Malacoherpesviridae*

Lineage: Picorna like
B: *Bromoviridae*
C: *Caliciviridae*
C2: *Comoviridae*
D: *Dicistroviridae*
L: *Luteoviridae*
M: *Microviridae*
N: *Nodaviridae*
P: *Papillomaviridae*
P1: *Parvoviridae*
P2: *Picornaviridae*
P3: *Polyomaviridae*
S: *Sequiviridae*
T: *Tetraviridae*
T1: *Tombusviridae*
T2: *Tymoviridae*

Despite their best efforts, most families are not yet assigned to orders.

In this review, we illustrate the shortcomings of these approaches to virus classification, especially at higher level (above family), and focus on instances where structure can help when sequence fails. Our fundamental argument is that because neither of the two classical pillars of virus classification is necessarily coupled to virion architecture they are unlikely to lead to a useful phylogeny. Having raised many of these issues previously (12, 27–30, 32), we focus here on clarifying the fundamental issues, present some recent results, and highlight areas where structure-based classification is problematic.

LESSONS FROM EARLY STRUCTURES

Virus crystals were first noted in the 1930s for *Tobacco mosaic virus* (TMV) and *Tomato bushy stunt virus* (33, 34). By 1941, Bernal & Fankuchen had showed that both gave clear X-ray diffraction images (35). In 1945, Dorothy Crowfoot Hodgkin (36) published diffraction patterns for *Tobacco necrosis virus*, and additional useful results emerged in the 1950s with work on TMV and poliovirus taken up by Rosalind Franklin (37) and carried on after her death by Aaron Klug (38). However, technical and methodological challenges precluded high-resolution X-ray analysis for another 20 years.

In 1978, the first high-resolution (2.9-Å) virion structure was determined—the small icosahedral ssRNA plant virus *Tomato bushy stunt virus* by the Harrison laboratory (39), followed by the structure of *Southern bean mosaic virus* by Rossmann's laboratory (40), each a tour de force of crystallography. These viruses showed similarity in the major part of their capsid subunit structures and a similar arrangement within the capsid. Bigger surprises came in 1985 when the Rossmann laboratory (41) published the structure of a human common cold virus and the Hogle laboratory published the structure of poliovirus (42), small icosahedral ssRNA animal viruses (picornaviruses) and similar to each other. What was unexpected was that the coat protein folds were closely related to those previously seen in the plant viruses (an eight-stranded β-barrel) (**Figure 3**). Later, an insect virus joined the structural family, determined in the Johnson laboratory (43). So a plant, an insect, and a human virus share a similar architecture! These observations inspired the idea that small ssRNA viruses may share a common ancestry (41, 44), a hypothesis supported by comparative genomics and phylogenetic analyses (45, 46), as well as by a comprehensive comparison of all currently available small icosahedral ssRNA virus coat protein structures (J. Ravantti, D.H. Bamford, & D.I. Stuart, unpublished data).

The small ssRNA viruses infect cells residing in different kingdoms but the same domain of life (eukarya). In the meantime, the Burnett laboratory (47) unwittingly took on the challenge of linking the domains of life

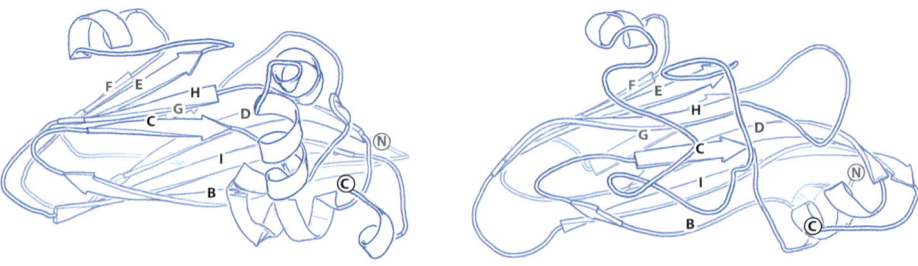

Figure 3

Side-by-side comparison of the structures of the jelly-roll capsid proteins of (*left*) *Southern bean mosaic virus* [Protein Data Bank identification (PDB ID) 4sbv] and (*right*) Rhinovirus VP2 (PDB ID 1nd2) represented as an outlined cartoon. Equivalent β-sheets correspond, respectively, to the BIDG and CHEF strands.

by extending their structural work from the dsDNA adenovirus (a human pathogen) to a bacterial virus (PRD1) infecting gram-negative bacterial hosts. The high-resolution PRD1 major coat protein structure was published in 1999 (48), and its closest structural relative was the corresponding protein of adenovirus! The simplest explanation for this observation was a common ancestry between the two viruses infecting hosts from two different domains of life (bacteria and eukarya). Unlike the small RNA viruses, where the building block is a single β-barrel lying (usually) parallel to the surface of the capsid, in these viruses, the building block is a trimeric arrangement of a polypeptide chain containing two concatenated β-barrels stacked against each other in a characteristic way and lying orthogonal to the capsid. Together with other independent similarities, including the presence of penton (49) and spike proteins (50) as well as a similar replication mechanism (51), this pointed strongly to divergence from a common origin. It was not long before this double β-barrel fold was observed in an archaeal virus, *Sulfolobus* turreted icosahedral virus (52), revealing that structurally similar viruses infect hosts in all three domains of cellular life. Since then, viruses belonging to a variety of different families have been shown to share this major coat protein fold (29, 53–55); so far this fold is utilized by 10 families of dsDNA viruses out of the 20 currently defined by the ICTV (56). The implication is that a common ancestor of these modern viruses infected cells before they diversified into the three domains of cellular life we know today.

These observations inspired us to develop a systematic approach to identifying lineages of viruses on the basis of virion architecture, with the expectation that these lineages may have separate but ancient origins. We aim to include all known viruses with enough structural information so that this analysis can cover the currently known viral structure space. Our underlying hypothesis is that the immense genetic diversity of viruses obscures the underlying structural bottleneck, dictated by the more-limited protein fold space and the complexity of the problem of forming a symmetrical functional capsid. Presumably, only a small subset of possible protein folds has the potential to make any kind of a virion, limiting available architectures. Furthermore, we propose that the invention of a self-assembling capsid is very difficult to achieve, so that, even where a structural fold (such as the β-barrel) is well suited to forming an isometric virus capsid, its formation, by variation and natural selection, is a very rare occurrence.

PROBLEMS OF COINCIDENCE AND ANALOGY

The argument above suggests that structure-based lineages may tend to reflect homology rather than structural convergence, but this question deserves careful analysis [a lucid exploration of many of the issues may be found in the paper by Williams (57)]. The basis of what we do was clearly explained by Woodger (58, pp. 98–99): "when we compare two things we set up a one-one relation or *correspondence* between the parts of the one and those of the other...choosing a pairing which will bring the maximum number of parts into identity correspondence." This principle applies equally well to the comparison of protein sequence, protein structure, or fossils. Inevitably, any comparison yields a result. In particular, when comparing atomic structures, the constraints of chemistry mean that it will always be possible to bring several Cα atoms into correspondence. Furthermore, it is very difficult to derive hard and fast rules to establish the point at which similarities become coincidental. Let us consider the common use of icosahedral symmetry by viruses. The key point, recognized by Crick & Watson (59), is that this reflects the fact that there are only five convex solids that satisfy the rules of space, and for reasons of parsimony, icosahedral symmetry is best suited to making a spherical virus. In this situation, there is no evolutionary significance to be attached to the common use of icosahedral symmetry by two families of virus. The question is thornier when applied to protein structure; as long ago as 1953, Astbury (60) proposed that protein structures would fall into a discreet number of families, and this has turned out to be true. Thus, the Structural Classification of Proteins currently recognizes just 1195 distinct folds (**http://scop.mrc-lmb.cam.ac.uk/scop/count.html#scop-1.75**). Furthermore, physicochemical rules appear to particularly favor certain protein wiring patterns, so that occurrences of folds, such as the β-barrel, might reflect analogy (convergence) rather than homology (divergence). Again, this is not a new problem, and the distinction between homology and analogy comes from Richard Owen (61) in the nineteenth century. For Owen the homology was with an archetype in the mind of god, whereas for us, it is with a real, but lost, common ancestor. We should, however, recognize that in the absence of historical data, such as a fossil record, we cannot formally distinguish the direction of evolution. As with classical anatomical comparisons, the guiding principle is that similarity in structure can be taken as evidence of homology, with analogy being the exception, indicated by finding inconsistencies in the affinities of differing characters for the two viruses. Overall, strong structural similarity enfolds within it a sufficient number of different characters (chain direction, secondary structure similarity, topological similarity, loop length similarity, and others) that the argument for homology is compelling. In situations where the structures are more divergent, commonality (or differences) in other key self (see below) characteristics of the virus, such as the mechanism of genome packaging, are taken into account, and comparisons need to be considered on a case-by-case basis. Ultimately, in difficult cases, judgment may have to be suspended in the hope that additional viruses will be identified as serial homologs (missing links).

TOOLS FOR STRUCTURE-BASED PHYLOGENY

Marked sequence similarity between the target proteins, allowing them to be aligned, can be translated into an evolutionary distance, and when a full pair wise comparison is performed between a series of sequences, the results can be represented as a phylogenetic tree. To this end several algorithms are available for sequence alignment (see, for example, **http://expasy.org** or **http://www.ebi.ac.uk**). To spot deeper relationships than those discernable by sequence analysis, structural comparison methods have been developed (62–66), but only recently has the three-dimensional structural similarity between viral structural proteins (coat proteins) been used to quantify evolutionary relationships (12, 27, 28, 49, 67). The human eye is remarkably good at picking out similarities in protein structures (if they are suitably displayed and similarly orientated); however, the reliable detection of these similarities by computational methods remains a challenging, and not fully solved, problem. Furthermore, once a superposition is obtained, one needs to infer from this an objective evolutionary distance, an even murkier issue. We have used the Structure Homology Program (SHP) (66), available for Linux and OS-X platforms upon request from the authors, for structural comparison. Based on ideas of Rossmann & Argos (65), the Structure Homology Program is rather sensitive to weak similarities, using the concept of "probability of equivalence." In brief the procedure compares one protein structure (A) with another (B), systematically exploring all possible relative orientations (conveniently defined in terms of three angles α, β, γ). For each relative orientation (α_h, β_h, γ_h) every residue (i) of A is compared with every residue (j) in B. The closeness in space of the Cα atoms of residues i and j, and the local shape of the two polypeptide chains around these residues, is then used to estimate the likelihood that these two residues are structurally equivalent to each other ($P_{i,j}$). Analysis of the matrix of $P_{i,j}$ values for all of i and j is then performed to provide a best set of matching residues between A and B, and the sum of the $P_{i,j}$ values for all residue pairs comprising this set ($\Sigma P_{i,j}$) provides an overall measure of structural similarity between these two structures in relative orientation α_h, β_h, γ_h. If A and B are indeed similar then by choosing the relative orientation for which $\Sigma P_{i,j}$ is the greatest, we have brought them to a common orientation, and $\Sigma P_{i,j}$ provides a measure of their similarity (65). It is then a simple matter to convert $\Sigma P_{i,j}$ into an indication of the evolutionary distance between A and B. This is achieved by the use of an empirical logarithmic function chosen by analogy with metrics for sequence alignment (67). If we have a set of protein structures that we wish to compare, then each member of the set should be compared with every other to provide a set of pair-wise evolutionary distances. We can then use these, in conjunction with well-established tools, such as PHYLYP (a package of programs for inferring phylogenies) (68), to generate and visualize the relative evolutionary distances through a phylogenetic tree. Because of the highly complex and recursive nature of this analysis (69), it is difficult to put the comparison on a simple statistical basis.

VIRION ARCHITECTURE COUPLES TO GENOME PACKAGING

A virus must not only construct a virion, it must also package the viral genome within it. This second fundamental process is a separate character that can help in addressing the question of analogy versus homology. There are possibly only two fundamental mechanisms: (*a*) translocation of the genome into a preformed procapsid using a NTP-driven motor and (*b*) genome condensation by specific nucleic acid/viral protein interactions during assembly. Although the mechanism of viral genome replication is irrelevant for the packaging (see below), packaging signals in the genome ensure specific genome/capsid recognition.

Examples of mechanism (*a*) can be found in icosahedral dsDNA viruses, such as bacteriophages T4, Φ29, P22, and PRD1, which use specific packaging ATPases (70, 71). The rather inflexible genome is translocated into the capsid through a special vertex equipped with a translocation apparatus, most often a dodecameric portal complex (72–75). The packaging NTPase works against an increasing pressure as the genome is condensed within the capsid, and the genome is usually packed as a tightly wound spool within the capsid (76).

Viral packaging NTPases cluster according to the architectural type of viruses where they function. Interestingly, the eukaryotic herpesvirus has been shown to possess a portal protein responsible for packaging the DNA, leading to the idea of strong functional and structural conservation between tailed phages and herpesvirus translocation machineries (77, 78). Recently, two other large eukaryotic viruses, mimivirus (79, 80) and *Paramecium bursaria* Chlorella virus 1 (81), have been also shown to have portals for DNA packaging, though the mimivirus portal is located centrally in a virus facet rather than at a unique vertex. Although the portals of some phages are reasonably well characterized (74, 82–84), only one component has been visualized in atomic detail for a eukaryotic virus (85), UL89-C.

Exceptions to this energy-driven packaging mechanism of large dsDNA viruses can be found in simian virus 40 (SV40) (and relatives) and in the lipid-containing bacteriophage PM2. In SV40, the viral dsDNA is packed by cellular histones into minichromosomes that act as a mold for capsid assembly (86), whereas the circular genome of PM2 is highly negatively supercoiled, facilitating direct genome condensation (53).

By contrast, studies on dsRNA bacteriophages have provided a mechanism of RNA packaging relevant to a number of eukaryotic dsRNA viruses, in particular those belonging to the *Reoviridae* family (87, 88). RNA internalization relies on a hexameric ring of NTPase packaging proteins located at the fivefold vertices of the virus. ssRNA versions of the segmented genome molecules selectively bind to the procapsid prior to coupling with the packaging NTPase for translocation to the particle interior (87, 89, 90).

For negative-sense ssRNA viruses, genome packaging utilizes nucleocapsid proteins, which bind the genome forming ribonucleoprotein complexes that act as scaffolds for virus assembly (91–95). Helical positive-sense ssRNA viruses, however, such as TMV, assemble via preformed disks, which nucleate RNA packaging (96). The icosahedral bacteriophage MS2, another positive ssRNA virus, also initiates RNA packaging using an RNA-binding site in the coat protein (97), although it remains unclear how genome packaging occurs in other positive ssRNA viruses, such as picornaviruses. A recent report (98) invoked interaction of a picornavirus nonstructural protein with the capsid protein, VP3, as essential, although other strategies may exist (99). Overall, the integration of the viral genome and the capsid is clearly a crucial step in virus assembly and explains why coat protein genes and packaging systems often coevolve.

CURRENT STATUS: DETECTING VIRAL LINEAGES

Major Lineages

On the basis of systematic structural comparisons, Abrescia et al. (12) identified four major lineages of viruses composed of several dozen separate virus families. Subsequent results have not fundamentally altered the overall picture (**Figure 4**). It seems likely that this classification will ultimately encompass the vast majority of icosahedral viruses, including those such as the tailed phages, which, although not icosahedral, have a major structural component built on an architecture closely derived from an icosahedron. We give updates of the four lineages before considering the outstanding problems of classification posed by the pleomorphic and helical viruses. We note that a further small number of viruses do not fall neatly within our classification scheme, for instance, the small positive ssRNA leviviruses, which have a distinct α/β fold and may be considered as a separate lineage with members infecting bacteria only. The current status of the mapping of the four major viral lineages onto the ICTV virosphere is shown in **Figure 2b**; note that almost half of the assigned families can be tentatively placed into one of these four lineages.

The picornavirus-like lineage contains 15 virus families. Work is required to establish whether viruses containing a DNA genome can be securely placed in this lineage (e.g., ϕX174) (100).

The HK97-like, tailed-phage lineage is well established, despite the differences in replication strategies across its members (discussed below). In addition there is a general consensus that herpesvirus belongs in this lineage (101, 102). This is the most successful lineage of self-replicating systems on Earth.

The PRD1/adenovirus lineage is also well established, although the status remains uncertain for some putative new recruits, for instance the SH1-like viruses (103, 104). These viruses may be related to the main lineage but have distinctive features. Thus, the building block of the PRD1/adenovirus capsid is a trimeric molecule, presumed to result from gene duplication because it comprises concatenated β-barrels arranged to form a pseudohexameric structure. Although at low resolution some of the SH1-type building blocks appear similar in shape and consistent in symmetry with this arrangement, others possess incompatible twofold symmetry. It remains to be seen if this apparent similarity is coincidental or if the viruses belong within the lineage as a distinct branch. The most remarkable recent addition to the lineage, however, is the D13 scaffolding protein from the massive vaccinia virus, a poxvirus (54). During morphogenesis, the vaccinia particle matures from roughly spherical to brick shaped, and a scaffolding protein, D13, forms a transitory, stabilizing lattice on the surface of the initial spherical particle. D13 comprises the characteristic double β-barrel subunit, arranged as pseudohexagonal trimers arranged in a honeycomb fashion. Standard structure-based phylogenetic analysis places the structure squarely within the lineage. Although D13 is larger than most members of this lineage, it is closer to the main block of viruses in the lineage than is the large coat protein of adenovirus (presumably reflecting the fact that, unlike all other known

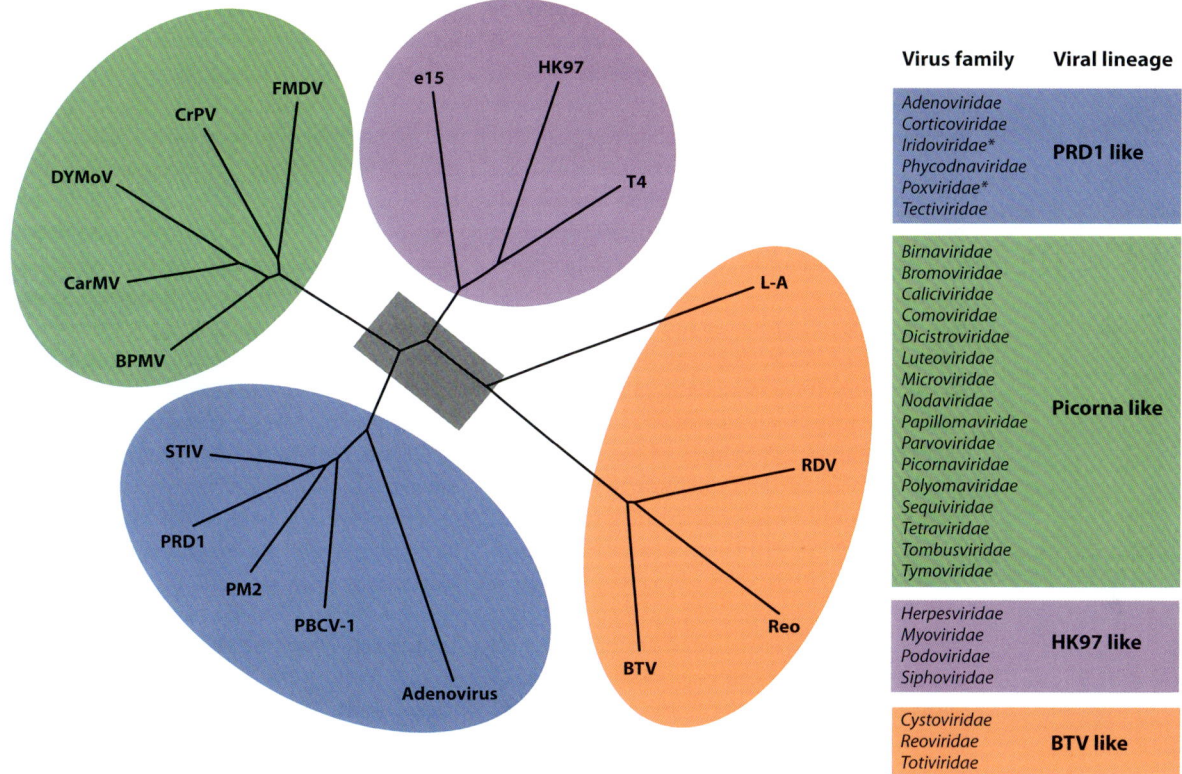

Figure 4
Structure-based phylogenetic tree showing the four different lineages so far detected by structural comparison of viral capsid proteins. The matrix of evolutionary distances was calculated as described in the text. Only some of the family members are represented in the tree, and the full list for each clade is given in the table (*right*). The gray rectangle at the center of the tree indicates that the lineages should not be considered as sharing a common ancestor, i.e., viruses are polyphyletic in origin. The Protein Data Bank codes used for each virus representative are shown in parentheses: FMDV, foot-and-mouth disease virus (1zba); BPMV, *Bean pod mottle virus* (1pgl); CarMV, *Carnation mottle virus* (1opo); DYMoV, *Desmodium yellow mottle virus* (1ddl); CrPV, cricket paralysis virus (1b35); HK97 bacteriophage (1ohg); T4 bacteriophage (1yue); ε15 bacteriophage (3c5b); STIV, *Sulfolobus* turreted icosahedral virus (2bbd); PRD1 bacteriophage (1cjd); PBCV-1, *Paramecium bursaria* Chlorella virus 1 (1j5q); PM2 bacteriophage (2vvf); adenovirus (1p2z); L-A virus (1m1c); RDV, *Rice dwarf* virus (1uf2); BTV, bluetongue virus (2btv); REO, Reovirus (1ej6). Reproduced with permission from Reference 12.

members of the lineage, adenovirus does not contain an internal membrane). This suggests that (analogous to higher organism embryogenesis) early poxvirus morphogenesis reflects their evolution from a lineage of viruses sharing a common icosahedral ancestor. The interesting question that this raises is whether this principle will apply to other large nonicosahedral viruses and allow them to be brought within the compass of the known major viral lineages.

Icosahedral dsRNA Viruses

For the fourth of the major viral lineages, the dsRNA viruses, our structural knowledge has expanded considerably over the last few years. These viruses are classified into nine families (*Birnaviridae, Chrysoviridae, Cystoviridae, Endornaviridae, Hypoviridae, Partitiviridae, Picobirnaviridae, Reoviridae,* and *Totiviridae*), infecting a broad spectrum of hosts ranging from bacteria to humans. The viruses also have varied genome compositions (from a single dsRNA molecule up to a maximum of 12 segments), capsid sizes (300–1,000 Å), and capsid morphologies. The structures of several dsRNA viruses have now been solved by X-ray crystallography and cryoelectron microscopy (cryo-EM), revealing the principles that underpin the virion architecture and providing a clearer picture of evolutionary relationships between these viruses (87, 105–113).

The first atomic-level structural data on a dsRNA virus came from the X-ray structure of the core particle of bluetongue virus (BTV) (an *Orbivirus* belonging to the *Reoviridae* family) (105), revealing 260 trimers of the major core protein VP7, cloaking a thin inner protein layer composed of 120 copies of the major core protein VP3 (we refer to this key protein as the capsid protein). VP3 is a flat, elongated molecule, and the BTV core structure showed how VP3 assembles into an icosahedral arrangement via conformational switching to yield a pattern of subunits not seen in other viruses and not predicted by the theory of Caspar & Klug (114). The nonequivalent arrangement of capsid proteins had not been seen previously in icosahedral viruses. The structure of orthoreovirus (106), another member of the *Reoviridae* family, showed an identical architectural arrangement of the equivalent capsid protein. Further structural studies of L-A virus, a totivirus (107), also revealed a similar disposition of the capsid protein. Subsequent work on other members of the *Reoviridae* family (108, 113, 115) and the *Pseudomonas* phage, phi6, in the *Cystoviridae* family (87) showed that, although the outer surfaces of the virus particles are very different, this unique quaternary arrangement of 120 copies of the inner capsid protein is a common signature among these diverse viruses. This feature seems to be shared by *Penicillium chrysogenum* virus, a new dsRNA virus infecting fungi, a member of the *Chrysoviridae* family whose cryo-EM study also provided low-resolution information on the capsid fold, putatively ascribed to the BTV-like lineage (112).

Very recently the crystal structures of two additional dsRNA viruses have been solved, a picobirnavirus and a partitivirus (110, 111). Picobirnaviruses are animal pathogens, whereas partitiviruses, such as *Penicillium stoloniferum* virus F, infect fungi. The capsid proteins of picobirnavirus and partitivirus are relatively small (55 kDa and 47 kDa, respectively) compared to those of the *Reoviridae* family, which can be over 100 kDa. The picobirnavirus and partitivirus capsid proteins are arranged as symmetric dimers within a capsid architecture that, at first glance, appears different to that observed for the *Reoviridae* and *Totiviridae*.

The folds of the capsid proteins of the members of the *Totiviridae* family, picobirnaviruses and partitiviruses, clearly differ from those of the *Reoviridae*, where there are strong structural similarities (**Figure 5**). Nevertheless, using the tools described above (66) provides superpositions that reveal common structural features. As shown in **Figure 5**, although the details of the fold differ, simply ramping the color of the capsid protein along the chain reveals obvious similarities in the general positioning of the structural elements within the proteins. In the case of the smaller picobirnaviruses and partitiviruses proteins, there are rearrangements in, for example, the C terminus (**Figure 5**), which once understood allow the structural similarities within domains to be seen more clearly. Because the building block of picobirnaviruses and partitiviruses is a symmetric dimer and so differs from that suggested for the members of *Reoviridae* and *Totiviridae* families, where the assembly of the equivalent shell occurs via clipping together capsid protein decamers, it has been proposed that the picobirnaviruses and partitiviruses are evolutionarily independent. However, although the path of assembly may have altered over time, as may be inferred from **Figure 5**, the underlying architecture and quaternary arrangement of capsid proteins in all these dsRNA viruses are in fact closely related. Taken with the similarities in the pattern of tertiary structure (albeit with some internal rigid body shifts) in the capsid proteins, this argues that all of these viruses originate from a common ancestor.

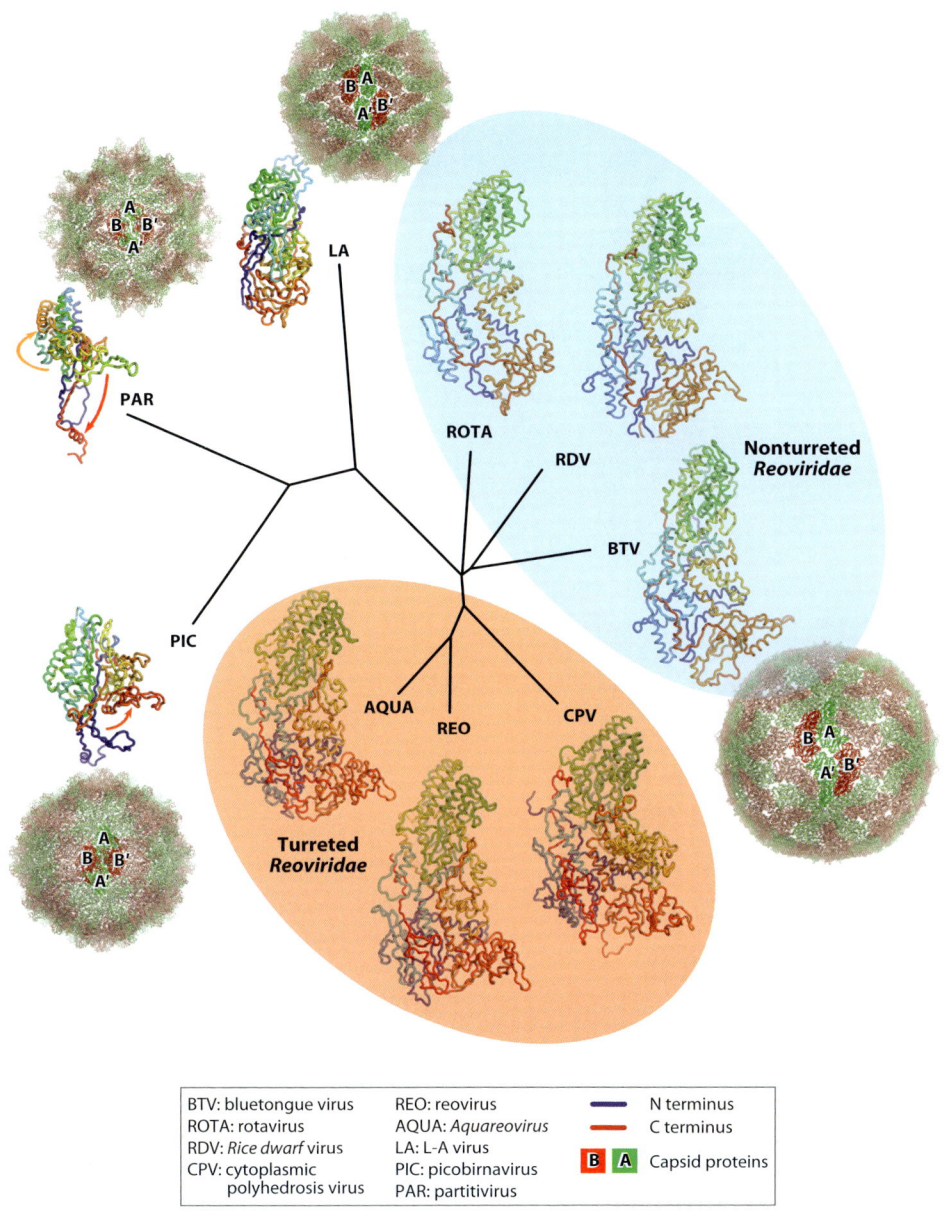

Figure 5

Structure-based phylogenetic tree for the dsRNA lineage of viruses. Using the Structure Homology Program for superposition, the tree shows capsid proteins (CPs) of the following viruses (with the protein and Protein Data Bank codes inside parentheses): BTV, bluetongue virus (VP3, 2btv); ROTA, rotavirus (VP2, 3kz4); RDV, *Rice dwarf* virus (P3, 1uf2); CPV, cytoplasmic polyhedrosis virus (VP1, 3izx); REO, reovirus (λ1, 1ej6); AQUA, *Aquareovirus* (VP3, 3iyl), LA, L-A virus (1m1c); PIC, picobirnavirus (2vf1); and PAR, partitivirus (3es5) (see Reference 66 for details). Equivalent CPs from the viruses are colored from the N terminus to the C terminus. Possible repositioning of structural elements in the picobirnavirus CP and the partitivirus CP to bring them into correspondence with L-A virus and the *Reoviridae* is represented by arrows. The "T2" shells (defined by the 120 copies of the CP) are shown for representatives of the clade, with the independent CP copies labeled A and B to highlight the identical quaternary arrangement.

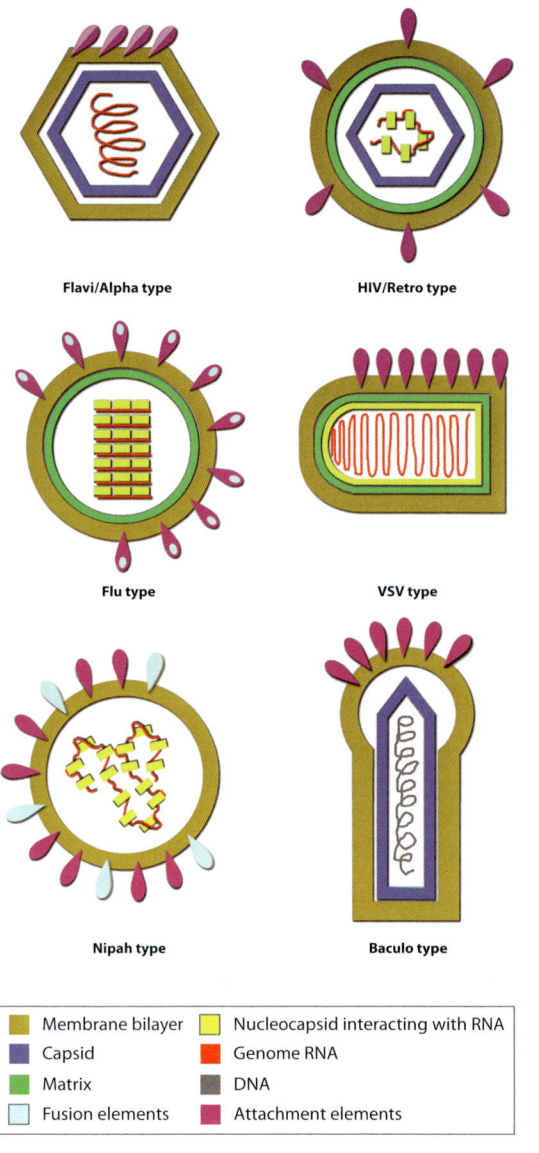

Figure 6

Schematic representation of the architecture of six major enveloped virus types. The elements shown in these drawings include the membrane bilayer, the capsid, the matrix, the nucleocapsid interacting with RNA, the genome RNA, and the DNA. The structural elements responsible for attachment (either as an individual protein or as a component of glycoprotein complex) and those for fusion appear on the outer layer. Abbreviations: Baculo type, baculovirus type; Flavi/Alpha virus type, flavivirus/alphavirus type; Flu type, influenza virus type; HIV/Retro type, human immunodeficiency virus/retrovirus type; Nipah type, Nipah and Hendra virus type; and VSV type, vesicular stomatitis virus type.

The final piece of structural information for a dsRNA virus is for the birnaviruses. Birnaviruses do not have a structural equivalent of the signature capsid protein of the BTV-like lineage. Indeed the major capsid protein, VP2, has structural similarities to capsid proteins of the picornavirus-like lineage (116), although the quaternary arrangement of VP2 within the capsid is identical to that seen for the trimeric outer capsid proteins of BTV and rotavirus (VP7 and VP6, respectively) (105, 113). One hypothesis is that the emergence of the *Reoviridae* is as a result of genome mixing between simpler toti-like and birna-like viruses, resulting in viruses that have a 120-copy inner core particle cloaked in a birnavirus-like T13 lattice (116). Alternatively, birnaviruses may have arisen from loss of an equivalent to the inner capsid protein seen in all other dsRNA viruses or even may have been derived from a separate lineage. It is clear that for now the birnaviruses cannot be securely placed into any of our major lineages and may even be a candidate for its own lineage.

Enveloped Viruses: Structural Links with Host Machines?

Enveloped viruses represent a conundrum for structure-based classification. The task of identifying possible lineages is made harder by the difficulty of defining the viral "self." Three architectural elements characterize many enveloped viruses: outer envelope proteins (usually trimeric, dimeric, or heterodimeric); a lipid bilayer (scavenged from the host); and a matrix/nucleocapsid below the membrane, which may contain more than one component in the mature virus. Thus, from the capsid-centric point of view if self elements exist, they must be recognized within the glycoproteins and/or nucleocapsid fold (see **Figure 6**). One of the most-studied enveloped viruses is dengue virus, a member of the *Flaviviridae* family. Cryo-EM on the entire virus and crystallographic studies of its glycoproteins have elucidated fundamental aspects of the flavivirus life cycle (117–119), although within this family is hepatitis

C virus whose morphogenesis and structure remain much more obscure (120, 121). Likewise no enveloped virus has been visualized by X-ray crystallography, but the West Nile virus, another flavivirus, has been crystallized (122).

The fold and domain organizations of the dengue virus envelope E protein are shared by the envelope protein E1 of Semliki Forest virus (a member of the *Alphaviridae* family), features taken to indicate homology between the two families (123). These envelope proteins belong to the class II fusion proteins and fuse the virus and host cell membranes to deliver the virus genome into the cytoplasm of the target cell. The recent emergence of the chikungunya virus, an alphavirus posing a major threat to human health, has stimulated structural work, resulting in X-ray structures of the precursor (p62-E1) and mature E1-E2-E3 glycoprotein complexes (124, 125). This latter study has shown for the first time the immunoglobulin-like fold of the domains composing the viral E2 protein, thought to be involved in attachment to the host cell.

Fusion machines exist in many different families of enveloped viruses and have been traditionally classified into three distinct types, I–III. They share a common function, although there are detailed mechanistic differences (126, 127). Recent structural studies (125, 128–130) have fleshed out our knowledge of these systems, and a detailed comparison of the flexible structural domains from which these proteins are assembled has led to the proposal that there may be unsuspected similarities between the three types (128). This raises a further problem for structure comparison, which works best when considering a rigid protein: For these proteins, it is necessary to recognize their inherent structural variation because their function is to perform structural gymnastics to manipulate membranes, applying asymmetric strain to simple hydrophobic anchors. This is quite different from the beautiful rigidity and symmetry that characterize the capsid components of the four major virus lineages. These problems mean that there is an urgent need to accrue more structural information, not only for viral fusion proteins, but also for relevant cellular proteins. Here, there is another confounding factor: Cellular life abounds in proteins that manipulate membranes, and it is quite conceivable that horizontal exchange might make the fusion machines unreliable markers for the history of the virus because they are probably acquired more easily than a functional icosahedral capsid. Furthermore, some of these viral glycoproteins are responsible for host cell recognition, and therefore, these glycoproteins are likely to change with changes in host range or even, by using membrane recognition for cell entry, confer a very broad host range (128).

Perhaps, are the internal viral proteins more likely to retain evolutionary signatures (131)? Certainly the nucleocapsid proteins of viruses of the *Bornaviridae* and *Rhabdoviridae* families show a strongly conserved α-helical topology (12). Recently, major efforts have been made to characterize additional nucleocapsid proteins from two pathogenic viruses: the human respiratory syncytial virus (*Paramyxoviridae* family) (95, 132) and the Rift Valley fever virus (*Bunyaviridae* family) (92, 133). The respiratory syncytial virus nucleoprotein is clearly related structurally to the nucleoproteins of rabies and vesicular stomatitis virus. By contrast, the Rift Valley fever virus nucleoprotein, crystallized in monomeric (133) and hexameric forms (92), is smaller and mainly α-helical with a bilobe domain organization, closer to the corresponding protein of *Influenza A*. Indeed a structure-based phylogeny clusters these together but separate from the Mononegavirales nucleoproteins (**Figure 7**). This suggests that there is a fundamental difference between the nucleoproteins of negative ssRNA viruses with segmented and nonsegmented genomes. Thus, the nucleocapsid might provide a useful fingerprint of evolutionary relationship, although we need to be cautious when working with proteins with close functional links to the replication machinery of the virus, an issue we explore below. In addition, it underlines the importance of increasing our database of structures solved to fill in

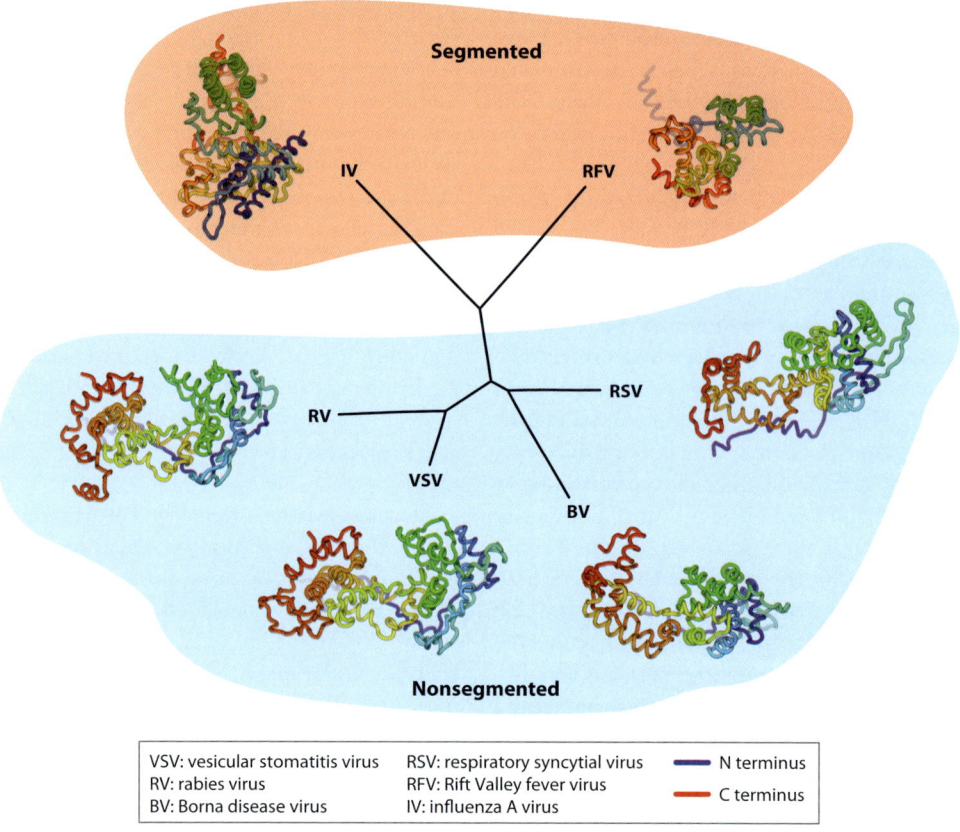

Figure 7

Structure-based phylogenetic tree of nucleocapsid proteins of the negative-sense RNA viruses. Nucleoproteins (Protein Data Bank codes are shown in parentheses) of VSV, vesicular stomatitis virus (2gic); RV, rabies virus (2gtt); BV, Borna disease virus (1n93); RSV, respiratory syncytial virus (2wj8); RFV, Rift Valley fever virus (3ov9); and IV, influenza A virus (2iqh) were superposed using the Structure Homology Program (see Reference 66 for details), revealing how members of the Mononegavirales (nonsegmented genomes) cluster separately from viruses with segmented genomes such as the influenza virus, suggesting that the nucleoprotein may be useful as a marker for the vertically inherited component defining the viral self within these groups of viruses. The structures of the nucleoproteins from the viruses are colored from the N terminus to the C terminus.

"missing links" to allow us to build a continuum of evolutionary relatedness.

Recent observations have revealed similar minimalist enveloped viruses infecting halophilic archaeal hosts (134, 135, 136). They consist of only a genome surrounded by a flexible lipid membrane with two major integral membrane proteins, a spike, and an internal small protein facing the genome. The virions are pleomorphic with a diameter of ∼50 nm. It may be that there are related viruses that infect bacteria, but more comparative work is needed to clarify this (134). An intriguing property of these closely related viruses is that in some the genome is ssDNA and in others dsDNA, challenging directly the usage of genome type as a classification criterion! In summary, although structure has provided unexpected insight into similarities (e.g., from the nucleoprotein structures), the question remains open as to whether a coherent phylogeny of the enveloped viruses even exists.

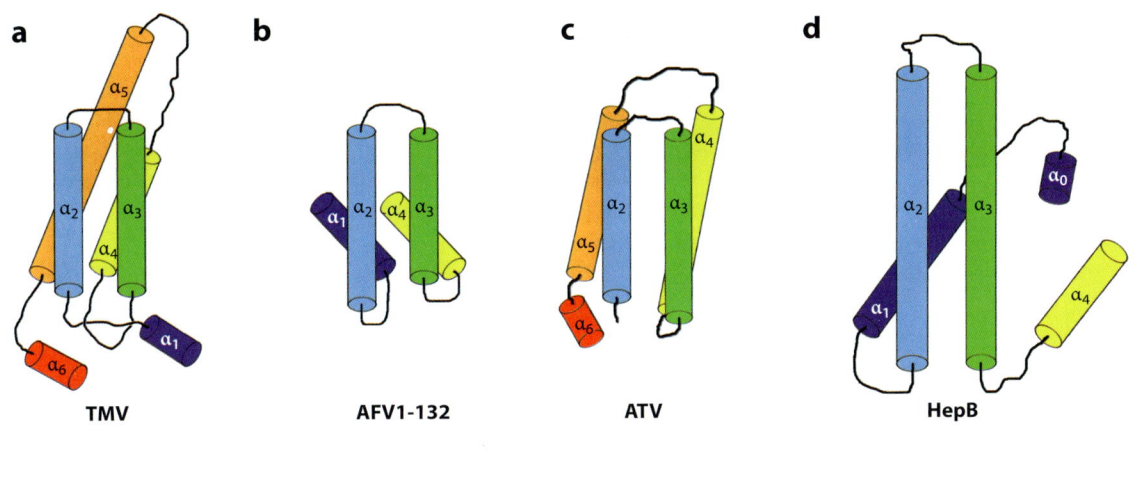
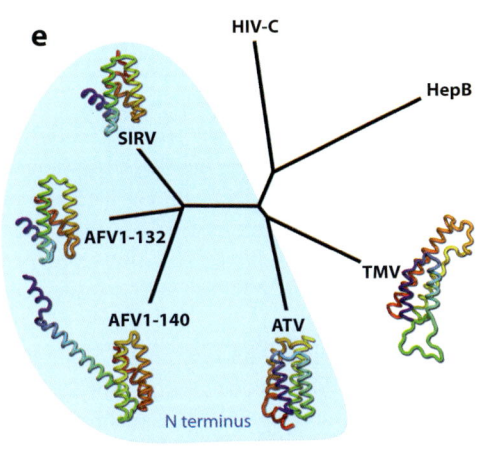

Figure 8

Schematic representation of the coat proteins of some helical viruses. The abbreviated virus name and Protein Data Bank codes are shown in parentheses in the following list. (*a*) *Tobacco mosaic virus* (TMV, 1ei7). (*b*) *Acidianus* filamentous virus 1 (AFV1-132, 3fbl). (*c*) *Acidianus* two-tailed virus (ATV, 3faj). (*d*) Hepatitis B antigen (HepB, 1qgt). Corresponding α-helices have been colored accordingly, with core α2 and α3 helices serving as a frame for the spatial arrangement of the additional N- and C-terminal α-helices. (*e*) A structure-based phylogenetic tree of all helical archaeal [AFV1-132, 3fbl; AFV1-140, 3fbz; ATV, 3faj; *Sulfolobus islandicus* rod-shaped virus (SIRV), 3f2e] and eukaryotic capsid proteins (TMV, 1ei7; HepB, 1qgt; HIV-C, 1a8o). Only the major capsid proteins for which satisfactory superimposition has been obtained are shown, colored from the N terminus to the C terminus. Although a single tree is shown, it is quite likely that the major branches are not related directly to a common ancestor.

Helical Viruses

Structural information on helical viruses is still very limited. For many years, the only high-resolution information was for TMV, a member of the *Virgaviridae* family. Its capsid subunit structure was solved in 1978 (96), revealing a four-helix bundle with additional short helices (∼2 turns) at the N and C termini (**Figure 8***a*). The bundle is organized as two sets of two antiparallel α-helices, packed against each other and tilted ∼30°. In the virus, the coat protein subunits are stacked in a single start helix. The N and C termini point outward, and the loops connecting the α-helices face inward, grabbing the viral ssRNA during virus morphogenesis (96, 137, 138). Recently, four archaeal virus structures have been determined (139–141). These are the two major coat proteins from *Acidianus* filamentous virus 1 (AFV1) (*Lipothrixviridae* family) (142), the coat protein of *Sulfolobus islandicus* rod-shaped virus (*Rudiviridae* family) (141), and the putative coat protein of the *Acidianus* two-tailed virus (**Figure 8b,c,e**) (140). The coat protein of AFV1 has the same overall structure as the coat protein of the *Sulfolobus islandicus* rod-shaped virus, despite little sequence identity. The strong structural similarity suggests that these helical archaeal viruses belong to a common lineage. Strikingly, the fold of the protein from the two-tailed phage (ATV) is also somewhat similar (**Figure 8e**). So seemingly diverse archaeal viruses may define a new lineage, characterized by this helical bundle. At first sight the TMV helical bundle is very different; however, a side-by-side schematic comparison of the fold of the AFV1-132 capsid protein and that of TMV is shown in **Figure 8a,b**, revealing that in both the helix-turn-helix motif formed by α_2 and α_3 serves as a framework against which helical extensions from the N and C termini organize. It is premature to take this similarity alone as compelling evidence for recruiting TMV into the archaeal helical bundle lineage.

One icosahedral viral shell has been found that is also built from a bundle of helices, hepatitis B virus (an enveloped virus belonging to the *Hepadnaviridae* family) (143). This structure is not formed from a four-helical bundle, although the arrangement could recapitulate the α_2 and α_3 frame in the context of the dimeric building block of the capsid shell (**Figure 8d**). We suggest that this similarity reflects analogy rather than homology. Indeed, this protein appears to share more common features with the capsid protein of the non-icosahedral retroviruses. Based on this similarity, a common evolutionary origin of the hepadnaviruses and retroviruses has been imputed (144).

A single α-helix forms the building block of filamentous bacteriophages belonging to the *Inoviridae* family. The capsid of these ssDNA viruses is composed of individual helices coiling with fivefold symmetry around the long helix axis (145), constituting a distinctive architectural arrangement, but the simplicity of the fold renders structural comparisons with more complex proteins worthless.

In summary, recent activity has identified a new lineage of viruses; some are rod-shaped helical structures, and all infect archaea. It is possible that there is an evolutionary connection with TMV, although the discovery of serial homologs would be required to make this compelling.

IMPLICATIONS AND FUTURE PERSPECTIVES

Viral genomes are composed of functional modules comprising one or several genes. Such modules are responsible for assembling the virion, replicating the genome, entering the host cell, modulating host defense systems, exiting the host, and other functions. By recombination, they can create new combinations and, via horizontal gene transfer, exchange material from unrelated sources. When a key module, such as the capsid, is lost, then the genome is no longer viral. This may be the origin of some plasmids that encode related proteins with no cellular counterparts, playing a crucial role in replication (146–148). More broadly, comparative genomics provides examples of the complexity of the evolution of these virus modules, such as where the replication module has decoupled from the virion assembly module, so that even the ICTV has classified the virus on the basis of the capsid. An elegant example is the bacterial virus PM2, where related prophages have acquired replication modules from unrelated sources, whereas the virion assembly module, containing the coat protein gene and the genome packaging ATPase encoding gene, is conserved (149). An even more striking example is the usage of a plethora of genome replication proteins by tailed dsDNA viruses infecting bacteria and archaea (see **Figure 9**), which nevertheless have very similar virion structure and assembly pathways. These and other examples clearly demonstrate that viral replication systems should not be used as the basis for virus classification, see also Reference 150.

The examples above demonstrate that there is no coherent viral evolutionary signal that includes all viral properties; rather, the viral genome is composed of a number of different modules or groups of genes, which may have quite distinct heritages. In fact, despite the apparent dominance of the genome-centric view of viruses, throughout the history of viral classification, virion architecture and genomic characteristics have been used as a guide to taxonomy. In practice, the current interpretation of the virus world is melded from many contributions, which, on a case-by-case basis, reflect whether a given person has a genomic or capsid-centric view of viruses. We argue that the gradual accumulation of structural information, although still woefully incomplete, has increasingly stacked the deck in favor of the view that the capsid-centered approach has the greatest power to provide a sensible and coherent phylogeny. Once again, we can see this with particular clarity for the tailed bacterial viruses. There is a general agreement that these are all related, even though they utilize several nonhomologous host recognition, replication, transcription, and lysis systems. Their virion assembly principle and coat protein fold (the hallmarks of their particular virus lineage) are the common denominator.

We may reformulate the capsid-centered view of viruses as: All virions are derived from their parent, creating a vertical evolutionary pathway that is tested for functionality with every generation. This is the basis for the term we previously put forward, the viral self (12, 27, 151), which includes those vertically inherited properties needed to assemble a virion. The huge selective pressure to keep the virus functional often preserves recognizable structural features between distantly related viruses. By contrast, functions that relate to interactions with the host exchange reasonably freely, allowing the virus to cope with the evolving host and changing environment. Such virus properties are consequently termed nonself. Furthermore, the genetic economy of viruses does

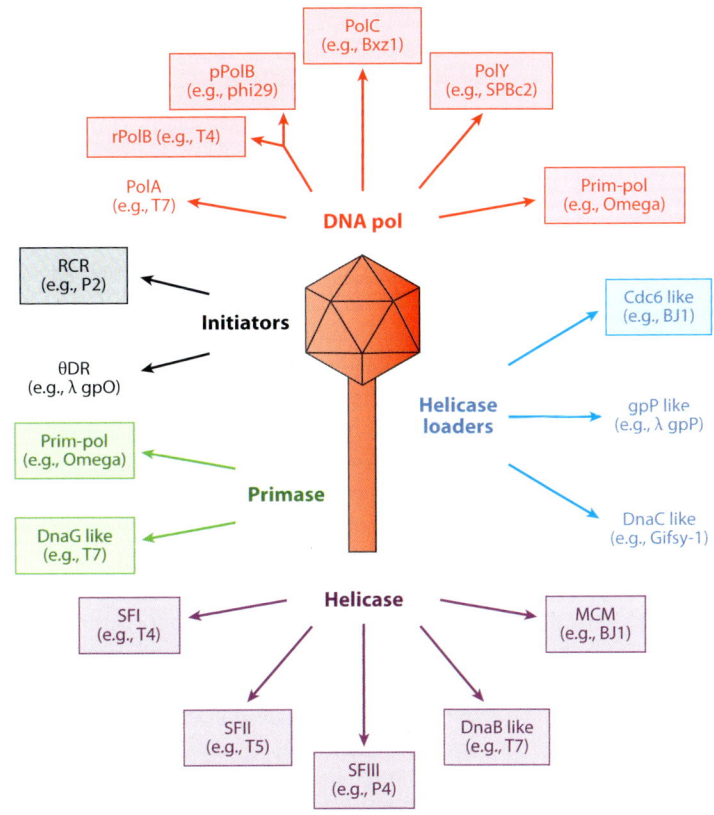

Figure 9

Diversity of genome replication proteins utilized by tailed dsDNA viruses infecting bacteria and archaea. Replication proteins that are also encoded by nontailed viruses and/or plasmids are boxed. Examples of viruses utilizing the indicated genome replication proteins are depicted. Abbreviations: Cdc6 like, cell division cycle 6 protein like; DnaC and DnaG like, helicase loader and primase like; gpP like, major tail protein like; MCM, minichromosome maintenance helicase like; PolA, -B, -C, and -Y, types A, B, C, and Y DNA polymerases, respectively; pPolB, protein-primed type B DNA polymerase; Prim-pol, primase-polymerase protein; RCR, rolling-circle DNA replication; rPolB, RNA/DNA-primed type B DNA polymerase; SFI–III, superfamily I to III helicases; θDR, theta-type DNA replication. Reproduced with permission from Reference 32.

not permit most viruses to carry nonfunctional copies of genes [rather, there are repeated examples of viruses inventing additional protein functions from within existing coding regions, for instance, by the use of alternative reading frames (152)]. This means that the mechanisms for radical protein structure invention available to viruses are more limited than for their cellular hosts, perhaps explaining why so few lineages of viruses have been found. If we look at the way complex functional bottlenecking has restricted all life to a broadly similar genetic code, we should probably not be too surprised. Perhaps, we should wonder instead whether the accumulation of virus structural information, which reaches far back, will eventually provide more insight into the melting pot that was life before the distinct domains of bacteria, eukarya, and archaea for cellular life and the major lineages of viral life settled out.

In the face of the sheer magnitude of the virus population, it is obvious that we need far more structural sampling of the virosphere, beyond the present 100 or so virus species we have characterized, if we are to properly test our hypothesis that there are only a few major virion architectures. For a rapid advance, we require new methods across a broad front. First, we need the ability to obtain information on single viruses from the environment without the need to go through the processes of isolation, cultivation, and purification. The first steps toward such technologies are emerging, with environmental bacterial samples being diluted with microfluidic techniques to provide single cells, which are then used for polymerase chain reactions to detect preselected viral markers (153, 154). Such methods would allow a systematic attack on the problem of virus ecology and diversity. In the absence of such techiques, there has been an attempt to achieve higher throughput using traditional approaches, with some 50 unique viruses characterized from high-salt environments. In this case, only one novel structural virus type was observed; the rest fell into known categories (136), very much in-line with our hypothesis that the number of fundamental structural types is small. A further obstacle is that the structure determination procedures become particularly difficult as the virion size increases. On both the cryo-EM and X-ray fronts, methods development has been rapid. Cryo-EM has now delivered atomic-level structures (109, 115, 155, 156), and for X-rays, new detectors, microbeams, and in situ diffraction are leading to a step change at synchrotron beam lines (157, 158). Beyond this, ideas are being developed to grow viral crystals in infected cells, bypassing the requirement for purification and crystallization (M. Pietila, M.-L. Parsey, E. Duke, G. Sutton, J. Huiskonen, et al., unpublished data). Such tiny crystals (typically sub-micrometer) might be amenable to analysis with advanced synchrotron microbeams or an X-ray free-electron laser (159). The X-ray free-electron laser might also offer possibilities for the analysis of single virus particles; however, at present, only very low resolution can be achieved (160). Once structures are available, we need accurate methods to validate and quantify structural similarities. To date, the sensitivity of these methods has not improved from the earliest methods, as discussed above. However, better methods are becoming available, as demonstrated by a complete, automatic, and reliable comparison of small icosahedral ssRNA virus coat proteins, which previously would have required manual intervention (J. Ravantti, D.H. Bamford, & D.I. Stuart, unpublished data). Perhaps, we are close to seriously addressing the organization of the viral universe, where the vast majority of the living entities reside. The insights that will flow from such an understanding will be crucial in combating viral diseases and may even shed light on such apparently remote issues as geochemical cycling and the global climate.

SUMMARY POINTS

1. In spite of the pervasive presence of viruses throughout the biosphere, conventional classification schemes struggle to provide a meaningful, comprehensive taxonomy.

2. The defining feature of a virus is its protein capsid, whose structure is strictly constrained by stereochemical rules and genetic parsimony. Capsid structure is therefore far more persistent than viral genome or protein sequences.

3. Other virus properties, such as the nature of the genome and the method by which it is replicated, are not so central to the virus "self."
4. Structures of virus capsids fall predominantly into a small number of lineages, with lineages tending to include viruses infecting all three domains of cellular life.
5. As an increasing number of structures have been determined, it has become possible to use objective computational procedures to suggest plausible phylogenetic relationships within these lineages, which otherwise might not be uncovered.
6. Non-icosahedral viruses present more challenges to this analysis; nevertheless, there are hints that similar methods will prove useful for helical viruses.
7. While the structural classification of enveloped viruses remains deeply problematic, the allocation of the immensely complex vaccina virus to a major icosahedral lineage suggests that certain enveloped viruses may betray their evolutionary roots in their morphogenesis.
8. The full diversity of the virus universe is still uncharted. We have especially sketchy knowledge of the viruses that infect archaea, and we have only rudimentary tools for more rapidly expanding our grasp of virus ecology; however, new methods are on the horizon.

FUTURE ISSUES

1. The methods of structure comparison aimed at providing an estimate of evolutionary distance between capsids are still primitive and are not yet on a robust statistical footing.
2. There is a need for novel high-throughput methods to enable the routine detection and characterization of viruses from individual cells.
3. If novel viruses can be detected and isolated more rapidly, there will be increasing pressure to quickly make the detailed structural characterization (at sufficiently high resolution to support meaningful structure comparison).
4. Exciting possibilities include new methods at conventional synchrotron sources and also the potential game changer of the X-ray free-electron laser, which may allow virus structure determination without extraction of the virus from infected cells.
5. The application of the knowledge gained and techniques developed in this endeavor to chart the virus universe is likely to have impact in the context of novel therapies. Perhaps whole viruses can be brought within the scope of structure-based therapy design, which is presently limited to individual viral protein targets.

DISCLOSURE STATEMENT

The authors are not aware of any affiliations, memberships, funding, or financial holdings that might be perceived as affecting the objectivity of this review.

ACKNOWLEDGMENTS

N.G.A.A. is supported by the Spanish Ministerio de Ciencia y Tecnología (BFU2009-08123) and the Basque Government (PI2010-20). The Finnish Center of Excellence Program (2006–2011) grant 1129684 supported the research of D.H.B. This work was enabled by the United Kingdom Medical Research Council and the European Commission contract number 031220FP6 (SPINE2-COMPLEXES). The Wellcome Trust provided administrative support (grant 075491/Z/04).

LITERATURE CITED

1. Plotkin SA. 2005. Vaccines: past, present and future. *Nat. Med.* 11:S5–11
2. Rappuoli R, Miller HI, Falkow S. 2002. Medicine. The intangible value of vaccination. *Science* 297:937–39
3. Halloran ME, Longini IM Jr, Nizam A, Yang Y. 2002. Containing bioterrorist smallpox. *Science* 298:1428–32
4. Bergh O, Borsheim KY, Bratbak G, Heldal M. 1989. High abundance of viruses found in aquatic environments. *Nature* 340:467–68
5. Srinivasiah S, Bhavsar J, Thapar K, Liles M, Schoenfeld T, Wommack KE. 2008. Phages across the biosphere: contrasts of viruses in soil and aquatic environments. *Res. Microbiol.* 159:349–57
6. Suttle CA. 2007. Marine viruses—major players in the global ecosystem. *Nat. Rev. Microbiol.* 5:801–12
7. Feschotte C. 2010. Virology: Bornavirus enters the genome. *Nature* 463:39–40
8. Bamford DH. 2003. Do viruses form lineages across different domains of life? *Res. Microbiol.* 154:231–36
9. Cairns J, Stent GS, Watson JD, eds. 1992. *Phage and the Origins of Molecular Biology—Expanded Edition*. Cold Spring Harbor, NY: Cold Spring Harb. Lab. Press
10. Watson JD, Crick FH. 1953. Molecular structure of nucleic acids; a structure for deoxyribose nucleic acid. *Nature* 171:737–38
11. Geslin C, Gaillard M, Flament D, Rouault K, Le Romancer M, et al. 2007. Analysis of the first genome of a hyperthermophilic marine virus-like particle, PAV1, isolated from *Pyrococcus abyssi*. *J. Bacteriol.* 189:4510–19
12. Abrescia NG, Grimes JM, Fry EE, Ravantti JJ, Bamford DH, Stuart DI. 2010. What does it take to make a virus: the concept of the viral "self." In *Emerging Topics in Physical Virology*, ed. PG Stockley, R Twarock, pp. 35–58. London: Imperial College Press
13. Prangishvili D, Forterre P, Garrett RA. 2006. Viruses of the Archaea: a unifying view. *Nat. Rev. Microbiol.* 4:837–48
14. Prangishvili D, Garrett RA, Koonin EV. 2006. Evolutionary genomics of archaeal viruses: unique viral genomes in the third domain of life. *Virus Res.* 117:52–67
15. Allan GM, Ellis JA. 2000. Porcine circoviruses: a review. *J. Vet. Diagn. Investig.* 12:3–14
16. Claverie JM, Ogata H, Audic S, Abergel C, Suhre K, Fournier PE. 2006. Mimivirus and the emerging concept of "giant" virus. *Virus Res.* 117:133–44
17. Suzan-Monti M, La Scola B, Raoult D. 2006. Genomic and evolutionary aspects of *Mimivirus*. *Virus Res.* 117:145–55
18. Raoult D, Audic S, Robert C, Abergel C, Renesto P, et al. 2004. The 1.2-megabase genome sequence of Mimivirus. *Science* 306:1344–50
19. Forterre P. 2006. The origin of viruses and their possible roles in major evolutionary transitions. *Virus Res.* 117:5–16
20. Forterre P, Prangishvili D. 2009. The origin of viruses. *Res. Microbiol.* 160:466–72
21. Koonin EV, Dolja VV. 2006. Evolution of complexity in the viral world: the dawn of a new vision. *Virus Res.* 117:1–4
22. Koonin EV, Wolf YI, Nagasaki K, Dolja VV. 2009. The complexity of the virus world. *Nat. Rev. Microbiol.* 7:250
23. Akita F, Chong KT, Tanaka H, Yamashita E, Miyazaki N, et al. 2007. The crystal structure of a virus-like particle from the hyperthermophilic archaeon *Pyrococcus furiosus* provides insight into the evolution of viruses. *J. Mol. Biol.* 368:1469–83

24. Sutter M, Boehringer D, Gutmann S, Gunther S, Prangishvili D, et al. 2008. Structural basis of enzyme encapsulation into a bacterial nanocompartment. *Nat. Struct. Mol. Biol.* 15:939–47
25. Tanaka S, Kerfeld CA, Sawaya MR, Cai F, Heinhorst S, et al. 2008. Atomic-level models of the bacterial carboxysome shell. *Science* 319:1083–86
26. Tanaka S, Sawaya MR, Yeates TO. 2010. Structure and mechanisms of a protein-based organelle in *Escherichia coli. Science* 327:81–84
27. Bamford DH, Burnett RM, Stuart DI. 2002. Evolution of viral structure. *Theor. Popul. Biol.* 61:461–70
28. Bamford DH, Grimes JM, Stuart DI. 2005. What does structure tell us about virus evolution? *Curr. Opin. Struct. Biol.* 15:655–63
29. Benson SD, Bamford JK, Bamford DH, Burnett RM. 2004. Does common architecture reveal a viral lineage spanning all three domains of life? *Mol. Cell* 16:673–85
30. Krupovic M, Bamford DH. 2008. Virus evolution: How far does the double beta-barrel viral lineage extend? *Nat. Rev. Microbiol.* 6:941–48
31. Baltimore D. 1971. Expression of animal virus genomes. *Bacteriol. Rev.* 35:235–41
32. Krupovic M, Bamford DH. 2010. Order to the viral universe. *J. Virol.* 84:12476–79
33. Stanley WM. 1935. Isolation of a crystalline protein possessing the properties of tobacco-mosaic virus. *Science* 81:644–45
34. Bawden FC, Pirie NW. 1943. Methods for the purification of tomato bushy stunt and tobacco mosaic viruses. *Biochem. J.* 37:66–70
35. Bernal JD, Fankuchen I. 1941. X-ray and crystallographic studies of plant virus preparations: I. Introduction and preparation of specimens II. Modes of aggregation of the virus particles. *J. Gen. Physiol.* 25:111–46
36. Crowfoot D, Schmidt GMJ. 1945. X-ray crystallographic measurements on a single crystal of a *Tobacco necrosis virus* derivative. *Nature* 155:504–5
37. Franklin RE. 1955. Structure of *Tobacco mosaic virus. Nature* 175:379–81
38. Klug A. 1999. The *Tobacco mosaic virus* particle: structure and assembly. *Philos. Trans. R. Soc. Lond. B* 354:531–35
39. Harrison SC, Olson AJ, Schutt CE, Winkler FK, Bricogne G. 1978. *Tomato bushy stunt virus* at 2.9 Å resolution. *Nature* 276:368–73
40. Abad-Zapatero C, Abdel-Meguid SS, Johnson JE, Leslie AGW, Rayment I, et al. 1980. Structure of *Southern bean mosaic virus* at 2.8 Å resolution. *Nature* 286:33–39
41. Rossmann MG, Arnold E, Erickson JW, Frankenberger EA, Griffith JP, et al. 1985. Structure of a human common cold virus and functional relationship to other picornaviruses. *Nature* 317:145–53
42. Hogle JM, Chow M, Filman DJ. 1985. Three-dimensional structure of poliovirus at 2.9 Å resolution. *Science* 229:1358–65
43. Tate J, Liljas L, Scotti P, Christian P, Lin T, Johnson JE. 1999. The crystal structure of cricket paralysis virus: the first view of a new virus family. *Nat. Struct. Biol.* 6:765–74
44. Chapman MS, Liljas L. 2003. Structural folds of viral proteins. *Adv. Protein Chem.* 64:125–96
45. Koonin EV, Wolf YI, Nagasaki K, Dolja VV. 2008. The big bang of picorna-like virus evolution antedates the radiation of eukaryotic supergroups. *Nat. Rev. Microbiol.* 6:925–39
46. Koonin EV, Dolja VV. 1993. Evolution and taxonomy of positive-strand RNA viruses: implications of comparative analysis of amino acid sequences. *Crit. Rev. Biochem. Mol. Biol.* 28:375–430
47. Athappilly FK, Murali R, Rux JJ, Cai Z, Burnett RM. 1994. The refined crystal structure of hexon, the major coat protein of adenovirus type 2, at 2.9 Å resolution. *J. Mol. Biol.* 242:430–55
48. Benson SD, Bamford JK, Bamford DH, Burnett RM. 1999. Viral evolution revealed by bacteriophage PRD1 and human adenovirus coat protein structures. *Cell* 98:825–33
49. Abrescia NG, Cockburn JJ, Grimes JM, Sutton GC, Diprose JM, et al. 2004. Insights into assembly from structural analysis of bacteriophage PRD1. *Nature* 432:68–74
50. Merckel MC, Huiskonen JT, Bamford DH, Goldman A, Tuma R. 2005. The structure of the bacteriophage PRD1 spike sheds light on the evolution of viral capsid architecture. *Mol. Cell* 18:161–70
51. Salas M. 1991. Protein-priming of DNA replication. *Annu. Rev. Biochem.* 60:39–71

52. Khayat R, Tang L, Larson ET, Lawrence CM, Young M, Johnson JE. 2005. Structure of an archaeal virus capsid protein reveals a common ancestry to eukaryotic and bacterial viruses. *Proc. Natl. Acad. Sci. USA* 102:18944–49
53. Abrescia NG, Grimes JM, Kivela HM, Assenberg R, Sutton GC, et al. 2008. Insights into virus evolution and membrane biogenesis from the structure of the marine lipid-containing bacteriophage PM2. *Mol. Cell* 31:749–61
54. Bahar MW, Graham SC, Stuart DI, Grimes JM. 2011. Insights into the evolution of a complex virus from the crystal structure of vaccinia virus D13. *Structure* 19:1011–20
55. Nandhagopal N, Simpson AA, Gurnon JR, Yan X, Baker TS, et al. 2002. The structure and evolution of the major capsid protein of a large, lipid-containing DNA virus. *Proc. Natl. Acad. Sci. USA* 99:14758–63
56. Krupovic M, Bamford DH. 2011. Double-stranded DNA viruses: 20 families and only five different architectural principles for virion assembly. *Curr. Opin. Virol.* 1:118–24
57. Williams J. 1977. The problem of homology in relation to protein structure. In *The Evolution of Metalloenzymes, Metalloproteins and Related Materials*, ed. GJ Leigh, pp. 17–38. London: Symp. Press
58. Woodger JH. 1945. On biological transformations. In *Essays on Growth and Form: Presented to D'Arcy Wentworth Thompson*, ed. WELG Clark, PB Medawar, pp. 95–120. Oxford: Clarendon
59. Crick FH, Watson JD. 1956. Structure of small viruses. *Nature* 177:473–75
60. Astbury WT. 1953. A discussion on the structure of proteins; introduction. *Proc. R. Soc. Lond. B* 141:1–9
61. Owen R. 1843. *Lectures on the Comparative Anatomy and Physiology of the Invertebrate Animals*. London: Longman Brown Green & Longmans
62. Holm L, Sander C. 1993. Protein structure comparison by alignment of distance matrices. *J. Mol. Biol.* 233:123–38
63. Krissinel E, Henrick K. 2004. Secondary-structure matching (SSM), a new tool for fast protein structure alignment in three dimensions. *Acta Crystallogr. Sect. D* 60:2256–68
64. Rao ST, Rossmann MG. 1973. Comparison of super-secondary structures in proteins. *J. Mol. Biol.* 76:241–56
65. Rossmann MG, Argos P. 1976. Exploring structural homology of proteins. *J. Mol. Biol.* 105:75–95
66. Stuart DI, Levine M, Muirhead H, Stammers DK. 1979. Crystal structure of cat muscle pyruvate kinase at a resolution of 2.6 Å. *J. Mol. Biol.* 134:109–42
67. Riffel N, Harlos K, Iourin O, Rao Z, Kingsman A, et al. 2002. Atomic resolution structure of Moloney murine leukemia virus matrix protein and its relationship to other retroviral matrix proteins. *Structure* 10:1627–36
68. Felsenstein J. 1989. PHYLIP—phylogeny inference package (version 3.2). *Cladistics* 5:164–66
69. Levine M, Stuart DI, Williams J. 1984. A method for the systematic comparison of the three-dimensional structures of proteins and some results. *Acta Crystallogr. Sect. A* 40:600
70. Stromsten NJ, Bamford DH, Bamford JK. 2005. In vitro DNA packaging of PRD1: a common mechanism for internal-membrane viruses. *J. Mol. Biol.* 348:617–29
71. Rao VB, Feiss M. 2008. The bacteriophage DNA packaging motor. *Annu. Rev. Genet.* 42:647–81
72. Sun S, Rao VB, Rossmann MG. 2010. Genome packaging in viruses. *Curr. Opin. Struct. Biol.* 20:114–20
73. Sun S, Kondabagil K, Gentz PM, Rossmann MG, Rao VB. 2007. The structure of the ATPase that powers DNA packaging into bacteriophage T4 procapsids. *Mol. Cell* 25:943–49
74. Olia AS, Prevelige PE Jr, Johnson JE, Cingolani G. 2011. Three-dimensional structure of a viral genome-delivery portal vertex. *Nat. Struct. Mol. Biol.* 18:597–603
75. Jardine PJ, Anderson DL. 2006. DNA packaging in double-stranded DNA phages. In *The Bacteriophages*, ed. R Calendar, pp. 49–65. New York: Oxford Univ. Press
76. Kindt J, Tzlil S, Ben-Shaul A, Gelbart WM. 2001. DNA packaging and ejection forces in bacteriophage. *Proc. Natl. Acad. Sci. USA* 98:13671–74
77. Nealon K, Newcomb WW, Pray TR, Craik CS, Brown JC, Kedes DH. 2001. Lytic replication of Kaposi's sarcoma-associated herpesvirus results in the formation of multiple capsid species: isolation and molecular characterization of A, B, and C capsids from a gammaherpesvirus. *J. Virol.* 75:2866–78
78. Newcomb WW, Juhas RM, Thomsen DR, Homa FL, Burch AD, et al. 2001. The UL6 gene product forms the portal for entry of DNA into the herpes simplex virus capsid. *J. Virol.* 75:10923–32

79. Zauberman N, Mutsafi Y, Halevy DB, Shimoni E, Klein E, et al. 2008. Distinct DNA exit and packaging portals in the virus *Acanthamoeba polyphaga mimivirus*. *PLoS Biol.* 6:e114
80. Xiao C, Kuznetsov YG, Sun S, Hafenstein SL, Kostyuchenko VA, et al. 2009. Structural studies of the giant Mimivirus. *PLoS Biol.* 7:e92
81. Cherrier MV, Kostyuchenko VA, Xiao C, Bowman VD, Battisti AJ, et al. 2009. An icosahedral algal virus has a complex unique vertex decorated by a spike. *Proc. Natl. Acad. Sci. USA* 106:11085–89
82. Simpson AA, Tao Y, Leiman PG, Badasso MO, He Y, et al. 2000. Structure of the bacteriophage phi29 DNA packaging motor. *Nature* 408:745–50
83. Orlova EV, Gowen B, Droge A, Stiege A, Weise F, et al. 2003. Structure of a viral DNA gatekeeper at 10 Å resolution by cryo-electron microscopy. *EMBO J.* 22:1255–62
84. Lebedev AA, Krause MH, Isidro AL, Vagin AA, Orlova EV, et al. 2007. Structural framework for DNA translocation via the viral portal protein. *EMBO J.* 26:1984–94
85. Nadal M, Mas PJ, Blanco AG, Arnan C, Sola M, et al. 2009. Structure and inhibition of herpesvirus DNA packaging terminase nuclease domain. *Proc. Natl. Acad. Sci. USA* 107:16078–83
86. Garcea RL, Liddington RC. 1997. Structural biology of polyomaviruses. In *Structural Biology of Viruses*, ed. W Chiu, RM Burnett, RL Garcea, pp. 187–208. New York: Oxford Univ. Press
87. Huiskonen JT, de Haas F, Bubeck D, Bamford DH, Fuller SD, Butcher SJ. 2006. Structure of the bacteriophage phi6 nucleocapsid suggests a mechanism for sequential RNA packaging. *Structure* 14:1039–48
88. Diprose JM, Burroughs JN, Sutton GC, Goldsmith A, Gouet P, et al. 2001. Translocation portals for the substrates and products of a viral transcription complex: the bluetongue virus core. *EMBO J.* 20:7229–39
89. Mancini EJ, Kainov DE, Grimes JM, Tuma R, Bamford DH, Stuart DI. 2004. Atomic snapshots of an RNA packaging motor reveal conformational changes linking ATP hydrolysis to RNA translocation. *Cell* 118:743–55
90. Kainov DE, Tuma R, Mancini EJ. 2006. Hexameric molecular motors: P4 packaging ATPase unravels the mechanism. *Cell Mol. Life Sci.* 63:1095–105
91. Albertini AA, Wernimont AK, Muziol T, Ravelli RB, Clapier CR, et al. 2006. Crystal structure of the rabies virus nucleoprotein-RNA complex. *Science* 313:360–63
92. Ferron F, Li Z, Danek EI, Luo D, Wong Y, et al. 2011. The hexamer structure of Rift Valley fever virus nucleoprotein suggests a mechanism for its assembly into ribonucleoprotein complexes. *PLoS Pathog.* 7:e1002030
93. Green TJ, Zhang X, Wertz GW, Luo M. 2006. Structure of the vesicular stomatitis virus nucleoprotein-RNA complex. *Science* 313:357–60
94. Rudolph MG, Kraus I, Dickmanns A, Eickmann M, Garten W, Ficner R. 2003. Crystal structure of the Borna disease virus nucleoprotein. *Structure* 11:1219–26
95. Tawar RG, Duquerroy S, Vonrhein C, Varela PF, Damier-Piolle L, et al. 2009. Crystal structure of a nucleocapsid-like nucleoprotein-RNA complex of respiratory syncytial virus. *Science* 326:1279–83
96. Bloomer AC, Champness JN, Bricogne G, Staden R, Klug A. 1978. Protein disk of *Tobacco mosaic virus* at 2.8 Å resolution showing the interactions within and between subunits. *Nature* 276:362–68
97. Pickett GG, Peabody DS. 1993. Encapsidation of heterologous RNAs by bacteriophage MS2 coat protein. *Nucleic Acids Res.* 21:4621–26
98. Liu Y, Wang C, Mueller S, Paul AV, Wimmer E, Jiang P. 2010. Direct interaction between two viral proteins, the nonstructural protein 2C and the capsid protein VP3, is required for enterovirus morphogenesis. *PLoS Pathog.* 6:e1001066
99. Sasaki J, Taniguchi K. 2003. The 5′-end sequence of the genome of Aichi virus, a picornavirus, contains an element critical for viral RNA encapsidation. *J. Virol.* 77:3542–48
100. Dokland T, McKenna R, Ilag LL, Bowman BR, Incardona NL, et al. 1997. Structure of a viral procapsid with molecular scaffolding. *Nature* 389:308–13
101. Baker ML, Jiang W, Rixon FJ, Chiu W. 2005. Common ancestry of herpesviruses and tailed DNA bacteriophages. *J. Virol.* 79:14967–70
102. Krupovic M, Forterre P, Bamford DH. 2010. Comparative analysis of the mosaic genomes of tailed archaeal viruses and proviruses suggests common themes for virion architecture and assembly with tailed viruses of bacteria. *J. Mol. Biol.* 397:144–60

103. Jaalinoja HT, Roine E, Laurinmaki P, Kivela HM, Bamford DH, Butcher SJ. 2008. Structure and host-cell interaction of SH1, a membrane-containing, halophilic euryarchaeal virus. *Proc. Natl. Acad. Sci. USA* 105:8008–13
104. Jaatinen ST, Happonen LJ, Laurinmäki P, Butcher SJ, Bamford DH. 2008. Biochemical and structural characterisation of membrane-containing icosahedral dsDNA bacteriophages infecting thermophilic *Thermus thermophilus*. *Virology* 379:10–19
105. Grimes JM, Burroughs JN, Gouet P, Diprose JM, Malby R, et al. 1998. The atomic structure of the bluetongue virus core. *Nature* 395:470–78
106. Reinisch KM, Nibert ML, Harrison SC. 2000. Structure of the reovirus core at 3.6 Å resolution. *Nature* 404:960–67
107. Naitow H, Tang J, Canady M, Wickner RB, Johnson JE. 2002. L-A virus at 3.4 Å resolution reveals particle architecture and mRNA decapping mechanism. *Nat. Struct. Biol.* 9:725–28
108. Nakagawa A, Miyazaki N, Taka J, Naitow H, Ogawa A, et al. 2003. The atomic structure of *Rice dwarf* virus reveals the self-assembly mechanism of component proteins. *Structure* 11:1227–38
109. Yu X, Jin L, Zhou ZH. 2008. 3.88 Å structure of cytoplasmic polyhedrosis virus by cryo-electron microscopy. *Nature* 453:415–19
110. Duquerroy S, Da Costa B, Henry C, Vigouroux A, Libersou S, et al. 2009. The picobirnavirus crystal structure provides functional insights into virion assembly and cell entry. *EMBO J.* 28:1655–65
111. Gertsman I, Gan L, Guttman M, Lee K, Speir JA, et al. 2009. An unexpected twist in viral capsid maturation. *Nature* 458:646–50
112. Luque D, González JM, Garriga D, Ghabrial SA, Havens WM, et al. 2010. The T = 1 capsid protein of *Penicillium chrysogenum* virus is formed by a repeated helix-rich core indicative of gene duplication. *J. Virol.* 84:7256–66
113. McClain B, Settembre E, Temple BR, Bellamy AR, Harrison SC. 2010. X-ray crystal structure of the rotavirus inner capsid particle at 3.8 Å resolution. *J. Mol. Biol.* 397:587–99
114. Caspar DL, Klug A. 1962. Physical principles in the construction of regular viruses. *Cold Spring Harb. Symp. Quant. Biol.* 27:1–24
115. Zhang X, Jin L, Fang Q, Hui WH, Zhou ZH. 2010. 3.3 A cryo-EM structure of a nonenveloped virus reveals a priming mechanism for cell entry. *Cell* 141:472–82
116. Coulibaly F, Chevalier C, Gutsche I, Pous J, Navaza J, et al. 2005. The birnavirus crystal structure reveals structural relationships among icosahedral viruses. *Cell* 120:761–72
117. Kuhn RJ, Zhang W, Rossmann MG, Pletnev SV, Corver J, et al. 2002. Structure of dengue virus: implications for flavivirus organization, maturation, and fusion. *Cell* 108:717–25
118. Zhang W, Chipman PR, Corver J, Johnson PR, Zhang Y, et al. 2003. Visualization of membrane protein domains by cryo-electron microscopy of dengue virus. *Nat. Struct. Biol.* 10:907–12
119. Yu IM, Zhang W, Holdaway HA, Li L, Kostyuchenko VA, et al. 2008. Structure of the immature dengue virus at low pH primes proteolytic maturation. *Science* 319:1834–37
120. Bartenschlager R, Penin F, Lohmann V, Andre P. 2011. Assembly of infectious hepatitis C virus particles. *Trends Microbiol.* 19:95–103
121. Yu X, Qiao M, Atanasov I, Hu Z, Kato T, et al. 2007. Cryo-electron microscopy and three-dimensional reconstructions of hepatitis C virus particles. *Virology* 367:126–34
122. Kaufmann B, Plevka P, Kuhn RJ, Rossmann MG. 2010. Crystallization and preliminary X-ray diffraction analysis of West Nile virus. *Acta Crystallogr. Sect. F* 66:558–62
123. Lescar J, Roussel A, Wien MW, Navaza J, Fuller SD, et al. 2001. The fusion glycoprotein shell of Semliki Forest virus: an icosahedral assembly primed for fusogenic activation at endosomal pH. *Cell* 105:137–48
124. Akahata W, Yang ZY, Andersen H, Sun S, Holdaway HA, et al. 2010. A virus-like particle vaccine for epidemic chikungunya virus protects nonhuman primates against infection. *Nat. Med.* 16:334–38
125. Voss JE, Vaney MC, Duquerroy S, Vonrhein C, Girard-Blanc C, et al. 2010. Glycoprotein organization of chikungunya virus particles revealed by X-ray crystallography. *Nature* 468:709–12
126. Harrison SC. 2005. Mechanism of membrane fusion by viral envelope proteins. *Adv. Virus Res.* 64:231–61
127. Harrison SC. 2008. Viral membrane fusion. *Nat. Struct. Mol. Biol.* 15:690–98
128. Kadlec J, Loureiro S, Abrescia NG, Stuart DI, Jones IM. 2008. The postfusion structure of baculovirus gp64 supports a unified view of viral fusion machines. *Nat. Struct. Mol. Biol.* 15:1024–30

129. Lee JE, Fusco ML, Hessell AJ, Oswald WB, Burton DR, Saphire EO. 2008. Structure of the Ebola virus glycoprotein bound to an antibody from a human survivor. *Nature* 454:177–82
130. Roche S, Rey FA, Gaudin Y, Bressanelli S. 2007. Structure of the prefusion form of the vesicular stomatitis virus glycoprotein G. *Science* 315:843–48
131. Bowden TA, Jones EY, Stuart DI. 2011. Cells under siege: viral glycoprotein interactions at the cell surface. *J. Struct. Biol.* 175:120–26
132. El Omari K, Dhaliwal B, Ren J, Abrescia NG, Lockyer M, et al. 2011. Structures of respiratory syncytial virus nucleocapsid protein from two crystal forms: details of potential packing interactions in the native helical form. *Acta Crystallogr. Sect. F.* 67:1179–83
133. Raymond DD, Piper ME, Gerrard SR, Smith JL. 2010. Structure of the Rift Valley fever virus nucleocapsid protein reveals another architecture for RNA encapsidation. *Proc. Natl. Acad. Sci. USA* 107:11769–74
134. Pietila MK, Roine E, Paulin L, Kalkkinen N, Bamford DH. 2009. An ssDNA virus infecting archaea: a new lineage of viruses with a membrane envelope. *Mol. Microbiol.* 72:307–19
135. Roine E, Kukkaro P, Paulin L, Laurinavicius S, Domanska A, et al. 2010. New, closely related haloarchaeal viral elements with different nucleic acid types. *J. Virol.* 84:3682–89
136. Atanasova NS, Roine E, Oren A, Bamford DH, Oksanen HM. 2012. Global network of specific virus-host interactions in hypersaline environments. *Environ. Microbiol.* 14:426–40
137. Sachse C, Chen JZ, Coureux PD, Stroupe ME, Fandrich M, Grigorieff N. 2007. High-resolution electron microscopy of helical specimens: a fresh look at *Tobacco mosaic virus*. *J. Mol. Biol.* 371:812–35
138. Klug A. 2010. From virus structure to chromatin: X-ray diffraction to three-dimensional electron microscopy. *Annu. Rev. Biochem.* 79:1–35
139. Goulet A, Pina M, Redder P, Prangishvili D, Vera L, et al. 2010. ORF157 from the archaeal virus *Acidianus* filamentous virus 1 defines a new class of nuclease. *J. Virol.* 84:5025–31
140. Goulet A, Vestergaard G, Felisberto-Rodrigues C, Campanacci V, Garrett RA, et al. 2010. Getting the best out of long-wavelength X-rays: de novo chlorine/sulfur SAD phasing of a structural protein from ATV. *Acta Crystallogr. D* 66:304–8
141. Szymczyna BR, Taurog RE, Young MJ, Snyder JC, Johnson JE, Williamson JR. 2009. Synergy of NMR, computation, and X-ray crystallography for structural biology. *Structure* 17:499–507
142. Goulet A, Blangy S, Redder P, Prangishvili D, Felisberto-Rodrigues C, et al. 2009. *Acidianus* filamentous virus 1 coat proteins display a helical fold spanning the filamentous archaeal viruses lineage. *Proc. Natl. Acad. Sci. USA* 106:21155–60
143. Wynne SA, Crowther RA, Leslie AG. 1999. The crystal structure of the human hepatitis B virus capsid. *Mol. Cell* 3:771–80
144. Zlotnick A, Stahl SJ, Wingfield PT, Conway JF, Cheng N, Steven AC. 1998. Shared motifs of the capsid proteins of hepadnaviruses and retroviruses suggest a common evolutionary origin. *FEBS Lett.* 431:301–4
145. Marvin DA, Welsh LC, Symmons MF, Scott WR, Straus SK. 2006. Molecular structure of fd (f1, M13) filamentous bacteriophage refined with respect to X-ray fibre diffraction and solid-state NMR data supports specific models of phage assembly at the bacterial membrane. *J. Mol. Biol.* 355:294–309
146. Koonin EV, Senkevich TG, Dolja VV. 2006. The ancient Virus World and evolution of cells. *Biol. Direct.* 1:29
147. Lipps G. 2004. The replication protein of the *Sulfolobus islandicus* plasmid pRN1. *Biochem. Soc. Trans.* 32:240–44
148. Forterre P. 2005. The two ages of the RNA world, and the transition to the DNA world: a story of viruses and cells. *Biochimie* 87:793–803
149. Krupovic M, Bamford DH. 2007. Putative prophages related to lytic tailless marine dsDNA phage PM2 are widespread in the genomes of aquatic bacteria. *BMC Genomics* 8:236
150. Krupovic M, Bamford DH. 2009. Does the evolution of viral polymerases reflect the origin and evolution of viruses? *Nat. Rev. Microbiol.* 7:250
151. Cockburn JJ, Bamford JK, Grimes JM, Bamford DH, Stuart DI. 2003. Crystallization of the membrane-containing bacteriophage PRD1 in quartz capillaries by vapour diffusion. *Acta Crystallogr. D* 59:538–40
152. Meier C, Aricescu AR, Assenberg R, Aplin RT, Gilbert RJ, et al. 2006. The crystal structure of ORF-9b, a lipid binding protein from the SARS coronavirus. *Structure* 14:1157–65

153. Krupovic M, Bamford DH. 2011. Virology. Revealing virus-host interplay. *Science* 333:45–46
154. Tadmor AD, Ottesen EA, Leadbetter JR, Phillips R. 2011. Probing individual environmental bacteria for viruses by using microfluidic digital PCR. *Science* 333:58–62
155. Liu H, Jin L, Koh SB, Atanasov I, Schein S, et al. 2010. Atomic structure of human adenovirus by cryo-EM reveals interactions among protein networks. *Science* 329:1038–43
156. Jiang W, Baker ML, Jakana J, Weigele PR, King J, Chiu W. 2008. Backbone structure of the infectious epsilon15 virus capsid revealed by electron cryomicroscopy. *Nature* 451:1130–34
157. Evans G, Axford D, Waterman D, Owen RL. 2011. Macromolecular microcrystallography. *Crystallogr. Rev.* 17:105–42
158. Wang X, Peng W, Ren J, Hu Z, Xu J, et al. 2012. A sensor/adaptor mechanism for enterovirus uncoating from structures of EV71. *Nat. Struct. Mol. Biol.* In press
159. Chapman HN, Fromme P, Barty A, White TA, Kirian RA, et al. 2011. Femtosecond X-ray protein nanocrystallography. *Nature* 470:73–77
160. Seibert MM, Ekeberg T, Maia FR, Svenda M, Andreasson J, et al. 2011. Single mimivirus particles intercepted and imaged with an X-ray laser. *Nature* 470:78–81

Cumulative Indexes

Contributing Authors, Volumes 77–81

A

Abrescia NGA, 81:795–822
Adler J, 80:42–70
Aebersold R, 81:379–405
Aitken CE, 77:177–203
Alber F, 77:443–77
Anckar J, 80:1,089–115
Antebi A, 80:885–916
Antonny B, 80:101–23

B

Back SH, 81:767–93
Bakal C, 79:37–64
Baker D, 77:363–82
Baker TA, 80:587–612
Balch WE, 78:959–91
Balci H, 77:51–76
Bamford DH, 81:795–822
Barlowe C, 79:777–802
Baron GS, 78:177–204
Barrera NP, 80:247–71
Bates M, 78:993–1,016
Begley TP, 78:569–603
Beltrao P, 77:415–41
Bensimon A, 81:379–405
Berg P, 77:14–44
Bergman Y, 81:97–117
Blobel G, 80:613–43
Block SM, 77:149–76
Bogan JS, 81:507–32
Bogdanov M, 78:515–40

Bolen DW, 77:339–62
Borgia A, 77:101–25
Bowen M, 78:903–28
Braakman I, 80:71–99
Breaker RR, 78:305–34
Broderick JB, 81:429–50
Brown S, 78:743–68
Brunger AT, 78:903–28
Bulleid NJ, 80:71–99
Buranachai C, 77:51–76
Bustamante C, 77:45–50, 205–28

C

Cairns BR, 78:273–304
Campelo F, 81:407–27
Carman GM, 80:859–83
Carrasco S, 80:973–1,000
Caughey B, 78:177–204
Cedar H, 81:97–117
Chait BT, 80:239–46
Chang HY, 78:245–71;
  81:145–66
Chapman ER, 77:615–41
Chemla YR, 77:205–28
Chen ZJ, 78:769–96
Cheng Y, 78:723–42
Chesebro B, 78:177–204
Chu S, 78:903–28
Ciccia A, 77:259–87
Clapier CR, 78:273–304
Clarke J, 77:101–25

Cole PA, 78:797–825
Colman PM, 78:95–118
Conaway JW, 81:61–64
Cooper JA, 81:661–86
Cotruvo JA Jr, 80:733–67
Cox J, 80:273–99
Crane BR, 79:445–70
Cravatt BF, 77:383–414
Crawford ED, 80:1,055–87

D

Dalbey RE, 80:161–87
Dancourt J, 79:777–802
Dar AC, 80:769–95
Das R, 77:363–82
Davies GJ, 77:521–55
Dean DR, 78:701–22
Debler EW, 80:613–43
DeGrado WF, 80:211–37
Dejosez M, 81:737–65
Dennis EA, 80:301–25
Derbyshire ER, 81:533–59
Deshaies RJ, 78:399–434
Dietrich JA, 79:563–90
Dikic I, 81:291–322
Dillin A, 77:727–54; 78:959–91
Dimroth P, 78:649–72
Doherty GJ, 78:857–902
Dorywalska M, 77:177–203
Dowhan W, 78:515–40
Driessen AJM, 77:643–67

823

E

Ealick SE, 78:569–603
Edwards A, 78:541–68
Elazar Z, 80:125–56
Ellenberger T, 77:313–38
Emr SD, 81:231–59
Endicott JA, 81:587–613
Engel A, 77:127–48

F

Fabian MR, 79:351–79
Farese RV Jr, 81:687–714
Fersht AR, 77:557–82
Filipowicz W, 79:351–79
Finley D, 78:477–513
Fontes CMGA, 79:655–81
Förster F, 77:443–77
Frank J, 79:381–412

G

Galletta BJ, 81:661–86
Gaub HE, 77:127–48
Geiduschek EP, 78:1–28
Gelb MH, 77:495–520
Geng F, 81:177–201
Gerton JL, 79:131–53
Gilbert HJ, 79:655–81
Goenrich M, 79:507–36
Gómez-García MR, 78:605–47
Gonzalez RL Jr, 79:381–412
Gonzalez-Cabrera PJ, 78:743–68
Gooptu B, 78:147–76
Graves BJ, 80:437–71
Greenleaf WJ, 77:149–76
Grigoryan G, 80:211–37
Grimes JM, 81:795–822
Guan K-L, 80:1,001–32

H

Ha T, 77:51–76
Hagan CL, 80:189–210
Halford RW, 78:1,017–40
Hamdan SM, 78:205–43
Han G, 80:859–83
Harkewicz R, 80:301–25
Hart GW, 80:825–58
Heck AJR, 81:379–405

Helenius A, 79:803–33
Henrissat B, 77:521–55
Herbert KM, 77:149–76
Herschlag D, 80:669–702
Hiromoto T, 79:507–36
Hoelz A, 80:613–43
Hoffman BM, 78:701–22
Hollenhorst PC, 80:437–71
Honig B, 79:233–69
Hsu P-C, 81:231–59
Huang B, 78:993–1,016
Hunter T, 78:435–75
Husnjak K, 81:291–322

I

Imlay JA, 77:755–76
Ishitsuka Y, 77:51–76

J

Jeffrey M, 78:177–204
Jensen GJ, 77:583–613
Jiang X, 78:769–96
Jin X, 79:233–69
Joazeiro CAP, 78:399–434
Joerger AC, 77:557–82
Johnson LN, 81:587–613
Joo C, 77:51–76
Joshi R, 79:233–69
Jurgenson CT, 78:569–603

K

Kahne D, 80:189–210
Kang JY, 80:917–41
Kaster A-K, 79:507–36
Kaufman RJ, 81:767–93
Kelly JW, 78:959–91
Keasling JD, 79:563–90
Khersonsky O, 79:471–505
Kiel C, 77:415–41
Kiessling LL, 79:619–53
Kim J, 80:1,001–32
Klein H, 77:229–57
Klug A, 79:1–35, 213–31
Knerr PJ, 81:479–505
Komander D, 81:203–29
Korkin D, 77:443–77
Kornberg A, 78:605–47
Kotti T, 78:1,017–40

Kozarich JW, 77:383–414; 78:55–63
Kuhn A, 80:161–87
Kwon YT, 81:261–89

L

Lagerlof O, 80:825–58
Lairson LL, 77:521–55
Lambeau G, 77:495–520
Lane MD, 81:715–36
Larsson N-G, 79:683–706
Lassila JK, 80:669–702
Lee J, 80:917–41
Leung EKY, 80:527–55
Li PTX, 77:77–100
Li T, 81:119–43
Lieber MR, 79:181–211
Lill R, 77:669–700
Lippard SJ, 80:333–55
Lippincott-Schwartz J, 80:327–32
Liu CC, 79:413–44
Liu Q, 79:295–319
Liu X, 81:561–85
Lomas DA, 78:147–76
Lovering AL, 81:451–78
Lu Z, 78:435–75
Lundeberg J, 81:359–78

M

MacGurn JA, 81:231–59
Mair W, 77:727–54
Malhotra V, 81:407–27
Mann M, 80:273–99
Mann RS, 79:233–69
Marletta MA, 81:533–59
Marshall RA, 77:177–203
Masai H, 79:89–130
Matsumoto S, 79:89–130
McDonald N, 77:259–87
McHenry CS, 80:403–36
McIntosh LP, 80:437–71
McKee AE, 79:563–90
McMahon HT, 78:857–902
Mehta S, 80:375–401
Mercer J, 79:803–33
Metcalf WW, 78:65–94
Meyer T, 80:973–1,000
Miyawaki A, 80:357–73
Mizuno-Yamasaki E, 81:637–59

Moffitt JR, 77:205–28
Mohr S, 79:37–64
Montal M, 79:591–617
Moore DT, 80:211–37
Mooren OL, 81:661–86
Morimoto RI, 78:959–91
Morris DM, 77:583–613
Mosammaparast N, 79:155–79
Mühlenhoff U, 77:669–700
Murphy MP, 77:777–98

N

Neufeld EF, 80:1–15
Neupert W, 81:1–33
Nikaido H, 78:119–46
Nishikura K, 79:321–49
Noble MEM, 81:587–613
Nocera DG, 78:673–99
Nomura M, 80:16–40
Nouwen N, 77:643–67
Novick P, 81:637–59
Nudler E, 78:335–61

O

Oda M, 79:89–130

P

Padrick SB, 79:707–35
Park KS, 81:261–89
Paroo Z, 79:295–319
Partridge L, 77:777–98
Patel BA, 79:445–70
Paulson JC, 80:797–823
Perrimon N, 79:37–64
Peters JW, 81:429–50
Piccirilli JA, 80:527–55
Ploegh HL, 81:323–57
Pluth MD, 80:333–55
Popot J-L, 79:737–75
Powers ET, 78:959–91
Price DH, 81:119–43
Puglisi JD, 77:177–203

R

Raines RT, 78:929–58
Ramirez DMO, 78:1,017–40
Ramirez-Correa G, 80:825–58

Rando OJ, 78:245–71
Rao NN, 78:605–47
Rape M, 81:203–29
Reece SY, 78:673–99
Reyes-Turcu FE, 78:363–97
Richardson CC, 78:205–43
Rillahan CD, 80:797–823
Rinn JL, 81:145–66
Riordan JR, 77:701–26
Rivera-Molina F, 81:637–59
Robinson CV, 80:247–71
Robinson NJ, 79:537–62
Rohs R, 79:233–69
Rose GD, 77:339–62
Rosen H, 78:743–68
Rosen MK, 79:707–35
Ross J, 77:479–94
Roth A, 78:305–34
Russell DW, 78:1,017–40

S

Safadi SS, 81:451–78
Sali A, 77:443–77
San Filippo J, 77:229–57
Sanna MG, 78:743–68
Sauer RT, 80:587–612
Schatz G, 81:34–59
Schelhaas M, 79:803–33
Schick M, 79:507–36
Schneider A, 80:1,033–53
Schramm VL, 80:703–32
Schultz PG, 79:413–44
Seefeldt LC, 78:701–22
Seeman NC, 79:65–87
Selinger Z, 77:1–13
Selth LA, 79:271–93
Sengupta R, 80:527–55
Serrano L, 77:415–41
Shah R, 78:1,017–40
Shajani Z, 80:501–26
Shi Y, 79:155–79
Shilatifard A, 81:65–95
Shima S, 79:507–36
Shokat KM, 80:769–95
Shoulders MD, 78:929–58
Shvets E, 80:125–56
Sigurdsson S, 79:271–93
Silhavy TJ, 80:189–210
Sistonen L, 80:1,089–115
Skaug B, 78:769–96
Slawson C, 80:825–58

Smith SB, 77:205–28
Sonenberg N, 79:351–79
Spiegelman BM, 77:289–312
Splain RA, 79:619–53
Sriram SM, 81:261–89
Ståhl PL, 81:359–78
Strynadka NCJ, 81:451–78
Stuart DI, 81:795–822
Stubbe J, 80:733–67
Sudhamsu J, 79:445–70
Suganuma T, 80:473–99
Sung P, 77:229–57
Suslov N, 80:527–55
Svejstrup JQ, 79:271–93
Sykes MT, 80:501–26

T

Tang QQ, 81:715–36
Tansey WP, 81:177–201
Tarrant MK, 78:797–825
Tasaki T, 81:261–89
Tawfik DS, 79:471–505
Thauer RK, 79:507–36
Tinoco I Jr, 77:77–100
Tomat E, 80:333–55
Tomkinson AE, 77:313–38
Tontonoz P, 77:289–312
Topf M, 77:443–77
Toyama BH, 80:557–85
Tuttle N, 80:527–55

V

van der Donk WA, 78:65–94;
    81:479–505
van der Veen AG, 81:323–57
Varshavsky A, 81:167–76
Ventii KH, 78:363–97
Vetter IR, 80:943–71
Vieregg J, 77:77–100
Voelker DR, 78:827–56
von Ballmoos C, 78:649–72
von Heijne G, 80:157–60

W

Walther TC, 81:687–714
Walz T, 78:723–42
Wang JC, 78:30–54
Wang L, 81:615–35

Wang P, 80:161–87
Weidberg H, 80:125–56
Weissman JS, 80:557–85
Wells JA, 80:1,055–87
Weninger K, 78:903–28
Wenzel S, 81:177–201
West SC, 77:259–87
West SM, 79:233–69
Wiedenmann A, 78:649–72
Wilkinson KD, 78:363–97
Williams PM, 77:101–25
Williamson JR, 80:501–26

Winey M, 81:561–85
Winge DR, 79:537–62
Withers SG, 77:521–55
Wittinghofer A, 80:943–71
Wolfenden R, 80:645–67
Wollam J, 80:885–916
Workman JL, 80:473–99
Wright AT, 77:383–414

X

Xiong B, 79:131–53

Y

Yoshizawa-Sugata N, 79:89–130
You Z, 79:89–130

Z

Zalatan JG, 80:669–702
Zhang J, 80:375–401
Zhou Q, 81:119–43
Zhuang X, 78:993–1,016
Zwaka TP, 81:737–65

Chapter Titles, Volumes 77–81

Prefatory

Discovery of G Protein Signaling	Z Selinger	77:1–13
Moments of Discovery	P Berg	77:14–44
Without a License, or Accidents Waiting to Happen	EP Geiduschek	78:1–28
A Journey in the World of DNA Rings and Beyond	JC Wang	78:31–54
From Virus Structure to Chromatin: X-ray Diffraction to Three-Dimensional Electron Microscopy	A Klug	79:1–35
From Serendipity to Therapy	EF Neufeld	80:1–15
Journey of a Molecular Biologist	M Nomura	80:16–40
My Life with Nature	J Adler	80:42–70
A Mitochondrial Odyssey	W Neupert	81:1–33
The Fires of Life	G Schatz	81:34–59

DNA

Chemistry and Structure

Nanomaterials Based on DNA	NC Seeman	79:65–87

Genomics

Genomic Screening with RNAi: Results and Challenges	S Mohr, C Bakal, N Perrimon	79:37–64

Methodology

Toward the Single-Hour High-Quality Genome	PL Ståhl, J Lundeberg	81:359–78

Replication

Motors, Switches, and Contacts in a Replisome	SM Hamdan, CC Richardson	78:205–43

Eukaryotic Chromosome DNA Replication: Where, When, and How?	H Masai, S Matsumoto, Z You, N Yoshizawa-Sugata, M Oda	79:89–130
DNA Replicases from a Bacterial Perspective	CS McHenry	80:403–36

Repair and Modifications

The COMPASS Family of Histone H3K4 Methylases: Mechanisms of Regulation in Development and Disease Pathogenesis	A Shilatifard	81:65–95
Programming of DNA Methylation Patterns	H Cedar, Y Bergman	81:97–117

Chromatin and Chromosomes

Genome-Wide Views of Chromatin Structure	OJ Rando, HY Chang	78:245–71
The Biology of Chromatin Remodeling Complexes	CR Clapier, BR Cairns	78:273–304
Regulators of the Cohesin Network	B Xiong, JL Gerton	79:131–53
Reversal of Histone Methylation: Biochemical and Molecular Mechanisms of Histone Demethylases	N Mosammaparast, Y Shi	79:155–79
Introduction to Theme "Chromatin, Epigenetics, and Transcription"	JW Conaway	81:61–64
Ubiquitin and Proteasomes in Transcription	F Geng, S Wenzel, WP Tansey	81:177–201

Recombination and Transposition

Mechanism of Eukaryotic Homologous Recombination	J San Filippo, P Sung, H Klein	77:229–57
The Mechanism of Double-Strand DNA Break Repair by the Nonhomologous DNA End-Joining Pathway	MR Lieber	79:181–211

Enzymes and Binding Proteins

Structural and Functional Relationships of the XPF/MUS81 Family of Proteins	A Ciccia, N McDonald, SC West	77:259–87
The Discovery of Zinc Fingers and Their Applications in Gene Regulation and Genome Manipulation	A Klug	79:213–31

RNA

Chemistry and Structure

How RNA Unfolds and Refolds	PTX Li, J Vieregg, I Tinoco Jr.	77:77–100
The Structural and Functional Diversity of Metabolite-Binding Riboswitches	A Roth, RR Breaker	78:305–34

Transcription and Gene Regulation

Single-Molecule Studies of RNA Polymerase: Motoring Along	KM Herbert, WJ Greenleaf, SM Block	77:149–76
Fat and Beyond: The Diverse Biology of PPARγ	P Tontonoz, BM Spiegelman	77:289–312
RNA Polymerase Active Center: The Molecular Engine of Transcription	E Nudler	78:335–61
Origins of Specificity in Protein-DNA Recognition	R Rohs, X Jin, SM West, R Joshi, B Honig, RS Mann	79:233–69
Transcript Elongation by RNA Polymerase II	LA Selth, S Sigurdsson, JQ Svejstrup	79:271–93
Genomic and Biochemical Insights into the Specificity of ETS Transcription Factors	PC Hollenhorst, LP McIntosh, BJ Graves	80:437–71
RNA Polymerase II Elongation Control	Q Zhou, T Li, DH Price	81:119–43
Genome Regulation by Long Noncoding RNAs	JL Rinn, HY Chang	81:145–66

Splicing, Posttranscriptional Processing, and Modifications

Biochemical Principles of Small RNA Pathways	Q Liu, Z Paroo	79:295–319
Functions and Regulation of RNA Editing by ADAR Deaminases	K Nishikura	79:321–49
Signals and Combinatorial Functions of Histone Modifications	T Suganuma, JL Workman	80:473–99

Translation

Regulation of mRNA Translation and Stability by microRNAs	MR Fabian, N Sonenberg, W Filipowicz	79:351–79
Structure and Dynamics of a Processive Brownian Motor: The Translating Ribosome	J Frank, RL Gonzalez Jr.	79:381–412
Assembly of Bacterial Ribosomes	Z Shajani, MT Sykes, JR Williamson	80:501–26
The Mechanism of Peptidyl Transfer Catalysis by the Ribosome	EKY Leung, N Suslov, N Tuttle, R Sengupta, JA Piccirilli	80:527–55

Proteins

Amino Acids and Their Chemistry

Adding New Chemistries to the Genetic Code	CC Liu, PG Schultz	79:413–44

Protein Chemistry and Structure

Eukaryotic DNA Ligases: Structural and Functional Insights	AE Tomkinson, T Ellenberger	77:313–38
Structure and Energetics of the Hydrogen-Bonded Backbone in Protein Folding	DW Bolen, GD Rose	77:339–62
Amyloid Structure: Conformational Diversity and Consequences	BH Toyama, JS Weissman	80:557–85

Methodology

Mass Spectrometry in the Postgenomic Era	BT Chait	80:239–46
Mass Spectrometry–Based Proteomics and Network Biology	A Bensimon, AJR Heck, R Aebersold	81:379–405

Folding and Design

Macromolecular Modeling with Rosetta	R Das, D Baker	77:363–82

Posttranslational Processing and Modifications

Regulation and Cellular Roles of Ubiquitin-Specific Deubiquitinating Enzymes	FE Reyes-Turcu, KH Ventii, KD Wilkinson	78:363–97
RING Domain E3 Ubiquitin Ligases	RJ Deshaies, CAP Joazeiro	78:399–434
Protein Folding and Modification in the Mammalian Endoplasmic Reticulum	I Braakman, NJ Bulleid	80:71-99
The Ubiquitin Code	D Komander, M Rape	81:203–29
Ubiquitin and Membrane Protein Turnover: From Cradle to Grave	JA MacGurn, P-C Hsu, SD Emr	81:231–59

Proteolysis and Turnover

Conformational Pathology of the Serpins: Themes, Variations, and Therapeutic Strategies	B Gooptu, DA Lomas	78:147–76
Degradation of Activated Protein Kinases by Ubiquitination	Z Lu, T Hunter	78:435–75
Recognition and Processing of Ubiquitin-Protein Conjugates by the Proteasome	D Finley	78:477–513

AAA+ Proteases: ATP-Fueled Machines of Protein Destruction	RT Sauer, TA Baker	80:587–612
The N-End Rule Pathway	T Tasaki, SM Sriram, KS Park, YT Kwon	81:261–89

Membrane Protein Structure and Function

Lipid-Dependent Membrane Protein Topogenesis	W Dowhan, M Bogdanov	78:515–40
Mechanisms of Membrane Curvature Sensing	B Antonny	80:101–23
Assembly of Bacterial Inner Membrane Proteins	RE Dalbey, P Wang, A Kuhn	80:161–87
β-Barrel Membrane Protein Assembly by the Bam Complex	CL Hagan, TJ Silhavy, D Kahne	80:189–210
Transmembrane Communication: General Principles and Lessons from the Structure and Function of the M2 Proton Channel, K+ Channels, and Integrin Receptors	G Grigoryan, DT Moore, WF DeGrado	80:211–37
The Structure of the Nuclear Pore Complex	A Hoelz, EW Debler, G Blobel	80:613–43
Membrane Fission: The Biogenesis of Transport Carriers	F Campelo, V Malhotra	81:407–27

Families and Evolution

Large-Scale Structural Biology of the Human Proteome	A Edwards	78:541–68
Bacterial Nitric Oxide Synthases	BR Crane, J Sudhamsu, BA Patel	79:445–70
Ubiquitin-Binding Proteins: Decoders of Ubiquitin-Mediated Cellular Functions	K Husnjak, I Dikic	81:291–322
Ubiquitin-Like Proteins	AG van der Veen, HL Ploegh	81:323–57

Proteomics

Activity-Based Protein Profiling: From Enzyme Chemistry to Proteomic Chemistry	BF Cravatt, AT Wright, JW Kozarich	77:383–414
Analyzing Protein Interaction Networks Using Structural Information	C Kiel, P Beltrao, L Serrano	77:415–41
Integrating Diverse Data for Structure Determination of Macromolecular Assemblies	F Alber, F Förster, D Korkin, M Topf, A Sali	77:443–77

Advances in the Mass Spectrometry of Membrane Proteins: From Individual Proteins to Intact Complexes	NP Barrera, CV Robinson	80:247–71
Quantitative, High-Resolution Proteomics for Data-Driven Systems Biology	J Cox, M Mann	80:273–99

Enzymology

Kinetics

From the Determination of Complex Reaction Mechanisms to Systems Biology	J Ross	77:479–94
Enzyme Promiscuity: A Mechanistic and Evolutionary Perspective	O Khersonsky, DS Tawfik	79:471–505

Catalytic Mechanisms

Biochemistry and Physiology of Mammalian Secreted Phospholipases A_2	G Lambeau, MH Gelb	77:495–520
Glycosyltransferases: Structures, Functions, and Mechanisms	LL Lairson, B Henrissat, GJ Davies, SG Withers	77:521–55
Benchmark Reaction Rates, the Stability of Biological Molecules in Water, and the Evolution of Catalytic Power in Enzymes	R Wolfenden	80:645–67
Biological Phosphoryl-Transfer Reactions: Understanding Mechanism and Catalysis	JK Lassila, JG Zalatan, D Herschlag	80:669–702
Enzymatic Transition States, Transition-State Analogs, Dynamics, Thermodynamics, and Lifetimes	VL Schramm	80:703–32

Cofactors and Prosthetic Groups

The Structural and Biochemical Foundations of Thiamin Biosynthesis	CT Jurgenson, TP Begley, SE Ealick	78:569–603
Hydrogenases from Methanogenic Archaea, Nickel, a Novel Cofactor, and H_2 Storage	RK Thauer, A-K Kaster, M Goenrich, M Schick, T Hiromoto, S Shima	79:507–36

Metalloenzymes

Copper Metallochaperones	NJ Robinson, DR Winge	79:537–62

Class I Ribonucleotide Reductases: Metallocofactor Assembly and Repair In Vitro and In Vivo	JA Cotruvo Jr, J Stubbe	80:733–67
Emerging Paradigms for Complex Iron-Sulfur Cofactor Assembly and Insertion	JW Peters, JB Broderick	81:429–50

Regulation and Metabolic Control

High-Throughput Metabolic Engineering: Advances in Small-Molecule Screening and Selection	JA Dietrich, AE McKee, JD Keasling	79:563–90

Inhibitors and Toxins

Botulinum Neurotoxin: A Marvel of Protein Design	M Montal	79:591–617
The Evolution of Protein Kinase Inhibitors from Antagonists to Agonists of Cellular Signaling	AC Dar, KM Shokat	80:769–95

Carbohydrates

Sugars and Their Chemistry

Chemical Approaches to Glycobiology	LL Kiessling, RA Splain	79:619–53

Methodology

Glycan Microarrays for Decoding the Glycome	CD Rillahan, JC Paulson	80:797–823

Glycoproteins and Protein Glycosylation

Cross Talk Between O-GlcNAcylation and Phosphorylation: Roles in Signaling, Transcription, and Chronic Disease	GW Hart, C Slawson, G Ramirez-Correa, O Lagerlof	80:825–58

Cell Walls, Extracellular Matrix, and Adhesion Molecules

Cellulosomes: Highly Efficient Nanomachines Designed to Deconstruct Plant Cell Wall Complex Carbohydrates	CMGA Fontes, HJ Gilbert	79:655–81
Structural Perspective of Peptidoglycan Biosynthesis and Assembly	AL Lovering, SS Safadi, NCJ Strynadka	81:451–78

Lipids

Methodology

Applications of Mass Spectrometry to Lipids and Membranes	R Harkewicz, EA Dennis,	80:301–25

Glycerophosolipids

Regulation of Phospholipid Synthesis in the Yeast *Saccharomyces cerevisiae*	GM Carman, G-S Han	80:859–83

Isoprenoids and Sterols

Sterol Regulation of Metabolism, Homeostasis, and Development	J Wollam, A Antebi	80:885–916

Other Biomolecules

Natural Products

Inorganic Polyphosphate: Essential for Growth and Survival	NN Rao, MR Gómez-García, A Kornberg	78:605–47

Antibiotics

Biosynthesis of Phosphonic and Phosphinic Acid Natural Products	WW Metcalf, WA van der Donk	78:65–94
Discovery, Biosynthesis, and Engineering of Lantipeptides	PJ Knerr, WA van der Donk	81:479–505

Drug Discovery and Combinatorial Chemistry

New Antivirals and Drug Resistance	PM Colman	78:95–118

Bioenergetics

Electron Transport and Oxidative Phosphorylation

Essentials for ATP Synthesis by F_1F_0 ATP Synthases	C von Ballmoos, A Wiedenmann, P Dimroth	78:649-72
Somatic Mitochondrial DNA Mutations in Mammalian Aging	N-G Larsson	79:683–706

Photosynthesis and Photobiology

Proton-Coupled Electron Transfer in Biology: Results from Synergistic Studies in Natural and Model Systems	SY Reece, DG Nocera	78:673–99

Nitrogen Fixation

Mechanism of Mo-Dependent Nitrogenase	LC Seefeldt, BM Hoffman, DR Dean	78:701–22

Permeases and Transporters

Multidrug Resistance in Bacteria	H Nikaido	78:119–46

Regulation of Glucose Transporter
    Translocation in Health and Diabetes   JS Bogan   81:507–32

Single-Molecule Biomechanics and Biological Nano-Devices

In singulo Biochemistry: When Less Is More   C Bustamante   77:45–50
Advances in Single-Molecule Fluorescence
    Methods for Molecular Biology   C Joo, H Balci, Y Ishitsuka, C Buranachai, T Ha   77:51–76
Single-Molecule Studies of Protein Folding   A Borgia, PM Williams, J Clarke   77:101–25
Structure and Mechanics of
    Membrane Proteins   A Engel, HE Gaub   77:127–48
Translation at the Single-Molecule Level   RA Marshall, CE Aitken, M Dorywalska, JD Puglisi   77:177–203
Recent Advances in Optical Tweezers   JR Moffitt, YR Chemla, SB Smith, C Bustamante   77:205–28
The Advent of Near-Atomic Resolution in
    Single-Particle Electron Microscopy   Y Cheng, T Walz   78:723–42
Emerging In Vivo Analyses of Cell Function
    Using Fluorescence Imaging   J Lippincott-Schwartz   80:327–32
Biochemistry of Mobile Zinc and Nitric Oxide
    Revealed by Fluorescent Sensors   MD Pluth, E Tomat, SJ Lippard   80:333–55
Development of Probes for Cellular Functions
    Using Fluorescent Proteins and
    Fluorescence Resonance Energy Transfer   A Miyawaki   80:357–73
Reporting from the Field: Genetically Encoded
    Fluorescent Reporters Uncover Signaling
    Dynamics in Living Biological Systems   S Mehta, J Zhang   80:375–401

Signal Transduction

Receptors and Adaptors

Sphingosine 1-Phosphate Receptor Signaling   H Rosen, PJ Gonzalez-Cabrera, MG Sanna, S Brown   78:743–68
The Role of Ubiquitin in the NF-κB
    Regulatory Pathways   B Skaug, X Jiang, ZJ Chen   78:769–96
Structural Biology of the Toll-Like
    Receptor Family   JY Kang, J-O Lee   80:917–41

Small GTPases and Heterotimeric G Proteins

Physical Mechanisms of Signal Integration
    by WASP Family Proteins   SB Padrick, MK Rosen   79:707–35

Structure-Function Relationships of the G Domain, A Canonical Switch Motif	A Wittinghofer, IR Vetter	80:943–71

Second Messengers

STIM Proteins and the Endoplasmic Reticulum-Plasma Membrane Junctions	S Carrasco, T Meyer	80:973–1000
Structure and Regulation of Soluble Guanylate Cyclase	ER Derbyshire, MA Marletta	81:533–59

Kinases, Phosphatases, and Phosphorylation Cascades

The Chemical Biology of Protein Phosphorylation	MK Tarrant, PA Cole	78:797–825
Amino Acid Signaling in TOR Activation	J Kim, K-L Guan	80:1001–32
The MPS1 Family of Protein Kinases	X Liu, M Winey	81:561–85
The Structural Basis for Control of Eukaryotic Protein Kinases	JA Endicott, MEM Noble, LN Johnson	81:587–613

Oncogenes and Tumor Suppressor Genes

Structural Biology of the Tumor Suppressor p53	AC Joerger, AR Fersht	77:557–82

Cellular Biochemistry

Biomembranes: Composition, Biology, Structure, and Function

Toward a Biomechanical Understanding of Whole Bacterial Cells	DM Morris, GJ Jensen	77:583–613
Genetic and Biochemical Analysis of Non-Vesicular Lipid Traffic	DR Voelker	78:827–56
Amphipols, Nanodiscs, and Fluorinated Surfactants: Three Nonconventional Approaches to Studying Membrane Proteins in Aqueous Solutions	J-L Popot	79:737–75
Introduction to Theme "Membrane Protein Folding and Insertion"	G von Heijne	80:157–60
Measurements and Implications of the Membrane Dipole Potential	L Wang	81:615–35

Vesicular Trafficking and Secretion

How Does Synaptotagmin Trigger Neurotransmitter Release?	ER Chapman	77:615–41
Protein Translocation Across the Bacterial Cytoplasmic Membrane	AJM Driessen, N Nouwen	77:643–67
Mechanisms of Endocytosis	GJ Doherty, HT McMahon	78:857–902

Single-Molecule Studies of the Neuronal SNARE Fusion Machinery	AT Brunger, K Weninger, M Bowen, S Chu	78:903–28
Protein Sorting Receptors in the Early Secretory Pathway	J Dancourt, C Barlowe	79:777–802
Biogenesis and Cargo Selectivity of Autophagosomes	H Weidberg, E Shvets, Z Elazar	80:125–56
GTPase Networks in Membrane Traffic	E Mizuno-Yamasaki, F Rivera-Molina, P Novick	81:637–59
Roles for Actin Assembly in Endocytosis	OL Mooren, BJ Galletta, JA Cooper	81:661–86

Intracellular Targeting and Localization

The Ubiquitin System, an Immense Realm	A Varshavsky	81:167–76

Organelles and Organelle Biogenesis

Maturation of Iron-Sulfur Proteins in Eukaryotes: Mechanisms, Connected Processes, and Diseases	R Lill, U Mühlenhoff	77:669–700
Mitochondrial tRNA Import and Its Consequences for Mitochondrial Translation	A Schneider	80:1033–53
Lipid Droplets and Cellular Lipid Metabolism	TC Walther, RV Farese Jr.	81:687–714

Apoptosis

Caspase Substrates and Cellular Remodeling	ED Crawford, JA Wells	80:1055–87

Organismal Biochemistry

Development and Differentiation

Collagen Structure and Stability	MD Shoulders, RT Raines	78:929–58
Adipogenesis: From Stem Cell to Adipocyte	QQ Tang, MD Lane	81:715–36
Pluripotency and Nuclear Reprogramming	M Dejosez, TP Zwaka	81:737–65

Biochemical Basis of Disease

CFTR Function and Prospects for Therapy	JR Riordan	77:701–26
The Biochemistry of Disease: Desperately Seeking Syzygy	JW Kozarich	78:55–63
Getting a Grip on Prions: Oligomers, Amyloids, and Pathological Membrane Interactions	B Caughey, GS Baron, B Chesebro, M Jeffrey	78:177–204

Biological and Chemical Approaches to Diseases of Proteostasis Deficiency	ET Powers, RI Morimoto, A Dillin, JW Kelly, WE Balch	78:959–91
Regulation of HSF1 Function in the Heat Stress Response: Implications in Aging and Disease	J Anckar, L Sistonen	80:1089–115
Endoplasmic Reticulum Stress and Type 2 Diabetes	SH Back, RJ Kaufman	81:767–93

Molecular Physiology and Nutritional Biochemistry

Aging and Survival: The Genetics of Life Span Extension by Dietary Restriction	W Mair, A Dillin	77:727–54
Cellular Defenses against Superoxide and Hydrogen Peroxide	JA Imlay	77:755–76
Toward a Control Theory Analysis of Aging	MP Murphy, L Partridge	77:777–98

Infectious Disease, Host-Pathogen, and Host-Symbiont Interactions

Super-Resolution Fluorescence Microscopy	B Huang, M Bates, X Zhuang	78:993–1016
Virus Entry by Endocytosis	J Mercer, M Schelhaas, A Helenius	79:803–33
Structure Unifies the Viral Universe	NGA Abrescia, DH Bamford, JM Grimes, DI Stuart	81:795–822

Neurochemistry

Cholesterol 24-Hydroxylase: An Enzyme of Cholesterol Turnover in the Brain	DW Russell, RW Halford, DMO Ramirez, R Shah, T Kotti	78:1017–40